Safe Science

Workplace Hazardous Materials Information System (WHMIS)

	compressed gas		dangerously reactive material
	flammable and combustible material		biohazardous infectious material
	oxidizing material		poisonous and infectious material causing immediate and serious toxic effects
	corrosive material		poisonous and infectious material causing other toxic effects

Hazardous Household Product Symbols (HHPS)

Symbol	Danger
	**Explosive** This container can explode if it is heated or punctured.
	**Corrosive** This product will burn skin or eyes on contact, or throat and stomach if swallowed.
	**Flammable** This product, or its fumes, will catch fire easily if exposed to heat, flames, or sparks.
	**Poisonous** Licking, eating, drinking, or sometimes smelling, this product is likely to cause illness or death.

Practise Safe Science in the Classroom

Be science ready.	Follow instructions.	Act responsibly.
• Come prepared with your textbook, notebook, pencil, and anything else you need. • Tell your teacher about any allergies or medical problems. • Keep yourself and your work area tidy and clean. Keep aisles clear. • Keep your clothing and hair out of the way. Roll up your sleeves, tuck in loose clothing, and tie back loose hair. Remove any loose jewellery. • Wear closed shoes (not sandals). • Do not wear contact lenses while doing investigations. • Read all written instructions carefully before you start an activity or investigation.	• Do not enter a laboratory unless a teacher is present, or you have permission to do so. • Listen to your teacher's directions. Read written instructions. Follow them carefully. • Ask your teacher for directions if you are not sure what to do. • Wear eye protection or other safety equipment when instructed by your teacher. • Never change anything, or start an activity or investigation on your own, without your teacher's approval. • Get your teacher's approval before you start an investigation that you have designed yourself.	• Pay attention to your own safety and the safety of others. • Know the location of MSDS (Material Safety Data Sheet) information, exits, and all safety equipment, such as the first aid kit, fire blanket, fire extinguisher, and eyewash station. • Alert your teacher immediately if you see a safety hazard, such as broken glass, a spill, or unsafe behaviour. • Stand while handling equipment and materials. • Avoid sudden or rapid motion in the laboratory, especially near chemicals or sharp instruments. • Never eat, drink, or chew gum in the laboratory. • Do not taste, touch, or smell any substance in the laboratory unless your teacher asks you to do so. • Clean up and put away any equipment after you are finished. • Wash your hands with soap and water at the end of each activity or investigation.

NELSON

SCIENCE PERSPECTIVES 10

Grade 10 Author Team

Christine Adam-Carr
Ottawa Catholic School Board

Martin Gabber
Formerly of Durham District School Board

Christy Hayhoe
Science Writer and Editor

Douglas Hayhoe, Ph.D.
Department of Education, Tyndale University College

Katharine Hayhoe, B.Sc., M.S.,
Professor, Department of Geosciences, Texas Tech University

Barry LeDrew
Curriculum and Educational Resources Consulting Ltd.

Milan Sanader, B.Sc., B.Ed., M.Ed
Dufferin-Peel Catholic District School Board

Senior Program Consultant

Maurice DiGiuseppe, Ph.D.
University of Ontario Institute of Technology (UOIT)
Formerly of Toronto Catholic District School Board

Program Consultants

Douglas Fraser
District School Board Ontario North East

Martin Gabber
Formerly of Durham District School Board

Douglas Hayhoe, Ph.D.
Department of Education, Tyndale University College

Jeffrey Major, M.Ed.
Thames Valley District School Board

NELSON EDUCATION

NELSON EDUCATION

Nelson Science Perspectives 10

Senior Program Consultant
Maurice DiGiuseppe

Program Consultants
Douglas Fraser
Martin Gabber
Douglas Hayhoe
Jeffrey Major

Authors
Christine Adam-Carr
Martin Gabber
Christy Hayhoe
Douglas Hayhoe
Katharine Hayhoe
Barry LeDrew
Milan Sanader

Contributing Authors
Douglas Fraser
Michael Stubitsch
Richard Towler
Judy Wearing

Vice President, Publishing
Janice Schoening

General Manager, Mathematics, Science, and Technology
Lenore Brooks

Publisher, Science
John Yip-Chuck

Associate Publisher, Science
David Spiegel

Managing Editor, Development
Susan Ball

National Director of Research and Teacher In-Service
Jennette MacKenzie

General Manager, Marketing, Math, Science, and Technology
Paul Masson

Product Manager
Lorraine Lue

Secondary Sales Specialist
Rhonda Sharp

Program Manager
Julia Lee

Project Managers
Christina D'Alimonte
Sarah Tanzini

Developmental Editors
Nancy Andraos
Vicki Austin
Barbara Booth
Jessica Fung
Julia Lee
Susan Skivington

Contributing Editor
Carmen Yu

Editorial Assistants
Vytas Mockus
Amy Rotman
Wally Zeisig

Senior Content Production Editor
Deborah Lonergan

Content Production Editors
Carolyn Pisani
Laurie Thomas

Copy Editor
Holly Dickinson

Proofreaders
Margaret Holmes
Linda Szostak

Indexer
Noeline Bridge

Senior Production Coordinator
Sharon Latta Paterson

Design Director
Ken Phipps

Interior Design
Bill Smith Studio
Greg Devitt Design

Feature Pages Design
Jarrel Breckon
Courtney Hellam
Julie Pawlowicz, InContext

Cover Design
Eugene Lo

Cover Image
Sebastian Kaulitzki/Shutterstock

Asset Coordinators
Renée Forde
Suzanne Peden

Illustrators
Steve Corrigan; Deborah Crowle; Steven Hall; Joel Harris; Sharon Harris; Stephen Hutching; Sam Laterza; Dave Mazierski; Dave McKay; Allan Moon; Nesbitt Graphics, Inc.; Jan-John Rivera; Theresa Sakno; Ann Sanderson; Bart Vallecoccia

Compositor
Nesbitt Graphics, Inc.

Photo Shoot Coordinator
Lynn McLeod

Photo/Permissions Researcher
David Strand

Printer
Transcontinental Printing, Ltd.

ISBN-13: 978-0-17-635528-9
ISBN-10: 0-17-635528-6

Printed and bound in Canada
5 6 12 11

For more information contact Nelson Education Ltd., 1120 Birchmount Road, Toronto, Ontario M1K 5G4. Or you can visit our Internet site at http://www.nelson.com.

REVIEWERS

Accuracy Reviewers

Andrew P. Dicks, Ph.D.
Senior Lecturer, Department of Chemistry, University of Toronto

Michelle French, B.Sc., M.Sc., Ph.D.
Lecturer, Department of Cell and Systems Biology, University of Toronto

William Gough,
Professor of Environmental Science, University of Toronto

Dr. Elizabeth L. Irving, O.D., Ph.D.
Canada Research Chair in Animal Biology
Associate Professor, School of Optometry, University of Waterloo

Meredith White-McMahon, Ph.D.
St. James-Assiniboia School Division

Assessment Consultants

Aaron Barry, M.B.A., B.Sc., B.Ed.
Sudbury Catholic DSB

Damian Cooper
Nelson Education Author

Mike Sipos, B.Ph.Ed., B.Ed.
Sudbury Catholic DSB

Catholicity Reviewer

Ted Laxton
Sacred Heart Catholic School, Wellington Catholic DSB

Environmental Education Consultant

Allan Foster, Ed.D., Ph.D.
Working Group on Environmental Education, Ontario
Former Director of Education, Kortright Centre for Conservation

ESL/Culture Consultant

Vicki Lucier, B.A., B.Ed., Adv. Ed.
ESL/Culture Consultant, Simcoe County DSB

Literacy Consultants

Jill Foster
English/Literacy Facilitator, Durham DSB

Jennette MacKenzie
National Director of Research and Teacher In-Service, Nelson Education Ltd.

Michael Stubitsch
Education Consultant

Numeracy Consultant

Justin DeWeerdt
Curriculum Consultant, Trillium Lakelands DSB

Safety Consultant

Jim Agban
Past Chair, Science Teachers' Association of Ontario (STAO) Safety Committee

STSE Consultant

Joanne Nazir
Ontario Institute for Studies in Education (OISE), University of Toronto

Technology/ICT Consultant

Luciano Lista, B.A. B.Ed., M.A.
Academic Information Communication Technology Consultant
Online Learning Principal, Toronto Catholic DSB

Advisory Panel and Teacher Reviewers

Christopher Bonner
Ottawa Catholic DSB

Charles J. Cohen
Community Hebrew Academy of Toronto

Jeff Crowell
Halton Catholic DSB

Tim Currie
Bruce Grey Catholic DSB

Lucille Davies
Limestone DSB

Greg Dick
Waterloo Region DSB

Matthew Di Fiore
Dufferin-Peel Catholic DSB

Ed Donato
Simcoe Muskoka Catholic DSB

Dave Doucette, B.Sc., B.Ed.
York Region DSB

Chantal D'Silva, B.Sc., M.Ed.
Toronto Catholic DSB

Naomi Epstein
Community Hebrew Academy of Toronto

Xavier Fazio
Faculty of Education, Brock University

Daniel Gajewski, Hon. B.Sc., B.Ed.
Ottawa Catholic DSB

Stephen Haberer
Kingston Collegiate and Vocational Institute, Limestone DSB
Faculty of Education, Queen's University

Shawna Hopkins, B.Sc., B.Ed., M.Ed.
Niagara DSB

Chris Howes, B.Sc., B.Ed.
Durham DSB

Janet Johns
Upper Canada DSB

Michelle Kane
York Region DSB

Dennis Karasek
Thames Valley DSB

Roche Kelly, B.Sc., B.Ed.
Durham DSB

Mark Kinoshita
Toronto DSB

Emma Kitchen, B.Sc., B.Ed.
Near North DSB

Stephanie Lobsinger
St. Clair Catholic DSB

Alistair MacLeod, B.Sc., P.G.C.E., M.B.A.
Limestone DSB

Doug McCallion, B.Sc., B.Ed., M.Sc.
Halton Catholic DSB

Nadine Morrison
Hamilton-Wentworth DSB

Dermot O'Hara, B.Sc., B.Ed., M.Sc.
Toronto Catholic DSB

Mike Pidgeon
Toronto DSB

William J.F. Prest
Rainbow DSB

Ron M. Ricci, B.E.Sc., B.Ed.
Greater Essex DSB

Charles Stewart, B.Sc., B.Ed.
Peel DSB

Richard Towler
Peel DSB

Carl Twiddy
Formerly of York Region DSB

Jim Young
Limestone DSB

Contents

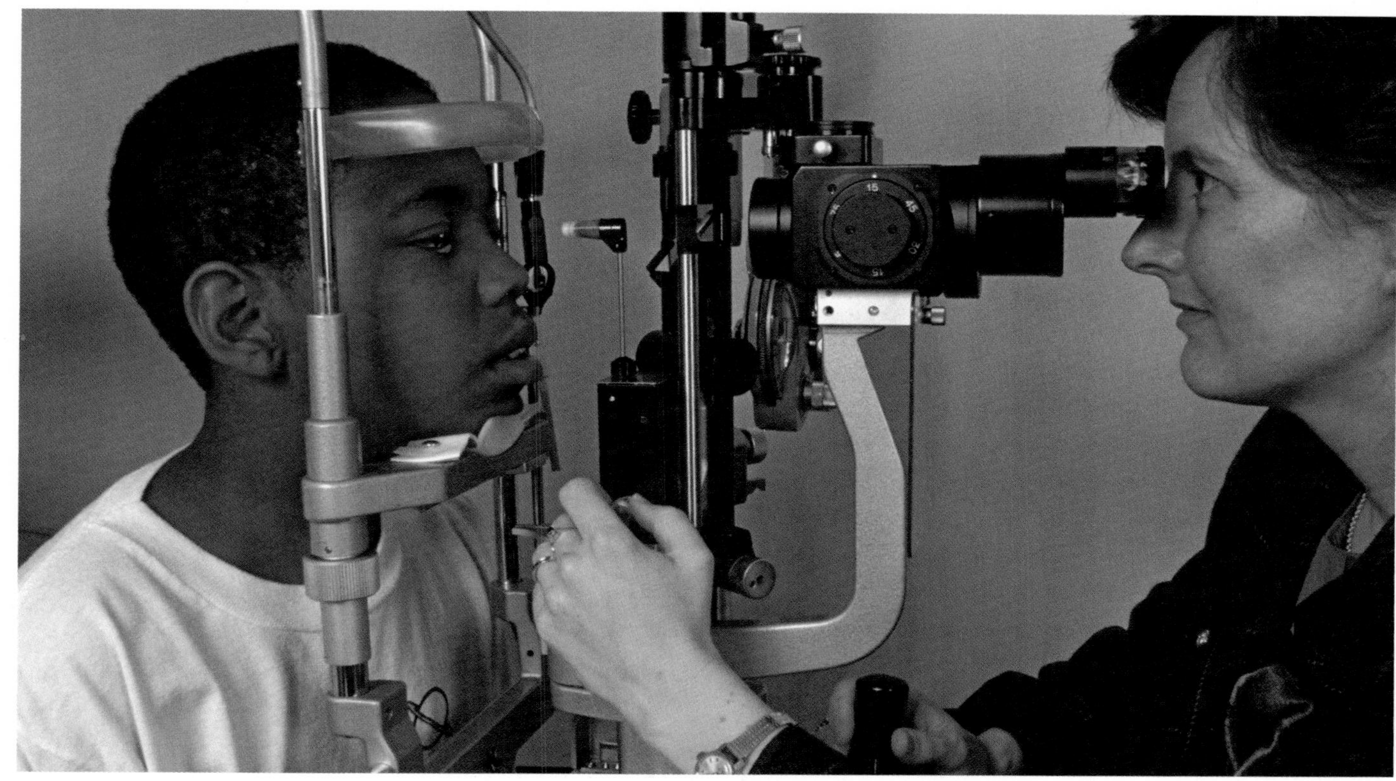

Discover Your Textbook

This textbook will be your guide to the exciting world of science. On the following pages is a tour of important features that you will find inside. **GET READY** includes all of the features of the introductory material that come before you begin each unit and chapter. **GET INTO IT** shows you all the features within each chapter. Finally, **WRAP IT UP** shows you the features at the end of each chapter and unit.

Get Ready

Unit Opener
Each of the five units has a letter and a title. Use the photo to help you predict what you might be learning in the unit.

Overall Expectations
The Overall Expectations describe what you should be able to do after completing the unit.

Big Ideas
The Big Ideas summarize the concepts you need to remember after you complete the unit.

Focus on STSE
These articles introduce real-world connections to the science topics you will be learning in the unit.

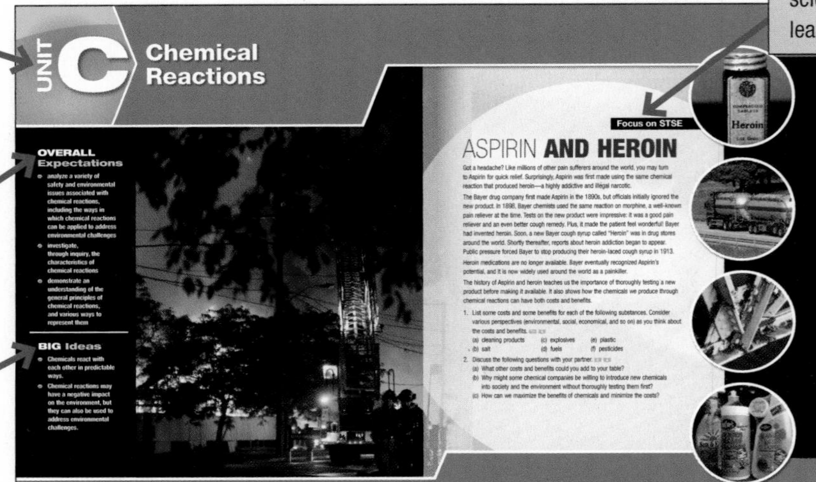

What Do You Already Know?
This feature lists the concepts and skills, developed in previous grades, that you will need to be successful as you work through the unit. Use the questions to see what you already know before you start the unit.

Concept Map
The Concept Map is a description of the topics, connected to picture clues, to help you predict what you will be learning in the unit.

Unit Task Preview
Find out about the Unit Task that you will complete at the end of each unit.

Unit Task Bookmark
When you see the Unit Task Bookmark, think about how the section relates to the Unit Task.

Assessment
The Assessment box tells you how you will demonstrate what you have learned by the end of the unit.

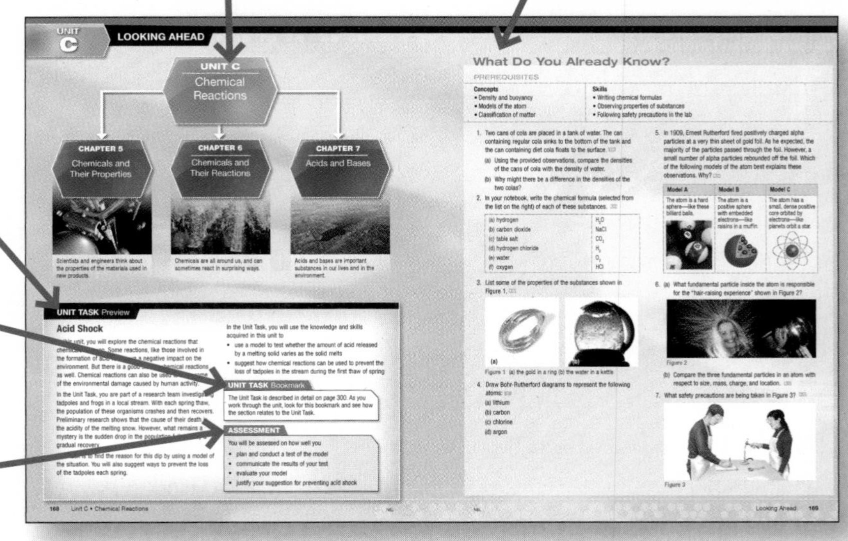

Chapter Opener

Each chapter has a number, a title, and a Key Question which you should be able to answer by the end of the chapter.

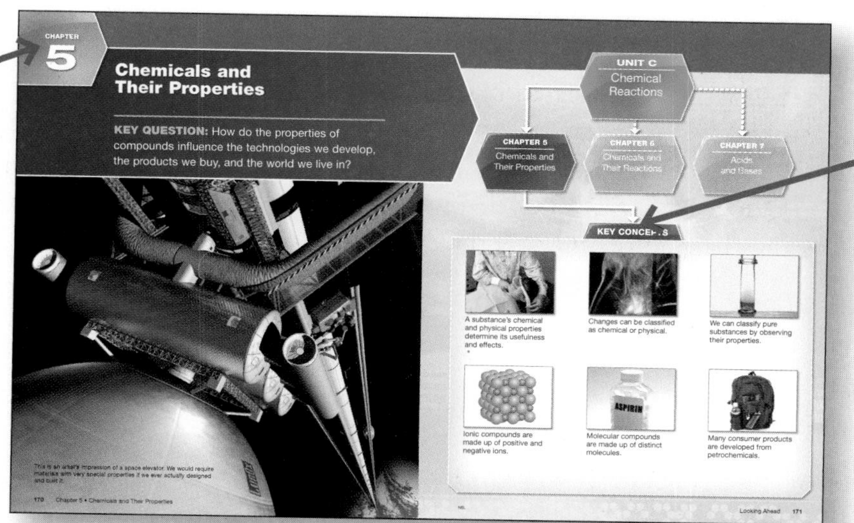

Key Concepts

The Key Concepts feature outlines the main ideas and skills you will learn in the chapter.

Engage in Science

These articles connect the topics you will learn in the chapter to interesting real-world developments in science.

What Do You Think?

Using what you already know, form an opinion by agreeing or disagreeing with statements that connect to ideas that will be introduced in the chapter.

Focus on Reading/ Focus on Writing

These reading and writing strategies help you learn science concepts and develop literacy skills in preparation for the OSSLT.

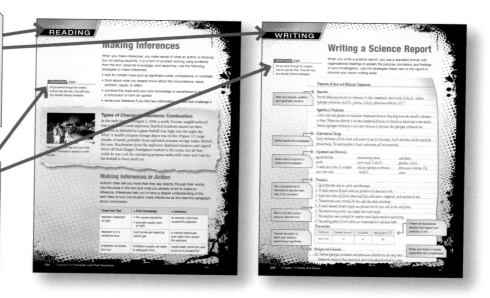

Reading/Writing Tip

Reading Tips suggest reading comprehension strategies to help you understand the science concepts presented in the text. Writing Tips provide suggestions to help you improve your writing skills.

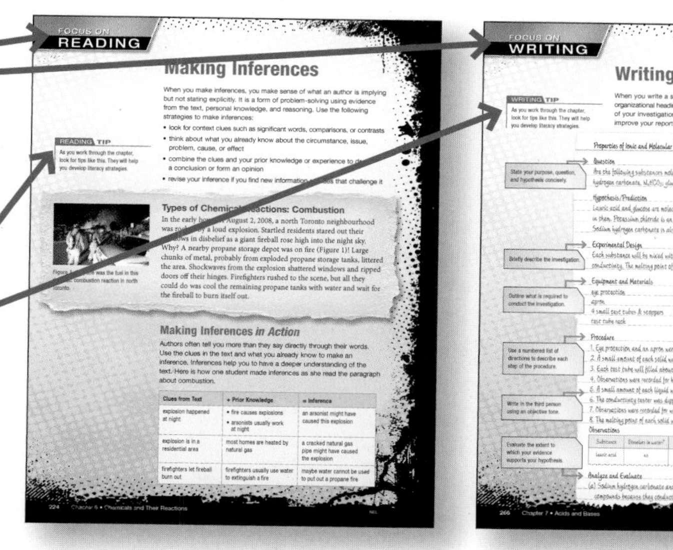

Get Into It

Vocabulary
You will learn many new terms as you work through the chapter. These key terms are in bold print. Their definitions can be found in the margins and in the Glossary at the back of the book.

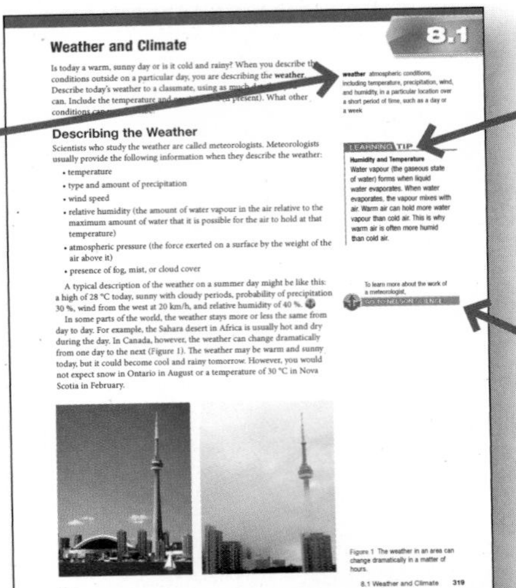

Learning Tip
Learning Tips are useful strategies to help you learn new ideas and make sense of what you are reading.

Career Link
The Career icon lets you know that you can visit the Nelson Science website to learn about science-related careers.

Sample Problems
This feature shows you how to solve numerical problems using the GRASS method. Make sure to check your learning by completing Practice problems.

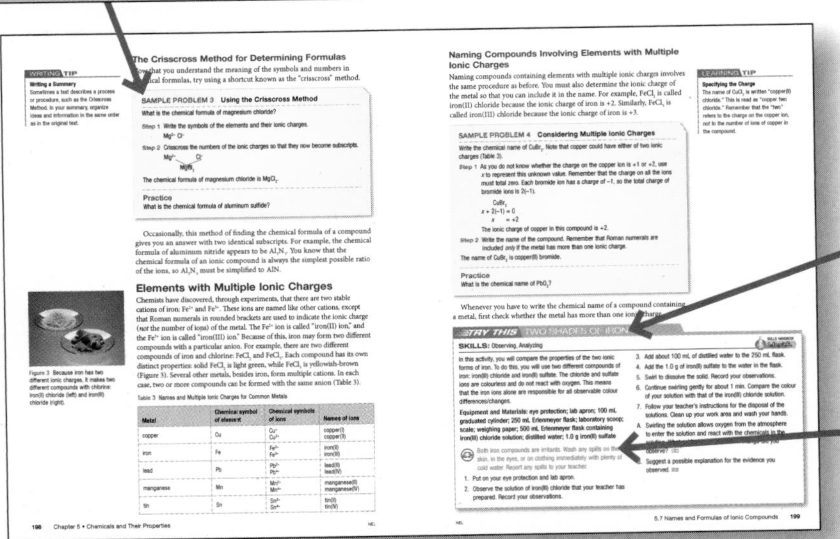

Try This
These are quick, fun activities designed to help you understand concepts and improve your science skills.

Safety Precautions
Look for these warnings about potential safety hazards in investigations and activities. They will be in red print with a safety icon.

Research This
These research-based activities will help you relate science and technology to the world around you and improve your critical thinking and decision-making skills.

Citizen Action
These activities encourage you to be a good citizen and a steward of the environment by taking action in the world around you.

Did You Know?
Read interesting facts about real-world events that relate to the topics you are learning.

Unit Task Bookmark
This icon lets you know that the concepts you learned in the section will help you to complete the Unit Task.

In Summary
At the end of each content section, this quick summary of the main ideas will help you review what you learned.

Check Your Learning
Complete these questions at the end of each content section to make sure you understand the concepts you have just learned.

Magazine Features
Look for these special features in each unit to learn about exciting developments in science, cool new technology, careers involving science, or how science relates to your everyday life.

OSSLT Icon
This icon lets you know that the material will help you develop literacy skills in preparation for the Ontario Secondary School Literacy Test.

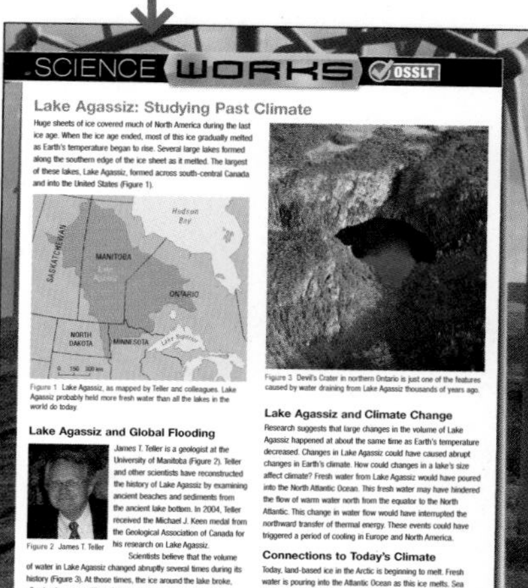
SCIENCE WORKS OSSLT

Lake Agassiz: Studying Past Climate

Lake Agassiz and Global Flooding

Lake Agassiz and Climate Change

Connections to Today's Climate

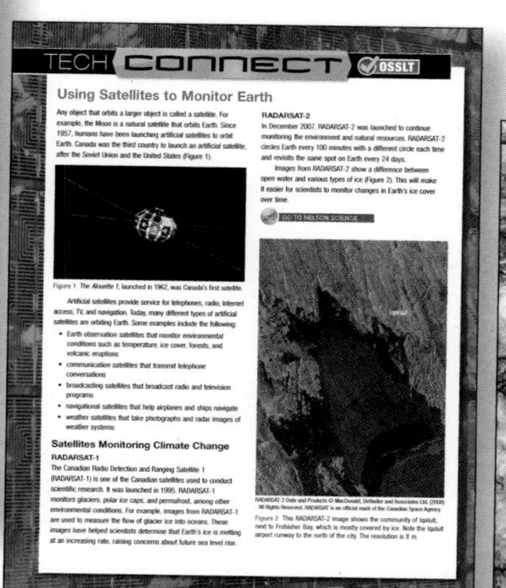
TECH CONNECT OSSLT

Using Satellites to Monitor Earth

RADARSAT-2

Satellites Monitoring Climate Change

AWESOME SCIENCE OSSLT

Geoengineering to Combat Climate Change?

Mirrors in Space

Farming Algae

Fertilizing the Oceans

Risks of Geoengineering

Perform an Activity
These are hands-on activities that allow you to observe the science that you are learning.

Skills Menu
The Skills Menu in each activity lists the skills that you will use to solve the problem or achieve the purpose of the activity.

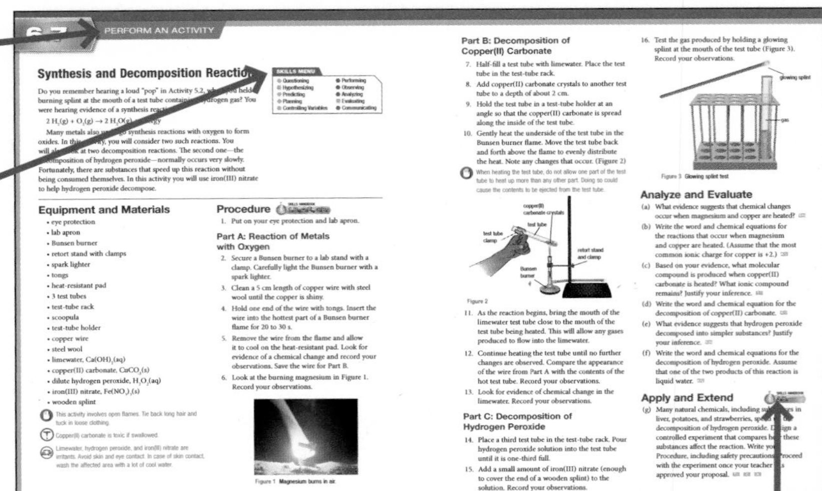

Skills Handbook Icon
This icon directs you to the section of the Skills Handbook that contains helpful information and tips.

Explore an Issue Critically
These activities allow you to examine social and environmental issues related to the unit. They often involve research, decision-making skills, and communication.

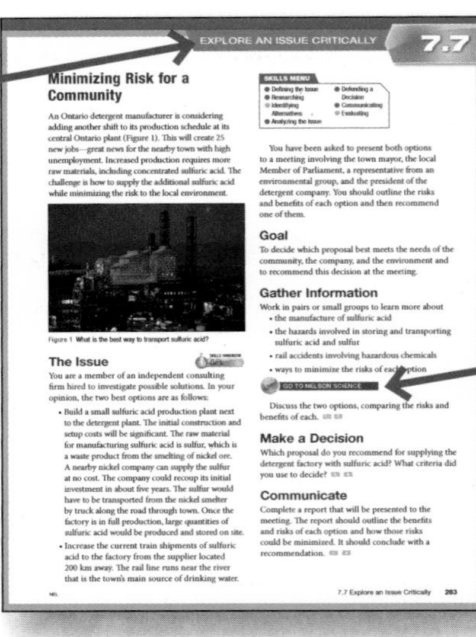

Weblink
When you see this weblink icon, you can visit the Nelson Science website to learn more about the topic, watch a video, do an online activity, or take a quiz.

Conduct an Investigation
These experimental investigations are an opportunity for you to develop science process skills.

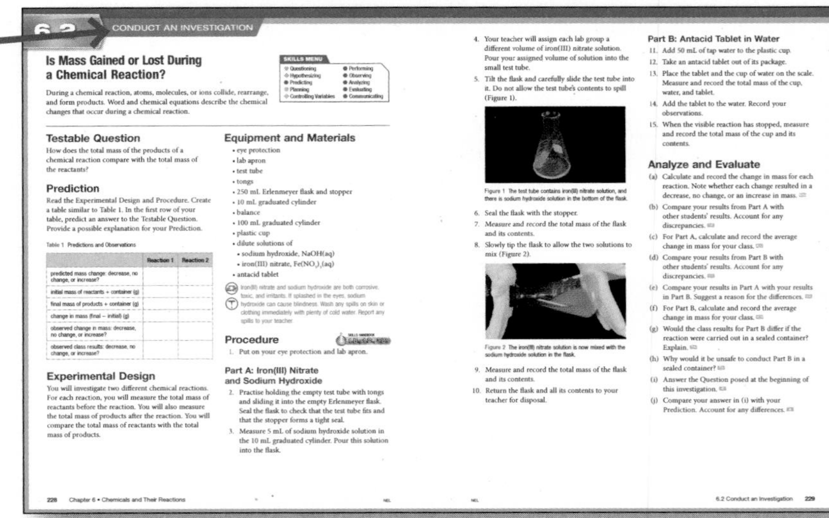

Wrap It Up

Key Concepts Summary

The Key Concepts Summary feature outlines the main ideas and skills that you learned in the chapter. The numbers in brackets indicate the section in which the concepts were taught.

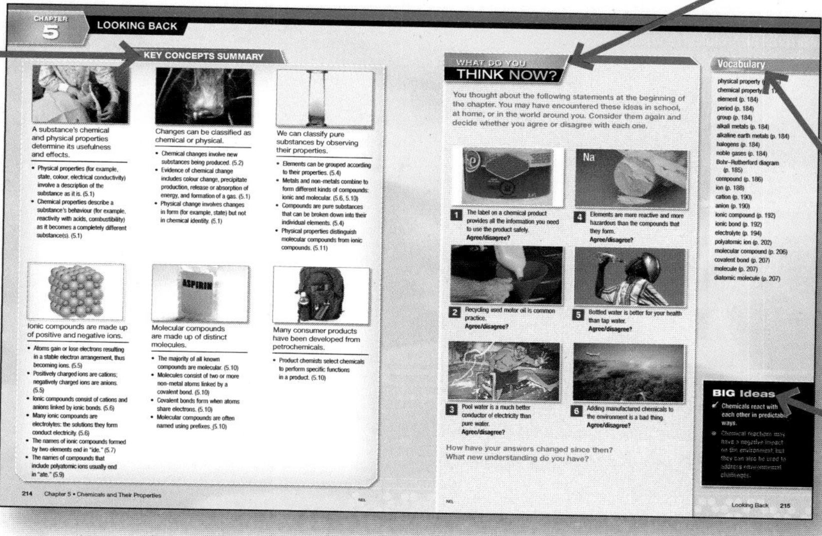

What Do You Think Now?

Think about what you learned in the chapter and consider whether you have changed your opinion by agreeing or disagreeing with the statements.

Vocabulary

This feature lists all the key terms you have learned and the page number where the term is defined.

Big Ideas

The checkmark indicates which Big Ideas were developed in the chapter.

Chapter Review

Complete these questions to check your learning and apply your new knowledge from the chapter.

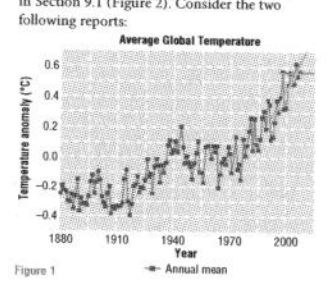

Achievement Chart Icons

All questions are tagged with icons that identify the types of knowledge and skills you must use to answer the question.

Online Quiz Icon

There is an online study tool for each chapter on the Nelson Science website.

Chapter Self-Quiz

The Chapter Self-Quiz is a helpful tool for you to make sure you understand all the concepts you learned in the chapter.

Master Concept Map
This feature brings together the Key Concepts from each chapter to summarize all the main ideas in the unit.

Make a Summary
Summarize what you have learned in the unit by completing the Make a Summary activity.

Career Links
Make connections between what you learned in the unit and future careers by completing the Career Links activity.

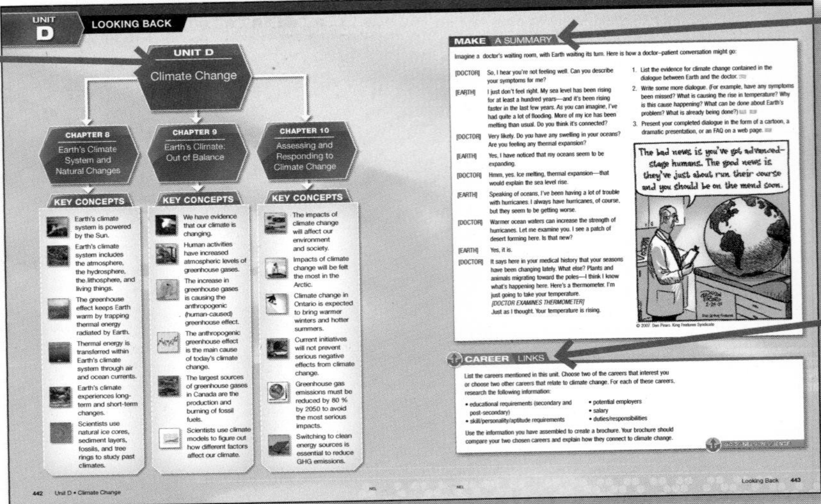

Unit Task
Demonstrate the skills and knowledge you developed in the unit by completing the challenge described in the Unit Task.

Skills Menu
The Skills Menu identifies the skills you will use to complete the Unit Task.

Assessment Checklist
This checklist lists the criteria that your teacher will use to evaluate your work on the Unit Task. Read this list carefully before completing the task.

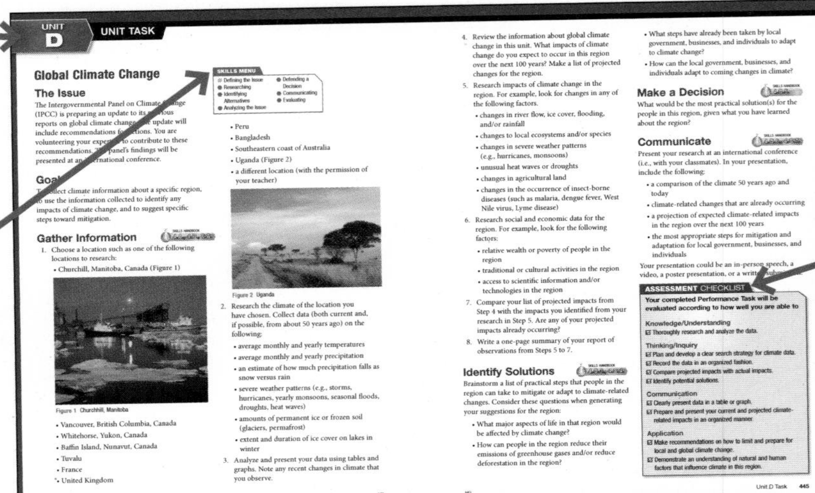

Unit Review
Complete the Unit Review questions to check your learning of all the concepts and skills in the unit.

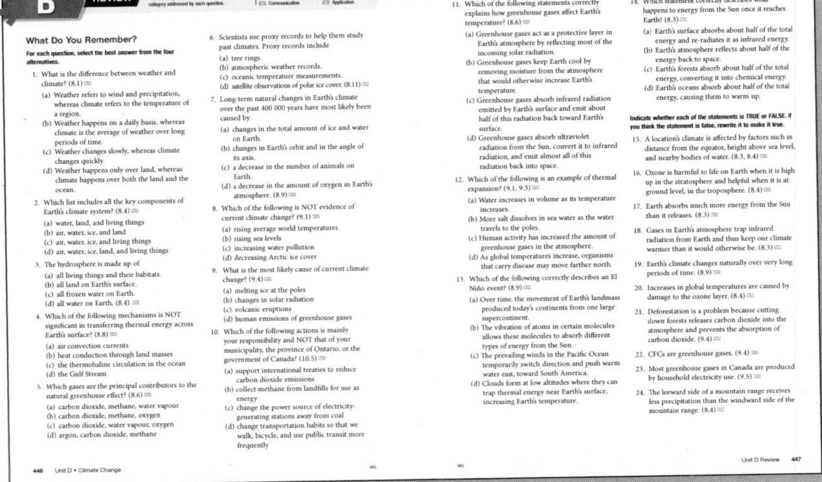

Unit Self-Quiz

The Unit Self-Quiz is an opportunity for you to make sure that you understand all the main ideas from the unit.

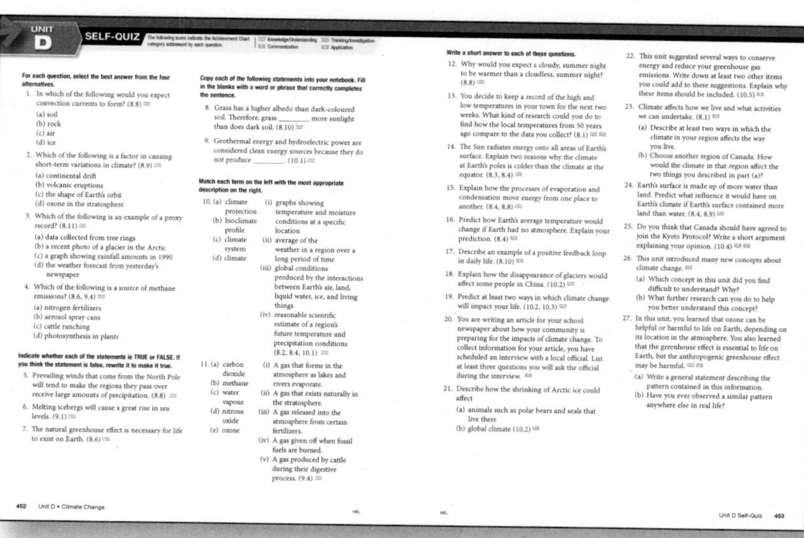

Skills Handbook

The Skills Handbook is your resource for useful science skills and information. It is divided into numbered sections. Whenever you see a Skills Handbook Icon, it will direct you to the relevant section of the Skills Handbook.

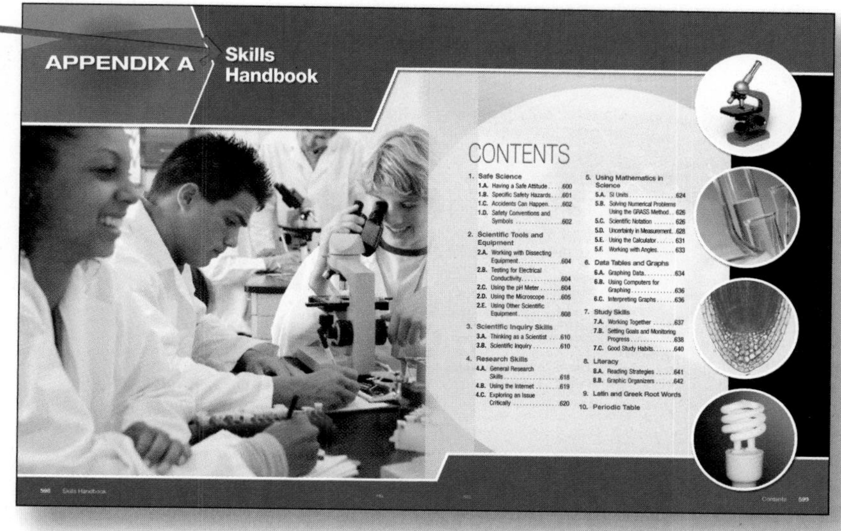

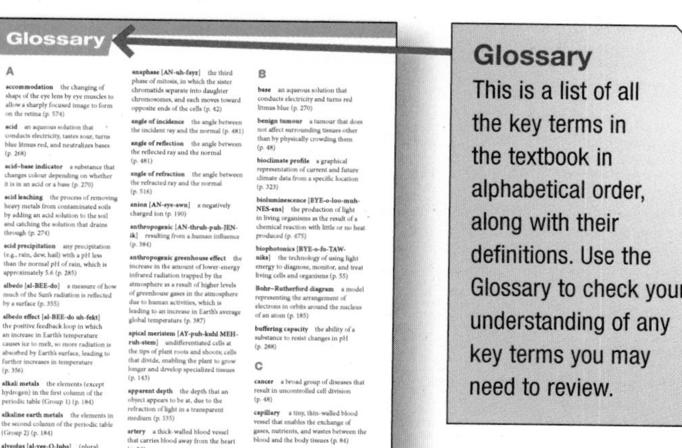

Glossary

This is a list of all the key terms in the textbook in alphabetical order, along with their definitions. Use the Glossary to check your understanding of any key terms you may need to review.

Introduction to Scientific Investigation Skills and Career Exploration

GOALS of the Science Program

- To relate science to technology, society, and the environment
- To develop the skills, strategies, and habits of mind required for scientific investigation
- To understand the basic concepts of science

Focus on STSE

SCIENCE AND YOUR LIFE

The goals of science education include much more than just the acquisition of scientific facts or knowledge. Your science program and this textbook are designed to help you understand the role of science in your everyday life and the impact of science and technology on society and the environment.

This introductory unit introduces the important scientific investigation skills that you will develop as you study biology, chemistry, Earth and space science, and physics in the following units.

In these four units, you will have many opportunities to learn through scientific inquiry. Through these inquiries, you will develop, practise, and refine the essential scientific investigation skills. These skills are useful not only in learning high school science but also in your post-secondary education and in your everyday life. You will also have opportunities to explore careers that are related to the various science topics.

The main purpose of learning science at this level is to make connections. As you progress through this course, you will develop an understanding of how science, technology, society, and the environment (STSE) are interrelated. You will connect the STSE relationships to your everyday life experiences, and you will develop scientific literacy.

Think, Pair, Share

1. List five examples where science or technology directly or indirectly influence your daily life. A
2. Pair up with a partner and share your ideas. Brainstorm to add further examples. C
3. Join another pair and share your lists. Eliminate duplicate examples and refine your list. Share your list with the whole class. C A

CHAPTER 1

Living and Working with Science

KEY QUESTION: What skills are required to carry out scientific investigations?

Science takes place in all kinds of places—not just in the lab.

UNIT A

Introduction to Scientific Investigation Skills and Career Exploration

CHAPTER 1

Living and Working with Science

KEY CONCEPTS

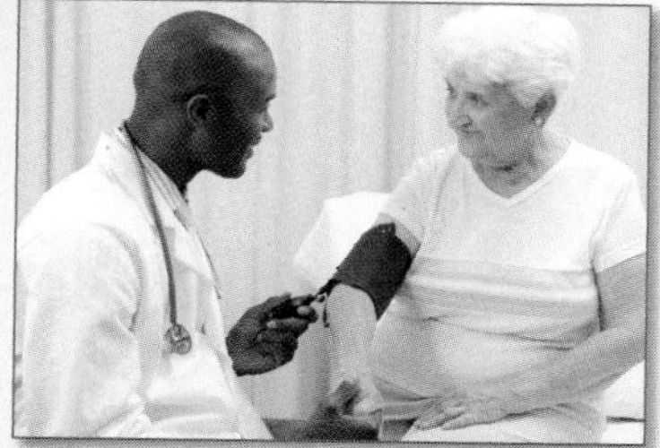

Science and technology are an important part of our everyday lives.

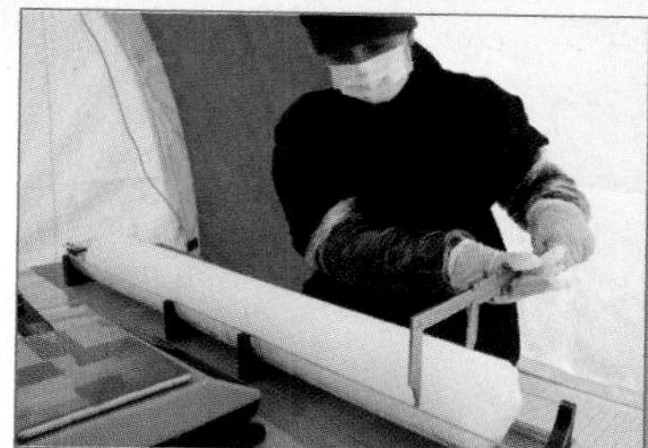

Scientific inquiry can be conducted in different ways, depending on the question to be answered.

All scientific inquiries rely on careful recording of accurate and repeated observations.

Careful analysis and interpretation of observations help make them meaningful.

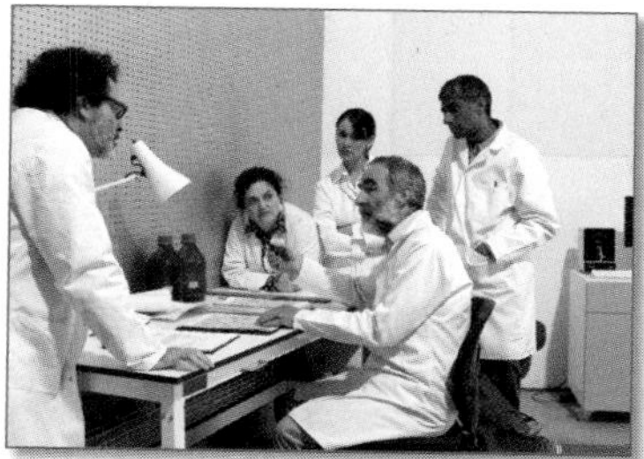

Clear communication is important for sharing scientific discoveries and ideas with others.

Scientific literacy is necessary for wise personal decisions, responsible citizenship, and careers.

ENGAGE IN SCIENCE

A BREATH OF *Fresh Air!*

Before Jonathan heads off to school, he checks the Air Quality Health Index (AQHI) online. The website, a service provided by the federal and provincial governments, provides an hourly update of the current air quality conditions. The weather forecast predicts that it's going to be a very hot day at 36 °C with little wind. These are perfect conditions for a bad smog day!

Jonathan is more aware of the air quality since his first asthma attack a few years ago. On days when the smog is bad, he experiences shortness of breath, wheezing, tiredness, and headaches. He has learned how to cope on those days when the air quality is poor by staying inside and limiting his physical activities. After the smog disappears, Jonathan's symptoms clear up, and he can enjoy his usual outdoor activities.

Jonathan uses the AQHI scale to help him determine the air quality and his outdoor activities for the day. The scale ranges from 1 to 10+. The lower numbers indicate good air quality and low health risk, while the higher numbers indicate poor air quality and high health risk.

Smog and traffic-related pollution cause major air quality problems in many large cities. A 2007 report by the City of Toronto suggests that traffic-related air pollution causes about 440 premature deaths and about 1700 hospitalizations per year in the city. The estimated direct and indirect costs related to these premature deaths are about $2.2 billion.

What is smog, and what causes it? Is everyone affected by smog or just those people who are at increased risk? Is climate change affecting our air quality? What can we do about it?

Figure 2 The AQHI scale indicates air quality and the level of risk to human health.

GO TO NELSON SCIENCE

How to Read Non-Fiction Text

Non-fiction text often includes a lot of information both in print and visuals. Using the following strategies will help you to better understand what you read:

Before Reading

- Preview the Text: Scan headings, bolded words, diagrams, photos, and captions.
- Identify how the text is organized.
- Think about what you already know about the topic.
- Set a Purpose for Reading: Change the title or heading into a question to help you determine your purpose for reading.

During Reading

- Read to answer the question you set as your purpose.
- Make connections to what you already know.
- Confirm, reject, or change your thinking based on new information.
- Ask questions.
- Pause from time to time to check your understanding.
- Identify main ideas.
- Make jot notes or use sticky notes to highlight key points.

After Reading

- Reflect on what you have learned.
- Check to see if you answered your purpose for reading question.
- Summarize what you learned in a graphic organizer or by remembering details about the text.
- Ask yourself how what you learned fits with what you already knew about the topic.

1.1 Skills of Scientific Investigation

Scientists assume that all events are caused by something. Hundreds of years ago, people believed that illnesses were caused by evil spirits or divine punishments. Some cultures still hold these beliefs today. Technological advances, such as the invention of the microscope, provide evidence that many diseases are the result of infections by micro-organisms. Scientists can now say that these micro-organisms "cause" disease (Figure 1).

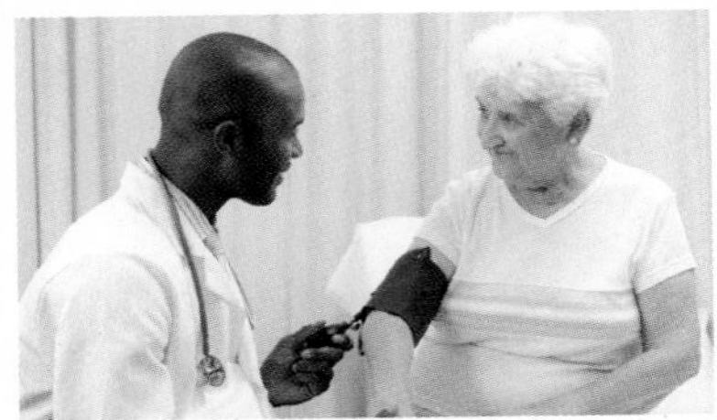

Figure 1 While blood pressure generally increases with age, high blood pressure is not directly caused by age—nor does high blood pressure cause old age.

Types of Scientific Inquiry

All scientific inquiry uses similar processes to find answers to questions. In most cases, these processes attempt to identify relationships between variables. A **variable** is any condition that could change in an inquiry. A variable that is deliberately changed or selected by the investigator is called the **independent variable**. A variable that changes in response to the independent variable but is not directly controlled by the investigator is called the **dependent variable**.

variable any condition that changes or varies the outcome of a scientific inquiry

independent variable a variable that is changed by the investigator

dependent variable a variable that changes in response to the change in the independent variable

There are three common types of scientific inquiry: (1) the controlled experiment, (2) the observational study, and (3) the correlational study.

Controlled Experiment

If the purpose of an inquiry is to determine whether one variable causes an effect on another variable, then you can carry out a **controlled experiment**. This is an experiment in which you control (change) the independent variable to determine if the change affects the dependent variable. For example, if you want to determine how temperature affects the rate of a chemical reaction, a controlled experiment is appropriate. In this case, you would change the temperature (independent variable) and observe any effect on the reaction rate (dependent variable).

controlled experiment an experiment in which the independent variable is purposely changed to find out what change, if any, occurs in the dependent variable

Observational Study

Often the purpose of a scientific inquiry is to gather information to answer a question about a natural phenomenon. **Observational studies** involve observing a subject or phenomenon without influencing it. Observational studies start with observations that lead to a question. Sometimes the researchers make a specific prediction about the answer to the question. Sometimes they also have an explanation for their prediction.

observational study the careful watching and recording of a subject or phenomenon to gather scientific information to answer a question

Sciences such as astronomy and ecology rely on observational studies. For example, if you wanted to determine the climatic conditions of a region, you would plan an observational study. You would make observations over many years, recording precipitation, temperature, and wind patterns. When sufficient data were collected, you could produce a complete description of the regional climate. Ongoing observations would make it possible to identify any changes in the climate.

Correlational Study

In a **correlational study**, a scientist tries to determine whether one variable is affecting another variable, without controlling any of the variables. Instead, the investigator simply observes variables that change naturally.

correlational study a study in which an investigator looks at the relationship between two variables

Correlation is the degree to which two sets of data vary together. A positive correlation indicates a direct relationship between variables: an increase in one variable corresponds to an increase in the other variable (Figure 2(a)). A negative correlation indicates an inverse relationship: an increase in one variable corresponds to a decrease in the other variable (Figure 2(b)). A line of best fit can be drawn through the points on the scatter graph. This line shows the relationship (positive or negative) between the two variables.

If there is no relationship between the two variables, we say that there is no correlation (Figure 2(c)).

Positive Correlation

y

(a)

x

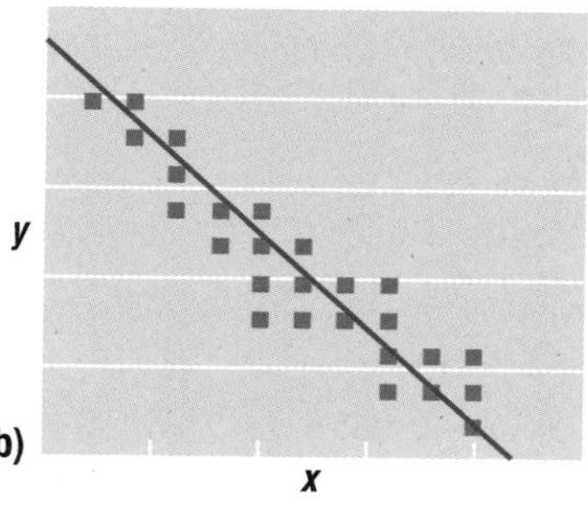

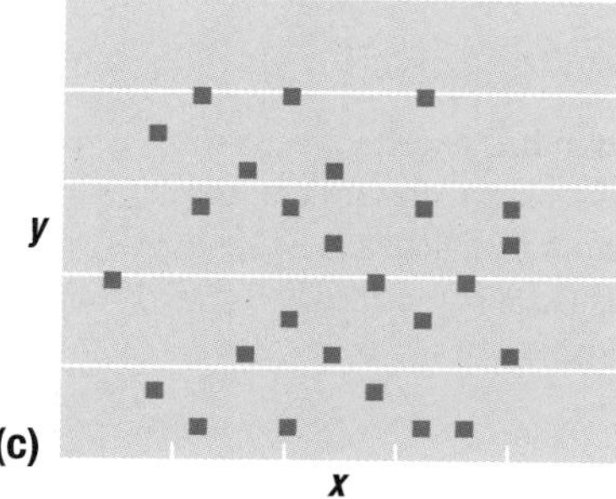

Figure 2 (a) In a positive correlation, variable *y* increases as variable *x* increases. (b) In a negative correlation, *y* decreases as *x* increases. (c) If there is no correlation, there is no pattern.

Table 1 Length of Education and Salary

Person	Education (years)	Annual salary ($000)	Person	Education (years)	Annual salary ($000)
1	12	45	14	12	50
2	19	110	15	15	75
3	11	40	16	12	55
4	15	65	17	14	70
5	14	60	18	17	65
6	20	140	19	12	55
7	18	100	20	16	80
8	11	50	21	15	145
9	18	95	22	13	60
10	18	90	23	16	55
11	13	50	24	10	40
12	19	100	25	17	95
13	10	50			

As an example, consider the relationship between average annual salary and amount of formal education (Table 1 and Figure 3). As you might expect, there is a positive correlation between these two variables. As the length of formal education increases, the average annual salary also increases. However, not everyone who has many years of formal education earns a high salary. Likewise, some people with little formal education earn a high salary. Also, two people with the same level of education could be earning different salaries.

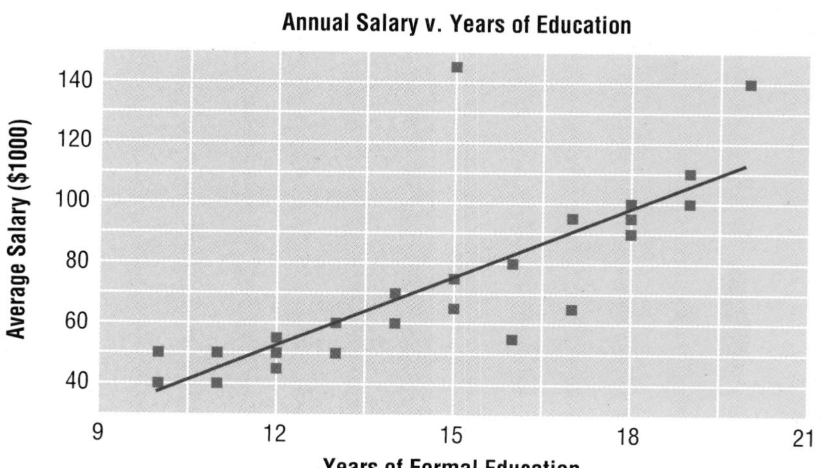

Figure 3 This graph shows a positive correlation between education and salary by plotting the data from Table 1.

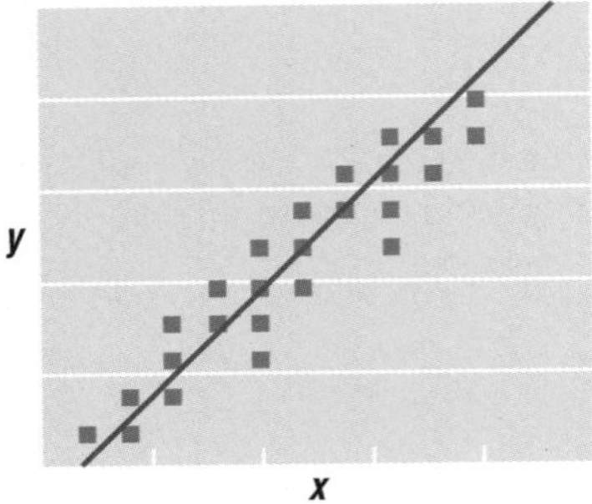

Figure 4 A strong positive correlation

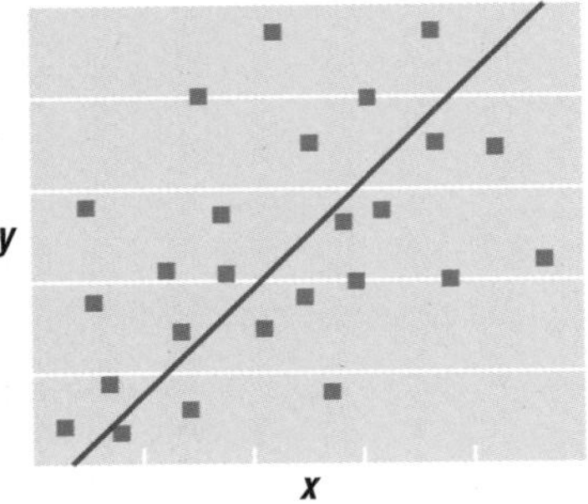

Figure 5 A weak positive correlation

When most of the data you collected is close to the line of best fit, there is a strong correlation between your variables (Figure 4). A weak correlation means that the relationship between your variables is less strong. The data points do not fall as close to your line of best fit (Figure 5).

When you know that two variables are correlated (either positively or negatively), you can predict one variable based on the other. Generally, the stronger the correlation (either positive or negative), the more probable it is that your prediction will be correct. In the example in Figure 3, you could predict with some certainty that a person who has a high level of education will earn a high salary.

Correlational studies require very large sample numbers and many replications to produce valid results. A correlation between two variables does not indicate that one variable causes an effect on the other. A correlation could simply be a coincidence. Consider a fictitious correlational study that shows that, in a given year, both the number of births increased and the number of earthquakes increased. It is highly unlikely that a reasonable link could be established between the two. Any correlation is totally due to coincidence.

Reports of correlations can be deceiving. For example, a newspaper headline reported that research showed a positive correlation between student height and mathematical ability. Taller students were better at solving mathematical problems. What the research failed to note (or the newspaper failed to report) was that the study included students of different ages. If the study had been done with students of the same age, there would have likely been no correlation between height and mathematical ability. The taller students—the better problem solvers—were probably older than the shorter students.

Researchers use correlational studies to further scientific understanding without performing experiments. They can make their own observations and measurements through fieldwork, interviews, and surveys. Alternatively, they can investigate relationships by using data from other researchers.

Scientific Investigation Skills

Regardless of the type of scientific investigation, certain skills are important in the process of conducting the investigation. These skills can be organized into four categories: (1) initiating and planning, (2) performing and recording, (3) analyzing and interpreting, and (4) communicating.

Initiating and Planning

All scientific investigations begin with a question. The question may have arisen from observations of a natural phenomenon or from an individual's curiosity (Figure 6). Often they come from previous experiments or studies.

Figure 6 Scientists who first observed unusual cave formations would have been curious so they proposed explanations for what they saw.

Some questions cannot be answered by scientific investigation, so it is important to ask the right questions. To lead to a scientific investigation, a question must be testable. Testable questions have certain characteristics:

- They must be about living things, non-living things, or events in the natural world.
- They must be answerable through scientific investigation—controlled experiments, observational studies, or correlational studies.
- They may be answered by collecting and analyzing data to produce evidence.

If a question suggests that a controlled experiment should be performed, then it is appropriate to propose a possible answer to the question. The tentative answer, which is based on existing scientific knowledge, is called a **hypothesis**. The hypothesis is directly related to the question. A hypothesis suggests a relationship between an independent variable and a dependent variable. A hypothesis serves two functions: (1) it proposes a possible explanation, and (2) it suggests a method of obtaining evidence that will support or reject the proposed explanation.

hypothesis a possible answer or untested explanation that relates to the initial question in an experiment

If you cannot make a hypothesis because you do not have a scientific explanation, then you can make a simple **prediction**. A prediction is a statement that predicts the outcome of a controlled experiment, without an explanation. A prediction is not a guess: it is based on prior knowledge and logical reasoning.

prediction a statement that predicts the outcome of a controlled experiment

A hypothesis usually includes a prediction. It is often written in the form "If ..., then ..., because ...." The "if ... then" part constitutes the prediction; the "because ..." part is the explanation.

TRY THIS IDENTIFYING VARIABLES AND MAKING PREDICTIONS

SKILLS: Predicting, Communicating

SKILLS HANDBOOK 3.B.2.

In this activity, you will identify dependent and independent variables and make predictions about the outcomes of scientific investigations.

Equipment and Materials: notebook or paper; pen

1. Each of the following questions could form the basis of a scientific investigation.
    - How is the volume of water related to its temperature?
    - How do phosphates affect the growth of aquatic plants?
    - What is the relationship between the size of the image in a mirror and the distance of the object from the mirror?
    - Is there a relationship between the average temperature and the number of pine trees in an area?
    - How does temperature affect the size of crystals that form in a solution?
    - How is the egg-laying ability of penguins affected by the availability of food?

A. For each statement, identify a possible independent variable and a possible dependent variable and make a prediction.

A hypothesis or a prediction provides the framework for the investigation. It identifies the variables and suggests which is the independent and which is the dependent variable. The hypothesis or prediction also suggests an experimental design by which the hypothesis can be tested fairly. The **experimental design** briefly describes the procedure. The value and success of the investigation depend on whether the experiment is fair, so careful planning at this stage is critical.

experimental design a brief description of the procedure in which the hypothesis is tested

Planning the investigation involves

- identifying the independent and dependent variables
- determining how the changes in the variables will be measured
- specifying how to control the variables not being tested
- selecting the appropriate equipment and materials (Figure 7)
- anticipating and addressing safety concerns
- deciding on a format for recording observations

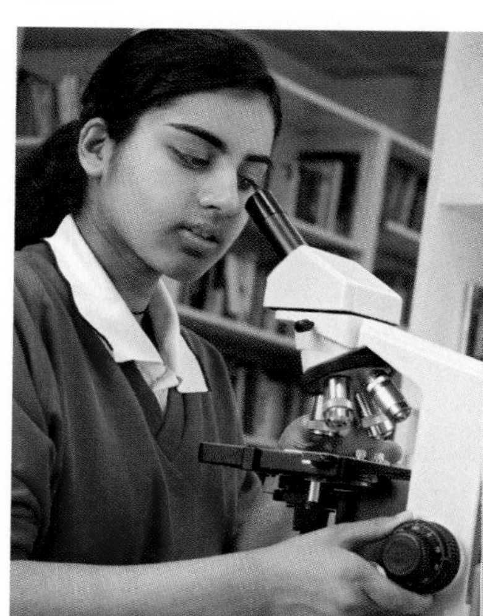

Figure 7 Specialized equipment is needed to conduct proper investigations.

Performing and Recording

After planning an investigation, it is important to follow the procedures described during the planning stage carefully (Figure 8). If the procedures present problems, they should be modified without changing the overall structure of the investigation. Record any modifications to the procedure in case you or someone else wants to repeat the investigation. If the problems cannot be overcome, you may have to go back to the planning stage and start again.

When performing the procedure, it is important to be constantly alert to potential safety concerns. Be sure to read the safety concerns in the Skills Handbook carefully before beginning an investigation and refer to them frequently if you have any concerns.

While performing an investigation, you will need to make accurate observations at regular intervals and record them carefully. Observations are any information that is obtained through the senses or by extension of the senses. Observations can be quantitative (numerical) or qualitative (non-numerical).

quantitative observation a numerical observation based on measurements or counting

qualitative observation a non-numerical observation that describes the qualities of objects or events

Quantitative observations are based on measurements or counting (Figure 9). Examples of quantitative measurements include length, mass, temperature, and population counts. Measuring is an important skill in making observations. It is important to select the right measuring tool to provide a precise and accurate measurement.

Qualitative observations are descriptions of the qualities of objects and events, without any reference to a measurement or a number. Common qualitative observations include the state of matter (solid, liquid, or gas), texture, and odour. These qualities cannot be measured directly.

The method or format of recording your observations depends on the type of observation. Quantitative observations are often recorded in an appropriate table. Qualitative observations can be written in words or recorded in pictures or sketches (Figure 10). Remember to record your observations clearly and accurately so that you do not have to rely on memory when you report your findings.

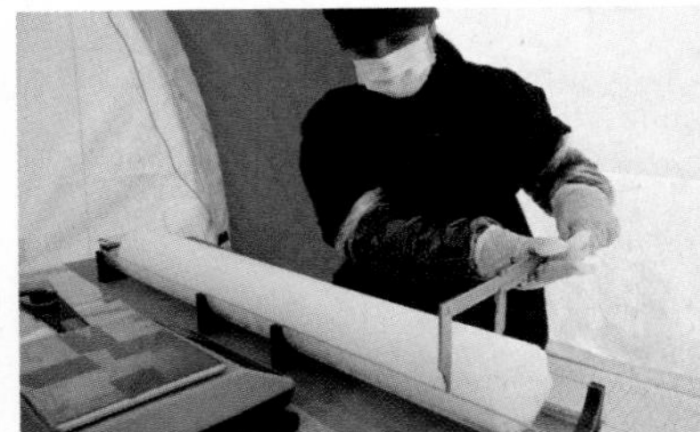

Figure 8 Scientists remove an ice core that has been cut from a glacier. These ice cores provide clues to the environmental conditions when the glacier was formed.

Figure 9 Two members of a ski patrol measure and record snow levels to forecast avalanches. Since this observation involves measuring, it is a quantitative observation. Noting that the lower layers of snow are more compacted is a qualitative observation.

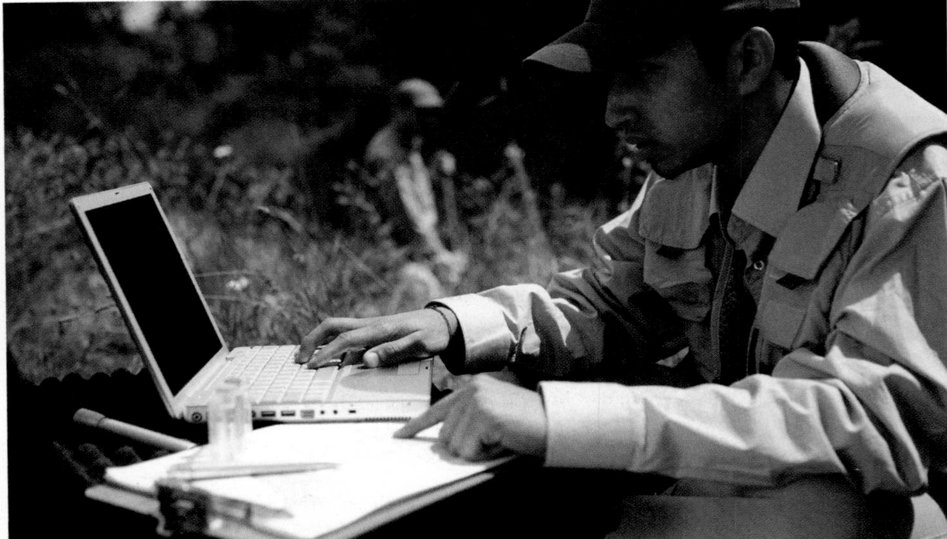

Figure 10 This field scientist transfers handwritten notes into electronic form.

Analyzing and Interpreting

Tables of observations are usually not the final product for the data collected during an investigation. Analyzing, or carefully studying, the observations usually provides more information than the raw data itself. In addition, you can plot graphs from the quantitative data to show up patterns and trends more clearly (Figure 11).

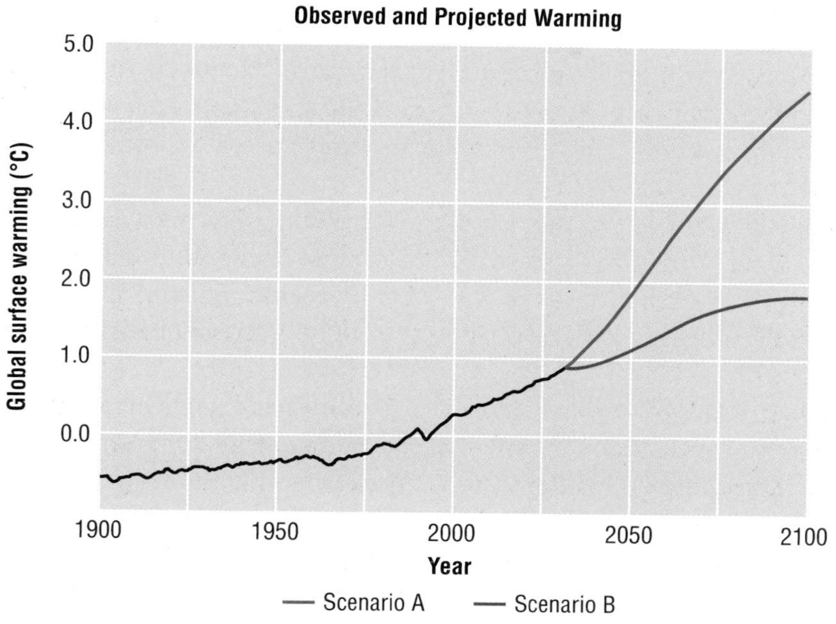

Figure 11 The two lines show the possible consequences of two scenarios. The red line shows a scenario in which humans continue to depend on fossil fuels. The blue line shows what is predicted to happen if humans switch to clean energy sources and conserve energy.

TRY THIS ANALYZING DATA

SKILLS: Analyzing, Communicating

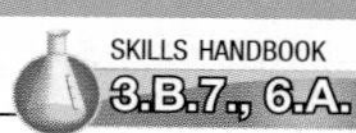

Data collection is the recording of information in an organized way. Analyzing data involves studying the data to uncover patterns and trends. Data interpretation involves explaining those patterns and trends. In this activity, you will analyze and interpret data.

Equipment and Materials: notebook; graph paper; pens or markers

1. Table 2 provides the estimated quarterly mouse population in a large field over six years. Study the data carefully.
2. Plot the provided data on a line graph using a coloured marker or pen.
3. Calculate the average population for each year by averaging the four quarterly numbers.
4. Plot the calculated yearly average on the same graph as well using a different coloured marker or pen.

A. Describe the patterns and trends that you observe in the data in Table 2.

B. What is happening to the population? Do you think this trend will continue indefinitely? Explain.

C. Propose a possible explanation for why the graph for the quarterly population data is a zigzagging line rather than a smooth line.

D. Is there a benefit to calculating and plotting the average population for each year? Explain.

E. Is it easier to see patterns and trends when the data are in the table or when the data are plotted on the graph? Explain.

Table 2 Mouse Population Data

Quarter	Mouse population	Quarter	Mouse population
2004 Q1	520	2007 Q1	790
2004 Q2	570	2007 Q2	870
2004 Q3	615	2007 Q3	930
2004 Q4	550	2007 Q4	860
2005 Q1	600	2008 Q1	875
2005 Q2	660	2008 Q2	940
2005 Q3	725	2008 Q3	1010
2005 Q4	675	2008 Q4	990
2006 Q1	705	2009 Q1	950
2006 Q2	780	2009 Q2	1040
2006 Q3	820	2009 Q3	1090
2006 Q4	780	2009 Q4	980

Analyzing observations also helps identify any errors in measurements. You should carefully check any measurement that is clearly very different from the others. If the very different measurement is caused by an error in measurement, record it but do not include it in the analysis.

A very important skill in a scientific investigation is evaluating the evidence that is obtained through observations. The quality of the evidence depends on the quality of other aspects of the investigation—the plan, the procedures, the equipment and materials, and the skills of the investigator. To evaluate the evidence, you need to evaluate all aspects of the investigation.

The whole purpose of analyzing and interpreting observations is to answer the question posed at the beginning. You may have evidence that you can use to confidently answer the question. You may conclude, however, that you do not have sufficient evidence to answer the question with any confidence. If your evidence confirms the prediction, then the hypothesis is supported. The evidence does not, however, *prove* the hypothesis to be true. If your evidence does not confirm the prediction, then the hypothesis may not be an acceptable explanation. Learning that the hypothesis is not supported is just as valuable as learning that it is supported. Rejection of a hypothesis is not a failure in a scientific investigation. It is simply another step along the path of finding an answer to the question.

Work in science seldom ends with a single experiment. Sometimes other investigators repeat the investigation to see if their evidence is the same. A question in science often sets off a chain reaction that leads to other questions, which then leads to other investigations and other questions. At the end of any investigation, the scientist asks questions such as What does this mean? Is the information of any practical value? How can the information be used? What other questions need to be answered? What new questions arose as a result of this investigation?

Communicating

One of the important characteristics of scientific investigation is that scientists share their information with the scientific community (Figure 12).

Figure 12 Sharing information is a key characteristic of scientific investigations.

Before scientific data can be published in a scientific journal, it must be examined by other scientists in a process called peer review. Other experts check that the data is valid and the science is correct. Clear and accurate communication is essential for sharing information. It is important to share not only the findings, but also the process by which evidence was obtained. If the investigation is to be repeated by others, sharing the design and procedures is just as important as sharing the findings.

By sharing their data and the techniques that they used to obtain, analyze, and interpret their data, scientists give others the opportunity to both review the data and use it in future research. The most common method for communicating with others about an investigation is by writing a lab report after the investigation is complete.

TRY THIS GETTING YOUR MESSAGE ACROSS

SKILLS: Planning, Evaluating, Communicating

SKILLS HANDBOOK 3.B.8., 3.B.9.

Accurate communication is as important in science as it is in everyday life. In this activity, you will work with a partner to demonstrate the importance of accurate communication and the need to develop and refine your skills in this area.

Equipment and Materials: notebook or paper; pen

1. You and your partner should each choose a different everyday task, such as tying a shoelace, preparing a meal, travelling from one location to another, or installing a piece of software on a computer. It must be a task that poses no significant hazards. Do not name the task or provide any additional instructions.
2. Write a set of detailed instructions that should enable someone else to complete the task.
3. Exchange instructions with your partner.
4. Use the instructions to complete the task without asking any questions. If necessary, complete the task at home.
5. Report back to your partner.

A. How successful were you in completing the task? Explain.

B. Why is clear, accurate communication difficult?

C. What skills do you need to communicate clearly? What do you need to do to develop these skills?

D. What strategy did you use to make your instructions as clear as possible?

E. What could you have done to communicate more clearly?

IN SUMMARY

- Observations of, and curiosity about, what we see around us often lead to questions that trigger scientific investigations.
- One type of scientific inquiry is the controlled experiment, in which the researcher keeps all but two variables constant, changes one (the independent variable), and observes the other (the dependent variable).
- A second type of inquiry is the observational study, in which the researcher collects data by observing a situation without affecting it.
- A third type of inquiry is the correlational study, in which the researcher analyzes data to see if there is a relationship between a pair of variables. The result may be a positive correlation, a negative correlation, or no correlation. The correlation could be strong or weak.
- Scientific investigation skills include initiating and planning (asking a question and deciding on the best way to find an answer); performing and recording (carrying out the procedure and making observations in an organized way); analyzing and interpreting (searching for patterns in the observations); and communicating (sharing findings with others).

1.2

Scientific Literacy for Living and Working in Canada

Scientific knowledge and technological innovations play an increasingly important role in everyday life. New technologies are designed so that the average person can use them without understanding how they work. Many people feel that science and technology are too complex for most of us to understand.

What Is Scientific Literacy?

Carl Sagan was a famous astronomer and author in the twentieth century. He recognized the need to be scientifically and technologically literate. In his book *The Demon-Haunted World*, Sagan said, "We've arranged a global civilization in which the most crucial elements profoundly depend on science and technology. We have also arranged things so that almost no one understands science and technology. This is a prescription for disaster. We might get away with it for a while, but sooner or later this combustible mixture of ignorance and power will blow up in our faces."

To make wise personal decisions and to act as a responsible citizen, it is necessary to be scientifically and technologically literate. The Science Teachers' Association of Ontario (STAO) defines a scientifically and technologically literate person as "one who can read and understand common media reports about science and technology, critically evaluate the information presented, and confidently engage in discussions and decision-making activities regarding issues that involve science and technology."

To read famous scientists' statements about the value of scientific literacy,
GO TO NELSON SCIENCE

Scientific Literacy for Careers in Science

Look around your classroom at your fellow science students. Some of you will probably pursue post-secondary education and a career in scientific research (Figure 1(a)). Others will find careers in medicine, geology, engineering, and environmental science Figure 1(b) and (c). Generally, employers hire individuals with strong critical-thinking and problem-solving skills and who have the ability to work as part of a team. These skills are emphasized throughout the entire science program.

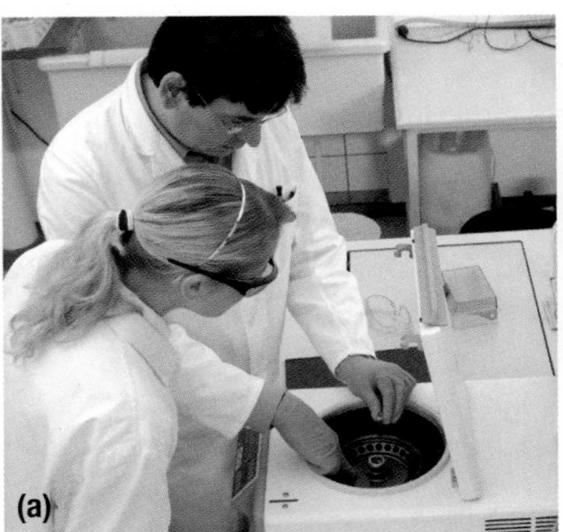
(a)

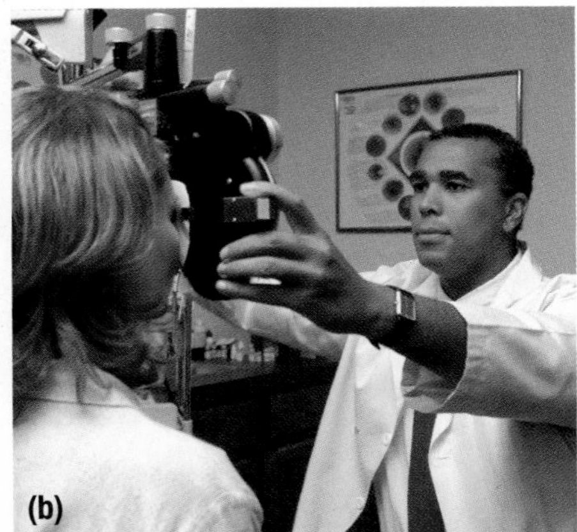
(b)

(c)

Figure 1 Some scientists conduct their research in laboratories. Others conduct their research in the natural world.

Canadians are actively working in a broad range of science disciplines. They are world leaders in such areas as astronomy, space exploration, medicine, genetics, environmental science, and information and communication technology. These Canadians are directly or indirectly responsible for a long list of scientific discoveries and technological inventions and innovations. The theory of plate tectonics, the discovery of insulin, the invention of the cardiac pacemaker, and the concept of standard time have made significant contributions to people and societies around the world.

Throughout this book, you will have opportunities to explore careers that are related to the area of science under study. Wherever you see a Career icon with the accompanying note in the margin of this book, you will be directed to the Nelson website. There you will be guided to research the education and training requirements for scientific careers and to find out about the roles and responsibilities of these careers.

Scientific Literacy for Life and Citizenship

A small number of people work in jobs that are directly related to science. However, every one of us is a citizen. Citizenship comes with certain rights and responsibilities. One of our basic rights is to have access to a full education. Along with that right comes the responsibility to use that education for the benefit of oneself and of society. Because science and technology influence our lives, it is important that we recognize and understand this influence. We can then make rational and ethical decisions about issues that affect us as individuals and as a society.

At some point in your life, you will have to make decisions about your own lifestyle. You will need to understand about the different types of medical diagnoses and the treatments available to you. It will also be in your best interest to know which products to buy (Figure 2). You will have to make decisions related to critical issues: climate change, environmental pollution, the depletion of natural resources, the protection of species, new medical technologies, space exploration, and world hunger.

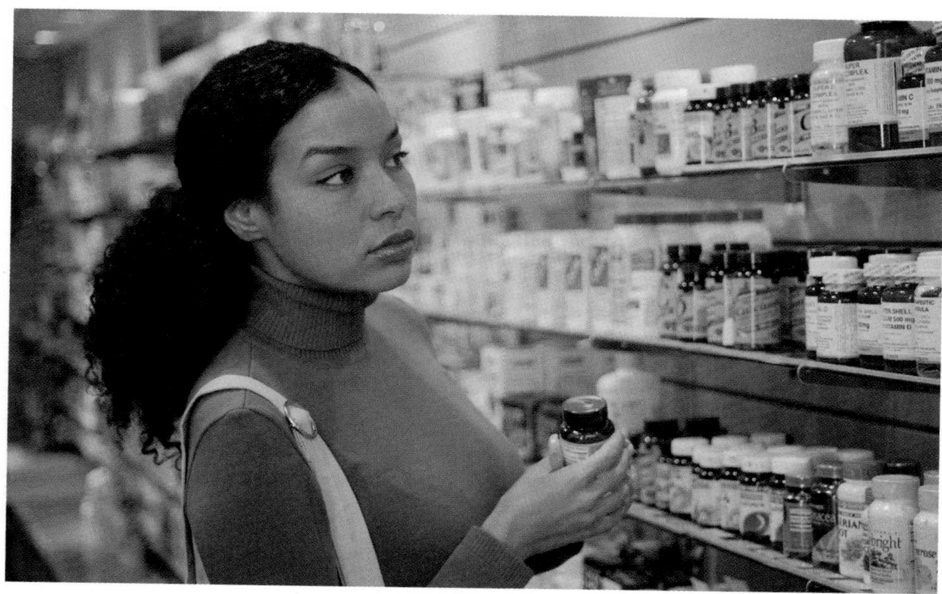

Figure 2 Science and technology affect the decisions we make every day.

As a society, we need to consider both the positive and the negative impacts of developing and using scientific knowledge and new technologies. We cannot foresee all the possible consequences of these achievements. However, we must be aware of the positive and the negative implications of new technologies. If we fail to do so, we may get some very unpleasant surprises after the technology has been adopted.

It is important to understand the basic concepts of science. However, it is impossible to know everything about science or to be aware of all the new scientific discoveries and technological advances. To achieve scientific literacy, it is just as important to learn about science as it is to learn science. It is important to know what science can do, to know that knowledge produced by science is reliable, and to know that science—despite its limitations—is the best way to learn about the world. It is just as important to be able to find and evaluate information and to use that information in making decisions.

A scientifically literate person understands that the future will be very different from the present. There are always new developments in science and technology. He or she also understands that society influences science and technology as much as science and technology influence society. Achieving scientific literacy is not the same as preparing to be a scientist. It is equally important to everyone, whether you are a a small-business person, a lawyer, a construction worker, a car mechanic, a travel agent, a doctor, an engineer, or a research scientist. Regardless of your plans and ambitions, achieving a level of scientific literacy is an important goal.

IN SUMMARY

- Scientific literacy helps people to understand and evaluate information relating to science and technology in the world so that they make better decisions.
- A general understanding of science is necessary to be an informed citizen.
- Specific scientific knowledge and skills are necessary for a wide range of careers.
- Canadians have made valuable contributions to the development of science and technology around the world.

KEY CONCEPTS SUMMARY

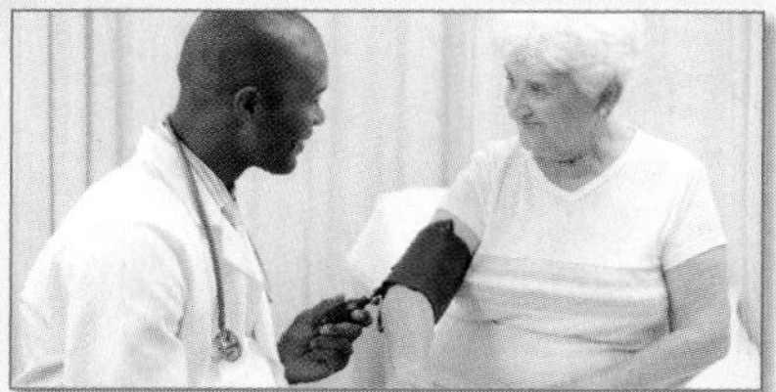

Science and technology are an important part of our everyday lives.

- All individuals are affected by science and technology every day of their lives.
- Most people do not understand the science and technology that they use in their everyday lives.
- Many major social issues, such as climate change and pollution, have a science and technology connection.

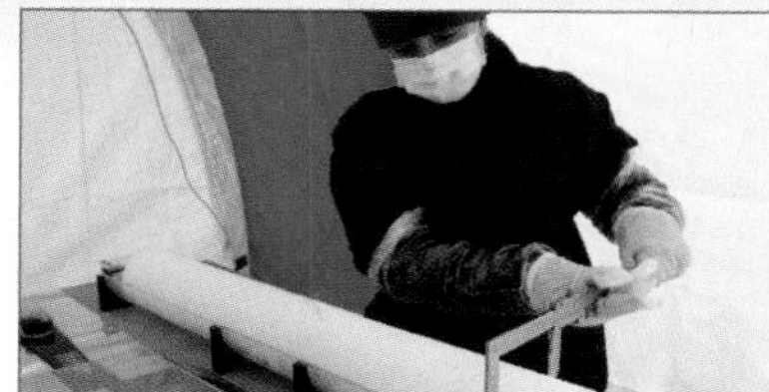

Scientific inquiry can be conducted in different ways, depending on the question to be answered.

- Observations often lead to questions that initiate scientific inquiry.
- The type of scientific inquiry depends on the nature of the question.
- Controlled experiments can be used to determine how an independent variable affects a dependent variable.
- A hypothesis is a possible answer and explanation for a scientific question. Hypotheses can be proposed at the beginning or the end of a scientific inquiry.

All scientific inquiries rely on careful recording of accurate and repeated observations.

- Observations are information obtained by using the senses or equipment that extends the senses.
- Observations can be quantitative (involving measuring or counting) or qualitative (involving descriptions).
- All observations should be recorded accurately.
- Repeating observations can eliminate errors and increase the value of the evidence.

Careful analysis and interpretation of observations help make them meaningful.

- Analysis of observations involves carefully studying data to identify patterns or trends.
- Interpretation of observations involves explaining patterns and trends in data.
- Analysis and interpretation of observations may lead to conclusions that answer the original scientific question.
- Interpretation of observations may raise additional questions and lead to further scientific inquiries.

Clear communication is important for sharing scientific discoveries and ideas with others.

- Scientists are expected to share their findings with other scientists and with the public.
- The results of scientific investigations should be reported clearly and honestly.
- Effective communication allows others to repeat scientific inquiries.

Scientific literacy is necessary for wise personal decisions, responsible citizenship, and careers.

- Careers in science and careers related to science require scientific knowledge and skills.
- Scientific knowledge and skills are essential for making logical and reasoned personal decisions.
- Citizens use their scientific knowledge and skills to make decisions that benefit society.
- A scientifically literate person is better able to understand issues related to science and technology.

Tissues, Organs, and Systems of Living Things

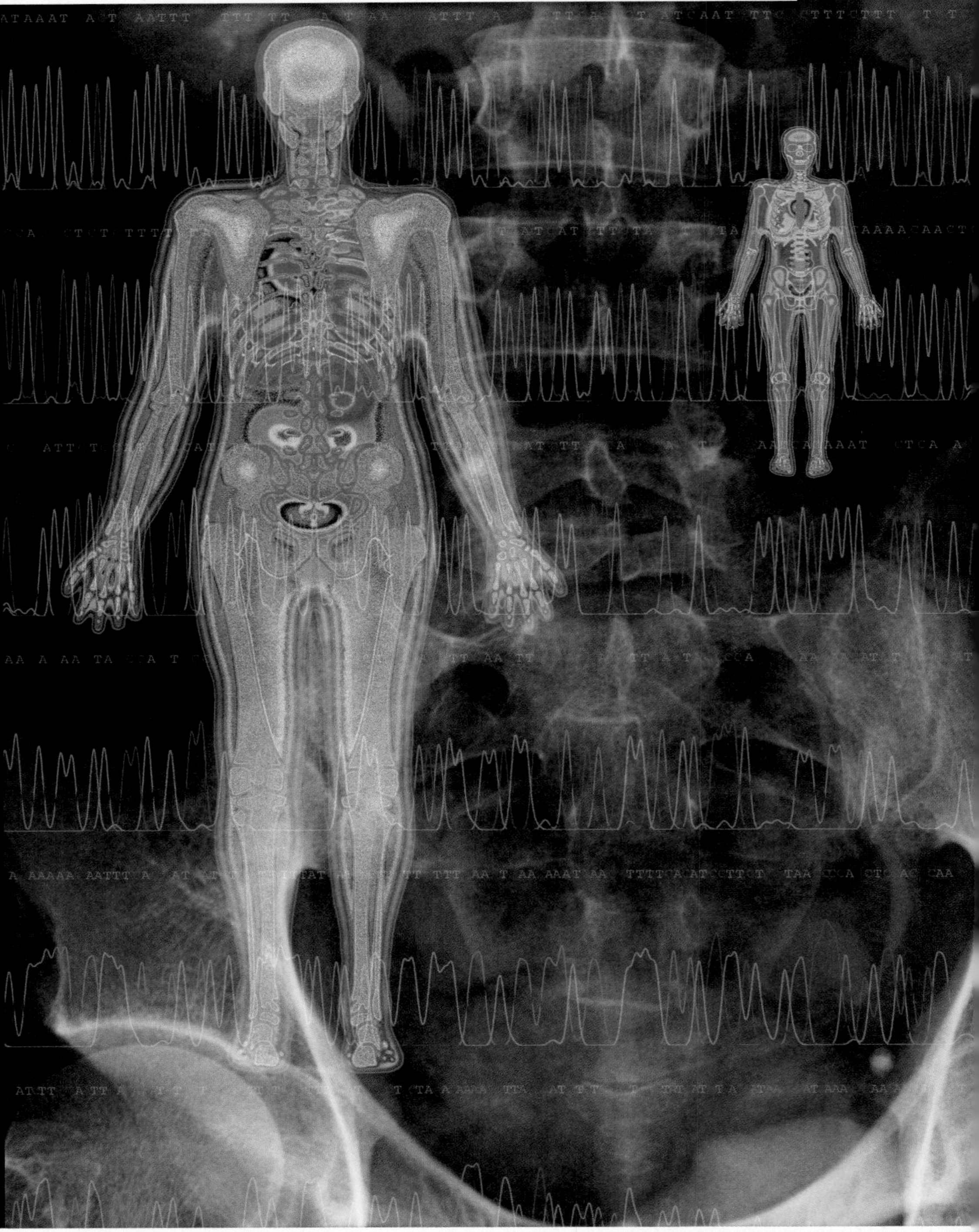

OVERALL Expectations

- evaluate the importance of medical and other technological developments related to systems biology and analyze their societal and ethical implications
- investigate cell division, cell specialization, organs, and systems in animals and plants, using research and inquiry skills, including various laboratory techniques
- demonstrate an understanding of the hierarchical organization of cells, from tissues, to organs, to systems in animals and plants

BIG Ideas

- Plants and animals, including humans, are made of specialized cells, tissues, and organs that are organized into systems.
- Developments in medicine and medical technology can have social and ethical implications.

Focus on STSE

MEDICAL TECHNOLOGY

Most of us have encountered some kind of medical technology in our lives. Even if we have not experienced it ourselves, someone we know probably has. If you have ever broken a bone, you have probably had an X-ray taken. If someone in your family has had kidney problems, he or she might have had to go on dialysis. Perhaps you know someone who has been through organ transplant surgery, heart surgery, or fertility treatment. There are many kinds of technology that help medical professionals diagnose and treat health problems.

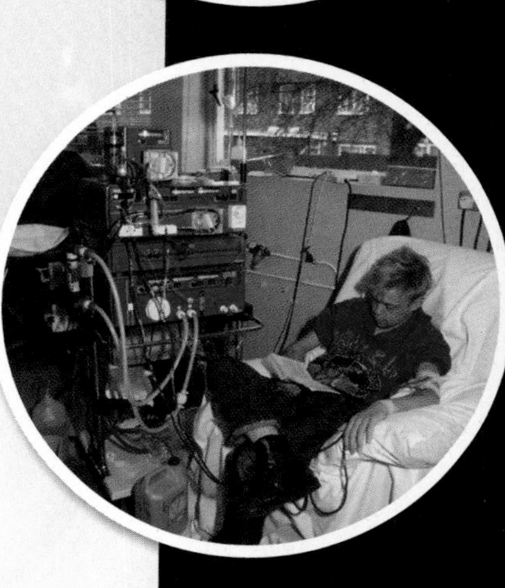

Think/Pair/Share

1. Write a list of medical technologies that you, or people you know, have experienced. Describe each technology and what it does in as much detail as possible. If possible, describe how you felt during the procedure. Did you have positive or negative feelings about the technology? C A
2. Share your list with a partner. C
3. Discuss your feelings about the technologies. Do any of them have ethical implications? C A
4. Have the technologies made a difference to the human lifespan and quality of life? Explain. C A

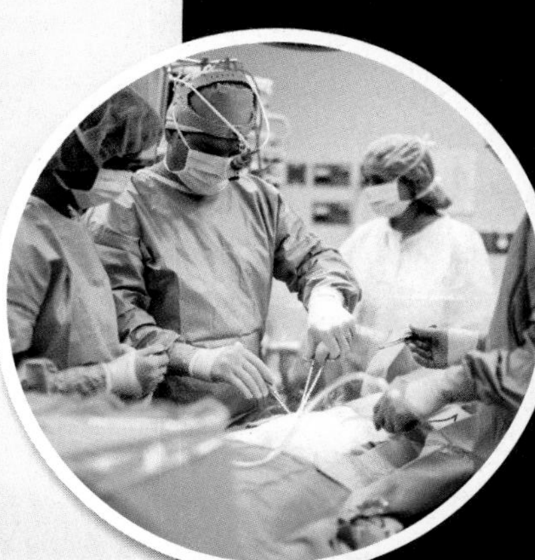

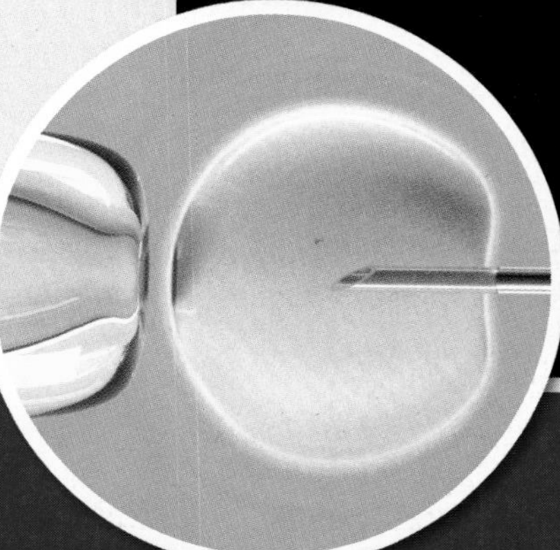

UNIT B

LOOKING AHEAD

UNIT B

Tissues, Organs, and Systems of Living Things

CHAPTER 2

Cells, Cell Division, and Cell Specialization

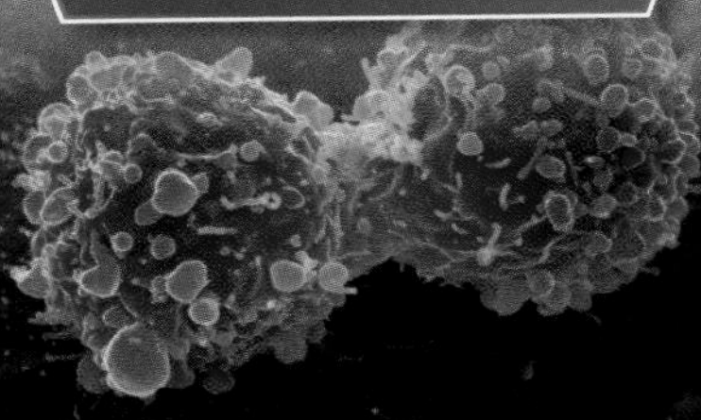

Cells divide so that the organism can grow, repair itself, and reproduce.

CHAPTER 3

Animal Systems

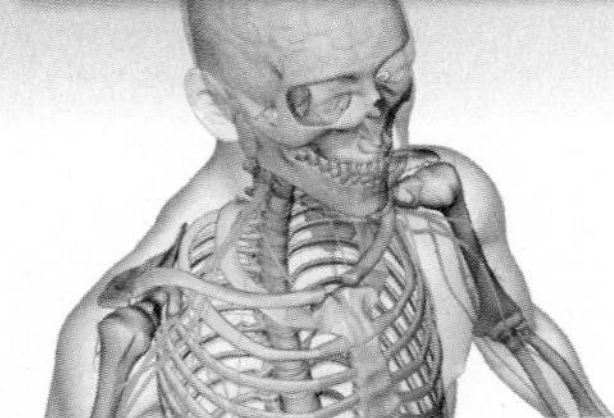

Animals have many organ systems that are interdependent.

CHAPTER 4

Plant Systems

Plants' shoots and roots contain specialized tissues organized into root, stems, leaves, and flowers.

UNIT TASK Preview

Family Health Supporter

Someone you know is ill or injured. Your support and knowledge are needed. You offer to help by researching and providing a package of information related to the particular health concern.

Working in a group, choose one of the suggested health scenarios. You will then consider the issue, research the illness or injury, collect information on how the functioning of a healthy organ system is affected, and find out what diagnostic tests and treatment options are available. You will analyze the information and decide what you think is the best course of action. You will then report your findings to the patient or the patient's family.

Remember that self-diagnosis is not reliable; only health professionals should make diagnoses and recommend treatments, in consultation with the patient or the patient's family.

As you work through this unit, you will be gaining the knowledge and skills to help you successfully complete your task.

UNIT TASK Bookmark

The Unit Task is described in detail on page 156. As you work through the unit, look for this bookmark and see how the section relates to the Unit Task.

ASSESSMENT

You will be assessed on how well you

- address the chosen health issue and present appropriate solutions
- develop a reasoned "best response" for the patient
- communicate this information effectively to the audience

What Do You Already Know?

PREREQUISITES

Concepts
- Cell structure
- Meeting needs for life
- Organization within an organism
- Human organ systems

Skills
- Using a light microscope
- Defining issues
- Identifying benefits and costs of scientific developments

1. All living things are made of cells. Explain how a single-celled organism can do all of the things that a multi-celled organism can do. K/U
2. Use a concept map to show how an *Amoeba* (single-celled organism), a worm, and a human obtain oxygen. K/U C
3. Construct a Venn diagram to list the key differences between plant and animal cells. K/U C
4. (a) What process is illustrated in Figure 1?
   (b) How does this process affect cells? K/U A

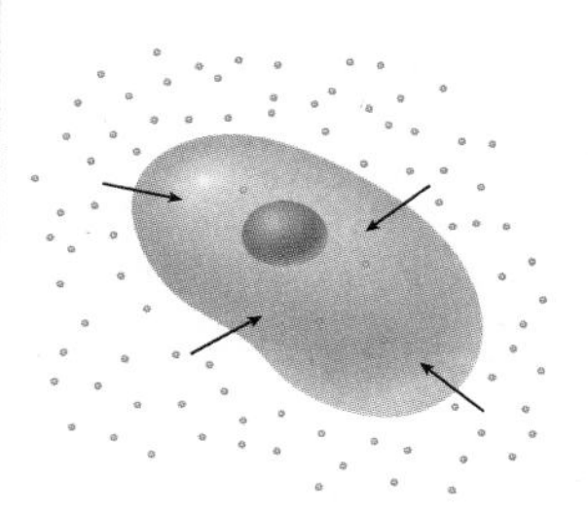
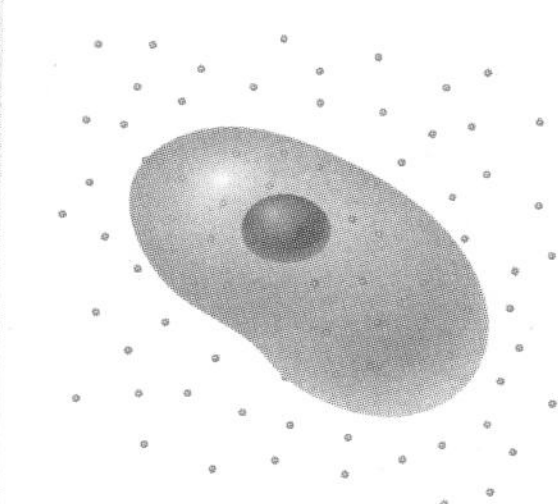

Figure 1 Particles moving into a cell

5. Copy Table 1 into your notebook. Complete the table to describe how the three organisms meet these basic needs (if they do). K/U

Table 1

Organism	Needs		
	Nutrition	Movement	Gas exchange
maple tree			
Amoeba			
human			

6. (a) List four structural differences between a water lily and a human.
   (b) List four structural differences between an *Amoeba* and a human. K/U
7. (a) Create a concept map that includes the following terms: cell; tissue; organ; organ system. Include linking words and examples.
   (b) Extend your concept map by adding other related terms. K/U
8. Match the following organs with their organ systems: K/U

Organ	System
(a) heart	respiratory
(b) stomach	circulatory
(c) lung	nervous
(d) spinal cord	digestive

9. (a) How did you observe cells in past science classes?
   (b) Describe any difficulties that you experienced while trying to examine cells. A
10. Microscopes have played a key role in our knowledge and understanding of cells. How has this technology improved our understanding of the structure of cells? A
11. Think about how you have used microscopes in your studies, and answer the following questions: K/U
    (a) What is the proper way to carry a microscope?
    (b) Which objective lens should you use to start viewing a slide?
    (c) When do you use the coarse-adjustment knob?
    (d) List the steps in preparing a wet-mount slide.
    (e) Describe the correct procedure for using the highest-power objective lens.
    (f) Which objective lens should you place over the stage when putting away the microscope?
12. Scientists have developed some pest-resistant crops by genetic engineering. These plants do not need as many chemical pesticides as normal plants. However, these pest-resistant crops may crossbreed with native plants and cause problems for natural populations. A
    (a) What is the main issue here?
    (b) List at least four possible stakeholders involved in this issue.
    (c) Name one positive impact and one negative impact of pest-resistant crops for each stakeholder.

CHAPTER 2

Cells, Cell Division, and Cell Specialization

KEY QUESTION: How and why do cells divide?

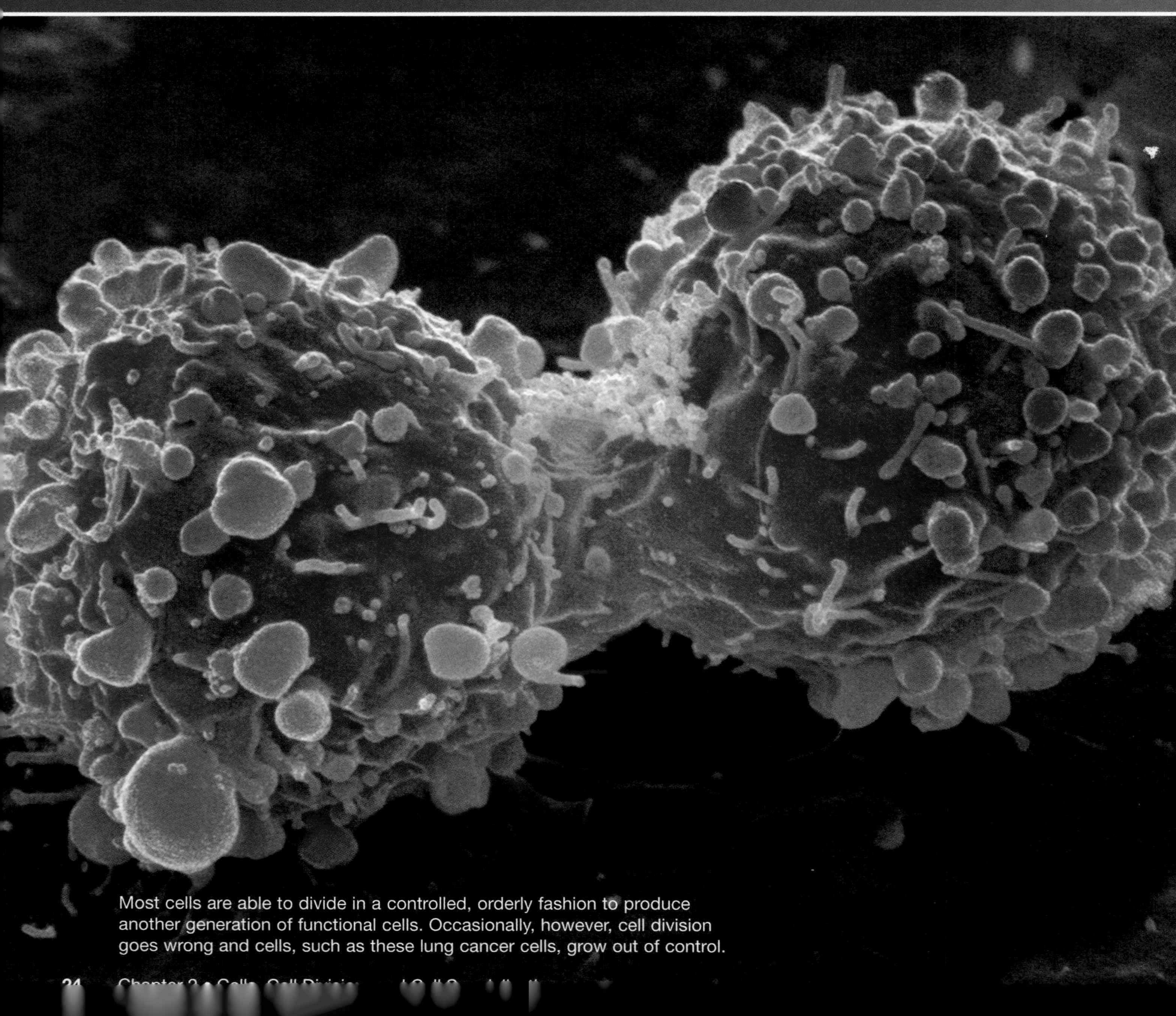

Most cells are able to divide in a controlled, orderly fashion to produce another generation of functional cells. Occasionally, however, cell division goes wrong and cells, such as these lung cancer cells, grow out of control.

UNIT B

Tissues, Organs, and Systems of Living Things

CHAPTER 2

Cells, Cell Division, and Cell Specialization

CHAPTER 3

Animal Systems

CHAPTER 4

Plant Systems

KEY CONCEPTS

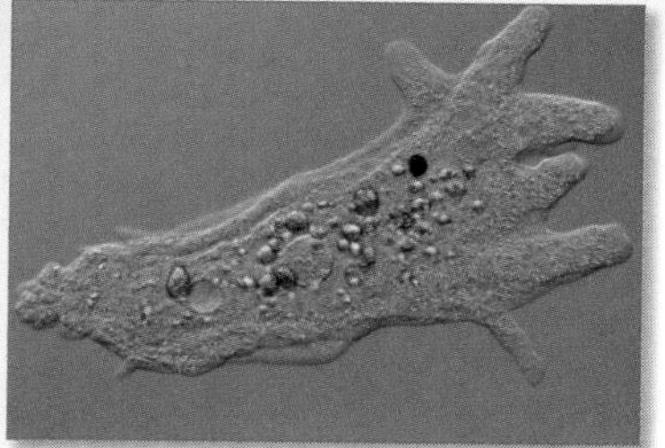

All organisms are made up of one or more cells.

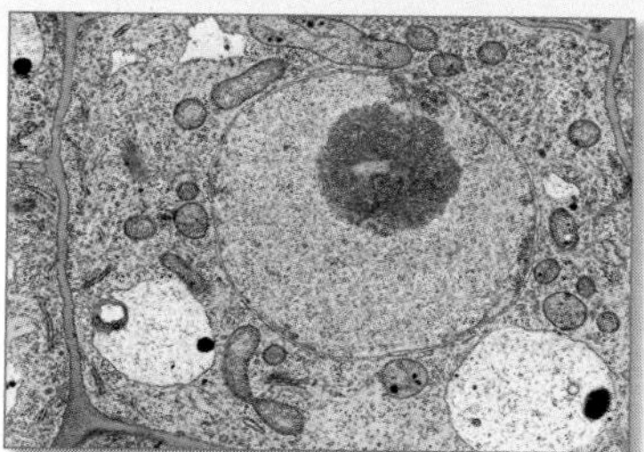

Microscopes enable us to examine cells in detail.

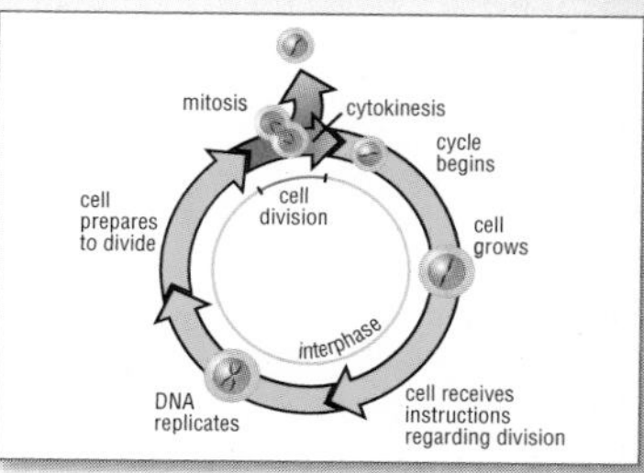

The cell cycle occurs in distinct stages.

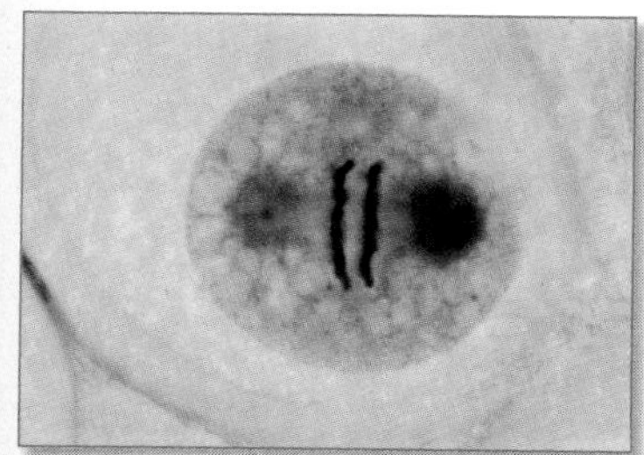

Cell division is important for growth, repair, and reproduction.

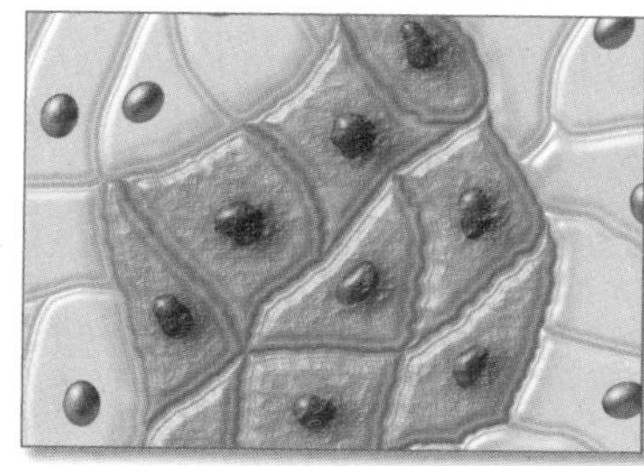

Cancer cells generally divide more rapidly than normal cells.

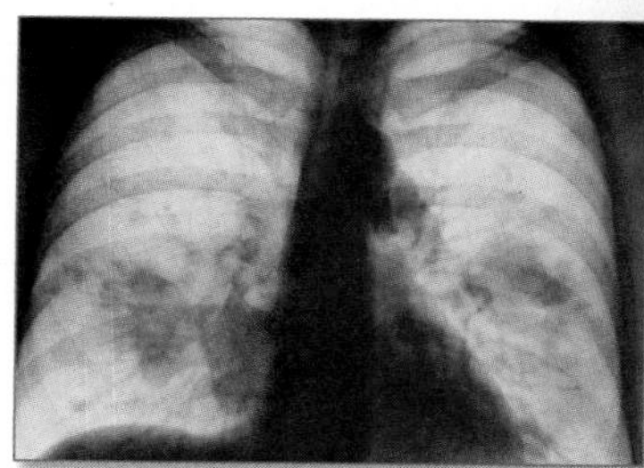

Medical imaging technologies are important in diagnosing and treating disease.

ENGAGE IN **SCIENCE**

You Can Make a Difference

Figure 1 The Canadian Cancer Society started selling daffodils as a fundraiser in 1957. Over 2 million daffodils are sold each year to support cancer research.

Every day, medical science is finding new treatments—and even cures—for some of the many types of cancer. Research pays, but research also costs. Lab space, chemicals, scientists, and hospitals all require money as research teams seek new treatments and cures.

To date, the Relay for Life has raised over $50 million toward cancer research. The Relay for Life is more than just a fundraiser, although fundraising is one of the goals of the Canadian Cancer Society's country-wide event. People all over Canada join the Relay for Life to celebrate cancer survivors, remember friends and family who have died of cancer, raise awareness of cancer research, and fight back. Participants believe that cancer can be beaten and are doing what they can to make that happen.

Every year in over one hundred communities in Ontario, people form relay teams and sign up sponsors. The teams take turns walking or jogging around a track through the night. Candles, music, banners, and visiting celebrities and entertainers add to the up-beat, hopeful mood. This is not a race; everybody wins. A sense of community and support develops. As the sun rises, participants celebrate with a renewed belief that we will find a cure for cancer.

No idea is too far-fetched for a fundraiser for medical research. A Toronto radio station sponsored a Ride for the Cure where bicyclists rode over 200 km from Toronto to Niagara Falls. A Waterloo archery shop held a Shoot for the Cure, in which sponsors rewarded good scores with bonus donations and raised over $6000 in their first year. The Canadian Breast Cancer Foundation and the Pink Ribbon campaign raise public awareness and money for breast cancer research.

Fundraising and awareness-raising can take place on any scale and do not have to involve athletic feats. Any activity can become a focus for change and can lead to a cure for cancer or for some other disease. Join an existing event or start one of your own. Knit for the Cure: make blankets for children with cancer while sponsors donate to your cause. Skate for the Cure: collect money for an ice-fun marathon. Ski down sponsored runs. Have a dance-a-thon. Sew, read, wash cars, ...

What is your school doing to help find a cure? What do you like to do? Could it become the next event to fund a cure for a disease? Is there a school club, team, group that you can organize? Every idea starts with one person who believes and inspires others.

You *can* make a difference.

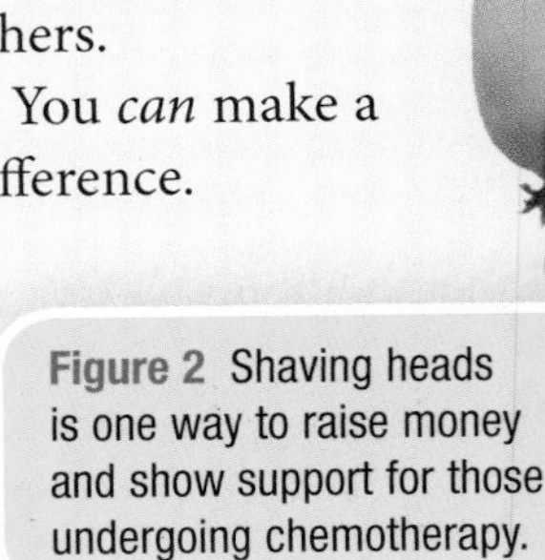

Figure 2 Shaving heads is one way to raise money and show support for those undergoing chemotherapy.

WHAT DO YOU
THINK?

Many of the ideas you will explore in this chapter are ideas that you have already encountered. You may have encountered these ideas in school, at home, or in the world around you. Not all of the following statements are true. Consider each statement and decide whether you agree or disagree with it.

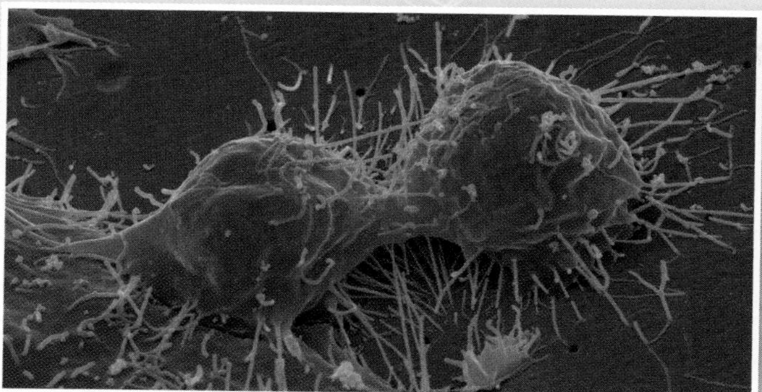

1 New cells come from previously existing cells.
Agree/disagree?

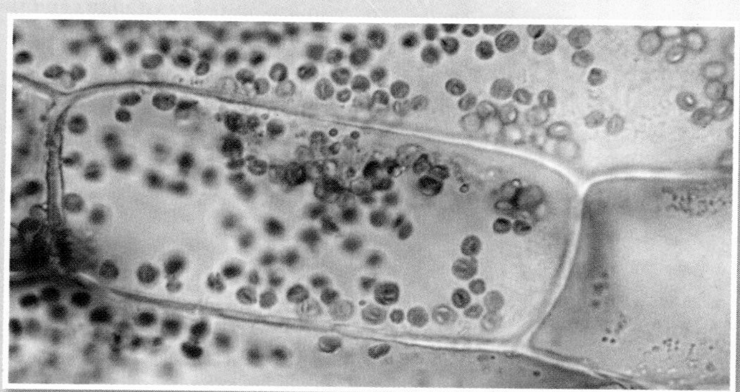

2 All plant cells are green.
Agree/disagree?

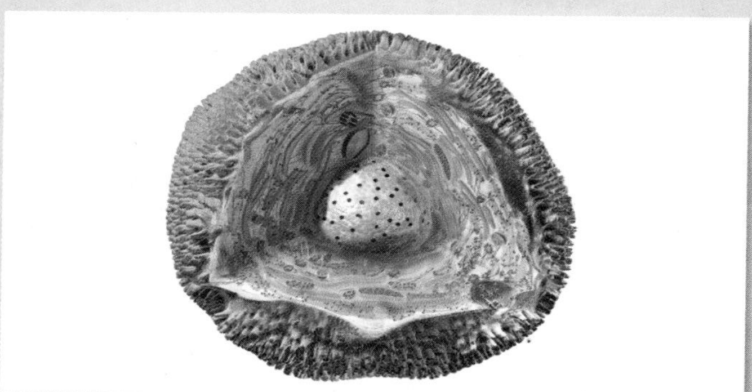

3 Plant and animal cells contain all the same organelles.
Agree/disagree?

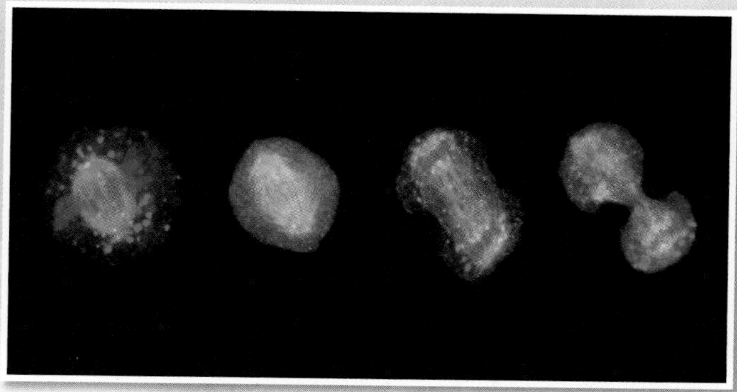

4 Mitosis and the cell cycle are identical processes.
Agree/disagree?

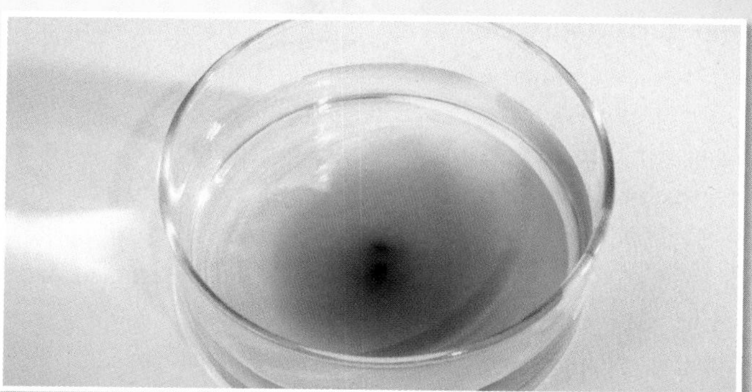

5 Diffusion is the movement of particles from an area of higher concentration to an area of lower concentration.
Agree/disagree?

6 All radiation is harmful to humans.
Agree/disagree?

Making Connections

When you make connections with a text, you are linking ideas and information in a new text with personal experiences or with your knowledge of other texts or world events. There are three main types of connections you can make to help you interpret a text.

- **Text-to-Self:** making connections between the text and your own experiences
- **Text-to-Text:** making connections between the text and other texts you have read or viewed.
- **Text-to-World:** making connections between events and issues in the new text and the world.

READING TIP

As you work through the chapter, look for tips like this. They will help you develop literacy strategies.

Cell Growth Rates and Cancer

A cancer cell is one that continues to divide despite messages from the nucleus or the surrounding cells to stop. The cell's checkpoints may fail to identify problems or kill off the cell. The uncontrolled growth and division may create a rapidly growing mass of cells that form a lump or tumour (Figure 1). The cells of the tumour may stay together and have no serious effect on surrounding tissues. This is called a benign tumour. Cells in a benign tumour are not cancerous. However, sometimes a benign tumour can grow so large that it physically crowds nearby cells and tissues. This can affect their normal function.

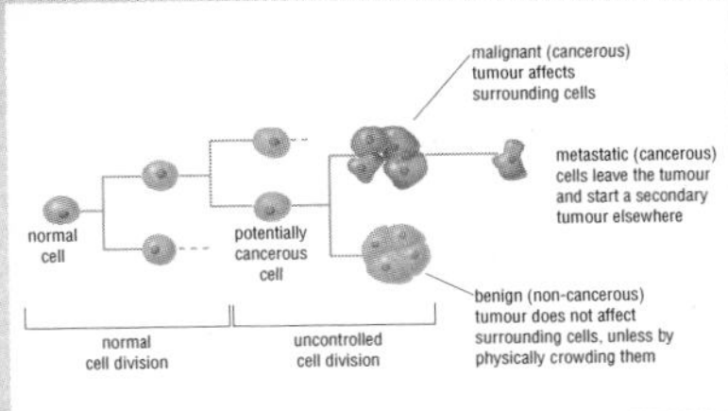

malignant (cancerous) tumour affects surrounding cells
metastatic (cancerous) cells leave the tumour and start a secondary tumour elsewhere
normal cell
potentially cancerous cell
benign (non-cancerous) tumour does not affect surrounding cells, unless by physically crowding them
normal cell division
uncontrolled cell division

Figure 1 A tumour is a mass of cells with no function. A tumour can remain benign, or it can become malignant. Tumour cells can metastasize, spreading to other areas of the body. Malignant and metastatic tumours are considered cancerous.

Making Connections *in Action*

Making connections can help you understand a text by visualizing, making inferences, forming opinions, or drawing conclusions. Here's how one student made connections with the text on cancer cells.

Connections I Made to the Text	How the Connection Helped Me Understand and Respond to the Text
Text-to-Self Connection: about my uncle whose prostate cancer was detected early and treated successfully	helps me form an opinion about the importance of regular medical check ups
Text-to-Text Connection: between graphic and *Invading Armies* video game	helps me visualize how cancer spreads through the body
Text-to-World Connection: between text and statistics about cancer as a leading cause of death in Canada	helps me draw conclusions about why almost 50 % of people with cancer die

Plant and Animal Cells

Biology as a science is built on three simple but very important ideas. These three ideas form the cell theory. The **cell theory** states that

1. All living things are made up of one or more cells and their products.
2. The cell is the simplest unit that can carry out all life processes.
3. All cells come from other cells; they do not come from non-living matter.

cell theory a theory that all living things are made up of one or more cells, that cells are the basic unit of life, and that all cells come from pre-existing cells

All living things are made up of cells, but these cells may be very simple or very complex. The simplest organisms are archaea and bacteria. These simple, single-celled life forms are called **prokaryotes** (Figure 1(a)). The cells do not have a nucleus. More complex cells can exist as single-celled organisms or multicellular organisms. The cells of these organisms, known as **eukaryotes**, have a more complex internal organization, including a nucleus. Eukaryotes include all protists, fungi, animals, and plants, from the tiniest *Amoeba* to the longest whale and the tallest tree (Figure 1(b) to (d)). The cells of eukaryotes are much larger than the cells of prokaryotes: tens to thousands of times larger. There are even some eukaryotes—beyond the scope of this unit—that are made up of one huge cell with very many nuclei.

prokaryote a cell that does not contain a nucleus or other membrane-bound organelles

eukaryote a cell that contains a nucleus and other organelles, each surrounded by a thin membrane

READING TIP

Making Connections

When trying to make connections with a text, use prompts or questions such as:

- This example reminds me of...
- This graph makes me ask why...
- Are these facts correct?
- Have I read about this before?

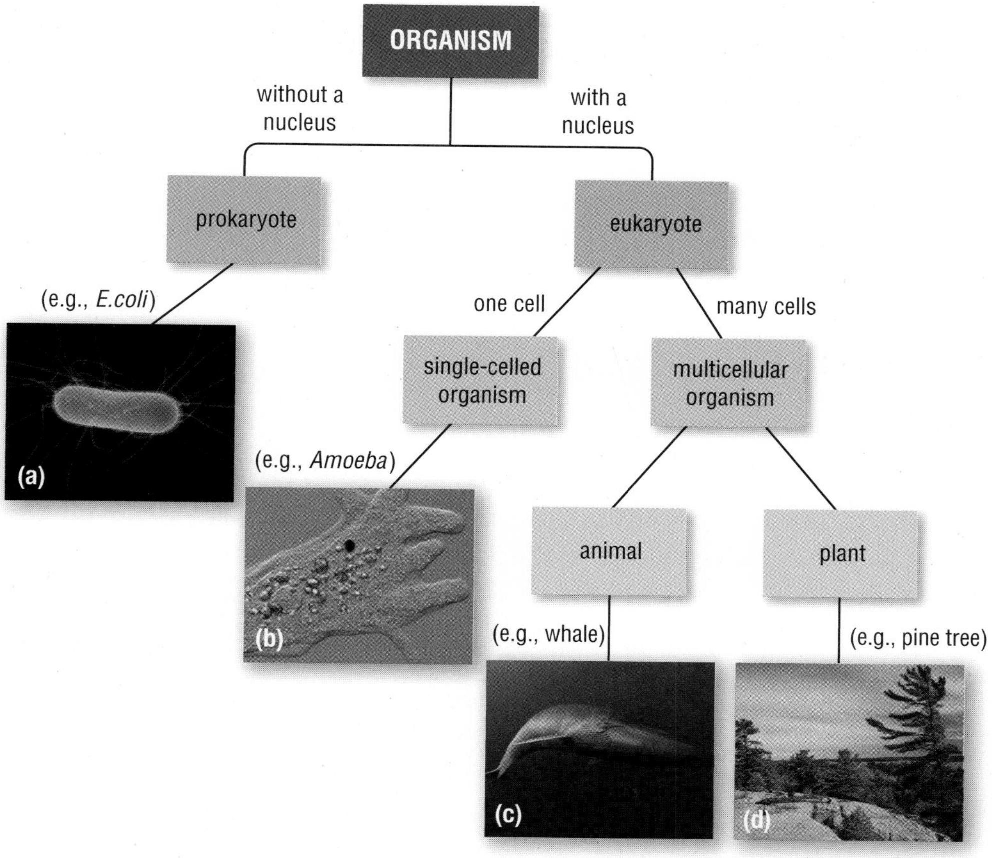

Figure 1 The relationship between prokaryotes and eukaryotes. The bacterium (a) is a prokaryote. The *Amoeba* (b), the whale (c), and the pine tree (d) are all eukaryotes.

Cell Structure

Your body is made up of many specialized organs that carry out all the processes needed to live. In the same way, a eukaryotic cell also has specialized parts, called **organelles**, that carry out specific functions necessary for life.

organelle a cell structure that performs a specific function for the cell

Structures Common to Plants and Animal Cells

All cells have to perform the same basic activities to stay alive: use energy, store materials, take materials from the environment, get rid of wastes, move substances to where they are needed, and reproduce. Each organelle has a specific function within the cell. Just as workers in a factory or a hospital coordinate their efforts to achieve a purpose, the various organelles of a cell work together to meet the needs of the cell—and the whole organism. Figure 2 shows the organelles in a typical plant cell and a typical animal cell.

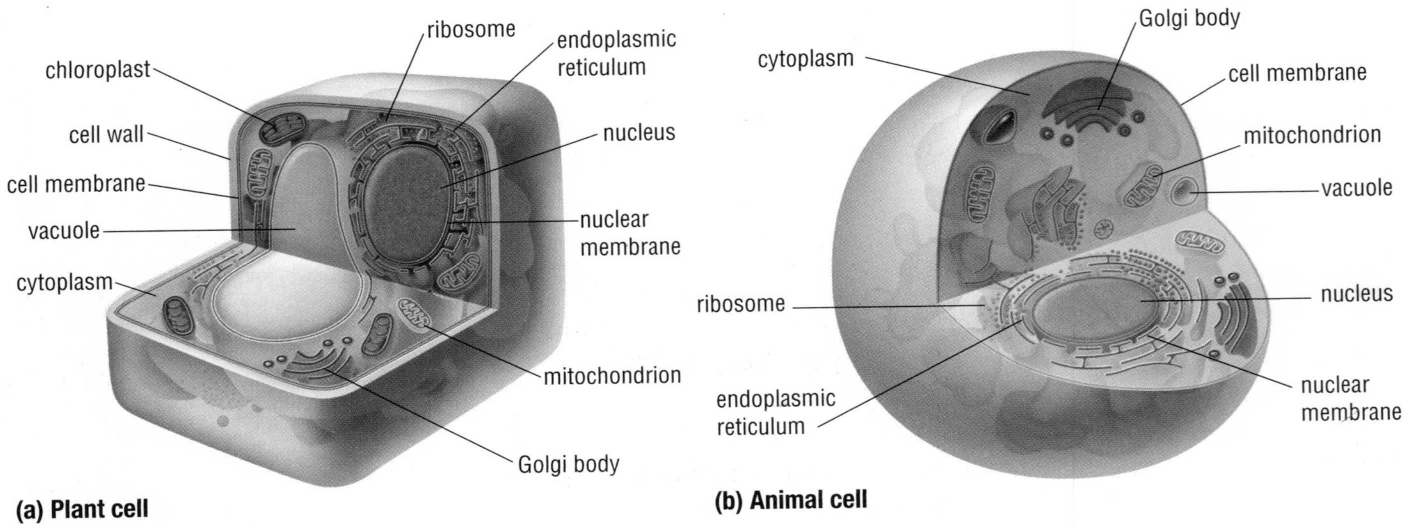

Figure 2 Plant and animal cells have many of the same organelles, but there are some differences.

CYTOPLASM

All the organelles inside the cell are suspended in the cytoplasm. The cytoplasm is mostly water, but it also contains many other substances that the cell stores until they are needed. Many chemical reactions take place within the cytoplasm, which can change from jelly-like to liquid, allowing organelles to be moved around.

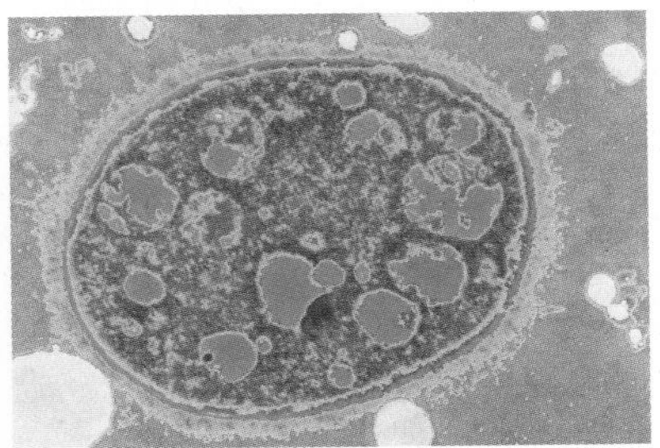

Figure 3 This TEM image of a cell highlights the cell membrane in green.

CELL MEMBRANE

The cell is surrounded by a flexible double-layered cell membrane (Figure 3). The function of the cell membrane is both to support the cell and to allow some substances to enter while keeping others out. For example, water and oxygen molecules can easily pass through the cell membrane, but larger molecules, such as proteins, cannot. Because of this ability, the cell membrane is called a "semi-permeable membrane."

A similar membrane also surrounds most organelles in a eukaryotic cell.

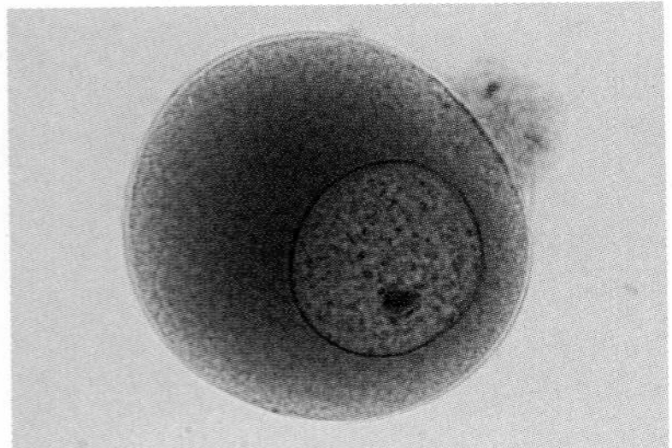

Figure 4 The large nucleus is easily visible inside this starfish cell.

NUCLEUS

The nucleus is a roughly spherical structure within the cell (Figure 4). The nucleus contains genetic information that controls all cell activities. This genetic information is stored on chromosomes. Chromosomes contain **DNA (deoxyribonucleic acid)**, the substance that carries the coded instructions for all cell activity. When a cell divides, the DNA is copied so that each new cell has a complete set of chromosmes.

DNA (deoxyribonucleic acid) the material in the nucleus of a cell that contains all of the cell's genetic information

MITOCHONDRIA

Cells contain many mitochondria (singular: mitochondrion) (Figure 5). Mitochondria are sometimes called the "power plants" of the cell because they make energy available to the cell. Active cells, such as muscle cells, have more mitochondria than less active cells, such as fat-storage cells. Cells store energy as a form of glucose (a sugar). The mitochondria contain enzymes that help to convert the stored energy into an easily usable form. This process is called cellular respiration and requires oxygen. The waste products of this reaction are carbon dioxide and water.

$$\text{glucose} + \text{oxygen} \rightarrow \text{carbon dioxide} + \text{water} + \text{usable energy}$$

Cells in which cellular respiration has to happen very fast, such as muscle cells and cells in the liver, have many mitochondria. In contrast, cells that are fairly inactive—that do not have to respire quickly—tend to have very few mitochondria. Fat cells may have only one or two mitochondria.

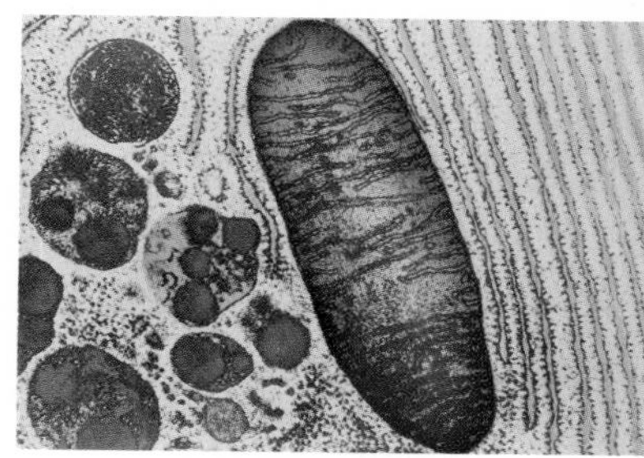

Figure 5 The mitochondrion (17 000 ×) is the large, reddish, oval structure in this TEM image.

ENDOPLASMIC RETICULUM

The endoplasmic reticulum is a three-dimensional network of branching tubes and pockets (Figure 6). It extends throughout the cytoplasm and is continuous from the nuclear membrane to the cell membrane. These fluid-filled tubes transport materials, such as proteins, through the cell.

Endoplasmic reticulum is important in many types of cells. In the brain it assists with the production and release of hormones. In the muscles the endoplasmic reticulum is involved with muscle contraction.

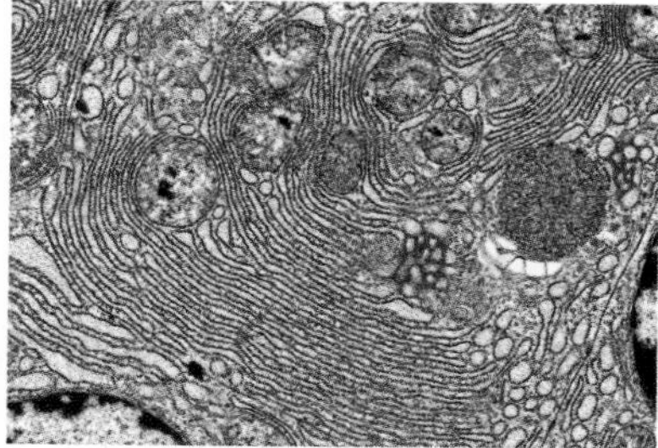

Figure 6 The endoplasmic reticulum (5 500x), coloured brown in this TEM, transports materials throughout the cell.

GOLGI BODIES

Golgi bodies collect and process materials to be removed from the cell (Figure 7). They also make and secrete mucus. Cells that secrete a lot of mucus, such as cells lining the intestine, have many Golgi bodies.

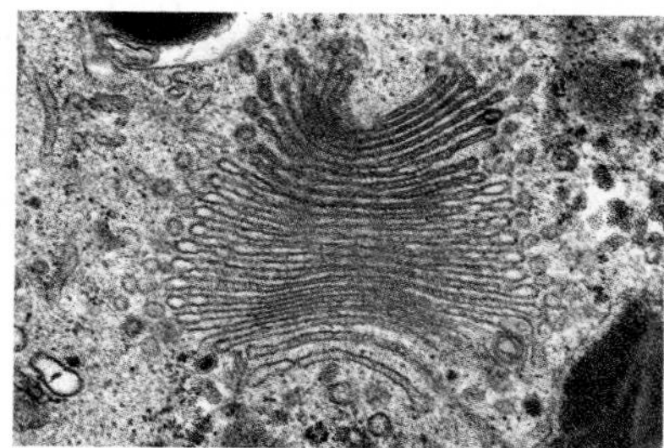

Figure 7 Golgi body (30 000×)

VACUOLES

A vacuole is a single layer of membrane enclosing fluid in a sac. The functions of vacuoles vary greatly, according to the type of cell. These functions include containing some substances, removing unwanted substances from the cell, and maintaining internal fluid pressure (turgor) within the cell. (The special role of plant vacuoles is explained below.) Animal cells may have many small vacuoles that are often not visible. Mature plant cells usually have one central vacuole that is visible under a light microscope.

Some animal cells can change their shape to wrap around and surround smaller objects to bring them inside the cell. *Amoeba* do this to obtain food. Some white blood cells engulf bacteria to kill them. During the engulfing process, a portion of the cell membrane turns inside out and forms a vacuole inside the cell until the engulfed object is digested. Then any waste material is ejected from the cell as the vacuole again joins up with the cell membrane.

Organelles in Plants Cells Only

Plant cells and animal cells have many structures in common, but there are also some differences. Plant cells have some organelles that animal cells do not have (Figure 8).

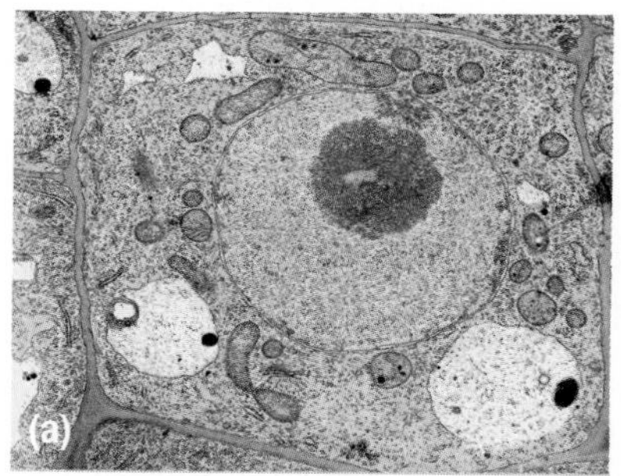

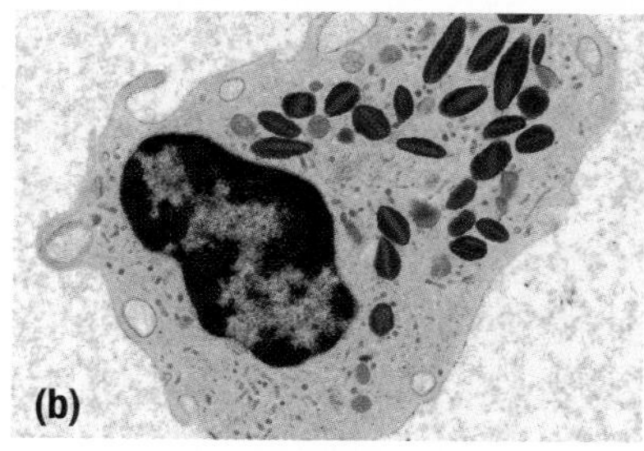

Figure 8 Plant cells (a) have a cell wall, large vacuoles, and chloroplasts (5 500×). Animal cells (b) do not (2 500×).

CELL WALL

The cell wall is found just outside the cell membrane of a plant cell. It is a rigid but porous structure made of cellulose. The cell wall provides support for the cell and protection from physical injury. The cellulose may hold together long after the plant has died. The paper in this book is composed mostly of cellulose from the cell walls of trees.

VACUOLE

Plant cells usually have one large vacuole, which takes up most of the space inside the cell. When these are full of water, turgor pressure keeps the cells plump, which keeps the plant's stems and leaves firm. If the water level drops, however, the vacuoles lose turgor pressure and the cells become soft. The plant stems and leaves become limp and droopy until the water is replaced.

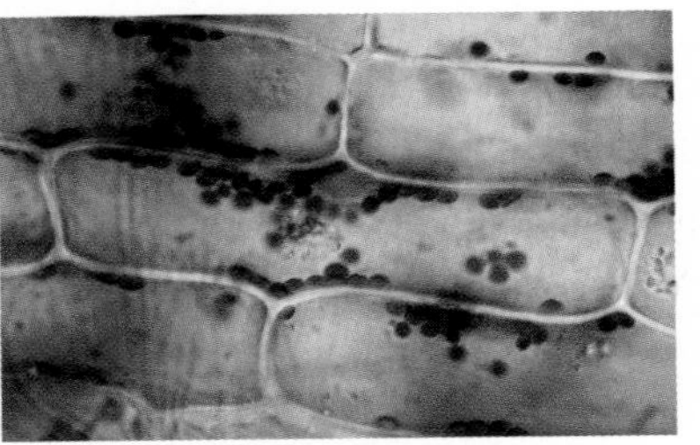

Figure 9 Chloroplasts in plant cells (250×)

CHLOROPLASTS

Many plant cells that are exposed to light, such as the cells of leaves, have structures called chloroplasts (Figure 9). Chloroplasts contain chlorophyll and give leaves their green colour. More importantly, chloroplasts absorb light energy. This light energy is used in photosynthesis—the process of converting carbon dioxide and water into glucose and oxygen.

carbon dioxide + water + energy (sunlight) → glucose + oxygen

Photosynthesis allows plants to obtain their energy from the Sun so that they can make their own food. Plant cells rely on mitochondria to metabolize glucose, just as animal cells do.

IN SUMMARY

- The cell theory states that all living things are made up of cells, the cell is the simplest unit that can carry out all life processes, and all cells are reproduced from other cells.
- The simplest single-celled organisms, including bacteria, are prokaryotes. More complex organisms, including multicellular organisms, are eukaryotes.
- Eukaryotic cells contain organelles that carry out specific life functions.
- The cell membrane, cytoplasm, nucleus, mitochondria, endoplasmic reticulum, Golgi bodies, and vacuoles occur in both plant and animal cells.
- Structures found only in plant cells are chloroplasts, a large vacuole, and the cell wall.

CHECK YOUR LEARNING

1. Summarize the cell theory in your own words. K/U
2. Are your cells prokaryotic or eukaryotic? Explain. K/U
3. What is the most obvious difference between prokaryotic and eukaryotic cells? K/U
4. How does the nucleus coordinate cell activities? K/U
5. When you exercise, you breathe harder and faster. Using your knowledge of organelles, explain why this happens. A
6. Not all plant cells contain chloroplasts. What is the most likely reason for this? K/U
7. Plant cells are surrounded by a cell wall. What is the function of this structure? K/U
8. Plant cells can make their own "food"—glucose. Why do plant cells have mitochondria? K/U

Seeing Inside

Scientists have been exploring the inner workings of our bodies with microscopes for centuries. Traditional light microscopes can only reveal so much. The very nature of light limits the size of objects that we can see. Also, it is not possible to look into solid tissues using traditional microscopy. However, new technology is extending our ability to see inside tissues and organs. The resulting images are revolutionizing how disease is diagnosed.

One development is called "confocal microscopy." This technique uses a process called fluorescence. A material fluoresces (glows) if it absorbs ultraviolet (UV) light and immediately emits visible light. In confocal microscopy, scientists incorporate fluorescent substances into the tissues or cells to be examined. These fluorescent markers glow when viewed under the microscope. Different fluorescent markers glow when they absorb different wavelengths of light. Confocal microscopy allows scientists to look at fluorescence on more than one plane. It does this by concentrating tiny beams of light on the specimen, instead of bathing the whole specimen in light from a single source. In this way, the viewer is able to see much more detail at high magnification, and in three dimensions (Figure 1). Confocal microscopy is used to diagnose a wide range of medical conditions including epilepsy, eye diseases, genetic disorders such as Alport syndrome, and skin cancer. We are still finding new uses for the technology.

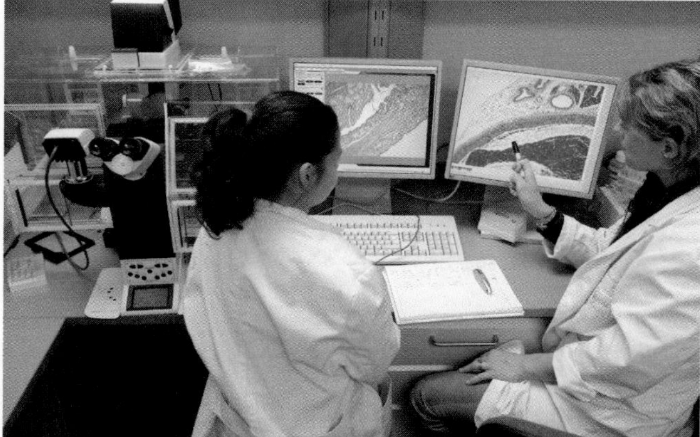

Figure 1 A confocal microscope

Multiphoton microscopy (MPM) is an extension of confocal microscopy. Mulitphoton microscopes deliver a precise infrared (IR) laser pulse to a desired point within a sample of cells. A fluorescent marker in the sample absorbs the IR light and glows at that specific point. Infrared light does not scatter as much as visible light when it passes through the specimen, so a scientist can "look" deeper into the sample (Figure 2). MPM is not damaging to living cells. In fact, it is so safe that doctors use MPM to diagnose eye conditions. Researchers are exploring whether this technology can also be used to understand brain diseases and cancer. It might be possible to find cancer in internal organs, such as the bladder and colon, by using a probe with a tiny camera to obtain an image. Then doctors could get enough information to make a diagnosis without performing surgery.

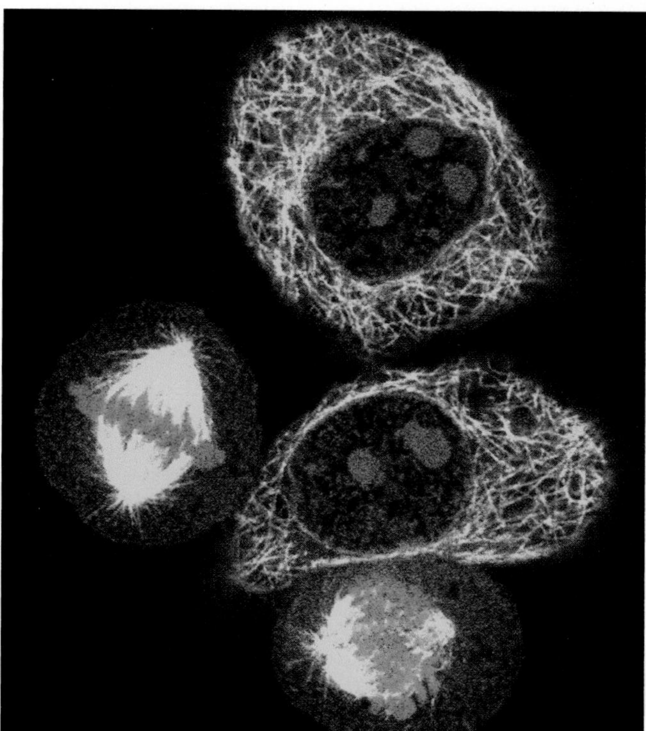

Figure 2 Recent developments in microscopy give scientists new insights into the structure and funtion of our bodies.

These technologies are just the beginning. Other emerging methods, such as digital holographic microscopy and three-dimensional structured illumination, might some day allow doctors to see everything going on inside our living cells, tissues, and organs up close and in detail.

GO TO NELSON SCIENCE

Observing Plant and Animal Cells

In this activity, you will observe and compare a typical plant cell and a typical animal cell. You will look at the organelles of each and identify the major differences. The plant cell will be taken from an onion; the animal cell will be prepared slides of human cheek cells. After observing these cells using a microscope, you will draw formal lab diagrams to record your observations.

SKILLS MENU

- Questioning
- Hypothesizing
- Predicting
- Planning
- Controlling Variables
- Performing
- Observing
- Analyzing
- Evaluating
- Communicating

Purpose

Part A: To prepare a wet mount of onion cells, to study one cell using a microscope, and to draw and label a diagram of the visible features.

Part B: To examine a prepared slide of human cheek cells, to study one cell using a microscope, and to draw and label a diagram of the visible features.

Equipment and Materials

- lab apron
- disposable gloves
- clean microscope slide and cover slip
- forceps or tweezers
- microscope
- dropper bottles of
  - distilled water
  - iodine stain
- onion
- paper towel
- prepared slide of human cheek cells

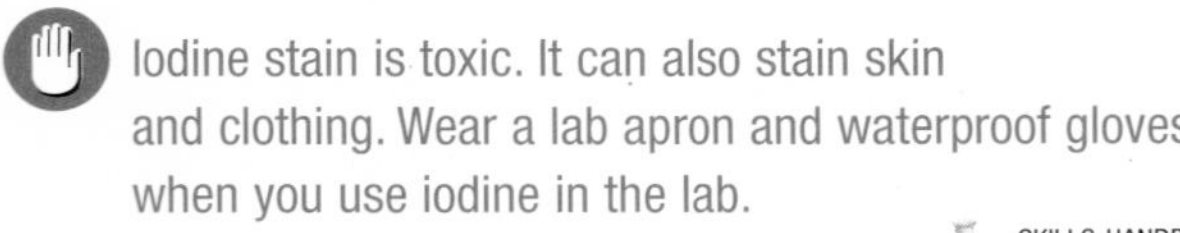

Iodine stain is toxic. It can also stain skin and clothing. Wear a lab apron and waterproof gloves when you use iodine in the lab.

SKILLS HANDBOOK 2.D., 3.B.6.

Procedure

Part A: Preparing a Wet Mount of Onion Cells

1. Put on your lab apron and the disposable gloves.
2. Put one drop of distilled water on a clean microscope slide.
3. Using the forceps, peel the inner translucent membrane from a small piece of onion.
4. Carefully place this membrane on the drop of water.
5. Carefully lower the cover slip from one side to cover the sample.
6. Observe the sample using low power. Focus using the coarse-adjustment knob.
7. Draw and label a sketch to show how the individual cells fit together.
8. Stain the specimen by putting one drop of iodine stain at one edge of the cover slip. Hold a piece of paper towel at the opposite edge to draw the stain across, under the cover slip (Figure 1).

Figure 1 Staining an onion skin wet mount with iodine

9. Observe the sample using medium power. Focus using the fine-adjustment knob.
10. Observe one cell. Draw and label a diagram of the visible structures.
11. Rinse the slide and clean it as instructed.
12. If your teacher provides prepared slides of other plant cells, look at these and record your observations.

Part B: Examining a Prepared Slide of Cheek Cells

13. Observe the provided slide of human cheek cells under the microscope using low power. Try to find an area of the slide where the cells do not overlap. Focus on one cell using the coarse-adjustment knob.
14. Observe the sample using medium power. Focus on one cell using the fine-adjustment knob.
15. Observe one cell. If necessary, use the high-power objective. Draw and label a diagram of the cell and any visible structures.

Analyze and Evaluate

(a) What structures were you able to identify in the onion cells? T/I

(b) What structures were you able to identify in the cheek cells? T/I

(c) What was the most obvious difference between the two types of cells? T/I

(d) Are onion cells typical plant cells? Explain. T/I

(e) Suggest why onions and humans were used as the source of cells in this activity. A

(f) What difference, if any, did you notice in the cells before and after they were stained with iodine? Why were you advised to stain the cells? T/I

Apply and Extend

(g) If you looked at other plant cells in this activity, describe how they differed from an onion cell. Give reasons for these differences. T/I

(h) Most books state that plant cells contain chloroplasts. Based on your observations, is this statement correct? Support and explain your answer. T/I

(i) Would you expect all animal cells to look the same? Explain your answer. T/I

2.3 The Importance of Cell Division

You started life as a single cell: a fertilized egg. Now your body is made up of trillions of cells. How does a single cell become a full-grown multicellular plant or animal? Cell division allows organisms to reproduce, to grow, and to repair damage.

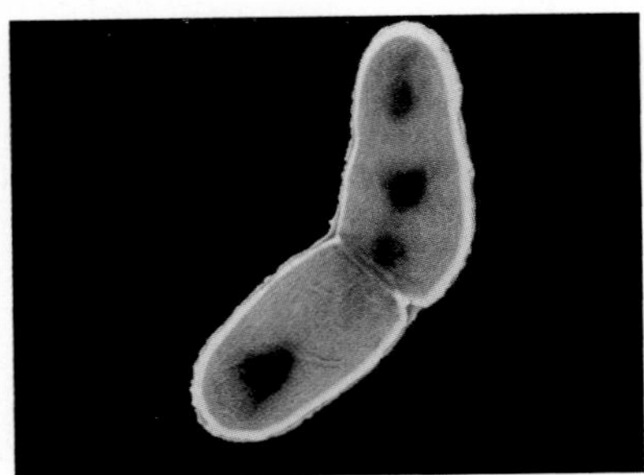

Figure 1 Bacteria produce new individuals simply by dividing in two. Under ideal conditions, bacteria can double their numbers every 20 minutes. (18 000×)

asexual reproduction the process of producing offspring from only one parent; the production of offspring that are genetically identical to the parent

sexual reproduction the process of producing offspring by the fusion of two gametes; the production of offspring that have genetic information from each parent

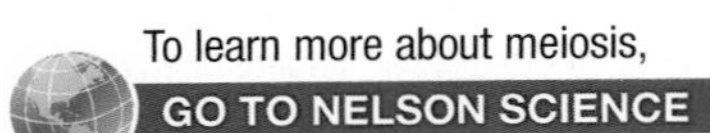

To learn more about meiosis, GO TO NELSON SCIENCE

Cell Division for Reproduction

The ability to reproduce is an important characteristic of all living things, from bacteria to elephants.

All cells, including single-celled organisms, use cell division to reproduce. Each time a parent cell divides, it results in two new organisms (Figure 1). Each organism inherits genetic information from its parent. It is very important that each new cell has a complete set of genetic information. This type of reproduction, called **asexual reproduction**, involves only one parent. The offspring are exact genetic copies of the parent.

Multicellular organisms also need to reproduce and pass their genetic information along to their offspring. Some multicellular organisms can produce offspring by asexual reproduction (Figure 2). Only one parent is involved, and the young will have exactly the same DNA as that parent.

We are more familiar with the idea of **sexual reproduction**, in which a cell from one parent joins with a cell from another parent. These two parental cells are different from normal body cells: they contain only half of the DNA usually found in a cell. These "half cells" are known as gametes. To produce gametes, some of the parents' cells undergo an additional cell division process called meiosis. When the two gametes combine, the offspring inherits characteristics from both parents (Figure 3).

Figure 2 This houseleek is reproducing asexually. Each rosette is an individual plant that is genetically identical to the parent.

Figure 3 Sexual reproduction results in offspring that have characteristics of both parents. This is because their cells contain DNA from both parents.

Cell Division for Growth

One of the main characteristics of life is that all organisms grow. As multicellular organisms grow, the number of cells increases. Why does the number of cells increase instead of simply increasing the size of the cells? This has to do with how a cell uses chemicals to function.

Plant and animal cells all need the same things: a source of energy, nutrients, water, and gases (Figure 4). Many chemicals need to be in solution (dissolved in water) so that they can be used in chemical reactions within the cell. It is therefore very important that cells contain plenty of water. The cell must excrete (get rid of) carbon dioxide and other waste products.

Why does the number of cells increase as an organism grows? Chemicals used during cell activity and growth enter the cell across the membrane and travel through the cell to where they are used. This movement of chemicals occurs by **diffusion**. Chemicals diffuse from an area of higher concentration to an area of lower concentration (Figure 5). **Concentration** is the amount of substance (the solute) in a given volume of solution.

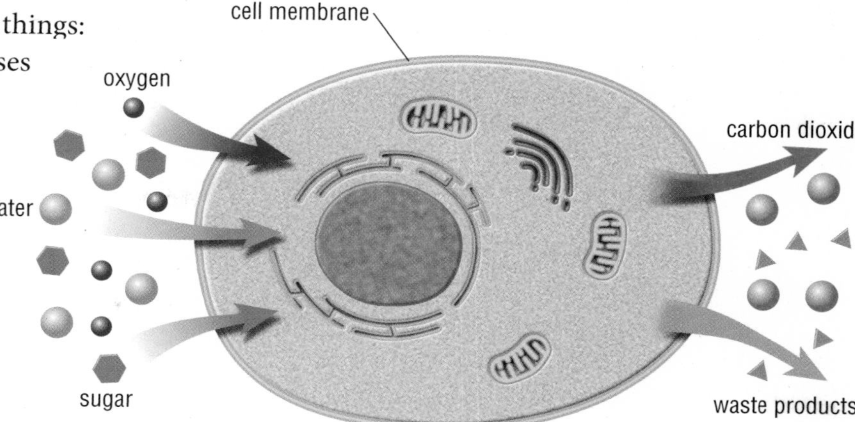

Figure 4 Substances entering and leaving a typical animal cell

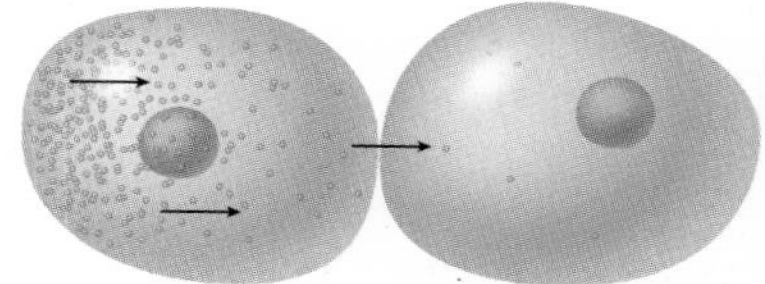

Figure 5 The red particles diffuse through the first cell and into the next one.

Water enters and leaves cells by a process called **osmosis**. In osmosis, water moves in the direction of greater solute concentration.

Diffusion and osmosis take time. Important chemicals must be available to all parts of the cell, in the right amount of water, for the cell to function properly. Waste products must also diffuse out of the cell quickly so they do not poison the cell. When a cell gets too large, chemicals and water cannot move through the cell fast enough.

diffusion a transport mechanism for moving chemicals into and out of the cell, from an area of higher concentration to an area of lower concentration

concentration the amount of a substance (solute) present in a given volume of solution

osmosis the movement of a fluid, usually water, across a membrane toward an area of high solute concentration

Cell Division for Repair

Every day, your body sheds millions of dead skin cells, all of which are replaced by new ones. Your body replaces each red blood cell about every 120 days. If you break a bone, cells divide to heal the break. Every cut and blister needs new cells to fill in the gaps. All organisms need to repair themselves to stay alive.

To learn more about osmosis,
GO TO NELSON SCIENCE

IN SUMMARY

- Cells undergo cell division for reproduction, growth, and repair.
- Reproduction involves the transfer of genetic information from the parent(s) to the offspring.
- As multicellular organisms grow, their cells duplicate their genetic information and divide.
- Chemicals diffuse into, throughout, and out of cells. This process must happen quickly enough for the cell to function.
- When part of an organism is damaged, the remaining cells divide to repair the injury.

CHECK YOUR LEARNING

1. Name three reasons for cell division. K/U
2. A cleaning product claims to kill "99.9 % of all bacteria." Will a cleaned surface stay bacteria-free forever? Explain your answer. A
3. List three differences between asexual reproduction and sexual reproduction. K/U
4. What processes are responsible for chemicals moving into, throughout, and out of cells? K/U
5. Why do cells divide instead of just getting bigger, as an organism grows? K/U
6. A minor wound heals over time. Explain how this happens. A

What Limits Cell Size?

SKILLS MENU

- Questioning
- Hypothesizing
- Predicting
- Planning
- Controlling Variables
- Performing
- Observing
- Analyzing
- Evaluating
- Communicating

Why do individual cells stop growing and start dividing when they reach a certain size?

In this investigation you will use models of cells to investigate factors that might limit cell size. The model cells contain a substance that changes from colourless to pink when it comes into contact with sodium hydroxide.

Testable Question

How does the size of a cell affect the distribution of chemicals throughout the cell?

Hypothesis/Prediction

Predict an answer to the Testable Question, using what you know about diffusion in cells.

Experimental Design

You will use different-sized agar cubes as models of living cells. You will submerge these cells in a solution of sodium hydroxide for 10 min. You will observe whether sodium hydroxide diffuses to the centre of a large cell as quickly as it diffuses to the centre of a small cell. You will also investigate the relationship between the surface area and the volume of different-sized cells and draw conclusions about maximum cell size.

Equipment and Materials

- eye protection
- lab apron
- disposable gloves
- 250 mL beaker or plastic cup
- 2 glass rods or stir sticks
- timing device
- scoopula
- ruler
- scalpel
- 3 different-sized cubes of phenolphthalein agar
- 100 mL dilute sodium hydroxide solution (0.1 mol/L)
- paper towels

Sodium hydroxide solution is corrosive and will irritate your skin or eyes. Avoid splashing this solution on your skin, in your eyes, or on clothing. Immediately rinse any spills with cold water and inform your teacher of the spill. Carefully follow your teacher's instructions for disposal.

Procedure

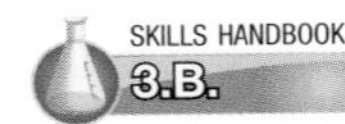

1. Put on your eye protection, lab apron, and gloves. Place your three agar "cells" into the beaker or cup.
2. Pour enough sodium hydroxide solution into the beaker to cover the cells completely.
3. Allow your cells to remain in the sodium hydroxide solution for 10 min. Use the glass rods or stir sticks to gently turn the cells often. Make sure that all sides of the cells come in contact with the solution. Avoid cutting or scratching the cells when you turn them.
4. After 10 min, remove the cells from the beaker with the scoopula. Blot each cell dry with the paper towels.
5. Copy Table 1 (on the next page) into your notebook.
6. Carefully measure the length of one side of one cell (cell A) in mm. Record this measurement in the appropriate place in your table.
7. Carefully cut cell A in half with the scalpel.

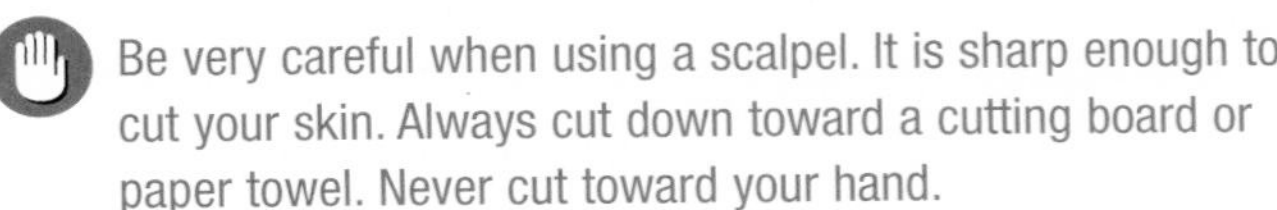

Be very careful when using a scalpel. It is sharp enough to cut your skin. Always cut down toward a cutting board or paper towel. Never cut toward your hand.

8. Where the sodium hydroxide diffused into the cells, the agar will have turned pink or purple. Measure how far the colour change extends into cell A. Record this measurement in your table.

Table 1 Observations of Three Cells in Sodium Hydroxide Solution

Cell	Length of one side (mm)	Area of one side (mm^2)	Total surface area (mm^2)	Volume of cell (mm^3)	Ratio of surface area: volume	Distance colour extended into cube (mm)
A						
B						
C						

9. Repeat Steps 6 to 8 with cell B and then with cell C.
10. Follow your teacher's instructions for proper disposal of all used materials.

Analyze and Evaluate

(a) What does the colour change in your cells indicate about the diffusion of the sodium hydroxide? T/I

(b) Did all of the cells change colour all the way through? Explain. T/I

(c) Calculate the area of one side of each cell. Record these values in your table. T/I

(d) Calculate the total surface area (for all six sides) of each cell. Record these values in your table. T/I

(e) Calculate the volume (length × width × height) of each cell. Record these values in your table. T/I

(f) Calculate the diffusion rate in millimetres per minute for each cell. Record this information in your table. T/I

(g) How does the surface area to volume ratio change as the cell gets larger? T/I

(h) Answer the Testable Question. T/I

(i) Compare your answer in (h) to your Hypothesis/ Prediction. Was your Prediction correct? T/I

(j) Imagine that sodium hydroxide represents a vital nutrient needed throughout a living cell. Did all parts of each cell obtain the "nutrient" within the 10 min time period? Which cell was able to have the nutrient diffuse into all (or most) of its volume? Explain, using the terms *surface area* and *volume*. T/I

(k) If all cell activity takes place in the cell's interior but all materials enter and exit through the cell's surface, explain the importance of the surface area to volume ratio. T/I

(l) If all parts of a cell need a supply of a substance quickly, would it be better for the cell to be large or small? T/I

(m) In what way(s) is a cube of agar a good model of a cell? In what way(s) is it not a good model? A

Apply and Extend

(n) Predict how temperature might affect the results of this activity. Explain your reasoning. T/I

(o) Predict how the concentration of sodium hydroxide might affect the results of this activity. Give reasons for your prediction. T/I

(p) In previous grades, you learned about the particle theory of matter. How does the particle theory relate to diffusion of substances throughout a cell? A

2.5 The Cell Cycle

cell cycle the three stages (interphase, mitosis, and cytokinesis) through which a cell passes as it grows and divides

To watch an animation of the cell cycle,
GO TO NELSON SCIENCE

As eukaryotic cells grow and divide, they move through three distinct stages. These stages make up what is known as the **cell cycle** (Figure 1). The stages of the cell cycle are interphase, mitosis, and cytokinesis. Cells grow and prepare to divide during interphase. Cell division occurs during mitosis and cytokinesis. You will learn about each of these three stages in more detail in this section.

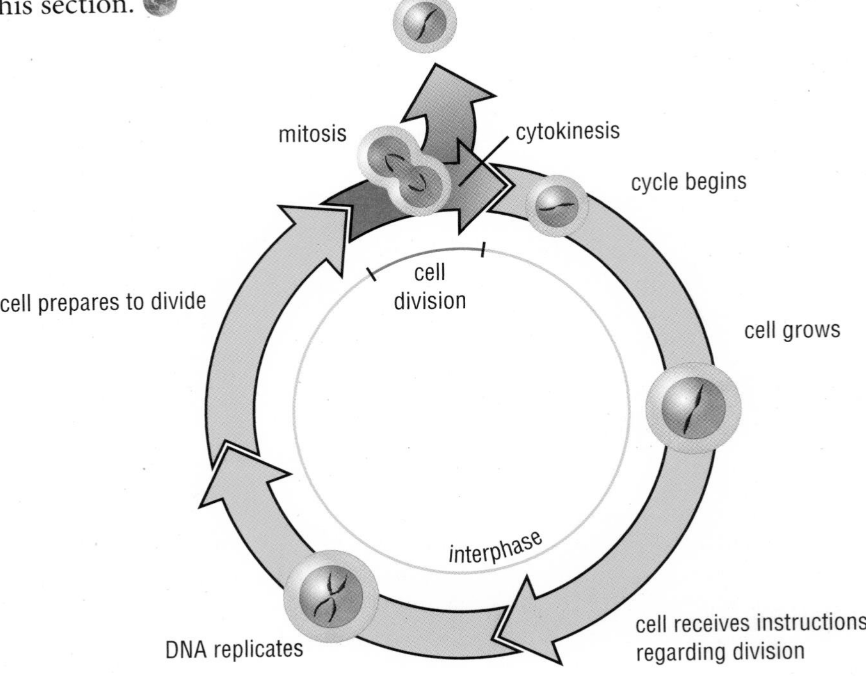

Figure 1 The cell cycle consists of three stages: interphase, mitosis, and cytokinesis. Interphase is the time between cell divisions when a cell grows.

The length of time it takes to complete one cycle varies. Embryonic cells divide rapidly. Some of your body cells may take as much as 30 h for a cycle. Very specialized cells, such as adult nerve cells, may never divide at all.

Interphase

interphase the phase of the cell cycle during which the cell performs its normal functions and its genetic material is copied in preparation for cell division

Interphase is the longest stage for most cells, but it is not a resting stage. During **interphase**, the cell is carrying out all life activities except division. These activities include growth, cellular respiration, and any specialized functions of that cell type. During this stage, the genetic material, DNA (deoxyribonucleic acid), is in very long, thin, invisible strands. When the cell prepares for cell division, the strands are duplicated so that there are two identical strands of the genetic material. More organelles are also formed.

Cell Division

mitosis the stage of the cell cycle in which the DNA in the nucleus is divided; the first part of cell division

cytokinesis the stage in the cell cycle when the cytoplasm divides to form two identical cells; the final part of cell division

daughter cell one of two genetically identical, new cells that result from the division of one parent cell

Cell division occurs in two stages: **mitosis**—the division of the contents of the nucleus—and **cytokinesis**—the division of the rest of the cell, such as cytoplasm, organelles, and cell membrane (Figure 2). Each cell division produces two genetically identical cells called **daughter cells**.

Mitosis is composed of four phases: **p**rophase, **m**etaphase, **a**naphase, and **t**elophase (PMAT). The cells move gradually from one phase to the next.

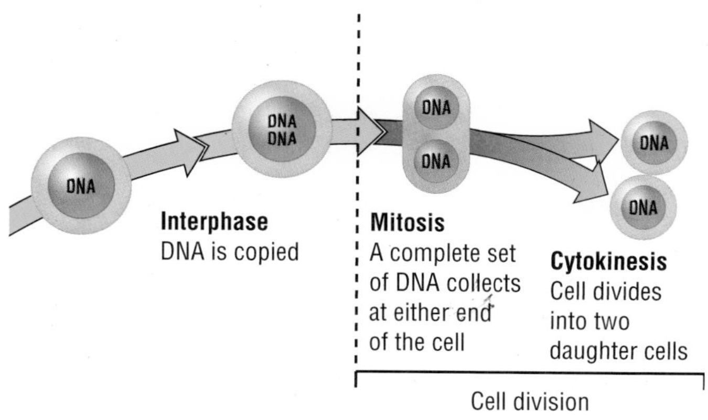

Figure 2 The three stages of the cell cycle in which a parent cell grows, duplicates its DNA, and divides into two daughter cells.

Prophase

As interphase ends, the cell enters the first phase of mitosis—**prophase**. The long strands of DNA condense into a compact form, becoming visible under a light microscope as **chromosomes** (Figure 3(a)). Because the DNA was copied during interphase, each chromosome consists of two identical strands called sister chromatids. An individual strand is called a **chromatid**. The sister chromatids are held together by a **centromere**. The nuclear membrane breaks down during prophase.

prophase the first stage of mitosis, in which the chromosomes become visible and the nuclear membrane dissolves

chromosome a structure in the cell nucleus made up of a portion of the cell's DNA, condensed into a structure that is visible under a light microscope

chromatid one of two identical strands of DNA that make up a chromosome

centromere the structure that holds chromatids together as chromosomes

Metaphase

During **metaphase**, the chromosomes line up in the middle of the cell (Figure 3(b)). This stage is easily recognized. All the chromosomes must be in the line for mitosis to continue.

metaphase the second stage of mitosis, in which the chromosomes line up in the middle of the cell

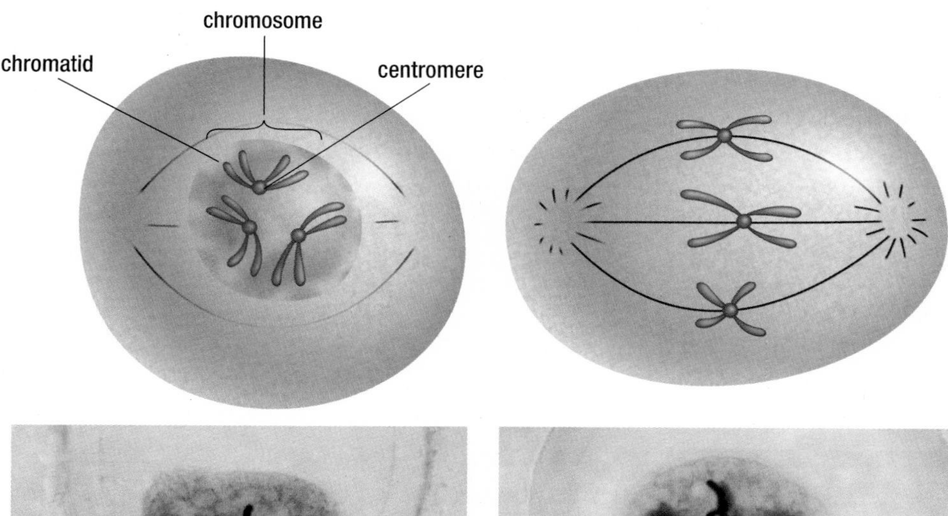

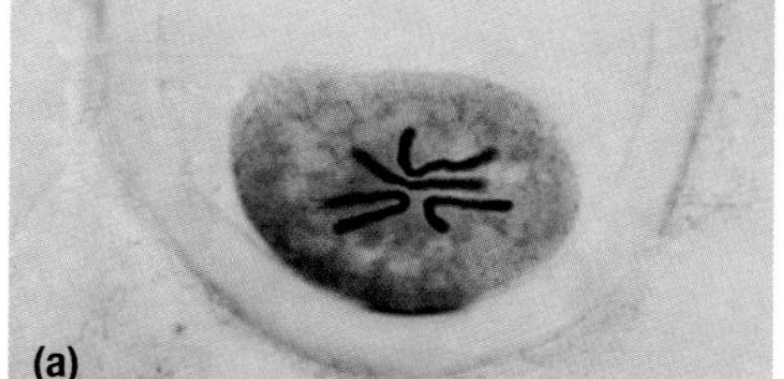

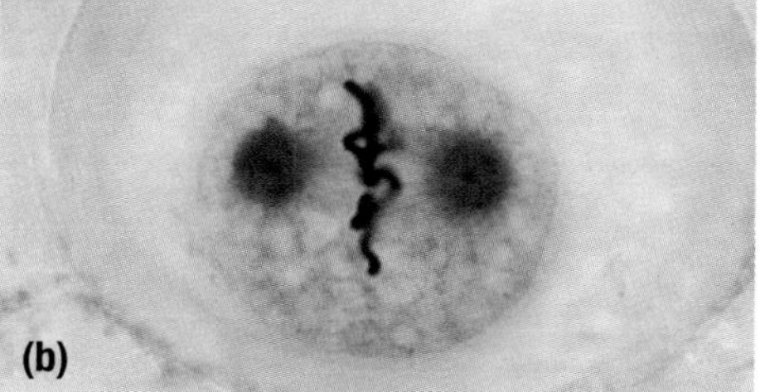

Figure 3 The first two phases of mitosis: (a) prophase and (b) metaphase

LEARNING TIP

Stages and Phases

There are three stages in the cell cycle: interphase, mitosis, and cytokinesis. Mitosis is divided into four phases. Even though "interphase" ends with *-phase*, do not get confused: it is a stage, not a phase!

Anaphase

anaphase the third phase of mitosis, in which the centromere splits and sister chromatids separate into daughter chromosomes, and each moves toward opposite ends of the cell

In **anaphase**, the centromere splits and the sister chromatids separate (Figure 4(a)). They are now called "daughter chromosomes." These daughter chromosomes move to opposite sides of the cell. Under the microscope, they appear to be pulled apart.

Telophase

telophase the final phase of mitosis, in which the chromatids unwind and a nuclear membrane reforms around the chromosomes at each end of the cell

To see an animation of mitosis,
GO TO NELSON SCIENCE

Telophase is the final stage of mitosis (Figure 4(b)). The daughter chromosomes stretch out, become thinner, and are no longer visible. A new nuclear membrane forms around each group of daughter chromosomes. At this stage, the cell appears to have two nuclei.

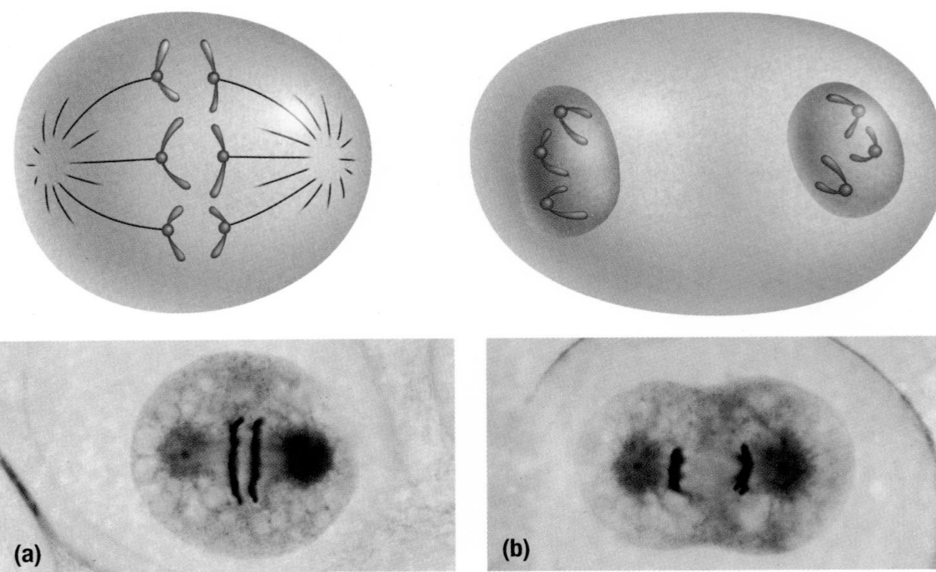

Figure 4 The last two phases of mitosis: (a) anaphase and (b) telophase

Cytokinesis

Cytokinesis is the final stage of cell division. The cytoplasm divides, producing two genetically identical daughter cells. The process of cytokinesis is slightly different in plant and animal cells. In a plant cell, a plate between the daughter cells develops into a new cell wall (Figure 5(a)). In an animal cell, the cell membrane is pinched off in the centre (Figure 5(b)).

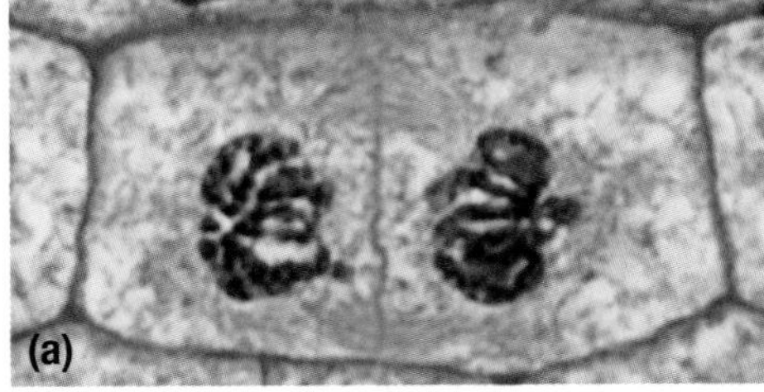

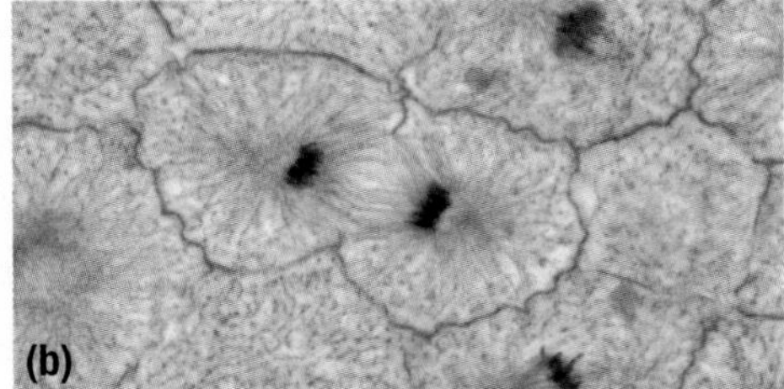

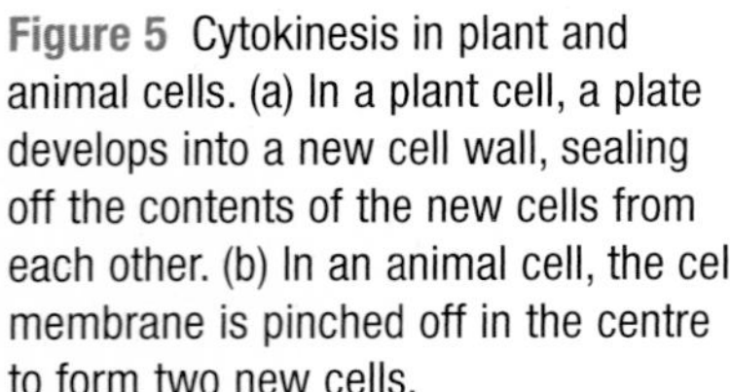

Figure 5 Cytokinesis in plant and animal cells. (a) In a plant cell, a plate develops into a new cell wall, sealing off the contents of the new cells from each other. (b) In an animal cell, the cell membrane is pinched off in the centre to form two new cells.

Moving the Chromosomes

During mitosis, the movements of chromosomes are controlled by spindle fibres: specialized structures that attach to the centromeres of each chromosome. They begin forming during late interphase. During prophase and metaphase, the spindle fibres pull the chromosomes into the middle of the cell. Finally, during anaphase they pull the daughter chromosomes toward opposite ends of the cell.

To learn more about the formation and action of spindle fibres,

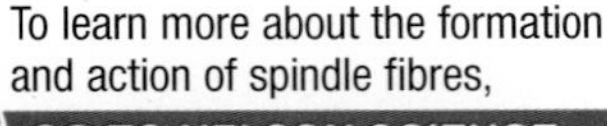

GO TO NELSON SCIENCE

Cell Division—The Big Picture

Figure 6 shows how the stages of cell division fit into the cell cycle.

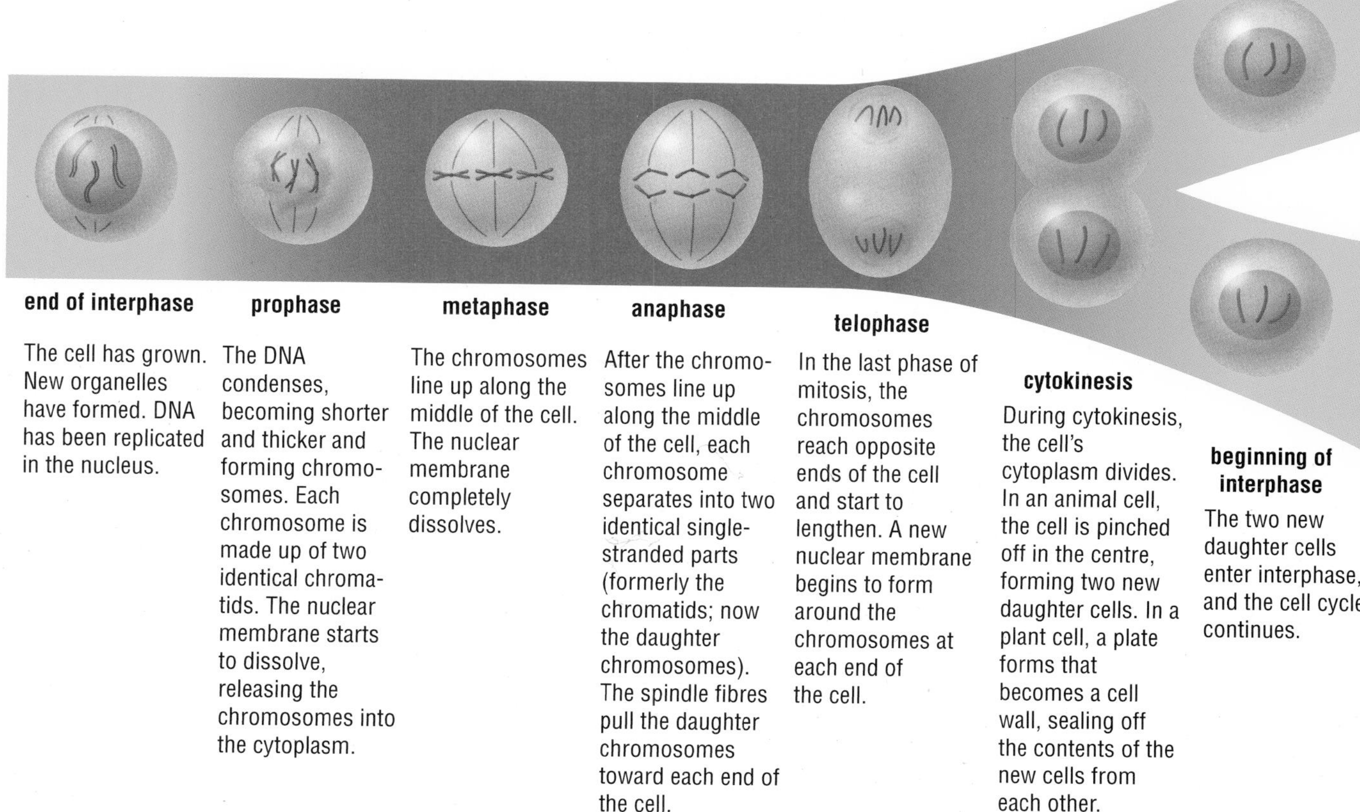

end of interphase
The cell has grown. New organelles have formed. DNA has been replicated in the nucleus.
prophase
The DNA condenses, becoming shorter and thicker and forming chromosomes. Each chromosome is made up of two identical chromatids. The nuclear membrane starts to dissolve, releasing the chromosomes into the cytoplasm.
metaphase
The chromosomes line up along the middle of the cell. The nuclear membrane completely dissolves.
anaphase
After the chromosomes line up along the middle of the cell, each chromosome separates into two identical single-stranded parts (formerly the chromatids; now the daughter chromosomes). The spindle fibres pull the daughter chromosomes toward each end of the cell.
telophase
In the last phase of mitosis, the chromosomes reach opposite ends of the cell and start to lengthen. A new nuclear membrane begins to form around the chromosomes at each end of the cell.
cytokinesis
During cytokinesis, the cell's cytoplasm divides. In an animal cell, the cell is pinched off in the centre, forming two new daughter cells. In a plant cell, a plate forms that becomes a cell wall, sealing off the contents of the new cells from each other.
beginning of interphase
The two new daughter cells enter interphase, and the cell cycle continues.

Figure 6 The stages of cell division

Checkpoints in the Cell Cycle

During the cell cycle, the cell's activities are controlled at specific points, or checkpoints. At each checkpoint, specialized proteins monitor cell activities and the cell's surroundings. These proteins send messages to the nucleus. The nucleus then instructs the cell whether or not to divide. A cell should remain in interphase and not divide if

- signals from surrounding cells tell the cell not to divide
- there are not enough nutrients to provide for cell growth
- the DNA within the nucleus has not been replicated
- the DNA is damaged

If the DNA is damaged and it is early enough in the cell cycle, there may be enough time for the cell to repair the damaged DNA. If there is too much damage to the DNA, the cell is usually destroyed. This is a vital process that helps keep organisms healthy.

DID YOU KNOW?

Discovering Cell Cycle Checkpoints

The discovery of the key proteins that regulate the cell cycle merited the Nobel Prize in Physiology or Medicine 2001. This prize was jointly awarded to three researchers: Leland H. Hartwell from the United States and Tim Hunt and Sir Paul M. Nurse from the United Kingdom.

GO TO NELSON SCIENCE

TRY THIS IDENTIFY THE PHASES OF MITOSIS

SKILLS: Observing, Evaluating, Communicating

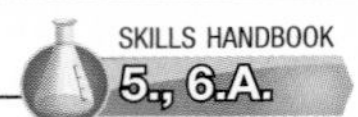

Explore an online resource to help you identify cells in different stages and phases of the cell cycle.

GO TO NELSON SCIENCE

1. Examine the images of cells at various stages of the cell cycle on the suggested websites. Study the descriptions of the different phases of mitosis. Identify which phases are illustrated.
2. Calculate the percentage of cells in each stage or phase.

A. Which phase of mitosis did you have the most trouble identifying? How will you identify this phase in the future? K/U
B. Compare the proportion of cells in the various stages and phases of the cell cycle. Write a statement summarizing the percentage of cells in the different stages and phases. K/U
C. Draw a circle graph showing the percentage of cells in each phase of mitosis. K/U
D. Which phase of mitosis takes the most time? Explain how you arrived at your answer. T/I

IN SUMMARY

- Cells follow a cell cycle that includes growth and preparation for division (interphase) followed by cell division (mitosis and cytokinesis).
- In interphase, the cell carries out all of the normal cell activities, including copying its DNA.
- Mitosis is the division of the nucleus into two identical nuclei. Mitosis has four phases: prophase, metaphase, anaphase, and telophase (PMAT).
- Mitosis is followed by cytokinesis, which results in the entire cell dividing into two new daughter cells.
- Animal cell cytokinesis features the cell pinching off to form two daughter cells.
- Plant cell cytokinesis features a new cell wall forming to separate the two daughter cells.

CHECK YOUR LEARNING

1. During which stage of the cell cycle does replication of the DNA occur? K/U
2. Why is it necessary that the cell copies its DNA? K/U
3. Why are chromosomes visible during mitosis but not at other times? K/U
4. Under a microscope, some cells can appear to be between metaphase and anaphase (Figure 7). Explain this observation. T/I

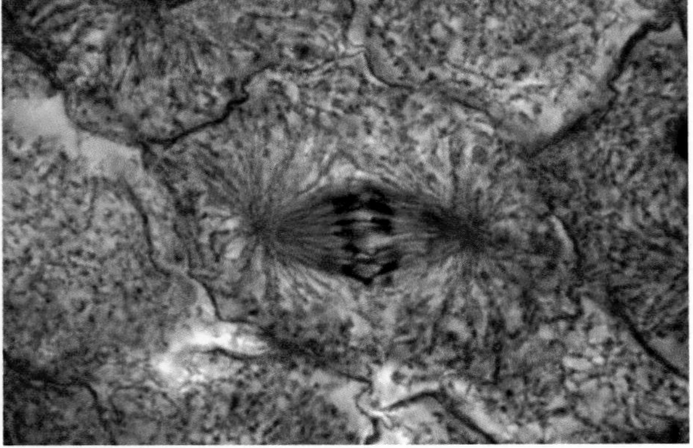

Figure 7

5. Which stage or phase of the cell cycle corresponds to each of the descriptions below? K/U
   (a) A new cell wall begins to form.
   (b) The membrane of the nucleus dissolves.
   (c) Daughter chromosomes begin to separate.
   (d) The cell begins to pinch together along its centre.
   (e) Thick chromosome threads are visible in two distinct regions of the cell.
   (f) The cell grows and copies its DNA.
6. Create a table to summarize what happens during the three stages of the cell cycle. K/U C
7. Biology books used to describe interphase as the "resting phase." Based on what we know now, what was wrong with this term? K/U

Aging: It Is in Our Cells

"Aging seems to be the only available way to live a long life." So said French composer Daniel-François-Esprit Auber. Why the body ages is one of the mysteries of science, but scientists all over the world are getting excited as the answer seems to be within our grasp: it is in the cells (Figure 1).

Figure 1 Why we grow old is one of the mysteries of science.

Aging is complicated, but scientists have recently discovered some clues as to what causes cells in our bodies to stop dividing and eventually die. One of these clues is that, starting in middle age, there is an increasing number of faulty cell divisions. This results in a gradual increase in cells throughout the body that do not function properly, which leads to problems such as Alzheimer's disease and osteoporosis. Sometimes errors in cell division directly affect one part of the body by damaging critical genetic information. For example, cells contain a gene called *COX-2*, which makes an essential protein. Errors during DNA duplication and cell division occasionally produce defective copies of this gene in daughter cells that can lead to heart and kidney failure. Sometimes genes that are related to general cellular function and maintenance stop functioning. In fact, scientists have identified hundreds of genes that are directly linked to development and aging. These are the "aging genes," featured in the online catalogue GenAge.

Chromosomes change as we age. Telomeres are regions of DNA located at the end of each chromosome (Figure 2). Their function is to protect the chromosomes from damage during cell division, much like the plastic covering on the end of a shoelace. As a cell ages, the telomeres on chromosomes get shorter. When telomeres get too short, cell division stops altogether. It is almost as though the telomeres are clocks, counting down the number of divisions that the chromosome can undergo.

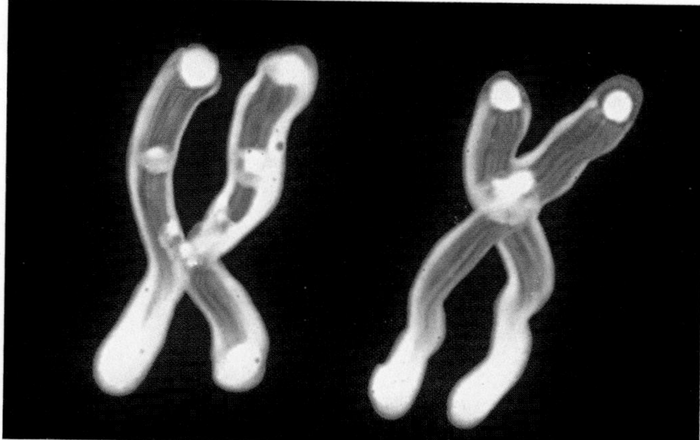

Figure 2 Telomeres, the region at the end of each chromosome, get shorter with age.

New research has revealed another possible link between cell division and aging: centromere location. In older fruit flies, the centromeres sometimes do not align properly during metaphase because the proteins responsible for alignment are missing or malfunctioning. Cell division is stalled until the centromeres line up properly (Figure 3). The proteins that "stand guard" over the orientation of centromeres could be key to the rate of cell division.

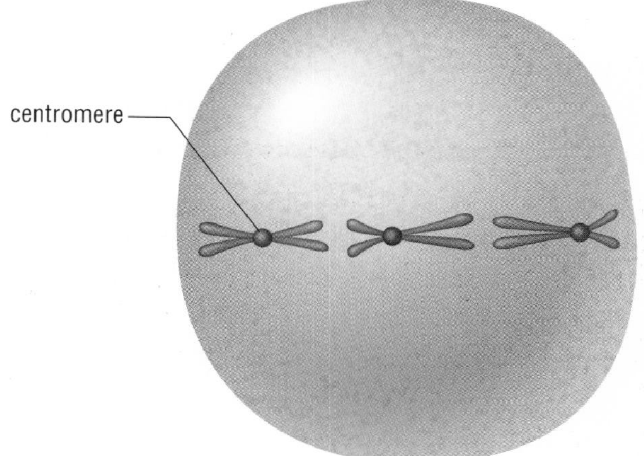

Figure 3 In aging cells, division may stall until the centromeres are lined up properly.

Scientists are exploring aging genes, telomeres, and centromere "guards" very closely. Not only do they hold the secrets of aging, they also hold the secrets of cancer. Aging and cancer are two sides of the same coin: not enough cell division or too much cell division. Solving one mystery may solve the other too.

GO TO NELSON SCIENCE

2.6 PERFORM AN ACTIVITY

Observing Cell Division

In this activity, you will observe prepared slides and photographs of cells that, when living, were dividing quickly. You will examine plant cells from a root tip and animal cells from a fish embryo. In Part A, you will identify characteristics of cells that are dividing quickly. In Part B, you will identify three stages of the cell cycle and the four phases of mitosis in each organism.

SKILLS MENU

Questioning	Performing
Hypothesizing	Observing
Predicting	Analyzing
Planning	Evaluating
Controlling Variables	Communicating

Purpose

Part A: To observe and identify onion root tip (plant) cells and whitefish embryo (animal) cells that are dividing quickly.
Part B: To identify the stages of the cell cycle and phases of mitosis in each organism.

Equipment and Materials

- microscope
- prepared microscope slide of onion root tip, longitudinal section
- photographs of whitefish embryo cells
- lens paper

Procedure

Part A: Observing Actively Dividing Cells

1. Clean the onion root tip slide with lens paper. Place the slide on the stage of your microscope, aiming for the cells just above the root cap (the U-shaped covering of tissue at the tip of the root) and slightly to one side. Use Figure 1 as your guide.

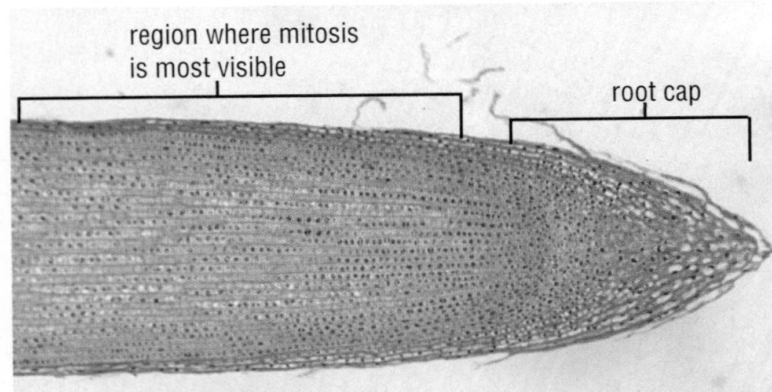

Figure 1 Cells above the root cap are the best ones for observing mitosis.

2. View the slide at low power. Bring it into focus using the coarse-adjustment knob. Begin a labelled scientific drawing that depicts the cell cycle stages that you observe. Include an appropriate title.
3. Move the slide until you can see an area of cells showing different stages of mitosis, similar to the view shown in Figure 2. Change to the medium-power objective lens. Use only the fine-adjustment knob to focus this view. Identify cells that are undergoing mitosis or cytokinesis and cells that are in interphase. Centre the slide so that you can see several cells that are dividing. Add details to your drawing.

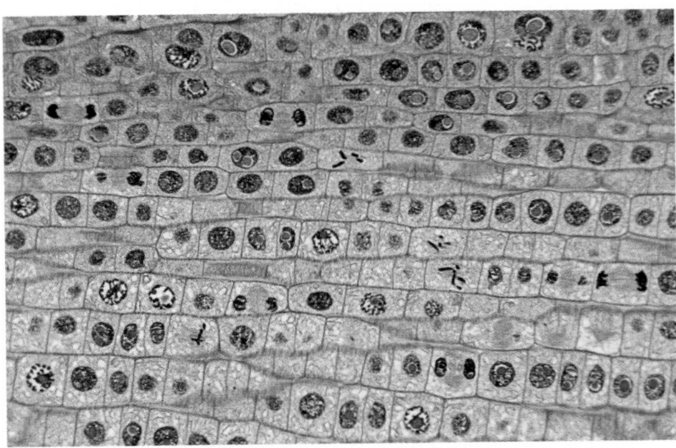

Figure 2 Onion root tip cells in various cell cycle stages, including some of the phases of mitosis

Part B: Identifying Phases of Mitosis

4. Move the high-power objective lens into place. Do not allow the objective lens to touch the slide or the stage. Locate cells in the different stages of the cell cycle: mitosis, cytokinesis, and interphase. Make labelled diagrams of these cells.

5. Look closely at the cells that are undergoing mitosis. Identify at least one cell in each of the four phases of mitosis: prophase, metaphase, anaphase, and telophase. Make labelled diagrams of these cells.
6. Return the microscope to low power and remove the onion root tip slide. Return the equipment to your teacher.
7. Observe the photographs of whitefish egg cells in Figure 3 (or similar photos provided by your teacher). Find cells in all stages of the cell cycle. (Note that these photographs are *not* in "PMAT" order.)
8. Draw and label diagrams of these cells in interphase, prophase, metaphase, anaphase, telophase, and cytokinesis.

Analyze and Evaluate

(a) Suggest why onion root tips and whitefish embryos were chosen as examples in this activity. T/I

(b) Do actively dividing cells look different from cells that divide less often? Explain your observations with reference to the onion root tip. T/I

(c) Given what you know about the cell cycle, give reasons for your answer in (b). T/I

(d) Compare the appearance of the animal cells with that of the plant cells during interphase, mitosis, and cytokinesis. You may use a graphic organizer to show your comparison. T/I C

(e) Describe the differences you observed between plant and animal cells during cytokinesis. T/I

(f) Which phases were the most difficult to distinguish between? Why? T/I

Apply and Extend

(g) Where were the actively dividing cells found in the root tip? Explain this observation. (Hint: What plant activity requires cell division?) T/I

(h) Would you expect whitefish embryo cells to continue to divide indefinitely? Explain why or why not. A

(i) Some herbicides, such as 2,4-D, are believed to make plant cells divide more rapidly than normal. Why would this kill plants? Why might this be dangerous to humans if it had the same effect on human cells? T/I A

(j) What do you think would happen if two daughter cells of an organism did not have identical chromosomes after division? T/I

GO TO NELSON SCIENCE

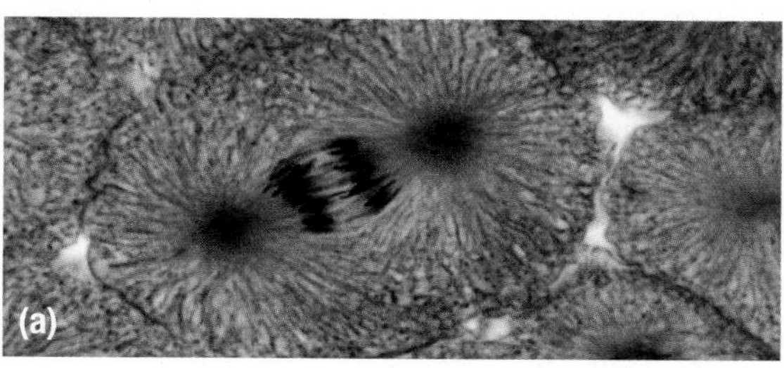

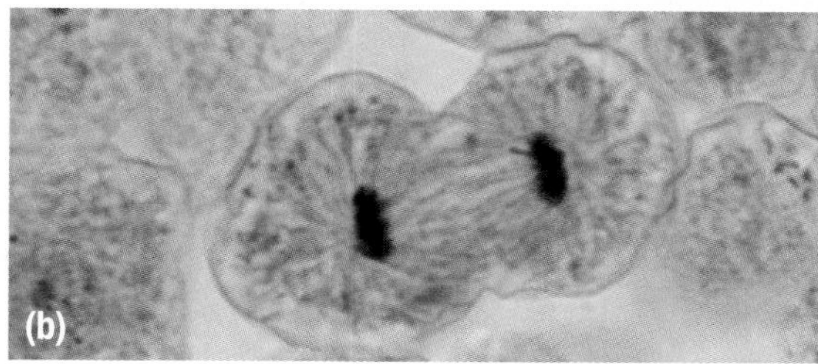

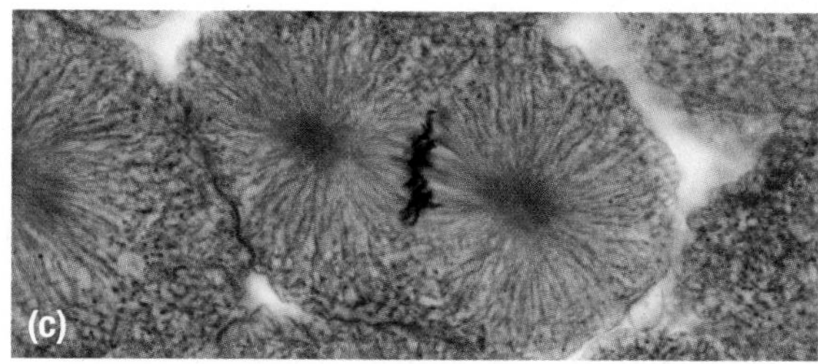

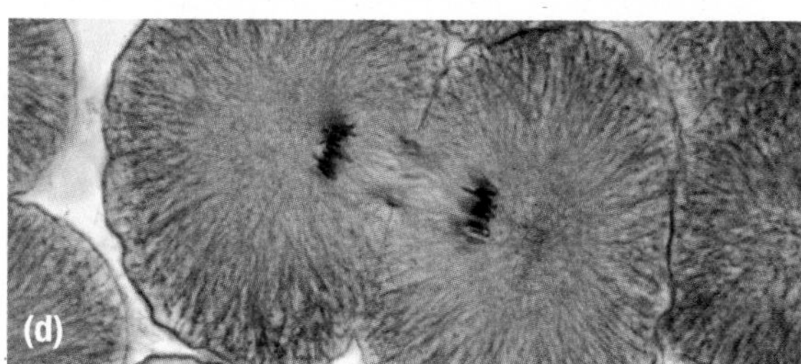

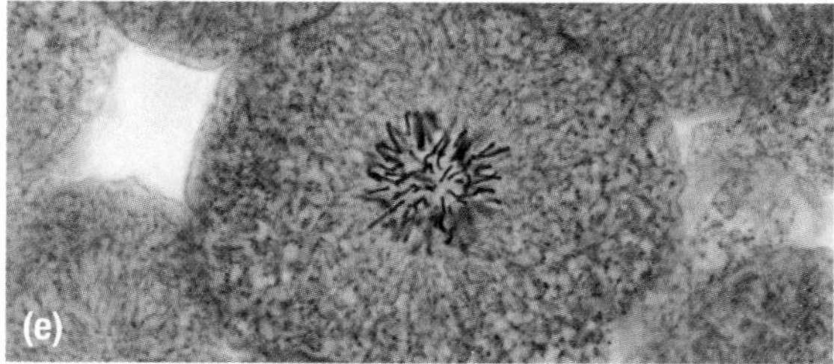

Figure 3 Whitefish embryo cells divide rapidly, so many of them are in the mitosis stage of the cell cycle.

2.7 Cell Division Going Wrong: Cancer

cancer a broad group of diseases that result in uncontrolled cell division

Cancer is a group of diseases in which cells grow and divide out of control. It results from a change in the DNA that controls the cell cycle. This change prevents the cells from staying in interphase for the normal amount of time. One or more of the checkpoints (addressed in Section 2.5) fails, so the cell and all of its subsequent daughter cells continue to divide uncontrollably.

Some types of cancer run in families, whereas others are triggered by environmental factors. Some cancers may have both hereditary and environmental causes. Cancer is not infectious: you cannot catch it from someone who has it.

Cancer is a serious concern for humans, but many other organisms, such as cats, dogs, fish, and even plants, can also develop cancer.

READING TIP

Making Connections

While reading a text, whenever you read a part that reminds you of a related experience that you, your family, or friends have had, jot it down on a sticky note and attach it to the text. After reading, you can review and synthesize your text-to-self connections.

Cell Growth Rates and Cancer

A cancer cell is one that continues to divide despite messages from the nucleus or the surrounding cells to stop growing and dividing. The cell's checkpoints may fail to identify problems or kill off the cell. The uncontrolled growth and division may create a rapidly growing mass of cells that form a lump or **tumour** (Figure 1). The cells of the tumour may stay together and have no serious effect on surrounding tissues. This is called a **benign tumour**. Cells in a benign tumour are not cancerous. However, sometimes a benign tumour can grow so large that it physically crowds nearby cells and tissues. This can affect their normal function.

tumour a mass of cells that continue to grow and divide without any obvious function in the body

benign tumour a tumour that does not affect surrounding tissues other than by physically crowding them

malignant tumour a tumour that interferes with the functioning of surrounding cells; a cancerous tumour

A mass of cells is a **malignant tumour** if it interferes with the function of neighbouring cells and tissues, such as the production of enzymes or hormones. Malignant tumours may even destroy surrounding tissues. The cells in a malignant tumour are considered cancerous.

To watch an animation of cancer cell growth over 24 hours,
GO TO NELSON SCIENCE

In some cases, cancer cells break away from the original (primary) tumour and move to a different part of the body. If they settle there and continue growing and dividing uncontrollably, they can start a new (secondary) tumour (Figure 2 on the next page). This process is known as **metastasis** and is one of the reasons why cancer is such a dangerous disease.

metastasis the process of cancer cells breaking away from the original (primary) tumour and establishing another (secondary) tumour elsewhere in the body

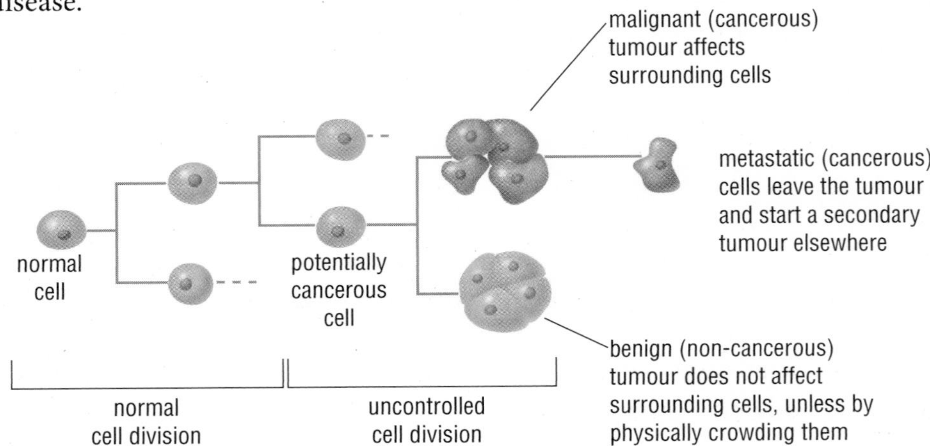
malignant (cancerous) tumour affects surrounding cells
metastatic (cancerous) cells leave the tumour and start a secondary tumour elsewhere
normal cell
potentially cancerous cell
benign (non-cancerous) tumour does not affect surrounding cells, unless by physically crowding them
normal cell division
uncontrolled cell division

Figure 1 A tumour is a mass of cells with no function. A tumour can remain benign, or it can become malignant. Tumour cells can metastasize, spreading to other areas of the body. Malignant and metastatic tumours are considered cancerous.

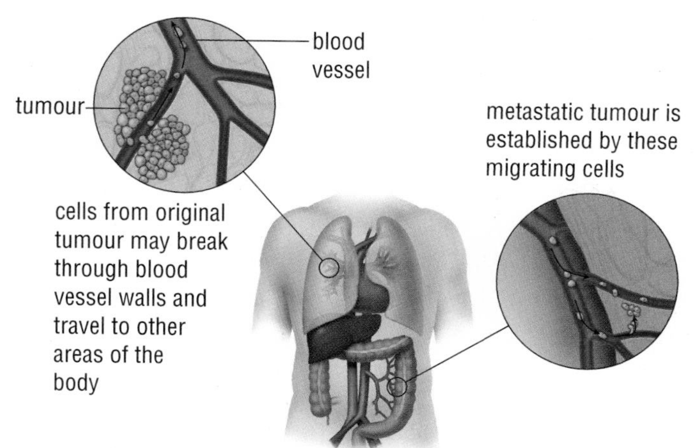

Figure 2 Cancer cells sometimes break free from the primary tumour site. These metastatic cells can then move through the blood vessels. Secondary tumours may develop at other sites.

Causes of Cancer

Every time a cell divides, its DNA is faithfully duplicated. Usually, this process is error-free and the genetic information in the daughter cells is exactly the same as that in the parent cell. Sometimes, however, random changes occur in DNA. These random changes are known as **mutations**. These changes may either result in the death of the cell or allow the cell to survive and continue to grow and divide. Very rarely, the change occurs in the DNA that controls cell division. Once this crucial cell cycle DNA starts behaving abnormally, the cells may become cancerous and proliferate wildly through repeated, uncontrolled mitosis and cytokinesis. They multiply until all nutrients are exhausted.

mutation a random change in the DNA

Some mutations are caused by **carcinogens**: environmental factors that cause cancer. Well-known carcinogens include tobacco smoke; radiation, such as X-rays and UV rays from tanning beds and sunlight; some viruses, such as human papillomavirus (HPV) and hepatitis B; certain chemicals in plastics; and many organic solvents. If a group of people is exposed to a carcinogen, some will develop cancer, but others will not. This is a major challenge for cancer researchers because they cannot predict who will develop cancer. Until this process is completely understood, it is best to limit your exposure to carcinogens.

carcinogen any environmental factor that causes cancer

To find out about some of the myths relating to things that cause, or cure, cancer,

GO TO NELSON SCIENCE

Some cancers appear to be at least partly hereditary. This means that the DNA passed from one generation to the next may contain information that leads to disease. These cancers include some breast cancers and some colon cancers. A genetic link makes it more likely that you will develop a particular type of cancer, but it does not guarantee that you will get cancer.

Smoking and Cancer

Lung cancer is one of the most common types of cancer in Canadians over 40. According to Health Canada, smoking currently causes 9 out of 10 cases of lung cancer. Carcinogens in tobacco affect more than just the lungs. Smoking also increases the risk of over a dozen other types of cancer (Figure 3). The good news is that most of these smoking-related cancers can be prevented by giving up smoking—or never starting in the first place—and staying away from second-hand smoke.

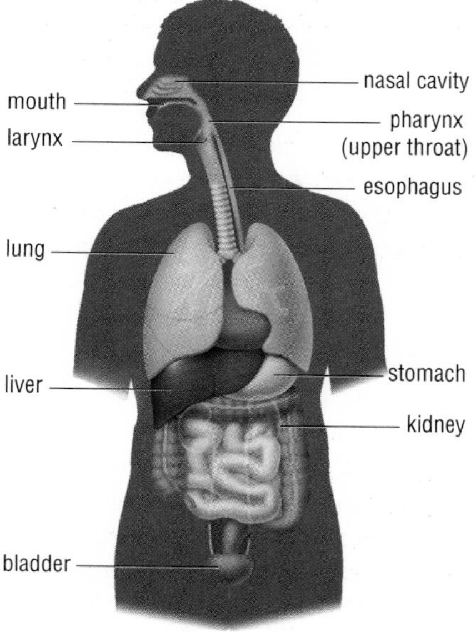

Figure 3 The carcinogens in tobacco smoke can affect all of these parts of the body.

Cancer Screening

READING TIP

Making Connections
Text-to-world connections involve facts about the world that you have collected during your lifetime. This type of connection can be very useful when reading a science text because it can help you take a more critical approach to the ideas and information in the text.

Cancer screening means checking for cancer even if there are no symptoms. Different types of cancer can be screened for in different ways. Cancer screening can be done at home, as part of a routine medical checkup, or with a special appointment. Screening is especially important for people who have a family history of certain cancers (such as breast cancer or colon cancer). If you have a family history of cancer, you may choose to go through genetic screening. This will determine if you have inherited DNA that is linked to cancer. Screening can also be valuable for people who are exposed to carcinogens at work or because of their lifestyle.

Screening does not prevent cancer, but it does increase the chance of detecting cancer early enough to successfully treat it. This is one important way to reduce your risk.

Many women take responsibility for their health by performing regular breast self-examinations to check for lumps that may indicate breast cancer. Women can also be screened for early signs of cervical cancer, starting around age 18, by getting a regular **Pap test**. This is a procedure in which a doctor takes a sample of cervical cells, which are checked for cancer.

Pap test a test that involves taking a sample of cervical cells to determine if they are growing abnormally

Men can detect testicular cancer early through testicular self-examination. There is also a blood test, called the PSA test, that a doctor can prescribe to screen for prostate cancer. This is not widely used for men under the age of 50 as the incidence of prostate cancer is lower for that age group.

To determine what your checkup checklist looks like,

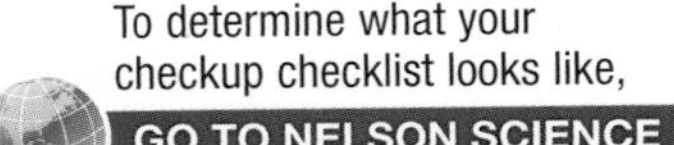

Other screening tests include a blood test for colon cancer and regular skin checks by a doctor or dermatologist to look for changes in moles, new growths, and sores. You can learn to check your own skin regularly for moles. Table 1 illustrates the "ABCD of moles." The letters stand for **A**symmetry, **B**order, **C**olour, and **D**iameter. If you see any suspicious-looking moles or growths, ask your doctor or dermatologist to check them out.

Table 1 The ABCD of Moles

	Asymmetry	Border	Colour	Diameter
Benign				
Malignant				

Reducing Your Cancer Risk

Cancer prevention and early detection are very important. Many factors can affect your risk of getting cancer. These risks include your personal and family medical history, carcinogens in your environment, and your lifestyle choices. Inform yourself about these factors so that you can minimize your exposure to known cancer risks. You cannot change your family history, and some aspects of your environment are beyond your control. You can, however, make lifestyle choices that have a big impact on your likelihood of getting cancer.

RESEARCH THIS CANCER SCREENING AND PREVENTION

SKILLS: Defining the Issue, Researching, Identifying Alternatives, Defending a Decision, Communicating

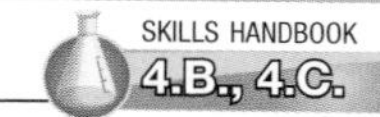
SKILLS HANDBOOK 4.B., 4.C.

Each year, approximately 150 women in Ontario die because of cervical cancer. There is no single cause of cervical cancer, but research has shown that the main risk factor is untreated infection with human papillomavirus (HPV). By preventing HPV infections, the risk of developing cervical cancer can be reduced.

1. Research cervical cancer, or another type of cancer, to discover why it is particularly dangerous.
2. Research information about ways to reduce the risk from your chosen type of cancer. Search a number of sources.

GO TO NELSON SCIENCE

A. Analyze the information that you have collected. Decide which of the cancer-reduction approaches you support. Write a statement of your decision, including your rationale. T/I C A
B. What questions do you still have regarding screening and prevention for this type of cancer? How do you think answers to these questions could be found? T/I
C. In your class, debate the medical and ethical issues surrounding cancer screening and prevention. C A

Lifestyle Choices

There are many lifestyle choices, besides avoiding tobacco smoke, that can help reduce your risk of developing cancer. A healthy diet including a lot of fruits and vegetables and less fatty meat may help. Research has shown that certain "super foods" contain substances that help your body protect itself from cancers (Figure 4). Even though vitamin supplements may include some of these substances, the best way to obtain them is by eating the food itself. These super foods do not prevent cancer; rather, they lower your cancer risk.

The risk of some cancers increases with the amount of body fat a person has. A healthy diet may aid weight loss, which could lower the risk of cancer.

To read about a Canadian researcher studying factors that affect prostate cancer,

GO TO NELSON SCIENCE

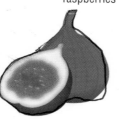

DID YOU KNOW?

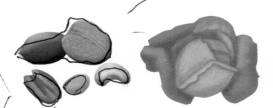

Antiperspirants and Breast Cancer—A Myth?
There is no known link between using antiperspirants and breast cancer. This is a myth that has been spread using e-mail and the Internet. A study in 2002 found no increase in breast cancer risk for women using antiperspirants or deodorants, even right after shaving their underarms.

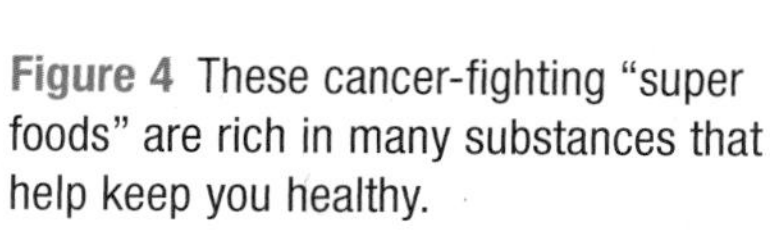

Figure 4 These cancer-fighting "super foods" are rich in many substances that help keep you healthy.

CITIZEN **ACTION**

Cancer Education and Research

SKILLS HANDBOOK 4.C.7.

The Canadian Cancer Society (CCS) works to educate people on the lifestyle factors that can lead to cancer, what we can do to prevent cancer, and how to find cancer in its earliest stages.

Cancer research is expensive and time-consuming. It can take years of work before a single research study is completed. There is only so much money to go around. Every year, many worthy research projects go unfunded.

What Can You Do to Help?

One person *can* make a difference—just look at Terry Fox! He was 18 years old when he was diagnosed with bone cancer in his right leg. The only treatment at that time was to amputate his leg just above the knee. This remarkable young man brought international awareness to the importance of cancer research by deciding to run a cross-country marathon. He ran more than 5 000 km from St. John's, Newfoundland, to Thunder Bay, Ontario, before discovering that the cancer had spread to his lungs. He died on June 28, 1981, at the age of 22. The efforts of this heroic young Canadian have inspired others to raise hundreds of millions of dollars in his name to fund cancer research.

There are many opportunities to volunteer in your local community to increase cancer awareness or raise funds.

Decide how you would like to get involved. You may make this decision by doing online research, talking to a cancer agency staff member or a researcher, or some other way. Prepare a written or verbal message to your fellow students to get them excited about your cause.

GO TO NELSON SCIENCE

Figure 5 Terry Fox on his Marathon of Hope

Diagnosing Cancer

In some cases, a growing tumour creates swelling or causes discomfort. In other cases, the patient may feel very tired or start losing weight for no apparent reason. The earlier a cancer is diagnosed, the better the chances of it being successfully treated. If cancer is suspected, the doctor will order medical tests to investigate further. These tests can include blood tests and special imaging techniques.

Imaging Technologies

Imaging techniques may include endoscopy, X-ray, ultrasound, CT scanning, and MRI.

An endoscope is commonly used to screen for colon cancer. The endoscope is made up of a fibre-optic cable to deliver light, a tiny camera, and a cable that sends the images to a screen (Figure 6). Tools, such as forceps, can also be attached. The patient may be given a sedative before the endoscope is inserted into the colon through the rectum. The camera allows the doctor to look for abnormal growths. Forceps can be used to remove a small sample (biopsy) of any suspicious-looking growth. The sample can then be studied under a microscope.

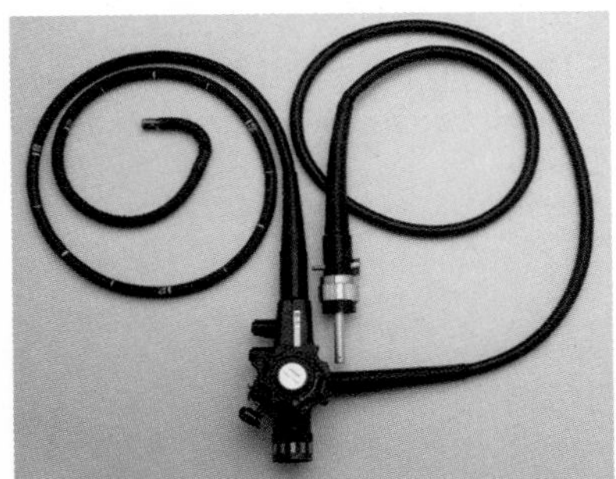

Figure 6 An endoscope

You may already be familiar with X-rays. Doctors use X-ray images to view parts of the body such as bones and lungs (Figure 7). A mammogram is a specialized X-ray technique for imaging breast tissue.

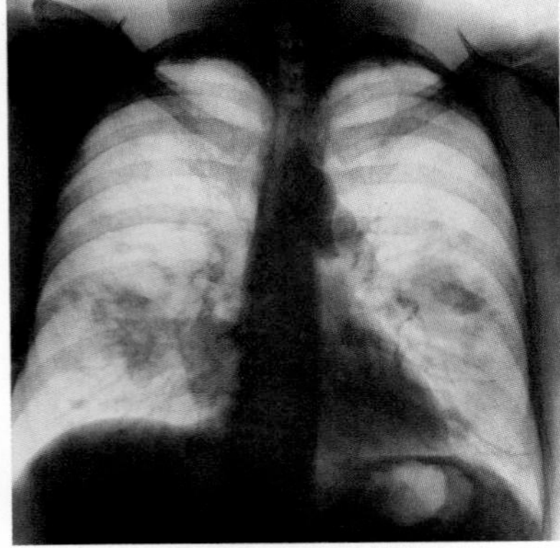

Figure 7 Chest X-ray of a patient with lung cancer (red areas).

X-rays can also cause DNA damage. They are particularly harmful to rapidly dividing cells such as those in a growing fetus. For this reason, women who are pregnant should not undergo X-ray examinations.

Another imaging technique that you may already be familiar with is ultrasound imaging. Ultrasound imaging uses ultra–high-frequency sound waves to create a digital image. The digital image allows doctors to view certain soft tissues, such as the heart or the liver (Figure 8).

Another commonly used imaging technique is a CT or CAT scan (computerized axial tomography). A CT scan allows the X-ray technician to take multiple X-rays of the body from many different angles. The images are then assembled by computer to form a series of detailed images. This technology allows doctors to view parts of the body that cannot be seen with a conventional X-ray scan (Figure 9).

A fourth imaging technique is MRI (magnetic resonance imaging) (Figure 10). In an MRI, radio waves and a strong magnetic field create images with more detail than a CT scan. Computers can assemble the information into three-dimensional models.

READING TIP

Recording Connections

Make a double-entry diary labelled "Connections" and "Understanding." Use the left side to record text-to-self, text-to-text, and text-to-world connections. On the right side, explain how the connections helped you to better understand the text by visualizing, making predictions, inferences, or judgements, forming opinions, drawing and supporting conclusions, or evaluating.

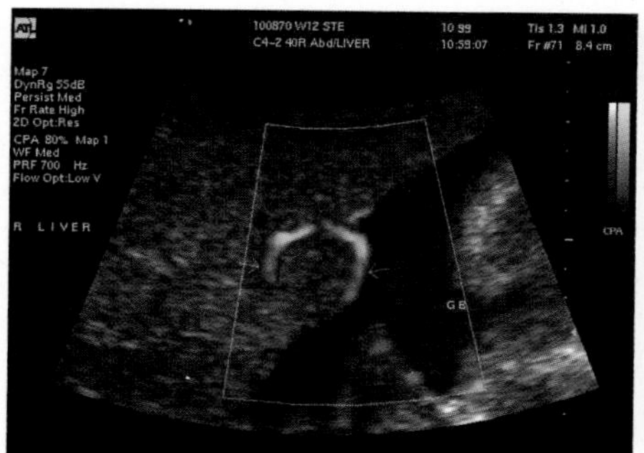

Figure 8 Ultrasound scan of the liver, showing increased blood flow (orange areas) due to malignant tumours

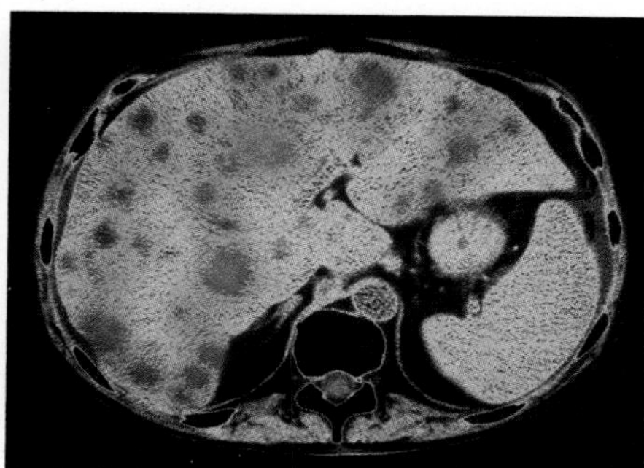

Figure 9 CT scan of a cross-section of the body, showing cancer tumours (darker red areas) in a liver

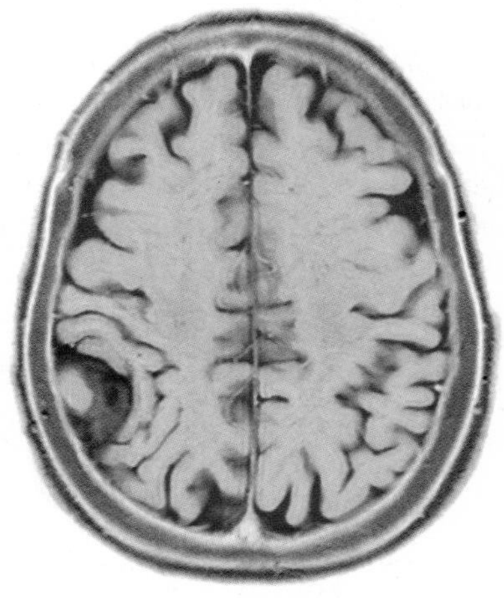

Figure 10 MRI scan showing brain cancer (green area)

Examining Cells

If any of the medical tests or images show abnormalities, the next step is to examine a sample of the suspected cancer cells under a microscope. This is the only way to confirm a diagnosis of cancer.

Certain cell samples can be obtained easily, such as blood cell samples. Leukemia is a cancer that affects the blood, often resulting in a high ratio of white blood cells to red blood cells. An experienced technician can identify this problem when looking at a sample of blood through a microscope.

A sample of tumour cells may have to be removed surgically. This technique is known as "taking a biopsy." The sample is then viewed under a microscope. It may also be tested for genetic abnormalities. If the tumour cells are determined to be non-malignant (not cancerous), then the tumour is diagnosed as benign. Cancer cells are often irregularly shaped and may be smaller or larger than the surrounding cells. Experienced medical professionals can identify cancer cells just by looking at them (Figure 11).

After diagnosis, the doctors must discover where the cancer originally began. They also need to find out how large the tumour is, how quickly it is growing, and whether the cancer has spread. This information helps to determine suitable treatments and to predict the outcome.

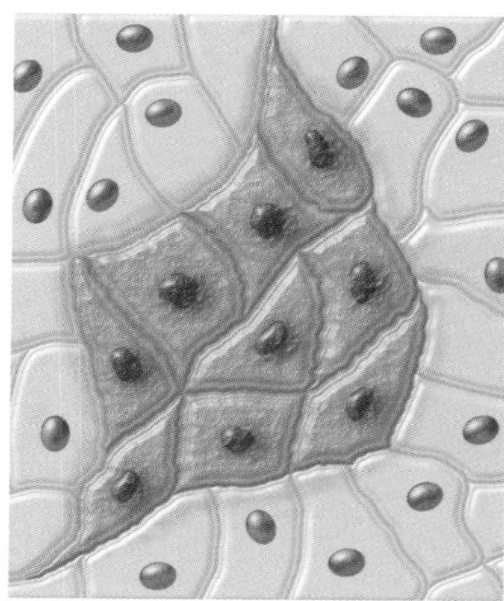

Figure 11 Illustration of cancer cells (shown in purple) among normal cells (pink). Notice that the cancer cells have irregular shapes.

Treatments for Cancer

The goal of cancer treatment is to slow down the growth of the tumours or destroy as many cancer cells as possible. Currently, there are three main conventional methods of treating cancer: surgery, chemotherapy, and radiation therapy. An emerging technique is biophotonics. A cancer treatment plan may consist of one or a combination of these methods.

Cancer treatment involves a team of medical specialists that may include surgeons, medical oncologists (doctors who specialize in cancer diagnosis and treatment), radiation oncologists, and oncology nurses.

To learn more about working in the field of medical oncology,
GO TO NELSON SCIENCE

Surgery

Surgery—physically removing the cancerous tissue—is sometimes the preferred way of treating cancer. If the tumour is easily accessible and fairly well defined, the doctors may recommend this option.

To find out the wait times for cancer surgery in your area,

GO TO NELSON SCIENCE

Chemotherapy

Chemotherapy is a method of treating cancer using drugs. These work by slowing or stopping the cancer cells from dividing and spreading to other parts of the body, and by killing the cells. The drugs can be injected or taken orally (by mouth). Side effects may include hair loss, nausea, and fatigue, but the benefits of the treatment generally outweigh the negative effects.

Chemotherapy is often one of the first stages of cancer treatment. Its aim is to shrink a tumour for surgical removal or for radiation treatments. A huge advantage of chemotherapy is that the drugs travel throughout the body and reach almost all tumours, even if they are much too small to be detected.

DID YOU KNOW?

The Blood-Brain Barrier
It is very difficult to deliver treatment drugs to tumours in the brain. This is because a barrier of dense cells prevents most chemicals from passing from the blood into the brain.

Radiation

Cancer cells are easily damaged by ionizing radiation because they divide rapidly. Radiation therapy takes advantage of this. The DNA of many of the daughter cells is damaged by the radiation, so the cells cannot divide further.

The radiation is directed at the tumour either by using a focused beam or by implanting a radioactive source into the tumour (Figure 12). This minimizes side effects.

READING TIP

Cross-Checking Connections
Use sticky notes while reading to make text-to-text connections with other texts you have read or viewed. After reading you can check these connections to see whether the new information matches or challenges what you already know.

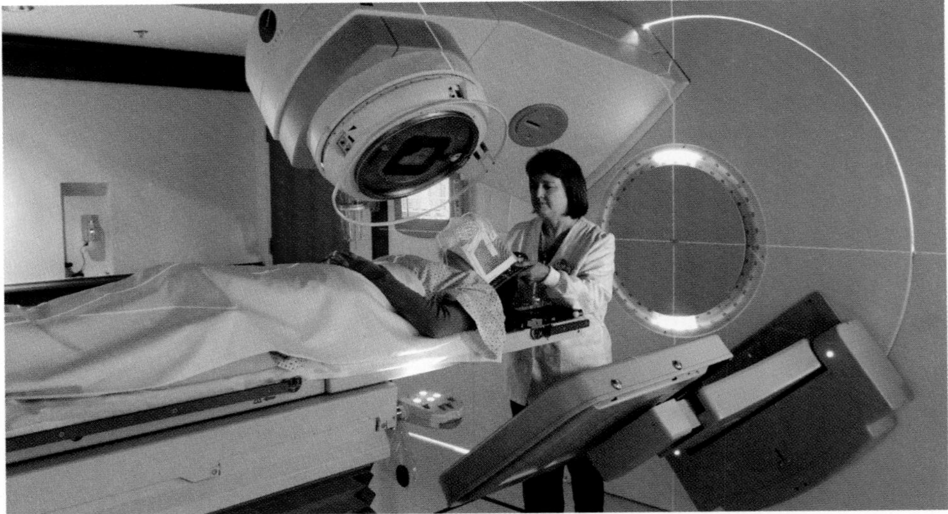

Figure 12 Radiation therapy is delivered by very technologically advanced machines.

Biophotonics

The newest weapon in the fight against cancer, **biophotonics**, uses beams of light to detect and treat cancer. It is a very sensitive diagnostic tool, allowing for early detection of cancer. It has fewer side effects than conventional radiation treatment as it can more accurately target the cancerous tissue. Much of this research is being pioneered at the University of Toronto.

Scientific research and technological advancements play an important part in understanding cell biology. Canadian researchers are at the forefront in discovering better ways to prevent, diagnose, and treat cancer.

biophotonics the technology of using light energy to diagnose, monitor, and treat living cells and organisms

To read more about the development of biophotonics at the University of Toronto,

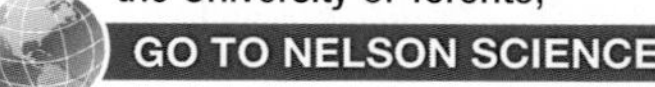

GO TO NELSON SCIENCE

UNIT TASK Bookmark

The information about cancer diagnosis and treatment may be useful to you as you work on the Unit Task described on page 156.

IN SUMMARY

- Cancer is a group of diseases that result from uncontrolled cell growth.
- Tumours can be benign (non-cancerous) or malignant (cancerous).
- Some cancer cells are able to move to new areas of the body in a process called metastasis.
- Many early stages of cancers have no noticeable symptoms.
- Cancer risks can be reduced through avoiding carcinogens and by making healthier lifestyle choices.
- Various imaging technologies, including endoscopy, X-ray, ultrasound, CT scanning, and MRI, can be used to identify abnormalities and diagnose cancer.
- Biopsy is a method of diagnosis in which a sample of cells is surgically removed and examined under a microscope.
- Many screening tests are available to diagnose cancer early and improve treatment success rates.
- The main treatment methods for cancer include surgery, chemotherapy, and radiation. A newer technology is biophotonics.

CHECK YOUR LEARNING

1. How is the behaviour of cancer cells different from that of normal cells? K/U
2. (a) Can a person inherit cancer genetically? Explain.
   (b) Can you catch cancer from a person who has cancer? Explain. K/U
3. (a) What is a carcinogen?
   (b) Give some examples of carcinogens that may be present in your everyday life. A
4. Why might it be easy to overlook cancer in its early stages? K/U
5. List at least five diagnostic techniques used to detect cancer. K/U
6. Briefly describe the three main conventional methods of treating cancer. K/U
7. Why might a doctor be concerned to find cancer cells in a patient's blood? K/U
8. Identify at least three simple lifestyle changes that could help reduce your risk of developing cancer. K/U
9. What cancer screening tests should young adults include in their cancer-fighting plan? K/U
10. Why might there be a risk of cancer recurring, even when surgery is performed to remove a malignant tumour? K/U

Comparing Cancer Cells and Normal Cells

In this activity, you will examine microslides or prepared slides to compare normal cells with cancer cells of the same type. You will look for evidence that the cells are dividing at different rates.

SKILLS MENU
- Questioning
- Hypothesizing
- Predicting
- Planning
- Controlling Variables
- Performing
- Observing
- Analyzing
- Evaluating
- Communicating

Purpose

To compare the rate of cell division in cancer cells and normal (non-cancer) cells.

Equipment and Materials

- microslide or prepared slides of normal cells and cancer cells
- microviewer or microscope
- lens paper

Procedure

SKILLS HANDBOOK 2.D., 3.B.6.

1. Carefully clean the microslide or slide with a sheet of lens paper.
2. Use the microviewer or microscope to observe a section of healthy tissue (Figure 1). If you are using a microscope, start at low power before moving on to medium and high power. Record your observations. Note the size and shape of the cells, their arrangement, and any significant structures in the cells.
3. Count the number of cells that are dividing. Include cells in both mitosis and cytokinesis. (Mitosis is the stage during which the chromosomes are visible as dark, threadlike structures. Cytokinesis is the stage during which the cell splits into two).
4. Estimate the percentage of cells that are dividing, using the formula

$$\left(\frac{\text{number of dividing cells}}{\text{total number of cells}}\right) \times 100\ \%$$

5. Make a scientific drawing of a small area showing about 10 typical cells.
6. Repeat Steps 3 to 5, this time looking at a section of cancerous tissue (Figure 2). Be sure to observe the same type of tissue as you observed in Step 3.
7. Return the equipment to your teacher.

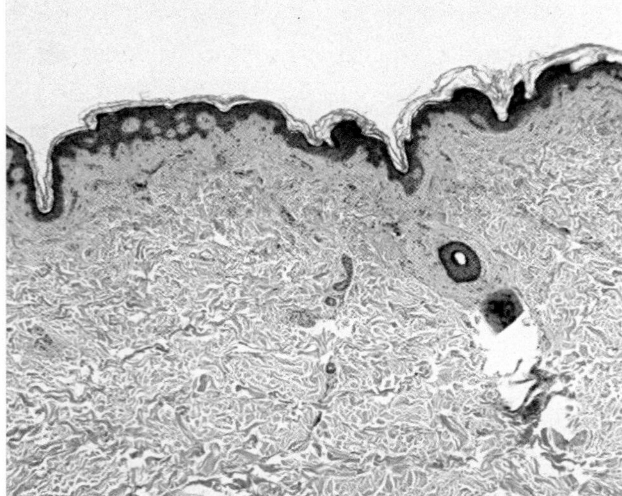

Figure 1 Sample of healthy skin cells

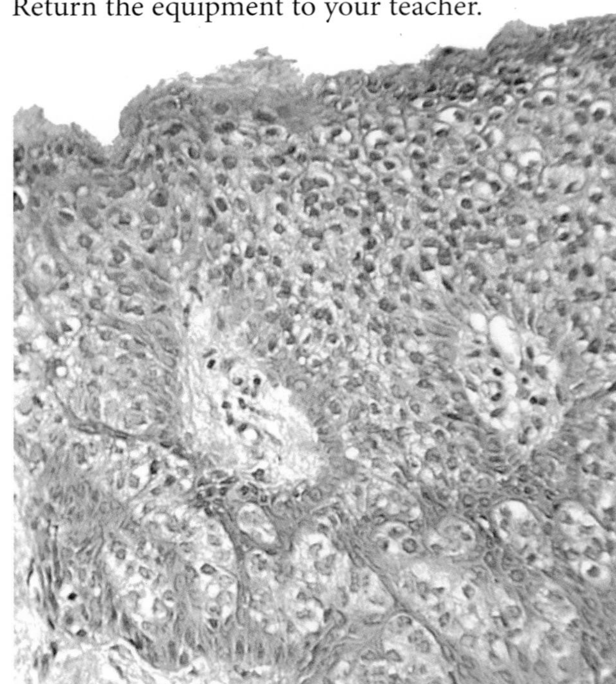

Figure 2 Sample of cancerous skin cells

Analyze and Evaluate

(a) What did you observe about the organization of the normal cells compared with the organization of cancer cells? T/I

(b) Describe any differences in the size and shape of the two types of cells. T/I

(c) What did you observe about the nuclei of normal cells compared with the nuclei of cancer cells? T/I

(d) (i) What percentage of normal cells were actively dividing?

(ii) What percentage of cancer cells were actively dividing?

(iii) Compare the rate of cell division in cancer cells to the rate in normal cells. T/I

(e) After normal cell division, daughter cells are the same type of specialized cell as the parent cell. Did the cancer cells appear to be specialized? Refer to your observations. T/I

(f) Normal cells have internal regulators that control division. This prevents cells from becoming overcrowded. From your observations, do you think that these internal regulators are working in cancer cells? Explain. T/I

(g) Summarize the differences that you observed between cancer cells and normal cells. K/U

Apply and Extend

(h) Cell division requires energy. Suggest how this may explain why cancer is harmful to surrounding cells. A

(i) Carcinogens sometimes cause cancer. What cell structure(s) is/are affected by carcinogens? Give reasons for your answer. A

2.9

Specialized Cells

In this chapter, you have looked at plant and animal cells as they grow and divide. You have seen that different cells have greater numbers, or fewer, of certain organelles depending on the function of the cell. You have also learned how cells sometimes grow out of control, forming tumours. Next you will look at some of the wide variety of different cell types—each with its own special function—that make up the bodies of plants and animals.

The cell theory states that every cell comes from a previously existing cell. You started life as a single fertilized egg. A towering maple tree also started as a single cell. The many cells in a complex organism are not all identical, however. Look at your best friend or any other living thing around you: they are made up of cells with different structures and different functions.

You can compare a multicellular organism to a large city. The city needs energy, transportation corridors, waste disposal facilities, a police force, and organization to keep everything operating efficiently. Different parts of the city have different purposes, and different businesses in the city meet different needs. Auto mechanics repair damaged cars. Dental hygienists help us keep our teeth healthy. Farmers grow and provide food. The list goes on. You do not expect the same person to repair your car, clean your teeth, and sell you food. Each of these jobs is performed by someone who is trained to do that job: a specialist.

Your body has needs similar to those of a city: energy, transportation, waste disposal, and so on. Every cell cannot digest food, fight disease, carry nutrients, and coordinate your body's movements. Similar to people working in a city, **specialized cells** have physical and chemical differences that allow them to perform one job very well. Figure 1 shows some of the different kinds of cells that line the trachea—the tube that carries air from the mouth to the lungs. Notice the orange-stained goblet cells. They contain many Golgi bodies to produce mucus.

specialized cell a cell that can perform a specific function

Cell specialization involves a change in form and in function. Specialized cells can look very different from each other.

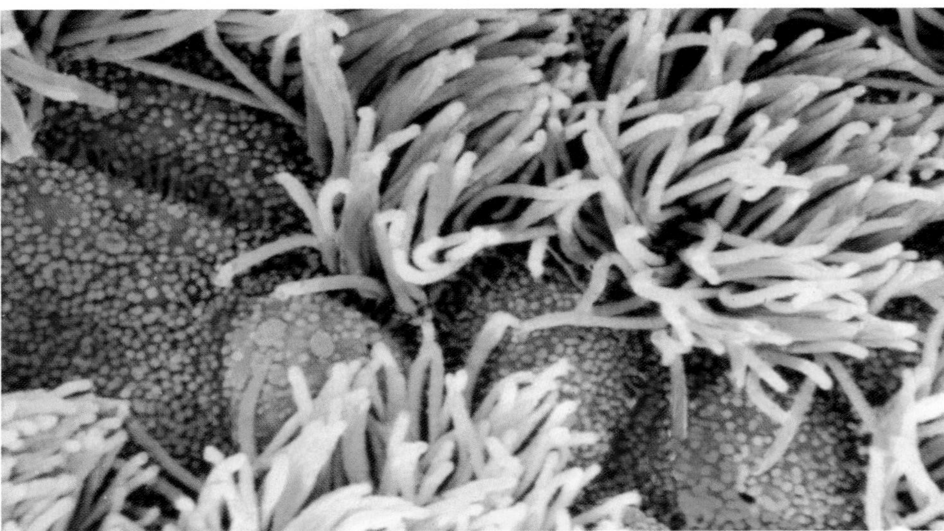

Figure 1 These specialized cells help to keep dirt out of the lungs. The orange cells are goblet cells. They secrete mucus. The parts that look like green grass are cilia—hair-like extensions of cells. The cilia can move. They move mucus along the trachea to trap and remove any inhaled dust and dirt.

Animal cells show a wide variety of specializations, as Figure 2 shows. They differ internally as well as externally. Cells such as muscle cells that use a lot of energy, for example, have a lot of mitochondria. Similarly, cells that produce mucus in the intestine have many Golgi bodies. These cells are specialized to perform particular functions.

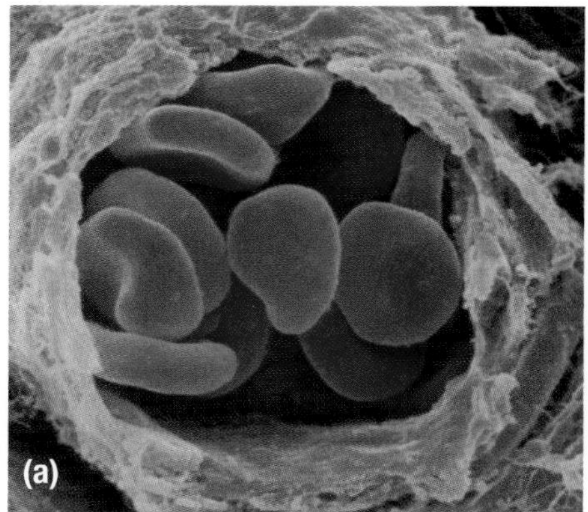

(a) Red blood cells contain hemoglobin that carries oxygen in blood. The cells are smooth so that they can easily pass through the blood vessels.

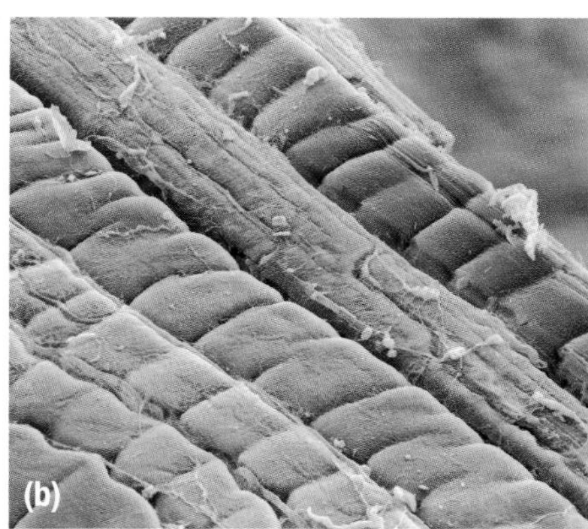

(b) Muscle cells are arranged in bundles called muscle fibres. Muscle cells can contract, which makes the fibre shorter and causes bones to move.

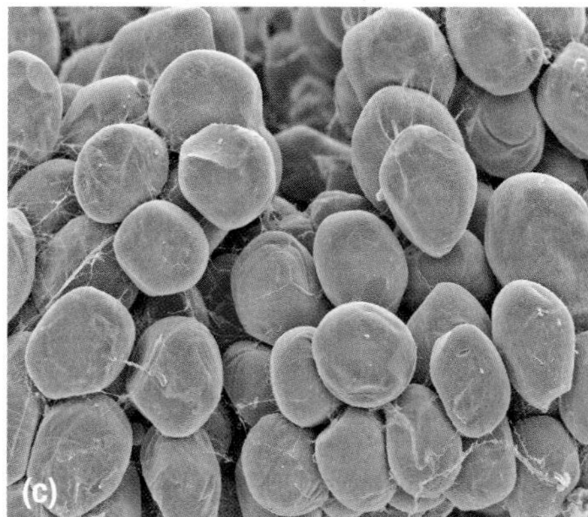

(c) Fat cells have a large vacuole in which to store fat molecules. This is how the cell stores chemical energy.

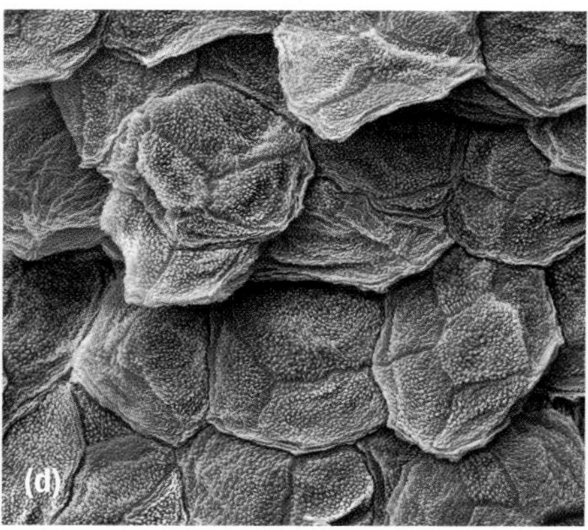

(d) Layers of skin cells fit together tightly, covering the outside of the body to protect the cells inside and to reduce water loss.

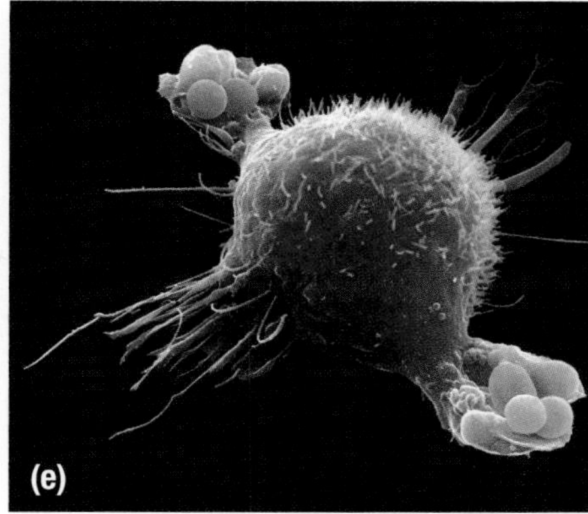

(e) White blood cells can move like an amoeba to engulf bacteria and fight infection.

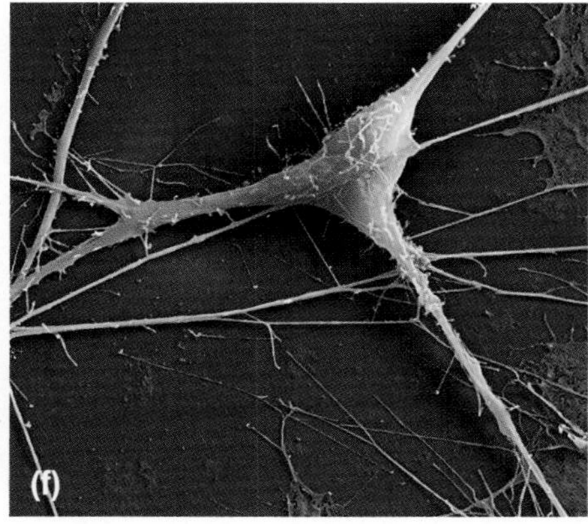

(f) Nerve cells are long, thin, and have many branches. They conduct electrical impulses to coordinate body activity.

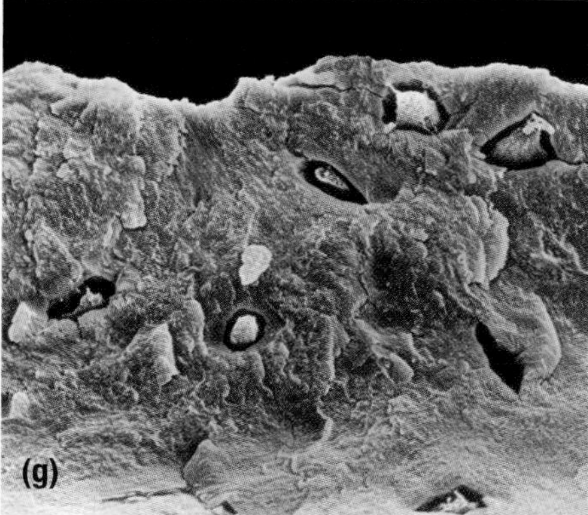

(g) Bone cells collect calcium from food and allow the growth and repair of bones. They build up bone around themselves, creating the body's skeleton.

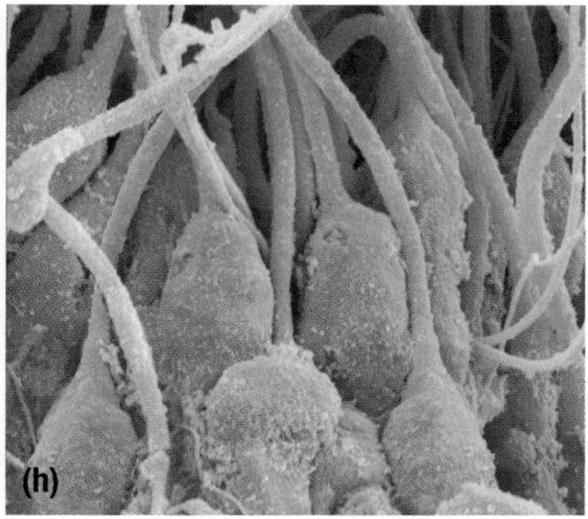

(h) Sperm cells are able to move independently, carrying DNA from the male parent to join with an egg cell from the female parent.

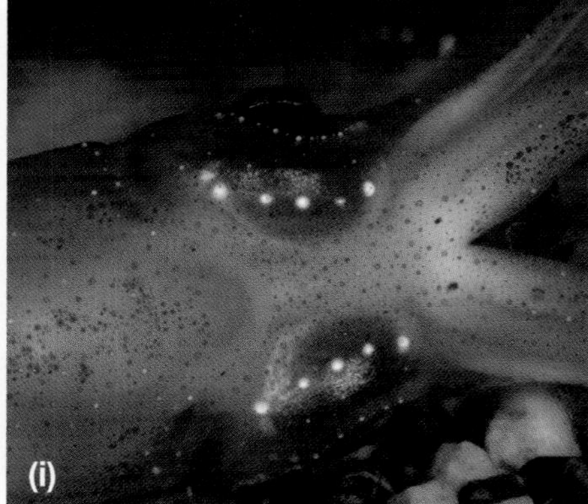

(i) Some animals that are active mainly at night, and others that live deep in the ocean, have cells that can emit light. These cells are called photophores.

Figure 2 A few of the many kinds of specialized cells found in animals

Plants also have specialized cells. The structure and function of the cells in a leaf are different from those of the cells in the trunk of a tree. Some of the specialized plant cells are shown in Figure 3.

In the next chapters, you will learn how cells work together in complex organisms. Chapter 3 focuses on animals and Chapter 4 on plants.

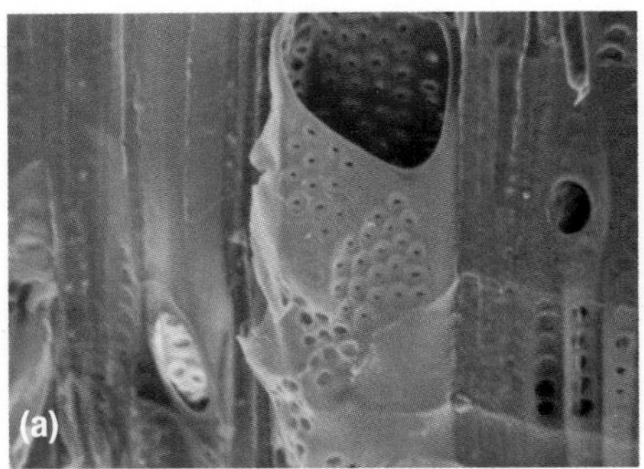

Some plant cells transport water and dissolved minerals throughout the plant.

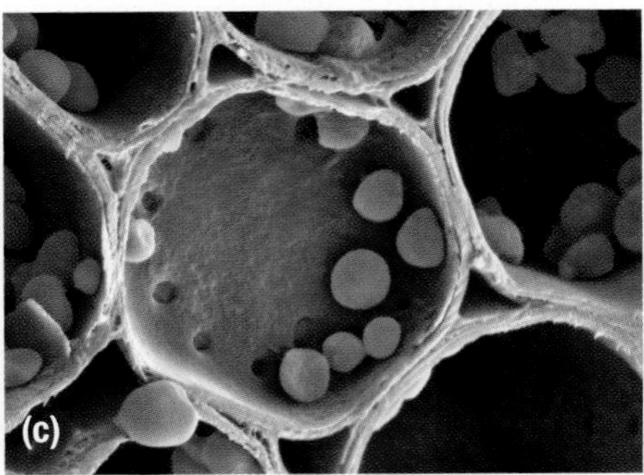

Storage cells contain special structures that store starch, a source of energy for the plant.

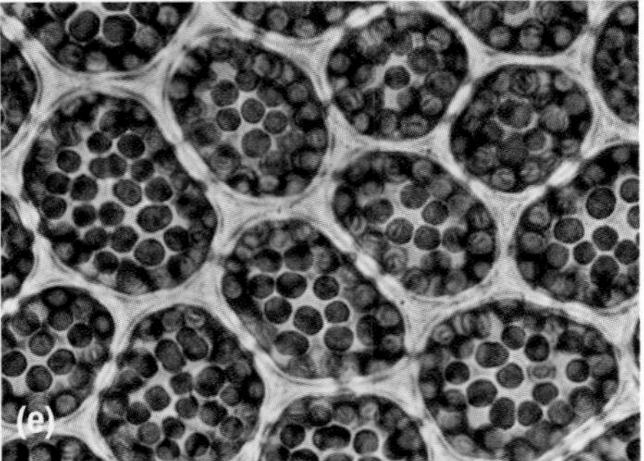

Photosynthetic cells contain many chloroplasts to collect energy from sunlight to make sugar for the plant.

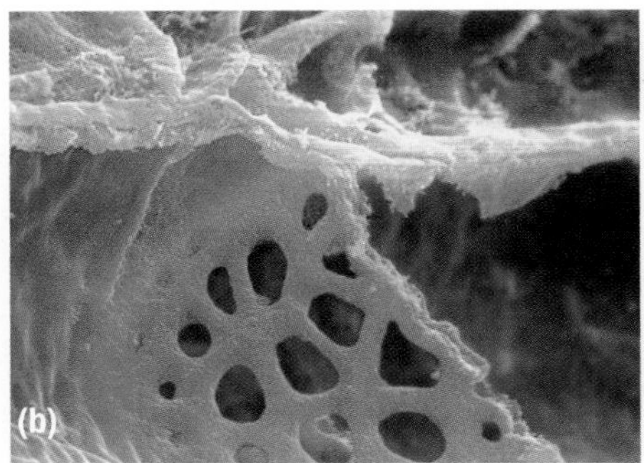

Other cells transport dissolved sugars around the plant.

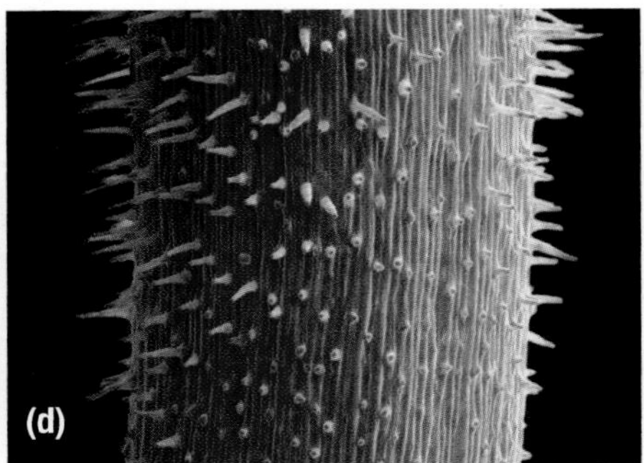

Epidermal cells on young roots have hairs that absorb water from the soil.

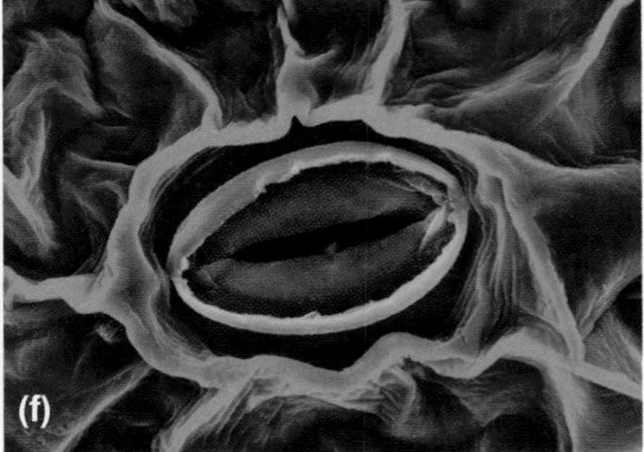

Guard cells in the surface of leaves control water loss.

Figure 3 Plants also have a wide variety of specialized cells.

IN SUMMARY

- All multicellular organisms are made mostly of specialized cells.
- Specialized cells have structures that allow them to perform specific functions.
- A specialized cell performs one primary function instead of doing everything an organism needs to stay alive.

CHECK YOUR LEARNING

1. Why are complex organisms made up of specialized cells? K/U
2. Think about your own body. List at least four activities that your body must do to keep you alive. K/U
3. Choose two specialized cells mentioned in this section. Compare their structure and function. K/U
4. Every cell in your body came from one fertilized egg cell. What does this tell you about the DNA differences between one body cell and another? K/U
5. Do plant cells specialize in the same way as animal cells? Use examples of each to illustrate your answer. A
6. Why do single-celled organisms not show specialization? K/U

Observing Specialized Cells

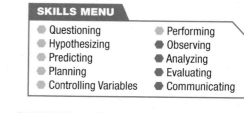

SKILLS MENU
- Questioning
- Hypothesizing
- Predicting
- Planning
- Controlling Variables
- Performing
- Observing
- Analyzing
- Evaluating
- Communicating

When you see your doctor for a routine examination, you are often asked to visit the lab to provide a blood sample. This blood sample is used to prepare a blood smear slide for examination. Slides are usually prepared with a stain that makes certain cells more visible (Figure 1).

The technicians who examine cell samples are very skilled at identifying cells that look different. Sometimes they count the proportions of different types of cells or look for abnormal shapes or clusters of cells. In this activity, you will have an opportunity to examine a variety of specialized cells from both plants and animals.

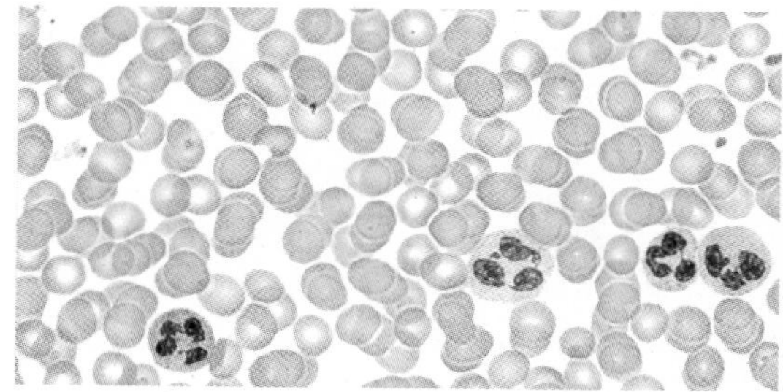

Figure 1 In this slide, the white blood cells are stained dark purple to make them easier to see.

Purpose

To examine a variety of specialized plant and animal cells.

Equipment and Materials

- prepared slides of specialized cells (such as epithelial cheek cells, skeletal muscle cells, lung tissue cells, onion root tip, leaf cross section)
- microscope
- lens paper

Procedure

1. Select one of the prepared slides and carefully clean it with lens paper. Using the correct method for viewing a slide, examine the cells on the slide. Record their shape, their colour, and how close together they are. Make a labelled scientific drawing of one or two cells.
2. Move the slide around to look for different types of cells. Record your observations.
3. Repeat Steps 1 and 2 with the other slides. Keep careful records and note differences between the types of cells.

Analyze and Evaluate

(a) How might the slide have been prepared to make the cells easier to see? T/I

(b) Think about each of the cell types that you examined. Summarize their structural differences. T/I

(c) For each cell type, infer how the structure of the cells is related to their function and give reasons for your inference. T/I

Apply and Extend

(d) Choose one of the cell types that you examined. Research a disease that results when these cells stop working properly. Relate the disease to the structure and function of the cell. Prepare and present a brief report on your chosen disease. Your report could be oral, written, or electronic. T/I C A

GO TO NELSON SCIENCE

KEY CONCEPTS SUMMARY

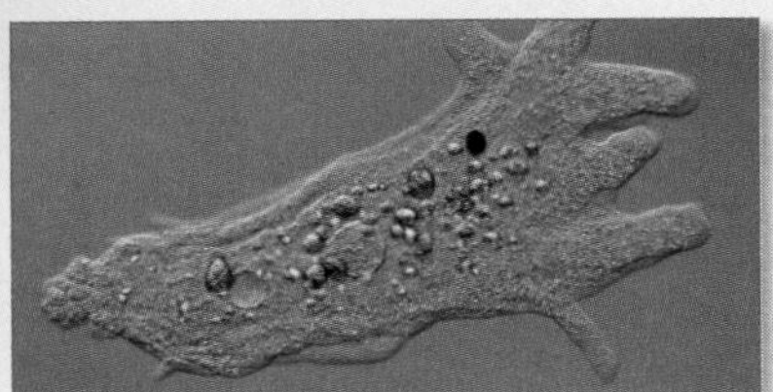

All organisms are made up of one or more cells.

- The cell theory states that the cell is the basic unit of life, that all organisms are made up of one or more cells, and that all cells come from pre-existing cells. (2.1)
- Single-celled organisms, such as bacteria, consist of just one cell. (2.1)
- Every plant and animal, including humans, is a multicellular organism and is classified as a eukaryote. (2.1)

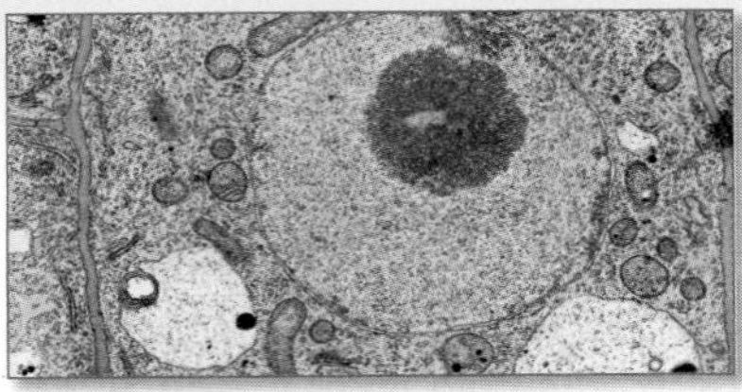

Microscopes enable us to examine cells in detail.

- Plant and animal cells have most of the major cell structures in common. (2.1, 2.2)
- Materials for cell activity pass through the cell membrane by diffusion and osmosis. (2.3)
- Plant cells contain a cell wall, a large central vacuole, and chloroplasts. (2.1, 2.2)
- Cells grow and divide to replace worn-out cells, to allow for growth in organisms, to repair damaged cells, and to reproduce. (2.3)

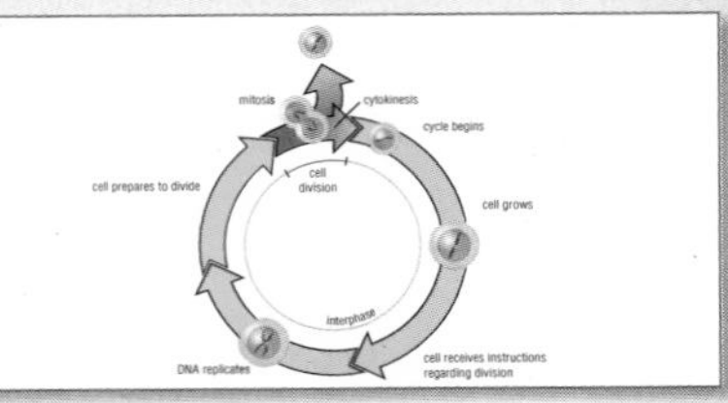

The cell cycle occurs in distinct stages.

- The cell cycle has three stages: interphase, mitosis, and cytokinesis. (2.5)
- Interphase is the stage when cells grow, perform their specific functions, produce more organelles, and replicate their DNA. (2.5)
- Mitosis is the division of the DNA in a cell's nucleus. (2.5)
- Cytokinesis is the division of the entire cell into two new identical daughter cells. (2.5)

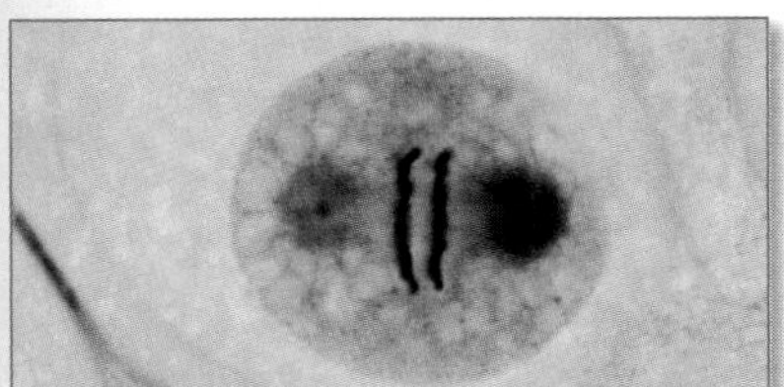

Cell division is important for growth, repair, and reproduction.

- Mitosis and cytokinesis together make up the cell division portion of the cell cycle. (2.5)
- Mitosis results in each daughter cell receiving an exact copy of the parent cell's DNA. (2.5)
- Cells go through four phases during mitosis: prophase, metaphase, anaphase, and telophase (PMAT). (2.5)

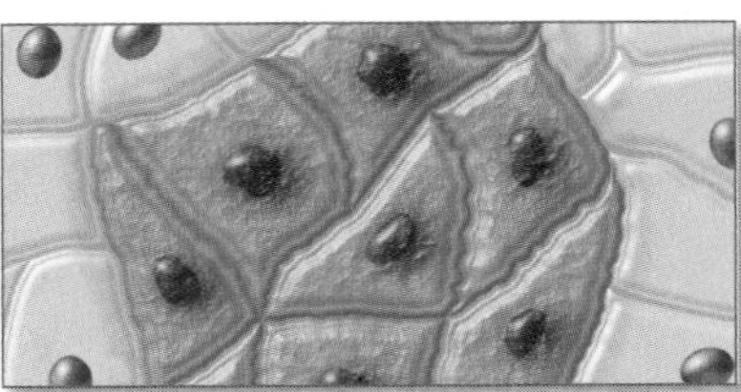

Cancer cells generally divide more rapidly than normal cells.

- Cancer is a broad group of diseases in which groups of cells grow and divide uncontrollably. (2.7)
- Uncontrolled cell growth and division create a mass of cells that may form a tumour. (2.7)
- A benign tumour does not seriously affect nearby cells and does not spread through the body. (2.7)
- A malignant tumour is made up of cancer cells; it may invade and damage surrounding tissues. (2.7)
- Cancer cells can metastasize to other parts of the body. (2.7)
- Prevention and screening minimize the risk of cancer. (2.7)

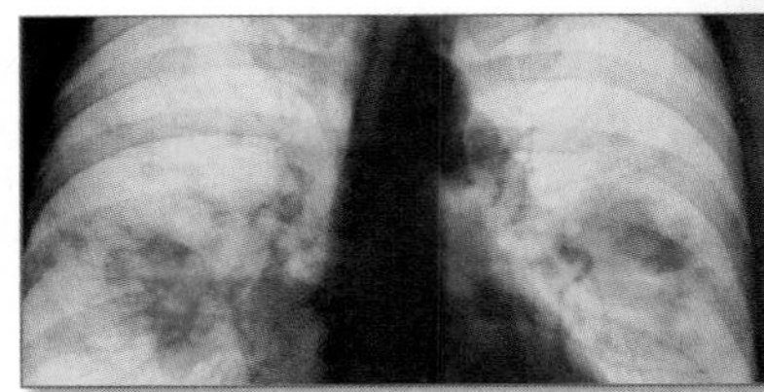

Medical imaging technologies are important in diagnosing and treating disease.

- Screening sometimes involves imaging technologies. (2.7)
- Medical imaging technologies include endoscopy, X-ray, ultrasound, CT scanning, and MRI. (2.7)
- Imaging technologies are widely used diagnostic tools that aid in the detection of cancer and other diseases. (2.7)
- Microscopic examination of cells is the only way to confirm a diagnosis of cancer. (2.7)

WHAT DO YOU THINK NOW?

You thought about the following statements at the beginning of the chapter. You may have encountered these ideas in school, at home, or in the world around you. Consider them again and decide whether you agree or disagree with each one.

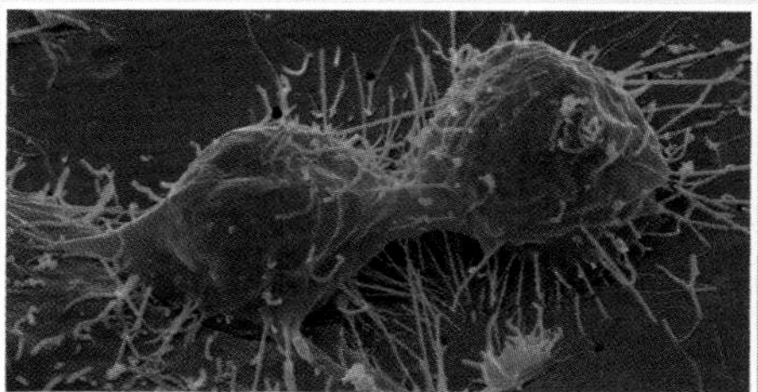

1 New cells come from previously existing cells.
Agree/disagree?

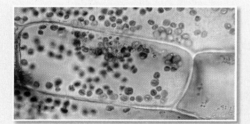

2 All plant cells are green.
Agree/disagree?

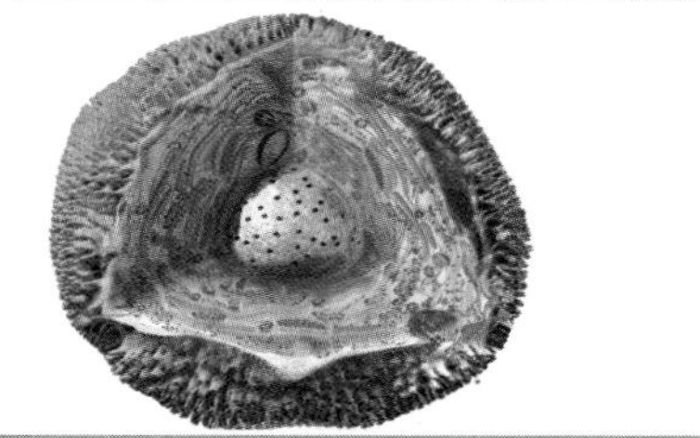

3 Plant and animal cells contain all the same organelles.
Agree/disagree?

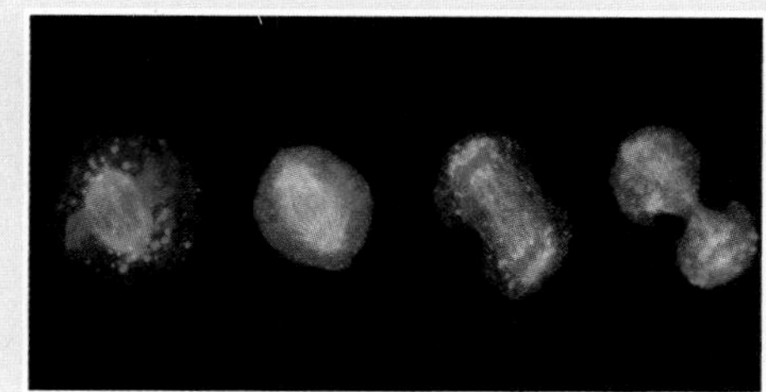

4 Mitosis and the cell cycle are identical processes.
Agree/disagree?

5 Diffusion is the movement of particles from an area of higher concentration to an area of lower concentration.
Agree/disagree?

6 All radiation is harmful to humans.
Agree/disagree?

How have your answers changed?
What new understanding do you have?

Vocabulary

cell theory (p. 29)
prokaryote (p. 29)
eukaryote (p. 29)
organelle (p. 29)
DNA (deoxyribonucleic acid) (p. 30)
asexual reproduction (p. 36)
sexual reproduction (p. 36)
diffusion (p. 37)
concentration (p. 37)
osmosis (p.37)
cell cycle (p. 40)
interphase (p. 40)
mitosis (p. 40)
cytokinesis (p. 40)
daughter cell (p. 40)
prophase (p. 41)
chromosome (p. 41)
chromatid (p. 41)
centromere (p.41)
metaphase (p. 41)
anaphase (p. 42)
telophase (p. 42)
cancer (p. 48)
tumour (p. 48)
benign tumour (p. 48)
malignant tumour (p. 48)
metastasis (p. 48)
mutation (p. 49)
carcinogen (p. 49)
Pap test (p. 50)
biophotonics (p. 55)
specialized cell (p. 58)

BIG Ideas

- ✓ **Plants and animals, including humans, are made of specialized cells, tissues, and organs that are organized into systems.**
- ✓ **Developments in medicine and medical technology can have social and ethical implications.**

The following icons indicate the Achievement Chart category addressed by each question.
K/U Knowledge/Understanding T/I Thinking/Investigation C Communication A Application

What Do You Remember?

1. What are the three stages in the cell cycle? (2.5) K/U
2. During which stage of the cell cycle does the replication of DNA take place? (2.5) K/U
3. List and briefly describe the phases of mitosis. (2.5) K/U
4. Look at Figure 1. (2.1–2.6) K/U
   (a) Identify whether each of the cells shown is an animal cell or a plant cell.
   (b) Which stage of the cell cycle (and, if appropriate, which phase of mitosis) is represented in each diagram?

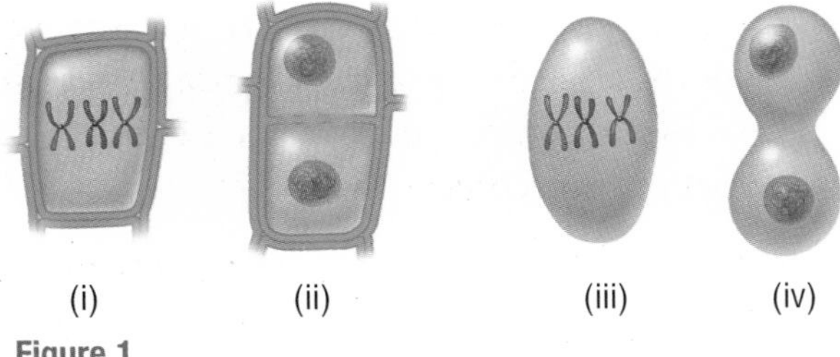

Figure 1

5. What are some features of cancer cells that make them dangerous? (2.7) K/U A

What Do You Understand?

6. You can often smell food cooking even though you may be several rooms away from the kitchen. What process is responsible for the odour spreading through the house? (2.3) K/U
7. Briefly describe the differences between plant and animal cell division. (2.5) K/U
8. Explain in your own words why mitosis is important to eukaryotes. (2.2, 2.5) T/I
9. How would the cell cycle of a cancer cell be different from that of a normal cell? (2.7, 2.8) T/I
10. What structures are responsible for the movement of chromosomes during mitosis? (2.5) K/U
11. Explain the difference between a chromosome and a chromatid. (2.5) K/U
12. (a) Explain why there is a maximum size for cells to be able to function efficiently.
    (b) Very active cells, such as muscle cells, tend to be smaller than fat storage cells. Explain why this is true. (2.3, 2.4, 2.9) K/U A

Solve a Problem

13. Why are skin cells so susceptible to cancer? (2.7) T/I A
14. Identify one major part of a plant where cells undergo frequent mitosis. (2.3, 2.6) K/U
15. List at least three "prevention and screening" steps that you can take to reduce your risk of premature death from cancer. (2.7) K/U A
16. Table 1 shows the number of cells in interphase, each phase of mitosis, and cytokinesis for two samples (Sample A and Sample B). For each sample, the percentage of cells in a stage or phase represents the length of time (as a percentage of the cell cycle) that that stage or phase takes. For example, if 50 % of the cells in a sample were in interphase, you would conclude that interphase takes up 50 % of the time of a complete cell cycle. (2.5, 2.6) K/U T/I
    (a) For each of the samples, calculate the percentage of time spent by the cells in each stage or phase of the cell cycle.
    (b) Draw a circle graph for each sample.
    (c) How do the two samples differ?
    (d) Suggest a reason, based on what you learned in this chapter, for the differences.

Table 1 Observations of Cells in Each Phase

	Number of cells in this stage or phase					
	Interphase	**Prophase**	**Metaphase**	**Anaphase**	**Telophase**	**Cytokinesis**
Sample A	320	10	3	2	1	1
Sample B	250	46	14	9	3	4

17. You notice that a mole on the back of your friend's arm seems to be getting bigger. Your friend does not think it is a big deal and is planning to ignore it. What would you advise? (2.7) K/U A

Create and Evaluate

18. What does sunlight have in common with environmental chemicals that are believed to cause cancer? (2.7) K/U T/I A

19. Why is it important for women who may be pregnant not to undergo X-rays? (2.7) K/U T/I A

20. Radiation is used in both X-rays and radiation therapy for treating cancer. Yet it also causes cancer. Explain how this is possible. (2.7) T/I A

Reflect On Your Learning

21. In this chapter, you learned about the cell cycle, including the phases of mitosis.
    (a) Was mitosis particularly difficult for you to understand?
    (b) If so, what strategies did you try to overcome this difficulty?
    (c) If not, what helped you to understand this concept?

Web Connections 

22. Since ancient times, people have searched for the "fountain of youth" to slow down or stop the aging process. Modern-day anti-aging research focuses on such areas as food, drugs, physical activity, and reduced calorie intake. T/I C A
    (a) Research a story related to the fountain of youth, such as the legend of Ponce de León.
    (b) Select one area of modern research and report on its potential as a method of prolonging life.
    (c) Write a story on the fountain of youth, using your area of research as the main theme.

23. Plants produce their own food from carbon dioxide and water by using the energy from sunlight in a process called photosynthesis. This process takes place in the chloroplasts. (2.1) T/I C
    (a) Research to find out which chemicals give chloroplasts their green colour and why these chemicals are necessary for photosynthesis.
    (b) Present your findings as a poster or in some other suitable format.

24. Research and report on the health aspects of using tanning salons. T/I C A

25. Plants sometimes develop growths or tumours called "galls." Conduct research to find the causes and effects of plant galls. Create an illustrated presentation of your research. T/I C

26. Research how scientists are working to prevent cancer. Pick a specific area of cancer research. Prepare a report on your research findings in a format of your choice. T/I C

27. How would applying sunscreen help reduce your risk of skin cancer? Conduct research to find out how effective different sunscreens are for preventing skin cancer. T/I A

28. Find out from a local veterinarian what your options are if your pet develops cancer. T/I A

29. Not all rapid cell growth is cancerous. A certain virus causes skin cells to divide quickly, producing a wart. Conduct research to find out how and why a virus can cause warts. T/I C A

30. With your teacher's approval, select any topic mentioned in this chapter. Research your topic and prepare a report of your findings in an appropriate format. T/I C

To do an online self-quiz or for all other Nelson Web Connections,
GO TO NELSON SCIENCE

CHAPTER 2

SELF-QUIZ

The following icons indicate the Achievement Chart category addressed by each question.

K/U Knowledge/Understanding
C Communication
T/I Thinking/Investigation
A Application

For each question, select the best answer from the four alternatives.

1. Chromosomes move toward opposite ends of the cell during (2.5) K/U
   (a) prophase.
   (b) metaphase.
   (c) anaphase.
   (d) telophase.
2. The job of mitochondria is to supply cells with (2.1) K/U
   (a) energy.
   (b) nutrients.
   (c) oxygen.
   (d) protein.
3. Plant cells and animal cells differ because plant cells have (2.1, 2.2) K/U
   (a) cell walls.
   (b) cell membranes.
   (c) nuclear membranes.
   (d) endoplasmic reticulum.
4. Prokaryotic cells differ from eukaryotic cells in that prokaryotic cells do not contain (2.1) K/U
   (a) cytoplasm.
   (b) DNA.
   (c) organelles.
   (d) a nucleus.

Indicate whether each of the statements is TRUE or FALSE. If you think the statement is false, rewrite it to make it true.

5. Cells of benign tumours can break away from the original tumour and move to different parts of the body. (2.7) K/U
6. Mitosis is the longest stage of the cell cycle. (2.5) K/U
7. Biophotonics is a technology that uses beams of light to detect and treat cancer. (2.7) K/U

Copy each of the following statements into your notebook. Fill in the blanks with a word or phrase that correctly completes the sentence.

8. In ________ reproduction, the offspring are exact genetic copies of the parent. (2.3) K/U
9. Environmental factors, such as tobacco smoke, X-rays, and UV rays, that cause cancer are ________. (2.7) K/U

Match each term on the left with the most appropriate description on the right.

10. (a) diffusion
    (b) osmosis
    (c) concentration
    (d) cancer
    (e) tumour

    (i) the amount of substance (solute) in a given volume of solution
    (ii) the disease resulting from uncontrolled cell division
    (iii) the process by which particles spread from areas of higher concentration to areas of lower concentration
    (iv) the movement of water from areas of higher water concentration to areas of lower water concentration
    (v) a mass of cells that divides uncontrollably without any function to the body (2.3, 2.7) K/U

Write a short answer to each of these questions.

11. The body cell of a horse has 60 chromosomes. How many chromosomes will each new horse cell have after mitosis? (2.3, 2.5) K/U
12. How do chloroplasts in individual plant cells contribute to the overall function of a plant? (2.1) K/U

13. Copy Table 1 into your notebook. In the second column, fill in information about the cell's function. In the third column, explain how the cell's structure suits its function. (2.9) K/U T/I

Table 1 Function and Structure of Cells

Type of Cell	What is the function?	How does the structure suit the function?
red blood cell		
nerve cell		
fat cell		
sperm cell		
epidermal cell of plant root		
photosynthetic cell		

14. The cell cycle varies among different types of cells. You are given samples of cells from a plant. Some of the cells have a cell cycle that is 24 h in length. Others have a cell cycle that is 72 h in length. (2.5, 2.6) T/I
    (a) Which set of cells would you predict came from the root tip of the plant? Explain your answer.
    (b) Where might the other set of cells have come from?

15. Is mitosis occurring in your body right now? Explain your answer. (2.3, 2.5) A

16. Draw a Venn diagram that illustrates the similarities and differences between plant and animal cells. (2.1, 2.2) C

17. During prophase, the nuclear membrane dissolves. It reforms during telophase. Explain why this action is important for cell division. (2.5) T/I

18. You are interested in comparing the current rates of various types of cancers with their rates of 50 years ago. (2.7) K/U T/I
    (a) Where could you find information to complete this research?
    (b) Based on your knowledge of cancer and cancer causes, what are some reasons that the rates may have changed over the last 50 years?

19. Some antibiotics interfere with a bacterial cell's ability to copy DNA. (2.3, 2.5) A
    (a) How would this type of antibiotic be able to stop a bacterial infection?
    (b) These antibiotics do not have any effect on the DNA replication of human cells. Why is this important?

20. Imagine you are writing an article for your school newspaper on cancer prevention. (2.7) C
    (a) Explain the causes of cancer that are most relevant to people your age.
    (b) Give three lifestyle choices students can make to decrease their odds of getting cancer.

21. Design an experiment to determine if temperature has an effect on the rate of mitosis in plant cells. Outline the procedure you will follow, including the independent and dependent variables, and controls necessary. (2.3, 2.5) T/I

22. The cells of human muscles and nerves rarely divide after they are formed. How does this characteristic of muscle and nerve cells affect an individual who has suffered from a spinal cord injury? (2.9) A

23. A friend comes to you with a mysterious mole on her arm. (2.7, 2.8) T/I C
    (a) What questions might you ask your friend about the mole?
    (b) What characteristics would you look for that would indicate whether she should see a dermatologist?

24. (a) Define asexual reproduction and sexual reproduction in your own words.
    (b) Which method of reproduction produces a population with more genetic variety? Explain your answer. (2.3) T/I

CHAPTER

3

Animal Systems

KEY QUESTION: What are the structures and functions of the various organ systems in animals?

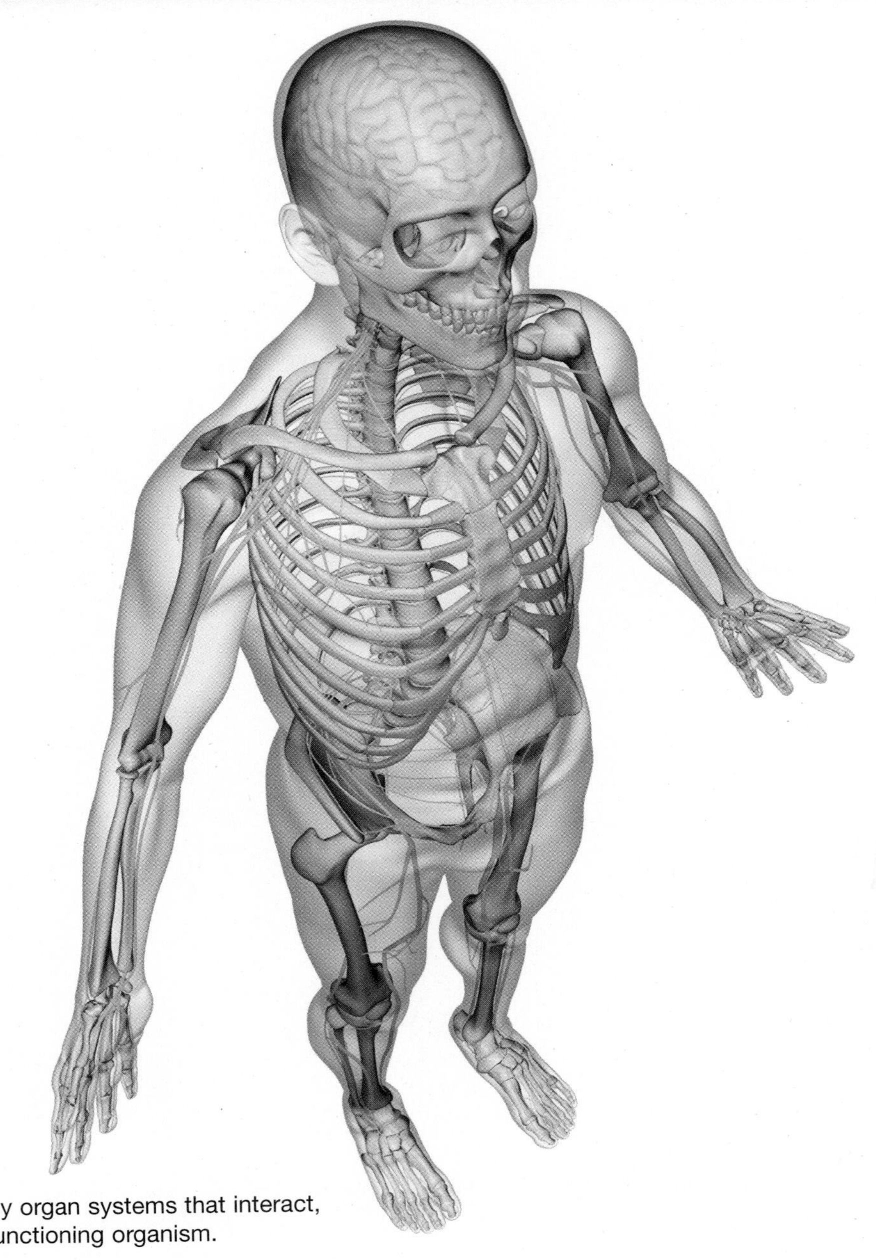

The human body has many organ systems that interact, resulting in an efficiently functioning organism.

UNIT B

Tissues, Organs, and Systems of Living Things

CHAPTER 2

Cells, Cell Division, and Cell Specialization

CHAPTER 3

Animal Systems

CHAPTER 4

Plant Systems

KEY CONCEPTS

Complex animals are made up of cells, tissues, organs, and organ systems.

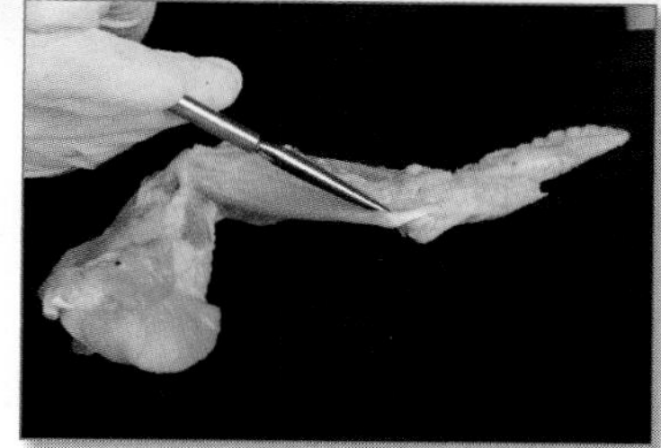

Scientists use laboratory techniques to explore the structures and functions within animals' bodies.

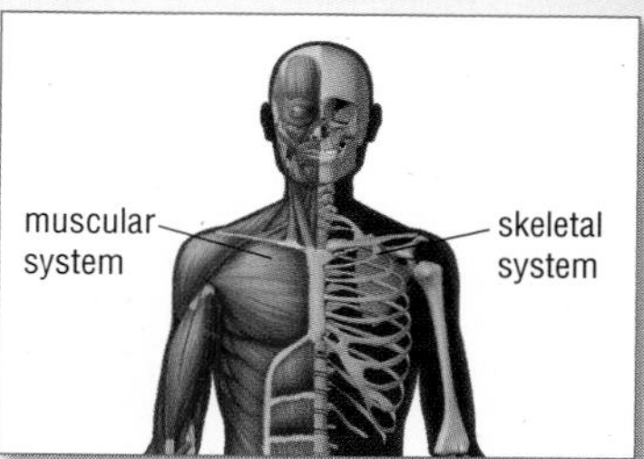

Each organ system has a specific function and corresponding specific structures.

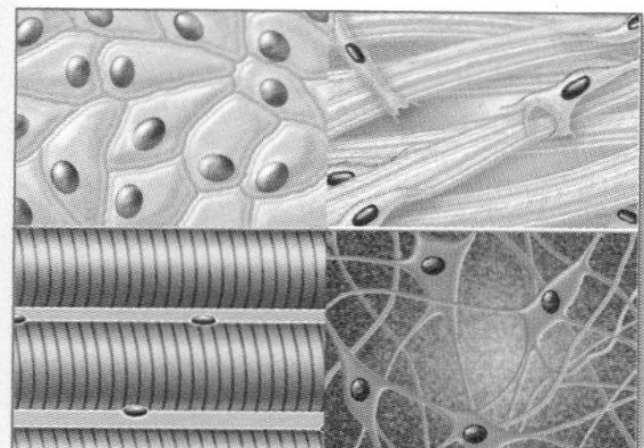

There are four main types of animal tissues.

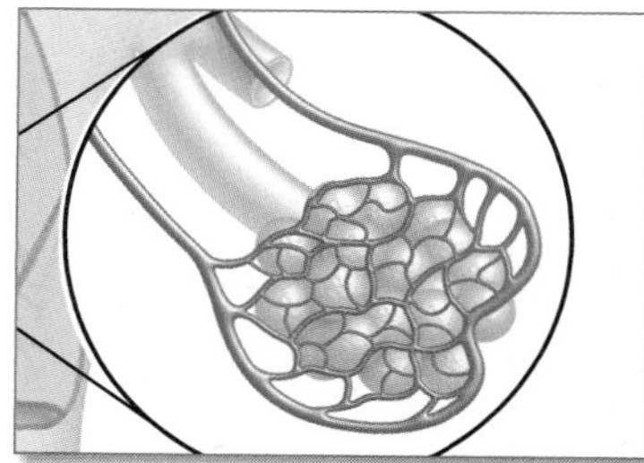

Organ systems interact to keep the organism functioning.

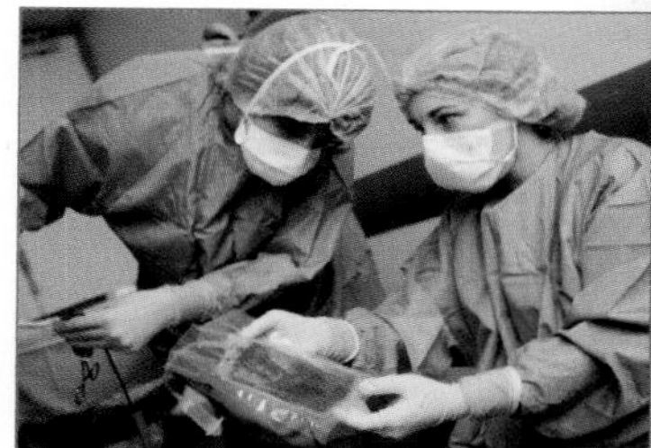

Research is helping people overcome illness and injury.

GROWING A NEW TRACHEA

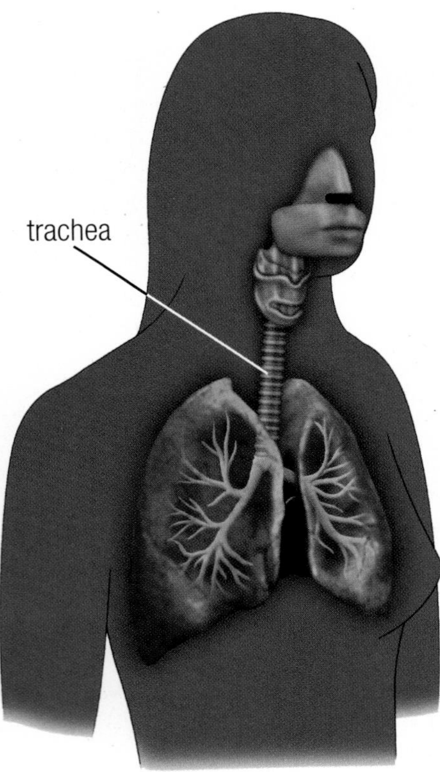

Thirty-year-old Claudia Castillo had a collapsed trachea (windpipe) as a result of tuberculosis. This left her short of breath, susceptible to infections, and unable to care for her children. Medical research offered a solution.

In spring 2008, doctors in Barcelona, Spain, used a trachea from a donor, who had died of a stroke, to help Claudia grow a new trachea. Doctors first stripped the living cells from the donor's trachea, leaving behind a supporting scaffold of non-living cartilage. Stem cells from Claudia's bone marrow were grafted over this trachea scaffold. This created a hybrid trachea, made up of Claudia's own cells and the donor scaffold. Because there were no living cells remaining in the donor scaffold, there was less chance of it being rejected by Claudia's body.

Since the transplantation, Claudia has not needed any antirejection medications. She has returned to regular activities with her children.

Are scientists able to produce entire organs in the laboratory from patients' own cells? Will stem cells eventually be used to delay the aging process or perhaps to avoid death altogether? What are your thoughts about these possibilities? Are these acceptable goals for science?

GO TO NELSON SCIENCE

WHAT DO YOU THINK?

Many of the ideas you will explore in this chapter are ideas that you have already encountered. You may have encountered these ideas in school, at home, or in the world around you. Not all of the following statements are true. Consider each statement and decide whether you agree or disagree with it.

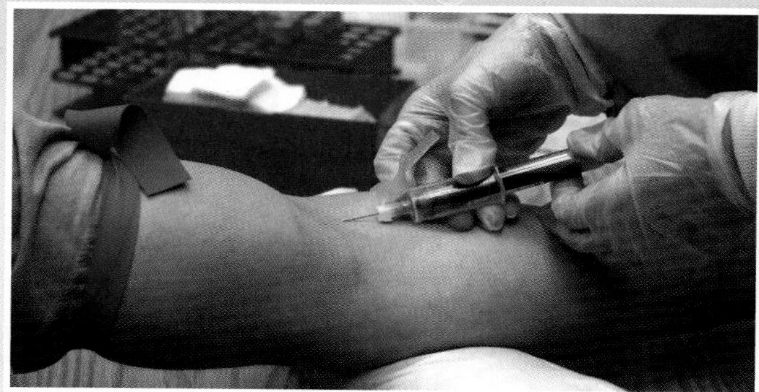

1 When blood is removed from the body, the remaining blood cells divide to make more blood.
Agree/disagree?

2 All organs in the body are made up of living cells.
Agree/disagree?

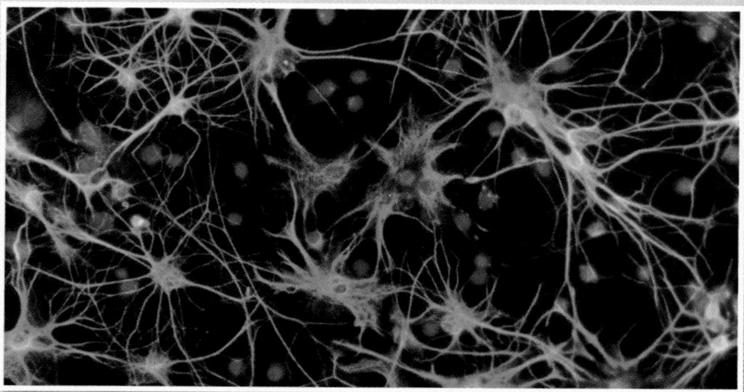

3 All animal cells look the same.
Agree/disagree?

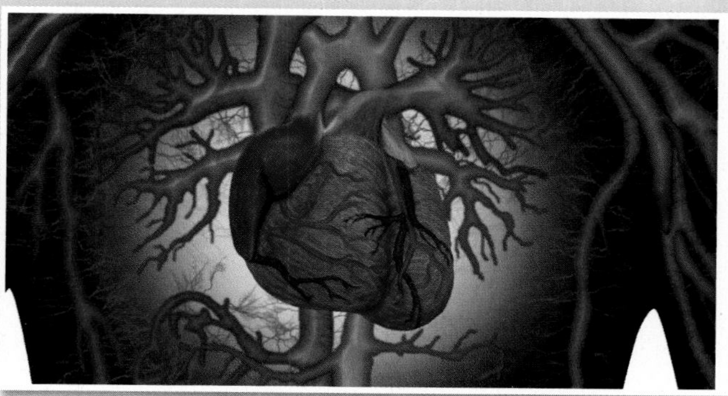

4 Our organs work independently of each other.
Agree/disagree?

5 The organ systems of a frog are the same as the organ systems of a human.
Agree/disagree?

6 Animals can grow replacement body parts.
Agree/disagree?

FOCUS ON
WRITING

Writing to Describe and Explain Observations

WRITING TIP

As you work through the chapter, look for tips like this. They will help you develop literacy strategies.

When you write to describe and explain observations, you note characteristics you observe with your senses such as shape, texture, or behaviour, and features you measure with instruments such as length, weight, or velocity. Focus on writing your observations very clearly and accurately. The following is an example of writing that describes and explains observations. Beside it are the strategies used to write effectively.

Observations of Tissues in a Chicken Leg

Parts of Chicken: The leg of a chicken is made up of two parts: the upper leg (thigh) and the lower leg (drumstick).

Mass: 227 g

Use scientific units correctly.

Record measurements (quantitative observations) accurately.

Skin: The skin on the chicken leg was bumpy, rough, and thick. The bumps were where the feathers used to grow.

Use short sentences.

Fat: There was fat under the skin. It felt soft and was white-yellow in colour. There was more fat in the thigh than in the drumstick.

Muscles: The muscles, or meat, of the chicken leg were pinkish-red in colour. The muscles were different shapes. A shiny transparent layer of epithelial tissue covered the muscles.

Use scientific terminology correctly.

Tendons: They were white. They looked like tough ropes. They connected the ends of the muscles to the bones.

Describe sensory (qualitative) observations clearly by using concrete words.

Joints: There was a joint between upper and lower legs where the two bones met. The joint was hinged, so it bent like a person's knee.

Use precise wording.

Ligaments and Cartilage: These tissues were shiny and white. Ligaments hold bone to bone. Cartilage is very smooth. It covers the surface of the bones at the joint.

Bone Marrow: In the centre of the bones is the soft, red, slimy marrow. This is where blood cells are made.

The Hierarchy of Structure in Animals

3.1

Multicellular organisms, such as animals, are made up of many different specialized types of cells. Each cell is specialized to perform a particular function. The stinging cells of jellyfish help it capture its prey, whereas the light-emitting cells of female fireflies can be used to attract a mate. Less unusual cells, but still highly specialized, include muscle and bone cells, blood cells, and sensory cells responsible for detecting sights, sounds, and odours.

Single-celled organisms, such as bacteria and blue-green algae, function independently. They do not directly depend on any other cells. In contrast, specialized animal cells cannot survive on their own. A single bone cell, hair cell, or stomach cell would quickly die if separated from its surrounding cells. These cells live and work as part of a much larger group of cells that collectively make up the body of the animal. In fact, the body of a large animal may be made up of trillions of individual cells. It is this entire collection of cells working together as a whole organism that is capable of survival and reproduction.

The complexity of animal bodies varies considerably. Some animals, such as sponges, have a simple body structure. Slugs and snails are more complex. Vertebrates (animals with backbones), such as birds, have highly complex bodies (Figure 1).

Figure 1 A sponge (a), a sea slug (b), and a cardinal (c) show increasing levels of complexity.

To understand how specialized cells work together in complex organisms, consider the many major tasks that must be performed by entire organisms, such as feeding, breathing, moving, and reproducing. In this section, you will examine how the animal body is organized for carrying out these functions.

The Animal Body—Levels of Organization

The bodies of animals look very different. A snail does not resemble a penguin. Yet all animals are made up of cells that are organized in a way that allows them to perform all of life's functions. There are levels of organization within each animal. These levels of organization form a **hierarchy**, with the "most complex" at the top and the "least complex" at the bottom.

hierarchy an organizational structure, with more complex or important things at the top and simpler or less important things below it

You are already familiar with some of these levels of organization. In earlier grades, you learned about some human organs and the digestive and circulatory systems, and you may even have used the word "tissue." These terms all refer to the hierarchy of structure within the animal body. How many levels are there? How are they arranged in terms of complexity?

tissue a collection of similar cells that perform a particular, but limited, function

organ a structure composed of different tissues working together to perform a complex body function

organ system a system of one or more organs and structures that work together to perform a major vital body function such as digestion or reproduction

Consider the hierarchy of organization within a specific animal: a white-tailed deer. Let's start with the simplest level of organization. Figure 2(a) shows a single muscle cell in the deer's heart. Each heart muscle cell is branched, allowing it to connect to other heart muscle cells. Together, these muscle cells make up the muscle **tissue** (Figure 2(b)). Figure 2(c) shows the heart itself, at the organ level of the hierarchy. An **organ** is made up of two or more types of tissues that work together to perform a complex function. In addition to muscle tissue, the heart includes two other types of tissue: nerve tissue and connective tissue. An **organ system** consists of one or more organs and other structures that work together to perform a vital body function. The heart, blood vessels, and blood are all parts of the circulatory system (Figure 2(d)). The organism—in this case, the deer in Figure 2(e)—is made up of many different organ systems working together.

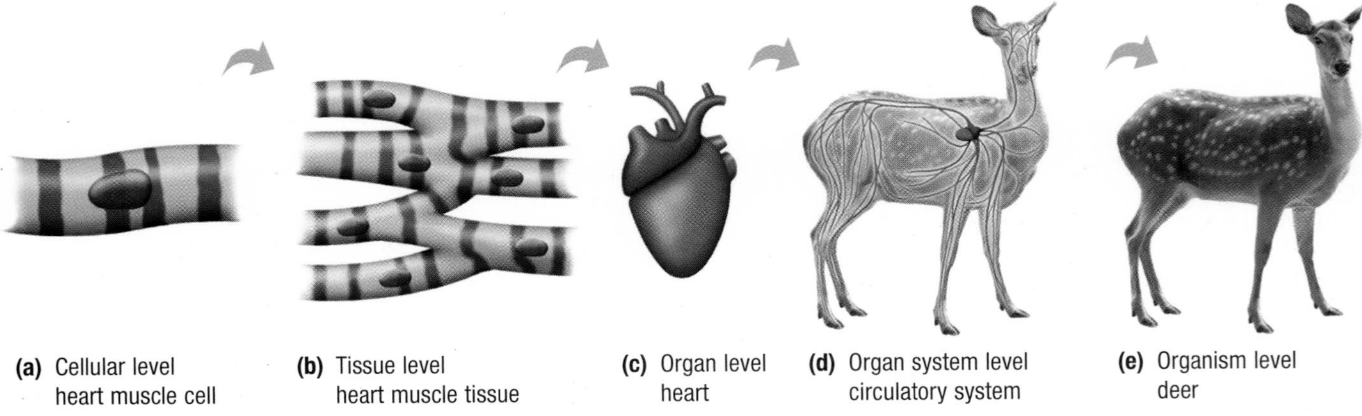

Figure 2 Levels of structural organization in an animal (a white-tailed deer)

The functioning of the whole organism depends on the hierarchy of organization within the animal. The deer needs a circulatory system to deliver nutrients and oxygen to its entire body. This system requires an organ, such as the heart, to pump the blood. It also needs a network of arteries and veins to distribute the blood throughout the body. In turn, the heart is itself made up of muscle tissue, which contracts, and nerve tissue, which keeps the heart beating regularly. The tissues are groups of specialized cells.

Living things are very complex. You might be wondering how many different organs and organ systems there are. Does each organ system have its own set of organs and associated tissues? How do these systems work together?

Organ Systems

All animals accomplish the same basic functions regardless of their appearance, behaviour, or where they live. They all obtain oxygen and nutrients and eliminate wastes. They all sense and respond to their environment, grow and repair damage, and reproduce. The task of organ systems is to perform these basic functions. Some well-known human organ systems are illustrated in Figure 3 on the next page. You will learn about some of these in more detail as you work through this chapter.

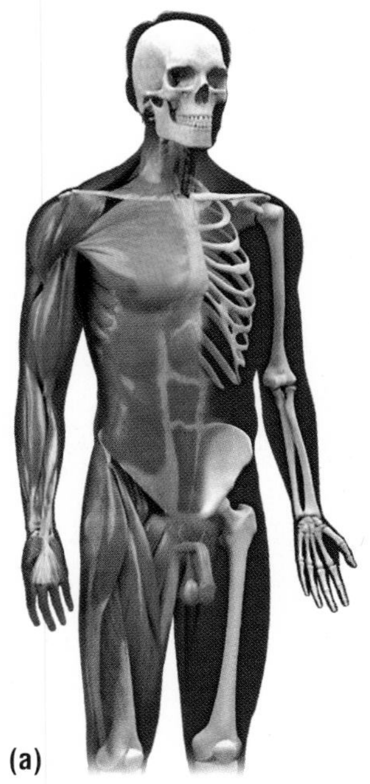

(a)

The musculoskeletal system supports the body and makes movement possible.
The reproductive system produces eggs (in females) and sperm (in males). In some animals, the female reproductive system also supports the growing fetus until it is born.

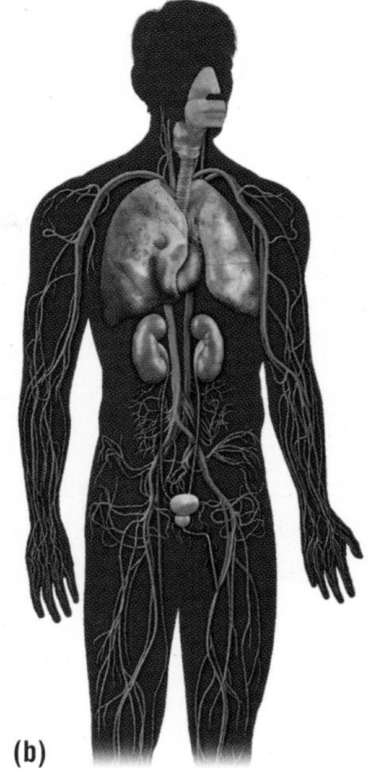

(b)

The respiratory system takes oxygen from the air and removes carbon dioxide from the body.
The circulatory system transports substances around the body.
The urinary system excretes waste and keeps the correct amount of water in the body.

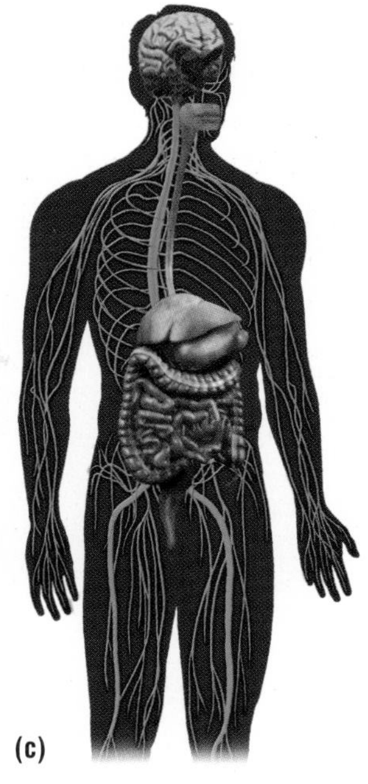

(c)

The nervous system sends messages around the body.
The digestive system breaks down the food you eat and makes it available to the body.

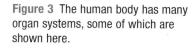

Figure 3 The human body has many organ systems, some of which are shown here.

WRITING TIP

Spell Check
Carefully check your spelling of scientific terms such as the names of tissues or organs. Since many of these terms are not used often in everyday speech or print, it is helpful to keep your own list of unfamiliar terms such as "musculoskeletal," "epithelial," and "cytokinesis" so you can easily check their meaning and correct spelling.

Organs

Each organ system is made up of highly specialized organs and other structures that work together to perform the overall function of the system. For example, the digestive system is made up of many organs, including the stomach, small and large intestines, liver, and pancreas. Most organs work within a single organ system. For example, the stomach is part of the digestive system and of no other system. Some organs, however, play a role in more than one system. For example, the pancreas is part of the digestive system and the endocrine system.

Tissues

Animals have four major types of tissue: **epithelial tissue**, **connective tissue**, **muscle tissue**, and **nerve tissue**. Each of these types of tissue contains many types of specialized cells, and each is found in most organ systems. Table 1, on the next page, summarizes these four types of tissue. Where do these tissues come from? How does an animal produce these different cell and tissue types, which are then organized into organs and organ systems? You will begin to answer these questions in the next section.

epithelial tissue (or **epithelium**) a thin sheet of tightly packed cells that covers body surfaces and lines internal organs and body cavities

connective tissue a specialized tissue that provides support and protection for various parts of the body

muscle tissue a group of specialized tissues containing proteins that can contract and enable the body to move

nerve tissue specialized tissue that conducts electrical signals from one part of the body to another

Table 1 Animal Tissue Types

Type	Example	Description	Function
epithelial tissue	• skin • lining of the digestive system	• thin sheets of tightly packed cells covering surfaces and lining internal organs	• protection from dehydration • low-friction surfaces
connective tissue	• bone • tendons • blood	• various types of cells and fibres held together by a liquid, a solid, or a gel, known as a matrix	• support • insulation
muscle tissue	• muscles that make bones move • muscles surrounding the digestive tract • heart	• bundles of long cells called muscle fibres that contain specialized proteins capable of shortening or contracting	• movement
nerve tissue	• brain • nerves in sensory organs	• long, thin cells with fine branches at the ends capable of conducting electrical impulses	• sensory • communication within the body • coordination of body functions

IN SUMMARY

- The bodies of animals are organized in a structural hierarchy.
- The levels of organization are organ systems, organs, tissues, and cells.
- Tissues are groups of similar cell types that perform a common function.
- There are four main tissue types: epithelial, connective, muscle, and nerve.

CHECK YOUR LEARNING

1. Create a concept map to illustrate the hierarchy of organization within an animal. Include examples in your concept map. K/U C
2. Give an example of an organ that is found in
   (a) only one organ system
   (b) more than one organ system K/U
3. In what way are organ systems more complex than highly specialized cells? K/U
4. Make a list of the main functions that must be performed by all living things. For each main function, name an organ system that is involved in performing that function. K/U
5. Most animals have the same kinds of organ systems. Why do you think there are not dozens or even hundreds of completely different kinds of organ systems? T/I
6. Why is there no hierarchy of organization within single-celled organisms? K/U

Stem Cells and Cellular Differentiation

3.2

In previous studies, you learned that all multicellular organisms start as a single cell. This single cell is sometimes called a zygote. A long process of development is needed to transform the zygote into a fully formed plant or animal. First, it goes through a series of divisions that generates many cells. As the cells of this early-stage organism, known as an embryo, continue to divide, the cells begin to show differences in their shapes, contents, and functions. In other words, the cells become specialized.

The process that produces specialized cells is **cellular differentiation**. Cellular differentiation is directed by the genetic information inside the cell. This genetic information is encoded in the cell's DNA. It is passed from parent to offspring in the eggs and sperm cells.

cellular differentiation the process by which a cell becomes specialized to perform a specific function

Stem Cells

In animals, a cell that can differentiate into many different cell types is called a **stem cell**. A stem cell divides into two daughter cells through the processes of mitosis and cytokinesis (Section 2.5). Each resulting daughter cell can develop into a different type of cell, based on which parts of its DNA are switched on (Figure 1). Stem cells generally occur in clumps that differentiate into different tissue layers, such as epithelial, muscle, and nerve tissues.

stem cell an undifferentiated cell that can divide to form specialized cells

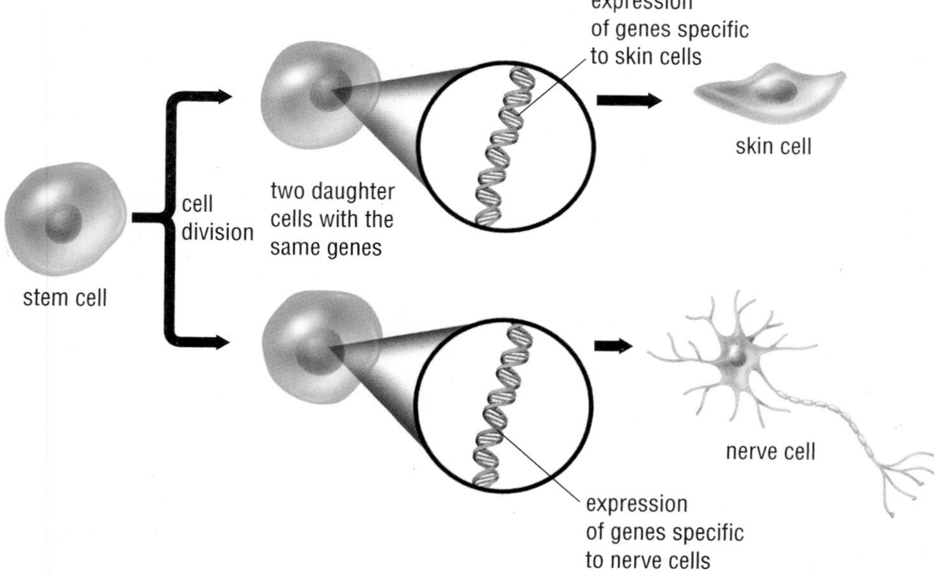

Figure 1 In this example, each daughter cell differentiates into two different types of specialized cells: a nerve cell and a skin cell.

There are two types of stem cells. Embryonic stem cells can differentiate into any kind of cell. Tissue stem cells (sometimes called "adult stem cells") exist within specialized tissue. They are only able to differentiate into certain types of cells. For example, tissue stem cells found in bone marrow can differentiate into white blood cells, red blood cells, or platelets (Figure 2). This is why bone marrow transplants are often a successful way to treat cancers that affect blood cells.

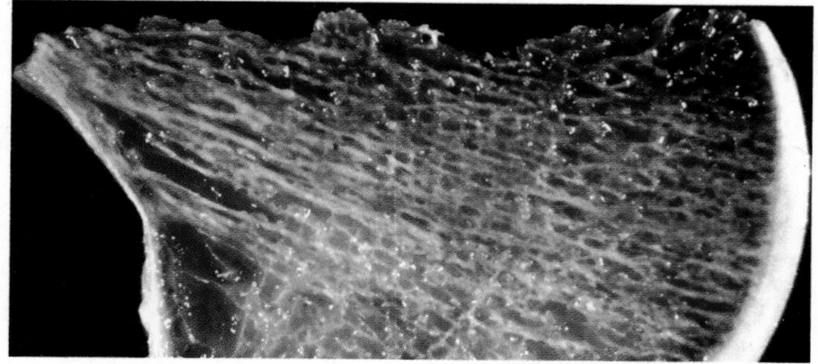

Figure 2 This cross section of bone shows the spongy bone tissue. The spaces are filled with bone marrow.

RESEARCH THIS STEM CELL RESEARCH IN CANADA

SKILLS: Researching, Analyzing the Issue, Communicating, Evaluating

SKILLS HANDBOOK 4.A., 4.C.

Many medical experts believe that stem cells can be used to treat a variety of injuries and illnesses. This field of medicine is called "stem cell research." Researchers are considering the use of stem cells in the treatment of spinal cord injuries, Parkinson's disease, Alzheimer's disease, diabetes, multiple sclerosis, and heart disease. There are many practical challenges in this line of work.

Human stem cell research is also full of legal, ethical, and social concerns. Many nations have developed ethical and legal guidelines for research that involves human stem cells.

1. Investigate Canada's current guidelines on stem cell research.
2. Identify and research one possible future medical advance that could result from stem cell research.
3. Research the ethical arguments for and against stem cell research.

GO TO NELSON SCIENCE

A. Summarize Canada's current guidelines for stem cell research. T/I
B. Write a short report on one possible application of stem cell research. T/I C A
C. Summarize the ethical arguments relating to this application of stem cell research. Present your position in an appropriate format, supporting your opinion with reasons. T/I C A

Cord Blood Cell Banking

The blood found in the umbilical cord immediately following birth is a rich source of stem cells. These are not embryonic stem cells; they are more similar to tissue stem cells. They can develop into any of the various kinds of blood cells. Umbilical cord blood has a high concentration of these tissue stem cells and is relatively easy to obtain (Figure 3). This collected blood could be "banked" (stored) in case it is needed later in the child's life. Current uses for such cord blood stem cells include treatment for childhood cancers such as leukemia.

DID YOU **KNOW?**

Canadians and Stem Cell Research
Research in the stem cell field grew out of findings by Canadian scientists Ernest A. McCulloch and James E. Till in the 1960s.

GO TO NELSON SCIENCE

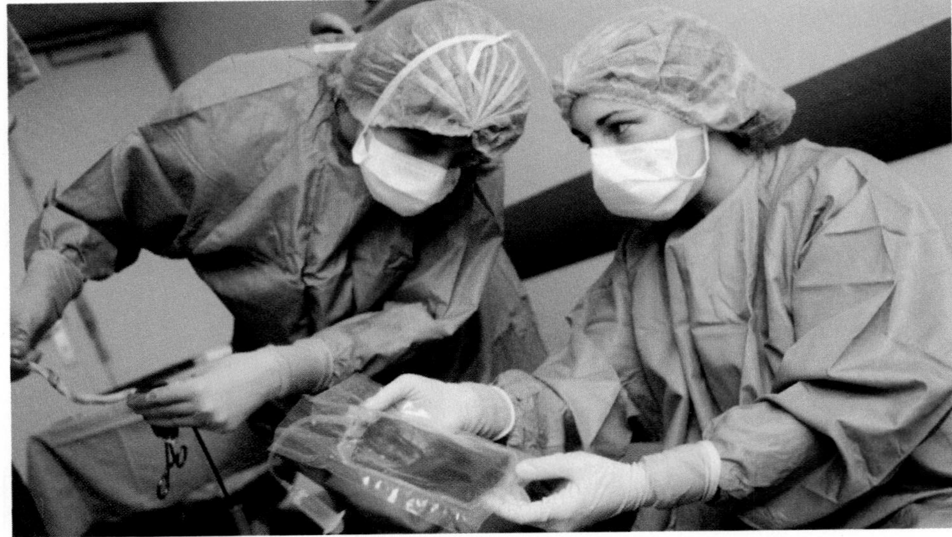

Figure 3 Umbilical cord blood collection

Some commercial companies sell the service of cord blood banking. They promote the idea that the child, or a sibling, may benefit from a future treatment using these tissue stem cells. Many new parents are asked whether they want to bank their new baby's cord blood.

Tissue Stem Cell Transplantation

Both cord blood and bone marrow stem cells are relatively easy to isolate. Both have been used to treat diseases such as leukemia. Leukemia is a cancer that occurs in the bone marrow. The stem cells that differentiate into blood cells divide too quickly, resulting in non-functioning blood cells. In leukemia treatment, all of the diseased white blood cells must first be removed and the bone marrow must be killed. Chemotherapy can target and kill bone marrow cells and white blood cells. Healthy bone marrow cells (or stem cells collected from the blood) are obtained from a carefully matched donor. The healthy stem cells are then injected into the patient's blood. In a successful stem cell transplant, the donor blood stem cells grow in the patient's own bone marrow. Eventually, they produce healthy, cancer-free blood cells.

Regeneration and Tissue Engineering

In complex animals, such as mammals, the term "regeneration" refers to the ability of a tissue to repair itself. Skin, muscle, and bone can regrow and heal after an injury. Not all cells regenerate, however. Nerve cells, for example, do not naturally regenerate completely.

In animals such as salamanders, sea stars (starfish), and flatworms, regeneration can sometimes replace lost limbs and even large portions of the body (Figure 4). Scientists are researching ways to regenerate human body tissues and parts that do not normally regenerate. This field of research is called tissue engineering. This is useful in treating spinal cord injuries and in different types of tissue grafting, which involves providing replacement tissues for patients. Tissue engineering could also provide a source of biological models for testing drugs and other potentially dangerous substances.

Figure 4 Some species of starfish have an extraordinary ability to regenerate. A single severed "arm" can grow into a whole new starfish.

To find out more about advances in tissue engineering,

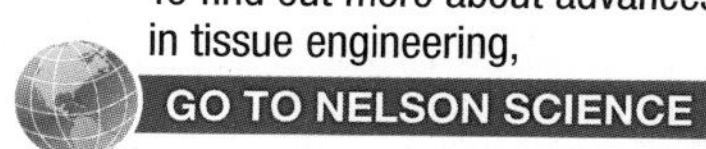

IN SUMMARY

- Cellular differentiation is the process by which a less specialized cell becomes a more specialized cell type.
- The two types of stem cells are embryonic stem cells and tissue stem cells.
- Stem cell research is discovering medical treatments for injuries and diseases.
- Regeneration results in the repair or replacement of tissues or body parts.

CHECK YOUR LEARNING

1. What is the meaning of the term "cellular differentiation"? K/U
2. What is the difference between tissue stem cells and embryonic stem cells? K/U
3. Why are stem cells from a newborn baby's umbilical cord blood considered to be tissue stem cells? K/U
4. What is the significance of being able to harvest stem cells that can specialize into any type of cell? K/U
5. Briefly describe the process of a bone marrow transplant to treat leukemia. K/U
6. How is regeneration beneficial to an animal? K/U
7. Can all animals regenerate? Explain. K/U

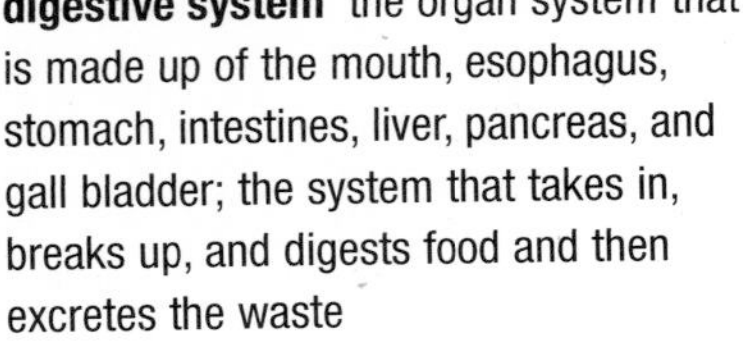

3.3 The Digestive System

digestive system the organ system that is made up of the mouth, esophagus, stomach, intestines, liver, pancreas, and gall bladder; the system that takes in, breaks up, and digests food and then excretes the waste

Your body needs food to survive. Each of your cells needs a supply of food (to provide chemical energy) and other nutrients. How do the necessary chemicals make their way from your mouth to your cells? The first part of the journey is through the digestive system. The second part is through the circulatory system, which you will learn about in Section 3.4.

The **digestive system** is the organ system that takes in food, digests it, and excretes the remaining waste. The digestive system is made up of the digestive tract and the accessory organs.

The Digestive Tract

The digestive tract in most animals is essentially one long tube with two openings, one at either end. This structure is very apparent in an earthworm, where the tube varies only slightly in diameter along its length (Figure 1(a)). In humans, the digestive tract is much more complex. It includes the mouth, esophagus, stomach, small intestine, large intestine, and anus. The accessory organs include the liver, gall bladder, and pancreas (Figure 1(b)).

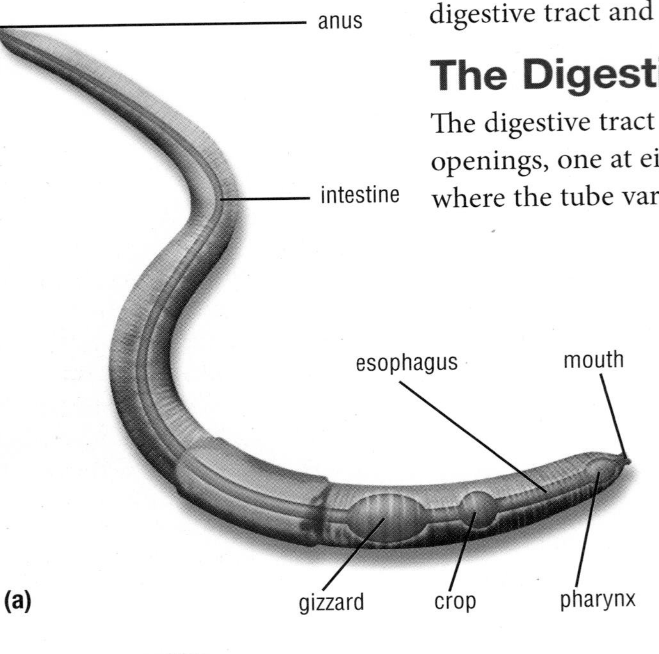

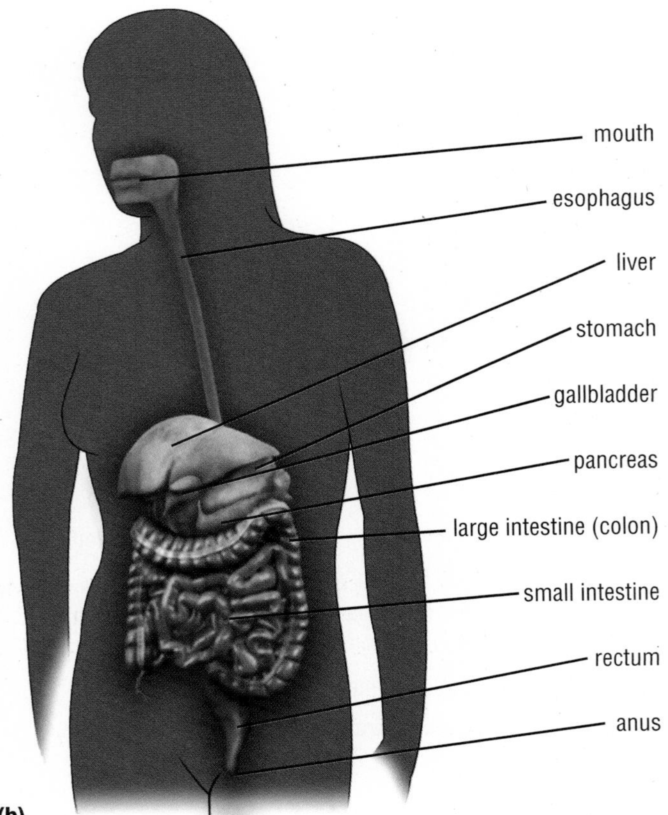

Figure 1 (a) In an earthworm, food passes through the esophagus and is stored in the crop. The gizzard grinds the food to break it into pieces. Nutrients are absorbed in the intestine. (b) In a human, the breakdown of food starts in the mouth and continues until nutrient absorption occurs in the small intestine.

The entire length of the digestive tract is lined with epithelial tissue. This tissue is made up of many different types of cells, including goblet cells that secrete mucus. The mucus serves two functions: it protects the digestive tube from digestive enzymes, and it allows the material to pass smoothly along the tube. The digestive tube also includes layers of muscle tissues and nerves (Figure 2).

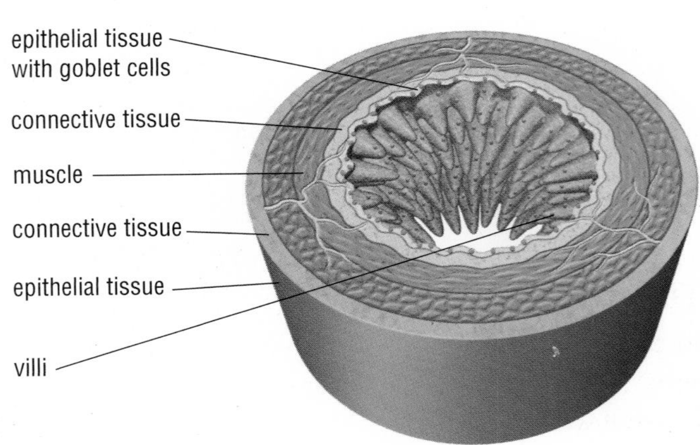

Figure 2 Tissues that make up the digestive tract

If you eat spoiled food, your body recognizes the presence of toxins produced by bacteria. Your digestive tract responds by attempting to remove the toxins rapidly, which you experience as vomiting or diarrhea. This can also happen as a result of too much alcohol or other poisonous substances.

The Mouth

The mouth starts the process of breaking down food. It does this in two ways: mechanically (with the teeth and tongue) and chemically (with chemicals called enzymes that break apart the molecules of food). The mouth adds saliva—a mixture of water and enzymes—to the food in the mouth. Saliva is produced by cells in the epithelial tissue that lines the mouth. Once the food is broken up and softened with saliva, it is swallowed and passed into the esophagus.

The Esophagus

The esophagus is a muscular tube connecting your mouth to your stomach. The muscles are a special type, called smooth muscle tissue, which can contract and relax without conscious thought. This movement is controlled by nerve tissue. The contractions slowly move the food along.

The Stomach

One of the major organs in the animal digestive system is the stomach (Figure 3). The main function of the stomach is to hold food and churn it to continue the process of digestion. The stomach lining contains cells that produce digestive enzymes and acids. Smooth muscle tissue contracts to mix the stomach contents.

The stomach is richly supplied with nerves that signal when we have had enough to eat.

DID YOU **KNOW?**

Heartburn Is a Real Pain

Sometimes if you eat too quickly or if you are under a lot of stress, acidic fluid from your stomach backs up into your esophagus. Your stomach has a thick protective mucous layer that protects it from your digestive acids. Your esophagus does not have the same protective layer to provide a barrier from the acid. This results in a burning sensation known as "heartburn." Heartburn treatment may include taking an antacid to neutralize and lessen the effects of the acid.

Interesting...

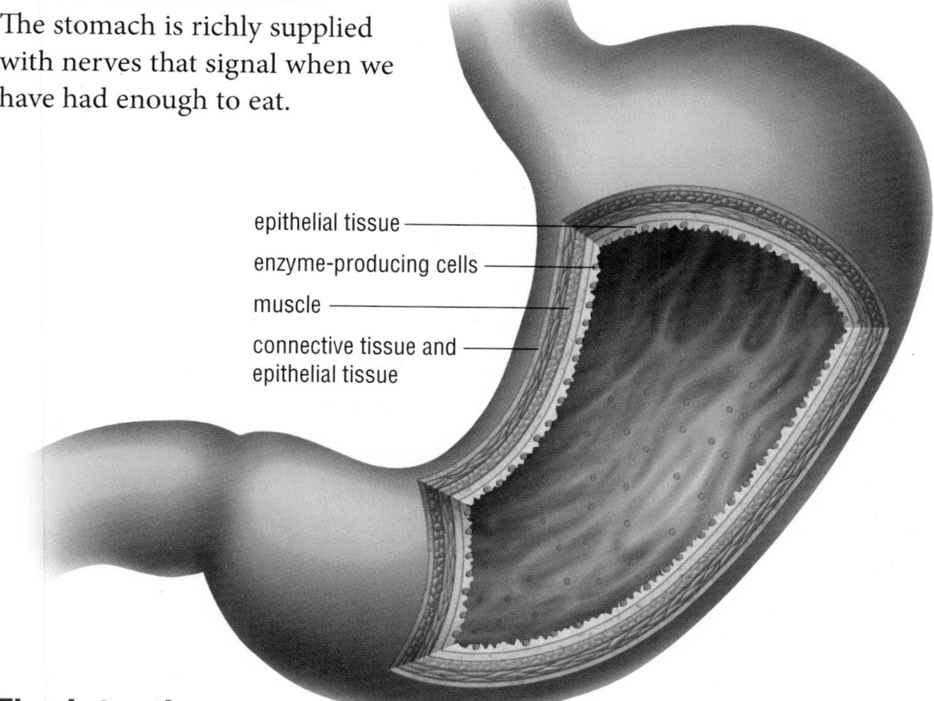

Figure 3 The stomach consists of many types of specialized tissues grouped together to function as an organ.

The Intestine

In mammals, the part of the digestive tract between the stomach and the anus is the intestine. The lining of the intestine has cells that produce mucus. It also has many fine blood vessels interlaced through the other tissues. Like the esophagus and stomach, the intestines contain smooth muscles that contract and relax without our conscious thought.

> **DID YOU KNOW?**
>
> **Expanding Mucus!**
> The mucus in the goblet cells is condensed but expands rapidly when released from the cells. Like shaving cream released from a pressurized can, the mucous gel increases in volume up to 500 times in only 20 ms!

There are two parts to the intestine: the small intestine and the large intestine. The small intestine, which is about 6 m long and relatively narrow, is where most digestion occurs. Goblet cells release mucus, and nutrients diffuse through the wall of the small intestine and enter the bloodstream.

The large intestine, or colon, is about 1.5 m long but larger in diameter than the small intestine. Its lining absorbs water from the indigestible food. The remaining solid matter is excreted as feces from the anus.

Sometimes the epithelial tissue lining the colon can become inflamed and stop working properly. This disease is known as colitis. There are several causes of colitis, including viruses, bacteria, narrowed blood vessels, and failure of the body's disease-fighting mechanism. Colitis is diagnosed with the aid of an endoscope and microscopic examination of tissue samples taken from the colon.

Accessory Organs

The liver, pancreas, and gall bladder all help with the digestion of food by supplying digestive enzymes. The liver also produces a fluid called bile, which helps in the breakdown of fats in our food. These substances are delivered into the digestive tract, where they mix with the partially digested food.

The pancreas produces an important enzyme called insulin. Insulin regulates the concentration of glucose (a sugar) in the blood. Diabetes is a disease in which the pancreas produces too much or too little insulin. A person with diabetes can experience weakness and dizziness due to too low or too high blood glucose levels. Some forms of diabetes can be controlled by diet.

> **UNIT TASK Bookmark**
>
> Consider this information about the digestive system as you work on the Unit Task on page 156.

IN SUMMARY

- The digestive system takes in food, digests it, absorbs nutrients and water, and excretes waste.
- The digestive system is made up of the digestive tract and accessory organs.
- The digestive tract in most animals contains one long tube with two openings, one at either end.
- The digestive tract in humans includes the mouth, esophagus, stomach, large intestine, small intestine, and anus.
- The digestive tract includes epithelial tissue, smooth muscle tissue, nerves, and connective tissue. The smooth muscle tissue can contract and relax without conscious thought.
- The accessory organs are the liver, the pancreas, and the gall bladder. They produce enzymes and other fluids that aid digestion.

CHECK YOUR LEARNING

1. List the main parts of the digestive tract and their major functions. K/U
2. Why is it necessary for food to be digested? K/U
3. Name at least four substances that are added to the food in the digestive tract to aid in digestion. K/U
4. Which kind of tissue contracts to push food through the digestive system? K/U
5. Briefly describe at least one disease or illness that can result from a problem in the digestive system. K/U

The Circulatory System

The human **circulatory system** is made up of the blood, the heart, and the blood vessels. The function of the circulatory system is to transport substances around the body. It moves nutrients absorbed from the intestine to all of the body's cells. Blood flows through the lungs (part of the respiratory system) to pick up oxygen and then flows through the body to deliver it to active cells. Blood also carries wastes from the body tissues for disposal. It carries carbon dioxide to the lungs, where it is released into the air. Other waste substances are carried to the kidneys (an organ of the urinary system), where the substances are filtered out and excreted.

Among the circulatory system's other vital functions are the regulation of body temperature and the transport of disease-fighting white blood cells to areas of the body where there are viruses or bacteria.

circulatory system the organ system that is made up of the heart, the blood, and the blood vessels; the system that transports oxygen and nutrients throughout the body and carries away wastes

To watch a dissection of the organ systems of a fish,
GO TO NELSON SCIENCE

Parts of the Circulatory System

The three main parts of the circulatory system are the blood, the heart, and the blood vessels. The heart pumps the blood through large blood vessels, called arteries, which branch into smaller and smaller blood vessels. The smallest blood vessels are called capillaries. In the capillaries, blood exchanges many substances with the surrounding tissues (Figure 1). After this exchange, blood flows into larger blood vessels called veins and eventually returns to the heart. Let's now look at these parts in detail.

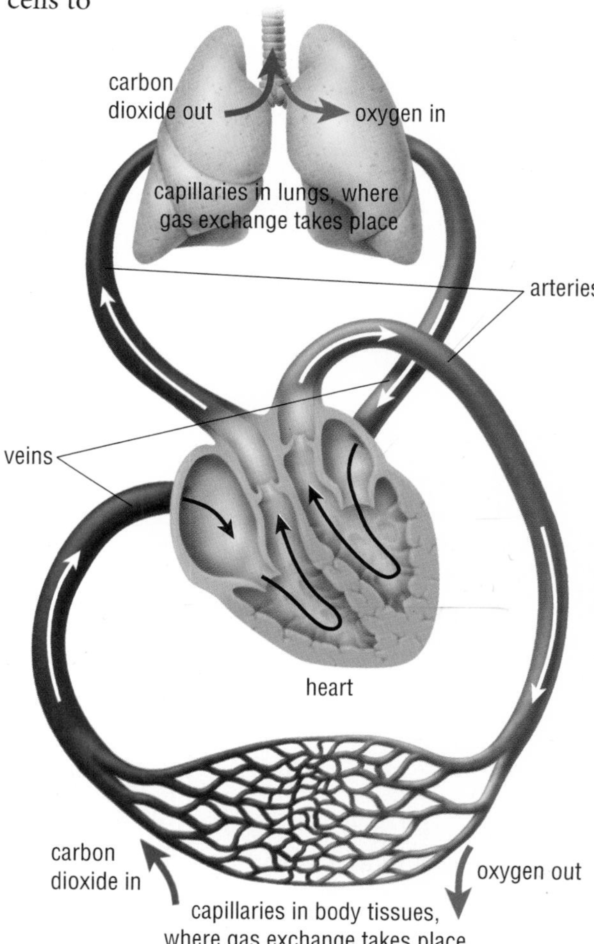

Figure 1 The circulatory system connects all parts of the body. In this diagram, the oxygenated blood is shown in red. The deoxygenated blood is shown in blue. Note that this diagram is not to scale.

Blood

Blood is a type of connective tissue that circulates throughout all parts of your body. The blood consists of four components (Figure 2):

- Red blood cells are the most plentiful of the body's blood cells. These cells make up almost half of the blood's volume. Red blood cells contain a protein called hemoglobin, which allows them to transport oxygen throughout the body. Hemoglobin makes the cells appear red.
- White blood cells are infection-fighting cells in the blood. They recognize and destroy invading bacteria and viruses. White blood cells make up less than 1 % of the volume of blood. They are the only blood cells to have a nucleus.
- Platelets are tiny cells that help in blood clotting. They also comprise less than 1 % of the blood.
- Plasma is a protein-rich liquid that carries the blood cells along. It makes up over half of blood's volume.

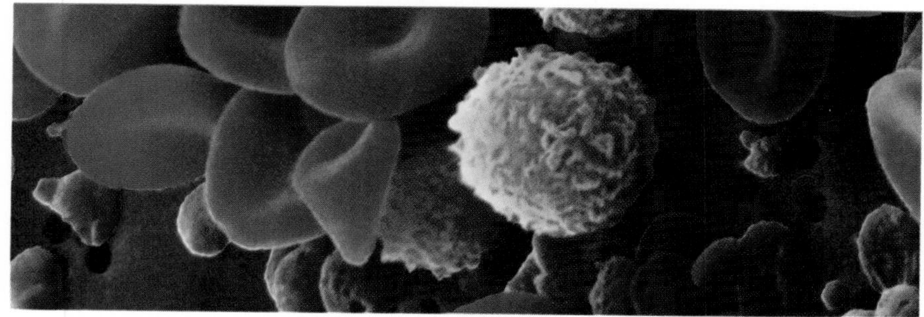

Figure 2 Blood cells

DID YOU **KNOW?**

Artificial Heart

Dr. Tofy Mussivand, of the Medical Devices Centre at the University of Ottawa Heart Institute, researches and designs artificial hearts. A patient needing a heart transplant may receive an artificial heart until a donor heart becomes available.

GO TO NELSON SCIENCE

The Heart

The heart is made up of three different types of tissue: cardiac muscle tissue, nerve tissue, and connective tissue. Cardiac muscle tissue is a special type of muscle found only in the heart (Figure 3). All of the cardiac muscle tissue in each part of the heart contracts at the same time. This makes the heart contract and moves the blood around the body.

Your heart pumps with a regular beat. The frequency of the beat (the heart rate) changes depending on your physical activity and other factors, such as stress, temperature, and your general health.

The muscles and nerves are covered by a smooth layer of epithelial tissues. This covering reduces friction and protects the heart from damage when the lungs expand and contract. The inner surface of the heart, where the blood flows, is also lined with smooth epithelial tissue to allow the blood to flow freely. Any hardening or roughening of this inner lining can lead to health problems.

Blood Vessels

artery a thick-walled blood vessel that carries blood away from the heart

vein a blood vessel that returns blood to the heart

capillary a tiny, thin-walled blood vessel that enables the exchange of gases, nutrients, and wastes between the blood and the body tissues

Three types of blood vessels form a network of tubes throughout the body to transport the blood. These three types of blood vessels are arteries, veins, and capillaries. **Arteries** carry blood away from the heart. Because the blood in the arteries is being pumped away from the heart, it is under greater pressure than the blood in other blood vessels. The walls of arteries are thicker than the walls of other blood vessels to withstand this pressure. **Veins** carry blood toward the heart. This blood is at lower pressure, so the walls of the veins are not as thick. Both arteries and veins can vary considerably in size. The largest are nearest the heart, where just a few blood vessels carry large volumes of blood. Further from the heart, the blood vessels are much smaller, and there are more of them, like twigs on a tree. Arteries and veins are linked together by the capillaries (Figure 4). **Capillaries** are tiny blood vessels with very thin walls that allow substances to diffuse between the blood and other body fluids and tissues. Oxygen and nutrients diffuse from the blood into the surrounding tissues. Carbon dioxide and other wastes pass from the body tissues into the blood to be carried away for disposal. Every part of the body is supplied with blood by a network of capillaries.

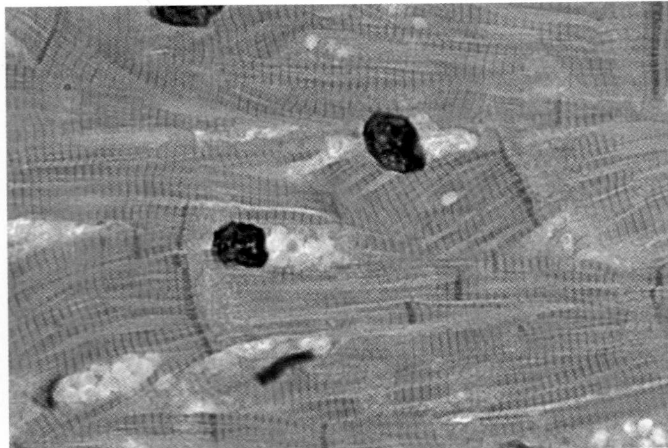

Figure 3 In this photo of healthy cardiac tissue, the fibres of cardiac muscle are stained pink. Their nuclei are purple.

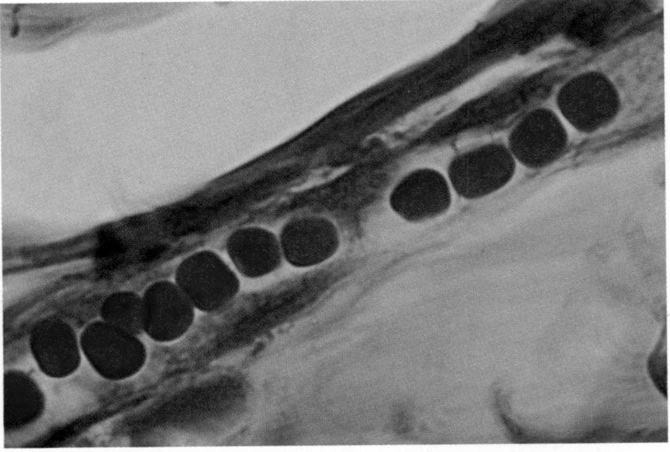

Figure 4 Capillaries can be so narrow that red blood cells can only pass through one at a time.

TRY THIS EXAMINING BLOOD VESSELS

SKILLS: Performing, Observing, Communicating

Arteries, veins, and capillaries are all blood vessels, but they have very different functions. In this activity, you will look at their structures and see how the structure is related to the function.

Equipment and Materials: microscope; lens paper; prepared slides of cross-sections through arteries, veins, and capillaries

1. Use your microscope to examine the slide showing an artery. Draw a diagram to illustrate what you see.
2. Repeat Step 1 with the other two slides.

A. How are the different functions of the three blood vessels reflected in their structures? T/I

Diseases and Disorders of the Circulatory System

There are many conditions that affect the function of the circulatory system. There are over a dozen types of heart disease alone that can affect people of all ages and all levels of fitness. The most common heart problems is coronary artery disease, which can lead to heart attack.

Coronary Artery Disease

Your heart is a hard-working organ, and the cardiac muscle tissue needs a steady supply of oxygen and nutrients. Coronary arteries are the blood vessels that provide blood to the heart muscle tissue itself. These arteries can become partially blocked with plaque—a deposit made of fat, cholesterol, calcium, and other substances that normally circulate in the blood. This plaque buildup can be caused by inherited genetic information or by poor lifestyle choices, such as a high-fat diet, smoking, and lack of exercise.

Symptoms of coronary artery disease include tiredness, dizziness, and pain or a burning sensation in the chest or arms. The problem can be diagnosed with the aid of a special X-ray called an angiogram, in which a fluorescent dye is injected into the bloodstream. This dye shows up on the X-ray image (Figure 5).

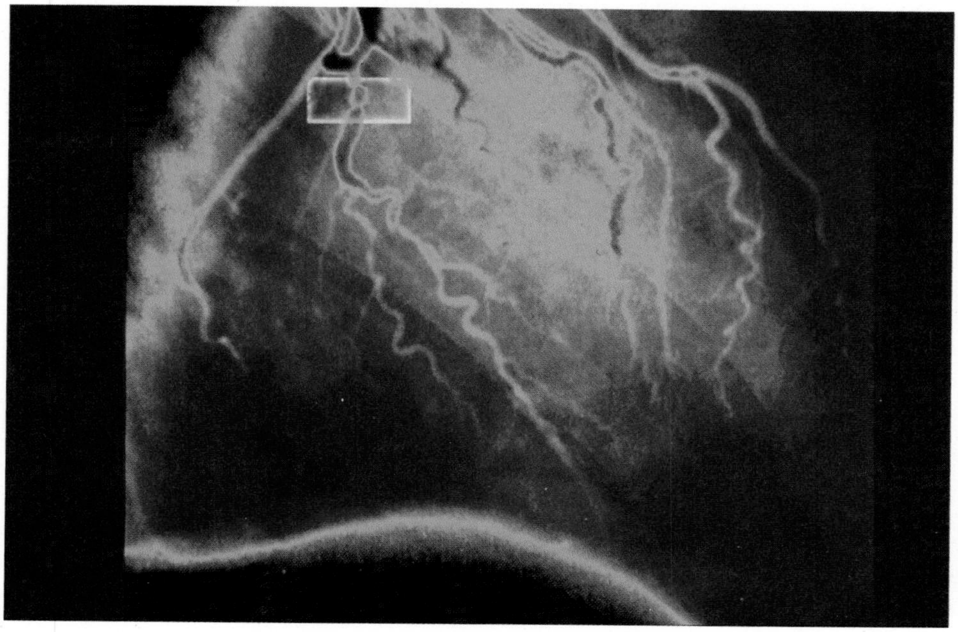

Figure 5 During an angiogram, a fluorescent dye is injected into the artery and X-ray scans are taken. This X-ray image has been colourized by a computer. A white rectangle highlights the blockage in the artery in a patient's heart.

WRITING TIP

Describing Observations

Use a thesaurus to find words that describe your observations as specifically and accurately as possible. Use words that help readers visualize your observations clearly. By using concrete nouns, adjectives, verbs, and adverbs, your description will give readers a clear picture of what you saw.

GO TO NELSON SCIENCE

DID YOU KNOW?

Blood Clotting

It is important that blood forms clots when the blood vessels are damaged by a cut or scrape. Some people have disorders that cause the blood to clot too easily, causing blockages, or not quickly enough, so they bleed uncontrollably. Both problems can be life-threatening.

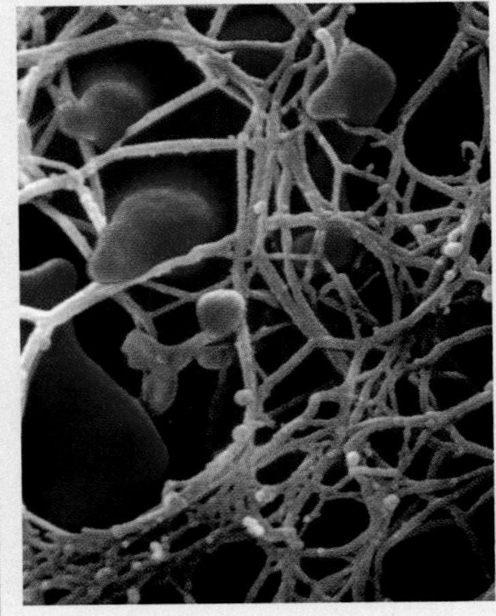

Heart Attack

Coronary arteries can become completely blocked, either with plaque or with a blood clot. When this happens, the heart muscle cells no longer receive the oxygen and nutrients they need to function. The heart stops pumping, and the heart tissue starts to die.

General symptoms of a possible heart attack include

- chest pain or pressure
- shortness of breath
- nausea
- anxiety
- upper body pain
- abdominal or stomach pain
- sweating
- dizziness
- unusual fatigue

The actual symptoms of a heart attack can vary widely between men and women and from person to person, but any suspicion of a heart attack requires immediate medical attention. A heart attack can be diagnosed with a blood test and an electrocardiogram. The blood test identifies certain proteins that are present only when cardiac muscle tissue dies. The **electrocardiogram** (or **ECG**) measures the electrical signals created by the heart as it beats (Figure 6). The electrical signals from damaged heart muscle tissue are not the same as those from healthy heart muscle.

electrocardiogram (ECG) a diagnostic test that measures the electrical activity pattern of the heart through its beat cycle

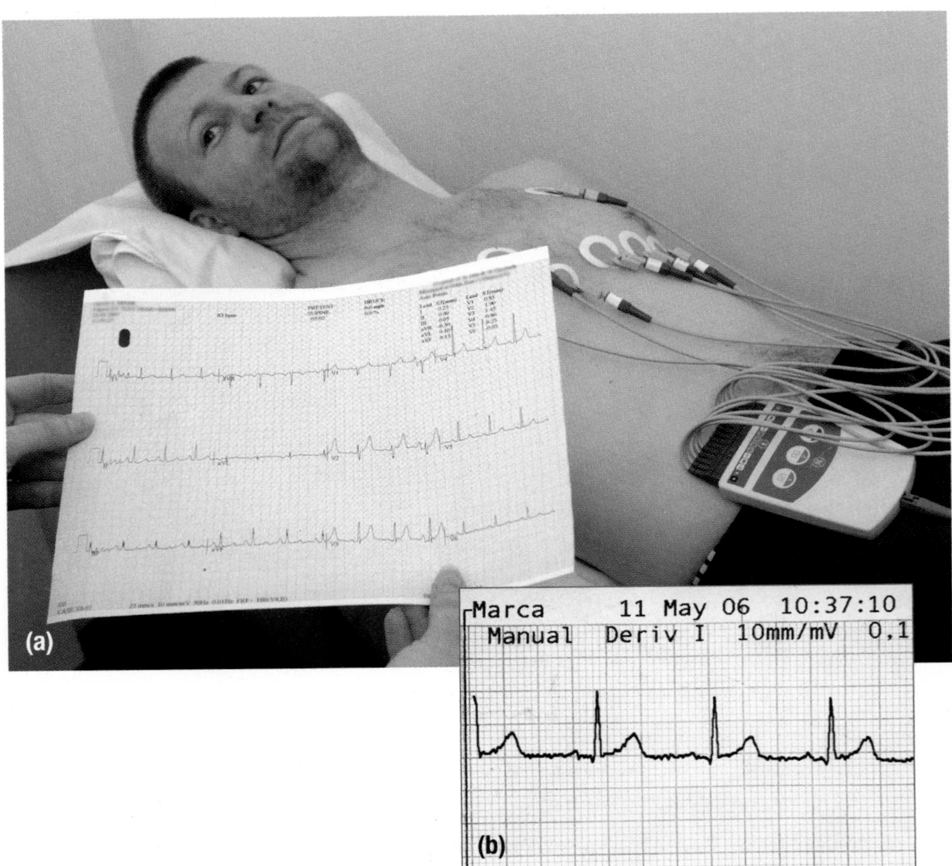

Figure 6 A patient having an electrocardiogram. The blue and white circles are electrodes placed on the skin. These electrodes detect electrical signals.

RESEARCH THIS PROBLEMS IN THE CIRCULATORY SYSTEM

SKILLS: Researching, Analyzing the Issue, Communicating

SKILLS HANDBOOK 4.A., 4.B.

The circulatory system is vitally important to our health. Blockages and other problems can occur in blood vessels in many parts of the body.

1. Chose one specific disease or disorder of the circulatory system to research, such as coronary artery disease, heart attack, stroke, deep vein thrombosis, or anemia.
2. Research your chosen disease. Find out the causes, symptoms, diagnostic technologies, treatment, and long-term effects. If you have time, you could also look at the social and economic impacts of your chosen disease.

GO TO NELSON SCIENCE

A. Think about how life would change for someone who discovers that he or she has this problem. Write a list of probable life changes. T/I A

B. Summarize your findings in an illustrated presentation or as a short dramatic performance. T/I C

C. As a class, discuss whether there are any drawbacks to the use of medical technologies. How expensive are they? Are they readily available to everybody? C A

UNIT TASK Bookmark

How could you use information about the symptoms of heart disease as you work on the Unit Task on page 156?

IN SUMMARY

- The circulatory system is an organ system made up of the blood, the heart, and the blood vessels.
- The function of the circulatory system is to move nutrients and gases to all of the cells of the body and to carry away wastes through the bloodstream.
- Heart disease is a group of conditions that affect the function of the heart.
- Angiograms and electrocardiograms are two medical technologies that are used to help diagnose abnormalities in the circulatory system.

CHECK YOUR LEARNING

1. Describe the function of the circulatory system. K/U
2. Name at least four substances that are carried by the circulatory system. K/U
3. Explain how the circulatory system interacts with the digestive system. K/U
4. How does an angiogram differ from a regular X-ray scan? K/U
5. Figure 7 shows cross-sections of three different blood vessels. Name each one and describe how its structure matches its function. K/U
6. Create a table that lists the main parts of the circulatory system and the tissue types in each part. K/U C
7. (a) Create a pie chart to illustrate the volumes of the various components of blood.
   (b) What challenges did you face when creating your chart? K/U C
8. How is cardiac muscle different from the smooth muscle that surrounds the digestive tract? K/U
9. Name and briefly describe two diseases or disorders of the circulatory system. K/U

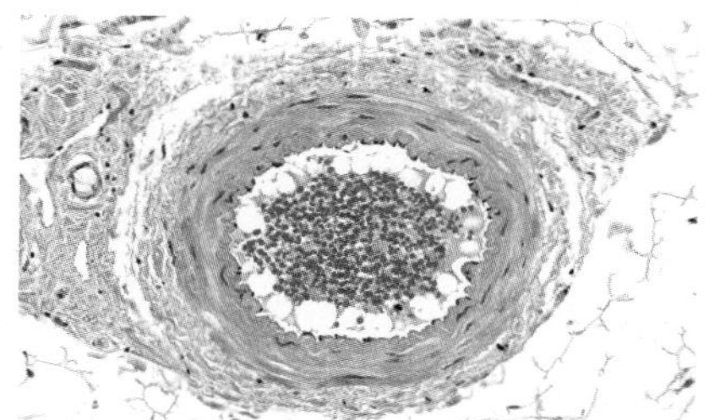

(a)

(b)

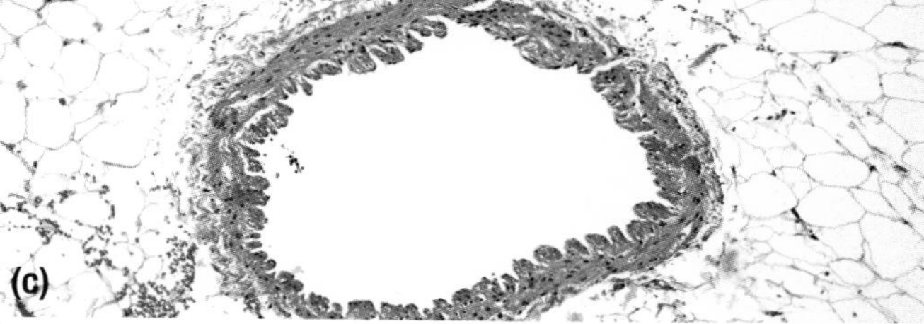

(c)

Figure 7

3.5 PERFORM AN ACTIVITY

Studying the Organ Systems of a Frog

SKILLS MENU	
Questioning	Performing
Hypothesizing	Observing
Predicting	Analyzing
Planning	Evaluating
Controlling Variables	Communicating

The organs and organ systems of a frog are similar to those of a human. Each organ in the frog has a specific function and plays a vital role as part of an organ system.

In this activity, you will work in a group to study the functions and organization of one of the following systems of the frog: digestive, circulatory, respiratory, nervous, reproductive. At the end of the activity, you and your classmates will come together to share your findings and discuss interrelationships between the systems studied. You will be using electronic resources for your research, including, when available, videos of dissections and computer-simulated dissection programs.

You will also have a chance to consider whether it is better to perform dissections on real animals, or to use simulations and photographs of dissections.

Purpose

To investigate the primary function and organization of organ systems in a particular animal system and to explore interrelationships between organ systems in an organism.

Equipment and Materials

- online access to electronic images

GO TO NELSON SCIENCE

Procedure

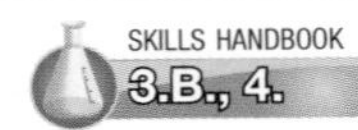
SKILLS HANDBOOK 3.B., 4.

1. Research the organ system of the frog that you have been assigned to study. Write a brief paragraph describing the overall purpose and function of this system in the organism. Be sure to discuss how this system allows the frog to survive in its environment.
2. Many photographs and diagrams of the frog organ system are available in print and on the Internet (Figures 1 and 2). Using various resources, draw a diagram of the organ system, labelling all organs and showing the connections between the organs. Be sure to include any external features that are part of the system.

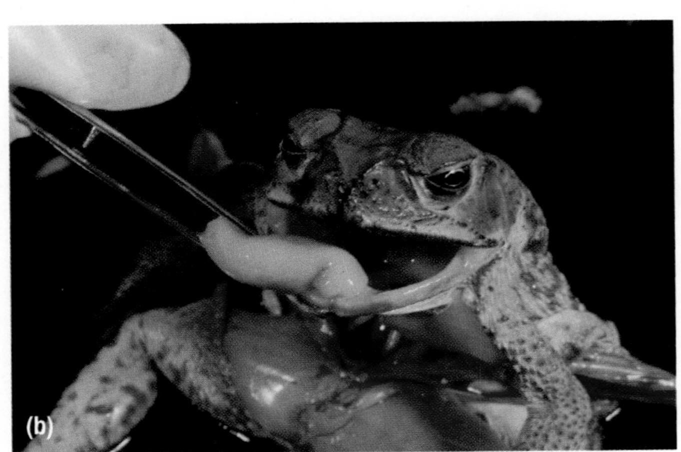

Figure 1 External features of a frog

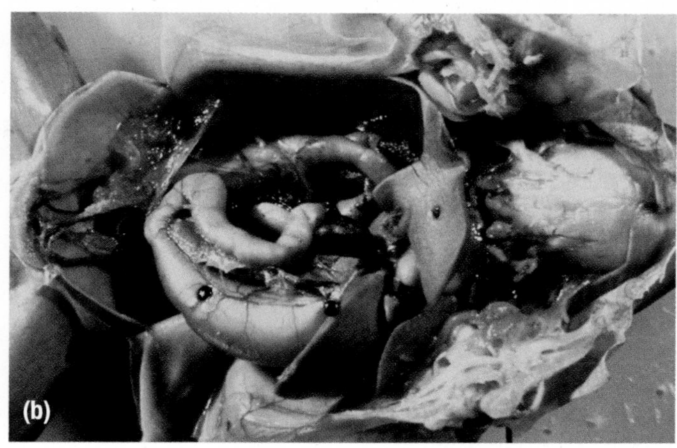

Figure 2 Internal features of a frog

3. Create a table similar to Table 1. Write a title that includes the name of your system. Complete the columns as you carry out your research. Locate and name each organ or structure in the first column. In column 2, summarize the functions of the individual organs. In column 3, list any interactions between organs and structures within that system. In column 4, write predictions of the interrelationships between your system and other systems. Your predictions can take the form of statements such as "The respiratory system brings oxygen into the body, and the circulatory system transports the oxygen to the cells."

Table 1 Organs of the __________ System of the Frog

Organ or structure	Function(s)	Interactions with other organs in the system	Interrelationships with other organ systems

4. As a group, prepare and deliver a short presentation to the class that describes your frog organ system. Assign different roles to different group members. Read the paragraph that you wrote in Step 1. Explain your diagram from Step 2, pointing out all organs and their functions. Emphasize the interaction of the organs within the system you studied.
5. While listening to the presentations of classmates, note any possible interrelationships between the function of their system and the function of yours. Add these to column 4 of your table.
6. Within your group, discuss the contents of column 4 of your table.
7. If you have time, research the interrelationships between systems.
8. As a class, discuss the interrelationships between the systems studied by each of the groups.

Analyze and Evaluate

(a) What is the role of the system you studied in the survival of the frog? T/I

(b) Did you learn about any organs that are unique to the frog? If so, what were they called and what was their function for the frog? T/I

(c) Describe the interrelationship between the systems. T/I

Apply and Extend

(d) Discuss as a class why you did not dissect a real frog in this activity. Research what alternatives are available. Debate the value of using preserved real specimens compared to alternatives such as virtual dissection programs. C A

GO TO NELSON SCIENCE

(e) Create a t-chart comparing the organ systems of a frog with those of a human. T/I C

West Nile Virus

In 1999, hundreds of people in New York City became ill with an illness that was new to North America. Medical researchers discovered that the cause was West Nile Virus (WNV).

WNV is mainly spread by mosquitoes (Figure 1). When an infected mosquito feeds on a bird's blood, the virus passes into the bird. Another mosquito then bites the same bird and the virus is transferred again. The virus also infects mammals, including horses and humans. One in five people with the virus become ill. In rare cases the illness causes death. WNV has now spread to Ontario and other provinces.

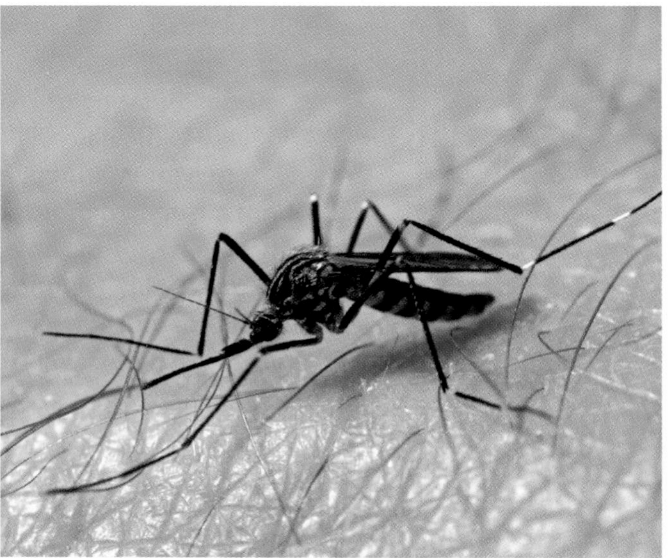

Figure 1 West Nile Virus is transferred to birds and humans by infected mosquitoes.

The Ontario government acted swiftly to implement control procedures. A Public Health Officer in each region is responsible for testing mosquitoes and birds determine the incidence of WNV in the area. These officers must decide if pesticides are needed to control the mosquito population. The Ministry of the Environment recommends pesticides and manages their use. Local governments are responsible for applying the pesticides. Currently, tablets with a fast-acting pesticide that kills mosquito larvae are put in drainage waters during spring months in many Ontario cities (Figure 2).

Figure 2 Chemicals that kill mosquito larvae are put in ponds and drainage water.

How has WNV affected us? People are more cautious about being outdoors and are using insect repellent more often (Figure 3). We do not yet know whether there are any long-term health effects of increased insect repellent use. There is also the possibility of mosquitoes becoming resistant to the chemicals.

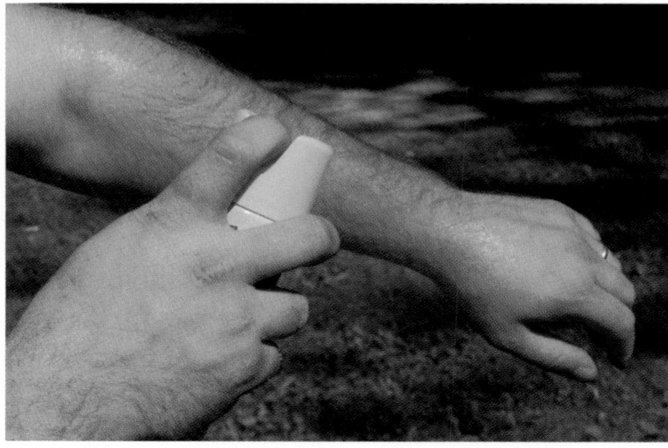

Figure 3 Ontarians are urged to protect themselves from West Nile Virus when out of doors by using insect repellent.

Monitoring and controlling WNV requires cooperation among a number of different professions. In addition to the Public Health Officers who make decisions about mosquito control, licensed applicators are required to apply the pesticides properly. Lab technicians test the birds and mosquitoes that are collected. The medical community is on alert for WNV symptoms. Public educators communicate ways to reduce the risk of WNV. They suggest cleaning up yards where mosquitoes might breed and protecting against mosquito bites. In the words of the Ontario Government, we must all "fight the bite."

GO TO NELSON SCIENCE

The Respiratory System

Whether you are aware of it or not, you breathe in and out 15 times each minute on average. This rate increases automatically if your physical activity increases. With normal breathing, the average person moves more than 10 000 L of air in and out of the lungs each day.

The **respiratory system** is responsible for providing the oxygen needed by the body and for removing the carbon dioxide produced as your body uses energy for growth, repair, and movement. The respiratory system works in close collaboration with the circulatory system (Figure 1). As you learned in Section 3.4, the circulatory system moves substances to all parts of the body.

respiratory system the organ system that is made up of the nose, mouth, trachea, bronchi, and lungs; the system that provides oxygen for the body and allows carbon dioxide to leave the body

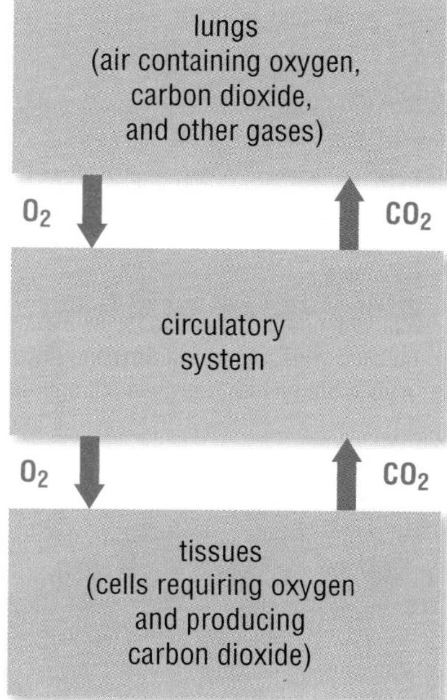

Figure 1 The respiratory system relies on the circulatory system to distribute oxygen to the cells and to remove carbon dioxide.

Structural Features

The respiratory system consists of the lungs and the other organs that connect the lungs to the outside (Figure 2(a)). Air enters through the mouth and the nose, passes through the pharynx (throat), and travels down the trachea (commonly known as the windpipe). The trachea separates into two branches called bronchi (singular: bronchus).

Some of the epithelial cells that line the trachea and bronchi produce mucus, similar to those in the digestive system. Many of the epithelial cells have cilia (hairlike projections). Cilia help move mucus and filter out any foreign material that might enter the system (Figure 2(b)). The bronchi deliver air into the lungs.

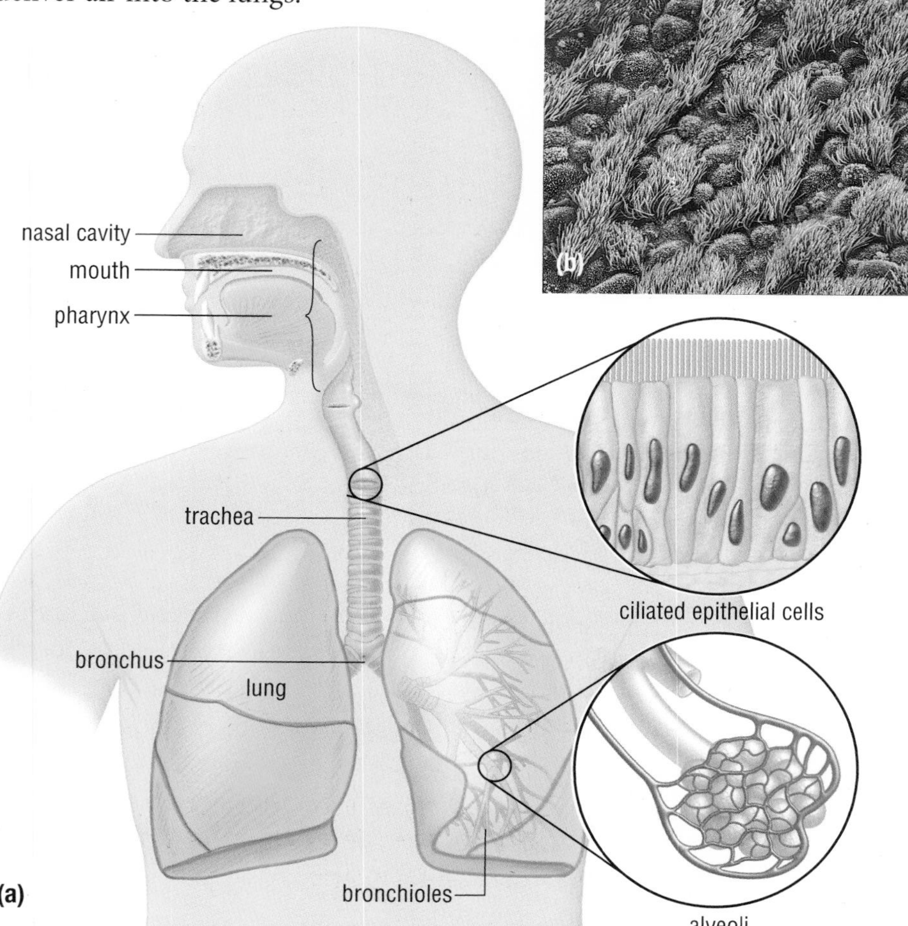

Figure 2 (a) The human respiratory system (b) Epithelial cells with cilia

DID YOU KNOW?

Keeping Food Out

Your mouth can contain both food and air. How does your body manage to send food to your stomach and air to your lungs? When you swallow food, a flap of tissue called the epiglottis covers the opening to the trachea. This prevents the food from going down the wrong tube. Occasionally, some food or liquid does sneak into the trachea. The resulting bout of coughing is your body's attempt to get it out of the trachea and into its rightful place.

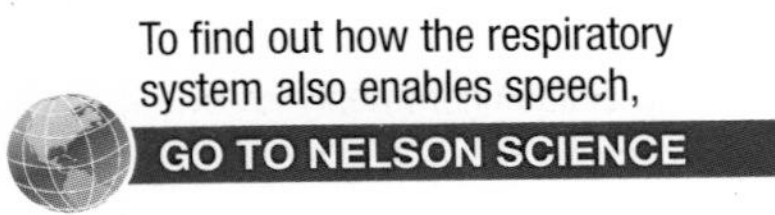
To find out how the respiratory system also enables speech,
GO TO NELSON SCIENCE

The trachea is supported by rings of cartilage. This keeps the trachea open and allows the air to flow freely. Cartilage is a special type of connective tissue consisting of specialized cells embedded in a matrix of strong but flexible fibres. This matrix was formed by cells but is not actually living material.

Gas Exchange

The main purpose of the respiratory system is gas exchange. Oxygen enters the bloodstream in the lungs by diffusion. Carbon dioxide leaves the blood in the same way. The respiratory system is adapted in several ways to make these processes as efficient as possible.

Each of the bronchi branch again and again, ending in tiny air sacs called **alveoli** (singular: alveolus) (Figure 3(a)). The alveoli have very thin walls. Each alveolus is surrounded by a network of capillaries. Oxygen and carbon dioxide have only to diffuse through two thin walls: the walls of the capillaries and the walls of the alveoli (Figure 3(b)).

alveolus (plural: alveoli) tiny sac of air in the lungs that is surrounded by a network of capillaries; where gas exchange takes place between air and blood

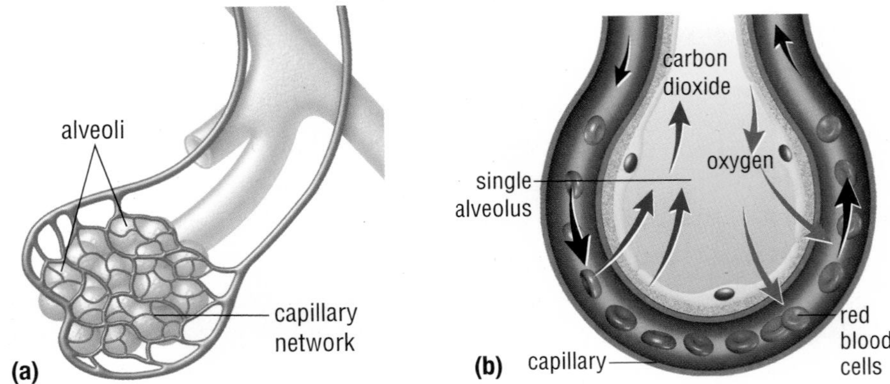

Figure 3 (a) Each alveolus is surrounded by a capillary network to ensure a good blood supply. (b) The alveoli provide a huge surface area in the lungs across which oxygen and carbon dioxide can diffuse.

LEARNING TIP

Diffusion

Substances always diffuse from where they are in higher concentration to where they are in lower concentration.

The circulatory system provides a good blood supply to the lungs. This helps make the respiratory system very efficient. The concentration of oxygen in the blood that flows through the lungs is always less than the concentration of oxygen in the air in the alveoli. This means that oxygen always diffuses into the blood. As the blood picks up oxygen, it is quickly carried away to other parts of the body, where the oxygen diffuses out of the blood and into the cells. At the same time, excess carbon dioxide diffuses from the cells into the blood. It is then carried by the blood to the lungs, where it diffuses out into the air in the alveoli and is expelled to the outside.

Breathing

The respiratory system includes a method of moving air into and out of the lungs. This process, which we call breathing, involves alternately drawing air into the lungs (inhalation) and then pushing air out (exhalation). This process involves muscles that move the ribs, making the rib cage expand and contract. Breathing also involves the diaphragm, a large sheet of muscle underneath the lungs. Together, the diaphragm and the muscles between the ribs increase or decrease the volume of the lungs (Figure 4). As the volume of the lungs changes, the pressure inside them also changes. In this way, fresh air flows into and out of the alveoli.

To see how the volume of the lungs increases and decreases as the diaphragm contracts and relaxes,

GO TO NELSON SCIENCE

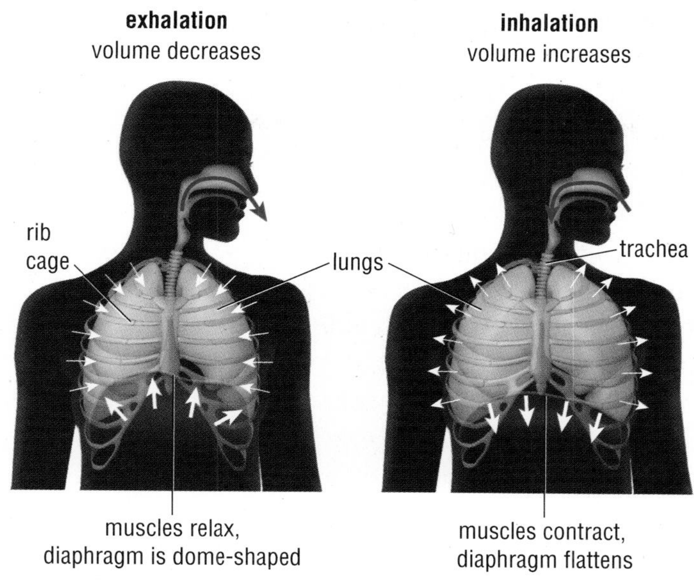

Figure 4 Inhalation involves drawing air into the lungs; exhalation involves pushing the air out.

DID YOU KNOW?

Hiccups!
The condition commonly known as hiccups is caused by the muscles of the diaphragm suddenly contracting. When the diaphragm quickly contracts, air rushes into the mouth and pharynx, causing the epiglottis to snap shut over the end of the trachea. This produces the "hic" sound. Numerous events can trigger hiccups, from eating too fast to coughing. Hiccups normally stop without intervention and rarely become a medical problem.

CONTROL OF BREATHING

The control over our breathing is involuntary; we do not generally have to think about breathing. We can override the involuntary system and stop breathing or control it while we talk, but this control is only temporary. The involuntary system soon takes over again after a short period of time. Try it and see: how long can you hold your breath?

Breathing is controlled by a part of the brain that detects the concentration of carbon dioxide in the blood. As the level of carbon dioxide increases, the brain sends signals to the diaphragm, the muscles between the ribs, and the heart. The breathing rate increases, and the heart beats faster. This has the double effect of decreasing the concentration of carbon dioxide in the blood and increasing the available oxygen.

The Respiratory System in Other Animals

The role of the respiratory system is to make oxygen available to all of the cells of the body and to get rid of waste carbon dioxide. In comparison to mammals, many other organisms have simple respiratory systems. Regardless of how simple or complex, all respiratory systems depend on the process of diffusion to move oxygen in and carbon dioxide out.

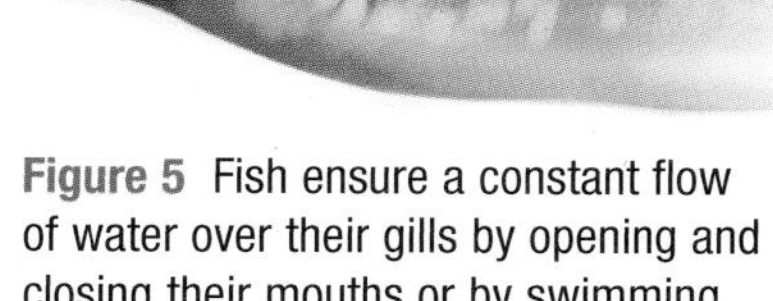

Figure 5 Fish ensure a constant flow of water over their gills by opening and closing their mouths or by swimming.

The Respiratory System in Fish

In fish, the gas exchange organs are the gills. The gills are exposed directly to the water (Figure 5). Like lungs, gills have many capillaries that bring blood very close to the water so that oxygen can diffuse from the water into the blood. Similarly, carbon dioxide can diffuse from the blood into the water. Fish do not actually breathe in the same way humans do, but you may have noticed fish opening and closing their mouths to create a flow of water over their gills. Some fish need to swim constantly to keep a supply of oxygenated water flowing over their gills.

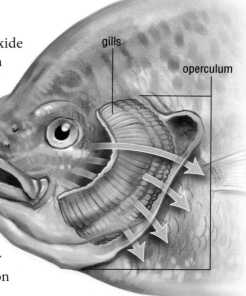

To see a video showing the respiratory organs of fish,
GO TO NELSON SCIENCE

Diseases of the Respiratory System

Because the respiratory system is constantly exposed to substances in the air, it is not surprising that is it affected by many different diseases.

Tuberculosis

Tuberculosis (TB) is an infectious disease, which means that it is easily passed between people. It is caused by bacteria that enter your body when you breathe. The bacteria grow in your lungs, although the disease can spread to other parts of your body, including your nervous system and your bones. TB has fairly general symptoms: fever, cough, weight loss, tiredness, and chest pain. If untreated, the disease can be fatal. A chest X-ray is one of the tests used to diagnose TB (Figure 6). However, other conditions such as pneumonia may show similar results on an X-ray.

DID YOU KNOW?

TB Around the World
TB is on the increase around the world, especially in developing countries. It has been estimated that more than 8 million new cases develop and 2 million people die from TB each year.

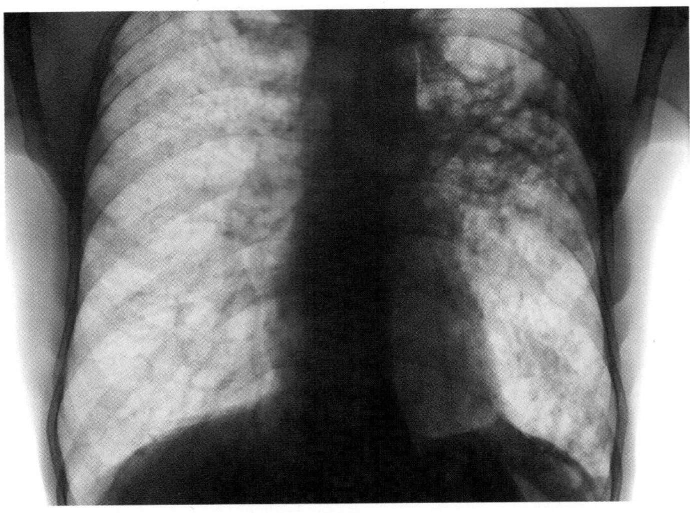

Figure 6 This chest X-ray shows evidence of TB in the upper part of the lung on the right.

To confirm a diagnosis of TB, medical technicians examine samples of stomach or lung secretions. One of the problems with TB is that, after the initial contact, the bacteria can remain dormant in the body for decades. Once diagnosed, however, the disease can be successfully treated with medicine and a few weeks of hospitalization.

Cancers

Tobacco smoke, both first-hand and second-hand, is a serious threat to the health of the respiratory system. As you read in Section 2.7, tobacco smoke contains many known carcinogens. These chemicals contribute to cancers not only of the lungs but also of the mouth, esophagus, larynx, pancreas, and bladder.

SARS

In early 2003, Canada—Toronto in particular—became gripped with fear of a new, deadly disease known as severe acute respiratory syndrome (SARS). SARS spread from a region of China to 37 countries around the world, threatening to become a global epidemic. In total, Canada identified 438 cases; 44 of these patients died. The symptoms of SARS are flulike and include high fever, shortness of breath, dry cough, sore throat, headache, muscle pain, and exhaustion. A diagnosis is made based on the above symptoms, a chest X-ray showing evidence of pneumonia, and positive lab results of cell samples taken from the patient.

RESEARCH THIS CANADIANS IN SCIENCE

SKILLS: Researching, Identifying Alternatives, Communicating, Evaluating

SKILLS HANDBOOK 4.A., 4.B.

Dr. Sheela Basrur was chief medical officer of health in Toronto during the 2003 SARS outbreak (Figure 7). Under her direction, Toronto implemented public health measures to control and reduce the spread of SARS. Her passion for medicine and public health began during her travels through India and Nepal after she graduated as a medical doctor.

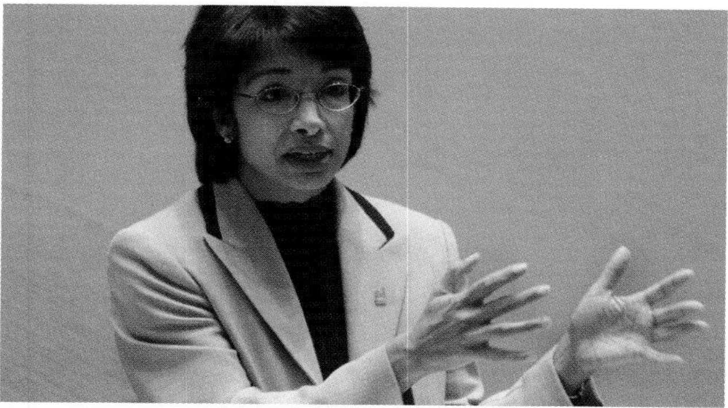

Figure 7 Dr. Sheela Basrur: the "SARS doctor."

Dr. Basrur was a tireless advocate for children, immigrants, and women. She died of a rare form of cancer in 2008 at the age of 51.

1. Research a Canadian scientist who has made a significant contribution to human health.
2. Research the scientist's specific contribution (e.g., research findings, development of a technology, treatment, advocacy).
3. Research the cultural and educational background of this individual and what motivated him/her to enter this field.

GO TO NELSON SCIENCE

A. Prepare an oral or written biography of this scientist to share with classmates. Include aspects from the research you performed in 1, 2, and 3 above. T/I C
B. Explain why you chose to profile this particular individual. C

UNIT TASK Bookmark

Think about how you could use information about the respiratory system as you work on the Unit Task on page 156.

IN SUMMARY

- The respiratory system exchanges gases between the body and the environment. Oxygen diffuses into the body, and carbon dioxide diffuses out.
- The main parts of the human respiratory system are the nose, mouth, trachea, bronchi, lungs, and diaphragm.
- Breathing brings air into and out of the lungs so that gas exchange can occur.
- Gas exchange takes place in the alveoli, which are surrounded by capillary networks containing blood.
- The circulatory system delivers oxygen to the cells and removes carbon dioxide from the cells.
- Many diseases such as tuberculosis, cancers, and SARS affect the respiratory system.
- The respiratory system in fish includes gills, which obtain oxygen from the surrounding water and get rid of carbon dioxide.

CHECK YOUR LEARNING

1. Name the main organs and structures of the respiratory system. K/U
2. What is the role of the epithelial tissue that lines the trachea and bronchi? K/U
3. How does an animal's respiratory system depend on its circulatory system? K/U
4. Explain the difference between breathing and gas exchange. K/U
5. (a) Why is an X-ray insufficient to make a positive diagnosis of tuberculosis?
   (b) What test is required to confirm a diagnosis of TB? K/U
6. Describe the similarities and differences between the respiratory systems of humans and fish. K/U

3.7 Organ Transplantation

The Engage in Science story at the beginning of this chapter introduced you to a young woman who received a replacement trachea. This replacement trachea was a combination of the cartilage from a donor and Claudia's own cells, grown onto the cartilage "scaffold." This new transplant technology was very successful for Claudia.

Tissue transplants have been performed since the early 1800s, when blood transfusions were first explored. The first successful organ transplant (a kidney) occurred in 1954. The living donor and the recipient were identical twins. Since that time, the science and the technology have advanced considerably.

The list of other organs that can be successfully transplanted now includes the heart, liver, lung, pancreas, and intestines. Transplantable tissues include the cornea, skin, bone, bone marrow, tendons, and blood vessels (Figure 1). Some organs and tissues can be successfully and safely transplanted from living donors. Other body parts can be taken only from deceased donors.

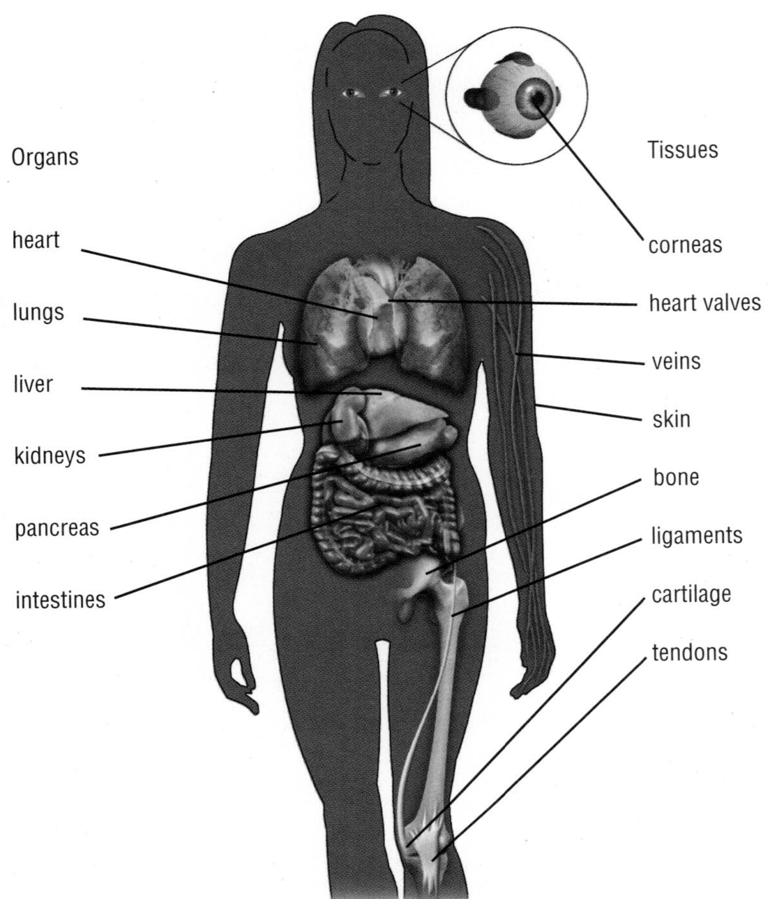

Figure 1 Many tissues and organs can be transplanted.

To find out more about tissue and organ transplantation in Canada,

Benefits and Risks

Both the recipient and the donor (or donor's family) benefit from the transplantation. The most obvious benefit is that the recipient can live a healthy, normal life. The benefit to the donor (or the deceased donor's family) is the satisfaction of knowing that the donated organ saved someone's life. In addition, medical researchers have gained a wealth of new knowledge about the human body as a result of the research in this area. Because of this, society as a whole benefits.

Unfortunately, there are also some risks associated with transplant surgery. The biggest risk is rejection. The recipient's immune system may recognize the new organ as foreign material and try to destroy it. The chance of rejection can be minimized by using tissues that are genetically similar to those of the recipient. However, even if the tissues are matched as closely as possible, most transplant patients will need to take drugs to prevent the immune system from rejecting the new tissue or organ. This solution also presents a risk, however. With the immune system suppressed, the body's ability to fight off infections is reduced.

DID YOU KNOW?

The Long Wait

There are far more people waiting for transplants than there are available organs, so there are usually long waiting lists. There are, on average, about 1700 people on waiting lists for organ transplants in Ontario. Up to 30 % of people on waiting lists die before a suitable organ is found.

Living Donor Organs

Living donor organs come from a living person who chooses to donate a kidney, a lobe of one of their lungs, or a part of their liver. A living donor lung transplantation requires two donors, each providing one lobe of a lung. These two lobes are then transplanted into the recipient (Figure 2). Kidney donation requires only one donor, who can lead a normal life with one remaining kidney after donating the other. In the case of a liver transplantation, doctors remove one lobe of the donor's liver and transplant it into the recipient. The liver has a unique ability to regenerate or grow again. The lobe that is transplanted will, over time, form new tissue and function as a complete liver. The donor's liver will also regenerate new tissue to replace the removed portion.

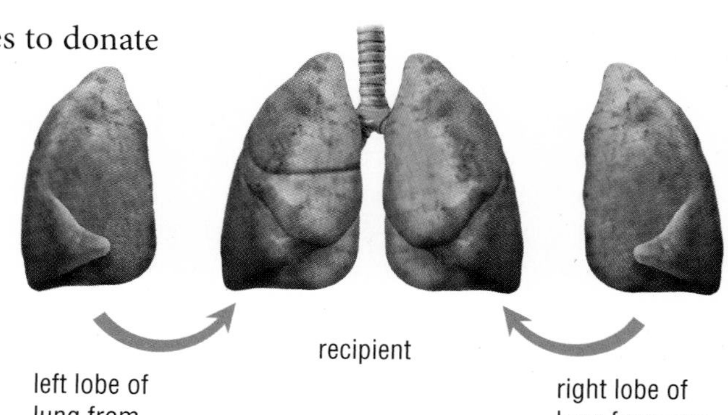

Figure 2 Removing part of one lung has minimal long-term effects on the donors. The two transplanted parts provide adequate lung function for the recipient.

In most cases, living donors are relatives of the recipient. This increases the chance that the organ offered for donation will be an appropriate genetic match, and the risk of rejection is minimized. An added advantage of a living donor is the reduced waiting time.

Obviously, there is some risk to donors. In the normal body, duplicate organs mean that if one organ fails, the other still functions. This backup system is lost or reduced after a donation. In addition, there are risks associated with any major surgery.

Deceased Donor Organs

The majority of organs for transplantation come from deceased persons. The decision to donate organs after death is usually made by the individual while the person is alive (Figure 3). However, family members can give consent to donate the organs after a person has died even if no donor card was signed.

Signing a donor card and informing family members of your decision makes organs and tissues available for donation.

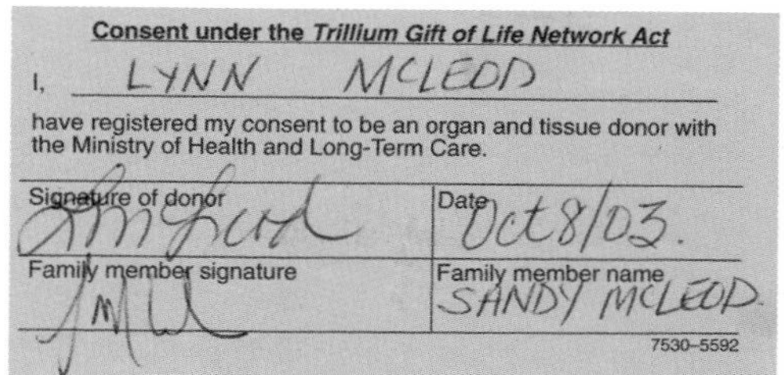

Consent under the *Trillium Gift of Life Network Act*

I, LYNN McLEOD

have registered my consent to be an organ and tissue donor with the Ministry of Health and Long-Term Care.

Signature of donor | Date Oct 8/03.

Family member signature | Family member name SANDY McLEOD

7530-5592

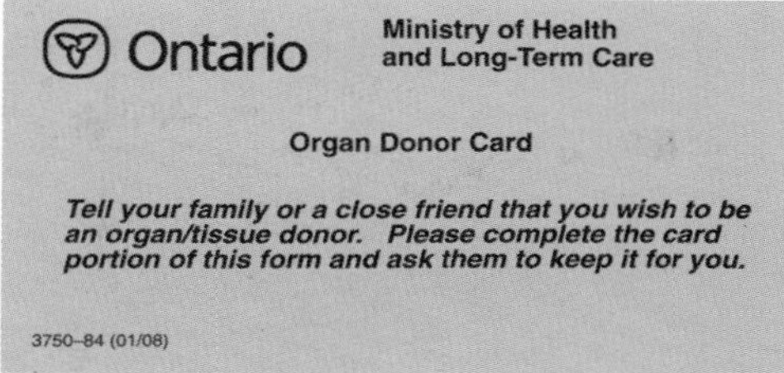

Ontario Ministry of Health and Long-Term Care

Organ Donor Card

Tell your family or a close friend that you wish to be an organ/tissue donor. Please complete the card portion of this form and ask them to keep it for you.

3750-84 (01/08)

Figure 3 Signing a donor card and informing family members of your decision makes organs and tissues available for donation.

When a potential organ donor dies, the organs must be checked to determine if they are healthy and undamaged. Then a search is done for potential recipients. Medical professionals ensure that the organs go to the most appropriate person. They take many factors into account, including blood and tissue types, the ages and locations of the donor and the recipient, and how long the recipient has been waiting for a transplant.

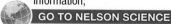

The Trillium Gift of Life Network encourages life-saving organ and tissue donations. For more information,

Xenotransplantation

Xenotransplantation is the transplanting of body parts from one species to another. This is not a new idea. Heart valves from pigs have already been used to replace damaged human heart valves. However, these valves have been chemically treated to kill the cells, so they are no longer considered living tissue. Rejection of living tissue is a major hurdle.

xenotransplantation the process of transplanting an organ or tissue from one species to another

RESEARCH THIS THE ETHICS OF XENOTRANSPLANTATION

SKILLS: Researching, Analyzing the Issue, Defending a Decision, Communicating, Evaluating

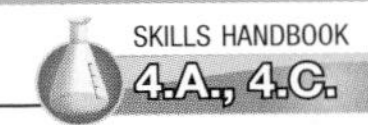
SKILLS HANDBOOK 4.A., 4.C.

The idea of xenotransplantation is a very controversial one. Using tissues or organs from other species could potentially save hundreds of human lives every year. Many people object to the idea of having part of another animal in their body. Others are opposed to killing animals to use their organs for transplantation. In Canada, there are no studies of xenotransplantation involving humans, but the Canadian Public Health Association has discussed the issue. Many questions have been raised.

- Is xenotransplantation needed?
- What are the risks to the public? Is this risk acceptable?
- What human and animal issues need to be considered?
- How should xenotransplantation be regulated and controlled?
- What are the alternatives to xenotransplantation?

1. Research Canada's position on xenotransplantation.
2. Research to determine if, and where, research on xenotransplantation is being carried out.

GO TO NELSON SCIENCE

A. Prepare a summary highlighting the potential benefits and risks of xenotransplantation. T/I A

B. Briefly describe the current status of xenotransplantation in Canada and elsewhere in the world. T/I

C. What is your position on the issue of xenotransplantation? Prepare and present a statement outlining your position in a format of your choice. Be sure to include information and arguments to support your position. T/I C A

IN SUMMARY

- Transplantation involves the transfer of living tissues or organs from one person to another.
- Donated tissues and organs can come from living or deceased donors. Transplants from deceased donors are much more common than from living donors.
- Rejection is the biggest risk for transplant patients. Time spent on a waiting list is a second risk factor.
- Xenotransplantation is the transfer of living tissues or organs from one species to another, usually from other animals to humans.

CHECK YOUR LEARNING

1. Distinguish between living donor transplants and deceased donor transplants. Which type is more common? K/U
2. Why is the list of living donor organs much shorter than the list of deceased donor organs? K/U
3. Briefly describe the procedure for a living organ donation. K/U
4. What are the two main risks for transplant patients? K/U
5. Define xenotransplantation. Why is it a controversial issue? K/U
6. It has been suggested that everyone should be required to be an organ donor. What are the arguments for and against this suggestion? Where do you stand? Explain your position. C A

The Musculoskeletal System

Imagine that all 206 of the bones in your body suddenly disappear! You are now just a big blob of soft tissues on the floor. Your arms and legs are like rubber. You cannot move from one place to another. Your brain could be damaged by the slightest bump. This may be a disturbing image, but it gives you a sense of how important our musculoskeletal system is (Figure 1). The **musculoskeletal system** is made up of all of the bones in your body and the muscles that make them move.

musculoskeletal system the organ system that is made up of bones and skeletal muscle; the system that supports the body, protects delicate organs, and makes movement possible

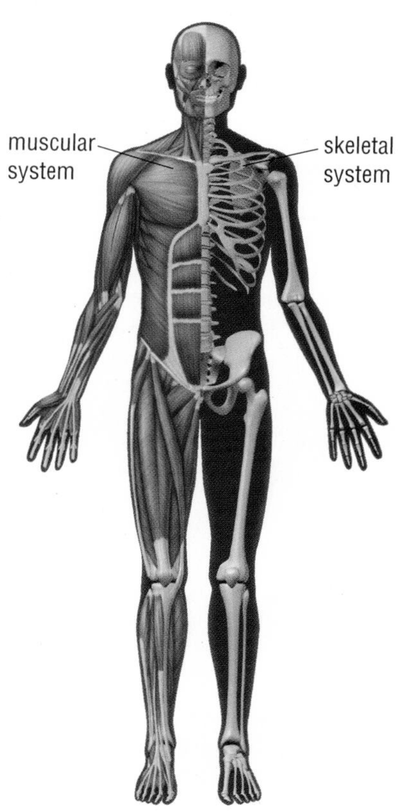

Figure 1 The bones and muscles work together to provide structure, support, protection, and movement.

Structural Features

The skeleton consists of three different types of connective tissue: bones, ligaments, and cartilage. Bone tissue is hard and dense. It consists of bone cells within a matrix of minerals (mainly calcium and phosphorus) and collagen fibres. Canals inside the bones contain nerves and blood vessels (Figure 2). Only a small percentage of bone tissue is actually living.

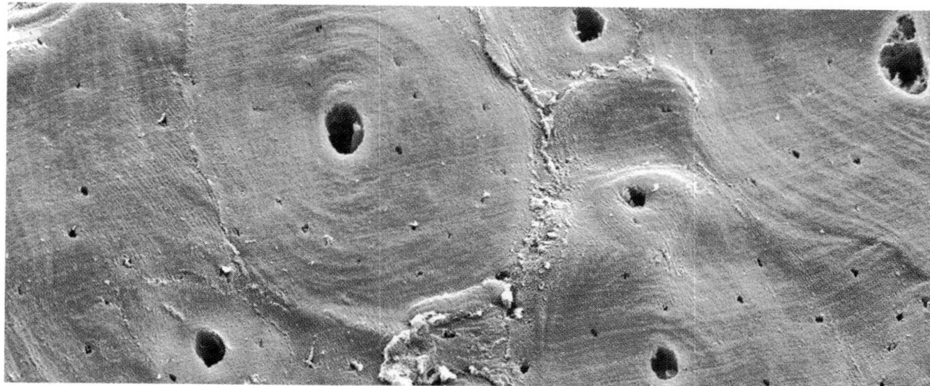
Figure 2 The canals in bone tissue provide space for blood vessels and nerve cells.

Ligaments are tough, elastic connective tissues that hold bones together at the joints. They are made up mostly of long fibres of collagen. Cartilage is a dense connective tissue found in the ear, nose, esophagus, the disks between our vertebrae, and joints (Figure 3). Cartilage is made up of special cells in a matrix of collagen fibres. It provides a strong, flexible, low-friction support for bones and other tissues.

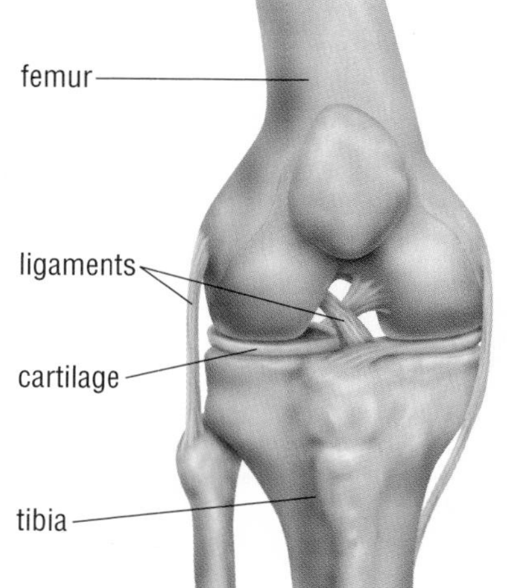

Figure 3 The knee joint is held together by several ligaments. There is a pad of cartilage that acts as a cushion between the ends of the bones (femur and tibia).

The other part of the musculoskeletal system is the muscle. Muscle tissue consists of bundles of long cells called muscle fibres that contain specialized proteins. These proteins cause the muscle to contract when signalled by nerve cells. When they contract, the muscles get shorter and thicker. Skeletal (voluntary) muscle tissue is one of the three types of muscle tissue. The others are smooth (involuntary) muscle, mostly located in the intestines, and cardiac muscle in the heart.

Skeletal muscle is attached to bones by tendons, allowing the movement of body parts (Figure 4).

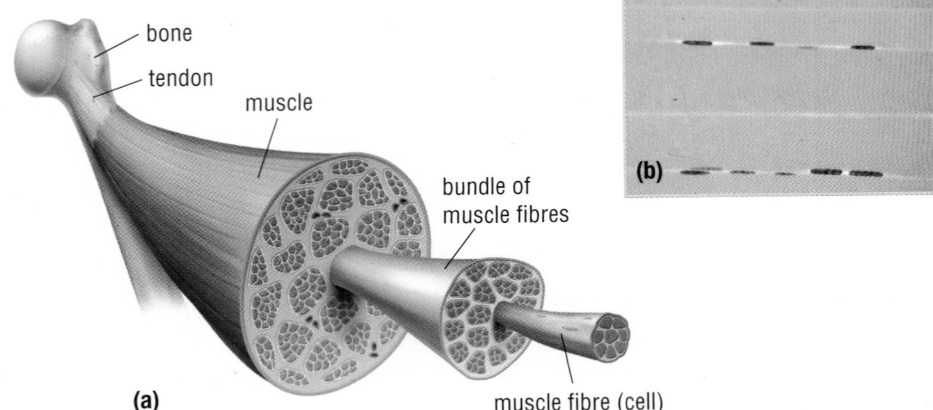

Figure 4 (a) Muscle consists of muscle fibres arranged in bundles. (b) Muscle cells have a unique structure that is visible under the microscope as stripes across the cell. This enables muscle cells to shorten when stimulated.

Support, Protection, and Movement

The main role of the skeleton is to provide structure and support for our bodies and anchor points for our muscles. Some bones also protect the soft internal organs and the brain. Bones also store calcium and other minerals needed by the organism, and some bones contain marrow, which produces red and white blood cells. Cartilage provides a smooth surface where bones come together at a joint, preventing damage to the ends of the bones. We use skeletal muscle for voluntary movements of the body, such as walking.

How Muscles Make Bones Move

Each end of a skeletal muscle is connected by tendons to one or more different bones in the skeleton. Tendons are similar to ligaments but are less elastic and connect muscles to bones. When muscles contract in response to signals from the nervous system, they exert a force. This force moves one or both of the bones to which the muscle is connected. Muscles can pull, but they cannot push, so skeletal muscles always work in opposing pairs or groups (Figure 5).

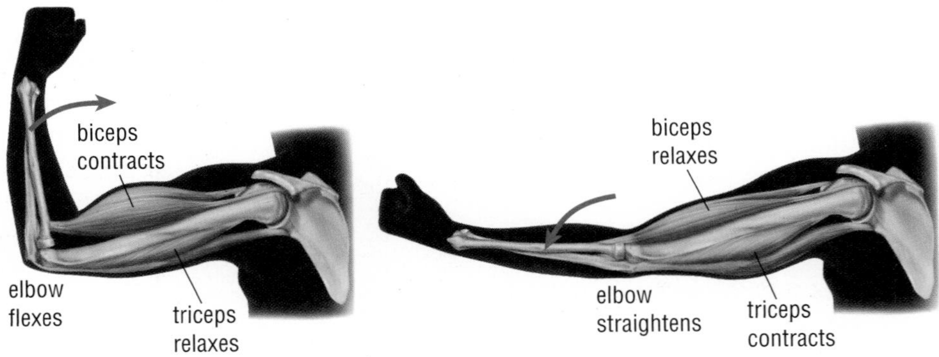

Figure 5 The triceps and the biceps work together to flex and straighten the elbow. Skeletal muscles commonly work in pairs.

Problems with the Musculoskeletal System

The musculoskeletal system, like other systems, is susceptible to diseases. Osteoporosis is a disease that can affect people of all ages but is a more common problem among older women. This disease involves loss of bone tissue, making the bones brittle and weak (Figure 6). Osteoporosis does not cause any pain, so only a bone density test can indicate the presence of the disease. It is linked to a loss of calcium in the bones, so women are encouraged to consume foods or supplements containing calcium and vitamin D. Physical exercise can also help increase bone mass and lessen the risks of osteoporosis.

Because of its role in support and protection, the musculoskeletal system experiences physical impacts and stresses. Extreme movements can tear ligaments, tendons, and muscle tissues; severe impacts can fracture bones. After a serious injury, X-rays are taken to determine if a bone is fractured and how the injury should be treated.

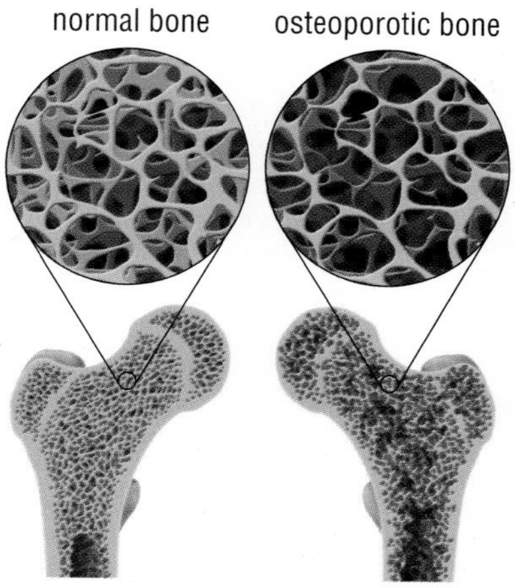

normal bone osteoporotic bone

Figure 6 Bone tissue loss in osteoporosis increases the risk of bone fractures.

The Skeletal System in Other Animals

All vertebrates (animals with backbones) have musculoskeletal systems similar to ours, with muscles attached to bones inside the skin. Invertebrates, however, have very different systems. Some, like worms and jellyfish, have no rigid frame to give them structure. Others—insects and arthropods—have their skeletal system on the outside. This hard external structure is called an exoskeleton (Figure 7). Muscles attached inside the exoskeleton enable the animal to walk, fly, eat, and so on.

Figure 7 The exoskeleton of this beetle protects its internal organs.

UNIT TASK Bookmark

Information about joints in this section could be useful as you prepare to answer the Unit Task on page 156.

IN SUMMARY

- The musculoskeletal system provides structure and support for the body and enables movement.
- The skeleton contains three types of connective tissue: bone, ligaments, and cartilage. Muscles contain muscle tissue and are connected to bones by special tissues called tendons.
- Muscle cells and tissues can contract, causing the bones to move.
- Loss of bone tissue due to osteoporosis makes bones more susceptible to fractures.
- Many invertebrates have exoskeletons to protect their internal organs.

CHECK YOUR LEARNING

1. List the main functions of the musculoskeletal system. K/U
2. Differentiate between a tendon and a ligament. K/U
3. Why is skeletal muscle tissue considered "voluntary" muscle? K/U
4. Use a simple diagram to describe how opposing muscle pairs produce movement of the lower leg. K/U C
5. Bone fractures are more common among senior citizens than among young people. Propose an explanation. A
6. Research has shown that there are basically two types of skeletal muscle fibres: fast twitch and slow twitch. The muscles of our neck and back have lots of slow twitch muscle fibres. These muscles are important in keeping our posture. Where in the body would you expect to find muscles with a lot of fast twitch muscle fibres? Explain your answer. A

Exploring the Structure and Function of Tissues in a Chicken Wing

SKILLS MENU
- Questioning
- Hypothesizing
- Predicting
- Planning
- Controlling Variables
- Performing
- Observing
- Analyzing
- Evaluating
- Communicating

In this activity, you and a partner will identify the tissues that make up a chicken wing. You will dissect the chicken wing in order to relate the structure of these tissues to their functions.

Purpose

To relate the structure of a bird's wing to its function.

Equipment and Materials

- lab apron
- disposable gloves
- dissection pan
- dissecting scissors
- forceps
- blunt probe
- pencil
- 5 different-coloured pencils
- fresh chicken wing

Wear disposable gloves for this activity. There could be bacteria on the chicken wing, or on any of the implements, that could make you very sick.

Procedure

Review the section in the Skills Handbook on using sharp instruments safely.

Always wash your hands thoroughly with hot water and soap after handling poultry products. They may contain *Salmonella* bacteria.

1. Decide which partner will perform the dissection and which will observe and record observations. The partner performing the dissection will put on the lab apron and disposable gloves.
2. Obtain one fresh chicken wing per pair. Place the wing in a dissection pan.
3. Compare the external features of your chicken wing with those shown in Figure 1.

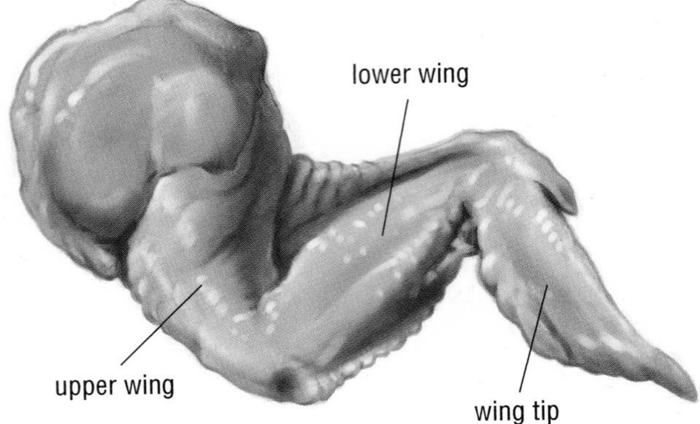

Figure 1

4. At the cut end of the upper wing, slip the tip of the dissecting scissors between the skin and the muscles underneath, as shown in Figure 2. Cut the skin lengthwise, stopping before you reach the lower wing. Be careful to cut only skin. Use the forceps to carefully remove the skin. Observe and describe the skin and any other tissue connected to it.

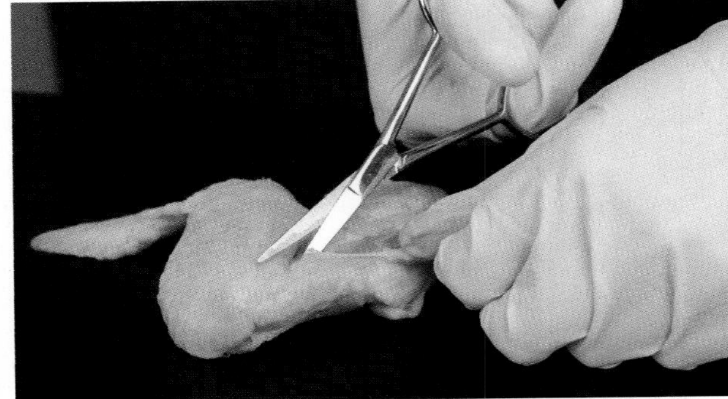

Figure 2 Step 4

5. Remove the skin from the lower wing in the same way, as shown in Figure 3 on the next page. Leave the skin on the wing tip. Using scissors, remove any tissues covering the muscle. Use the blunt probe to separate the individual muscles from each other. Be careful not to tear any muscles.

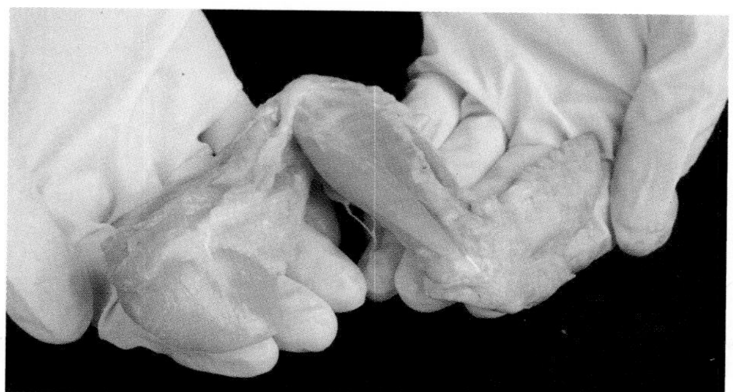

Figure 3 Step 5

6. Examine the muscle tissue. Describe its appearance and texture. Draw the chicken wing, showing all of its separate muscles.
7. Examine the bone structure of the wing. Depict in your drawing each of the bones and where they meet at joints. Bend and straighten the joint and observe how the bones fit together. Look at the "shoulder" part of the wing. The shiny, white covering of the joint surface is cartilage. Use a coloured pencil to draw the cartilage. Start a key to identify what each colour represents.
8. Figure 4 shows a ligament. Using the forceps, find as many ligaments as possible in your chicken wing. Use a second coloured pencil to show these ligaments in your drawing.

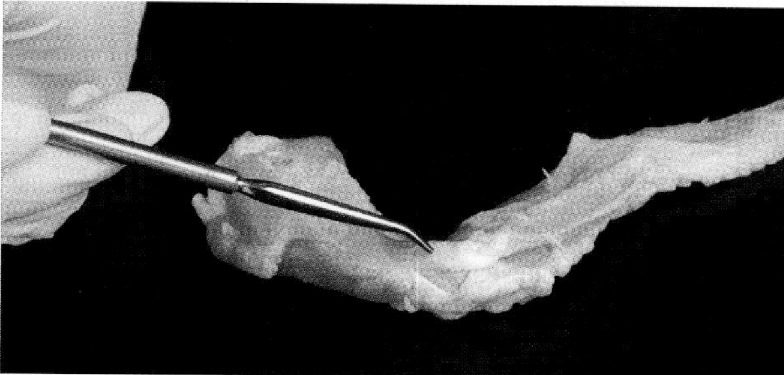

Figure 4 Step 8

9. Find as many tendons as possible in your chicken wing (Figure 5). Use a third coloured pencil to show these tendons in your drawing.

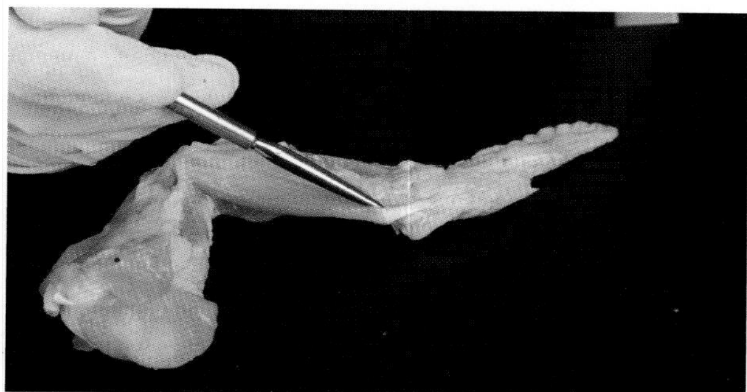

Figure 5 Step 9

10. Straighten the chicken wing and hold it horizontally above the tray. Pull on each muscle and note the movement that results. Turn the wing upside down and bend the joints. Pull on each muscle and note the movement that results. On your drawing, use a fourth colour to indicate each muscle that *flexes* (bends) a joint. These muscles are called "flexors." Use a fifth colour to indicate each muscle that *extends* (straightens) a joint. These muscles are called "extensors."
11. Use the scissors to cut through the middle of a flexor muscle for the lower wing. Record what happens to the wing.
12. Cut through the middle of an extensor muscle for the lower wing. Record what happens to the wing.
13. Dispose of the chicken wing as directed by your teacher. Place your dissection equipment in the collection bin as directed.

Analyze and Evaluate

(a) Complete and label your diagram. C

(b) Consider all of the tissues you observed during your dissection of the chicken wing. T/I

  (i) Name as many of these tissues as possible.

  (ii) Categorize them as epithelial, connective, nerve, or muscle tissue.

  (iii) Describe their functions in a live bird.

(c) What tissue did you *not* observe that is necessary for a live bird to move its wing? T/I

(d) Relate the functions of cartilage, ligaments, and tendons to their appearance in a chicken wing. T/I

Apply and Extend

(e) Refer to your drawing. Identify the muscles, bones, and joints in the chicken wing that would correspond to the muscles, bones, and joints in your own arm. A

(f) Compare the range of motion of a bird's wing with that of the lower part of a human arm. Why can you not compare the range of motion of the entire limb from this activity? A

3.10 The Nervous System

nervous system the organ system that is made up of the brain, the spinal cord, and the peripheral nerves; the system that senses the environment and coordinates appropriate responses

central nervous system the part of the nervous system consisting of the brain and the spinal cord

peripheral nervous system the part of the nervous system consisting of the nerves that connect the body to the central nervous system

There are some organs and tissues that we can live without. We could lose one or both kidneys and still survive with the help of a machine. We could even live with a mechanical heart. But the brain is essential! No one can live without a brain. The brain controls almost everything that happens in your body. Because it is so crucial to our survival, it is not surprising, then, that the brain is protected inside a very hard skull. However, this protection does not completely guarantee its safety; the brain is a very fragile organ.

The brain is just part of the **nervous system**: that delicate network of nerves that carries messages around the body, allowing us to interact successfully and safely with our environment.

Structural Features

The core of the nervous system—the **central nervous system**—consists of the brain and spinal cord. The nerves that carry signals between the central nervous system and the body make up the **peripheral nervous system** (Figure 1). The peripheral nervous system relays information about the internal and external environments to the brain. It also relays instructions from the brain to other parts of the body to control many of the body's functions and responses. The peripheral nervous system can be further divided into three groups of nerves:

- nerves that control the voluntary muscles
- nerves that carry information from the sensory organs, such as the eyes, ears, taste buds, and touch receptors, to the brain
- nerves that regulate involuntary functions such as breathing, heartbeat, and digestion

To protect it from physical damage, the central nervous system is shielded by bones. The skull protects the brain, and the spine guards the spinal cord (Figure 2). The brain and the spinal cord are surrounded by cerebrospinal fluid. This fluid helps cushion the brain and spinal cord from injury, transports chemicals, and removes wastes that are produced in the brain.

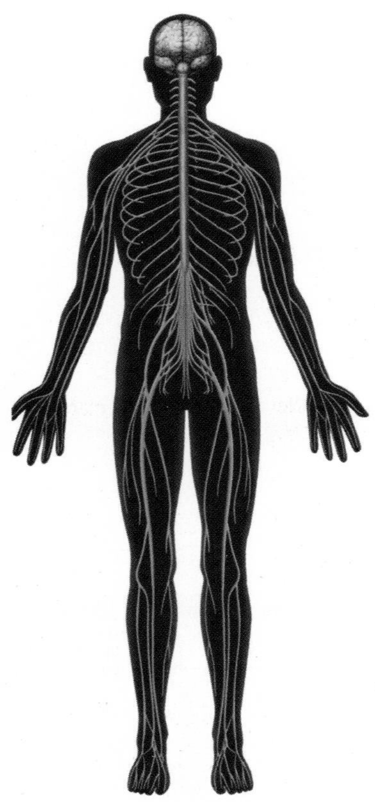

Figure 1 The peripheral nervous system (shown in brown) brings information from the body to the central nervous system (pink).

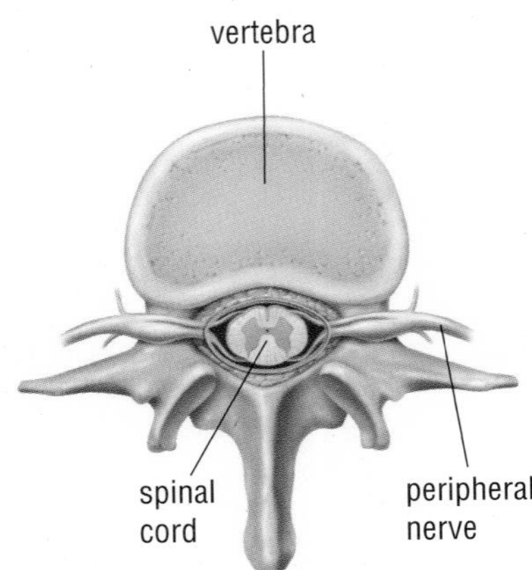

Figure 2 Each vertebra has a space in the middle for the spinal cord and grooves on either side to accommodate the peripheral nerves.

Nerve Tissue

neuron a nerve cell

Nerve tissue is made up of special cells called **neurons**. Nerve tissue is found in the brain, spinal cord, and nerves. There are an estimated 100 billion neurons in the human brain. Neurons are communication specialists. Their structure enables them to send information around your body. They do this by conducting electrical signals—nerve impulses—from one area of the body to another (Figure 3, next page). The axons of some neurons are covered by a fatty material called myelin. The myelin sheath acts like the insulation on an electrical wire, preventing electrical impulses from passing to the wrong neuron.

WRITING TIP

Recording Measurements

When measuring using an instrument, measure three times to ensure it is correct. Use a table to record quantitative measurements and double-check that numbers are accurate and in the correct cell of the table.

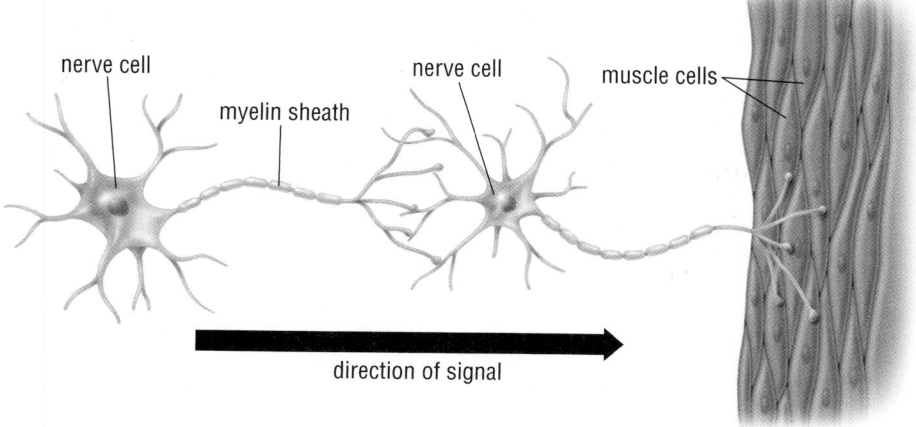

Figure 3 Electrical signals pass through neurons in only one direction.

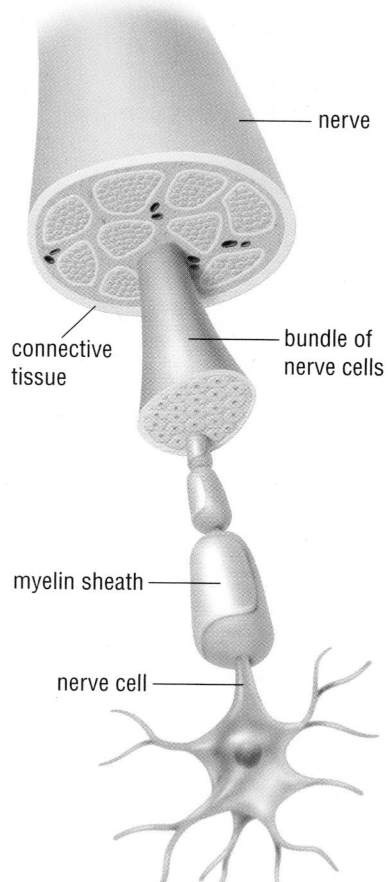

Figure 4 Nerves can transmit many signals at the same time.

Nerves are bundles of neurons that are surrounded by connective tissue (Figure 4). Nerves allow a two-way flow of information even though each neuron transmits information only in one direction.

Many cells in the body undergo cell division to repair an injury. Injured neurons in the central nervous system, however, do not easily regenerate. Some neurons in the peripheral nervous system can regrow to repair a small gap (a few millimetres) between the ends of severed nerves.

Sensory Receptors

Sensory receptors are special cells or tissues that receive input from our external environment and send signals along the peripheral nerves to our central nervous system. Our eyes have receptors that are sensitive to light. The ears, mouth, nose, muscles, and skin have other sensory receptors.

TRY THIS MAPPING SENSORY RECEPTORS

SKILLS: Controlling Variables, Performing, Observing, Analyzing, Communicating

SKILLS HANDBOOK 3.B.

In this activity, you will explore the touch sensitivity of the skin. You will work with a partner to discover the minimum distance between "touch points" that can be felt on different areas of skin.

Equipment and Materials: caliper or paper clip; ruler; blindfold (optional)

1. Create a table in which to record observations for point distances of 5 mm, 10 mm, and 15 mm to be tested at the following locations: fingertip, palm, inner arm, knee cap, behind the knee, back of the neck.
2. Adjust the caliper so that the distance between the points matches one of the values in your table.
3. Touch the subject's skin firmly, at one of the locations listed, with both points of the caliper at the same time (Figure 5). The subject should not know the distance. Perform several tests, using different distances, at each location. The subject should indicate, each time, whether he or she feels one point of contact or two. Record the responses.

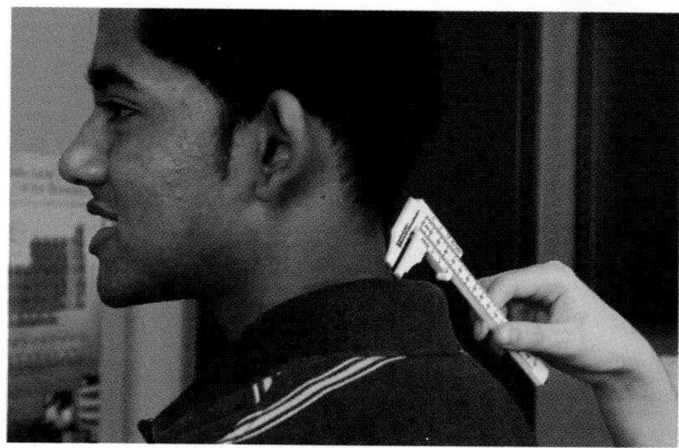

Figure 5 Do not hurt your partner with the caliper.

A. What minimum distance is required to discriminate between two points of contact at each location? What does this tell you about the concentrations of touch receptors? T/I

B. Based on your observations, predict which locations have the highest and lowest concentrations of touch receptors. T/I

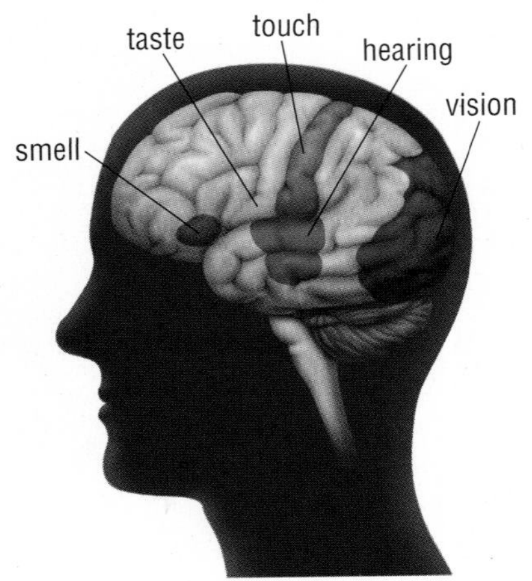

Figure 6 Different parts of the brain receive information from our various senses.

In addition to the familiar senses (sight, hearing, taste, smell, and touch), we have receptors in our muscles and skin that are sensitive to pressure, temperature, and pain and receptors that make us aware of our balance, position, and motion. All of the sensory receptors around the body send information to the brain, where it is processed. Information from different sensory receptors goes to specific parts of the brain (Figure 6).

Communication, Coordination, and Perception

The overall function of the nervous system is to transmit signals in both directions between your brain and the rest of your body. This allows your body to respond both to the outside world and to the internal environment. For example, the nervous system tells the respiratory system when to increase the breathing rate and tells the circulatory system when the heart should beat faster. The nervous system lets us know when we should eat or drink and when we should stop. Another function of the brain is perception: interpreting or making sense of all the information we receive from our environment.

The spinal cord has another important function: it acts as a short cut for reflexes. Reflexes are actions that do not require the involvement of the brain; they occur without conscious thought. Quickly moving your hand away from a hot surface is an example of a reflex action.

Nerve Impulses and the Bionic Arm

Figure 7 The bionic arm is still in development, but already it is having a positive impact on people's lives.

The human arm is made up of many specialized tissues with different functions. These tissues work together to accomplish the tasks that we perform using our arms, such as brushing teeth, carrying books, writing assignments, and so on. As you may have noticed in Activity 3.9, there are many similarities in limb structure and function between animal species.

Scientists and engineers used everything they know about the human arm to develop an artificial arm that is controlled by nerve impulses (Figure 7). It is designed for people who have had an arm amputated as a result of an accident or disease. Nerves that once served the amputated arm are rerouted and connected to healthy muscle in the chest and other surrounding muscles. The rerouted nerves grow into these muscles and direct impulses, originally intended for the amputated arm, to the robotic arm. This lets the wearer move the prosthetic arm simply by thinking about it.

DID YOU **KNOW?**

Gene Hunter

Dr. Michael Hayden was named "researcher of the year" for 2008 by the Canadian Institutes of Health Research. Dr. Hayden is the Director of the Centre for Molecular Medicine and Therapeutics at the University of British Columbia. Among many accomplishments, this world-renowned geneticist identified the DNA responsible for Huntington disease and other diseases of the nervous system.

GO TO NELSON SCIENCE

Diseases and Disorders of the Nervous System

Problems with the nervous system can be very serious. The brain can be permanently damaged by viruses or bacteria. Diseases can also be caused by problems in other body systems. For example, multiple sclerosis is caused by a malfunction of the immune system. This disease destroys the myelin sheaths of neurons in the central nervous system. Symptoms, which worsen as the disease progresses, include muscle weakness, slurred speech, and difficulty walking.

Physical trauma, such as a fall or a blow, can cause severe damage to the spinal cord, often resulting in paralysis. One of the most serious—and possibly avoidable—sports injuries is damage to the brain. We hear about hockey players suffering from head traumas, ranging from concussions to death, when their heads hit the ice. Helmets can reduce the effect of head trauma. Helmets are mandatory in Canada for motorcyclists, hockey players, and competitors in many other sporting events. They are strongly recommended for any activity that has a risk of head injury.

If someone is hit on the head and a doctor suspects brain injury, the person is likely to be sent for a CT or MRI scan. The CT scan in Figure 8 shows an area where blood has collected between the brain and its protective membrane, causing pressure on the brain tissue.

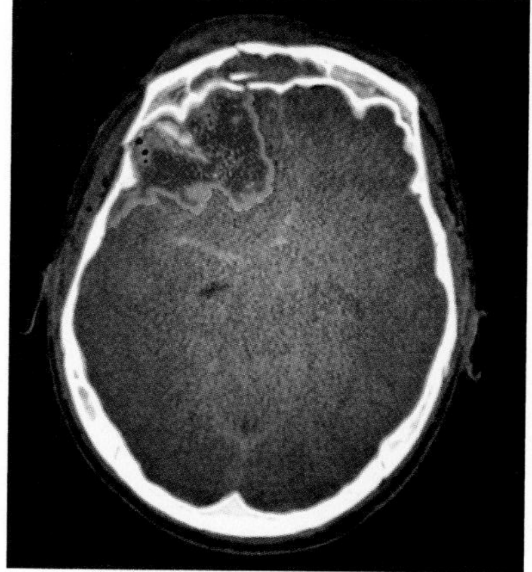

Figure 8 CT head scan

RESEARCH THIS DNA SCREENING

SKILLS: Defining the Issue, Researching, Analyzing the Issue, Defending a Decision

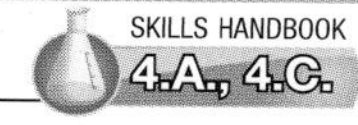

Some nervous system disorders are linked to certain genes. This means that there is a chance that the disorder could be passed via the DNA from parent to child. One such disorder is known as Huntington's disease (sometimes called Huntington's chorea).

1. Research Huntington's disease, including how it affects the nervous system and how it is treated.
2. Research the option of genetic screening for this disease.

GO TO NELSON SCIENCE

A. Imagine that a relative has been diagnosed with Huntington's disease. Should you and your family get tested for the Huntington's disease gene? Defend your answer. T/I C A

UNIT TASK Bookmark

How can you use this information about the central nervous system as you work on the Unit Task on page 156?

IN SUMMARY

- The nervous system is made up of the central nervous system and the peripheral nervous system.
- The central nervous system consists of the brain and the spinal cord. The peripheral nervous system consists of nerves that connect all parts of the body with the central nervous system.
- Nerves are made up of bundles of neurons, each surrounded by connective tissue.
- The body has millions of sensory receptors that receive information from the environment. The peripheral nervous system sends this information to the central nervous system.
- The main functions of the nervous system are communication and coordination of body activities.
- Diseases and injuries can cause serious damage to the nervous system.

CHECK YOUR LEARNING

1. Briefly describe the structure and function of the nervous system. K/U
2. Sketch a neuron and label its structures. K/U C
3. Create a flow chart of the events in the nervous system when you swing at a baseball with a bat. Start with the sight of the ball coming toward you. K/U C
4. Give three examples of the nervous system coordinating activities in other body systems. K/U
5. What is the function of our sensory receptors? Mention at least two different sensory receptors in your answer. K/U
6. What medical technology can be used in the diagnosis of head injuries or injuries of the central nervous system? What is the advantage of this type of technology? K/U A
7. After a car accident, Jila lost the hearing in one ear. Examinations indicate no damage to the eardrum. Suggest a reason for the loss of hearing. T/I

3.11

Interactions of Systems

All cells have to perform the same basic activities to stay alive: they use energy and materials from the environment, store materials, get rid of wastes, move substances to where they are needed, grow, and reproduce. Similarly, most living things have to perform similar basic functions: they obtain food, transport necessary substances (such as food, other nutrients, and oxygen) to the cells, remove unwanted substances from the cells, grow, and reproduce. Very simple animals can perform these functions with fairly simple arrangements of cells and tissues. Larger, more complex animals, however, need organ systems like the ones you have learned about in this chapter. Even these organ systems do not work independently. They interact with each other to allow the animal to carry out all the processes necessary for life.

In this chapter you have looked at five organ systems in detail: the digestive system, the circulatory system, the respiratory system, the musculoskeletal system, and the nervous system. Other organ systems include the urinary system, the reproductive system, the integumentary system, and the endocrine system. How do all these systems interact? Obviously it is a very complicated arrangement. Medical professionals and researchers around the world are still trying to puzzle out some of the details. We can, however, consider some of the system interactions within a simplified version of an organism (Figure 1).

To find out more about other organ systems,
GO TO NELSON SCIENCE

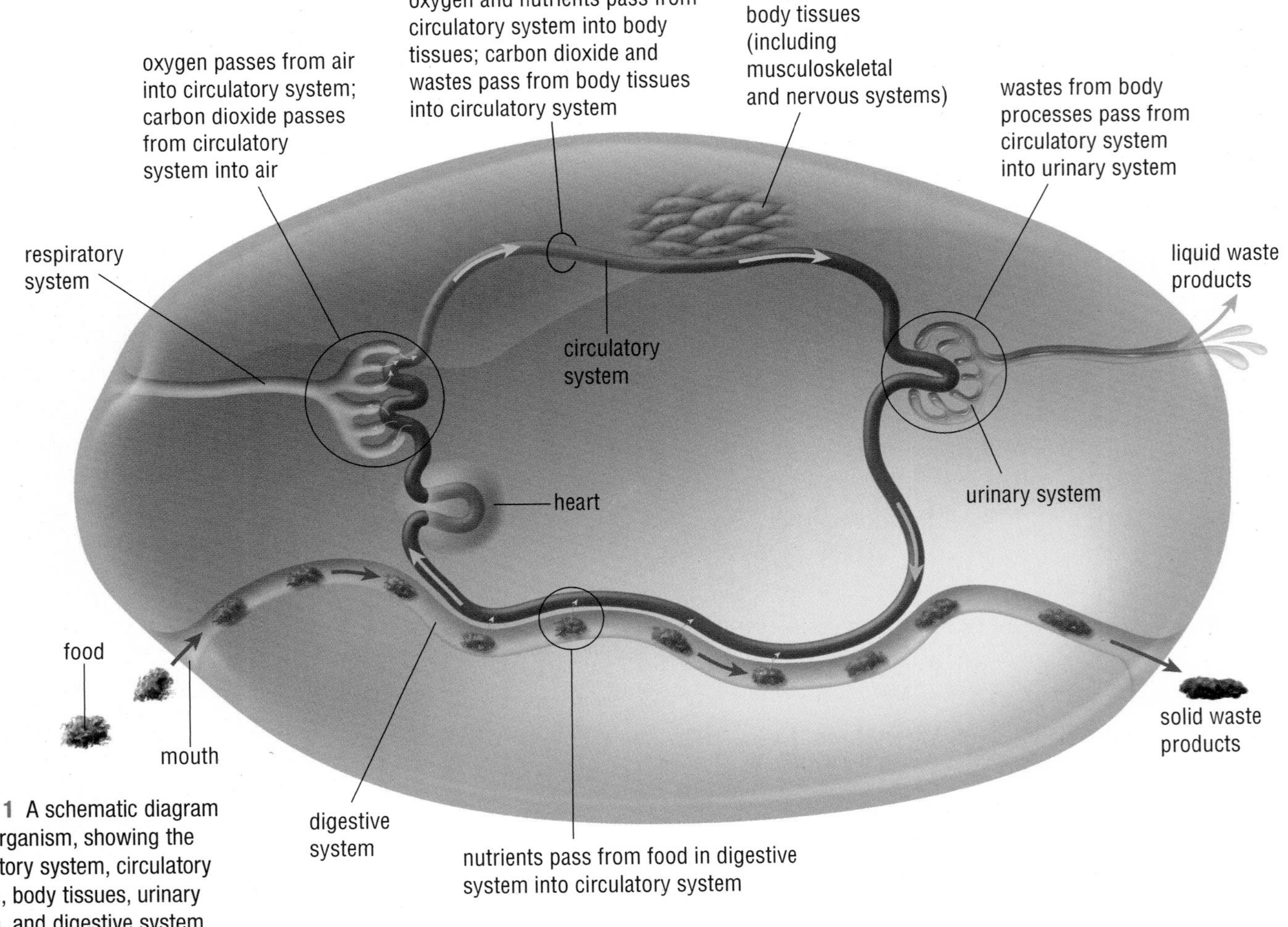
oxygen passes from air into circulatory system; carbon dioxide passes from circulatory system into air
oxygen and nutrients pass from circulatory system into body tissues; carbon dioxide and wastes pass from body tissues into circulatory system
body tissues (including musculoskeletal and nervous systems)
wastes from body processes pass from circulatory system into urinary system
respiratory system
liquid waste products
circulatory system
heart
urinary system
food
mouth
solid waste products
digestive system
nutrients pass from food in digestive system into circulatory system

Figure 1 A schematic diagram of an organism, showing the respiratory system, circulatory system, body tissues, urinary system, and digestive system

Let's look at some points of interaction between these systems. Consider the digestive system and the circulatory system. The digestive system breaks down food into small molecules that can pass through the walls of the digestive tract. Without a circulatory system, only the tissues right next to the digestive tract would receive nutrients. The circulatory system provides a way to transport nutrients to all tissues in the organism. Thousands of capillaries surround the digestive tract, carrying blood that absorbs the nutrients (Figure 2). Materials move from a region of higher concentration to a region of lower concentration by diffusion. The circulatory system continuously delivers low-nutrient blood to the capillaries surrounding the digestive system, and carries away blood that now contains a high concentration of nutrients. This nutrient-rich blood then enters larger blood vessels and travels to every part of the body, where nutrients diffuse from the blood into the cells.

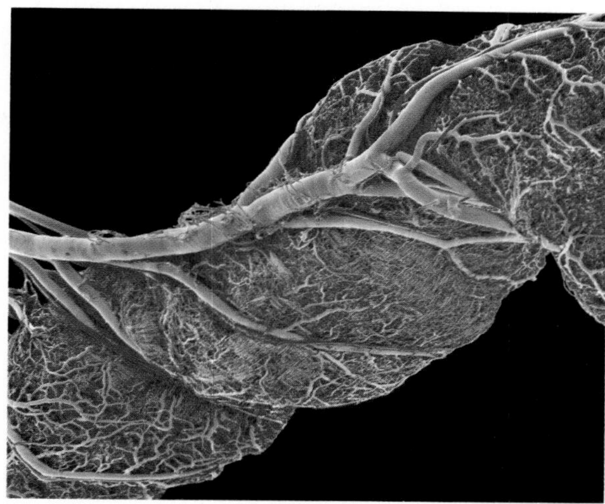

Figure 2 The digestive tract is surrounded by blood vessels. The blood absorbs nutrients for delivery around the body.

The musculoskeletal system—particularly the skeletal muscle tissue—uses oxygen and nutrients to make the body move. Every time a muscle contracts, its rate of cellular respiration increases. To fuel this active system, the circulatory system must deliver a constant supply of oxygen and nutrients. In addition, it must remove waste products such as carbon dioxide and, during strenuous exercise, lactic acid. These wastes enter the blood by diffusion through capillary walls and are carried away for "disposal." Carbon dioxide, as you know, passes into the lungs and leaves the body through the respiratory system. The liver—an accessory organ to the digestive system—removes lactic acid from the blood.

How are other waste products of cell activity removed from the body? Not surprisingly, the circulatory system is involved. The blood collects waste products as it travels through all the body's tissues. In particular, it collects nitrogen-containing waste products of protein breakdown. As the blood travels through the kidneys (part of the urinary system), the unwanted and toxic substances are removed (Figure 3). These substances, dissolved in water, drain into the bladder for temporary storage. The mixed solution is called urine, and is periodically excreted from the body.

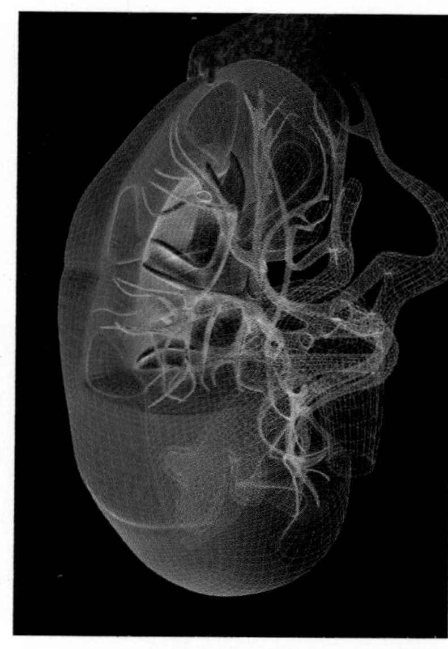

Figure 3 Each kidney processes almost 100 L of blood each day, producing approximately 1 L of urine.

Other Interactions Between Systems

Figure 4 A chameleon's tongue can stretch out to be longer than its body. The tongue is sticky at the end to catch insects.

The examples discussed so far apply to humans and other mammals, and most vertebrates. Some animals, depending on their environment and needs, have developed interesting alternatives to the familiar organ systems. For instance, a jellyfish does not have a circulatory system. Its digestive system extends into the animal's fluid-filled body cavity and exchanges nutrients directly with this fluid. The animal has so few cells, and such a small demand for food, that this arrangement is adequate to distribute nutrients. Even animals that have familiar digestive tracts may have developed interesting adaptations to acquire food. Two examples of muscular adaptations are the elephant's trunk and the chameleon's tongue (Figure 4).

The integumentary (skin) system and muscles interact to provide information to the nervous system. There are sensors in our skin that detect temperature, pressure, pain, and so on. The ears of many mammals are mobile—thanks to muscles—so they can be directed toward a sound (Figure 5). The shape of the ears amplifies the sound, increasing the amount of information that the nervous system can collect. Similarly, the eyes of many animals include muscles that control the amount of light entering the eye, improving vision at a variety of light levels.

In birds the musculoskeletal system and the integumentary (skin) system interact to make flight possible. Specialized flight feathers grow from the skin, but they would be useless without the bird's light, hollow bones, and powerful flight muscles (Figure 6).

Respiratory systems are quite different in air-breathing animals than in aquatic animals that get their oxygen from the water. Recall from Section 3.6 that fish have gills that are richly supplied with blood. Oxygen and carbon dioxide diffuse between the blood and water that constantly flows over the gills. The circulatory system then carries oxygenated blood around the body. Amphibians such as newts and frogs have lungs, but can also exchange gases through their thin skin when moist (Figure 7). In effect, the entire surface of the body becomes part of the respiratory system.

Different groups of animals have different ways of excreting cellular wastes. Freshwater fish carry nitrogen compounds (from protein breakdown) in the blood until it reaches the gills. The compounds then diffuse through the gills into the water. Thus, the fish's respiratory system serves double duty as an excretory system, aided by the circulatory system.

DID YOU KNOW?

Tapeworms: Just the Essentials

Because of where they live—inside the digestive systems of vertebrate animals—tapeworms do not need to find, capture, and digest food. They do not need to protect themselves from predators or worry about the weather. Tapeworms have almost no organs or organ systems. They do not need these structures; they can just absorb food that has been digested for them! Tapeworms are really little more than a reproductive system. Most other functions are performed for them by their host.

Figure 5 A jackrabbit has very large ears so that it can hear and avoid predators.

Figure 6 Most of the surface of the wing is made up of feathers—part of the integumentary system—not bone and muscle.

Figure 7 This young Eastern newt obtains oxygen through its skin.

There are many ways in which organ systems interact. For example, the urinary system and the reproductive system are very closely connected in mammals—particularly in males. The nervous system works very closely with the endocrine system. Each organ system interacts with at least one other organ system. It is through the interaction and coordination of organ systems that complex organisms can carry out all life functions to survive.

RESEARCH THIS SYSTEMS WORKING TOGETHER

SKILLS: Researching, Communicating

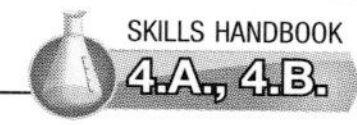

In this activity you will research how a specific animal performs an essential function: providing its cells with nutrients. For example, you might consider how a cat hunts, eats, and digests (Figure 8).

1. Choose an animal to investigate. It can be as familiar or as unusual as you like.
2. Research how your animal performs this function. Pay particular attention to which parts of the body are involved, and to which organ systems these parts belong.

GO TO NELSON SCIENCE

A. Create a concept map or any other graphic organizer to communicate what you have discovered. Be sure to clearly indicate the interactions between the various organ systems. C A

UNIT TASK Bookmark

Consider the interactions among organ systems as you are preparing to complete the Unit Task on page 156.

IN SUMMARY

- Organ systems work together to accomplish specific functions.
- All organ systems in the body interact with at least one other organ system.
- Animals have different ways of meeting their needs so organ systems vary greatly. Not all animals have all organ systems.
- In complex animals, the circulatory system connects all other systems in the body.

CHECK YOUR LEARNING

1. List at least three interactions of organ systems within the bodies of mammals. K/U
2. (a) Which organ system interacts with most other systems in the body?
   (b) Explain why it is advantageous for the system named in (a) to be integrated with so many other systems. K/U
3. What two systems do amphibians use for gas exchange? K/U
4. How is it possible for an animal like a tapeworm to live without having a digestive system? K/U
5. Frogs and ducks have webbed feet. How does this illustrate the interactions of systems? A
6. Briefly describe one interaction (not necessarily in humans) between the nervous system and K/U A
   (a) the integumentary (skin) system.
   (b) the musculoskeletal system.
   (c) the respiratory system.
   (d) the urinary system.

Monitoring the Health of an Unborn Baby

Ultrasound scanning is a technique very commonly used in North America to monitor the health of a baby before birth. An ultrasound technician uses a device that contains a small transmitter and a receiver. This device is moved repeatedly across the pregnant mother's abdomen. The transmitter produces sound waves at a higher frequency than the human ear can detect. These sound waves pass through some tissue but bounce back from other tissue. The reflected sound waves are detected by the receiver. They are interpreted by a computer and converted into a picture of the tissues in their path. This picture can be viewed on a television screen (Figure 1). An ultrasound image of a fetus (unborn baby) can provide a lot of valuable information about the baby's size, position, and development. In particular, this technology can help doctors detect serious health problems in the baby, such as spina bifida.

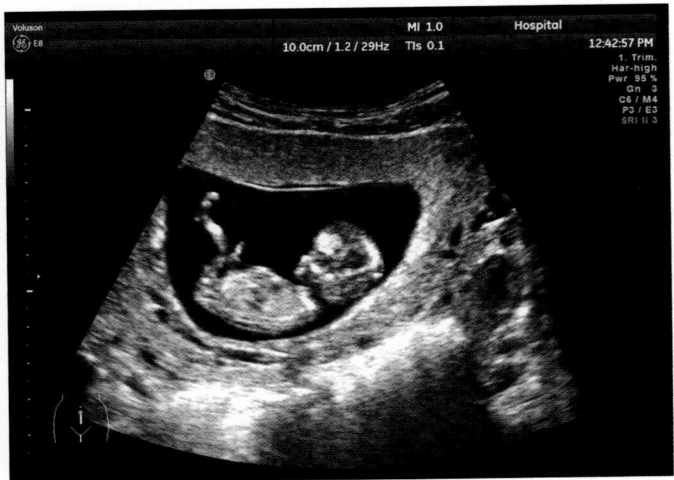

Figure 1 An ultrasound image of a fetus in the uterus

Ultrasound scans can provide a lot of information, but they do not reveal everything about the fetus's development. This technology can, however, help doctors perform another test that can reveal a great deal more: amniocentesis. Amniocentesis involves inserting a long needle into the uterus and taking a sample of the amniotic fluid that surrounds the fetus. With the help of ultrasound as the "eyes," an obstetrician can safely direct the needle to the right part of the uterus (Figure 2). This fluid contains fetal cells, which contain DNA. The DNA is checked for genes known to cause certain diseases.

Knowing what health problems exist for a child before it is born can be of great benefit. One example is phenylketonuria (PKU). People born with this genetic disorder lack an enzyme that helps digest certain foods.

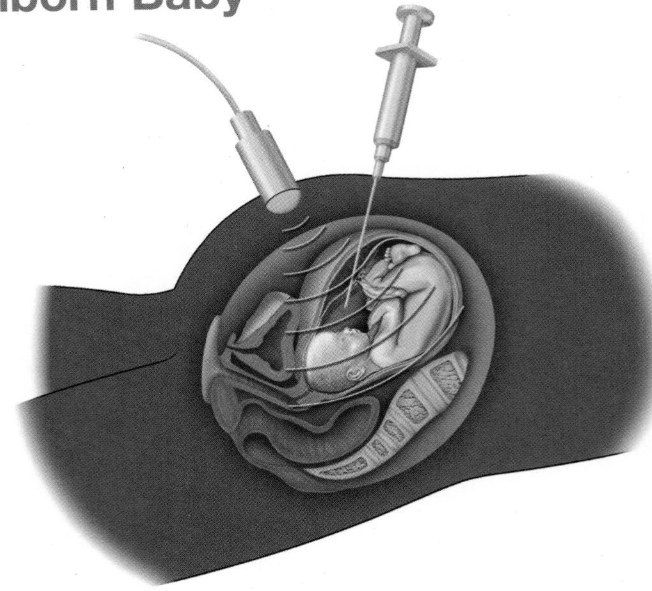

Figure 2 Ultrasound technology is used to help guide the placement of the needle for amniocentesis.

Some proteins in our food contain a substance called phenylalanine. High concentrations of phenylalanine in the body can cause damage to the nerves and brain. Normally, phenylalanine is digested in the digestive tract. People with PKU do not have the enzyme to digest phenylalanine, so it builds up in the body to toxic levels. Fortunately, the consequences of PKU can be avoided by controlling phenylalanine in the diet. PKU is a genetic disorder, so if either of the parents has DNA associated with PKU, there is a chance that the unborn baby might also have this DNA. Doctors can use amniocentesis to collect a sample of fetal cells and then test these cells for PKU. This gives a mother early warning if her baby has the condition. She can change her diet to avoid food containing phenylalanine. She can also feed the baby a special diet after birth. These changes will ensure that phenylalanine does not build up and harm the baby.

Ultrasound is a remarkable tool for monitoring health. However, like many technologies, there are some risks associated with it. There is some evidence that ultrasonic scans lasting 30 minutes or more cause brain damage in fetal mice. However, the long-term effects of ultrasound are unknown. Medical use must balance the benefits of this technology with its potential risks.

GO TO NELSON SCIENCE

To Immunize or Not to Immunize?

Your parents probably had you immunized against a range of diseases when you were a child. Or perhaps they chose not to. They may be among the portion of the population who believe that the health risks of vaccinating children outweigh the benefits. Individual decisions on whether or not to immunize are the basis of a current public health controversy.

On one side of the argument are the parents who believe that public immunization programs are a benefit to the individual and to society. They might remember a time when pregnant mothers caught rubella (German measles) and miscarried their babies. Perhaps they know someone who barely survived polio. Or maybe they came from a place where deadly smallpox epidemics were a regular occurrence. Early-childhood vaccination programs have eliminated many of the diseases that once killed hundreds of thousands worldwide (Figure 1).

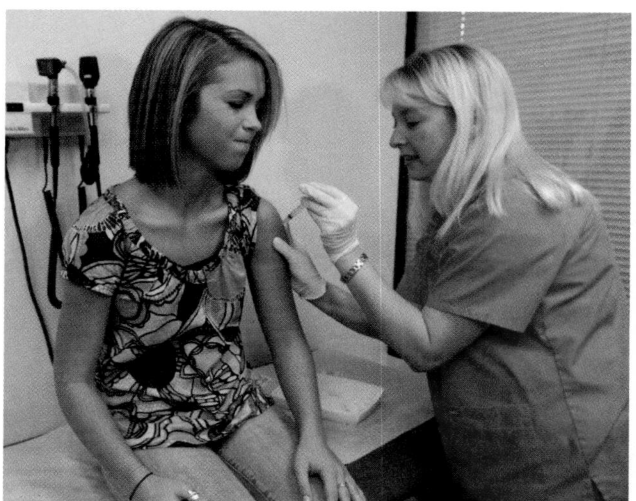

Figure 1 Vaccinations given to children and young adults can reduce their chances of catching certain life-threatening infections.

Parents who oppose childhood vaccination may point out that vaccinations do not provide total protection from disease. Some parents believe that an increased number of vaccinations can leave a child vulnerable to diseases such as asthma and diabetes. They argue that studies into the risks of vaccines are not done or are biased because the big drug companies make so much money from vaccines.

SKILLS MENU

- Defining the Issue
- Researching
- Identifying Alternatives
- Analyzing the Issue
- Defending a Decision
- Communicating
- Evaluating

The Issue

Do the benefits of immunizing children outweigh the risks? You are part of a committee that will be collecting evidence from both sides of the issue. Your committee will present this evidence to a community forum and recommend whether or not children should be immunized.

Goal

To research public immunization programs and present recommendations to parents regarding immunization for their children.

Gather Information

Work in small groups to research how immunization works, the risks and benefits of public immunization programs, and any available alternatives to immunization.

GO TO NELSON SCIENCE

Identify Solutions

You may want to consider the following ideas:

- The question of the rights of the individual versus the rights of society in general
- Respect for individual values and opinions
- Alternatives to vaccination

Make a Decision

What will you recommend to parents regarding whether or not to immunize their children? What evidence supports your recommendations? T/I C

Communicate

Complete a written report that will be distributed to parents. Your report should include the benefits and risks of public immunization programs and a final recommendation based on the evidence. T/I C A

KEY CONCEPTS SUMMARY

Complex animals are made up of cells, tissues, organs, and organ systems.

- Groups of similar cells that perform a common function form tissues. (3.1)
- Organs are made up of several types of tissue. (3.1, 3.3, 3.4, 3.6, 3.8, 3.10)
- Organ systems are made up of organs and tissues. (3.1, 3.3, 3.4, 3.6, 3.8, 3.10)

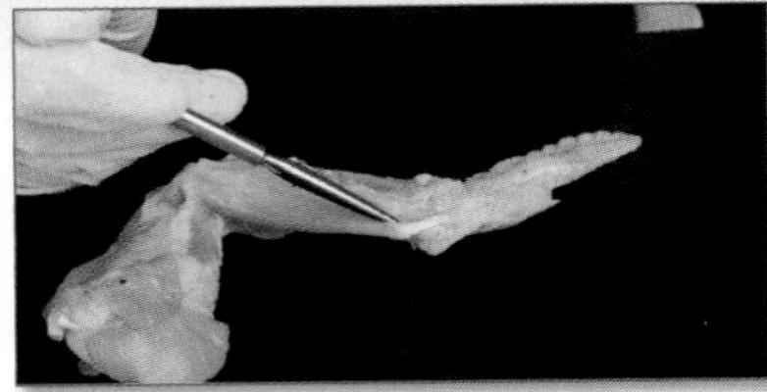

Scientists use laboratory techniques to explore the structures and functions of animals' bodies.

- The structures and functions of tissues can be explored using laboratory techniques such as dissection. (3.5, 3.9)
- Observations of tissue structures can be recorded in scientific drawings, which help relate structures to functions. (3.5, 3.9)
- Models of cells and tissues can be used to simulate structures and functions of living things. (3.5, 3.9)

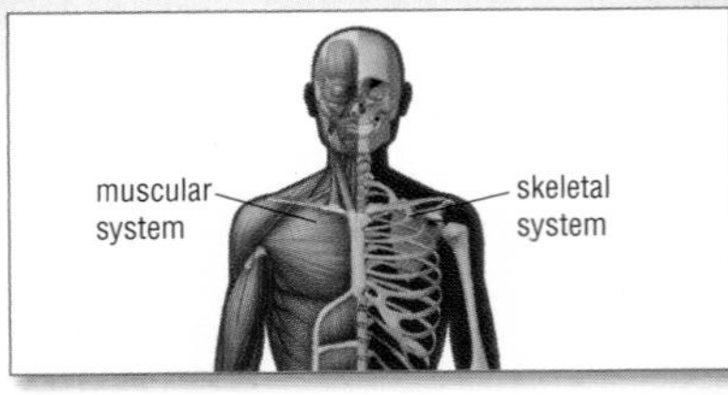

Each organ system has a specific function and corresponding specific structures.

- Organ systems have specific functions within the body. (3.1, 3.3, 3.4, 3.6, 3.8, 3.10)
- The structure of each organ system reflects its function. (3.1, 3.3, 3.4, 3.6, 3.8, 3.10)

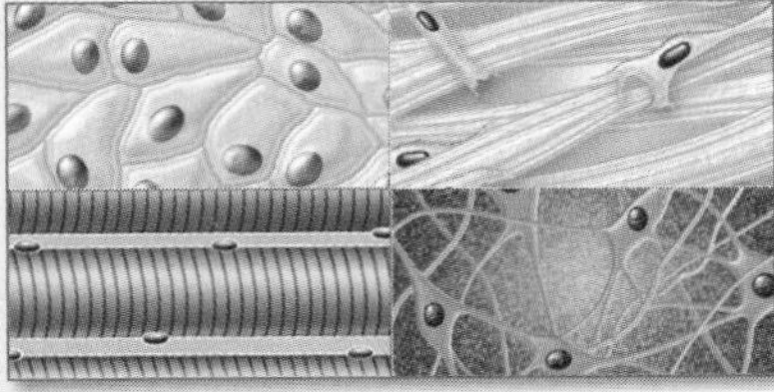

There are four main types of animal tissues.

- Epithelial tissue covers the outside of the body and lines the respiratory system and the digestive system. (3.1, 3.3, 3.6)
- Nerve tissue reaches every part of the body, carrying messages to and from the central nervous system. (3.1, 3.10)
- There are three types of muscle tissue: skeletal (for voluntary movement), smooth (for involuntary movement), and cardiac (for keeping the heart beating regularly). (3.1, 3.3, 3.8)
- Connective tissue includes blood, bones, and cartilage. (3.1, 3.4, 3.8)

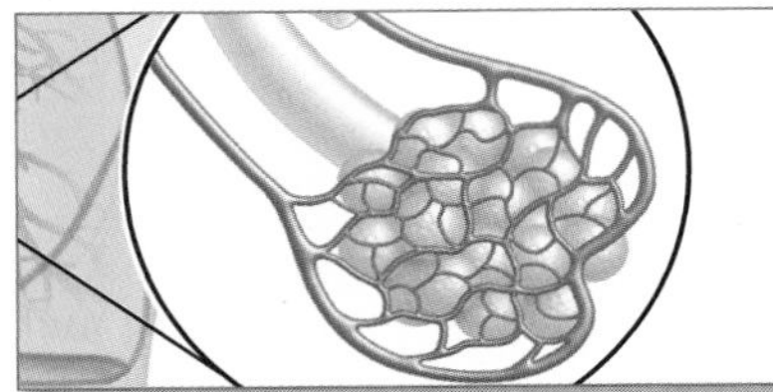

Organ systems interact to keep the organism functioning.

- Organ systems are dependent on each other: none can function for long without the others. (3.1, 3.11)
- The circulatory system, for example, transports oxygen around the body from the respiratory system and nutrients from the digestive system. (3.1, 3.4, 3.6, 3.11)

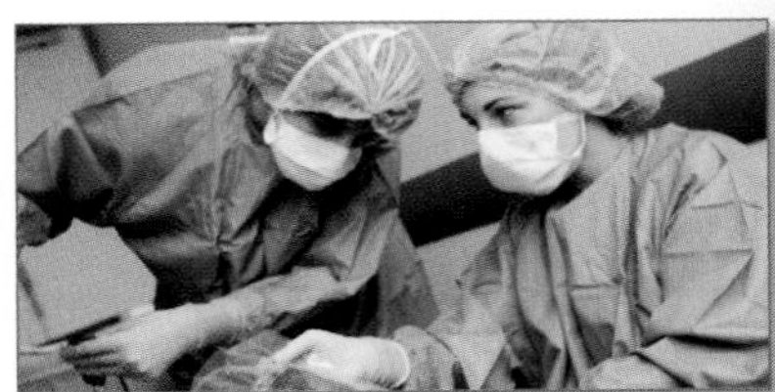

Research is helping people overcome illness and injury.

- Diagnostic and treatment technologies help health professionals detect and treat problems. (3.4, 3.6, 3.8, 3.10)
- There are legal, ethical, and social concerns associated with many technological and medical advances, such as stem cell research, DNA screening, and immunization. (3.2, 3.6, 3.10, 3.12)

WHAT DO YOU THINK NOW?

You thought about the following statements at the beginning of the chapter. You may have encountered these ideas in school, at home, or in the world around you. Consider them again and decide whether you agree or disagree with each one.

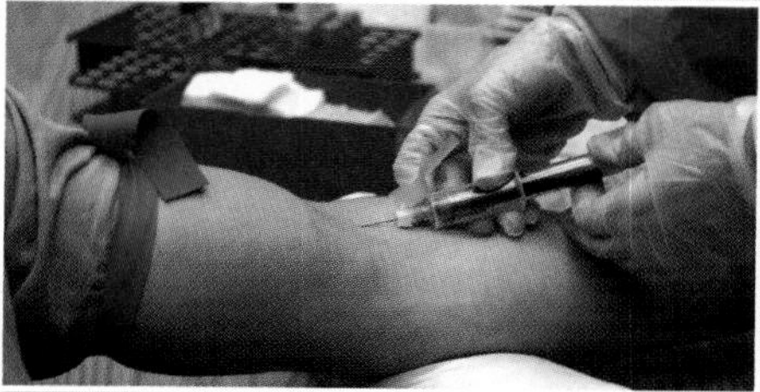

1 When blood is removed from the body, the remaining blood cells divide to make more blood.
Agree/disagree?

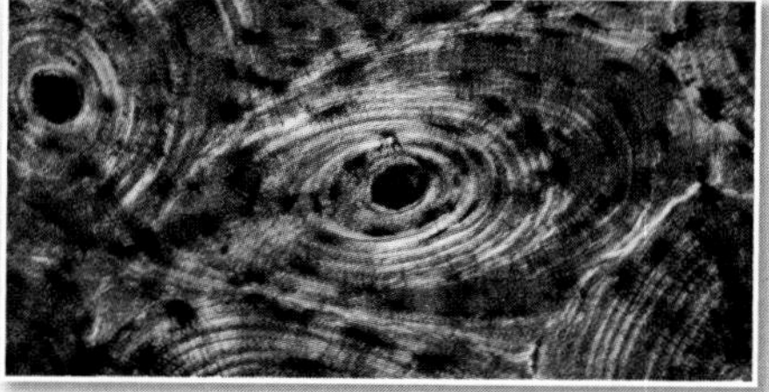

2 All organs in the body are made up of living cells.
Agree/disagree?

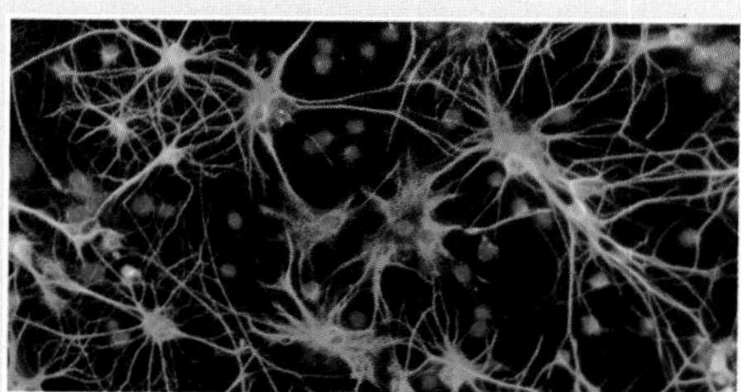

3 All animal cells look the same.
Agree/disagree?

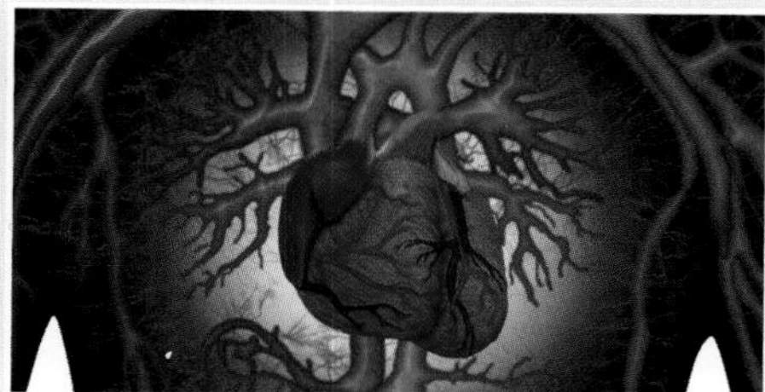

4 Our organs work independently of each other.
Agree/disagree?

5 The organ systems of a frog are the same as the organ systems of a human.
Agree/disagree?

6 Animals can grow replacement body parts.
Agree/disagree?

How have your answers changed?
What new understanding do you have?

Vocabulary

hierarchy (p. 73)
tissue (p. 74)
organ (p. 74)
organ system (p. 74)
epithelial tissue (p. 75)
connective tissue (p. 75)
muscle tissue (p. 75)
nerve tissue (p. 75)
cellular differentiation (p. 77)
stem cell (p. 77)
digestive system (p. 80)
circulatory system (p. 83)
artery (p. 84)
vein (p. 84)
capillary (p. 84)
electrocardiogram (ECG) (p. 86)
respiratory system (p. 91)
alveolus (p. 92)
xenotransplantation (p. 97)
musculoskeletal system (p. 99)
nervous system (p. 104)
central nervous system (p. 104)
peripheral nervous system (p. 104)
neuron (p. 104)

BIG Ideas

- ✓ Plants and animals, including humans, are made of specialized cells, tissues, and organs that are organized into systems.
- ✓ Developments in medicine and medical technology can have social and ethical implications.

CHAPTER 3 REVIEW

The following icons indicate the Achievement Chart category addressed by each question.

K/U Knowledge/Understanding T/I Thinking/Investigation
C Communication A Application

What Do You Remember?

1. Which of these levels includes all of the others: cell, organ system, tissue, organ, organism? (3.1) K/U

2. Copy Table 1 into your notebook. Complete the table by filling in the appropriate information. (3.1) K/U C

Table 1 Tissue Types in Animals

Tissue type	Structure	Function
	Made up of cells that are able to contract	
		Transmitting information around the body
Connective		
		Protection and reduction of water loss

3. Briefly describe the main function of
   (a) the digestive system.
   (b) the brain.
   (c) blood. (3.3, 3.4, 3.10) K/U

4. List the main organs in the nervous system, along with their functions. (3.10) K/U

5. (a) Which type of muscle moves food through the digestive tract?
   (b) Why is this type of muscle suitable for this task? (3.3, 3.8) K/U

6. (a) Name the system that provides support and structure for the body.
   (b) What is another function of this system?
   (c) Name the tissue types that make up this system. (3.1, 3.8, 3.11) K/U

7. (a) Which system is responsible for transporting nutrients to all parts of the body?
   (b) What is the main organ in this system? (3.4) K/U

What Do You Understand?

8. (a) What is the difference between embryonic stem cells and tissue stem cells?
   (b) Briefly describe how tissue stem cells can be used to cure a disease. (3.2) K/U

9. (a) What does the term "regeneration" mean?
   (b) Can humans regenerate? Explain. (3.2) K/U

10. (a) What are the functions of the human arm?
    (b) What are the functions of a bird's wing?
    (c) What are the similarities and differences in these functions?
    (d) How are these similarities and differences reflected in their structures? (3.9) K/U T/I

11. Name three systems that are interdependent. Describe how they depend on each other. (3.1, 3.3, 3.4, 3.6, 3.8, 3.10, 3.11) K/U T/I

12. Create and complete a table that lists and describes four technologies that are used to diagnose or treat damage or disease of the human body. Include a column to describe briefly how each technology is used. (3.4, 3.6, 3.8, 3.10) K/U C

13. Where do all of the different cells in our bodies come from? In your answer, use the terms "stem cell," "specialized tissue," "differentiate," "organ," and "adult." (3.1, 3.2) K/U

14. Use a concept map to link the four types of animal tissues to the five organ systems described in this chapter. (3.1, 3.3, 3.4, 3.6, 3.8, 3.10, 3.11) T/I C

Solve a Problem

15. Create a t-chart listing the pros and cons of organ transplantation. (3.7) A C

16. In sports, the illegal practice of "blood doping" involves removing some blood cells from an athlete about two weeks before competition and storing them. Just before the competition, the blood cells are injected back into the athlete. Explain why this practice might give the athlete an advantage. (3.4, 3.6) T/I

17. Your body is said to be in "homeostasis" when there is a healthy balance in its internal conditions and processes (body temperature, blood pressure, heart rate, breathing rate). Explain how the circulatory, respiratory, digestive, and nervous systems contribute to homeostasis. (3.3, 3.4, 3.6, 3.10) K/U T/I

18. Your friend tells you that she has had special training that helps her hold her breath for long periods of time. She claims that she can swim underwater for almost 30 minutes. What is your reaction? Explain why. (3.6, 3.10) T/I

19. In a developing fetus, the mother's blood provides all of the necessary nutrients and removes waste. Why, then, is it necessary for an expectant mother to pay careful attention to her diet and lifestyle choices? (3.3, 3.4) A

Create and Evaluate

20. What is DNA screening? Why might someone choose not to use this technology? (3.10) K/U A

21. Imagine that a drug company has developed a new drug to treat coronary artery disease. What ethical issues might arise if the company wants to run and fund its own trials of this drug? (3.4) A

22. Are there any "downsides" to the advances in diagnostic technologies? Explain. (3.4, 3.6, 3.8, 3.10) C A

Reflect On Your Learning

23. Explain how your understanding of your body as "a system of interdependent systems" has changed as a result of studying this chapter.

24. In this chapter, you have learned about many medical technologies and developments.
    (a) What did you think about medical technologies before reading about them in this chapter?
    (b) How has your understanding of medical technologies changed?

Web Connections 

25. There are other systems in the body, in addition to the seven covered in this chapter. One is the immune system. Research how this system works, what tissues are involved, and what an autoimmune disease is. Write a short magazine article summarizing your research. T/I C

26. Adipose tissue is a type of connective tissue that contains fat cells. At a certain stage of development, individuals have essentially all of the fat cells they will ever have. Research the relationship between the number and size of fat cells and obesity. Report on the latest findings regarding attempts to prevent young people from developing obesity. T/I C A

27. West Nile virus is a disease transmitted through blood. Find out how it is transmitted. How have health authorities responded to the threat of West Nile virus? What effect does this response have on people's lifestyles? Create a skit or video to share your findings. T/I C A

28. (a) Research ophthalmology. What system of the body does it relate to? What types of imaging technologies are used in ophthalmology?
    (b) Create an illustrated poster to communicate your findings. T/I C

29. What is transgenic research? Research this topic, including the discussions surrounding the ownership and use of the findings. Prepare a brief presentation in any format of your choice. T/I C A

30. Cloning animals is currently being explored for use in agriculture and medical research.
    (a) Research the cloning techniques of nuclear transfer or embryo splitting. Investigate the arguments of people who support cloning and those who oppose it. Assemble your arguments to identify the main issues.
    (b) Decide on your position regarding the cloning of agricultural animals. Write a one-page report that states your position and explains your rationale. T/I C A

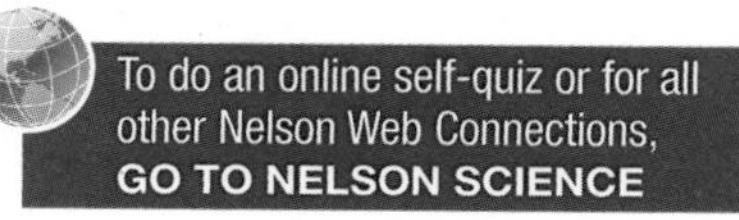
To do an online self-quiz or for all other Nelson Web Connections,
GO TO NELSON SCIENCE

CHAPTER 3 SELF-QUIZ

The following icons indicate the Achievement Chart category addressed by each question.

K/U Knowledge/Understanding · T/I Thinking/Investigation · C Communication · A Application

For each question, select the best answer from the four alternatives.

1. Which tissue type covers and protects the human body? (3.1) K/U
   (a) connective
   (b) nerve
   (c) muscle
   (d) epithelial

2. Which organ systems work together to absorb nutrients from food? (3.1–3.11) K/U
   (a) musculoskeletal system and digestive system
   (b) nervous system and circulatory system
   (c) digestive system and circulatory system
   (d) respiratory system and musculoskeletal system

3. The cells that carry oxygen are
   (a) white blood cells.
   (b) red blood cells.
   (c) platelets.
   (d) plasma. (3.4) K/U

4. Which of the following statements about organ donation and organ transplants is true? (3.7) K/U
   (a) The recipient of an organ donation is the only person who benefits.
   (b) Doctors are not sure how to minimize the risk of organ transplant rejection.
   (c) Organ transplants from deceased donors are more common than from living donors.
   (d) The person who has been waiting the longest automatically receives the first available organ.

Indicate whether each of the statements is TRUE or FALSE. If you think the statement is false, rewrite it to make it true.

5. The process of digestion begins in the stomach. (3.3) K/U

6. All tissues are able to regenerate themselves. (3.2) K/U

Copy each of the following statements into your notebook. Fill in the blanks with a word or phrase that correctly completes the sentence.

7. The ______________ nervous system is made up of the brain and spinal cord, and the ______________ nervous system is made up of nerves that connect the rest of the body to the brain and spinal cord. (3.10) K/U

8. The exchange of oxygen and carbon dioxide between the blood and the lungs occurs in tiny sacs called ______________. (3.6) K/U

Match each term on the left with the most appropriate description on the right.

9. (a) epithelial tissue
   (b) connective tissue
   (c) muscle tissue
   (d) nerve tissue
   (e) stem cells

   (i) cells that can differentiate to form specialized cells
   (ii) cells that can contract to move bones
   (iii) cells arranged in thin sheets to cover surfaces
   (iv) long, thin cells that conduct electrical impulses
   (v) various types of cells that provide support or insulation (3.1, 3.2) K/U

Write a short answer to each of these questions.

10. Name two systems that interact with the circulatory system. (3.3–3.11) K/U

11. Explain how the musculoskeletal system carries out each of the following functions:
    (a) support
    (b) protection
    (c) movement (3.8) K/U

12. Which of the following terms does not belong with the others? Explain your answer.
    esophagus, stomach, small intestine, lungs
    (3.3, 3.6) K/U

13. (a) What is a neuron?
    (b) How does the structure of a neuron help it to carry out its function? (3.10) K/U

14. Write a paragraph describing the passage of a slice of pizza through the digestive tract. (3.3) K/U C

15. Copy Table 1 into your notebook. Complete your table with information about each disease and the system or systems involved. (3.3, 3.4, 3.8, 3.10) K/U

Table 1 Diseases and Systems

Disease	Description of the disease	System(s) involved
osteoporosis		
diabetes		
coronary artery disease		
multiple sclerosis		

16. Explain how the structure of each blood vessel suits its function. (3.4) T/I
    (a) arteries
    (b) veins
    (c) capillaries

17. Imagine that doctors identified a disease that causes the smooth muscles in the human body to stop working. What consequences would this disease have for the person suffering from it? (3.3, 3.4) T/I

18. A person who takes many rapid, deep breaths may become dizzy. (3.4) T/I A
    (a) What has happened to the balance of oxygen and carbon dioxide in the person's bloodstream?
    (b) Why is breathing into a paper bag sometimes recommended during a situation like this?

19. As you travel to higher altitudes, the concentration of gases in the air, including oxygen, decreases. (3.6, 3.10) T/I A
    (a) What effect does this environment have on your breathing?
    (b) How might a runner who trained at sea level perform in a race taking place at 2 000 m above sea level? Explain.

20. Write a letter to a female friend about osteoporosis and its consequences. In the letter, explain how the proper diet and exercise regimen can help to prevent osteoporosis. (3.8) C

21. Surface area is a term used to describe how much exposed area an object has. (3.6) T/I
    (a) In situations where materials need to be transported across surfaces, what is the advantage of having a large surface area for transport?
    (b) In the lining of the small intestine, there are thousands of little projections called microvilli that provide a large surface area for nutrients to diffuse into the bloodstream. Name another place in the body where increased surface area helps the transfer of materials.

22. Many hospitals have their own cord blood donation programs. When women give birth, they can choose to donate the cord blood to the hospital. The hospital stores the blood and keeps it available for patients needing transplants. Imagine that you are responsible for talking to pregnant women about this program. How could you convince them that it is a good idea? (3.2) C

23. (a) Describe how the muscle tissue and the nerve tissue work together to control the respiratory system.
    (b) What parts do connective tissues and epithelial tissues play in the respiratory system? (3.11) T/I

CHAPTER 4

Plant Systems

KEY QUESTION: How do plants use the processes of cell division and differentiation to grow?

A water lily, like other plants, has a hierarchy of structures that perform various functions.

UNIT B
Tissues, Organs, and Systems of Living Things

CHAPTER 2
Cells, Cell Division, and Cell Specialization

CHAPTER 3
Animal Systems

CHAPTER 4
Plant Systems

KEY CONCEPTS

Plants have a shoot system and a root system.

The organization of tissues, organs, and systems is different in plants to that in animals.

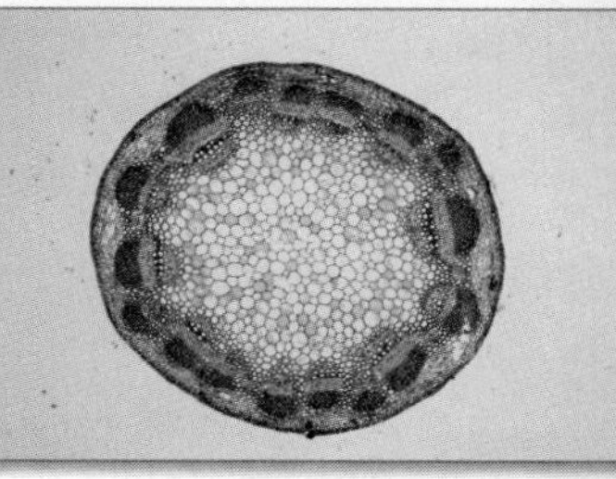
Plants have dermal, vascular, and ground tissue systems.

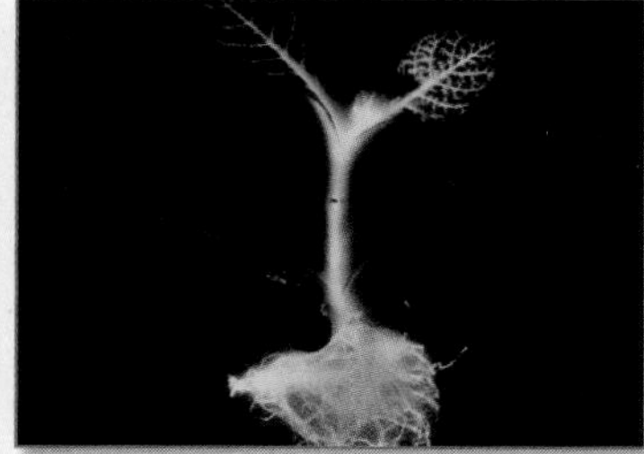
Scientists can change the genetic makeup of plants.

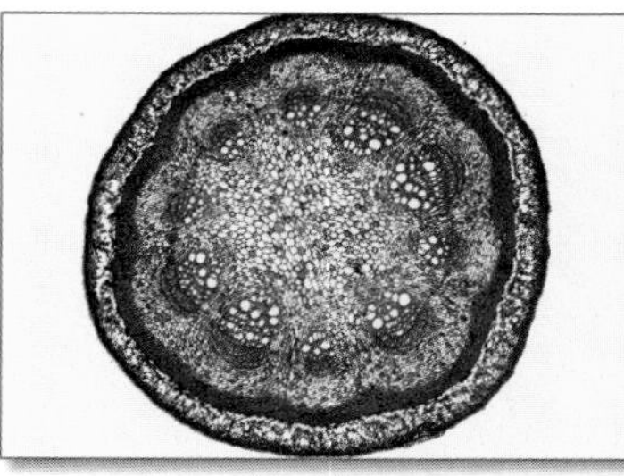
Plant tissue systems interact to perform complex tasks.

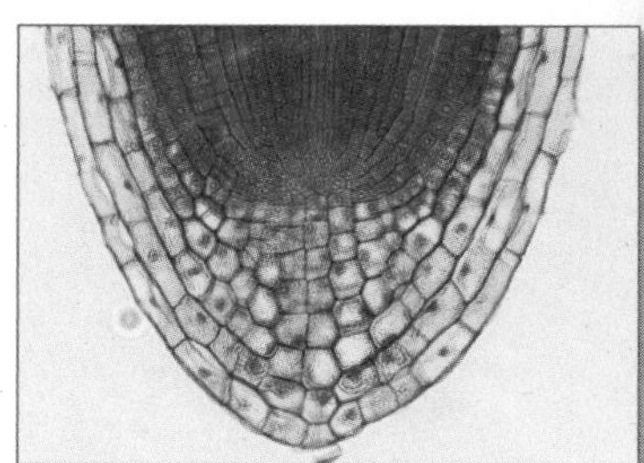
Meristems determine the pattern of plant growth.

ENGAGE IN
SCIENCE

PLANT TISSUES: SOMETHING TO CHEW ON

How would you answer if asked to explain why plants are important to us? Most people would reply that plants are important because they produce oxygen and make food. Others might mention how plants influence the climate and provide habitats for many other living things. Few people, however, understand just how important plants are in our daily lives.

We eat plants. This is our most direct use of plant tissues. Three of the most widely consumed food plants are the seeds of types of grass: wheat, rice, and corn. Of course, we also eat potatoes, lettuce, apples, peanuts, olive oil, and soy sauce. All of these food basics come from plants. Luxury plant foods include chocolate, chicle (tree sap used to make chewing gum), and coffee. Coffee beans are second only to petroleum in total dollar trading value.

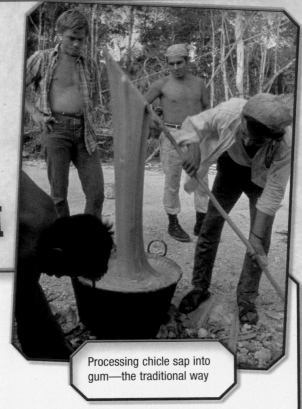

Processing chicle sap into gum—the traditional way

But what about non-food uses? Without wood from trees, this book would not exist and most homes would collapse. If the soft fibres from cotton seeds suddenly disappeared, most people on Earth would have no clothes.

How do you use plants? Consider favourite or unusual foods, musical instruments, sports equipment, and works of art. Try to write a “top 10 list” of the most unusual or most important uses of plants.

A lot of wood goes into the construction of the average Canadian house.

WHAT DO YOU THINK?

Many of the ideas you will explore in this chapter are ideas that you have already encountered. You may have encountered these ideas in school, at home, or in the world around you. Not all of the following statements are true. Consider each statement and decide whether you agree or disagree with it.

1 Plants have organs, just like animals.
Agree/disagree?

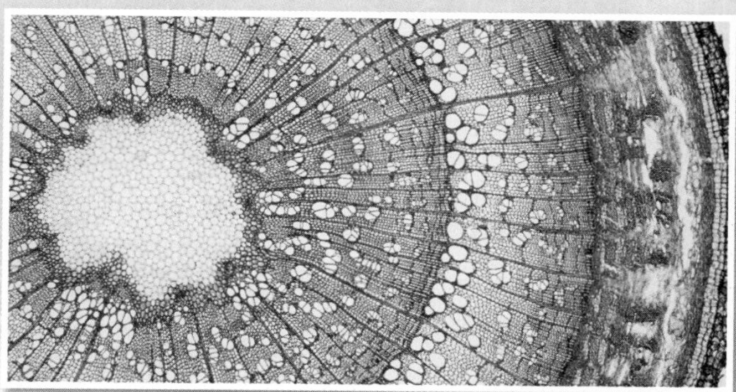

4 Plants have a circulatory system.
Agree/disagree?

2 Respiration is the same as breathing.
Agree/disagree?

5 Cell division occurs in all parts of a plant.
Agree/disagree?

3 All plants grow from seeds.
Agree/disagree?

6 Plants use water in the process of photosynthesis.
Agree/disagree?

Asking Questions

You can ask yourself questions about a text before reading to think about what you already know about the topic, during reading to work out what a text means, and after reading to think critically. You can ask and answer three types of questions:

- **Literal ("On the Lines"):** These questions are about information directly stated on the page.
- **Inferential ("Between the Lines"):** These questions require that you connect information in the text with information you already know.
- **Evaluative ("Beyond the Lines"):** These questions ask you to make a judgment based on evidence from the text.

READING TIP

As you work through the chapter, look for tips like this. They will help you develop literacy strategies.

Figure 1 Cootes Paradise is a valuable wetland at Royal Botanical Gardens. It is part of Hamilton Harbour.

For the Love of Plants

Scientists are actively engaged in restoring a large wetland called Cootes Paradise at RBG (Figure 1). They are interested in how a habitat changes over time. They also conduct research on invasive species that come from other countries and damage Canadian ecosystems. RBG works with many other conservation organizations, including McMaster University and Environment Canada. Their aim is to preserve Canada's native plant species, wild habitats, and ecosystems.

Asking Questions *in Action*

Asking questions before, during, and after reading can help you understand the text. Asking questions helps you focus on what you already know, find factual information, make inferences and form opinions. Here's how one student asked questions to understand the text:

Questions I asked before, during, and after reading the text	How that question helps me understand the text
Literal: What is RBG?	the caption for Figure 1 helps me understand that RBG stands for Royal Botanical Gardens
Inferential: Why do scientists want to study how a habitat changes?	helps me understand that scientists do research to figure out how they can improve the environment
Evaluative: How does this text compare with other texts?	helps me form an opinion about the importance of restoring wetlands because I saw a documentary about how salt water from Hurricane Katrina destroyed vegetation in marshes

Systems in Plants

Plants—including mosses, ferns, conifers, and flowering plants—are multicellular organisms. Plants have two obvious distinguishing features: they are typically green in colour, and they cannot move from place to place. Their green colour is caused by chlorophyll—a chemical that plants use to photosynthesize. They all have structures—usually roots—to anchor them firmly in one place. These two features have a profound influence on the overall structure and functioning of plants.

In this chapter we will focus on the structures in flowering plants. Other plants, such as mosses, ferns, and coniferous trees, have different structures.

Figure 1 illustrates the main features in a flowering plant. The plant body is divided into two main " body systems." This contrasts strikingly with animals, which are made up of many systems.

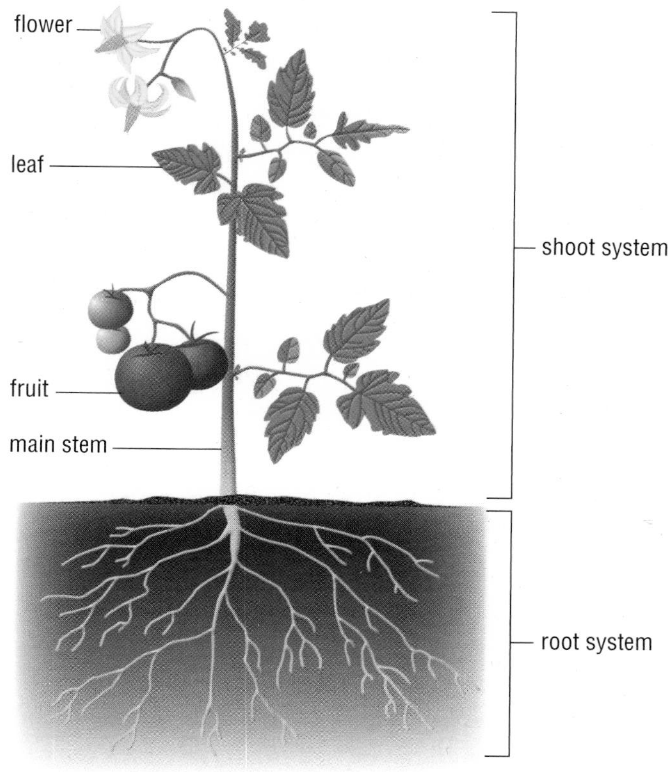

Figure 1 A typical flowering plant has two body systems: the root system and the shoot system.

Plants perform photosynthesis to make their own food. This means that they do not need to move around in search of food. As a direct consequence, they have no need for the complex and coordinated organ systems found in animals. A plant does not require a digestive or a musculoskeletal system, or a nervous system, to sense its surroundings and coordinate movements. However, plants do have to perform many of the same functions as animals.

- Plants need to exchange gases with their surroundings.
- They require an internal transportation system to move water and nutrients around within their bodies.
- They must have a way of reproducing.

DID YOU KNOW?

Canadian Researcher

Approximately 75 % of modern medicines come from plants. Dr. Memory Lewis studies the medicines used by native peoples in South America. Her research into traditional medicines has uncovered a tree sap that speeds up healing of wounds. Another plant contains compounds used to treat malaria, which affects millions of people worldwide. There is still much to be discovered about traditional plant medicines. Unfortunately, tropical jungles are being cut down for agriculture, and many plant species are disappearing before they can be properly studied.

We will now look at the organization of plant bodies and the terminology used to describe them. Because plants are so dramatically different from animals, scientists do not apply the same terminology for describing the hierarchy of organization in the plant body.

Hierarchy of Organization in the Plant Body

As Figure 1 on the previous page shows, the flowering plant body has two main body systems: the root system and the shoot system. For simplicity, scientists often just refer to them as the root and the shoot. These systems consist of a number of structures.

The root system is made up of one or more separate roots, whereas the shoot system consists of the stem, leaves, and flowers (when present). Although roots, stems, leaves, and flowers are sometimes referred to as plant "organs," scientists rarely use this terminology. They generally refer simply to plant parts.

LEARNING TIP

Systems in Plants

Plant terminology can be confusing. Plants are described as having two body systems and three tissue systems. Scientists sometimes use the term "organ" but not the term "organ system" when describing plants.

Plant parts are made up of a wide variety of specialized cells. As in animals, groups of similar specialized cells are called tissues. In addition to the plant's two main body systems, scientists also refer to their tissues as belonging to separate systems. There are three main types of plant tissue systems: the dermal tissue system, the vascular tissue system, and the ground tissue system. The **dermal tissue system** is made up of tissues that form the outer surfaces of plant parts. The **vascular tissue system** is made up of tissues specialized for the transportation of water, minerals, and nutrients throughout the plant. The tissues in the **ground tissue system** make up all of the other structures within a plant. You will learn about these tissue systems in more detail in the next section.

dermal tissue system the tissues covering the outer surface of the plant

vascular tissue system the tissues responsible for conducting materials within a plant

ground tissue system all plant tissues other than those that make up the dermal and vascular tissue systems

root system the system in a flowering plant, fern, or conifer that anchors the plant, absorbs water and minerals, and stores food

The Root System

The **root system** is the part of the plant that typically grows below ground. The root system anchors the plant, absorbs water and minerals from the soil, and stores food (Figure 2). Most of the water and minerals obtained by the plant are absorbed by root hairs: fine extensions of dermal tissue cells.

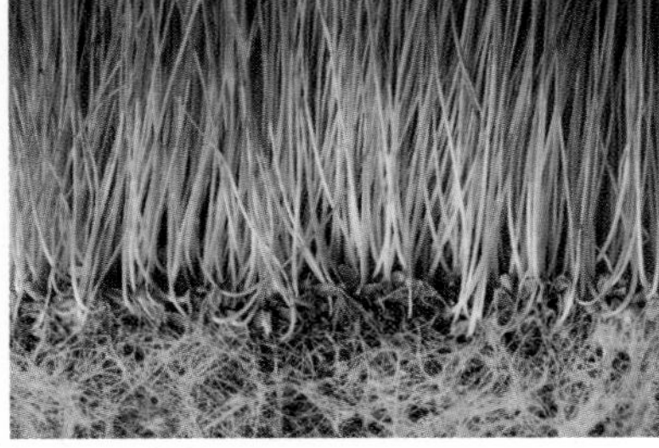

Figure 2 Roots spread down into the soil to anchor the plant and collect water.

A plant's root system can spread underground to cover a very large area. Some roots even appear above ground or above water (Figure 3(a) and (b)). Other roots, such as radishes and carrots, are specialized for nutrient storage (Figure 3(c)). Different types of plants have tremendous variation in their tissues and organs according to the environment in which they live.

Roots have long been a useful source of foods (sweet potatoes, carrots, sugar beet), flavourings (liquorice, ginger), fibres (used in basketry), and a variety of natural remedies.

Figure 3 (a) This banyan tree has aerial roots that grow downward from its branches to anchor and support the tree. (b) Some plants have specialized root extensions that grow up out of the water to obtain oxygen for the root cells below. (c) Carrot roots store food for the plant.

The Shoot System

The **shoot system** is the system that is specialized for two main functions: to conduct photosynthesis and to produce flowers for sexual reproduction. The shoot systems of flowering plants are made up of three parts—the leaf, the flower, and the stem—which all have distinctive functions.

Let us look at the parts of the shoot system in detail.

shoot system the system in a flowering plant that is specialized to conduct photosynthesis and reproduce sexually; it consists of the leaf, the flower, and the stem

The Leaf

The leaf is the main photosynthetic structure of the plant. In photosynthesis, tissues in the leaf use carbon dioxide, water, and light energy to produce glucose (a form of sugar) and oxygen. The glucose is used for plant growth, cellular respiration, and energy storage.

$$\text{light energy} + \text{carbon dioxide} + \text{water} \xrightarrow{\text{chlorophyll}} \text{glucose} + \text{oxygen}$$

The cell structure that actually performs photosynthesis is an organelle called a chloroplast. (You learned in Section 2.1 that chloroplasts are organelles that occur in plant cells but not in animal cells.) Chloroplasts contain flat, disc-like structures called thylakoids. Thylakoids are arranged in stacks called grana. These stacks act as solar collectors, using chlorophyll in the membranes of thylakoids (Figure 4).

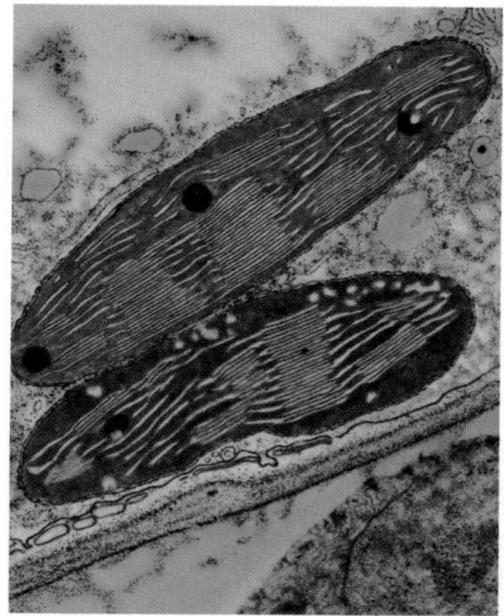

Figure 4 This light micrograph (magnification 12 000X) shows two chloroplasts (dark green) within a plant leaf cell. The grana appear light yellow in this picture.

Some leaves are also adapted for support, protection, reproduction, and attraction. Figure 5(a) shows leaves that are specialized to provide support. Can you recognize the leaves of the plant in Figure 5(b) and what appear to be flower petals of the plant in Figure 5(c)?

We use leaves in a variety of ways. Many are edible, such as lettuce, spinach, onion, tea, and herbs. Others are sources of waxes and medicines. In the agriculture industry, leaves are a major source of nutrition for livestock.

You will have a chance to examine leaves in greater detail in Section 4.4.

Figure 5 (a) The tendrils on this cucumber plant are modified leaves that help support the growing plant. (b) The sharp spines on this cactus are actually modified leaves that protect the plant from herbivorous animals. (c) Some poinsettia leaves are red, but these are not part of the flower. The true flower consists of the yellow-green parts in the centre of the red leafy region.

The Flower

Flowers are specialized structures developed for sexual reproduction. They contain male or female reproductive structures, or sometimes both. Male reproductive structures produce pollen grains; female reproductive structures produce eggs. Eggs are fertilized by pollen. After pollination, the female flower parts form seeds. In most flowering plants, the seeds are contained within a specialized structure called a fruit.

READING TIP

Asking Questions

Sometimes information in a text makes you ask an evaluative question. For example, you might wonder about plants that are pollinated by the wind and ask, "If there is little wind when the pollen is ready to be spread, will grasses and trees not be pollinated that year?" Your evaluative question might prompt you to conduct further research on the issue.

Figure 6 (a) Grasses rely on wind for pollination. (b) Other plants produce nectar and colourful flowers to attract insects that help with pollination.

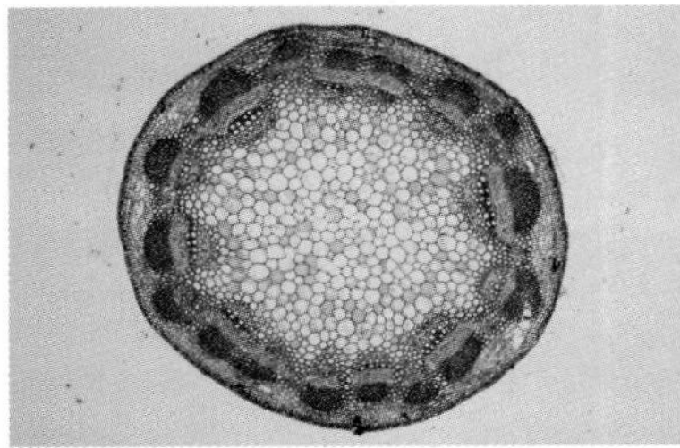

Figure 7 Cross-section through a sunflower stem. The brownish clusters of cells are the plant's transportation system.

Pollination occurs with the help of wind or animals. Flowering plants such as grasses and many common tree species are wind pollinated. The flowers are small and drab but produce large amounts of pollen (Figure 6(a)). In contrast, other flowering plants are pollinated by animals, such as insects, bats, and birds. These plants often have large, colourful, and fragrant flowers to attract their would-be pollinators (Figure 6(b)). Most also produce nectar as an added form of attraction.

Flowers, and the seeds and fruits that come from them, are very important sources of food and flavourings. Rice, wheat, corn, vanilla, chocolate, coffee, bananas, apples, mangos, cotton, and even some medicines all originate from flowers.

Mosses, ferns, and coniferous trees are not flowering plants. They have different sexual reproductive systems. Conifers are all wind pollinated and, instead of having flowers, produce pollen and seeds in specialized cones. In this chapter we will concentrate on flowering plants.

The Stem

The flowering plant stem (or trunk, in the case of trees) has several functions. It supports the branches, leaves, and flowers and provides a way to transport materials. The stem contains significant amounts of vascular tissue for carrying substances to and from roots, leaves, flowers, and fruits (Figure 7).

Some stems are specialized for food storage, protection, photosynthesis, and reproduction. Plant stems provide us with sugar cane, potatoes, wood and paper products, cork, linen, and a variety of medicines.

IN SUMMARY

- Flowering plants have two body systems: the root system and the shoot system.
- The function of the root system is to anchor the plant, absorb water and minerals from the soil, and store food.
- The shoot system of flowering plants is made up of the stem, leaves, and flowers (when present).
- The two main functions of the shoot system of flowering plants are to conduct photosynthesis and produce flowers for sexual reproduction.
- Leaves are mainly responsible for photosynthesis.
- The various parts of plants work together to perform all of the functions necessary to keep the plant alive.

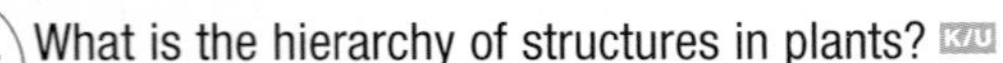

CHECK YOUR LEARNING

1. What is the hierarchy of structures in plants? K/U
2. (a) What are the two main differences between plants and animals?
   (b) How do these differences suggest the need or lack of need for various organ systems? K/U
3. Name the three tissue types in plants and briefly describe them. K/U
4. Compare the functions of the leaf and the stem in plants. K/U
5. Compare the functions of the stem and the root in plants. K/U
6. (a) Describe the primary function of a flower.
   (b) Describe two ways in which different plants carry out this function. K/U
7. How can roots, stems, and leaves all be involved in food storage? Give examples in your answer. T/I

Plant Tissue Systems

Plant tissues are classified into three tissue systems. Each tissue includes a variety of specialized cell types that enable them to carry out specific functions within the plant body. As you will learn, some cells are specialized to perform photosynthesis just below the surface of leaves, whereas others are specialized to absorb water. Some specialized plant cells even act as doors, literally opening and closing to control the entry and exit of gases! Just as in animals, these specialized cells develop from unspecialized cells during the process of cellular differentiation.

Cellular Differentiation and Specialization in Plants

When a seed starts to grow, the cells divide very quickly. As the seed develops into an embryo, many of the cells start to differentiate into specific tissues (Figure 1). The regions of growth are located at the tips of the plant's roots and shoots. Plants with woody stems, such as trees, also have a region of growth just below the surface of their stems. Plant growth occurs because undifferentiated cells are actively growing and dividing in these regions. As these new cells mature, some of them specialize and develop different features according to their location and future function. For example, cells in leaves are very different from those in the roots and stem of the plant.

DID YOU **KNOW?**

Clean as a Lotus Leaf

The surfaces of some plants are of great interest to scientists and engineers. The leaves of the lotus plant have slippery surfaces that materials simply cannot stick to. As a result, they are always extremely clean. Engineers are trying to learn how to use this living lotus technology on solar panels and other exposed surfaces so they never need cleaning.

Meristematic Cells

Just as animals contain some unspecialized cells called stem cells, plants have unspecialized cells that are called **meristematic cells**. Meristematic cells can differentiate into specialized tissue types (Figure 2).

meristematic cell an undifferentiated plant cell that can divide and differentiate to form specialized cells

Figure 1 A germinating runner bean seed, showing the shoot and root

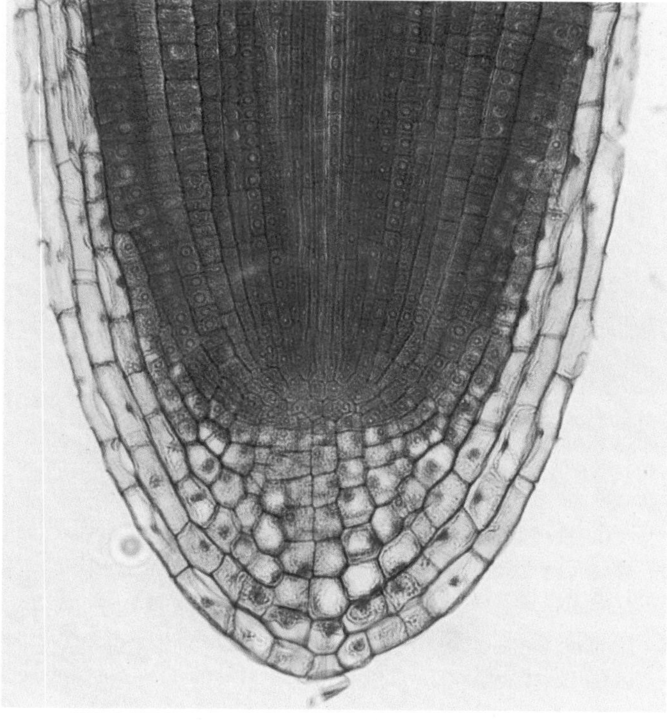

Figure 2 There are regions in all plant roots and stems, near the tips where rapid growth occurs, that contain meristematic cells.

Think back to Activity 2.6, in which you looked at the phases of mitosis in plant and animal cells. The plant cells you examined were from the meristem region in an onion root tip. The meristematic cells in this region divide and differentiate into specialized tissues in the roots. Similarly, the tissues of new stalks, leaves, and flowers arise from meristem regions in the shoot. As cells in the meristem regions continue to divide, some of the new cells always remain in the undifferentiated state.

Tissue Systems in Plants

As you learned in Section 4.1, plants have three major tissue systems: dermal, ground, and vascular. Each of these three tissue systems contains different types of specialized cells. As you can see in Figure 3, the tissues of the three systems are found in all parts of the body and have a distinct arrangement. The dermal tissues cover the entire outer surface of the plant but do not occur within the plant. The vascular tissues are found within every root, shoot, and leaf of the plant. The vascular tissue system is continuous, so all plant parts are joined together by vascular tissues. The ground tissues account for all other internal tissues in the root, stem, and leaves.

READING TIP

Asking Questions

Use sticky notes while reading to record your questions beside the part of the text you find confusing. After reading you can check to see if the rest of the selection, including the graphs, illustrations, and captions, answered your questions.

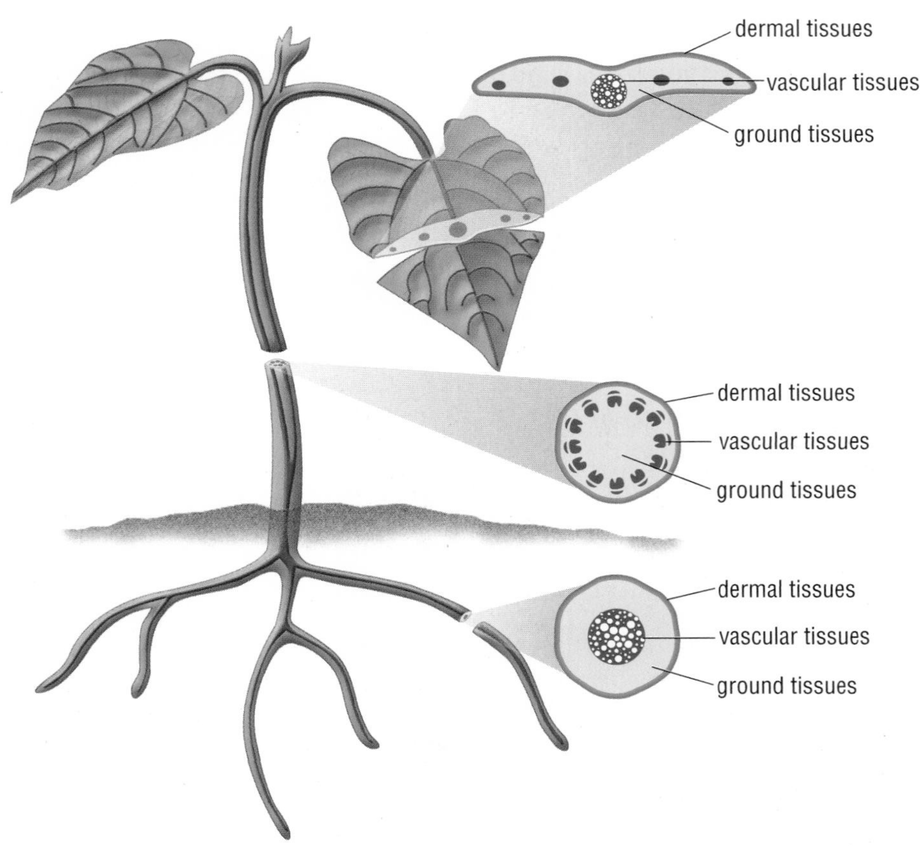

Figure 3 General arrangement of plant tissues. The vascular tissues are shown in red, the dermal tissues in green, and the ground tissues in light blue.

Dermal Tissue System

The dermal tissue system forms the outermost layer of a plant. The dermal tissue system includes both epidermal and periderm tissues. **Epidermal tissue (epidermis)** is a thin layer of cells that covers the surfaces of leaves, stems, and roots. In woody plants, the epidermal tissue is replaced by **periderm tissue** that forms bark on stems and large roots.

The cells of the dermal tissue system are specialized to perform a wide variety of unique functions. Some epidermal root cells have long extensions called root hairs to help absorb water and minerals from the surrounding soil. Most epidermal leaf cells produce a layer of wax, called the cuticle, that helps waterproof the leaf's surface (Figure 4). Some epidermal leaf cells are adapted for defence: they have hairlike structures that contain chemical irritants. Anyone who has experienced the discomfort of touching a plant called stinging nettle knows just how irritating it can be (Figure 5).

epidermal tissue (epidermis) a thin layer of cells covering all non-woody surfaces of the plant

periderm tissue tissue on the surface of a plant that produces bark on stems and roots

DID YOU KNOW?

Dental Hygiene and the Rainforest
Chewing sticks are used by many people instead of toothbrushes. The neem tree grows in many Asian, African, and Middle Eastern countries. Indigenous people in these areas know that chewing neem twigs is good for their teeth. The epidermal tissues contain chemicals that kill bacteria and reduce inflammation. The study of how indigenous (native) plants are used by different cultures is known as ethnobotany.

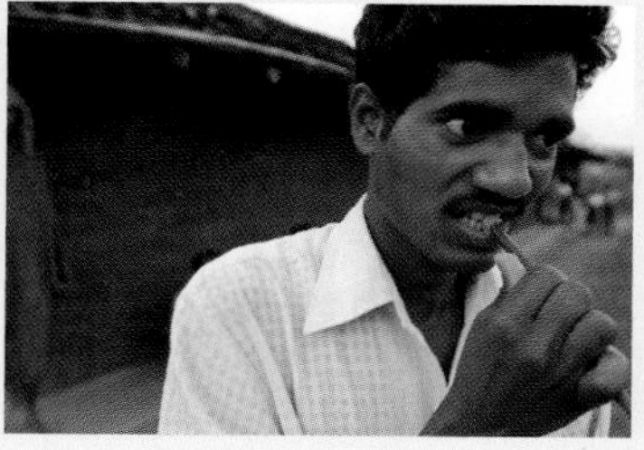

GO TO NELSON SCIENCE

Figure 4 The wax produced by the epidermis keeps a plant from losing too much water. It can also be harvested and used by humans. Carnauba wax, from the carnauba palm tree, is the hardest known wax.

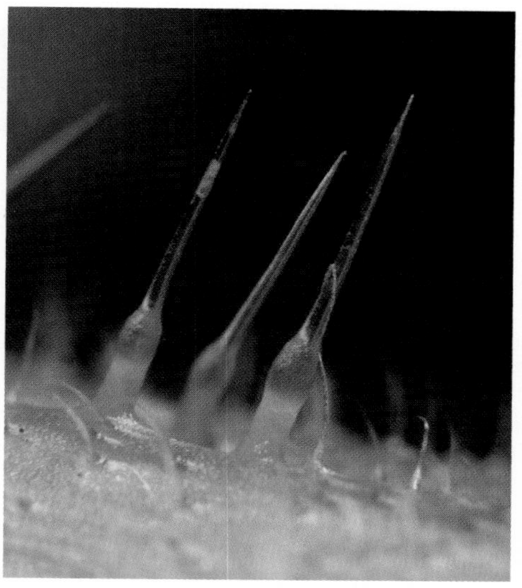

Figure 5 Fine hairlike structures on this stinging nettle leaf inject an irritating chemical into the skin of any curious animal. This adaptation helps protect the plant from being eaten.

Vascular Tissue System

A plant must obtain from the soil all the water and minerals that it needs for growth. These substances are absorbed by the plant's roots. The water and nutrients then have to be distributed to all of the plant's cells. In addition, plants must be able to distribute the sugars and other chemicals, produced during photosynthesis, to all of its parts. The plant's vascular tissue system is the transportation system that moves water, minerals, and other chemicals around the plant. It is like a network of tubes that reaches from the roots, up the stalk, and through the leaves. As the plant grows, new vascular tissues differentiate in the growing tips of the plant. The new tissues are lined up with existing vascular tissues, maintaining the tubes' connections throughout the plant.

xylem vascular tissue in plants that transports water and dissolved minerals from the roots to the leaves and stems of the plant

There are two types of vascular tissue: xylem and phloem. **Xylem** (pronounced *zye-luhm*) is made up of elongated cells. Xylem transports water and dissolved minerals upward from the roots. This solution travels along an interconnected network to the stems and leaves.

Once they are fully formed, xylem cells are little more than hollow tubes with rigid walls (Figure 6). They have no cytoplasm, nucleus, or other organelles. This allows water to move easily through the tubes. Mature xylem cells are no longer living tissue.

phloem vascular tissue in plants that transports dissolved food materials and hormones throughout the plant

Phloem (pronounced *flo-uhm*) is a specialized tissue that transports solutions of sugars produced by photosynthesis. Phloem also transports other dissolved nutrients and hormones throughout the plant. Under certain conditions, food materials are transported downward from the photosynthesizing leaves to the stem and roots. Under other conditions, the food materials are carried upward from the root and stem to the leaves. Phloem is made up of elongated cells that, unlike xylem cells, are alive when mature and functioning.

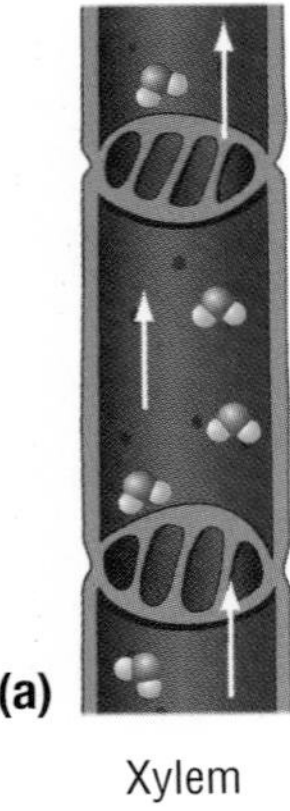

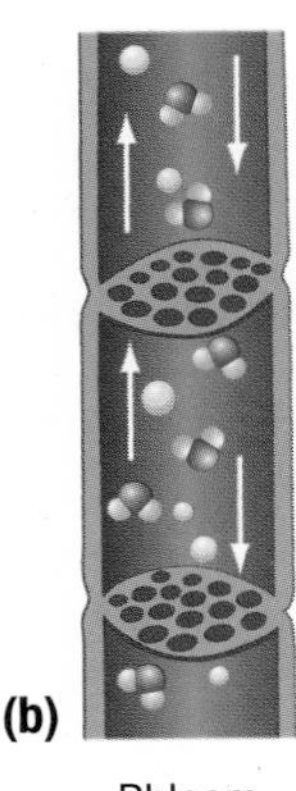

Figure 6 (a) In xylem tissue, water carries dissolved minerals upward from the root to the shoot through dead, hollow cells. (b) In phloem tissue, dissolved sugar and other nutrients are moved around the plant through living cells.

The arrangement of vascular tissues within the stems of plants differs dramatically between woody and non-woody plants. In non-woody plants, the vascular tissues are arranged in bundles and accompanied by ground tissues (Figure 7). You will find out about the unique arrangement and growth of vascular tissues in woody plants in Section 4.6.

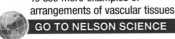

To see more examples of arrangements of vascular tissues,

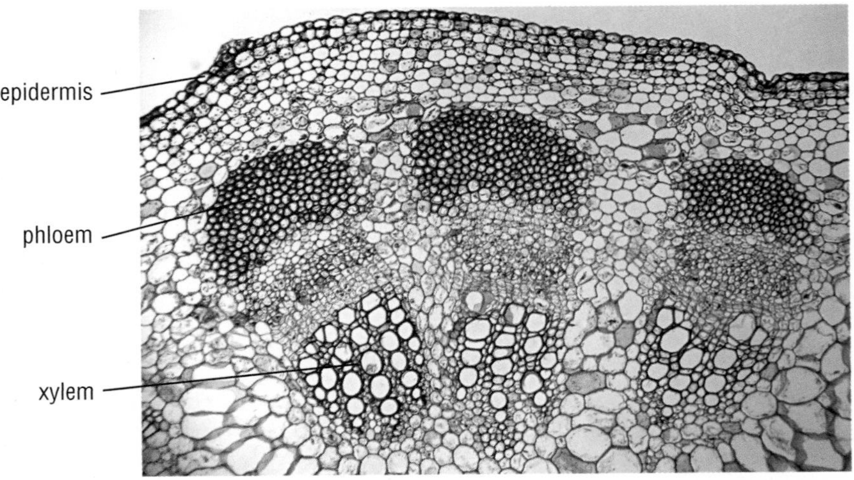

Figure 7 Vascular tissues in the stems of non-woody plants are arranged in bundles.

All plants and animals use sugars to provide the energy necessary for cell processes. This is called cellular respiration, and requires oxygen. Unlike animals, however, most plant tissues do not need a special supply of oxygen. Leaves produce their own oxygen during photosynthesis. Most stem and root tissues get enough oxygen by diffusion from the surrounding air and air spaces in the soil. One exception to this is the water lily (see page 120). This plant actually pumps oxygen from leaves floating on the surface of the water down through vascular tissues to supply the roots buried in the mud.

Ground Tissue System

Most of the cells of young plants are ground tissue cells. Ground tissues are the filler between the dermal and the vascular tissues. Ground tissues perform a variety of functions: in the green parts of the plant, they manufacture nutrients by the process of photosynthesis; in the roots, they store carbohydrates; and in the stems, they provide storage and support.

You will explore some of the functions of specialized ground tissues when you examine the leaf in more detail in Activity 4.5.

RESEARCH THIS WHEN PLANTS GET SICK

SKILLS: Defining the Issue, Researching, Identifying Alternatives

SKILLS HANDBOOK 4.A., 4.C.

Plants, like all living things, are subject to a wide variety of diseases. Plant diseases can kill or injure a plant, reducing its value in a number of ways. Injured or dying plants may have less economic value, less aesthetic value, or less ecological value.

GO TO NELSON SCIENCE

1. Find out the meanings of the terms "economic value," "aesthetic value," and "ecological value."
2. Select a plant disease and research answers to the following questions.
   - What is the cause of the disease?
   - How does the disease affect the plant?
   - How does the disease affect the economic, aesthetic, and/or ecological value of the plant?
   - How can the disease be controlled, prevented, and/or treated?

A. Present your findings in an informative and creative way. You could make a brochure, audiovisual presentation, or a public information "news story" video about the disease. T/I C A

IN SUMMARY

- Meristematic cells are undifferentiated plant cells that can form any kind of specialized tissue.
- Meristematic cells are located at the tips of roots and shoots and in the stems of woody plants.
- Plants have three tissue systems: dermal, vascular, and ground.
- Xylem and phloem are vascular tissues that are responsible for transporting water, minerals, hormones, and nutrients within the plant.

CHECK YOUR LEARNING

1. What is the purpose of cell division in plants? K/U
2. What are the main functions of the three plant tissue systems? K/U
3. Compare the structure and function of xylem and phloem. K/U
4. Give some examples to show how specialized plant cells perform specific functions. K/U
5. Explain why it is important for plant leaves to be waterproof. K/U
6. Explain why the movement of water and minerals in xylem is always upward. K/U

Transgenic Plant Products

SKILLS MENU

- Defining the Issue
- Researching
- Identifying Alternatives
- Analyzing the Issue
- Defending a Decision
- Communicating
- Evaluating

We use plants in many different ways. On the most basic level, we need plants to produce food and oxygen and to remove carbon dioxide from the air. We rely on plants to maintain the right conditions on Earth for us to survive. We also need them to provide the wood pulp for our textbooks, the timber used to build our homes, the cotton in our clothing, and even many of the medicines we use.

Humans have always harvested the natural products of plants. A few thousand years ago, humans began selecting which plants they grew, choosing the strongest, the most drought-resistant, or those that produced the sweetest fruit or the most grains (Figure 1). Over the centuries, this artificial selection has changed the plants. Modern apples, for example, are much bigger and sweeter than the apples that grew wild thousands of years ago. Until very recently, plants only produced their own naturally occurring sugars and chemicals. Now, however, with the aid of genetic engineering, we can use plants to produce substances that are normally found in entirely different organisms.

Figure 1 By selecting and growing seeds from the best plants, humans have gradually changed how the plants grow. Corn has been bred selectively for thousands of years.

Genetic engineering is a technology in which genetic material is taken from one organism and inserted into the genetic material of a different organism. For example, DNA might be taken from a fish and introduced into a plant. This transplanted genetic material contains the information needed to produce substances normally made by the original organism. The aim is that the new "transgenic" plant will use this genetic information to produce the desired compound (Figure 2). Organisms that contain DNA from other species are sometimes called genetically modified organisms: GMOs.

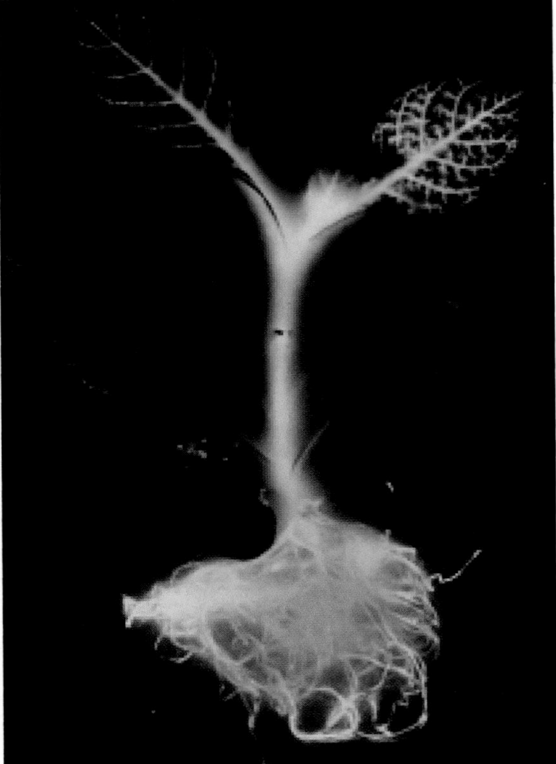

Figure 2 Because this plant contains some genetic information from a firefly, it is able to produce GFP (green fluorescent protein). The protein glows and helps scientists uncover the inner workings of plant cells.

Research scientists are working on a wide range of transgenic modifications. The research may produce plants that:

- are resistant to pests, diseases, and adverse growing conditions such as drought or frost
- manufacture vaccines and other valuable medicines
- have greater nutritional value
- are able to absorb more carbon dioxide from the air
- manufacture valuable biological materials such as spider web protein

We already grow transgenic plants in Canada (Figure 3). There are several varieties of crop plants (such as corn, soybeans, and canola) that have been modified to make them resistant to certain herbicides. This allows the farmer to spray the entire field with a chemical that will kill all the weeds but leave the crop plants unharmed.

Figure 3 A field of transgenic canola

Other plants (such as corn and potatoes) have been given bacterial DNA that kills caterpillars. This reduces the need for pesticides, thus saving farmers money and reducing the quantity of pesticides added to the environment.

The Issue

The creation of transgenic plants is controversial. Many people believe that genetically engineered plants could alter the natural ecology of the planet. Others argue that this technology has the potential to provide huge benefits to society and the environment.

Goal

To assess the potential risks and benefits of one type of transgenic plant and prepare a set of recommendations based on your findings.

Gather Information

Use the library and online resources to gather information on the possible uses of transgenic plants. Choose a specific transgenic plant that interests you. Research to find out what organism the transplanted DNA was taken from. How might the transgenic plant benefit people and the environment? What are the possible risks of growing this transgenic plant? What should farmers consider if they are thinking of growing this plant? As you collect information, evaluate its reliability and bias.

GO TO NELSON SCIENCE

READING TIP

Asking Questions

Sometimes missing information in a text makes you ask an inferential question. For example, you might wonder about whether transplanting DNA from a fish to a plant is safe. The text does not say that genetic engineering is foolproof. But you have read an article about possible negative consequences of genetic engineering. By combining what the text does not say with what you already know, you end up asking an inferential question.

Identify Solutions

Consider the challenges and risks associated with developing the transgenic plant you selected. Identify possible ways of overcoming these challenges and reducing or eliminating these risks.

Make a Decision

Prepare a set of recommendations regarding your chosen plant. Should this plant be approved for use? If so, should there be any regulations on how, when, or where it is grown? Who should establish the regulations?

Communicate

Discuss your presentation options with your teacher. Present your recommendations to the class.

4.4

Tissues Working Together

The leaves of most plants are highly specialized structures with one primary function: performing photosynthesis. In this section, we will consider what leaves need to photosynthesize. You will look at the various ways in which the tissue systems of the leaf work together to perform this vital function.

Photosynthesis is the process that enables plants to build glucose from simple molecules using light energy, usually sunlight. Let us begin by reviewing the process of photosynthesis. The word equation for this reaction is

$$\text{light energy} + \text{carbon dioxide} + \text{water} \xrightarrow{\text{chlorophyll}} \text{glucose} + \text{oxygen}$$

This equation tells us that the leaf uses light energy, carbon dioxide, and water to produce glucose and oxygen. Glucose can then be converted into other carbohydrates, including complex sugars, in the leaf. These carbohydrates in the form of starch are used as a source of stored chemical energy. Alternatively, the plant can use them as the basic building blocks for all of the complex chemicals and structures that it needs to make.

Sugar is a valuable product needed by all parts of the plant. The plant therefore has to have some system to transfer sugars from the leaf to the rest of the plant. Both glucose (a form of sugar) and oxygen are required in all plant cells for cellular respiration. (Recall that cellular respiraton is the release of energy from glucose.) However, leaves produce more oxygen than the plant needs, so excess oxygen is released as a waste product.

Let's examine in more detail how plants obtain the three factors required for photosynthesis: energy (sunlight), carbon dioxide, and water.

READING TIP

Asking Questions

If you do not understand something while reading a text, stop and try to ask yourself a question about it. Try rereading the previous sentence to see if you missed some information that might answer your question. Another strategy is to continue reading to see if the information in the next few sentences or paragraphs answers your question.

Absorbing Light

As you know, most leaves are green and thin (Figure 1). Both of these features are ideal for absorbing light. Wide, thin leaves have a much greater surface area than thick, narrow leaves. This enables them to absorb more light and accounts for why this leaf shape is extremely common. The green colour is produced by chlorophyll—the pigment that actually absorbs light to begin the process of photosynthesis.

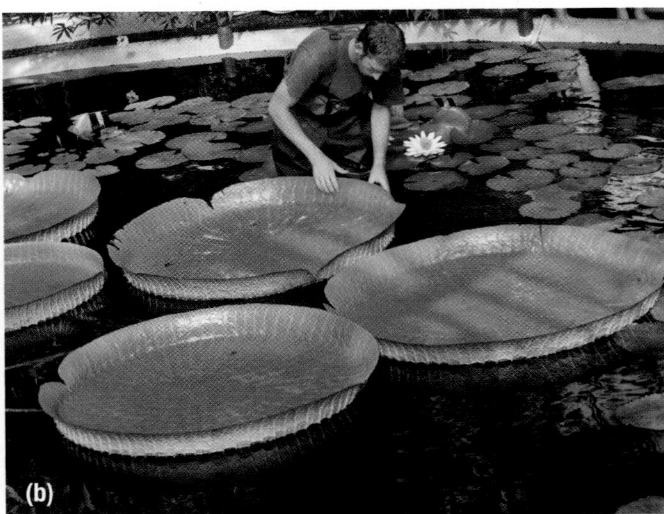

Figure 1 (a) This plant is sometimes called the "poor man's umbrella." It grows in the rainforests of Panama. (b) The leaves of the Victoria water lily are well over a metre in diameter.

As you learned in Section 4.1, chlorophyll is contained in cell organelles called chloroplasts. It is these organelles that actually conduct the chemical process of photosynthesis. In a typical leaf (Figure 2), the chloroplasts are located mostly in the **palisade layer** and the **spongy mesophyll**. The palisade cells are located where there is the maximum amount of light: just below the leaf's upper surface. Spongy mesophyll cells are located throughout the interior of the leaf. The palisade and spongy mesophyll cells are part of the ground tissue system. You may have noticed in Figure 2 that the palisade cells are packed closely together, whereas the spongy mesophyll cells are more loosely packed. The close packing helps palisade cells capture as much incoming light as possible, whereas the spaces between the spongy mesophyll cells permit gases to move around within the leaf. All of these photosynthesizing cells require a supply of carbon dioxide and water.

palisade layer a layer of tall, closely packed cells containing chloroplasts, just below the upper surface of a leaf; a type of ground tissue

spongy mesophyll a region of loosely packed cells containing chloroplasts, in the middle of a leaf; a type of ground tissue

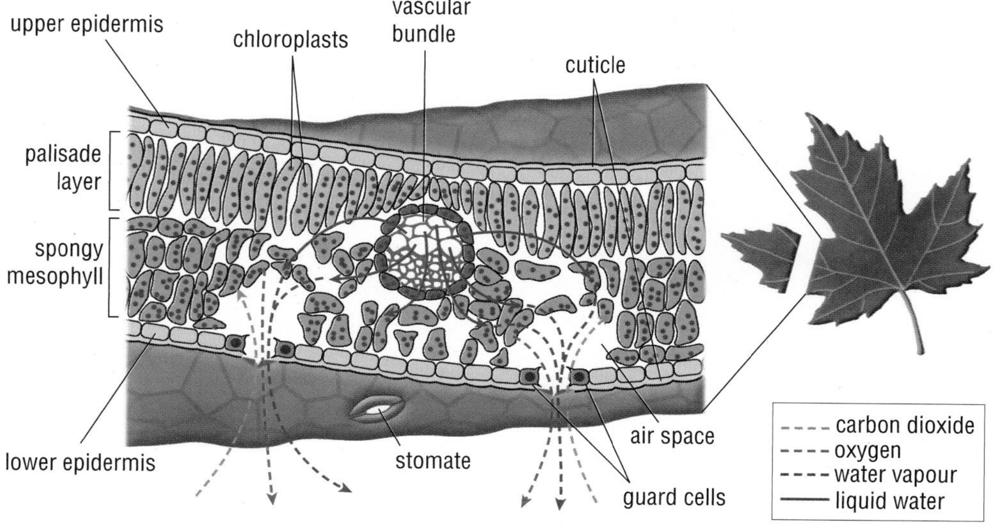

Figure 2 The specialized tissues and cells in a leaf

Obtaining Carbon Dioxide

How does carbon dioxide enter leaves? Carbon dioxide gas is present in the atmosphere in low concentrations. As you can see in Figure 2, the entire upper and lower leaf surfaces are covered by a layer of epidermal tissue (epidermis). The cells that make up this tissue produce a thin layer of wax called the **cuticle**. This waxy layer keeps the leaf from drying out. This same cuticle also prevents gases from entering the leaf by direct diffusion through the surface cells. Instead, gases enter and exit through special openings in the leaf surface called **stomata** (singular: stomate). These openings are surrounded and controlled by pairs of special epidermal **guard cells** (Figure 3). The guard cells can either bend outward, causing stomata to open, or they can collapse inward, causing stomata to close.

The majority of plants have most or all of their stomata in the lower surface of the leaf. This location reduces water loss, provides more surface area for photosynthesis, and reduces the chances of airborne viruses, bacteria, and fungal spores from entering the leaf.

cuticle a layer of wax on the upper and lower surfaces of a leaf that blocks the diffusion of water and gases

stomate (plural: stomata) an opening in the surface of a leaf that allows the exchange of gases

guard cell one of a pair of special cells in the epidermis that surround and control the opening and closing of each stomate

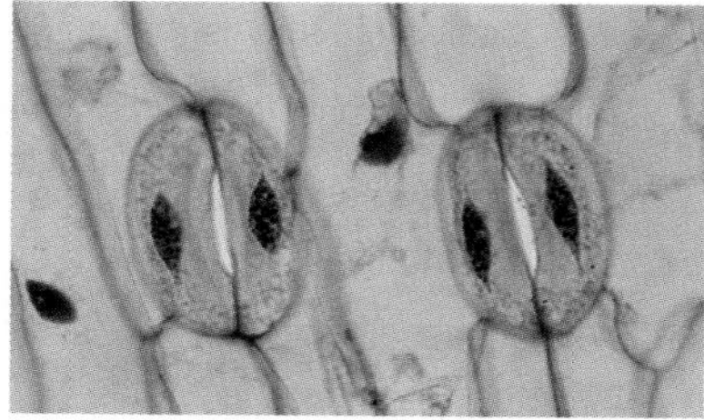

Figure 3 Two stomata, each surrounded by a pair of guard cells

Once carbon dioxide enters the leaf through the stomata, it can move around through the air spaces in the spongy mesophyll. This enables carbon dioxide to more easily reach the photosynthetic cells within the leaf.

Controlling Stomata

When and why do plants open and close their stomata?

When stomata are open, carbon dioxide can enter the leaf and oxygen can escape. This helps the plant photosynthesize. Ideally, plants would open their stomata whenever it was sunny. However, when stomata are open, water vapour can also escape. A very thin leaf could dry out and die very quickly on a sunny or windy day. Preventing too much water loss is therefore a major concern for many plants.

Guard cells are an adaptation to help the plant conserve water by altering their shape in response to water levels in the leaf. If there is a good supply of water within the leaf, the guard cells expand and bend apart—opening the stomata. If there is a shortage of water, the guard cells become soft and collapse—closing the stomata. Guard cells also have a mechanism that responds to light levels. This lets them close the stomata at night when carbon dioxide is not needed because there is no light for photosynthesis.

LEARNING TIP

Door Cells

It may help you to remember the roles of guard cells and stomata if you think about guard cells as doors and stomata as doorways. You walk through doorways just like gas moves through stomata. You close these openings with doors, like a leaf closes the openings with guard cells.

TRY THIS YOUR OWN GUARD CELLS

SKILLS: Performing, Observing, Analyzing, Evaluating, Communicating

SKILLS HANDBOOK 3.B.

In this activity you will make a model of guard cells to see how they affect the size of the stomate.

Students who are allergic or sensitive to latex should avoid all contact with the balloons.

Equipment and Materials: 2 long balloons; masking tape

1. Partially inflate two long balloons. Hold the ends of the balloons shut.
2. Have a partner place several strips of tape lengthwise along one side of each balloon.
3. Blow some more air into each balloon. Hold the two balloons side by side, with the two taped portions facing one another.
4. Release some air from the balloons until the two balloons lie flat against one another.

A. Complete two labelled diagrams of your balloons at Step 3 and Step 4. C

B. Which diagram represents two guard cells filled with water, and which represents the guard cells with less water in them? T/I

C. What does the space between the balloons represent? T/I

D. What effect does the masking tape have on the shape of the completely filled balloons? T/I

E. The balloons are a model of the guard cells in a leaf. How good of a model are they? A

Obtaining Water

Leaves reduce water loss with a waxy cuticle and by closing their stomata when water levels are low. How do leaves obtain water in the first place? Vascular tissues made up of xylem and phloem are arranged as bundles and run through the plant from root to leaf. Long, thin epidermal cells on the roots, called root hairs, grow into the surrounding soil and absorb water by the process of osmosis. The root hairs greatly increase the surface available for water absorption, allowing the osmosis to happen very quickly. The water is then transported by xylem from the roots, up the stem, and through the leaves.

Comparing Plant and Animal Systems

Plants and animals are very different organisms. At the cellular level, however, there are similarities. Plant and animal cells both need to carry out some of the same processes. Plants cells, like animals cells, respire. In this process, they use oxygen to "burn" glucose, releasing its energy for other chemical processes in the cells. As well as producing energy, this reaction produces carbon dioxide and water.

In some ways, the fluid circulation systems in plants and animals are similar. Vascular bundles are similar to the veins and arteries in animals (Figure 4). There is one important difference, however. In animals, the blood is actively pumped around the body by the heart; in plants, liquids move around passively. Plants have no organ equivalent to a heart.

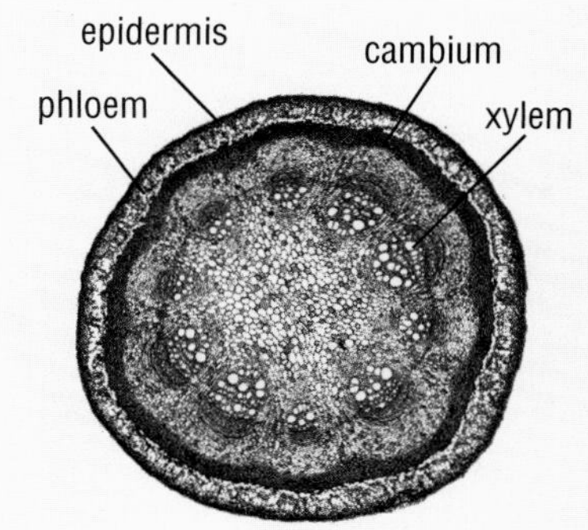

Figure 4 Vascular bundles transport water and other substances around the plant. They can be compared to veins and arteries in an animal.

In Chapter 3, you learned how different animal organ systems work together to accomplish complex tasks. Table 1 compares how animal and plant systems accomplish a similar task.

Table 1 A Comparison of Animal and Plant Tissue Systems

Task	Animal organ systems	Plant tissue systems
Obtaining food and transporting it within the body	• The nervous and musculoskeletal systems are involved with obtaining food. • The digestive system processes food. • In the circulatory system, blood vessels transport all digested food nutrients.	• The entry of necessary carbon dioxide is controlled by cells in the dermal tissue system. • The ground tissue system in leaves produces the plant's own food (sugars). • The vascular system transports sugars and other complex compounds in phloem.

IN SUMMARY

- A plant leaf is made up of a system of tissues, each with its own specific structure and function.
- Ground tissue in a leaf is largely responsible for photosynthesis and fills the space between the dermal layers and the vascular tissues.
- The leaf epidermis contains many tiny openings called stomata, which allow gas exchange and the release of water vapour.
- Special dermal cells known as guard cells surround and control each stomate in a leaf.
- Both animal and plant cells use sugar and oxygen in the process of respiration.
- Animals and plants both possess systems that must work together to accomplish complex tasks.

CHECK YOUR LEARNING

1. (a) Do you find it confusing to compare plant structures with animal structures?
   (b) Determine how your classmates feel about this comparison.
   (c) Discuss the matter as a class with your teacher. K/U T/I C
2. Explain how each region of the leaf contributes to the task of photosynthesis. K/U
3. Why does the middle of the leaf (the spongy region) have loosely arranged cells? K/U
4. (a) In what ways do the cuticle and guard cells perform the same function?
   (b) In what ways do their roles differ? K/U
5. Compare the functions and arrangements of cells in the palisade and spongy mesophyll layers. K/U
6. How do the shape and colour of the leaves in Figure 1 on page 136 help the leaf accomplish its primary function? K/U
7. Briefly describe how guard cells are able to control stomata. K/U

Plant Cells and Tissues

In this activity, you will make your own slide of a plant tissue. You will also observe prepared slides of other plant tissues. For each slide, you will complete a labelled diagram highlighting the key features of the cells and tissues that you see. You will consider the differences between the tissue samples.

SKILLS MENU

- Questioning
- Hypothesizing
- Predicting
- Planning
- Controlling Variables
- Performing
- Observing
- Analyzing
- Evaluating
- Communicating

Purpose

Part A and B: To examine different plant tissues.
Part C: To compare the differences between the structures of the tissues and relate these structures to their functions.

Equipment and Materials

- lab apron
- disposable gloves
- scalpel or sharp knife
- cutting board or tray
- hand lens (magnifying glass)
- microscope slide
- light microscope
- celery or prepared slide of vascular bundles in a celery stalk
- dropper bottle of iodine solution
- prepared slide of leaf cross-section
- complete plant with roots, stems, and leaves, such as a radish or beet

Be careful when using sharp objects such as scalpels or knives. Always cut downward on a cutting board or tray on the lab bench.

Iodine solution is toxic and can stain skin and clothing. Use disposable gloves, handle the solution with care, and wash your hands afterwards.

Procedure

SKILLS HANDBOOK
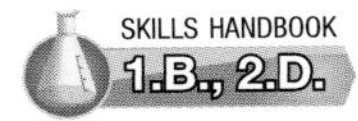

Part A: Cross-Section of a Celery Stalk

1. Put on your lab apron and gloves.
2. Obtain a piece of celery.
3. On the cutting board, use a scalpel or sharp knife to cut across the stalk (Figure 1).

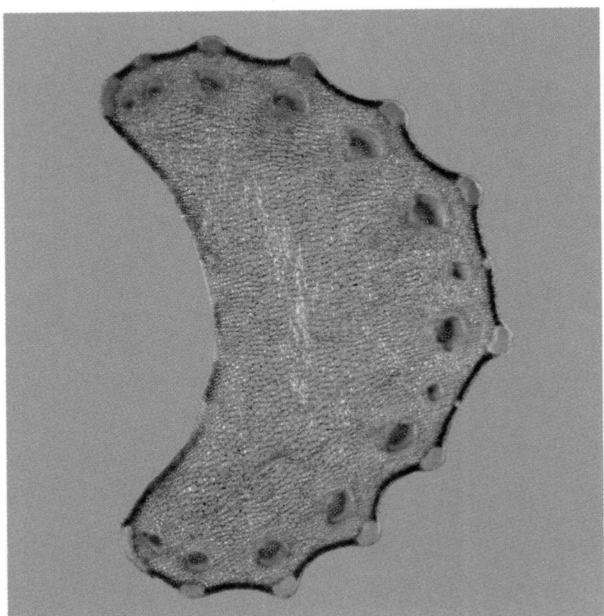

Figure 1 Thin slice of celery stalk

4. Use the hand lens to observe the cut section. Draw a diagram of your observations.
5. Carefully add a drop of iodine stain to the cut section and use the hand lens to carefully observe the vascular bundles.
6. Cut the thinnest slice possible from the stained celery stalk. Place this on a clean microscope slide and observe using low power. Draw a diagram of your observations indicating the arrangement of the vascular bundles.

Part B: Cross-Section of a Leaf

7. Obtain a prepared slide of a leaf cross-section.
8. Using low power, observe the organization of cells in the leaf. Centre the slide and focus using the coarse-adjustment knob. Rotate the nosepiece to medium power and focus using the fine-adjustment knob. Compare what you observe with Figure 2.

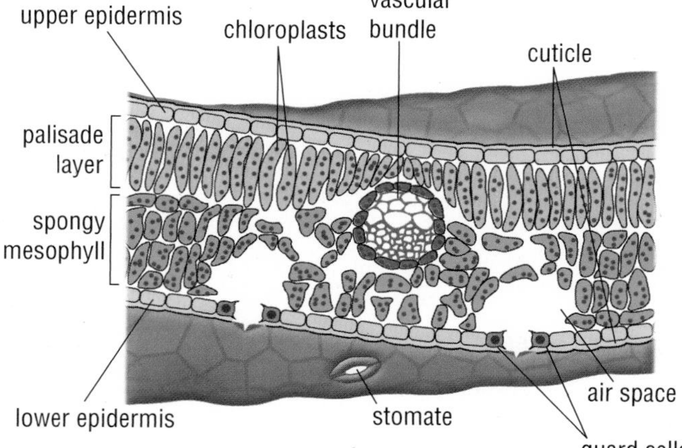

Figure 2 Part of a leaf in cross-section

9. Make a labelled scientific drawing of the cells, noting their shape and arrangement.
10. Carefully return the microscope to low power and return the slide to the teacher.

Part C: Whole Plant

11. Obtain a whole beet or radish seedling. Carefully remove any soil on the roots so that the root structure can be seen.
12. Draw a diagram of this plant. Label the root system and the shoot system. Identify the two parts of the shoot system that can be seen.

Analyze and Evaluate

(a) What are the "strings" in a celery stalk? What is their function? T/I

(b) On your leaf cross-section diagram from Step 8, identify the regions where each of these events take place: T/I

(i) light absorption

(ii) water supply

(iii) carbon dioxide absorption

(c) The root system of the plant observed in Part C stores carbohydrates for the plant's use. Roots are usually underground and are rarely photosynthetic. Explain how the carbohydrates are transported to the root for storage. T/I

Apply and Extend

(d) Which unique cellular structures did you see that could be linked to specific functions? T/I

(e) Create a table to summarize the major functions of the root, stem, and leaf of a plant. C

(f) Why does a plant need light, water, and carbon dioxide? K/U

(g) Explain how water is transported to the leaf of a plant. K/U

For the Love of Plants: The Royal Botanical Gardens

Things are blooming at the largest public garden in Canada. Royal Botanical Gardens (RBG) near Hamilton, Ontario, covers 11 km^2: about the size of 2500 football fields. The natural landscapes and manicured gardens are home to more than 1100 species of plants. Over 20 professional plant specialists care for the plants and grounds. Many enthusiastic volunteers also help out. RBG workers guard against damage from pests and encourage pollination by bees and butterflies. Every year, RBG botanists and their helpers start thousands of plants from seeds, bulbs, and cuttings (Figure 1). They also grow some woody plants, such as rose bushes and fruit trees, from grafts (Figure 2). In grafting, a branch cut from a tree or shrub is attached to a parent tree, called the root stock. The cut area heals against the parent tree, fusing the two together. The plant grows like a single organism, but the two parts keep their genetic differences.

Figure 1 The young plants need care and attention from the greenhouse staff.

Figure 2 A twig of an apple tree is grafted onto a root stock.

The result of this hard work is a stunning show of flowers, shrubs, and trees for visitors to enjoy.

RBG is much more than a tourist attraction. It also helps conserve Canada's natural habitats and threatened species. RBG is home to some extremely rare plants, including the largest Canadian population of the endangered red mulberry tree. The bashful bulrush grows naturally in the Gardens and in only one other location in Canada.

Scientists are actively engaged in restoring a large wetland called Cootes Paradise at RBG (Figure 3). They are interested in how a habitat changes over time. They also conduct research on invasive species that come from other countries and damage Canadian ecosystems. RBG works with many other conservation organizations, including McMaster University and Environment Canada. Their aim is to preserve Canada's native plant species, wild habitats, and ecosystems.

Figure 3 Cootes Paradise is a valuable wetland at Royal Botanical Gardens. It is part of Hamilton Harbour.

RBG places great importance on public education. From gardening tips to information about species survival, RBG educators teach visitors about how plants affect our lives. The work of the Gardens even extends well beyond its garden walls. RBG is Canada's national focal point for the Global Strategy for Plant Conservation, which is an international agreement to protect Earth's plants. RBG staff present travelling workshops across Ontario to train health care professionals to use plants in therapy with the elderly and those with physical or mental challenges.

GO TO NELSON SCIENCE

Plant Growth

The pattern of growth in plants is dramatically different from what we observe in animals. In animals, cell division occurs throughout the body with many kinds of differentiated cells being able to divide. In plants, cell division occurs only in certain parts of the plant. Most differentiated plant cells cannot divide further. In addition, most animals grow to a certain maximum size. Your arms and legs do not continue to grow throughout your life. In contrast, plants continue growing for as long as they live.

How can some plants grow to be massive trees if cell division does not occur throughout the entire plant?

Plant Meristems

You learned in Section 4.2 that specialized plant cells grow from unspecialized meristematic cells. These meristematic cells are located near the tips of the roots, in the growing parts of the shoots, and just beneath the outside layer of the stem of woody plants (Figure 1). The meristematic cells at the tips of roots and shoots form **apical meristems**. Meristematic cells around the stem and roots form **lateral meristems**. As the cells in these regions divide and grow, these parts of the plant become longer and wider.

apical meristem undifferentiated cells at the tips of plant roots and shoots; cells that divide, enabling the plant to grow longer and develop specialized tissues

lateral meristem undifferentiated cells under the bark in the stems and roots of woody plants; cells that divide, enabling the plant to grow wider and develop specialized tissues in the stem

Apical Meristems

In Activity 2.6, when you examined the phases of mitosis in an onion root, you were actually looking at an apical meristem region. You may have noticed that many of the cells were quite elongated. The growing tips of roots actually have three distinct regions: the meristem region of cell division, the region of elongation, and the region of maturation (Figure 2).

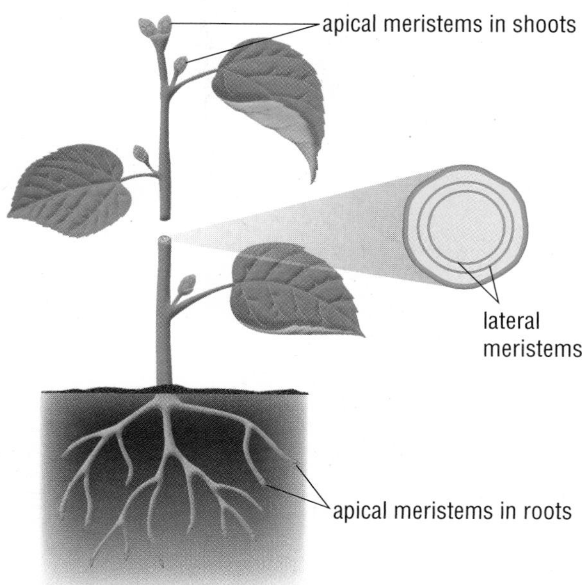

Figure 1 Meristem regions in a woody plant

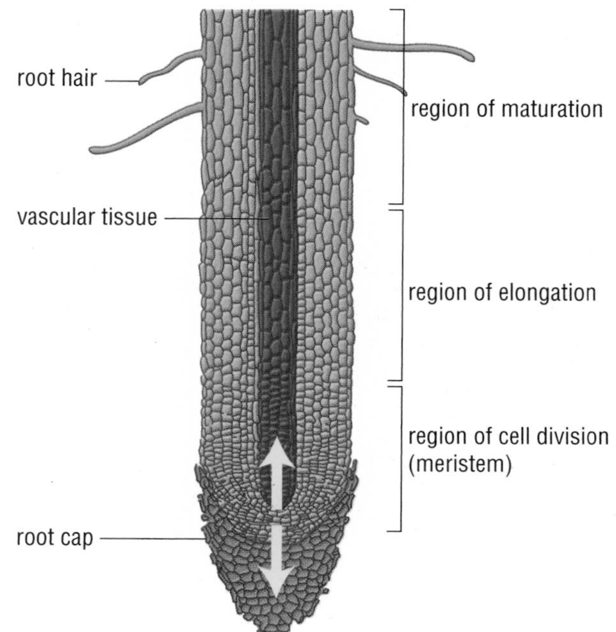

Figure 2 The only cell division in a root takes place in the apical meristem. Then the cells elongate, differentiate into various kinds of specialized tissue, and mature.

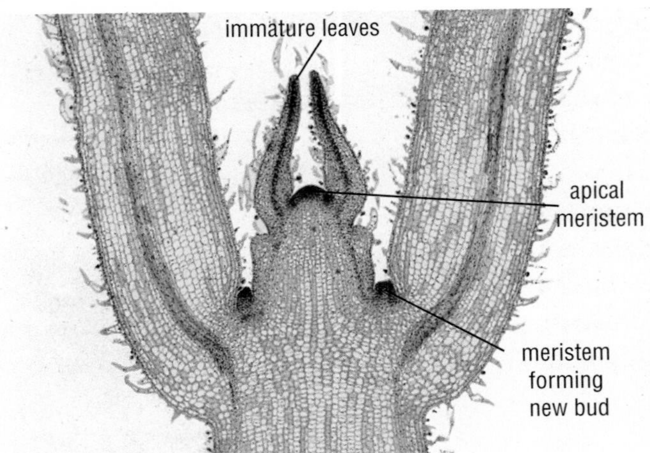

Figure 3 In the shoot meristem, cell division forms tissues that will mature into leaves, stems, and perhaps flowers.

When meristem cells first divide in the tips of roots, many of them begin to elongate. This makes the root longer, so it pushes its way through the soil. The cells in root tips may increase in length by more than 10 times. As they elongate, they begin to differentiate into specialized cells of the dermal, ground, and vascular tissue systems. They complete this process in the region of maturation. After this point, most cells cannot continue to grow or divide.

Growth in the apical meristems of shoots is more complex. The apical meristems are located in buds at the very tips of the growing stem, as you saw in Figure 1 on the previous page. Apical meristems also occur at points along the stem, giving many plants the ability to grow side branches off the main stem. The buds contain the meristem as well as immature stem, leaf, and sometimes flower parts (Figure 3). Cells produced by the apical meristems divide and specialize as they form the new tissue systems of the stems, leaves, and flowers.

DID YOU **KNOW?**

Growing a Witch's Broom

In medieval Europe, some people thought that magic caused dense growths of branches on the side of a tree. More recently, scientific study has shown that a fungus alters chemical signals in the apical meristem cells, causing them to form numerous side branches. Some plant growers intentionally infect young ornamental trees with the same fungus to keep them small and densely branched.

Lateral Meristems

Our most familiar long-lived plants are trees. These woody plants grow in both height and diameter. In addition to having apical meristems, woody plants contain lateral meristems within their stems and roots. Lateral meristems form two cylinders, one inside the other, that run the full length of the shoots and roots. As the plant grows in diameter, the outer lateral meristem produces new dermal tissue, called cork, to replace the old epidermal cells. The inner lateral meristem produces new phloem tissue on its outer surface and new xylem tissue toward its interior (Figure 4).

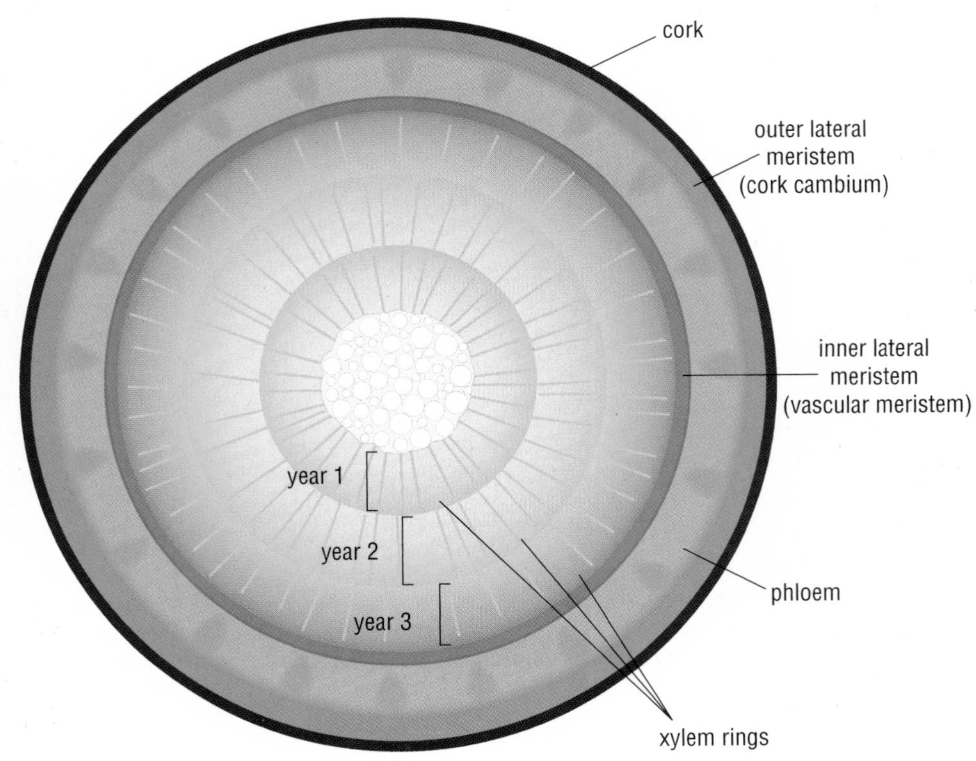

Figure 4 In the roots and shoots, lateral meristem differentiates into vascular tissue (xylem and phloem) and epidermal tissue. Each year's growth of new vascular tissue forms a visible ring of cells.

As you see in Figure 4, the phloem and cork form the bark of the growing tree, whereas the rings of xylem tissues form the interior of the tree trunk. The xylem keeps accumulating and causes the trunk to increase in diameter year after year. We can read the age of trees because each year's growth produces a noticeable ring of new xylem.

Many plants do not contain lateral meristems. As a result, their stems and roots cannot grow any thicker once the first tissues formed by the apical meristems have matured. These plants cannot produce any woody tissue and therefore usually remain quite small. They are generally short-lived annuals, completing their entire life cycle in one year.

LEARNING TIP

Other Names for Lateral Meristem

As you read other sources for information about plants, you may come across new scientific terminology. The innermost ring of lateral meristem is sometimes called the vascular cambium. The outer ring of lateral meristem is called the cork cambium.

Vegetative Reproduction

Clones are individuals that are genetically identical to each other. Clones can occur naturally in both plants and animals. Identical twins are a familiar example of natural animal clones.

Many plants produce clones naturally all the time. In this process, called **vegetative reproduction**, a plant puts out special shoots or roots that develop into new plants. These new plants have exactly the same genetic information as the parent plant. There are many examples of vegetative reproduction occurring naturally. A well-known example is that of strawberry plants that reproduce vegetatively by sending out shoots called runners across the surface of the soil. When the meristems in the tips of these shoots make contact with the ground, they begin growing roots and shoots that develop into a new plant (Figure 5(a)). A large clump of poplar trees are often all descended from a single tree, so they, too, are all clones (Figure 5(b)). Like the strawberry plants, the poplar trees usually remain connected to the parent plant. In this case, it might be more accurate to refer to the cluster as a single tree with many stems rather than as many clones of the original tree.

vegetative reproduction the process in which a plant produces genetically identical offspring from its roots or shoots

READING TIP

Checking for Meaning

You might come across a term that you are unfamiliar with and ask, "What does this mean?" Try looking in the margin to see if a definition is given. You can also check the text surrounding the unfamiliar term. A definition might be given in parentheses after the term, or an example might be used to help you figure out the meaning of the unknown word.

Figure 5 Plant clones (a) Strawberry plants produce new plants by putting out long stems called "runners" that can grow new roots. (b) Clusters of poplar trees are actually cloned stems of the same plant. We can often identify these "super-organisms" in the fall because they all change colour at exactly the same time!

Mass-Producing Plants

The natural cloning ability of plants to produce new young plants has been used by farmers, gardeners, and agricultural scientists for centuries. Different parts of a plant can be used to grow whole new plants. Every time you grow a new plant from a stem, root, or leaf cutting, you are cloning the original parent plant. Plant growers use these and other artificial forms of vegetative reproduction to mass-produce clones of individual plants that have desirable characteristics. Figure 6 shows two types of vegetative reproduction commonly used in Canada.

Figure 6 Vegetative reproduction can be used to create new plants from (a) "seed" potatoes, which are not seeds at all but cut-up potatoes that sprout, and (b) leaf cuttings of pelargonium shoots.

Tissue Culture Propagation

tissue culture propagation a method of growing many identical offspring by obtaining individual plant cells from one parent plant, growing these cells into calluses, and then into whole plants

One of the most fascinating techniques for plant cloning is called **tissue culture propagation**. This process involves taking pieces of root or another plant organ and breaking them down into individual cells. These cells are grown in a container with special chemicals that cause them to revert back into undifferentiated cells and to begin dividing, similar to meristem cells. As they continue to divide, each forms a clump of cells called a callus (Figure 7(a)). These growing calluses can be separated over and over again, producing a virtually unlimited number of growing cell masses. At any point, new chemicals can be added to each callus. These chemicals act like plant hormones and trigger the callus cells to begin the processes of growth and differentiation. The end result is a new plant (Figure 7(b)).

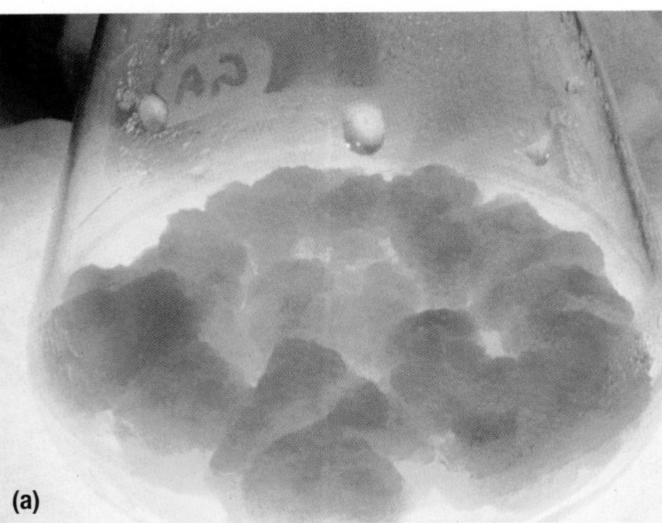

Figure 7 (a) Plant cells are specially treated to make them form calluses. (b) New plants grow from the calluses.

Plants grown by tissue culture propagation are genetically identical to their original parent plant. Thousands of clones can be produced from a single parent plant using this method. This is very useful for the mass growing of special high-yield varieties of crop plants or to produce uniform ornamental plants (Figure 8).

Figure 8 Tissue culture propagation allows thousands of clones to be developed from a single parent plant.

IN SUMMARY

- Apical and lateral meristems are responsible for plant growth.
- Apical meristems result in the growth in length of roots and shoots and the production of leaves and flowers.
- Lateral meristems, which produce cork and vascular tissues, are found only in woody plants and result in the growth in width of stems.
- Many plants use vegetative reproduction to produce new young plants with exactly the same genetic information as the parent plant.
- Humans take advantage of vegetative reproduction to produce many identical new plants from one parent plant.
- Tissue culture propagation is a modern technique used to mass-produce plants.

CHECK YOUR LEARNING

1. In what ways is plant growth fundamentally different from animal growth? K/U
2. Where are the two main types of meristem in plants? Explain the role of each type. K/U
3. What enables roots to push through the soil and trees to grow tall? K/U
4. Describe the location and arrangement of new phloem and xylem tissues in a growing tree. K/U
5. Only woody plants have lateral meristems, which produce vascular tissues. Where do the vascular tissues in non-woody plants come from? K/U
6. In a small group, brainstorm a list of the advantages a plant has when it is able to produce woody tissue. C A
7. In what way is a callus similar to embryonic stem cells? A

CHAPTER 4

LOOKING BACK

KEY CONCEPTS SUMMARY

Plants have a shoot system and a root system.

- The root system anchors the plant in the ground and absorbs water and nutrients from the soil. Some roots store food for the plant. (4.1, 4.2)
- The shoot system is primarily responsible for photosynthesis: using solar energy to make food from carbon dioxide and water. (4.1, 4.4)
- In most plants, the shoot system also produces flowers for sexual reproduction. (4.1)

The organization of tissues, organs, and systems is different in plants to that in animals.

- Plant organs are generally referred to as plant parts. (4.1)
- Plant parts are made up of one or more of three different types of tissue: dermal, vascular, and ground tissue. (4.1, 4.2)

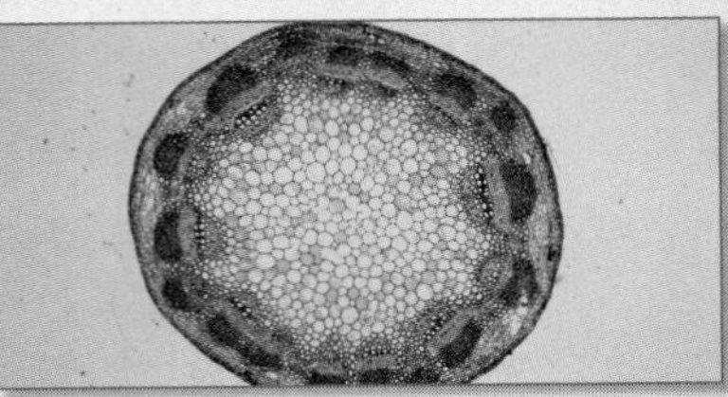

Plants have dermal, vascular, and ground tissue systems.

- Dermal tissues form the outermost layer of plants. (4.2)
- There are two types of vascular tissue: xylem, which transports water and dissolved minerals up from the roots, and phloem, which transports the food produced by photosynthesis throughout the plant. (4.2)
- Ground tissue, found between the dermal and vascular tissues, is responsible for photosynthesis, food storage, and support. (4.2)

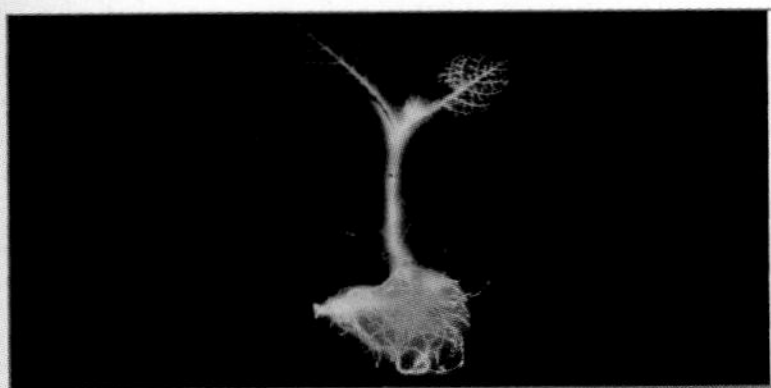

Scientists can change the genetic makeup of plants.

- For centuries, people have selected and grown plants that have the best characteristics. (4.3)
- Genetic engineering allows scientists to grow plants that produce substances that normally occur in other organisms. (4.3)
- Genetic engineering involves transferring selected sections of DNA from one organism to another. The organisms with the new genetic material are genetically modified organisms (GMOs). (4.3)

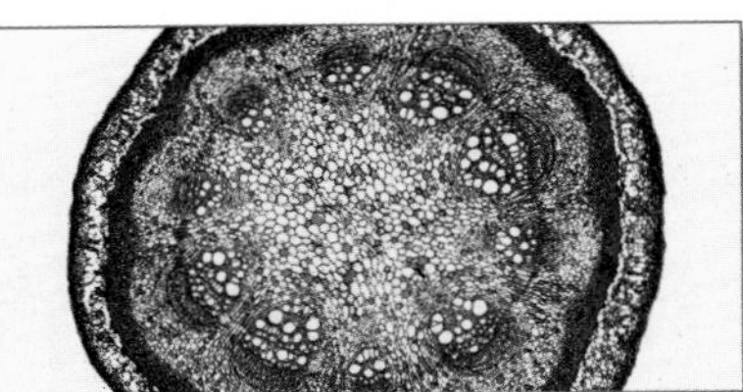

Plant tissue systems interact to perform complex tasks.

- Specialized cells in the tissues of a leaf have functions related to photosynthesis. (4.1, 4.4)
- The palisade layer and the spongy mesophyll contain the photosynthesizing cells. (4.4)
- Vascular tissue brings water and nutrients to the leaf and carries the manufactured food to other parts of the plant. (4.2, 4.4)
- Guard cells in the epidermis create openings (stomata) that allow gas exchange and regulate water loss from the leaf. (4.4)

Meristems determine the pattern of plant growth.

- Meristematic cells are unspecialized cells, similar to animal stem cells. (4.2, 4.6)
- Meristematic cells are found in apical meristems at the root and shoot tips of all plants and in lateral meristems in the stems and roots of woody plants. (4.6)
- Cell division, growth, and differentiation in apical meristems make roots and shoots longer. (4.6)
- Cell division, growth, and differentiation in lateral meristems form new vascular tissue and epidermal tissue, making the stems wider. (4.6)

WHAT DO YOU THINK NOW?

You thought about the following statements at the beginning of the chapter. You may have encountered these ideas in school, at home, or in the world around you. Consider them again and decide whether you agree or disagree with each one.

1 Plants have organs, just like animals.
Agree/disagree?

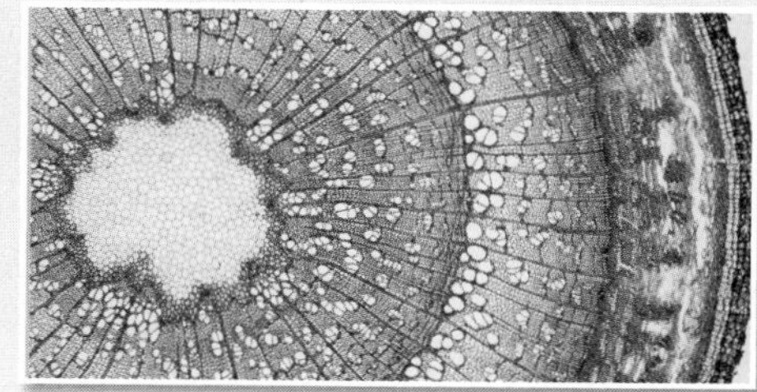

4 Plants have a circulatory system.
Agree/disagree?

2 Respiration is the same as breathing.
Agree/disagree?

5 Cell division occurs in all parts of a plant.
Agree/disagree?

3 All plants grow from seeds.
Agree/disagree?

6 Plants use water in the process of photosynthesis.
Agree/disagree?

How have your answers changed? What new understanding do you have?

Vocabulary

dermal tissue system (p. 126)
vascular tissue system (p. 126)
ground tissue system (p. 126)
root system (p. 126)
shoot system (p. 127)
meristematic cell (p. 129)
epidermal tissue (epidermis) (p. 131)
periderm tissue (p. 131)
xylem (p. 132)
phloem (p. 132)
palisade layer (p. 137)
spongy mesophyll (p. 137)
cuticle (p. 137)
stomate (p. 137)
guard cell (p. 137)
apical meristem (p. 143)
lateral meristem (p. 143)
vegetative reproduction (p. 145)
tissue culture propagation (p. 146)

BIG Ideas

- ✓ Plants and animals, including humans, are made of specialized cells, tissues, and organs that are organized into systems.
- Developments in medicine and medical technology can have social and ethical implications.

The following icons indicate the Achievement Chart category addressed by each question.

K/U Knowledge/Understanding
C Communication
T/I Thinking/Investigation
A Application

What Do You Remember?

1. Copy the left-hand column of Table 1 into your notebook. Complete the right-hand column as indicated. (4.1, 4.2) K/U

Table 1 Levels of Organization in Plants

Level of organization	List
Systems	(a) List two systems.
Tissue systems	(b) List three tissue systems.
Parts of the plant	(c) List four parts of the plant.
Tissues used for transportation	(d) List two tissues used for transportation.

2. Draw a simple diagram of a plant. Label the main parts. (4.1) K/U
3. State the main function for each part of a plant. (4.1) K/U
4. Identify the parts of a plant and the specialized cells involved in gas exchange. (4.4) K/U
5. What is the name given to undifferentiated plant cells, and where are these cells located? (4.2, 4.6) K/U
6. Plants and animals must perform some of the same functions. Name three functions that plants and animals both carry out. (4.1, 4.4) K/U
7. What is plant cloning? (4.6) K/U
8. Plants have many adaptations to stop them from drying out. Briefly describe two of these adaptations. (4.2, 4.4) K/U
9. List the functions of a plant's roots. (4.1, 4.2, 4.4) K/U
10. What are the structures inside chloroplasts that collect light energy? (4.1) K/U

What Do You Understand?

11. Why do plants not need to eat? Describe how they obtain their nutrients. (4.1, 4.2, 4.4) K/U
12. Draw a simple diagram of a plant root tip. Indicate where the three types of plant tissue are located. (4.6) K/U C
13. Briefly describe the general function of each tissue type in a plant root tip. (4.1, 4.2, 4.6) K/U
14. Why are the spikes of a cactus considered to be modified leaves? (4.1) K/U
15. Potato plants store starch. We use this starch as a food. In which part of the plant do potatoes store their starch? Where does this starch come from? (4.1, 4.2) K/U
16. How is the growth of plants different from the growth of animals? (4.6) K/U
17. Name and briefly describe the two types of reproduction in most plants. (4.1, 4.6) K/U
18. Why are non-woody plants limited in how large they can grow? (4.6) K/U
19. Name the plant tissues that are specialized to provide strength and support. (4.2) K/U
20. Indicate where water enters and exits a plant. Name and describe the specialized cells involved at both the entry and exit points. (4.1, 4.2, 4.4) K/U
21. What substances does a plant require for respiration? (4.1) K/U
22. How are meristems in plants similar to and different from stem cells in animals? (4.2, 4.6) K/U

Solve a Problem

23. Identify what parts of a plant are shown in Figure 1(a) to (d). (4.1, 4.2, 4.4–4.6) K/U

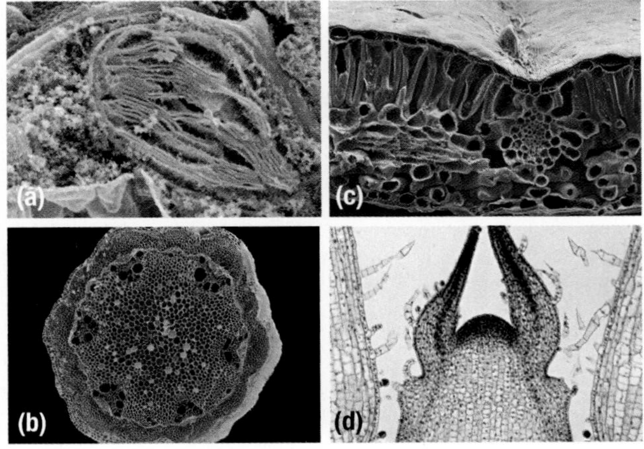

Figure 1

24. Identify one location in plants where the cells are loosely arranged. What is the function of these cells, and why are they loosely arranged? (4.4) K/U

25. How can a plant easily regrow a damaged stem or leaf? (4.2, 4.6) K/U

26. Explain how stripping the bark from a tree might kill the tree. (4.6) A

27. Why do you think wind-pollinated plants generally produce more pollen than animal-pollinated plants? (4.1) A

28. Mowing your lawn could result in the top half of the leaves of grass being cut off, yet the grass keeps on growing. Using your knowledge of meristems and plant growth, explain why the grass keeps on growing. (4.6) A

29. Write the word equation for photosynthesis and then, considering each component in the equation, describe how the structure of a leaf has developed to allow photosynthesis to occur. (4.4) K/U T/I

30. How could you distinguish between the upper surface and the lower surface of a leaf? (4.4) K/U

Create and Evaluate

31. Imagine you are a water molecule. Write a short story describing your travel route after being absorbed through a root hair and ending up in the process of photosynthesis. (4.1, 4.2, 4.4) K/U T/I C

32. Compare the advantages of asexual reproduction and sexual reproduction in plants. (4.1, 4.6) K/U

33. Design an experiment in which you might determine which fertilizer formula promotes the best plant growth. Be sure to identify the independent and dependent variables and suggest a suitable control for the experiment. T/I A

34. Jack and Jill carved their initials in the bark of a tree trunk 1 m above ground level. The tree was 5 m tall at the time. Twenty years later, the tree is 15 m tall. How far above ground level are their initials now? Explain your answer. (4.6) T/I

35. Climatologists use the growth rings of trees to infer what the climate was like during the lifetime of the tree. Explain how the growth rings might provide information to make this kind of inference. (4.6) K/U T/I

Reflect On Your Learning

36. In this chapter, you learned about specialized cells and tissues in plants. What did you find most interesting or most surprising? Write two questions that you would like to have answered through further study.

37. Plants are often considered to be less important in the ecosystem. At-risk or threatened plants generally receive less public attention than at-risk or threatened animals. In a paragraph, present an argument that plants are more important and should receive more attention than animals. How has this chapter helped you prepare your argument?

Web Connections 

38. The McIntosh apple (Figure 2) is one of the most popular apple varieties in Canada. Research the origins of this type of apple. Present your findings as a paragraph to be printed in an advertising brochure for an apple pie bakery. T/I C A

Figure 2

39. Investigate why poinsettia leaves turn red around midwinter. How could you use this information to encourage your poinsettia to turn red in time for the next winter holiday? Communicate your findings in an e-mail to your grandparents. T/I C A

40. Find out what structures are present in a plant seed, such as a bean. Create a visual presentation showing how these structures grow to form an adult plant. T/I C

41. The leaves of a plant growing in a darkened room will bend toward any source of light. This is referred to as phototropism. Research why plants behave this way. Also research other plant tropisms and what causes them. T/I

To do an online self-quiz or for all other Nelson Web Connections,
GO TO NELSON SCIENCE

CHAPTER 4

SELF-QUIZ

The following icons indicate the Achievement Chart category addressed by each question.

K/U Knowledge/Understanding T/I Thinking/Investigation C Communication A Application

For each question, select the best answer from the four alternatives.

1. Which part of the plant carries out the process of reproduction? (4.1) K/U
   (a) flower
   (b) leaf
   (c) root
   (d) stem
2. In which of the following plants would you expect to find periderm tissue? (4.2) K/U
   (a) a water lily
   (b) an oak tree
   (c) a cactus
   (d) a poinsettia
3. What are the products of photosynthesis? (4.1, 4.4) K/U
   (a) carbon dioxide and water
   (b) water and oxygen
   (c) oxygen and sugar
   (d) sugar and carbon dioxide
4. Plant cells that can differentiate into specialized tissues are called
   (a) stomata cells.
   (b) transgenetic cells.
   (c) epidermal cells.
   (d) meristematic cells. (4.2, 4.6) K/U
5. What material forms the tube-shaped structures that carry water from the roots to the leaves? (4.2) K/U
   (a) chlorophyll
   (b) cuticle
   (c) stomata
   (d) xylem
6. Which of these plants stores the most starch in its roots? (4.1) K/U
   (a) yam
   (b) tomato
   (c) pumpkin
   (d) apple tree

Indicate whether each of the statements is TRUE or FALSE. If you think the statement is false, rewrite it to make it true.

7. The shoot system is another name for the stem of a plant. (4.1) K/U
8. Phloem transports food produced by photosynthesis throughout the plant. (4.2) K/U
9. The spongy mesophyll is an inner region of a leaf with loosely packed cells. (4.4) K/U

Copy each of the following statements into your notebook. Fill in the blanks with a word or phrase that correctly completes the sentence.

10. The organelle where photosynthesis takes place is called a(n) ________________. (4.1) K/U
11. Thylakoids are arranged in stacks called ________________. (4.1) K/U
12. Plant eggs produced by the female reproductive organ are fertilized by ____________ produced by the male reproductive organ. (4.1) K/U

Match each term on the left with the most appropriate description on the right.

13. (a) stomata
    (b) cuticle
    (c) chloroplast
    (d) guard cell

    (i) a layer of wax on the surface of leaves
    (ii) a structure that controls the size of openings
    (iii) an opening that allows for the exchange of gases
    (iv) an organelle that carries out photosynthesis
    (4.1, 4.2, 4.4) . K/U

Write a short answer to each of these questions.

14. Briefly describe one difference between plants and animals that applies to all plants and animals. (4.1, 4.4) K/U
15. You are planning a study of the cells in the stem of a tomato plant. (4.1, 4.2) T/I
    (a) Identify two tools you would need for this study and describe how you would use them.
    (b) Identify three types of cells you would look for and describe their functions.

16. (a) Identify the two products of photosynthesis, and describe how animals take in each of these products.

(b) Explain why each product is essential to the survival of animals. (4.1, 4.4) K/U

17. Plants need water, carbon dioxide, and sunlight to produce food. Design an experiment to study the effect of each of these resources on plant growth. Describe how you would first deprive the plant of each resource and then observe the effect of restoring the resource. (4.1, 4.4) T/I

18. A student's backyard contains two maple trees. One maple tree has a trunk that is 2 m in diameter, and one maple tree has a trunk that is 3 m in diameter. Which tree is older? Explain your answer. (4.6) T/I

19. Draw a diagram of the cells in a root tip. Label the root cap, the region of cell division, the region of elongation, and the region of maturation. (4.6) C

20. Explain how each plant tissue has a similar function to the organ or organ system in the human body.

(a) dermal tissue and human skin

(b) vascular tissue and the circulatory system

(c) ground tissue and the skeletal system (4.2, 4.4) A

21. You are planning to study the reproduction and propagation methods of several flowering plants in your area.

(a) Name four questions you would ask during your study to determine each plant's methods of reproduction.

(b) Name four questions you would ask during your study to determine each plant's methods of spreading its seeds. (4.1, 4.6) T/I C

22. Complete the diagrams below by labelling th tissues. Each space should be labelled as derma vascular, or ground. (4.2, 4.4) C

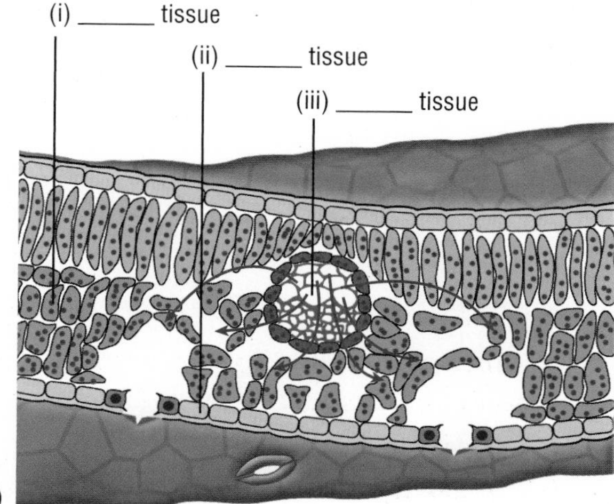

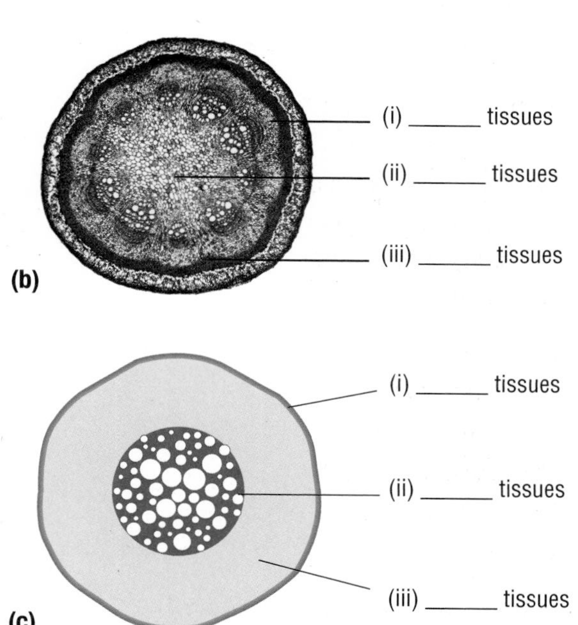

23. A hedge always grows new leaves after it is pruned. However, a human cannot grow a new finger if it is lost in an accident. Explain why animals and plants respond differently to the loss of a body part. (4.2, 4.6) A

24. Imagine you are a carbon atom that is part of a compound floating in the air. Describe your journey into a plant and the changes that take place there. Then describe how you become part of an animal and how you return to the air. In each phase tell what chemical compound you are part of. (4.1, 4.2, 4.4) K/U T/I C

LOOKING BACK

UNIT B
Tissues, Organs, and Systems of Living Things

CHAPTER 2
Cells, Cell Division, and Cell Specialization

KEY CONCEPTS

All organisms are made up of one or more cells.

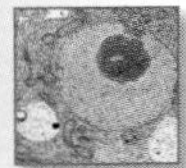
Microscopes enable us to examine cells in detail.

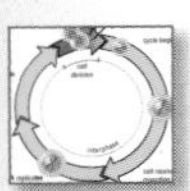
The cell cycle occurs in distinct stages.

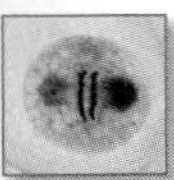
Cell division is important for growth, repair, and reproduction.

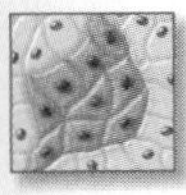
Cancer cells generally divide more rapidly than normal cells.

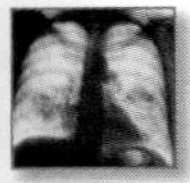
Medical imaging technologies are important in diagnosing and treating disease.

CHAPTER 3
Animal Systems

KEY CONCEPTS

Complex animals are made up of cells, tissues, organs, and organ systems.

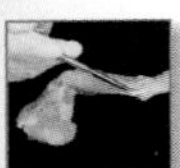
Scientists use laboratory techniques to explore structures and functions within animals' bodies.

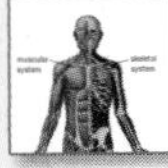
Each organ system has a specific function and corresponding specific structures.

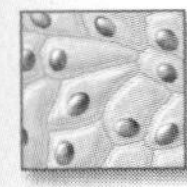
There are four main types of animal tissues.

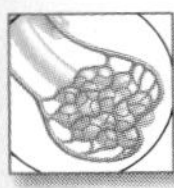
Organ systems interact to keep the organism functioning.

Research is helping people overcome illness and injury.

CHAPTER 4
Plant Systems

KEY CONCEPTS

Plants have a shoot system and a root system.

The organization of tissues, organs, and systems is different in plants to that in animals.

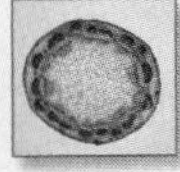
Plants have dermal, vascular, and ground tissue systems.

Scientists can change the genetic makeup of plants.

Plant tissue systems interact to perform complex tasks.

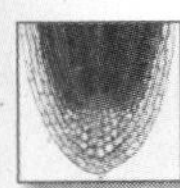
Meristems determine the pattern of plant growth.

MAKE A SUMMARY

Imagine that you are attending a conference entitled "Tissues, Organs, and Systems of Living Things." The whole class will attend the conference to share "best practices" in a jigsaw-style learning structure. Each person will have two important roles:

(1) as a member in one home Trio Group

(2) as a delegate in an Expert Group

Part A: Selecting Areas of Expertise within the Trio Group

1. In your Trio Group, cooperatively choose one of the following roles for each member.
    - Implications Expert (Specialty: the importance of medical and technological developments related to systems biology; analysis of social, ethical, and environmental implications)
    - Animal Expert (Specialty: animal cell division and cell specialization; tissues, organs, and systems in animals; research skills)
    - Plant Expert (Specialty: plant cell division and cell specialization; tissues and tissue systems in plants; research skills)

Part B: Expert Group Workshops

2. Experts with the same specialties from each Trio Group will meet as delegates at a "workshop." For example, all of the Implications Experts will get together.
3. The conference organizer (your teacher) will post a list of topics to be covered. Sign up to present a sample of your work that best covers one of the topics.
4. Take turns with the other delegates to present your contribution at the workshop. Hand out a printed summary of your presentation to the other delegates in the workshop.
5. During or between the presentations, you may ask questions of other delegates to help clarify the work and to develop your expertise.

Part C: Trio Group Sharing

6. Reassemble with your Trio Group. Share what was learned at the expert workshops.
7. Working together with your Trio write point-form notes for each of the Overall Expectations listed at the beginning of this unit (page 20).

CAREER LINKS

List the careers mentioned in this unit. Choose two of the careers that interest you, or choose two other careers that relate to tissues, organs, and systems of living things. For each of these careers, research the following information:

- educational requirements (secondary and post-secondary)
- skill/personality/aptitude requirements
- duties/responsibilities
- potential employers
- salary

Assemble the information you have discovered into a poster. Your poster should compare your two chosen careers and explain how they connect to this unit.

GO TO NELSON SCIENCE

Family Health Supporter

A friend or family member has come to you for support in coping with a health care concern. You offer to help by researching and providing a package of information related to the particular health concern. Remember that self-diagnosis is not reliable; only health professionals should make diagnoses and recommend treatments, in consultation with the patient or the patient's family.

With your group, read the scenarios that follow. Choose one of the scenarios as your health issue.

The Issue

SCENARIO 1 Your uncle had abdominal pain and was mysteriously losing weight. Four weeks later, he was diagnosed with Crohn's disease. His family needs to understand the origins and effects of the disease and how it can be treated or managed.

SCENARIO 2 During a hockey game, one of your teammates was knocked down onto the ice and slid headfirst into the boards. Your teammate was taken to the hospital for tests. You will be visiting your friend in hospital and want to share as much information as you can about the possible injuries, what to expect, and how to avoid further injury.

SCENARIO 3 Your elderly cat has been unwell. The veterinarian says that the cat's kidneys are not working effectively. The vet mentions several treatment options: changing to a low-ash cat food, a drug to help the kidneys work better, and peritoneal dialysis. You need to help your family come up with an informed decision on the cat's treatment.

SCENARIO 4 Your best friend's mother has had a chronic cough for several months. She visited her doctor last week to discuss her chest pain, weight loss, and severe fatigue. Her doctor ordered tests to look for lung cancer. She is anxious about going through the tests. You would like to prepare information for her so that she understands the types and purposes of the diagnostic tests she may be undergoing.

SKILLS MENU

- Defining the Issue
- Researching
- Identifying Alternatives
- Analyzing the Issue
- Defending a Decision
- Communicating
- Evaluating

SCENARIO 5 Your dog was running after a ball and suddenly started limping. The vet said that your dog tore her anterior cruciate ligament and may require surgery to repair the ligament. Before your family commits to the surgery, you want to find out about this injury and alternative treatment options.

SCENARIO 6 Your tennis partner has developed a very sore shoulder. Initial online research suggests that the cause might be rotator cuff tendonitis. Your partner thinks there is no point in going to the doctor. You decide to assemble some information for your friend about long-term effects of untreated tendonitis and treatment options.

Goal

To provide information and support to enable your friend or family member to take the appropriate actions toward diagnosis, treatment, or management of the condition.

Gather Information

Working in your group, review your chosen scenario. Research the functioning of the organ in the healthy state and how this functioning is affected by the disease or injury. Find out what specialized cells, tissues, organs, or organ systems are affected by the condition.

Find out about the disease or injury's possible causes, diagnosis, and treatment. Investigate how the diagnostic technologies and treatments relate to the "biology" of the disease or injury. Decide what information you need to gather in order to best address the issue you selected.

Distribute the various research tasks among the members of your group. Determine what resources you will use to gather your information. Use a variety of resources. Write a brief plan for conducting your research.

Assemble information on

- how that part of the body functions when healthy
- how the illness or injury affects that part of the body
- what diagnostic technology is likely to be used
- what possible treatments might be available
- pros and cons of the possible treatments
- recommended follow-up care

As you research, keep these points in mind:

- Keep detailed notes and sketches.
- Record your sources of information.
- Analyze your information for reliability and bias.
- Remember that your goal is to provide information and support to your friend or family member. You are not qualified to act as a medical expert.
- Focus on the positive.

Identify Solutions

Decide what is the main issue raised in your chosen issue. As you put together the information package, make sure that you address this issue and present some possible solutions. Consider the following questions:

- What is needed: information about diagnostic tests, treatment options, or disease management strategies? This should determine the focus of the information you provide.
- What recent technological developments are there in the diagnosis, treatment, or management of the patient's condition?
- What studies have been done showing the risks and benefits of certain tests or treatments?
- Are there any controversial technologies or ethical questions regarding the treatments that you have identified?

Make a Decision

You are providing information that will help someone else make a decision regarding the main issue that you have identified. You must maintain an objective stance and present all sides of the issue. However, when a person or family is looking for information, they are sometimes also looking for opinions. Consider what you feel would be the best response to the issue for the patient.

Communicate

Communicate the information you have gathered as

- an informal chat with the person or family about your research findings, including the main issue and what diagnostic and treatment technologies might be available;
- a written summary of the information you have discovered, with diagrams, plus a list of resources (websites, community agencies) for individuals who would like more detailed information;
- a written list of questions for the person (or the animal's owner) to ask the doctor or vet; and
- your response to the main issue in the form of recommended "next steps" (accompanied by an acknowledgement that you are not a medical professional) in a format of your choice.

ASSESSMENT CHECKLIST

Your completed Performance Task will be assessed according to the following criteria:

Knowledge/Understanding

- ☑ Address diagnostic and treatment technologies in your informal chat.
- ☑ Use appropriate terminology in your written summary and diagrams.
- ☑ Display a clear understanding of the tissues and organs involved in the questions that you suggest.
- ☑ Recommend reasonable "next steps."

Thinking/Investigation

- ☑ Identify the main issue you need to address.
- ☑ Use research skills to gather information about the health concern.
- ☑ Analyze the sources for reliability and bias.
- ☑ Use critical thinking to recommend "next steps."

Communication

- ☑ Communicate orally in a clear and encouraging manner.
- ☑ Prepare organized, logical, and factually accurate written information.
- ☑ Acknowledge, when communicating possible "next steps," where to go for professional medical advice.

Application

- ☑ Assess the importance of medical technological developments for human or animal health.
- ☑ Suggest a course of action to deal with the specific health issue.

UNIT B REVIEW

The following icons indicate the Achievement Chart category addressed by each question.

K/U Knowledge/Understanding | T/I Thinking/Investigation | C Communication | A Application

What Do You Remember?

For each question, select the best answer from the four alternatives.

1. All living things contain
   (a) cells.
   (b) tissues.
   (c) organs.
   (d) organ systems. (2.1) K/U

2. All living things are able to
   (a) photosynthesize.
   (b) reproduce.
   (c) move.
   (d) breathe. (2.1, 2.3) K/U

3. Identify the type(s) of tissue found in the heart. (3.4, 3.8) K/U
   (a) epithelial
   (b) muscle
   (c) nerve
   (d) all of the above

4. Which one of the following tissue systems is not part of a plant? (4.2) K/U
   (a) vascular
   (b) ground
   (c) respiratory
   (d) dermal

5. Which phase of mitosis is marked by chromosomes lining up along the middle of the cell? (2.5) K/U
   (a) prophase
   (b) metaphase
   (c) anaphase
   (d) telophase

6. Which of the following organs cannot be transplanted from one person to another? (3.7) K/U
   (a) cornea
   (b) lung
   (c) spinal cord
   (d) blood

7. In plants, the cells that can develop into any kind of tissue are called
   (a) meristem
   (b) xylem cells
   (c) the root cap
   (d) palisade cells (4.2, 4.6) K/U

8. Cells that contain only half of the DNA from each parent are called
   (a) clones.
   (b) zygotes.
   (c) gametes.
   (d) chromosomes. (2.3) K/U

9. Which of the following best describes connective tissue? (3.1) K/U
   (a) tissue that provides support and protection for various parts of the body
   (b) tissue that conducts electrical signals from one part of the body to another
   (c) tissue that contains proteins that can contract and enable the body to move
   (d) tissue made of tightly packed cells that covers body surfaces and lines internal organs

10. Which of the following is a product of photosynthesis? (2.1, 4.1, 4.4) K/U
    (a) water
    (b) oxygen
    (c) solar energy
    (d) carbon dioxide

11. During which part of the cell cycle are the DNA strands replicated? (2.5) K/U
    (a) anaphase
    (b) prophase
    (c) metaphase
    (d) interphase

Indicate whether each of the statements is TRUE or FALSE. If you think the statement is false, rewrite it to make it true.

12. Animals have levels of organization (a hierarchy) for structure and function, including cells, tissues, tissue systems, and organ systems. (3.1) K/U

13. The tissue system responsible for transporting materials around a plant is called the dermal tissue system. (4.2) K/U

14. The phase of mitosis during which DNA replicates is called prophase. (2.5) K/U

15. The organ system in humans responsible for sending information to and from the heart is called the nervous system. (3.4, 3.10) K/U

16. The part of a plant that contains only the male sex cells is the seed. (4.1) K/U

17. Cancerous tumours may form when cells stop dividing. (2.7) K/U

18. The main function of leaves on a plant is protection. (4.1, 4.2) K/U

19. When there is a good supply of water within a plant leaf, the guard cells close the stomata to prevent the water from escaping. (4.4) K/U

20. Veins carry oxygenated blood from the lungs to other parts of the body. (3.4) K/U

Copy each of the following statements into your notebook. Fill in the blanks with a word or phrase that correctly completes the sentence.

21. Two main sources of blood stem cells are ____________ and ____________. (3.2) K/U

22. __________ tissue transmits signals to the __________ tissue, instructing it to contract. (3.8, 3.10) K/U

23. The organ systems that help remove waste substances from your body are the ____________, the ____________, the ____________, and the ____________ systems. (3.3, 3.4, 3.6, 3.8) K/U

24. The organ system that enables you to detect changes in your environment is the ____________ system. (3.10) K/U

25. Guard cells close to conserve ____________ at night. (4.4) K/U

26. Skeletal muscles always work in opposing ____________: one muscle causes the joint to ____________; the other muscle causes the joint to ____________. (3.8) K/U

27. During the process of diffusion, substances move from an area of ____________ concentration to an area of ____________ concentration. (2.3) K/U

28. The process by which a cell becomes specialized to perform a specific function is called ________________. (3.2) K/U

Match each term on the left with the appropriate description on the right.

29. (a) diabetes
(b) tuberculosis
(c) osteoporosis
(d) multiple sclerosis
(e) colitis

(i) an inflammation of the lining of the large intestine
(ii) a disease that destroys the myelin sheaths of neurons
(iii) a condition in which the pancreas produces too much or too little insulin
(iv) a disease that involves the loss of bone tissue, making bones brittle and weak
(v) an infectious disease caused by bacteria growing in the lungs (3.3–3.10) K/U

Write a short answer to each of these questions.

30. Give three reasons why cells divide. (2.3) K/U

31. Define the term *interphase*. Explain what occurs during this part of the cell cycle. (2.5) K/U

32. Why must a cell's nucleus replicate during mitosis before cell division proceeds? (2.5) K/U

33. When looking at two samples of cells, how can you tell that cell division is more rapid in one sample than in the other? (2.6, 2.8) K/U

34. Identify several factors that affect the growth and cell cycle rate of healthy cells. (2.5, 2.7) K/U

35. List three factors that are known to increase the risk of cancer in humans. (2.7) K/U

36. Which body systems work together to provide nutrients to all of your cells? Explain your answer. (3.3, 3.4) K/U

37. One way that plants differ from animals is that plants cannot move from place to place. Explain why plants do not have to move from place to place in order to survive. (4.1, 4.2, 4.4) K/U

38. Choose one type of animal tissue and explain how the structure of that tissue supports its function. (3.1, 3.3, 3.4, 3.6, 3.8, 3.10) K/U

What Do You Understand?

39. Why are the roots of a carrot plant so much larger than the roots of a grass plant? (4.1, 4.2) K/U

40. Describe an example of regeneration that you or someone you know has experienced. (3.2) A

41. (a) Describe at least two similarities between the process of photosynthesis and the process of cellular respiration.
    (b) Describe at least two differences between the process of photosynthesis and the process of cellular respiration. (2.1, 4.1, 4.4) T/I

42. Draw the stages of mitosis in animal and plant cells. Use these diagrams to compare mitosis in plant and animal cells. Identify differences between animal and plant cell mitosis. (2.5, 2.6) K/U C

43. Create a table to compare healthy cells with cancer cells. Include the following headings in your table: Rate of cell division; Level of specialization; Length of mitosis; Appearance of cell; and Ability to move. (2.6–2.8) K/U C

44. Briefly describe at least three medical imaging technologies. Explain how each technology is used in the diagnosis of injuries, diseases, or disorders. (2.7, 3.4, 3.6, 3.8, 3.9) K/U

45. Explain in your own words how an increase in the volume of a cell affects its ability to meet its needs. (2.3, 2.4) K/U

46. Differentiate between the three types of muscle. (3.8) K/U

47. What is the advantage of treating specialized cells to make them behave as stem cells? (3.2) K/U A

48. Describe the main difference between xenotransplantation and regular organ transplantation. (3.7) K/U A

49. Some kinds of pollen are very light and are blown by the wind. Explain how this helps a plant reproduce. (4.1) K/U

50. In what type of climate would you be likely to find a plant with a very thick cuticle around its leaves and stem? (4.2, 4.4) K/U

51. A frog uses its skin as a respiratory surface. Explain how this can be both an advantage and a disadvantage. (3.4) K/U

52. In addition to food and medicine, list four ways in which plants are used in your everyday life. (4.1) A

53. (a) What are the functions of animal epithelial tissues?
    (b) What are the functions of plant dermal tissues?
    (c) What are the similarities and differences in these functions? (3.1, 4.2, 4.4) K/U

54. Compare the transport system in a plant with that in an animal. (3.4, 4.2, 4.4) K/U

55. Compare how a plant and an animal obtain nutrients. (3.3, 4.2, 4.4) K/U

56. Create a table to compare the hierarchy of organization in plant and animal bodies. (3.1, 4.1) K/U C

57. Create and complete a table that compares asexual reproduction in plants with asexual reproduction in animals. (2.3, 4.6) K/U C

58. Several animal organ systems were mentioned in Chapter 3. Create a concept map to show how these organ systems work together to keep the human body functioning. (3.3, 3.4, 3.6, 3.8, 3.10, 3.11) K/U C

59. Create a concept map that shows the four components of blood. Include a brief description of each component's function in your map. (3.4) C

60. Some organelles in a single-celled organism perform functions similar to those of organ systems in a human body. For each of the organelles listed below, choose an organ or organ system that performs a similar function. Explain each of your choices. (2.1, 3.3, 3.4, 3.8, 4.2) T/I
    (a) endoplasmic reticulum
    (b) Golgi bodies
    (c) cytoplasm

Solve a Problem

61. Some aquatic plants grow on the surface of a pond, while others grow at the bottom of a pond. Predict what would happen to the plants growing at the bottom if the plants on the surface grew and covered the entire surface of the pond. (4.4) T/I

62. What is one advantage and one disadvantage of having a body made mostly of specialized cells rather than a body made of one type of cell that performs many functions? (2.9) T/I

63. People with the blood disorder anemia have a deficiency of hemoglobin in their red blood cells. Explain how this disorder could affect the function of the circulatory system. (3.4) T/I A

64. During a lab activity, a researcher counts cells in two areas of a sample. Cells in all stages of the cell cycle are present in the sample. The observations are recorded in Table 1. These cells normally take 15 hours to complete one cell cycle. T/I C

Table 1 Cell phase data

Cell stage or phase	Area 1	Area 2	Percentage in each stage/phase	Time for each phase (h)
interphase	85	78		
prophase	14	9		
metaphase	3	1		
anaphase	2	3		
telophase	3	5		
cytokinesis	9	6		
Total cell count				

(a) Copy Table 1 into your notebook. Calculate the percentage of cells in each phase and how much time is spent in each stage or phase of the cell cycle.
(b) Create a circle graph for either Area 1 or Area 2, indicating the percentage of cells in each stage/phase.

65. Your red blood cells and your outermost layer of skin cells do not contain a nucleus. (2.5, 3.2) K/U T/I A
(a) What happens to these cells, in terms of the cell cycle?
(b) If a skin cream claimed to restore cells in your skin, could your cells produce new skin cells?
(c) Design a test for the claim that new skin cells are being created by a skin cream.

66. Table 2 shows the approximate lifespan of different human body cells. T/I A

Table 2 Human Cell Lifespans

Cell type	Lifespan
stomach and intestinal cells	2 to 5 days
skin cells	2 weeks
red blood cells	3 months
bone	10 years
brain	30 to 50 years

(a) Why do you think different cells have such a range of lifespans?
(b) How would this affect injuries in different regions of the body?

67. A new sunblock product claims to have an SPF rating of 60. This means that you get 60 times your body's natural defence against the Sun's UVB rays. T/I C A
(a) How could you test this claim? Comment on the ethics of your test.
(b) If many skin cancers are associated with UVA rays, what impact could this product have on cancer rates?

68. Following the Chernobyl nuclear disaster in 1986, many people developed cancers. The radiation damaged chromosomes, so they could not control their cells in the normal way. (2.7) K/U T/I A
   (a) How might this affect cell division?
   (b) How could you detect this effect on cells?

69. Most flowering plants will produce fruit only if the flower has been pollinated. How might this affect a fruit grower's decision to use pesticides? T/I A

70. When fruits mature, they begin to release a gas called ethene. Ethene will cause nearby fruits to ripen as well. T/I A
   (a) How could you test this effect using a ripe banana?
   (b) How could you design a method to reduce this effect so that you could make fruit last longer in your home?

71. How could you demonstrate that the vascular tissue connects all parts of a plant? T/I

72. Choose two related organ systems and explain how they work together. (3.3, 3.4, 3.6, 3.8, 3.10, 3.11, 4.2, 4.4) K/U A

73. Suggest what causes solutions to move water through the xylem and phloem of plants. Explain your answer. (4.2) K/U

Create and Evaluate

74. Using students in your class, act out the phases of the cell cycle, including mitosis. (2.5) K/U C

75. Choose one local fundraising activity associated with one of the diseases described in this unit. Prepare a public relations campaign for the students in your class or school to promote this event. (2.7, 3.2, 3.3, 3.4, 3.6, 3.8, 3.10) K/U C A

76. For plants, what are the advantages and disadvantages of growing tall? (4.1, 4.2, 4.4, 4.6) A

77. A friend tells you that you should not have a plant in your bedroom because at night the plant takes oxygen from the air that you need to breathe. Is your friend correct? Explain your answer. (4.4) K/U T/I

78. Write a brief paragraph describing the journey of an oxygen molecule as it enters the nose and finally ends up in a muscle cell. (3.3, 3.6, 3.8) K/U C

79. Grafting is a technique sometimes used to grow plants and trees. For example, to produce an apple tree by grafting, the stem from one tree is inserted into a cut in the stump of another tree. If the graft is successful, these two parts will grow together and form a new apple tree. Is grafting an example of sexual or asexual reproduction? How do you know? (2.3, 4.1, 4.6) A

Reflect On Your Learning

80. Which phase of the cell cycle is the most difficult to clearly identify? Explain why you had difficulty with this particular phase. Ask your classmates for tips in helping to identify this phase.

81. Prior to completing this unit, what were your thoughts and ideas about transgenic plants? How has your understanding of GMOs changed?

82. Before starting this unit, you may have had certain beliefs about the causes of cancer and heart disease. Is there anything you can do now to reduce your own risk of developing cancer or heart disease? Explain.

83. In this unit, you learned that plants and animals have both transport and gas exchange systems.
   (a) What relationship is there between the transport and gas exchange systems of plants and animals?
   (b) Does this relationship make sense to you? Explain why or why not.
   (c) What other comparisons between plants and animals can you make?

84. In this unit, you have read about many uses for imaging technology. Has this information made you interested in a career in imaging technology? Briefly explain your answer.

85. (a) Before studying this unit, what were your opinions about organ donation?
    (b) How has your understanding of organ donation changed?

Web Connections 

86. Copy or trace Figure 1 into your notebook. Find out what each sections of the ECG corresponds to the different parts of a heartbeat cycle. Use the Internet and other resources to help you. Use your research to add descriptive labels to your ECG trace. T/I C

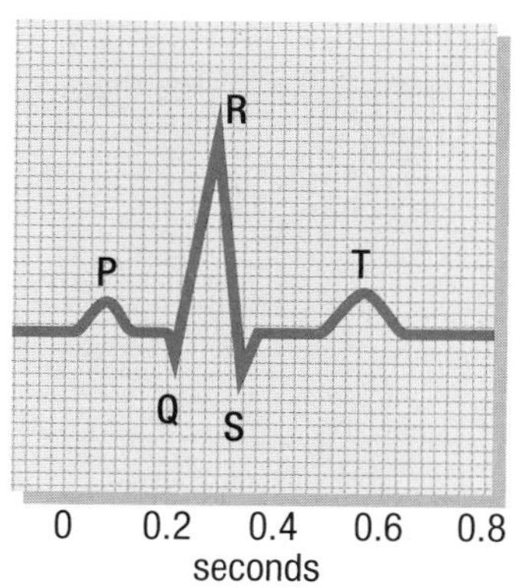

Figure 1 A portion of an electrocardiogram (ECG)

87. Research the differences in the digestive systems of mammals that eat only meat and mammals that eat only plants. Create two labelled models or diagrams to communicate your finding. T/I C

88. (a) What is tissue engineering? Summarize your findings in a brief report.
    (b) Name at least one organ that should be considered for future tissue engineering research. Explain why and how this organ should be considered for this process. T/I C A

89. Choose a type of cancer that you are interested in learning more about. Research this disease to determine whether there are any environmental causes, lifestyle factors, or other contributing factors. Create a short radio "spot" telling listeners how to minimize their risk of contracting this disease. T/I C A

90. Investigate the phenomenon of colony collapse disorder. This disorder has been reported in North America and Europe. T/I C A
    (a) Conduct research to identify what this is and where it has been observed.
    (b) Predict what impact this might have on our food supply. Write up your prediction as a letter to the editor of an agricultural magazine.

SELF-QUIZ

The following icons indicate the Achievement Chart category addressed by each question.

K/U Knowledge/Understanding | T/I Thinking/Investigation | C Communication | A Application

For each question, select the best answer from the four alternatives.

1. Which of the following organelles is present in plant cells but not in animal cells? (2.1) K/U
   (a) nucleus
   (b) cell wall
   (c) cytoplasm
   (d) Golgi body

2. Which of the following lists the levels of organization within an animal in order from simplest to most complex? (3.1, 4.1, 4.2) K/U
   (a) tissue, organ, cell, organ system, organism
   (b) organism, tissue, organ system, cell, organ
   (c) organ, organism, organ system, tissue, cell
   (d) cell, tissue, organ, organ system, organism

3. Which phrase correctly describes the vascular tissue system in a plant body? (4.2) K/U
   (a) tissues that store carbohydrates
   (b) tissues that manufacture nutrients
   (c) tissues that form the outer surfaces of plant parts
   (d) tissues that transport materials throughout the plant

4. Which type of animal cell is able to divide into different types of specialized cells? (3.2) K/U
   (a) stem cell
   (b) skin cell
   (c) nerve cell
   (d) blood cell

Indicate whether each of the statements is TRUE or FALSE. If you think the statement is false, rewrite it to make it true.

5. Both malignant and benign tumours are made up of cells that have undergone uncontrolled growth and division. (2.7) K/U

6. Ligaments attach skeletal muscle to bones. (3.8) K/U

7. Most digestion occurs in the small intestine. (3.3) K/U

Copy each of the following statements into your notebook. Fill in the blanks with a word or phrase that correctly completes the sentence.

8. Woody plants, such as trees, contain both apical and ____________ meristems. (4.6) K/U

9. The respiratory system and the ____________ system work together to provide oxygen to the body and to remove carbon dioxide from the body. (3.4, 3.6) K/U

Match each term on the left with the most appropriate description on the right.

10. (a) mitochondrion
    (b) cytoplasm
    (c) nucleus
    (d) cell membrane
    (e) chloroplast

    (i) holds the genetic information that controls all cell activities
    (ii) absorbs light energy and allows photosynthesis to occur
    (iii) contains enzymes that help convert stored energy into an easily usable form
    (iv) consists of a fluid in which all the organelles are suspended
    (v) allows some substances to enter a cell while keeping other substances out (2.1) K/U

Write a short answer to each of these questions.

11. Why do guard cells generally close the stomata on plant leaves at night? (4.4) K/U

12. Name two places meristematic cells are found in plants. (4.2, 4.6) K/U

13. Identify the main functions of each of the organ systems below. (3.3, 3.4, 3.10) K/U
    (a) digestive system
    (b) circulatory system
    (c) nervous system

14. Do all single-celled organisms have the same internal organization? Explain your answer using examples. (2.1, 2.9) K/U

15. You explain to your family that your body sheds millions of dead skin cells every day, and that every day your body replaces these dead cells with new ones. Your little sister wants to know why she does not see piles of dead skin cells all over the floor of the house. How would you answer her question? (2.3, 2.5) T/I C

16. Muscle cells and white blood cells are two different kinds of specialized cells. (2.9, 3.4, 3.8) K/U T/I
    (a) Compare the movement and arrangement of these two kinds of cells.
    (b) Describe how each of these cells' arrangement and movement supports their functions.

17. List four choices you can make to reduce your risk of developing cancer. (2.7) C

18. Explain how the circulatory system uses diffusion to move the following gases through the body. (2.3, 3.4) T/I
    (a) oxygen
    (b) carbon dioxide

19. You have learned that plants with colourful flowers, such as dandelions, depend on insects and other animals to spread their pollen, while plants with small, drab flowers, such as grasses, rely on wind to spread their pollen. Which type of plant, a dandelion or a grass, would you expect to produce more pollen? Justify your answer. (4.1) T/I

20. Your friend tells you that you should always "warm up" your muscles before going for a long run. (3.4, 3.8) A
    (a) What do you think it means to "warm up" muscles?
    (b) Do you agree with your friend? Why or why not?

21. Scientists have developed tests to screen for a number of diseases. Some people say we should use these tests to screen for as many diseases as possible. Others believe overuse of these tests leads to unnecessary worry and expense because many tests result in false positives. A false positive result shows that a condition exists when in reality it does not. Do you think we should or should not test for as many diseases as possible? Justify your answer. T/I

22. Describe the process of cell division in your own words. Use the following terms in your description: *mitosis*, *interphase*, *cytokinesis*, *prophase*, *anaphase*, *metaphase*, and *telophase*. (2.5) K/U C

23. Some plants produce flowers that bloom at night. These plants still need to reproduce by pollination. Explain how these night-blooming flowers could attract animal pollinators in the dark. (4.1) A

24. You are writing an article about the health benefits and risks of X-rays for your school newspaper. For your research, you will interview a dental technician. What are three questions you could ask him or her about the issue? (2.7) C

UNIT C Chemical Reactions

OVERALL Expectations

- analyze a variety of safety and environmental issues associated with chemical reactions, including the ways in which chemical reactions can be applied to address environmental challenges
- investigate, through inquiry, the characteristics of chemical reactions
- demonstrate an understanding of the general principles of chemical reactions, and various ways to represent them

BIG Ideas

- Chemicals react with each other in predictable ways.
- Chemical reactions may have a negative impact on the environment, but they can also be used to address environmental challenges.

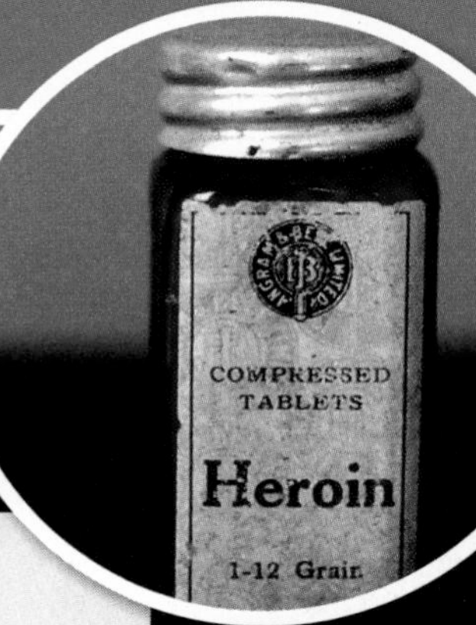

Focus on STSE

ASPIRIN **AND HEROIN**

Got a headache? Like millions of other pain sufferers around the world, you may turn to Aspirin for quick relief. Surprisingly, Aspirin was first made using the same chemical reaction that produced heroin—a highly addictive and illegal narcotic.

The Bayer drug company first made Aspirin in the 1890s, but officials initially ignored the new product. In 1898, Bayer chemists used the same reaction on morphine, a well-known pain reliever at the time. Tests on the new product were impressive: it was a good pain reliever and an even better cough remedy. Plus, it made the patient feel wonderful! Bayer had invented heroin. Soon, a new Bayer cough syrup called "Heroin" was in drug stores around the world. Shortly thereafter, reports about heroin addiction began to appear. Public pressure forced Bayer to stop producing their heroin-laced cough syrup in 1913.

Heroin medications are no longer available. Bayer eventually recognized Aspirin's potential, and it is now widely used around the world as a painkiller.

The history of Aspirin and heroin teaches us the importance of thoroughly testing a new product before making it available. It also shows how the chemicals we produce through chemical reactions can have both costs and benefits.

1. List some costs and some benefits for each of the following substances. Consider various perspectives (environmental, social, economical, and so on) as you think about the costs and benefits. T/I A
   (a) cleaning products
   (b) salt
   (c) explosives
   (d) fuels
   (e) plastic
   (f) pesticides
2. Discuss the following questions with your partner. C A
   (a) What other costs and benefits could you add to your table?
   (b) Why might some chemical companies be willing to introduce new chemicals into society and the environment without thoroughly testing them first?
   (c) How can we maximize the benefits of chemicals and minimize the costs?

UNIT C

LOOKING AHEAD

UNIT C
Chemical Reactions

CHAPTER 5
Chemicals and Their Properties

Scientists and engineers think about the properties of the materials used in new products.

CHAPTER 6
Chemicals and Their Reactions

Chemicals are all around us, and can sometimes react in surprising ways.

CHAPTER 7
Acids and Bases

Acids and bases are important substances in our lives and in the environment.

UNIT TASK Preview

Acid Shock

In this unit, you will explore the chemical reactions that chemicals undergo. Some reactions, like those involved in the formation of acid rain, have a negative impact on the environment. But there is a good side to chemical reactions as well. Chemical reactions can also be used to undo some of the environmental damage caused by human activity.

In the Unit Task, you are part of a research team investigating tadpoles and frogs in a local stream. With each spring thaw, the population of these organisms crashes and then recovers. Preliminary research shows that the cause of their death is the acidity of the melting snow. However, what remains a mystery is the sudden drop in the population followed by a gradual recovery.

Your task is to find the reason for this dip by using a model of the situation. You will also suggest ways to prevent the loss of the tadpoles each spring.

In the Unit Task, you will use the knowledge and skills acquired in this unit to

- use a model to test whether the amount of acid released by a melting solid varies as the solid melts
- suggest how chemical reactions can be used to prevent the loss of tadpoles in the stream during the first thaw of spring

UNIT TASK Bookmark

The Unit Task is described in detail on page 300. As you work through the unit, look for this bookmark and see how the section relates to the Unit Task.

ASSESSMENT

You will be assessed on how well you

- plan and conduct a test of the model
- communicate the results of your test
- evaluate your model
- justify your suggestion for preventing acid shock

What Do You Already Know?

PREREQUISITES

Concepts	Skills
• Density and buoyancy • Models of the atom • Classification of matter	• Writing chemical formulas • Observing properties of substances • Following safety precautions in the lab

1. Two cans of cola are placed in a tank of water. The can containing regular cola sinks to the bottom of the tank and the can containing diet cola floats to the surface. T/I
   (a) Using the provided observations, compare the densities of the cans of cola with the density of water.
   (b) Why might there be a difference in the densities of the two colas?

2. In your notebook, write the chemical formula (selected from the list on the right) of each of these substances. K/U

(a) hydrogen	H_2O
(b) carbon dioxide	NaCl
(c) table salt	CO_2
(d) hydrogen chloride	H_2
(e) water	O_2
(f) oxygen	HCl

3. List some of the properties of the substances shown in Figure 1. K/U

Figure 1 (a) the gold in a ring (b) the water in a kettle

4. Draw Bohr-Rutherford diagrams to represent the following atoms: C
   (a) lithium
   (b) carbon
   (c) chlorine
   (d) argon

5. In 1909, Ernest Rutherford fired positively charged alpha particles at a very thin sheet of gold foil. As he expected, the majority of the particles passed through the foil. However, a small number of alpha particles rebounded off the foil. Which of the following models of the atom best explains these observations. Why? K/U

Model A	Model B	Model C
The atom is a hard sphere—like these billiard balls.	The atom is a positive sphere with embedded electrons—like raisins in a muffin.	The atom has a small, dense positive core orbited by electrons—like planets orbit a star.

6. (a) What fundamental particle inside the atom is responsible for the "hair-raising experience" shown in Figure 2?

Figure 2

   (b) Compare the three fundamental particles in an atom with respect to size, mass, charge, and location. K/U

7. What safety precautions are being taken in Figure 3? K/U

Figure 3

CHAPTER 5

Chemicals and Their Properties

KEY QUESTION: How do the properties of compounds influence the technologies we develop, the products we buy, and the world we live in?

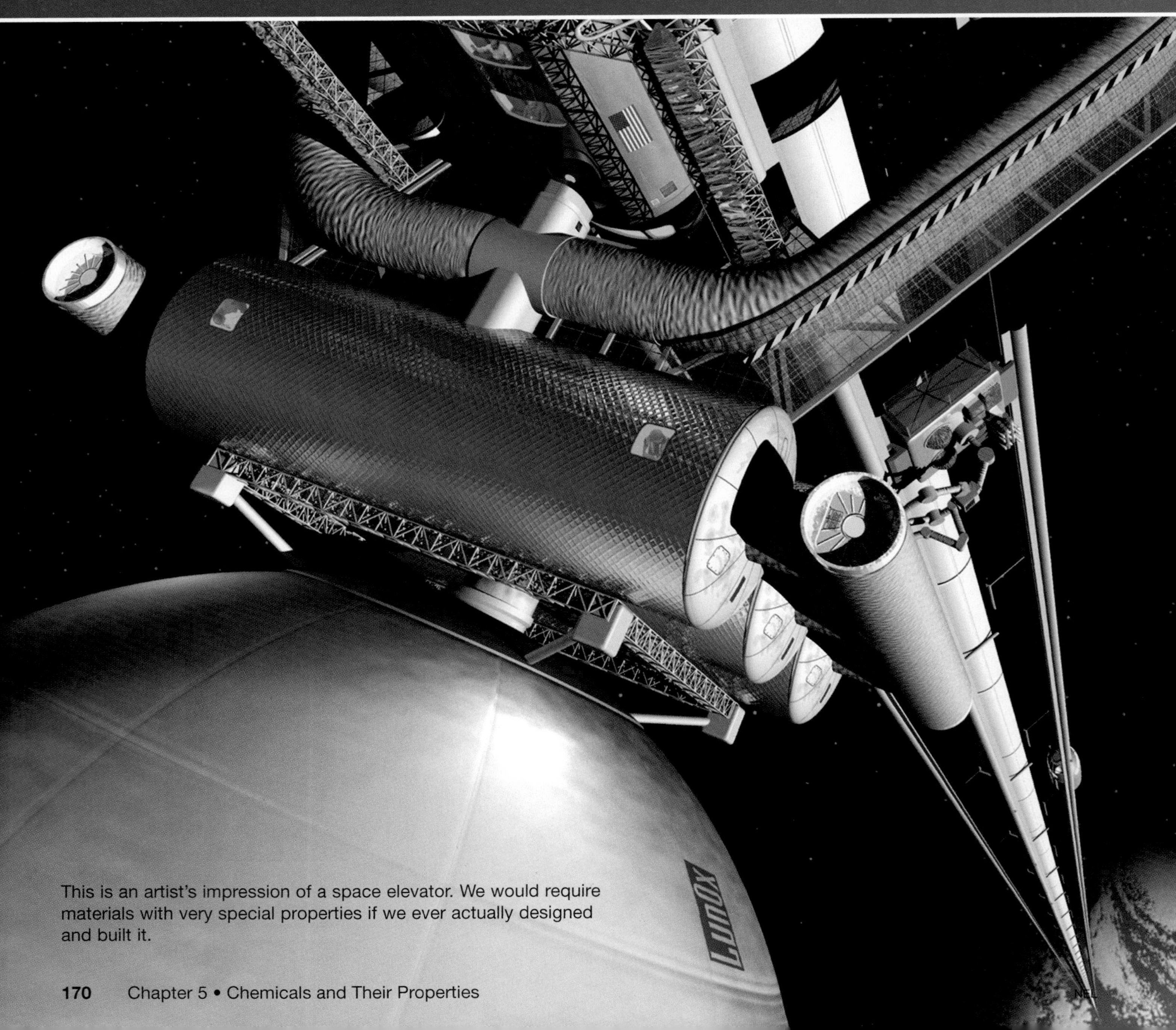

This is an artist's impression of a space elevator. We would require materials with very special properties if we ever actually designed and built it.

UNIT C
Chemical Reactions

CHAPTER 5
Chemicals and Their Properties

CHAPTER 6
Chemicals and Their Reactions

CHAPTER 7
Acids and Bases

KEY CONCEPTS

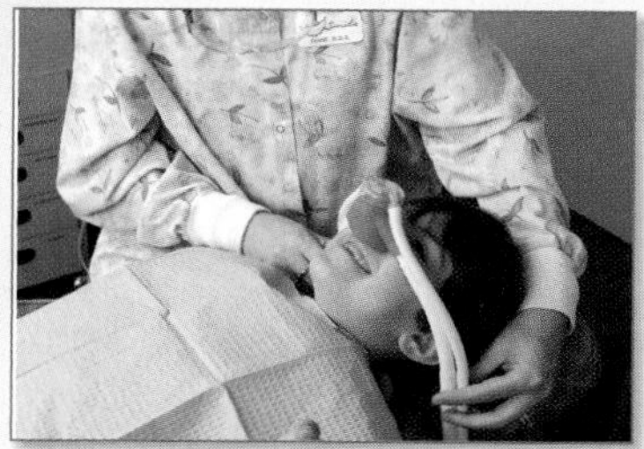
A substance's chemical and physical properties determine its usefulness and effects.

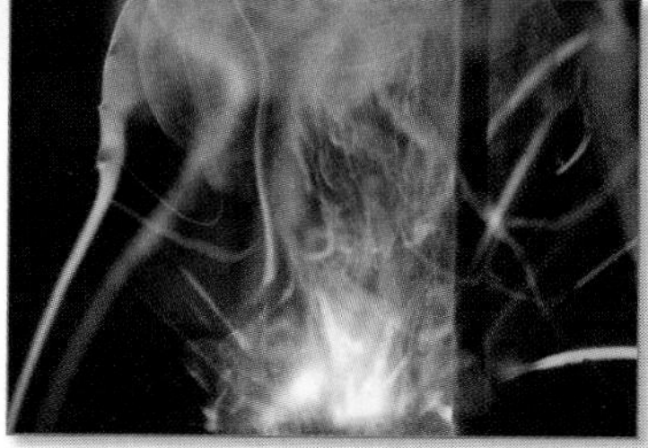
Changes can be classified as chemical or physical.

We can classify pure substances by observing their properties.

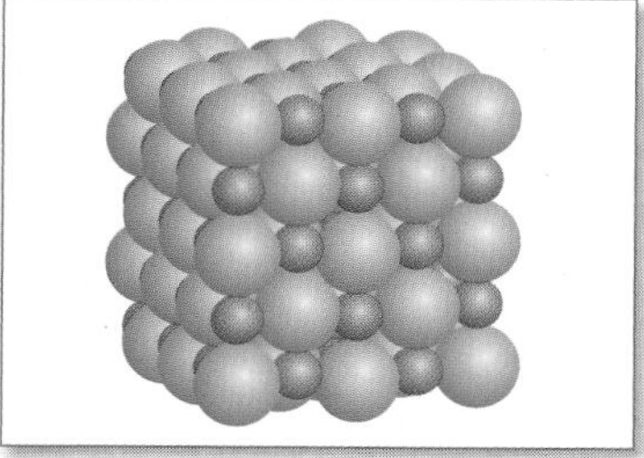
Ionic compounds are made up of positive and negative ions.

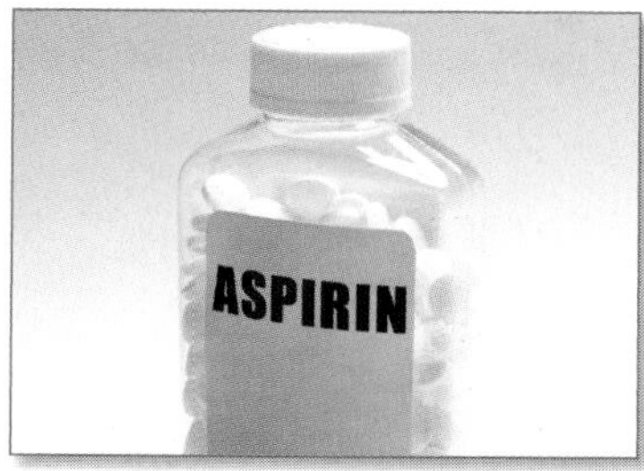

Molecular compounds are made up of distinct molecules.

Many consumer products are developed from petrochemicals.

ENGAGE IN
SCIENCE

SPACE ELEVATOR

Imagine stepping onto an elevator, pressing the button for the 12 millionth floor, and going for the ride of your life. Within an hour or two, the land beneath your feet and the warmth of the Sun are replaced by the cold darkness of space.

Getting people and cargo into space the old-fashioned way—by using rockets—is expensive, dangerous, and unreliable, so NASA engineers are looking for a better way. Some believe that a space elevator may be the answer.

The space elevator would consist of a long cable—more than three times as long as the diameter of Earth—firmly attached to a point on Earth's equator. At the other end, about 40 000 km straight up, would be a large mass. As long as this mass orbits Earth at the same rate as Earth is rotating, the cable connecting it to Earth remains taut. This would allow a "space-proof" elevator car to climb the cable. But plenty of questions need to be answered before we start putting this plan into action. For example, what does this cable have to do, and where must it operate? What physical and chemical properties should the cable have? Are there any common substances that have at least some of these properties? Do these common substances have any properties that make them unsuitable? Could the substances be changed to overcome this problem? And, perhaps most importantly, how would this device impact our lives and our planetary home?

WHAT DO YOU THINK?

Many of the ideas you will explore in this chapter are ideas that you have already encountered. You may have encountered these ideas in school, at home, or in the world around you. Not all of the following statements are true. Consider each statement and decide whether you agree or disagree with it.

1 The label on a chemical product provides all the information you need to use the product safely.
Agree/disagree?

2 Recycling used motor oil is common practice.
Agree/disagree?

3 Pool water is a much better conductor of electricity than pure water.
Agree/disagree?

4 Elements are more reactive and more hazardous than the compounds that they form.
Agree/disagree?

5 Bottled water is better for your health than tap water.
Agree/disagree?

6 Adding manufactured chemicals to the environment is a bad thing.
Agree/disagree?

FOCUS ON
WRITING

Writing a Summary

When you write a summary you condense a text by restating only the main idea and key points in your own words. Personal opinions or interpretations are not included. Use the summary below and the information boxes to help you improve your writing.

WRITING TIP

As you work through the chapter, look for tips like this. They will help you develop literacy strategies.

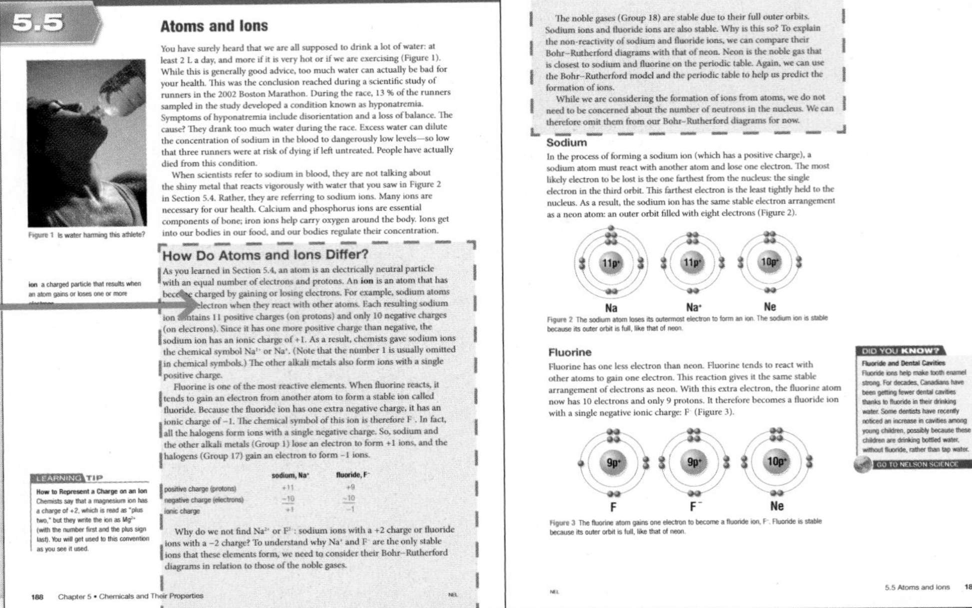

How Do Atoms and Ions Differ?

State the general topic of the text.

Atoms are electrically neutral particles with an equal number of electrons and protons. Ions are atoms that have been charged by gaining or losing electrons. The chemical formula for ions depends upon the number of positive and negative charges of each.

Organize ideas and information in the same order as in the original text.

Notice that the examples, descriptions, and explanations that were in the original text above have been left out.

Tools such as the Bohr-Rutherford model and the Periodic table can help a scientist make predictions about the formation of ions. The number of neutrons in a nucleus is not considered when looking at the formation of ions from atoms.

Write one clear sentence for each major point.

Write a closing sentence that explains how the ideas are connected together.

Ions are atoms that have been charged positively or negatively by subtracting or adding electrons.

Properties and Changes

Chemistry is the study of the substances around us: what is in them, what they do, and what they are used for. These substances can be the laboratory chemicals that you will use in this unit. They can also be common things like air, water, the food you eat, and the products you buy. An understanding of chemistry teaches us how to change substances into new and useful products. It also teaches us how to carry out these changes in socially and environmentally responsible ways.

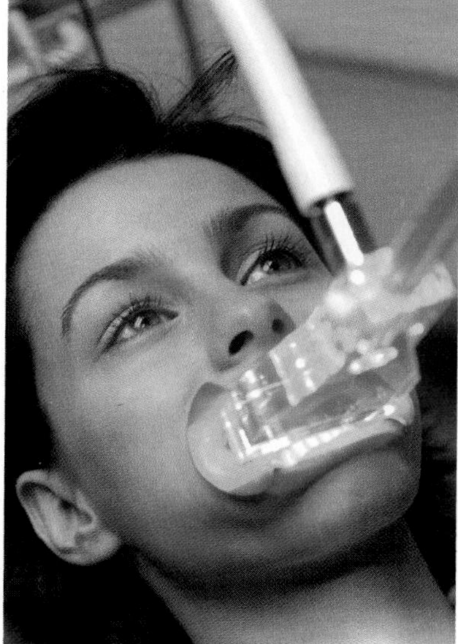

Figure 1 Should we change the natural colour of our teeth?

Physical and Chemical Properties

Does a "whiter than white" Hollywood smile appeal to you (Figure 1)? What are the benefits of a whiter smile? Before you jump on the "whiter smile" bandwagon, check your facts: research the pros and cons using reliable sources. Consider these factors:

- Teeth are not naturally pure white. They can range in colour from off-white to yellow. Furthermore, teeth naturally darken with age.
- Whitening your teeth does not make them healthier. Perfectly white teeth are still subject to decay and gingivitis.
- Whitening procedures are not permanent and must be repeated periodically to keep the teeth white.
- According to the Canadian Dental Association, the long-term effects of whitening have not been thoroughly researched.

There are two common ways of whitening teeth: surface whitening and bleaching. In the case of surface whitening, a hard, abrasive material such as baking soda is used to scrape surface stains off your teeth. Bleaching is a chemical process that uses a reactive substance such as hydrogen peroxide, and sometimes light.

We have just described baking soda as being a hard, abrasive material. Hardness is an example of a physical property of matter. A **physical property** is a characteristic or description of a substance. Colour, texture, density, smell, solubility, taste, melting point, and physical state are other common physical properties.

physical property a description of a substance that does not involve forming a new substance; for example, colour, texture, density, smell, solubility, taste, melting point, and physical state

chemical property a description of what a substance does as it changes into one or more new substance(s)

Conversely, a **chemical property** is a characteristic behaviour that occurs when the substance changes into something new. A useful chemical property of hydrogen peroxide is that it bleaches coloured substances.

Other common examples of chemical properties are listed in Table 1. You will learn more about these properties in this unit.

Table 1 Examples of Chemical Properties

Chemical property	Example
reaction of an acid with a base	Vinegar reacts with baking soda to produce carbon dioxide gas.
flammability	Gasoline burns easily if ignited.
bleaching ability	Hydrogen peroxide breaks down the pigment (colour) in hair.
corrosion	Discarded batteries in landfill sites break down readily when they come in contact with groundwater.

CITIZEN **ACTION**

Cadmium Cleanout

Nickel–cadmium or NiCd batteries were the first popular rechargeable batteries for small appliances (Figure 2). They cannot be used indefinitely, though. They lose their ability to hold their charge after many discharge-recharge cycles. Then the batteries have to be replaced and disposed of. Some are recycled, but many NiCd batteries go to landfill sites. Over time, the batteries' toxic contents leak out. It is estimated that over 50 % of the cadmium leaching from landfills into groundwater comes from NiCd batteries. This is an environmental concern because cadmium is highly toxic. It is a human carcinogen and has been linked to lung, liver, and kidney disorders. Furthermore, cadmium cannot be broken down into a safer substance. The best we can do to reduce the danger is ensure that cadmium does not leak into our air and water.

GO TO NELSON SCIENCE

What Can You Do to Help?

Search your home for small battery-powered appliances that are no longer in use: old cellphones, portable phones, laptop computers, cordless power tools, and electric toothbrushes. Check the battery for the "NiCd" symbol. If it is a NiCd battery, remove it and bring it to an authorized recycling centre or hazardous waste collection depot.

Figure 2 NiCd battery

WRITING **TIP**

Summarizing Text
You can condense a text by combining sentences and omitting examples. For example, "A change that does not produce a new substance is called a physical change. Changes of state (including melting, evaporation, condensation, sublimation, and dissolving) are examples of physical changes." can be shortened to "Physical changes such as changes of state do not make new substances."

Physical and Chemical Changes

A change that does not produce a new substance is called a physical change. Changes of state (including melting, evaporation, condensation, sublimation, and dissolving) are examples of physical changes. Many physical changes can be reversed, while others cannot. Dissolving sugar in water can easily be reversed by evaporating off the water. The process of cutting logs into lumber, however, cannot be reversed. Think of the physical changes that you have encountered today. Are they reversible?

The change that a substance goes through to produce a new substance (or more than one new substance) is called a chemical change. Table 2 shows some of the clues indicating that a chemical change may have taken place. Note, however, that these clues are not conclusive evidence that a chemical change took place. For example, boiling water produces a gas; so does mixing baking soda with vinegar. One of these events is a chemical change; the other is not. Which is which? The only way to tell for sure that a chemical change has taken place is to conduct further tests on the products. If the products are different from the starting materials, then a chemical change has taken place.

Many chemical changes, like a forest fire, cannot be reversed. However, some chemical changes *can* be reversed. For example, the design of rechargeable batteries is based on a reversible chemical change. As you use the battery, the chemical change that generates electricity produces new substances. As the battery is recharged, these substances are changed back to the original chemicals.

LEARNING **TIP**

What It's Like and What It Does
A ***property*** is like an adjective: it describes what the substance is like. A ***change*** is like a verb: it describes what the substance is doing.

Table 2 Possible Evidence of Chemical Changes

Visible change	Example	Visible change	Example
A new colour appears.		A solid material (a precipitate) forms in a liquid.	
Heat or light is produced or absorbed.		The change is (generally) difficult to reverse.	
Bubbles of gas are formed.			

RESEARCH THIS CHEMICALS FOR YOUR HAIR

SKILLS: Researching, Analyzing the Issue, Communicating, Evaluating

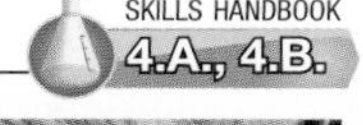

Hair stylists consider the properties of hair dyes when advising clients on the best product for them (Figure 3). Some hair dyes are temporary, while others are permanent.

1. Research how temporary hair dyes work.
2. Research how permanent hair dyes work.

A. Using your researched information, classify the action of temporary and permanent hair dyes as either chemical or physical changes. Justify your answer. K/U T/I

B. Identify at least one disadvantage of each process. T/I A

C. If you were to dye your hair, which process would you use? What concerns would you have about using this process? A

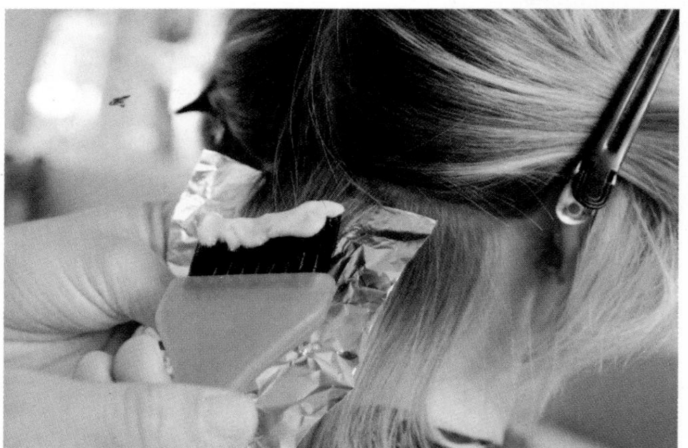

Figure 3 This workplace uses chemistry.

IN SUMMARY

- Properties are either chemical properties (which describe the ability of a substance to form one or more different substances) or physical properties (which describe a substance when it is not in the process of forming a new substance).
- Changes are either chemical changes (in which a substance changes into one or more different substances) or physical changes (in which a substance remains the same substance but changes its physical properties in some way).

CHECK YOUR LEARNING

1. (a) Describe an idea in the reading that is new to you.
   (b) How does this idea add to your understanding? K/U
2. Classify each of the following observations as an example of either a chemical or a physical property. K/U
   (a) Liquid nitrogen boils at –196 °C.
   (b) Propane, leaking from a damaged tank, ignites easily.
   (c) Silver jewellery tarnishes (darkens) in air.
   (d) Spilled oil generally floats on the surface of water.
   (e) Meat darkens when it is heated on a grill.
   (f) Sulfur trioxide changes to sulfuric acid in the atmosphere.
3. What kind of change is described in each of the following situations? Justify your answer. K/U
   (a) Air is often blended into ice cream to give it a lighter texture.
   (b) When popping corn is heated, water inside the kernels becomes a gas and expands. This creates enough pressure to explode the kernel.
   (c) A loud pop is heard when a lit match is placed at the mouth of a test tube containing hydrogen gas.
   (d) Ethanol is used as an alternative source of energy to power vehicles.
   (e) Geothermal energy from underground hot springs is used to heat water to turn turbines and produce electricity.
   (f) Some silver rings leave a green stain on your finger.
4. Figure 4 shows a warning label commonly found on chlorine bleaches. Is this hazard a result of a chemical property or a physical property of chlorine bleach? Justify your answer. K/U

Figure 4 Warning label from a bottle of chlorine bleach

5. Over time, crusty scales form on the heating coils inside a kettle. These can be removed by covering the coils with vinegar. As the scales disappear, bubbles of gas are observed. Does this method of cleaning a kettle represent a physical or chemical change? Explain. K/U A
6. The solvents in house paint allow the paint to flow smoothly onto a surface. Once paint is exposed to the air, these solvents evaporate and the paint dries. It is the job of a paint chemist to select the most suitable solvents. A
   (a) What chemical and physical properties should these solvents have?
   (b) What other characteristics should the ideal solvents have?
7. Auto mechanics sometimes use cola to remove the crusty solid that forms around battery terminals. When the cola comes in contact with this solid, bubbles of carbon dioxide gas are observed. Is the cleaning of the battery terminal a physical or chemical change? Explain. K/U A
8. Drain cleaners often produce a great deal of heat as they unclog drains. Is the action of drain cleaners a physical or chemical change? Explain. K/U A
9. Adding vinegar to milk causes small lumps called curds to form in the milk. Is this a physical or chemical change? Explain. K/U A
10. Describe two physical properties and one chemical property of the materials used for dental braces. K/U
11. After having read this section, has your opinion of teeth whitening procedures changed? Why or why not? What other information should be considered before having your teeth whitened? T/I

Processing Hazardous Waste

Crusty jars of cleaners, unlabelled bottles of solvents, clogged spray cans of insecticides, and rusted cans of paint lurk in the basements and garages of many Canadian homes. It is easy to recycle or neutralize many of these substances at a hazardous waste collection depot. Let's follow a typical box of household waste as it is processed.

When hazardous waste arrives at a hazardous waste collection depot, it is first sorted into different classes of materials. The workers separate paints, oils, solvents, pesticides, batteries, medicines, and so on (Figure 1).

Figure 1 About 50 % of the material that comes to a hazardous waste transfer station is paint and motor oil.

The sorted materials are then packaged and shipped to a hazardous waste processing company. There, the hazardous materials are emptied from their original containers into large drums of similar materials. Different types of waste motor oils, for example, are mixed together into one large container. The oil is then shipped elsewhere for treatment and recycling. This process is so effective that almost all the motor oil sold in Ontario is recycled oil.

Almost 85 % of the waste paint that you bring to the hazardous waste depot can be recycled. Hotz Environmental Services Inc. receives most of the waste paint in Ontario. At their recycling facility in Hamilton, Hotz sorts waste paint first into oil-based or latex paints. Each type of paint is then further sorted and blended into eight different colour groups. A master technician, affectionately known as the "brewmaster," controls the colour of each batch. Hotz then sells the paints to large institutional users such as Canadian and foreign governments.

Even single-use propane canisters can be recycled using a process developed by Hotz Environmental. First, they remove any remaining propane in the canister using a vacuum system. Propane recovered from discarded propane canisters can be used to heat the facility. Then they safely puncture, cut, and recycle the empty propane canisters as scrap metal.

We obviously need to recycle or neutralize hazardous waste. The technologies required to perform these tasks are well developed and in place. Table 1 shows the recycling and disposal processes for various kinds of hazardous waste. Hazardous household wastes do not have to be an environmental threat. How many Ontario households routinely bring their hazardous waste to waste collection depots for disposal? Unfortunately, only about 10 %. What can you do to make a difference?

Table 1 Treatment of Hazardous Waste Products

Hazardous waste	Process
solvents (e.g., paint thinner, adhesives, antifreeze)	• Flammable solvents are burned as fuel in high-temperature kilns used to produce cement.
automobile lead acid batteries	• Acids are neutralized. • Plastic cases are recycled into new cases. • Lead is refined and used to make new batteries.
propane barbecue tanks	• Tanks are emptied and refurbished into new tanks or converted to scrap metal and recycled.
pesticides	• Substances are chemically treated to make them harmless.
fluorescent lights (Figure 2)	• Toxic mercury in the lights is collected, purified, and reused.

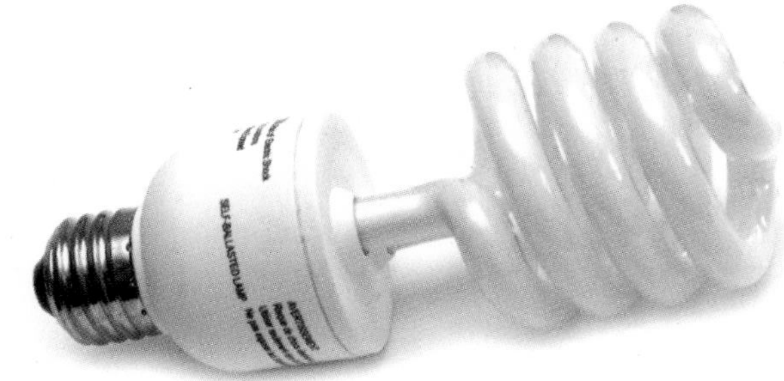

Figure 2 Fluorescent light

Identifying Physical and Chemical Changes

SKILLS MENU

Questioning	Performing
Hypothesizing	Observing
Predicting	Analyzing
Planning	Evaluating
Controlling Variables	Communicating

As you have already learned, a chemical property describes the ability of a substance to react to form something new. When this happens, we say that a chemical change has occurred.

In this activity, you will observe a number of physical and chemical changes. As you classify these changes, you will identify specific evidence of chemical change.

Purpose

To collect and use evidence to identify physical and chemical changes.

Equipment and Materials

- eye protection
- lab apron
- 2 test tubes
- test-tube rack
- Bunsen burner
- utility stand with clamp
- spark lighter
- test-tube stopper
- laboratory scoop
- warm water bath
- thermometer (optional)
- dropper bottles of
  - dilute hydrochloric acid, $HCl(aq)$
  - distilled water
  - dilute sodium hydroxide, $NaOH(aq)$
- magnesium ribbon, $Mg(s)$
- wooden splint
- copper(II) sulfate, $CuSO_4(s)$
- steel wool
- prepared test tube of lauric acid, $C_{12}H_{24}O_2(s)$

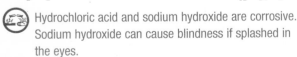

Hydrochloric acid and sodium hydroxide are corrosive. Sodium hydroxide can cause blindness if splashed in the eyes.

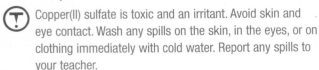

Copper(II) sulfate is toxic and an irritant. Avoid skin and eye contact. Wash any spills on the skin, in the eyes, or on clothing immediately with cold water. Report any spills to your teacher.

Use caution around the hotplate and water bath. Do not touch surfaces that might be hot. This activity involves open flames. Long hair should be tied back and loose clothing tucked in.

Procedure

1. Prepare a data table in which to record your observations during this activity.
2. Put on your eye protection and lab apron.

Part A

CHANGE 1

3. Add hydrochloric acid to a test tube to a depth of about 2 cm.
4. Add two 1 cm strips of magnesium ribbon to the test tube. Check for evidence of change occurring. Test the bottom of the test tube with your hand for temperature changes. Record your observations.
5. Place the test tube in a test-tube rack and wait 30 s for the gas produced to push any air out of the test tube.

CHANGE 2

6. Clamp the Bunsen burner to the utility stand for stability.
7. Light the Bunsen burner with a spark lighter; then light a splint from the burner flame.
8. Hold the burning splint near the mouth of the "acid + magnesium" test tube. Record your observations.
9. Dispose of the contents of the test tube as directed by your teacher.
10. Rinse the test tube with tap water.

Part B

CHANGE 3

11. Pour distilled water into a test tube to a depth of about 3 cm.
12. Add about 0.5 g of copper(II) sulfate (as much as about half an Aspirin tablet) to the test tube.
13. Stopper and invert the test tube several times to mix its contents well. Record your observations.

CHANGE 4

14. Remove the stopper. Add a small ball of steel wool (about the size of an Aspirin tablet) to the test tube (Figure 1).

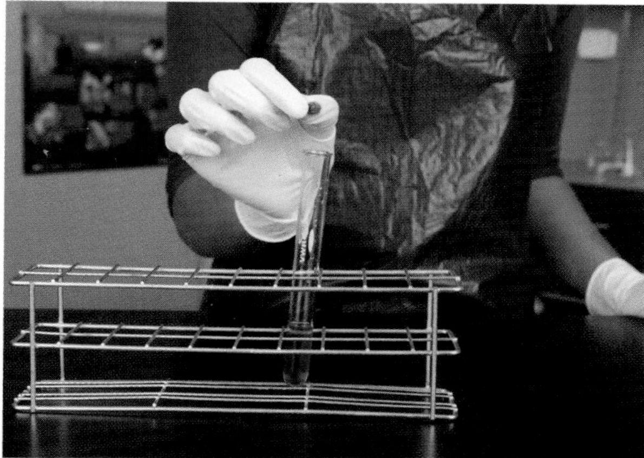

Figure 1 Adding steel wool to copper(II) sulfate solution

15. Stopper the test tube again and mix.
16. Allow the solids in the test tube to settle to the bottom. Record your observations.

CHANGE 5

17. Remove the stopper. Add about 5 drops of sodium hydroxide solution to the test tube.
18. Slowly add drops of hydrochloric acid to the test tube. Gently swirl the test tube after every couple of drops. Continue adding dropwise until the solid disappears. Record your observations.
19. Dispose of the contents of the test tube as directed by your teacher.

Part C

CHANGE 6

20. Examine a test tube of lauric acid.
21. Place the test tube in a warm water bath. Wait until the substance in the test tube completely liquefies.
22. Remove the test tube and cool it in a stream of tap water until its contents solidify again. Record your observations.

Analyze and Evaluate

(a) Classify each of the changes that you observed as either chemical or physical. Use specific evidence from your observation table to justify your inference. T/I

(b) Which changes were the most difficult to classify? Why? T/I

(c) Give one example, from your everyday life, of a physical change that is

  (i) reversible. Justify your inference.

  (ii) not reversible. Justify your inference. A

Apply and Extend

(d) Identify one chemical change in this activity that was reversible. What chemical could you add to reverse this change again? T/I

(e) In Change 2, you may have heard a "pop" when the burning splint was inserted into the mouth of the test tube. Name the gas produced in the test tube. K/U

(f) Plan an experiment to determine the factors that could make the "pop" louder. Once your teacher approves your plan, conduct the experiment. T/I

5.3 Hazardous Products and Workplace Safety

While going about her normal cleaning duties one day, a school custodian unintentionally mixed two cleaning solvents. Soon, irritating fumes of toxic chlorine gas filled the room, making it difficult for her to breathe. She was rushed to hospital, where she died later that evening.

This tragic accident prompted an investigation. School board officials asked whether any other products used in schools were hazardous. One of the first places they looked for this information was the WHMIS information for each product.

What Is WHMIS?

Workplace Hazardous Materials Information System (WHMIS) provides Canadian workers with information on the safe use of hazardous products in their workplace. Employers must, by law, provide this information. The information is conveyed in three ways: WHMIS product labels, materials safety data sheets (MSDS), and worker training.

WHMIS Product Labels

A WHMIS product label is your first alert that the product may be hazardous. There are two common types of product labels in most workplaces: supplier labels and workplace labels. A supplier label is required on any hazardous material that is sold or imported to a workplace in Canada (Figure 1). A supplier label must always have a hatched border and must be written in both English and French. It must also include the name of the product, any relevant hazard symbol, the supplier's contact information, and a reference to the MSDS. A workplace label must appear on all hazardous materials produced in a workplace or transferred to other containers within a workplace (Figure 2). Workplace labels do not have to have a hatched border, but must show the name of the product, information on safe handling, an MSDS reference, and any relevant hazard symbol.

WRITING TIP

Writing a Summary

Use the topic sentence to restate the main idea of the text in your own words. For example, if the text says "A WHMIS product label is your first alert that the product may be hazardous.", you might paraphrase this main idea by saying "WHMIS labels are warnings about dangerous products."

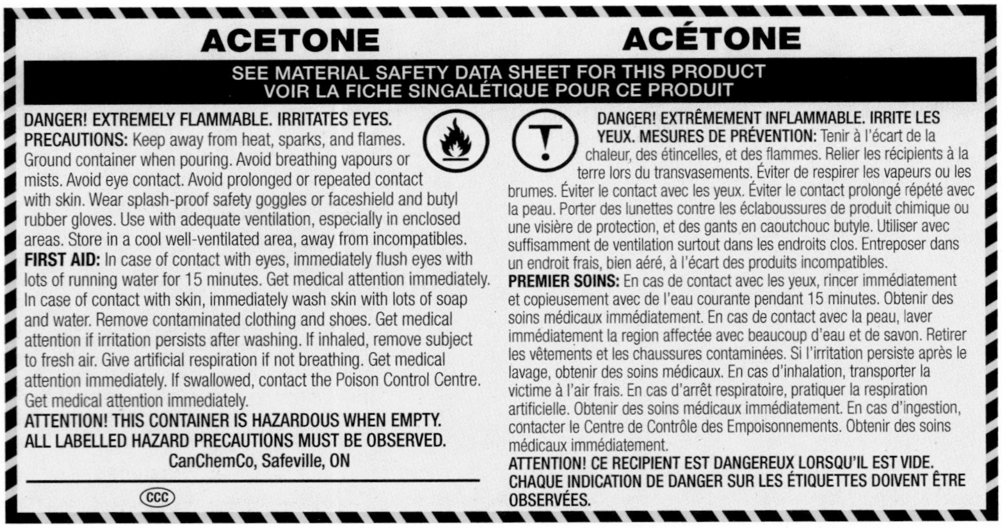

ACETONE **ACÉTONE**

SEE MATERIAL SAFETY DATA SHEET FOR THIS PRODUCT
VOIR LA FICHE SINGALÉTIQUE POUR CE PRODUIT

DANGER! EXTREMELY FLAMMABLE. IRRITATES EYES.
PRECAUTIONS: Keep away from heat, sparks, and flames. Ground container when pouring. Avoid breathing vapours or mists. Avoid eye contact. Avoid prolonged or repeated contact with skin. Wear splash-proof safety goggles or faceshield and butyl rubber gloves. Use with adequate ventilation, especially in enclosed areas. Store in a cool well-ventilated area, away from incompatibles.
FIRST AID: In case of contact with eyes, immediately flush eyes with lots of running water for 15 minutes. Get medical attention immediately. In case of contact with skin, immediately wash skin with lots of soap and water. Remove contaminated clothing and shoes. Get medical attention if irritation persists after washing. If inhaled, remove subject to fresh air. Give artificial respiration if not breathing. Get medical attention immediately. If swallowed, contact the Poison Control Centre. Get medical attention immediately.
ATTENTION! THIS CONTAINER IS HAZARDOUS WHEN EMPTY. ALL LABELLED HAZARD PRECAUTIONS MUST BE OBSERVED.
CanChemCo, Safeville, ON

CCC

DANGER! EXTRÊMEMENT INFLAMMABLE. IRRITE LES YEUX. MESURES DE PRÉVENTION: Tenir à l'écart de la chaleur, des étincelles, et des flammes. Relier les récipients à la terre lors du transvasements. Éviter de respirer les vapeurs ou les brumes. Éviter le contact avec les yeux. Éviter le contact prolongé répété avec la peau. Porter des lunettes contre les éclaboussures de produit chimique ou une visière de protection, et des gants en caoutchouc butyle. Utiliser avec suffisamment de ventilation surtout dans les endroits clos. Entreposer dans un endroit frais, bien aéré, à l'écart des produits incompatibles.
PREMIER SOINS: En cas de contact avec les yeux, rincer immédiatement et copieusement avec de l'eau courante pendant 15 minutes. Obtenir des soins médicaux immédiatement. En cas de contact avec la peau, laver immédiatement la region affectée avec beaucoup d'eau et de savon. Retirer les vêtements et les chaussures contaminées. Si l'irritation persiste après le lavage, obtenir des soins médicaux. En cas d'inhalation, transporter la victime à l'air frais. En cas d'arrêt respiratoire, pratiquer la respiration artificielle. Obtenir des soins médicaux immédiatement. En cas d'ingestion, contacter le Centre de Contrôle des Empoisonnements. Obtenir des soins médicaux immédiatement.
ATTENTION! CE RECIPIENT EST DANGEREUX LORSQU'IL EST VIDE. CHAQUE INDICATION DE DANGER SUR LES ÉTIQUETTES DOIVENT ÊTRE OBSERVÉES.

Figure 1 This supplier label is on the container of acetone when it is purchased.

Flammable

- Keep away from heat, sparks, and flames
- Wear butyl rubber gloves and safety goggles
- Use with local exhaust ventilation

Material Safety Data Sheet Available

Figure 2 A workplace label is required when acetone is transferred to other containers that the workers use on site.

Materials Safety Data Sheet (MSDS)

A product label provides only a limited amount of safety information. For more details, consult the materials safety data sheet (MSDS) that comes with the product. This includes information about any hazardous properties, safe handling and storage procedures, and what to do in case of an emergency. Many manufacturers post MSDS information online. Workers should consult the MSDS *before* using the product.

To look at online MSDS information,
GO TO NELSON SCIENCE

Worker Training

There are potentially hazardous products in almost every workplace. It is important for everybody to understand the hazards. People who frequently work with hazardous products must have special training. They must know how to handle the products and what to do if something goes wrong.

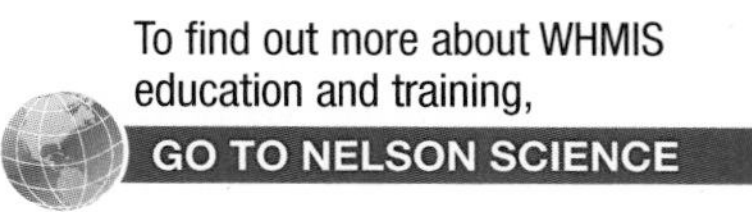
To find out more about WHMIS education and training,
GO TO NELSON SCIENCE

RESEARCH THIS WHICH BLEACH IS BEST?

SKILLS: Researching, Analyzing the Issue, Communicating, Evaluating

SKILLS HANDBOOK 4.B., 4.C.8.

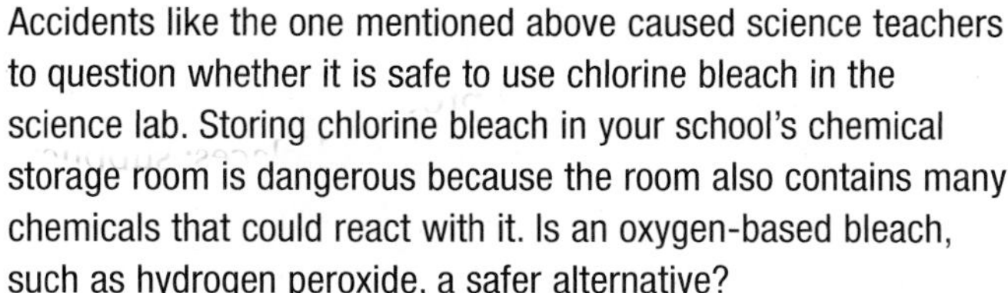

Accidents like the one mentioned above caused science teachers to question whether it is safe to use chlorine bleach in the science lab. Storing chlorine bleach in your school's chemical storage room is dangerous because the room also contains many chemicals that could react with it. Is an oxygen-based bleach, such as hydrogen peroxide, a safer alternative?

1. Research the MSDS of 6 % hydrogen peroxide bleach.
2. Research the MSDS for chlorine bleach.

GO TO NELSON SCIENCE

A. Compare the hazards of the two types of bleach. T/I
B. Based on what you have read, is hydrogen peroxide safer to use? Why? T/I
C. What extra information, if any, do you need before you can say which is better overall? T/I A

IN SUMMARY

- WHMIS legislation requires employers to provide information on the safe use of hazardous products used in the workplace.
- Safety information is provided through product labels, MSDS, and worker training.

CHECK YOUR LEARNING

1. Why is worker training an essential part of WHMIS? K/U
2. When is a workplace label required? K/U
3. In which section(s) of the MSDS are the following properties listed? K/U
   (a) physical properties
   (b) chemical properties
4. Why do you think it is necessary to provide the preparation date on MSDS? A
5. Where might a consumer find MSDS information for consumer products such as household bleach and paint thinner? A
6. Consumers who buy cleaning products at the supermarket rely on the product label for safety information. What additional information is required if this product is to be used in a workplace? A
7. Give one example of a hazardous chemical that can be used safely. K/U
8. Give one example of a "safe" chemical that could, under the right conditions, be hazardous. K/U
9. The phrase "the danger is in the dose" is sometimes used when describing poisonous (toxic) chemicals. What does this mean? Include an example in your answer. T/I

5.4 Patterns and the Periodic Table

element a pure substance that cannot be broken down into simpler substances

You have already learned that **elements** are pure substances that cannot be broken down into simpler substances. You probably also know that the periodic table is a powerful tool that chemists use to explain and predict the properties of the elements (Figure 1).

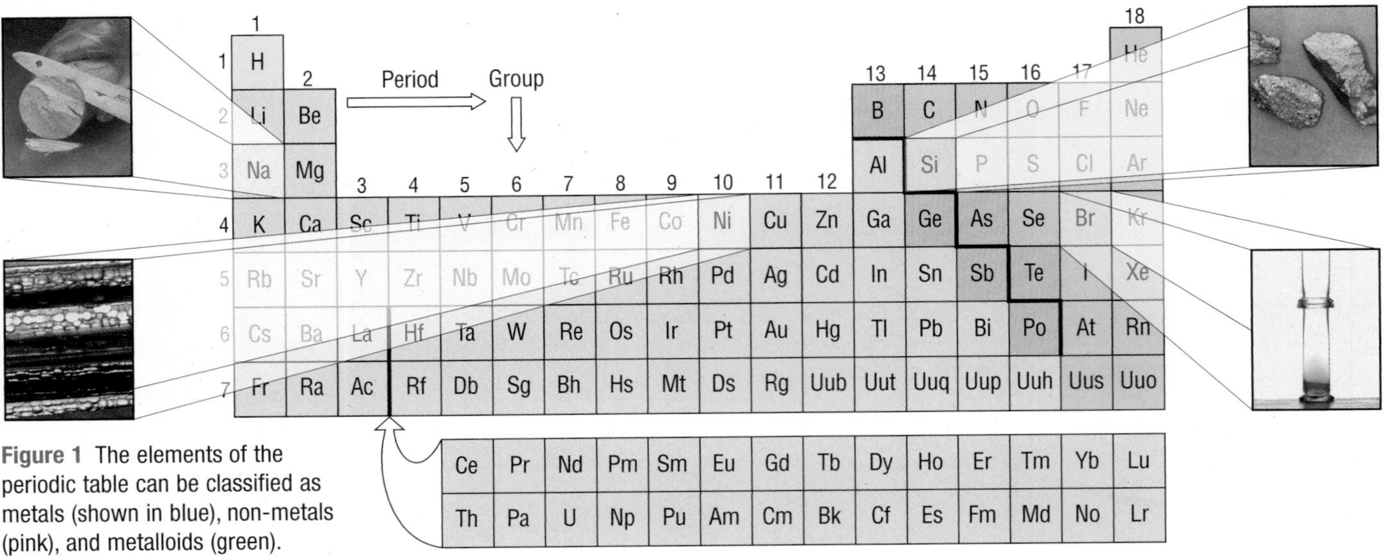

Figure 1 The elements of the periodic table can be classified as metals (shown in blue), non-metals (pink), and metalloids (green).

Table 1 Summary of Properties of Metals and Non-Metals

Property	Metals	Non-metals
example	nickel, Ni	bromine, Br
state at room temperature	solid	solid, liquid, or gas
lustre	shiny	dull
malleability	generally malleable	brittle (if solid)
electrical conductivity	conductors	insulators

Table 1 provides a summary of the general properties of metals and non-metals. Note that hydrogen, H, has its own unique colour. This is because it has some properties in common with the metals in the first column. However, it lacks many of the characteristic physical properties of metals at room temperature. We cannot really classify hydrogen as a metal, so we group it on its own.

Chemical Periods and Groups

period a row of elements in the periodic table

group a column of elements in the periodic table with similar properties

alkali metals the elements (except hydrogen) in the first column of the periodic table (Group 1)

alkaline earth metals the elements in the second column of the periodic table (Group 2)

halogens the elements in the seventeenth column of the periodic table (Group 17)

noble gases the elements in the eighteenth column of the periodic table (Group 18)

The periodic table also categorizes elements into periods and groups. Each row of elements on the periodic table is called a **period**. Each column is a **group** of elements with similar properties. Four of the best-known groups of elements are listed below:

- The Group 1 elements (with the exception of hydrogen) are the **alkali metals**. These elements are soft, highly reactive metals (Figure 2).
- The Group 2 elements are light, reactive **alkaline earth metals**.
- The Group 17 elements are the **halogens**. They are one of the most reactive groups on the periodic table.
- The Group 18 elements are **noble gases**. Unlike the halogens, the noble gases are so stable that they rarely react with any other chemical.

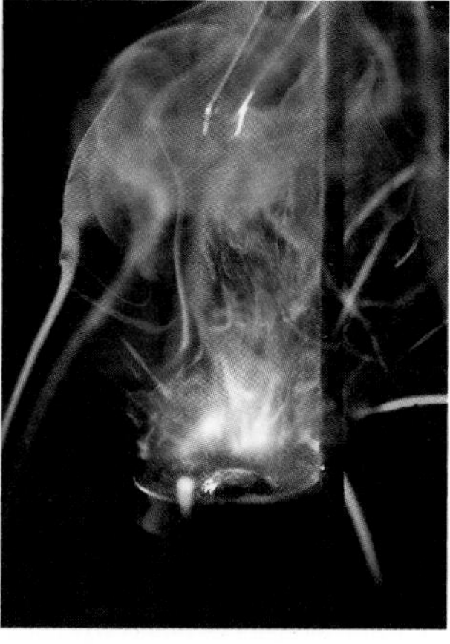

Figure 2 Lithium, sodium, and potassium react at different rates with water to produce flammable hydrogen gas. The potassium reaction is so vigorous that the hydrogen gas ignites.

Atomic Structure

Why do elements behave so differently? The answer to this question lies in the structure of atoms. Scientists developed a simple model of the atom to explain the properties of elements. In this model, most of the mass of the atom is concentrated in an extremely small, dense, positively charged core called the nucleus (Figure 3).

Atoms are made up of three kinds of subatomic particles (Table 2).

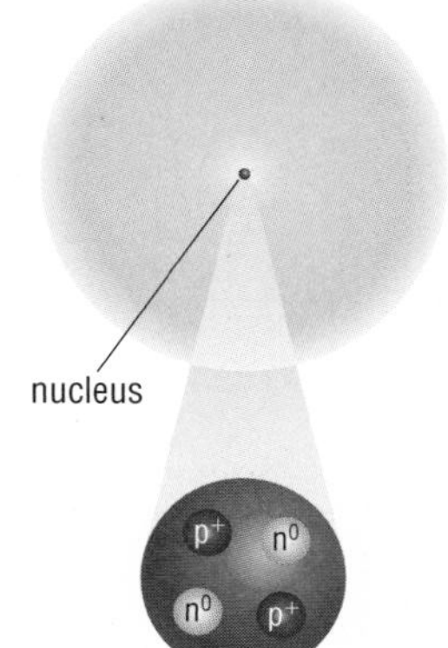

Figure 3 Most of the atom's volume is empty space. Most of the atom's mass is concentrated in the nucleus. The nucleus takes up only about 1/100 000 of the volume of an atom.

Table 2 Subatomic Particles

	Proton	Neutron	Electron
electrical charge	positive	neutral	negative
symbol	p^+	n^0	e^-
location	nucleus	nucleus	orbit around the nucleus

The number of protons in the nucleus is called the atomic number of the element. For example, because carbon contains six protons in its nucleus, its atomic number is 6. The elements of the periodic table are arranged in order of *increasing* atomic number. Atoms are electrically neutral, with equal numbers of protons and electrons.

DID YOU KNOW?

Feeling Dense?
If you could somehow squeeze all the space out of your body's atoms you would be smaller than a penny. However, that "penny" would be as heavy as you are right now.

Electron Arrangements and the Bohr–Rutherford Diagram

The **Bohr–Rutherford diagram** of the atom is a useful way of representing the arrangement of electrons around the nucleus for the first 20 elements. In Bohr–Rutherford diagrams, each electron orbit is shown as a ring around the nucleus. Evidence indicates that only a limited number of electrons can occupy each orbit. The first orbit can hold up to a maximum of two electrons. The second and third orbits can each hold a maximum of 8 electrons. Elements with atomic numbers above 18 must have some electrons in the fourth orbit. Because all atoms are electrically neutral, the total number of electrons in these orbits must match the number of protons in the nucleus. Recall that this is only a model of the atom. The actual behaviour of electrons is much more complicated, but this model meets our needs for now.

Bohr–Rutherford diagram a model representing the arrangement of electrons in orbits around the nucleus of an atom

We can use this model and the periodic table to help us predict the atomic structures and properties of elements. Let's use Bohr–Rutherford diagrams for hydrogen, helium, lithium, and fluorine to illustrate this idea of electron orbits.

WRITING TIP

Writing a Summary
Write one clear sentence for each major point. For example, if summarizing the Bohr–Rutherford diagrams, use separate sentences to explain the orbit of each electron.

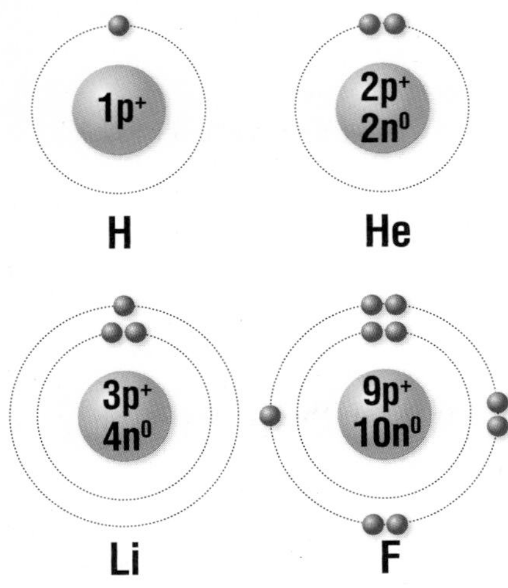

Figure 4 Bohr–Rutherford diagrams of hydrogen (H-1), helium (He-4), lithium (Li-7), and fluorine (F-19)

An atom of lithium is sometimes represented as Li-7. This means that this particular atom has a mass number of 7. Recall that the mass number is the total number of protons and neutrons. Lithium atoms always have three protons, so this atom must also have four neutrons in its nucleus. To balance the charge of the three protons, three electrons must be orbiting the nucleus. The first orbit can hold up to two electrons. Since the third electron cannot fit in the first orbit, it has to go in a second orbit (Figure 4).

Our atomic model suggests that the second orbit can accommodate up to eight electrons. So, as you proceed from element to element in the second period, the number of electrons in the second orbit increases by one until there are eight: the maximum number. Therefore, fluorine has seven electrons in its second orbit and neon has eight in its second orbit. After neon, another orbit is required to accommodate the next set of eight electrons. The outermost electron of sodium, therefore, is in the third orbit (Figure 5). Notice that the period, or row, number of an element tells you how many electron orbits the atoms have. You can predict that elements 19 and 20, which are in the fourth row, have electrons in a fourth orbit. Beyond this orbit, the model of the atom gets more complicated, but you need not be concerned about these larger atoms for this course.

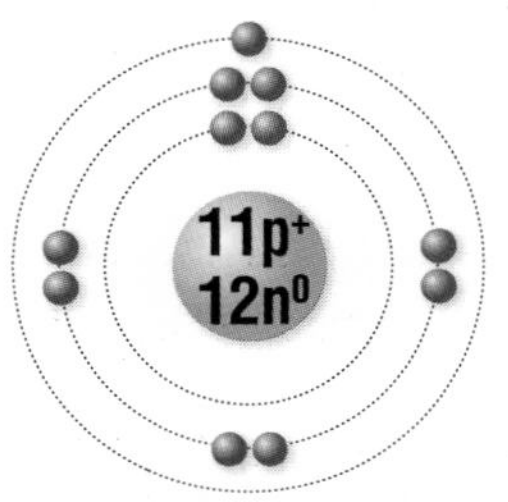

Figure 5 Bohr–Rutherford diagram of sodium (Na-23).

Electron Arrangements and Reactivity

The noble gases are known for their stability. They are so stable that they almost never react with other elements. Why are they so stable? From experimental evidence, chemists infer that the outer electrons of an element are responsible for the element's reactivity. Since the noble gases all have completely filled outer orbits, we can conclude that there is something particularly stable about full outer orbits (Figure 6). As you will see in the following sections, the stability of filled outer electron orbits is important to understanding how elements combine to form compounds. **Compounds** are substances made up of two or more elements in a fixed ratio.

MATH TIP

Ratio

A ratio is a comparison of quantities. The ratio of boys to girls in Ontario is approximately 1:1. The ratio of hydrogen atoms to oxygen atoms in a water molecule is 2:1. Ratios are usually expressed using the smallest possible whole numbers.

compound a pure substance composed of two or more elements in a fixed ratio

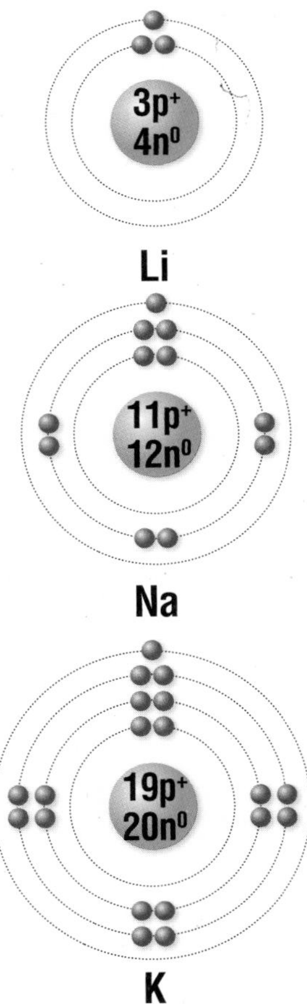

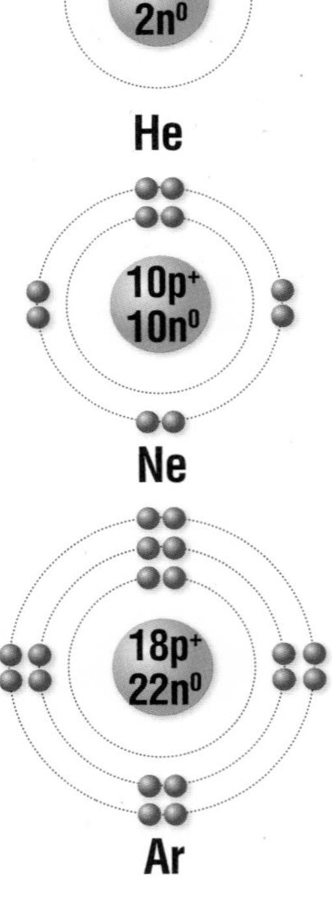

Figure 6 Lithium (Li-7), sodium (Na-23), and potassium (K-39) are reactive because of their single outer electron. Helium (He-4), neon (Ne-20), and argon (Ar-40) are stable because they have filled outer electron orbits.

While the noble gases are very stable, the elements at the other side of the periodic table—the alkali metals—are very reactive. Observations show that every alkali metal reacts with water (Figure 2 on page 185). Chemists theorize that alkali metals are highly reactive because each of their atoms contains one electron in its outer orbit.

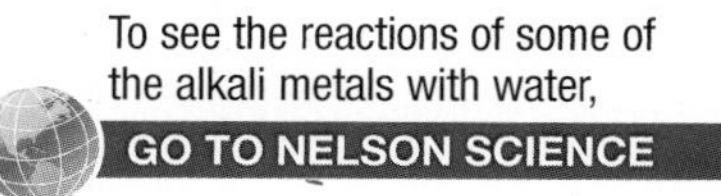

To see the reactions of some of the alkali metals with water,
GO TO NELSON SCIENCE

IN SUMMARY

- Elements are arranged in the periodic table in order of atomic number (the number of protons in the nucleus).
- Electrically neutral elements have the same number of electrons as protons in each atom.
- Elements in vertical columns (groups) in the periodic table all have the same number of electrons in their outer orbits.
- The number of electrons in the outer orbit affects the reactivity of an element.
- Bohr–Rutherford diagrams illustrate the numbers of protons, neutrons, and electrons in an atom and the arrangement of the electrons.

CHECK YOUR LEARNING

1. What information on the periodic table lets you predict the number of electrons in an atom? K/U
2. Compare metals and non-metals in terms of their
   (a) state at room temperature
   (b) electrical conductivity
   (c) lustre
   (d) number of electrons in their outermost orbit K/U
3. Refer to the periodic table to name and write the symbols for the following elements: K/U
   (a) the halogen of the second period
   (b) the alkaline earth metal in the fifth period
   (c) the noble gas with the smallest atomic number
   (d) the non-metal in the fifth period with seven outermost electrons
   (e) the alkali metal of the fourth period
   (f) the metal of the third period with three outermost electrons
   (g) the unreactive gas of the second period
4. Sketch the Bohr–Rutherford diagrams for the following elements: nitrogen (N-14), aluminum (Al-27), chlorine (Cl-35), and magnesium (Mg-24). T/I C
5. Imagine that chemists discovered a new element with atomic number 119. T/I
   (a) Use the periodic table to predict what chemical family this element would belong to.
   (b) How many outer electrons would an atom of this element have?
   (c) Predict one physical property and one chemical property of this element.
6. Look at the physical appearance of the elements in Figure 7. K/U
   (a) Classify each element as either a metal or a non-metal.
   (b) Identify one unusual physical property of element (iv).
   (c) Which of the elements are likely to conduct electricity?

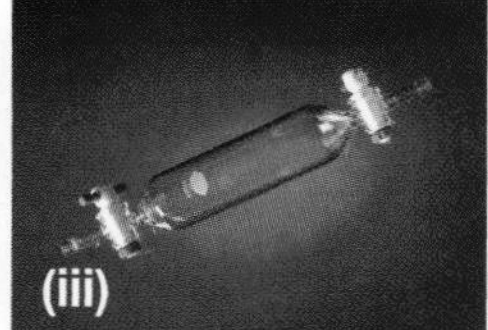

Figure 7 Elements at room temperature

7. Compare the number of outermost electrons
   (a) within a period
   (b) within a group K/U
8. Why are atoms electrically neutral? K/U
9. Many high schools have banned the use of potassium. What property of potassium may have led to this ban? T/I

5.5 Atoms and Ions

Figure 1 Is water harming this athlete?

You have surely heard that we are all supposed to drink a lot of water: at least 2 L a day, and more if it is very hot or if we are exercising (Figure 1). While this is generally good advice, too much water can actually be bad for your health. This was the conclusion reached during a scientific study of runners in the 2002 Boston Marathon. During the race, 13 % of the runners sampled in the study developed a condition known as hyponatremia. Symptoms of hyponatremia include disorientation and a loss of balance. The cause? They drank too much water during the race. Excess water can dilute the concentration of sodium in the blood to dangerously low levels—so low that three runners were at risk of dying if left untreated. People have actually died from this condition.

When scientists refer to sodium in blood, they are not talking about the shiny metal that reacts vigorously with water that you saw in Figure 2 in Section 5.4. Rather, they are referring to sodium ions. Many ions are necessary for our health. Calcium and phosphorus ions are essential components of bone; iron ions help carry oxygen around the body. Ions get into our bodies in our food, and our bodies regulate their concentration.

How Do Atoms and Ions Differ?

ion a charged particle that results when an atom gains or loses one or more electrons

As you learned in Section 5.4, an atom is an electrically neutral particle with an equal number of electrons and protons. An **ion** is an atom that has become charged by gaining or losing electrons. For example, sodium atoms lose one electron when they react with other atoms. Each resulting sodium ion contains 11 positive charges (on protons) and only 10 negative charges (on electrons). Since it has one more positive charge than negative, the sodium ion has an ionic charge of +1. As a result, chemists gave sodium ions the chemical symbol Na^{1+} or Na^{+}. (Note that the number 1 is usually omitted in chemical symbols.) The other alkali metals also form ions with a single positive charge.

Fluorine is one of the most reactive elements. When fluorine reacts, it tends to gain an electron from another atom to form a stable ion called fluoride. Because the fluoride ion has one extra negative charge, it has an ionic charge of –1. The chemical symbol of this ion is therefore F^{-}. In fact, all the halogens form ions with a single negative charge. So, sodium and the other alkali metals (Group 1) lose an electron to form +1 ions, and the halogens (Group 17) gain an electron to form –1 ions.

	sodium, Na^{+}	fluoride, F^{-}
positive charge (protons)	+11	+9
negative charge (electrons)	–10	–10
ionic charge	+1	–1

LEARNING TIP

How to Represent a Charge on an Ion
Chemists say that a magnesium ion has a charge of +2, which is read as "plus two," but they write the ion as Mg^{2+} (with the number first and the plus sign last). You will get used to this convention as you see it used.

Why do we not find Na^{2+} or F^{2-}: sodium ions with a +2 charge or fluoride ions with a –2 charge? To understand why Na^{+} and F^{-} are the only stable ions that these elements form, we need to consider their Bohr–Rutherford diagrams in relation to those of the noble gases.

The noble gases (Group 18) are stable due to their full outer orbits. Sodium ions and fluoride ions are also stable. Why is this so? To explain the non-reactivity of sodium and fluoride ions, we can compare their Bohr–Rutherford diagrams with that of neon. Neon is the noble gas that is closest to sodium and fluorine on the periodic table. Again, we can use the Bohr–Rutherford model and the periodic table to help us predict the formation of ions.

While we are considering the formation of ions from atoms, we do not need to be concerned about the number of neutrons in the nucleus. We can therefore omit them from our Bohr–Rutherford diagrams for now.

Sodium

In the process of forming a sodium ion (which has a positive charge), a sodium atom must react with another atom and lose one electron. The most likely electron to be lost is the one farthest from the nucleus: the single electron in the third orbit. This farthest electron is the least tightly held to the nucleus. As a result, the sodium ion has the same stable electron arrangement as a neon atom: an outer orbit filled with eight electrons (Figure 2).

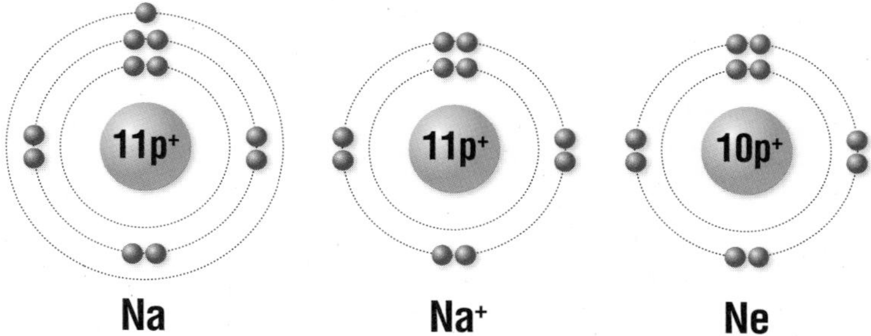

Figure 2 The sodium atom loses its outermost electron to form an ion. The sodium ion is stable because its outer orbit is full, like that of neon.

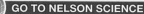

Fluorine

Fluorine has one less electron than neon. Fluorine tends to react with other atoms to gain one electron. This reaction gives it the same stable arrangement of electrons as neon. With this extra electron, the fluorine atom now has 10 electrons and only 9 protons. It therefore becomes a fluoride ion with a single negative ionic charge: F^- (Figure 3).

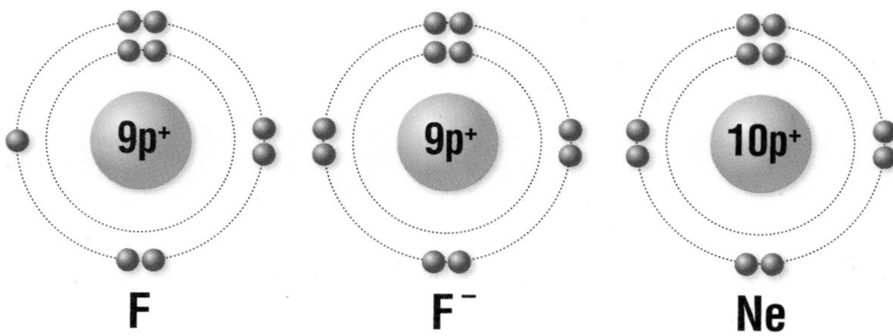

Figure 3 The fluorine atom gains one electron to become a fluoride ion, F^-. Fluoride is stable because its outer orbit is full, like that of neon.

DID YOU KNOW?

Fluoride and Dental Cavities

Fluoride ions help make tooth enamel strong. For decades, Canadians have been getting fewer dental cavities thanks to fluoride in their drinking water. Some dentists have recently noticed an increase in cavities among young children, possibly because these children are drinking bottled water, without fluoride, rather than tap water.

GO TO NELSON SCIENCE

Aluminum

The Bohr–Rutherford diagram of aluminum shows that aluminum has three outer electrons (Figure 4(a)). To have a stable outer orbit (like a noble gas) aluminum could—in theory—either gain five electrons or lose three. Experimental evidence shows that metals tend to lose electrons, while non-metals tend to gain them. The result is an aluminum ion with ionic charge +3: Al^{3+} (Figure 4(b)).

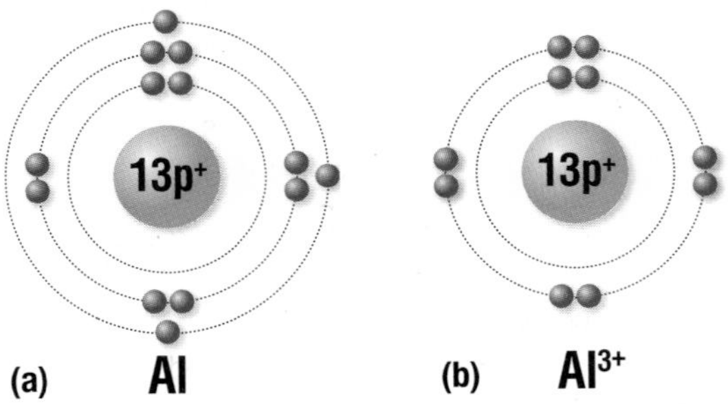

Figure 4 (a) An aluminum atom has three electrons in its outer orbit. (b) Losing these electrons leaves the aluminum ion positively charged.

Sulfur

Sulfur has six electrons in its third orbit (Figure 5(a)). To achieve a stable electron arrangement, a sulfur atom reacts with other atoms and gains two electrons. When it does, sulfur forms an ion with the chemical symbol S^{2-} (Figure 5(b)). This is called a sulfide ion.

Sulfur can also form compounds without forming ions. You will learn more about these compounds in Section 5.11.

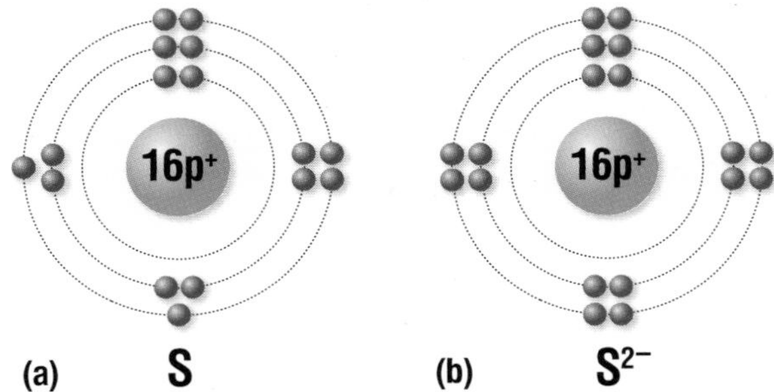

Figure 5 (a) A sulfur atom has six electrons in its outer orbit. (b) Filling the outer orbit with electrons makes the sulfide ion negatively charged.

DID YOU KNOW?

Hydrogen
Hydrogen can form both positive and negative ions. It can gain one electron to fill its only orbit, forming an ion with a charge of −1. More often, though, hydrogen loses its only electron to form an ion with a charge of +1.

Naming Ions

We can classify ions as **cations**—those that have positive charges, and **anions**—those that have negative charges.

The name of a positive ion is the same as the name of the element: sodium forms sodium ions, for example. The name of a negative ion is determined by adding "ide" to the stem of the name. For example, **ox**ygen forms **ox**ide ions and **phosph**orus forms **phosph**ide ions.

cation a positively charged ion

anion a negatively charged ion

LEARNING TIP

Cations and Anions
Remember "cation" contains the letter "t," which looks like a + sign; **an**ions **a**re **n**egatively charged.

TRY THIS IONS AND THE PERIODIC TABLE

SKILLS: Analyzing, Communicating

SKILLS HANDBOOK 3.B., 10.

Some elements gain or lose electrons to form stable ions. Is there a pattern to how elements form ions? In this activity you will explore how some of the first 20 elements in the periodic table form ions. In the process, you will learn how to predict the ionic charge of elements based on their location on the periodic table.

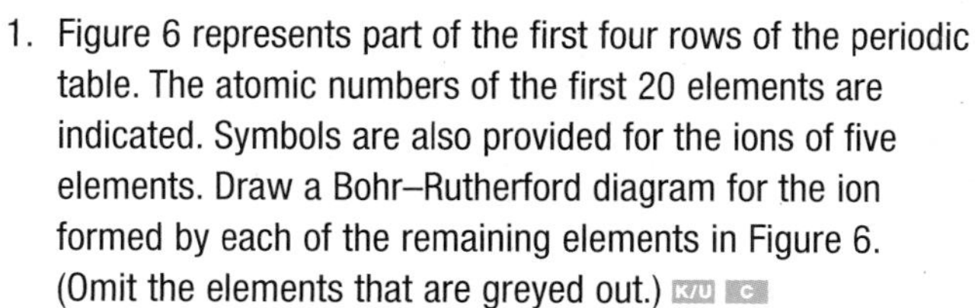

1. Figure 6 represents part of the first four rows of the periodic table. The atomic numbers of the first 20 elements are indicated. Symbols are also provided for the ions of five elements. Draw a Bohr–Rutherford diagram for the ion formed by each of the remaining elements in Figure 6. (Omit the elements that are greyed out.) K/U C

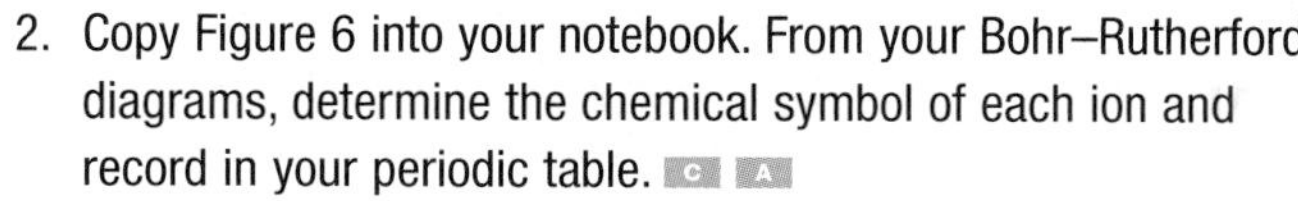

2. Copy Figure 6 into your notebook. From your Bohr–Rutherford diagrams, determine the chemical symbol of each ion and record in your periodic table. C A

A. Describe the patterns or similarities that exist within a period and within a group for the Bohr–Rutherford diagrams. T/I A

B. Describe the patterns or similarities that exist within a period and within a group for the ionic charges. T/I A

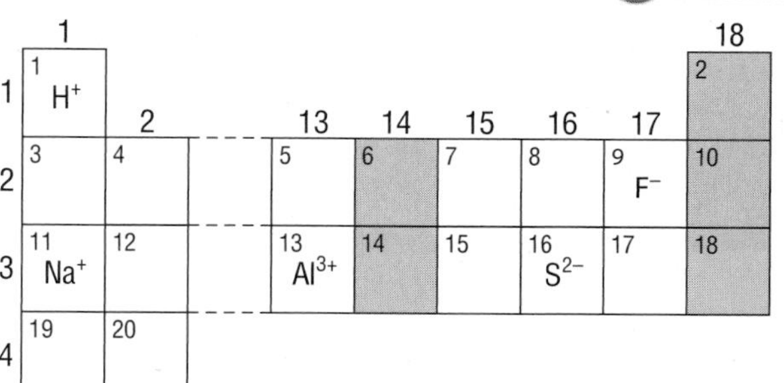

Figure 6

C. How can the ionic charge be predicted from the location of the element on the periodic table? T/I

D. Use your answer in C to predict the chemical symbol of the ion of each of the following elements: T/I

(a) barium, Ba (b) iodine, I (c) rubidium, Rb (d) arsenic, As

E. You were not required to determine the ions for the greyed-out cells in Figure 6. Why? T/I

IN SUMMARY

- Ions are atoms that have gained or lost electrons. Many ions have complete outer orbits, so they are stable.
- Anions have more electrons than protons and therefore have a negative charge. Anions often have "ide" at the end of their names.
- Cations have fewer electrons than protons and therefore have a positive charge.
- Atoms and ions can be represented by Bohr–Rutherford diagrams.
- Some ions—in the appropriate concentrations—are necessary for good health.

CHECK YOUR LEARNING

1. Compare a sodium ion to
   (a) a sodium atom (b) a neon atom K/U
2. (a) Draw the Bohr–Rutherford diagram (without neutrons) for an atom of each of the following elements: lithium, oxygen, calcium, and phosphorus.
   (b) Draw the Bohr–Rutherford diagram (without neutrons) for the ion formed by each of the elements in (a).
   (c) Write the chemical symbol for each ion.
   (d) Name the noble gas with the same electron arrangement as each ion. T/I C
3. Distinguish between a cation and an anion. K/U
4. Name these ions. K/U
   (a) Mg^{2+} (b) S^{2-} (c) Fe^{3+} (d) Br^- (e) N^{3-}
5. List three atoms or ions that have the same number of electrons as each of the following:
   (a) S^{2-} (b) Al^{3+} (c) P^{3-} (d) Kr (e) Cs^+ T/I
6. Suppose that a new element has been made. Chemical tests show that it is an alkaline earth metal. T/I
   (a) Predict how many electrons there will be in the outer orbit.
   (b) Predict the ionic charge of the ion that this element forms.
7. Justify why these ions do not exist under normal conditions. A
   (a) K^{2+} (b) O^-
8. (a) What is the trend in the ionic charges of the elements in Groups 1, 2, and 13 of the periodic table?
   (b) What is the trend in ionic charges of the elements in Groups 15 to 17? K/U
9. What type of drink would you recommend for endurance runners who suffer from hyponatremia? Why? K/U

5.6

Ionic Compounds

As you have already learned, sodium is a very reactive metal. Chlorine is a poisonous gas. When these elements mix, a violent reaction occurs. The compound that results, however, is a safe and familiar one: sodium chloride or table salt (Figure 1). What happens to the sodium and chlorine atoms as this reaction takes place?

To see a video of this reaction, GO TO NELSON SCIENCE

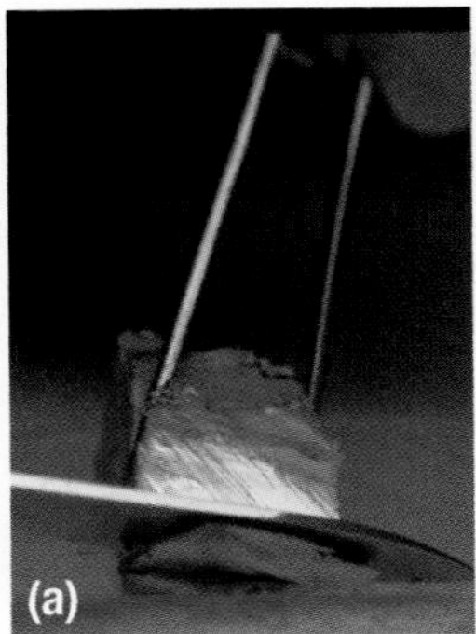

Figure 1 This series of photographs shows what we can actually see during the formation of sodium chloride. (a) Sodium metal (b) Chlorine gas (c) Sodium and chlorine reacting vigorously (d) The product: sodium chloride (e) We recognize sodium chloride, in a more familiar container, as table salt.

Making Ionic Compounds from Elements

In the previous section, you saw that metals tend to lose electrons to form positive ions called cations. Conversely, non-metals tend to gain electrons to form negatively charged ions called anions. When a metal such as sodium reacts with a non-metal such as chlorine, both processes occur (Figure 2). The non-metal atoms take electrons from the metal atoms. This electron transfer is possible because the metal's hold of its outermost electrons is weak. At the same time, the non-metal attracts the metal's electrons strongly. The resulting ions all have the same stable, filled outer electron arrangements as the nearest noble gas.

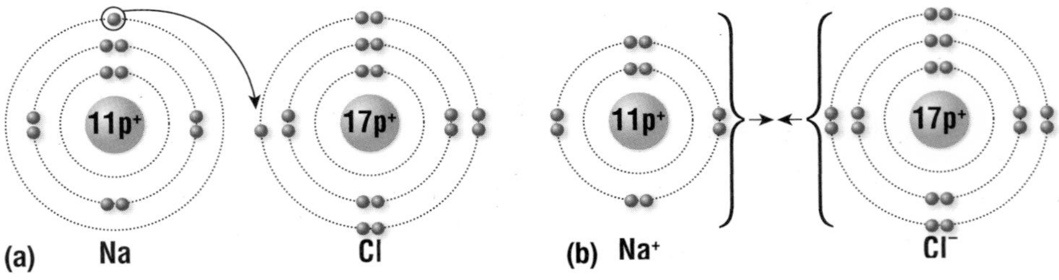

Figure 2 A model of the formation of sodium chloride. (a) A chlorine atom fills its outermost orbit by taking the third-orbit electron from a sodium atom. (b) The stable sodium and chloride ions attract each other to form sodium chloride.

Once they form, positive and negative ions from different elements attract each other to form compounds. Compounds that are made up of positive and negative ions are called **ionic compounds**. For example, sodium chloride (table salt) is an ionic compound made up of sodium ions, Na^+, and chloride ions, Cl^-. For ionic compounds that contain only two elements, one element is always a metal and the other is a non-metal. The attraction that holds oppositely charged ions together in a compound is called an **ionic bond**.

ionic compound a compound made up of one or more positive metal ions (cations) and one or more negative non-metal ions (anions)

ionic bond the simultaneous strong attraction of positive and negative ions in an ionic compound

Large numbers of sodium and chloride ions join together to form an ionic crystal. This crystal consists of alternating sodium and chloride ions, in a ratio of 1:1, extending in three dimensions (Figure 3). This is why the chemical formula for sodium chloride is NaCl. There is no individual "NaCl" particle: the compound always consists of many sodium ions and chloride ions held together in a crystal.

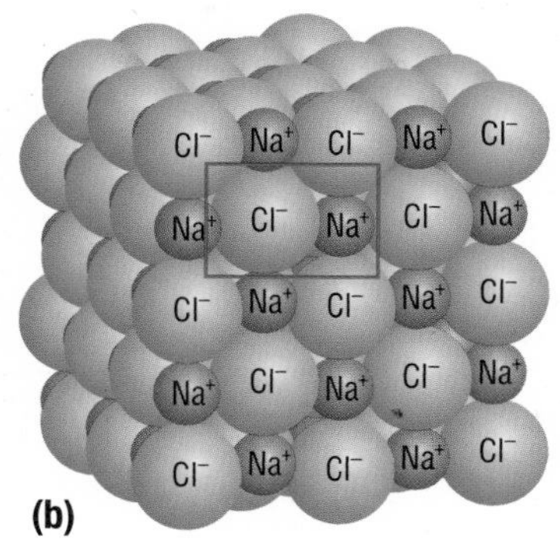

Figure 3 (a) Under the microscope, sodium chloride appears as cubes. (b) A crystal of sodium chloride could contain billions of alternating sodium and chloride ions. However, the number of sodium ions is always equal to the number of chloride ions, so their ratio is 1:1.

Some ionic compounds are soluble in water. When they dissolve, they separate into ions. Water molecules surround each ion as it leaves the crystal (Figure 4). This prevents ions from rejoining the crystal.

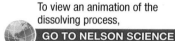

To view an animation of the dissolving process,
GO TO NELSON SCIENCE

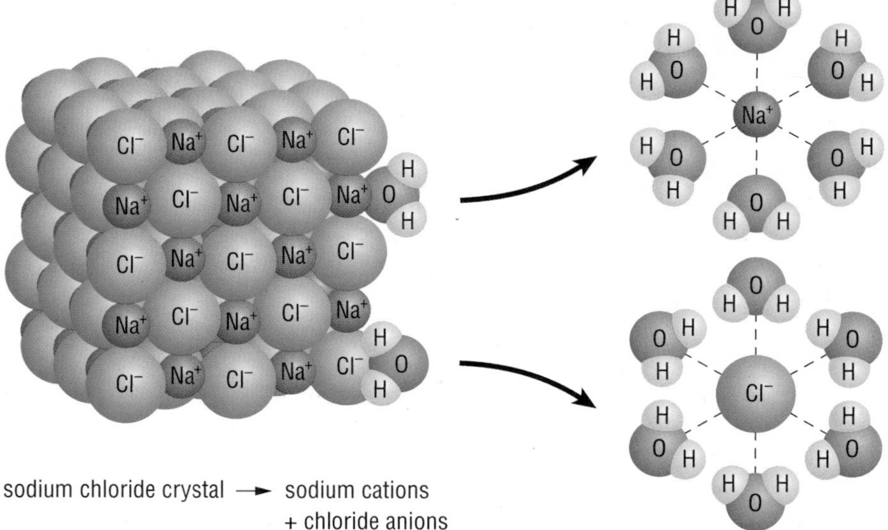

Figure 4 When ionic substances dissolve, their positive and negative ions are pulled away from the crystal by water molecules. The water molecules arrange themselves around ions in particular ways: the oxygen atoms of water molecules are attracted to positive ions and the hydrogen atoms are attracted to negative ions.

The element aluminum can also react with chlorine gas. Each aluminum atom, however, has three electrons to lose, while each chlorine atom can accommodate only one extra electron. How can this be resolved? Each aluminum atom reacts with three chlorine atoms (Figure 5). The result is an ionic compound called aluminum chloride, a common ingredient in many antiperspirants. When dissolved in water (or sweat), the aluminum ions and the chloride ions separate, just like the ions in sodium chloride.

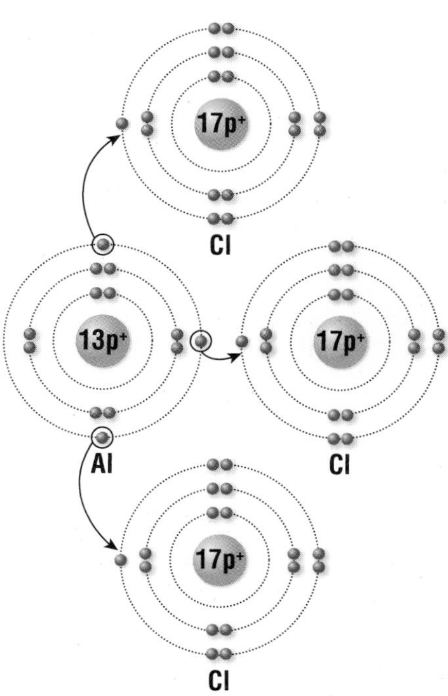

Figure 5 Aluminum transfers its three outer electrons to chlorine atoms to form aluminum chloride.

Properties of Ionic Compounds

electrolyte a compound that separates into ions when it dissolves in water, producing a solution that conducts electricity

Due to the strength of the ionic bond, ionic compounds are hard, brittle solids with high melting points. Most ionic compounds are also **electrolytes**, which means that they dissolve in water to produce a solution that conducts electricity. As ionic compounds dissolve, their ions are pulled apart by water molecules. The presence of these ions improves the electrical conductivity of water (Figure 6). Pure water is a poor conductor of electricity, but tap water, lake water, and seawater are good conductors because they contain ions from a variety of sources, such as minerals. That is why it is essential to stay out of swimming pools or lakes during a lightning storm.

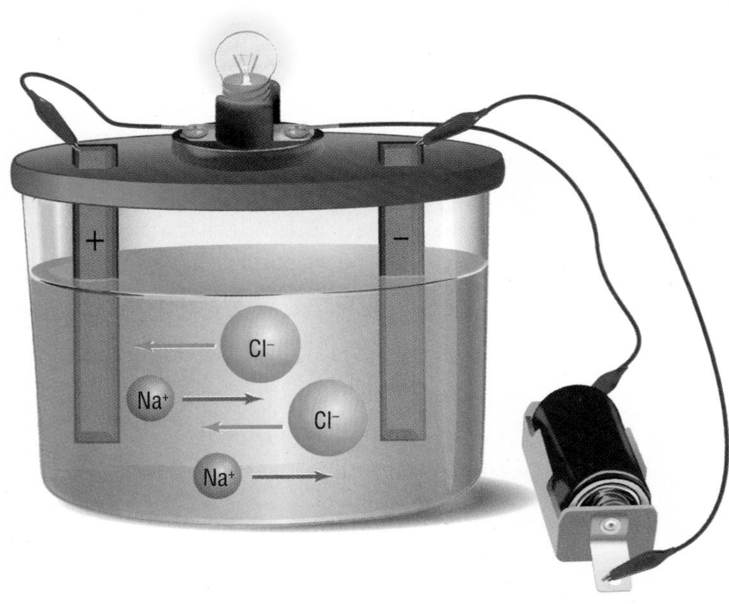

Figure 6 Sodium chloride is an electrolyte because it separates into ions when it dissolves. A solution can conduct electricity only if it contains ions that are free to move.

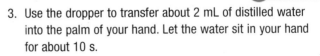

TRY THIS TESTING FOR ELECTROLYTES

SKILLS: Observing, Analyzing, Communicating

SKILLS HANDBOOK 1.B., 2.B.

Electrolytes in our water come from a variety of sources. Some are added naturally as water flows over rocks containing minerals. Others, such as compounds containing fluoride ions, are artificially added to drinking water to help prevent tooth decay. In this activity, you will compare the electrical conductivity of distilled water before and after it contacts your skin.

Equipment and Materials: low-voltage conductivity tester; small beaker; dropper; distilled water

 Use only low-voltage conductivity testers in this activity.

1. Add 5 to 10 mL of distilled water to a clean, dry beaker.
2. Test the electrical conductivity of distilled water.
3. Use the dropper to transfer about 2 mL of distilled water into the palm of your hand. Let the water sit in your hand for about 10 s.
4. Test the electrical conductivity of the liquid in your hand.

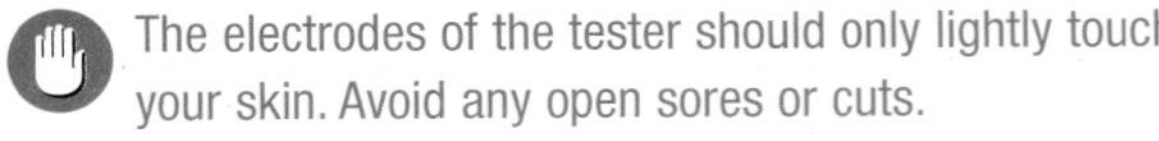

The electrodes of the tester should only lightly touch your skin. Avoid any open sores or cuts.

5. Wash your hands.

A. Why does the conductivity of distilled water change after it has been in contact with your skin? T/I

B. Why is there a risk of electrical shock when handling electrical equipment with wet hands? T/I

UNIT TASK Bookmark

You can apply what you learned in this section about the properties of ionic compounds to the Unit Task described on page 300.

IN SUMMARY

- Elements (a metal and a non-metal) can react to form an ionic compound.
- During the reaction, the non-metal atoms pull electrons away from the metal atoms.
- The ratio of metal ions to non-metal ions in an ionic compound depends on the number of electrons each ion gains or loses.
- Most ionic compounds have high melting points and are hard, brittle electrolytes: they dissolve in water to form solutions that conduct electricity.
- Most ionic compounds form three-dimensional crystals, with many of each kind of ion in a fixed ratio held together by their opposite charges.

CHECK YOUR LEARNING

1. What kinds of elements combine to form ionic compounds? K/U
2. Look at each of the following pairs of elements. Predict whether each pair would form ionic bonds. Explain your reasoning. K/U
   (a) Mg, O (b) Zn, Cl (c) C, F (d) H, F
3. Magnesium and chlorine react to form an ionic compound. K/U C
   (a) Which element is the metal and which is the non-metal?
   (b) Draw a Bohr–Rutherford diagram (without neutrons) for each element.
   (c) How many electrons must the atoms of each element gain or lose to become a stable ion?
   (d) Sketch a diagram to show the transfer of electrons that occurs when these two elements react. Your diagram should be similar to Figure 2 on page 192.
4. Repeat question 3 for lithium and oxygen. K/U C
5. Explain why two non-metallic elements are not likely to form ionic bonds. K/U
6. When each of these compounds dissolves in water, what ions are released and in what ratio? T/I
   (a) NaF (b) Li_3N (c) $FeCl_3$ (d) K_2O
7. Element X has three electrons in its outermost orbit. Element Y has seven electrons in its outermost orbit. K/U T/I
   (a) Classify each of these elements as metal or a non-metal.
   (b) What is the chemical formula of the ionic compound formed by these elements when they react together? (Use X and Y as the chemical symbols.)
8. Dissolved ions are surrounded by water molecules. Explain how the water prevents the ions from forming a solid again. K/U
9. Silver reacts with sulfur to form a compound that has two silver ions for every sulfide ion. Write the chemical formula for the compound. K/U
10. Producing sodium chloride (table salt) using the reaction in Figure 1 on page 192 is impractical and expensive. Sodium chloride can be extracted from seawater. How could this be done using a common renewable energy source? A
11. (a) Compare the electrical conductivity of pure water, tap water, and seawater.
    (b) Why is there a difference? T/I
12. A penny and a dime inserted into a pickle and connected in an electric circuit can generate enough electricity to power a small electric buzzer. Suggest why a pickle is better for this activity than a raw cucumber. A
13. Table 1 shows the melting points of four Group 1 chlorides. T/I C

Table 1 Melting Point Data

Compound	Melting point (°C)	Period number of metal ion
NaCl	801	
KCl	775	
RbCl	718	
CsCl	645	

   (a) Copy and complete Table 1 in your notebook.
   (b) Draw a line graph of the melting point against the period number of each metal ion.
   (c) Chemists do not know much about the very rare Group 1 element francium, or its compounds. Extend the line of your graph in (b) to predict the melting point of francium chloride.

5.7 Names and Formulas of Ionic Compounds

Figure 1 Some chemicals have been known for centuries, such as (a) blue vitriol, (b) cinnabar, and (c) Glauber's salt. At the time of their discovery, there was no organized way of naming them.

Blue vitriol, cinnabar, Glauber's salt (Figure 1). These unusual names may sound like the ingredients in a magical potion. However, they are actually the traditional names of three chemicals commonly found in chemistry labs. These names were developed centuries ago, when few chemicals were known. Today, the number of known chemicals has grown to over 10 million! To keep track of them all, chemists have developed a systematic method of naming chemicals. The International Union of Pure and Applied Chemistry (IUPAC) is the organization that decides how chemicals will be named. A common naming system allows scientists around the world to communicate with each other without misunderstandings.

Naming Ionic Compounds

Many ionic compounds are made up of two elements: a metal and a non-metal. It is therefore logical that the names of ionic compounds have two parts. The first part refers to the metal ion in the compound and the second part to the non-metal ion. Remember that the name of the metal ion remains the same as the name of the neutral metal atom (Table 1) but that the ending of the name of the second ion—the non-metal—changes to "ide."

Table 1 Examples of Naming Ionic Compounds

Metal	Metal ion	Non-metal	Non-metal ion	Compound
magnesium	magnesium ion	chlorine	chloride ion	magnesium chloride
aluminum	aluminum ion	oxygen	oxide ion	aluminum oxide

Table 2 gives the ion names of common non-metals found in ionic compounds. Remember, from Section 5.5, that all non-metals form negatively charged ions: anions.

Table 2 Names and Charges of Common Anions

Name of element	Name of ion	Ionic charge	Ion symbol
fluorine	fluoride ion	–1	F^-
chlorine	chloride ion	–1	Cl^-
bromine	bromide ion	–1	Br^-
iodine	iodide ion	–1	I^-
oxygen	oxide ion	–2	O^{2-}
sulfur	sulfide ion	–2	S^{2-}
nitrogen	nitride ion	–3	N^{3-}
phosphorus	phosphide ion	–3	P^{3-}

Writing Chemical Formulas of Ionic Compounds

When elements form ionic compounds, electrons move from metal atoms to non-metal atoms. The resulting charged ions attract other ions of the opposite charge until the charges balance out. The compound that forms is electrically neutral. In other words, in the compound, the total number of positive charges must equal the total number of negative charges. This basic idea helps us determine the chemical formulas of ionic compounds.

To determine the chemical formula of an ionic compound, you must first figure out the correct number of ions required to produce an electrically neutral compound. The compound's total ion charge (the negative and positive ion charges added together) must equal zero. Here is a strategy that will help.

LEARNING TIP

The Zero-Sum Rule

The sum of all charges in the chemical formula of the compound must equal zero. This model may help you understand this idea. It shows that two chloride ions (triangles) are needed to "complete" the rectangle. The rectangle represents the smallest number of ions that must combine to give an overall charge of zero. The ratio of ions in this rectangle is the same as the ratio of ions in the chemical formula of the compound.

$MgCl_2$

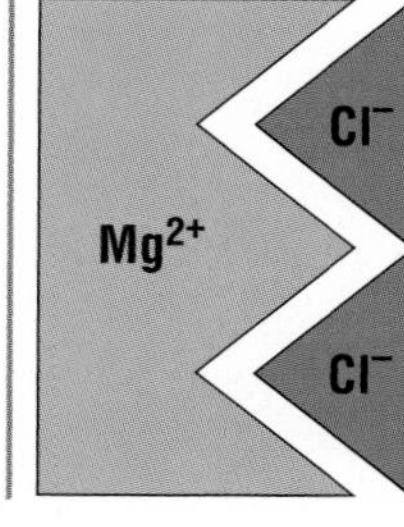

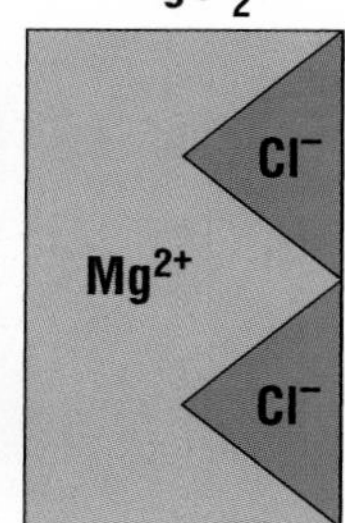

SAMPLE PROBLEM 1 Chemical Formula of an Ionic Compound

What is the chemical formula of magnesium chloride?

Step 1 Write the symbols of the elements, with the metal on the left-hand side and the non-metal on the right-hand side.

Mg Cl

Step 2 Add the ionic charge of each ion above the symbol.

+2 −1

Mg Cl

Step 3 Determine how many ions of each type are required to bring the total charge to zero. The sum of all charges in the compound must equal zero.

Total ionic charge: 1(+2) + 2(−1) = 0

Mg Cl

Step 4 Write the chemical formula using the (red) coefficients in front of each bracket as subscripts.

Mg_1Cl_2

Step 5 Do not write the subscript "1" in chemical formulas because the symbol itself represents one ion.

The chemical formula for magnesium chloride is $MgCl_2$

SAMPLE PROBLEM 2 Chemical Formula of an Ionic Compound

What is the chemical formula of aluminum oxide (Figure 2)?

Step 1 Write the symbols of the metal and non-metal elements.

Al O

Step 2 Add the ionic charge of each ion above the symbol.

+3 −2

Al O

Step 3 Determine the number of ions required to bring the total charge to zero.

2(+3) + 3(−2) = 0

Al O

The chemical formula of aluminum oxide is Al_2O_3.

Figure 2 A layer of aluminum oxide forms a protective coating over the metal underneath.

The Crisscross Method for Determining Formulas

WRITING TIP

Writing a Summary

Sometimes a text describes a process or procedure, such as the Crisscross Method. In your summary, organize ideas and information in the same order as in the original text.

Now that you understand the meaning of the symbols and numbers in chemical formulas, try using a shortcut known as the "crisscross" method.

SAMPLE PROBLEM 3 Using the Crisscross Method

What is the chemical formula of magnesium chloride?

Step 1 Write the symbols of the elements and their ionic charges.

Mg^{2+} Cl^{-}

Step 2 Crisscross the numbers of the ionic charges so that they now become subscripts.

Mg^{2+} Cl^{-}

$MgCl_2$

The chemical formula of magnesium chloride is $MgCl_2$.

Practice

What is the chemical formula of aluminum sulfide?

Occasionally, this method of finding the chemical formula of a compound gives you an answer with two identical subscripts. For example, the chemical formula of aluminum nitride appears to be Al_3N_3. You know that the chemical formula of an ionic compound is always the simplest possible ratio of the ions, so Al_3N_3 must be simplified to AlN.

Elements with Multiple Ionic Charges

Figure 3 Because iron has two different ionic charges, it makes two different compounds with chlorine: iron(II) chloride (left) and iron(III) chloride (right).

Chemists have discovered, through experiments, that there are two stable cations of iron: Fe^{2+} and Fe^{3+}. These ions are named like other cations, except that Roman numerals in rounded brackets are used to indicate the ionic charge (*not* the number of ions) of the metal. The Fe^{2+} ion is called "iron(II) ion," and the Fe^{3+} ion is called "iron(III) ion." Because of this, iron may form two different compounds with a particular anion. For example, there are two different compounds of iron and chlorine: $FeCl_2$ and $FeCl_3$. Each compound has its own distinct properties: solid $FeCl_2$ is light green, while $FeCl_3$ is yellowish-brown (Figure 3). Several other metals, besides iron, form multiple cations. In each case, two or more compounds can be formed with the same anion (Table 3).

Table 3 Names and Multiple Ionic Charges for Common Metals

Metal	Chemical symbol of element	Chemical symbols of ions	Names of ions
copper	Cu	Cu^{+} Cu^{2+}	copper(I) copper(II)
iron	Fe	Fe^{2+} Fe^{3+}	iron(II) iron(III)
lead	Pb	Pb^{2+} Pb^{4+}	lead(II) lead(IV)
manganese	Mn	Mn^{2+} Mn^{4+}	manganese(II) manganese(IV)
tin	Sn	Sn^{2+} Sn^{4+}	tin(II) tin(IV)

Naming Compounds Involving Elements with Multiple Ionic Charges

Naming compounds containing elements with multiple ionic charges involves the same procedure as before. You must also determine the ionic charge of the metal so that you can include it in the name. For example, $FeCl_2$ is called iron(II) chloride because the ionic charge of iron is +2. Similarly, $FeCl_3$ is called iron(III) chloride because the ionic charge of iron is +3.

LEARNING TIP

Specifying the Charge

The name of $CuCl_2$ is written "copper(II) chloride." This is read as "copper two chloride." Remember that the "two" refers to the charge on the copper ion, not to the number of ions of copper in the compound.

SAMPLE PROBLEM 4 Considering Multiple Ionic Charges

Write the chemical name of $CuBr_2$. Note that copper could have either of two ionic charges (Table 3, previous page).

Step 1 As you do not know whether the charge on the copper ion is +1 or +2, use *x* to represent this unknown value. Remember that the charge on all the ions must total zero. Each bromide ion has a charge of –1, so the total charge of bromide ions is 2(–1).

$$CuBr_2$$
$$x + 2(-1) = 0$$
$$x = +2$$

The ionic charge of copper in this compound is +2.

Step 2 Write the name of the compound. Remember that Roman numerals are included *only* if the metal has more than one ionic charge.

The name of $CuBr_2$ is copper(II) bromide.

Practice

What is the chemical name of PbO_2?

Whenever you have to write the chemical name of a compound containing a metal, first check whether the metal has more than one ionic charge.

TRY THIS TWO SHADES OF IRON

SKILLS: Observing, Analyzing

SKILLS HANDBOOK 1.B., 3.B.

In this activity, you will compare the properties of the two ionic forms of iron. To do this, you will use two different compounds of iron: iron(III) chloride and iron(II) sulfate. The chloride and sulfate ions are colourless and do not react with oxygen. This means that the iron ions alone are responsible for all observable colour differences/changes.

Equipment and Materials: eye protection; lab apron; 100 mL graduated cylinder; 250 mL Erlenmeyer flask; laboratory scoop; scale; weighing paper; 500 mL Erlenmeyer flask containing iron(III) chloride solution; distilled water; 1.0 g iron(II) sulfate

Both iron compounds are irritants. Wash any spills on the skin, in the eyes, or on clothing immediately with plenty of cold water. Report any spills to your teacher.

1. Put on your eye protection and lab apron.
2. Observe the solution of iron(III) chloride that your teacher has prepared. Record your observations.
3. Add about 100 mL of distilled water to the 250 mL flask.
4. Add the 1.0 g of iron(II) sulfate to the water in the flask.
5. Swirl to dissolve the solid. Record your observations.
6. Continue swirling gently for about 1 min. Compare the colour of your solution with that of the iron(III) chloride solution.
7. Follow your teacher's instructions for the disposal of the solutions. Clean up your work area and wash your hands.

A. Swirling the solution allows oxygen from the atmosphere to enter the solution and react with the chemicals in the solution. What evidence of a chemical change did you observe? K/U

B. Suggest a possible explanation for the evidence you observed. C

IN SUMMARY

- When writing chemical formulas of ionic compounds, the number of each ion must balance the positive and negative charges so that the overall charge is zero.
- The numbers of the ions in the chemical formula of an ionic compound must be in the simplest ratio. The subscript "1" is not included in a chemical formula.
- Ionic compounds are named with the metal first and then the non-metal. The ending of the non-metal's name is changed to "ide."
- Some metals can form ions with different charges. In the name of an ionic compound containing one of these metals, the charge is written as Roman numerals in brackets after the name of the metal.

CHECK YOUR LEARNING

1. (a) Describe at least one idea or skill in the reading that you think you will need to practise.
   (b) How do you plan to improve with this idea or skill?
   (c) Discuss your plans with your teacher. C
2. Name each of these compounds: K/U
   (a) CaF_2
   (b) K_2S
   (c) Al_2O_3
   (d) LiBr
   (e) Ca_3P_2
3. Determine the chemical formula of the ionic compound that forms when each of these pairs of elements react. K/U
   (a) K and Br (b) Ca and O (c) Na and S
4. Which is the correct chemical formula of tin(IV) oxide: SnO_2 or Sn_2O_4? Why? K/U
5. Copper forms two different compounds with bromine. One compound contains one copper ion for every bromide ion. The other contains two bromide ions for every copper ion. Name and write the chemical formulas of these two compounds. K/U
6. Why must the net charge on any ionic compound always equal zero? K/U
7. Write the chemical formulas for the following compounds: K/U
   (a) calcium chloride
   (b) aluminum bromide
   (c) magnesium sulfide
   (d) lithium nitride
   (e) calcium nitride
8. The blue compound shown in Figure 1(a) was traditionally called blue vitriol or bluestone. The systematic name for this compound is copper(II) sulfate. What is the advantage of a systematic naming system for chemicals? A
9. Copy and complete Table 4. K/U

Table 4 Chemical Names and Formulas of Ionic Compounds

	Name	Formula
(a)	iron(II) bromide	
(b)	manganese(IV) oxide	
(c)	tin(IV) chloride	
(d)	copper(I) sulfide	
(e)	iron(III) nitride	
(f)	copper(II) oxide	
(g)		$PbCl_2$
(h)		Fe_2O_3
(i)		SnS
(j)		Cu_3P_2
(k)		$CaBr_2$
(l)		CuF_2
(m)		K_3P
(n)		Cu_3P

10. Magnetite is a substance composed of only iron ions and oxygen ions. It has the chemical formula Fe_3O_4 (Figure 4). Given that iron has only two common ionic charges, suggest an explanation for this formula. T/I

Figure 4

Chlorine Conclusions

What could be more refreshing than a cool dip in a backyard pool on a hot summer's day? Ideal pool water contains just the right amount of disinfecting chemicals: enough to kill any micro-organisms in the water but not enough to cause skin or eye irritation or harm the environment.

For many pool owners, chlorine is the disinfectant of choice. It is inexpensive, readily available, and very effective. However, chlorine has a downside: it smells unpleasant, bleaches hair, and dries skin. High concentrations can cause respiratory problems.

The Issue

The homes in a new housing development will all have backyard pools. The development backs onto an environmentally sensitive wetland (Figure 1).

Figure 1 The disinfection system for these pools should be friendly to both the environment and the wallet.

Your environmental engineering firm has been contracted to recommend the best means of keeping the new backyard pools clean and free of micro-organisms. Your solution must satisfy at least three criteria: it should be affordable, safe, and have as little impact on the environment as possible. You will present your recommendation to a joint meeting of the local community council and the developer.

Goal

To recommend a swimming pool disinfection system.

SKILLS MENU
- Defining the Issue
- Researching
- Identifying Alternatives
- Analyzing the Issue
- Defending a Decision
- Communicating
- Evaluating

Gather Information

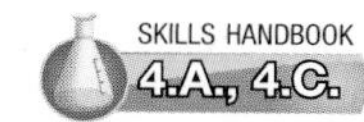

Work in pairs or small groups to brainstorm and research answers to the following questions. C A

- What are the health risks of a poorly maintained pool?
- What are the most popular methods of keeping outdoor pools clean?
- What are the potential impacts of discarded pool water on the sensitive ecosystem?

Consider where you can find more information. If you are doing an Internet search, what keywords can you use? Do you know somebody who has first-hand experience of maintaining a pool?

Identify Solutions

Two of the most common methods of disinfecting pools involve introducing chlorine to the water. These two methods are

- adding chlorine-containing solutions or soluble solids directly to the water
- producing small amounts of chlorine directly in the water using a process called electrolysis. This technology is commonly used in "saltwater pools."

Decide whether there is another alternative that should be considered. T/I

Make a Decision

What disinfection method would you recommend for the pools in this development? What criteria did you use to decide? T/I C A

Communicate

Complete a report that will be presented to a meeting of the developer and the town council. The report should examine the advantages and disadvantages of at least two systems of disinfection. The report should conclude with your recommendation. C A

5.9 Polyatomic Ions

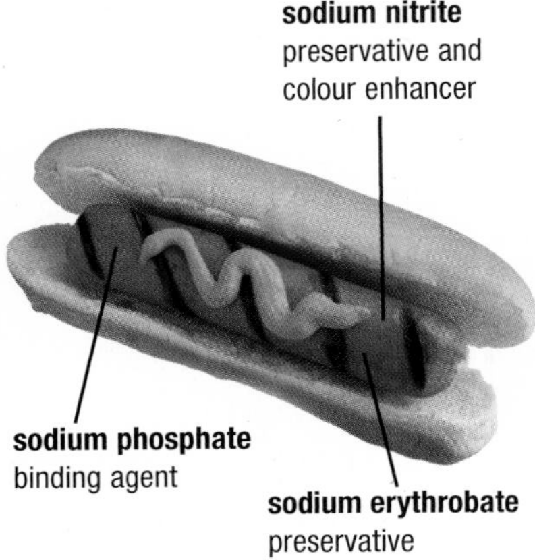

Figure 1 Hot dogs contain many chemical additives.

Processed foods contain a lot of sodium, mostly from sodium chloride, NaCl (table salt). Sodium chloride enhances the flavour and extends the shelf-life of food.

Other additives in processed foods also contribute to your daily sodium intake (Figure 1). For example, sodium phosphate helps bind together the meat in a hot dog. Sodium nitrite acts as a preservative and flavour enhancer and gives hot dogs their pinkish colour. There is some concern about adding compounds containing nitrite ions to foods. Nitrite ions react with substances in the digestive tract to form nitrosamines in the body. Nitrosamines are chemicals that have been linked to certain types of cancer in laboratory animals. An occasional hot dog is unlikely to make you ill, but a steady diet of processed meats is not recommended.

Nitrites are not always bad news, however. Research suggests that sodium nitrite may help protect heart tissue after a heart attack or transplant. Also, nitrites and a related group of compounds called nitrates are important plant nutrients. They occur naturally in soil and are manufactured and spread as fertilizer on many crops. You can see how difficult it is to make a simple judgment about whether a chemical is good or bad.

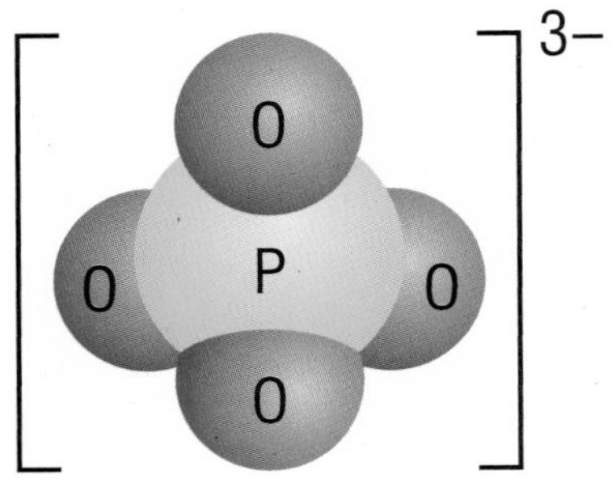

Figure 2 The phosphate ion is made up of four oxygen atoms bonded to a central phosphorus atom.

The sodium compounds in Figure 1 are ionic compounds similar to the others you have learned about in this chapter. Sodium phosphate, for example, is a white solid, relatively stable, and an electrolyte. The chemical formula for this compound is Na_3PO_4. Its cation is sodium, but its anion, phosphate, $(PO_4)^{3-}$, is an example of a polyatomic ion (Figure 2). A **polyatomic ion** is an ion that consists of a stable group of several atoms acting together as a single charged particle. The ionic charge of a polyatomic ion is shared over the entire ion rather than being on just one atom.

polyatomic ion an ion made up of more than one atom that acts as a single particle

Table 1 lists some of the most common polyatomic ions and their ionic charges. Note that all the ions are anions except ammonium. Also, note that all the anion names end in "ate" except nitrite and hydroxide.

Table 1 Formulas and Charges of Common Polyatomic Ions

Name of polyatomic ion	Ion formula	Ionic charge
nitrate ion	NO_3^-	−1
nitrite ion	NO_2^-	−1
hydroxide ion	OH^-	−1
hydrogen carbonate ion (also called bicarbonate ion)	HCO_3^-	−1
chlorate ion	ClO_3^-	−1
carbonate ion	CO_3^{2-}	−2
sulfate ion	SO_4^{2-}	−2
phosphate ion	PO_4^{3-}	−3
ammonium	NH_4^+	+1

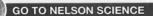

DID YOU **KNOW?**

Not Just for Baking

Sodium hydrogen carbonate (baking soda) is an extremely versatile substance. It can substitute for toothpaste if you run out, it soaks up odours from your shoes and refrigerator, it is a good abrasive cleaner, and it is even useful to help relieve insect bites and sunburn.

GO TO NELSON SCIENCE

Naming Compounds Involving Polyatomic Ions

The strategy for naming polyatomic compounds uses the same steps that you learned in the previous section. The only difference is that the anion is named according to the polyatomic ion rather than the names of the individual elements.

SAMPLE PROBLEM 1 Naming Compounds Involving Polyatomic Ions

Write the name of the compound Na_2CO_3. (Sodium does not form multiple ions.)

Step 1 Write the name of the metal and check whether it has more than one ionic charge. Sodium always has a charge of +1.

Step 2 Write the name of the compound.

The name of Na_2CO_3 is sodium carbonate.

Practice

What is the name of $Ca(OH)_2$?

As the Practice example indicates, several polyatomic ions can be associated with each cation. This is indicated by placing brackets around the polyatomic ion and writing the subscript outside the brackets.

Some metal ions have more than one charge. Check this before naming the compound. In compounds containing these ions, use a Roman numeral to indicate the charge.

SAMPLE PROBLEM 2 Naming Compounds Involving Polyatomic Ions

Write the name of the compound $Fe(NO_3)_3$.

Step 1 Write the name of the metal and check whether it has more than one possible ionic charge. If it does, proceed to step 2. If not, proceed to step 3.
Iron has two possible ionic charges: +2 and +3.

Step 2 Determine the ionic charge of the metal.

$$\begin{aligned} &Fe(NO_3)_3 \\ &x + 3(-1) = 0 \\ &x = +3 \end{aligned}$$

The ionic charge of iron in this compound is +3, indicated by the Roman numeral "III."

Step 3 Write the name of the compound, with the Roman numeral if necessary.

The name of $Fe(NO_3)_3$ is iron(III) nitrate.

Practice

What is the name of $CuSO_4$?

Writing Formulas Involving Polyatomic Ions

You can apply the rules that you have already learned for writing ionic compound formulas to polyatomic compounds. Remember to treat the polyatomic ion as one unit. Consider the following Sample Problem.

SAMPLE PROBLEM 3 Writing Formulas for Compounds Containing Polyatomic Ions

What is the chemical formula of sodium phosphate?

Step 1 Write the symbols of each ion, beginning with the cation (metal).

Na PO_4

Step 2 Write the ionic charges above each ion.

+1 −3
Na PO_4

Step 3 Determine how many ions of each type are required to give a total charge of zero.

Total ionic charge: 3(+1) + 1(−3) = 0
Na PO_4

Three sodium ions balance the −3 charge of the phosphate ion.

Step 4 Write the chemical formula using the coefficients as subscripts.

Na_3PO_4

The formula of sodium phosphate is Na_3PO_4.

DID YOU KNOW?

The Phosphate Fiasco
In the 1960s, laundry detergents included phosphates. The phosphate-laden wastewater was flushed into lakes and streams. Phosphates are an important plant nutrient. The increased dose of nutrients caused unnaturally rapid growth in aquatic plant populations. As these plants died and decomposed, much of the oxygen in the water was used up, killing many aquatic animals. To avoid the problem, chemical engineers redesigned detergents to work effectively without phosphates.

Note that sodium phosphate contains three sodium ions and one phosphate ion (Figure 3).

The "crisscross method" also works for compounds with polyatomic ions. Sample Problem 4 shows how. Note that, if a coefficient is required for the polyatomic ion, first put curved brackets around it and then write the subscript outside of the final bracket.

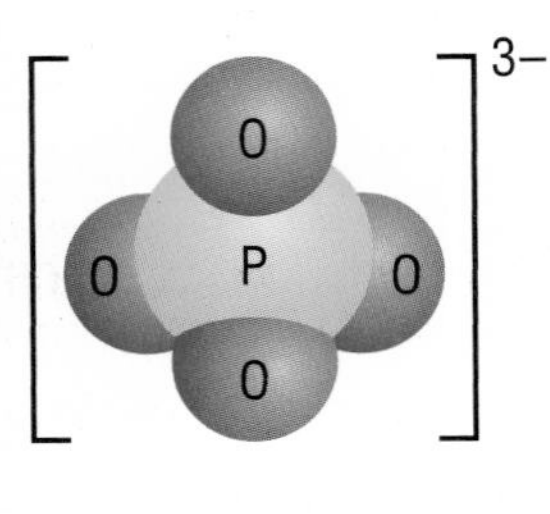

Figure 3 Sodium phosphate, Na_3PO_4, consists of three sodium ions and one phosphate ion, PO_4^{3-}. When this compound dissolves, these four ions separate. However, the phosphate ion remains intact.

SAMPLE PROBLEM 4 Writing Formulas for Compounds Containing Polyatomic Ions

What is the chemical formula of copper(II) nitrate?

Step 1 Write the symbol of each ion, with its charge.

Cu^{2+} $(NO_3)^-$

Step 2 Crisscross the numbers of the ionic charges so that they now become subscripts.

Cu^{2+} $(NO_3)^-$

Step 3 Write any necessary subscripts *after* the brackets around each ion. (Remember, you do not need to write "1" as a subscript, so no brackets are necessary.)

Cu $(NO_3)_2$

The formula of copper(II) nitrate is $Cu(NO_3)_2$.

Practice
What is the chemical formula of ammonium carbonate?

IN SUMMARY

- Polyatomic ions are made up of more than one atom, with a charge spread over the entire ion.
- Polyatomic ions are present in many natural and artificial compounds. They are used as food additives, fertilizers, and cleaners.
- Consider the polyatomic ion as a single unit when writing the chemical formula for a compound that includes a polyatomic ion. If the compound contains more than one of the polyatomic ion, write curved brackets around the ion and write a subscript outside the brackets.

CHECK YOUR LEARNING

1. Name the polyatomic ion in each of the following compounds and name the compound. (Watch for metals with more than one possible ionic charge.) K/U
   (a) KNO_3 (found in gun powder)
   (b) $Ca(OH)_2$ (an ingredient in plaster)
   (c) $CaCO_3$ (in chalk, limestone, and antacid medicines)
   (d) $CuSO_4$ (a fungicide)
   (e) KOH (used to make soap)
   (f) $Fe(NO_3)_3$ (used in water treatment)
   (g) $Cu(ClO_3)_2$ (used to colour fireworks)
   (h) $(NH_4)_3PO_4$ (an ingredient in bread dough)
2. Write the chemical formula for each of the following compounds: K/U
   (a) potassium nitrate (used to colour fireworks purple)
   (b) barium sulfate (given prior to an X-ray of the intestine)
   (c) ammonium nitrate (a common ingredient in fertilizer)
   (d) aluminum sulfate (used in preparing pickles)
   (e) potassium chlorate (an explosive)
   (f) copper(II) nitrate (used to colour ceramics)
   (g) lead(II) sulfate (found in car batteries)
   (h) tin(II) phosphate (used in the dyeing of silk)
3. What is the most common ending for the name of
   (a) a polyatomic anion?
   (b) an anion made up of only one element? K/U
4. Well water in agricultural areas is often monitored for nitrate ions because of possible health effects. Where can nitrate contamination come from on a farm? A
5. Write the names of the following compounds. Note that some of them contain polyatomic ions. K/U
   (a) $SnCO_3$
   (b) $CaCl_2$
   (c) $Fe(OH)_3$
   (d) MnO_2
   (e) K_2S
   (f) $(NH_4)_2SO_4$
   (g) $Mn(ClO_3)_2$
   (h) PbI_2
6. Write the chemical formulas of the following compounds. Note that some of them contain polyatomic ions. K/U
   (a) calcium sulfate
   (b) ammonium chloride
   (c) copper(I) carbonate
   (d) barium sulfide
   (e) calcium chlorate
   (f) tin(II) hydroxide
   (g) iron(II) phosphate
   (h) aluminum nitride
7. Explain why the chemical formula for calcium hydroxide, $Ca(OH)_2$, is not written as CaO_2H_2. K/U
8. Most ionic compounds are made up of a metal cation and a non-metal anion. Look at Table 1 on page 202 and identify an exception to this rule. K/U
9. When writing the chemical formula of an ionic compound, which ion is always written first? K/U
10. Copy Table 2 into your notebook. Complete your table using the example provided as a guide. K/U

Table 2 Identifying Ions

Compound	Cation(s)	Anion(s)
$Fe(OH)_3$	1 Fe^{3+}	3 OH^-
$Cu(NO_3)_2$		
$Al_2(SO_4)_3$		
$(NH_4)_2CO_3$		
K_3PO_4		

11. The names "sodium chloride" and "sodium chlorate" sound similar. However, these compounds are very different. Sodium chloride enhances the flavour of food while sodium chlorate is a toxic herbicide. For each compound, write
    (a) the chemical formula
    (b) the chemical formula of the anion
    (c) the chemical formula of a compound that this anion makes with calcium K/U
12. Describe one strategy you can use to reduce the amount of salt in your diet. A

5.10 Molecules and Covalent Bonding

Nitrous oxide, N_2O, is a colourless, sweet-smelling gas that dentists sometimes use to relax nervous patients (Figure 1(a)). You would not, however, want to inhale nitrogen dioxide, NO_2. This reddish-brown toxic gas is produced in the atmosphere from pollutants emitted in automobile exhaust. Nitrogen dioxide is a dangerous part of the smog that hangs over large cities in the summer (Figure 1(b)). People with asthma and other respiratory problems often find it more difficult to breathe on smoggy days.

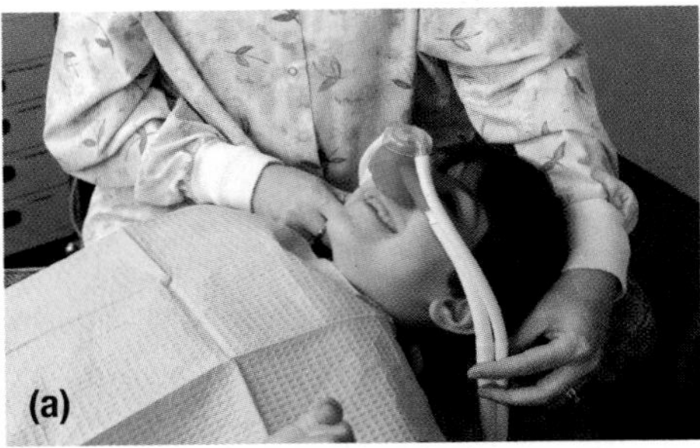

Figure 1 (a) Inhaling nitrous oxide, laughing gas, helps dental patients relax. (b) Nitrogen dioxide is the gas responsible for the reddish-brown colour of smog over large cities during the summer.

Pollution control technology in modern cars helps reduce nitrogen dioxide emissions. A catalytic converter, attached to the car's exhaust system, converts nitrogen dioxide into harmless nitrogen and oxygen. The Ontario Drive Clean Program requires that most cars over five years old pass an emissions test every two years. This helps reduce pollution in Ontario.

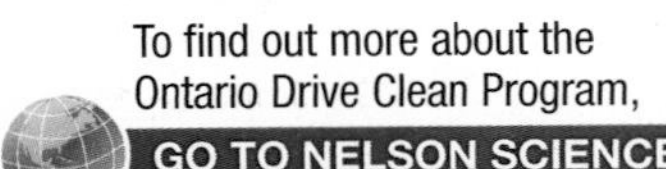

To find out more about the Ontario Drive Clean Program,
GO TO NELSON SCIENCE

Nitrous oxide and nitrogen dioxide are both **molecular compounds**. As the name implies, molecular compounds are made up of individual particles called molecules (Figure 2(a)). (Ionic compounds, as Figure 2(b) shows, consist of many ions in a crystal.) The chemical formula of a molecular compound gives the exact numbers of atoms in each molecule. The elements that make up molecular compounds are all non-metals.

molecular compound a pure substance formed from two or more non-metals

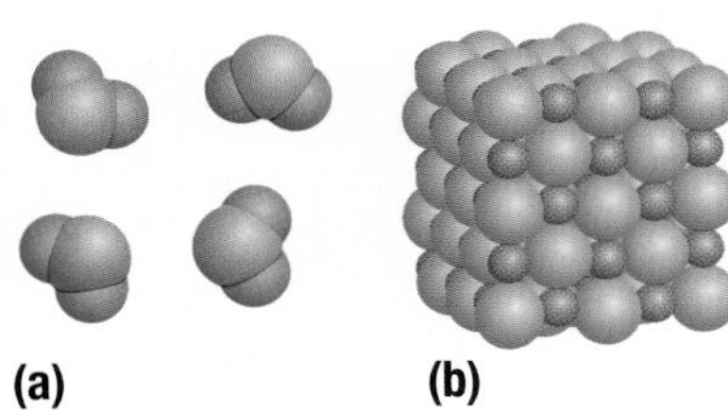

Figure 2 (a) Each nitrogen dioxide particle is a molecule made up of one nitrogen atom and two oxygen atoms. (b) An ionic compound does not exist as individual particles. Instead, millions of ions are tightly held together in a crystal.

Molecular compounds are all around you: in the air that you breathe and the substances that you eat and drink (Figure 3). A soft drink, for example, contains water molecules, H_2O, and sugar molecules, $C_{12}H_{22}O_{11}$, as well as flavouring and colouring molecules. Carbonated soft drinks, like colas, also contain a large quantity of dissolved carbon dioxide molecules, CO_2. In fact, the majority of all known compounds are molecular. Living organisms make thousands of different kinds of molecular compounds. Sugars, fats, and proteins are all molecular compounds. Some of them are very large, containing thousands of atoms in a single molecule! Like nitrogen dioxide, many molecular compounds affect our environment.

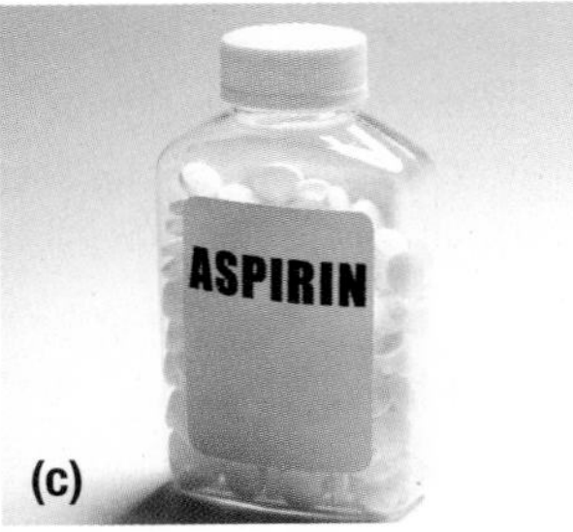

Figure 3 Most of the chemicals that we encounter are molecular compounds, including (a) sugar, (b) water, and (c) acetylsalicylic acid, or Aspirin.

Bonding in Molecules

In Section 5.4, you learned that ions form when metallic elements lose electrons to non-metal elements. An electron transfer occurs because

- the metal's hold on its outer electrons is weak
- the attraction of the non-metal for the metal's electrons is strong
- a full outer electron orbit is very stable

Remember that non-metals all have almost-full outer electron orbits with "spaces" available that attract other electrons. This allows non-metal atoms to get relatively close to each other.

When two non-metals bond with each other, both nuclei form strong attractions for the other's electrons. However, neither atom attracts the other's electrons strongly enough to pull them away completely. What results is a "tug of war" for electrons that neither atom ever wins. The net effect is that the two atoms share each other's electrons, resulting in a bond that holds the atoms together. A chemical bond that results from atoms sharing electrons is called a **covalent bond**. The bonded atoms form a **molecule**. Molecules that consist of two atoms joined with a covalent bond are called **diatomic molecules.**

covalent bond a bond that results from the sharing of outer electrons between non-metal atoms

molecule a particle in which atoms are joined by covalent bonds

diatomic molecule a molecule consisting of only two atoms of either the same or different elements

Covalent bonds can form between two identical atoms or between atoms of different elements. We will start by looking at two identical hydrogen atoms. A hydrogen atom has one outer electron. To achieve a stable outer orbit like that of the nearest noble gas (helium), hydrogen must acquire one more electron. When two hydrogen atoms collide, the proton of one atom attracts the electron of the other and vice versa (Figure 4). Since the atoms are identical, they have the same ability to attract electrons. As a result, both electrons are shared equally between the two atoms. This results in a covalent bond between the atoms. The resulting hydrogen molecule has the chemical formula H_2. A solid line linking the atoms, H—H, represents a covalent bond.

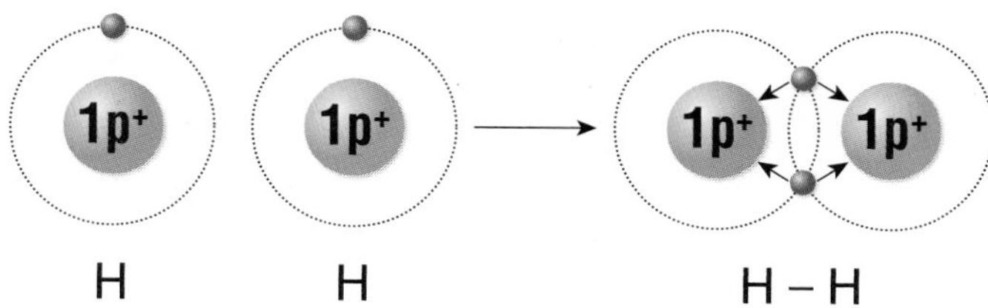

Figure 4 A covalent bond results from the sharing of a pair of electrons represented by a dash.

Fluorine is another example of a diatomic molecule. Fluorine has seven outer electrons—one electron short of a stable electron arrangement. When two fluorine atoms share a pair of electrons (one from each atom) to form a covalent bond, they form a relatively stable fluorine molecule with the chemical formula F_2.

There are other kinds of diatomic molecules. Some are made up of atoms that share two pairs of electrons. Oxygen is an example of this. There is a double covalent bond joining the two atoms: O═O. Other diatomic molecules are made up of two different elements (such as hydrogen fluoride, HF, shown in Figure 5).

Other molecules are made up of three or more atoms. A water molecule, for example, consists of one oxygen atom and two hydrogen atoms: H_2O.

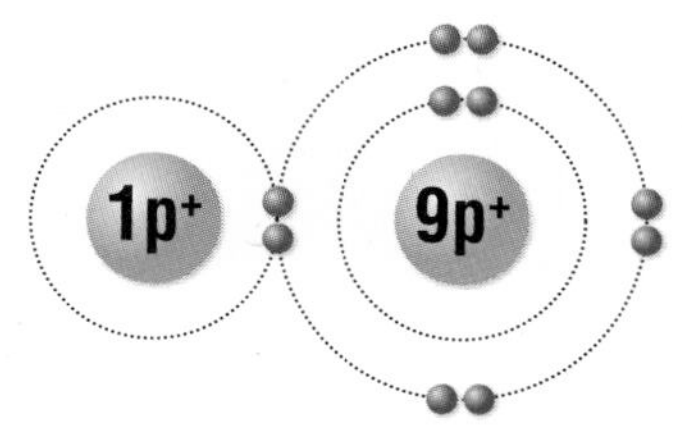

Figure 5 A hydrogen atom and a fluorine atom form a hydrogen fluoride molecule.

LEARNING TIP

Who Is Hofbrincl?

Hofbrincl, more correctly written as HOFBrINCl, is a name made up using the chemical symbols of the diatomic elements. It might help you remember this list of elements.

Table 1 lists common elements that exist as diatomic molecules. This list will be useful when you write chemical equations in Chapter 6.

Table 1 Common Diatomic Elements

Name of element	Chemical symbol	Formula of molecule	State at room temperature
hydrogen	H	H_2	gas
oxygen	O	O_2	gas
fluorine	F	F_2	gas
bromine	Br	Br_2	liquid
iodine	I	I_2	solid
nitrogen	N	N_2	gas
chlorine	Cl	Cl_2	gas

TRY THIS MOLECULAR MODELS

SKILLS: Observing, Communicating

Making models of molecules might help you understand how some elements form compounds. Each sphere, representing an atom, has a number of connection sites. This number represents the number of bonds that the atom can make with another atom. Each different colour represents a different element: white = hydrogen; red = oxygen; green = a halogen (e.g., chlorine); black = carbon.

Equipment and Materials: molecular model kit

1. Select two white spheres and connect them together to represent a hydrogen molecule. Sketch your model. C
2. Select two red spheres and connect them together to represent a molecule of oxygen. Sketch your model. C
3. Connect one black and four white spheres. Sketch your model. C
4. Build a model of a molecule with one oxygen atom and two hydrogen atoms. Sketch your model. C
5. Build a molecule of hydrogen chloride. Sketch your model. C
6. Build any other molecule using the molecular model kit. Sketch your model. C

A. Beside each sketch, write the chemical formula and, if possible, the name of the molecule. K/U

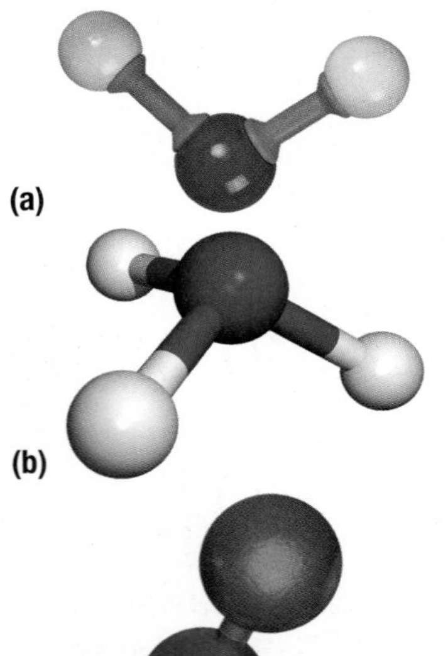

Figure 6 Molecular models of (a) water, (b) ammonia, and (c) nitric oxide

Naming Molecular Compounds

Unfortunately, the naming of molecular compounds is not as straightforward as the naming of ionic compounds. Many molecular compounds have been known for centuries and have common names that are still in use today (Figure 6). Some of these common names are given in Table 2.

Table 2 Common Names of Some Molecular Compounds

Common name	Chemical formula	Use/Occurrence
water	H_2O	the most commonly available molecular compound on Earth; the "universal solvent"
ammonia	NH_3	used in window cleaners and in the production of fertilizers
nitric oxide	NO	an air pollutant produced in the automobile engine when gasoline is burned
hydrogen sulfide	H_2S	an invisible gas with a distinctive "rotten eggs" odour

Chemists have established a system for naming molecular compounds that involves using prefixes to specify the number of atoms. The prefix is attached to the name of the element to which it refers (Table 3). For example, the name "dinitrogen pentoxide" tells us that there are two nitrogen atoms (*di* means two) and five oxygen atoms (*penta* means five) in the compound. The chemical formula for this compound, therefore, is N_2O_5. The prefix *mono* is used only for the second element in the compound, so CO_2 is carbon dioxide. Similarly, the name "carbon monoxide" states that there is one carbon and one oxygen atom in the molecule: CO.

LEARNING TIP

Using Prefixes
Note that *mono* is used only for the second element in the compound. Further, the second "o" in mono is dropped when used with oxide to become "monoxide" rather than "monooxide."

Table 3 Prefixes Used for Molecular Compounds

Prefix	Number of atoms	Sample molecular compound
mon(o)-	1	carbon monoxide, CO
di-	2	carbon dioxide, CO_2
tri-	3	sulfur trioxide, SO_3
tetra-	4	carbon tetrachloride, CCl_4
penta-	5	phosphorus pentafluoride, PF_5

When you are asked to write the name of a compound, first check the formula to see if it includes a metal. If the first element is a metal, the substance is an ionic compound and should be named accordingly (with no prefixes). If the compound consists only of non-metals, it is a molecular compound, and you should follow these steps to name it.

SAMPLE PROBLEM 1 Naming Molecular Compounds

Name the molecular compound with chemical formula PCl_3.

Step 1 Write the names of both elements in the same order as in the formula. Replace the ending of the second element with "ide."
phosphorus chlor~~ine~~ ide

Step 2 Add prefixes. Remember that the prefix "mono" is never used for the first element.
phosphorus *tri*chloride

The compound with chemical formula PCl_3 is called phosphorus trichloride.

Practice

Name the compound with chemical formula N_2O.

WRITING TIP

Concluding Your Summary
Write a closing sentence that connects the main idea and key points. For example, "The use of prefixes helps students remember the number of atoms in a molecular compound."

Writing Chemical Formulas of Molecular Compounds

Given its name, writing the formula of a molecular compound is relatively simple. The prefixes in the name become the subscripts in the formula. For example, the molecular compound called sulfur dioxide has the chemical formula SO_2 (Figure 7).

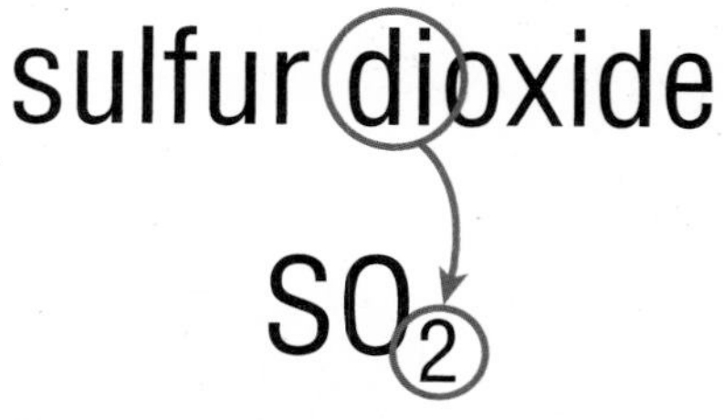

Figure 7 In the chemical name, the prefix specifying the number of atoms comes *before* the element's name. In the chemical formula, however, the number of atoms is specified by a subscript *after* the element's chemical symbol.

Figure 8 Even something as ordinary as a backpack could contain hundreds of different molecular compounds, most of them made from fossil fuels.

Molecular Compounds from Fossil Fuels

Most compounds are molecular. Living things make a huge variety of different molecular compounds. Another source of thousands of different molecules is fossil fuels. Coal, oil, and natural gas are the most common fossil fuels. These substances take millions of years to form from the partially decayed remains of ancient plants and animals. Fossil fuels are called a non-renewable resource because they are not formed as quickly as we are using them.

Fossil fuels have become very important to our way of life. When we burn fossil fuels, the energy stored within them heats our homes, powers our vehicles, and can be harnessed to generate electricity. Without the energy released from fossil fuels, our lives would be very different. But there is more to fossil fuels than just their energy. Compounds extracted from fossil fuels are processed into petrochemicals. We use these compounds to make important consumer products and industrial chemicals, including plastics, pharmaceuticals, and synthetic fabrics (Figures 8 and 9).

Try to imagine life without the products from petrochemicals. Half of your clothing is now gone! Personal products like toiletries, cosmetics, and their containers no longer exist. The paint on the walls and the synthetic carpet on the floor—gone. Do you need medication regularly? Too bad! Many drugs are made using petrochemicals. And what about communication? Computers, phones, and all portable electronic devices cannot function without their plastic cases and the insulation covering their electrical parts.

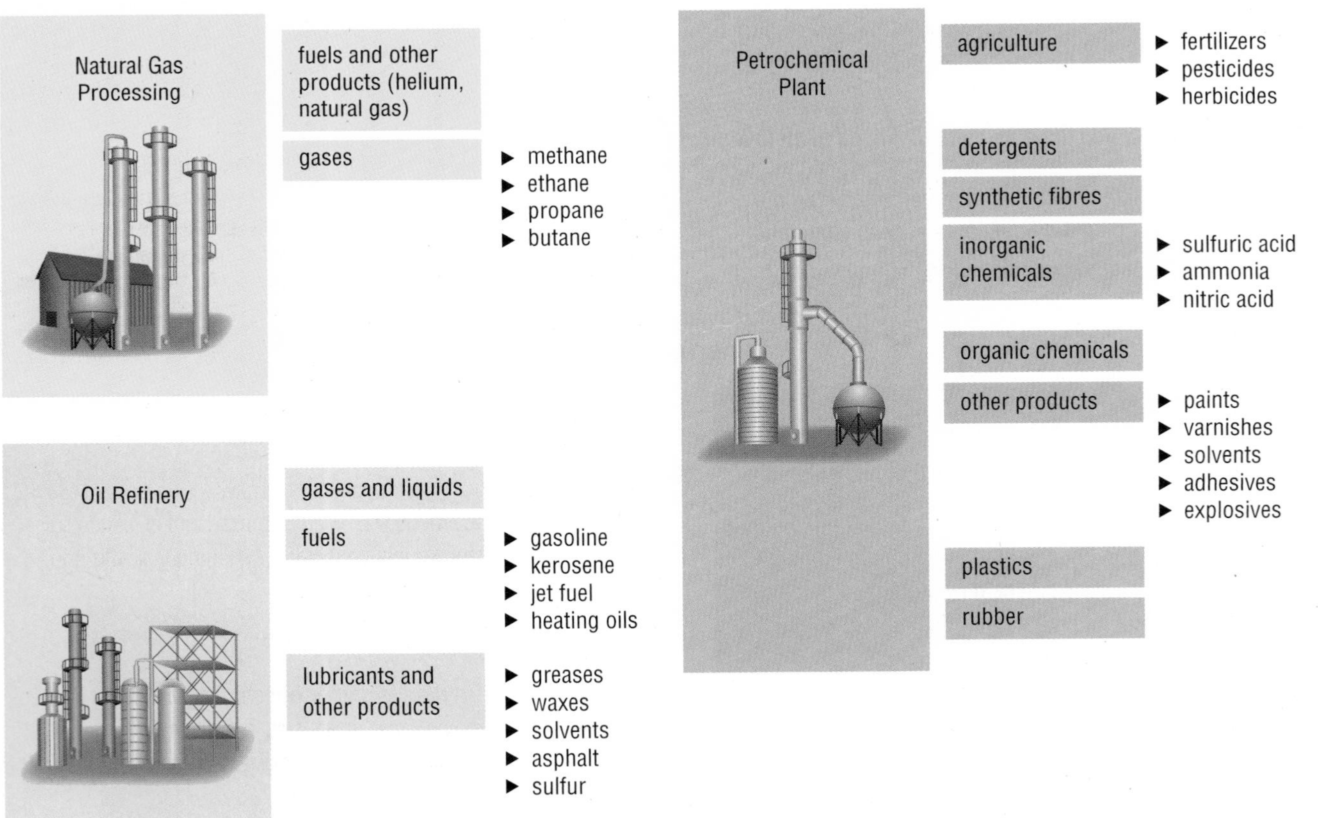

Figure 9 Crude oil and natural gas are the raw materials for a vast range of chemical products.

Spills and Leaks of Molecular Compounds

The world's oil and natural gas deposits are concentrated in a few places, far from the places where they are sold and used. Most of Canada's oil is in the west and north of the country, while most Canadian consumers live in the south and east. As a result, huge quantities of oil and natural gas are carried across North America and around the world. Oil is transported by rail, ship, or pipeline. Natural gas is transported by pipeline or in specially designed tankers. Accidents are inevitable. Loaded oil tankers travel across oceans and through the Great Lakes. Spills from oil tankers can have a devastating effect on local ecosystems, contaminating water and shorelines and killing birds and other aquatic organisms (Figure 10). A lot of research is done on the best way to deal with the leaked chemicals.

Accidental spills and poor waste disposal methods leak toxins into the ground, where they pollute groundwater. One such toxin is a molecular compound called trichloroethene. Trichloroethene is widely used in industry as a degreaser for cleaning metal and glass. Spills of this compound are a serious problem in Canada because they can contaminate the groundwater. Nearly 9 million Canadians rely on groundwater for their drinking water. Canadian researchers are working on innovative ways to clean groundwater. Dr. Elizabeth Edwards, a professor of chemical engineering at the University of Toronto, has found that some pollutants are food for microbes. Dr. Edwards has successfully used microbes to remove trichloroethene from soil. As the microbes "dine," they convert the pollutant into ethene, a relatively harmless gas. These microbes are now being used to remove trichloroethene from polluted sites around the world. 

DID YOU KNOW?

Petrodollars

Annual sales from the global petrochemical industry are estimated to be over \$1 trillion. With a global population of almost 7 billion, that is about \$150 for every person in the world.

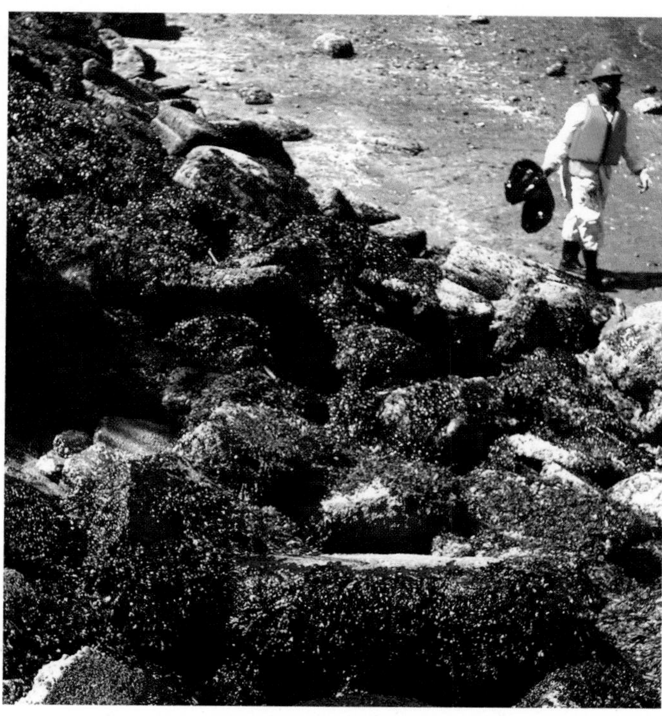

Figure 10 Spilled oil is a frequent environmental problem.

University professors are often involved in solving real-world problems. To find out more about Dr. Edwards' research,

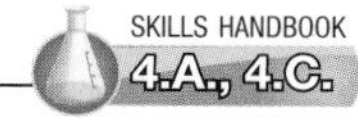

GO TO NELSON SCIENCE

RESEARCH THIS SLICKS FROM SHIPS

SKILLS: Researching, Identifying Alternatives, Analyzing the Issue, Defending a Decision, Evaluating

SKILLS HANDBOOK 4.A., 4.C.

Accidental spills of oil can be devastating for the environment. Environmental engineers develop strategies to clean up oil spills, using their knowledge of oil's chemical and physical properties.

GO TO NELSON SCIENCE

Recently, biological agents have also been successfully used to clean up oil spills. Naturally occurring micro-organisms such as algae and bacteria help break down the spilled oil. This process is very slow but can be sped up by adding fertilizer to the contaminated area.

1. Research the role of chemical methods, physical methods, and biological agents in cleaning an oil spill.

GO TO NELSON SCIENCE

A. Analyze the advantages and disadvantages of each clean-up strategy. Which strategies are the most economically viable? Which strategies are the most environmentally friendly? T/I A

B. Which method can be used to clean up the majority of a spill on calm water? T/I

IN SUMMARY

- Molecular compounds both occur naturally and are produced synthetically. Some are beneficial; others are not.
- Molecular compounds are made up of molecules. A molecule is a group of two or more atoms joined by covalent bonds.
- A covalent bond forms when two non-metallic atoms share electrons.
- A variety of strategies are used to reduce the environmental damage resulting from spilled chemicals.
- The name of a molecular compound includes prefixes to indicate how many atoms of each element are present (for example, dinitrogen pentoxide). If there is only one atom of the first element, the prefix "mono" is omitted. The name of the last element ends in "ide."
- Fossil fuels provide valuable energy and petrochemicals. Many important industrial chemicals and consumer products are made from petrochemicals.

CHECK YOUR LEARNING

1. (a) Name these compounds: NI_3, CCl_4, OF_2, P_2O_5, and N_2O_3.
   (b) Describe how each compound name indicates the ratio of elements. K/U
2. Write the chemical formula for each of the following molecular compounds: K/U
   (a) carbon monoxide
   (b) sulfur tetrafluoride
   (c) dinitrogen tetroxide
   (d) nitrogen tribromide
   (e) carbon disulfide
3. For each of the following compounds, classify the elements as metal or non-metal, classify the compound as ionic or molecular, and name the compound. K/U
   (a) SO_2
   (b) PbO_2
   (c) $AlCl_3$
   (d) N_2O
   (e) $KClO_3$
   (f) SnO_2
   (g) $FePO_4$
   (h) N_2O_4
4. (a) How many electrons do atoms of hydrogen and oxygen have in their outer orbits?
   (b) How many electrons will these elements gain before they become stable?
   (c) Sketch a diagram to show how hydrogen and oxygen could bond to form a stable molecule. K/U
5. Explain, with diagrams, why the term "molecule" is appropriate for hydrogen chloride but not for sodium chloride. K/U
6. Contrast the way in which the elements in ionic and molecular compounds achieve stability. K/U
7. (a) Why are fossil fuels a non-renewable resource?
   (b) What are the two main benefits that we get from fossil fuels?
   (c) What are two disadvantages of our dependence on fossil fuels? K/U A
8. Explain why chlorine occurs as diatomic molecules in nature, rather than individual atoms. K/U
9. Hydrogen peroxide, H_2O_2, is a molecular compound used to disinfect cuts (Figure 11). Why is the formula of this compound not written as HO? A

Figure 11

10. How can molecular compounds be distinguished from ionic compounds
    (a) by looking at their chemical formulas?
    (b) by testing them in the lab? K/U
11. What effect could a disruption in the supply of oil have on the cost of goods you purchase? Why? A

Properties of Ionic and Molecular Compounds

SKILLS MENU	
Questioning	Performing
Hypothesizing	Observing
Predicting	Analyzing
Planning	Evaluating
Controlling Variables	Communicating

As you learned earlier in this chapter, ionic compounds are solids made up of positive and negative ions. Molecular compounds are made up of tiny individual molecules. Ionic and molecular compounds have quite different properties.

In this investigation, you will compare some physical properties of ionic and molecular compounds: solubility in water, electrical conductivity of the mixtures they form with water, and melting point.

Testable Question

Which of the following substances are molecular compounds and which are ionic compounds: lauric acid, $C_{12}H_{24}O_2$; sodium hydrogen carbonate, $NaHCO_3$; glucose, $C_6H_{12}O_6$; potassium chloride, KCl?

Hypothesis/Prediction

Use the information in the chemical formulas to write a hypothesis regarding the classification of the compounds. Give reasons for your hypothesis. Use your hypothesis to predict an answer to the Testable Question.

Experimental Design

You will use solubility and conductivity tests, along with researched melting point data, to determine whether four solids are ionic or molecular compounds.

Equipment and Materials

- eye protection
- lab apron
- 4 small test tubes and stoppers
- test-tube rack
- low-voltage conductivity tester
- well plate
- samples of
  - lauric acid, $C_{12}H_{24}O_2(s)$
  - sodium hydrogen carbonate, $NaHCO_3(s)$
  - glucose, $C_6H_{12}O_6(s)$
  - potassium chloride, KCl(s)

Procedure

1. Design a procedure to compare
   - the solubility of the compounds in water
   - the conductivity of a solution of each of the compounds with water

   Figure 1 shows the maximum quantity of each compound that you should use. Check the MSDS for each compound. Be sure to include all necessary safety precautions.

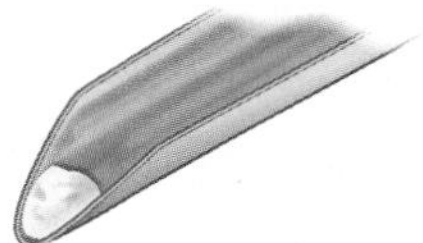

Figure 1 Only a small quantity of each compound is required.

2. With your teacher's approval, perform your procedure. Record your observations.
3. Look up the melting point of each substance in a reference book or online. Add these data to your experimental information.

GO TO NELSON SCIENCE

Analyze and Evaluate

(a) Use your evidence to answer the Testable Question. Compare your answer with your Prediction. Account for any differences. T/I

(b) Which of the three tests was most useful in classifying the compounds as ionic or molecular? Explain. T/I

Apply and Extend

(c) Create a table to compare the typical properties of ionic and molecular compounds. K/U C

CHAPTER 5

LOOKING BACK

KEY CONCEPTS SUMMARY

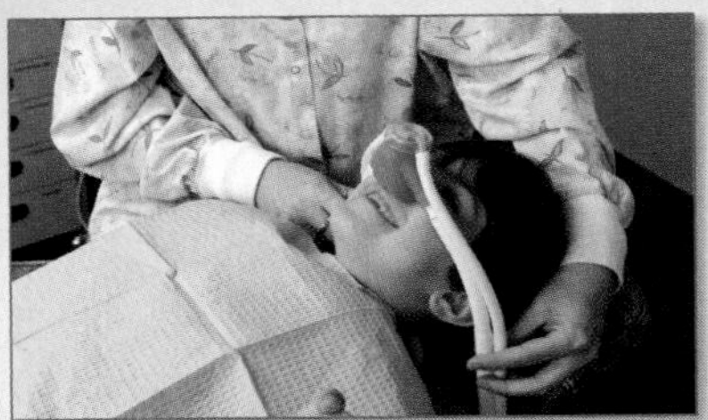

A substance's chemical and physical properties determine its usefulness and effects.

- Physical properties (for example, state, colour, electrical conductivity) involve a description of the substance as it is. (5.1)
- Chemical properties describe a substance's behaviour (for example, reactivity with acids, combustibility) as it becomes a completely different substance(s). (5.1)

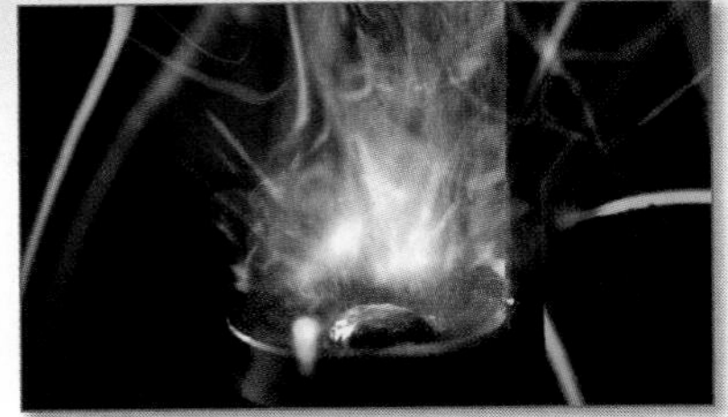

Changes can be classified as chemical or physical.

- Chemical changes involve new substances being produced. (5.2)
- Evidence of chemical change includes colour change, precipitate production, release or absorption of energy, and formation of a gas. (5.1)
- Physical change involves changes in form (for example, state) but not in chemical identity. (5.1)

We can classify pure substances by observing their properties.

- Elements can be grouped according to their properties. (5.4)
- Metals and non-metals combine to form different kinds of compounds: ionic and molecular. (5.6, 5.10)
- Compounds are pure substances that can be broken down into their individual elements. (5.4)
- Physical properties distinguish molecular compounds from ionic compounds. (5.11)

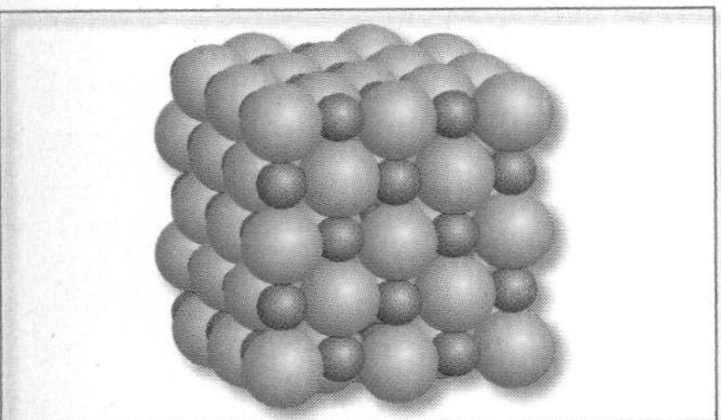

Ionic compounds are made up of positive and negative ions.

- Atoms gain or lose electrons resulting in a stable electron arrangement, thus becoming ions. (5.5)
- Positively charged ions are cations; negatively charged ions are anions. (5.5)
- Ionic compounds consist of cations and anions linked by ionic bonds. (5.6)
- Many ionic compounds are electrolytes: the solutions they form conduct electricity. (5.6)
- The names of ionic compounds formed by two elements end in "ide." (5.7)
- The names of compounds that include polyatomic ions usually end in "ate." (5.9)

Molecular compounds are made up of distinct molecules.

- The majority of all known compounds are molecular. (5.10)
- Molecules consist of two or more non-metal atoms linked by a covalent bond. (5.10)
- Covalent bonds form when atoms share electrons. (5.10)
- Molecular compounds are often named using prefixes. (5.10)

Many consumer products have been developed from petrochemicals.

- Product chemists select chemicals to perform specific functions in a product. (5.10)

WHAT DO YOU THINK NOW?

You thought about the following statements at the beginning of the chapter. You may have encountered these ideas in school, at home, or in the world around you. Consider them again and decide whether you agree or disagree with each one.

1 The label on a chemical product provides all the information you need to use the product safely.
Agree/disagree?

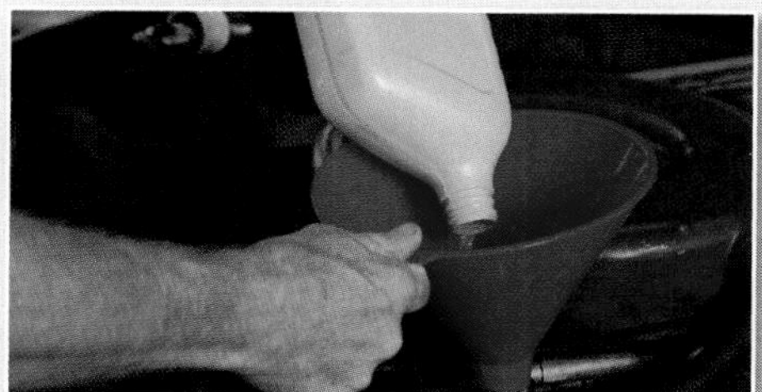

2 Recycling used motor oil is common practice.
Agree/disagree?

3 Pool water is a much better conductor of electricity than pure water.
Agree/disagree?

4 Elements are more reactive and more hazardous than the compounds that they form.
Agree/disagree?

5 Bottled water is better for your health than tap water.
Agree/disagree?

6 Adding manufactured chemicals to the environment is a bad thing.
Agree/disagree?

How have your answers changed since then? What new understanding do you have?

Vocabulary

physical property (p. 175)
chemical property (p. 175)
element (p. 184)
period (p. 184)
group (p. 184)
alkali metals (p. 184)
alkaline earth metals (p. 184)
halogens (p. 184)
noble gases (p. 184)
Bohr–Rutherford diagram (p. 185)
compound (p. 186)
ion (p. 188)
cation (p. 190)
anion (p. 190)
ionic compound (p. 192)
ionic bond (p. 192)
electrolyte (p. 194)
polyatomic ion (p. 202)
molecular compound (p. 206)
covalent bond (p. 207)
molecule (p. 207)
diatomic molecule (p. 207)

BIG Ideas

- ✓ **Chemicals react with each other in predictable ways.**
- **Chemical reactions may have a negative impact on the environment, but they can also be used to address environmental challenges.**

CHAPTER 5 REVIEW

The following icons indicate the Achievement Chart category addressed by each question.

K/U Knowledge/Understanding T/I Thinking/Investigation C Communication A Application

What Do You Remember?

1. Describe, using an example, how you would recognize the chemical formula of
   (a) an element
   (b) a compound (5.4) K/U
2. Distinguish between the following terms, using specific examples. (5.1–5.11) K/U
   (a) physical and chemical properties
   (b) ionic and molecular compounds
   (c) ionic and covalent bonds
3. Explain your answers to the following questions. (5.1–5.10) K/U
   (a) Is tap water a pure substance?
   (b) Is the ability to burn a physical property?
   (c) What type of property describes the formation of a new substance?
   (d) What kind of elements make up molecular compounds?
   (e) What is special about the outer electron orbits of helium, neon, and argon?
   (f) What kind of ion has more protons than electrons?
   (g) Which are polyatomic ions: hydroxide, chloride, ammonium, and carbonate?
4. Copy and complete Table 1 in your notebook. (5.5) K/U

Table 1 Ion Formation for Three Elements

Element	Bohr–Rutherford diagram of the atom	Bohr–Rutherford diagram of the ion	Chemical symbol of the ion
(a) Na			
(b) S			
(c) Cl			

What Do You Understand?

5. List the numbers and names of atoms in the following molecules. (5.10) K/U
   (a) CO_2
   (b) N_2
   (c) CCl_4
   (d) HBr
6. Name or write the chemical formula for each of the following compounds and classify each one as an ionic or a molecular compound. (5.7, 5.10) K/U
   (a) $FeCl_3$
   (b) $CuSO_4$
   (c) NI_3
   (d) PbO_2
   (e) P_2O_3
   (f) $Sn(NO_3)_2$
   (g) carbon tetrabromide
   (h) calcium carbonate
   (i) nitrogen monoxide
   (j) hydrogen sulfide
7. In which of the following groupings do each of the three ions have the same number of electrons? How many electrons do they have? (5.5) K/U
   (a) O^{2-}, F^-, N^{3-}
   (b) Na^+, K^+, Li^+
   (c) K^+, P^{3-}, Ar
   (d) F^-, Cl^-, Br^-
8. Classify each of the following compounds as ionic or molecular and write their chemical formulas. (5.7, 5.10) K/U
   (a) potassium chloride
   (b) carbon monoxide
   (c) carbon tetrafluoride
   (d) calcium iodide
   (e) sulfur dioxide
   (f) lithium oxide
9. Look at your answers to question 8. Explain why Roman numerals are not required to name these compounds. (5.7) K/U
10. Write the name and chemical formula of the compound that forms when the following pairs of elements combine. (5.7) K/U
    (a) calcium and sulfur
    (b) aluminum and chlorine
    (c) sodium and phosphorus
    (d) aluminum and sulfur
11. Name or write the chemical formula for each of the following compounds involving polyatomic ions. (5.9) K/U
    (a) calcium nitrate
    (b) silver carbonate
    (c) $Fe(OH)_3$
    (d) $Cu(ClO_3)_2$
    (e) lead(II) phosphate
12. Imagine the chemical formula of a compound formed from each of these pairs of elements. What are the mostly likely subscripts in each chemical formula? (5.7) K/U T/I
    (a) an alkali metal and a halogen
    (b) an alkaline earth metal and a member of Group 16
    (c) an alkali metal and a member of Group 16

13. Hazardous household products used in the home have different safety symbols than those used in the workplace. For example, Figure 1 shows symbols found on a spray can of furniture polish.
   (a) What are the risks involved in using this product?
   (b) What are the equivalent WHMIS symbols for these risks?
   (c) What would be the advantage of having the same labelling system for home and work? (5.3) K/U A

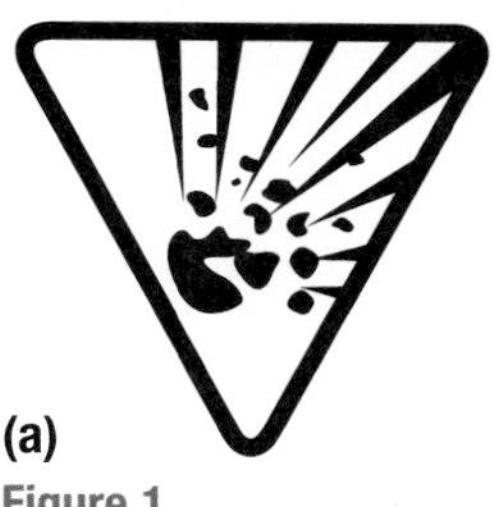
(a)

(b)

Figure 1

Solve a Problem

14. An unknown element X forms a compound with chlorine: XCl_2. Predict the chemical formula of the compound that element X makes with oxygen. Justify your answer. (5.7) T/I

15. As you have seen in this chapter, hydrogen is very different from the other elements in Group 1. (5.4) T/I
   (a) Chemists have put hydrogen and the alkali metals in the same column of the periodic table. Why?
   (b) Hydrogen is sometimes shown above fluorine on the periodic table? Why?

16. Figure 2 shows two properties of oil and water. What impact do these properties have on attempts to clean up an oil spill on water? (5.10) K/U A

(a)

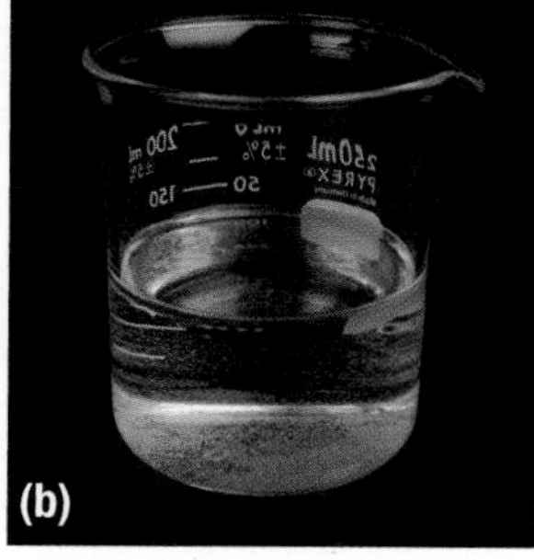
(b)

Figure 2 (a) Oil and water do not mix. The oil stays in droplets. (b) Oil floats on water.

17. The price of gasoline changes a great deal depending on the price of oil. Some environmentalists argue that high gas prices are good for the environment. Why? (5.10) A

18. Why is it important to understand the chemical composition of chlorinating agents used in swimming pools before using them? (5.8) A

Create and Evaluate

19. Look back to the Engage in Science page about the space elevator. Think about the cable that might some day anchor the space elevator to Earth. (5.1–5.11) T/I C A
   (a) What physical and chemical properties should the cable have? Think of properties that were mentioned in the chapter and others that would be relevant. Explain.
   (b) Choose one of these properties. Suggest a way to test that the cable has this property.
   (c) Think about how the space environment might affect the cable. How should this affect the selection of a material for the cable?
   (d) What might be some environmental and social impacts of building a space elevator?
   (e) Write a paragraph about the space elevator from the perspective of someone living when the space elevator is being built. Refer to a chemical and its properties in your writing. Your paragraph could be a letter to a newspaper editor, a blog, a news flash, or any other format of your choice.

Reflect on Your Learning

20. (a) Complete this sentence: "One idea that I found particularly interesting in this chapter and would like to explore further is..."
   (b) Why did this idea catch your interest? Share this idea with a classmate or your teacher.

Web Connections 

21. The two most common household bleaches are chlorine bleach and oxygen bleach. Research their advantages and disadvantages. Which would you recommend for home use? Why? (5.3) T/I A

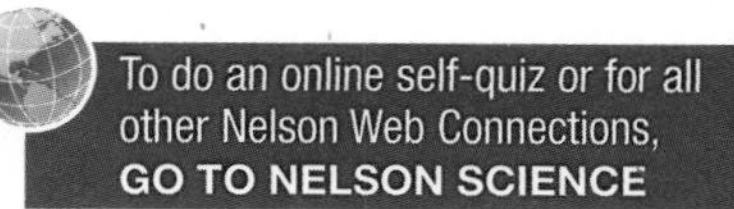
To do an online self-quiz or for all other Nelson Web Connections, GO TO NELSON SCIENCE

CHAPTER 5 SELF-QUIZ

The following icons indicate the Achievement Chart category addressed by each question.

K/U Knowledge/Understanding
T/I Thinking/Investigation
C Communication
A Application

For each question, select the best answer from the four alternatives.

1. A physical property is a
   (a) characteristic that does not involve forming a new substance.
   (b) behaviour that occurs when a substance becomes something new.
   (c) change that does not produce a new substance.
   (d) characteristic of a substance that allows it to take part in a reaction. (5.1) K/U
2. Which statement correctly describes the relationship between electron arrangements and reactivity? (5.6) K/U
   (a) Atoms with completely filled outer orbits usually react with other atoms.
   (b) Atoms frequently react with other atoms in order to have one outer electron.
   (c) Atoms tend to react with other atoms in order to have a full outer electron orbit.
   (d) Stable atoms readily react with other atoms of other elements to form compounds.
3. In reactions with non-metals, metals tend to
   (a) lose electrons to become positively charged anions.
   (b) gain electrons to become positively charged cations.
   (c) gain electrons to become negatively charged anions.
   (d) lose electrons to become positively charged cations. (5.5) K/U
4. Which of the following is the correct formula of diphosphorus pentoxide? (5.10) K/U
   (a) N_2O_5
   (b) P_2O_5
   (c) PO_4^{3-}
   (d) P_5O_2

Indicate whether each of the statements is TRUE or FALSE. If you think the statement is false, rewrite it to make it true.

5. The number of protons in the nucleus of an atom is its atomic number. (5.4) K/U
6. The elements in the periodic table are arranged in order of decreasing atomic number. (5.4) K/U

Copy each of the following statements into your notebook. Fill in the blanks with a word or phrase that correctly completes the sentence.

7. A(n) _______ is a horizontal row of elements on the periodic table. (5.4) K/U
8. Most of an atom's mass is concentrated in an extremely small, dense, positively charged core called the _______. (5.5) K/U

Match each term on the left with the most appropriate term on the right.

9. (a) Ca — (i) ionic charge of −2
   (b) S — (ii) ionic charge of +3
   (c) K — (iii) ionic charge of +1
   (d) Al — (iv) ionic charge of +2 (5.5) K/U
10. (a) $Fe(NO_3)_2$ — (i) iron(III) nitride
    (b) Fe_3N_2 — (ii) iron(II) nitrate
    (c) FeN — (iii) iron(II) nitride
    (d) $Fe(NO_3)_3$ — (iv) iron(III) nitrate (5.7) K/U

Write a short answer to each of these questions.

11. What has to happen to a substance in order for you to observe its chemical properties? (5.1) K/U
12. Explain why ionic compounds are electrically neutral. (5.6) K/U
13. Why are groups of elements on the periodic table sometimes referred to as "families"? (5.4) K/U

14. An element is a substance that cannot be broken down into simpler substances. Does this mean that elements cannot undergo chemical change? Explain. (5.1, 5.4) K/U T/I

15. Compare metals and non-metals in terms of their
(a) chemical properties.
(b) physical properties. (5.1, 5.4) K/U

16. Briefly describe the trend in how elements form ions, as you look across the periodic table from Group 1 to Group 17. (5.4) K/U

17. What are some characteristics of a substance that could be discovered by studying its chemical properties? (5.1–5.11) T/I

18. The production of a gas is evidence that a chemical change *may* have occurred. However, the boiling of water also produces a gas. (5.1–5.3) T/I
(a) Why is boiling not a chemical change?
(b) Give an example of a chemical change that produces a gas. Explain why this example is a chemical change.

19. Write a paragraph describing the relationship between atoms and ions. Use each of the following words at least once in your paragraph: atom, ion, neutral, cation, anion, electron, and proton. (5.5) K/U C

20. For each property described below, identify the property and state whether it is a physical or a chemical property. (5.1, 5.2) T/I
(a) liquid nitrogen boils at -196 °C
(b) silver jewellery tarnishes in air
(c) propane ignites easily

21. Many people know water by its common name, but not by its systematic name. Write a one-paragraph advertisement warning people of the potential dangers of dihydrogen monoxide (water). (5.10) C

22. (a) The term "polyatomic ion" can be broken into three parts: poly, atomic, and ion. Explain the meaning of each part of this term.
(b) Use the meaning of each part to write a definition of "polyatomic ion." (5.9) T/I C

23. (a) Describe a jewellery-cleaning method that requires a chemical change.
(b) Describe a jewellery-cleaning process that involves a physical change. (5.1) A

24. Describe, using an example, how a substance being added to water can cause
(a) a chemical change.
(b) a physical change. (5.1, 5.2) T/I

CHAPTER

6

Chemicals and Their Reactions

KEY QUESTION: What are chemical reactions and how do they impact our lives?

What chemical reaction could draw this many people to a world-record-setting event in Belgium, in 2008?

UNIT C

Chemical Reactions

CHAPTER 5
Chemicals and Their Properties

CHAPTER 6
Chemicals and Their Reactions

CHAPTER 7
Acids and Bases

KEY CONCEPTS

Chemical reactions involve the change of one or more substances into one or more different substances.

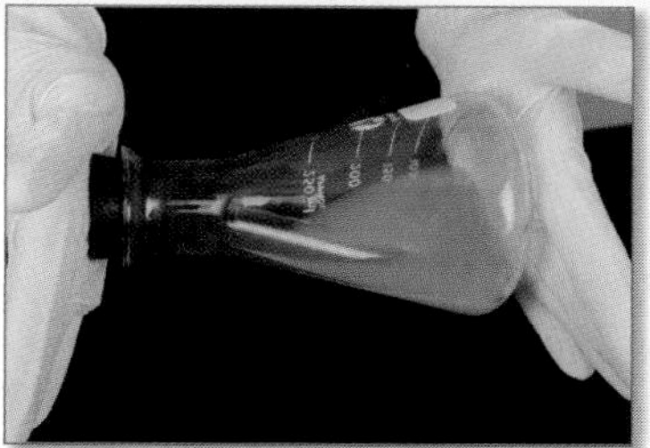

Chemical reactions obey the law of conservation of mass and can be represented by balanced chemical equations.

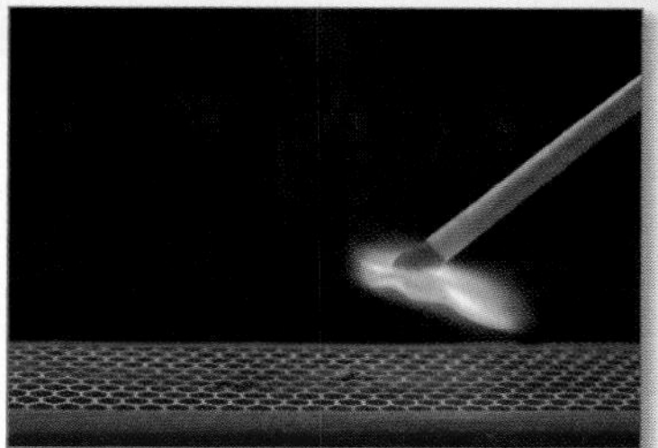

Chemical reactions involving consumer products can be useful or harmful.

We can use patterns in chemical properties to classify chemical reactions.

Corrosion is the reaction of metals with substances in the environment.

Chemical reactions affect us and our environment.

THE MINT-COLA FOUNTAIN

The mint–cola fountain experiment has fascinated me ever since I saw a demonstration of it at a science fair. I searched the Internet and found thousands of hits and every possible variation of this experiment you can imagine! I tried it myself with exciting results, as the photo shows.

Then I started wondering which mint–cola mixture produces the highest fountain. Some friends and I did our own experiment, playing around with some of the variables and measuring how high the fountain blew. We compared diet and regular cola (freshly opened 2 L bottles each time) and always added 30 g of mint candy. We tried crushing the mints before adding them to the cola. Then we tried adding either table salt, rock salt, or detergent to the cola before adding the mints. The results of the tests are summarized in the diagram. (In case you're wondering, we didn't drink any of the cola that had been used in the experiment!)

We have a couple of things to follow up on: we plan to find out how high the fountain will go if we add detergent to regular cola. Any predictions? I've also read that it works with tonic water…

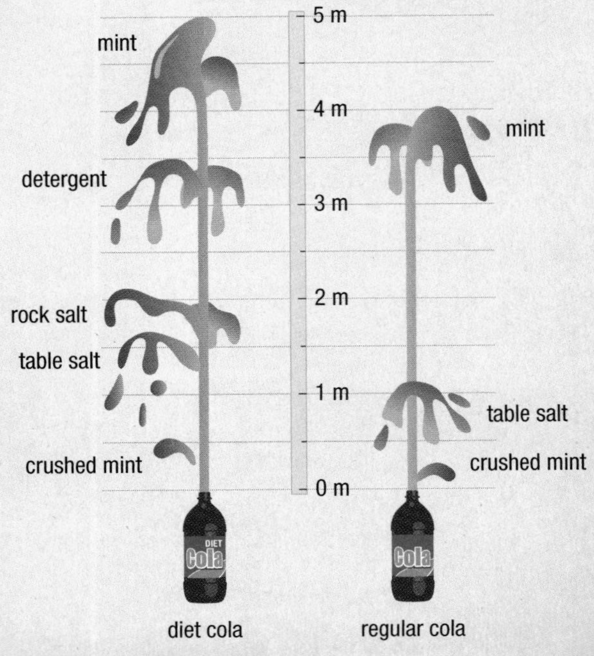

GO TO NELSON SCIENCE

WHAT DO YOU

THINK?

Many of the ideas you will explore in this chapter are ideas that you have already encountered. You may have encountered these ideas in school, at home, or in the world around you. Not all of the following statements are true. Consider each statement and decide whether you agree or disagree with it.

1 Chemical reactions are bad for the environment.
Agree/disagree?

2 Chemical reactions are reversible.
Agree/disagree?

3 The total mass of substances involved in a chemical change remains constant.
Agree/disagree?

4 Carbon dioxide in your classroom makes you sleepy.
Agree/disagree?

5 Some people are allergic to cellphones.
Agree/disagree?

6 Hydrogen will replace gasoline as the fuel of the twenty-first century.
Agree/disagree?

Making Inferences

When you make inferences, you make sense of what an author is implying but not stating explicitly. It is a form of problem-solving using evidence from the text, personal knowledge, and reasoning. Use the following strategies to make inferences:

- look for context clues such as significant words, comparisons, or contrasts
- think about what you already know about the circumstance, issue, problem, cause, or effect
- combine the clues and your prior knowledge or experience to draw a conclusion or form an opinion
- revise your inference if you find new information or clues that challenge it

READING TIP

As you work through the chapter, look for tips like this. They will help you develop literacy strategies.

Types of Chemical Reactions: Combustion

In the early hours of August 2, 2008, a north Toronto neighbourhood was rocked by a loud explosion. Startled residents stared out their windows in disbelief as a giant fireball rose high into the night sky. Why? A nearby propane storage depot was on fire (Figure 1)! Large chunks of metal, probably from exploded propane storage tanks, littered the area. Shockwaves from the explosion shattered windows and ripped doors off their hinges. Firefighters rushed to the scene, but all they could do was cool the remaining propane tanks with water and wait for the fireball to burn itself out.

Figure 1 Propane was the fuel in this dramatic combustion reaction in north Toronto.

Making Inferences *in Action*

Authors often tell you more than they say directly through their words. Use the clues in the text and what you already know to make an inference. Inferences help you to have a deeper understanding of the text. Here is how one student made inferences as she read the paragraph about combustion.

Clues from Text	+ Prior Knowledge	= Inference
explosion happened at night	• fire causes explosions • arsonists usually work at night	an arsonist might have caused this explosion
explosion is in a residential area	most homes are heated by natural gas	a cracked natural gas pipe might have caused the explosion
firefighters let fireball burn out	firefighters usually use water to extinguish a fire	maybe water cannot be used to put out a propane fire

Describing Chemical Reactions

6.1

Chemical reactions are everywhere, resulting in changes that we see all around us: fires burn, grass grows, joggers run, milk sours, autumn leaves change colour, and marshmallows toast to a golden brown (Figure 1).

In a **chemical reaction**, one or more substances change into different substances. One of the most familiar types of chemical reaction is combustion (burning). We use combustion reactions to cook our food, heat our homes, and travel long distances. Combustion reactions can also lead to many health and environmental problems. Chemists have developed a set of rules that enable people around the world to communicate and share information about chemical reactions. In this chapter, you will learn some of these rules of communication.

Figure 1 Combustion is a chemical reaction that releases energy.

chemical reaction a process in which substances interact, causing the formation of new substances with new properties

word equation a way of describing a chemical reaction using the names of the reactants and products

chemical equation a way of describing a chemical reaction using the chemical formulas of the reactants and products

reactant a chemical, present at the start of a chemical reaction, that is used up during the reaction

product a chemical that is produced during a chemical reaction

Describing Chemical Reactions with Equations

Chemists use equations to describe chemical reactions. You will learn how to write and interpret two kinds of equations. In **word equations**, the names of the chemicals are written out in full. In **chemical equations**, chemical formulas are used to represent the chemicals.

Examples of Word and Chemical Equations

During a chemical reaction, reactant particles collide, allowing their atoms (or ions) to rearrange and form products. **Reactants** are the substances that are used up during the reaction. **Products** are the substances that are produced during the reaction. Figure 2 shows the reaction that occurs when a mixture of powdered iron and sulfur is heated. The reactants in this case are iron and sulfur. The product is iron(II) sulfide. Chemical reactions can either absorb or release energy. In this reaction, more energy is released than is absorbed. The word "energy" is therefore written on the right side of the equation, with the products. If energy is absorbed, it is written with the reactants.

We can communicate the reaction between iron and sulfur using either a word equation or a chemical equation. Since energy is released in this reaction, "energy" is written on the right side of the chemical equation.

	Reactants	yields	Products
Word equation:	iron + sulfur	→	iron(II) sulfide + energy
Chemical equation:	$Fe + S$	→	FeS + energy

LEARNING TIP

Clues of a Chemical Reaction
Look back to Table 2 in Section 5.1. These clues indicate that a chemical reaction may be taking place.

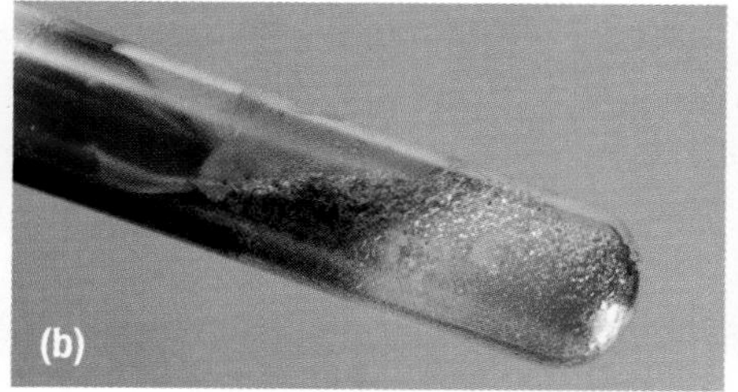

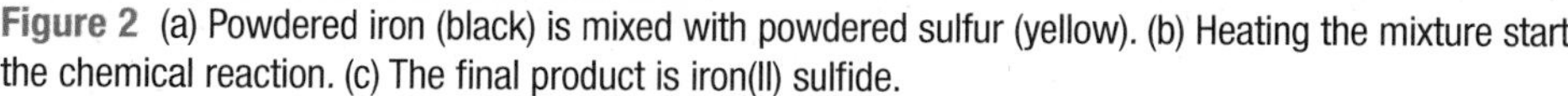

Figure 2 (a) Powdered iron (black) is mixed with powdered sulfur (yellow). (b) Heating the mixture starts the chemical reaction. (c) The final product is iron(II) sulfide.

Word equations and chemical equations have a lot in common:

- An arrow indicates the direction in which the chemical reaction is going. The arrow is read as "yields," "forms," or "produces."
- Substances to the left of the arrow are called reactants.
- Substances to the right of the arrow are called products.
- If several reactants are involved, "+" signs are placed between the reactants. This indicates that the reactants must be in contact with each other.
- If several products are formed, "+" signs are placed between the products.

Both equations list the reactants and products of the reaction. The chemical equation, however, provides far more detail: it gives the chemical formulas of the reactants and products as well as their state. **State symbols** tell us the state, or form, of each substance in a chemical equation. For example, the state symbol (s) means "solid." All the chemicals involved in this reaction are solids. Table 1 summarizes the most common state symbols.

state symbol a symbol indicating the physical state of the chemical at room temperature (i.e., solid (s), liquid (l), gas (g), or aqueous (aq))

Table 1 Common State Symbols in Chemical Equations

State symbol	Meaning
(s)	solid
(l)	liquid
(g)	gaseous
(aq)	aqueous (dissolved in water)

Now look at Figure 3. Pale green copper(II) carbonate absorbs energy to produce carbon dioxide gas and copper(II) oxide. The equations for this reaction are:

Word equation:

energy + copper(II) carbonate → carbon dioxide + copper(II) oxide

Chemical equation:

$$\text{energy} + CuCO_3(s) \rightarrow CO_2(g) + CuO(s)$$

The state symbols in this case tell you that the reactant is a solid and that the products are a gas and a solid. Since energy must be absorbed for this reaction to occur, the energy term is written on the left side of the equation.

One final example: if a piece of zinc metal is placed in a solution of copper(II) sulfate, a fuzzy reddish-brown coating forms on the zinc (Figure 4). After about 20 min, the entire zinc strip is covered by this new solid.

Word equation:

zinc + copper sulfate → zinc sulfate + copper + energy

Chemical equation:

$$Zn(s) + CuSO_4(aq) \rightarrow ZnSO_4(aq) + Cu(s) + \text{energy}$$

The state symbol (aq) tells us that the chemical is dissolved in water. Both reactants and products can be aqueous.

READING TIP

Making Inferences

Look for context clues such as significant words, comparisons, or contrasts. For example, in the comparison of word and chemical equations, you notice an arrow pointing right. From the direction of the arrow, you infer that the reactants on the left of the arrow *lead to* the products on the right of the arrow. You infer that the direction of the arrow is a visual symbol to help you understand these equations.

Figure 3 Pale green copper(II) carbonate reacts to become black copper(II) oxide when it is heated.

Figure 4 The first test tube contains copper(II) sulfate solution. The second test tube contains the same solution and also a strip of zinc. The blue colour of the solution fades as a reddish-brown solid forms on the zinc. What is this solid?

IN SUMMARY

- Chemical reactions always involve one or more reactants changing to give one or more products.
- State symbols are often written after a chemical formula to indicate the state of the substance.
- We can use word equations or chemical equations to describe chemical reactions. In both, the reactants are written on the left and an arrow points right, toward the products.

CHECK YOUR LEARNING

1. What is the purpose of the arrow in a chemical equation? K/U
2. Write word equations for the following reactions: K/U
   (a) Acetic acid (vinegar) and sodium hydrogen carbonate (baking soda) react to form water, carbon dioxide, and sodium acetate.
   (b) Aluminum metal reacts with oxygen from the air to form a protective coating called aluminum oxide.
   (c) Water and carbon dioxide are produced when propane burns in oxygen.
3. Some barbecues cook food by burning charcoal. (Charcoal is mostly carbon.) The chemical equation for this reaction is
   $C(s) + O_2(g) \rightarrow CO_2(g)$ K/U T/I
   (a) Write the word equation, including an energy term, for this reaction.
   (b) Write the state of each substance in the reaction.
   (c) What evidence suggests that a chemical change is taking place?
   (d) What would you expect to see when this reaction is complete?
4. Consider the reaction in Figure 5:
   $AgNO_3(aq) + NaCl(aq) \rightarrow AgCl(s) + NaNO_3(aq)$ K/U
   (a) Name the reactants and products in this reaction.
   (b) Name the chemicals that are dissolved in water.
   (c) Name the white solid.
   (d) What physical property do both reactants have in common?

Figure 5 When two aqueous reactants mix, they sometimes form a solid product.

5. Consider the following chemical equation:
   $Zn(s) + H_2SO_4(aq) \rightarrow H_2(g) + ZnSO_4(aq) + \text{energy}$ K/U
   (a) Name the products of this reaction.
   (b) What liquid is also present in the reaction vessel, along with the reactants and products?
   (c) What evidence would indicate that this reaction is occurring?
   (d) Will the test tube in which this reaction is occurring become warmer or cooler during the reaction? Why?
   (e) What evidence would indicate that the reaction has stopped?
   (f) How does the quantity of zinc metal change as the reaction proceeds?
6. Take another look at the figures in this section. For each figure, what evidence suggests that a chemical change has taken place? K/U
7. In a burning marshmallow (Figure 1 on page 225), sugar breaks down into carbon (the black residue) and water vapour. K/U T/I
   (a) Write a word equation for this reaction.
   (b) What evidence tells you that the reaction is complete?
   (c) Write a hypothesis predicting how the mass of a marshmallow will change from before to after burning. Include an explanation in your prediction.
8. Bread rises due to the action of a single-celled organism called yeast. Yeast converts some glucose molecules in bread dough into carbon dioxide and ethanol. Carbon dioxide and ethanol then bubble through the dough, making it rise. K/U
   (a) Write a word equation for this reaction.
   (b) Is the action of carbon dioxide a physical or chemical change? Explain.
9. Under the right conditions, some chemical reactions can be reversed. For example, K/U A
   (a) An important step in bottling carbonated soft drinks is bubbling carbon dioxide gas into cold water and then sealing the bottle. Inside, a solution of hydrogen carbonate (also know as carbonic acid), $H_2CO_3(aq)$, soon forms. Write the chemical equation for this reaction.
   (b) Describe two things you could do to quickly reverse this reaction.

6.2 CONDUCT AN INVESTIGATION

Is Mass Gained or Lost During a Chemical Reaction?

SKILLS MENU

- Questioning
- Hypothesizing
- Predicting
- Planning
- Controlling Variables
- Performing
- Observing
- Analyzing
- Evaluating
- Communicating

During a chemical reaction, atoms, molecules, or ions collide, rearrange, and form products. Word and chemical equations describe the chemical changes that occur during a chemical reaction.

Testable Question

How does the total mass of the products of a chemical reaction compare with the total mass of the reactants?

Prediction

Read the Experimental Design and Procedure. Create a table similar to Table 1. In the first row of your table, predict an answer to the Testable Question. Provide a possible explanation for your Prediction.

Table 1 Predictions and Observations

	Reaction 1	Reaction 2
predicted mass change: decrease, no change, or increase?		
initial mass of reactants + container (g)		
final mass of products + container (g)		
change in mass (final − initial) (g)		
observed change in mass: decrease, no change, or increase?		
observed class results: decrease, no change, or increase?		

Experimental Design

You will investigate two different chemical reactions. For each reaction, you will measure the total mass of reactants before the reaction. You will also measure the total mass of products after the reaction. You will compare the total mass of reactants with the total mass of products.

Equipment and Materials

- eye protection
- lab apron
- test tube
- tongs
- 250 mL Erlenmeyer flask and stopper
- 10 mL graduated cylinder
- balance
- 100 mL graduated cylinder
- plastic cup
- dilute solutions of
  - sodium hydroxide, $NaOH(aq)$
  - iron(III) nitrate, $Fe(NO_3)_3(aq)$
- antacid tablet

Iron(III) nitrate and sodium hydroxide are both corrosive, toxic, and irritants. If splashed in the eyes, sodium hydroxide can cause blindness. Wash any spills on skin or clothing immediately with plenty of cold water. Report any spills to your teacher.

Procedure

SKILLS HANDBOOK 1.B., 1.D., 3.B.

1. Put on your eye protection and lab apron.

Part A: Iron(III) Nitrate and Sodium Hydroxide

2. Practise holding the empty test tube with tongs and sliding it into the empty Erlenmeyer flask. Seal the flask to check that the test tube fits and that the stopper forms a tight seal.
3. Measure 5 mL of sodium hydroxide solution in the 10 mL graduated cylinder. Pour this solution into the flask.

4. Your teacher will assign each lab group a different volume of iron(III) nitrate solution. Pour your assigned volume of solution into the small test tube.
5. Tilt the flask and carefully slide the test tube into it. Do not allow the test tube's contents to spill (Figure 1).

Figure 1 The test tube contains iron(III) nitrate solution, and there is sodium hydroxide solution in the bottom of the flask.

6. Seal the flask with the stopper.
7. Measure and record the total mass of the flask and its contents.
8. Slowly tip the flask to allow the two solutions to mix (Figure 2).

Figure 2 The iron(III) nitrate solution is now mixed with the sodium hydroxide solution in the flask.

9. Measure and record the total mass of the flask and its contents.
10. Return the flask and all its contents to your teacher for disposal.

Part B: Antacid Tablet in Water

11. Add 50 mL of tap water to the plastic cup.
12. Take an antacid tablet out of its package.
13. Place the tablet and the cup of water on the scale. Measure and record the total mass of the cup, water, and tablet.
14. Add the tablet to the water. Record your observations.
15. When the visible reaction has stopped, measure and record the total mass of the cup and its contents.

Analyze and Evaluate

(a) Calculate and record the change in mass for each reaction. Note whether each change resulted in a decrease, no change, or an increase in mass. K/U

(b) Compare your results from Part A with other students' results. Account for any discrepancies. T/I

(c) For Part A, calculate and record the average change in mass for your class. K/U

(d) Compare your results from Part B with other students' results. Account for any discrepancies. T/I

(e) Compare your results in Part A with your results in Part B. Suggest a reason for the differences. T/I

(f) For Part B, calculate and record the average change in mass for your class. K/U

(g) Would the class results for Part B differ if the reaction were carried out in a sealed container? Explain. T/I

(h) Why would it be unsafe to conduct Part B in a sealed container? T/I

(i) Answer the Question posed at the beginning of this investigation. C

(j) Compare your answer in (i) with your Prediction. Account for any differences. A

6.3

Conserving Mass in Chemical Reactions

Think again about Investigation 6.2: Is Mass Gained or Lost During a Chemical Reaction? In Part A, you saw that when two solutions are combined to form a product in a sealed container, the total mass of reactants equals the total mass of products. No mass is gained; none is lost. In other words, mass is conserved during the chemical reaction.

Is mass conserved in other chemical reactions as well? The answer to this question may not be immediately obvious. After all, everyday experience may suggest that mass does change during a chemical reaction. For example, a campfire burns down to a pile of fluffy ashes that have much less mass than the original wood (Figure 1). One of the first scientists to study this question was the eighteenth-century French chemist Antoine Lavoisier. Until this time, chemists had not considered the possibility that some reactions could involve gases. Gases, like other forms of matter, have mass. In his experiments, Lavoisier used equipment designed to trap all the reactants and products, including any gases produced during the reactions. Lavoisier found that mass was always conserved if all the reactants and products were considered.

Figure 1 Wood disappears to almost nothing when it burns. How does this reaction obey the law of conservation of mass?

Lavoisier's conclusion has been supported by the work of many other scientists and is now considered to be a scientific law. It is known as the **law of conservation of mass**:

law of conservation of mass the statement that, in any given chemical reaction, the total mass of the reactants equals the total mass of the products

In any given chemical reaction, the total mass of the reactants equals the total mass of the products.

The law of conservation of mass makes sense when you consider what is occurring at the atomic level. (Recall the Bohr–Rutherford model in Section 5.4.) Experiments have shown that, during a chemical reaction, the atoms in reactant molecules are rearranged to form products. Therefore, all the atoms that existed in the reactants are still present in the products of the reaction. Atoms cannot be created or destroyed. That explains why the total mass of reactants is equal to the total mass of products.

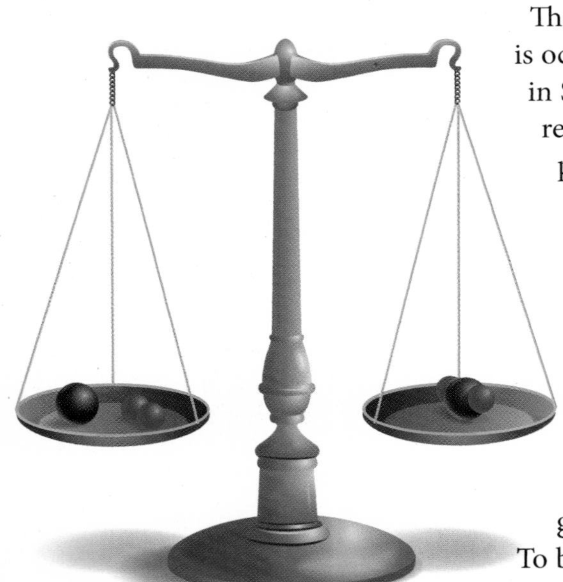

Figure 2 There are two (red) oxygen atoms and one (black) carbon atom on either side of the scale. Similarly, there are two oxygen atoms and one carbon atom on either side of the arrow in the chemical equation.

Reactions, Equations, and the Conservation of Mass

In Section 6.1, you learned that chemical equations can provide a lot of information about chemical reaction. Equations give the formulas and physical state of the reactants and products. To be a completely accurate description of the chemical reaction, an equation must also follow the law of conservation of mass. It must show an equal number of each kind of atom on both sides of the equation. This indicates that there are equal numbers of each kind of atom *before and after* the chemical reaction takes place.

Let's consider the reaction in which carbon and oxygen react to form carbon dioxide (Figure 2):

$$C(s) + O_2(g) \rightarrow CO_2(g)$$

Note that the same numbers and kinds of atoms are present on both sides of the equation. The product (carbon dioxide) contains exactly the same numbers and kinds of atoms as the reactants (carbon and oxygen). Therefore, the chemical equation obeys the law of conservation of mass.

Let's look at another chemical equation:

$H_2(g) + Cl_2(g) \rightarrow HCl(g)$

This equation does not accurately describe the reaction between hydrogen and chlorine. Can you tell why? Count the atoms on each side of the arrow. In this reaction, two atoms of hydrogen react with two atoms of chlorine. (Remember that hydrogen and chlorine both exist as diatomic molecules.) The product, however, contains only one atom of hydrogen and one of chlorine (Figure 3(a)). One atom of hydrogen and one atom of chlorine remain unaccounted for. An equation in which the reactants and products are not balanced is sometimes called a "skeleton equation." Since atoms cannot just vanish, we can assume that two molecules of HCl(g) are produced in this reaction. Laboratory investigations indicate that this is the case (Figure 3(b)).

To show that two hydrogen chloride molecules are produced in this reaction, the coefficient "2" is placed before HCl(g) in the chemical equation:

$H_2(g) + Cl_2(g) \rightarrow 2\ HCl(g)$

The coefficient in a chemical equation applies to all the atoms in the molecule. "2 HCl" means that there are two molecules of hydrogen chloride, each containing one hydrogen atom and one chlorine atom.

Now the chemical equation obeys the law of conservation of mass. Chemical equations must always be balanced, with the same kinds and same numbers of atoms on both sides of the arrow.

LEARNING TIP

Diatomic Molecules

Remember that some elements exist as diatomic molecules. Refer back to Table 1 in Section 5.10.

LEARNING TIP

Coefficients versus Subscripts

Do not confuse coefficients with subscripts in chemical formulas. Coefficients give the ratio of reactants and products in a reaction. Subscripts give the ratio of elements in a chemical formula; they cannot change in a given chemical. Thus, chemical equations can be balanced only by using coefficients.

READING TIP

Adjusting Inferences

Sometimes as you continue reading, you come across information that conflicts with an inference that you have already made. For example, you read that an equation must obey the law of conservation of mass. Later you read a chemical equation that appears to break this fundamental law. You infer that exceptions are possible. Later you read that the equation as stated is incorrect. You revise your inference and conclude that the law of conservation of mass is unbreakable.

Figure 3 In (a), there are two (white) hydrogen atoms and two (green) chlorine atoms on the left side of the scale, but only one hydrogen atom and one chlorine atom on the right side. In (b), there are two hydrogen atoms and two chlorine atoms on each side of the scale. Reactants and products are balanced.

TRY THIS MODELLING BALANCED CHEMICAL EQUATIONS

SKILLS: Predicting, Observing, Analyzing

In this activity, you will use molecular models to visualize the law of conservation of mass.

Equipment and Materials: molecular model kit

1. Build one molecule of hydrogen, H_2. Build one molecule of bromine, Br_2. Predict how many molecules of hydrogen bromide, HBr, can be made from these two models. Verify your prediction by making the "product" from the "reactants."
2. Build two molecules of hydrogen and one molecule of oxygen. Predict how many molecules of water, H_2O, can be made from them. Verify your prediction by making the product.
3. Build two molecules of hydrogen peroxide, H_2O_2. Predict how many molecules of oxygen and water can be made from them. Verify your prediction.
4. Build two molecules of ammonia, NH_3. Imagine that this is the product of a reaction between hydrogen and nitrogen. Predict how many molecules of hydrogen and nitrogen are required to make two molecules of ammonia. Check your prediction.

A. Write the word equations and chemical equations for each of these four reactions. K/U

B. Explain how the results of this activity illustrate the law of conservation of mass. T/I

IN SUMMARY

- The law of conservation of mass states that in any given chemical reaction, the total mass of the reactants equals the total mass of the products.
- Chemical equations obey the law of conservation of mass. They show that all the atoms in the reactants are still present in the products.
- Coefficients are added before chemical formulas in a chemical equation to ensure that the numbers of atoms on each side of the arrow are equal (balanced).

CHECK YOUR LEARNING

1. (a) The idea that gases have mass can be difficult to accept. How has this reading helped your understanding of this concept?
   (b) Identify what remains unclear. Discuss this with your teacher. C
2. (a) State the law of conservation of mass.
   (b) Explain the law of conservation of mass by referring to the atoms involved in a chemical reaction.
   (c) Which best represents the law of conservation of mass: a skeleton equation or a balanced chemical equation? Explain. K/U
3. Are the following situations exceptions to the law of conservation of mass? Justify your answer in each case. K/U
   (a) The mass of a hamburger decreases as it is barbecued.
   (b) A tree's mass is continually increasing as the tree grows.
   (c) The mass of a copper penny increases if it is heated in a Bunsen burner flame.
   (d) You are often lighter in the morning than you were when you went to bed.
4. You might have noticed that new copper roofs turn green over time. This occurs because copper reacts with substances in the air to form a hard, protective coating. Will the mass of the new copper roof increase or decrease over time? Explain. Does this prediction violate the law of conservation of mass? Explain. T/I A
5. Design an experiment involving the reaction of vinegar and baking soda to test the law of conservation of mass. T/I
6. A 20 g sample of compound A is mixed with 45 g of compound B. A chemical reaction occurs in which a gas is produced. Once the reaction is complete, the final mixture has a mass of 55 g. T/I
   (a) What is the mass of the gas?
   (b) What assumption did you make in (a)?
7. Soon after learning about the work of Lavoisier, John Dalton proposed that atoms are never created or destroyed in chemical reactions, only rearranged. Explain how this statement applies to the law of conservation of mass. A

Information in Chemical Equations

6.4

What is a cookie recipe doing in a science textbook (Figure 1)? Believe it or not, recipes and balanced chemical equations have a lot in common. A recipe provides more than just a list of required ingredients. It provides the quantities of each ingredient. It also tells us the conditions under which the ingredients are combined. Recipes also tell you how much product to expect. Chemical equations give us similar information (Table 1).

Chocolate Chip Oatmeal Cookies

Turn on the oven to 170 °C.
Place the first five ingredients in the mixing bowl, one at a time, mixing well after each addition. In a cup, dissolve the baking soda in the hot water, then add it to the mixing bowl. Add the flour, oats, and chocolate chips. Mix well. Drop in spoonfuls onto cookie sheets. Bake in the oven until starting to turn golden. Makes about 50 cookies.

1 cup butter at room temperature
1 ½ cups brown sugar
2 eggs
1 teaspoon vanilla essence
1 teaspoon salt
½ teaspoon baking soda
1 tablespoon hot water
1 ½ cups unbleached flour
1 ½ cups rolled oats
1 ½ cups chocolate chips

Figure 1 A cookie recipe has some features in common with a chemical equation.

Table 1 Comparison of Recipes and Chemical Equations

Information communicated	Recipe	Chemical equation
	See Figure 1.	$2\ H_2(g) + O_2(g) \rightarrow 2\ H_2O(g)$
starting materials	ingredients list (e.g., eggs and butter)	chemical symbols for the reactants (to the left of the arrow) (e.g., $2\ H_2(g)$ and $O_2(g)$)
conditions of starting materials	directions (e.g., butter at room temperature)	state symbols: (s), (l), (g), (aq) (e.g., $2\ H_2(g)$)
proportions of starting materials	quantities in ingredients list (e.g., 2 eggs, 1 cup butter)	coefficients of reactants (e.g., $2\ H_2(g)$)
instructions for combining materials	directions (e.g., mixing well after each addition)	plus sign (+) between the formulas of the reactants, indicating that reactants must come into contact (e.g., $2\ H_2(g) + O_2(g)$)
resulting product	title (e.g., Chocolate Chip Oatmeal Cookies)	chemical symbols for the products (to the right of the arrow) (e.g., $2\ H_2O\ (g)$)
proportions/ quantities of product	final sentence (e.g., Makes about 50 cookies.)	coefficients of products (e.g., $2\ H_2O\ (g)$)

Now you know what the various parts of a balanced chemical equation represent. The next stage is to learn *how* to balance chemical equations.

Balancing Chemical Equations

To learn how to balance a chemical equation, let's work through some Sample Problems. To make things simpler when we are learning to balance equations, we will not use state subscripts for now.

SAMPLE PROBLEM 1 Balancing a Chemical Equation

Write the balanced chemical equation for the reaction of magnesium with oxygen.

Step 1 Write the word equation for the reaction.

magnesium + oxygen → magnesium oxide

Step 2 Replace each chemical name with the correct chemical formula. (This is the skeleton equation.)

$Mg + O_2 \rightarrow MgO$

Step 3 Count the number of atoms of each type on either side of the arrow.

$Mg + O_2 \rightarrow MgO$

1 Mg atom 1 Mg atom

2 O atoms 1 O atom

Step 4 Multiply the formulas by an appropriate coefficient until all the atoms are balanced. Keep checking whether the numbers of each type of atom on both sides are balanced.

- MgO (on the right) must be multiplied by the coefficient 2 to balance the oxygen atoms.

  $Mg + O_2 \rightarrow 2\ MgO$

- Mg (on the left) must now be multiplied by the coefficient 2 so that there are two Mg atoms on both sides.

The final balanced chemical equation is

$2\ Mg + O_2 \rightarrow 2\ MgO$

Practice

Write a balanced chemical equation for the reaction between potassium and bromine to make potassium bromide.

LEARNING TIP

Better Balancing

When balancing a chemical equation, start by balancing the element(s) that occurs only once on both sides of the equation. Also, leave any substances appearing as elements in the equation until the end since they can be balanced without affecting any other atom types. Keep checking as you go along because balancing one element may "unbalance" another.

SAMPLE PROBLEM 2 Balancing a Chemical Equation

Methane gas, CH_4, burns in oxygen to produce carbon dioxide and water (Figure 2). Write a balanced chemical equation for this reaction.

Step 1 methane + oxygen → carbon dioxide + water

Step 2 $CH_4 + O_2 \rightarrow CO_2 + H_2O$

Step 3 $CH_4 + O_2 \rightarrow CO_2 + H_2O$

1 C atom 1 C atom

4 H atoms 2 H atoms

2 O atoms 3 O atoms (2 + 1)

Step 4 Check whether the numbers of each type of atom on both sides are balanced.

- Carbon is already balanced.
- H_2O must be multiplied by 2 to balance hydrogen atoms.

  $CH_4 + O_2 \rightarrow CO_2 + 2\ H_2O$

  1 C atom 1 C atom

  4 H atoms 4 H atoms

  2 O atoms 4 O atoms (2 + 2)

Figure 2 Methane makes up about 80 % of the natural gas burned in a natural gas furnace.

- O_2 must be multiplied by 2 to balance oxygen atoms.

  $CH_4 + 2\ O_2 \rightarrow CO_2 + 2\ H_2O$

  1 C atom 1 C atom

  4 H atoms 4 H atoms

  4 O atoms 4 O atoms

All the atoms balance, so the final balanced chemical equation is

$CH_4 + 2\ O_2 \rightarrow CO_2 + 2\ H_2O$ (Figure 3)

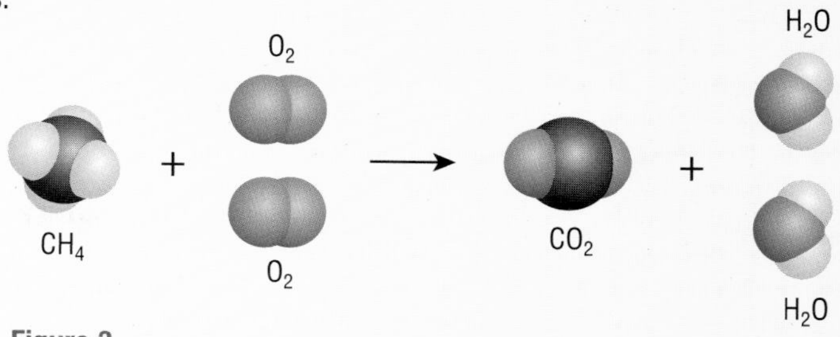

CH_4 + O_2 O_2 → CO_2 + H_2O H_2O

Figure 3

Practice

Write a balanced chemical equation for the reaction between oxygen and a hydrocarbon called pentane, C_5H_{12}, to produce carbon dioxide and water.

Our third example shows how to balance equations involving polyatomic ions. In these cases, consider the polyatomic ion as one unit rather than as individual atoms.

SAMPLE PROBLEM 3 Balancing an Equation Involving Polyatomic Ions

Zinc metal reacts in a silver nitrate solution to produce zinc nitrate and silver metal. Write a chemical equation for this reaction. (See Table 1 in Section 5.9 for a list of polyatomic ions.)

Step 1 zinc + silver nitrate → zinc nitrate + silver

Step 2 $Zn + AgNO_3 \rightarrow Zn(NO_3)_2 + Ag$

Step 3 $Zn + AgNO_3 \rightarrow Zn(NO_3)_2 + Ag$

1 Zn atom 1 Zn^{2+} ion

1 Ag^+ ion 1 Ag atom

1 NO_3^- ion 2 NO_3^- ion

Step 4 Check whether the numbers of each type of atom on both sides are balanced. Because each polyatomic ion generally stays intact, you can count polyatomic ions in the same way as you count atoms.

- Zinc and silver are already balanced.
- $AgNO_3$ must be multiplied by 2 to balance nitrate ions.
- $Zn + 2\ AgNO_3 \rightarrow Zn(NO_3)_2 + Ag$

  1 Zn atom 1 Zn^{2+} ion

  2 Ag^+ ions 1 Ag atom

  2 NO_3^- ions 2 NO_3^- ions
- Ag^+ must be multiplied by 2 to balance the number of silver atoms.

The final balanced chemical equation is

$Zn + 2\ AgNO_3 \rightarrow Zn(NO_3)_2 + 2\ Ag$

Practice

Write a balanced chemical equation for the reaction between iron(III) nitrate and sodium hydroxide to produce iron(III) hydroxide and sodium nitrate.

RESEARCH THIS GAS FURNACE TECHNICIAN

SKILLS: Researching, Evaluating

SKILLS HANDBOOK 4.A.

Many careers involve chemical reactions and the delicate balance between the reactants. Consider the responsibilities of the people who install, service, and repair gas furnaces. If the reactants (natural gas and oxygen) are not available in the appropriate ratios, the outcome could be deadly.

1. Research this career, paying particular attention to the training required.
2. Research the main dangers of badly installed or poorly maintained gas furnaces.

A. List the educational background and training required to be a gas furnace technician. T/I

B. Write balanced chemical equations for at least two chemical reactions that furnace technicians must know about. K/U T/I

C. Write a paragraph explaining why this would, or would not, be a good career for you. C A

GO TO NELSON SCIENCE

IN SUMMARY

- Chemical equations contain information about which substances are reactants and which are products, and the ratios of these substances.
- Chemical equations obey the law of conservation of mass.
- Coefficients can be added before chemical formulas in a chemical equation to make the numbers of atoms in reactants and products balance.

CHECK YOUR LEARNING

1. What strategy for balancing chemical equations did you find particularly useful? Share this strategy with a classmate who is having difficulty with balancing equations. K/U C
2. What is the difference between a skeleton equation and a balanced chemical equation? K/U
3. Consider this skeleton equation: $HI \rightarrow H_2 + I_2$ K/U
   (a) Which of the following two equations is the correct balanced chemical equation?
   $2\ HI \rightarrow H_2 + I_2$ or $H_2I_2 \rightarrow H_2 + I_2$
   (b) Explain your answer.
4. (a) What is the difference between a subscript and a coefficient in a chemical equation?
   (b) Which is the only one that can be changed as you balance a skeleton equation? Why? K/U
5. Write a balanced chemical equation for each of the following reactions taking place in water: K/U
   (a) potassium iodide → potassium and iodine
   (b) magnesium + silver nitrate → silver + magnesium nitrate
   (c) sodium + water → hydrogen + sodium hydroxide
   (d) lead(II) nitrate + sodium chloride → lead(II) chloride + sodium nitrate
6. Octane, C_8H_{18}, is a compound in gasoline. Octane burns in oxygen to produce carbon dioxide gas and water vapour. T/I
   (a) Write a balanced chemical equation for this reaction
   (b) How many carbon dioxide molecules are produced for every octane molecule that burns?
7. Balance the following skeleton equations if they are not already balanced. (It does not matter if you do not know the names of all the compounds, but you must recognize the polyatomic ions. You do not need to add state symbols.) T/I
   (a) $Ca + Cl_2 \rightarrow CaCl_2$
   (b) $K + Br_2 \rightarrow KBr$
   (c) $H_2O_2 \rightarrow H_2O + O_2$
   (d) $Na + O_2 \rightarrow Na_2O$
   (e) $N_2 + H_2 \rightarrow NH_3$
   (f) $NH_4OH + HBr \rightarrow H_2O + NH_4Br$
   (g) $CaSO_4 + KOH \rightarrow Ca(OH)_2 + K_2SO_4$
   (h) $Ba + HNO_3 \rightarrow H_2 + Ba(NO_3)_2$
   (i) $H_3PO_4 + NaOH \rightarrow H_2O + Na_3PO_4$
   (j) $C_3H_8 + O_2 \rightarrow CO_2 + H_2O$
   (k) $Al_4C_3 + H_2O \rightarrow CH_4 + Al(OH)_3$
   (l) $FeBr_3 + Na \rightarrow Fe + NaBr$
   (m) $Fe + H_2SO_4 \rightarrow H_2 + Fe_2(SO_4)_3$
   (n) $C_2H_6 + O_2 \rightarrow CO_2 + H_2O$
8. Ammonium dichromate, $(NH_4)_2Cr_2O_7$, is an orange solid that releases nitrogen gas and water vapour when it is heated. The green solid produced in this reaction is a toxic form of chromium oxide. K/U T/I
   (a) Write a word equation for this reaction.
   (b) When 2.5 g of ammonium dichromate is heated, the mass of nitrogen and water vapour released is 1.0 g. What is the final mass of the solid product?

Types of Chemical Reactions: Synthesis and Decomposition

6.5

There are currently about 10 million known compounds. Each compound can react in many different ways. It is impossible for anyone to memorize all the reactions. To make it easier to predict what reactions will take place, chemists have grouped similar reactions into categories.

One way of grouping chemical reactions is based on recognizing patterns in the chemical formulas. For example, Figure 1 shows three chemical reactions that at first seem unrelated. But when you compare the chemical equations for these reactions, a pattern emerges. Can you tell what it is?

LEARNING TIP

State Symbols

It is good practice to write state symbols after chemical formulas to clarify the states of the chemicals during the reaction. For example, HCl(g) tells us that the hydrogen chloride exists as a gas.

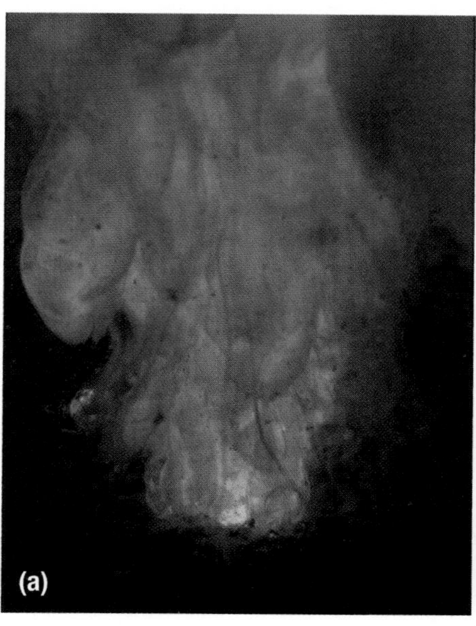

(a)

$Zn(s) + S(s) \longrightarrow ZnS(s)$

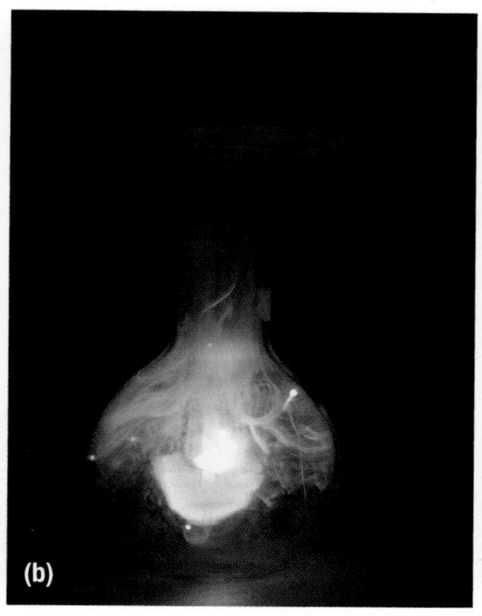

(b)

$2\ Na(s) + Cl_2(g) \longrightarrow 2\ NaCl(s)$

(c)

$HCl(g) + NH_3(g) \longrightarrow NH_4Cl(s)$

Figure 1 (a) Powdered zinc metal reacts with sulfur to produce zinc sulfide powder. (b) A small piece of sodium metal ignites as it is placed in a flask of chlorine gas (yellowish gas). (c) Hydrogen chloride gas and ammonia gas both diffuse out of their aqueous solutions. When the two gases come in contact, they react to produce a cloudy white powder of ammonium chloride.

Synthesis Reactions

The three reactions shown in Figure 1 are examples of synthesis reactions. In a **synthesis reaction**, two simple reactants combine to make a larger or more complex product (Figure 2). The chemical equations for synthesis reactions follow the general pattern:

$$A + B \rightarrow AB$$

synthesis reaction a reaction in which two reactants combine to make a larger or more complex product; general pattern: $A + B \rightarrow AB$

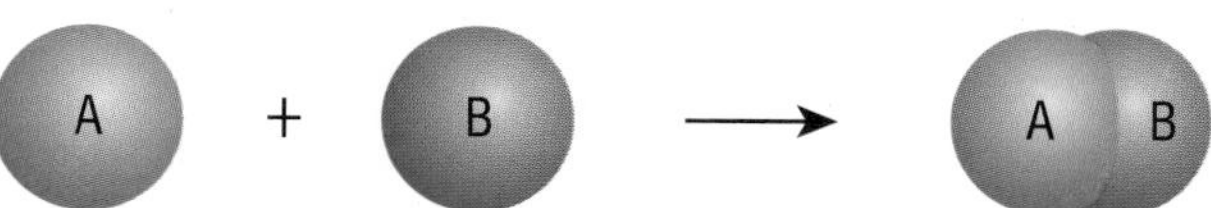

Figure 2 In some cases, the reactants are atoms (elements), while in others, they are molecules (elements or compounds).

Table 1 shows how the three synthesis reactions in Figure 1, on the previous page, follow the general pattern.

Table 1 Examples of Synthesis Reactions

Synthesis reaction	Equation
zinc sulfide (Figure 1(a))	zinc + sulfur → zinc sulfide $Zn(s) + S(s) \rightarrow ZnS(s)$
sodium chloride (Figure 1(b))	sodium + chlorine → sodium chloride $2\ Na(s) + Cl_2(g) \rightarrow 2\ NaCl(s)$
ammonium chloride (Figure 1(c))	ammonia + hydrogen chloride → ammonium chloride $NH_3(g) + HCl(g) \rightarrow NH_4Cl(s)$
General pattern	A + B → AB

Decomposition Reactions

decomposition reaction a reaction in which a large or more complex molecule breaks down to form two (or more) simpler products; general pattern: AB → A + B

We can think of decomposition reactions as being the opposite of synthesis reactions. During a **decomposition reaction**, large compounds are broken down into smaller compounds or elements (Figure 3). The general pattern for decomposition reactions is

$$AB \rightarrow A + B$$

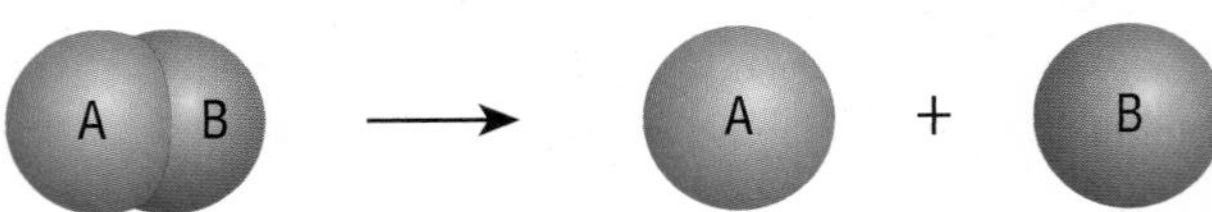

Figure 3 In a decomposition reaction, a complex molecule breaks down, or decomposes, into simpler products. The products can be elements or compounds.

Just the Opposite
Decomposition reactions (AB → A + B) are the reverse of synthesis reactions (A + B → AB).

Decomposition reactions usually absorb energy (such as thermal or electrical energy) from an external source. This energy is then used to convert reactants into products. For example, water can be decomposed into its elements using electricity as its energy source.

Table 2 shows how two decomposition reactions follow the general pattern.

Table 2 Examples of Decomposition Reactions

Decomposition reaction	Equation
water	energy + water → hydrogen + oxygen energy + $2\ H_2O(l) \rightarrow 2\ H_2(g) + O_2(g)$
sodium azide	energy + sodium azide → sodium + nitrogen $2\ NaN_3(s) \rightarrow 2\ Na(s) + 3\ N_2(g)$
General pattern	AB → A + B

Figure 4 Airbags are designed to slow your forward motion during a collision. An airbag inflates in about 1/20th of a second and stays fully inflated for only about 1/10th of a second.

Many nitrogen compounds undergo important decomposition reactions. For example, the use of airbags in cars is a lifesaving application of a decomposition reaction (Figure 4). Airbags contain a nitrogen compound called sodium azide, $NaN_3(s)$. During a collision, a sudden flow of electricity is automatically sent to the airbag. This electrical energy triggers the rapid decomposition of sodium azide to produce nitrogen gas and sodium metal.

RESEARCH THIS PROPOSAL TO BAN FERTILIZERS

SKILLS: Researching, Analyzing the Issue, Defending a Decision, Communicating

SKILLS HANDBOOK 4.A., 4.C.

Ammonium nitrate is one of the cheapest and most widely used fertilizers. However, it can also be used to make explosives (Figure 5):

ammonium nitrate → water + nitrogen + oxygen + energy

$$2\ NH_4NO_3(s) \rightarrow 4\ H_2O(g) + 2\ N_2(g) + O_2(g) + \text{energy}$$

Because of this, some politicians have proposed that the sale of ammonium nitrate should be restricted or banned.

1. Research the facts regarding the use of ammonium nitrate as a fertilizer and in the manufacture of explosives.
2. Research the arguments for and against the proposed ban.
3. Organize your findings in a "pros" and "cons" chart. T/I C

A. Do you think the proposed ban is fair? Defend your decision in a letter to a local politician. C

Figure 5 A few well-placed explosive charges can demolish old, vacant buildings quickly and safely. Many explosives release their destructive energy through decomposition reactions.

IN SUMMARY

- Chemical reactions are grouped into categories. Within each category, the reactions follow the same pattern.
- In a synthesis reaction, two simple reactants combine to make a larger or more complex product and follow the general pattern $A + B \rightarrow AB$.
- In a decomposition reaction, a complex reactant breaks down to make two or more simpler product and follow the general pattern $AB \rightarrow A + B$.

CHECK YOUR LEARNING

1. Classify each of the following as either a synthesis or a decomposition reaction: K/U
   (a) zinc chloride → zinc + chlorine
   (b) potassium + iodine → potassium iodide
   (c) potassium oxide + water → potassium hydroxide
   (d) calcium carbonate → calcium oxide + carbon dioxide
2. Write a balanced chemical equation for each of the word equations given in question 1. K/U T/I
3. Copper metal was first made over 3 000 years ago by heating a mineral containing copper(II) oxide. The other product of this reaction is oxygen. K/U T/I
   (a) Write a word equation for this reaction.
   (b) Is this a synthesis reaction or a decomposition reaction? Explain.
   (c) Write a balanced chemical equation for the reaction.
4. Write a balanced chemical equation for each of the following reactions. Include state symbols. Classify each one as either synthesis or decomposition. K/U T/I
   (a) Hydrogen gas reacts explosively with chlorine gas to form hydrogen chloride gas.
   (b) A solution of hydrogen peroxide, H_2O_2, breaks down into water and oxygen.
   (c) Solid potassium chlorate breaks down into solid potassium chloride and oxygen when heated.
   (d) Ammonia gas, NH_3, can be made by combining hydrogen gas and nitrogen gas.
   (e) Aluminum metal reacts with oxygen from the air to form a hard coating of aluminum oxide. This coating prevents aluminum objects from corroding.

6.6 Types of Chemical Reactions: Single and Double Displacement

Figure 1 The beaker contains solid carbon dioxide that quickly sublimates to release carbon dioxide gas. This gas then reacts vigorously with hot magnesium metal.

Magnesium metal is one of the most combustible substances in your school's chemical storeroom. Most fires can be extinguished with a standard carbon dioxide fire extinguisher. A magnesium fire is particularly dangerous because spraying carbon dioxide onto it only makes it worse (Figure 1).

Single Displacement Reactions

The word and chemical equations for the reaction of magnesium with carbon dioxide are

magnesium + carbon dioxide → magnesium oxide + carbon

$2\ Mg(s) + CO_2(g) \rightarrow 2\ MgO(s) + C(s)$

This reaction is an example of a single displacement reaction. In a **single displacement reaction**, one element displaces or replaces an element in a compound (Figure 2). The general pattern for this type of reaction is

$$A + BC \rightarrow AC + B$$

A represents an element; BC represents a compound.

single displacement reaction a reaction in which an element displaces another element in a compound, producing a new compound and a new element

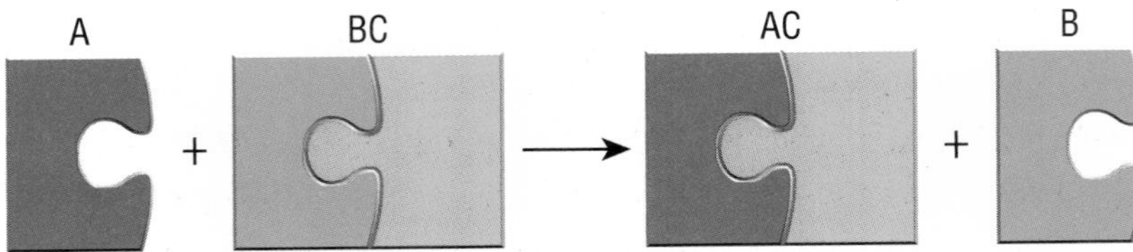

Figure 2 In a single displacement reaction, one element, A, displaces element B in a compound, BC. The new compound, AC, is one product. The displaced element, B, is the second product.

To see a dramatic video of the reaction of magnesium and carbon dioxide,

LEARNING TIP

Non-Metals Place Second

Remember that the second element in a chemical compound is almost always a non-metal. That is why, in Figure 2, element C is written second in both compounds: BC and AC.

Notice how the chemical equation for the reaction of magnesium with carbon dioxide is similar to this general pattern. In single displacement reactions involving an ionic compound and a metal, it is always the positive ion (cation) that is replaced in the compound.

Single displacement reactions often occur in aqueous solution. Figure 3 shows what happens when a coiled copper wire is placed into a solution of silver nitrate:

copper + silver nitrate → copper(II) nitrate + silver

$Cu(s) + 2\ AgNO_3(aq) \rightarrow Cu(NO_3)_2(aq) + 2\ Ag(s)$

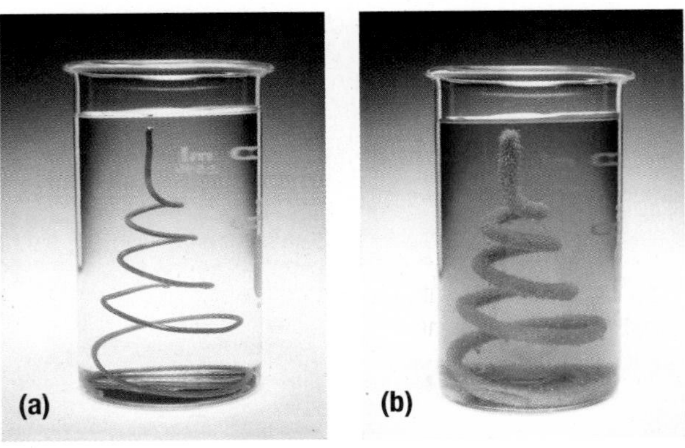

Figure 3 (a) A coil of copper wire is placed into a solution of the silver nitrate. (b) The fuzzy coating on the copper wire is silver metal. The solution is blue because of dissolved $Cu^{2+}(aq)$ ions.

Single displacement reactions also occur when metals are placed into acids. The chemical formula of any acid includes one or more hydrogen atoms. (For example, the chemical formula for hydrochloric acid is HCl(aq).) In these reactions, metal atoms displace the hydrogen atoms in the compound. Figure 4 shows the reaction of zinc metal and hydrochloric acid:

$$\text{zinc} + \text{hydrochloric acid} \rightarrow \text{hydrogen} + \text{zinc chloride}$$

$$Zn(s) + 2\,HCl(aq) \rightarrow H_2(g) + ZnCl_2(aq)$$

Hydrogen, you will recall, forms diatomic molecules. That is why the chemical formula for hydrogen is H_2 rather than H.

Figure 4 Zinc reacts in hydrochloric acid. As zinc displaces the hydrogen in the acid, bubbles of hydrogen gas appear on the surface of the zinc.

Displacement Reactions in Mining

Metals rarely occur naturally as pure elements. Instead, they combine with other elements to form rock deposits called ores. Nickel, for example, occurs in rock as nickel sulfide. The processing of nickel ore is called smelting. The first step in smelting is converting nickel sulfide into nickel oxide. Nickel oxide is then burned with coke (carbon) to produce pure nickel and poisonous carbon monoxide:

$$C(s) + NiO(s) \rightarrow Ni(s) + CO(g)$$

Note that the chemical equation for smelting fits the general pattern of single displacement reactions.

The factory in which this process occurs is called a smelter. Another of the products of processing nickel is sulfur dioxide. As you will learn in the next chapter, sulfur dioxide emissions from nickel smelters are responsible for some of the damage caused by acid precipitation.

RESEARCH THIS WHEN GOLD LOSES ITS GLITTER

SKILLS: Defining the Issue, Researching, Communicating

SKILLS HANDBOOK 4.A., 4.C.

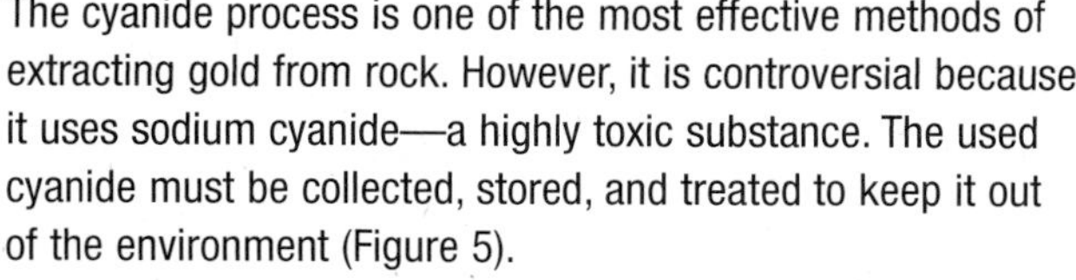

The cyanide process is one of the most effective methods of extracting gold from rock. However, it is controversial because it uses sodium cyanide—a highly toxic substance. The used cyanide must be collected, stored, and treated to keep it out of the environment (Figure 5).

Figure 5 Water from gold mines is treated to remove the cyanide before being released into the environment.

Consider this scenario. A company wants to mine a newly discovered gold deposit near a remote town in northern Ontario. The president of the company has invited the following people to discuss the project:

- the mayor of the town
- a representative from the Ministry of the Environment
- the leader of a First Nations group
- a member of a local non-governmental environmental group

1. Take the role of one of the four people at the meeting.
2. Research background information to support your role, including the history of cyanide use in gold processing.

A. In your role, summarize your perspective on the development of the mine. Present your perspective as an opening argument. T/I C

Double Displacement Reactions

double displacement reaction a reaction that occurs when elements in different compounds displace each other or exchange places, producing two new compounds

Double displacement reactions occur when two elements in different compounds trade places (Figure 6). The general pattern for these reactions is:

$$AB + CD \rightarrow AD + CB$$

The symbols A, B, C, and D represent atoms, single ions, or polyatomic ions.

LEARNING TIP

Polyatomic Ions

Treat polyatomic ions as a single unit in a chemical equation. In a double displacement reaction, a polyatomic ion (such as the nitrate ion, NO_3^-) can change places with an ion composed of only one atom (such as chloride, Cl^-).

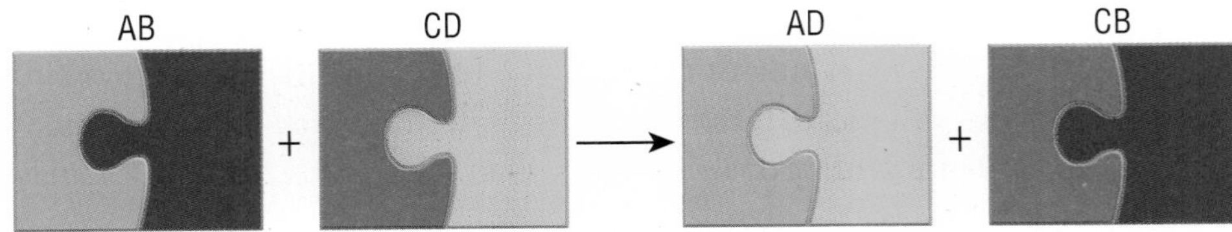

Figure 6 In a double displacement reaction, the two non-metals, B and D, trade places. Alternatively, you could think of it as the two metals, A and C, switching over.

DID YOU KNOW?

Heavy Metals Banned

Heavy metals such as lead, cadmium, and mercury are very toxic. Solutions containing cations of these metals are also dangerous. For this reason, many school boards have banned their use. Even in schools where they are not banned, they are generally only used in very small quantities. After use, they are collected in special containers so that as little as possible is released into the environment. What is your school's policy on heavy metals?

Many double displacement reactions occur between two ionic compounds in solution. For example, Figure 7 shows the reaction of a solution of silver nitrate with a solution of sodium chloride. In this reaction, nitrate ions and chloride ions trade places. The word and chemical equations for this reaction are as follows:

silver nitrate + sodium chloride → silver chloride + sodium nitrate

$$AgNO_3(aq) + NaCl(aq) \rightarrow AgCl(s) + NaNO_3(aq)$$

Notice that both the word and chemical equations fit the general pattern for this reaction:

$$AB + CD \rightarrow AD + CB$$

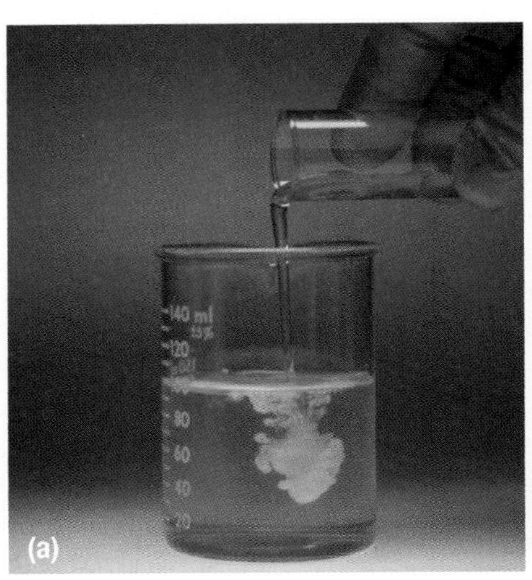

Figure 7 (a) When silver nitrate solution is added to sodium chloride solution, specks of silver chloride appear. (b) When silver nitrate solution is added to tap water, a faint haze appears. This haze indicates that chloride ions are present in the water—perhaps from road salt.

Forming a Precipitate

precipitate a solid formed from the reaction of two solutions

Look closely at the chemical equation for the reaction in Figure 7. Note that the reactants are both in solution (aq), and so is one of the products: sodium nitrate. The other product, silver chloride, is a solid (s). Chemists have discovered, through experimentation, that some ionic compounds do not dissolve in water. If these insoluble compounds are formed during a reaction, they become visible as a **precipitate**: tiny specks of solid material in the solution. The silver chloride formed in Figure 7 is a precipitate.

Figure 8 The bright yellow precipitate is insoluble lead(II) iodide. It is produced when a solution containing Pb^{2+} ions is mixed with a solution containing I^- ions.

Not all double displacement reactions result in the formation of a precipitate, but many do. Lead(II) nitrate and potassium iodide are both soluble in water. When their solutions are mixed, a bright yellow precipitate of lead(II) iodide appears (Figure 8).

$$AB + CD \rightarrow AD + CB$$

lead(II) nitrate + potassium iodide → lead(II) iodide + potassium nitrate

$$Pb(NO_3)_2(aq) + 2\ KI(aq) \rightarrow PbI_2(s) + 2\ KNO_3(aq)$$

Potassium iodide can be used to test for Pb^{2+} ions in water: a yellow precipitate indicates that lead ions are present.

IN SUMMARY

- In a single displacement reaction, an element and a compound react to produce a different element and compound and have the general pattern $A + BC \rightarrow AC + B$.
- In a double displacement reaction, two compounds react to produce two different compounds and have the general pattern $AB + CD \rightarrow AD + CB$.
- Sometimes, in a reaction of aqueous reactants, one of the products is insoluble. This product, called a precipitate, appears as a solid in the solution.

CHECK YOUR LEARNING

1. Compare single and double displacement reactions. K/U
2. What types of reactants are likely to be involved in
   (a) a single displacement reaction?
   (b) a double displacement reaction? K/U
3. Classify the following word equations as representing either single or double displacement reactions: K/U
   (a) aluminum + iron(III) oxide → aluminum oxide + iron
   (b) barium chloride + sodium sulfate → barium sulfate + sodium chloride
   (c) zinc + copper(II) sulfate → zinc sulfate + copper
   (d) silver nitrate + sodium phosphate → silver phosphate + sodium nitrate
   (e) calcium + water → hydrogen + calcium hydroxide
4. Rewrite the word equations in question 3 as balanced chemical equations (without state symbols). K/U T/I
5. Consider the chemical equation $CuSO_4(aq) + Fe(s) \rightarrow FeSO_4(aq) + Cu(s)$. K/U A
   (a) Classify the reaction as a single displacement or a double displacement reaction.
   (b) Copper compounds such as copper(II) sulfate are toxic. Before disposal, these compounds must be treated to reduce their toxicity. Describe how to use steel wool (which is made mostly of iron) to remove the Cu^{2+} ions from an aqueous solution of copper(II) sulfate.
6. Firefighters suggest that the best way to put out a magnesium fire is to pour sand or salt over it. Why is this better than using a carbon dioxide fire extinguisher? A
7. The dark tarnish that sometimes forms on silver is silver sulfide, Ag_2S. A common home remedy for tarnish is represented by the chemical equation $3\ Ag_2S(s) + 2\ Al(s) \rightarrow 6\ Ag(s) + Al_2S_3(s)$ (Figure 9) K/U A
   (a) Classify the reaction as a single displacement or a double displacement reaction.
   (b) Which method do you think is better for cleaning silverware: the method described in (a) or scrubbing and polishing? Why?

Figure 9 To remove the tarnish from silverware, soak it in a hot solution of baking soda in an aluminum pan.

8. The fuzzy silver coating in Figure 3 of this section is impure silver. It can be converted back into silver nitrate by reacting it with nitric acid, $HNO_3(aq)$, as shown: $Ag(s) + HNO_3\ (aq) \rightarrow AgNO_3(aq) + NO_2(g) + H_2O(l)$ K/U T/I A
   (a) Balance the equation.
   (b) What has to be done to the reaction mixture to recover solid silver nitrate?
   (c) Why must this process be done in a well-ventilated area? (Hint: see Figure 1 in Section 5.10.)

Synthesis and Decomposition Reactions

SKILLS MENU
- Questioning
- Hypothesizing
- Predicting
- Planning
- Controlling Variables
- Performing
- Observing
- Analyzing
- Evaluating
- Communicating

Do you remember hearing a loud "pop" in Activity 5.2, when you held a burning splint at the mouth of a test tube containing hydrogen gas? You were hearing evidence of a synthesis reaction:

$$2\ H_2(g) + O_2(g) \rightarrow 2\ H_2O(g) + \text{energy}$$

Many metals also undergo synthesis reactions with oxygen to form oxides. In this activity, you will consider two such reactions. You will also look at two decomposition reactions. The second one—the decomposition of hydrogen peroxide—normally occurs very slowly. Fortunately, there are substances that speed up this reaction without being consumed themselves. In this activity you will use iron(III) nitrate to help hydrogen peroxide decompose.

Equipment and Materials

- eye protection
- lab apron
- Bunsen burner
- retort stand with clamps
- spark lighter
- tongs
- heat-resistant pad
- 3 test tubes
- test-tube rack
- scoopula
- test-tube holder
- copper wire
- steel wool
- limewater, $Ca(OH)_2(aq)$
- copper(II) carbonate, $CuCO_3(s)$
- dilute hydrogen peroxide, $H_2O_2(aq)$
- iron(III) nitrate, $Fe(NO_3)_3(s)$
- wooden splint

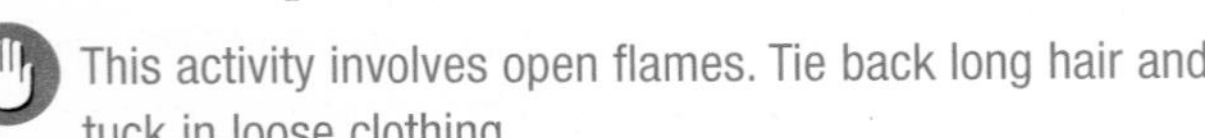
This activity involves open flames. Tie back long hair and tuck in loose clothing.

Copper(II) carbonate is toxic if swallowed.

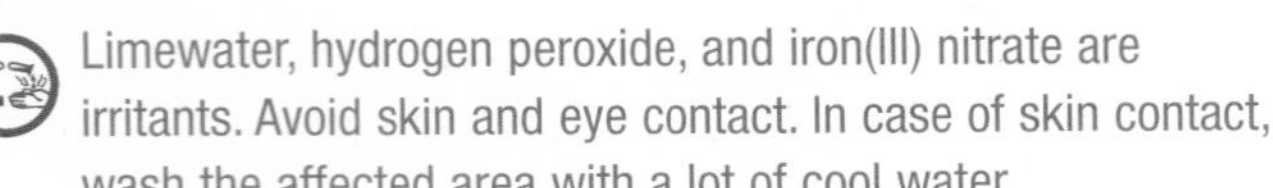
Limewater, hydrogen peroxide, and iron(III) nitrate are irritants. Avoid skin and eye contact. In case of skin contact, wash the affected area with a lot of cool water.

Procedure

SKILLS HANDBOOK 1.B., 2.E., 3.B.

1. Put on your eye protection and lab apron.

Part A: Reaction of Metals with Oxygen

2. Secure a Bunsen burner to a lab stand with a clamp. Carefully light the Bunsen burner with a spark lighter.
3. Clean a 5 cm length of copper wire with steel wool until the copper is shiny.
4. Hold one end of the wire with tongs. Insert the wire into the hottest part of a Bunsen burner flame for 20 to 30 s.
5. Remove the wire from the flame and allow it to cool on the heat-resistant pad. Look for evidence of a chemical change and record your observations. Save the wire for Part B.
6. Look at the burning magnesium in Figure 1. Record your observations.

Figure 1 Magnesium burns in air.

Part B: Decomposition of Copper(II) Carbonate

7. Half-fill a test tube with limewater. Place the test tube in the test-tube rack.
8. Add copper(II) carbonate crystals to another test tube to a depth of about 2 cm.
9. Hold the test tube in a test-tube holder at an angle so that the copper(II) carbonate is spread along the inside of the test tube.
10. Gently heat the underside of the test tube in the Bunsen burner flame. Move the test tube back and forth above the flame to evenly distribute the heat. Note any changes that occur. (Figure 2)

When heating the test tube, do not allow one part of the test tube to heat up more than any other part. Doing so could cause the contents to be ejected from the test tube.

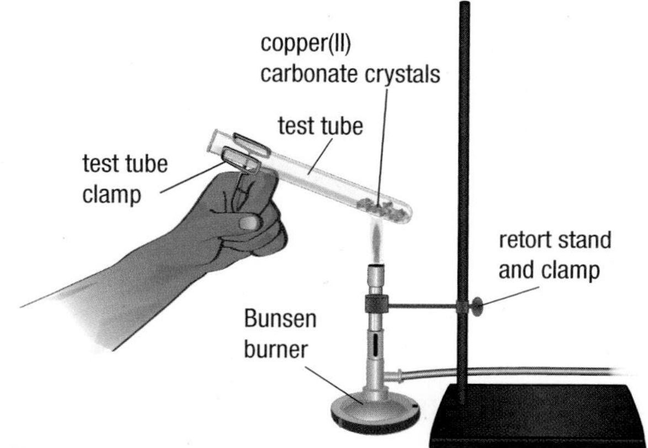

Figure 2

11. As the reaction begins, bring the mouth of the limewater test tube close to the mouth of the test tube being heated. This will allow any gases produced to flow into the limewater.
12. Continue heating the test tube until no further changes are observed. Compare the appearance of the wire from Part A with the contents of the hot test tube. Record your observations.
13. Look for evidence of chemical change in the limewater. Record your observations.

Part C: Decomposition of Hydrogen Peroxide

14. Place a third test tube in the test-tube rack. Pour hydrogen peroxide solution into the test tube until it is one-third full.
15. Add a small amount of iron(III) nitrate (enough to cover the end of a wooden splint) to the solution. Record your observations.
16. Test the gas produced by holding a glowing splint at the mouth of the test tube (Figure 3). Record your observations.

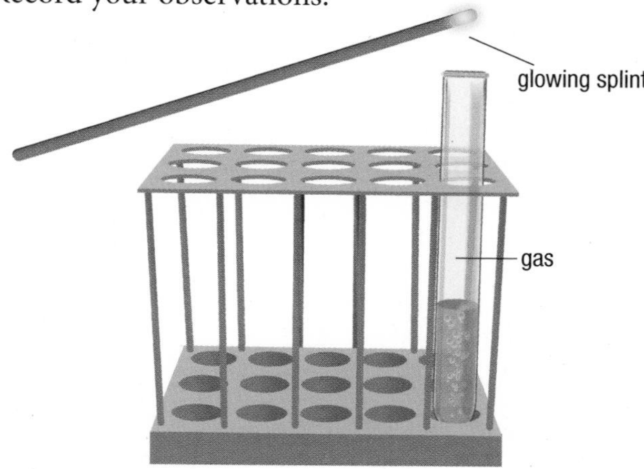

Figure 3 Glowing splint test

Analyze and Evaluate

(a) What evidence suggests that chemical changes occur when magnesium and copper are heated? K/U

(b) Write the word and chemical equations for the reactions that occur when magnesium and copper are heated. (Assume that the most common ionic charge for copper is +2.) K/U

(c) Based on your evidence, what molecular compound is produced when copper(II) carbonate is heated? What ionic compound remains? Justify your inference. T/I

(d) Write the word and chemical equation for the decomposition of copper(II) carbonate. K/U

(e) What evidence suggests that hydrogen peroxide decomposed into simpler substances? Justify your inference. K/U

(f) Write the word and chemical equations for the decomposition of hydrogen peroxide. Assume that one of the two products of this reaction is liquid water. K/U

Apply and Extend

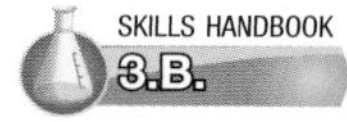

(g) Many natural chemicals, including substances in liver, potatoes, and strawberries, speed up the decomposition of hydrogen peroxide. Design a controlled experiment that compares how these substances affect the reaction. Write your Procedure, including safety precautions. Proceed with the experiment once your teacher has approved your proposal. T/I C A

Displacement Reactions

SKILLS MENU

- Questioning
- Hypothesizing
- Predicting
- Planning
- Controlling Variables
- Performing
- Observing
- Analyzing
- Evaluating
- Communicating

Displacement reactions involve elements displacing other elements from their compounds. Table 1 summarizes single and double displacement reactions. How are these reactions similar? How are they different?

Table 1 Summary of Single and Double Displacement Reactions

Type of reaction	Single displacement	Double displacement
general pattern	A + BC → AC + B	AB + CD → AD + CB
example	$Zn(s) + Pb(NO_3)_2(aq) \rightarrow Zn(NO_3)_2(aq) + Pb(s)$	$Fe(NO_3)_3(aq) + 3\ NaOH(aq) \rightarrow Fe(OH)_3(s) + 3\ NaNO_3(aq)$

Purpose

To observe and compare single and double displacement reactions.

Equipment and Materials

- lab apron
- eye protection
- large well plate
- steel wool or sandpaper
- 3 strips of magnesium ribbon
- 3 small pieces of zinc
- 3 strips of copper wire
- dropper bottles of
  - dilute copper(II) nitrate solution
  - dilute zinc nitrate solution
  - dilute magnesium nitrate solution
  - dilute sodium carbonate solution
- 3 toothpicks

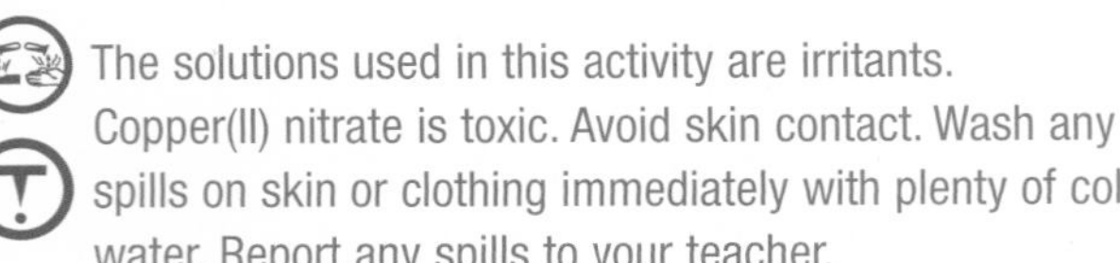

The solutions used in this activity are irritants. Copper(II) nitrate is toxic. Avoid skin contact. Wash any spills on skin or clothing immediately with plenty of cold water. Report any spills to your teacher.

Procedure

1. Read the following steps carefully. Prepare a table in which to record your observations.
2. Put on your lab apron and eye protection.

Part A: Single Displacement Reactions

3. Clean three strips of magnesium ribbon, zinc metal, and copper wire with steel wool or sandpaper until they are shiny.
4. Half-fill one well of the well plate with the copper(II) nitrate solution.
5. Half-fill a second well with the zinc nitrate solution.
6. Half-fill a third well with the magnesium nitrate solution.
7. Place one strip of each metal (magnesium, zinc, and copper) in each of the wells (Figure 1). Use the toothpick to keep the metals submerged and not touching each other.
8. Observe the wells for several minutes. Record your observations in your table, noting the properties of the reaction products.
9. Save the contents of the wells for Part B.

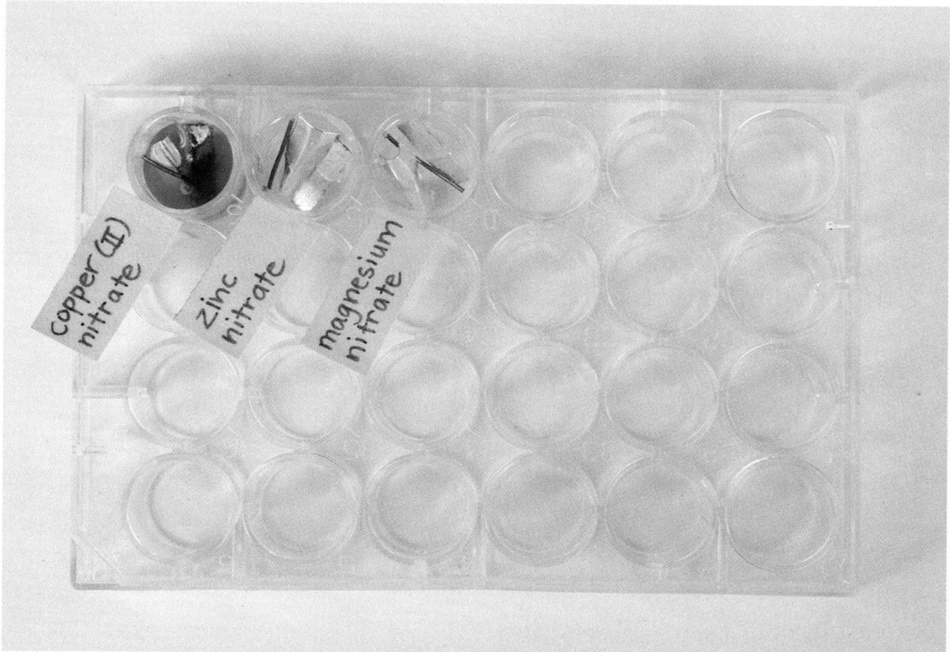

Figure 1 Each metal is placed in each of the three solutions.

Part B: Double Displacement Reactions

10. Add three drops of sodium carbonate solution to each of the three wells from Part A. (Note: Avoid cross-contamination of solutions. Never let the tip of the dropper touch or enter another solution.) Record your observations in your table, noting the properties of the reaction products.
11. Dispose of the contents of the well plate as directed by your teacher.
12. Clean your workstation and wash your hands.

Analyze and Evaluate

(a) What evidence suggests that chemical reactions occurred? K/U

(b) Rank the three metals in Part A in order of decreasing reactivity. (Rank the most reactive as #1 and the least reactive as #3.) T/I

(c) Rank the three solutions in Part A in order of decreasing reactivity. T/I

(d) What pattern do you notice when you compare your answers to (b) and (c)? T/I

(e) Write the chemical equations for three reactions that occurred in Part A. K/U

(f) Write the chemical equations for three reactions that occurred in Part B. K/U

(g) Why was it necessary to clean the metal surfaces before putting them in the solutions? T/I

Apply and Extend

(h) Table 1 shows the chemical equation for the reaction of zinc in a solution of lead(II) nitrate. Given this information, predict whether magnesium will also react in a lead(II) nitrate solution. Explain your prediction. T/I

(i) Water is considered to be "hard" if it contains high concentrations of calcium and magnesium ions. Research and prepare a brief report summarizing
- why hard water can be a nuisance.
- how the sodium carbonate solution from this activity "softens" water. T/I C A

6.9 Types of Chemical Reactions: Combustion

Figure 1 Propane was the fuel in this dramatic combustion reaction in north Toronto.

In the early hours of August 2, 2008, a north Toronto neighbourhood was rocked by a loud explosion. Startled residents stared out their windows in disbelief as a giant fireball rose high into the night sky. Why? A nearby propane storage depot was on fire (Figure 1)! Large chunks of metal, probably from exploded propane storage tanks, littered the area. Shockwaves from the explosion shattered windows and ripped doors off their hinges. Firefighters rushed to the scene, but all they could do was cool the remaining propane tanks with water and wait for the fireball to burn itself out. Extinguishing a propane blaze is almost impossible. People were shocked and angry that propane—a highly explosive fuel—had been stored and handled in a way that led to such a damaging explosion.

What Is Combustion?

combustion the rapid reaction of a substance with oxygen to produce oxides and energy; burning

Combustion is a chemical reaction in which a fuel "burns" or reacts quickly with oxygen. The products of this reaction are usually an oxide and energy. Propane, C_3H_8, is one of a group of molecular compounds called hydrocarbons. As their name implies, these compounds contain only the elements hydrogen and carbon. Most hydrocarbons originate from fossil fuels. The combustion of hydrocarbons powers cars and buses, warms homes, generates electricity, and even lights up the candles on your birthday cake.

Complete Combustion of Hydrocarbons

complete combustion a combustion reaction of hydrocarbons that uses all the available fuel and produces only carbon dioxide, water, and energy; occurs when the supply of oxygen is plentiful

The products of a hydrocarbon combustion reaction can vary. They depend on the availability of oxygen. If oxygen is plentiful, hydrocarbons burn completely to release the energy they contain. The only products of complete combustion are carbon dioxide and water. The word equation for the **complete combustion** of a hydrocarbon is

hydrocarbon + oxygen → carbon dioxide + water + energy

Carbon dioxide is an important greenhouse gas. You will learn more about this product of combustion in Unit B: Climate Change.

Methane, $CH_4(g)$, is a typical hydrocarbon. Natural gas is mostly composed of methane. The balanced chemical equation for the complete combustion of methane is

$$CH_4(g) + 2\ O_2(g) \rightarrow CO_2(g) + 2\ H_2O(g) + \text{energy (complete combustion)}$$

The complete combustion of hydrocarbons can be represented by the general equation

$$C_xH_y + O_2 \rightarrow CO_2 + H_2O + \text{energy}$$

During complete combustion, fuels burn cleanly with no sooty residue.

DID YOU **KNOW?**

Is Your Classroom Putting You to Sleep?

Carbon dioxide is exhaled by students and produced by the school heating system. This gas accumulates in a classroom over the course of a day if ventilation is inadequate. Excess carbon dioxide can cause headaches and drowsiness.

Incomplete Combustion of Hydrocarbons

If the oxygen supply is limited, **incomplete combustion** may occur, releasing carbon monoxide gas and carbon (soot), in addition to carbon dioxide and water. An orange, flickering flame often indicates incomplete combustion.

incomplete combustion a combustion reaction of hydrocarbons that may produce carbon monoxide, carbon, carbon dioxide, soot, water, and energy; occurs when the oxygen supply is limited

Butane gas, $C_4H_{10}(g)$, is burned as fuel in some portable stoves. If the stove burner is not adjusted properly, or if there is not enough oxygen, incomplete combustion could occur.

$$C_4H_{10}(g) + 5\ O_2(g) \rightarrow 2\ CO_2(g) + 5\ H_2O(g) + CO(g) + C(s) + \text{energy}$$

Carbon Monoxide

Carbon monoxide, $CO(g)$, is an odourless, colourless gas that is highly toxic. Symptoms of carbon monoxide poisoning include headache, dizziness, nausea, and respiratory problems. These are fairly general symptoms, so carbon monoxide may not immediately be identified as the cause. Many people have died from inhaling carbon monoxide. It is often produced as a result of the incomplete combustion of fuels in a confined space. Carbon monoxide is typically found in a home with a poorly ventilated furnace or in a closed garage in which a vehicle is running. 

To learn more about the effects of carbon monoxide on the body,

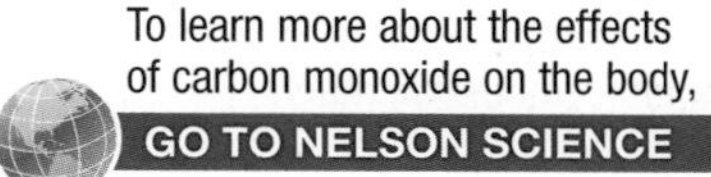

CITIZEN **ACTION**

Carbon Monoxide Detectors

Carbon monoxide in homes is particularly dangerous at night because people who are asleep are totally unaware of the symptoms of carbon monoxide poisoning. To prevent deaths from faulty home furnaces or stoves, we are advised to install carbon monoxide detectors in our homes. These devices save lives every year.

GO TO NELSON SCIENCE

What Can You Do to Help?

Contact your local fire department. Find out what they recommend regarding carbon monoxide detectors. How many detectors should you have in your home? Where should they be placed? Why? Consider protecting your family from carbon monoxide poisoning.

Soot

Soot is made up of particles of carbon. Soot is evidence of incomplete combustion, which causes pollution and wastes energy. Soot is common in older vehicles with poorly maintained engines (Figure 2(a)). Forest fires also produce huge quantities of soot that travel far downwind (Figure 2(b)).

Figure 2 Two signs of incomplete combustion are (a) soot production and (b) orange flames.

Other Combustion Reactions

Many other substances—besides hydrocarbons—undergo combustion reactions. Elements, for example, react with oxygen to form oxides. Magnesium burns to produce magnesium oxide just as carbon burns to produce carbon dioxide.

General word equation:	element + oxygen → oxide	+ energy
General chemical equation:	$A + O_2 \rightarrow AO$	+ energy
Example:	$2\ Mg(s) + O_2(g) \rightarrow 2\ MgO(s)$	+ energy

You might have noticed that combustion reactions involving elements are also synthesis reactions: they follow the pattern $A + B \rightarrow AB$.

Combustion of Hydrogen

Hydrogen reacts (burns) with oxygen to form water:

$$2\ H_2(g) + O_2(g) \rightarrow 2\ H_2O(g) + \text{energy}$$

Hydrogen is already being used as a fuel in a few technologies. The source of hydrogen is usually water. The decomposition reaction that produces hydrogen from water is the exact reverse of the hydrogen combustion reaction:

$$2\ H_2O(g) + \text{energy} \rightarrow 2\ H_2(g) + O_2(g)$$

The energy on the reactant side of this equation usually comes from electricity.

At first glance, hydrogen is an ideal fuel because

- it burns cleanly, producing only water and energy, and
- there is an almost endless supply of water to produce hydrogen. As long as you have energy to decompose the water, you have a source of hydrogen.

However, some technical problems have to be overcome before hydrogen becomes a common vehicle fuel.

- Making hydrogen requires energy. What non-polluting source of energy can be used?
- The engines for hydrogen-fuelled cars are currently very expensive to make.
- Hydrogen is an explosive gas. It is difficult to transport and store.

DID YOU KNOW?

Iceland: A Hydrogen Powerhouse
Iceland has ambitious plans. It intends to be the first country to completely switch from fossil fuels to hydrogen. Iceland was formed from volcanoes, so it has abundant stores of geothermal energy. Icelandic scientists plan to use this energy to extract hydrogen from seawater.

Combustion of Phosphorus

The combustion of phosphorus has particularly interesting applications. Phosphorus comes in two forms: white and red phosphorus (Figure 3(a)). You might have seen red phosphorus on the striking strip of a package of safety matches (Figure 3(b)).

DID YOU KNOW?

Phosphorus from Urine
Like many alchemists of his time, Hennig Brandt was preoccupied with trying to turn objects into gold. In 1669, Brandt chose urine as his starting material because it is gold coloured. Brandt boiled urine until it became a thick paste. He continued heating this paste, collecting and condensing the vapours produced. Much to his disappointment, instead of gold, he had produced a white solid that glowed in the dark. Brandt named this substance "phosphorus," which means "light-giving" in Greek.

Figure 3 (a) White phosphorus is so reactive that it has to be stored in oil to prevent it from coming into contact with the air. Red phosphorus is relatively unreactive. (b) Safety matches will ignite only if they are scraped against the red phosphorus on the match box.

When you rub a match against the striking strip, the friction releases heat energy. This energy converts red phosphorus into white phosphorus, which instantly burns in the air:

$$P_4(s) + 5\ O_2(g) \rightarrow P_4O_{10}(g) + \text{energy}$$

The heat from this reaction ignites the chemicals in the head of the match.

The combustion of non-metals is an important first step in the formation of acid precipitation. You will learn more about these reactions in Chapter 7.

RESEARCH THIS FIREFIGHTING WITH AN MSDS

SKILLS: Researching, Identifying Alternatives, Defending a Decision, Communicating

SKILLS HANDBOOK 4.A.

How you put out a fire depends on the properties of the fuel. An MSDS can provide firefighters with valuable tips on how to put out a fire, particularly with chemical fires.

1. Research the MSDS for the following substances: propane, olive oil, and magnesium.

GO TO NELSON SCIENCE

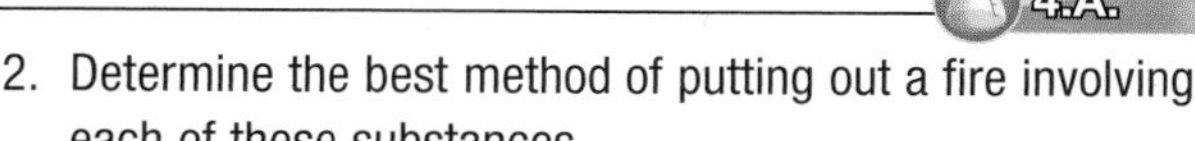

2. Determine the best method of putting out a fire involving each of these substances.

A. Create a poster or other information campaign to share your discoveries with others. Explain, in your campaign, why the specific methods are effective. T/I C A

IN SUMMARY

- Hydrocarbons often react with oxygen in combustion reactions. Complete combustion produces only carbon dioxide and water; incomplete combustion may produce carbon (soot), carbon monoxide, carbon dioxide, and water.
- Some metals react with oxygen in combustion reactions, producing oxides of the metal (e.g., magnesium oxide, MgO).
- Hydrogen reacts with oxygen in combustion reactions, producing water. This is a possible future source of energy for vehicles.

CHECK YOUR LEARNING

1. (a) Describe an idea in this section that could affect your life.
   (b) Why is this idea important? K/U A
2. Complete these skeleton equations. Remember to balance the equations, where necessary, by adding coefficients before the chemical symbols. K/U T/I
   (a) ________ (s) + ________ (g) → SO_2(g) + energy
   (b) __ Ca(s) + __ ______ → __ CaO(s) + energy
   (c) __ C_3H_8(g) + __ O_2(g) → __ ________ + __ ________ + energy
   (d) __ C_2H_4(g) + __ O_2(g) → __ ________ + __ ________ + energy
3. Propane is used as a fuel in camping stoves (Figure 4). It is a hydrocarbon with the chemical formula C_3H_8. Propane is a gas at room temperature and pressure but becomes a liquid when compressed. K/U T/I A
   (a) Write the general equation for the complete combustion of a hydrocarbon.
   (b) Write the balanced chemical equation for the complete combustion of propane.
   (c) Examine the Hazardous Household Products Symbols (HHPS) on the label. Outline the precautions you should take when using this product.
   (d) Why is it unwise to use a camping stove inside a tent?

Figure 4 Propane is highly flammable. Note the warnings on the container.

4. Explain why you can save money on home heating fuel if you keep your gas furnace clean and operating at peak efficiency. A
5. (a) Give at least two reasons why the use of hydrogen fuel is potentially better for the environment than gasoline.
   (b) What is meant by the statement "hydrogen is only as environmentally clean as the energy used to make it"? K/U A
6. List the five types of reactions that have been discussed in this chapter so far. K/U
7. (a) Use specific examples to show that some combustion reactions are also synthesis reactions.
   (b) Under what conditions does this occur? A

6.10 Corrosion

corrosion the breakdown of a metal resulting from reactions with chemicals in its environment

With just a few exceptions, such as gold and platinum, most metal elements corrode. **Corrosion** is the breakdown of a metal as a result of chemical reactions with its environment. Metalworkers have, over the centuries, given different names to corrosion reactions involving different metals. Silver, for example, tarnishes when it comes in contact with sulfur compounds in air.

Beneficial Corrosion

In some cases, the corrosion of a metal is beneficial. For example, when aluminum is exposed to air, it quickly corrodes to form aluminum oxide—one of the hardest substances known. Aluminum oxide tightly coats the underlying aluminum metal, preventing any further corrosion from occurring. This explains why aluminum camping pans can be safely left outside in the rain, while a cast iron pan rusts in a matter of days. Zinc and copper are other common metals that form protective coatings when they corrode. Copper develops an attractive greenish patina after being exposed to the atmosphere for several months (Figure 1). This patina is so corrosion resistant that a copper roof remains weatherproof for up to 75 years.

Figure 1 A colorful patina develops on copper roofs over several years

Rust

Rust is the familiar reddish-brown flaky material produced when metals containing iron corrode. Unlike the corrosion products of aluminum and copper, rust does not stick well to the underlying steel. Instead, rust is very porous and readily flakes away from the surface of steel. As it does, fresh steel is exposed for further corrosion. This process continues until the steel is completely corroded or "eaten away." All that remains is a trail of rust flakes!

LEARNING TIP

Corrosion and Rusting

The terms *corrosion* and *rusting* are often used interchangeably. Strictly speaking, corrosion is a general term that can be applied to any metal that reacts with chemicals in the environment. Rusting, however, refers specifically to the corrosion of metals that contain iron, such as steel.

Causes of Rust

The corrosion of iron or rusting is a complex process that is affected by many things: the presence of air, water, and electrolytes, along with acidity and mechanical stress.

OXYGEN AND WATER

The most obvious factors necessary for the corrosion of iron are oxygen (in air) and water. Steel will not corrode if it is kept away from water and oxygen. This is why steel lasts much longer in dry climates than in Ontario.

ELECTROLYTES

Contrary to popular belief, salt (sodium chloride) does not actually cause corrosion of iron. It does, however, speed up corrosion once it starts. This is because salt is an electrolyte that helps the rusting process along. The combination of road salt and saltwater spray off the ocean affects both the bodies of cars and the metal supports of bridges.

LEARNING TIP

Electrolytes

In Chapter 5, you learned that electrolytes are compounds that, when dissolved in water, conduct electricity.

Preventing Corrosion

Several strategies are used to prevent corrosion in various situations. Some are more effective than others, but none are perfect. These strategies can be divided into three categories: using corrosion-resistant materials, protective coatings, and galvanizing.

DID YOU KNOW?

The Eiffel Tower

The world-famous icon of France is an iron tower. Preventing the Eiffel Tower from rusting requires the application of 50 to 60 tonnes of paint every seven years.

Protective Coatings

A simple way to prevent corrosion is to cover the metal with a rust-inhibiting paint, chrome, or plastic coating. This strategy works well on above-ground structures provided that the metal remains completely covered. However, once the coating is chipped or scratched, corrosion is inevitable (Figure 2).

Figure 2 This bridge, spanning Halifax Harbour, is exposed to all the factors that speed up corrosion. It requires constant care to keep it corrosion-free.

Corrosion-Resistant Materials

A straightforward way of preventing corrosion is to use materials that do not rust. For example, decades ago, car bumpers were made out of steel, which tended to rust if they became dented or scratched. Today, most bumpers are made of plastic. Plastic does not corrode and is lighter than steel. This helps reduce the overall weight of the car and improve fuel efficiency.

If steel is the only appropriate material for a specific object, improving its corrosion resistance will be an advantage. For example, the steel used by the auto industry today has more corrosion-resistant additives than ever before. New cars and bridges remain rust-free longer, even in Ontario's challenging winter conditions.

Many other corrosion-resistant alloys have been developed as well. An alloy is a metal produced by blending metals (and sometimes non-metals) in specific proportions. For example, most cutlery is made from stainless steel, an alloy of various elements, including iron, carbon, nickel, and chromium. Surgical-grade stainless steel, which is used to make medical tools and implants, contains enough chromium to make the steel corrosion-proof almost indefinitely (Figure 3).

Figure 3 This surgical implant is made of stainless steel, an alloy designed to resist corrosion inside the body.

Galvanizing

galvanized steel steel that has been coated with a protective layer of zinc, which forms a hard, insoluble oxide

Galvanized steel is steel that has been coated with a thin layer of zinc. Galvanizing protects steel because zinc corrodes before the iron in the steel does. As zinc corrodes, it forms a protective oxide layer that sticks to both the zinc layer and any steel that may be exposed. The corrosion protection remains intact even if there are nicks or scratches in the zinc layer. That is why galvanizing steel provides better rust protection than painting it (Figure 4).

READING TIP

Making Inferences

Think about what you already know to help you make inferences. For example, you might recall seeing rusted nails in an old wooden fence. You might also have seen a galvanized chain-link fence that has not rusted. Consequently, you infer that galvanized metal resists corrosion better than non-galvanized metal.

Figure 4 Galvanized steel is corrosion resistant and requires no maintenance.

IN SUMMARY

- Corrosion is the breakdown of a metal as it reacts with chemicals in the environment.
- Corrosion of some metals forms a tough protective layer that prevents further corrosion.
- Rusting is the corrosion of iron and steel. Rust does not form a protective layer but continues flaking away until the metal is severely damaged.
- Rusting occurs in the presence of oxygen and water and is made worse by electrolytes such as salt.
- Corrosion can be slowed or avoided by using corrosion-resistant materials, covering the metal with a protective layer (for example, paint), or galvanizing with zinc.

CHECK YOUR LEARNING TREATMENT

1. (a) Before reading about it in this section, what did you think rust was?
   (b) Describe similarities and differences between what you already knew about rust and what you learned about it as you read this section. K/U T/I
2. (a) In your own words, define "corrosion."
   (b) Describe the difference between corrosion and rusting. K/U
3. (a) What two substances react to form rust?
   (b) What other factors help rust form quickly? K/U
4. Consider an experiment in which an aluminum soft drink can and a steel soup can are left outside for a few days. Use your knowledge of the corrosion of steel and aluminum to predict how they would look different after a week exposed to rainy weather. Explain your prediction. K/U T/I
5. Why is it important for a car to be clean and dry before being treated with a rust-proofing product? A
6. A car design company predicts that the bodies of automobiles will last much longer in the islands of the Caribbean than in Canada. Explain this prediction. K/U A
7. Why is galvanized steel preferred for outdoor uses? K/U T/A A

Poisonous Jewellery

Does your jewellery stain your skin? Do you wear 18-karat yellow gold with no ill effects but break out in a rash if you wear white gold? The culprit in both situations may be nickel, the shiny, silvery metal that is a part of so many common metallic objects: earrings, coins, zippers, cellphones (Figure 1). For some people, prolonged skin contact with a nickel-containing object (like that stud through your eyebrow) is like rubbing your skin in poison ivy, except the effects are less immediate. In both cases, a chemical is transferred to your skin, sensitizing the skin to the chemical and potentially triggering an allergic response known as allergic contact dermatitis. Each year, hundreds of Canadians become sensitized to nickel, meaning that even brief exposure to nickel can trigger an allergic response. And once you are sensitized, there is no cure!

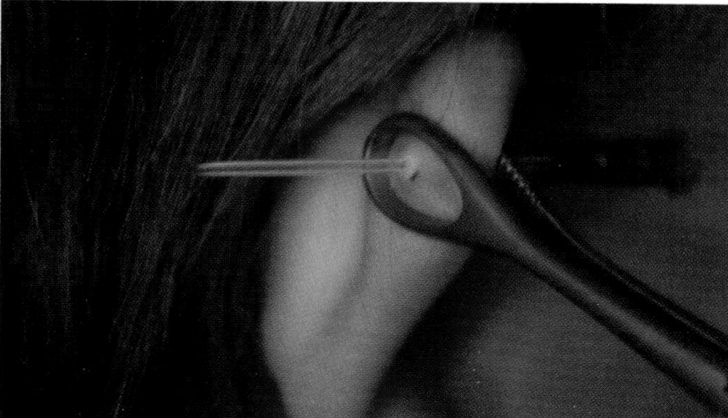

Figure 1 If you have your ears pierced, give some thought to what kind of studs or hoops you want inserted.

The real cause of nickel allergies is nickel(II) ions, $Ni^{2+}(aq)$. These ions are produced when nickel metal is corroded by the acidity of body fluids, such as sweat. The reaction involved is

$$Ni(s) + 2\ H^{+}(aq) \longrightarrow Ni^{2+}(aq) + H_2(g)$$

Nickel Allergy: Contributing Factors

Three factors contribute to a nickel allergy: prolonged, direct skin contact; the presence of electrolytes, which occur in sweat; and the type of nickel-containing metal involved.

The type and composition of a metal often determine how quickly it corrodes. Stainless steel is an alloy of iron, nickel, and chromium. The resistance of stainless steel to corrosion varies depending on how much chromium it contains. The stainless steel used for dental braces and surgical implants does not corrode at all, unlike the steel in inexpensive jewellery.

So You Want to Pierce Your . . .

The number of nickel allergies has recently increased dramatically because of the popularity of body piercing. The reason is obvious: a nickel-containing object is in constant contact with the body. Piercing any body part results in some bleeding. Blood contains electrolytes, which can corrode nickel in the studs or hoops, releasing nickel ions. The ions can easily enter the body through the new wound. Soft tissue, like that in the earlobes, heals quickly because it does not have a lot of blood flow.

Tongue Piercing

Getting your tongue pierced is far riskier than getting your ears pierced (Figure 2). A tongue is much thicker and contains more blood vessels than an ear lobe. Consequently, a newly pierced tongue takes much longer to heal than a newly pierced ear lobe. It is also more prone to infection than most other body parts. An infected tongue can swell large enough to block the airway to the throat. Some deaths have resulted from complications due to tongue piercing. Even without complications, a newly pierced tongue can take from four to six weeks to heal. During the normal healing period, a stud through the tongue is bathed with enough blood and saliva to trigger a nickel allergy.

Figure 2 Tongue piercings are more likely than other types of piercings to result in infection or nickel allergy.

Making an Informed Decision

If you are considering a body piercing, carefully weigh the risks involved. Consult only with the most reputable business or health care professional. Insist that sterile procedures and high-grade surgical stainless steel be used.

LOOKING BACK

KEY CONCEPTS SUMMARY

Chemical reactions involve the change of one or more substances into one or more different substances.

- Reactants are substances present at the beginning of a reaction; products are substances present at the end of the reaction. (6.1)
- Reactions can involve energy as an input or as an output. (6.1)
- Different reactions occur at different rates: combustion reactions are generally fast, while corrosion reactions are much slower. (6.1–6.9)

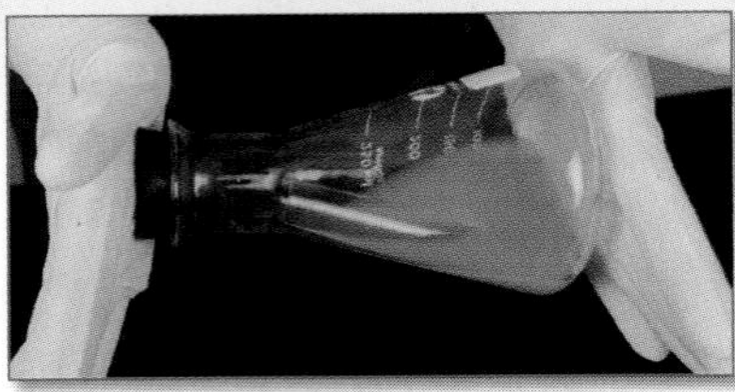

Chemical reactions obey the law of conservation of mass and can be represented by balanced chemical equations.

- The law of conservation of mass states that in any given chemical reaction, the total mass of reactants equals the total mass of products. (6.3)
- A balanced chemical equation describes the reactants and products involved in the reaction, their ratios, and their states. (6.4)
- Balanced chemical equations obey the law of conservation of mass. (6.3, 6.4)

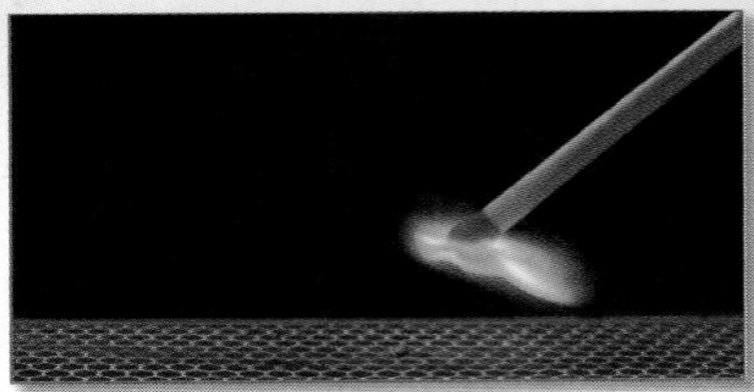

Chemical reactions involving consumer products can be useful or harmful.

- The usefulness of many consumer products depends on the chemical reactions of their ingredients. (6.4–6.9)
- Not all chemical reactions involving consumer products are beneficial. Some are dangerous to people and to the environment. (6.4–6.10)
- Reactive consumer products should be stored, used, and handled with care. (6.4–6.9)

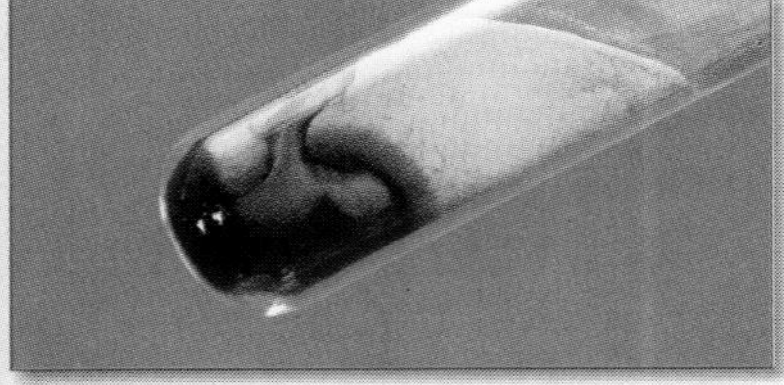

We can use patterns in chemical properties to classify chemical reactions.

- Synthesis reactions can be represented by $A + B \rightarrow AB$. (6.5)
- Decomposition reactions can be represented by $AB \rightarrow A + B$. (6.5)
- Single displacement reactions can be represented by $A + BC \rightarrow AC + B$. (6.6)
- Double displacement reactions can be represented by $AB + CD \rightarrow AD + CB$. (6.6)
- Combustion reactions of elements can be represented by $A + O_2 \rightarrow AO +$ energy; complete combustion of hydrocarbons can be represented by $C_xH_y + O_2 \rightarrow CO_2 + H_2O +$ energy. (6.9)
- Some reactions fit into more than one class. (6.8, 6.10)

Corrosion is the reaction of metals with substances in the environment.

- Most metals corrode. (6.10)
- Some metals, such as aluminum, form hard, protective coatings when they corrode. (6.10)
- Rust is produced when iron or steel corrode. (6.10)
- Rusting can be prevented by using corrosion-resistant materials, applying protective coatings, or galvanizing. (6.10)

Chemical reactions affect us and our environment.

- Some combustion reactions form products that are damaging to the environment and human health. (6.9)
- The complete combustion of hydrocarbons releases carbon dioxide—a greenhouse gas. (6.9)
- The incomplete combustion of hydrocarbons releases soot and carbon monoxide. (6.9)
- Corrosion can weaken iron and steel, resulting in damage to cars and structures. (6.10)

WHAT DO YOU THINK NOW?

You thought about the following statements at the beginning of the chapter. You may have encountered these ideas in school, at home, or in the world around you. Consider them again and decide whether you agree or disagree with each one.

1 Chemical reactions are bad for the environment.
Agree/disagree?

4 Carbon dioxide in your classroom makes you sleepy.
Agree/disagree?

2 Chemical reactions are reversible.
Agree/disagree?

5 Some people are allergic to cellphones.
Agree/disagree?

3 The total mass of substances involved in a chemical change remains constant.
Agree/disagree?

6 Hydrogen will replace gasoline as the fuel of the twenty-first century.
Agree/disagree?

How have your answers changed since then? What new understanding do you have?

Vocabulary

chemical reaction (p. 225)
word equation (p. 225)
chemical equation (p. 225)
reactant (p. 225)
product (p. 225)
state symbol (p. 226)
law of conservation of mass (p. 230)
synthesis reaction (p. 237)
decomposition reaction (p. 238)
single displacement reaction (p. 240)
double replacement reaction (p. 242)
precipitate (p. 242)
combustion (p. 248)
complete combustion (p. 248)
incomplete combustion (p. 249)
corrosion (p. 252)
galvanized steel (p. 254)

BIG Ideas

- ✓ **Chemicals react with each other in predictable ways.**
- ✓ **Chemical reactions may have a negative impact on the environment, but they can also be used to address environmental challenges.**

The following icons indicate the Achievement Chart category addressed by each question.

K/U Knowledge/Understanding T/I Thinking/Investigation
C Communication A Application

What Do You Remember?

1. Zinc oxide, ZnO, is an active ingredient in some sunscreens. Zinc oxide can be made by heating zinc sulfide strongly in air:

   $2\ ZnS(s) + 3\ O_2(g) \rightarrow 2\ ZnO(s) + 2\ SO_2(g)$

   Referring to this chemical equation, write
   (a) the coefficient for zinc sulfide
   (b) the number of sulfur atoms in sulfur dioxide
   (c) the coefficient for oxygen (the element)
   (d) the number of different reactants
   (e) the total number of reactant molecules
   (f) the states of matter present in this reaction (6.1) K/U

2. The combustion of hydrocarbons sometimes produces a yellow flame and a sooty residue. (6.9) K/U
   (a) Describe the conditions when this might occur.
   (b) Name this type of combustion.

What Do You Understand?

3. Consider the six types of reactions discussed in this chapter. All involve elements and compounds. Identify the type(s) of reaction that has the following reactants:
   (a) two elements
   (b) two compounds
   (c) oxygen and a fuel
   (d) only one compound
   (e) an element and a compound (6.5–6.10) K/U

4. Use the law of conservation of mass to explain why chemical equations should be balanced. (6.2, 6.3) K/U

5. Hydrogen peroxide is a clear, colourless liquid used to disinfect cuts. The chemical equation for the decomposition of hydrogen peroxide is

   $2\ H_2O_2(aq) \rightarrow 2\ H_2O(l) + O_2(g)$ (6.1, 6.3) K/U T/I

   (a) Explain the difference between the state symbols (aq) and (l).
   (b) Predict what you would expect to see in a test tube of hydrogen peroxide undergoing this reaction.
   (c) Describe how to tell when the reaction is complete.
   (d) Predict how the mass might change as the reaction proceeds. Explain.

6. People can choose to have the underside of their vehicle sprayed with oil once a year. How does a coating of oil help prevent rusting? (6.10) K/U A

7. Balance and classify each of the following chemical equations. (Note that some may fit into more than one category.) (6.3–6.9) K/U T/I
   (a) $K_2O \rightarrow K + O_2$
   (b) $Na + I_2 \rightarrow NaI$
   (c) $Cu(NO_3)_2 + NaOH \rightarrow Cu(OH)_2 + NaNO_3$
   (d) $KClO_3 \rightarrow KCl + O_2$
   (e) $Ca(NO_3)_2 + HBr \rightarrow CaBr_2 + HNO_3$
   (f) $Sn(OH)_2 \rightarrow SnO + H_2O$
   (g) $P_4 + N_2O \rightarrow P_4O_6 + N_2$
   (h) $Fe + Al_2(SO_4)_3 \rightarrow FeSO_4 + Al$
   (i) $AlCl_3 + Na_2CO_3 \rightarrow Al_2(CO_3)_3 + NaCl$
   (j) $C_3H_6 + O_2 \rightarrow CO_2 + H_2O$

8. Explain why you should not cook on a barbecue in an enclosed space. (6.9) T/I A

9. The following chemical reactions take place, one after another, when an automobile airbag inflates. Balance and classify each reaction. (6.3–6.6) K/U
   (a) The gas needed to inflate the bag comes from the reaction of sodium azide, NaN_3.

   $NaN_3(s) \rightarrow N_2(g) + Na(s)$

   (b) The sodium produced in reaction (a) is dangerous. It is removed by a reaction with iron(III) oxide, Fe_2O_3, in the airbag.

   $Na(s) + Fe_2O_3(s) \rightarrow Na_2O(s) + Fe(s)$

   (c) The sodium oxide quickly reacts with carbon dioxide and moisture from the air to form sodium hydrogen carbonate.

   $Na_2O(s) + CO_2(g) + H_2O(g) \rightarrow NaHCO_3(s)$

Solve a Problem

10. Fuels burn faster as the concentration of oxygen is increased. For example, iron wool burns in pure oxygen to produce iron(III) oxide and a great deal of energy. (6.1, 6.3, 6.9) K/U T/I
    (a) Write the word and chemical equations for this reaction.
    (b) Design an experiment using this reaction to confirm the law of conservation of mass.

11. A student placed a piece of zinc metal in a solution of hydrochloric acid. The chemical equation for the reaction that occurred is

$Zn(s) + 2\ HCl(aq) \rightarrow H_2(g) + ZnCl_2(aq)$

The following data were collected from the experiment:

initial mass of zinc reacted:	2.5 g
initial mass of hydrochloric acid:	52.6 g
final mass of solution:	54.8 g

(6.3, 6.5) T/I

(a) Calculate the mass of hydrogen produced.
(b) Do the results of this experiment violate the law of conservation of mass? Explain.

12. Two properties that make gasoline potentially dangerous are that it evaporates quickly at room temperature and that its vapour ignites readily. Despite these hazards, many homeowners keep a small amount of gasoline on hand to fuel small devices like lawn mowers and snowblowers. What must be considered when handling or storing gasoline? Explain your answer. (6.9) A

Create and Evaluate

13. A routine test of the drinking water at Edgevale High School revealed slightly higher than normal concentrations of lead. Exposure to lead is known to slow brain development, particularly in children. The lead was coming from the corrosion of the solder metal used to join lengths of copper pipe together. Edgevale school officials were faced with a difficult decision. Closing the school was not an option. The copper pipes could not be replaced until the following summer. However, staff and students needed a safe supply of drinking water now. Two options being considered were
    - providing bottled water for all 1500 staff and students or
    - turning on all drinking fountains for at least 20 minutes each morning before students arrive. (After 20 minutes, the lead concentrations decreased to within the "normal" range.) (6.10) A

(a) List the risks and benefits of both options.
(b) If you were the school principal, which option would you choose? Why?

14. (a) Develop a visual organizer, such as a chart or a consequence map, to show the risks and benefits of burning fossil fuels.
(b) Write two paragraphs, each written from a different perspective, outlining why we should (or should not) continue to burn fossil fuels for energy. (6.9) A

Reflect on Your Learning

15. (a) How relevant was the idea of a chemical reaction to you before reading this chapter?
(b) Outline how your understanding of the role of chemical reactions in your life has changed.

Web Connections 

16. What is the "fire triangle"? How does it relate to the work of firefighters? (6.10) K/U A

17. The "carbonyl process" is a method of refining nickel that involves passing carbon monoxide gas over impure nickel at high temperature. This produces a compound called nickel carbonyl:

$Ni(s) + CO(g) \rightarrow Ni(CO)_4(g)$

Nickel carbonyl is then strongly heated to produce pure nickel and carbon monoxide gas:

$Ni(CO)_4(g) \rightarrow Ni(s) + CO(g)$ (6.5) K/U T/A

(a) Classify each of these reactions.
(b) Research the properties of nickel carbonyl and carbon monoxide. Why must the reactions be conducted in an airtight chamber?

18. Dinitrogen monoxide is also known as nitrous oxide, $N_2O(g)$. Dentists sometimes use this gas to relax their patients. Nitrous oxide can be made from solid ammonium nitrate. Water is also produced in this reaction. (6.3, 6.5, 6.9) K/U C A

(a) Write the chemical equation for this reaction.
(b) Classify this reaction.
(c) Research how nitrous oxide can be used to improve the performance of racing cars. Summarize your findings in a short web article for car enthusiasts. Include safety advice regarding the use of nitrous oxide in car engines.

To do an online self-quiz or for all other Nelson Web Connections,
GO TO NELSON SCIENCE

CHAPTER 6

SELF-QUIZ

The following icons indicate the Achievement Chart category addressed by each question.

K/U Knowledge/Understanding T/I Thinking/Investigation C Communication A Application

For each question, select the best answer from the four alternatives.

1. Which of the following equations is balanced? (6.4) K/U
   (a) $H_2 + O_2 \rightarrow 2\ H_2O$
   (b) $Zn + 2\ AgNO_3 \rightarrow Zn(NO_3)_2 + 2\ Ag$
   (c) $N_2 + H_2 \rightarrow NH_3$
   (d) $PbCl_2 + Li_2SO_4 \rightarrow LiCl + 2\ PbSO_4$
2. Which of the following is a double displacement reaction? (6.6) K/U
   (a) $2\ PbO_2 \rightarrow 2\ PbO + O_2$
   (b) $2\ Al + Fe_2O_3 \rightarrow 2\ Fe + Al_2O_3$
   (c) $N_2 + 3\ H_2 \rightarrow 2\ NH_3$
   (d) $ZnBr_2 + 2\ AgNO_3 \rightarrow Zn(NO_3)_2 + 2\ AgBr$
3. The decomposition of calcium carbonate is given by the equation:

   $CaCO_3(s) \rightarrow CaO(s) + CO_2(g)$

   If 25 g of $CaCO_3$ is heated to give 15g of CaO, what mass of CO_2 is also produced? (6.3) K/U
   (a) 5 g
   (b) 10 g
   (c) 15 g
   (d) 25 g
4. Carbon disulfide is produced by the reaction of carbon with sulfur dioxide. Carbon monoxide is also formed. What is the balanced chemical equation for this reaction? (6.4) K/U
   (a) $C + SO_2 \rightarrow CS_2 + CO$
   (b) $5\ C + 2\ SO_2 \rightarrow CS_2 + 4\ CO$
   (c) $4\ C + SO \rightarrow CS_2 + 3\ CO$
   (d) $C + SO \rightarrow CS + CO$

Indicate whether each of the statements is TRUE or FALSE. If you think the statement is false, rewrite it to make it true.

5. In a balanced chemical equation, each side of the equation has the same number of atoms of each element. (6.3) K/U
6. Equations are balanced by changing the subscripts in the chemical formula of a substance. (6.3) K/U

Copy each of the following statements into your notebook. Fill in the blanks with a word or phrase that correctly completes the sentence.

7. A _______ is a whole number that appears in front of a chemical formula in an equation. (6.3) K/U
8. In an experiment combining aluminum with oxygen to form aluminum oxide, the aluminum oxide is the _______ formed from the reactants. (6.1) K/U

Match the general pattern on the left with the appropriate type of chemical reaction on the right.

9. (a) $A + B \rightarrow AB$ (i) decomposition
   (b) $A + BC \rightarrow AC + B$ (ii) double displacement
   (c) $AB + CD \rightarrow AD + CB$ (iii) single displacement
   (d) $AB \rightarrow A + B$ (iv) synthesis
   (e) $A + O_2 \rightarrow AO$ + energy (v) combustion
   (6.5, 6.6, 6.9) K/U

Write a short answer to each of these questions.

10. Consider the following chemical reaction:

    $CaO(s) + H_2O(l) \rightarrow Ca(OH)_2(s)$ + energy
    (6.1, 6.6) K/U T/I
    (a) What are the reactants in this reaction? How do you know?
    (b) What substance is produced in this reaction?
    (c) Why is the final reaction mixture a cloudy liquid?
    (d) Will the test tube in which this reaction is occurring become warmer or cooler during the reaction? Explain.

11. When baking soda (sodium hydrogen carbonate, $NaHCO_3(s)$) is heated, it decomposes to form sodium carbonate, $Na_2CO_3(s)$, carbon dioxide, and water. (6.1, 6.4) K/U C
    (a) Write the word equation for this reaction.
    (b) What evidence suggests that a chemical change is taking place?
    (c) Write a balanced chemical equation for this reaction.
12. The following unbalanced equation represents the formation of magnesium oxide from magnesium and oxygen.

    $Mg + O_2 \rightarrow MgO$

    Can this equation be balanced by changing the formula of the product to MgO_2? Explain. (6.3, 6.4) K/U
13. List three pieces of information that can be obtained from a balanced chemical equation. (6.4) K/U
14. Compare corrosion and rusting, using specific examples. (6.10) K/U C
15. A synthesis reaction is defined as "a reaction in which two reactants combine to make a larger or more complex product." (6.5) C
    (a) Write a definition of "synthesis reaction" in your own words.
    (b) Draw a diagram representing a typical synthesis reaction.
16. Describe one way you could protect a bicycle from corrosion. (6.10) T/I
17. (a) How does the name "hydrocarbon" give clues about the composition of hydrocarbons?
    (b) Give the names and formulas of three hydrocarbons. (6.9) A
18. (a) Briefly describe five chemical reactions that you may encounter in a typical day.
    (b) Choose one of the reactions and identify the names of the products and reactants. (6.1–6.10) A
19. The Eiffel Tower is an iron monument located in Paris, over 150 km from the ocean. Predict how the rate of corrosion would change if the Eiffel Tower were located in a seaside town. (6.10) T/I
20. Your family is planning to grill hamburgers in the backyard, but it starts to rain. Someone suggests grilling in the garage instead. Explain why it would be risky to grill in the garage. (6.9) C
21. Give an example of how one chemical reaction can be beneficial in some applications and harmful in others. (6.1–6.10) A

CHAPTER 7

Acids and Bases

KEY QUESTION: How do acids and bases affect our lives and the environment?

Sudbury has experienced the effects of acids on its environment, and is now recovering from those effects.

UNIT C
Chemical Reactions

CHAPTER 5
Chemicals and Their Properties

CHAPTER 6
Chemicals and Their Reactions

CHAPTER 7
Acids and Bases

KEY CONCEPTS

Acids are aqueous solutions that have characteristic properties.

Bases are aqueous solutions that have characteristic properties.

Acids and bases have a significant impact—both good and bad—on society and the environment.

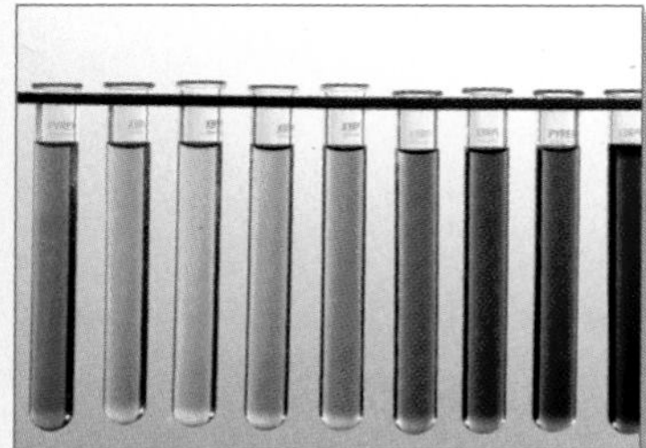
The acidity of solutions is measured using the pH scale.

Acids and bases react together in neutralization reactions.

Industrial and vehicle emissions that cause acid precipitation can be reduced by technology.

ENGAGE IN SCIENCE

THE GREENING OF SUDBURY

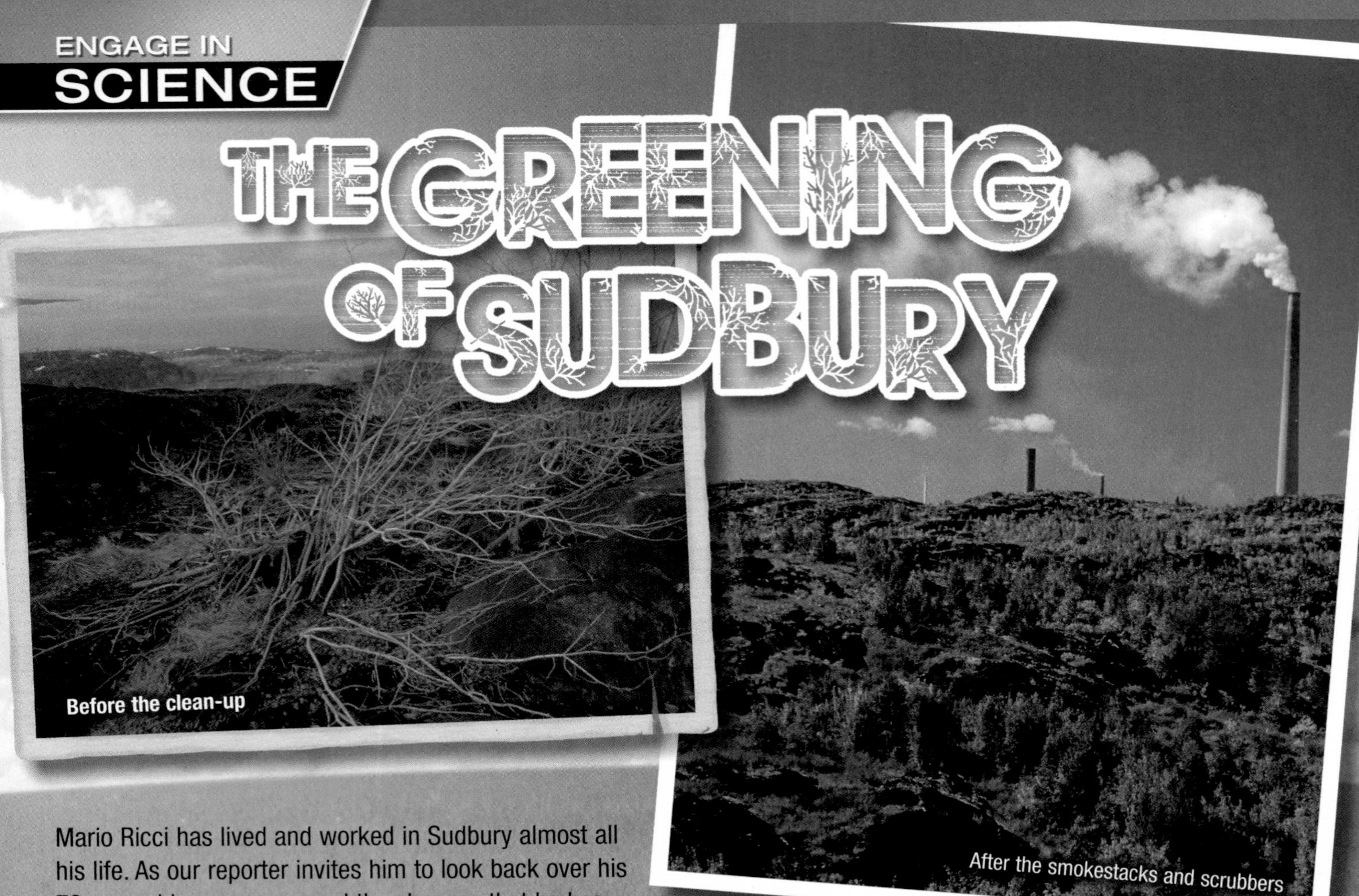
Before the clean-up

After the smokestacks and scrubbers

Mario Ricci has lived and worked in Sudbury almost all his life. As our reporter invites him to look back over his 70 years, his answers reveal the changes that he has witnessed in the community.

Reporter: How old were you when you left school and started work?

MR: I left after Grade 8—most of us did—and went to work at the nickel mine.

Reporter: What was it like in Sudbury at that time?

MR: It was much smaller, but growing fast. The nickel mines and smelter brought people in from all over. My parents came here for the work when I was about six. I remember it was pretty dirty. Most of the trees were dead. There wasn't much grass. My parents were disappointed that they couldn't grow vegetables like they could back home in Italy.

Reporter: What was causing that?

MR: The stuff that came out of the factory chimneys. When you smelt nickel, you have to heat it up and add some chemicals. Some of the chemicals come out of the chimneys with the smoke. Then they fall on the ground and make it so that nothing can grow. There was also a lot of copper and nickel in the soil. Some of the streams were bright orange, with the metal in them. And a lot of people had bad coughs and bronchitis from the pollution.

Reporter: When did things start to change?

MR: When they put up the Superstack in 1972. It's the second highest smokestack in the world, you know! And they installed scrubbers in the chimneys, too, to take some of the chemicals out of the smoke.

Reporter: How did that make a difference?

MR: It seems that less poisonous stuff fell on the ground. It took a while, but after a few years things started to grow again. People also planted millions of trees, but some people in other towns complained that the smoke just blew over to their area.

Reporter: What do you think of the Sudbury area now?

MR: Parts of it are beautiful now. There are trees and grass in the summer. It's much nicer to walk around and sit outside.

How have acids affected the environment where you live? Has it been a positive or a negative change? Where did the chemicals come from? Was their source anthropogenic or natural?

WHAT DO YOU THINK?

Many of the ideas you will explore in this chapter are ideas that you have already encountered. You may have encountered these ideas in school, at home, or in the world around you. Not all of the following statements are true. Consider each statement and decide whether you agree or disagree with it.

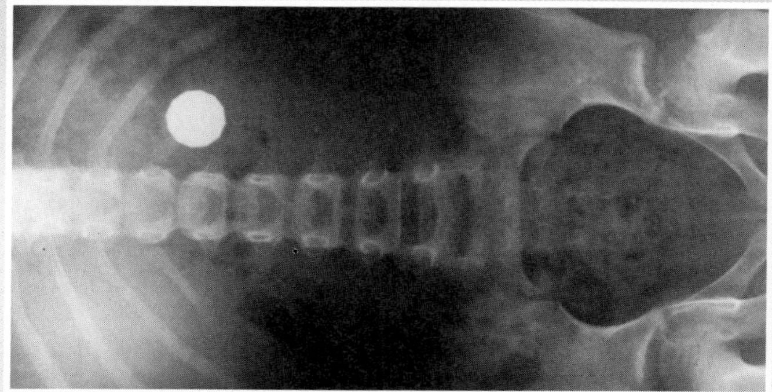

1 Stomach acid can dissolve metals like the coin in this child's stomach.
Agree/disagree?

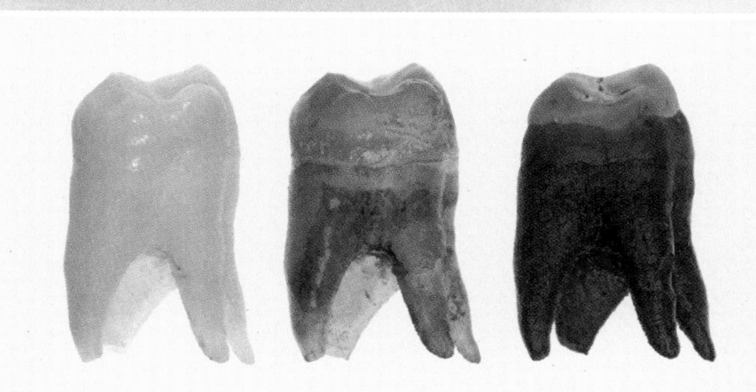

2 Even diet soft drinks can cause tooth decay.
Agree/disagree?

3 All acids are dangerous.
Agree/disagree?

4 We all contribute to the production of acid precipitation.
Agree/disagree?

5 Medications for heartburn work by cooling the stomach.
Agree/disagree?

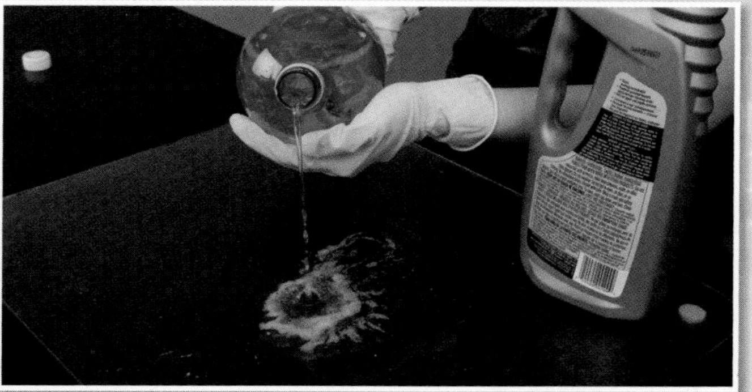

6 Soft drinks can be used to make spilled drain cleaner safer.
Agree/disagree?

FOCUS ON WRITING

Writing a Science Report

When you write a science report, you use a standard format with organizational headings to explain the purpose, procedure, and findings of your investigation. Use the strategies listed next to the report to improve your report-writing skills.

WRITING TIP

As you work through the chapter, look for tips like this. They will help you develop literacy strategies.

Properties of Ionic and Molecular Compounds

State your purpose, question, and hypothesis concisely.

Question

Are the following substances molecular or ionic compounds: lauric acid, $C_{12}H_{24}O_2$; sodium hydrogen carbonate, $NaHCO_3$; glucose, $C_6H_{12}O_6$; potassium chloride, KCl?

Hypothesis/Prediction

Lauric acid and glucose are molecular compounds because they only have non-metallic elements in them. Potassium chloride is an ionic compound because it contains a metal and a non-metal. Sodium hydrogen carbonate is also ionic because it contains the hydrogen carbonate ion.

Briefly describe the investigation.

Experimental Design

Each substance will be mixed with water to see if it dissolves. Each solution will be tested for conductivity. The melting point of each substance will be researched.

Outline what is required to conduct the investigation.

Equipment and Materials

- eye protection
- lab apron
- 4 small test tubes & stoppers
- test tube rack
- conductivity tester
- lauric acid, $C_{12}H_{24}O_2$
- sodium hydrogen carbonate, $NaHCO_3$
- well plate
- glucose, $C_6H_{12}O_6$
- potassium chloride, KCl
- water

Use a numbered list of directions to describe each step of the procedure.

Write in the third person using an objective tone.

Procedure

1. Eye protection and an apron were obtained.
2. A small amount of each solid was placed in its own test tube.
3. Each test tube will filled about half full with water, stoppered, and inverted to mix.
4. Observations were recorded for how well the solids dissolved.
5. A small amount of each liquid was poured into its own well in the well plate.
6. The conductivity tester was dipped into each liquid.
7. Observations were recorded for whether each liquid conducts electricity.
8. The melting point of each solid was researched in a reference book.

Present all observations, whether they support your prediction or not.

Observations

Substance	Dissolves in water?	Conductive	Melting point (°C)
lauric acid	no	no	45

Evaluate the extent to which your evidence supports your hypothesis.

Revise your drafts to improve organization and completeness.

Analyze and Evaluate

(a) Sodium hydrogen carbonate and potassium chloride are the only ionic compounds because they conducted electricity when dissolved in water.

Classifying Acids and Bases

Acids and bases are two important classes of chemicals that play an important role in many consumer products and environmental problems. In this activity, you will plan and conduct five tests on two typical acids (hydrochloric acid and acetic acid) and two typical bases (sodium hydroxide and calcium hydroxide). This will help you to recognize some typical properties of acids and bases.

SKILLS MENU

- Questioning
- Hypothesizing
- Predicting
- Planning
- Controlling Variables
- Performing
- Observing
- Analyzing
- Evaluating
- Communicating

Purpose

To observe the typical properties of acids and bases

Equipment and Materials

- eye protection
- lab apron
- well plate
- tweezers
- conductivity tester
- magnesium ribbon (0.5 cm)
- toothpick
- sodium hydrogen carbonate (baking soda), $NaHCO_3(s)$
- red litmus paper
- blue litmus paper
- dropper bottles of
  - distilled water
  - bromothymol blue indicator
  - dilute hydrochloric acid, $HCl(aq)$
  - dilute acetic acid, $HC_2H_3O_2(aq)$
  - dilute sodium hydroxide solution, $NaOH(aq)$
  - dilute calcium hydroxide solution, $Ca(OH)_2$
  - unknown acids and bases

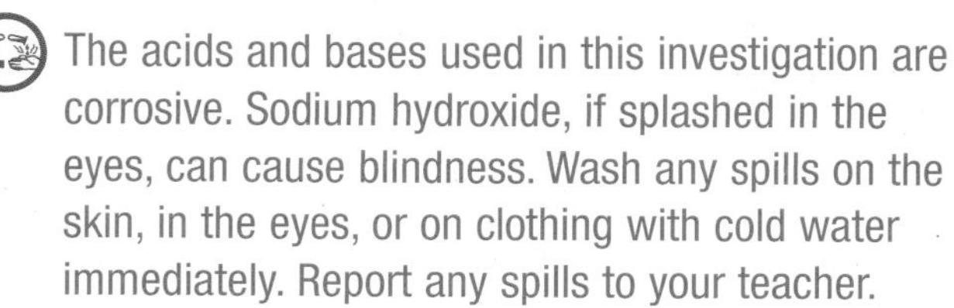

The acids and bases used in this investigation are corrosive. Sodium hydroxide, if splashed in the eyes, can cause blindness. Wash any spills on the skin, in the eyes, or on clothing with cold water immediately. Report any spills to your teacher.

Procedure

Part A: Known Acids and Bases

1. Write a detailed, step-by-step Procedure for each of the following tests:
   (i) electrical conductivity
   (ii) reaction with magnesium
   (iii) reaction with sodium hydrogen carbonate
   (iv) colour with bromothymol blue indicator
   (v) effect on red and blue litmus paper

   You will test hydrochloric acid, acetic acid, sodium hydroxide, and calcium hydroxide. Use the Equipment and Materials list to help you. Include safety precautions as necessary. All tests should be done in the well plate.
2. Prepare a table for your observations.
3. With your teacher's approval, proceed with the tests. Record your observations as you go along.

Part B: Unknown Acids and Bases

4. Your teacher will provide you with several unknown solutions.
5. Select two chemical tests that you think are best for distinguishing between acids and bases. Plan and use these tests to classify each of the unknown solutions as an acid or a base. Record your observations.
6. Dispose of all substances as directed.

Analyze and Evaluate

(a) Summarize the properties of acids and bases. K/U

(b) Classify your unknowns (in Part B) as acids or bases. Explain your classification. T/I

7.2 Properties, Names, and Formulas

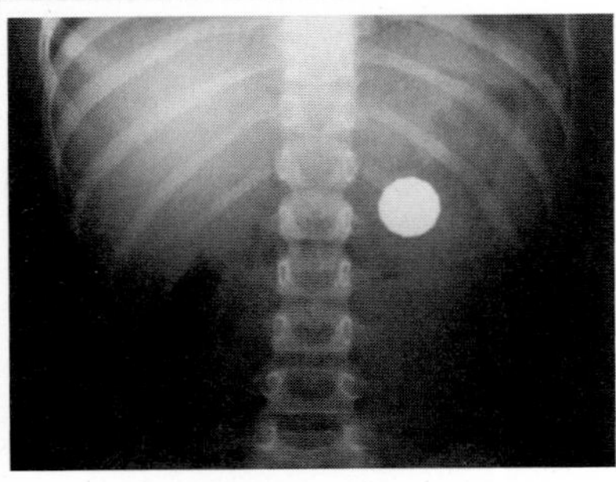

Figure 1 The bright white disc in the centre of this X-ray image is a penny in a young child's stomach.

acid an aqueous solution that conducts electricity, tastes sour, turns blue litmus red, and neutralizes bases

Young children like to explore objects by putting things in their mouths. Occasionally, to the horror of their parents, the interesting object gets swallowed! Coins are the most common foreign objects swallowed by young children (Figure 1). Once in the stomach, a coin is bathed in a corrosive mixture that includes hydrochloric acid, HCl(aq), and other digestive juices.

The concentration of hydrochloric acid in your stomach is about the same as the hydrochloric acid used in Activity 7.1. What do you think happens to the coin?

Properties of Acids

Chemists often classify substances by their properties. This is also the case with acids. **Acids** are substances that react with metals and carbonates, conduct electricity, turn blue litmus red, and neutralize bases. Acids also taste sour, but you should never taste chemicals in the chemistry lab.

Acids are very useful in the food industry: they act as a preservative. Harmful microorganisms cannot survive in acid. Acids such as vinegar and lemon juice act as preservatives. Pickles, barbecue sauce, and ketchup all have long shelf-lives because they contain a lot of vinegar.

Food scientists are very concerned with improving the shelf-life of foods. To find out more about this career,

WRITING TIP

Writing a Science Report
When you are writing a science report, ask yourself what question you will answer. For example, if the purpose of the report is to identify the gas produced by the reaction of an acid with a metal, you might ask "What gas is produced when hydrochloric acid reacts with magnesium?"

Reaction with Metals

In Activity 7.1, you discovered that acids react with magnesium to produce bubbles of gas. If you were to test the gas produced with a burning splint, you would discover that it is hydrogen. Acids typically react with metals to produce hydrogen. For example, the word and chemical equations for the reaction of hydrochloric acid, HCl(aq), with zinc are

hydrochloric acid + zinc → hydrogen gas + zinc chloride

$$2\ HCl(aq) + Zn(s) \rightarrow H_2(g) + ZnCl_2(aq)$$

Reaction with Carbonates

Acids have another typical reaction: they react with carbonate compounds to produce bubbles of carbon dioxide gas. You have witnessed this reaction if you have ever used vinegar and baking soda to unblock a sink (Figure 2). As you know, vinegar's chemical name is acetic acid and baking soda is sodium hydrogen carbonate. The equations for this reaction are

acetic acid + sodium hydrogen carbonate → carbon dioxide + water + sodium acetate

$$HC_2H_3O_2(aq) + NaHCO_3(aq) \rightarrow CO_2(g) + H_2O(l) + NaC_2H_3O_2(aq)$$

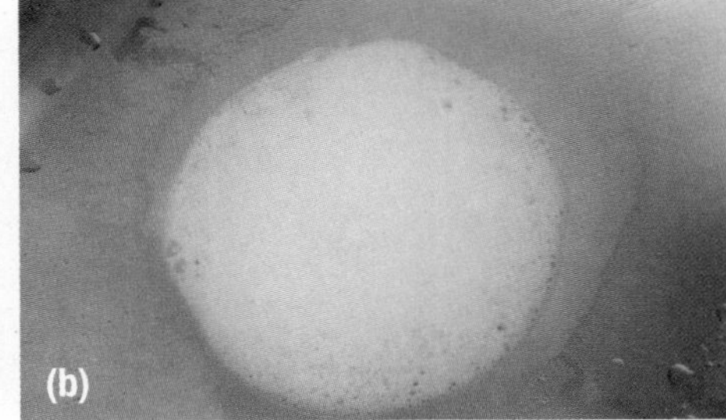

Figure 2 (a) Acids react with compounds that contain carbonate ions. (b) The bubbles are produced by the reaction of vinegar (acetic acid) and baking soda (sodium hydrogen carbonate).

Electrical Conductivity

Many acids are good conductors of electricity. A solution can conduct electricity only if it contains ions. Since acids are molecular compounds, they do not contain ions. However, collisions with water molecules break acid molecules apart to form cations (hydrogen ions) and anions. Hydrochloric acid, for example, forms hydrogen ions and chloride ions:

$$HCl(aq) \rightarrow H^{+}(aq) + Cl^{-}(aq)$$

Chemical Formulas of Acids

All acids release at least one hydrogen ion when they dissolve in water. Hydrofluoric acid, $HF(aq)$, for example, forms one hydrogen ion and one fluoride ion:

$$HF(aq) \rightarrow H^{+}(aq) + F^{-}(aq)$$

Because all acids form hydrogen ions when dissolved in water, chemists deduce that these hydrogen ions give acids their properties.

You will also notice that the chemical formula of an acid begins with an H and is usually followed by "(aq)." This is because an acid shows its properties only when it is dissolved in water. For example, hydrogen chloride is a gas at room temperature. However, if you accidentally inhaled hydrogen chloride, it would dissolve in the water in your throat and lungs to form highly corrosive hydrochloric acid. This is one reason why caution is important when you smell a substance.

Names of Acids

There are two common groups of acids. One group is called binary acids because these compounds contain only two elements (Table 1).

Table 1 Common Binary Acids

Acid name	Chemical formula	Use
hydrofluoric acid	$HF(aq)$	etching glass
hydrochloric acid	$HCl(aq)$	cleaning concrete
hydrobromic acid	$HBr(aq)$	to make cleaning compounds
hydrosulfuric acid	$H_2S(aq)$	purifying metals

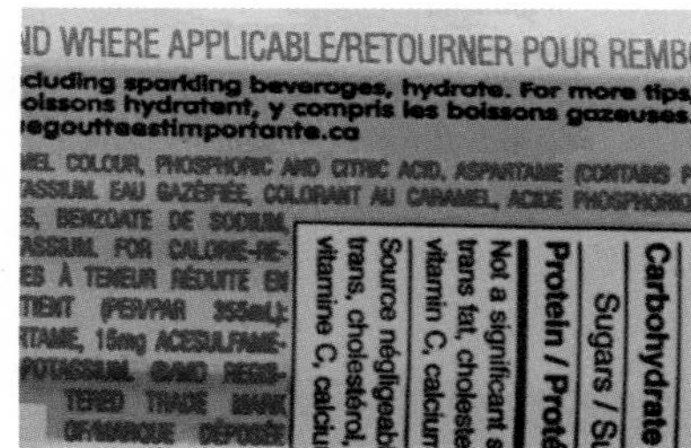

Figure 3 Phosphoric acid helps give colas their tart taste. Cola would be very sour if it did not also contain so much sweetener.

Most acids that you encounter are in a group called oxyacids. They are related to polyatomic ions. Their chemical formulas differ only by one or more hydrogen ions (Table 2). Phosphoric acid, $H_3PO_4(aq)$, for example, is the oxyacid related to the phosphate ion, $PO_4^{3-}(aq)$ (Figure 3).

Table 2 Common Oxyacids and Their Related Polyatomic Ions

Acid	Chemical formula	Related polyatomic ion	Polyatomic ion name
acetic acid	$HC_2H_3O_2(aq)$	$C_2H_3O_2^{-}(aq)$	acetate
nitric acid	$HNO_3(aq)$	$NO_3^{-}(aq)$	nitrate
carbonic acid	$H_2CO_3(aq)$	$CO_3^{2-}(aq)$	carbonate
sulfuric acid	$H_2SO_4(aq)$	$SO_4^{2-}(aq)$	sulfate
phosphoric acid	$H_3PO_4(aq)$	$PO_4^{3-}(aq)$	phosphate

LEARNING TIP

Formulas of Oxyacids

The number of hydrogen atoms in the formula of an oxyacid is equal to the value of the charge of its related polyatomic ion. For example, the phosphate ion, PO_4^{3-}, has an ionic charge of −3. Therefore, the chemical formula of phosphoric acid is $H_3PO_4(aq)$.

Properties of Bases

base an aqueous solution that conducts electricity and turns red litmus blue

Just as acids are classified by their properties, bases also have a common set of properties. Like acids, **bases** conduct electricity and change the colour of acid–base indicators; unlike acids, bases feel slippery and taste bitter. (Again, you should *never* taste chemicals in the chemistry lab.)

DID YOU **KNOW?**

Alkaline Ashes
The word "alkaline" is sometimes used to describe a basic solution. This word comes from *al-qaliy*, an Arabic word meaning "the ashes." The ashes left over from burning plants were an early source of basic compounds. Early Canadian pioneers used ashes from burning wood as their source of basic compounds to make soap.

Electrical Conductivity

Bases are electrolytes, meaning that their solutions are good conductors of electricity. For example, sodium hydroxide, NaOH, is an electrolyte because it completely separates into its ions as it dissolves in water:

$$NaOH(s) \rightarrow Na^{+}(aq) + OH^{-}(aq)$$

Many common bases are ionic compounds; they are made up of ions (unlike acids, which are molecular compounds made up of molecules). As these compounds dissolve in water, their ions are released.

Barium hydroxide, $Ba(OH)_2$, is another base. As a pure compound, it is a solid, but when placed in water, it releases one barium ion and *two* hydroxide ions:

$$Ba(OH)_2(s) \rightarrow Ba^{2+}(aq) + 2\ OH^{-}(aq)$$

Colour with Acid–Base Indicators

acid–base indicator a substance that changes colour depending on whether it is in an acid or a base

Have you ever noticed that clear tea changes colour when lemon juice is added to it? This is because the acidity of the lemon juice slightly changes the chemicals that give tea its distinctive colour. Several natural chemicals and many synthetic (human-made) chemicals change colour when they are placed in acidic or basic solutions (Figure 4). A substance that changes its colour depending on the acidity or basicity of the solution is known as an **acid–base indicator** (Table 3).

To see videos of acid–base indicators in action,

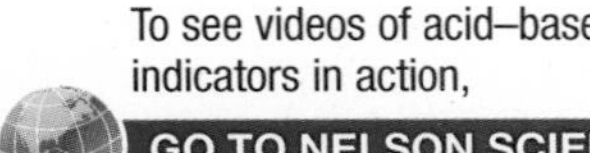

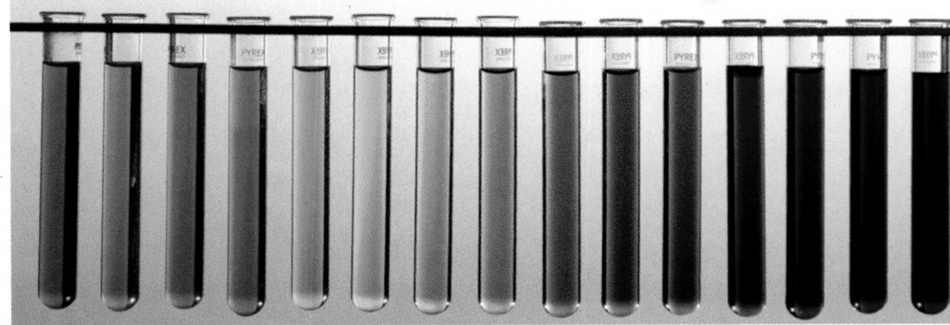

Figure 4 An acid–base indicator has a different colour depending on the acidity or basicity of the solution.

Table 3 Colours of Common Synthetic Acid–Base Indicators

Indicator	Colour in acid	Colour in base
bromothymol blue	yellow	blue
phenolphthalein	colourless	pink
phenol red	yellow	red/pink
litmus	red	blue
methyl orange	red	orange/yellow

LEARNING **TIP**

Indicators to Remember
You do not need to memorize the colour changes of most indicators, but you will find it useful to remember the colours of phenolphthalein and litmus.

Names and Chemical Formulas of Bases

Table 4 lists some common bases. Many bases are ionic compounds containing hydroxide or carbonate ions. Remember that an ionic compound contains a positive ion (usually a metal) and a non-metal or polyatomic anion.

Table 4 Common Bases and Their Uses

Base	Chemical formula	Uses
sodium hydroxide	$NaOH(aq)$	making paper
calcium hydroxide	$Ca(OH)_2(aq)$	decreasing the acidity of lakes and soil
ammonium hydroxide	$NH_4OH(aq)$	window cleaners
magnesium hydroxide	$Mg(OH)_2(aq)$	antacids
aluminum hydroxide	$Al(OH)_3(aq)$	heartburn medications
sodium hydrogen carbonate (baking soda)	$NaHCO_3(aq)$	making baked goods rise an abrasive cleaner

UNIT TASK Bookmark

You can apply what you learned about the properties of acids and bases to the Unit Task described on page 300.

IN SUMMARY

- Acids are molecular compounds. In solutions, acids react with metals, conduct electricity, and change the colours of acid–base indicators.
- Acids may be binary acids ($HCl(aq)$, $HBr(aq)$) or oxyacids ($HNO_3(aq)$, $H_2SO_4(aq)$).
- Bases are ionic compounds. Many are hydroxides. In aqueous solutions, they conduct electricity and change the colours of acid–base indicators.
- Acid–base indicators show whether a solution is acidic or basic.

CHECK YOUR LEARNING

1. Why are solutions of acids and bases often good conductors of electricity? K/U
2. Classify the following substances as being acidic or basic. K/U
   (a) $KOH(aq)$
   (b) $HNO_3(aq)$
   (c) barium hydroxide solution
   (d) $KHCO_3(aq)$
   (e) sodium hydrogen carbonate solution
3. Write the name or chemical formula for each compound in question 2. K/U
4. What part of an acid's chemical formula is responsible for its acidic properties? K/U
5. What polyatomic ion most commonly appears in bases? K/U
6. People who suffer from bulimia sometimes self-induce vomiting to prevent weight gain. Consider the properties of the chemicals in your stomach. Why do the teeth of people with bulimia often appear worn or eroded? A
7. The fluid inside an alkaline battery can be corrosive. K/U T/I
   (a) Describe a chemical test that you could perform to determine if the substance is an acid or a base.
   (b) What safety precautions must you take when conducting this test?
8. Teeth can be damaged by acid erosion (Figure 5). What foods could contribute to this problem? How can it be avoided? A

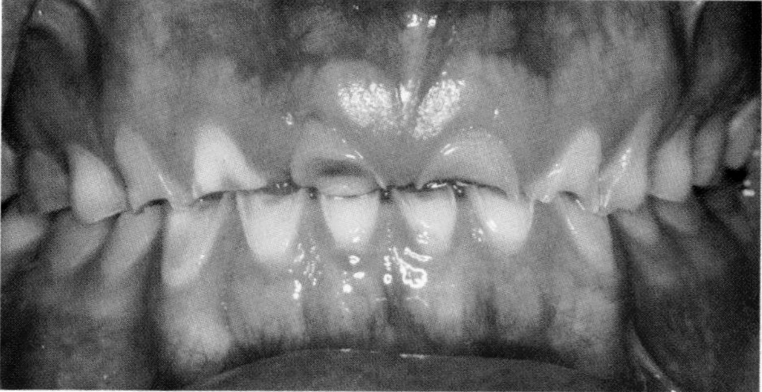

Figure 5

7.3 The pH Scale

Propionibacterium acnes is the species of bacteria that causes acne (Figure 1). All adults, regardless of whether or not they have acne, have colonies of *P. acnes* on their skin. But why do only some people get acne? One factor that helps prevent acne is the natural acidity of your skin. While this acidity is not harmful to you, it is toxic to *P. acnes* colonies and helps limit their growth.

DID YOU KNOW?

Too Clean?
The acids on your skin are produced by "good" bacteria. Frequent washing may kill some of these bacteria, making your skin less acidic and more inviting to disease-causing bacteria. Washing with soap is important for proper hygiene, but it is also important to allow your skin time to recover to its normal acidity between washings.

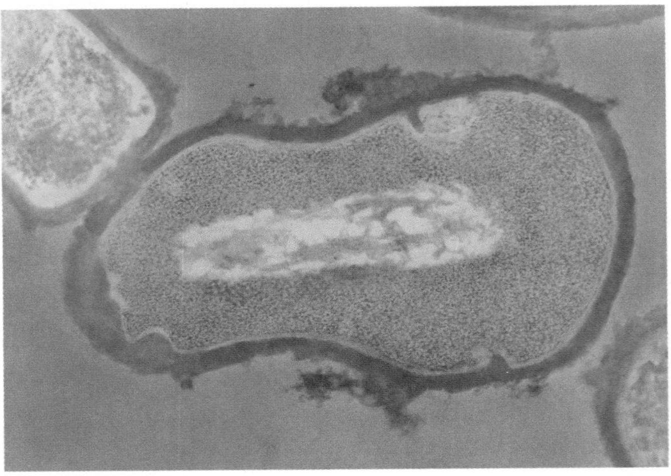

Figure 1 *Propionibacterium acnes* bacteria on the skin can cause acne.

Chemists measure the acidity of a solution using a scale called "pH." You may have heard the term "pH" in the media—perhaps in a skin or hair product commercial. But what exactly is pH? **pH** is a measure of how acidic or basic a solution is. The **pH scale** is a numerical scale of all the possible numerical values of pH from 0 to 14 (Figure 2). The pH of a solution can range from 0 (for example, car battery acid) to 14 (some industrial-strength drain cleaners). A solution with a pH of 7 is considered **neutral**: neither acidic nor basic. A solution whose pH is less than 7 is acidic; a solution whose pH is greater than 7 is basic. The surface of healthy skin is slightly acidic, with a pH of 5.5: enough to keep acne bacteria under control.

pH a measure of how acidic or basic a solution is

pH scale a numerical scale ranging from 0 to 14 that is used to compare the acidity of solutions

neutral neither acidic or basic; with a pH of 7

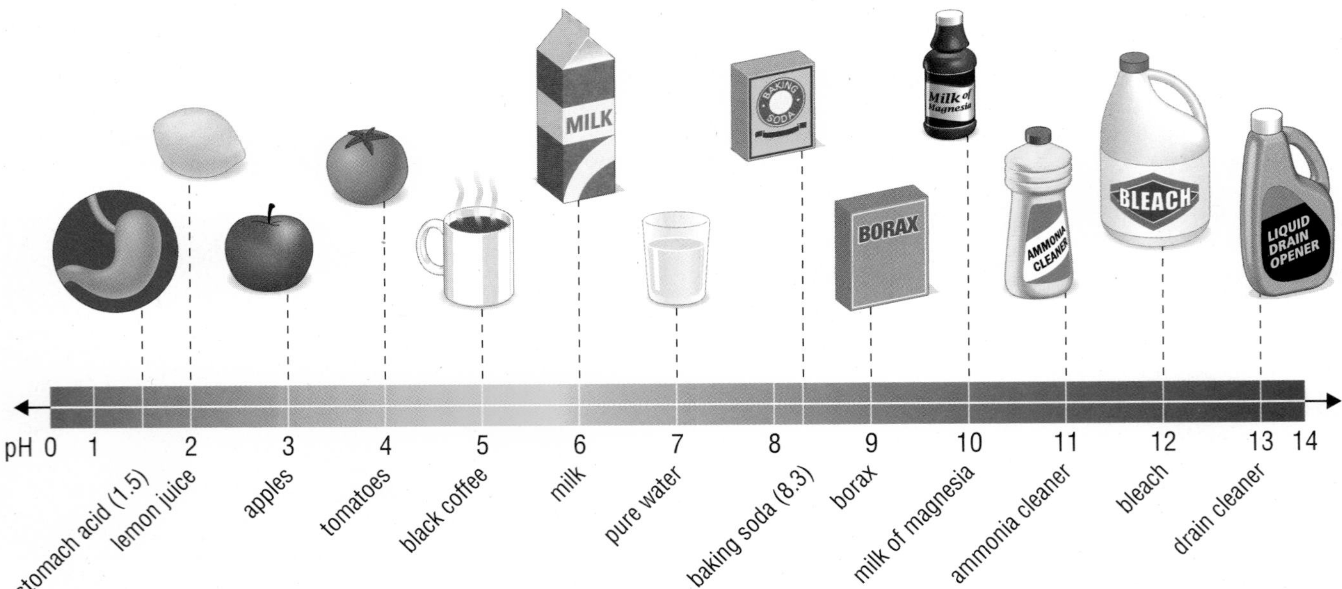

Figure 2 The pH scale is used to compare the hydrogen ion concentration of a broad range of substances. Consumer products that are on opposite ends of the scale are corrosive and have Household Hazardous Product Symbols indicating this on their labels.

You know that acids form hydrogen ions in a solution. The concentration of these hydrogen ions is what determines the solution's pH. (Concentration is a measure of the quantity of a substance dissolved in a given volume of a solution.) The greater the concentration of hydrogen ions in a solution, the stronger the acidic properties. Solutions with the highest hydrogen ion concentration are positioned near the "zero" end of the pH scale. In other words, highly acidic solutions have a low pH. Highly acidic solutions are very corrosive and reactive, and must be handled with extreme care.

The pH scale also indicates the concentration of hydroxide ions. Solutions with a pH greater than 7 have a higher concentration of hydroxide ions than of hydrogen ions. Consequently, solutions with distinctly basic properties, like some drain cleaners, are positioned near the "14" end of the pH scale. In other words, solutions that are highly basic have a high pH. Highly basic solutions are also corrosive and reactive, and require very careful handling. We can use the word "basicity" to describe a property of a base, just as we use "acidity" to describe a property of an acid.

Solutions with pH 7 are said to be neutral because their hydrogen and hydroxide concentrations are equal and balance each other out.

WRITING TIP

Using the Right Tone

Use a serious and formal tone when writing a science report. The report should focus on the details of the investigation, not on details about the writer. Avoid first-person and second-person pronouns such as "I," "me," "we," and "you," and expressions of personal feelings, reactions, or opinions.

TRY THIS VISUALIZE THE pH SCALE

SKILLS: Observing, Communicating

SKILLS HANDBOOK 3.B.

In this activity, you will use colour to simulate the changes in pH that occur as an acid is diluted.

Equipment and Materials: well plate; micropipette; toothpick; dropper bottle of food colouring; water

1. Add one drop of food colouring into the first two wells of the well plate. (The colour of this liquid represents hydrochloric acid with pH 0.)
2. Using the micropipette, add 9 drops of tap water to the second well. The total volume of liquid in the second well is now 10 drops.
3. Use a toothpick to mix the contents of the second well. The concentration of food colouring in this mixture is one-tenth the concentration of the original. (The colour of this liquid represents hydrochloric acid with pH 1.)
4. Use a new, clean micropipette to transfer one drop of the mixture in well #2 to well #3.
5. Add 9 drops of tap water to well #3 and stir to mix. (The colour of this liquid represents a hydrochloric acid solution with a pH of 2.)
6. Repeat steps 4 and 5 three more times, transferring a drop of the mixture from well #3 to #4, and so on.

A. What is the pH corresponding to the solutions in wells #4, #5, and #6? T/I

B. As you dilute an acid solution, what happens to its concentration? T/I

C. As you dilute an acid solution, what happens to its pH? T/I

pH in Our Lives

We may not be aware of it, but the pH of solutions can have a huge effect on us: in our bodies, in our homes, and in the wider world. Once we understand the effects of acids and bases, we can use them to solve all kinds of problems.

pH and Soil

The pH of soil can vary considerably, depending on the type of rock in the area, the kinds of plants growing there, and the materials that people have added—intentionally or not. Different plants grow best in different conditions of soil acidity. Some, such as legumes (beans and clover, for example), grow best in soil that is slightly basic (pH 7 to 10). Corn thrives in mildly acidic soil (pH 5 to 6) (Figure 3). Potatoes, however, prefer acidic soil that is below pH 5.

Figure 3 Farmers monitor and adjust the pH of the soil to get the best possible yield from their crop plants.

DID YOU **KNOW?**

Acidic Oceans

Carbon dioxide reacts with falling rain to form very dilute carbonic acid. In recent years, the concentration of carbon dioxide in the atmosphere has increased. This means that rain is more acidic. Gradually, the oceans are also becoming more acidic. This is becoming a serious ecological problem.

The pH of soil can sometimes be altered to improve growing conditions. For example, the addition of compost (decaying organic matter) or aluminum sulfate makes soils more acidic. Soil pH can be raised by mixing in calcium oxide (lime).

METAL TOXINS IN SOIL

There are thousands of sites across Canada where once-clean soil is now contaminated with toxic chemicals. Many of these contaminants are metals. Some, like cadmium, come from batteries buried in the ground. Others come from industrial processes. For example, the soil near a factory that once processed lead is likely to have abnormally high levels of lead. One way of treating the soil involves a technology called **acid leaching**. In this process, soil is first removed from the site. Then it is acidified to dissolve its metal contaminants and washed. The cleaned soil can be returned to its original site. However, this process is very expensive and disruptive to local ecosystems.

acid leaching the process of removing heavy metals from contaminated soils by adding an acid solution to the soil and catching the solution that drains through

Fortunately, there is a better way! Scientists have discovered that certain plants are natural "sponges" for metal toxins. As they grow, these plants absorb metals from the soil (Figure 4). Later, the plants are harvested and burned. The metals are then recycled from the ashes. The process of using plants to remove toxins from an area is called "phytoremediation."

Figure 4 Scientists at the University of Guelph have found that geraniums are useful for phytoremediation.

Figure 5 This household product is a very corrosive substance. It should be used with great care. The label cautions that this product must not be mixed with any other cleaning products.

pH and Consumer Products

Many of your shampoo and skin care products were probably designed to have a pH that is close to neutral. Other products, though, are far from neutral. Many cleaning products contain high concentrations of hydroxide ions. These cleaners are very corrosive. This makes them effective at cleaning but also potentially quite damaging to skin and eyes. To warn consumers of the danger, products with a very high or very low pH are labelled with Hazardous Household Products Symbols (Figure 5).

To learn more about Hazardous Household Product Symbols,

GO TO NELSON SCIENCE

pH and Swimming Pools

Maintaining a backyard swimming pool requires keeping a close watch on the pH of the water. Ideally, the pH of the pool water should be maintained within a narrow range of 7.2 to 7.8 (Figure 6). If the pH falls too much below 7, pool water irritates the eyes. If the pH increases above 8, the pool water becomes cloudy and irritating to the eyes, and the chlorine compounds used to disinfect the pool begin to lose their effectiveness. Many pool owners use a pH test kit to monitor the pH of their pools on a regular basis. Hydrochloric acid (also called "muriatic acid") is often added to pool water to reduce the pH if it is too high. Conversely, products containing sodium carbonate can be added to raise the pH when it is too low.

Figure 6 The pH of pool water must be monitored and adjusted regularly.

IN SUMMARY

- pH describes the acidity or basicity of a solution.
- Solutions with a pH of 7 are neutral. The lower the pH, the more acidic the solution. The higher the pH, the more basic the solution.
- Solutions that are highly basic or highly acidic are corrosive and reactive. They must be handled with caution.
- Living things are sensitive to small pH changes in their environment.

CHECK YOUR LEARNING

1. (a) Describe an idea related to pH that you were aware of before you started reading this section.
   (b) When and where did you first learn about this idea?
   (c) How has your understanding of this idea changed? K/U
2. Refer to Figure 2 to write the pH of each of the following substances: K/U
   (a) lemon juice
   (b) milk of magnesia
   (c) borax
3. Classify the following solutions as highly acidic, slightly acidic, neutral, slightly basic, or highly basic: K/U
   (a) a solution with a pH of 13
   (b) a solution with a pH of 6
   (c) a solution with a pH of 1
   (d) the moisture on your skin
   (e) water in a swimming pool (ideally)
4. (a) Arrange the following list of substances in order of increasing acidity: baking soda; tomatoes; stomach acid; bleach; black coffee; pure water.
   (b) Describe the pH of the substances in their new order. K/U
5. How can we make use of acids or bases to
   (a) remove heavy metals from soils?
   (b) improve crop yields?
   (c) prevent the growth of micro-organisms in swimming pools? A
6. Why should household products be stored in their original containers? T/I A
7. Look around your home for consumer products with a Household Hazardous Product Symbol warning that the contents are corrosive (Figure 5). K/U A
   (a) In which room(s) were most of these products found?
   (b) Using only the information given on the product label, try to classify these substances as acidic or basic. Caution: Do not open any of these products.
8. Some skin creams claim that they are "pH balanced" and yet do not have a pH of 7. What do the manufacturers mean when they say "pH balanced"? A
9. What are the advantages of phytoremediation over acid leaching in treating contaminated soils? A
10. Experiments show that teeth begin to lose minerals at pH 5.5 or less. How could you adjust your diet to minimize mineral loss? A

The pH of Household Substances

SKILLS MENU

- Questioning
- Hypothesizing
- Predicting
- Planning
- Controlling Variables
- Performing
- Observing
- Analyzing
- Evaluating
- Communicating

In Activity 7.1, Classifying Acids and Bases, you used the colour changes of the bromothymol blue indicator to distinguish between acids and bases. Bromothymol blue is an example of an acid–base indicator. These substances change colour when the pH of a solution changes. Some, such as bromothymol blue, are synthetic. Others, such as the juice from red cabbage, are natural.

Some acid–base indicators change colour at a specific pH. For example, bromothymol blue is yellow in acid, and changes to green at pH 7. Phenolphthalein changes from colourless (in acid) to pink at pH 9. Some, like universal indicator and red cabbage extract, go through a series of colour changes over a wide range of pH values. This is because they are made up of many substances, each of which changes colour at a different pH.

In this activity, you will compare the effectiveness of universal indicator and red cabbage juice at determining the pH of a variety of household substances. To do this, you will test solutions of pH 1, 3, 5, 7, 9, 11, and 13 with both universal and red cabbage indicators, observing the resulting colours. You will then use these colours as standards to determine the pH of solutions of common household products.

WRITING TIP

Recording Observations
Prepare a table with clearly worded labels and accurate units of measurement or scientific notation at the top of the columns. If appropriate, start with the most important observations and end with the least important.

Purpose

To compare the effectiveness of a natural indicator (red cabbage extract) with that of universal indicator at determining the pH of household substances.

Equipment and Materials

- eye protection
- lab apron
- two 24-well plates
- droppers
- pH meter (optional)
- swimming pool pH testing kit (optional)
- prepared red cabbage extract
- dropper bottles of
  - prepared solutions of pH 1, 3, 5, 7, 9, 11, 13
  - universal indicator
- 100 mL beakers of household products (each with its own dropper), labelled I, II, III, etc.

Some of the acids and bases used in this investigation are corrosive. Any spills on the skin, in the eyes, or on clothing should be washed immediately with cold water. Report any spills to your teacher.

Procedure

SKILLS HANDBOOK 3.B.6., 6.B.

1. Read Parts A and B. Prepare a table in which to record your observations.
2. Put on your eye protection and lab apron.

Part A: Colour Changes of Universal and Red Cabbage Indicators

3. Label one set of seven wells "red cabbage" 1, 3, 5, 7, 9, 11, 13. Label another set of seven wells "universal" 1, 3, 5, 7, 9, 11, 13. Each pH level should have its own well for each indicator.
4. Add two drops of red cabbage extract to each of the "red cabbage" wells. Use a clean dropper for each of the pH solutions being tested.
5. Add two drops of universal indicator to each of the "universal" set of wells (Figure 1, next page).
6. Add two drops of the pH 1 solution to the first well of both indicators. Record your observations.

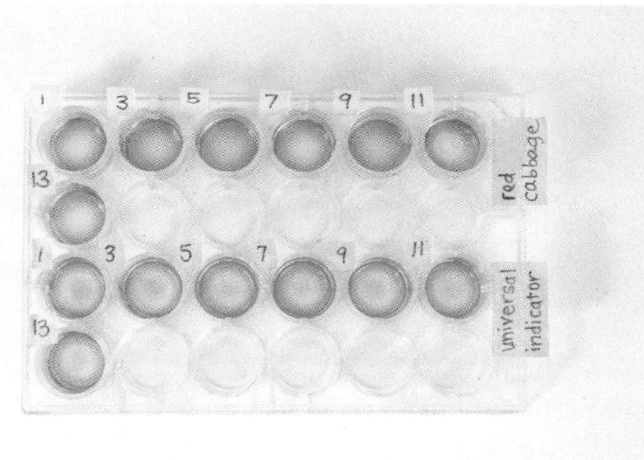

Figure 1 First the indicators are added to the wells.

7. Add two drops of the pH 3 solution to the second well of both indicators. Again, record your observations.
8. Continue with solutions of pH 5, 7, 9, 11, and 13. Note the colour changes and the pH values at which they occur.
9. Save the contents of the well plate for Part B.

Part B: Testing the pH of Household Products

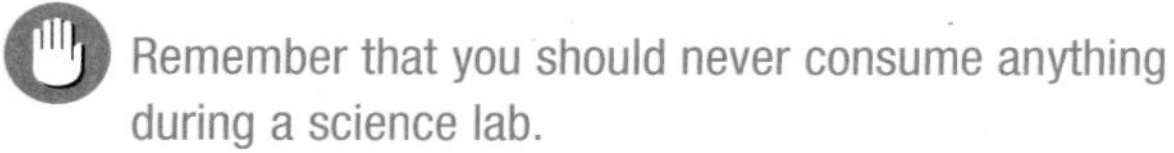
Remember that you should never consume anything during a science lab.

10. Obtain a second well plate.
11. Add two drops of red cabbage extract to each well of the first column of the second well plate.
12. Add two drops of universal indicator to each well of the second column.
13. Add two drops of the first household product (I) to the red cabbage indicator in the first row. Note any colour change that occurs.
14. Now add two drops of the household product (I) to the universal indicator in the first row. Note any colour change that occurs and compare the colours with those obtained in Part A.
15. Repeat steps 12 and 13 for the other assigned household products: II, III, and so on. Use different wells each time.
16. Dispose of the contents of both well plates as directed by your teacher. Clean up your workstation. Wash your hands.

Analyze and Evaluate

(a) Determine the pH of each of the household products, according to the red cabbage indicator, using the well plate from Part A as your guide. Indicate the position of each product on a pH scale (similar to Figure 2 in Section 7.3). K/U T/I

(b) Determine the pH of each of the household products, according to the universal indicator, using the well plate from Part A as your guide. T/I

(c) Classify each of the substances as acidic or basic. K/U

(d) Which household substances were difficult to classify? Why? T/I

(e) Compared with universal indicator, how effective is red cabbage juice as an indicator? Explain. T/I

(f) Of the two indicators you tested, which do you think was most accurate at determining pH? Explain why you think this. T/I C

(g) How well were you able to compare the effectiveness of a natural indicator (red cabbage extract) with that of universal indicator? What could you do to make a better comparison? T/I

Apply and Extend

(h) Use a pH meter to determine the pH of the household products. Use these results to evaluate the effectiveness of acid–base indicators to determine pH. T/I

(i) Measure the pH of a sample of tap water using a swimming pool pH test kit. Use small quantities of the appropriate chemicals to adjust the pH of the water upward and downward within the pH range of the test kit. A

(j) Research other natural acid–base indicators. Select a set of natural indicators whose colour changes cover a large range of pH values. Blend these indicators together to make your own version of universal indicator. If time permits, test the colour changes using the pH solutions from this investigation. T/I

(k) In a small group, look at some Hazardous Household Product Symbols. Discuss the necessity of adequate accurate labelling of products. If the containers are not empty, handle them with the appropriate caution. C A

7.5 Neutralization Reactions

On June 6, 2007, firefighters in Niagara Falls were called to the scene of a truck accident. A truck carrying potassium hydroxide, KOH, slammed into a telephone pole and was leaking some of its corrosive cargo. Once the firefighters realized what they were dealing with, they went to a local grocery store to purchase what they needed to treat the spill: 40 large bottles of cola. Remembering their science training, they knew that cola is mildly acidic. The firefighters neutralized the spill with the carbonated beverage and then soaked up the mess with an absorbent material, much like cat litter.

Firefighters have to know a lot of chemistry to deal with the emergencies they encounter. To find out more about being a firefighter,

GO TO NELSON SCIENCE

The reaction of potassium hydroxide (base) and the acid in cola is an example of a neutralization reaction. **Neutralization reactions** occur when an acid and a base react to form products that have a pH closer to 7 than either of the reactants. These products are usually an ionic compound (sometimes called a "salt") and water.

neutralization reaction a chemical reaction in which an acid and a base react to form an ionic compound (a salt) and water. The resulting pH is closer to 7.

You have seen that acids form hydrogen ions in water, and most bases release hydroxide ions. Think what happens when an acid and a base are mixed. The hydrogen ions and the hydroxide ions quickly react to produce water (Figure 1). As you know, water is neutral, with a pH of around 7.

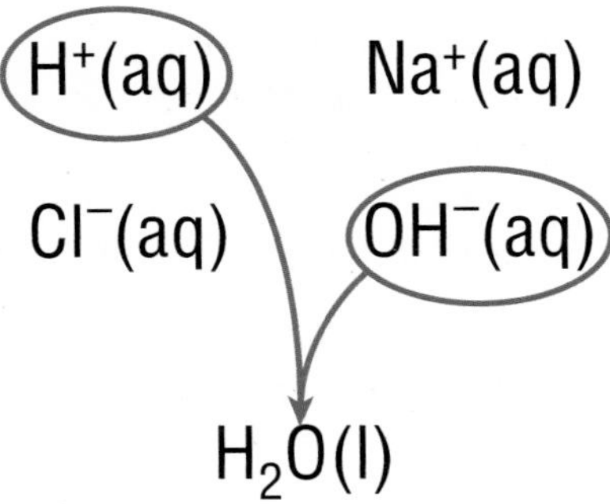

Figure 1 The hydrogen ion from the acid and the hydroxide ion from the base react to form water.

Not all bases contain hydroxide ions. Some, such as sodium hydrogen carbonate, have carbonate as their anion. A similar, but more complex, neutralization reaction takes place between acids and carbonate bases. You do not need to know the equations for these reactions at this stage of your chemistry education.

The neutralization reaction is very important in chemistry. It has many useful applications in everyday life and affects our society and the environment.

Predicting the Products of Neutralization Reactions

If we know the reactants of a neutralization reaction, we can predict the products. Let's look at the neutralization of hydrochloric acid with sodium hydroxide to see how this is done. The pattern of the reaction might look familiar. The chemical equation for this reaction is

$$HCl(aq) + NaOH(aq) \rightarrow H_2O(l) + NaCl(aq)$$

You can see that the sodium ion in NaOH(aq) and the hydrogen ion in HCl(aq) switch places: this is a double displacement reaction. The sodium and chloride ions remain dissolved in water. If we evaporated away the water, we would be left with solid sodium chloride crystals.

Let's apply the general pattern for double displacement reactions to the hydrochloric acid–sodium hydroxide example:

$$\begin{array}{ccccccc} HCl(aq) & + & NaOH(aq) & \rightarrow & H_2O(l) & + & NaCl(aq) \\ AB & + & CD & \rightarrow & AD & + & CB \end{array}$$

The general formula "AD" fits for water if you think of the chemical formula H_2O as being the same as HOH.

All acid–base neutralization reactions are double displacement reactions. The resulting ionic compound may remain dissolved in water or may form a precipitate. The general chemical equation for a neutralization reaction is

$$\text{acid} + \text{base} \rightarrow \text{water} + \text{ionic compound}$$

Remember the firefighters' use of cola to neutralize the chemical spill? Cola contains both carbonic acid, $H_2CO_3(aq)$, and phosphoric acid, $H_3PO_4(aq)$. The chemical equation for the neutralization of potassium hydroxide with phosphoric acid is

$$H_3PO_4(aq) + 3\ KOH(aq) \rightarrow 3\ H_2O(l) + K_3PO_4(aq)$$

In this example, the ionic compound produced is potassium phosphate, K_3PO_4. Note that three "KOH units" are needed for each H_3PO_4 molecule. Why? Because phosphoric acid can release three H^+ ions into the solution.

LEARNING TIP

Salty Product

Traditionally, chemists have referred to ionic compounds as "salts." You might therefore see the general equation for a neutralization reaction written as

acid + base $\rightarrow$ water + salt

This does not necessarily mean that table salt, NaCl, is produced.

Applications of Neutralization Reactions

There are a lot of acids and bases around. Their properties make them useful but can also make them hazardous. As we benefit from the positive aspects of these valuable chemicals, we also have to manage the negative aspects.

Chemical Spills

Sulfuric acid and sodium hydroxide (also called caustic soda) are two of the most widely used industrial chemicals. Huge quantities of these corrosive chemicals are transported annually by truck, rail, or ship. Even with strict laws controlling how industrial chemicals are transported, accidents happen.

In March 31, 2007, a train was hauling 150 000 L of sulfuric acid near Englehart in northern Ontario. The train suddenly derailed and spilled some of its cargo into the Blanche River (Figure 2). An ecological disaster unfolded. Dead fish washed up on shore, and local residents were warned not to allow their livestock to drink river water. Luckily, the town of Englehart gets its drinking water from wells.

The emergency response crew added calcium oxide (lime), CaO(s), to the river slightly upstream from the spill site. They wanted to neutralize the acid leaking from the containers.

DID YOU KNOW?

The Sour Spike Theory of Acids

In the seventeenth century, some scientists believed that acids were sour because the particles in an acid were in the shape of tiny spikes that pierced the tongue. Base particles were thought to contain holes on their surfaces. During neutralization, the scientists hypothesized, the spiky acid particles fit into the holes of the bases, locking the two particles together.

Figure 2 Derailed tanker cars spilled sulfuric acid near Englehart, Ontario.

DID YOU KNOW?

Train Derailments

According to the Transport Safety Board of Canada, about 570 train derailments occur each year. More than a quarter of these trains haul hazardous chemicals, so the chances are quite high of a derailment resulting in a chemical spill.

DID YOU **KNOW?**

Large-Scale Chemical Reactions
Calcium oxide (lime) is also widely used to raise the pH of lakes acidified by acid precipitation.

This spill cleanup strategy relied on a two-step neutralization reaction:

1. Calcium oxide, CaO, reacts in water to form basic calcium hydroxide:

   $CaO(s) + H_2O(l) \rightarrow Ca(OH)_2(aq)$

2. Calcium hydroxide neutralizes sulfuric acid in the river, producing a harmless solution of water and calcium sulfate:

   $Ca(OH)_2(s) + H_2SO_4(aq) \rightarrow 2\ H_2O(l) + CaSO_4(aq)$

During this procedure, the pH of the river water downstream was carefully monitored to ensure that just the right amount of lime was being added.

Antacids

Your stomach produces a solution of hydrochloric acid with pH as low as 1.5. The acid is needed for digestion. Sometimes, however, stomach acid can irritate the lining of the stomach, leading to discomfort or pain. Antacids provide relief by neutralizing stomach acid. The two most common active ingredients in antacids are hydroxide and carbonate compounds.

Figure 3 The active ingredient in milk of magnesia is magnesium hydroxide.

Milk of magnesia is a typical antacid containing magnesium hydroxide. It is a thick paste because magnesium hydroxide (a base) is not very soluble in water. As it is not very soluble, it releases only a relatively low concentration of hydroxide ions. This explains why you can swallow milk of magnesia without suffering a chemical burn (Figure 3). However, once in your stomach, magnesium hydroxide effectively neutralizes stomach acid:

$2\ HCl(aq) + Mg(OH)_2(aq) \rightarrow 2\ H_2O(l) + MgCl_2(aq)$

Powdered antacids usually contain sodium hydrogen carbonate (baking soda), $NaHCO_3(s)$, as the active ingredient:

$HCl(aq) + NaHCO_3(aq) \rightarrow CO_2(g) + H_2O(l) + NaCl(aq)$

As you can see, the neutralization reaction between an acid and a carbonate base is not a simple double displacement reaction. You do not have to write equations for this type of reaction.

WRITING **TIP**

Analyzing and Evaluating
When you write the Analysis and Evaluation for an activity or investigation, explain how they connect with the reason for doing the lab.

TRY THIS NEUTRALIZE IT

SKILLS: Questioning, Observing, Analyzing, Communicating

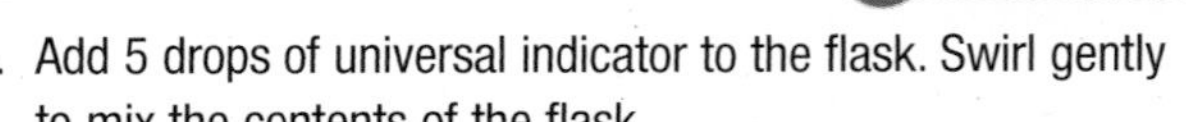

SKILLS HANDBOOK 3.B.7.

Is milk of magnesia effective at neutralizing hydrochloric acid? You will use universal indicator in this activity. The colours of universal indicator are red—very acidic; orange/yellow—acidic; green—neutral; blue—basic; purple—very basic.

Equipment and Materials: **eye protection; lab apron; 250 mL Erlenmeyer flask; teaspoon; water; milk of magnesia; dropper bottle of universal indicator; dropper bottle of dilute hydrochloric acid, HCl(aq)**

Hydrochloric acid is corrosive. Avoid spills on skin, clothing, or in the eyes. Wash spills immediately with cold water and alert your teacher.

1. Put on your eye protection and lab apron.
2. Add about 150 mL of tap water to the flask.
3. Add about a quarter of a teaspoon of milk of magnesia.
4. Add 5 drops of universal indicator to the flask. Swirl gently to mix the contents of the flask.
5. While continuously swirling the flask, slowly add the hydrochloric acid, about 10 drops at a time.
6. Continue to add acid until the mixture turns green.
7. Add a slight excess of acid (about 10 drops) until the solution is no longer cloudy.

A. What observation told you that the magnesium hydroxide was completely neutralized? K/U

B. Suggest why the solution was no longer cloudy after step 7. T/I

C. A homemade remedy for heartburn is to drink a glass of water containing half a teaspoon of baking soda. Do you think this remedy is effective? Why or why not? T/I

UNIT TASK Bookmark

You can apply what you learned about neutralizing acidic solutions to the Unit Task described on page 300.

IN SUMMARY

- Acids and bases react together in neutralization reactions.
- At the end of a neutralization reaction, the pH is closer to 7 (neutral) than it was for either of the reactants.
- The products of a neutralization reaction are water and an ionic compound (salt).
- Neutralization reactions have environmental and consumer applications, including restoring lakes affected by acid precipitation.

CHECK YOUR LEARNING

1. (a) What do we mean when we describe a solution as being neutral?
   (b) Describe the pH changes expected if an acid is used to neutralize a base with pH 12. K/U
2. Write the chemical equations for each of these acid–base reactions. (The formulas of acids and bases are given in Tables 1, 2, and 4 in Section 7.2.) K/U
   (a) hydrochloric acid and potassium hydroxide
   (b) sulfuric acid and potassium hydroxide
3. (a) The cola that the firefighters used in the opening example contains both phosphoric acid and carbonic acid. Write the chemical equation for the neutralization of potassium hydroxide solution with carbonic acid (from cola).
   (b) Predict the approximate pH change that occurs during this reaction. K/U T/I
4. Which would be the best to neutralize a large acid spill in your school lab: sodium hydroxide or baking soda? Explain. A
5. Figure 4(a) shows a pink mixture of sodium hydroxide and phenolphthalein indicator being injected into a lemon. The lemon was then cut in half (Figure 4(b)). Explain why the inside of the lemon was not pink. T/I

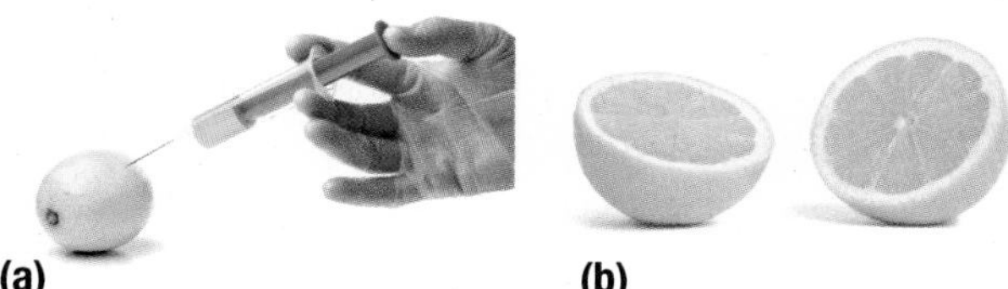

(a) (b)

Figure 4

6. Calcium oxide, CaO, is a common ingredient in cement. Based on what you have learned in this section, what precautions should a bricklayer take when mixing cement? Why? A
7. The hard parts of coral and shellfish are mostly made up of calcium carbonate. Evidence suggests that air pollution is causing the oceans to become slightly acidic.
   (a) What kind of chemical reaction occurs when acids are in contact with calcium carbonate? T/I A
   (b) Write the chemical equation for the reaction of sulfuric acid with calcium carbonate.
   (c) What effect would this reaction have on coral ecosystems? Explain.
8. Lead acid batteries used in cars contain a concentrated solution of sulfuric acid. When these batteries are recycled, their sulfuric acid is drained and neutralized to produce water and sodium sulfate. Suggest a compound that could be used to neutralize sulfuric acid and produce sodium sulfate. Justify your choice by writing the chemical equation for the reaction. T/I
9. Fish muscle contains bases that give cooked fish their "fishy" odour. Suggest why a squirt of lemon juice often makes this odour disappear. A
10. Over time, a crusty deposit of calcium carbonate forms on the heating element inside a kettle. Use your understanding of acids and bases to suggest a way to remove this deposit. A

Analyzing an Acid Spill

SKILLS MENU

- Questioning
- Hypothesizing
- Predicting
- Planning
- Controlling Variables
- Performing
- Observing
- Analyzing
- Evaluating
- Communicating

You are an environmental scientist involved in monitoring the effect of a sulfuric acid spill near a river. Each day for a week, you collect a sample of river water from a point just downstream of the spill site. To compare the acidity of the samples, you count the number of drops of sodium hydroxide solution needed to neutralize each sample. The chemical equation for this reaction is

$$H_2SO_4(aq) + 2\ NaOH(aq) \rightarrow 2\ H_2O(l) + Na_2SO_4(aq)$$

You use bromothymol blue indicator to show when the samples are neutralized (Figure 1).

Figure 1 Bromothymol blue indicator is yellow in an acidic solution (on the left), green in a neutral solution, and blue in a basic solution (on the right).

Equipment and Materials

- eye protection
- lab apron
- 125 mL Erlenmeyer flask
- pH meter (optional)
- dropper bottles of
  - dilute sodium hydroxide, $NaOH(aq)$
  - bromothymol blue indicator
- seven contaminated river-water samples labelled "Day 1," "Day 2," etc.

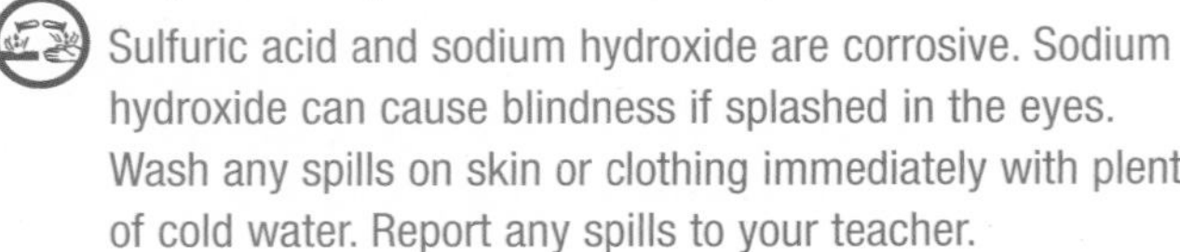

Sulfuric acid and sodium hydroxide are corrosive. Sodium hydroxide can cause blindness if splashed in the eyes. Wash any spills on skin or clothing immediately with plenty of cold water. Report any spills to your teacher.

Procedure

1. Write a series of detailed steps for your Procedure, including safety precautions.
2. Create a table in which to record your observations.
3. With your teacher's approval, carry out your Procedure.

Analyze and Evaluate

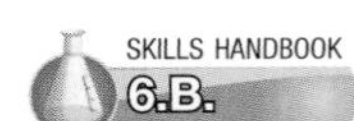

SKILLS HANDBOOK 6.B.

(a) What did you have to keep constant to ensure that your test was fair? T/I

(b) What is the relationship between acidity and the number of drops of base required? T/I

(c) Plot a graph of the number of drops of sodium hydroxide required to neutralize the acid against time (Day 1, Day 2, etc.). C

(d) Describe the shape of your graph. What information does it communicate? K/U T/I

(e) How many days did it take for the river to return to a neutral pH? T/I

(f) How did adding sodium hydroxide to the samples help you determine their acidity? K/U T/I

(g) Evaluate this method of determining acidity. T/I

Apply and Extend

(h) The following chemical equation represents the neutralization of sulfuric acid with calcium hydroxide:

$$H_2SO_4(aq) + Ca(OH)_2(aq) \rightarrow 2\ H_2O(l) + CaSO_4(s)$$

How would your observations differ if calcium hydroxide were used in the analysis rather than sodium hydroxide? Why? Refer to chemical equations for both reactions in your answer. T/I

(i) A pH meter can measure small changes in pH (Figure 2). If your school has a pH meter, use it to measure the pH of the water samples. Record your observations and plot a graph. How does this graph differ from the graph you created in question (c)? T/I

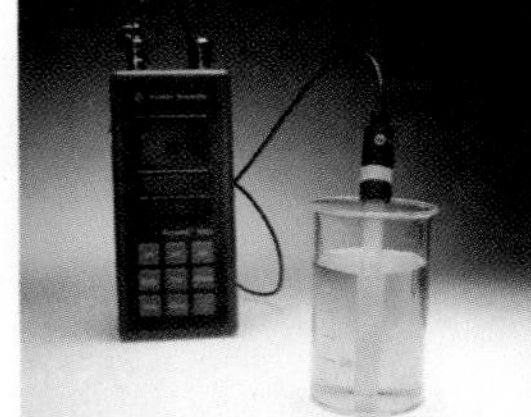

Figure 2 A pH meter

UNIT TASK Bookmark

You can apply what you learned about measuring pH to the Unit Task described on page 300.

Minimizing Risk for a Community

SKILLS MENU

- Defining the Issue
- Researching
- Identifying Alternatives
- Analyzing the Issue
- Defending a Decision
- Communicating
- Evaluating

An Ontario detergent manufacturer is considering adding another shift to its production schedule at its central Ontario plant (Figure 1). This will create 25 new jobs—great news for the nearby town with high unemployment. Increased production requires more raw materials, including concentrated sulfuric acid. The challenge is how to supply the additional sulfuric acid while minimizing the risk to the local environment.

Figure 1 What is the best way to transport sulfuric acid?

The Issue

You are a member of an independent consulting firm hired to investigate possible solutions. In your opinion, the two best options are as follows:

- Build a small sulfuric acid production plant next to the detergent plant. The initial construction and setup costs will be significant. The raw material for manufacturing sulfuric acid is sulfur, which is a waste product from the smelting of nickel ore. A nearby nickel company can supply the sulfur at no cost. The company could recoup its initial investment in about five years. The sulfur would have to be transported from the nickel smelter by truck along the road through town. Once the factory is in full production, large quantities of sulfuric acid would be produced and stored on site.
- Increase the current train shipments of sulfuric acid to the factory from the supplier located 200 km away. The rail line runs near the river that is the town's main source of drinking water.

You have been asked to present both options to a meeting involving the town mayor, the local Member of Parliament, a representative from an environmental group, and the president of the detergent company. You should outline the risks and benefits of each option and then recommend one of them.

Goal

To decide which proposal best meets the needs of the community, the company, and the environment and to recommend this decision at the meeting.

Gather Information

Work in pairs or small groups to learn more about

- the manufacture of sulfuric acid
- the hazards involved in storing and transporting sulfuric acid and sulfur
- rail accidents involving hazardous chemicals
- ways to minimize the risks of each option

Discuss the two options, comparing the risks and benefits of each. T/I C

Make a Decision

Which proposal do you recommend for supplying the detergent factory with sulfuric acid? What criteria did you use to decide? C A

Communicate

Complete a report that will be presented to the meeting. The report should outline the benefits and risks of each option and how those risks could be minimized. It should conclude with a recommendation. C A

Painting Out Pollution

Imagine a city whose roads and buildings neutralize acidic air pollutants on contact. Does this sound too good to be true? Believe it or not, the technology already exists. Researchers in Italy and Hong Kong have developed a pollution-fighting coating that can be painted onto pavement and the walls of buildings. The active compounds in this coating are titanium dioxide, TiO_2, and calcium carbonate, $CaCO_3$. You probably brushed your teeth with titanium dioxide this morning: it is a common whitening agent in toothpaste. Calcium carbonate is found in limestone rock, blackboard chalk, and antacid tablets. It is widely available, inexpensive, and safe.

The pollution-fighting technology relies on a unique property of titanium dioxide: it can absorb ultraviolet radiation from the Sun (Figure 1). This is why titanium dioxide is used in many sunblock lotions.

The energy that the titanium dioxide absorbs is transferred to airborne water molecules that collide with the surface. It turns out that the quantity of energy transferred is just the right amount to make water react with nitrogen dioxide, a serious air pollutant. The reaction of nitrogen dioxide with water produces nitric acid. The acid is then neutralized by calcium carbonate particles in the coating.

The carbon dioxide and water from this reaction return to the air. Calcium nitrate is washed harmlessly to the ground by rainwater. The titanium oxide is not used up in this reaction. It remains available to absorb and transfer energy indefinitely.

Roadways and buildings in cities across Europe have been covered with this coating (Figure 2). Designers have developed decorative panels that can be retro-fitted onto existing buildings or surfaces in polluted areas. Where it is in use, area residents immediately reported a noticeable improvement in the air quality. They were right: analysis of the air at street level showed that nitrogen dioxide concentrations were reduced to about 40 % of their previous levels!

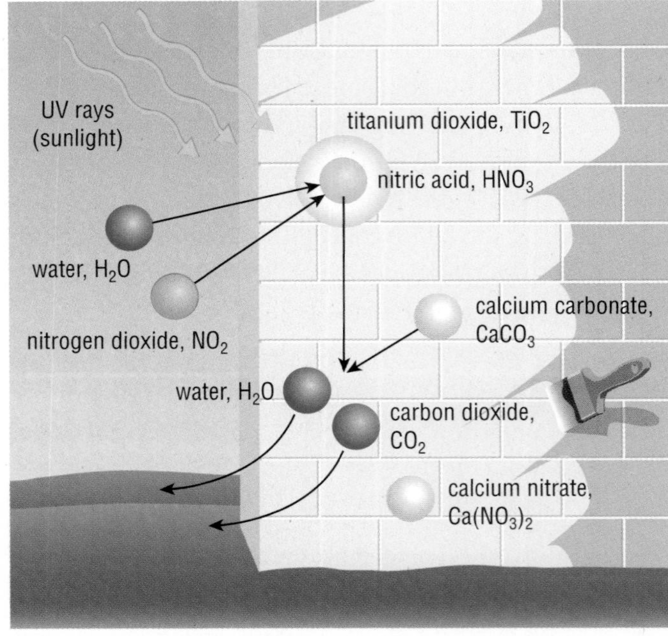

Figure 1 A series of reactions converts toxic nitrogen dioxide gas to relatively harmless compounds.

Figure 2 This church in Rome is coated with white cement that contains titanium dioxide. The titanium oxide helps the surface stay white, as well as fight pollution.

Acid Precipitation

The air currently filling your lungs is made up of millions of molecules. These molecules have probably already been in the lungs or bloodstreams of the people around you or even the people who live on the other side of the world. You share air with everyone and everything in your environment. This includes wildlife, grass, automobiles, lakes, factories, and even your pet dog.

Air is constantly on the move and carries some rather interesting cargo: bacteria, dust particles, moisture, traces of pollutants, and so on. Some of these pollutants may be natural, and some may be the result of human activity. Some air pollutants concern scientists, environmentalists, and the general population because they contribute to acid precipitation (Figure 1). The twentieth century was a good time in North America for industrial and economic growth but a bad time for the natural environment. All that industrial growth resulted in the production of vast quantities of chemicals, many of which polluted the air, ground, and water.

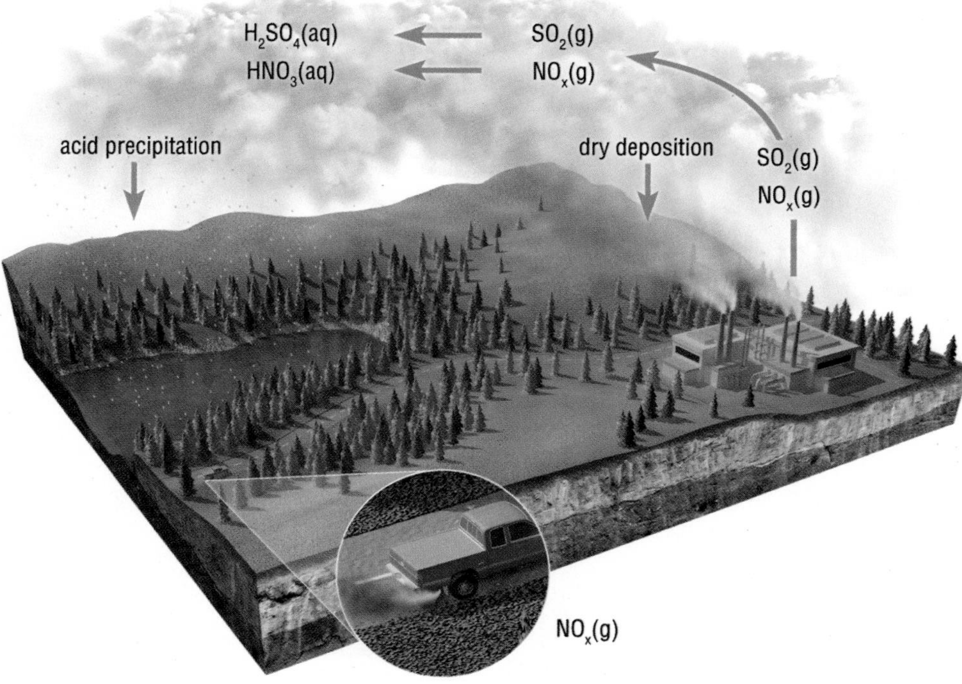

Figure 1 Emissions of sulfur dioxide and nitrogen oxides combine with water in the atmosphere to form acids that fall back to Earth as acid precipitation (rain, snow, and sleet). Acid-forming pollutants can also fall to Earth directly as dry deposition (particles or gases).

What Is Acid Precipitation?

Acid precipitation is a term used to describe any precipitation (rain, snow, fog) that has become acidic from reacting with compounds in the atmosphere. Acid precipitation forms when certain pollutants—most importantly sulfur dioxide and nitrogen oxides—combine with water in the atmosphere before falling to Earth. Acid precipitation has a pH less than about 5.6—the normal pH of rain.

acid precipitation any precipitation (e.g., rain, dew, hail) with a pH less than the normal pH of rain, which is approximately 5.6

dry deposition the process in which acid-forming pollutants fall directly to Earth in the dry state

There is another way that acidic substances can reach Earth: **dry deposition** occurs when acid-forming chemicals fall directly onto surfaces like soil or leaves. These particles form acids when they come in contact with water on these surfaces.

Sources of Acid-Forming Pollutants

Several pollutants lead to acid precipitation. Two that are of particular concern are sulfur dioxide, SO_2, and nitrogen oxides, NO_x.

Sulfur Dioxide, SO_2

DID YOU KNOW?

Venus: Extreme Acidic Rain
There is so much volcanic activity on Venus that much of the planet is permanently covered in a dense, swirling cloud of sulfuric acid. This is just one of the challenges faced by engineers designing interplanetary probes to investigate Venus.

Sulfur dioxide is a clear, colourless gas that has a strong, choking odour. As Figure 2 shows, most of Canada's production of sulfur dioxide comes from industry. This includes burning coal to generate electricity and the mining and refining of metals. Smelting is a process used to separate a metal from the ore extracted from the ground. Smelting involves heating the ore to high temperatures and collecting the molten metal. During the process, any sulfur present in the ore reacts with oxygen in the air, forming sulfur dioxide gas:

$$S(s) + O_2(g) \rightarrow SO_2(g)$$

The same reaction takes place during the combustion of fossil fuels—which generally contain some sulfur. Over half of the electricity generated in the United States comes from the combustion of coal. As a result, over 50 % of the sulfur dioxide emissions that reach Eastern Canada come from the United States (Figure 3).

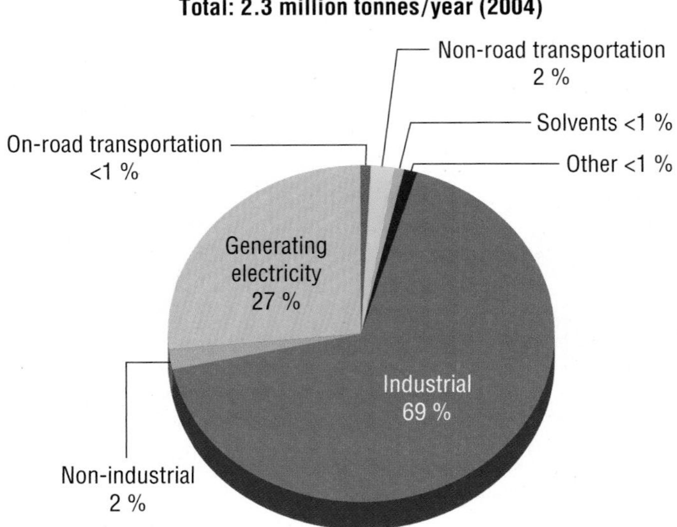

Figure 2 Sources of sulfur dioxide emissions in Canada

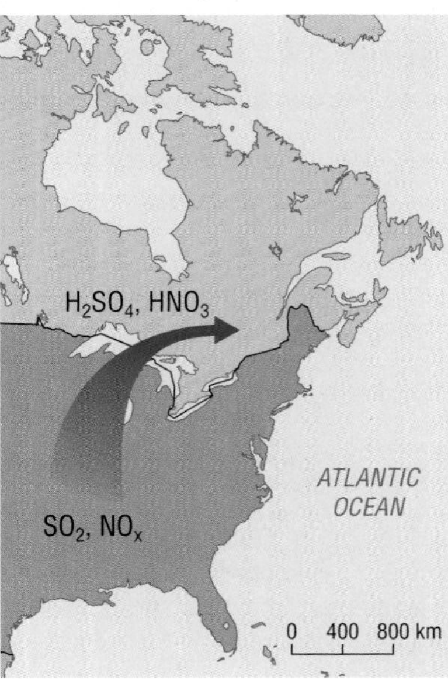

Figure 3 Most of the acid-causing pollutants falling on Ontario come from the United States.

Once in the atmosphere, sulfur dioxide reacts with more oxygen to produce sulfur trioxide:

$$2\ SO_2(g) + O_2(g) \rightarrow 2\ SO_3(g)$$

Then sulfur trioxide combines with water droplets in the atmosphere to form sulfuric acid:

$$SO_3(g) + H_2O(l) \rightarrow H_2SO_4(aq)$$

Nitrogen Oxides, NO_x

NO_x is a general chemical formula used to represent a number of nitrogen oxides, including nitrogen monoxide, NO, and nitrogen dioxide, NO_2. Most of the nitrogen oxide emissions in North America are produced by vehicles that burn fossil fuels—mostly gasoline (Figure 4).

The temperatures inside the internal combustion engine of a car or train are high enough for atmospheric nitrogen and oxygen to react to form nitrogen monoxide:

$N_2(g) + O_2(g) \rightarrow 2\ NO(g)$

Pollution-control technology in today's vehicles reverses this chemical reaction, converting nitrogen monoxide back into harmless nitrogen and oxygen. However, some nitrogen monoxide still escapes into the atmosphere. There it reacts with more oxygen to produce toxic, reddish-brown nitrogen dioxide:

$2\ NO(g) + O_2(g) \rightarrow 2\ NO_2(g)$

In Section 5.11, you learned that nitrogen dioxide is one of the gases that make up smog. In the atmosphere, nitrogen dioxide combines with water to produce nitric acid and more nitrogen monoxide:

$3\ NO_2(g) + H_2O(l) \rightarrow 2\ HNO_3\ (aq) + NO(g)$

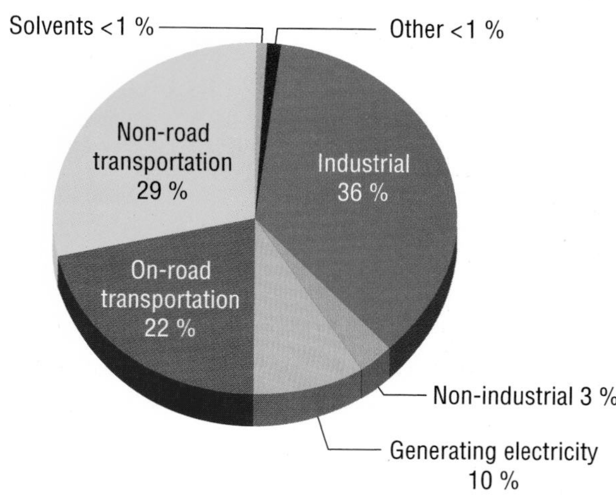

Figure 4 Sources of nitrogen oxide emissions in Canada

Environmental Impact of Acid Precipitation

Acid precipitation has a serious effect on almost every part of Ontario's environment, from the soil to the Great Lakes.

Aquatic Ecosystems

Aquatic life can tolerate only minor changes in the pH of the water in their environments (Figure 5). As the pH decreases, the youngest and most fragile organisms die first (Figure 6). The loss of fish affects organisms that are higher on the food chain. Predators such as loons, ospreys, otters, and larger fish must find other sources of food or move elsewhere to survive.

Figure 5 As the acidity of a lake increases, loons have less available food.

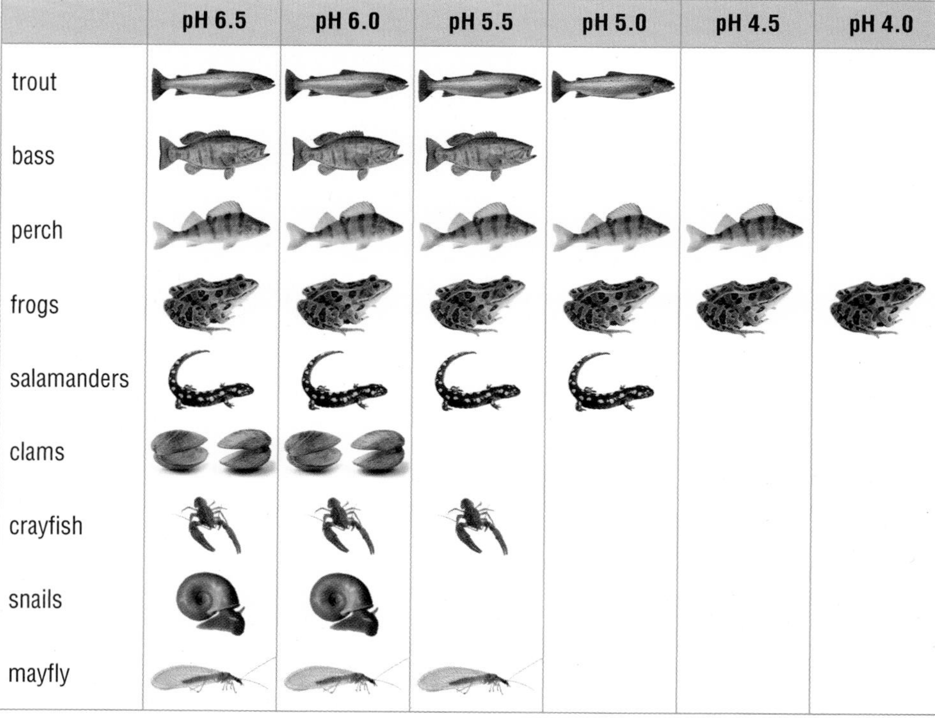

pH 6.5
pH 6.0
pH 5.5
pH 5.0
pH 4.5
pH 4.0
trout
bass
perch
frogs
salamanders
clams
crayfish
snails
mayfly

Figure 6 At pH 5, many young fish fail to develop properly and many invertebrates are killed. Some adult fish may be able to tolerate pH levels as low as 4, but they cannot survive without food.

DID YOU **KNOW?**

Fish Heart Attacks

An increase in the acidity of lake water can increase the viscosity of a fish's blood. The thickened blood can become so difficult to pump that the fish dies of a heart attack.

Acid precipitation can indirectly cause another problem for aquatic life: aluminum ions can be washed from acidified soil into the streams and lakes, where they damage the gills of fish, causing them to suffocate.

Soils

Acid precipitation can have a drastic effect on soil. Acidic groundwater can dissolve and wash away metal ions. Some ions, such as calcium, magnesium, and potassium, are essential nutrients for plant growth. Others, including aluminum, are toxic to plants and to aquatic life.

buffering capacity the ability of a substance to resist changes in pH

All soils have some ability to resist changes in acidity. This ability, called **buffering capacity**, depends on substances in the soil that act like bases, neutralizing the acid. If a soil's buffering capacity is high, acids passing through the soil are neutralized before reaching nearby streams or lakes. Soils in areas where the rock is mostly limestone have high buffering capacities. Limestone is mostly made up of calcium carbonate, $CaCO_3$, which neutralizes acids:

$$CaCO_3(s) + H_2SO_4(aq) \rightarrow CaSO_4(aq) + H_2O(l) + CO_2(g)$$

However, neutralization reactions use up buffering minerals in the soil, so the buffering capacity eventually decreases.

Forests

The buffering capacity of soil partially protects forests from the effects of acid precipitation. If the soil has only a small buffering capacity, however, it can only neutralize a small "dose" of acid. Then the soil loses valuable nutrients needed for trees to grow properly. Acid precipitation rarely kills trees directly, but it weakens the trees. They are then more vulnerable to diseases, strong winds, or extreme cold (Figure 7).

Figure 7 Acid precipitation kills mature trees and prevents young trees from growing.

The Economic Impact of Acid Precipitation

We know that acid precipitation can harm the environment, but it can also harm the Canadian economy. For example,

- Environment Canada estimates that poor growing conditions result, every year, in the loss of wood valued at billions of dollars.
- A reduction in fish stocks has already affected Ontario's multibillion dollar recreational fishing industry.
- Acid precipitation damages steel structures, limestone buildings, and stone monuments (Figure 8, next page).

Figure 8 The first photograph of this statue of George Washington (a) was taken in 1935. The second (b), was taken in 1994. The effect of acid precipitation on the face of the statue over those 59 years is quite obvious.

Cleaning Up Acid Precipitation

Since 1980, there have been reductions in both $SO_2(g)$ and $NO_x(g)$ emissions in North America (Table 1). Scientists predict that emissions of both pollutants will continue to decline until 2020. This improvement has been achieved by

- switching to low-sulfur fossil fuels to generate electricity
- the installation of scrubbers to remove sulfur from the emissions of smelting operations and fossil fuel-burning power plants
- improvements in the pollution control equipment on vehicles
- stricter laws governing vehicle emissions (such as Ontario's Drive Clean Program)

Table 1 Trends in the Emission of Acid-Forming Pollutants

	1980 to 2000		Predicted 2000 to 2020	
	Eastern Canada	U.S.	Eastern Canada	U.S.
SO_2 reduction	53 %	40 %	21 %	38 %
NO_x reduction	17 %	little	39 %	47 %

Source: Environment Canada 2004 Canadian Acid Deposition Science Assessment

These reductions are bringing some encouraging signs. For example, the pH of some lakes has increased to more normal levels. This is especially noticeable near smelters that have been fitted with technologies that significantly reduce emissions. There has also been some improvement in the wildlife observed in these lakes. For example, populations of loons have increased. The food chains in these lakes remain fragile, however. For improvements to continue, further reductions are needed.

Not all of the news is good either, however. Despite these reductions, many of Ontario's ecosystems are still far from healthy. According to a 2004 Environment Canada assessment, ecosystems in Ontario have not yet fully recovered from the effects of acid precipitation in the twentieth century. Furthermore, many regions in Ontario are still receiving more acidic precipitation than their ecosystems can safely absorb.

It will take a long time for Ontario's ecosystems to recover completely from the effects of acid precipitation—much longer than it took to pollute them!

UNIT TASK Bookmark

You can apply what you learned about acid precipitation and its effects to the Unit Task described on page 300.

IN SUMMARY

- Acid precipitation is any precipitation that has a pH less than the normal pH of rain.
- Acid precipitation is caused by pollutants (mostly SO_2 from burning fossil fuels and NO_x from vehicle engines) reacting with water in the air to form acids.
- Acid precipitation can have serious environmental and economic impacts.
- Steps taken to reduce the effects of acid precipitation include using low-sulfur fuels, adding scrubbers to smokestacks, installing catalytic converters to vehicles, and enacting stricter anti-pollution laws.
- The production of pollutants is decreasing but is still causing damage in some areas. Other areas are starting to recover.

CHECK YOUR LEARNING

1. In your own words, define K/U
   (a) acid precipitation
   (b) buffering capacity
2. (a) How do acid rain and acid precipitation differ?
   (b) Acid precipitation is sometimes called "wet deposition." Why is this term appropriate?
   (c) How do wet and dry deposition differ? K/U
3. Rain is naturally acidic. Under what conditions does "normal" rain become "acid" rain? K/U
4. (a) What two groups of compounds are largely responsible for causing acid precipitation?
   (b) Write a chemical equation(s) to illustrate how one of these compounds reacts to produce acid precipitation. K/U
5. Look at Figures 2 and 4 in this section. K/U A
   (a) What percentage of sulfur dioxide and nitrogen oxide emissions came from transportation in 2004?
   (b) Suggest two things that you could do to reduce the amount of acid pollutants resulting from transportation.
6. Look at Figure 6 in this section. K/U
   (a) Which aquatic animals are most tolerant of acidic conditions?
   (b) Which animals are most affected by acidic conditions?
7. Briefly outline at least three damaging effects that acid precipitation may have on the environment. A
8. List two economic impacts of acid precipitation. A
9. What is the prediction for acid precipitation over the next decade? K/U
10. What evidence is there that the environment is beginning to recover from the effects of acid precipitation? K/U
11. (a) Calcium oxide, CaO, reacts with water to form calcium hydroxide. Write the chemical equation for this reaction.
    (b) Why is calcium oxide capable of neutralizing acids despite not having the hydroxide ion in its chemical formula?
    (c) Explain why adding calcium oxide (sometimes called "lime") to lakes polluted with acid precipitation solves the problem only for the short term.
    (d) Suggest a permanent solution to the problem. K/U T/I A
12. Why is global co-operation essential in our battle against the effects of acid precipitation? A
13. Consider two neighbouring lakes in an isolated region of Ontario. One lake is on limestone (calcium carbonate) while the other is lined with granite. Both receive the same amount of acid precipitation. A
    (a) Explain why the limestone lake has a higher pH than the granite lake.
    (b) Which lake is likely to have a healthier aquatic ecosystem? Explain your answer.
14. Why are the populations of young aquatic animals in an ecosystem often a good indicator of the overall health of the ecosystem? A

Scrubbers: Antacids for Smokestacks

Over the past 30 years, there have been significant reductions in the emissions of pollutants that cause acid precipitation. Much of this success is due to advancements in technologies, including the installation of sulfur dioxide scrubbers into smokestacks.

The Problem

The smelting of nickel ore is a major source of sulfur dioxide emissions. Smelting involves heating the ore and collecting the molten metal. During the process, sulfur in the ore reacts with oxygen in the air, forming sulfur dioxide gas. The sulfur dioxide mixes with the products of combustion: water and carbon dioxide. This is the mixture that used to billow from the smokestacks of nickel smelting plants. Once in the atmosphere, sulfur dioxide reacts with water to form sulfuric acid. The acid then falls to Earth as acid precipitation.

In the past, this reaction caused serious environmental damage in nickel-producing areas like Sudbury. It remains a problem in countries with less strict environmental laws than Canada's. As long as there is a demand for nickel, we will be faced with the problem of cleaning up smelter emissions.

The Technological Solution

To remove the sulfur dioxide, combustion gases—including sulfur dioxide—are first fed into a large scrubber tower (Figure 1). There, these gases are showered with a fine paste-like mixture of limestone and water. The limestone paste absorbs sulfur dioxide from the combustion gases and converts it into calcium sulfite, $CaSO_3$. The chemical equation for this reaction is

$$CaCO_3(s) + SO_2(g) \longrightarrow CaSO_3(s) + CO_2(g)$$

The remaining combustion gases continue up the scrubber tower and are released. Calcium sulfite particles fall to the bottom of the tower, where they can be removed. The waste calcium sulfite can then be chemically converted into gypsum, which can be used to make drywall.

There are other technologies that can reduce the release of sulfur dioxide into the atmosphere. These include capturing sulfur dioxide and turning it into a valuable product: sulfuric acid.

Making products that can be sold turns pollution reduction into a profitable process.

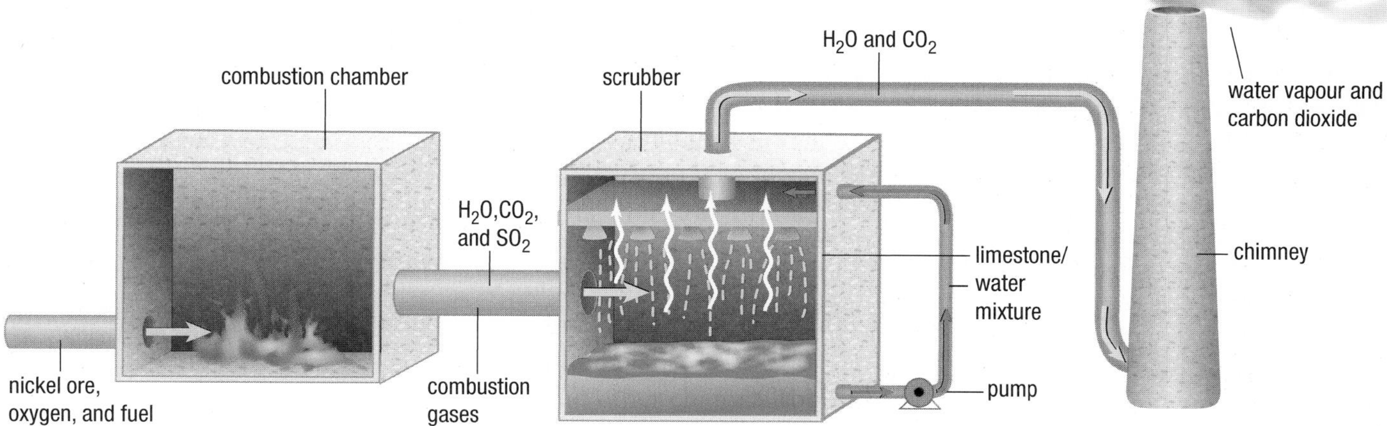

Figure 1 Companies that produce scrubbers claim that their products remove up to 95 % of the sulfur dioxide in combustion emissions.

LOOKING BACK

KEY CONCEPTS SUMMARY

Acids are aqueous solutions that have characteristic properties.

- Acids turn indicators specific colours (e.g., bromothymol blue becomes yellow; litmus becomes red). (7.2)
- Acids have a pH less than 7. (A pH of 7 is neutral.) (7.3)
- Acidic solutions result when certain molecular compounds form hydrogen ions in the solution. (7.2)
- Acids neutralize bases. (7.5)

Bases are aqueous solutions that have characteristic properties.

- Bases turn indicators specific colours (e.g., bromothymol blue becomes blue; litmus becomes blue). (7.2)
- Bases have a pH greater than 7. (7.3)
- Basic solutions result when certain ionic compounds release hydroxide ions into the solution. (7.2)
- Bases neutralize acids. (7.5)

Acids and bases have a significant impact—both good and bad—on society and the environment.

- Chemical spills can seriously affect natural ecosystems. (7.3, 7.6)
- Acids can be used to remove heavy metal contaminants from soil. (7.3)
- Acids are added to processed foods for taste or as a preservative. (7.3)
- Bases are widely used in cleaning products. (7.3, 7.4)
- Products have been developed to modify the pH of swimming pools, skin, soil, stomach contents, and corrosive chemical spills. (7.3–7.5)

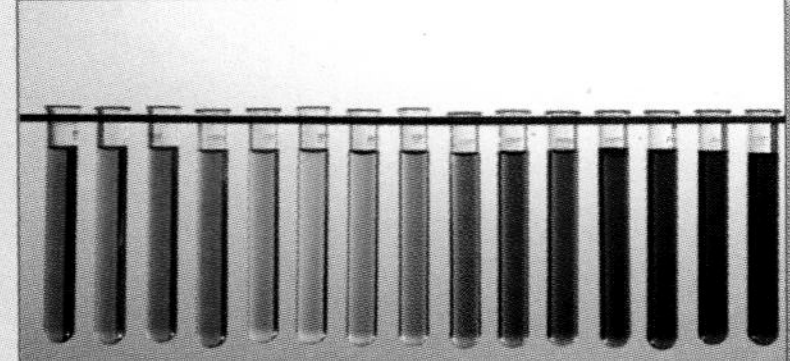

The acidity of solutions is measured using the pH scale.

- pH is a term used to describe how acidic or basic a solution is. (7.3)
- The pH of a solution can be altered by adding an acid or base. (7.5)
- Most ecosystems function best when the pH is near 7. (7.3, 7.6–7.8)

Acids and bases react together in neutralization reactions.

- Acids and bases react to produce water and an ionic compound. (7.5)
- When an acid and a base mix, the pH moves closer to 7 (neutral). (7.5)
- Neutralization reactions can be used to offset the effects of acid precipitation in terrestrial and aquatic ecosystems. (7.5, 7.8)
- Acid spills can be neutralized by adding a base and vice versa. (7.5, 7.8)
- Consumer products rely on neutralization reactions to neutralize acid. (7.5, 7.6)

Industrial and vehicle emissions that cause acid precipitation can be reduced by technology.

- Precipitation with pH below 5.6 is considered acid precipitation (7.3, 7.8)
- Sulfur dioxide, SO_2, and nitrogen oxides, NO_x, are the two main causes of acid precipitation (7.8)
- Pollution reduction technologies (e.g., scrubbers) use chemical reactions to remove acid-forming pollutants. (7.8)

WHAT DO YOU THINK NOW?

You thought about the following statements at the beginning of the chapter. You may have encountered these ideas in school, at home, or in the world around you. Consider them again and decide whether you agree or disagree with each one.

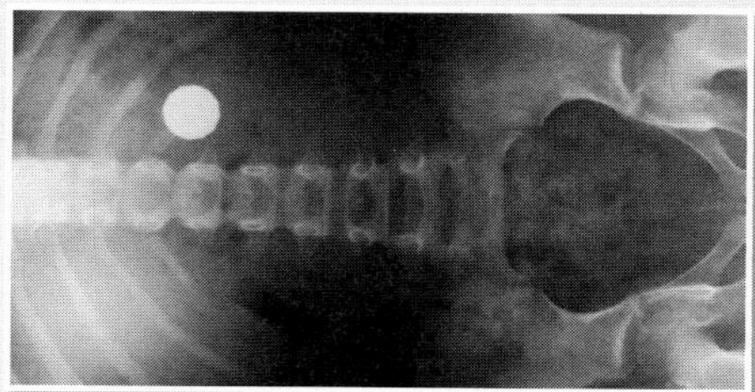

1 Stomach acid can dissolve metals like the coin in this child's stomach.
Agree/disagree?

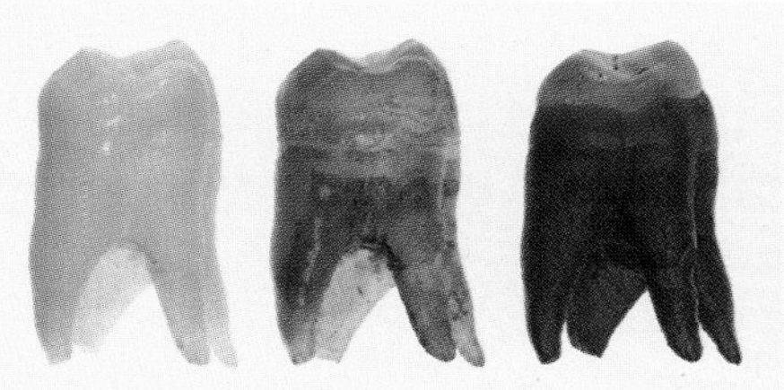

2 Even diet soft drinks can cause tooth decay.
Agree/disagree?

3 All acids are dangerous.
Agree/disagree?

4 We all contribute to the production of acid precipitation.
Agree/disagree?

5 Medications for heartburn work by cooling the stomach.
Agree/disagree?

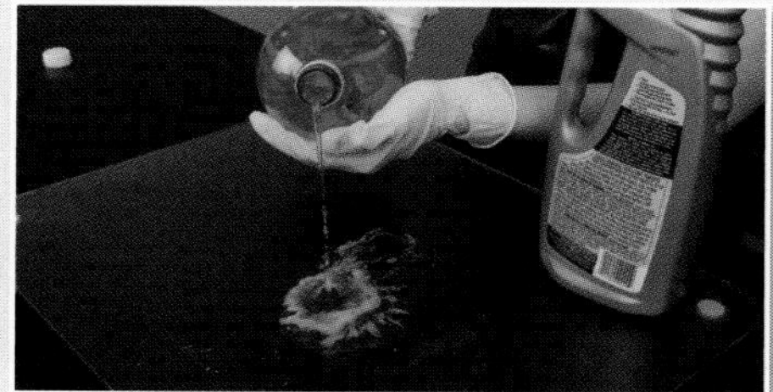

6 Soft drinks can be used to make spilled drain cleaner safer.
Agree/disagree?

How have your answers changed?
What new understanding do you have?

Vocabulary

acid (p. 268)
base (p. 270)
acid–base indicator (p. 270)
pH (p. 272)
pH scale (p. 272)
neutral (p. 272)
acid leaching (p. 274)
neutralization reaction (p. 278)
acid precipitation (p. 285)
dry deposition (p. 286)
buffering capacity (p. 288)

BIG Ideas

- ✓ **Chemicals react with each other in predictable ways.**
- ✓ **Chemical reactions may have a negative impact on the environment, but they can also be used to address environmental challenges.**

The following icons indicate the Achievement Chart category addressed by each question.

K/U Knowledge/Understanding T/I Thinking/Investigation C Communication A Application

What Do You Remember?

1. Name the following compounds. (7.2) K/U
   (a) $H_3PO_4(aq)$
   (b) $HBr(aq)$
   (c) $Fe(OH)_3$
   (d) $H_2SO_4(aq)$
   (e) $Ca(HCO_3)_2$
   (f) KNO_3
2. In your own words, define the terms "acid" and "base." (7.2) K/U
3. (a) What gas is produced when acids react with metals like magnesium?
   (b) What atoms cause the acidity of an acid?
   (c) What colour is bromothymol blue indicator in an acid and in a base?
   (d) What are the typical products of a neutralization reaction?
   (e) How is pH related to acidity? (7.2, 7.3, 7.5) K/U
4. (a) What two pollutants are primarily responsible for acid precipitation?
   (b) Which human activities are the major sources of these pollutants?
   (c) Identify two technologies used to decrease the emissions of these pollutants. (7.8) K/U

What Do You Understand?

5. Consider these seven compounds: $H_3PO_4(aq)$, $HBr(aq)$, $Fe(OH)_3$, $H_2SO_4(aq)$, $Ca(HCO_3)_2$, and KNO_3. Predict which of these compounds will produce a solution with a pH
   (a) less than 7.
   (b) greater than 7.
   (c) equal to 7. (7.2, 7.3) K/U T/I
6. Drinking water standards in Canada require that the pH of drinking water be between 6.5 and 8.5. Name a household substance that could, if incorrectly discarded,
   (a) lower the pH of water leaving your home
   (b) increase the pH of water leaving your home (7.3, 7.4) K/U A
7. (a) How does acid precipitation affect aquatic life in a lake? (7.3, 7.4)
   (b) How can the addition of lime help renew lakes that have been affected by acid precipitation? (7.3, 7.8) A
8. Complete the following word equations. (7.5) K/U
   (a) hydrochloric acid + ______________ → water + potassium chloride
   (b) sulfuric acid + calcium hydroxide → ______________ + ______________
   (c) phosphoric acid + sodium hydroxide → water + ______________
9. Write balanced chemical equations for the word equations in question 8. (6.3, 7.5) K/U T/I
10. Describe, in your own words, the process of acid–base neutralization. (7.5) K/U
11. (a) Write the word and chemical equations for the reaction of nitric acid with calcium hydroxide.
    (b) This reaction can be classified in two ways. Name these two types of reactions. (6.6, 7.5) K/U
12. (a) When fossils fuels are burned, a chemical is released that causes acid precipitation. Describe this process, including chemical equations.
    (b) Name one technology that can reduce this problem. Briefly outline how this technology works. (7.8) K/U A
13. Farmers sometimes add calcium hydroxide to soil to increase its pH.
    (a) Why does calcium hydroxide raise soil pH?
    (b) Calcium hydroxide dissolves very slowly in water. Is this property an advantage or a disadvantage for a soil additive? Explain. (7.3) K/U A
14. Summarize your understanding of acids and bases in a table or graphic organizer. (Your graphic organizer might look something like Figure 1.) Include the following terms: synthesis; properties; neutralization; indicator; uses; pH; typical ion; environmental impact; metal; non-metal. (7.1–7.8) K/U

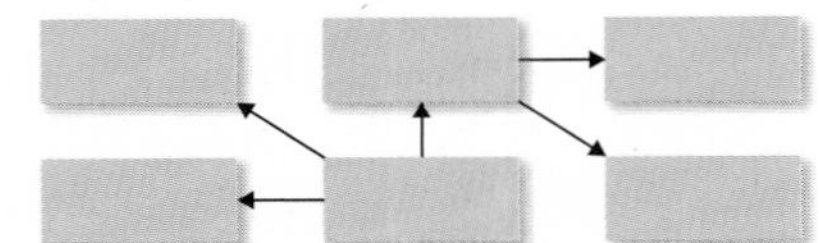

Figure 1

Solve a Problem

15. Phenol red is an acid–base indicator that is commonly used to test the pH of swimming pools. (7.3–7.5) K/U T/I

Figure 2

    (a) What is the pH of the water in Figure 2?
    (b) Swimming pools should have a pH between 7.2 and 7.6. What can the owner of this pool do to adjust the pH to return it to within the ideal range?

16. Three unlabelled bottles containing solutions were found in the chemical storeroom. The labels had fallen off the bottles and were found on the shelf. The solutions contained hydrochloric acid, sodium chloride, and sugar. A series of tests was carried out to identify the chemicals (Table 1). Identify the three solutions. (If necessary refer to Table 3, Colours of Common Synthetic Acid–Base Indicators, in Section 7.2.) (7.1, 7.2) T/I

Table 1 Summary of Evidence from Tests on Three Unknown Solutions

Solution	Colour with bromothymol blue	Reaction with magnesium	Electrical conductivity
A	blue to yellow	bubbles	high
B	remains blue	no reaction	high
C	remains blue	no reaction	none

17. An environmental chemist tested two different 500 g soil samples, taken from different sites, to compare their buffering capacity. Two identical volumes of acidic water were passed through the two soil samples and collected. The chemist then slowly added a sodium hydroxide solution to neutralize each sample of acidic water. Sample 1 required 5.2 mL and Sample 2 required 6.4 mL of basic solution to neutralize the acid. (7.5–7.8) K/U T/I
    (a) Define "buffering capacity."
    (b) Why are different volumes of NaOH required?
    (c) Which soil sample has the highest buffering capacity? Give your reasoning.

18. What information must be on the labels of chemical containers? (5.3) K/U T/I

Create and Evaluate

19. Some parents let their infants go to sleep with a bottle of apple juice. Discuss, with reasons, whether this is a good or a bad idea. (7.2) C A

20. (a) Design an acid spill kit or a base spill kit for home use. Include in your plans
    - the names and active ingredients of the substances to be included
    - any safety information that should appear on the label of each product
    - instructions for using the kit
    - a way of knowing when a spill has been neutralized

    (b) Evaluate the usefulness of your kit. (7.5, 7.6) T/I

Reflect on Your Learning

21. (a) Before you started this chapter, what did the terms "acid" and "base" mean to you?
    (b) How did your understanding of these terms change as you worked through this chapter?
    (c) What questions do you still have about acids and bases?
    (d) Suggest two resources where you could perhaps find answers to your questions.

22. When hearing about accidents involving the transportation or use of acids and bases, many people wonder, "Why don't they ban the use of these dangerous chemicals?"
    (a) Was this your opinion before reading this chapter? Explain.
    (b) Has your opinion changed as a result of reading this chapter? Explain.

Web Connections 

23. Research and identify two examples of acids commonly found in skin care products. What are these products used for? T/A A

24. Research the use of acids or bases in an industrial or environmental process. Assemble your findings into a presentation of your choice. T/A A C

CHAPTER 7

SELF-QUIZ

The following icons indicate the Achievement Chart category addressed by each question.

K/U Knowledge/Understanding T/I Thinking/Investigation C Communication A Application

For each question, select the best answer from the four alternatives.

1. Which of the following is a property of acids? (7.1, 7.2) K/U
   (a) slippery feel
   (b) bitter taste
   (c) ability to conduct electricity
   (d) ability to turn red litmus blue

2. Substances that change color depending on the acidity or basicity of a solution are called
   (a) indicators.
   (b) buffers.
   (c) electrolytes.
   (d) antacids. (7.2) K/U

3. Which liquid in Table 1 is the most acidic? (7.3) K/U

Table 1 The pH of Four Liquids

	Liquid	pH
(a)	milk	6
(d)	lemon juice	2
(c)	vinegar	3
(d)	ammonia	11

4. What is the pH of a neutral solution? (7.3) K/U
   (a) 0
   (b) 1
   (c) 7
   (d) 14

Indicate whether each of the statements is TRUE or FALSE. If you think the statement is false, rewrite it to make it true.

5. An oxyacid is an acid that contains only two elements. (7.2) K/U

6. In an acidic solution, the hydrogen ion and hydroxide ion concentrations are equal. (7.3, 7.5) K/U

7. Acid–base neutralization reactions are single displacement reactions. (7.5) K/U

Copy each of the following statements into your notebook. Fill in the blanks with a word or phrase that correctly completes the sentence.

8. The products of a neutralization reaction are an ionic compound and ______. (7.5) K/U

9. The ______ scale is a numerical scale ranging from 0 to 14 that is used to compare the acidity of solutions. (7.3) K/U

Match the term on the left with the appropriate description on the right.

10. (a) acid — (i) produces hydroxide ions in solution
    (b) base — (ii) is produced from hydrogen and hydroxide ions
    (c) ionic compound — (iii) produces hydrogen ions in solution
    (d) water — (iv) also known as a salt
    (7.2, 7.5) K/U

Write a short answer to each of these questions.

11. Why are acids and bases considered to be electrolytes? (7.2) K/U

12. Write the correct names and formulas of three acids and three bases. (7.2) K/U

13. Complete and balance the following equations: (7.3) K/U C
    (a) $HNO_3 + KOH \rightarrow$
    (b) $Ca(OH)_2 + H_2SO_4 \rightarrow$
    (c) $H_2SO_4 + NaOH \rightarrow$

14. (a) Identify four uses of acids.
    (b) Identify four uses of bases.
    (7.2, 7.3, 7.5, 7.7) A

15. (a) What two ions typically combine when an acid and a base react?
    (b) What is the product of this reaction?
    (c) What is this type of reaction called? (7.5) K/U

16. You test the pH of your backyard pool and find that the pH is 8.2. (7.3, 7.5) K/U A
    (a) Is the pool water acidic or basic?
    (b) At this pH, how is the water quality affected?
    (c) How could you adjust the pH of your swimming pool to be in the acceptable range?

17. Your friend wants to determine whether a household product is an acid or a base. Write a note to your friend explaining how she could accomplish this task. Include safety precautions. (7.1, 7.2) T/I C

18. Where would you expect the pH of the most corrosive substances to lie on the pH scale? (7.3) T/I

19. Write a paragraph describing neutralization reactions. Include the following terms at least once in the paragraph: acid, base, pH, hydrogen ion, hydroxide ion, neutralization, and neutral. (7.2, 7.3, 7.5) C

20. (a) How could turning out the light when leaving a room reduce your contribution to acid rain formation?
    (b) Name two other practices that could reduce your contribution to acid rain formation. (7.8) A

21. (a) Many soft drinks and fruit drinks have a pH of less than 5. Why do dentists discourage people from consuming too many of these beverages?
    (b) What beverages could you substitute for soft drinks that would be less damaging? Explain your answer. (7.3) T/I A

22. (a) Give two environmental applications of neutralization reactions.
    (b) Give two consumer applications of neutralization reactions. (7.5, 7.7, 7.8) A

23. Describe how you could use chalk and vinegar to demonstrate the effect of acid rain on statues and buildings. (7.8) T/I

LOOKING BACK

UNIT C Chemical Reactions

CHAPTER 5 Chemicals and Their Properties

KEY CONCEPTS

A substance's chemical and physical properties determine its usefulness and effects.

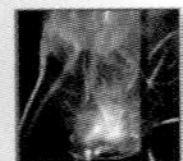
Changes can be classified as chemical or physical.

We can classify pure substances by observing their properties.

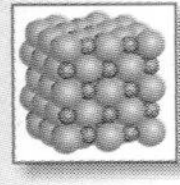
Ionic compounds are made up of positive and negative ions.

Molecular compounds are made up of distinct molecules.

Many consumer products are developed from petrochemicals.

CHAPTER 6 Chemicals and Their Reactions

KEY CONCEPTS

Chemical reactions involve the change of one or more substances into one or more different substances.

Chemical reactions obey the law of conservation of mass and can be represented by balanced chemical equations.

Chemical reactions involving consumer products can be useful or harmful.

We can use patterns in chemical properties to classify chemical reactions.

Corrosion is the reaction of metals with substances in the environment.

Chemical reactions affect us and our environment.

CHAPTER 7 Acids and Bases

KEY CONCEPTS

Acids are aqueous solutions that have characteristic properties.

Bases are aqueous solutions that have characteristic properties.

Acids and bases have a significant impact—both good and bad—on society and the environment.

The acidity of solutions is measured using the pH scale.

Acids and bases react together in neutralization reactions.

Industrial and vehicle emissions that cause acid precipitation can be reduced by technology.

MAKE A SUMMARY

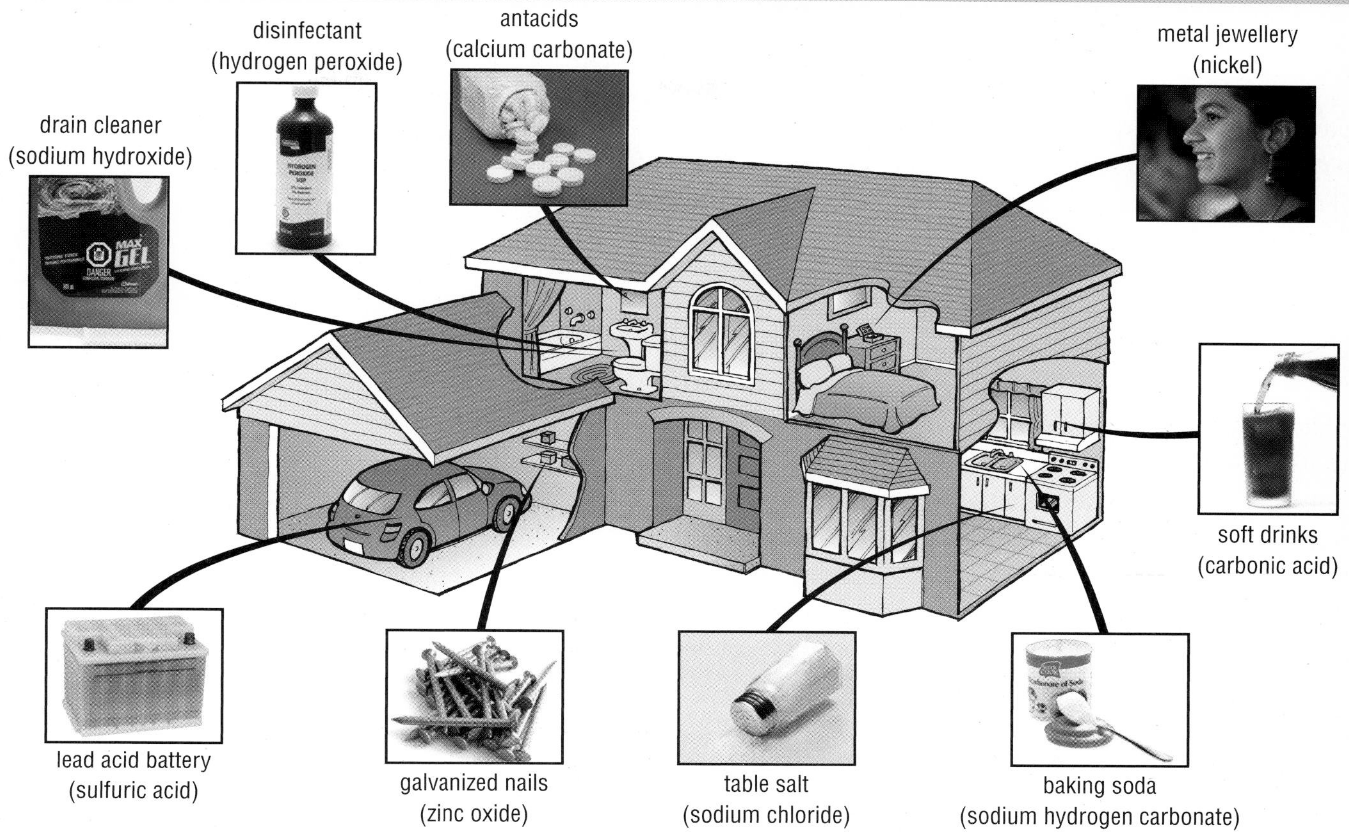

drain cleaner (sodium hydroxide)
disinfectant (hydrogen peroxide)
antacids (calcium carbonate)
metal jewellery (nickel)
soft drinks (carbonic acid)
lead acid battery (sulfuric acid)
galvanized nails (zinc oxide)
table salt (sodium chloride)
baking soda (sodium hydrogen carbonate)

Figure 1

Figure 1 shows nine common products found in our homes. An important ingredient in each product is named. Provide the following information for each named compound:

1. Write the chemical formula of the named ingredient. K/U
2. Classify it as being either ionic or molecular. K/U
3. Describe its most important physical and chemical properties. K/U
4. Write the balanced chemical equation for at least one chemical reaction involving this substance (either as a reactant or product). K/U T/I
5. Identify at least one common use of the compound. A
6. Identify at least one risk or benefit related to this compound. A

CAREER LINKS

List the careers mentioned in this unit. Choose two of the careers that interest you, or choose two other careers that relate to chemical reactions. For each of these careers, research the following information:

- educational requirements (secondary and post-secondary)
- skill/personality/aptitude requirements
- potential employers
- salary
- duties/responsibilities

Assemble the information you have discovered into a job application. Your application should mention both careers, but should focus on the one you are applying for, giving your reasons. In your application, explain how your chosen job relates to chemical reactions.

GO TO NELSON SCIENCE

Acid Shock: A Silent Killer

During a study of the wildlife in a local stream, biologists began to notice a disturbing pattern (Figure 1). Each spring, the population of young fish and amphibians plummeted and then recovered (Figure 2). The population crash seemed to happen soon after the first warm days in spring, when the snow started to melt.

The biologists knew that falling snow collects air pollutants, so they began to suspect that air pollutants were being trapped and stored in the snow over winter. The biologists wondered whether the melting snow released its stored pollutants gradually or dumped a sudden dose of pollutants into the waterways. The biologists wondered whether sulfur dioxide was the culprit. It has all the right properties: it is soluble in water and forms an acid when it reacts with water. Did the pH of meltwater from snow remain constantly acidic, they wondered, or could the first snow melt contain a big dose of acid? Rather than wait for spring, the biologists designed a model to test their idea.

You will now take the role of one of the scientists on the team. You will use the model they have designed, along with the skills you have learned throughout this unit, to test the "big dose of acid" idea. You will then use your findings and your understanding of chemical reactions to recommend ways to fix the problem.

Testable Question

Does the pH of fluid from a melting acidic mixture change over time?

SKILLS MENU

- Questioning
- Hypothesizing
- Predicting
- Planning
- Controlling Variables
- Performing
- Observing
- Analyzing
- Evaluating
- Communicating

Prediction

Predict the pH changes that occur, if any, as frozen vinegar melts.

Experimental Design

You will freeze a sample of vinegar to simulate frozen snow contaminated with acidic pollutants. As the vinegar melts, you will collect at least five fluid (meltwater) samples. You will measure the pH of these samples with a pH meter, recording the dependent and independent variables.

You will check that your observations are reproducible.

Equipment and Materials

- eye protection
- lab apron
- 100 mL graduated cylinder
- plastic cup
- pH meter
- hot water bath
- 6 small plastic vials or beakers
- vinegar
- buffer solution (for calibrating the pH meter)

Figure 1 Environmental biologists track the populations of aquatic animals through the year.

Figure 2 Leopard frogs, and their tadpoles, are one of the species that the team is monitoring.

Procedure

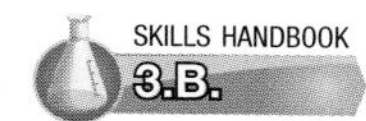

1. Freeze 100 mL of vinegar in a plastic cup overnight.
2. Read the Experimental Design, the Equipment and Materials list, and these procedural steps. Write your own procedure for collecting and testing meltwater from the frozen sample. Your procedure should include the following points:
   - Calibrate the pH meter with buffer solution, according to the manufacturer's instructions (Figure 3).
   - Practise using the pH meter.
   - Examine the length of the electrode portion of the pH meter. Determine the minimum depth of fluid you will need to take a reading.

   A warm water bath can be used to speed up the melting process.

Figure 3 A pH meter can detect small changes in pH.

Analyze and Evaluate

(a) What are the controlled, dependent, and independent variables in this investigation?

(b) Plot a graph of the pH changes occurring during the experiment.

(c) Answer the Testable Question. Justify your answer.

(d) Compare your answer in (c) to your Prediction. Comment on any differences.

(e) Evaluate the Experimental Design and your skill at conducting the investigation. Are there any parts of the investigation that might cast some doubt on your results? What improvements would you recommend?

(f) The melting block of vinegar was a model of thawing snow contaminated with acidic pollutants. Evaluate the strengths and limitations of the model. How could the model be made more realistic?

Apply and Extend

(g) Make recommendations to prevent the population crash of tadpoles in the spring. Assume that it is not possible, in the short term, to eliminate pollutants at their source. Include a discussion of the specific chemicals you recommend and the chemical reactions they undergo. Justify your choices. Consider the impact of your recommended chemicals on the environment. How could further unwanted environmental impacts be minimized?

ASSESSMENT CHECKLIST

Your completed Performance Task will be assessed according to the following criteria:

Knowledge/Understanding

- ☑ Understand the concepts of solutions and changes in state.
- ☑ Write balanced chemical equations for the chemical reactions involved.
- ☑ Demonstrate an understanding of the pH scale and the process of acid–base neutralization.

Thinking/Investigation

- ☑ Plan an investigation.
- ☑ Conduct fair test safely.
- ☑ Record observations accurately and in an organized manner.
- ☑ Analyze the results.
- ☑ Evaluate the Experimental Design and skills.

Communication

- ☑ Prepare a suitable lab report that includes a complete Procedure, summary of the observations, graph, any necessary analysis, and an evaluation of the investigation.
- ☑ Demonstrate an understanding of the language and symbols used to represent chemical reactions.

Application

- ☑ Use a model to represent an environmental challenge.
- ☑ Demonstrate an understanding of how chemical reactions can be used to address an environmental challenge such as acid precipitation.

The following icons indicate the Achievement Chart category addressed by each question.

K/U Knowledge/Understanding T/I Thinking/Investigation C Communication A Application

What Do You Remember?

For each question, select the best answer from the four alternatives.

1. Which of the following is evidence of a chemical change occurring? (5.1) K/U
   (a) Energy is released or absorbed.
   (b) A precipitate forms.
   (c) A gas is produced.
   (d) All of the above.
2. Which of the following compounds has a name that ends in *-ate*? (5.9) K/U
   (a) $NaNO_3$
   (b) KOH
   (c) CaO
   (d) N_2O
3. Which of the following is an ionic compound? (5.6) K/U
   (a) NH_3
   (b) $PbSO_4$
   (c) $C_6H_{12}O_6$
   (d) OCl_2
4. Which of the following compounds has a prefix in its name? (5.7, 5.10) K/U
   (a) PbO_2
   (b) $Ba(OH)_2$
   (c) P_2O_3
   (d) Al_2O_3
5. Which of the following causes rust? (6.10) K/U
   (a) water only
   (b) oxygen only
   (c) water and oxygen
   (d) salt only
6. WHMIS requires employers to provide safety information through
   (a) product labels
   (b) MSDS
   (c) employee training
   (d) all of the above (5.3) K/U
7. Which of these observations is the best evidence that a chemical reaction is taking place? (5.1, 5.2) K/U
   (a) An open flame appears.
   (b) A liquid changes to gas.
   (c) An object changes volume.
   (d) A solid disappears into a liquid.
8. What is the ending of the name of the compound with the formula Na_2SO_4? (5.9) K/U
   (a) -ane
   (b) -ate
   (c) -ide
   (d) -ite
9. Based on their locations in the periodic table, which combination of elements most often forms ionic compounds? (5.4–5.6) K/U
   (a) an element from period 1 combined with an element from period 2
   (b) an element from period 1 combined with an element from period 7
   (c) an element from group 1 combined with an element from group 17
   (d) an element from group 17 combined with an element from group 18

Indicate whether each of the statements is TRUE or FALSE. If you think the statement is false, rewrite it to make it true.

10. An ionic bond will form between two elements that are non-metals. (5.6, 5.10) K/U
11. Non-metallic elements form anions, while metallic elements form cations. (5.5) K/U
12. The element arsenic (atomic number 33) forms an anion with an ionic charge of +3. (5.5) K/U
13. Covalent bonds result from the transfer of electrons from one element to another. (5.6, 5.10) K/U

14. The law of conservation of mass states that the total number of reactant molecules and total number of product molecules in a chemical reaction are equal. (6.3) K/U K/U

15. The reaction of hydrochloric acid and potassium hydroxide is an example of both a neutralization reaction and a double displacement reaction. (6.6, 7.5) K/U

16. The products of the complete combustion of a hydrocarbon are carbon monoxide and water. (6.9) K/U

17. The reactants of a single displacement reaction are both elements. (6.6) K/U

18. Dissolving sulfur dioxide gas in water produces a basic solution. (7.8) K/U

19. Metals form positively-charged ions by gaining electrons. (5.5) K/U

20. Atoms of carbon dioxide are held together by ionic bonds. (5.10) K/U

21. Sulfur dioxide gas released from some power plants is one of the causes of acid precipitation. (7.8) K/U

22. When magnesium chloride ($MgCl_2$) is dissolved in water, it releases Mg^+ and Cl_2^- ions. (5.6) K/U

23. When a hydrocarbon such as methane, CH_4, burns where there is very little oxygen, one of the products is carbon monoxide, CO. (6.9) K/U

Copy each of the following statements into your notebook. Fill in the blanks with a word or phrase that correctly completes the sentence.

24. A(n) ____________ change is one that involves the formation of a new substance. (5.1) K/U

25. The substances found on the right-hand side of a chemical equation are called ____________ . (6.4) K/U

26. A(n) ____________ cannot be chemically broken down into simpler substances. (5.1) K/U

27. Atoms gain, lose, or share electrons to get electron arrangements like their nearest ____________ . (5.5) K/U

28. Scrubbing smokestack emissions is effective at removing ____________ emissions from smelting operations. (7.8) K/U

29. ____________ is the corrosion of iron. (6.10) K/U

30. A substance that changes colour in acids or bases is called a(n) ____________ . (7.2) K/U

31. Acid leaching can be used to treat soils contaminated with ____________ . (7.3) K/U

32. A change of state, such as melting, is a ____________ change. (5.1, 5.2) K/U

33. A pure substance made up of molecules that contain atoms of more than one element is called a ____________. (5.10) K/U

34. Atoms tend to form bonds to achieve electron arrangements like those of Group ____________ elements. (5.5, 5.10) K/U

35. Acid–base indicators change colour because of changes in the ____________ ion concentration of a solution. (7.2) K/U

Write a short answer to each of these questions.

36. The compounds $BaSO_4$ and KCl are the products of a chemical reaction. Write a balanced chemical equation for this reaction. (6.1–6.4) K/U T/I

37. Copper turns black and becomes heavier when it reacts with oxygen in a Bunsen burner flame. What type of chemical reaction is this? (6.5) K/U

38. How does the addition of calcium oxide, CaO, reduce the acidification of water? Use specific chemical equations to support your answer. (7.5) K/U C

39. Describe how the periodic table can be used to predict the ionic charges of the following elements:

    (a) Ca
    (b) S
    (c) K
    (d) Al (5.4, 5.5) K/U

40. Which of the four basic types of reactions occurs as rust forms on iron? (6.5, 6.9) K/U

41. Write the chemical equation for the complete combustion of ethyne, C_2H_2, also known as acetylene. (6.9) K/U C

What Do You Understand?

42. Use a Bohr–Rutherford diagram to explain why the phosphide ion has an ionic charge of –3. (5.5) K/U

43. List three elements or ions that have the same number of electrons as the sulfide ion, S^{2-}. (5.5) K/U

44. Write the chemical formula for each of the following compounds. (5.6–5.10) K/U
    (a) magnesium chloride
    (b) aluminum sulfide
    (c) tin(II) sulfate
    (d) iron(III) oxide
    (e) lead(II) nitrate
    (f) silver phosphate
    (g) sulfuric acid
    (h) hydrochloric acid
    (i) chlorine dioxide
    (j) dinitrogen monoxide

45. Name these compounds. (5.6–5.10) K/U
    (a) K_2O
    (b) CuS
    (c) Na_3PO_4
    (d) $Pb(OH)_2$
    (e) $HNO_3(aq)$
    (f) CO
    (g) NO

46. Describe how a balanced chemical equation is like a "recipe" for the chemical reaction. (6.4) K/U

47. Classify the following chemical equations as representing synthesis, decomposition, single displacement, double displacement, or combustion reactions. (6.5, 6.6, 6.9) K/U
    (a) ammonia + sulfuric acid → ammonium sulfate
    (b) aluminum + copper(II) chloride → aluminum chloride + copper
    (c) phosphoric acid + sodium hydroxide → water + sodium phosphate
    (d) aluminum sulfate → aluminum oxide + sulfur trioxide
    (e) ethane (C_2H_6) + oxygen → carbon dioxide + water

48. Write a balanced chemical equation for each word equation in question 32. (6.3–6.9) K/U C

49. The refining of iron ore into iron in a blast furnace is an important industrial chemical process (Figure 1). During this reaction, carbon reacts with iron(III) oxide to produce iron and carbon dioxide. (6.1, 6.3, 6.6) K/U T/I
    (a) Write a word equation for this reaction.
    (b) Write a balanced chemical equation for this reaction.
    (c) Classify this reaction.

Figure 1 A blast furnace

50. Carbon dioxide reacts with water to form a weak acid called carbonic acid. (7.3, 7.8) A
   (a) What effect is excess carbon dioxide in the atmosphere likely to have on the pH of the world's oceans?
   (b) Why is this a concern for us?

51. Draw Bohr–Rutherford diagrams of specific examples to explain why some atoms form cations while others form anions. (5.5) K/U C

52. Elements in the same group (column) of the periodic table tend to have similar chemical properties. Explain the reason for this pattern in terms of the element's electrons. (5.4, 5.5) K/U C

53. Write the names of the compounds represented by these formulas: (5.6–5.10) K/U C
   (a) K_2S
   (b) CBr_4
   (c) FeO
   (d) $CuSO_4$
   (e) $AgNO_3$
   (f) PbO_2
   (g) N_2O

54. Write a balanced chemical equation for the following word equation.

   iron(III) chloride + tin(II) chloride →
   iron(II) chloride + tin(IV) chloride
   (6.3, 6.6) C

Solve a Problem

55. A shiny, silver-coloured substance X is a good conductor of electricity. The substance is burned in oxygen to produce a white solid. The white solid is then added to water to produce a solution with a pH of 10. (5.4, 5.7, 7.5) K/U T/I
   (a) What type of substance is X? Give reasons for your answer.
   (b) If substance X acquires an ionic charge of +2 when it reacts, what is the likely chemical formula of the white solid?
   (c) Suggest a compound that can be used to neutralize the solution so that it can be safely disposed of.

56. The following chemical equation represents the combustion of propane in a sealed container:

   $C_3H_8(g) + 5\ O_2(g) \rightarrow 3\ CO_2(g) + 4\ H_2O(g)$

   Use the data in Table 1 to determine the mass of oxygen consumed in the reaction. (6.3, 6.9) T/I

Table 1

Item	Mass (g)
container only	50.0
propane and container	61.8
carbon dioxide, water, and container	104.8

57. The following chemical equation describes the chemical reaction that occurs when calcium metal is added to an open test tube containing water:

   $Ca(s) + 2\ H_2O(l) \rightarrow 2\ H_2(g) + Ca(OH)_2(aq)$
   (5.1, 6.4, 7.2) T/I

   (a) Explain the difference between the symbols (aq) and (l).
   (b) Predict what you would see in the test tube as the reaction proceeds.
   (c) How can you tell that the reaction is complete?
   (d) Predict how the mass of the contents of the test tube changes as the reaction proceeds. Explain your prediction.
   (e) Predict what you would observe if a few drops of phenolphthalein indicator are added to the test tube after the reaction is complete. Explain your prediction.

58. What can you and your family do immediately and in the long-term to help reduce emissions that cause acid precipitation? (7.8) A

59. An experiment was carried out in which sodium hydrogen carbonate (baking soda) reacted with an acetic acid solution (vinegar). The products are sodium acetate, water, and carbon dioxide gas. The reaction is represented by the equation

$$NaHCO_3 + CH_3COOH \rightarrow NaCH_3COO + H_2O + CO_2$$

In the experiment, 42 g of $NaHCO_3$ reacted completely with 30 g of CH_3COOH. Analysis of the products showed that 41 g of $NaCH_3COO$ and 9 g of H_2O had been produced. The mass of the CO_2 gas was unknown because it bubbled off into the air. Calculate much CO_2 was produced. (6.3, 6.4) T/I

60. Element X forms a compound with magnesium with the formula Mg_3X_2. What would be the formula of the compound that element X forms with hydrogen? (5.6, 5.7) T/I

Create and Evaluate

61. Since the 1960s, the number of cars on Canadian roads has increased faster than the Canadian population. (7.8) A
    (a) How do you think this trend has affected acid precipitation? Justify your speculation.
    (b) Propose a plan of action that the planning committee of a new fast-growing town should adopt to minimize the acid-causing emissions from cars in their town.

62. (a) Describe how you would design a new community in order to minimize the vehicle emissions that cause acid precipitation.
    (b) How does the design of your community compare to the community where you live now? (7.8) A

63. "Tadpoles and snails are indicator species of the health of an ecosystem, just as bromothymol blue is an indicator of the acidity of a solution." Compare the uses of the word "indicator" in this statement. (7.2) K/U

64. Acid precipitation affects the health of Ontario's forests. (7.8) A
    (a) How does this affect you (directly or indirectly)?
    (b) Identify two sectors of the Ontario economy that feel the impact of this directly.
    (c) What can you or your family do to help reduce acid precipitation?

65. Only vehicles with two or more occupants can use high-occupancy vehicle or HOV lanes (Figure 2). Explain and evaluate the environmental benefits of this practice. A

Figure 2

66. A slowdown in the global economy will likely result in less demand for metals mined in Ontario. How might an economic slowdown impact acid precipitation in Ontario? Why? (7.8) M1

67. Two possible products of fossil fuel combustion are carbon monoxide, CO, and carbon dioxide, CO_2. For each of these products, describe
    (a) one way it is produced,
    (b) one hazard it creates, and
    (c) one way to reduce the amount that enters the atmosphere. (6.9) T/I

Reflect on Your Learning

68. How has your attitude toward the waste materials that you pour down the drain changed after completing this unit? Which ideas influenced you the most? Why?

69. How have your ideas about air pollution changed because of this unit?

Web Connections 

70. Research and compare two related consumer products or processes that involve chemical reactions (e.g., teeth whitening procedures, types of fuels). Evaluate the two products using a rubric or rating scale of your own design. Make a recommendation for the best product or process, with your reasons. A T/A

71. Research and prepare a report describing two opposing positions on a current environmental issue involving chemicals. Choose one of the sample issues below or another important issue affecting your community.
    - Are "green alternatives" as effective as commercial household chemical products (e.g., cleaners, polishes, disinfectants)?
    - Are biofuels the solution to our future energy needs?
    - Is salt a better disinfectant for pools than chlorine?
    - Should the transport of oil by ship through the Great Lakes be permitted?

    One position must be from an environmental point of view. The other position can be from one or more different points of view, such as moral/ethical, political, economic, social, or cultural.

    Present your report in an innovative format approved by your teacher. T/I C A

72. Ontario generates about a quarter of its electricity by burning coal. This process is a significant source of acid emissions. Research Ontario's plans to reduce or phase out coal-burning generating stations. Find out how this target date has changed over the last few years. Write a letter to your MPP to express your opinion on the use of coal for electricity generation. C A

For all Nelson Web Connections,
GO TO NELSON SCIENCE

SELF-QUIZ

The following icons indicate the Achievement Chart category addressed by each question.

K/U Knowledge/Understanding T/I Thinking/Investigation C Communication A Application

For each question, select the best answer from the four alternatives.

1. Which of the following properties of a substance is a chemical property? (5.1) K/U
   (a) density
   (b) flammability
   (c) hardness
   (d) solubility
2. Which of the following chemical formulas represents a compound? (5.10) K/U
   (a) CO
   (b) Na
   (c) P_4
   (d) Br_2
3. In which of the following pairs do the atoms or ions have the same total number of electrons? (5.5) K/U
   (a) N and P
   (b) F^- and O^{2-}
   (c) Na and K^+
   (d) Fe^{2+} and Fe^{3+}
4. The equation for the reaction of silver nitrate and copper is
   $2\ AgNO_3 + Cu \rightarrow 2\ Ag + Cu(NO_3)_2$
   Which type of reaction is this? (6.6) K/U
   (a) synthesis
   (b) decomposition
   (c) single displacement
   (d) double displacement
5. What is the name of the compound with the formula H_3PO_4? (7.2) K/U
   (a) phosphoric acid
   (b) hydrophoric oxide
   (c) hydrogen phosphate
   (d) phosphorus hydroxide
6. What is the pH of pure water? (7.3) K/U
   (a) 0
   (b) 1
   (c) 7
   (d) 14

Indicate whether each of the statements is TRUE or FALSE. If you think the statement is false, rewrite it to make it true.

7. Covalent bonds are formed by sharing electrons. (5.10) K/U
8. The chemical equation shown below is balanced. (6.3) K/U
   $CH_4 + 3\ O_2 \rightarrow CO_2 + 4\ H_2O$
9. Highly acidic solutions have low pH. (7.3)

Copy each of the following statements into your notebook. Fill in the blanks with a word or phrase that correctly completes the sentence.

10. The equation below represents a __________ reaction. (6.6) K/U
    $KCl + AgNO_3 \rightarrow KNO_3 + AgCl$
11. Some coal-fired power plants release oxides of sulfur and nitrogen into the atmosphere, which can lead to environmental damage in the form of __________ precipitation. (7.8) K/U
12. Complete the following word equation.
    nitric acid + potassium hydroxide $\rightarrow$
    water + ______________ (7.5) K/U

Match the formula on the left with the chemical name on the right.

13. (a) $FeCl_3$ (i) carbon tetrachloride
    (b) $SnCl_4$ (ii) tin(IV) chloride
    (c) $(NH_4)_3PO_4$ (iii) iron(III) chloride
    (d) CCl_4 (iv) carbon monoxide
    (e) CO (v) ammonium phosphate
    (5.7, 5.9, 5.10) K/U

Write a short answer to each of these questions.

14. Write the chemical equation for the following word equation.

    hydrochloric acid + potassium hydroxide →
    potassium chloride + water (6.3, 7.5) K/U C

15. Describe some of distinguishing the characteristics of elements in the following groups of the periodic table. (5.4)
    (a) Group 1 (alkali metals)
    (b) Group 2 (alkaline earth metals)
    (c) Group 18 (noble gases) K/U

16. Write the chemical formulas for the following compounds.
    (a) aluminum oxide
    (b) iron(II) chloride
    (c) ammonium sulfate (5.7, 5.9) K/U C

17. (a) Describe two physical properties of the element aluminum.
    (b) Describe one chemical property of the element aluminum. (5.4, 5.6, 6.10) A

18. An unknown element, X, combines with aluminum to form a compound with the empirical formula Al_2X_3. What would be the most likely formula for the compound that element X forms with calcium, Ca? Explain your answer. (5.4–5.6) T/I

19. An iron nail left exposed to the weather increased in mass slightly as it became coated with rust. Explain why this does not violate the law of conservation of mass. (6.3, 6.5, 6.10) T/I

20. When zinc, Zn, was added to sulphuric acid, H_2SO_4, a gas bubbled out of the reaction mixture. Analysis of the reaction products showed that zinc sulphate had been produced.
    (a) What was the gas that bubbled off?
    (b) Write a balanced equation for the reaction. (6.3, 6.6) T/I

21. Hydrogen peroxide, H_2O_2, is an unstable compound that readily decomposes in light to produce water and oxygen. Bottles of 3% hydrogen peroxide are sold in pharmacies for home use.
    (a) Why are hydrogen peroxide bottles dark brown rather that clear glass or plastic?
    (b) Write a balanced equation for the decomposition of hydrogen peroxide. (6.3, 6.5) K/U C A

22. Consider the following household liquids: tap water, whole milk, lemon juice, and drain cleaner.
    (a) Predict which liquid would have the highest concentration of hydrogen ions. Explain your prediction.
    (b) Predict which liquid would have the highest pH. Explain your prediction. (7.3, 7.4) A

23. A janitor used ammonia cleaner (active ingredient, NH_4OH) to clean and disinfect a school bathroom. He thought it would be best to neutralize the leftover ammonia before discarding it. The chemistry teacher told him he could use dilute hydrochloric acid and phenolphthalein. Describe the procedure the janitor should use, including all necessary safety precautions. (7.2, 7.4, 7.5) T/I A

24. In this unit, you have learned that reactions take place all around you. You should now be able to look at some everyday chemical changes and see them in a new way.
    (a) Choose two common chemical changes and explain how you understood them before you knew about chemical reactions.
    (b) Explain how you understand the same two chemical changes now. (6.1–6.10) A

UNIT D Climate Change

OVERALL Expectations

- analyze some of the effects of climate change around the world and assess the effectiveness of initiatives that attempt to address the issues of climate change
- investigate various natural and human factors that influence Earth's climate and climate change
- demonstrate an understanding of natural and human factors, including the greenhouse effect, that influence Earth's climate and contribute to climate change

BIG Ideas

- Earth's climate is dynamic and is the result of interacting systems and processes.
- Global climate change is influenced by both natural and human factors.
- Climate change affects living things and natural systems in a variety of ways.
- People have the responsibility to assess their impact on climate change and to identify effective courses of action to reduce this impact.

Focus on STSE

WHO **IS TO BLAME?**

A group of students is meeting around a large conference table. This is the First International Student Climate Change Congress. Students from around the world have decided to meet by themselves, since they are the ones who will inherit the Earth. The Canadian delegation is on edge because Canada's production of greenhouse gases is under discussion.

Per person, Canadians produce the most carbon dioxide

Americans used to be the largest per capita producers of carbon dioxide. However, the additional carbon dioxide released by Canada's forests because of the pine beetle has now put Canadians in first place.

1. Do you think it is fair to single out Canadians as the biggest producers of CO_2? C A
2. List reasons why you think Canadians are ranked as the top consumers of fossil fuels. T/I C

Per person, Canadians are one of the top ten producers of greenhouse gases

Countries that produce a lot of petroleum and natural gas emit huge amounts of methane as well as carbon dioxide. Small amounts of methane have the same effect as larger amounts of carbon dioxide. Many oil-producing countries with tiny populations, such as Kuwait and Brunei, produce more greenhouse gases per person than Canada.

3. If you came from Kuwait, how would you defend your production of greenhouse gases? A T/I

Canada as a country is the eighth largest producer of carbon dioxide

China produces the most carbon dioxide at 22 % of the world's total. The United States is second at 20 %. Canada produces about 2 %.

4. If you came from China, how would you defend your production of greenhouse gases? T/I A

We are all in it together

After a heated discussion, the student leaders realize that they will inherit the world together. It will not help to blame any one country. We all share the same atmosphere.

5. Do you agree? Or should it be up to just a few countries to solve the problem? C A

UNIT D

LOOKING AHEAD

UNIT D

Climate Change

CHAPTER 8

Earth's Climate System and Natural Change

CHAPTER 9

Earth's Climate: Out of Balance

CHAPTER 10

Assessing and Responding to Climate Change

What factors cause the two different climates visible in this photograph?

The Angel Glacier in Jasper National Park has retreated since 1939.

The construction of wind farms is one possible response to climate change.

UNIT TASK Preview

Global Climate Change

The Intergovernmental Panel on Climate Change (IPCC) is preparing to update its report on global climate change. You are part of a task force requested by the IPCC to examine international impacts of climate change.

In this task, you will select a location somewhere in the world and learn about its climate. You will collect data on the present climate and on the climate from about 50 years ago. You will then compare the two sets of data to identify any past changes in climate.

You will use these conclusions, along with information from this unit, to predict what climate-related changes may be happening now. You will also predict what will happen over the next 100 years. Next, you will conduct further research on the impacts of climate change in your location to see if your prediction agrees with that of climate experts.

Finally, you will compose a list of practical recommendations on what people in that region can do to prepare for the impacts of climate change. You will also compose a list of recommendations for international changes that can be made to limit the extent of climate change.

UNIT TASK Bookmark

The Unit Task is described in detail on page 444. As you work through the unit, look for this bookmark and see how the section relates to the Unit Task.

ASSESSMENT

You will be assessed on how well you

- research and analyze your data
- present your current and predicted climate-related changes
- make local and global recommendations on how to limit and prepare for climate change

What Do You Already Know?

PREREQUISITES

Concepts	Skills
• Radiation from the Sun • Greenhouse gases and the greenhouse effect • Effects of human activities on the environment	• Graphing data and analyzing graphs • Researching and collecting information • Planning and conducting experiments • Communicating scientific information appropriately

1. (a) Where do we obtain most of the energy we use to power our technologies?
   (b) What are the risks and benefits of using this energy source?
   (c) What other sources of energy are available? K/U

2. Water can occur in three different states. All three states of water are found naturally on Earth. K/U
   (a) List the three states of water.
   (b) Give a natural example of water in each state.

3. Explain how the climate of each region in Figure 1 is affected by its location. K/U C

Figure 1 (a) Tobermory is beside a large lake. (b) Vancouver Island is on the coast and has a warm ocean current nearby.

4. Describe and explain, as if to a younger student, what is happening in Figure 2. K/U C

Figure 2

5. List at least 10 ways in which human activities affect the environment. A

6. Draw a concept map similar to Figure 3 that includes the terms listed below. Draw lines to connect all the terms that you think may be related to each other. Write words or expressions on the lines to explain the relationships. K/U C

climate	humans	atmosphere
Sun	Earth	greenhouse gases

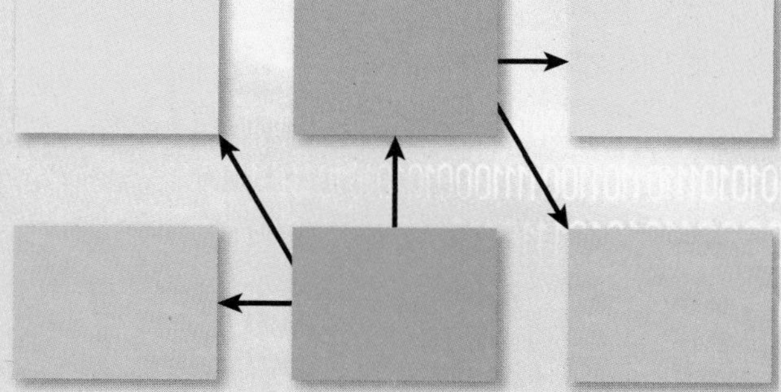

Figure 3

7. (a) Is all the information you find on the Internet reliable?
   (b) What do you look for on a website to be sure the site is unbiased and accurate? T/I C

CHAPTER

8

Earth's Climate System and Natural Change

KEY QUESTION: How does Earth's climate system keep Earth's temperature in balance, and what natural events cause changes in Earth's climate?

What factors cause the two different climates visible in this photograph?

UNIT D

Climate Change

CHAPTER 8

Earth's Climate System and Natural Change

CHAPTER 9

Earth's Climate: Out of Balance

CHAPTER 10

Assessing and Responding to Climate Change

KEY CONCEPTS

Earth's climate system is powered by the Sun.

Earth's climate system includes the atmosphere, the hydrosphere, the lithosphere, and living things.

The greenhouse effect keeps Earth warm by trapping thermal energy radiated by Earth.

Thermal energy is transferred within Earth's climate system through air and ocean currents.

Earth's climate experiences long-term and short-term changes.

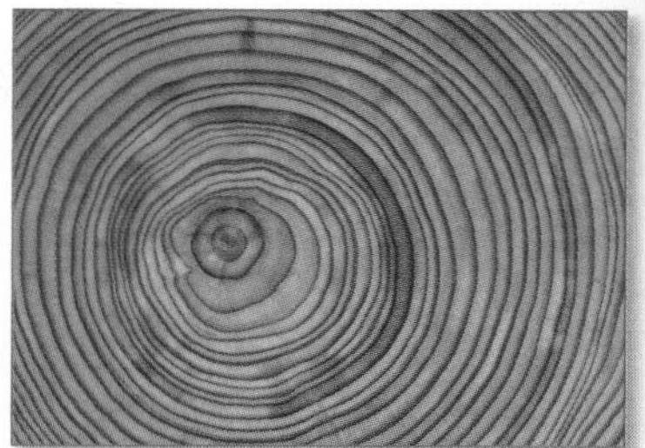

Scientists use natural ice cores, sediment layers, fossils, and tree rings to study past climates.

ENGAGE IN SCIENCE

Evidence of a Different Climate

This summer, Manuel Ramirez of Cardston, Alberta, made an exciting discovery. While scrambling up the slope of a hill, 15-year old Manuel noticed a dull brown object sticking out of the side of the hill. On taking a closer look, he saw that it looked like an enormous bone.

Manuel went back the next day, taking a small shovel with him. He carefully dug away the soil to reveal more bones embedded in the ground. Manuel notified the museum in Calgary of his discovery.

Paleontologists from the museum worked with Manuel to uncover the rest of the bones. The scientists took the bones back to the museum, where they washed and assembled them. The final skeleton looks like an enormous elephant.

Alberta's climate was very different 10 000 years ago. Some species from that time, such as woolly mammoths, are now extinct. Others, such as caribou, still exist in Alberta today.

"I thought elephants lived in warm climates like in India and Africa," said Manuel. "But the paleontologists told me that it's not an elephant—it's a woolly mammoth. Mammoths existed more than 10 000 years ago. They lived in very cold climates where much of the ground was covered in permanent ice. On the Internet, I've seen pictures of a baby mammoth that was found frozen in the ice."

"I still have some questions," says Manuel. "Alberta doesn't have an icy climate, so why did mammoths live here? How has Alberta's climate changed?"

1. From the information in the article, how has Alberta's climate changed?
2. What evidence is there that Ontario's climate has changed over time?

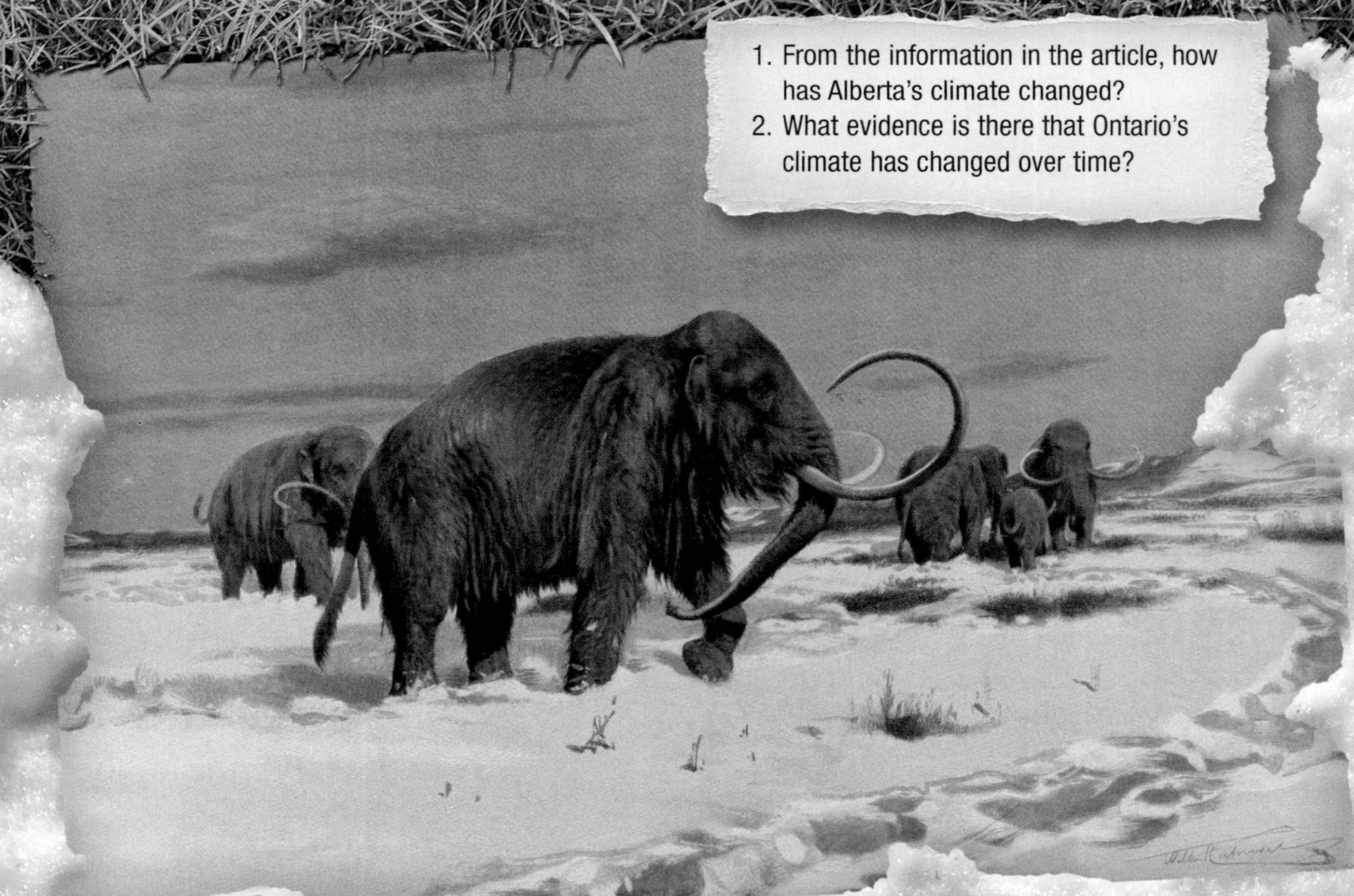

WHAT DO YOU THINK?

Many of the ideas you will explore in this chapter are ideas that you have already encountered. You may have encountered these ideas in school, at home, or in the world around you. Not all of the following statements are true. Consider each statement and decide whether you agree or disagree with it.

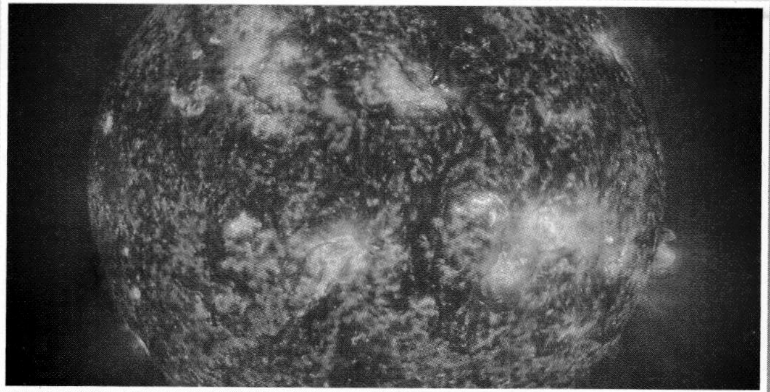

1 About half the energy on Earth comes from the Sun.
Agree/disagree?

2 The greenhouse effect is a natural phenomenon.
Agree/disagree?

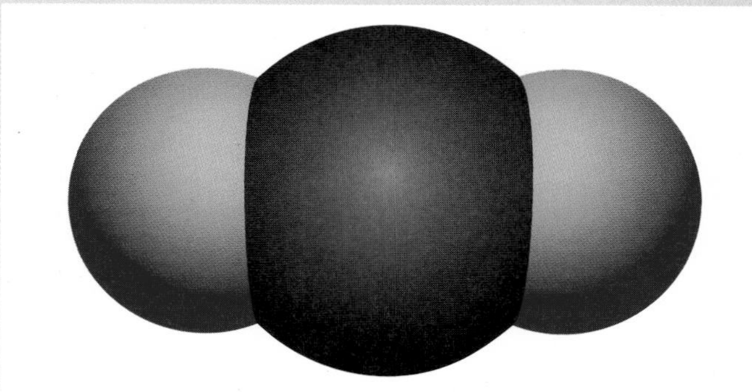

3 Carbon dioxide is an important part of Earth's climate system.
Agree/disagree?

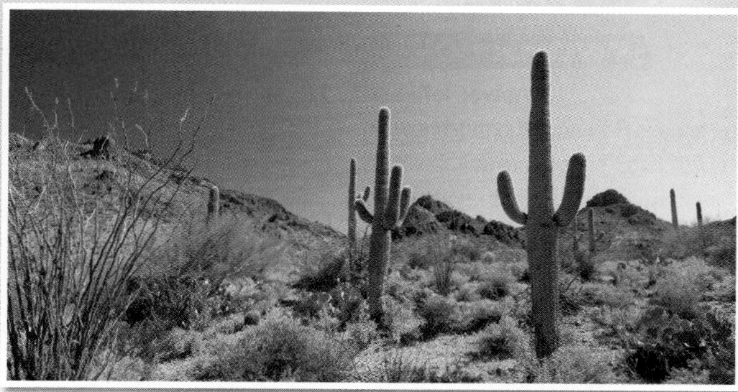

4 Earth's climate has remained stable for several thousands of years.
Agree/disagree?

5 Volcanic eruptions cause the climate to change.
Agree/disagree?

6 Weather and climate are the same thing.
Agree/disagree?

Finding the Main Idea

When you look for the main idea of a text, you identify a general topic (climate change) and a related key concept (*humans add to the greenhouse effect*). Details are usually given to support the main idea (*fossil fuels, unsustainable farming*). Use the following strategies to find the main idea:

- use clues in the title and the headings
- check the first, second, and last sentences
- look for repeated words and bolded words
- look for signal words (*therefore, consequently, for this reason*)
- change the heading or main idea into a question to see if other sentences explain it
- make a concept map to help separate details from the main idea

READING TIP

As you work through the chapter, look for tips like this. They will help you develop literacy strategies.

Components of Earth's Climate System

Would it be possible to live on another planet? Mercury has no atmosphere. During the day, Mercury's temperature is scorching hot at 450 °C, and at night, it drops to a freezing –170 °C. Venus's atmosphere is 100 times more dense than Earth's, and its surface temperature is over 700 °C. In contrast, Mars has an atmosphere 100 times less dense than that of Earth, and its average surface temperature is –63 °C. In fact, Earth may be the only planet in our solar system that can support life as we know it.

Finding the Main Idea *in Action*

A main idea narrows a topic to a specific point or opinion. Details in the text support the main idea by explaining it with facts or examples. Here is how one student found the main idea in the paragraph above.

Strategy	Information in the Text
Read the heading	I predicted that the text would be about the Earth's climate.
Reread the first sentence	The first sentence was a question: "Would it be possible to live on other planets?" This narrowed the topic down.
Looked for repeated words	The word atmosphere and temperature were repeated.
Looked for examples and facts	The examples told me why other planets were not suitable for life as we know it.
Reread the last sentence	The last sentence answered the question: "Would it be possible to live on other planets?"
Main Idea: The atmosphere and temperature of planets, other than Earth, are not able to support life as we know it.	

Weather and Climate

8.1

Is today a warm, sunny day or is it cold and rainy? When you describe the conditions outside on a particular day, you are describing the **weather**. Describe today's weather to a classmate, using as much detail as you can. Include the temperature and precipitation (if present). What other conditions can you describe?

weather atmospheric conditions, including temperature, precipitation, wind, and humidity, in a particular location over a short period of time, such as a day or a week

Describing the Weather

Scientists who study the weather are called meteorologists. Meteorologists usually provide the following information when they describe the weather:

- temperature
- type and amount of precipitation
- wind speed
- relative humidity (the amount of water vapour in the air relative to the maximum amount of water that it is possible for the air to hold at that temperature)
- atmospheric pressure (the force exerted on a surface by the weight of the air above it)
- presence of fog, mist, or cloud cover

A typical description of the weather on a summer day might be like this: a high of 28 °C today, sunny with cloudy periods, probability of precipitation 30 %, wind from the west at 20 km/h, and relative humidity of 40 %.

In some parts of the world, the weather stays more or less the same from day to day. For example, the Sahara desert in Africa is usually hot and dry during the day. In Canada, however, the weather can change dramatically from one day to the next (Figure 1). The weather may be warm and sunny today, but it could become cool and rainy tomorrow. However, you would not expect snow in Ontario in August or a temperature of 30 °C in Nova Scotia in February.

LEARNING TIP

Humidity and Temperature

Water vapour (the gaseous state of water) forms when liquid water evaporates. When water evaporates, the vapour mixes with air. Warm air can hold more water vapour than cold air. This is why warm air is often more humid than cold air.

To learn more about the work of a meteorologist,

GO TO NELSON SCIENCE

Figure 1 The weather in an area can change dramatically in a matter of hours.

Predicting the Weather

We often want to know what the weather will be like this afternoon, tomorrow, or on the weekend. Meteorologists gather information on weather around the world and use this information to forecast the weather for specific areas. Environment Canada maintains thousands of weather stations across the country. In some parts of the world, weather information has been collected and recorded daily since the 1800s. Methods of collecting weather data include weather stations, weather balloons, aircraft, and satellites (Figure 2).

To see weather predictions in your region,

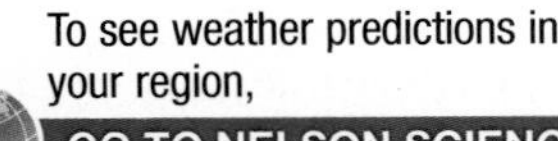

GO TO NELSON SCIENCE

Figure 2 This automatic weather station is located on Alexander Island, Antarctica. It records the daily temperature, precipitation, wind speed and direction, and humidity.

DID YOU **KNOW?**

Earth's Tilt Causes Seasons

Earth is presently tilted at 23.5°. Because of the tilt, during part of Earth's orbit the northern hemisphere is tilted away from the Sun (winter), and during another part it is tilted toward the Sun (summer). This tilt accounts for our distinct summer and winter seasons.

What causes weather? Interactions between water and air on Earth and energy from the Sun contribute to our weather. Energy originating from the Sun heats Earth's atmosphere, creating winds and other air movement. Water in oceans, lakes, and rivers evaporates, cools, and condenses. This process forms clouds that can produce rain or snow. Ocean water moves in currents from the poles to the equator and back again. Together, air movement and water movement create weather.

What Is Climate?

climate the average of the weather in a region over a long period of time

Climate is the usual pattern of weather in a region over a long period of time. To determine the climate of a region, climatologists collect weather measurements made over 30 years or more and average the results. Scientists who study climate include climatologists, paleoclimatologists, atmospheric scientists, and climate modellers.

To find out more about the differences between these groups of scientists,

GO TO NELSON SCIENCE

The climate of a region gives a range of temperatures that you might expect at a certain time of year. It also tells you whether to expect rain, snow, or high winds in certain seasons. For example, the climate in southern Ontario is warm and humid during the summer and cold with snow during the winter.

What is the difference between weather and climate? Weather describes the atmospheric conditions over a short period of time—an hour, a day, or even a week. Climate describes the typical weather you can expect in a region based on weather data gathered over many years. As the science fiction author Robert Heinlein wrote, "Climate is what you expect, but weather is what you get."

The climate of a region determines the types of plants and animals that live there. Animals, such as polar bears and seals, which live in the Arctic, must be able to survive the cold, dark winters (Figure 3). The few plants that live in the Arctic grow only during the summer months. In contrast, the climate of southern Ontario encourages the lush growth of a variety of trees, bushes, and other plants. Many insects, birds, and mammals live in central and southern Ontario forests.

Figure 3 (a) Polar bears have thick layers of fat under their fur to help them survive the winters. (b) Many plant and animal species, such as the red fox, live in Ontario forests.

IN SUMMARY

- Weather is a description of the atmospheric conditions, including temperature, precipitation, wind, and humidity, in a particular location over a short period of time.
- Climate is the average weather in a region over a long period of time, usually 30 years.
- The climate of a region determines the types of plants and animals that live there.

CHECK YOUR LEARNING

1. Classify the following as *either* a weather observation for a specific location and day *or* an aspect of a location's climate. Explain each choice with a short sentence. K/U
   (a) temperature highs and lows
   (b) precipitation
   (c) hours of sunshine
   (d) wind speed
   (e) humidity
2. List three different technologies that a meteorologist might use to collect weather data. K/U
3. (a) What factors should you include in a description of weather?
   (b) Describe today's weather in your area using as much detail as you can. K/U T/I
4. What is the difference between weather and climate? K/U
5. Describe the climate where you live. K/U
6. How does a region's climate determine its plant and animal life? Explain. K/U
7. Suppose that accurate weather forecasts could be extended from five days to seven days, but a government investment of $10 billion over 10 years would be necessary. Would this be a worthwhile investment? Create a chart that lists the pros and cons from the point of view of a ski hill operator, municipal works department, and private citizens who will pay the additional taxes as well as enjoy the additional benefits. T/I A C
8. Two students were discussing the difference between weather and climate. Sandeep said, "Since Ottawa's climate is characterized by a cold winter, and we're moving into November, the weather next week is definitely going to be colder than it was this week." Rodrigo replied, "I would agree with you if you changed the word 'week' to 'month.'" Do you agree with Rodrigo or with Sandeep? Explain in terms of weather and climate. Be prepared to share your answer with the class. K/U C
9. Suppose you were marooned on an island. You started collecting weather data so that you could come to a conclusion about the climate of the island. How long would you need to keep weather records before you could make a definite conclusion about the climate? T/I

8.2 Classifying Climate

The climate in your region affects the way you live. For example, houses in hot climates are often built on one level, with white walls and inner courtyards (Figure 1(a)). These features help keep the houses cool. Houses in colder climates are often smaller. They may have basements or more than one level and steep roofs to prevent snow from accumulating. These features make the houses easier to heat in the winter (Figure 1(b)). The climate also determines what you wear and what activities you do outside. How would your life be different if you lived in a different climate?

Figure 1 (a) Houses in hot climates are usually designed to stay cool. (b) Houses in colder climates are designed to stay warm during the winter.

Climate Zones

In the early 1900s, a scientist named Vladimir Köppen (1846–1940) used temperature, precipitation, and plant communities to identify climate zones. As you might expect, polar regions such as the Arctic and Antarctic are placed in the same climate zone. Northern Canada's cold spruce forests are classified with Russia's similar forest regions. Köppen's system has been revised since then, but current systems are similar to Köppen's original zones.

Figure 2 shows a map of Earth's main climate zones. In which climate zone do you live? Can you find another country containing the same climate zone?

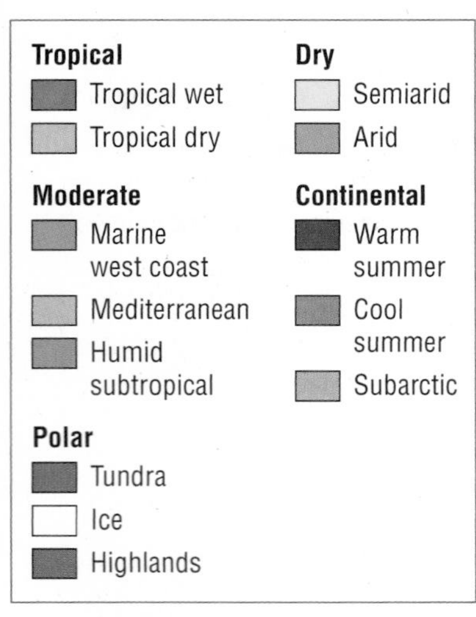

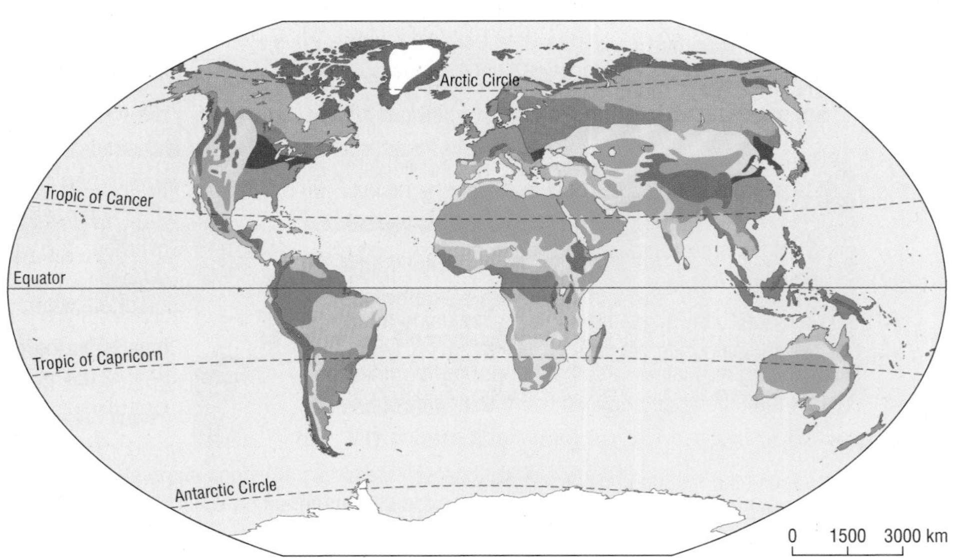

Figure 2 World climate zones

Ecoregions

Over the last 30 years, people have become more concerned about the survival of ecosystems. A new method of classifying climate has been developed to reflect this concern. New climate zones, which focus on the ecology of the region, are called ecoregions.

Ecoregions are based on landforms, soil, plants, and animals, as well as climate. They also consider human factors such as crops and urban centres. Figure 3 shows 867 distinct land-based ecoregions around the globe.

To access the Terrestrial Ecoregions of the World map and examine any ecoregion in detail,

GO TO NELSON SCIENCE

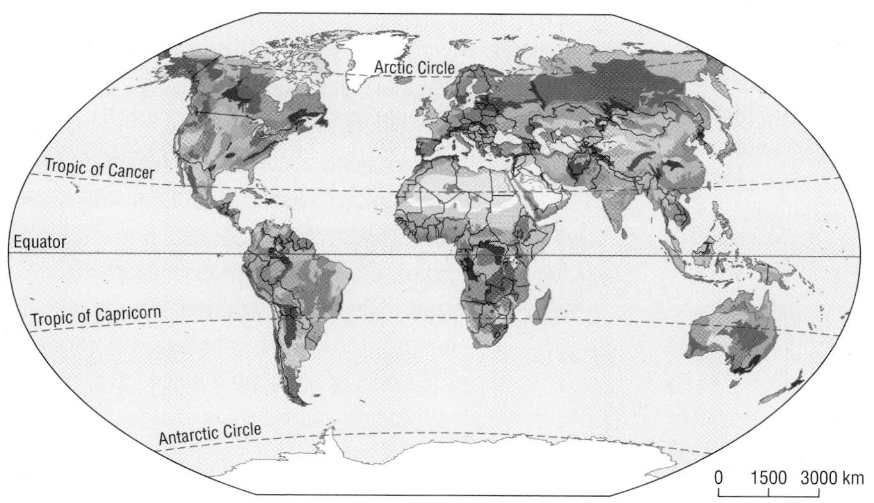

Ontario Ecoregions

- Western Great Lakes forest
- Midwestern Canadian Shield forests
- Southern Hudson Bay taiga
- Central Canadian Shield forests
- Eastern forest-boreal transition
- Eastern Great Lakes lowland forest
- Southern Great Lakes forests

Figure 3 Terrestrial ecoregions of the world. Which ecoregion do you live in?

Canada recently developed its own system for mapping ecoregions. In Canada's system, major ecozones are divided into smaller ecoregions. For example, the community of Timmins is in the Abitibi Plains ecoregion, which is within the larger Boreal Shield ecozone. Remember, Canada's ecozones are similar to global ecoregions.

LEARNING TIP

Ecoregion Versus Ecozone?

The global ecoregion classification is equivalent to Canada's ecozone classification. In the Canadian system, an ecozone is divided into smaller ecoregions. For example, the boreal ecozone contains 30 ecoregions. In the international system, the boreal forest is considered an ecoregion. Remember: Canadian Ecozone = Global Ecoregion.

Bioclimate Profiles

An ecoregion describes the climate and ecosystem of a region in its current state. A **bioclimate profile** is a series of graphs that show temperature and moisture conditions at a given location. An important difference between ecoregions and bioclimate profiles is that bioclimate profiles only describe climate. Also, bioclimate profiles display the location's projected climate 40 to 80 years into the future. Over 500 sites in Canada are described by bioclimatic profiles.

bioclimate profile a graphical representation of current and future climate data from a specific location

To learn more about bioclimate profiles,

GO TO NELSON SCIENCE

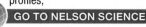

Factors Affecting Climate

Why do climate zones exist? The climate in a region is caused by a variety of factor. These factors include:

- the distance from the equator (latitude)
- the presence of large bodies of water
- the presence of ocean or air currents
- land formations
- the height above sea level (altitude).

As you read the rest of this chapter, you will revisit each of these factors to see how they affect climate.

RESEARCH THIS CLASSIFY YOUR CLIMATE

SKILLS: Researching, Evaluating

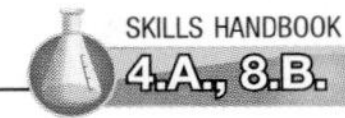

SKILLS HANDBOOK 4.A., 8.B.

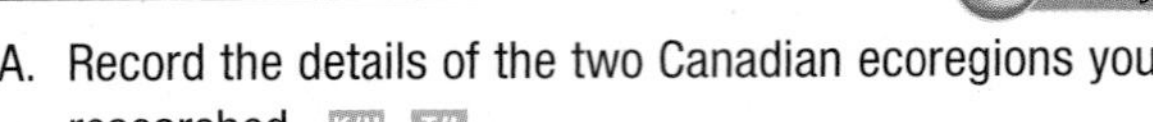

1. Research the Canadian ecozone and ecoregion that you live in. Research another Canadian ecozone and ecoregion that is very different from the one you live in.
2. Research (a) another country with an ecoregion similar to the Canadian ecozone you live in and (b) another country with an ecoregion that is different from the Canadian ecozone where you live.
3. Examine the Bioclimate Profiles website. Find a weather station close to your home and compare the maximum April to October temperatures for 1971–2000 with the maximum temperatures projected for the same months in 2070–2099.

GO TO NELSON SCIENCE

A. Record the details of the two Canadian ecoregions you researched. K/U T/I

B. Use a graphic organizer such as a Venn diagram or a compare and contrast chart to organize and display similarities and differences in the climate of the two Canadian locations. C

C. Which country with a similar ecoregion did you research? Describe how it is like your ecozone. T/I

D. Which country with a different ecoregion did you research? Describe how it differs from your ecozone. T/I

E. What does the bioclimate profile tell you about the climate where you live? How is this different from the description of an ecoregion? A

UNIT TASK Bookmark

You can apply what you learned in this section about climate zones and bioclimate profiles to the Unit Task described on page 444.

IN SUMMARY

- Traditional climate zones are usually classified based on temperature, precipitation, and vegetation.
- Ecoregions and ecozones classify climate regions based on a region's landforms, soil, vegetation, and human factors, in addition to temperature and precipitation variables.
- Bioclimate profiles display observed climate data and allow us to compare a region's current climate variables with those projected 40 to 80 years into the future.

CHECK YOUR LEARNING

1. (a) What three factors did Köppen take into account in his climate classification system?
   (b) What additional factors do we take into account when we classify climate into ecoregions? K/U
2. The average daily maximum temperature in central Egypt reaches 40 °C in the summer. There is very little rain—less than 10 mm per year. The lack of rain means that there is little vegetation in this part of the country. K/U
   (a) In what climate zone does central Egypt belong?
   (b) Is there any part of North America that belongs in the same climate zone?
3. Find another country far away that has climate zones similar to those in Canada. Predict the kind of global ecoregion this country might have. Use the Web to find out if your hypothesis is accurate. K/U T/I
4. Ontario has three ecozones, the largest of which is the Boreal Shield. The Boreal Shield subdivides into many smaller ecoregions. How many ecoregions in Ontario belong to the Boreal Shield? K/U
5. Bioclimate profiles are restricted to traditional climate variables such as temperature and precipitation. Why are they more useful than ecoregion descriptions for studying the future of Ontario's climate? T/I

The Sun Powers Earth's Climate System

8.3

So far, you have learned about climate zones around the world. To understand Earth's climate, you need to look at the bigger picture. Earth has a global **climate system**, which includes air, land, liquid water, ice, and living things. The climate system is powered by the Sun (Figure 1). The interactions between these components and the Sun produce climate zones. In this section, you will learn about the energy Earth receives from the Sun and what happens to this energy on Earth.

climate system the complex set of components that interact with each other to produce Earth's climate

Figure 1 The Sun powers the climate system.

The Balance of Energy on Earth

Almost all energy on Earth comes from the Sun. The Sun emits a number of different types of radiation (Figure 2). The types of radiation include **ultraviolet radiation** (invisible short wavelength, higher-energy radiation), visible light, and **infrared radiation** (invisible long wavelength, lower-energy radiation).

ultraviolet radiation a form of invisible higher-energy radiation

infrared radiation a form of invisible lower-energy radiation

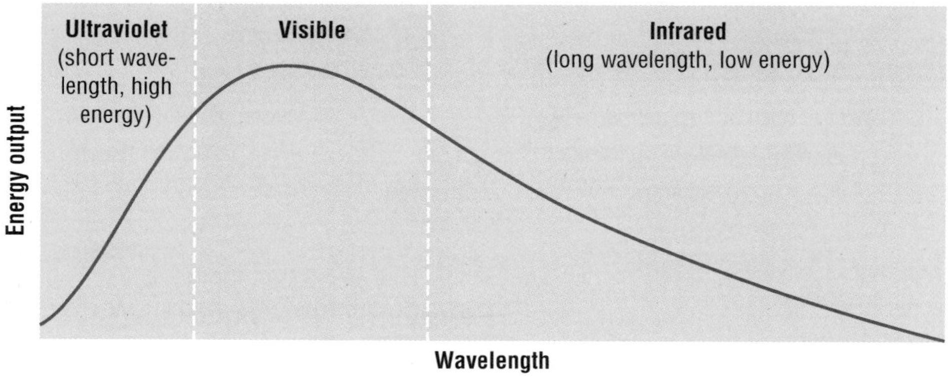

Figure 2 The Sun's electromagnetic spectrum at the top of the atmosphere

Earth Absorbs Energy from the Sun

When radiation contacts a particle of matter, one of three things happens.

1. The radiation may be absorbed by the particle, causing the particle to gain energy.
2. The radiation may be transmitted through the particle.
3. The radiation may be reflected off the particle.

READING TIP

Finding the Main Idea
The first place to look for the main idea is the first two sentences of a paragraph. The first sentence of a paragraph is called the topic sentence and this is where the main idea usually is stated. Sometimes the first sentence gives an introduction to the topic, so check the second sentence to see if it states the main idea.

What happens to radiation from the Sun once it reaches Earth? About 30 % of the energy is reflected back to space by clouds, particles in the atmosphere, and Earth's surface (Figure 3). The remaining 70 % is absorbed by Earth's surface, clouds, and certain gases in the atmosphere. This accounts for 100 % of the incoming radiation from the Sun.

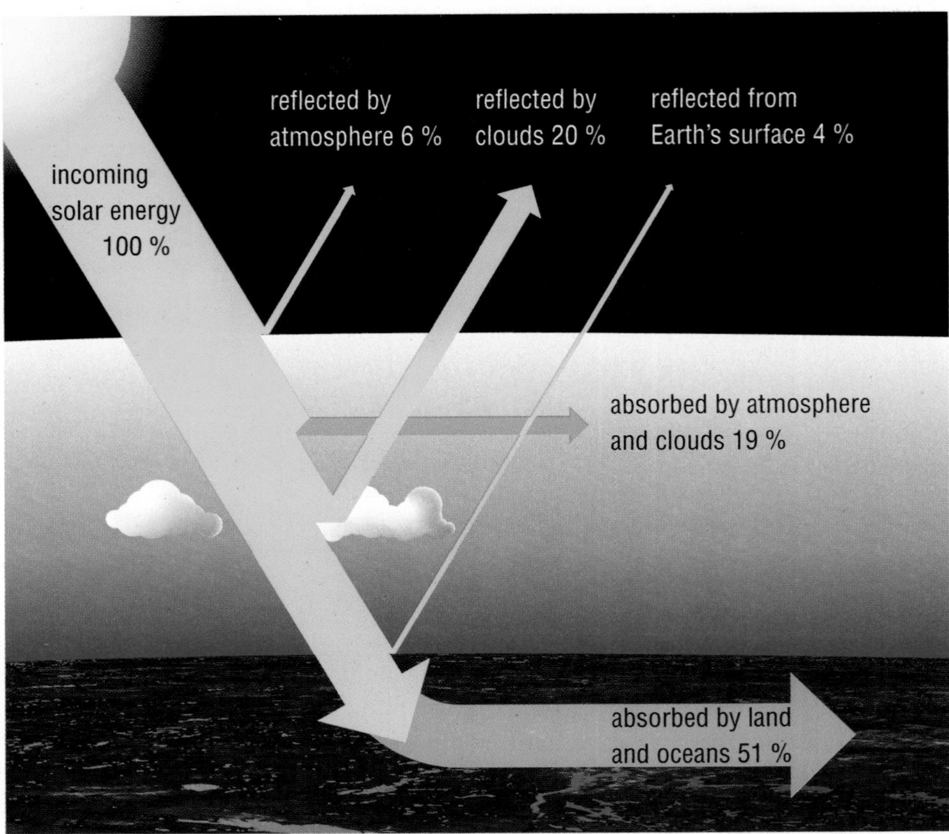

Adapted from original material from NASA, available through the Atmospheric Science Data Center

Figure 3 Incoming energy from the Sun is reflected (30 %) and absorbed (70 %) by Earth's surface, clouds, and the atmosphere.

What happens to the energy absorbed by Earth's surface? Plants trap a small proportion of the energy (<1 %) and use it to power the process of photosynthesis. Rocks and water absorb the Sun's energy, causing them to heat up. As Earth's surface temperature increases, it heats the air above.

How Does Earth Maintain a Balance?

Your home probably has a thermostat to control the indoor temperature. A thermostat switches off the heating system once your home has reached a set temperature. What would happen to the temperature in your home if the thermostat broke and the heating system continued to heat the house?

Earth is constantly absorbing energy from the Sun because the Sun has no "off" switch or thermostat. In the next Try This activity, you will construct a simple model to help answer this question: If the Sun continuously shines on Earth, why does Earth's average temperature remain relatively constant?

Earth's Surface Emits Energy

Energy can be converted from one form to another. Earth's surface absorbs energy from the Sun at different wavelengths (ultraviolet, visible light, and infrared). As this energy is absorbed, Earth's surface gains **thermal energy** and its temperature rises. Earth's warm surface then emits mostly lower-energy infrared radiation back out. In the following Try This activity, you will feel the radiation resulting from this increased thermal energy.

thermal energy the energy present in the motion of particles at a particular temperature

TRY THIS TESTING A MODEL OF THE EARTH–SUN ENERGY SYSTEM

SKILLS: Questioning, Hypothesizing, Predicting, Controlling Variables, Performing, Observing, Analyzing

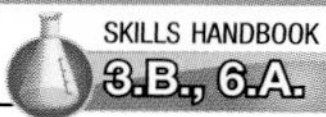

Equipment and Materials: desk lamp with 60 W incandescent bulb; paper or plastic plate; gravel; thermometer; white paper; ruler

Do not touch the light bulb, even after it has been turned off. It could cause a burn.

Do not unplug the desk lamp by pulling on the cord. Pull the plug out.

1. Prepare a model of Earth's surface by placing a layer of gravel on the plate. Put the thermometer bulb under the layer of stones so that the light from the lamp will not shine directly onto the thermometer (Figure 4).

Figure 4

2. Record the initial temperature.
3. Bend the lamp so that the bulb is approximately 10 cm above the stones. Measure and record the exact distance.
4. Put white paper over the exposed part of the thermometer to block it from the light. (You will remove the paper to take readings.) Turn on the lamp. Record the temperature every 10 min for 60 min.
5. After 60 min, turn off the lamp. Place your hand over the gravel. Record your observations.
6. Plot a graph of temperature versus time. C

A. Did the temperature of model Earth rise without stopping when the lamp was on? T/I

B. (i) What did you notice when you put your hand over the stones after the lamp was turned off? Try to explain what you felt.

   (ii) Do you think this was happening when the lamp was still on? T/I

C. Predict what you would feel if you placed your hand over the stones 30 min after the lamp has been turned off. T/I

D. Use your answers to questions A, B, and C to write a general statement about what happens as the Sun's light continuously shines on Earth. C

E. Predict what would happen if you used an equal mass of water instead of stones in this experiment. T/I

F. What were the limitations of this model in terms of Earth, the Sun, and the transfer of thermal energy? T/I

Now you can see that Figure 3 does not show the whole picture. Earth's surface both absorbs energy and emits energy.

Figure 5 shows both incoming energy (yellow arrows) and outgoing energy (red arrows). The amount of energy radiated by Earth's system is equal to the amount of energy Earth's system absorbs from the Sun. The amounts of energy balance. As a result of this balance of energy, Earth's global temperature stays fairly constant.

MATH TIP

Percentages

All the different types of energy reflected or radiated must add up to 100% of the incoming solar energy.

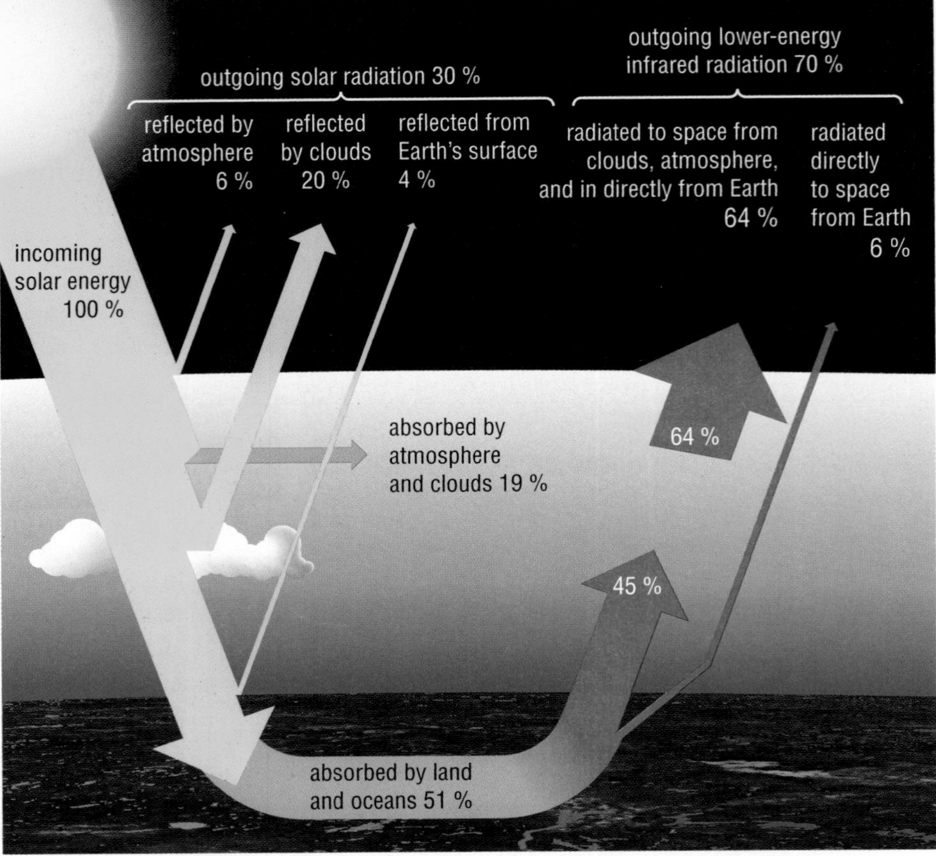

Figure 5 The energy absorbed by Earth (yellow arrows) is equal to the energy radiated by Earth (red arrows). As a result of this balance of energy, Earth's global temperature stays fairly constant.

Figure 6 Because Nigeria is close to Earth's equator, it has a hot climate.

Latitude and Climate Zones

In Section 8.2, you learned about climate zones on Earth. Which country would you expect to have a warmer climate: Nigeria, located near the equator, or Greenland, located near the North Pole? You probably chose Nigeria, and you are correct. The climate is warmer at lower latitudes and colder at higher latitudes near the North and South Poles (Figure 6). Latitude is a measure of distance from the equator. For example, Nigeria has a lower latitude than Greenland because it is closer to the equator.

Why is the climate colder at higher latitudes? Near the equator, the Sun shines directly overhead. There, the energy from the Sun is spread over a small area and feels very strong. Closer to the North and South Poles, the Sun is not directly overhead. As a result, the Sun's energy is spread over a larger area and feels weaker (Figure 7, next page).

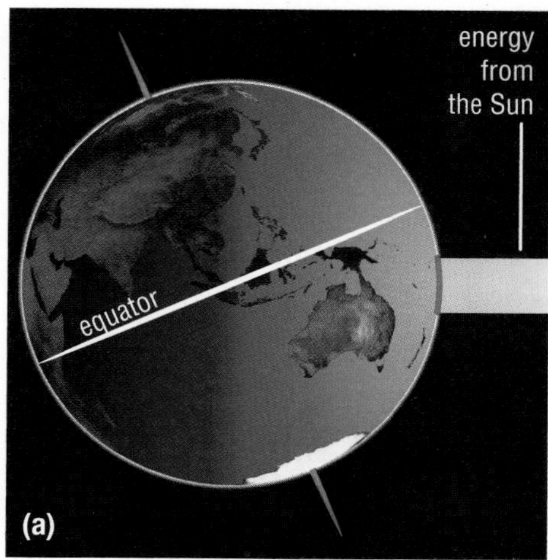

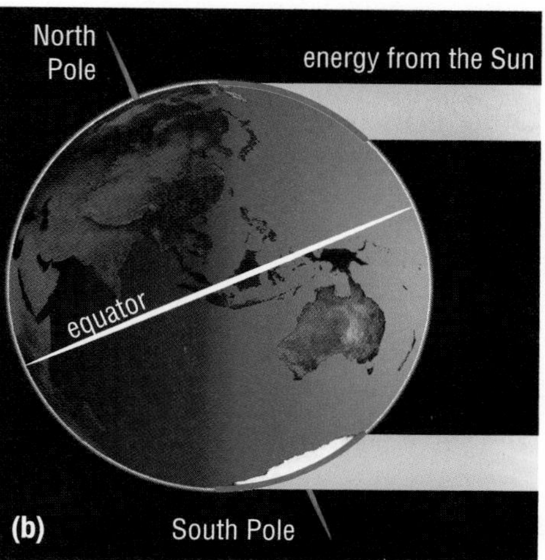

Figure 7 (a) The energy from the Sun is more intense near Earth's equator since it hits Earth's surface directly. (b) Energy from the Sun is less intense near the two poles since energy hits Earth's surface at an angle and spreads over a larger area.

There is another reason why climate changes with latitude. From Figure 7 you can see that radiation from the Sun hits higher latitudes at an angle. Because of this, the Sun's radiation must pass through more of the atmosphere before it strikes Earth's surface (Figure 8). Thus, at high latitudes more radiation is absorbed and reflected by the atmosphere, so less reaches the ground. At lower latitudes, the Sun shines directly on Earth's surface. Radiation from the Sun passes through less of the atmosphere, and less radiation is absorbed and reflected.

Figure 8 The Arctic is cold because energy received by Earth at high latitudes is spread over a larger area and because the radiation must pass through more atmosphere.

IN SUMMARY

- Earth receives radiation from the Sun, causing Earth to warm up.
- Regions near the equator are usually warmer because they receive more energy from the Sun per area than regions near the North and South Poles.
- Earth's warm surface emits energy in the form of infrared radiation.
- The balance between energy absorbed from the Sun and energy emitted from Earth ensures that Earth's global temperature remains fairly constant.

CHECK YOUR LEARNING

1. Explain how the ocean interacts with the Sun's radiation. K/U
2. The amount of energy Earth absorbs is equal to the amount of energy it radiates. Why is this important? Explain your answer. K/U
3. The Sun shines continuously on Earth. Explain why Earth does not keep warming up. K/U
4. (a) Describe what happens to the Sun's energy once it reaches Earth.
   (b) Describe what happens to the energy Earth emits. K/U
5. What would you expect to happen to the temperature on Earth if
   (a) the amount of energy radiated by Earth increased but the amount of energy coming from the Sun stayed the same?
   (b) the amount of energy radiated by Earth decreased but the amount of energy coming from the Sun stayed the same? T/I
6. State two reasons why you would expect the climate in Nigeria (near the equator) to be warmer than the climate in Greenland (closer to the North Pole). K/U

8.4 Components of Earth's Climate System

Other planets in our solar system have a balance of energy just as Earth does. Would it be possible to live on another planet? Mercury has no atmosphere. During the day, Mercury's temperature is scorching hot at 450 °C, and at night, it drops to a freezing −170 °C. Venus's atmosphere is 100 times more dense than Earth's, and its surface temperature is over 700 °C. In contrast, Mars has an atmosphere 100 times less dense than that of Earth, and its average surface temperature is −63 °C. In fact, Earth may be the only planet in our solar system that can support life as we know it. What makes Earth so special?

The Climate System

Earth's climate system makes Earth unique among the planets. It keeps the global temperature constant and maintains the conditions needed for life. There are four main components of Earth's climate system: the atmosphere, the hydrosphere, the lithosphere, and living things (Table 1). Each of these components receives the Sun's energy and interacts with the other components. Individual parts of the climate system are continually changing as organisms grow and die, clouds form, and wind and ocean currents flow.

Table 1 Components of Earth's Climate System

Atmosphere	Hydrosphere	Lithosphere	Living Things
The atmosphere is made of layers of gases surrounding Earth.	The hydrosphere includes liquid water in lakes and oceans, water vapour in the atmosphere, and ice in glaciers and at the poles.	The lithosphere is Earth's rock crust, including land surfaces.	All living things on Earth are part of the climate system.

As you learned in Section 8.3, Earth absorbs energy from the Sun. The climate system traps, stores, and transports this energy from one place to another and eventually radiates all of it back out to space. These complex processes are what keep Earth's global temperature stable.

The Atmosphere

atmosphere the layers of gases surrounding Earth

Earth is wrapped in layers of mixed gases. Together, these layers of gases make up the **atmosphere**. Although the atmosphere is thin compared with the radius of Earth—like the skin on a tomato—the gases in the atmosphere reach more than 100 km above Earth's surface (Figure 1). Above this height they are present only in very low concentrations.

Air is a mixture of gases. In the troposphere, the air we breathe is 78 % nitrogen gas and 21 % oxygen gas. The remaining 1 % is a combination of gases, including argon, carbon dioxide, and traces of helium, hydrogen, and ozone. Some water vapour and dust are also present. The proportion of gases changes at different levels in the atmosphere.

The atmosphere reflects some of the Sun's energy, absorbs and radiates some of the energy, and transmits some of it to Earth's surface. Once the energy from the Sun reaches Earth's surface, the atmosphere traps much of it, warming Earth. You can think of the atmosphere as a layer of blankets wrapped around Earth, conserving thermal energy to keep Earth warm. The atmosphere also shields Earth from dangerous radiation. How does it do this?

Ozone in the Stratosphere

Although life could not exist on Earth without the Sun, the Sun's energy is sometimes dangerous. The Sun causes damage, such as sunburn and skin cancer, to living things. Ozone, O_3, in the atmosphere prevents most of the harmful energy from reaching us.

There is more naturally occurring ozone gas in the stratosphere than in the rest of the atmosphere. In the stratosphere, ozone absorbs high-energy ultraviolet (UV) radiation from the Sun, preventing it from reaching Earth's surface. UV radiation damages plants and causes cancer in animals and people. Therefore, stratospheric ozone protects human health, as well as that of plants and other animals.

In the 1970s, scientists noticed that the ozone layer over Antarctica was thinning. In the 1990s, a similar ozone "hole" began to form over the Arctic (Figure 2). Ozone depletion in the stratosphere is caused by human-made compounds called chlorofluorocarbons (CFCs). CFCs belong to a family of chemical compounds called halocarbons. Halocarbons are molecules made up of carbon atoms linked by chemical bonds to fluorine, chlorine, bromine, or iodine. In the case of CFCs, chlorine and fluorine are linked to the carbon atoms. This is why they are called chlorofluorocarbons.

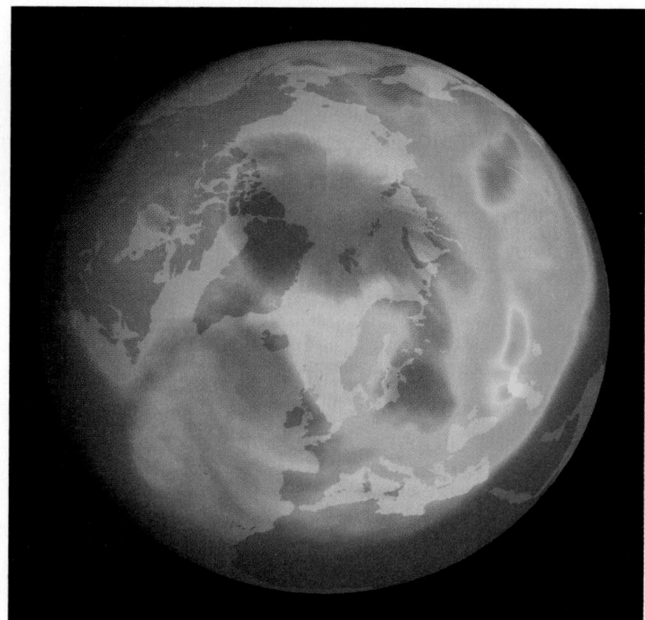

Figure 2 This is a view of Earth, looking directly down on the North Pole. The dark blue region over the Arctic has an ozone layer that is about 40 % thinner than normal.

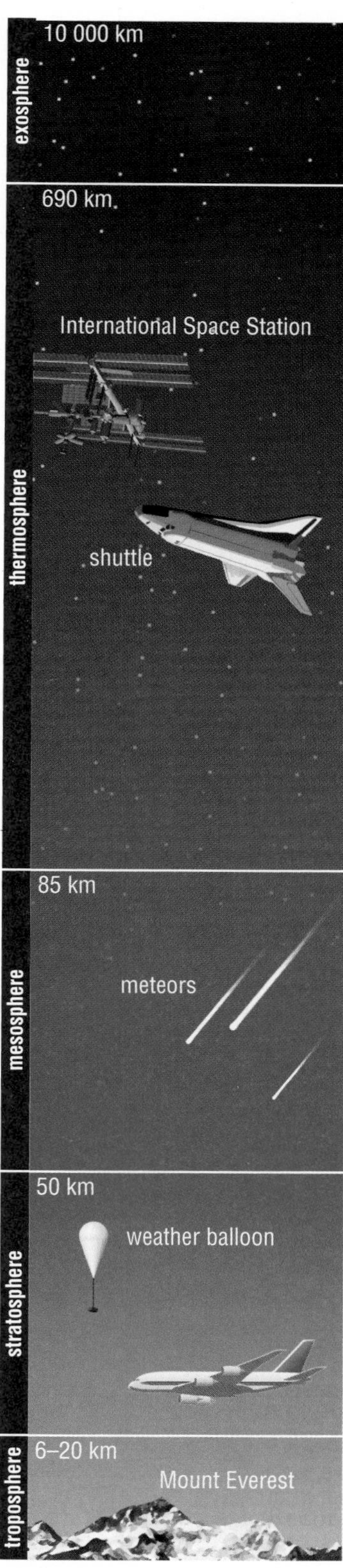

Figure 1 Weather occurs in the troposphere. Airplanes often travel in the lower stratosphere because the air is calmer there. Meteors from space burn up as they reach the mesosphere, becoming "shooting stars." The space shuttle orbits in the thermosphere, and most satellites orbit in the exosphere.

READING TIP

Finding the Main Idea

The main idea is what the author thinks about a topic or key concept. Start by identifying the topic or key concept of the text (*Ozone Layer*) and then look for the author's perpective on it (*chlorofluorocarbons, or CFCs, are harmful to the ozone layer*).

For many years, CFCs were used in products such as pressurized spray cans, refrigerators, and air conditioners. Because they are very stable and last for a long time, CFCs gradually travel up to the stratosphere. There, the chlorine atoms from CFCs react with ozone molecules, destroying the protective ozone layer. Each CFC molecule can destroy hundreds or thousands of ozone molecules.

In 1987, governments around the world signed an agreement in Montréal called the Montréal Protocol on Substances That Deplete the Ozone Layer. They agreed to stop the production and use of CFCs. Other short-lived halocarbons were not banned and are still used today. The Montréal Protocol was successful, and the ozone layer is beginning to recover. However, scientists estimate that it will take at least 50 more years for the ozone layer to return to its original thickness.

Figure 3 The brown haze over the city is photochemical smog.

To find out more about the work of atmospheric scientists,

GO TO NELSON SCIENCE

To learn more about Ontario's Drive Clean program,

GO TO NELSON SCIENCE

Ozone in the Troposphere

Although ozone has a protective role in the stratosphere, it has a toxic and corrosive effect in the lower troposphere. UV radiation from the Sun combines with the exhaust from cars to produce toxic chemicals and ozone gas at ground level. This mixture of gases and particles is called photochemical smog (Figure 3). (You learned about smog in Chapter 5.)

Photochemical smog is harmful to human health, damages buildings, and affects plants and animals. Unfortunately, the ozone being released into the troposphere does not move up into the stratosphere and offers no significant UV protection.

To help deal with the smog problem in Ontario, the government has introduced the Drive Clean program. The aim of this program is to reduce smog-causing emissions produced by vehicles. All vehicles over five years old must be tested every two years to make sure that their emissions are within limits set by the government.

RESEARCH THIS SMOG DAYS

SKILLS: Researching, Analyzing the Issue, Communicating, Evaluating

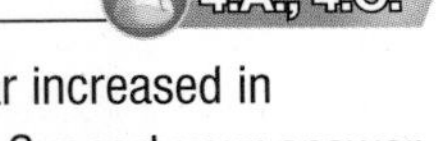

SKILLS HANDBOOK 4.A., 4.C.

Changes that we are making to the lower level of the atmosphere are causing health problems. Ground-level ozone, for example, is a principal component of smog. Many people consider it to be the most difficult air pollution problem in large cities.

1. Research the causes of ground-level ozone.
2. Research the health, agricultural, and environmental effects of ground-level ozone.
3. Research what technology we are using to reduce ground-level ozone.
4. Research the Great Smog of London, England, in December 1952.
5. Find out whether smog over cities is only a problem in the summer months.

GO TO NELSON SCIENCE

A. Have the number of smog days per year increased in Ontario's cities over the past 30 years? Support your answer with data. T/I
B. What is a catalytic converter? Who do you think should pay the extra cost to put catalytic converters on old cars: individual owners or the government? K/U T/I A
C. What evidence is there that increased ground-level ozone is related to people's health? T/I A
D. Will climate warming increase ground-level ozone? T/I
E. Why do you think the problem of ozone smog is especially associated with large cities? T/I A
F. Why is ozone smog considered to be the most difficult air pollution problem? T/I
G. How many people died of respiratory illnesses related to the Great Smog? K/U T/I

The Hydrosphere

The **hydrosphere** includes liquid water, water vapour, and ice. Liquid water absorbs energy from warm air and the Sun and then releases energy back. It also reflects some of the energy from the Sun. Water vapour and clouds in the atmosphere also reflect, absorb, and transmit energy from the Sun.

hydrosphere the part of the climate system that includes all water on and around Earth

The Water Cycle

The water cycle is an important part of the climate system (Figure 4). Energy is absorbed when water evaporates from oceans and lakes. This process has the effect of cooling its surroundings. Energy is given off when water vapour condenses into clouds in the atmosphere. This process warms the surroundings. As a result, the water cycle is one way that the climate system moves energy from one place to another.

READING TIP

Finding the Main Idea

The last sentence usually is the concluding sentence of a paragraph. Its job is to remind the reader of the main idea and key reasons for accepting it. Sometimes an author waits until the last sentence to state the main idea. The last sentence of a paragraph is a good place to check for the main idea.

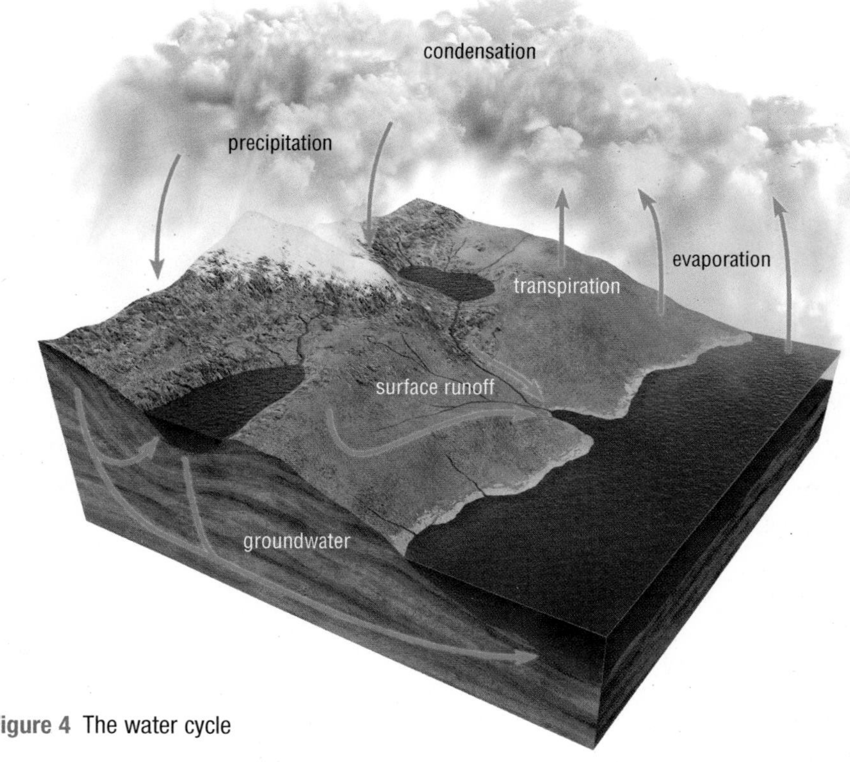

Figure 4 The water cycle

Large Bodies of Water and Climate Zones

Large bodies of water have an effect on the climate of nearby regions. Water absorbs and stores more thermal energy than land. Consequently, water heats up and cools down more slowly than land. Regions near an ocean or large lake tend to be cooler in the summer than inland locations because the water takes a long time to warm up as it absorbs thermal energy. These regions also tend to be warmer in the fall as the water slowly emits stored thermal energy.

Regions that are downwind from a large body of water have more snowfall in the winter. If the water is not covered with ice, air passing over the water can absorb water vapour. Once the air reaches the colder land, the water vapour condenses as snow.

Ice and the Climate System

Figure 5 Far more fresh water is frozen at the poles and in glaciers than exists in all the freshwater lakes in the world.

About 2 % of all Earth's water is frozen. Most of this ice is located at the two poles: as sea ice in the Arctic and as land-based ice sheets in the Antarctic. Sea ice, or pack ice, is relatively thin ice—only a few meters thick—formed from frozen sea water. It floats in the ocean near the North and South Poles. Ice sheets are enormous areas of permanent ice stretching over land of the Antarctic and Greenland. They are often several kilometres thick.

Permanent ice is also found in glaciers, on mountaintops, and in permafrost (Figure 5). Glaciers are permanent ice fields found in mountainous areas. Icebergs are large pieces of glaciers that have broken off and are floating in the ocean. Permafrost is ground that remains frozen all year round.

Earth's permanent ice plays a vital role in the climate system. Surfaces covered in ice and snow reflect more radiant energy than surfaces covered in soil, rock, or vegetation. Since most of Earth's polar regions are covered in ice, these regions reflect back a great deal of the Sun's energy. This reflection is another reason why polar regions are so cold. (The other reason is latitude, as you learned in Section 8.3.) In Section 8.10, you will learn more about the effect that Earth's ice has on climate.

The Lithosphere

lithosphere the part of the climate system made up of the solid rock, soil, and minerals of Earth's crust

The **lithosphere** is the Earth's crust. It includes all the solid rock, soil, and minerals on land and extends under the oceans as well. Together with the hydrosphere, the exposed lithosphere absorbs higher-energy radiation from the Sun, converts it into thermal energy, and then emits the energy back as lower-energy infrared radiation.

Land Formations and Climate Zones

Land formations also affect climate zones. Mountains and other land formations affect how air moves over an area. As clouds are blown upward over mountains, they lose their moisture as rainfall on the windward side. The leeward side of the mountain receives little rain. This process is called the rain shadow effect (Figure 6).

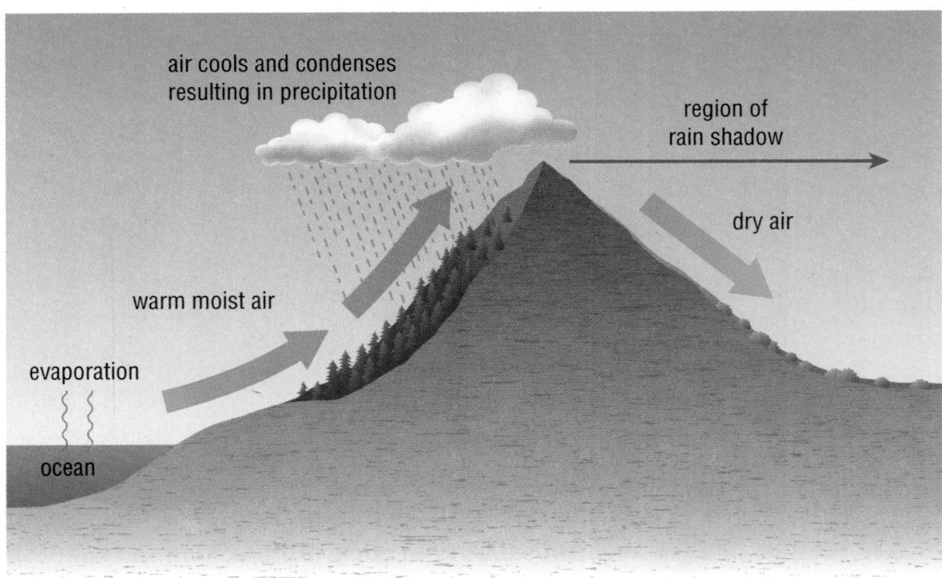

Figure 6 As air is forced up the windward side of a mountain, the air cools and condenses. Precipitation occurs. On the leeward side of the mountain, there is a rain shadow effect because the air has lost its moisture.

Altitude and Climate Zones

At high altitudes, atmospheric pressure is lower because there is less air above pushing down. This means that as the air from lower altitudes rises to high altitudes, it expands and cools down. Therefore, at high altitudes, such as in the mountains, air is cooler than at low altitudes. The "alpine climate" has a strong effect on the ecosystems found at high altitudes.

Living Things

All organisms are part of the climate system (Figure 7). Through various processes, plants and animals change the relative amounts of gases in the atmosphere. Through photosynthesis, plants take in carbon dioxide and release oxygen. Through cellular respiration, plants, animals, and other organisms take in oxygen and release carbon dioxide. Some animals, such as cows and sheep, produce methane gas as they digest their food. Tiny organisms such as termites, and bacteria that live in wetlands and shallow bodies of water, also produce methane.

Some gases in the atmosphere, such as carbon dioxide and methane, absorb infrared radiation emitted by Earth. Therefore, if the amount of carbon dioxide or methane in the atmosphere changes, it affects how much radiation the atmosphere can absorb. Later in this unit, you will learn more about how changes in the composition of the atmosphere affect Earth's climate.

READING TIP

Signal Words

Signal words are words or phrases that signal a connection between ideas. "For example" and "for instance" signal that what follows is an example, not a main idea. "Therefore," "thus," and "for this reason" signal a conclusion is coming. Signal words like these can provide clues to the main idea.

Figure 7 Living things are part of the climate system.

IN SUMMARY

- The climate system is composed of the atmosphere, the hydrosphere, the lithosphere, and living things.
- Earth's climate system moves energy around the globe.
- Ozone in the stratosphere absorbs harmful UV radiation. Human activities have caused a decrease in ozone in the stratosphere.
- Ozone in the troposphere, which forms from pollution at ground level, is harmful to human health, plants, and other animals.
- Large bodies of water and ice and land formations such as mountains influence climate.
- Living things affect the amount of carbon dioxide, oxygen, and methane in the atmosphere.

CHECK YOUR LEARNING

1. a) List the four main components of the climate system on Earth.
   (b) Describe one way in which each component is important to the climate system. K/U
2. Name the different layers of the atmosphere from lowest to highest altitude. K/U
3. Scientists and governments are working to reduce ozone levels in the troposphere, but they are trying to protect ozone levels in the stratosphere. Explain why this is not a contradiction. K/U
4. How does permanent ice on Earth's surface affect Earth's climate? K/U
5. What role do large bodies of water play in Earth's climate system and the flow of thermal energy? K/U
6. How does each of the following affect a region's climate? K/U
   (a) altitude
   (b) nearness to a mountain range
   (c) which side of a large lake a location is on (i.e., upwind or downwind from the lake)
7. Predict what might happen if
   (a) plants were able to absorb more carbon dioxide.
   (b) microorganisms began releasing more methane. T/I A

Comparing Canadian Climates

In this investigation, you will use climate data from Environment Canada weather stations to compare the temperatures and precipitation of three Canadian locations.

SKILLS MENU

- Questioning
- Hypothesizing
- Predicting
- Planning
- Controlling Variables
- Performing
- Observing
- Analyzing
- Evaluating
- Communicating

Purpose

- To determine the effect that a large body of water, such as an ocean, has on climate.
- To determine the effect that latitude has on climate.

Equipment and Materials

- detailed map of Canada
- access to the Internet or a Canadian climate data source
- electronic spreadsheet or graph paper

Procedure

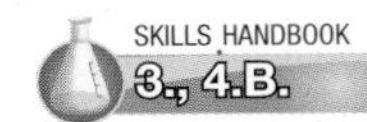

1. Refer to a detailed map of Canada while you look at the weather stations on the "Canadian Climate Normals or Averages 1971-2000" website.
2. Choose two Canadian locations on the website that have similar latitudes. One should be by the ocean and the other should be inland, far from the ocean.

GO TO NELSON SCIENCE

3. Choose a third Canadian location that is
   - a similar distance from the ocean as your other inland site, and
   - at least 4° farther north or south (i.e., more than 300 km) from your first inland site.
4. Predict an answer for each of the following questions. Give reasons for your predictions.
   - Which of the three locations will have the most precipitation?
   - Which of the three locations will have the highest summer temperatures?
   - Which of the three locations will have the most moderate climate?
5. Use the Internet to obtain monthly data for the three locations. Find the following data:
   - average daily temperature for each month of the year and for the entire year
   - average precipitation (mm) for each month and for the entire year
6. Record your data in a spreadsheet or table similar to Table 1. This data is from Windsor, Ontario (Figure 1). Remember to include data for all three locations.

Figure 1 Windsor sits beside the Detroit River, between Lake St. Clair and Lake Erie. Does the presence of these bodies of water impact Windsor's climate?

Table 1 Canadian Climate Normals for Windsor Airport (42°16′)

Month	J	F	M	A	M	J	J	A	S	O	N	D	Year
Precipitation (mm)	57.6	57.3	75	85.1	80.8	89.8	81.8	79.7	96.2	64.9	75.5	74.7	918.0
Average temperature (°C)	−4.5	−3.2	2.0	8.2	14.9	20.1	22.7	21.6	17.4	11.0	4.6	−1.5	9.4

Source: Environment Canada

Analyze and Evaluate

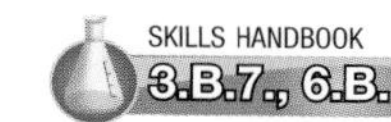

(a) Using a graphing program (or pencil and paper), plot a graph for each location similar to Figure 1. The graph should be a composite, showing months on the x-axis, precipitation (mm) on the left y-axis, and average temperature (°C) on the right y-axis. Use a bar graph to show precipitation and a line graph to show the average temperature. To make the graphs easy to compare, use the same scales for the three locations. C

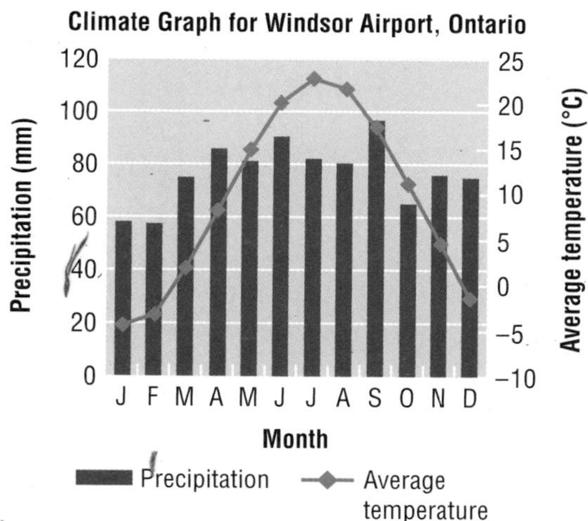

Figure 1 Produce a graph similar to this for each of your three locations.

(b) Compare the yearly average temperature and precipitation for each of your locations. What do you observe? Suggest explanations for the differences. T/I

(c) Compare the average temperatures in January and July for each location. Which location gets the coldest in the winter? Which gets the hottest in the summer? Suggest an explanation for your observations. T/I

(d) Compare your answers to (b) with those of your classmates who examined different locations.

(e) Compare your observations with your predictions. Suggest explanations for any differences. T/I C

(f) Does latitude affect the climate of a location? Does distance from a large body of water affect the climate of a location? Explain your answers. T/I

Apply and Extend

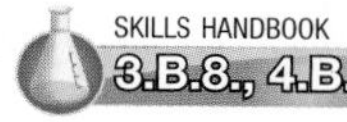

(g) What other factors may play a large role in determining climate? T/I

(h) Predict how the size of a city affects its climate, directly or indirectly. T/I

(i) Analyze other climate factors for Canadian locations (e.g., population, altitude). A

(j) What ecozone and ecoregion were your three locations in? K/U

(k) Find the bioclimatic profiles located nearest your three locations. Compare the information you find at the weather site and at the bioclimate site. You could use a t-chart for your comparison. T/I C

GO TO NELSON SCIENCE

(l) Using the Internet, find the average January and July temperatures and the annual precipitation for three different locations around the world. Account for temperature and precipitation data in terms of latitude, proximity to the ocean, or other climate factors. T/I

GO TO NELSON SCIENCE

UNIT TASK Bookmark

You can apply what you learned in this activity on researching and summarizing climate data to the Unit Task described on page 444.

8.6 The Greenhouse Effect

Even without the climate system, Earth would reach a balance of energy. The amount of energy absorbed would be equal to the amount of energy emitted. However, without a climate system, Earth would be much colder than it is. How much of a difference does the climate system make to Earth?

The climate system moderates Earth's temperature by trapping and storing energy from the Sun and distributing it around the world. As a result, the air temperature remains relatively constant day and night and across large regions of Earth.

DID YOU **KNOW?**

Not Quite Like a Greenhouse
Scientists once thought that greenhouses warmed up following the same principles as in the atmosphere. In fact, greenhouses warm up because colder air currents on the outside are prevented from interacting with the air inside. Therefore, the greenhouse effect is misnamed.

What Is the Greenhouse Effect?

How does the climate system trap energy to keep Earth warm? The atmosphere allows much of the higher-energy radiation from the Sun to pass through it. This radiation is absorbed by Earth's surface, becoming thermal energy. As a result, Earth's surface warms up. Earth's warm surface then emits lower-energy infrared (IR) radiation. Gases in Earth's atmosphere trap much of the IR radiation. These gases then radiate the energy equally in all directions, which means that about half of the radiation gets sent back toward Earth's surface. This warms Earth even more. The trapped energy keeps Earth's global temperature much higher than it would otherwise be. The energy-trapping process is called the **greenhouse effect** (Figure 1).

greenhouse effect a natural process whereby gases and clouds absorb infrared radiation emitted from Earth's surface and radiate it, heating the atmosphere and Earth's surface

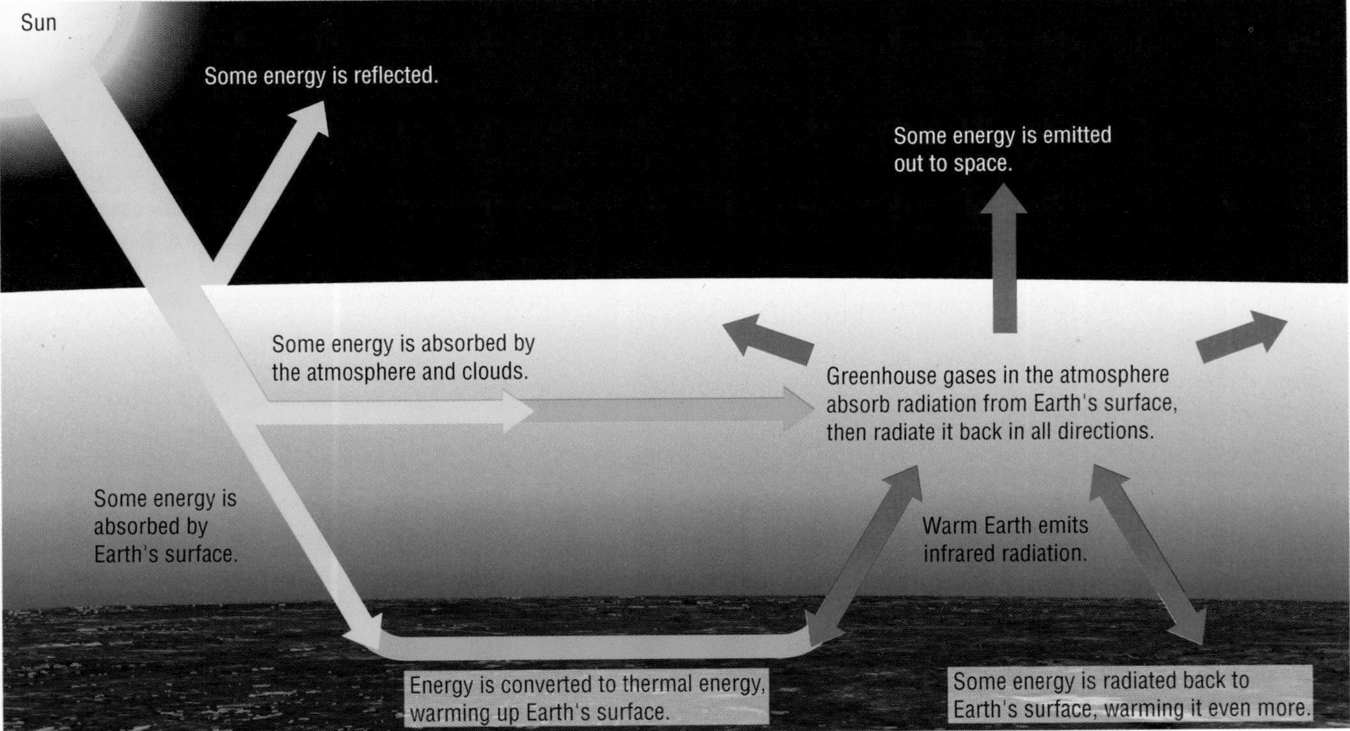
Sun
Some energy is reflected.
Some energy is emitted out to space.
Some energy is absorbed by the atmosphere and clouds.
Greenhouse gases in the atmosphere absorb radiation from Earth's surface, then radiate it back in all directions.
Some energy is absorbed by Earth's surface.
Warm Earth emits infrared radiation.
Energy is converted to thermal energy, warming up Earth's surface.
Some energy is radiated back to Earth's surface, warming it even more.

Figure 1 High-energy radiation from the Sun enters the atmosphere. Gases and clouds in the atmosphere trap some of the infrared radiation from Earth's surface and radiate it back. This is the greenhouse effect.

If Earth did not have a climate system, the average global temperature would be about −18 °C. Because of the climate system's greenhouse effect, Earth's actual average temperature is around 15 °C. The greenhouse effect is a natural process that has been happening for millions of years.

Greenhouse Gases

Most of the air in the atmosphere is made up of nitrogen and oxygen gases. These gases do not absorb radiation from Earth's surface. In fact, the greenhouse effect is caused by gases that exist in very low concentrations in the atmosphere. These gases are called **greenhouse gases**. The most important greenhouse gases are water vapour, H_2O, and carbon dioxide, CO_2. Other, less significant, greenhouse gases are methane, CH_4, tropospheric ozone, O_3, and nitrous oxide, N_2O. Their contribution to the greenhouse effect is determined by their concentration in the atmosphere and by how much thermal energy each molecule of gas can absorb.

greenhouse gas any gas in the atmosphere (such as water vapour, carbon dioxide, and methane) that absorbs lower-energy infrared radiation

Carbon Dioxide

Earth's atmosphere contains only 385 ppm (parts per million) carbon dioxide, or 0.0385 %. This is just a small percentage of all the gases in the atmosphere. However, carbon dioxide is estimated to cause up to a quarter of the natural greenhouse effect on Earth.

Before the industrial age, the concentration of carbon dioxide in the atmosphere was 280 ppm. Natural sources of atmospheric carbon dioxide include volcanic eruptions, the burning of organic matter, and cellular respiration of plants and animals (Figure 2). In Section 9.4, we will look at how human activity has increased atmospheric carbon dioxide and other greenhouse gases.

The carbon cycle is the movement of carbon through living things, the lithosphere, the atmosphere, and the hydrosphere (Figure 3). Living things and oceans are important **carbon sinks**. This means that they remove carbon dioxide from the atmosphere and store the carbon atoms in a different form. Trees and other plants capture carbon dioxide during photosynthesis and use it to grow. When trees decompose or burn, the carbon is released back into the atmosphere as carbon dioxide. In the ocean, carbon dioxide dissolves and some forms solid calcium carbonate, which sinks to the bottom of the ocean.

Figure 2 Volcanic eruptions release carbon dioxide into the atmosphere.

carbon sink a reservoir, such as an ocean or a forest, that absorbs carbon dioxide from the atmosphere and stores the carbon in another form

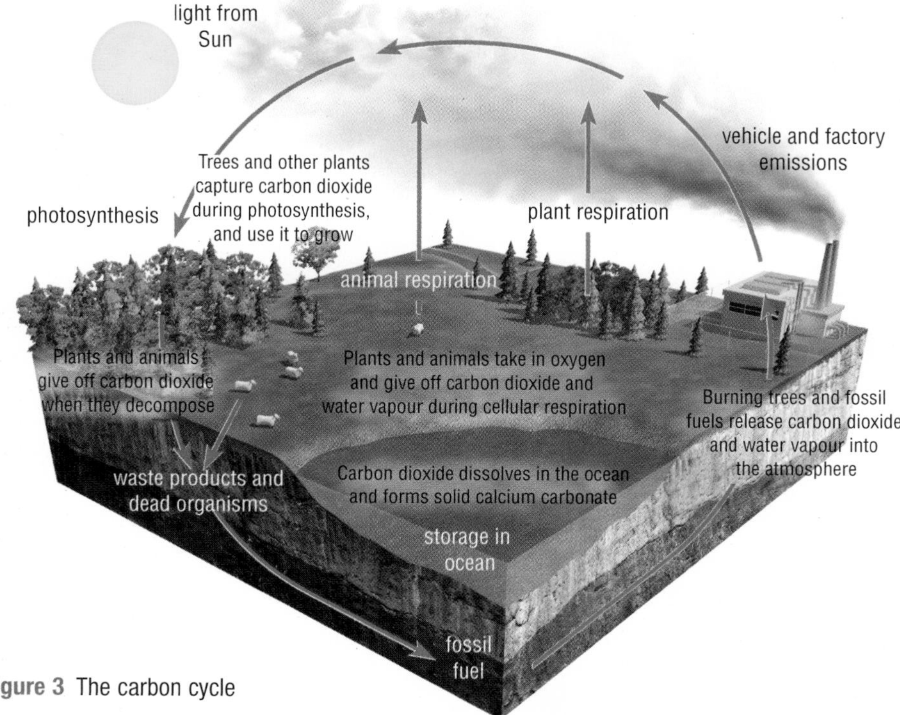

Figure 3 The carbon cycle

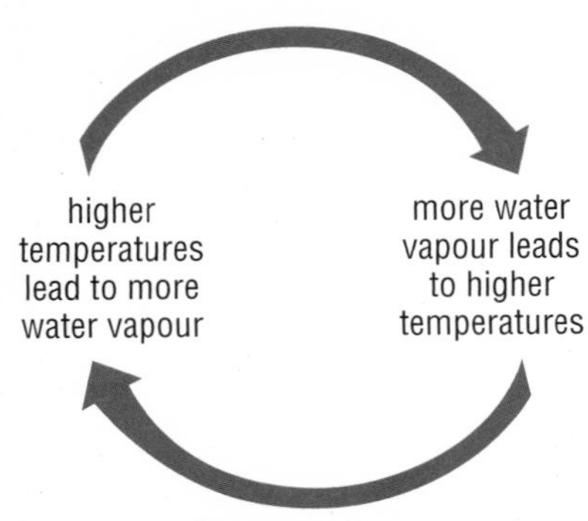

Figure 4 Higher temperatures cause more water to evaporate and form water vapour. Since water vapour traps heat in the atmosphere, more water vapour increases the temperature further.

feedback loop a process in which the result acts to influence the original process

Water Vapour

About two-thirds of Earth's natural greenhouse effect is caused by water vapour in the atmosphere. The quantity of atmospheric water vapour depends on the temperature of the atmosphere. It varies from trace amounts to about 4 %.

How are water vapour and temperature related? Water evaporates more readily when it is heated. Also, warmer air can hold more water vapour. Thus, as Earth's temperature increases, more liquid water becomes water vapour. Because water vapour traps energy, the more water vapour there is in the atmosphere, the warmer Earth becomes (Figure 4).

This type of relationship is called a **feedback loop**. In a feedback loop, the cause (here, high temperature) creates an effect (more water vapour in the air) that affects the original cause (warming Earth further). In a positive feedback loop, the effect enhances the original cause. Water vapour and temperature are related in a positive feedback loop. In a negative feedback loop, the effect decreases the original cause. You will encounter more feedback loops in Section 8.10.

MATH TIP

Very Small Units

"Parts per million" (ppm) is similar to percent, only while "percent" is out of 100, "ppm" is out of 1 million. "Part per billion" (ppb) means "out of 1 billion." 750 ppb means there are 750 methane molecules out of a billion molecules, or 750/1 000 000 000.

Methane

There is much less methane in the atmosphere than there is carbon dioxide. However, a molecule of methane can absorb much more thermal energy than a molecule of carbon dioxide. As a result, a molecule of methane is about 23 times more powerful as a greenhouse gas than a molecule of carbon dioxide.

Methane, like carbon dioxide, comes from both natural and human sources. It is produced naturally by biological processes such as plant decomposition in swamps and animal digestion (Figure 5). Before the industrial age, the concentration of methane in the atmosphere was 0.700 ppm (or 700 ppb). It has now risen to 1.785 ppm (or 1785 ppb).

Figure 5 Bacteria that live in swamps and other wetlands, and in animal digestive systems, produce methane as a by-product.

Ozone

You learned in Section 8.4 that ozone gas exists naturally in the stratosphere where it forms a layer protecting Earth's surface from the Sun's higher-energy UV radiation. Lower down in the troposphere, ozone acts as a greenhouse gas. Scientists do not have a clear picture of what the average concentration of tropospheric ozone is because it changes rapidly. However, they do know that it contributes to the greenhouse effect.

Nitrous Oxide

A molecule of nitrous oxide, N_2O, is almost 300 times more effective than a molecule of carbon dioxide as a greenhouse gas. However, a much smaller concentration of nitrous oxide is present in the atmosphere. Before the industrial age, the concentration of nitrous oxide in the atmosphere was 270 ppb (0.270 ppm). It has since risen to 321 ppb (0.321 ppm). Like carbon dioxide and methane, nitrous oxide is produced from both natural and human sources. Nitrous oxide is produced naturally by the reactions of bacteria in soil and water (Figure 6).

Figure 6 Tropical soils are a significant source of nitrous oxide.

Carbon dioxide, methane, nitrous oxide, and other greenhouse gases are present in the atmosphere in minute quantities. The following Try This activity models the effect of a very small concentration of a substance. You will find out whether such a tiny concentration can have any effect on the absorption of radiation. Could such low concentrations of greenhouse gases really affect the amount of energy trapped by the atmosphere?

TRY THIS HOW TINY CONCENTRATIONS CAN MAKE A DIFFERENCE

SKILLS: Planning, Controlling Variables, Performing, Observing, Evaluating, Communicating

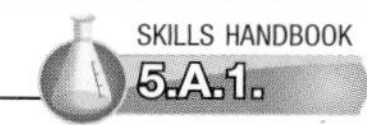

The concentration of carbon dioxide in the atmosphere at the present time is approximately 0.04 %. In this activity, you will see if a concentration that small can make a difference to how radiation, such as light, travels through a fluid.

Equipment and Materials: lab apron; water; transparent glass; measuring cup; eyedropper; water-soluble acrylic black ink (or green or blue food colouring)

1. Put on your lab apron.
2. Fill a transparent glass with 250 mL of water (approximately 1 cup).
3. Calculate the volume (in mL) of black ink you need to add to the 250 mL of water to produce a concentration similar to that of carbon dioxide in the atmosphere.
4. Use your answer to step 3 to calculate how many *drops* of black ink you need to add to the water. Assume 1 mL is equal to approximately 20 drops. Since acrylic black ink is mostly water and only 10-15 % ink, you should multiply the number of drops you calculated by a factor of 7 so that you actually use the correct amount of ink. This also applies to food colouring.
5. Using the eyedropper, add the number of drops of black ink or food colouring that you calculated in step 4 to the water. Stir until solution looks uniform.

A. What difference did the addition of this tiny amount of black ink have on the visibility of the water in the glass? T/I

B. How does this simple model compare with how tiny concentrations of greenhouse gases in the atmosphere trap infrared radiation emitted by Earth? A

C. Carbon dioxide is very difficult to remove from the atmosphere. T/I A

(i) Can you think of a way of removing the black ink from the glass of water?

(ii) Could this method be applied to the removal of carbon dioxide in the atmosphere?

How Do Greenhouse Gases Trap Infrared Radiation?

Nitrogen gas and oxygen gas each consist of two identical atoms (Figure 7(a)). The two atoms in these molecules can only vibrate one way: back and forth. This limits the type of energy the molecules can absorb. When infrared radiation reaches these molecules, they cannot absorb it.

Water, carbon dioxide, and methane consist of three or more atoms, and have different types of atoms (Figure 7(b)). Nitrous oxide also has three atoms. The atoms in these molecules can vibrate and wiggle in many ways, and can absorb different types of energy. Thus, when infrared radiation reaches water vapour, carbon dioxide, or methane, these molecules trap the infrared energy and re-radiate it back out in every direction.

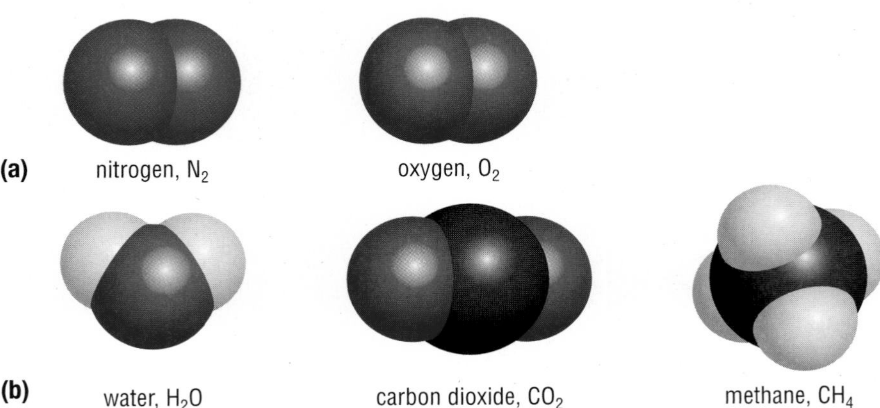

Figure 7 (a) Nitrogen and oxygen are very poor absorbers of infrared radiation because they consist of only two atoms. (b) Water, carbon dioxide, and methane molecules contain several atoms as well as different types of atoms. These molecules can absorb different types of energy, including infrared radiation.

IN SUMMARY

- The climate system traps and stores energy through the greenhouse effect.
- The greenhouse effect is caused by gases in the atmosphere absorbing the infrared radiation that is emitted from Earth's surface and radiating it back again.
- The greenhouse effect warms the atmosphere and Earth's surface so that life can exist on Earth.
- Water vapour, carbon dioxide, methane, ozone, and nitrous oxide are important greenhouse gases because they trap Earth's infrared radiation.
- Living things, especially forests and oceans, are carbon sinks because they remove carbon dioxide from the atmosphere and store the carbon atoms in a different form.

CHECK YOUR LEARNING

1. Explain why the greenhouse effect is important to life on Earth. K/U
2. Describe how the greenhouse effect in the atmosphere works and draw a diagram illustrating it. K/U C
3. (a) Write a definition, in your own words, of "greenhouse gas."
   (b) Name the two most important greenhouse gases that occur naturally in the atmosphere. K/U
4. List two factors that affect how important a particular greenhouse gas is in contributing to the greenhouse effect. K/U
5. If forests serve as important sinks for greenhouse gases, describe how past ice ages might have affected the concentration levels of carbon dioxide in the atmosphere. A
6. Give one natural source for each of the following greenhouse gases: K/U
   (a) carbon dioxide
   (b) methane
   (c) nitrous oxide
   (d) water vapour
7. Explain how greenhouse gas molecules, in contrast to oxygen and nitrogen molecules, trap infrared radiation. Use a diagram and provide a sentence description. K/U C

Modelling the Greenhouse Effect

Scientists and engineers build physical models of systems that they are studying. Models help us demonstrate our ideas in a hands-on way. They also help us study the different variables in a system. Models help us answer scientific questions.

SKILLS MENU

- Questioning
- Hypothesizing
- Predicting
- Planning
- Controlling Variables
- Performing
- Observing
- Analyzing
- Evaluating
- Communicating

Purpose

To construct and test a physical model of the greenhouse effect

Equipment and Materials

List the Equipment and Materials you will need. Include any necessary safety equipment.

Procedure

In this activity, you will design and build a physical model of the greenhouse effect. Your model should include the following functions:

- allow visible light to enter the system
- absorb that light energy
- emit the light energy in the form of lower-energy infrared radiation (You will feel this infrared radiation as thermal energy.)
- prevent thermal energy from leaving the system
- allow you to monitor the temperature of your system

Part A: Designing Your Model

1. With your group, brainstorm some materials that you could use in your model.
2. Decide how you will test your materials to see if they work before you build your model. Test your materials.
3. Decide how you will control the variables in your investigation. Also, decide how you will compare the temperature inside your system with the temperature outside your system.
4. Plan a way to reduce the amount of thermal energy leaving your system.
5. Sketch a diagram of your model.
6. List any safety precautions you will take.

Part B: Building Your Model

7. Describe the design of your model to another student or to your teacher. Incorporate any appropriate changes that they suggest.
8. Write a paragraph describing what you plan to build and how you will do it. Include any materials you plan to use.
9. Build your model.

Part C: Testing Your Model

10. Plan a procedure to test your greenhouse model in the presence of a strong light. Remember to account for and control all the variables, except the ones you are measuring. What measurements will be made and for how long?

Analyze and Reflect

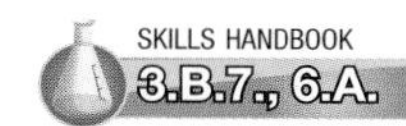

(a) Graph your data. Compare it with data from previous activities in this chapter. C

(b) Thermal energy can leak out of a system in three ways: conduction, convection, and radiation. (See the Learning Tip on page 345.) Explain how your model blocked thermal energy from escaping in each of these three ways. T/I

(c) Did your model give the results that you expected? Describe why it did or did not. T/I

(d) Use your model to explain the greenhouse effect. A

(e) How is your model similar to the greenhouse effect? A

(f) How does your model differ from the greenhouse effect? A

Apply and Extend

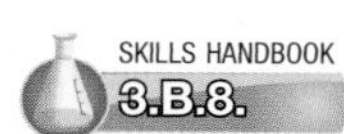

(g) How could you improve your model to make it a more effective illustration of the greenhouse effect? T/I A

8.8

Energy Transfer within the Climate System: Air and Ocean Circulation

In Section 8.3, you learned that the Sun's radiation reaches Earth's surface with different intensities at different latitudes. In addition, water and land absorb energy at different rates. As a result, Earth is unevenly heated. The climate system transports thermal energy from areas that receive a lot of radiation to areas that receive less radiation. This reduces the temperature difference over Earth, keeping tropical regions cooler and polar regions warmer than they would otherwise be.

heat sink a reservoir, such as the ocean, that absorbs and stores thermal energy

Figure 1 The absorption and release of thermal energy from vast oceans has a large effect on the climate system.

The atmosphere and the hydrosphere are essential parts of the climate system. Both are able to absorb and store thermal energy, so they act as **heat sinks**. The ocean is particularly important as a heat sink because water can absorb much more thermal energy than air (Figure 1). When the air is warmer than the ocean surface, the ocean absorbs energy from the air. When the air is cooler than the ocean surface, the ocean releases energy back into the air. This phenomenon is the reason why large bodies of water affect the climate of nearby regions (Section 8.3).

Most of Earth's thermal energy circulation occurs in the atmosphere and the hydrosphere. The following Try This activity will help you understand why.

TRY THIS EXAMINE AIR AND WATER CURRENTS

SKILLS: Hypothesizing, Observing, Analyzing, Evaluating, Communicating

SKILLS HANDBOOK 3.B.6., 3.B.7.

In this activity, you will make air and water currents.

Equipment and Materials: heavy cardstock; ruler; pencil; scissors; thread; lamp with incandescent bulb; clear glass container; warm water; 2 ice cubes coloured with food colouring; 250 mL salt water (3 %)

Do not touch the bulb after it is turned off; it could still be hot.

To unplug the lamp, pull on the plug, not the cord.

Part A: Air Currents

1. From cardstock, cut a circle with a diameter of 15 cm. Cut a continuous strip 1.5 cm wide around the circle, making a spiral.
2. Tie a piece of thread to the centre of the card spiral.
3. Remove the shade from a lamp with an incandescent bulb. Turn on the lamp. Hold the spiral by its thread above the light bulb, without allowing the card to touch the light bulb (Figure 2). Record your observations.

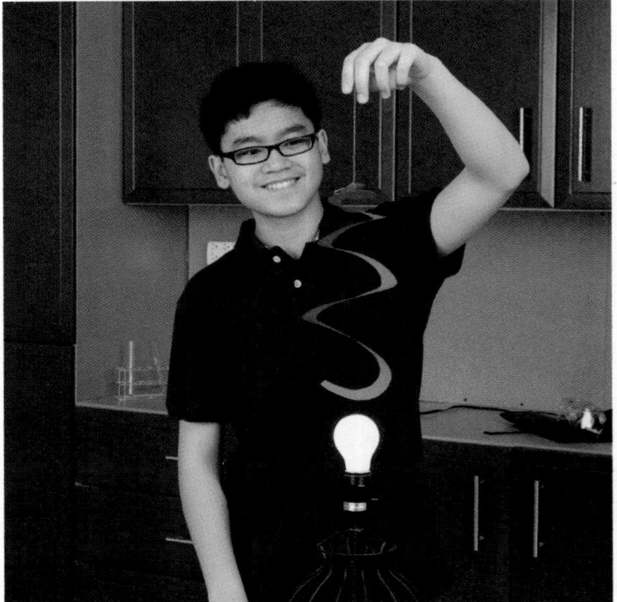

Figure 2 Spiral of card held above an incandescent bulb

Part B: Water Currents

4. Fill a clear glass container such as a drinking glass with warm tap water. This represents the ocean in your model. Carefully lower a coloured ice cube into the water. This represents a freshwater ice sheet. Watch it closely as it melts. Record your observations in a diagram.
5. Repeat the process using 3 % salt water to more accurately represent the ocean.

A. Explain your observations in Part A. T/I

B. Explain your observations in Part B. T/I

In the Try This activity, you observed the card moving when you placed it over the lamp. The card moved because of a current in the air. This current formed because energy from the lamp heated the air around it. The air particles began to move faster and farther apart. The warm air around the lamp became less dense than the colder air above the lamp. The colder, denser air above the lamp sank toward the lamp, causing the warmer, less dense air to rise. This movement created a continuous current in which colder air above the lamp moved down to the lamp and warmed up, and warmer air near the lamp moved up and cooled down.

The same principle applies to water: warm water is less dense than cold water. A current forms when water is unevenly heated. Colder, denser water falls and pushes the warmer, less dense water up. We can apply this principle to the atmosphere and the hydrosphere to explain how thermal energy is moved around the world.

LEARNING TIP

Thermal Energy Transfer

In convection, the movement of heated matter transfers thermal energy. For example, ocean currents transfer thermal energy. In conduction, thermal energy is transferred through matter as particles bump against each other. For example, air in contact with warm water will heat up. In radiation, electromagnetic waves transport energy through space. For example, the Sun radiates energy.

Energy Transfer in the Atmosphere

Near the equator, the Sun's rays reach Earth's surface with the greatest intensity (Section 8.3). Air at the equator heats up rapidly and becomes less dense. Colder, more dense air above it drops, pushing the warm air up into the atmosphere. As the warm air moves up, it creates an area of low pressure below it. Once the warm air is high in the troposphere, it spreads out toward the poles and cools down. The cooler air sinks back to Earth's surface, resulting in an area of high pressure (Figure 3). This movement of warm and cold air creates a circular current called a **convection current**.

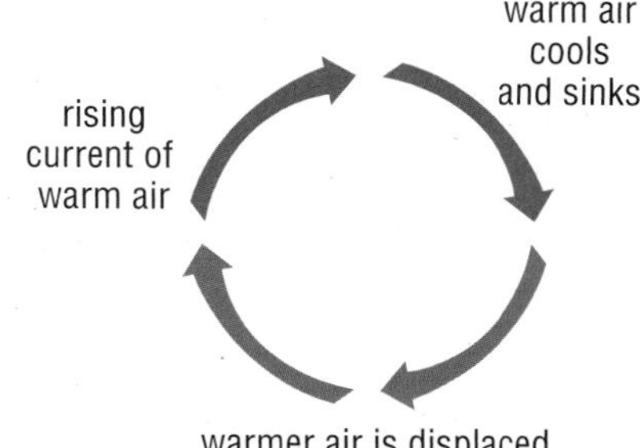

Figure 3 The movement of warm and cold air creates a convection current.

convection current a circular current in air and other fluids caused by the rising of warm fluid as cold fluid sinks

The pattern of convection currents at the equator is repeated closer to the poles. As a result, Earth has permanent bands of high and low air pressure, parallel with the equator (Figure 4). Convection currents are one of the main ways that energy is transported in the atmosphere. They move thermal energy from the equator toward the North and South Poles.

Air tends to flow from areas of high pressure to areas of low pressure. This causes air currents, which we recognize as wind. Since Earth has permanent bands of high and low pressure, there are prevailing winds that blow in the same direction almost all the time. Because Earth rotates, these winds curve around the globe instead of moving directly north or south. Prevailing winds eventually move warm air from the equator toward the poles. Prevailing winds also push warm ocean water toward the North and South Poles.

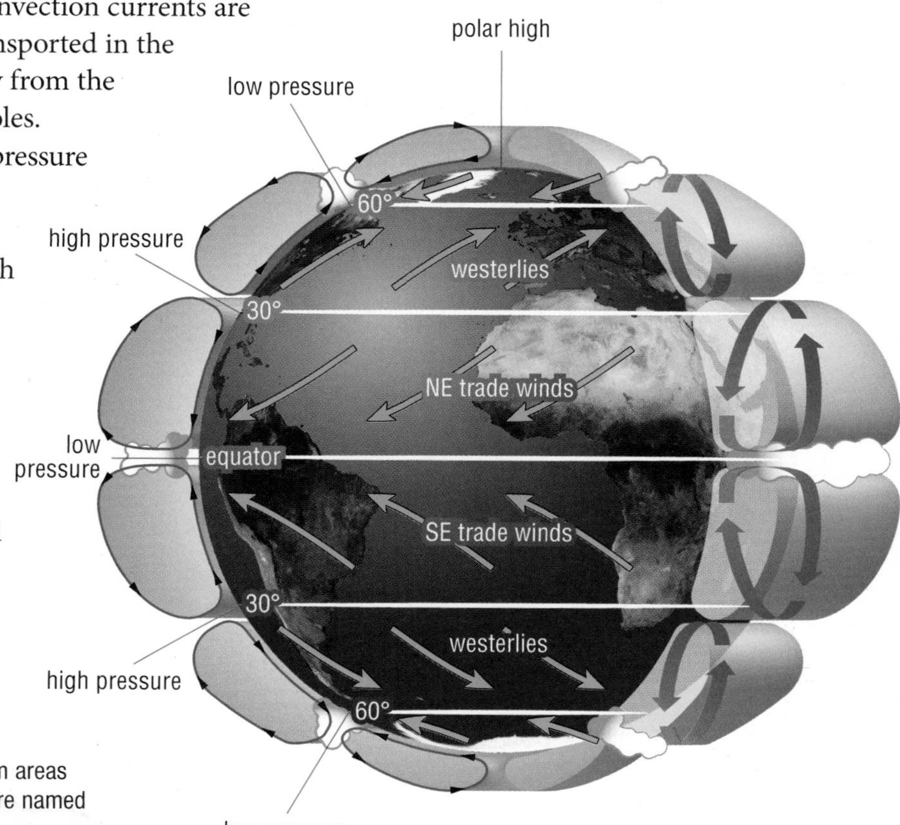

Figure 4 Bands of high and low pressure around the globe create air currents (winds) that blow from areas of high pressure to areas of low pressure. Winds are named based on the direction they originate from.

Prevailing Winds and Climate Zones

Prevailing winds are yet another factor in determining climate zones. As a prevailing wind passes over the ocean, it picks up water vapour. When the wind reaches land, the water vapour eventually condenses, bringing rain. Regions where the prevailing winds pass over water before reaching land have higher amounts of precipitation. If a prevailing wind comes from the North Pole, it will be cold and dry. The regions this wind passes over may become colder and drier because of the wind.

READING TIP

Clues to the Main Idea
Sometimes an author hints at a main idea rather than spelling it out clearly and directly. You can infer (make a reasonable guess at) an implied (indirectly stated) main idea. Look for clues in the title, headings, examples, facts, and reasons. Ask yourself what the clues add up to.

Energy Transfer in the Oceans

As water travels toward the poles, it gets colder. It also becomes more salty as surface water evaporates and sea ice forms. Sea ice is mostly fresh water because it rejects the salt when it freezes. This leaves the remaining water saltier. Both of these factors—the low temperature and saltiness of the water—make the water at the poles more dense. As a result, the water sinks to the ocean floor.

Warmer surface water from the equator then flows toward the poles to take its place. This process is called the **thermohaline circulation** of the oceans. *Thermo* means heat and *haline* means salt in Greek. Thermohaline circulation includes all ocean currents caused by changes in temperature and salinity.

thermohaline circulation the continuous flow of water around the world's oceans driven by differences in water temperatures and salinity

Ocean currents around the globe act like an enormous conveyor belt, slowly moving water (and the thermal energy it carries) from the equator to the poles. Figure 5 shows the main ocean currents around the world.

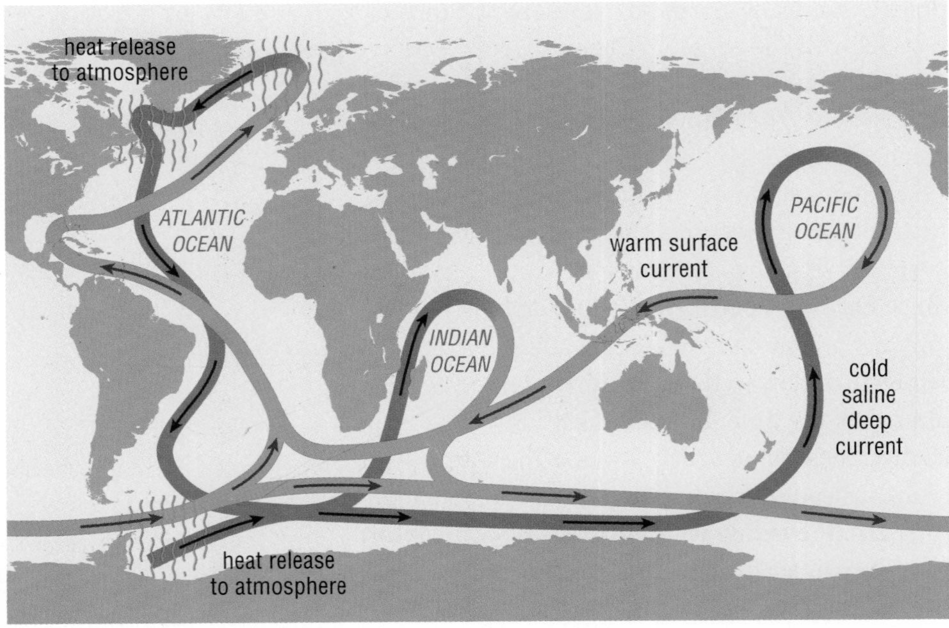

Figure 5 Red lines show warm ocean currents and blue lines show cold ones. Warm water currents travel on the surface, but cold water currents travel deep down in the ocean.

Ocean currents can also be caused by winds. Winds are the main cause of the Gulf Stream that transports warm water from the tropics up the eastern coast of North America and across to Europe (Figure 6).

Ocean Currents and Climate Zones

Ocean currents have a strong effect on the climates of nearby land. Warm ocean currents heat the air above them. When this warm, moist air reaches land, it warms the land and produces rain, affecting the climate of that area. The warm Gulf Stream current gives the northwest coast of Europe a warmer, damper climate than it would otherwise have at those latitudes.

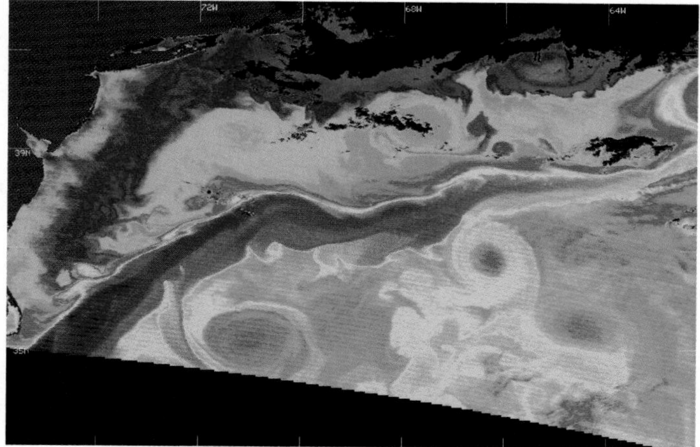

Figure 6 Warm ocean currents such as the Gulf Stream transport heat energy from the equator to higher latitudes. This is shown by this satellite photograph of sea surface temperatures. Red tones are warmer currents and blue tones are cooler currents. The east coast of North America is the dark grey area in the upper left.

Cold ocean currents cool the air above them. When the cold, dry air reaches land, it cools the land and creates desert areas. California and parts of Mexico are cooler and drier because of a cold ocean current along their west coasts. You can feel the effect of a cold ocean current when visiting Newfoundland and Labrador. The water at beaches in Newfoundland and Labrador is colder than the water near Prince Edward Island because of the cold Labrador current.

IN SUMMARY

- The climate system transports thermal energy from areas that have more energy to areas that have less energy.
- Water, air, and land heat up at different rates when they absorb energy.
- When a fluid such as air or water is unevenly heated, convection currents form as colder, denser fluid sinks, pushing up the warmer, less dense fluid.
- Uneven heating of Earth causes the convection currents that create prevailing winds and ocean currents.
- Air and ocean currents are the main ways that energy is transported around Earth.
- Prevailing winds and ocean currents affect climate in nearby regions.

CHECK YOUR LEARNING

1. (a) Use the image of ocean currents in Figure 5 of this section to identify an area in the world that is likely to have a warm, damp climate due to ocean currents.
   (b) Use the map to identify a region that is likely to have a cold, dry climate due to ocean currents. T/I
2. Describe an idea in this section that is new to you. How did this idea change how you picture the climate system? K/U
3. Draw a diagram to show the movement of cold and warm air in the atmosphere around the equator. K/U C
4. What effect would each type of prevailing wind have on the land it crosses? T/I
   (a) a prevailing wind coming from the North Pole
   (b) a prevailing wind coming from the ocean
5. Define "thermohaline." K/U
6. Earth's thermohaline circulation is like an "enormous conveyor belt." What is being moved on this "global conveyor belt" that affects Earth's climate? K/U
7. How are convection currents formed? K/U

8.9 Long-Term and Short-Term Changes in Climate

Figure 1 Boulders such as this one, called erratics, triggered scientists to think about the movement of glaciers, leading to research into past climates. Have you ever seen a boulder similar to this one?

ice age a time in Earth's history when Earth is colder and much of the planet is covered in ice

Large, strangely placed boulders can be seen in parts of Canada (Figure 1). These boulders do not match the rock in the surrounding landscape. Where did these boulders come from? Glaciers move very slowly over long periods of time, scraping away soil and rock and carrying large boulders for miles. The boulders are left behind when the glacier eventually melts.

About 200 years ago, scientists began to study land formations caused by moving glaciers. They hypothesized that there was a time when much more of Earth's surface was covered in ice—in other words, an **ice age**. Since then, scientists have found more evidence of past climate change.

Scientists have discovered that Earth's climate goes through a variety of natural changes. Changes in climate are triggered by changes in Earth's energy balance. If something causes Earth's surface and atmosphere to absorb the Sun's energy differently, the climate will change. Or, if the amount of energy received from the Sun changes, Earth's climate will also change.

Over millions of years, the movement of Earth's crust triggers changes in the climate by affecting how much of the Sun's energy is absorbed. Over hundreds of thousands of years, Earth's climate also undergoes cyclic changes from warmer to colder climates. These cycles are a result of variations in Earth's orbit. Changes like these are sometimes called long-term changes because they happen over very long periods of time. Within long-term changes, there are also shorter periods of climate change caused by natural events such as volcanic eruptions. These are known as short-term changes.

plate tectonics the theory explaining the slow movement of the large plates of Earth's crust

continental drift the theory that Earth's continents used to be one supercontinent named Pangaea

Long-Term Changes Due to Continental Drift

According to the theory of **plate tectonics**, Earth's continents have moved over the surface of the globe for hundreds of millions of years. This movement of the continents is called **continental drift**. Over time, one large supercontinent split up to form the continents we see today, 225 million years later (Figure 2).

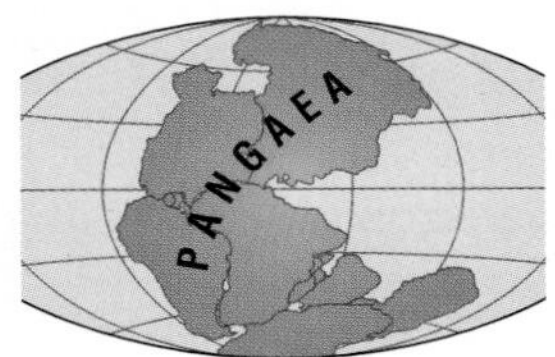

PERMIAN
225 million years ago

TRIASSIC
200 million years ago

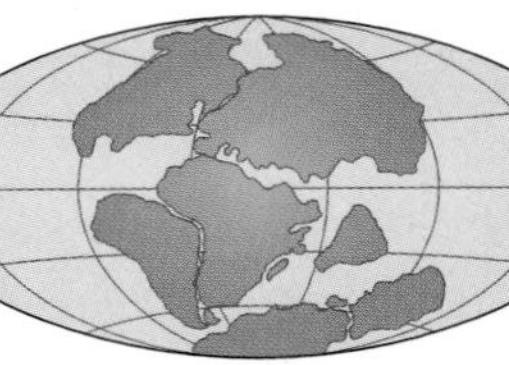

JURASSIC
135 million years ago

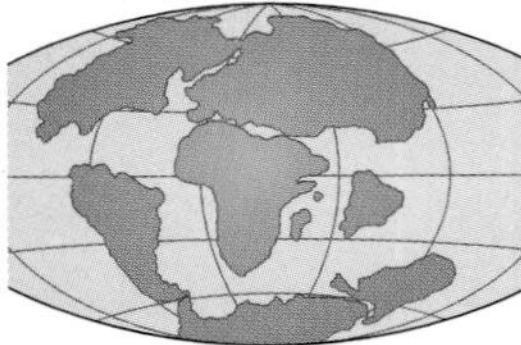

CRETACEOUS
65 million years ago

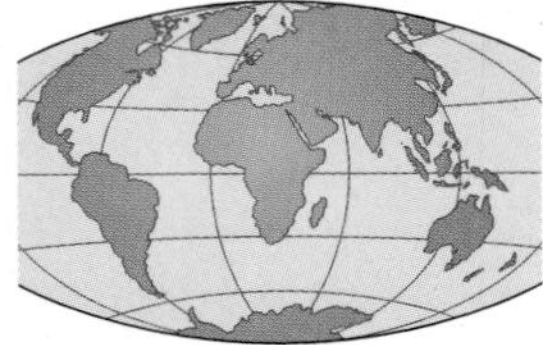

PRESENT DAY

Figure 2 Continental drift affects climate because it changes the distribution of land around the globe.

Continental drift influences Earth's climate in many ways. When continents move, ocean currents and wind patterns change. This affects heat transfer. Millions of years ago, major air and ocean currents were not the same as they are now.

Continental drift also affects the distribution of land mass. At present, the northern hemisphere, which includes Canada, has the most land mass.

Because there are fewer large bodies of water, the northern hemisphere has the coldest winters and the warmest summers. The greater amount of ocean in the southern hemisphere produces a more moderate climate. This was not true, however, in the Permian period (225 million years ago).

To find out more about the movements of the continents,
GO TO NELSON SCIENCE

The uplifting of new mountain ranges, caused by the movement of Earth's plates, affects local and regional climates. The weathering of old mountain ranges over millions of years also results in regional climate changes. These two effects have changed the regional climate in different parts of Canada. Canada contains younger mountain ranges, such as the Canadian Rockies, and older ranges, such as the Appalachians.

Scientists have identified other factors that may affect climate over very long or "geological" periods of time. For example, the amount of energy the Sun produces can gradually change over time.

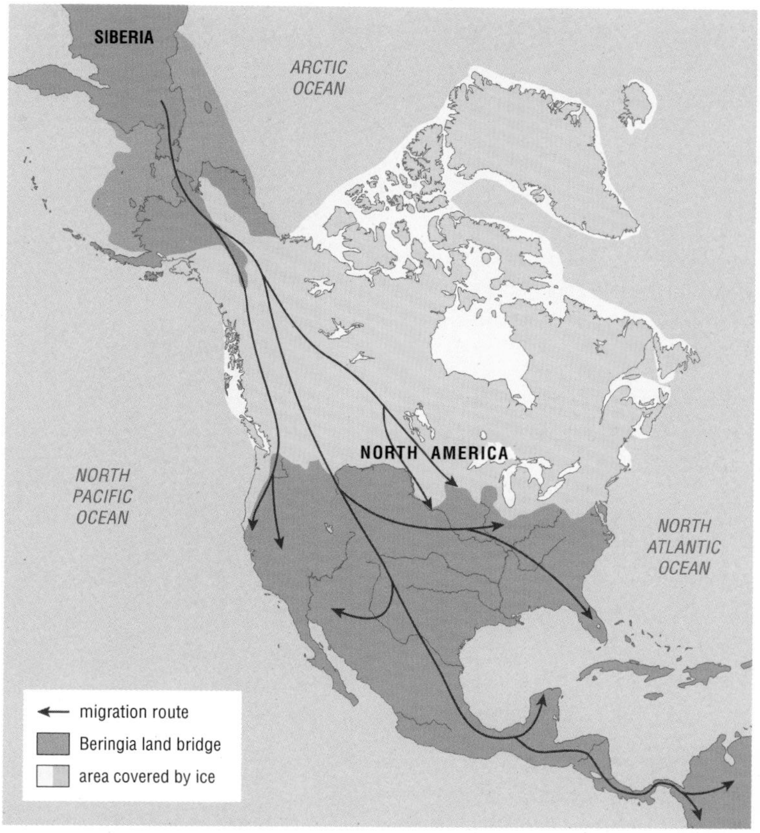

Figure 3 The Beringia land bridge was a temporary formation of land between Alaska and northern Europe. One theory suggests that humans walked over the land bridge to settle in North America about 25 000 years ago.

Long-Term Cycles in Climate

About 20 000 years ago, Earth experienced its last ice age. The average temperature was almost 10 °C lower than it is today. Ice sheets about 3 km thick covered much of Canada. As the water in the oceans froze, sea levels dropped. Land that was normally under the ocean became exposed, forming links between continents. Plants and animals, including humans, migrated across these bridges and settled in new lands (Figure 3).

For the last 800 000 years or more, Earth's climate has cycled between freezing ice ages and warmer **interglacial periods** (Figure 4).

Archaeologists study what life was like hundreds or thousands of years ago. To find out more about the work of archaeologists,

GO TO NELSON SCIENCE

interglacial period a time between ice ages when Earth warms up

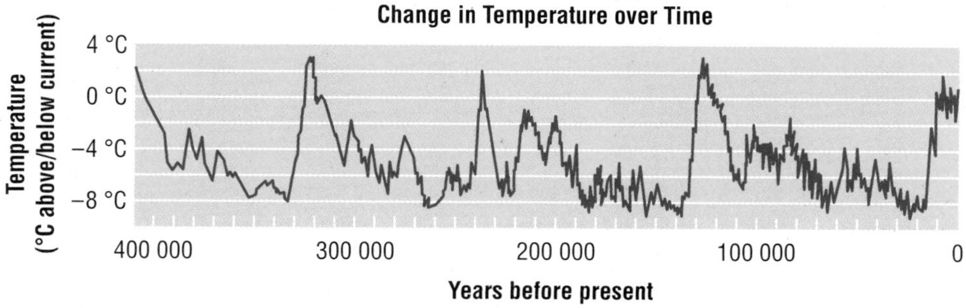

Figure 4 Graph of changes in Earth's average temperature over the past 400 000 years. The values on the *y*-axis represent deviations from Earth's average temperature today. Notice that major changes in temperature happen in regular cycles. Warm interglacial periods occur about every 100 000 years.

Why Do Interglacial Periods and Ice Ages Keep Happening?

In 1941, an engineer and amateur astronomer named Milutin Milankovitch studied long-term cycles of climate change. He developed a theory on the causes for these changes. Milankovitch calculated that Earth's orbit around the Sun changes in three main ways (Figure 5).

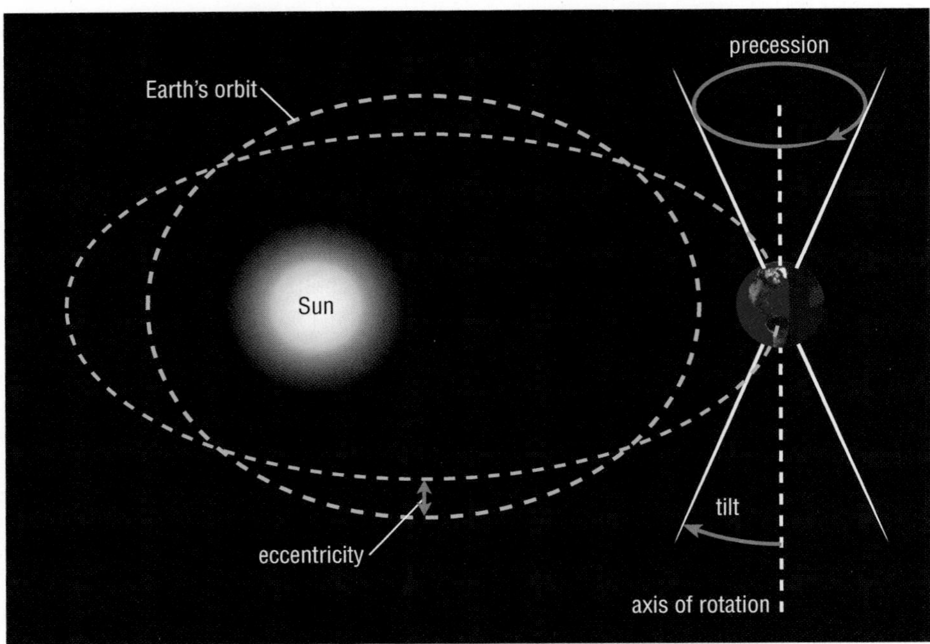

Figure 5 Milankovitch cycles

- **Eccentricity (shape of orbit):** The shape of Earth's orbit around the Sun varies from being nearly circular to being more elliptical (like a flattened circle). This variation is caused by the influence of Jupiter's and Saturn's gravities. It has several components, which combine to give an approximate cycle of 100 000 years. Earth's orbit is currently more elliptical.
- **Tilt:** Over a cycle of about 41 000 years, Earth tilts back and forth on its axis from 22.1° to 24.5°. As the angle increases, seasonal differences increase. Earth's axis is currently at 23.5°. The angle is slowly decreasing.
- **Precession of tilt (wobble):** As Earth spins on its axis, it slowly wobbles in a cycle over 26 000 years. The angle of tilt remains approximately the same, but the direction of tilt changes. This is similar to how a top behaves when you spin it (Figure 6). Earth's axis is currently pointing toward Polaris, which we call the North Star. More than 5000 years ago, the North Star was Thuban. In another thousand years, Airai will be the new North Star.

Figure 6 The axis of a spinning toy gyroscope or top slowly wobbles around the vertical.

Together, these changes add up to cause regular cycles of ice ages and interglacial periods. These regular cycles have taken place for more than 400 000 years.

These small changes in Earth's orbit happen very slowly, over tens and hundreds of thousands of years. However, most climate scientists think that, combined, they are the main trigger of the 100 000-year cycles in Earth's climate seen in Figure 3 of this section. They are the most likely reason that astronomers have identified so far.

To learn more about the work of an astronomer,
GO TO NELSON SCIENCE

The small changes in the amount of energy Earth receives from the Sun temporarily unbalance the climate system. Positive feedback effects then enhance this small change. The climate system rebalances again, but at a different global temperature.

A decrease in energy from the Sun, causing lower temperatures, creates an ice age. An increase in energy, causing warmer temperatures, creates a warm interglacial period. Today, we are in a relatively warm period.

Short-Term Variations in Climate

Even though Earth's climate is in a stable warm period right now, we still see small variations in climate over tens of years to hundreds of years. Short-term variations in climate are caused by a variety of factors, including volcanic eruptions, small changes in the Sun's radiation, and changes in the circulation of air and ocean currents. In the rest of this section, you will learn how relatively small changes in these factors can affect the entire climate system.

Volcanic Eruptions

Volcanic eruptions spew rocks, dust, and gases high into the atmosphere (Figure 7). In particular, ejected particles of sulfur dioxide reflect the Sun's energy back out to space. This has the effect of shading Earth's surface. As a result, less energy is present in the climate system and Earth temporarily cools down.

DID YOU KNOW?

The Year with No Summer

In 1815, Mount Tambora in Indonesia erupted, sending 150 cubic kilometres of dust into the stratosphere. The next year was called "the year with no summer." There was widespread famine in North America following summer snowfalls and frost.

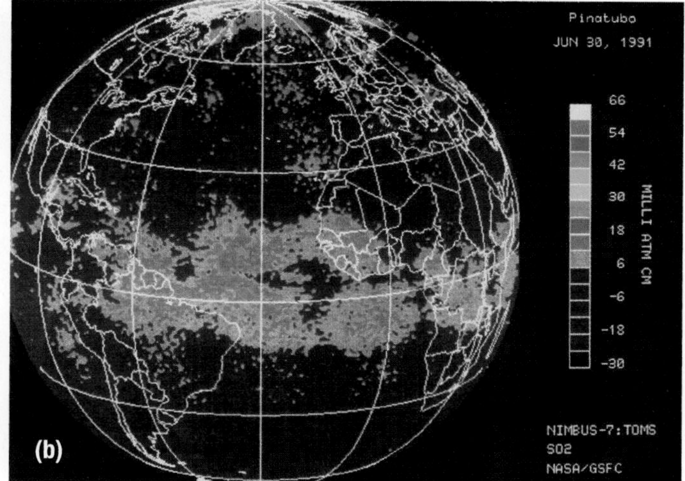

Figure 7 (a) Mount Pinatubo, in the Philippines, erupted in June 1991. (b) This map, based on satellite images taken 18 days after the eruption of Mount Pinatubo, shows the distribution of 17 Mt (megatonnes) of sulfur dioxide that was produced during the eruption.

Air and Ocean Currents

Air and ocean currents affect Earth's climate in many different ways. Changes to the ocean's thermohaline circulation may cause abrupt changes in climate, although this phenomenon is not fully understood.

Earlier, you learned that thermohaline circulation is an important part of the "conveyor belt" of ocean currents. About 12 000 years ago, Earth's climate was in transition from the last ice age into the present warm period. Enormous ice sheets across Earth's surface melted, dumping vast quantities of fresh water into the oceans. This fresh water was less dense than the salty ocean water, so it stayed near the surface of the ocean.

Scientists believe that this fresh water may have disrupted the thermohaline circulation. Earth's climate abruptly became colder again for a while, possibly due to this disruption.

Some changes in air and ocean currents occur regularly. A dramatic change takes place in the Pacific Ocean every three to seven years. The prevailing winds temporarily switch direction, changing the ocean currents in that area. Instead of pushing warm surface water toward the west Pacific, the prevailing winds push the warm water east, toward South America. This periodic shift in Pacific winds and ocean currents is called **El Niño** (Figure 8).

El Niño a recurring change in the Pacific winds and ocean currents that brings warm, moist air to the west coast of South America

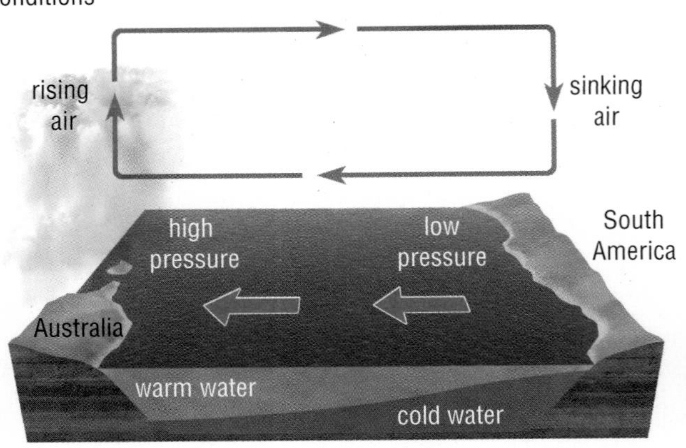

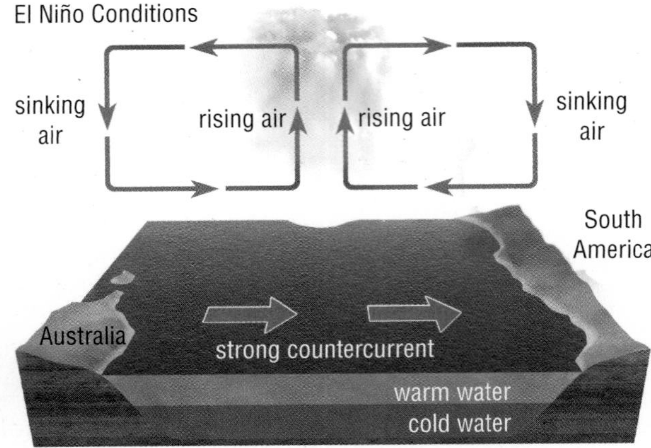

Figure 8 Red arrows represent warm water currents. (a) Normally, the west coast of South America is cold and dry, due to a cold ocean current nearby. (b) During an El Niño event, changes to prevailing winds affect the movement of ocean water. The west coast of South America receives warmer, wetter weather.

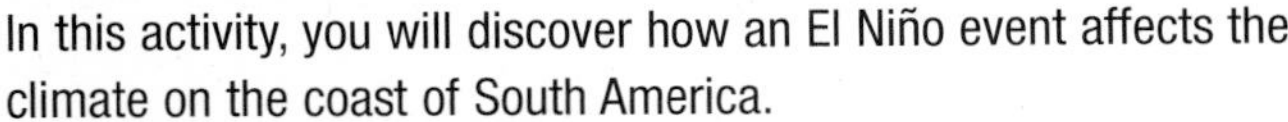

SKILLS: Researching, Analyzing the Issue, Communicating

SKILLS HANDBOOK 4.A.1., 4.A.2.

In this activity, you will discover how an El Niño event affects the climate on the coast of South America.

1. Research the normal climate in Peru and Ecuador. Record your findings.
2. Research the effect of El Niño on temperature and precipitation in Peru and Ecuador. Record your findings.
3. Research the relationship between El Niño and severe weather like droughts, storms, and floods in California, Australia, and southeast Asia.

GO TO NELSON SCIENCE

A. The cold Peru Current, also called the Humboldt Current, flows north along the west coast of South America. What effect does the Peru Current have on the climate of Peru and Ecuador? T/I
B. During an El Niño year, winds push warm water toward the coast of Peru and Ecuador. What effect does this warm water have on the temperature and precipitation in Peru and Ecuador? T/I
C. Warm water and air contain more energy than cold water or air. Also, storms and hurricanes need a lot of energy to develop. How does El Niño affect the number of storms and hurricanes in Peru and Ecuador? T/I
D. Summarize the effects of El Niño on countries bordering the east side of the Pacific Ocean. Include environmental, economic, and social impacts. T/I A
E. What effect does El Niño have on the west side of the Pacific Ocean? T/I
F. What impacts does El Niño have on Canada? Think about direct and indirect effects. T/I A

Changes in the Sun's Radiation

Even small changes in the Sun's radiation can affect climate. If the amount of radiation from the Sun drops, Earth receives less energy. Earth's climate cools down. If the amount of radiation from the Sun increases, our climate warms up. Scientists do not yet fully understand the reasons why the Sun's energy decreases or increases over shorter time scales.

TRY THIS EXPLORING THE SUN'S ENERGY OUTPUT

SKILLS: Analyzing, Evaluating

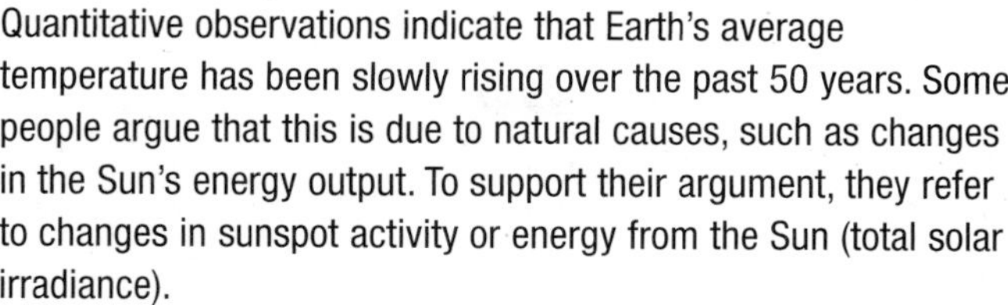

Quantitative observations indicate that Earth's average temperature has been slowly rising over the past 50 years. Some people argue that this is due to natural causes, such as changes in the Sun's energy output. To support their argument, they refer to changes in sunspot activity or energy from the Sun (total solar irradiance).

1. Use the Internet to find a graph that plots sunspot number, cosmic ray intensity, and total solar irradiance since 1950. Notice the cycles.

GO TO NELSON SCIENCE

A. Using the graph, examine the "smoothed sunspot number" from 1950 to 2005. Since 1950, has the number of sunspots in each peak year increased significantly? T/I

B. Examine the total solar irradiance. This satellite data only goes back to 1975. T/I
   (i) Compare solar irradiance variation with the variation in sunspots.
   (ii) How has total solar irradiance varied from 1975 to 2005?
   (iii) What values have the peaks varied between? (Be careful to choose the correct vertical scale.)
   (iv) Would you conclude that the total solar irradiance has been increasing or decreasing since 1975?

C. What do you conclude about the Sun's activity since 1975? Since 1950? T/I

IN SUMMARY

- Continental drift and other natural factors have profoundly affected Earth's climate over the past hundreds of millions of years.
- Over the last 400 000 years or more, climate has continually cycled from ice ages to warmer interglacial periods about every 100 000 years.
- Long-term cycles in Earth's climate correspond to changes in the shape of Earth's orbit, changes in Earth's tilt, and the precession of Earth's axis.
- Short-term variations in climate can be caused by volcanic eruptions, changes in the Sun's radiation, and changes in the circulation of air and ocean currents.

CHECK YOUR LEARNING

1. Describe three ways in which plate tectonics and continental drift might have affected global climate patterns in the past. K/U
2. (a) What is an ice age?
   (b) How did scientists infer the possibility of an ice age by examining boulders and other marks left by glaciers? K/U
3. (a) What is an interglacial period?
   (b) How frequently do warm interglacial periods occur?
   (c) Are we currently in an ice age or an interglacial period?
   (d) Has Earth spent more time in cold ice age periods or in mild interglacial periods over the past 400 000 years? K/U A
4. List at least three possible causes of short-term climate change. K/U
5. How do volcanic eruptions affect climate? Use a labelled diagram in your answer. K/U C
6. Describe one way in which changing air or ocean currents can affect climate. K/U
7. How do we know that the long cycle of ice ages is not caused by some other factor, such as
   (a) volcanic activity?
   (b) continental drift and plate tectonics?
   (c) long cycle changes in the Sun's energy output? K/U T/I
8. Explain the changes that scientists think are responsible for the 100 000-year climate cycles. K/U

Lake Agassiz: Studying Past Climate

Huge sheets of ice covered much of North America during the last ice age. When the ice age ended, most of this ice gradually melted as Earth's temperature began to rise. Several large lakes formed along the southern edge of the ice sheet as it melted. The largest of these lakes, Lake Agassiz, formed across south-central Canada and into the United States (Figure 1).

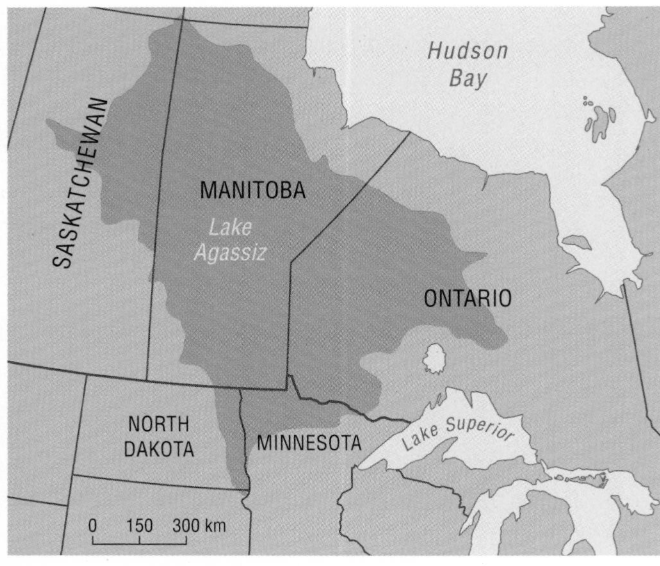

Figure 1 Lake Agassiz, as mapped by Teller and colleagues. Lake Agassiz probably held more fresh water than all the lakes in the world do today.

Lake Agassiz and Global Flooding

Figure 2 James T. Teller

James T. Teller is a geologist at the University of Manitoba (Figure 2). Teller and other scientists have reconstructed the history of Lake Agassiz by examining ancient beaches and sediments from the ancient lake bottom. In 2004, Teller received the Michael J. Keen medal from the Geological Association of Canada for his research on Lake Agassiz.

Scientists believe that the volume of water in Lake Agassiz changed abruptly several times during its history (Figure 3). At those times, the ice around the lake broke, allowing huge quantities of water to drain into the ocean. The additional water caused sea levels around the world to rise slightly. Teller believes flooding from Lake Agassiz may be the source of flood stories found in the Bible, First Nation legends, and other ancient stories.

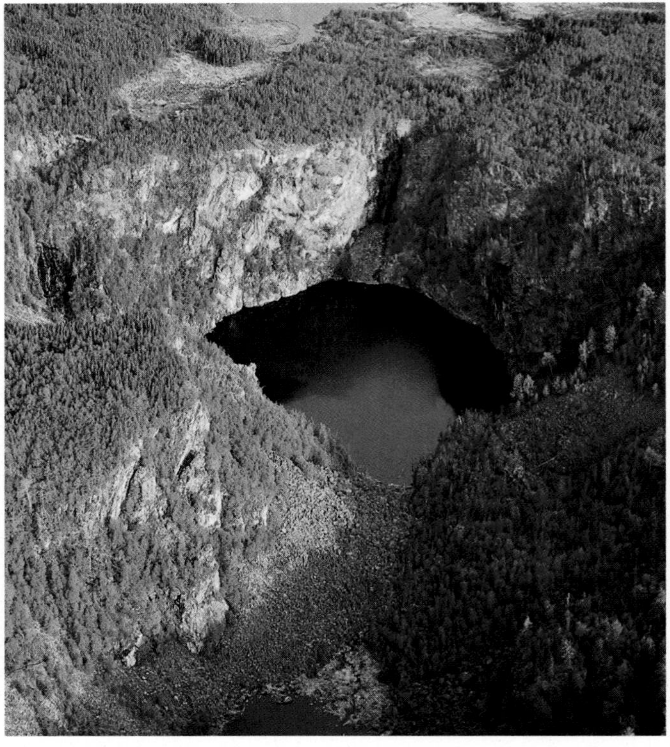

Figure 3 Devil's Crater in northern Ontario is just one of the features caused by water draining from Lake Agassiz thousands of years ago.

Lake Agassiz and Climate Change

Research suggests that large changes in the volume of Lake Agassiz happened at about the same time as Earth's temperature decreased. Changes in Lake Agassiz could have caused abrupt changes in Earth's climate. How could changes in a lake's size affect climate? Fresh water from Lake Agassiz would have poured into the North Atlantic Ocean. This fresh water may have hindered the flow of warm water north from the equator to the North Atlantic. This change in water flow would have interrupted the northward transfer of thermal energy. These events could have triggered a period of cooling in Europe and North America.

Connections to Today's Climate

Today, land-based ice in the Arctic is beginning to melt. Fresh water is pouring into the Atlantic Ocean as this ice melts. Sea levels are expected to rise, just as they did when Lake Agassiz drained. Scientists want to know whether the flow of fresh water will interrupt ocean currents and affect world climate. Information from Lake Agassiz may help scientists determine how Earth's climate will change over the next century.

Feedback Loops and Climate

Small changes, such as a decrease in snow cover, can have a very large effect on Earth's climate. This is because small changes are sometimes enhanced, or made bigger, by feedback loops (Section 8.6). Feedback loops can also act in the opposite way and cancel out changes.

Recall that, in a feedback loop, the cause creates an effect that impacts the original cause. In a positive feedback loop, the effect increases the original cause. In a negative feedback loop, the effect decreases the original cause. Feedback loops make it difficult for climatologists to predict the effects of changes.

The Water Vapour Feedback Loop

You probably know that more water vapour enters the atmosphere when the climate warms up, due to increased evaporation from Earth's lakes and oceans. (You studied the water cycle in Section 8.4.) This causes the climate to warm up even more because water vapour is a greenhouse gas and traps infrared radiation emitted by Earth. Conversely, if the climate cools down, less water vapour forms and the climate cools further. These are positive feedback loops.

The water vapour feedback loop becomes more complex when you consider clouds (Figure 1). More water vapour usually means more clouds. If the clouds form relatively low in the atmosphere, then they trap thermal energy near Earth's surface. You may have noticed that cloudy nights are usually warmer than clear nights. Low clouds create a positive feedback loop:

warmer temperatures → more (low) clouds → even warmer temperatures

If the clouds form at high altitudes, however, then they reflect the Sun's radiation back out to space. This creates negative feedback:

warmer temperatures → more (high) clouds → cooler temperatures

The Albedo Effect

Different surfaces reflect different amounts of the Sun's radiation. The proportion of radiation reflected by a surface is called its **albedo**. Ice and snow have high albedos because they reflect more radiation than grass or trees (Figure 2). This is why you squint when you go outside after a snowfall.

albedo a measure of how much of the Sun's radiation is reflected by a surface

Figure 1 Low clouds are involved in positive feedback loops. High clouds are involved in negative feedback.

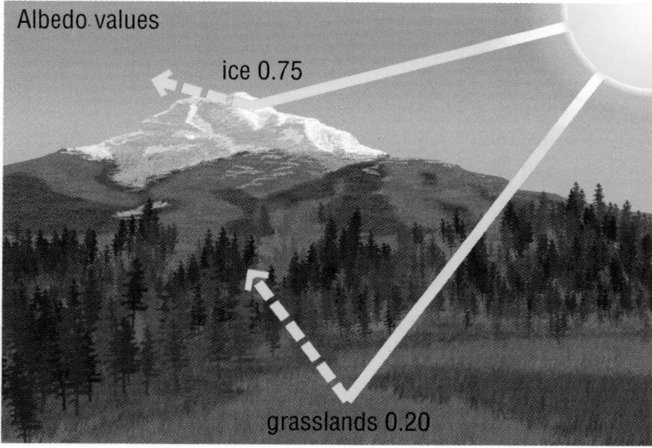

Figure 2 Ice reflects about 75 % of the Sun's radiation; its albedo is 0.75. Grass reflects about 20 % of the Sun's radiation; its albedo is 0.20.

DID YOU **KNOW?**

Venus, the Bright Planet
Venus has the highest albedo of the planets in our solar system: 0.65. Venus's high albedo is one of the reasons why this planet is often the brightest object in the night sky. Venus appears sometimes as the brightest object in the western sky after sunset, and sometimes as the brightest object in the eastern sky before sunrise.

albedo effect the positive feedback loop in which an increase in Earth's temperature causes ice to melt, so more radiation is absorbed by Earth's surface, leading to further increases in temperature

A planet's albedo is a measure of the amount of radiation that is reflected back when the Sun shines on it. Different parts of Earth's surface have different albedos: water (8 %), forest (10 %), sand and desert (25 %), fresh snow (85 %), and clouds (40 to 70 %). On average, Earth reflects 30 to 40 % of the Sun's radiation, so Earth's average albedo is between 0.30 and 0.40.

One of the most important feedback loops in Earth's climate system is called the **albedo effect**. This is the positive feedback loop between ice on Earth's surface and Earth's average temperature.

If Earth's average temperature drops slightly, more ice forms. This ice reflects more of the Sun's radiation, and Earth's temperature decreases even more. If Earth's average temperature increases slightly, more ice melts (Figure 3). More of the Sun's radiation is absorbed, and Earth's temperature increases even more.

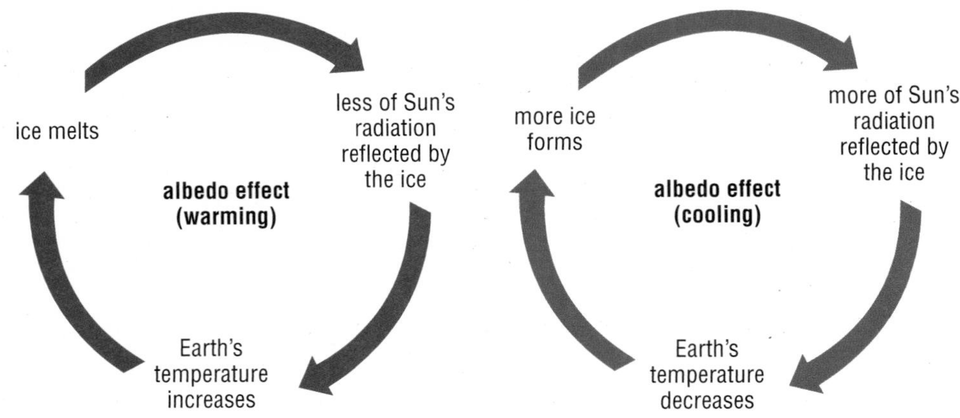

Figure 3 The albedo effect is the relationship between ice and Earth's temperature.

TRY THIS TESTING THE ALBEDO EFFECT

SKILLS: Predicting, Planning, Controlling Variables, Performing, Observing, Analyzing, Evaluating, Communicating

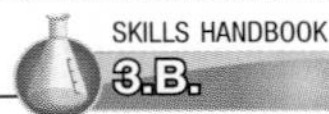

In this activity, you will design an experiment in which you model the albedo effect. You will plan and create a simple system with a source of light shining on a closed container with a transparent side and then measure the temperature of your system over time. Next, you will change the albedo of some part of your system and repeat your experiment. To draw meaningful conclusions, change only one variable at a time.

Equipment and Materials: thermometers or temperature sensors; light source; surfaces with different albedos (i.e., white or black construction paper or paint); sealable glass or plastic container

1. Design and write up a procedure that you will follow. Consider the following questions.
   - How often and for how long will you measure the temperature of your system?
   - How will you change the albedo of your system?
   - How will you record your observations?
2. Submit your procedure to your teacher for approval.
3. Carry out your investigation and record your observations.

A. What variable did you change? How do you know that changing this variable gave your system a different albedo? T/I

B. List all the variables that you kept constant. How confident are you that all variables except the one you tested were constant? T/I

C. The ice caps at Earth's poles are becoming smaller as the ice melts. Based on your investigation, explain what effect the melting of the polar ice caps could have on Earth's climate. Is this a positive feedback loop or a negative feedback loop? A

In Figure 4 you can see that Earth's climate shifts relatively quickly from an ice age into the warm, interglacial period that follows it. This rapid change can be partly explained by the albedo effect.

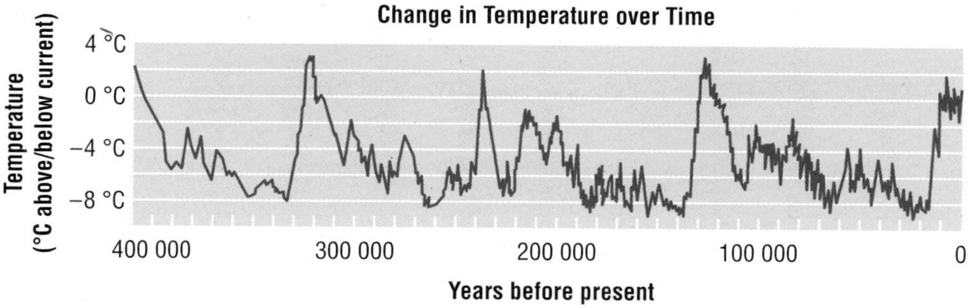

Figure 4 Notice how average temperatures swing quite rapidly between warm and cold periods.

IN SUMMARY

- In a positive feedback loop, the effect increases the original cause.
- In a negative feedback loop, the effect decreases the original cause.
- Feedback loops can enhance small changes in the climate system.
- Low clouds are involved in a positive feedback loop between Earth's temperature and water vapour. High clouds are involved in a negative feedback loop between temperature and water vapour.
- The albedo effect is a positive feedback loop linking the area of permanent ice on Earth's surface and Earth's average temperature.

CHECK YOUR LEARNING

1. What is a feedback loop? K/U
2. What is the difference between a positive feedback loop and a negative feedback loop? Give an example of each. K/U
3. Grass has an albedo of about 0.20 and rock has an albedo of about 0.30. Which will reflect more of the Sun's radiation, grass or rock? Explain why. K/U
4. Explain, using a concept map or flow diagram, how an increase in water vapour might result in a positive feedback loop. Include in your explanation the fact that Earth gives off infrared radiation. T/I C
5. Suppose that an increase in water vapour resulted in the increase in the number of cloudy days. These clouds kept the Sun's light from reaching Earth, thus lowering the average temperature. K/U T/I
   (a) What would happen to the amount of water evaporating from lakes and oceans in this scenario?
   (b) Would this be a positive or a negative feedback loop? Explain with a diagram.
6. (a) Write out a definition of a planet's albedo using your own words.
   (b) If a planet's albedo is affected by ice, would it contribute to a positive or a negative feedback loop in the planet's climate? Explain. K/U T/I
7. During the last ice age, more than half of Earth's surface (both land and sea) was covered in ice. K/U T/I C
   (a) Create a hypothesis about the albedo of Earth during the last ice age.
   (b) Based on your answer to (a), did Earth absorb more or less of the Sun's radiation during the last ice age than it does today? How would this have affected Earth's climate?
   (c) Draw a diagram for the feedback loop caused by the albedo effect during the last ice age.

8.11 Studying Clues to Past Climates

Scientists have recorded temperature, rainfall, and other climate data over the last 200 years and more. These data give us a good record of the world's climate during this period. Before that, people kept informal climate records in journals, paintings, farming records, and oral histories (Figure 1).

Figure 1 Paintings and drawings from the 1500s to 1850s show snow and people skating on rivers that do not freeze over today. This period is known as the Little Ice Age. It was likely caused by small decreases in the Sun's radiation, along with several large volcanic eruptions at that time.

Paleoclimatologists study past climate. How can they determine what the climate was like thousands of years ago? Natural materials such as rocks and ice preserve clues to past climates. **Proxy records** are stores of natural information that we can measure today that tell us what the climate was like in the distant past. For example, climatologists study fossils, tree rings, layers of ice, and coral reefs. Scientists are interested in the structures of these materials as well as in their chemical makeup.

proxy record stores of information in tree rings, ice cores, and fossils that can be measured to give clues to what the climate was like in the past

Proxy records are indirect records of climate. This means that they are not quantitative measurements of temperature or precipitation taken directly at the time. Scientists compare proxy records with quantitative historical records over the last several hundred years to determine what the proxy observations represent. This technique allows scientists to extend proxy records back into the distant past.

Ice Cores

The ice in Greenland and Antarctica contains air bubbles that have been trapped for thousands of centuries. Scientists drill deep into the ice and extract long cylinders of ice called ice cores. The ice at the top of an ice core is very recent, whereas the bottom of an ice core may be up to 800 000 years old. Ice cores provide our longest record of conditions in the atmosphere.

Scientists cut the ice cores into very thin slices and test the air bubbles in each slice for various gases (Figure 2). The tests establish how much carbon dioxide, methane, and nitrous oxide was in the air when the air bubble formed. Ice cores show paleoclimatologists that concentrations of these greenhouse gases have changed dramatically over Earth's history.

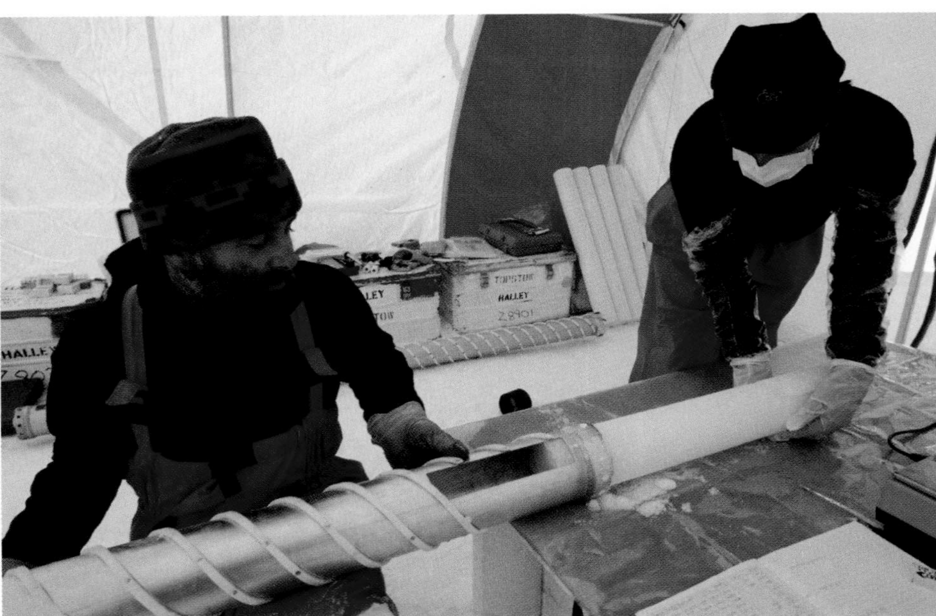

Figure 2 Scientists drill and examine ice cores to study gases trapped inside the ice.

Scientists also test ice cores for oxygen. Different types of atoms of oxygen exist. Some types are heavier than others. By measuring the ratio of light to heavy oxygen atoms, scientists can obtain information about air temperature. The colder the air when the bubble was formed, the more light oxygen is present in the bubble.

Ice cores also give information on precipitation and on volcanic eruptions through preserving layers of dust.

Ice core records show scientists that Earth has gone through many changes in climate, from ice ages to interglacial periods and back to ice ages (Section 8.9). We also know that temperature and greenhouse gas concentrations increase and decrease at the same time. During warmer periods, levels of greenhouse gases were higher. When it was cooler, levels were lower.

Tree Rings and Coral Reefs

Trees create one growth ring per year. Tree rings are thickest in years with good growing conditions. For example, a warm, wet year will produce a thick growth ring, whereas a cold, dry year will produce a thin ring (Figure 3). Some trees, such as the bristlecone pine and the California redwood, live for thousands of years. Scientists assemble clues from both living and dead trees to collect records of climate going back as far as 10 000 years. In Ontario, trees have provided proxy climate data for the past 2 767 years.

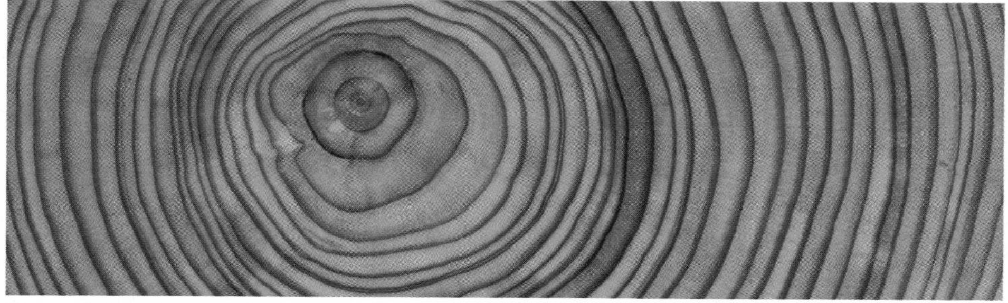

Figure 3 Tree rings are wider in good growing years and narrower in poor growing years.

DID YOU KNOW?

Old Trees in Ontario

Ontario has some very ancient trees. Some Eastern white cedar trees growing on the Niagara Escarpment are over 1 500 years old. Scientists have collected data from living and dead trees to construct a climate record of southern Ontario going back 2 767 years.

Records are also preserved in coral reefs (Figure 4). Like trees, corals add layers of growth each season. Scientists drill cylinders of coral and study their layers. Information from coral layers helps determine the temperature of the surface ocean water when each layer was growing.

To learn more about using proxy records such as tree rings and coral reefs,

GO TO NELSON SCIENCE

Figure 4 Coral reefs grow near the surface of the ocean. Layers of coral grow at different rates in warm water than in cold water.

Rock, Ocean Sediment, and Caves

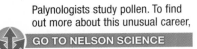

Palynologists study pollen. To find out more about this unusual career,

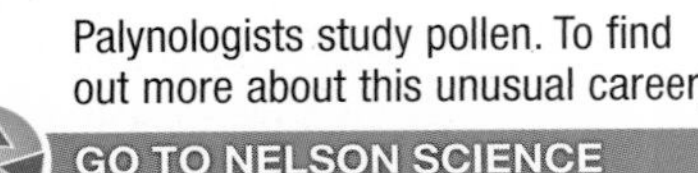

Layers of soil and rock build up on Earth's surface over time. Each layer may contain clues, such as plant pollen or fossils, to the climate at that time in that location. Fossils of pollen grains can be used to identify plants that grew thousands of years previously. Scientists who study pollen are called palynologists. They use the size, shape, and presence of pores, furrows, and air sacs to identify the species of the plants (Figure 5).

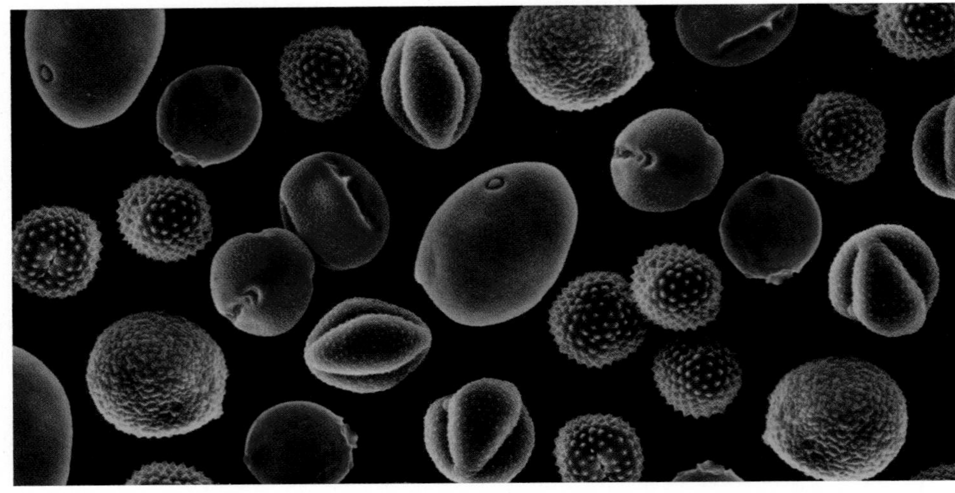

Figure 5 How many types of pollen can you see?

DID YOU KNOW?

Alice Wilson: A Canadian Scientist

Alice Evelyn Wilson (1881–1964) was the first female geologist at the Geological Survey of Canada. She worked there for 54 years, cataloguing the geological formations in the Ottawa region. Wilson covered about 16 000 km^2 of land—much of it by bicycle—identifying, classifying, and writing about fossil formations.

In the ocean, layers of sediment drift to the ocean floor and form layers of rock. Scientists drill cores of sediment from the ocean floor (Figure 6). Sometimes scientists find fossils of marine plants and animals that lived in warmer water than the location where they are found today. This is evidence that the layer containing these fossils formed during a warmer climate. Information from these sediment cores has allowed scientists to build a picture of Earth's climate over thousands of years in the past.

In caves, rock formations grow as the minerals that are dissolved in dripping water solidify into rock (Figure 7). Scientists can measure and date layers from these rocks. These rock formations grow faster in rainy weather, so analysis of the layers helps determine how much precipitation occurred at specific times in the past.

Figure 6 This sediment core from the ocean floor shows layers containing fossils from the past.

Figure 7 Stalactites (from the roof) and stalagmites (from the floor) show indirect evidence of precipitation patterns.

TRY THIS ANCIENT CLIMATES IN ARCTIC LANDS

SKILLS: Predicting, Analyzing, Communicating

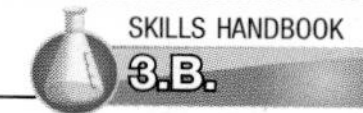

In Section 8.9, you learned that the continents have moved over the past millions of years. This means that some of the islands in the Canadian Arctic might not always have been so far north.

1. Locate on a map some of the northern islands in Canada's Arctic, such as Axel Heilberg, Baffin, Devon, Ellesmere, and Victoria Islands. What kind of plant and animal life do you think lived on these islands millions of years ago? Give reasons for your predictions. (Refer to Figure 2 in Section 8.9.)
2. Over the past 25 years, scientists have discovered fossils on some of these islands. Locate and analyze the fossil data reported from these sites.

GO TO NELSON SCIENCE

A. If these islands were located in another place on Earth's surface millions of years ago, what do you think their climate might have been like? Do you think the ocean currents that flowed past them millions of years ago might have been different? (Refer to Figure 2 in Section 8.9.) T/I

B. What kind of fossils have been found on or around these Arctic islands? Was your prediction correct? T/I

C. What kind of climate did the islands have at the time when the plants or animals represented by these fossils were active? Contrast this with the current climate of the region. T/I

D. Summarize your findings in a short paragraph of several sentences, describing what kind of life existed in the Arctic millions of years ago and how we know that the climate then was quite different from the climate today. T/I A

UNIT TASK Bookmark

You can apply what you learned in this section about records of past climates to the Unit Task described on page 444.

IN SUMMARY

- Proxy records are indirect records of past climates contained in natural materials.
- Analyzing air bubbles in ice cores provides data on greenhouse gases and temperature from the past.
- Analyzing the growth rings on ancient trees and coral provides data on temperature and precipitation from the past.
- Sediment cores from the ocean floor contain clues such as fossils that provide data about past climates.

CHECK YOUR LEARNING

1. (a) How do old paintings inform us that there was a Little Ice Age from 1500 to 1850 in Europe?
   (b) Are paintings proxy records or direct records of climate? K/U
2. (a) How do scientists know what Earth's climate was like over the past 200 years?
   (b) How do scientists know what Earth's climate was like thousands of years in the past?
   (c) How do scientists know that these records of Earth's climate thousands of years ago are accurate? K/U
3. Suppose you are a climatologist who is studying climate in Ontario 300 years ago. List three different sources you could examine for information on the climate at that time. K/U
4. Give one example of information scientists can obtain from each of the following proxy records: K/U
   (a) ice cores
   (b) tree rings
   (c) sediment cores from the ocean floor
5. Explain why the following statement is incorrect: Climatologists use the data from ice cores to directly measure past temperatures. K/U
6. In Ontario, our historical records of temperature and precipitation only go back 200 years. However, climatologists have been able to establish a climate record for southern Ontario that goes back 2 767 years. K/U
   (a) What was the main proxy record that they used to do this?
   (b) What can they do to show that this proxy record is accurate?
7. How do fossils tell us about the climate that existed in the past? (Hint: Use the idea of ecozones and ecoregions that you studied in Section 8.2.) T/I A

CHAPTER 8 LOOKING BACK

KEY CONCEPTS SUMMARY

Earth's climate system is powered by the Sun.

- Climate describes the weather that you can expect in a region. (8.1)
- The climate system is the complex set of components that interact with each other to produce Earth's climate. (8.3, 8.4)
- About 30 % of the energy from the Sun is reflected back to space. The remaining 70 % is absorbed by Earth's surface, by clouds, and by some gases in the atmosphere before being all re-emitted back out into space. (8.3)
- The amount of energy per unit of area reaching Earth is more intense near the equator than near the poles. (8.3)

Earth's climate system includes the atmosphere, the hydrosphere, the lithosphere, and living things.

- The atmosphere is composed of 78 % nitrogen gas, 21 % oxygen gas, and trace amounts of other gases, including carbon dioxide, methane, and ozone. (8.4)
- The hydrosphere and lithosphere absorb higher-energy radiation from the Sun, convert it into thermal energy, and then emit lower-energy infrared radiation. (8.4)
- Landforms and large bodies of water influence climate. (8.4)
- Living things affect the composition of gases in the atmosphere. (8.4)

The greenhouse effect keeps Earth warm by trapping thermal energy radiated by Earth.

- Water vapour is the most important greenhouse gas contributing to the natural greenhouse effect, followed by carbon dioxide and methane. (8.6)
- Greenhouse gases absorb lower-energy infrared radiation, preventing it from escaping into space. (8.6)
- The greenhouse effect is a natural process that keeps Earth's surface and atmosphere warmer than they would otherwise be. (8.6)

Thermal energy is transferred within Earth's climate system through air and ocean currents.

- The atmosphere and the oceans are heat sinks. (8.8)
- Convection currents cause zones of high and low pressure, thereby circulating heat in the atmosphere. (8.8)
- Thermohaline circulation is the main part of the conveyor belt that circulates thermal energy in the oceans. (8.8)

Earth's climate experiences long-term and short-term changes.

- Continental drift influences global circulation patterns in the atmosphere and oceans. (8.9)
- The shape of Earth's orbit, the tilt of Earth on its axis, and the wobble as Earth spins all influence Earth's climate. These are thought to cause ice ages approximately every 100 000 years. (8.9)
- Volcanic eruptions and variations in air and ocean currents cause short-term changes in climate. (8.9)

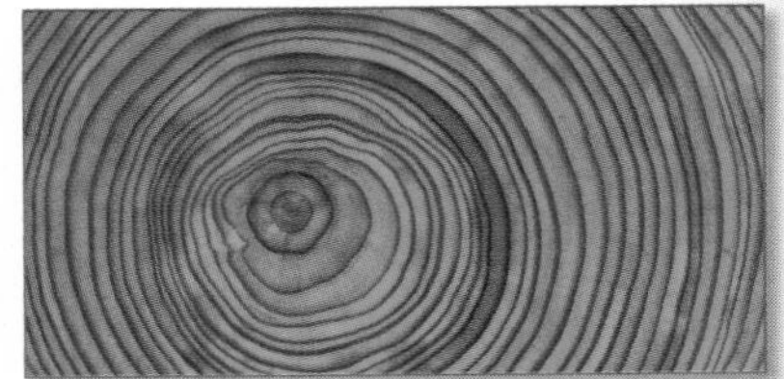

Scientists use natural ice cores, sediment layers, and tree rings to study past climates.

- Proxy records are indirect measures of Earth's past climate. (8.11)
- Ice cores record temperature data by trapping gases such as oxygen, carbon dioxide, methane, and nitrous oxide. (8.11)
- Tree rings and coral reefs grow annual layers in proportion to how favourable the climate is. (8.11)
- Sediment may contain evidence, such as fossils and plant pollen, of past climates. (8.11)

WHAT DO YOU THINK NOW?

You thought about the following statements at the beginning of the chapter. You may have encountered these ideas at school, at home, or in the world around you. Consider them again and decide whether you agree or disagree with each one.

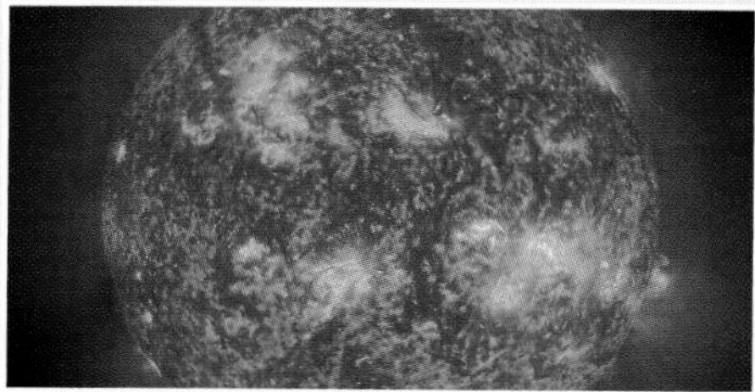

1 About half the energy on Earth comes from the Sun.
Agree/disagree

2 The greenhouse effect is a natural phenomenon.
Agree/disagree

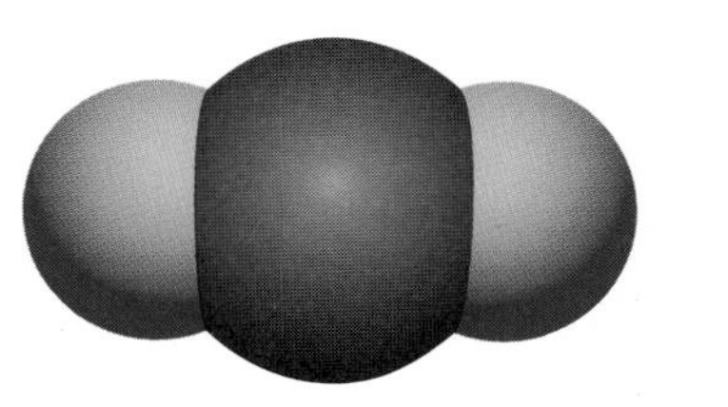

3 Carbon dioxide is an important part of Earth's climate system.
Agree/disagree

4 Earth's climate has remained very stable for thousands of years.
Agree/disagree

5 Volcanic eruptions cause the climate to change.
Agree/disagree

6 Weather and climate are the same thing.
Agree/disagree.

How have your answers changed since then? What new understanding do you have?

Vocabulary

weather (p. 319)
climate (p. 320)
bioclimate profile (p. 323)
climate system (p. 325)
ultraviolet radiation (p. 325)
infrared radiation (p. 325)
thermal energy (p. 327)
atmosphere (p. 330)
hydrosphere (p. 333)
lithosphere (p. 334)
greenhouse effect (p. 338)
greenhouse gas (p. 339)
carbon sink (p. 339)
feedback loop (p. 340)
heat sink (p. 344)
convection current (p. 345)
thermohaline circulation (p. 346)
ice age (p. 348)
plate tectonics (p. 348)
continental drift (p. 348)
interglacial period (p. 349)
El Niño (p. 352)
albedo (p. 355)
albedo effect (p. 356)
proxy record (p. 358)

BIG Ideas

- ✓ **Earth's climate is dynamic and is the result of interacting systems and processes.**
- ✓ **Global climate change is influenced by both natural and human factors.**
- **Climate change affects living things and natural systems in a variety of ways.**
- **People have the responsibility to assess their impact on climate change and to identify effective courses of action to reduce this impact.**

CHAPTER 8 REVIEW

The following icons indicate the Achievement Chart category addressed by each question. K/U Knowledge/Understanding T/I Thinking/Investigation C Communication A Application

What Do You Remember?

1. (a) Define climate.
   (b) List three climate zones. (8.1) K/U
2. (a) How is climate similar to weather?
   (b) How is climate different from weather? (8.1) K/U
3. What are the four main components of the climate system on Earth? (8.3, 8.4) K/U
4. (a) What is a greenhouse gas?
   (b) List three greenhouse gases in addition to water vapour. (8.6) K/U
5. What are some of the causes of long-term climate change? (8.9) K/U
6. What are some of the causes of short-term climate variations? (8.9) K/U
7. (a) What is a feedback loop?
   (b) Give three examples of feedback loops in Earth's climate system. (8.10) K/U
8. (a) What are proxy records?
   (b) How are they useful to climatologists? (8.11) K/U
9. Describe two methods a climatologist could use to collect climate data from the distant past. (8.11) K/U
10. (a) What is albedo?
    (b) What is the albedo effect? (8.10) K/U

What Do You Understand?

11. Describe one way in which each of the factors below affects the climate of a region. (8.4, 8.8) K/U
    (a) distance from a large body of water
    (b) prevailing winds
    (c) land formations
12. Draw a diagram to explain why the climate is colder close to the North and South Poles than it is at the equator. (8.3) K/U
13. Earth constantly absorbs energy from the Sun. Explain why Earth's surface does not continue heating up. (8.3) K/U
14. Explain how the greenhouse effect works. (8.6) K/U
15. How can relatively small changes, such as a small drop in the Sun's radiation, cause large changes to Earth's climate? (8.10) K/U
16. (a) When Earth begins to warm up from an ice age, the ice begins to melt. Describe how the albedo of ice becomes important at this stage.
    (b) Is the albedo effect a positive or a negative feedback loop? Use a diagram to explain. (8.10) K/U C
17. Explain how tree rings are used as proxy records. (8.11) K/U
18. How do volcanoes influence the climate? (8.9) K/U
19. Explain how clouds can be part of both positive and negative feedback loops. (8.10) K/U

Solve a Problem

20. Choose a city on the north shore of Lake Ontario (Figure 1). (8.4) K/U T/I

Figure 1

    (a) Describe how Lake Ontario affects the climate in your chosen city.
    (b) Explain how the Great Lakes can influence the amount of snowfall in nearby regions downwind.
21. Write a paragraph summarizing the role of Earth's climate system and how it affects conditions on Earth. (8.3, 8.4) K/U
22. (a) Compare how thermal energy circulates in the ocean with how thermal energy circulates in the atmosphere.
    (b) Why is the circulation of thermal energy around Earth important for living things? (8.8) K/U
23. The Gulf Stream carries warm water past the west coast of Europe. How would you expect this current to affect the climate of the west coast of Europe? (8.8) T/I

24. Greenhouse gases trap energy in the atmosphere, causing Earth to be warmer than it would be without greenhouse gases. Use this fact to predict what might happen to Earth's average temperature in each of the following scenarios: (8.6) T/I
    (a) Greenhouse gas levels in the atmosphere increase.
    (b) Greenhouse gas levels in the atmosphere decrease.
25. (a) Suggest two difficulties that climatologists might encounter when interpreting proxy records to reconstruct climate.
    (b) Identify one type of proxy record. Suggest specific examples of difficulties scientists might have. (8.11) K/U T/I
26. Scientists sometimes use measurements of types of oxygen in coral to obtain climate data. The graphs in Figure 2 compare data of heavy and light oxygen ratios in two coral reefs (Tarawa Coral and Galápagos Coral) with the rainfall index. (8.11) T/I

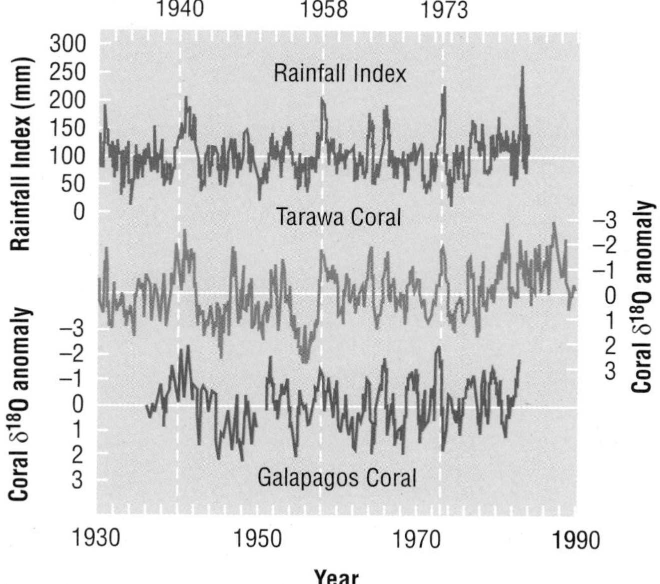

Figure 2

    (a) Do the data from the two coral reefs correspond? If they do, explain how.
    (b) Do the coral reef data correspond to the rainfall index? If they do, explain how. (Hint: Concentrate on the high and low points of the data.)
    (c) Are these data useful as proxy data?
    (d) Did you find it difficult to detect signals from the data? Explain why or why not.

Create and Evaluate

27. You are part of a team that has a long-term assignment to determine the climate of an isolated region in the Himalayas. What measurements will you take, and for how long? (8.1) K/U T/I
28. "Ozone is both a helpful and a harmful gas." Explain this statement. (8.4) K/U A
29. Yukon Territory has a dry climate with fewer trees than other parts of Canada. You are a climatologist working in Yukon. You want to find out what the local climate was like 6000 years ago. Develop a plan for collecting evidence. What clues might you look for? (8.11) K/U T/I C

Reflect on Your Learning

30. (a) What information in this chapter did you already know before reading it?
    (b) What information in this chapter was completely new to you?
    (c) How might the new information that you learned affect how you think about Earth?
31. (a) Before reading this chapter, how much did you know about previous periods of global warming and cooling?
    (b) How has your knowledge changed since reading this chapter?

Web Connections 

32. Suppose you want to move to a European city with a warm, moist climate. (8.4, 8.8) T/I
    (a) Use maps and the information in this chapter to identify a suitable city.
    (b) Explain why you think the climate of that city or town is warm and moist.
    (c) Use the Internet to research the climate of that city or town. Was your prediction correct? Explain.
33. Winter in the northern hemisphere has lower average temperatures than winter in the southern hemisphere. Why might this be so? Examine a globe to see how the two hemispheres differ and how this might affect their respective climates and average winter temperatures. Explain. (8.4, 8.9) T/I A

SELF-QUIZ

The following icons indicate the Achievement Chart category addressed by each question.

K/U Knowledge/Understanding T/I Thinking/Investigation C Communication A Application

For each question, select the best answer from the four alternatives.

1. Why is carbon dioxide considered a greenhouse gas? (8.6) K/U
   (a) It destroys the ozone layer.
   (b) It traps radiation from Earth's surface.
   (c) It is released by living things.
   (d) It is needed for photosynthesis.
2. Which of the following phrases best describes a bioclimate profile? (8.2) K/U
   (a) a graph of the atmospheric conditions in a location over a short period of time
   (b) an average of the weather in a certain region over a long period of time
   (c) a series of graphs that show present and future climate at a given location
   (d) a climate zone based on landforms, soil, plants, and animals, as well as climate
3. Which of the following series orders greenhouse gases from highest to lowest in effectiveness? (8.6) K/U
   (a) nitrous oxide, methane, carbon dioxide
   (b) carbon dioxide, methane, nitrogen dioxide
   (c) carbon dioxide, methane, nitrous oxide
   (d) methane, carbon dioxide, nitrous oxide
4. A weather description states: "a high of 35 °C today, sunny with cloudy periods, a 30 % chance of precipitation, wind from the west at 25 km/h, and relative humidity of 45 %." Which of the following describes the amount of water in the air in relation to the maximum amount that the air can possibly hold at that temperature? (8.1) K/U
   (a) 30 % chance of precipitation
   (b) wind from the west at 25 km/h
   (c) relative humidity of 45 %
   (d) a high of 35 °C
5. Scientists believe that Earth's slow wobble on its axis, its orbital shape, and changes in the tilt of its axis result in
   (a) the slow movement of plates in the lithosphere.
   (b) the uplift of new mountain ranges.
   (c) cyclical changes in the Sun's energy output.
   (d) long-term cycles in climate. (8.9) K/U
6. A positive feedback loop is defined as a process in which
   (a) the result influences the original process.
   (b) the result amplifies the original process.
   (c) the result diminishes the original process.
   (d) the result stops the process from proceeding. (8.10) K/U

Indicate whether each of the statements is TRUE or FALSE. If you think the statement is false, rewrite it to make it true.

7. Land near an ocean or large lake tends to be cooler in the summer than inland locations at the same altitude. (8.4) K/U
8. More energy is absorbed by Earth than is radiated by Earth. (8.3) K/U
9. Scientists use proxy records to directly measure past temperatures. (8.11) K/U
10. Convection currents form because warm air tends to rise and cold air tends to sink. (8.8) K/U

Copy each of the following statements into your notebook. Fill in the blanks with a word or phrase that correctly completes the sentence.

11. An ecoregion describes the ______ and ecology of a region in its current state. (8.2) K/U
12. Climate is the average of the ______ in a region over a long period of time. (8.1) K/U
13. Ice cores, tree rings, fossils, and coral reefs are examples of ______. (8.11) K/U
14. The greenhouse effect occurs when _______ absorb _______ emitted from Earth's surface and radiate it, heating Earth's surface and the atmosphere. (8.6) K/U
15. The continuous flow of water in the world's oceans is called _________. It is caused by differences in water ______ and _______. (8.8) K/U

Match each term on the left with the most appropriate description on the right.

16. (a) atmosphere (i) includes solid rock, soil, and minerals of Earth's crust
    (b) hydrosphere (ii) layer of the atmosphere where ozone absorbs ultraviolet radiation
    (c) stratosphere (iii) includes all water on and around Earth
    (d) lithosphere (iv) layers of gases surrounding Earth
    (e) troposphere (v) layer of the atmosphere where ozone has a toxic effect (8.4) K/U

Write a short answer to each of these questions.

17. Why is the greenhouse effect described as an "energy trapping" process? (8.6) K/U

18. In addition to the atmosphere, hydrosphere, and lithosphere, living things are important components of Earth's climate system. (8.4) K/U T/I
    (a) How do living things affect Earth's climate system?
    (b) Name two processes that organisms use to absorb gases from and release gases to the atmosphere.

19. Paraphrase the following statement: "Climate is what you expect, but weather is what you get." (8.1) C

20. The climate system transfers energy around the globe. (8.8) K/U
    (a) Explain why energy transfer is important.
    (b) How is energy transferred in the atmosphere?
    (c) How is energy transferred in the oceans?

21. Many parts of the climate systems are controlled and influenced by feedback loops. (8.10) K/U
    (a) Water vapour and temperature are involved in both positive and negative feedback loops. Using a diagram, explain how this can be so.
    (b) Explain how the albedo effect is part of a feedback loop.

22. Explain how each of the following would affect the land near which it passes: (8.8) T/I
    (a) a warm Gulf Stream current
    (b) a cold ocean current

23. Climate system is defined as the complex set of components that interact with each other to produce Earth's climate. Write a definition of "climate system" in your own words. (8.3, 8.4) C

24. Ozone in the stratosphere protects us from dangerous ultraviolet radiation. (8.4) T/I
    (a) Explain how the ozone is affected by the use of products that release CFCs?
    (b) Is stratospheric ozone concentration increasing or decreasing? What brought about this change?

25. Earth continuously absorbs energy from the Sun and yet Earth stays at a relative constant temperature. Explain how this occurs. (8.3) T/I

26 Would a significant increase in carbon dioxide in the atmosphere have a positive or negative effect on the populations of polar bears and seals that live in the Arctic? Explain why. (8.6) A

27. Scientists are concerned about current changes in climate. (In Chapters 9 and 10, you will learn about current climate change.) Use your understanding of past climate to explain why a change in climate would affect all living things on Earth. (8.9) A

CHAPTER 9

Earth's Climate: Out of Balance

KEY QUESTION: What is causing the changes in Earth's global climate?

Now the glacier
ends here.
1939

The Angel Glacier in Jasper National Park has retreated since 1939 (inset).

UNIT D

Climate Change

CHAPTER 8

Earth's Climate System and Natural Changes

CHAPTER 9

Earth's Climate: Out of Balance

CHAPTER 10

Addressing and Responding to Climate Change

KEY CONCEPTS

We have evidence that our climate is changing.

Human activities have increased atmospheric levels of greenhouse gases.

The increase in greenhouse gases is causing the anthropogenic (human-caused) greenhouse effect.

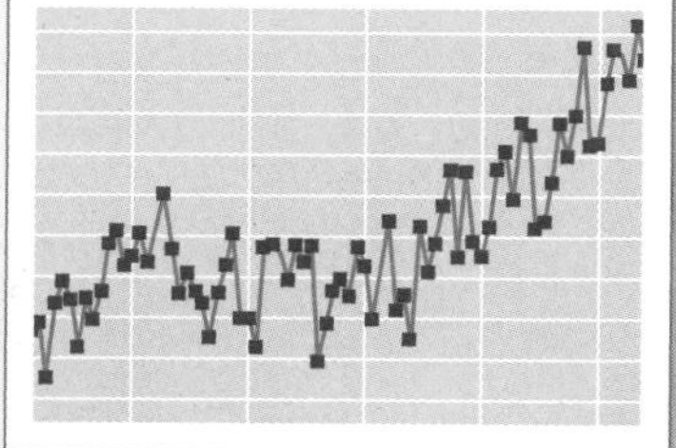

The anthropogenic greenhouse effect is the main cause of today's climate change.

The largest sources of greenhouse gases in Canada are the production and burning of fossil fuels.

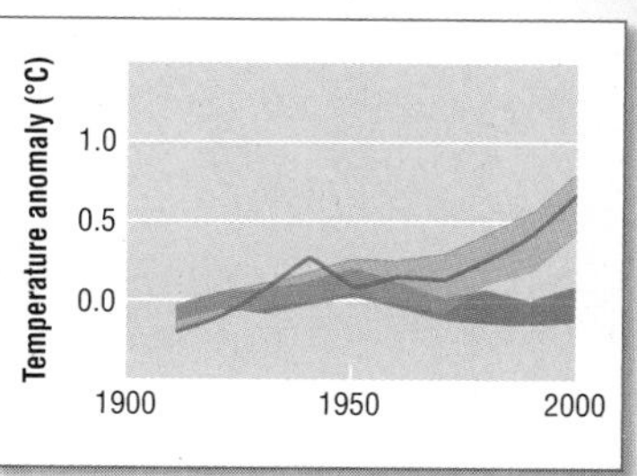

Scientists use climate models to figure out how different factors affect our climate.

CHANGING OUR LIVES

Figure 1 Banks Island is in the Northwest Territories.

Amaruq lives in a small community located on the coast of Banks Island, in the Arctic (Figure 1). He and his family live in a small house overlooking the ocean (Figure 2). In the spring, Amaruq and his family travel to their summer camp to hunt, fish, and gather berries. When Amaruq's family gathers with other families, the elders compare this year's conditions with those of past seasons (Figure 3).

"It's not as cold as it was when I was younger," says Tulugaq, speaking in Inuktitut. "It wasn't so long ago that the sea ice filled the harbour. All we had to do was paddle a few strokes out to hunt seals at the blow holes! There isn't as much sea ice now."

"That's right," his sister Elisapee agrees. "This fall, there was less ice, and it was far out in the ocean. It took a long time for my son and the other hunters to reach the seals."

Amaruq's father agrees with the elders. "There isn't as much ice this winter either," he says. "I was almost trapped by thin ice just a bit north of town. It's getting dangerous to travel. You can't trust the ice to be thick enough anymore."

Amaruq's mother reminds everyone about last spring: "We usually drive on the frozen river north of town, but it had already melted!"

After a while, Amaruq wanders off to talk to a couple of his friends. "Do you really think the climate is different now?" he asks his best friend, Meeka. "The elders always say things were better when they were young!"

"Yes, I've noticed changes in the last few years," agrees Meeka. "Remember when the schoolhouse had to be moved? The permafrost under the building melted, and the foundations started to sink. It's never been warm enough for that to happen before. Last month a scientist came to talk to the elders about our climate."

"Yeah!" chimes in Irniq. "It's definitely getting warmer. We're starting to get animals from the south on the island! My dad caught a salmon in the river this summer. My grandpa says salmon have never been this far north before."

On the way home, Amaruq thinks about what he has heard. He wonders whether the climate is really different now from what it used to be.

1. What evidence did Amaruq's family and friends use to infer that the climate is changing? Is this evidence scientific?
2. How are Inuit lifestyles changing? Are the changes in lifestyle related to the changes in Arctic ice?

Figure 2 Amaruq lives in a typical Arctic community.

Figure 3 The elders remember a time when the snow cover lasted much longer.

WHAT DO YOU THINK?

Many of the ideas you will explore in this chapter are ideas that you have already encountered. You may have encountered these ideas in school, at home, or in the world around you. Not all of the following statements are true. Consider each statement and decide whether you agree or disagree with it.

1 Today's changes in climate are similar to changes that occurred in the past.
Agree/disagree?

2 Increased atmospheric carbon dioxide is mainly due to human agriculture practices around the world.
Agree/disagree?

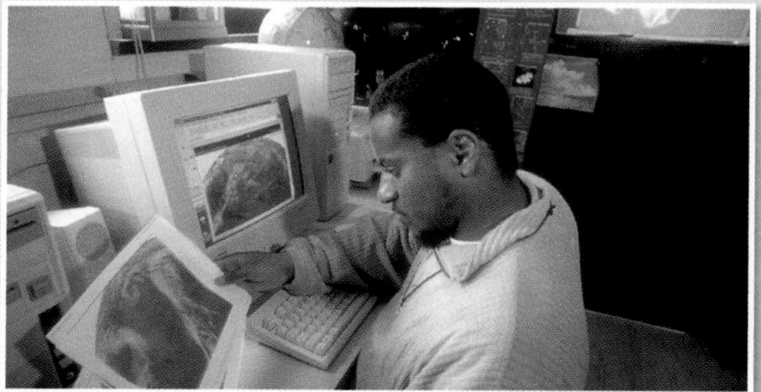

3 Scientists can predict exactly how the climate will change in the future.
Agree/disagree?

4 Climate change will have both negative and positive effects.
Agree/disagree?

5 Changes to Earth's climate can be reversed by changing human activities.
Agree/disagree?

6 The burning of fossil fuels in heating, transportation, and industry is the greatest cause of greenhouse gas production in Canada.
Agree/disagree?

FOCUS ON
READING

Summarizing

When you summarize a text, you shorten it by restating only the main idea and key points in your own words. Specific facts, examples, and questions are not included. Use the following strategies when summarizing a text:

- determine the main idea and key points using text features such as the title, headings, topic sentences, and signal words such as “thus,” “therefore,” and “in other words”
- ignore words or details that do not expand your understanding of the main idea
- use the same organizational pattern as the text uses (cause/effect, concept/definition)
- replace several specific words with a general word

READING TIP

As you work through the chapter, look for tips like this. They will help you develop literacy strategies.

Melting Glaciers, Ice Sheets, and Sea Ice

Over the last few decades, the average size of glaciers all over the world has begun to decrease as global temperature has risen (Figure 1). As you saw in the chapter opening, we can compare old photographs of glaciers with what they look like now. Many glaciers are much smaller now than they used to be.

The water from melting glaciers runs through rivers and lakes to the ocean. Nearly half the people around the world, including people in China and south Asia, depend on glaciers for their water. If glaciers disappear, serious water shortages could occur around the globe.

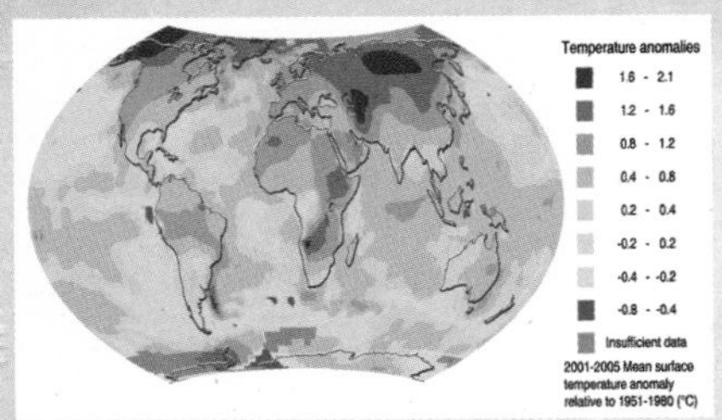

Figure 1 Global increases in average temperature for 2001 to 2005, compared with the 1951 to 1980 average. Temperatures are continuing to rise.

Summarizing *in Action*

A summary is a shorter version of a longer text. The key to summarizing is paraphrasing, which means using your own words to restate the main idea in the original text. Rewording the original text is crucial. Identify the main idea for the summary and briefly highlight the key points that support the main idea. Eliminate unimportant or redundant information. Here is how one student used the strategies to summarize the selection about melting glaciers.

Clues	Unimportant Words	Text Pattern	Replacement Words
The title suggests heat is causing ice to melt.	“Over the last few decades”	Cause/Effect	Use “Many people” to replace “including people in China and south Asia.”
Glaciers supply water for half the people on Earth.	“As you saw in the chapter opening”		
No glaciers = water shortages			
Summary: Global warming has caused glaciers to melt. Many people depend on glaciers for their water supply. Reduced glaciers means reduced water supply globally.			

Evidence of a Changing Climate

9.1

In Chapter 8, you learned how Earth's climate works and how scientists study past climate. Today, Earth's climate is being discussed around the world. Why? Scientists who study climate believe that Earth's climate is changing again—and that this time the change is happening because of human activities.

Scientists have observed changes in Earth's climate system. Most of these changes are things that have not happened for thousands of years. For example, ice that has remained frozen on Earth's surface for thousands of years is beginning to melt (Figure 1). Changes like this are symptoms of global changes in Earth's climate. Many climatologists, glaciologists, and biologists are measuring and analyzing evidence of climate change. In this section, you will examine the evidence we have that Earth's climate is changing.

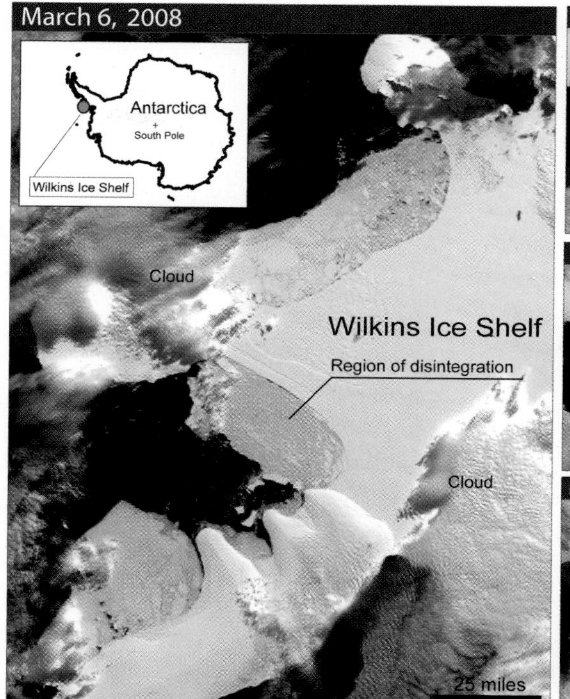

Figure 1 Between February 28 and March 6, 2008, a 405 km² section of the Wilkins Ice Shelf collapsed. The rest of the 14 000 km² ice shelf is showing signs that it could disintegrate and pull away from Antarctica.

Rising Temperatures

One of the main ways scientists can tell that the climate is changing is by studying past temperature records. Weather stations have recorded daily temperatures and other weather data around the world since the late 1800s. Scientists use these records to calculate Earth's average temperature each year going back more than 100 years.

This historical record shows that the average temperature goes up and down from year to year but also shows long-term trends. Earth's average temperature rose from 1910 to 1940, stayed relatively even from 1940 to 1970, and has continued to rise since then (Figure 2). By 2006, 11 of the previous 12 years were the warmest ever recorded.

Average Global Temperature

Temperature anomaly (°C) vs. Year (1880–2000); —■— Annual mean

Figure 2 Earth's average temperature has generally increased from 1880 to 2006. The data were collected from all over Earth's surface.

LEARNING TIP

Temperature Anomaly

Climate change graphs show "temperature anomaly" on the *y*-axis. Temperature anomaly is the difference between a long-term average temperature and the data point. A temperature anomaly of +0.1 °C means that the data point is 0.1 °C above the average temperature.

In Canada, average national temperatures have increased by 1 °C in the last 55 years. Over the western and northern parts of the country, average temperatures have risen even more. Average temperatures have increased by as much as 2.5 °C in some parts of Yukon.

Figure 3 shows changes in global temperatures by comparing the long-term average temperatures (between 1951 and 1980) with temperatures between 2001 and 2005. The dark red areas have the highest increases in temperature. Notice that these areas are concentrated over land in the northern hemisphere. A few places in the southern hemisphere, particularly over the oceans, show temperature decreases.

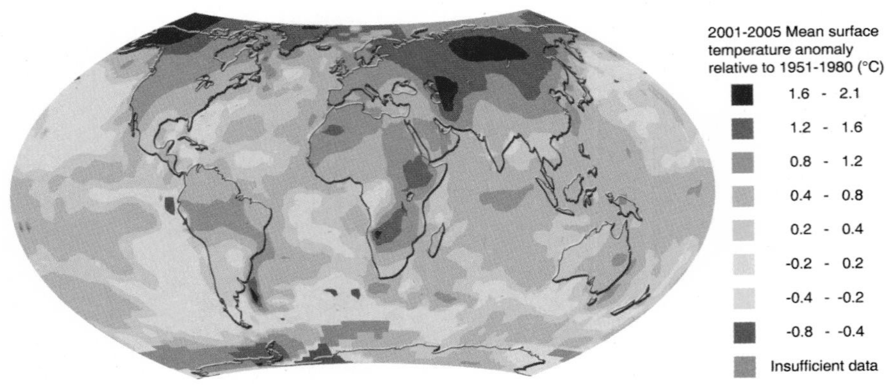

Figure 3 Global increases in average temperature for 2001 to 2005, compared with the 1951 to 1980 average. Temperatures are continuing to rise.

Melting Glaciers, Ice Sheets, and Sea Ice

Over the last few decades, the average size of glaciers all over the world has begun to decrease as the global temperature has risen. As you saw in the chapter opening, we can compare old photographs of glaciers with what they look like now. Many glaciers are much smaller now than they used to be.

The scientists who study glaciers are called "glaciologists." To find out more about glaciologists' work,

GO TO NELSON SCIENCE

The water from melting glaciers runs through rivers and lakes to the ocean. Nearly half the people around the world, including people in China and south Asia, depend on glaciers for their water. If glaciers disappear, serious water shortages could occur around the globe.

READING TIP

Paraphrasing

Paraphrasing is restating an idea in your own words in your own way. Avoid reusing the author's words. Pretend you are explaining the idea to someone who does not understand the original text. Use words that you understand to capture the meaning of the original text.

Ice sheets covering the expanses of Greenland and Antarctica are also melting. Twice as much of Greenland's surface ice sheet is now melting each summer, compared with 15 years ago. In Antarctica, snow is now melting much farther inland and at higher altitudes than ever recorded. In Figure 1 (on the previous page), you saw that an ice shelf in the Antarctic is beginning to disappear.

Arctic sea ice is also disappearing. An area of sea ice equivalent to Ontario and Quebec together disappeared from the Arctic Ocean in September 2007 (Figure 4, next page). Much of the ice re-formed during the next winter, but summer ice continues to decrease dramatically. Scientists estimate that the Arctic could be entirely ice-free in the summer within just a few years. You will learn more about the extent of sea ice in Activity 9.2.

To view an animation of the change in sea ice coverage in the Arctic,

GO TO NELSON SCIENCE

Figure 4 These satellite images show decreases in sea ice between (a) September 1979 and (b) September 2007. This month holds the record for the smallest area of sea ice in the Arctic, at the time of publication, since records were first started in 1979.

Rising Sea Level

Global sea level has risen significantly over the past 120 years. Since 1993, the sea level has been rising almost twice as fast as during the previous 30 years. Even small increases in sea level can result in devastating floods in low-lying countries such as the islands of Tuvalu in the South Pacific.

TRY THIS CALCULATE SEA LEVEL RISE

SKILLS: Analyzing, Evaluating

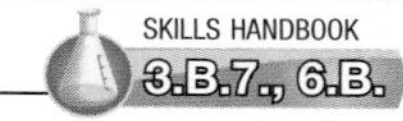

Global average temperature and sea level have changed since the late 1800s (Figure 5). Use these data to answer the questions.

Equipment and Materials: calculator

1. Examine the two graphs in Figure 5.

A. Estimate by how much the global average temperature has risen over the past 150 years. T/I

B. Estimate the rise of global average sea level over the same time period. T/I

C. Can you use these two graphs to show that sea level rise may be connected to climate change? Explain why or why not. What other data do you think you might need to prove that the two are linked? T/I C

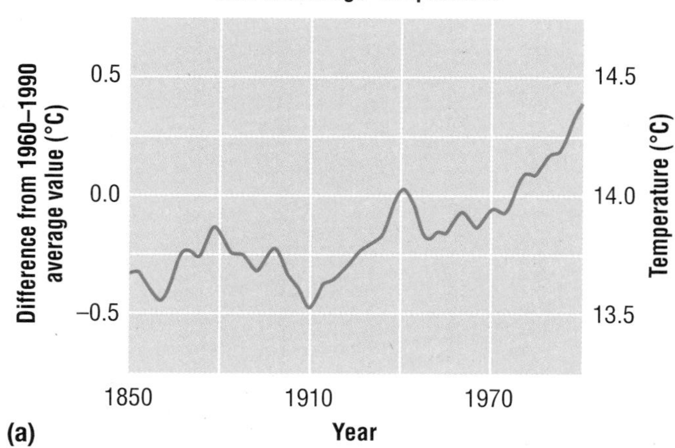

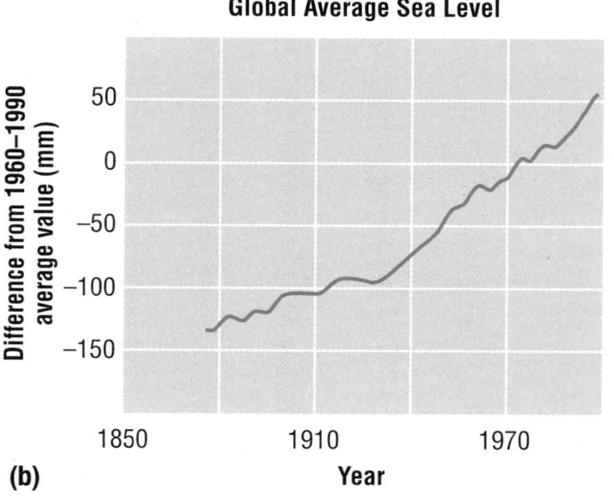

Figure 5 Changes in (a) global average temperature and (b) global average sea level.

LEARNING TIP

What Happens When Floating Ice Melts?

Floating sea ice and icebergs have little effect on sea level when they melt because they already displace water when they float. If you float an ice cube in a glass of water, the water level will not change after it melts.

When glaciers and ice sheets on land melt, water runs into the oceans, causing the sea level to rise. This melting may explain why sea level rise is accelerating. If the Greenland Ice Sheet melted completely, it would raise the global sea level by about 7 m (Figure 6). This is unlikely to happen in the near future, however. So far, not enough glaciers and ice sheets on land have melted to account for most of the observed rises in sea level. What other factor can be causing sea levels to rise?

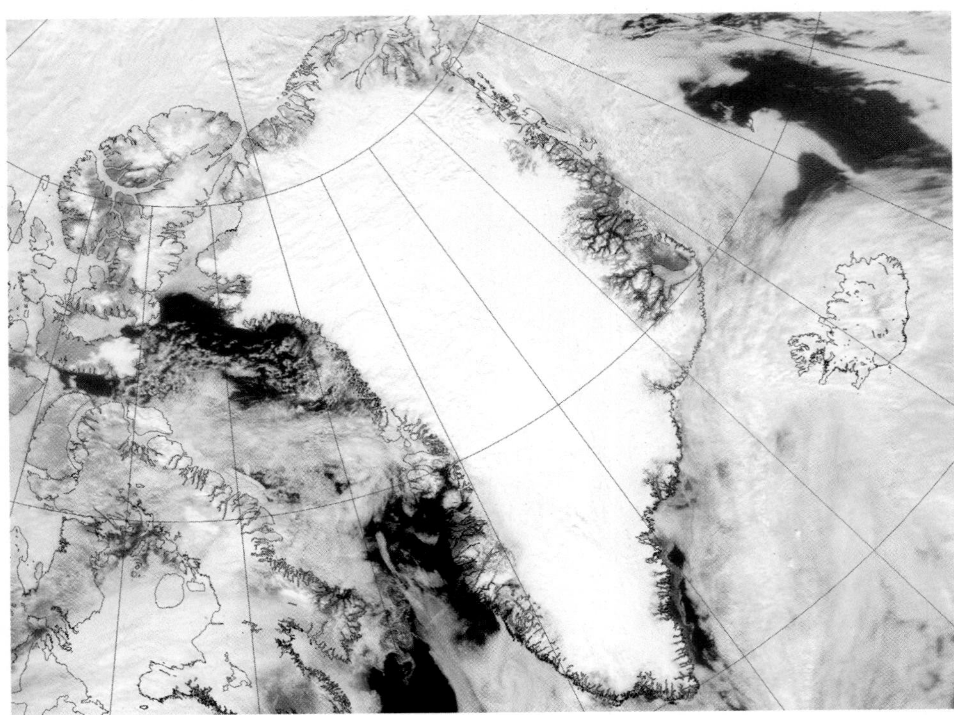

Figure 6 A huge volume of water is stored as ice that covers most of Greenland.

Thermal Expansion and Sea Level Rise

thermal expansion the increase in the volume of matter as its temperature increases

Water expands slightly when it warms up. This is known as **thermal expansion**. Scientists believe that thermal expansion has been responsible for much of the rise in sea level over the past 120 years. The thermal expansion of water is very small, but since the oceans are so deep, it is enough to make a difference. You will see this phenomenon in Activity 9.3.

Another Reason for Sea Level Rise?

Canadian scientist Richard Peltier is one of the world's pioneers in understanding the effect of climate change on sea level rise (Figure 7). Peltier continues to be puzzled by rising sea levels around the world. Scientists know that sea levels are rising due to thermal expansion and melting land ice. However, these two explanations do not fully account for all sea level rise. Professor Peltier thinks the additional water may be coming from North America, where the levels of groundwater have dropped substantially. Alternatively, ice in Greenland or Antarctica may be melting faster than we thought.

In 2008, Peltier was awarded the Milutin Milankovic Medal by the European Geosciences Union in recognition of his work. He presently directs the Centre for Global Change Science at the University of Toronto. His team of scientists uses a supercomputer system to study past climate changes.

Figure 7 Richard Peltier

Changes in Severe Weather

Certain types of severe weather events, such as heat waves and hurricanes, are becoming more intense (Figure 8). A heat wave swept across Europe in the summer of 2003. That summer was one of the hottest ever recorded. Thousands of people died from heat-related causes. Hurricanes have also become stronger over the past 50 years, fuelled by warmer ocean temperatures. The number of category 4 or 5 hurricanes (winds of 178 to 249 km/hr) per year has nearly doubled over the last 40 years.

READING TIP

Supporting Details

Details like facts, reasons, and examples are used to support the main idea. They help the reader understand the main idea by saying something specific about it that the reader may already know or be able to visualize. An example of supporting details is "such as heat waves and hurricanes." When summarizing a text, it is better not to include supporting details.

Changes in Precipitation Patterns

In the northern hemisphere, more precipitation is falling as rain and less as snow. There are more heavy precipitation events such as rainstorms and snowstorms. The total annual precipitation is increasing in northern Canada. Southern Africa, the Mediterranean, and southern Asia, however, are becoming drier (Figure 9). As you know, precipitation is one of the main factors used to identify the climate of a region. Changes in precipitation point to changes in climate.

Figure 9 The water level in the Ganjiang River in China is dropping.

Figure 8 The heat wave in Melbourne, Australia, in early 2009, caused rail lines to expand and bend.

Changing Seasons

Seasons in Canada and in other parts of the world are gradually changing. The amount of snow that remains on the ground in winter is decreasing throughout the northern hemisphere. The frequency of very cold days has been decreasing worldwide. Very cold days and frosty nights are coming later in the year and ending earlier in the spring. As a result, many regions are experiencing longer growing seasons.

READING TIP

Summarizing

When summarizing, find ways to condense the original text. "Very cold days and frosty nights" can be reduced from six words to two: "Cold weather." Sometimes several specific words can be replaced by a general word or phrase. For example, "telephone, radio, and Internet access" can be changed to "communication media."

Changes in Ecosystems

Plants and animals are responding to the changes in temperature and precipitation. Trees, shrubs, and other plants across North America are flowering earlier in the spring. Animals such as squirrels are breeding earlier in the year. Climate scientists can track these changes as evidence that climate change is occurring.

Animal and plant communities are slowly migrating toward the poles and to higher altitudes as their regions warm up. Unfortunately, this means that undesirable insects and plants are also moving north into new regions. The mountain pine beetle has moved into areas of British Columbia where, in the past, it has been too cold for the beetle to survive.

RESEARCH THIS IS CLIMATE CHANGE ALWAYS BAD?

SKILLS: Researching, Analyzing the Issue, Defending a Position, Communicating, Evaluating

SKILLS HANDBOOK  4.A., 4.C.

Most discussions of climate change give the impression that the impacts of climate change will always be negative. However, there may also be some positive impacts.

1. Select one aspect of climate change from the list below. Brainstorm negative and positive impacts that could result. Consider social, environmental, and economic impacts. Think short-term and long-term impacts. Research the issue on the Internet and in the library. Discuss it with friends and family.
   - warmer temperatures
   - reduction of Arctic sea ice
   - rising sea level
   - changing rainfall patterns

 GO TO NELSON SCIENCE

A. Analyze and prioritize each item on your list of positive and negative impacts as to its importance and potential effect. T/I

B. Write a conclusion to your research. Describe the positive and the negative impacts of the aspect of climate change that you selected. T/I A C

C. Make a recommendation to "The Organization of Students Concerned About Climate Change." You could present your recommendation as a written paper, as a video submission, in person, or in another format approved by your teacher. Defend your recommendation with good arguments. Evaluate whether the impact you studied is likely to be very important, moderately important, or not important for the future of people and ecosystems. C A

UNIT TASK Bookmark

You can apply what you learned in this section about evidence for climate change to the Unit Task described on page 444.

IN SUMMARY

- Most of the changes that we are currently seeing in the climate system have not happened for thousands of years.
- Earth's average temperature has been rising steadily for the past 50 years and overall for the past 100 years.
- The extent of Earth's glaciers, sea ice, and ice sheets has decreased over the past few decades.
- The rate of sea level rise is increasing.
- Temperature and precipitation patterns are changing, as well as the distribution of many plant and animal species.

CHECK YOUR LEARNING

1. For each of the following topics, write one observed symptom of a changing climate. Which symptoms do you think are the most convincing that Earth's climate is definitely changing? K/U
   (a) ecosystems
   (b) sea level
   (c) growing seasons
   (d) glaciers and ice sheets
   (e) hurricanes
   (f) precipitation
2. Some people think that observed changes are just the usual year-to-year variations. They remember the 1930s and 40s, when winters were so mild and dry that they rode scooters on the streets of Toronto. How would you respond to them? K/U C
3. People who live in valleys below melting glaciers have more drinking water available than 30 years ago. In the future, however, they will have less drinking water available than 30 years ago. Explain why this may happen. T/I
4. (a) What are two possible causes of sea level rise?
   (b) If the sea level has risen by 20 cm over the past century on the coast of North America, by how much do you think it has risen over the past century on the coast of India? Explain. K/U T/I
5. (a) State one positive and one negative effect of climate change for Canada.
   (b) State one positive and one negative effect of climate change for a country closer to the equator, such as India. K/U A
6. Why are insect and plant pests becoming more of a problem in northern latitudes? K/U
7. Describe two problems that are caused by melting glaciers and polar ice. K/U

Using Satellites to Monitor Earth

Any object that orbits a larger object is called a satellite. For example, the Moon is a natural satellite that orbits Earth. Since 1957, humans have been launching artificial satellites to orbit Earth. Canada was the third country to launch an artificial satellite, after the Soviet Union and the United States (Figure 1).

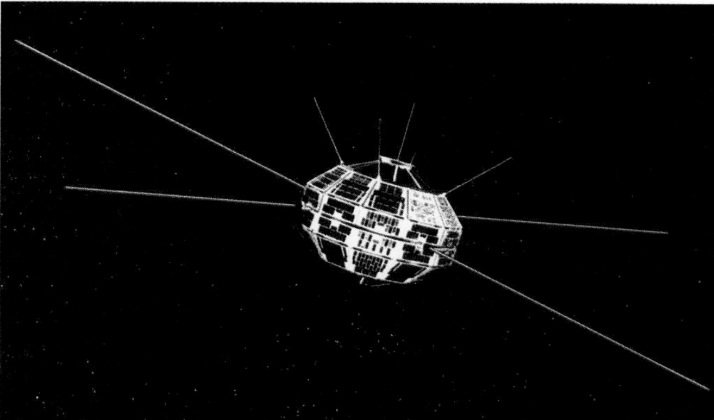

Figure 1 The *Alouette 1*, launched in 1962, was Canada's first satellite.

Artificial satellites provide service for telephones, radio, Internet access, TV, and navigation. Today, many different types of artificial satellites are orbiting Earth. Some examples include the following:

- Earth observation satellites that monitor environmental conditions such as temperature, ice cover, forests, and volcanic eruptions
- communication satellites that transmit telephone conversations
- broadcasting satellites that broadcast radio and television programs
- navigational satellites that help airplanes and ships navigate
- weather satellites that take photographs and radar images of weather systems

Satellites Monitoring Climate Change

RADARSAT-1

The Canadian Radio Detection and Ranging Satellite 1 (RADARSAT-1) is one of the Canadian satellites used to conduct scientific research. It was launched in 1995. RADARSAT-1 monitors glaciers, polar ice caps, and permafrost, among other environmental conditions. For example, images from RADARSAT-1 are used to measure the flow of glacier ice into oceans. These images have helped scientists determine that Earth's ice is melting at an increasing rate, raising concerns about future sea level rise.

RADARSAT-2

In December 2007, RADARSAT-2 was launched to continue monitoring the environment and natural resources. RADARSAT-2 circles Earth every 100 minutes with a different circle each time and revisits the same spot on Earth every 24 days.

Images from RADARSAT-2 show a difference between open water and various types of ice (Figure 2). This will make it easier for scientists to monitor changes in Earth's ice cover over time.

GO TO NELSON SCIENCE

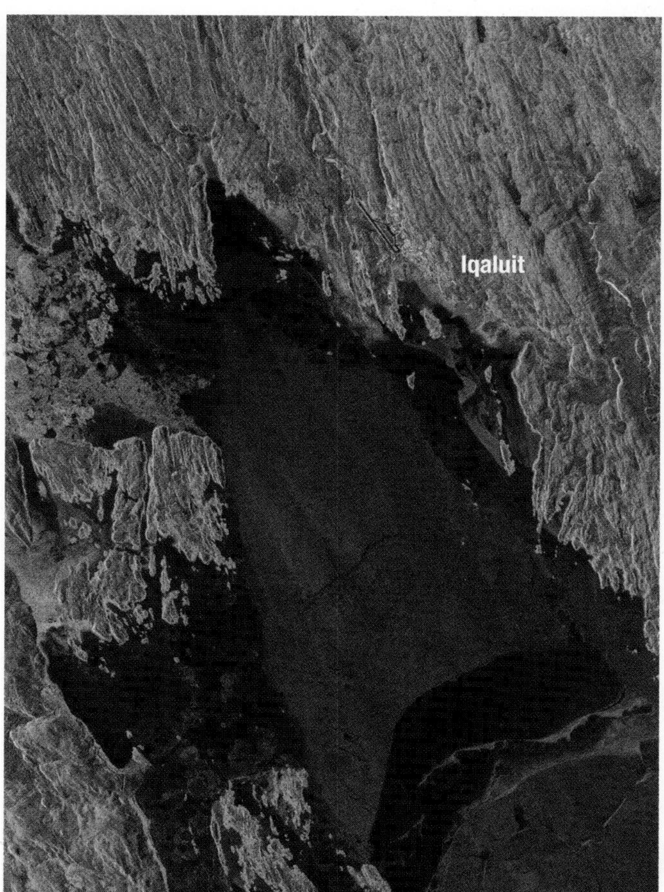

Figure 2 This RADARSAT-2 image shows the community of Iqaluit, next to Frobisher Bay, which is mostly covered by ice. Note the Iqaluit airport runway to the north of the city. The resolution is 8 m.

9.2 PERFORM AN ACTIVITY

Analyzing Sea Ice Extent for Evidence of Climate Change

Canadians are asking questions about changes in the amount (or extent) of sea ice in the Arctic (Figure 1). Will it become easier for ships to travel along Canada's north coast? How will changes in sea ice extent affect the polar bear population?

Every year, sea ice increases (maximum sea ice extent) and decreases (minimum sea ice extent) depending on the season (Figure 2). In this activity, you will analyze data to see whether the maximum or minimum sea ice extent is changing.

SKILLS MENU

- Questioning
- Hypothesizing
- Predicting
- Planning
- Controlling Variables
- Performing
- Observing
- Analyzing
- Evaluating
- Communicating

Figure 1 Sea ice in the Antarctic, as well as in the Arctic, is vulnerable to climate change. It is an important habitat for many Antarctic species, such as the emperor penguin.

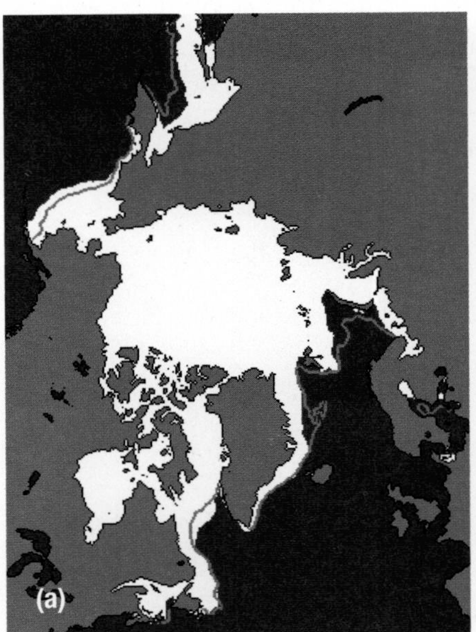

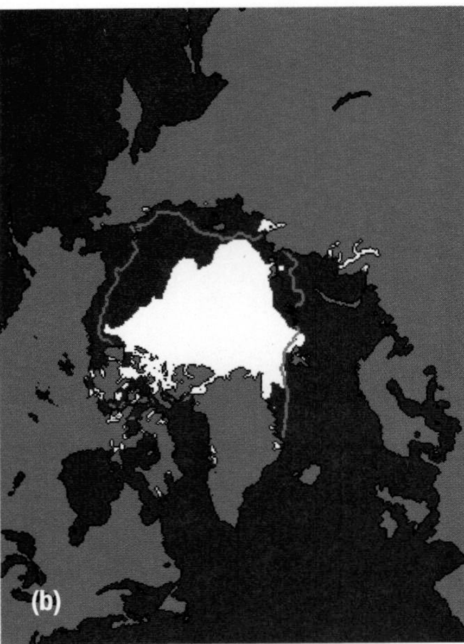

Figure 2 These images, created in (a) March and (b) September, 2008, show how Arctic sea ice extent changes with the season. The pink line is the median ice edge between 1979 and 2000.

Purpose

To determine whether sea ice extent in the Arctic or Antarctic is changing over time.

Equipment and Materials

- Internet access or data for sea ice extent
- electronic spreadsheet and graphing calculator (optional)
- graph paper (2 pieces)

Procedure

1. Prepare a data table similar to Table 1. There should be enough rows to collect data up to the present year.

Table 1 Maximum and Minimum Sea Ice Extent in the Arctic and Antarctic

	Arctic		Antarctic	
	March	September	February	September
1979				
1980				
1981				

2. Examine online maps of sea ice extent in the Arctic and the Antarctic since 1979. Your teacher will give you instructions for locating the latest data at the National Sea Ice Data Centre.

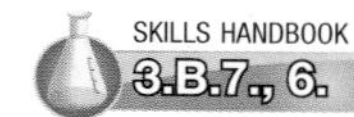

GO TO NELSON SCIENCE

3. Record the data for sea ice extent from the maps onto your data table.

Analyze and Evaluate

SKILLS HANDBOOK 3.B.7., 6.

(a) On a single graph, plot your maximum and minimum sea ice extent data for March and September in the Arctic and for September and February in the Antarctic. Put the year across the x-axis and the sea ice extent (millions of km^2) on the y-axis. C

(b) How has the maximum sea ice extent in the Arctic changed since 1979? T/I

(c) How has the minimum sea ice extent in the Arctic changed since 1979? T/I

(d) Write a short summary describing how sea ice extent in the Arctic and the Antarctic has changed since 1979. T/I C

(e) Do the results of this activity support the idea that Earth has been warming up over the past few decades? T/I

(f) If you were skeptical about climate change, how might you use these data to support your position that we cannot be sure that long-range warming of Earth is taking place? T/I A

Apply and Extend

SKILLS HANDBOOK 3.B.8., 4.B.

(g) In this activity, you worked with maps made from satellite photographs. What other data would you want to know to further support or test your conclusions? T/I

(h) How might the continued reduction of the polar ice cap influence transportation and shipping? A C

(i) The Environment Canada website provides further data on Arctic ice. To obtain ice thickness data to answer the following questions, go to the Nelson Science website. It will direct you to the Environment Canada website. T/I

GO TO NELSON SCIENCE

(i) What is the average thickness of the sea ice in Canada's Arctic?

(ii) What has been happening to the thickness of the Arctic sea ice?

(j) How does ice in the Arctic and Antarctic help keep Earth cool? K/U

(k) In September, 2008, a group of Canadian high school students travelled to the Arctic. They were part of an expedition looking for evidence of climate change (Figure 3). Go online to learn about their experiences and discoveries and how they are using what they learned. Write a one-page summary of their activities. T/I C

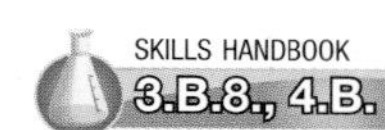

GO TO NELSON SCIENCE

Figure 3 Canadian students exploring the Arctic

UNIT TASK Bookmark

You can apply what you learned about changing ice cover to the Unit Task described on page 444.

Thermal Expansion and Sea Level

SKILLS MENU
- Questioning
- Hypothesizing
- Predicting
- Planning
- Controlling Variables
- Performing
- Observing
- Analyzing
- Evaluating
- Communicating

The sea level around the world has been slowly rising over the past century. (Remember the Try This: Calculate Sea Level Rise in Section 9.1.) Sea level rise is caused by runoff from melting glaciers, ice caps, and ice sheets and from thermal expansion of the oceans as they warm. Most of the media's attention is focused on the first reason. The second reason, however, may be more important in explaining what has happened over the last hundred years. In this activity, you will measure the thermal expansion of water.

Purpose

To model the effect of thermal expansion of water on sea level.

Equipment and Materials

- eye protection
- lab apron
- two-hole stopper with an alcohol thermometer pushed through one hole and the other hole left open
- clear plastic tubing
- 500 mL beaker or large cup
- Erlenmeyer flask (125 or 250 mL)
- ruler
- hot plate
- wire gauze
- utility stand and clamp
- safety gloves
- 50 or 100 mL graduated cylinder
- 100 cm length of tubing, identical in diameter to that used in the above flask
- 10 mL graduated cylinder
- large second thermometer (if necessary)
- water
- 2 ice cubes
- food colouring (optional)

Be careful not to touch any hot pieces of equipment. Never touch the hot plate when it is on.

Wear eye protection to protect against spills of hot water.

When you plug the hot plate into the electric socket, make sure everything is dry.

When you unplug the hot plate, do not pull on the cord; pull the plug itself.

Procedure

1. Create a table similar to Table 1 (next page) in your notebook.
2. Put on your eye protection and lab apron.
3. Push the clear plastic tubing into the unused hole of the stopper. Leave about 20 cm of tubing outside the flask for the water to expand into.
4. Fill a large cup or 500 mL beaker with enough cold water to fill the flask. Use ice cubes to adjust the temperature to approximately 10 °C. (To make it easier to see the water when it expands into the clear tubing, you could add a few drops of food colouring.)
5. Fill the Erlenmeyer flask with the cold water (no ice) almost to the brim. Push the rubber stopper firmly into the neck of the flask until it is watertight. Some water may spill out, which is not a problem. Water will rise partway up the tubing. Make sure that the initial level of the water in the tubing is visible above the stopper.
6. Read the initial temperature of the water in the flask. Record all observations in your table.
7. With a ruler, measure the initial height of the water level in the tubing outside the flask.

Table 1 Measurements Taken During the Activity

Change in water temperature	Initial temperature: ____ °C	Final temperature: ____ °C	Temperature change: ____ °C
change of water level in tubing (above stopper)	initial level: _____ cm	final level: _____ cm	change in level: _____ cm
volume of water in flask	_____ mL		
volume of water needed to fill 100 cm of tubing	_____ mL		

8. Place the flask on a wire gauze on a hot plate (Figure 1). Clamp the flask to the utility stand. When the temperature has risen by approximately 10 °C (i.e., from 10 °C to 20 °C), remove the flask from the hot plate using safety gloves. Measure the new level of water in the clear tubing. At approximately the same time, measure the final temperature of the water in the flask.

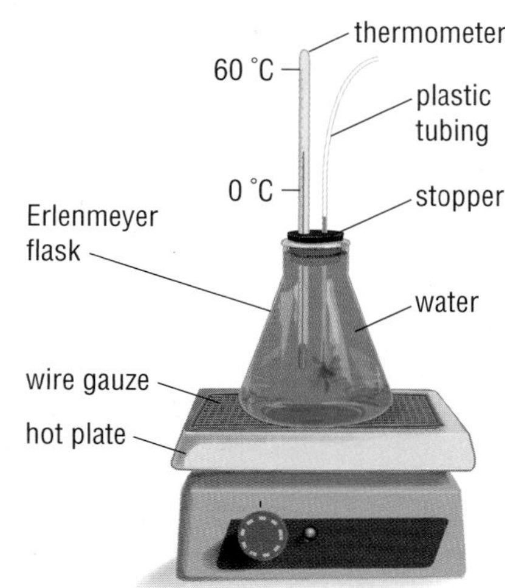

Figure 1 Experimental set-up

9. Carefully remove the stopper from the top of the flask without spilling water. Use a 50 mL graduated cylinder to measure the volume of water that was in the flask. (This volume should be a little more than the volume printed on the side of the flask.)
10. To find out how much water the tubing can hold per centimetre of length, take 100 cm of tubing and fill it with water. Then empty the water from the tubing into a 10 mL graduated cylinder. Record the volume in Table 1.

Analyze and Evaluate

(a) Did the water in the flask change in volume when you heated it? How do you know? T/I C

(b) Estimate the thermal expansion of the ocean by following these steps: T/I
   (i) By how many centimetres did the water expand up the tube when it was heated?
   (ii) Use your measurement of the volume of water in 100 cm of tubing to calculate how much the water expanded in mL.
   (iii) Use the total volume of the flask to express the volume of expansion as a percentage. By what percentage did the water expand?
   (iv) To find the percentage expansion per degree Celsius of temperature change, divide your answer to (iii) by the temperature change.
   (v) The ocean's average temperature has risen by approximately 0.375 °C over the last hundred years. Use this value to estimate by how much the ocean water might have expanded in items of percentage increase.
   (vi) Use the percentage increase calculated in (v) to estimate by how much the sea level might have risen with a 0.375 °C temperature increase. The average depth of the ocean is 3 740 m.

(c) Note any obvious sources of error in your calculation of thermal expansion. T/I

Apply and Extend

(d) What can you conclude about the effect of the thermal expansion of water on sea level? T/I

(e) Refer back to Figure 5(b) on page 375. T/I
   (i) By how many centimetres has the sea level risen over the past 100 years?
   (ii) According to your results in this activity, how much of this rise can be accounted for by thermal expansion?

9.4 Greenhouse Gases: Changing the Climate

Greenhouse gases have been part of our atmosphere for hundreds of thousands of years. Over the past 200 years, however, the atmospheric levels of most greenhouse gases have increased (Figure 1).

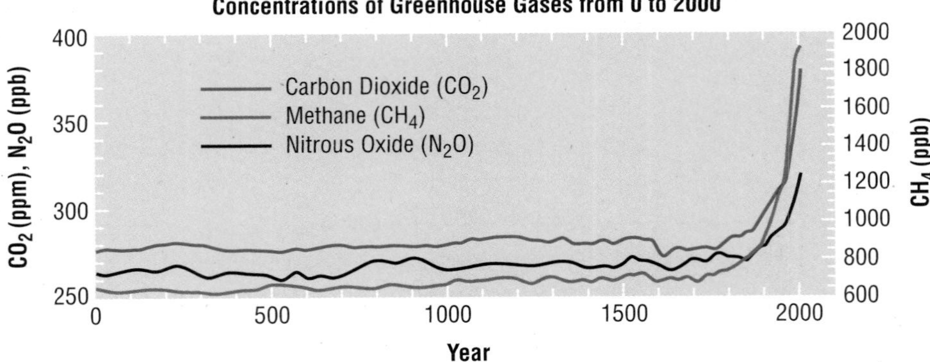

Figure 1 Atmospheric concentrations of important long-lived greenhouse gases over the last 2000 years. Increases since about 1750 are attributed to human activities in the industrial era.

We know that humans produce greenhouse gases by the combustion of fossil fuels in energy-production, transportation, and industry. But how do we know that the gases we produce end up in the atmosphere, and are not absorbed by oceans and forests?

Scientists have studied both natural and human sources of greenhouse gases. For example, they have studied the flow of carbon through the carbon cycle to determine where human-produced carbon dioxide ends up. These studies show that about half of human-produced carbon dioxide ends up in the atmosphere, causing an increase in atmospheric carbon dioxide.

Based on studies like these, scientists conclude that the increase in greenhouse gases in the atmosphere is due to human activities. The additional amount of human-produced greenhouse gases is called "**anthropogenic** greenhouse gases."

anthropogenic resulting from a human influence

Anthropogenic Sources of Greenhouse Gases

The principal anthropogenic greenhouse gases are carbon dioxide, methane, nitrous oxide, and chlorofluorocarbons (CFCs).

Carbon Dioxide (CO_2)

Carbon dioxide is the most significant greenhouse gas produced by humans today. Burning fossil fuels (coal, gasoline, and natural gas) produces carbon dioxide (Figure 2 on the next page). Fossil fuels are used for energy in transportation, heating, electricity generation, and industry. In addition, fossil fuels are used to provide energy for manufacturing.

Forests play an important role in determining how much carbon dioxide will remain in the atmosphere. Trees take in carbon dioxide from the air and use it in photosynthesis. The chemical equation for photosynthesis is shown below:

$6\ CO_2$	+	$6\ H_2O$	+	light energy	→	$C_6H_{12}O_6$	+	$6\ O_2$
carbon dioxide	+	water	+	light energy	→	glucose	+	oxygen

Figure 2 The combustion of fossil fuels produces greenhouse gases.

A carbon dioxide molecule contains one atom of carbon and two atoms of oxygen. The carbon atoms become part of new glucose molecules during photosynthesis. (Glucose is a simple sugar.) The glucose is then made into other compounds that form wood, leaves, and roots. Trees take in carbon atoms from the atmosphere and store them in this new form. Therefore, as long as a tree is alive, it is a carbon sink because it removes carbon from the atmosphere and stores it.

About 10 % of our carbon dioxide emissions are due to deforestation, mostly in tropical countries (Figure 3). Deforestation has two unwanted effects. It stops the forest from absorbing carbon and releases some of the previously absorbed carbon back into the atmosphere as carbon dioxide. How does this occur?

When forests are cut down, the leftover forest waste decomposes. The process of decomposition produces greenhouse gases, including methane and carbon dioxide. Cutting down a forest causes the forest to become a *source* of carbon instead of a sink. Furthermore, deforestation means that there are fewer trees to absorb atmospheric carbon dioxide through photosynthesis.

Figure 3 Deforestation releases greenhouse gases.

Methane (CH_4)

Methane emissions come from many sources. Agricultural activities such as rice farming and cattle ranching produce methane (Figure 4). Methane is also produced from the decay of organic material in landfills and sewage treatment plants. Coal mining and natural gas extraction release methane gas that was trapped underground in fossil fuel deposits.

DID YOU **KNOW?**

Carbon Gluttons

The number of carbon atoms stored in Earth's forests, smaller plants, soil, and dead organic matter is greater than the number of carbon atoms present in the atmosphere as carbon dioxide gas.

Figure 4 The digestive systems of cattle release methane.

Deforestation also results in methane emissions. Forests are often cleared by burning. If the trees smoulder slowly, the combustion reaction produces methane as well as carbon dioxide and water.

Nitrous Oxide (N_2O)

About two-thirds of nitrous oxide emissions comes from the management of livestock feed and waste. The rest comes from farmers using nitrogen fertilizers, certain industrial processes, and, to a much smaller degree, fossil fuel use.

Chlorofluorocarbons (CFCs)

CFCs are commonly used as refrigeration agents. There are no natural sources of CFCs. These gases leak out of refrigerators and air conditioners or are released by industrial processes. Atmospheric levels of CFCs are now decreasing because of international treaties, such as the Montréal Protocol on Substances that Deplete the Ozone Layer.

Summary of Greenhouse Gas Concentrations

Concentrations of the major greenhouse gases have been increasing since the start of the industrial age. You can see in Table 1 that atmospheric concentrations of methane and nitrous oxide are much smaller than those of carbon dioxide. However, methane and nitrous oxide molecules are much more powerful than a carbon dioxide molecule in contributing to the greenhouse effect.

LEARNING TIP

Tiny Concentrations

Concentrations of greenhouse gases in the atmosphere are very low compared with concentrations of oxygen and nitrogen. Scientists therefore use different units for them: parts per million (ppm), parts per billion (ppb), and parts per trillion (ppt).

Table 1 Concentrations and Lifetimes of Atmospheric Greenhouse Gases

	Concentration in the atmosphere		
Greenhouse gas	**Pre-industrial level**	**2008 level**	**Approximate atmospheric lifetime**
carbon dioxide	280 ppm	384 ppm	100 to 1 000 years
methane	0.700 ppm (700 ppb)	1.785 ppm (1785 ppb)	12 years
nitrous oxide	0.270 ppm (270 ppb)	0.321 ppm (321 ppb)	114 years
CFC-11	trace amounts	0.000251 ppm (251 ppt)	45 years
CFC-12	trace amounts	0.000525 ppm (525 ppt)	100 years

Data from *IPCC Climate Change 2007: The Physical Science Basis* and *The NOAA Annual Greenhouse Gas Index*

Anthropogenic Greenhouse Gases and Global Temperature

We have evidence that concentrations of greenhouse gases are increasing. In previous sections, you learned that global temperatures are also increasing. But how do we know whether these changes are related?

Scientists compare data on Earth's past climate with current data to determine how anthropogenic greenhouse gases are affecting Earth's climate. (In Section 9.6, you will take a closer look at the scientific reasoning used to conclude that humans are affecting Earth's climate.) As part of these studies, scientists have examined ice cores to collect data on greenhouse gases stretching back nearly 800 000 years (Figure 5). In the following Try This, you will use ice core data to compare changes in greenhouse gases with changes in Earth's average temperature over 400 000 years.

Figure 5 Data from ice cores show how carbon dioxide concentrations have changed over the past 800 000 years.

TRY THIS COMPARING CARBON DIOXIDE CONCENTRATIONS WITH GLOBAL TEMPERATURE

SKILLS: Hypothesizing, Predicting, Analyzing, Communicating

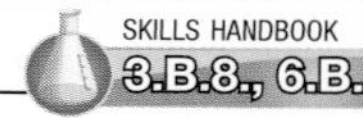

In this activity, you will compare ice core data on carbon dioxide with global temperatures over the past 400 000 years.

Equipment and Materials: computer with Internet access and a graphing program

1. Write a hypothesis describing the relationship (if any) between carbon dioxide concentration and global temperature. K/U C
2. Your teacher will direct you to the carbon dioxide data and global temperature data online. Use a graphing program to plot the two sets of data against time. If necessary, adjust the *y*-axis scales so that the two graphs are about the same size. Make sure the *x*-axis scales are the same on both graphs.

GO TO NELSON SCIENCE

A. What is the highest level of carbon dioxide in the atmosphere over the past 400 000 years? T/I

B. What is the lowest level of carbon dioxide in the atmosphere over the past 400 000 years? T/I

C. Today, the concentration of carbon dioxide in the atmosphere is 385 ppm and rising. Compare this value with your values from parts A and B. What do you conclude? T/I

D. Compare your two graphs. Identify three times over the past 400 000 years when the temperature changed rapidly. T/I

E. What happened to the concentration of carbon dioxide at the same three times identified above? T/I

F. What is happening to the carbon dioxide concentrations today? What is happening to global temperature? T/I

G. Summarize your findings about carbon dioxide concentration, global temperature, and any correlation between them. T/I C

In the Try This activity, you observed that changes in carbon dioxide concentration closely match changes in global temperatures over the past 400 000 years. Your graphs do not prove that increases in carbon dioxide concentrations *cause* global temperature to rise. They also do not show that global temperature rise causes increases in carbon dioxide concentrations. However, your graphs do show a strong correlation between carbon dioxide and global temperature. What might be one cause of this correlation? Recall what you learned about the greenhouse effect in Chapter 8. You already know about a connection between greenhouse gases and Earth's temperature.

LEARNING TIP

Relationships between Variables
Two variables are correlated when they change at the same time. The fact that the two variables are correlated, however, does not mean that one causes the other. It is up to scientists to determine the nature of the relationship.

The Anthropogenic Greenhouse Effect

What do anthropogenic greenhouse gases have to do with the current changes in Earth's climate? You already know that Earth absorbs radiation from the Sun, converts it to thermal energy, and then radiates lower-energy infrared radiation. More and more of this lower-energy infrared radiation is being absorbed by Earth's atmosphere as concentrations of greenhouse gases in the atmosphere increase. This human-caused increase is known as the **anthropogenic greenhouse effect**. Carbon dioxide accounts for about 30 % of the anthropogenic greenhouse effect. Methane and nitrous oxide contribute just under 20 % and 10 % respectively.

The anthropogenic greenhouse effect involves the same process as the natural greenhouse effect. What is happening is that humans are *enhancing* the natural greenhouse effect. As humans release more greenhouse gases into the atmosphere, Earth's energy balance changes. More thermal energy is trapped inside the atmosphere, raising global temperatures beyond what they would be from the natural greenhouse effect alone.

anthropogenic greenhouse effect the increase in the amount of lower-energy infrared radiation trapped by the atmosphere as a result of higher levels of greenhouse gases in the atmosphere due to human activities, which is leading to an increase in Earth's average global temperature

The Feedback Loop between Carbon Dioxide and Global Temperature

Remember that even small changes in Earth's climate can be made much larger by feedback loops. We have already looked at the water vapour and albedo feedback loops in Section 8.10. Another important feedback loop involves carbon dioxide levels in the atmosphere and Earth's temperature. You already know that increases in carbon dioxide levels result in increases in global temperature due to the greenhouse effect. However, increases in temperature can also cause increases in carbon dioxide levels.

Over the past 400 000 years, rising temperatures increased the release of carbon dioxide stored in plants and oceans (carbon sinks). As you know, carbon dioxide is a greenhouse gas that traps thermal energy inside the atmosphere. Thus, as rising temperatures caused carbon dioxide levels in the atmosphere to increase, Earth's average temperature rose even further (Figure 6).

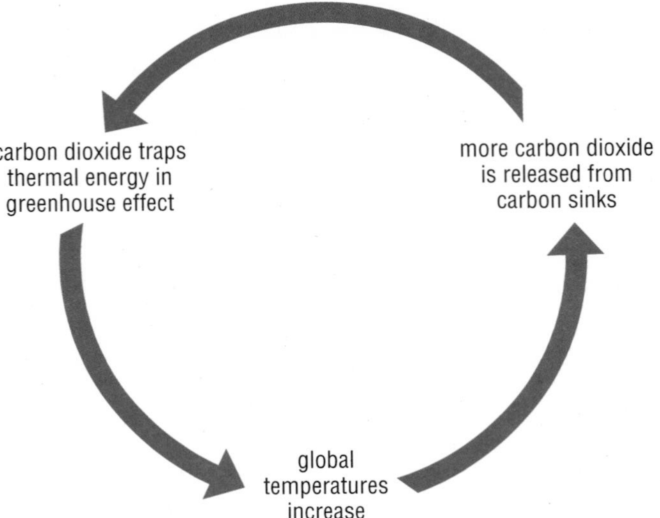
carbon dioxide traps thermal energy in greenhouse effect
more carbon dioxide is released from carbon sinks
global temperatures increase

Figure 6 Carbon dioxide concentrations and global temperatures are connected in a positive feedback loop.

The ocean is a carbon sink, storing vast amounts of carbon. Much of the carbon is dissolved carbon dioxide and carbonic acid. Scientists are concerned that this stored carbon might be released as global temperatures increase. The release of carbon would further increase levels of atmospheric carbon dioxide and add to the anthropogenic greenhouse effect (Figure 7).

Figure 7 As global temperatures rise, the ocean will likely absorb less and less carbon from the atmosphere. In addition, the ocean may release increasing amounts of stored carbon.

Increases to Earth's Average Temperature

Your carbon dioxide graph from the Try This activity showed that the concentration of atmospheric carbon dioxide is now higher than it has been for hundreds of thousands of years. The same is true for other greenhouse gases, such as methane and nitrous oxide.

If humans continue producing greenhouse gases at a similar rate, Earth's average temperature will likely increase by 2 °C to 6 °C by the end of the century. This increase may not sound extreme. However, it could cause devastating impacts, such as more severe heat waves, flooding and droughts, and species extinctions. You will learn more about future impacts of climate change in Chapter 10.

IN SUMMARY

- The concentrations of carbon dioxide and other greenhouse gases in the atmosphere are higher now than they have been for hundreds of thousands of years.
- The increase in greenhouse gases is caused by human activities such as the burning of fossil fuels, deforestation, agricultural practices, and industrial processes.
- Some of the carbon dioxide produced by human activities ends up in carbon sinks such as the oceans and forests. However, about half the carbon dioxide and most other greenhouse gases end up in the atmosphere.
- As the concentrations of greenhouse gases increase, more energy is trapped and absorbed by the atmosphere. This process is called the anthropogenic greenhouse effect.
- The anthropogenic greenhouse effect is causing Earth's temperature to increase.
- Carbon dioxide concentrations and global temperatures are connected in a positive feedback loop.
- Positive feedback loops can increase the anthropogenic greenhouse effect.

CHECK YOUR LEARNING

1. What is the anthropogenic greenhouse effect? K/U
2. You learned in this section that the greenhouse gas concentrations have all increased over the past 100 years. T/I
   (a) Using the data in Table 1, calculate the percentage increase for each of the following: carbon dioxide, methane, nitrous oxide.
   (b) Are scientists most concerned about the gases that have increased by the greatest percentage? Why or why not?
3. How do trends in the atmospheric concentration of carbon dioxide compare with trends in Earth's average temperature over the past 400 000 years? Explain. K/U
4. (a) What is a carbon sink?
   (b) How can Earth's forests affect Earth's temperature? K/U
5. The atmospheric concentration of carbon dioxide has varied over the past 400 000 years. For most of this time, there was no industrial activity. Why, then, do we think that the present increase in atmospheric carbon dioxide is due to human activities? State two reasons. A
6. Computer models of Earth's climate project increases in the average temperature from 2 °C to 6 °C by the end of the century. Have these high temperatures ever occurred before in the past 400 000 years? K/U
7. In Section 8.7, you designed and built a model of the greenhouse effect. How might you redesign it to model the additional anthropogenic greenhouse effect? T/I

9.5

Canadian Emissions of Greenhouse Gases

In this chapter, you have learned that humans are producing increasing amounts of greenhouse gases. Now it is time to bring the issue closer to home. How do Canadians produce greenhouse gases, and how much do we produce? Canada contains vast forests that act as carbon sinks (Figure 1). Do our forests soak up the carbon dioxide we produce?

Figure 1 About 30 % of the world's boreal forest is in Canada.

Canadian Sources of Greenhouse Gases

On average, individual Canadians emit more greenhouse gases than most people in the world, ranking in the top ten worldwide. Alberta is Canada's largest provincial producer of greenhouse gases. Ontario is the second largest. Ontario alone releases more than 200 million tonnes of greenhouse gases per year (Figure 2). If we do not take into account tiny oil producing countries such as Kuwait, Qatar, and Brunei, Canadians rank in the top two or three emitters of greenhouse gases per person.

Figure 2 Individual Canadians emit more greenhouse gases than most other people in the world. Heating our homes in the winter contributes to our emissions.

Why do Canadians produce so much greenhouse gas? Table 1 lists some of Canada's main sources of greenhouse gases. Most industrialized countries produce greenhouse gases from similar sources.

Table 1 Some Canadian Sources of Greenhouse Gases

Source	Examples	Quantity per year based on data from 2006 (Mt (megatonnes) CO_2-eq)
producing and using energy	generating electricity and heat, fossil fuel industries, mining, lighting and heating buildings, manufacturing	324
transportation	exhaust from cars, trucks, airplanes, trains	190
fugitive emissions	gases released during fossil fuel extraction and processing	67
agriculture	production of nitrogen fertilizers, farm machinery exhausts	62
industrial processes	mineral and metal production, chemical industry	54
waste management	sewage treatment, landfills	21
land use and forestry	forests, crops, grassland, wetlands, settlements	20

Data from *Sectoral Greenhouse Gas Emission Summary*, Environment Canada

DID YOU **KNOW?**

A Natural Resource

There are more fossil fuel deposits in Canada than in any other country except Saudi Arabia. Companies in Alberta are world leaders in developing technologies to process oil sands from open-pit mines.

In Table 1, the unit for the numbers in the third column is "Mt CO_2-eq." This is read as "megatonnes of carbon dioxide equivalent." This unit is used because different greenhouse gases have different abilities to trap thermal energy. Scientists consider the relative ability of each gas to absorb lower-energy infrared radiation and convert it to thermal energy. They compare this ability to the effect of carbon dioxide gas: "CO_2 equivalents" (CO_2-eq). Therefore, 1 Mt of a gas that is 10 times more efficient at trapping thermal energy than carbon dioxide is the same as 10 Mt CO_2-eq.

Many Canadian sources of greenhouse gases come from burning fossil fuels (Figure 3).

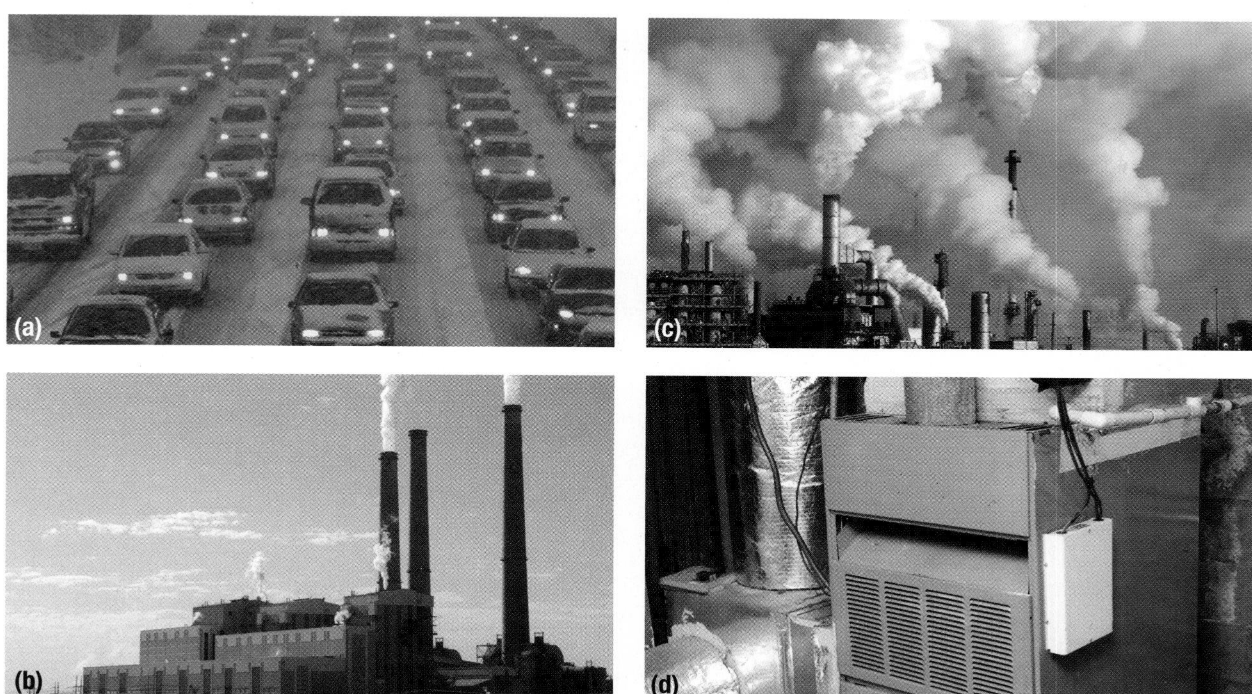

Figure 3 Sources of greenhouse gases include (a) gasoline burned in car and truck engines, (b) coal-based power plants, (c) manufacturing and other industrial processes, and (d) natural gas heating systems in homes.

The main component of natural gas is methane (Figure 4). Methane is a greenhouse gas, so any emissions of methane also increase concentrations of greenhouse gases.

DID YOU **KNOW?**

Generating Trouble

Passenger vehicles in Ontario emit, all together, about 30 Mt CO_2-eq of greenhouse gases per year. Driving an SUV releases 1.5 t more carbon dioxide per year, on average, than driving a car. Coal-generating stations produce about 20 Mt CO_2-eq of greenhouse gases each per year—equivalent to the emissions of six million cars.

Figure 4 Methane and other waste gases are often "flared off" from oil rigs to prevent a dangerous buildup of pressure.

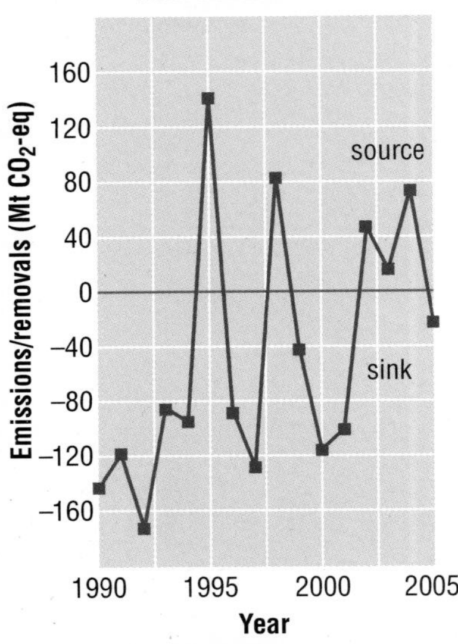

Figure 5 Are Canada's forests more often a source or a sink for greenhouse gases?

Canadian Forests: Sources or Sinks?

Canada has about 400 million hectares of forest. Our forests are among our most valuable natural resources. When healthy, Canadian forests are a carbon sink. Our forests, however, are sometimes a source of carbon due to the combination of insect damage, wildfires, and deforestation (Figure 5). Insect infestations kill trees, speeding up decomposition and increasing carbon dioxide emissions. Forest fires release large quantities of carbon dioxide and other greenhouse gases into the atmosphere. When forests are cut down, greenhouse gases, including methane and carbon dioxide, are released.

An International Issue

Greenhouse gases produced in Canada enter the atmosphere and travel around the world. Likewise, greenhouse gases produced elsewhere affect our climate. This is truly an international issue.

Reducing our use of fossil fuels is the most important thing we can do to help limit future changes in climate. Canadian governments and industries are already making some changes to reduce our dependence on fossil fuels. You will examine these changes in Section 10.4.

IN SUMMARY

- On average, individual Canadians emit more greenhouse gases than most other people in the world.
- Ontario is the second largest provincial emitter.
- The greatest causes of greenhouse gases in Canada include burning fossil fuels for transportation and to produce electricity, and processing fossil fuels.
- Canadian forests may be a source or a sink of carbon.

CHECK YOUR LEARNING

1. Canadians produce more greenhouse gases per capita than most other countries in the world. K/U C A
   (a) List six Canadian sources of greenhouse gases in order of their contribution, beginning with the source that produces the most.
   (b) Think carefully about the list in part (a). Which of these sources should be a priority for Canadians, in terms of reducing greenhouse gas production? Explain.
2. (a) What other greenhouse gases does Canada produce in a significant amount in addition to carbon dioxide?
   (b) How are these greenhouse gases produced? K/U
3. (a) The pine beetle is currently destroying large areas of forest in British Columbia. Some scientists fear that it will travel across Canada and damage forests in Ontario and other provinces. How will this affect Canada's output of greenhouse gases into the atmosphere?
   (b) If forests destroyed by the pine beetle are then burned, how will this affect Canada's greenhouse gas contribution? T/I A
4. List three factors that affect Canada's forests and determine whether they are a carbon sink or a source of carbon dioxide. K/U
5. The ocean and living things (including forests) are Earth's two main carbon sinks. Draw a flow chart (similar to Figure 6) or other diagram that shows the relationship between these carbon sinks, greenhouse gases, and Earth's temperature. C

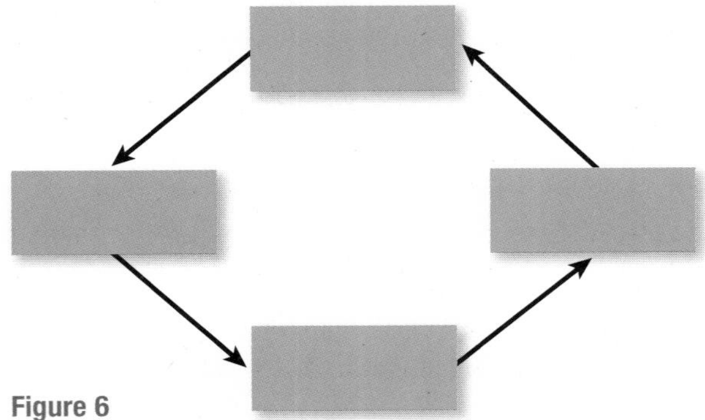
Figure 6

Computer Modelling: Evidence that Human Activity Is Causing Current Changes

9.6

When we look at Earth's history, we see that the climate has always been changing. Even over the past thousand years, the climate has changed because of natural causes such as volcanic eruptions and changes in oceanic currents (Figure 1). More recently, however, Earth's climate has changed because of human activities. Today, scientists are more than 90 % certain that most of the change observed over the last century is due to human emissions of greenhouse gases. A greater than 90 % certainty means that something is very likely to be true.

Figure 1 The eruptions of (a) Mount St. Helens in 1980 and (b) El Chichón in 1982 caused a temporary cooling as millions of tonnes of particles were shot into the atmosphere. The photograph of El Chichón was taken two years after the volcano erupted.

When scientists study the effect of a variable, they try to hold all other variables constant. Scientists would prefer to conduct controlled experiments with a second identical Earth. They could then compare our Earth with a second Earth with no people. This way, they could measure exactly how humans are affecting Earth's climate system. As this is not possible, scientists use past observations and complex computer models of the climate system to determine what is causing the changes we see today.

Modelling the Climate

Scientists have developed detailed computer models that represent important components of the climate system. These models are used to create simulations of Earth's climate under different circumstances. Scientists can identify which factors are affecting the climate by comparing these simulations with real-life observations. Thus, the models can "predict" what the climate would be like under certain circumstances.

Two scenarios are described on the next page. In the first, the model includes only natural factors acting on the climate system. In the second scenario, scientists include both natural and anthropogenic influences.

Climate scientists often work closely with computer programmers. To find out more about being a computer programmer,

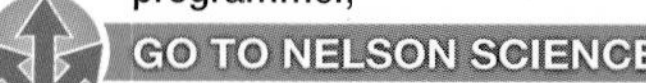

Figure 2 Swamps are a natural source of methane and carbon dioxide gases. Climate scientists take natural emissions like these into account when modelling factors affecting climate.

Scenario 1: Natural Changes Only

This scenario shows what Earth's climate would be like if no humans existed. This scenario considers the following:

- changes in energy from the Sun
- volcanic eruptions
- natural processes and variability that are part of Earth's climate system, including natural emissions of greenhouse gases (Figure 2)

The scenario does not include any human-related emissions of greenhouse gases.

In this scenario, Earth's average global temperature stays about the same, even decreasing a little from the 1950s to today. The thick blue line in Figure 3 shows the range of expected global temperature due to natural changes only, as predicted by this scenario.

Global Average Temperatures during the 20th Century

Temperature anomaly (°C): 1.0, 0.5, 0.0

Year: 1900, 1950, 2000

- Scenario 1: Expected temperature range if only natural changes are occurring
- Scenario 2: Expected temperature range if human activity is influencing climate
- Actual temperatures observed

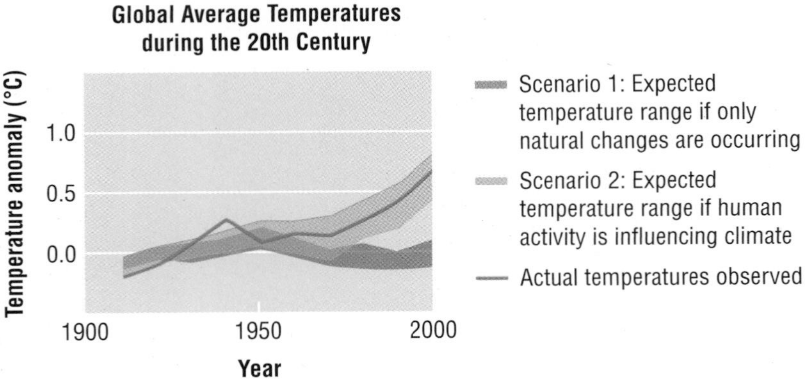

Figure 3 This graph compares predicted values for two scenarios with actual global temperature (green line) changes over the past century. The *y*-axis gives the "temperature anomaly," or changes above and below the average temperature (0.0).

Scenario 2: Natural and Anthropogenic Factors

Figure 4 Car exhaust is one of many human sources of greenhouse gases.

This scenario models an Earth that includes both natural changes and changes caused by human activity (Figure 4). This scenario considers the natural changes listed above *and* anthropogenic greenhouse gas emissions.

In this scenario, Earth's average global temperature increases, particularly over the last 50 years. The thick pink line in Figure 3 shows the range of expected global temperatures if human activity is affecting climate, as predicted by this scenario.

Comparing Scenarios: Which Model Fits?

The two scenarios present two different pictures: a picture of Earth's climate with only natural changes and a picture of Earth's climate with both natural changes and human influences. Which matches the real world best? The green line in Figure 3 represents actual data on Earth's global temperature. As you can see, over the past 50 years, the green line matches the pink line, not the blue line. In other words, the real world matches the scenario in which human activity is affecting Earth's climate.

From these computer simulations, scientists have concluded that human activity is affecting the climate in a noticeable way, particularly over the last 50 years.

To view a simulation of how climate will change over time,
GO TO NELSON SCIENCE

Conclusion: Human Factors Have Affected the Climate Over the Past 50 Years

Are the changes that we are currently experiencing within the range of natural changes in the past? To answer this question, scientists look at proxy records such as ice cores to see what variations in climate happened in the past. Is it possible that today's changes are just part of these natural cycles?

Greenhouse gas levels in the atmosphere are higher today than they have been at any time in the past 800 000 years. However, Earth's average temperature is not yet warmer than it has been in past cycles. If human emissions of greenhouse gases continue to increase at the same rate, by the year 2100, Earth could be warmer than at any time in the past 800 000 years.

Look back at Figure 3 in Section 8.9. This graph shows Earth's average temperatures over the last 400 000 years. There have been times when Earth's temperature was just as warm or 1 °C or 2 °C warmer than it is currently. Why, then, are scientists worried about the fact that Earth's temperature will probably increase in the next century? They are worried for three reasons. First, this temperature change will happen much more quickly than in the past. Second, the temperature change will be higher than anything seen over the last 800 000 years. Third, human civilization is adapted to recent climate conditions. Any major change in these conditions will have drastic effects. In Chapter 10, we will consider how scientists make projections about Earth's future climate and what actions we should take.

READING TIP

Summarizing

When you are summarizing a text, use the same organizational pattern that the text uses. If the text uses a problem-solution pattern to give the problem and then several possible solutions, do the same. If the text describes a process, summarize the steps in the same order as the original text.

IN SUMMARY

- Complex computer models of the climate system are used to understand factors that have affected climate in the past.
- Using models, scientists have concluded that human activity is affecting the climate in a noticeable way, particularly over the last 50 years.
- At the current rate of human emissions of greenhouse gases, Earth could be warmer by the year 2100 than at any time in the previous 800 000 years.

CHECK YOUR LEARNING

1. A computer model compares two worlds: one with humans and one without humans. All variables are the same in both worlds, except for the effects of human activities. In the computer world with humans, temperatures are rising. In the computer world without humans, temperatures are staying the same. K/U T/I
   (a) What variable does the computer model study?
   (b) Why are all other variables kept the same?
   (c) Suppose temperatures in the real world had stayed the same over the past 50 years. What would this show about the computer model and its predictions regarding the effect of humans on the world's climate?
   (d) Temperatures around the world have, in fact, been rising over the past 50 years. What does the computer model show about the effect of humans on the world's climate?
2. In your own words, interpret the graph in Figure 3 in this section. T/I C
3. Earth's average temperature is not yet as warm as it has been in past cycles of natural climate change (Figure 3 in Section 8.9). T/I A
   (a) Does this mean that today's climate change might be natural? Explain your answer.
   (b) Why are scientists concerned about the change in climate that we are currently experiencing?
4. (a) How well do you understand the idea of using percentages to describe scientists' certainty?
   (b) What could you do to improve your understanding? T/I

CHAPTER 9

LOOKING BACK

KEY CONCEPTS SUMMARY

We have evidence that our climate is changing.

- The average temperature of Earth is rising. (9.1)
- Glaciers, ice sheets, and sea ice are melting. (9.1, 9.2)
- The sea level is rising and ecosystems are changing. (9.1)

Human activities have increased atmospheric levels of greenhouse gases.

- Human activities produce carbon dioxide, methane, nitrous oxide, and CFCs. (9.4)
- Most of our greenhouse gas emissions end up in the atmosphere. (9.4)
- Carbon dioxide concentrations in the atmosphere have risen to the highest level in at least 800 000 years. (9.6)
- Positive feedback loops increase the effect of anthropogenic greenhouse gases. (9.4)

The increase in greenhouse gases is causing the anthropogenic (human-caused) greenhouse effect.

- As the concentration of greenhouse gases in the atmosphere increases, more energy is trapped. (9.4)
- The trapped energy leads to an increase in Earth's global temperature. (9.4)

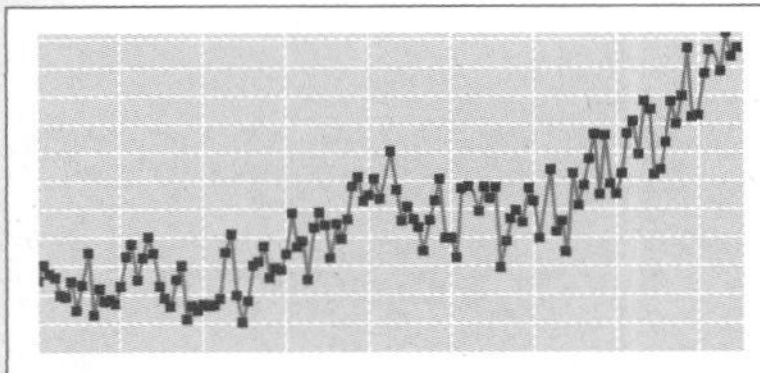

The anthropogenic greenhouse effect is the main cause of today's climate change.

- There is a strong relationship between atmospheric carbon dioxide levels and global temperature. (9.4, 9.6)
- Scientists have studied both natural and anthropogenic causes of climate change. They have concluded that human activities are very likely to be the cause of observed climate change. (9.4, 9.6)

The largest sources of greenhouse gases in Canada are the production and burning of fossil fuels.

- On average, individual Canadians are in the top ten highest emitters of greenhouse gases. (9.5)
- Fossil fuels are used for transportation, heating, and the production of electrical energy. (9.5)
- The production and use of energy produce about 324 Mt CO_2-eq in Canada each year. (9.5)
- Canada's forests are carbon sinks and carbon sources. (9.5)

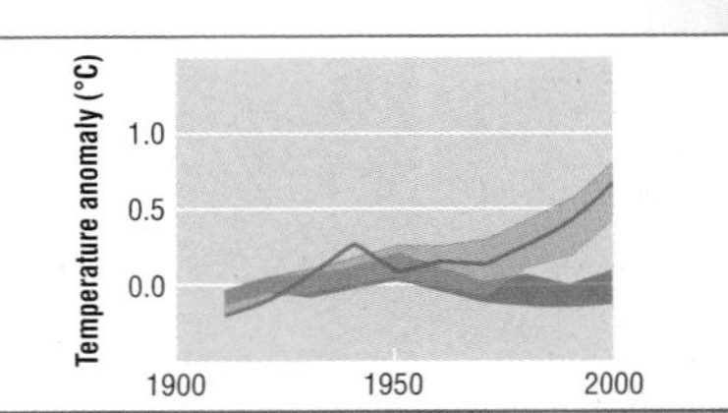

Scientists use climate models to figure out how different factors affect our climate.

- Climate models show that human activities are affecting the climate, particularly over the last 50 years. (9.6)

WHAT DO YOU THINK NOW?

You thought about the following statements at the beginning of the chapter. You may have encountered these ideas in school, at home, or in the world around you. Consider them again and decide whether you agree or disagree with each one.

1 Today's changes in climate are similar to changes that occurred in the past.
Agree/disagree?

2 Increased atmospheric carbon dioxide is mainly due to human agriculture practices around the world.
Agree/disagree?

3 Scientists can predict exactly how the climate will change in the future.
Agree/disagree?

4 Climate change will have both negative and positive effects.
Agree/disagree?

5 Changes to Earth's climate can be reversed by changing human activities.
Agree/disagree?

6 The burning of fossil fuels in heating, transportation, and industry is the greatest cause of greenhouse gas production in Canada.
Agree/disagree?

How have your answers changed since then? What new understanding do you have?

Vocabulary

thermal expansion (p. 376)
anthropogenic (p. 384)
anthropogenic greenhouse effect (p. 387)

BIG Ideas

- ✓ **Earth's climate is dynamic and is the result of interacting systems and processes.**
- ✓ **Global climate change is influenced by both natural and human factors.**
- ✓ **Climate change affects living things and natural systems in a variety of ways.**
- **People have the responsibility to assess their impact on climate change and to identify effective courses of action to reduce this impact.**

CHAPTER 9 REVIEW

The following icons indicate the Achievement Chart category addressed by each question.

K/U Knowledge/Understanding T/I Thinking/Investigation C Communication A Application

What Do You Remember?

1. Describe five different signs that indicate that climate change is already affecting the environment we live in. (9.1) K/U
2. List three Canadian sources of greenhouse gases. (9.5) K/U
3. What is the connection between fossil fuels and greenhouse gases? (9.4) K/U
4. List three different purposes for burning fossil fuels in Canada. (9.5) K/U
5. The term "anthropogenic greenhouse gases" includes carbon dioxide and other gases. Name three of these other gases. (9.4) K/U
6. (a) In your own words, define "carbon sink."
   (b) Give an example of a carbon sink.
   (c) Explain why carbon sinks are important. (9.4, 9.5) K/U C
7. List three factors that affect whether Canadian forests act as a source or a sink for greenhouse gases. (9.5) K/U

What Do You Understand?

8. Scientists believe that today's climate change is very likely caused by human activity. Summarize the evidence that leads scientists to this conclusion. (9.6) K/U
9. Draw a diagram that summarizes the relationship between using fossil fuels for transportation and Earth's global temperature. (9.4) K/U C
10. The greenhouse effect is natural and important to life on Earth. Scientists today are concerned about something called the "anthropogenic greenhouse effect." (9.4) K/U
    (a) Distinguish between the anthropogenic greenhouse effect and the natural greenhouse effect.
    (b) Explain why scientists consider the anthropogenic greenhouse effect to be a problem.
11. Describe the feedback loops illustrating the connection between each variable below and climate change: (9.4, 9.5) K/U
    (a) Raising the temperature of the ocean results in more melting of the ice.
    (b) Higher temperatures result in increased evaporation of moisture from the soil, making it more likely that forest fires will occur, adding carbon dioxide to the atmosphere.
    (c) Increasing atmospheric carbon dioxide raises the temperature of the ocean, reducing the amount of carbon dioxide that can be absorbed by the ocean.
12. (a) When two variables, such as average world temperature and sea level, change at the same time, does this mean that one causes the other? Explain.
    (b) Think about several reasons or a piece of evidence that strongly suggests that these two things are related. Write an argument of several sentences that you can use to convince friends and family members. (9.4) K/U T/I C
13. In Chapter 5 you learned that nitrous oxide is a harmless gas administered to dental patients. In this chapter you learned about a different aspect of nitrous oxide. Comment on the description of nitrous oxide as a "harmless" gas. (9.4) T/I

Solve a Problem

14. A scientist wants to examine the relationship between sea level rise and carbon dioxide concentrations in the atmosphere. Describe one method the scientist could use. (9.1, 9.3) K/U C
15. (a) Brainstorm with some friends to create a list of extreme natural events related to climate. Ask your parents and other people for additional examples.
    (b) Which, if any, of the events on your list had positive impacts?
    (c) Outline what kind of research you could do to find out if extreme events have increased in frequency and/or intensity over the past 50 years where you live. (9.1) T/I A

Create and Evaluate

16. Examine Figure 1. You saw this previously in Section 9.1 (Figure 2). Consider the two following reports:

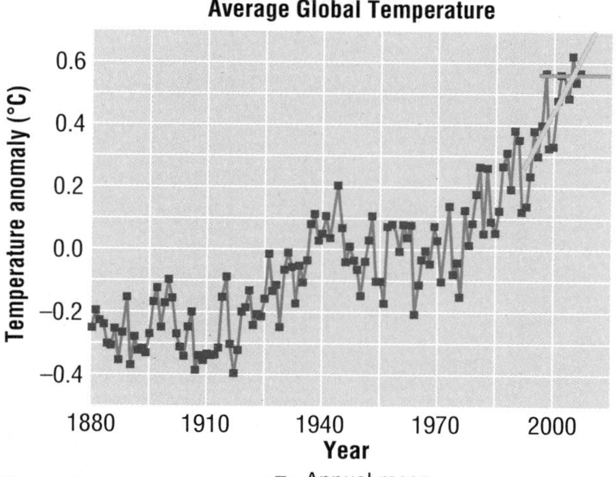

Figure 1

- In a 2007 report, an independent researcher drew a line between the temperature data points for 1998 and for 2006 (orange line). Looking at the horizontal line, the researcher concluded that Earth stopped warming after 1998. He concluded that we do not have anything to worry about.
- Another 2007 report drew a trend line through the annual temperature data points from 1996 to 2006 (yellow line). The writers of this report concluded that since the line of best fit had a positive slope, Earth's temperature increased continuously over the past century. Since human use of fossil fuel increased continuously over the past century, we do not have to consider any other factors. (9.1) T/I C A

(a) For each report, write a sentence or two describing a possible bias behind the report. How might the bias be related to the editor's knowledge? Belief? Values?

(b) Find an approximate trend line for other time periods besides the ones above. Place a straight edge over the data points for a 10-year period (1996–2006), a 20-year period (1986–2006), a 30-year period (1976–2006), and another 30-year period (1940–1970). What do you conclude about each time period?

(c) How would you interpret the reports summarized above?

(d) If you were asked to draw your own line of best fit, would it be more similar to the orange line or the yellow line?

Reflect on Your Learning

17. In Chapter 9, you learned that increases in carbon dioxide concentrations in the atmosphere cause increases in global temperatures.

(a) Did you find this concept difficult to understand? Why or why not?

(b) How did your learning in Chapter 8 prepare you for this concept?

(c) Did you find that the evidence used to support this concept was reasonable and convincing? Explain.

(d) What further research can you do to help yourself understand this concept?

Web Connect 

18. The National Snow Information System for Water website provides projections of what might happen to the extent of snow over Canada in the future. (9.2) K/U T/I

(a) Is snow cover increasing or decreasing in Canada?

(b) How does snow help keep Earth cool?

(c) What effect might a reduction in snow cover have on Earth's climate?

(d) How is this an example of a feedback loop in climate change?

19. Some people refuse to believe that climate change is happening, or that its effects will be mainly negative. T/I C A

(a) Research three common arguments skeptics present that disagree with the IPCC's findings. Summarize your findings.

(b) Come up with your own replies to these three arguments, based on the information in this book.

(c) Research responses to these three common arguments that have been written by climate change experts. Summarize your findings.

(d) Compare the experts' responses with your own. What new information did you learn from the experts' responses?

CHAPTER 9

SELF-QUIZ

The following icons indicate the Achievement Chart category addressed by each question.

K/U Knowledge/Understanding · T/I Thinking/Investigation · C Communication · A Application

For each question, select the best answer from the four alternatives.

1. Which statement best describes thermal expansion? (9.1) K/U
   (a) The volume of water increases as its temperature decreases.
   (b) The mass of water increases as its temperature increases.
   (c) The volume of matter increases as its temperature increases.
   (d) The volume of a substance increases as its concentration increases.
2. What is the most abundant human-produced greenhouse gas? (9.4) K/U
   (a) water vapour
   (b) ammonia
   (c) carbon dioxide
   (d) methane
3. Which of the following may be attributed to an increase in greenhouse gases in Earth's atmosphere? (9.1) K/U
   (a) rising sea level
   (b) falling temperatures
   (c) formation of new ice sheets
   (d) expansion of Arctic ice
4. Which of the following series orders Canadian greenhouse gases from highest to lowest emissions? (9.5) K/U
   (a) transportation; agriculture; land use and forestry
   (b) agriculture; transportation; land use and forestry
   (c) agriculture; land use and forestry; transportation
   (d) land use and forestry; agriculture; transportation

Indicate whether each of the statements is TRUE or FALSE. If you think the statement is false, rewrite it to make it true.

5. Refrigeration agents are a significant source of anthropogenic carbon dioxide in the atmosphere. (9.4) K/U
6. Earth's average temperature has been continuously increasing since 1880. (9.1) K/U
7. Carbon dioxide stored in the oceans might be released as global temperatures increase. (9.4) K/U
8. Melting glaciers and continental ice sheets cause sea levels to rise. (9.1) K/U

Copy each of the following statements into your notebook. Fill in the blanks with a word or phrase that correctly completes the sentence.

9. Reservoirs such as oceans and forests that absorb carbon dioxide from the air are known as carbon __________. (9.4) K/U
10. Greenhouse gases are emitted during the burning of __________ for transportation, heating, and industrial use. (9.4) K/U
11. The concentration of carbon dioxide in the atmosphere is higher now than it has been for __________ years. (9.4) K/U

Match each substance on the left with the most appropriate source on the right.

12. (a) carbon dioxide — (i) rice farming and cattle ranching
    (b) methane — (ii) deforestation and burning fossil fuels
    (c) nitrous oxide — (iii) refrigeration agents
    (d) chlorofluorocarbons — (iv) nitrogen fertilizers and livestock feed
    (9.4) K/U

Write a short answer to each of these questions.

13. Describe the two main factors that contribute to rising sea levels on Earth. (9.1) K/U
14. How does deforestation affect the amount of carbon dioxide in the atmosphere? (9.4) K/U

15. The word "anthropogenic" means "human caused." Write a phrase using the word "anthropogenic" that describes the relationship between humans, greenhouse gases, and global climate change. (9.4) C

16. Explain how a tree on your lawn could be a
    (a) carbon sink.
    (b) carbon source. (9.4) A

17. How could the melting of glaciers have a
    (a) positive effect on human populations?
    (b) negative effect on human populations? (9.1) T/I

18. Feedback loops can make small changes in Earth's climate much larger. (9.4) K/U
    (a) Is the feedback loop between carbon dioxide concentration and global temperature positive or negative?
    (b) Describe the feedback loop using a diagram.

19. The Kyoto Protocol is an international agreement that aims to reduce greenhouse gas emissions. It places restrictions on human activities that produce greenhouse gases. Why might a country be reluctant to participate in this agreement? (9.1) T/I

20. The choices we make about our personal transportation will influence how much greenhouse gases are produced. (9.4, 9.5) A T/I
    (a) An SUV releases roughly 1.5 t more carbon dioxide per year than a car does. How might this information affect your decisions about transportation?
    (b) Most buses release even more carbon dioxide than SUVs do. Do you think people should stop riding buses?
    (c) Respond to this statement: "Some countries have forests that soak up significant amounts of carbon dioxide. People in these countries shouldn't have to worry about what they drive."

21. When two variables change at the same time, they are considered to be correlated. (9.4) T/I
    (a) When two variables are correlated, does this mean that one causes the other?
    (b) How have scientists concluded that human activity is affecting climate in a noticeable way?

22. Scientists believe that anthropogenic sources of greenhouse gases are causing the climate to change. (9.1) K/U C
    (a) What evidence of anthropogenic climate change do you think is most convincing?
    (b) Write an argument that you can use to communicate this to friends and family members.

23. Why is climate change considered a truly global issue? (9.5) K/U

24. (a) Describe the two main causes of sea level rise.
    (b) What additional causes might contribute to the sea level rise? (9.1, 9.3) K/U

25. If all countries in the world stopped their emissions of greenhouse gases tomorrow, could we stop climate change from happening? Explain your answer. (9.4) K/U C

26. Scientists use computer models to understand factors that have affected the climate in the past. Explain how they use these models to understand the effect humans have had on Earth's climate system. (9.6) K/U C

27. Imagine that you are taking part in a debate about Canadian emissions of greenhouse gases. Your opponent says, "Canada's forests absorb carbon dioxide, so we don't have to worry about our emissions. The more greenhouse gases we emit, the more our forests will absorb." Write a short paragraph critiquing your opponent's argument. (9.5) C A

28. Use a table or graphic organizer to compare the natural greenhouse effect with the anthropogenic greenhouse effect. (9.4) K/U C

CHAPTER 10

Assessing and Responding to Climate Change

KEY QUESTION: What can we do to reduce the impacts of climate change and to prepare for coming changes?

Constructing wind farms is one possible response to climate change.

UNIT D

Climate Change

CHAPTER 8
Earth's Climate System and Natural Changes

CHAPTER 9
Earth's Climate: Out of Balance

CHAPTER 10
Assessing and Responding to Climate Change

KEY CONCEPTS

The impacts of climate change will affect our environment and society.

Impacts of climate change will be felt the most in the Arctic.

Climate change in Ontario is expected to bring warmer winters and hotter summers.

Current initiatives will not prevent serious negative effects from climate change.

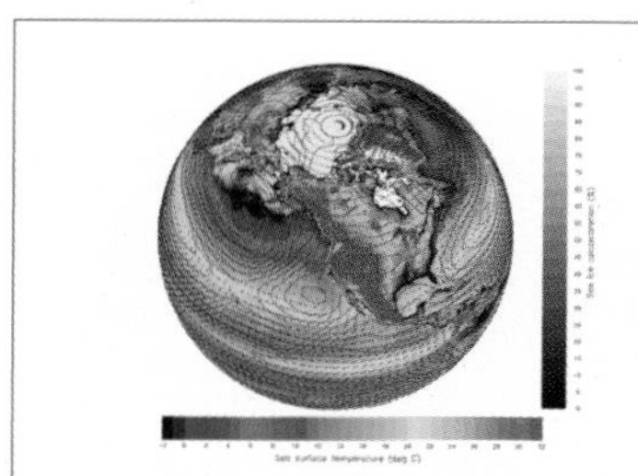

Greenhouse gas emissions must be reduced by 80 % by 2050 to avoid the most serious impacts.

Switching to clean energy sources is essential to reduce greenhouse gas emissions.

ENGAGE IN SCIENCE

GREEN TECHNOLOGIES

In 2010, the Vancouver Convention & Exhibition Centre (VCEC) will be the international broadcast centre for the Vancouver Olympics. The centre will become one of the buildings that people around the world associate with Canada. Fortunately for Canadians, the newly expanded VCEC will be a worthy representative of our country. The centre is environmentally friendly, with many green technologies that help reduce emissions of greenhouse gases.

Green Roof Technology

The centre boasts an enormous green roof, made up of 400 000 plants that are native to the West Coast. Although the plants started off small, they are quickly growing into a lush living carpet. On a green roof, plants absorb the Sun's energy and use it for photosynthesis, instead of emitting thermal energy back into the atmosphere as an ordinary roof does. Green roofs also provide better insulation for buildings. As a result, buildings with green roofs use less energy for heating and cooling. Buildings with green roofs stay cooler in hot weather.

Green roofs absorb stormwater, reducing the amount of water going into storm sewers. They also improve air quality and reduce noise pollution. Other major cities, including London, Chicago, and Toronto, are looking into green roofs as one way to prepare for changes in climate.

Other Green Technologies

Built partly on water, the new VCEC is designed to provide an underwater habitat for algae, mussels, and other marine life. The building is energy efficient, resulting in fewer greenhouse gas emissions than a traditional building of the same size. It also contains an on-site water treatment plant and is built to last a long time. The VCEC provides a good example of how green technologies help prepare us for a changing climate.

How could green technologies be used in your home or neighbourhood? Will these technologies be enough to reduce the impacts of climate change? Can you suggest a further course of action related to the green technologies described above?

WHAT DO YOU THINK?

Many of the ideas you will explore in this chapter are ideas that you have already encountered. You may have encountered these ideas in school, at home, or in the world around you. Not all of the following statements are true. Consider each statement and decide whether you agree or disagree with it.

1 We can stop climate change from happening if we reduce our consumption of fossil fuels.
Agree/disagree?

2 Scientists are always objective and unbiased in their data collection, analyses, and conclusions.
Agree/disagree?

3 Since the impact of climate change will be most felt in the Arctic, people in countries near the Equator do not need to be concerned.
Agree/disagree?

4 The best way to decrease the impact of human activity on climate is to switch to clean energy sources.
Agree/disagree?

5 Canadians' use of fossil fuels is wasteful.
Agree/disagree?

6 We need to take action on climate change at individual, local, regional, and international levels.
Agree/disagree?

Synthesizing

When you synthesize, you combine different sources of information in a way that brings you to a new understanding. Use the following strategies when synthesizing a text:

- make connections between new information in the text and what you already know
- explain relationships among ideas in the text and the ideas in other texts
- monitor how the new ideas and information change your perspectives
- consider how features such as graphs and illustrations combine with the text
- draw conclusions about what you have learned

READING TIP

As you work through the chapter, look for tips like this. They will help you develop literacy strategies.

Electricity Use

Figure 1 Blackouts could occur more often as Canadians use more electricity to cool their homes and offices.

As summers get hotter, we will need to use more electricity for air conditioning. Generating electricity from coal or natural gas produces greenhouse gases, which makes the problem of climate change worse. In the winter, energy use may decrease due to warmer weather.

In Ontario, about a quarter of our electricity is produced using hydroelectric power, which does not produce greenhouse gases. During heat waves, however, people might try to use more electricity than can be produced, leading to blackouts (Figure 1). In addition, less hydroelectricity will be available if the lake levels drop because of climate change. This could increase our use of fossil fuels.

Synthesizing *in Action*

Synthesizing is a powerful way to use what you already know as a context for thinking about new information and drawing your own conclusions. Here is an example of how one student used the strategies to synthesize the selection about electricity use.

Information in the Text	What I Already Know	What I Think Now
More air conditioning will be needed if summers get hotter.	The power outage of 2003 happened in the summer.	We will need to generate more electricity.
Hydroelectric power is better than coal or gas.	Greenhouse gases cause climate change.	We need to generate more electricity that is clean.
Lower lake levels will mean less hydroelectric energy.	Wind and solar energy are clean.	We need to use more wind and solar technology.

Climate Models and Clean Energy

10.1

Science fiction movies and books contain many interesting ideas about the future. Depending on the climate, Earth in 100 years could be very different from Earth today (Figure 1). At the same time, the way in which human technologies and society develop will affect Earth's climate.

Figure 1 The climate will affect our future way of life—and our present way of life will affect future climate. How do you think human technologies and society will develop?

Uncertainties in Predicting Future Climate

Think back on what you have learned about Earth's complex climate system. Among other things, the climate system depends on the following:

- the carbon and water cycles
- concentrations of greenhouse gases (GHGs) in the atmosphere
- positive and negative feedback loops, including the albedo effect
- ocean currents, including thermohaline circulation

With this many variables, scientists who study the climate system cannot be sure how quickly Earth's climate will respond to changes in greenhouse gas concentrations. For example, Arctic sea ice is melting faster than the climate models predicted. This suggests that Earth's climate system might be more sensitive to our greenhouse gas emissions than previously thought.

Also, scientists have discovered that the rate at which the ocean and living things absorb carbon is slower than previously thought (Figure 2). This means that the concentration of atmospheric carbon dioxide may increase faster than we expect, even if our emissions remain constant.

Figure 2 Oceans are absorbing carbon at a slower rate than scientists previously thought.

In spite of its complexity, computer models are still able to model the climate system with reasonable accuracy. In fact, the main reason why predicting climate is difficult is that scientists do not know what choices people will make. How much the climate changes over the next 100 years depends on the choices we make now. For example, if we continue to use fossil fuels at the same rate, we can expect the climate to change drastically. Conversely, if we quickly switch to **clean energy sources** that produce few or no greenhouse gases, Earth's climate may change less dramatically. We have a good chance of preventing the most severe impacts of climate change if we act soon.

clean energy source a source of energy that produces no significant greenhouse gases

Climate Projections

climate projection a scientific forecast of future climate based on observations and computer models

A **climate projection** is a reasonable scientific estimate of future climate conditions. Climate projections are based on simulations by complex computer models. These models, called climate models, consider future changes that will affect greenhouse gas production (Figure 3).

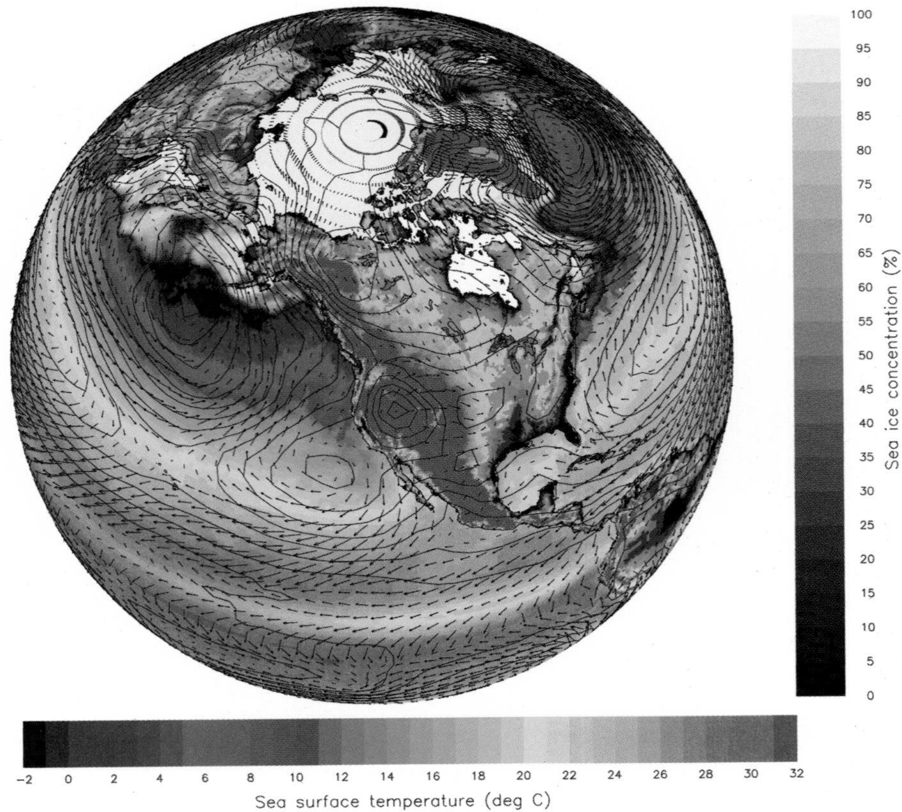

Figure 3 This image, showing changes in sea ice cover and sea surface temperature, was generated using output from climate models.

We do not know what our future GHG emissions will be or how quickly Earth will respond. However, scientists can develop projections that forecast what Earth's future climate will probably be like under specific conditions.

To generate climate projections, scientists make assumptions about future human behaviour. This kind of assumption is called a scenario. One scenario (A) is that humans continue to use fossil fuels at the same rate as today (Figure 4). A different scenario (B) is that humans switch from fossil fuels to clean energy sources within 20 years. For each scenario, scientists use climate models to make projections of how the climate would change under those circumstances.

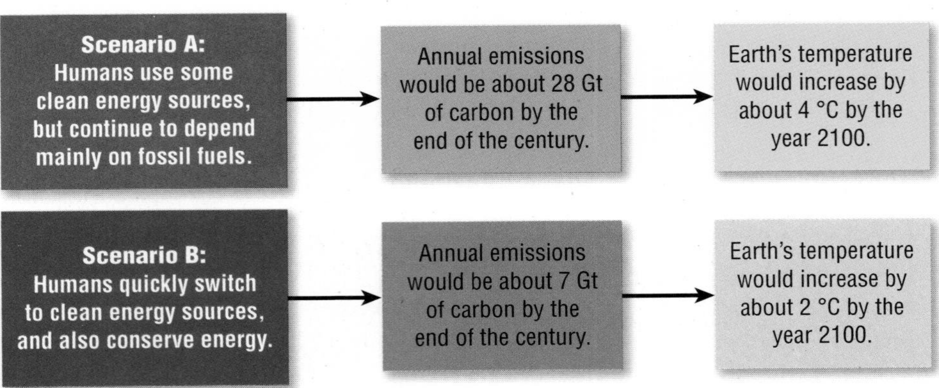

Scenario A:
Humans use some clean energy sources, but continue to depend mainly on fossil fuels.
Annual emissions would be about 28 Gt of carbon by the end of the century.
Earth's temperature would increase by about 4 °C by the year 2100.
Scenario B:
Humans quickly switch to clean energy sources, and also conserve energy.
Annual emissions would be about 7 Gt of carbon by the end of the century.
Earth's temperature would increase by about 2 °C by the year 2100.

Figure 4 To make climate projections, scientists first develop possible scenarios. Next, they determine the amount of greenhouse gases that would be produced under each scenario. Finally, they input these values into climate models to calculate how Earth's climate would change under those conditions.

To develop future climate scenarios, scientists think about the following:

- How fast will the world's population increase?
- What kinds of technologies will we be using in 10 years? 50 years? 100 years?
- What energy sources will we use in 10 years? 50 years? 100 years?

Scientists who study the climate have developed projections for many scenarios. The projections have something in common. Scenarios with higher greenhouse gas emissions result in larger temperature increases than scenarios with lower emissions. Under a higher-emissions scenario, such as Scenario A, the temperature changes by the end of the century are double those expected under a lower-emissions scenario, such as Scenario B (Figure 5).

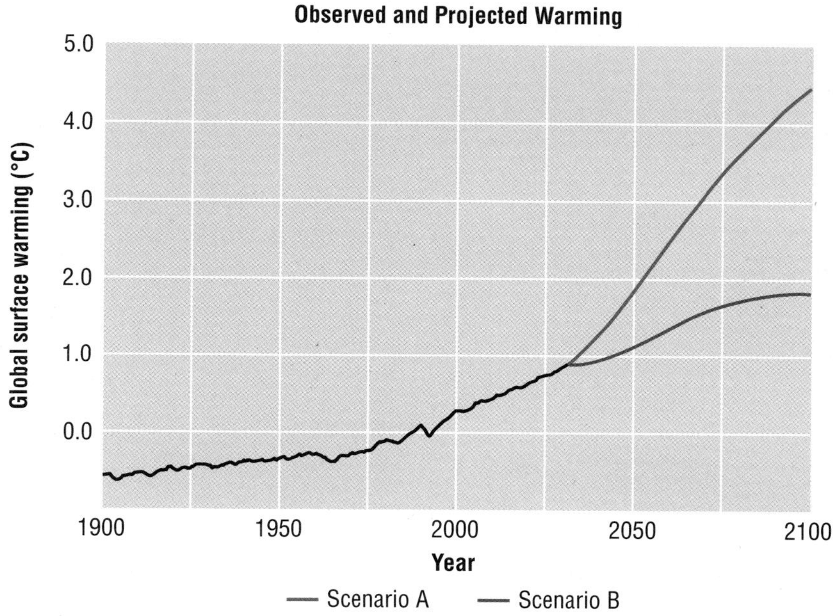

Figure 5 The two coloured lines show the possible consequences of two scenarios. In Scenario A (red), humans continue to depend on fossil fuels. In Scenario B (blue), humans switch to clean energy sources and conserve energy.

READING TIP

Visual Information

If you are struggling to understand the main idea in a text, you can use the information in an illustration or photo to help you. Alternatively, the information in a graph or table could give specific details to help you synthesize.

TRY THIS ESTIMATE ONTARIO'S FUTURE CLIMATE

SKILLS: Predicting, Controlling Variables, Performing, Evaluating

SKILLS HANDBOOK 3.B.

In this activity, you will run a computer simulation and discover how your chosen scenario might affect Ontario's climate.

Equipment and Materials: computer with Internet access

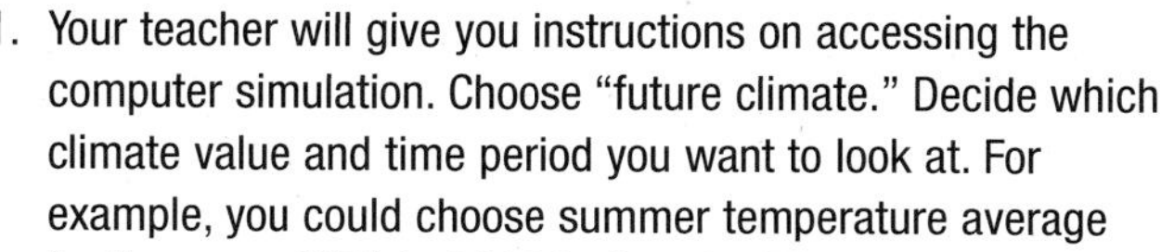

1. Your teacher will give you instructions on accessing the computer simulation. Choose "future climate." Decide which climate value and time period you want to look at. For example, you could choose summer temperature average for the years 2071 to 2100 in Ontario. Click on your choices.
2. Choose either the higher greenhouse gas scenario or the lower greenhouse gas scenario.
3. Choose a map type (i.e., province) and a location.
4. Read the map produced by the program. Record your observations.
5. Repeat the projection for a different set of values.

A. Compare your results from the two sets of values. Ask yourself questions such as the following: T/I
   - What is the greatest change that will happen in Ontario? What is the least amount of change?
   - Which cities or towns will see the greatest change? Which will see the least amount of change?

B. Explain how a climate model like this one can help scientists estimate changes in climate. T/I

C. Identify any drawbacks of using a climate model such as this one. For example, what is not shown? What incorrect assumptions (if any) could be made from a climate model like this one? T/I

Changing Our Energy Sources

Energy use is one of the main variables needed to generate climate projections. We depend on fossil fuels to power the many technological devices we are accustomed to using. Extracting, refining, and using fossil fuels release greenhouse gases.

To limit the extent of climate change, we will have to change from using fossil fuels to using clean energy sources. The most common sources of clean energy are summarized in Table 1.

Table 1 Clean Energy Sources

Clean energy source	Description
wind power	Wind causes the blades of wind turbines to turn, powering generators that produce electricity.
geothermal energy	Thermal energy below Earth's surface is used to heat homes and other buildings.
solar power	Solar panels absorb radiation from the Sun and convert it into electricity. The Sun's radiation can also be used to heat water.
hydroelectricity	The energy of moving water (e.g., a waterfall) turns turbines to power generators that produce electricity.
biofuels	Biofuels use plant-based fuels to produce energy.
nuclear power	Nuclear energy is created by splitting the nuclei of atoms. Although nuclear energy plants do not emit greenhouse gases, they do produce radioactive waste, thus creating a different set of problems.

DID YOU **KNOW?**

Biofuels

Biofuels produce greenhouse gases, but these emissions have much less effect on the climate system. Why? The carbon stored in biofuels was part of the atmosphere only a few years ago, so releasing it does not change the composition of the atmosphere. Thus, biofuels are considered to be carbon neutral as long as fossil fuels are not used to produce them.

Even if we stopped all use of fossil fuels today, climate change would not stop entirely. Earth would continue to warm over the next century due to greenhouse gases that have already been added to the atmosphere. However, by switching from fossil fuels to clean energy sources quickly, we can prevent the most serious consequences of climate change. Serious consequences include the following:

- the disintegration of the Greenland and West Antarctic Ice Sheets, which would raise sea level by almost 15 m and displace over a billion people
- the extinction of up to half of the world's species, including many that have not even been identified yet
- increases in extreme weather events such as heat waves, floods, and droughts

Figure 6 Geothermal plants mine thermal energy from the Earth's crust.

Iceland once depended on imported coal for heating. In the past 50 years, however, Iceland has switched to renewable energy sources, primarily geothermal and hydroelectricity (Figure 6). How long will it take other countries to make the change to clean energy sources? The answer to this question will likely have a major impact on the future of our civilization.

DID YOU KNOW?

Arctic Bananas

Long summer days and cheap geothermal power make it possible to grow tomatoes, cucumbers, roses, and even bananas in Iceland. About 90 % of houses are heated using geothermal power.

UNIT TASK Bookmark

This information about changing our energy sources might be useful as you work on the Unit Task described on page 444.

IN SUMMARY

- To make climate projections, scientists develop scenarios by specifying the amount of greenhouse gases that would be produced under each scenario. These amounts are used to calculate how Earth's climate would change under those conditions.
- Switching from fossil fuels to clean energy sources is one of the main ways to reduce the impacts of human activity on climate.
- Clean energy sources include wind power, geothermal energy, solar power, hydroelectricity, biofuels, and nuclear power.

CHECK YOUR LEARNING

1. Give two reasons why it is difficult for scientists to predict exactly how the climate will change in the future. K/U
2. Describe two complexities about Earth's climate system that result in changes happening faster than expected. K/U
3. (a) What is a clean energy source?
   (b) Give examples of two clean energy sources that you have seen being used successfully in Ontario. K/U
4. How do climate models take human uncertainty into account? A
5. Clean energy sources also have their problems. For each of the following, suggest one way that it might have a negative effect on the environment or on people. K/U A
   (a) hydroelectric dams
   (b) nuclear power stations
   (c) biofuel production from corn
6. Suppose your family is making a five-year plan for switching to clean energy sources. Where will you start? What steps will you take? Create a timeline to show when you would implement the steps in your plan. C A

10.2

Global Impacts of Climate Change

For thousands of years, the Arctic tundra supported only low-growing flowering plants, mosses, and lichens because of its cold temperatures and short growing season. Over the last century, forests of spruce trees have begun to spring up in this land, where no trees used to grow (Figure 1). Due to the changing climate, the summer season is getting longer and spruce tree populations are migrating north.

Changes that are caused by rising global temperatures, by changing precipitation patterns, or by other changes in climate are called **impacts of climate change**. The changes that you learned about in Section 9.1 are all impacts of climate change. There will be many more impacts of climate change in the years to come.

impacts of climate change effects on human society and our natural environment that are caused by changes in climate, such as rises in Earth's global temperature

DID YOU **KNOW?**

Disappearing Country

The tiny country of Tuvalu is in danger of disappearing. Tuvalu is a series of islands in the Pacific Ocean. At its highest, Tuvalu is only 4.5 m above sea level. This country could disappear within decades if the sea level continues to rise. Where will the 11 000 people from Tuvalu go?

Figure 1 Spruce trees began growing in the tundra around 80 years ago.

Expected Changes

The **Intergovernmental Panel on Climate Change (IPCC)** was formed in 1988 to evaluate the risks of human-caused climate change. As part of the IPCC, several thousand climate scientists participate voluntarily in sharing and synthesizing their work. The IPCC represents a consensus among scientists internationally (Figure 2). This means that the majority of scientists support the IPCC's conclusions regarding the future impacts of climate change.

Intergovernmental Panel on Climate Change (IPCC) a group of several thousand climate scientists who have summarized the latest scientific research on climate change

Figure 2 In 2007, the IPCC, represented here by its chairman Dr. Rajendra Pachauri, and Al Gore were awarded the Nobel Peace Prize for their work on climate change.

Not all scientists are in complete agreement with the IPCC, however. Some scientists believe that the IPCC reports do not capture the full severity of the impacts. They argue that humans will probably not reduce their emissions quickly enough, so the impacts of climate change will be even greater than the IPCC expects. A small minority of scientists believe that most of the impacts from climate change will be less severe than predicted, or even positive. However, the majority of scientists support the IPCC's conclusions regarding the future impacts of climate change.

The IPCC reports have identified many potential impacts of climate change. Let's look at four of the main global impacts in more detail.

To read the IPCC's reports,

GO TO NELSON SCIENCE

Rising Sea Level

As glaciers and ice sheets melt and the oceans warm up and expand, the sea level of the oceans will rise. Low-lying coastal areas, home to millions of people, will be at increased risk of flooding. Some island states including Tuvalu and Federated States of Micronesia may end up below sea level (Figure 3). Bigger countries such as Bangladesh and the Netherlands could lose large areas of land. Much of southern Florida is also at risk.

Figure 3 Low coastal areas, such as Bhola Island, Bangladesh, are vulnerable to sea level rise. Half of this island (more than 3000 km^2) has eroded away, and half a million people are homeless.

Impacts on Agriculture

Dry regions of the world, such as parts of Africa, may get even less rainfall (Figure 4). Crops may be less productive, and millions of people could experience famine. Other areas, such as the southern United States and Japan, could have more rainfall, leading to flooding. Warm, wet weather will probably lead to more damage from insects and other pests.

Figure 4 Corn crops during a drought

Impacts on Ecosystems

Some plants and animals are likely to migrate toward the poles as their current habitats become unsuitable. As a result, ecosystems around the world will change. Biodiversity may be lost. About 30 % of species could become extinct by 2050. Fragile coastal wetlands could be drowned by the rising seas. Changes in one population will have repercussions throughout the food web. For example, climate fluctuations reduce plankton populations. Plankton is the main food of the North Atlantic right whale (Figure 5). Less plankton leads to higher mortality in the whale population.

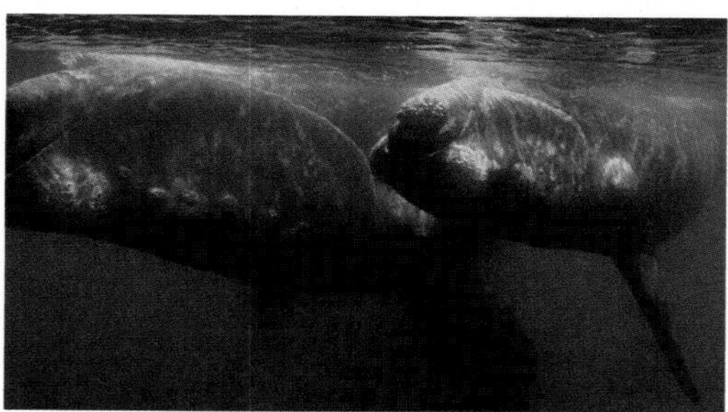

Figure 5 Fewer than 350 North Atlantic right whales remain.

Impacts on Human, Plant, and Animal Health

Pests, diseases, and disease carriers that inhabit warmer climates could spread toward the poles. This includes human diseases, such as malaria and dengue fever, that are transmitted by mosquitoes (Figure 6). Climate change also affects plant diseases and pests that infest crops and forests. For example, the mountain pine beetle that has destroyed forests in British Columbia is expanding farther north and higher into the mountains.

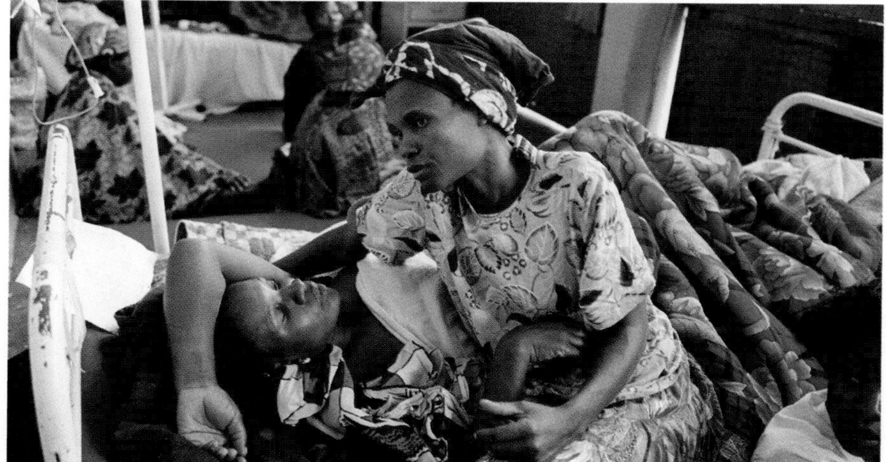

Figure 6 Malaria and dengue fever are common in the equatorial regions of Africa and Asia.

Continental Changes

Many impacts are already happening. They will likely become even more severe in decades to come. Figure 7 describes some possible impacts of climate change in different regions of the world.

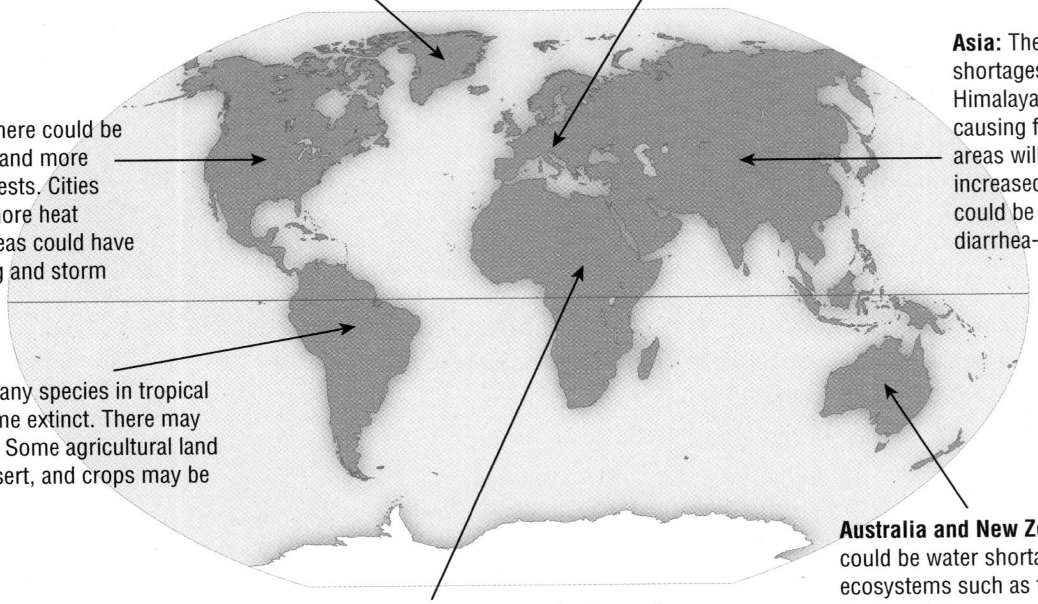

Arctic, Greenland, Antarctic: Glaciers and ice sheets are projected to melt, and sea ice and permafrost to decrease. Species such as polar bears will have trouble adapting. Traditional ways of life in the Arctic may be lost.

Europe: Glaciers are projected to gradually melt, resulting in flooding. There may be more heat waves and more forest fires.

Asia: There could be water shortages. Glaciers in the Himalayas could melt, causing flooding. Coastal areas will be at risk for increased flooding. There could be more diarrhea-type illnesses.

North America: There could be more forest fires, and more insect and plant pests. Cities may experience more heat waves. Coastal areas could have increased flooding and storm damage.

Latin America: Many species in tropical areas could become extinct. There may be more flooding. Some agricultural land could become desert, and crops may be less productive.

Australia and New Zealand: There could be water shortages. Important ecosystems such as the Great Barrier Reef could become endangered.

Africa: There could be more water shortages. Some agricultural land could become desert. Crops may suffer. Low coastal areas where many people live could have increased flooding. Mangrove and coral reef ecosystems could be damaged.

Figure 7 Possible global impacts of climate change projected by the IPCC

RESEARCH THIS CONTROVERSY ABOUT CLIMATE CHANGE

SKILLS: Researching, Analyzing the Issue, Communicating, Evaluating

Most climate scientists agree about the causes and possible consequences of climate change. However, a few skeptics believe that changes in Earth's climate are just part of natural climate cycles. Others agree that human activity is changing Earth's climate but think that the impacts will not be as severe as generally believed.

1. Research what climate change skeptics say about the evidence for climate change and about what is causing this change. This evidence includes retreating glaciers, increasing average temperatures, and rising sea levels.
2. Find responses on the Internet to the claims of these skeptics. Look particularly for responses made by professional, currently practising climate scientists.

GO TO NELSON SCIENCE

A. What do you think about the skeptics' arguments? T/I

B. How do climate scientists counter the claims of skeptics? T/I

C. When scientific studies first linked lung cancer to cigarette smoke, in the 1950s, some scientists argued against these new discoveries. Over the years, it has become accepted knowledge that smoking affects health. T/I A

(a) Comment on whether we should be skeptical about new scientific results when they are first published.

(b) How much evidence should we expect before we make major and costly decisions?

(c) What are the dangers of waiting another 10 years before making decisions about anthropogenic greenhouse gas emissions and climate change?

Special Concern for the Canadian Arctic

Scientists have observed that climate change is happening more rapidly in the Arctic than anywhere else. Under climate projections, this is what is expected due to the albedo effect.

Regional Problems Due to Climate Change in the Arctic

Sea ice is melting and habitats are changing as temperatures across the Arctic rise. These changes have important ecological repercussions for all species. For example, with less ice, it is harder for polar bears to reach their food, the ringed seals that live on the ice (Figure 8). In addition, many people in the Arctic depend on hunting for food and will be affected by changes in animal populations.

Figure 8 Animal species in the Arctic are being affected by climate change.

Other problems are occurring as well. The coastline is eroding as the sea ice melts. Communities that were protected by the ice are now more vulnerable to autumn storms coming in from the ocean. Traditional ways of life are being affected because it is dangerous to travel on melting ice. Permanently frozen soil (permafrost) is beginning to melt, creating sinkholes and shifting the foundations of buildings and other structures.

Possible Benefits of Climate Change in the Arctic

Less sea ice means that it will become easier for ships to reach the Arctic and the valuable natural resources there. In addition, ships could follow much shorter routes by travelling across the Arctic through the Northwest Passage rather than taking longer, more southern routes (Figure 9).

Figure 9 The Northwest Passage has, in the past, been blocked by ice for much of the year. Now it is becoming passable for longer periods.

Trees are beginning to grow in the warmer Arctic climate, helping to absorb carbon dioxide from the atmosphere. In the future, it may become possible for farmers to grow crops at higher latitudes, depending on the soil. However, having trees and crops in the Arctic will reduce the albedo of the tundra (especially in winter) and may result in a net increase in warming.

RESEARCH THIS COMPETITION FOR THE ARCTIC

SKILLS: Researching, Analyzing the Issue, Communicating, Evaluating

SKILLS HANDBOOK
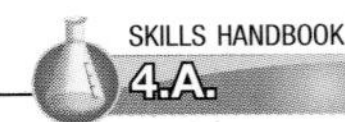

As polar ice melts, many countries are beginning to see possibilities for profit. Rich natural resources may be available to Aboriginal peoples to claim. In addition, new routes for shipping may open up across the Arctic Ocean.

Equipment and Materials: globe; measuring tape

1. Using a globe, research which countries are likely to have a political or economic interest in the Arctic.
2. Measure the distance for a ship to travel from St. John's, NL, Canada, to Magadan, Russia. Assume that the ship can travel through the Panama Canal.
3. Measure the distance for a ship to travel from St. John's to Magadan if there were no polar ice.
4. Research the history of the Northwest Passage.

A. Which countries do you think could claim rights to natural resources found in the Arctic? Explain why. T/I A

B. How could the melting polar ice affect shipping and transportation for Canada and other countries ? T/I

C. (a) How much shorter would the shipping route be between Halifax and Magadan if the Arctic route could be used instead of the Panama Canal route?

(b) What would be the potential impact on trade between Canada and Russia? T/I A

D. Explain why so many explorers in the past attempted to find the Northwest Passage. A

E. What impacts might development of the Northwest Passage have on traditional ways of life? A

GO TO NELSON SCIENCE

How Can Changes in the Arctic Affect the Rest of the World?

Climate change in the Arctic will have major impacts for the rest of the world.

- **Albedo effect:** As Arctic ice melts, the ocean and land will reflect less of the Sun's energy and absorb more. As a result, the Arctic will warm up faster than it would otherwise. Energy absorbed by the Arctic surface will be spread around the world by the climate system.
- **Release of carbon dioxide:** Earth's permafrost may contain more stored carbon dioxide and methane than exists in the atmosphere today. The permafrost in Canada, Alaska, and other parts of the world has already started to melt. If large amounts of carbon dioxide and methane are released by melting permafrost, the greenhouse effect would be further enhanced, and Earth's climate could change much faster than expected.
- **Sea level rise:** More water will flow into the oceans as the Greenland ice sheet and glaciers in Canada, Alaska, and Russia melt.
- **Ocean currents:** Fresh water flowing into the Arctic Ocean from melting ice may slow or even stop ocean currents that transport thermal energy around the globe (thermohaline circulation). Ocean currents affect the climate in many countries.
- **Biodiversity:** Many migratory species have breeding grounds in the Arctic. If the Arctic ecosystems change, this could affect species around the world (Figure 10).
- **Changes in shipping and transportation:** As the polar ice cap melts, ships will be able to travel through the Arctic en route to other locations. The shorter route will save thousands of kilometres and reduce transportation costs and energy use.

DID YOU KNOW?

Climate Change Boomtown
The port of Churchill, Manitoba, on the west coast of Hudson's Bay, may benefit from climate change. The port is likely to see increased shipping with the possible opening of the Northwest Passage. Unfortunately, the reduced ice also means that polar bears wander into town looking for food.

Figure 10 The Arctic is an important habitat for many species of migratory birds.

CITIZEN ACTION

How Can We Protect the Arctic?

Climate change cannot be stopped entirely. The greenhouse gases that we have already emitted will affect Earth's climate for many years to come. However, the impacts of climate change can be reduced.

Think about things you can do in your own community to protect the Arctic from climate change. Put your suggestions into a short article for your local newspaper or for an e-zine. Your article should persuade others that the Arctic should be protected. The suggestions should be practical actions that the average citizen can do.

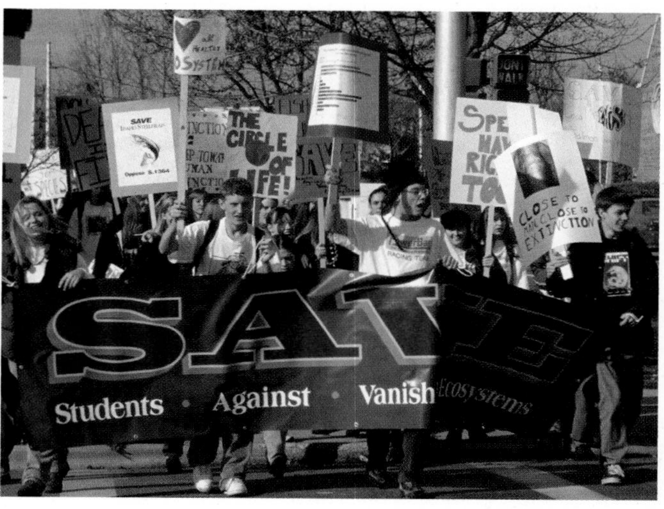

UNIT TASK Bookmark

Think about the impacts of climate change that are mentioned in this section as you work on the Unit Task, page 444.

IN SUMMARY

- The Intergovernmental Panel on Climate Change (IPCC) has summarized the latest scientific research on climate change.
- Changes expected around the world include increased temperatures, shifting precipitation patterns, and a rise in sea level.
- Impacts of climate change will affect human society and the natural environment, including agriculture, ecosystems, and the spread of pests and diseases.
- Climate change is occurring more rapidly in the Arctic than anywhere else.
- Climate change in the Arctic will have economic and ecological repercussions worldwide.

CHECK YOUR LEARNING

1. Choose one of the following impact areas of climate change.
   - rising sea level
   - changes to agriculture
   - changes to ecosystems
   - increased spread of diseases

   Describe how your chosen impact could affect Earth and its people over the next century. A
2. For each of the four climate change impact areas listed in question 1, name a country that might suffer severely. Briefly describe how each country would be affected. T/I
3. Explain how the albedo effect of polar ice means that the largest climate change in Canada will occur in the Arctic and not in southern Ontario. You may need to refer back to Section 8.10. K/U
4. Why is it important to have a strong consensus among climate scientists regarding climate change? A
5. Explain why the melting of permafrost in the Arctic may start another positive feedback loop that further contributes to climate change. K/U
6. Climate change in the Arctic will have major impacts on the rest of the world. Choose one of the six areas below. Describe in several sentences why it will be negative or positive (or both) for the rest of the world. Be prepared to share your thoughts with your peers. A
   - albedo effect and ice
   - release of carbon dioxide
   - ocean currents
   - sea level rise
   - biodiversity
   - navigation

Geoengineering to Combat Climate Change?

Some scientists are trying to figure out ways to use technology on a global scale to address the issue of climate change. This kind of problem-solving is called *geoengineering*: the use of technology to modify Earth's environment. Here are just three of the many geoengineering ideas out there. Do you think they would work?

Mirrors in Space

Mirrors could be used to reflect some of the Sun's radiation back into space to decrease Earth's temperature (Figure 1). To balance out the effect of climate change, it would take 55 000 mirrors orbiting Earth, each 100 km^2 in size. Alternatively, we could imitate a volcano by spraying millions of tonnes of sulfur into the atmosphere. The sulfate droplets would act like tiny mirrors.

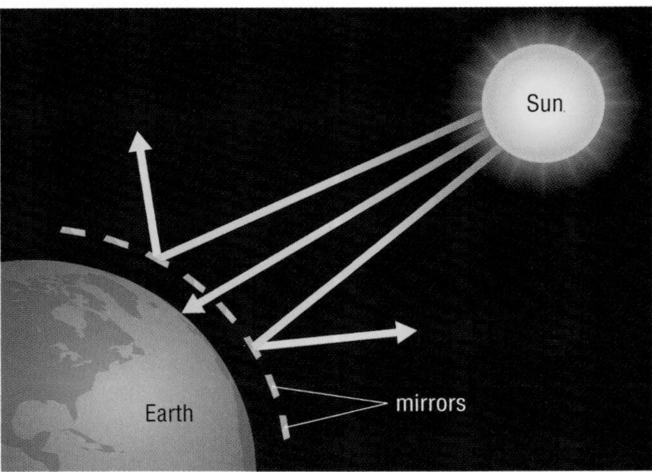

Figure 1 Mirrors in space

Fertilizing the Oceans

During photosynthesis, plants capture carbon dioxide from the atmosphere. Scientists are investigating the effects of adding fertilizers to the ocean to increase algal growth (Figure 2). The algae would then absorb carbon dioxide from the atmosphere.

Figure 2 Fertilizing the oceans

Farming Algae

Bags, vats, or tubes of algae can be used to absorb the carbon dioxide produced by power plants and other factories (Figure 3). The algae can be processed into fuel.

Figure 3 Farming algae

Risks of Geoengineering

There are many ways of using technology to counter climate change. However, intentionally changing Earth's climate system could be a very bad idea. Why? Here are just a few of the reasons why geoengineering is a big risk:

- We cannot predict all the consequences of changing Earth's climate. A geoengineering project could have enormous side effects and cause great harm. For example, sulfate droplets in the atmosphere are known to damage the ozone layer and to cause acid rain.
- The projects may not work out as expected. For example, research shows that dumping fertilizer into the ocean may not result in as much carbon dioxide removal as hoped. And as soon as you stop putting in the fertilizer, the ocean stops absorbing carbon dioxide.
- Geoengineering might cause people to assume that climate change is "being taken care of." People might stop reducing their emissions of greenhouse gases.

Geoengineering cannot fix the problems of climate change all by itself. However, it could provide us with a backup plan if Earth's climate begins to change even more suddenly than expected.

GO TO NELSON SCIENCE

Impacts of Climate Change on Ontario

Canada and other high-latitude countries have already experienced greater temperature increases than low-latitude countries. Canada could therefore expect greater warming in the future than many other countries. For example, with moderate greenhouse gas emissions by 2100, Ontario's average temperature could increase by 3 to 6 °C in the winter and 4 to 8 °C in the summer.

What would this feel like? A recent study shows that the climate of southern Ontario has already become similar to the climate of upstate New York just 20 years ago. By the end of the century, Ontario's winters might match those in Pennsylvania today, and our summers might be as hot and humid as northern Virginia's (Figure 1).

DID YOU KNOW?

Small Difference—Big Effect
Small changes in Earth's average temperature can cause large changes in climate. The difference between today's average temperature and that of the last ice age is estimated to be less than 10 °C. If we continue consuming fossil fuels, Earth's temperature could increase by a similar amount within the next 100 years.

Figure 1 In 100 years, Ontario's summers might be as warm as Virginia's summers are today.

Temperature and Precipitation

Some of the changes to Ontario's climate could be considered positive. Our winters will probably become warmer, with fewer extremely cold days and less snow (Figure 2). As a result, heating costs may go down, and it will be easier to keep the roads clear of snow and ice. There may be less ice coverage on the Great Lakes, leading to longer shipping seasons.

There are likely to be more negative than positive changes, however. For example, there would be more extremely hot, humid days in the summer. There will also be many more heat waves. Both of these affect human health (see page 421).

Figure 2 Warmer winters may be good for some people, but will affect winter sports such as snowboarding and skiing.

Precipitation patterns in Ontario are likely to shift in the future. There will likely be more rainfall overall, but some areas will become drier, whereas others will become wetter. The frequency of heavy rainstorms will probably increase. Rain events will probably be heavier, with long dry spells between rainstorms.

Changing Lake Levels

The level of Lake Superior has dropped significantly over the last few years. Scientists have not yet positively identified the cause, but climate change is likely to be a contributing factor. The other Great Lakes have not changed as significantly, but changes may occur as temperatures increase. Higher temperatures mean that there will be less ice cover on lakes in the winter. More water will evaporate year-round as the air warms. However, increases in precipitation may counter some of the losses due to evaporation.

As the lake water warms, fish that live in cold water, such as trout, may migrate north or die out. Algae would grow faster in warm weather, resulting in foul-smelling water and beaches when the algae die and decompose. Invasive species such as zebra mussels and lamprey could increase in numbers and disrupt local ecosystems.

Ecosystems

Studies are now examining how Canadian ecosystems are adapting to current changes. Some tundra plants are flowering earlier and reproducing faster in northwest Newfoundland and Labrador. Figure 3 shows how ecoregions in Canada would change if atmospheric carbon dioxide concentrations were doubled and if climate were the only factor involved. Factors such as soil type also determine where plants and animals can live. This makes it difficult to predict which species will survive and which will become extinct.

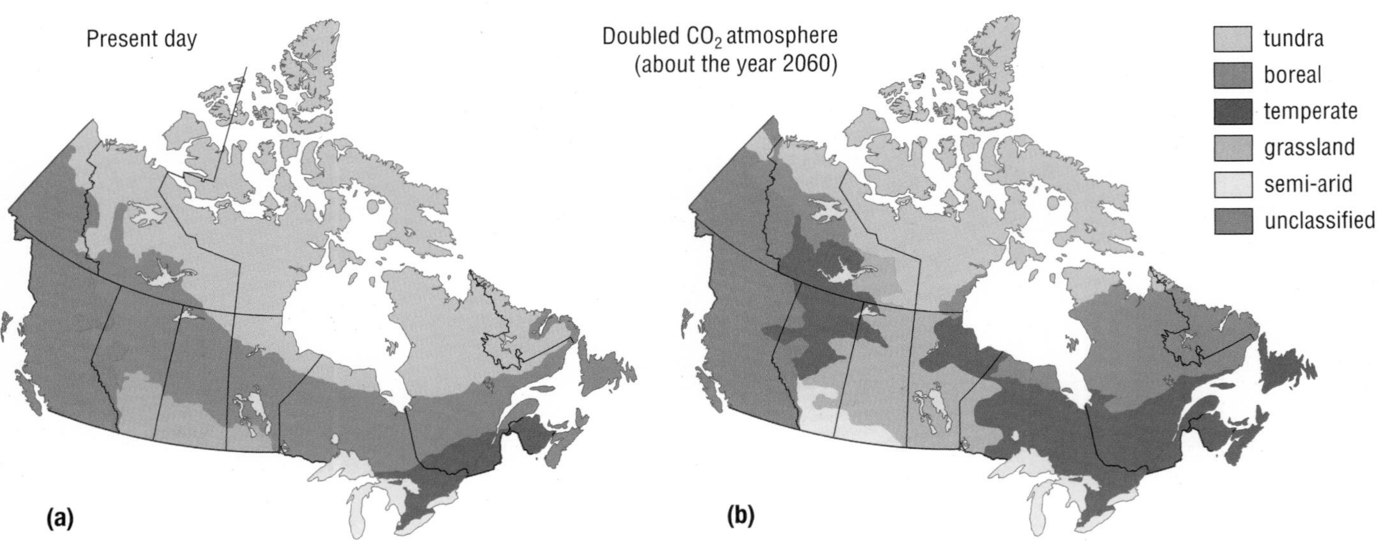

Figure 3 Ecoregions will move northward as carbon dioxide concentrations and temperatures increase.

Figure 4 Kudzu is an invasive plant that grows over almost anything. It may arrive in Ontario as the average winter temperature rises.

Some animals in southern Ontario, such as white-tailed deer and cardinals, can adapt to or even benefit from higher temperatures. Plants and animals that thrive in colder temperatures, including black spruce and moose, are likely to migrate farther north over time.

Plants and animals from the United States could move north into Canada as temperatures rise. These new species would disrupt existing ecosystems. Some local species could become endangered or extinct.

Kudzu is a fast-growing vine that spreads over buildings, trees, power lines, and anything else in its way (Figure 4). The southern U.S. states spend millions of dollars each year tearing kudzu off buildings. Currently, kudzu is limited to the United States because it cannot survive the cold Canadian winters. If winters become warmer, kudzu could move northward into Canada.

Disease and Illness

Disease-carrying organisms may increase as average temperatures rise and precipitation patterns change. The risk of West Nile virus, already appearing in Ontario, could increase. Lyme disease, which is carried by deer ticks and causes fever and a skin rash, could become a worse problem in Ontario (Figure 5).

Epidemiology is a field of medicine that deals with the incidence and distribution of diseases. To learn more about the work of an epidemiologist,

GO TO NELSON SCIENCE

Figure 5 (a) Lyme disease is caused by a bacteria spread by deer ticks. (b) The ticks are carried by deer, squirrels, and other small mammals.

Worsening heat waves are likely to increase heat-related illnesses and deaths. For example, heat stress often sends people to the hospital and can cause strokes. In addition, car exhaust reacts with sunlight to produce ground-level ozone, a component of smog. This reaction happens faster in warmer air, so smog will worsen as temperatures increase. Increasing air pollution would be bad for people who have asthma or other respiratory illnesses.

DID YOU KNOW?

Smog Gets Worse

Although smog is not caused by climate change, increased temperatures will make it worse. Smog is a serious concern in cities around the world. The annual "cost" of smog for Ontario is $10.8 billion. In 2005, smog contributed to the premature death of 5800 people in Ontario, and resulted in 17 000 additional hospital visits. By 2015, smog could cause 10 000 premature deaths per year.

Agriculture

Spring would come earlier as the climate warms, and the growing season for crops and other plants would lengthen. Some crops, such as soybeans and corn, could benefit from warmer temperatures and increased carbon dioxide concentration in the atmosphere. However, these factors could also encourage the growth of unwanted plants. Farmers may need to use more herbicides than they do now. An increase in smog would also damage agricultural crops.

Southern Ontario farmers might be able to grow fruits and vegetables that normally grow farther south. In Yukon and the Northwest Territories, farming may become more viable, depending on the soil. Land in the Arctic could become useful for agriculture as the permafrost melts.

To learn more about how climate will impact Ontario's agriculture,

GO TO NELSON SCIENCE

Forests

Studies show that the rain in Ontario could occur in shorter, heavy bursts, with long dry spells in between. Summers are expected to be hotter and drier, resulting in more forest fires. Insect pests could migrate northward, attacking the southern fringes of our forests. Southern plants could survive the warmer winters and increase in numbers.

Close to 50 % of the land area of Ontario is currently covered by boreal forest (Figure 3). This area could shrink as the ideal climate for Canadian forests shifts northward. As our forests become less healthy, they may become a carbon source rather than a carbon sink. The invasion of the pine beetle has already caused the forests in British Columbia to become carbon sources.

READING TIP

Synthesizing

After reading a text, a powerful way of synthesizing is to draw a conclusion about what you have just read:
What I already know about the concept + What I read in the text = What I think now about the concept.

Electricity Use

As summers get hotter, we will use more electricity for air conditioning. Generating electricity from coal or natural gas produces greenhouse gases, which makes the problem of climate change worse. In the winter, energy use may decrease due to warmer weather.

In Ontario, about a quarter of our electricity is produced using hydroelectric power, which does not produce greenhouse gases. During heat waves, however, people might try to use more electricity than can be produced, leading to blackouts (Figure 6). In addition, less hydroelectricity will be available if the lake levels drop because of climate change. This could increase our use of fossil fuels.

Figure 6 Blackouts could occur more often as Canadians use more electricity to cool their homes and offices.

In the next section, you will learn how actions by individuals, businesses, cities, and governments can help prevent the most severe impacts of climate change from occurring in the Arctic and elsewhere.

IN SUMMARY

- Changes expected to Ontario's climate include warmer winters and more extremely hot days in the summer. Precipitation patterns are also likely to change.
- Ontario's boreal forest will likely shrink, and plant and animal species will migrate northward.
- Other possible effects include changing lake levels, higher risk of insect-borne diseases and heat-related illnesses, a longer growing season, increased energy use in summer, and decreased energy use in winter.

CHECK YOUR LEARNING

1. Suppose you are a climate researcher in northern Ontario. You want to study the effects of climate change on local ecosystems. List three factors you could study in the area to monitor the effects of climate change. K/U
2. (a) Create a table headed "Risks" and "Benefits" to compare the positive and negative impacts of climate change for Ontario.
   (b) Some people say that Canada could benefit from a warming climate. Write a short paragraph giving your opinion on this. Refer to some of the points in your table. C A
3. How is climate change likely to affect Ontario's agriculture? K/U
4. List several species that might migrate away from southern Ontario as our climate changes and several species that might migrate into southern Ontario. K/U
5. List four factors, related to climate change, that might negatively affect Ontario's forests. K/U
6. A warmer climate may lower lake levels, which would reduce the capacity to generate hydroelectricity, which would increase electricity production from fossil fuels, which would lead to a warmer climate. K/U A
   (a) Is this a negative or a positive feedback loop?
   (b) What can the Ontario government do to break this loop?

Taking Action to Limit Climate Change

Climate change is already occurring, and scientists know that some additional climate change is inevitable. Our climate will continue to change this century because of greenhouse gases that we have already emitted. How much the climate changes will depend on decisions that we make now and in the next few decades. There are many actions we can take to keep our climate from changing too drastically. Actions taken to reduce unwanted change are referred to as **mitigation**.

mitigation reducing an unwanted change by deliberate decisions and actions

Deciding How Quickly to Reduce Greenhouse Gas Emissions

Global temperatures have already risen by 0.74 °C over the last 100 years. Most scientists agree that we should limit increases to 2 °C to avoid the most dangerous impacts of climate change, including drastic sea level rise and species extinctions. How do we limit temperature increases to 2 °C? Climate models suggest that atmospheric concentrations of greenhouse gases must stabilize at no more than the equivalent of 450 ppm of carbon dioxide. Even staying at this concentration gives us only a 50 % chance of limiting warming to 2 °C.

How much would we have to cut our pollution emissions to reach this goal? To have a chance of stabilizing greenhouse gas concentrations at 450 ppm by 2050, all industrialized nations would have to cut their annual greenhouse gas emissions by 80 % relative to their emissions in 1990 (Figure 1). Developing countries would need to begin reducing their emissions 10 to 20 years after industrialized nations began their reductions. Ontario's provincial government has already committed to this 80 % reduction. The Canadian federal government has not.

READING TIP

Synthesizing

Synthesizing means joining different things together to make something new. Think of combining 400 individual pieces to complete a jigsaw puzzle. Or imagine being a crime scene investigator and combining a variety of clues to solve a crime. When synthesizing a text, look for ways to combine clues to understand the main idea.

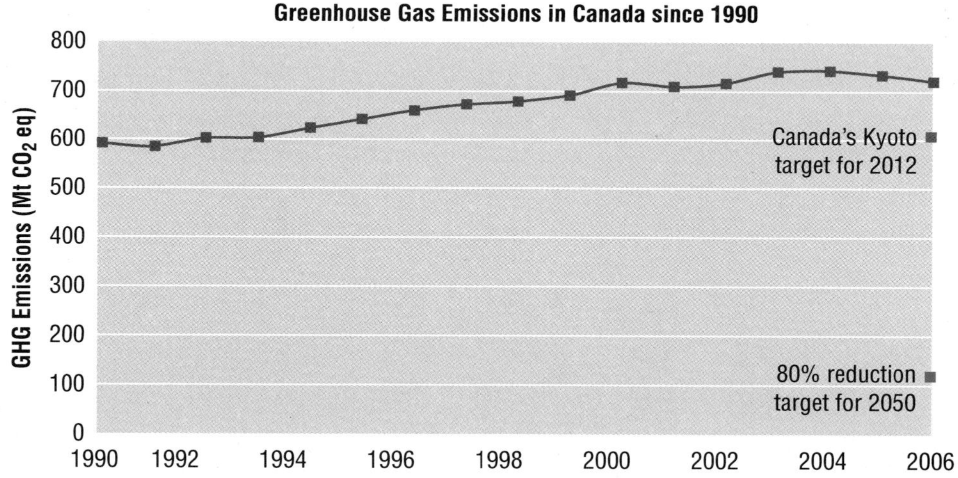

Figure 1 Canada's greenhouse gas (GHG) emissions since 1990

The Canadian Government and Climate Change

The Canadian government is working with many governments around the world to reduce greenhouse gas emissions. In 2002, Canada agreed to join the **Kyoto Protocol**. This international treaty set short-term mitigation goals. It requires industrialized countries to reduce their emissions of greenhouse gases to specific concentrations by 2012.

Kyoto Protocol a plan within the United Nations for controlling greenhouse gas emissions

In December 2007, the United Nations Climate Change Conference in Bali announced a plan. They decided to develop a new international agreement to limit the impacts of climate change. This new agreement was the topic of the Framework Convention on Climate Change in Copenhagen, Denmark, in December 2009. About 170 nations participated. The Copenhagen Agreement will take effect in 2013, the year after the Kyoto Protocol expires.

About 70 % of the Canadian public supported the Kyoto Protocol. There was also some vocal opposition. The government of Alberta protested that thousands of jobs in the fossil fuel industries would be lost (Figure 2). Other industries also have objections.

In 2006, Canada's prime minister announced that we would not meet our Kyoto targets by 2012. Instead, the government issued a Notice of Intent in which it promised to develop a plan to regulate industrial greenhouse gas and air pollutant emissions. Some of the opposition parties, however, saw this as an excuse for not keeping Canada's Kyoto commitments.

In April 2007, the government unveiled its Turning the Corner Plan to reduce greenhouse gases and air pollution. In March 2008, after consulting with environmental and industrial groups, the federal government announced further details of its greenhouse gas regulations. The government claims that it is now putting into place one of the strongest climate change programs in the world. The focus of its program is an absolute 20 % reduction in greenhouse gases from the year 2006 to 2020. The plan also mentions a reduction of 60 % to 70 % by 2050, although this is not emphasized.

The international community is criticizing Canada for not living up to our Kyoto commitments. How do Canada's efforts to improve compare with the efforts of other countries? In 2008, a comparison of the 57 largest greenhouse gas producers was released. Canada's efforts at reduction ranked second-last.

In June 2008, the opposition parties in the Canadian government passed legislation (opposed by the minority government) requiring Canada to cut greenhouse gas emissions by 80 % by 2050 (Figure 3). This legislation requires Canada to do its part to mitigate the most serious impacts of climate change. As you learned in this section, scientists consider that a reduction of 80 % by 2050 is necessary to keep Earth from warming up by more than 2 °C. As this textbook is being written, the new legislation has not yet been implemented.

Figure 2 The Alberta oil sands provide a large financial return and many jobs. Extracting, processing, and using fossil fuels from the oil sands also produce enormous quantities of greenhouse gases.

Figure 3 Opposition parties supported the *Climate Change Accountability Act.*

The European Union is aiming to limit global temperature increases to 2 °C. Canada's target, an 80 % reduction, only gives a 50 % chance of keeping the global increase below 2 °C. To ensure that warming is limited to 2 °C, the European Union may need to reduce its emissions by even more than 80 %.

To find out more about Canada's current position on reducing greenhouse gas emissions,

Provincial and Municipal Governments and Climate Change

The government of Ontario introduced Go Green: Ontario's Action Plan on Climate Change in August 2007. This action plan includes the following:

- the decision to stop burning coal at Ontario's four remaining coal-fired generating stations by 2014
- targets to reduce Ontario's greenhouse gas emissions by 6 % by 2014, 15 % by 2020, and 80 % by 2050
- a public transportation plan for 902 km of new or improved rapid transit routes in the Greater Toronto Area and Hamilton (MoveOntario 2020)
- a fund to support green technologies and businesses in Ontario
- the planting of 50 million trees in southern Ontario by 2020
- a plan to work with a panel of leading scientists and environmental experts who will make recommendations on how Ontario can adapt to climate change
- legislation to fast-track the approval of renewable energy projects, such as wind turbines, rather than putting them through long municipal processes for approval

Between 2004 and 2006, Ontario's greenhouse gas emissions decreased in most sectors of the economy, but overall they were still 7 % higher than in 1990 (Figure 4).

To learn more about Ontario's Action Plan on Climate Change,

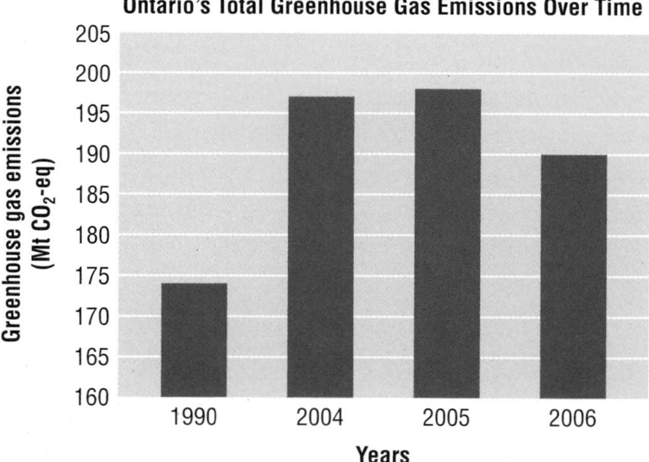

Figure 4 The decrease in greenhouse gas emissions between 2004 and 2006 is primarily due to a reduction in the use of coal-fired generating plants and the mild winter in 2006 that reduced natural gas use.

Local governments in Canada have started to conserve energy and reduce greenhouse gas emissions.

- The City of Toronto is planning a 6 % cut in greenhouse gas emissions (relative to 1990 levels) by 2012. This amounts to a 30 % cut by 2020, and an 80 % cut by 2050.
- The City of Calgary is achieving its target of 6 % below 1990 concentrations ahead of time. It is committed to reducing emissions by 50 % by 2012.
- The City of Halifax has introduced a composting program to reduce greenhouse gas emissions (Figure 5).

Figure 5 The Hatch composting facility in Halifax, Nova Scotia.

RESEARCH THIS GREENHOUSE GAS EMISSIONS IN YOUR COMMUNITY

SKILLS: Researching, Identifying Alternatives, Analyzing the Issue, Defending a Decision, Communicating

1. Research what projects your municipal government is engaged in to reduce greenhouse gas emissions. Examine the government website and look at their publications.
2. Find out how people in your municipality feel about the local projects and whether they have any other ideas.

A. Describe three projects being undertaken by your municipal government to reduce greeenhouse gas emissions.

B. What are the costs of these projects to the average person or residence in your community? T/I A

C. Do people in your community agree with what is being done? Do they feel that the projects are effective? Are they willing to pay the additional costs? A

D. Imagine you were running for office in your municipality in the next local election. Develop a platform regarding greenhouse gas emissions in your community. Would you expand what is already happening, reduce it, or continue it as is? A

E. Present your platform to the class as a short skit, in which the candidate presents their platform to a "citizen" at the door. Alternatively, create a short campaign video, song, or poster. C

What Canadian Businesses, Industries, and Governments Can Do

Some specific actions that can be done in Canada to reduce our emissions of greenhouse gases are listed below. Compare this list with Table 1 in Section 9.5.

TRANSPORTATION

- Use less fuel by driving more efficiently (less idling, less harsh acceleration and braking).
- Use fuels that produce fewer or no greenhouse gases.
- Use hybrid (part gasoline-powered, part electric) or electric vehicles (Figure 6).
- Drive less and increase travel by rail, public transit, cycling, and walking.
- Support government-imposed restrictions on pollution levels or incentives for lower emissions and better fuel efficiency.

Figure 6 Several car companies now make hybrid cars.

PRODUCING ENERGY

- Improve energy efficiency through new technologies.
- Use clean energy sources, such as wind and solar (Figure 7).

Figure 7 Wind turbines do not produce greenhouse gases.

INDUSTRIES

- Use more efficient equipment to consume less energy.
- Recycle energy (e.g., by capturing thermal energy produced by industrial processes and using it to power other processes).
- Capture and store carbon dioxide released by smokestacks (Figure 8).
- Impose taxes and limits on fossil fuel use.

Figure 8 Devices in smokestacks can remove carbon dioxide from industrial emissions.

BUILDING AND CONTRACTING

- Increase energy efficiency (Figure 9).
- Install better insulation.
- Introduce rebates or tax incentives for insulating and energy efficiency.

AGRICULTURE

- Restore polluted land so that it absorbs more carbon than it emits (Figure 10).
- Study and implement ways to reduce methane emissions in rice and cattle farming.
- Use less nitrogen fertilizer to reduce nitrous oxide emissions.

WASTE MANAGEMENT

- Collect methane from landfills and use it for energy.
- Compost all organic waste so it does not go to a landfill.
- Reduce the volume of waste generated.
- Reduce consumption by avoiding the purchase of unnecessary materials.
- Recycle (Figure 11).

FORESTS

- Plant more forests and replace trees that are cut down (Figure 12).
- Reduce deforestation.
- Manage forests carefully so that they are carbon sinks, not sources.
- Produce and use fuels from waste forestry products (biofuel) instead of fossil fuels.

Industries and businesses in Canada are also taking steps to reduce their emissions of greenhouse gases. Among others, Alcan, DuPont, General Motors, IBM, the Mining Association of Canada, and the Canadian Chemical Producers' Association have all made substantial cuts in their greenhouse gas emissions.

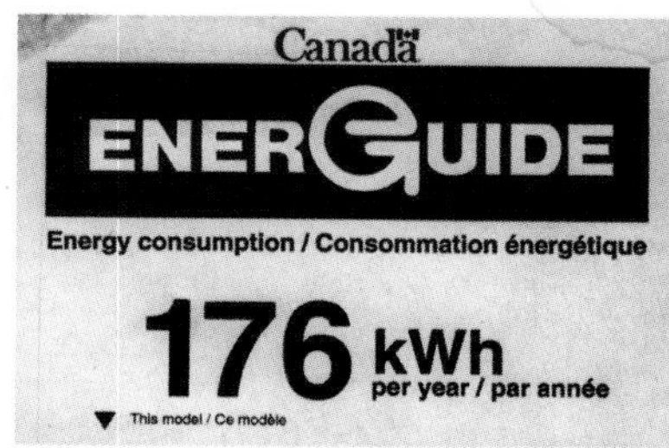

Figure 9 New appliances have an energy consumption label.

Figure 10 Planting native plants can help restore polluted land while removing carbon dioxide from the air.

Figure 11 Recycling reduces the amount of waste going to landfills.

Figure 12 Replacing trees when they are cut down helps maintain the amount of carbon being absorbed.

Adapting to a Changing Climate

In this chapter you have encountered many ways in which you can help reduce the most severe impacts of climate change. Even if we take all these steps, some changes in climate will still happen. It is important to plan for climate change that is going to happen. Planning how to deal with future climate changes is called "adaptation." It is a sensible approach for any person or organization to take.

City governments are deciding how to deal with increased heat waves and flooding in the future. Forest managers are planning how to handle more frequent wildfires and insect infestations. People who work with protected species are considering ways to help species migrate and cope with higher temperatures. Researchers are studying how farmers can grow crops that survive higher temperatures.

We can limit the most severe impacts of climate change by taking action to reduce our emissions of greenhouse gases. In the next two sections you will learn more about appropriate actions and when we should take these actions. However, we must also prepare for expected changes in climate.

UNIT TASK Bookmark

How can you use this information on large-scale steps to reduce climate change as you work on the Unit Task described on page 444?

IN SUMMARY

- To avoid the most dangerous impacts of climate change, we should aim to limit increases in global temperature to 2 °C. This requires limiting greenhouse gas concentrations to 450 ppm.
- Industrialized nations will have to cut their greenhouse gas emissions at least 80 % by 2050 to meet the limit of 450 ppm.
- In 2002, the Canadian government joined the Kyoto Protocol, agreeing to reduce emissions by 6 % of 1990 levels. Since then, our emissions have continued to increase, and Canada will not meet its 2012 targets.
- In 2007, legislation was introduced that calls for reducing greenhouse gas emissions 80 % from 1990 levels by 2050.
- Canadian businesses, industries, and governments can reduce emissions by switching to clean energy sources, reducing emissions from cars and airplanes, conserving energy, and managing farms and forests better.
- We can adapt to climate change by planning and preparing for changes.

CHECK YOUR LEARNING

1. Why is some change in our climate inevitable? K/U
2. Why is it important that the world not put all of its resources into adapting to climate change but use a large part of its resources for mitigation of climate change? T/I
3. For each of the following categories, state one action that will help reduce the impacts of climate change: K/U
   (a) transportation
   (b) energy production
   (c) energy conservation
   (d) forests
4. Scientists believe there is a maximum temperature increase that we must not allow the Earth to pass. K/U
   (a) Earth's temperature has already risen by 0.74 °C. To what maximum temperature rise do these scientists hope to limit Earth?
   (b) To keep Earth within this temperature rise, what is the maximum concentration of greenhouse gases we must allow by the year 2050?
   (c) To keep the maximum concentration to the number you mentioned in part (b), by how much must the world reduce its annual production of greenhouse gases by the year 2050?
5. Adapting to climate change and mitigating further climate change will be very costly and controversial. Nevertheless, it needs to be done. C A
   (a) Suggest several actions that our federal, provincial, and municipal governments can take before they bring in any further, costly legislation to deal with climate change.
   (b) When the time comes to act, will governments be able to do it in a way that satisfies everyone? Include people who depend on fossil fuel production and consumption for their livelihood. Give your own suggestions.
6. Some changes in climate are inevitable. Therefore, people are beginning to plan on how to adapt to a changing climate. K/U A
   (a) List four ways in which people and governments are preparing to adapt to climate change.
   (b) Think of two additional examples, not mentioned here, of things people might have to do to adapt to a changing climate in the next 30 years. (Hint: Think about recreation, tourism, outdoor work, etc., in summer and winter.)

What Can Individuals Do?

10.5

Most of the impacts of climate change are expected to occur to some degree even if humans quickly switch to clean energy sources. The risk of dangerous impacts such as extreme sea level rise and massive species extinctions increases if we take longer to make changes. The main reason to take action is that not taking action will have dangerous consequences for us all. Not taking action could harm millions of people. It is easy to make lists about what actions large groups should be taking. We should also look at what each of us can do. What is our role, as individuals, in reducing climate change?

Reducing Your Emissions

More than one-third of Canada's greenhouse gas emissions come from the activities of individuals. Most of the greenhouse gas emissions come from transportation and home heating and cooling (Figure 1). This is good news because it means that you, your friends, and your family can make a genuine difference by reducing your own emissions of greenhouse gases.

There are many ways to conserve energy and reduce your greenhouse gas emissions. Here are just a few. Many of them will even save you money.

- Walk, bike, take public transport, or carpool to school.
- If buying a car, choose the most fuel-efficient one you can afford. Maintain it regularly to keep it running efficiently.
- Switch off lights and unplug appliances, such as computers and televisions, which use electricity on standby when they are not in use.
- Fly less, to reduce the amount of pollution (including greenhouse gases) emitted by airplanes and by the production of airplane fuel.
- Use air conditioners and heating only when necessary. Use a programmable thermostat to turn down the heat when you do not need it.
- Plant trees that are native to your area. Trees absorb carbon dioxide, and native trees will not harm local ecosystems.
- Turn off the water when you brush your teeth or shave. Take short showers. Conserving water saves energy because fossil fuels are used to heat the water.

What else could you add to this list?

Personal Greenhouse Gas Emissions from Energy Use in Canada

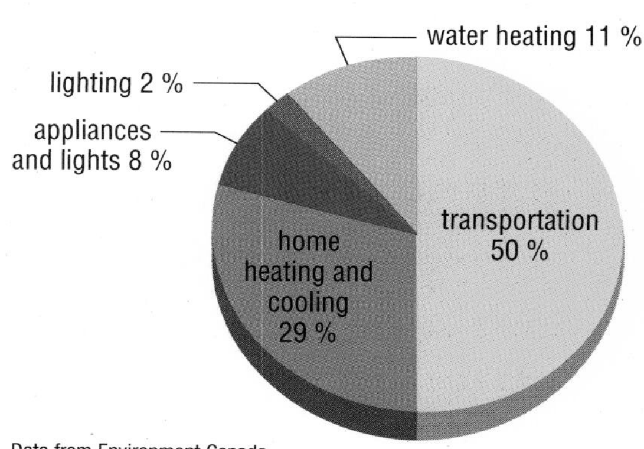

Data from Environment Canada

Figure 1 Personal greenhouse gas emissions from energy use in Canada

READING TIP

Existing Knowledge

You have a treasure chest of knowledge stored in your memory. Use it to make connections with ideas or information in a text. Ask yourself what you already know that relates to the text. For example, for the topic of energy conservation, has your family made a change, recently, to reduce their energy use?

RESEARCH THIS BUYING ENERGY STAR® APPLIANCES

SKILLS: Researching, Identifying Alternatives, Communicating

SKILLS HANDBOOK 4.A.7.

Buying energy efficient appliances can save money and reduce greenhouse gas emissions. In Canada, qualified ENERGY STAR® products must meet technical specification to ensure that products are energy efficient.

1. Research the ENERGY STAR® program.
2. Research what problems the disposal of old appliances might cause and where old appliances are disposed of.

GO TO NELSON SCIENCE

A. Write a brief (one page) description of the ENERGY STAR® program. T/I C
B. What ENERGY STAR® rebates and incentives are available to Ontario residents? T/I
C. What options are available to dispose of old appliances? What environmental problems might they cause? T/I A
D. What ENERGY STAR® appliances do you have in your home? K/U

RESEARCH THIS EATING LOCAL

SKILLS: Defining the Issue, Researching, Communicating, Evaluating

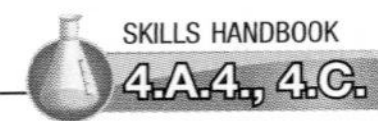

Preparing and transporting foods consumes a great deal of energy. Eating local produce and eating fewer processed foods can help save energy.

1. Research the benefits of eating local produce. Look for information on the methods and energy used to transport produce and other foods from other countries.
2. Research the benefits to the environment of growing and eating organic food.
3. Go to a grocery store and look at a frozen packaged meal. Examine the components and the packaging. See if the package says what city the meal comes from.
4. Research Community Supported Agriculture (CSA). Locate at least one CSA farm near you.

A. Some people try to "eat local." What issue are they addressing when they make this choice? T/I

B. Write a brief description of the energy that may have gone into processing, packaging, and transporting the frozen packaged meal you examined. Compare it with the energy used in cooking a similar meal at home, using local foods. T/I

C. Evaluate the decision to "eat local." Explain your evaluation. A

D. How might switching from conventional farming to organic farming give local ecosystems a better chance of adapting to changes in climate? K/U A

E. Would you consider buying produce from a CSA farm? Explain why or why not. T/I A

GO TO NELSON SCIENCE

Carbon Offset Credits

Even committed environmentalists such as David Suzuki say that it is impossible to reduce our greenhouse gas emissions to zero, no matter how hard we try. However, we can achieve "carbon neutrality" by purchasing carbon offset credits. When we purchase credits in clean energy programs such as new wind energy farms or solar power, we reduce the burning of fossil fuels (Figure 2). In this way, we *offset* carbon dioxide (and other greenhouse gas) emissions. Similarly, when we purchase credits for reforestation in Brazil, we are increasing carbon sinks; thus, we help offset the increase in atmospheric carbon dioxide.

To learn more about carbon offset credits,

GO TO NELSON SCIENCE

Figure 2 Buying credits in renewable energy sources such as (a) solar power or (b) reforestation projects can offset your greenhouse gas emissions.

You may hear criticisms of the carbon offset program. Some people feel it is a way for rich westerners to ease their conscience and continue their fossil fuel–intensive lifestyle. Many organizations suggest that we should purchase carbon offset credits only after we have made a strong effort to minimize our carbon footprint. Others believe that organizations receiving carbon offset credit payments do little with the money. For these reasons, it is important to deal with certified carbon offset organizations.

CITIZEN **ACTION**

Everyone Can Make a Difference

Work with a partner or in a small group to brainstorm specific actions that you, your school, or your community can take to mitigate or adapt to climate change. Your suggestions must be practical. Present your actions on a poster or as a presentation in your school or in a local library or shopping mall.

Try to persuade others to join you in taking action. Consider putting on a skit, developing a performance art piece, or creating a video and posting it online.

GO TO NELSON SCIENCE

Climate Change and Stewardship

Stewardship is an old concept that is coming back into focus in Ontario. This term means the careful management of something that one does not own. We now realize that Earth has not been given to us to exploit for ourselves but to manage responsibly for our children. One of the worst inheritances we could leave to future generations would be an Earth spoiled by our extravagance.

The Ontario Ministry of Natural Resources (MNR) has an interesting stewardship program for 17-year-old students. Every summer, they employ "Ontario Stewardship Rangers" to work on a variety of projects. These include conducting forest research, planting trees, and studying species at risk (Figure 3). The MNR also supports a volunteer organization for adults to get involved in caring for Ontario's environment and dealing with climate change. This organization is called Ontario Stewardship.

To learn more about Ontario stewardship programs,

GO TO NELSON SCIENCE

Figure 3 Ontario Stewardship Rangers may rehabilitate wetlands, carry out ecological research, or create wildlife habitats.

You read above that stewardship is not a new idea. In fact, it has been part of many cultural, traditional, and religious beliefs throughout human history. If you or your family is affiliated with a particular tradition, you might like to find out what viewpoint it takes with regard to climate change and stewardship.

Ultimately, how we think about our environment concerns values. Ontario's science curriculum says, "Values that are central to responsible stewardship are: using non-renewable resources with care; reusing and recycling what we can; switching to renewable resources where possible" (*The Ontario Curriculum: Grades 9 and 10 Science, 2008 (revised)*, page 5).

DID YOU **KNOW?**

Deforestation in the Amazon Rainforest

Much of Earth's deforestation is taking place in the Amazon rainforest as people clear land for farming. We could protect the rainforest and reduce further emissions by providing support to developing nations in the Amazon region.

To take the One Less Tonne challenge and learn how you can reduce your greenhouse gas emissions,

GO TO NELSON SCIENCE

The impacts of climate change may seem too large or global to be affected by your personal choices, but they are not. Choices made by you, your family, your school, and your community can make a big difference. You can help limit further climate change by reducing your emissions of greenhouse gases. There are important reasons why you should take action. Some are discussed below.

Figure 4

TO PROTECT PEOPLE'S HEALTH

A warming climate would increase smog levels in cities. By reducing emissions of greenhouse gases, you can improve air quality and prevent further pollution (Figure 4). Switching to non-motorized activities (canoeing instead of power boating, and walking or biking instead of driving) is good for the planet and for your health. In addition, eating less meat and fewer packaged foods is a good choice.

Figure 5

TO SAVE MONEY

Adapting to a different climate is likely to be very expensive for cities, governments, and individuals. Reducing emissions of greenhouse gases can be costly in the short term but could save money in the long term by preventing extreme climate change (Figure 5). Also, being energy efficient saves money and reduces emissions.

Figure 6

TO IMPROVE YOUR CITY OR TOWN

Many of the actions that reduce greenhouse gases have additional positive results for our cities and our health. For example, reducing energy consumption—perhaps by developing communities that require less car travel—could result in less traffic, less smog, and less risk of electricity blackouts (Figure 6).

Figure 7

TO PROTECT TRADITIONAL ACTIVITIES

As plant and animal species migrate into new areas, cultural activities such as hunting and fishing will change (Figure 7). Reducing climate change protects these traditional activities. It also protects other winter activities, such as skiing, and even some summer activities since it could become much hotter in the summer.

Figure 8

TO PROTECT THE ENVIRONMENT

Reducing climate change will discourage invading species that prefer warmer temperatures. This will help protect native plants and animals in your region (Figure 8).

UNIT TASK Bookmark

How could you use information about individual reductions in greenhouse gas emissions as you work on the Unit Task, described on page 444?

IN SUMMARY

- There are many actions that individuals can take to help limit the impacts of climate change.
- You can change your mode of transportation, alter your buying habits, or take action, such as planting trees in your neighbourhood.
- Some reasons to reduce the impacts of climate change include protecting the environment, protecting human health, saving money, improving your community, and protecting traditional activities.

CHECK YOUR LEARNING

1. (a) What one activity does the average Canadian do regularly that produces half of the individual greenhouse gas emissions in Canada?
   (b) Think of four ways that you and your family can reduce greenhouse gas emissions in this activity. K/U A
2. (a) What activity at the individual level produces the second greatest amount of greenhouse gases for Canadians?
   (b) Describe how you could reduce your greenhouse gas emissions in this activity in both summer and winter. K/U A
3. In a journal, record all the activities you engage in or see in your daily life that cause greenhouse gas emissions. With your friends, make a list of suggestions on how to reduce emissions in each of these ways. Prioritize them. A C
4. (a) If you or your family is affiliated with a particular cultural, traditional, or religious perspective, find out what its viewpoint is with regard to climate change and stewardship. Do you agree with this point of view?
   (b) Ask your parents, family members, or guardians what the concept of stewardship means to them. Is stewardship a useful idea as regards their attitude toward reducing greenhouse gas emissions? A C
5. Describe how reducing greenhouse gas emissions could improve your standard of living. A
6. (a) How does tree planting reduce the impact of climate change?
   (b) What trees are native to the area where you live? How much would it cost to plant a young native tree around your house? K/U T/I
7. (a) Write a few sentences describing what carbon offset credits are and why we should use them.
   (b) What are two concerns with carbon offset credits? K/U A
8. Suppose that your family is purchasing a new appliance, such as a refrigerator, stove, or dish washer. You can purchase one model for \$400, or you can purchase a second model almost identical to the first, but much more energy efficient, for \$600. The energy-efficient model will save your family \$25 in electricity each year. How many years would it take to recover the additional cost of purchasing the energy-efficient model, assuming that the additional capital (\$200) is loaned to you interest-free for these years? T/I
   (a) if the cost of electricity stayed the same?
   (b) if the cost of electricity went up 6 % a year on average?
   (c) Should cost be the only consideration when buying an appliance? Justify your answer.
9. Use a table, such as Table 1, or a diagram to summarize the information in this section, and your response to it. C

Table 1

Section says . . .	I think . . .

10. In your opinion, would it solve the problem of Canadian and U.S. emissions if everyone bought enough carbon offset credits to offset their emissions? Explain why or why not. T/I A
11. Explain the meaning of the phrase, "responsible stewardship of Earth." K/U
12. What is the biggest reason for you, personally, to take action against climate change? A

Taking Action on Climate Change Now or Later?

In 2006, the Conservative government proposed a new *Clean Air Act* for Canada. The act was presented as a positive step toward reducing greenhouse gas emissions. Some people argued, however, that the act was not strong enough. They pointed out that following the act meant that Canada would not meet the Kyoto Protocol emission limits. Canada's *Clean Air Act* is being revised to address these concerns. The key question at stake is: How quickly should Canada act to reduce the impacts of climate change?

The Issue

The Canadian government, provincial legislatures, and city governments have taken some steps toward reducing greenhouse gas emissions. Nevertheless, Canada's emissions continue to increase. Scientists and some concerned citizens believe that we are moving too slowly; others believe we should proceed cautiously. In this activity, you will take the role of a stakeholder at a conference on climate change. The Canadian government has set up the conference to discuss how to respond to climate change. Several stakeholders have been asked to present their views.

Roles

- A climate scientist points out that most scientists agree that climate change is a result of human activities. She reminds the conference that humans must reduce their emissions by more than is being done now to save us from disastrous impacts (Figure 1).
- A politician argues that taking quick action to reduce emissions will harm the Canadian economy. The government should move more carefully, making sure that our economy remains stable.
- A scientist funded by an oil company suggests that since we are not certain about all the causes and impacts of climate change, it makes no sense to take action. He wants the government to fund studies on causes and impacts rather than pay for new technologies and incentives to reduce emissions.

SKILLS MENU

- Defining the Issue
- Researching
- Identifying Alternatives
- Analyzing the Issue
- Defending a Decision
- Communicating
- Evaluating

- An economist points out that there are great economic benefits to promoting energy conservation.
- An oil executive says he is willing to fund energy-saving technologies only after we wait a few more years to make sure that climate change is as bad as some people claim.
- An environmental activist says that we are taking too little action, too late. Climate change is already harming people and ecosystems around the world. We are being very irresponsible in delaying action.

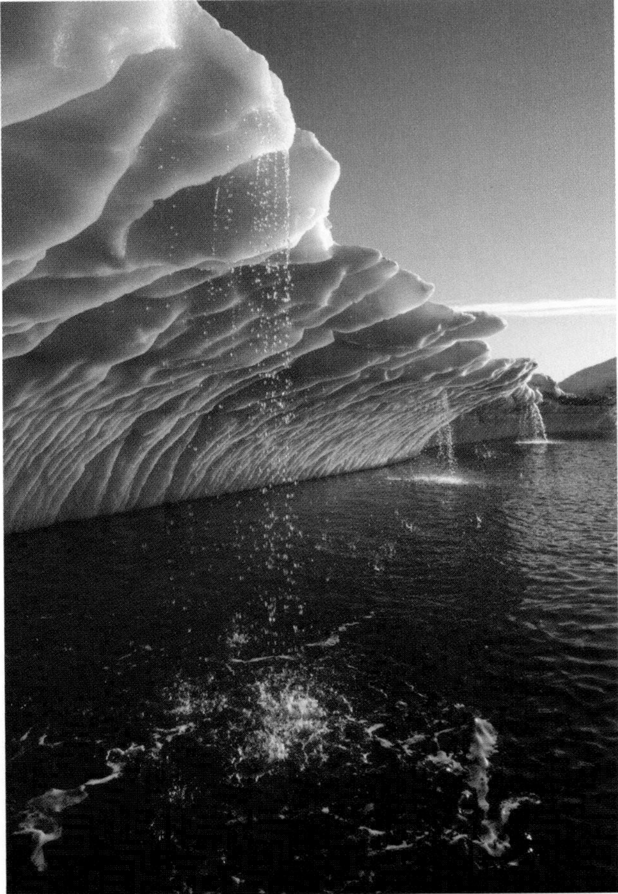

Figure 1 The melting of Arctic ice is a dramatic impact of climate change.

- A political scientist examined 900 scientific papers on climate change. She found that all agreed with the IPCC consensus that climate change is happening and is caused by humans. She points out that skeptics do not usually publish papers in scientific journals. Instead, they present their views on the Internet and to the media (Figure 2).

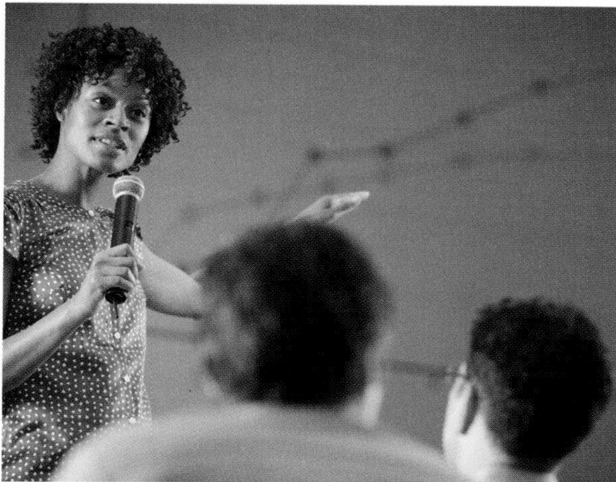

Figure 2 The political scientist explains the value of peer-reviewed research.

- The owner of a manufacturing plant argues that laws restricting her company's actions will harm her business. She is willing to voluntarily reduce some of her emissions but objects to laws that force her to do so.
- An industrial designer explains how reducing greenhouse gas emissions can benefit many industries. Canada could become a leader in green technology. We could export our expertise around the world.
- A citizen is concerned that not enough is being done by government and industry to reduce greenhouse gas emissions. He wonders if the actions he is taking to reduce his personal emissions are having any effect.

Goal

To debate the statement: The Canadian government should take immediate and extreme action to reduce emissions of greenhouse gases and limit the impacts of climate change.

Gather Information

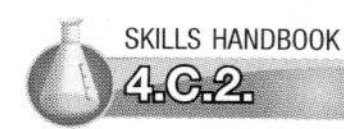

1. As a group, choose a stakeholder whom you will represent. Using a variety of resources, find information that supports the perspective of your stakeholder.

GO TO NELSON SCIENCE

Identify Solutions

Consider the following questions as you prepare to defend your position.

- Why should the government follow your suggestions?
- What evidence do you have to support your position?
- Are there ways to reduce emissions without harming businesses and industries?
- Will reducing emissions save money or cost money overall?
- Does the stakeholder have a personal bias that is affecting his or her judgment? How might this bias be addressed in a different way?
- Can a general agreement be reached that is acceptable to all stakeholders? Why or why not?
- Should all stakeholders have equal say, or should the opinion of some be considered more weighty than the opinion of others, since they may have a better understanding of the situation?

Make a Decision

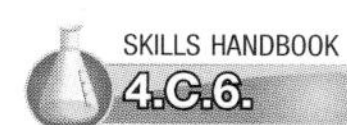

After the presentations have been made and the questions have been answered, discuss in your group whether or not your position has changed. Decide on the final position of your stakeholder.

Communicate

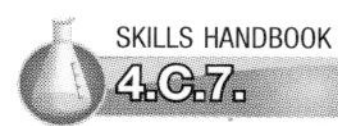

Using your position, prepare a list of recommendations to the government. Make sure that your recommendations are supported by facts and research.

Write up your recommendations as a formal letter to the minister of the environment. Alternatively, you could present your information as a news report, a television or newspaper advertisement, or a convincing work of art. T/I C A

CHAPTER 10

LOOKING BACK

KEY CONCEPTS SUMMARY

The impacts of climate change will affect our environment and society.

- The Intergovernmental Panel on Climate Change (IPCC) has summarized the latest scientific research on climate change. (10.2)
- Impacts of climate change are affecting people's health, where they can live, and what crops they can grow. (10.2)

Impacts of climate change will be felt most in the Arctic.

- The climate is changing more rapidly in the Arctic than anywhere else. (10.2)
- Climate change in the Arctic will have repercussions worldwide. (10.2)

Climate change in Ontario is expected to bring warmer winters and hotter summers.

- Ontario's ecology and economy will be impacted by climate change. (10.3)
- Summer days will get hotter and more humid, affecting people's health and increasing the demand for air conditioning. (10.3)
- Many plant and animal species native to southern Ontario will move northward. New species will move up into Ontario from the south. (10.3)

Current initiatives will not prevent serious negative effects from climate change.

- All levels of government should consider how to adapt to inevitable climate change, as well as how to mitigate more severe impacts. (10.4)
- Governments, business, and industry have roles to play in reducing climate change. (10.4)

Greenhouse gas emissions must be reduced by 80 % by 2050 to avoid the most serious impacts.

- Scientists develop climate scenarios for different amounts of greenhouse gases. (10.1)
- Climate models make projections about how Earth's climate would change under these scenarios. (10.1)
- To limit increases in global temperature to 2 °C, we must limit atmospheric greenhouse gas concentrations to 450 ppm. (10.4)

Switching to clean energy sources is essential to reduce greenhouse gas emissions.

- Clean energy sources produce little or no greenhouse gases. (10.1)
- Examples of clean energy sources are wind power, geothermal energy, solar power, hydroelectricity, biomass, and nuclear power. (10.1)
- Individuals can and should take action to reduce climate change. (10.5)

WHAT DO YOU THINK NOW?

You thought about the following statements at the beginning of the chapter. You may have encountered these ideas at school, at home, or in the world around you. Consider them again and decide whether you agree or disagree with each one.

1 We can stop climate change from happening if we reduce our consumption of fossil fuels.
Agree/disagree?

2 Scientists are always objective and unbiased in their data collection, analyses, and conclusions.
Agree/disagree?

3 Since the impact of climate change will be most felt in the Arctic, countries situated near the Equator do not need to be concerned."
Agree/disagree?

4 The best way to decrease the impact of human activity on climate is to switch to clean energy sources.
Agree/disagree?

5 Canadians' use of fossil fuels is wasteful.
Agree/disagree?

6 We need to take action on climate change at individual, local, regional, and international levels.
Agree/disagree?

How have your answers changed since then? What new understanding do you have?

Vocabulary

clean energy source (p. 407)
climate projection (p. 408)
impacts of climate change (p. 412)
Intergovernmental Panel on Climate Change (IPCC) (p. 412)
mitigation (p. 423)
Kyoto Protocol (p. 423)

BIG Ideas

- ✓ **Earth's climate is dynamic and is the result of interacting systems and processes.**
- ✓ **Global climate change is influenced by both natural and human factors.**
- ✓ **Climate change affects living things and natural systems in a variety of ways.**
- ✓ **People have the responsibility to assess their impact on climate change and to identify effective courses of action to reduce this impact.**

CHAPTER 10 REVIEW

The following icons indicate the Achievement Chart category addressed by each question.

K/U Knowledge/Understanding T/I Thinking/Investigation C Communication A Application

What Do You Remember?

1. Describe four possible global impacts of climate change. (10.2) K/U
2. List three reasons why it is important to take action to address the issue of climate change. (10.4) K/U
3. About how much of Ontario's greenhouse gas emissions comes from individuals' activities? (10.5) K/U
4. Give four possible impacts of climate change on Ontario. (10.3) K/U

What Do You Understand?

5. It is difficult for climate models to predict exactly how Earth's climate will change over the next century. Explain this difficulty in terms of
   (a) the complexity of the climate system.
   (b) human choices. (10.1) K/U A
6. Specify a possible impact of climate change for each of the following parts of the world. (10.2) K/U
   (a) Europe
   (b) Asia
   (c) Australia and New Zealand
   (d) Africa
   (e) Latin America
   (f) North America
   (g) Arctic, Greenland, Antarctica
7. In this chapter, you learned about several possible impacts of climate change on Ontario. Which impact do you think will affect you personally the most? Explain why. (10.3) C A
8. Climate projections are based on complex computer model simulations. These simulations consider scenarios of future changes in factors that affect human production of greenhouse gases.
   (a) What factors are included in the future scenarios?
   (b) What factors do the models consider? (10.1) K/U A
9. The cost of switching from fossil fuels would be high, but the cost of doing nothing might be even higher. Thus, people talk about a balanced approach. Name several components of a balanced approach. (10.4–10.6) C A

Solve a Problem!

10. (a) List three things you do or participate in that result in greenhouse gas emissions.
    (b) For each of the three things you listed, suggest two alternatives that would result in lower greenhouse gas emissions.
    (c) Assess how likely it is that you will act on these six suggestions. Write a paragraph explaining why you probably will, or probably will not, carry out your own suggestions. (10.5) C A
11. The issue of sea level rise is attracting a great deal of media attention. Why do you think sea level rise is such an important issue? (10.2) A
12. If a world government required everyone to reduce their consumption of fossil fuels by 80 % over the next 10 years, what negative effects might this have on the following sectors? (10.4, 10.6) A
    (a) the economy of oil-producing nations such as Canada
    (b) global transportation
    (c) food production and distribution
    (d) tourism

Create and Evaluate!

13. Why do you think the world is more concerned about climate change in the Arctic than in the Antarctic? Suggest several reasons. (10.2) K/U
14. Evaluate at least two positive and two negative impacts that Canada might experience due to climate change. (10.2) K/U A
15. Anticipate how a changing climate might affect your daily life. In a blog, a short magazine article, or a video diary, describe a day in your new life. Include at least four examples of changes. (10.3) C A
16. Brainstorm a list of actions that your school could take to reduce greenhouse gas emissions. Prioritize the actions and explain your priorities. Write a summary explaining why it is important that these actions be taken. Present your proposal to the student council and/or the principal. (10.5) T/I C A

17. The Canadian government is making important decisions about climate change that will affect your life. If the government asked you for your opinion on the top three priorities it should focus on, what would you say, and why? (10.4) C A

18. "A change in climate would affect plants and animals but would not affect our economy." Evaluate this statement. Explain why you agree or disagree with the statement (Figure 1). (10.2) C A

Figure 1

19. Find out what students in your school think about the issue of climate change. T/I C
    (a) Develop a questionnaire with at least five opinion-based questions on the topic of climate change. For example, you might ask whether students know what climate change means and/or whether they agree that climate change is caused by human activities. You may want to write your questions so the answers can be scaled from 1 (strongly disagree) to 5 (strongly agree).
    (b) Decide how you will survey a good sample of your school. For example, will you include staff members? Will you include students from different grades?
    (c) Carry out the survey.
    (d) Compile and analyze your data. What trends do you observe? Use a chart and/or graph to display your data.

20. (a) What types of composting and recycling programs are available in your community?
    (b) What proportion of locally generated garbage do they divert from landfill sites?
    (c) How do these programs reduce climate change? (10.5) T/I K/U

21. Do you have family relatives, friends, or neighbours who work in any of the following sectors of our economy: transportation, producing energy, industries, building and contracting, agriculture, waste management, and forests? If so, notice the recommendations made for that sector in Section 10.4. Contact your relative or friend and ask them to share any initiatives they know of in their workplace. Take notes to share with your class. (10.4) C A

Reflect On Your Learning

22. Reread Section 10.2 on global impacts. What was your initial response after reading the list of possible effects of global warming on the environment? How do you think your emotions might affect your critical thinking about the issue of climate change?

23. Which impacts of climate change had you heard about before starting this chapter? Which impacts came as a surprise to you?

Web Connections 

24. Research what the government of Ontario is doing to reduce climate change. (10.4) T/I C
    (a) Make a list of three initiatives of the government of Ontario in this area.
    (b) How does the government of Ontario assess its progress on these initiatives?
    (c) Decide on your personal assessment of each of the three areas listed in (a). Which do you think is likely to be the most successful? The least successful? Why?

25. Suggest a cost-efficient way to build a house on permafrost so that it will not be affected by any permafrost thaw that might occur unexpectedly. Conduct research and write a short paragraph with some ideas and a design. T/I C

To do an online self-quiz or for all other Nelson Web Connections, **GO TO NELSON SCIENCE**

CHAPTER 10

SELF-QUIZ

The following icons indicate the Achievement Chart category addressed by each question.

K/U Knowledge/Understanding T/I Thinking/Investigation
C Communication A Application

For each question, select the best answer from the four alternatives.

1. What is one reason that explains why scientists have trouble making exact projections about the rate of climate change? (10.3) K/U
   (a) It is difficult to determine the relative amounts of atmospheric gases.
   (b) It is difficult to measure how the energy radiated by the Sun varies over time.
   (c) It is difficult to predict how quickly people will switch from fossil fuels to other energy sources.
   (d) It is difficult to calculate how much carbon dioxide is produced by the burning of fossil fuels.
2. Which of the following is most likely to be a direct result of rising global temperatures? (10.2) K/U
   (a) more sea ice
   (b) more forest fires
   (c) more freshwater
   (d) more biodiversity
3. Which of the following projections best describes how climate change may affect precipitation? (10.2) K/U
   (a) Climate change will cause the patterns of precipitation to vary.
   (b) Climate change will increase precipitation everywhere.
   (c) Climate change will decrease precipitation everywhere.
   (d) Climate change will increase snowfall and decrease rainfall.
4. Which of these has been proposed as a way to absorb excess carbon dioxide from the atmosphere? (10.5) K/U
   (a) fish farms
   (b) large reservoirs
   (c) cattle herds
   (d) reforestation
5. Approximately what fraction of Ontario's greenhouse gas emissions are caused by the activities of individuals? (10.5) K/U
   (a) 10 %
   (b) 30 %
   (c) 75 %
   (d) 85 %
6. Which of the following is likely to be a positive effect of rising global temperatures? (10.2) K/U
   (a) fewer crop pests
   (b) higher lake levels
   (c) longer growing season
   (d) less spread of tropical diseases

Indicate whether each of the statements is true or false. If you think the statement is false, rewrite it to make it true.

7. If nations drastically reduce greenhouse gas emissions, global temperatures will immediately stop rising. (10.4) K/U
8. The rate of climate change can be reduced by relying less on power plants that burn fossil fuels and more on nuclear power plants. (10.1) K/U

Copy each of the following statements into your notebook. Fill in the blanks with a word or phrase that correctly completes the sentence.

9. The ____________ has summarized the latest research on climate change. (10.2) K/U
10. Carbon dioxide and methane will be released into the atmosphere by melting ____________ in the Arctic. (10.2) K/U
11. To make climate ____________, scientists develop ____________ which specify the amount of greenhouse gases produced. (10.1) K/U

Match each effect of climate change on the left with the region most likely to be affected on the right.

12. (a) forest fires (i) Canada
    (b) species loss (ii) Pacific islands
    (c) sea level rise (iii) northern Africa
    (d) expanding deserts (iv) Amazon rainforest
    (e) increasing storm intensity (v) United States Gulf Coast (10.2) K/U

Write a short answer to each of these questions.

13. Describe two ways in which planting grass on a roof can reduce greenhouse gases. (10.5) K/U
14. Describe the relationship between global population and climate change. (10.1) K/U

15. Describe a series of global events that connect fossil-fuel-driven transportation activities to the rise of sea level that could flood low-lying islands and coastal areas. (10.2) T/I

16. Most scientists agree that humans must rapidly reduce their dependence on fossil fuels and convert to energy sources that do not emit greenhouse gases. Some people think all nations should cut back emissions by the same percentage. However, given that switching to alternative energy sources will be expensive, other people think that developing nations should not have to change their energy usage as much as the more industrialized nations. Write a short paragraph supporting one of these viewpoints. (10.4) K/U C

17. The warming of Canada's far north and the Arctic Ocean may lead to the following changes:
    - less sea ice
    - habitat loss for some species
    - rising sea level
    - melting permafrost
    - growth of trees farther north
    - more open water for shipping
    - lowering of heating costs

    Write a short paragraph explaining whether you think these changes will be an overall benefit or problem for Canada. (10.2) T/I

18. Imagine you are planning a survey of local businesses that advertise themselves as "green." Make a list of three questions you would ask the business owners to help you evaluate just how green their businesses are. (10.4, 10.5) C

19. A certain species of migratory songbird that once spent its summers in southern Ontario has not been seen there for several years. (10.3) T/I
    (a) State a hypothesis that could explain the birds' disappearance based on your understanding of climate change.
    (b) How would you go about testing this hypothesis?

20. Explain how the planting of trees can reduce the impact of climate change. Be sure to include the cause-and-effect steps between the act of planting trees and the effect on climate. (10.5) K/U A

21. Figure 1 shows glaciers on the sides of mountains in the Canadian Rockies. (10.2) A

Figure 1

    (a) Describe how climate change is likely to change the appearance of these glaciers.
    (b) Describe one global effect produced by the melting of these glaciers.

22. The area of the Arctic Ocean covered by sea ice is decreasing at an accelerating rate because open water absorbs more solar energy than ice does. Imagine you are a climatologist beginning to create a computer model that will predict changes in Arctic sea ice. Describe two physical properties you would need to know in order to begin calculations for your model. (10.1) T/I

23. Identify two ways in which you add to atmospheric carbon dioxide or other greenhouse gases while at home. Explain how each of these activities leads to the release of greenhouse gases. In each case, explain how you could reduce the amount of greenhouse gas emissions. (10.5) A

24. Describe three ways that people can change their shopping habits and dietary choices to reduce greenhouse gas emissions. In each case, explain how the change reduces emissions. (10.5) A

25. Describe the role of the Intergovernmental Panel on Climate Change (IPCC). (10.1) K/U

UNIT D

Climate Change

CHAPTER 8
Earth's Climate System and Natural Changes

KEY CONCEPTS

Earth's climate system is powered by the Sun.

Earth's climate system includes the atmosphere, the hydrosphere, the lithosphere, and living things.

The greenhouse effect keeps Earth warm by trapping thermal energy radiated by Earth.

Thermal energy is transferred within Earth's climate system through air and ocean currents.

Earth's climate experiences long-term and short-term changes.

Scientists use natural ice cores, sediment layers, fossils, and tree rings to study past climates.

CHAPTER 9
Earth's Climate: Out of Balance

KEY CONCEPTS

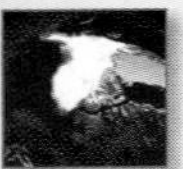
We have evidence that our climate is changing.

Human activities have increased atmospheric levels of greenhouse gases.

The increase in greenhouse gases is causing the anthropogenic (human-caused) greenhouse effect.

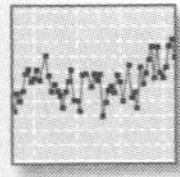
The anthropogenic greenhouse effect is the main cause of today's climate change.

The largest sources of greenhouse gases in Canada are the production and burning of fossil fuels.

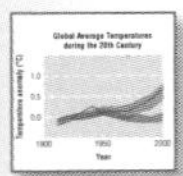
Scientists use climate models to figure out how different factors affect our climate.

CHAPTER 10
Assessing and Responding to Climate Change

KEY CONCEPTS

The impacts of climate change will affect our environment and society.

Impacts of climate change will be felt the most in the Arctic.

Climate change in Ontario is expected to bring warmer winters and hotter summers.

Current initiatives will not prevent serious negative effects from climate change.

Greenhouse gas emissions must be reduced by 80 % by 2050 to avoid the most serious impacts.

Switching to clean energy sources is essential to reduce GHG emissions.

MAKE A SUMMARY

Imagine a doctor's waiting room, with Earth waiting its turn. Here is how a doctor–patient conversation might go:

[DOCTOR] So, I hear you're not feeling well. Can you describe your symptoms for me?

[EARTH] I just don't feel right. My sea level has been rising for at least a hundred years—and it's been rising faster in the last few years. As you can imagine, I've had quite a lot of flooding. More of my ice has been melting than usual. Do you think it's connected?

[DOCTOR] Very likely. Do you have any swelling in your oceans? Are you feeling any thermal expansion?

[EARTH] Yes, I have noticed that my oceans seem to be expanding.

[DOCTOR] Hmm, yes. Ice melting, thermal expansion—that would explain the sea level rise.

[EARTH] Speaking of oceans, I've been having a lot of trouble with hurricanes. I always have hurricanes, of course, but they seem to be getting worse.

[DOCTOR] Warmer ocean waters can increase the strength of hurricanes. Let me examine you. I see a patch of desert forming here. Is that new?

[EARTH] Yes, it is.

[DOCTOR] It says here in your medical history that your seasons have been changing lately. What else? Plants and animals migrating toward the poles—I think I know what's happening here. Here's a thermometer. I'm just going to take your temperature.
[DOCTOR EXAMINES THERMOMETER]
Just as I thought. Your temperature is rising.

1. List the evidence for climate change contained in the dialogue between Earth and the doctor. K/U
2. Write some more dialogue. (For example, have any symptoms been missed? What is causing the rise in temperature? Why is this cause happening? What can be done about Earth's problem? What is already being done?) T/I C
3. Present your completed dialogue in the form of a cartoon, a dramatic presentation, or an FAQ on a web page. C

CAREER LINKS

List the careers mentioned in this unit. Choose two of the careers that interest you or choose two other careers that relate to climate change. For each of these careers, research the following information:

- educational requirements (secondary and post-secondary)
- skill/personality/aptitude requirements
- potential employers
- salary
- duties/responsibilities

Use the information you have assembled to create a brochure. Your brochure should compare your two chosen careers and explain how they connect to climate change.

GO TO NELSON SCIENCE

Global Climate Change

The Issue

The Intergovernmental Panel on Climate Change (IPCC) is preparing an update to its previous reports on global climate change. The update will include recommendations for actions. You are volunteering your expertise to contribute to these recommendations. The panel's findings will be presented at an international conference.

SKILLS MENU
- Defining the Issue
- Researching
- Identifying Alternatives
- Analyzing the Issue
- Defending a Decision
- Communicating
- Evaluating

Goal

To collect climate information about a specific region, to use the information collected to identify any impacts of climate change, and to suggest specific steps toward mitigation.

Gather Information

SKILLS HANDBOOK 4.A., 4.B., 6.A.

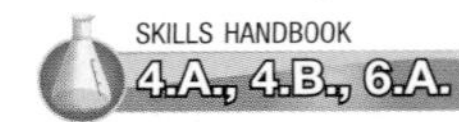

1. Choose a location such as one of the following locations to research:
   - Churchill, Manitoba, Canada (Figure 1)

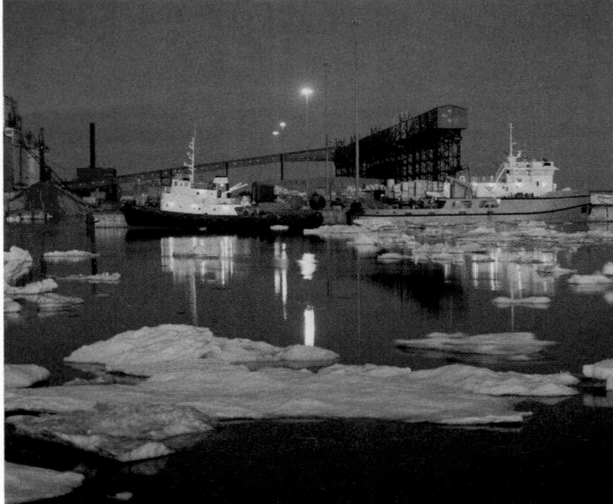

Figure 1 Churchhill, Manitoba

   - Vancouver, British Columbia, Canada
   - Whitehorse, Yukon, Canada
   - Baffin Island, Nunavut, Canada
   - Tuvalu
   - France
   - United Kingdom
   - Peru
   - Bangladesh
   - Southeastern coast of Australia
   - Uganda (Figure 2)
   - a different location (with the permission of your teacher)

Figure 2 Uganda

2. Research the climate of the location you have chosen. Collect data (both current and, if possible, from about 50 years ago) on the following:
   - average monthly and yearly temperatures
   - average monthly and yearly precipitation
   - an estimate of how much precipitation falls as snow versus rain
   - severe weather patterns (e.g., storms, hurricanes, yearly monsoons, seasonal floods, droughts, heat waves)
   - amounts of permanent ice or frozen soil (glaciers, permafrost)
   - extent and duration of ice cover on lakes in winter
3. Analyze and present your data using tables and graphs. Note any recent changes in climate that you observe.

4. Review the information about global climate change in this unit. What impacts of climate change do you expect to occur in this region over the next 100 years? Make a list of projected changes for the region.
5. Research impacts of climate change in the region. For example, look for changes in any of the following factors.
    - changes in river flow, ice cover, flooding, and/or rainfall
    - changes to local ecosystems and/or species
    - changes in severe weather patterns (e.g., hurricanes, monsoons)
    - unusual heat waves or droughts
    - changes in agricultural land
    - changes in the occurrence of insect-borne diseases (such as malaria, dengue fever, West Nile virus, Lyme disease)
6. Research social and economic data for the region. For example, look for the following factors:
    - relative wealth or poverty of people in the region
    - traditional or cultural activities in the region
    - access to scientific information and/or technologies in the region
7. Compare your list of projected impacts from Step 4 with the impacts you identified from your research in Step 5. Are any of your projected impacts already occurring?
8. Write a one-page summary of your report of observations from Steps 5 to 7.

Identify Solutions

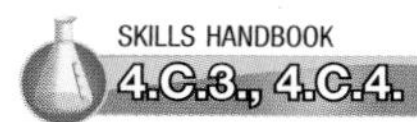

Brainstorm a list of practical steps that people in the region can take to mitigate or adapt to climate-related changes. Consider these questions when generating your suggestions for the region:

- What major aspects of life in that region would be affected by climate change?
- How can people in the region reduce their emissions of greenhouse gases and/or reduce deforestation in the region?
- What steps have already been taken by local government, businesses, and individuals to adapt to climate change?
- How can the local government, businesses, and individuals adapt to coming changes in climate?

Make a Decision

What would be the most practical solution(s) for the people in this region, given what you have learned about the region?

Communicate

Present your research at an international conference (i.e., with your classmates). In your presentation, include the following:

- a comparison of the climate 50 years ago and today
- climate-related changes that are already occurring
- a projection of expected climate-related impacts in the region over the next 100 years
- the most appropriate steps for mitigation and adaptation for local government, businesses, and individuals

Your presentation could be an in-person speech, a video, a poster presentation, or a written submission.

ASSESSMENT CHECKLIST

Your completed Performance Task will be evaluated according to how well you are able to

Knowledge/Understanding

☑ Thoroughly research and analyze the data.

Thinking/Inquiry

☑ Plan and develop a clear search strategy for climate data.
☑ Record the data in an organized fashion.
☑ Compare projected impacts with actual impacts.
☑ Identify potential solutions.

Communication

☑ Clearly present data in a table or graph.
☑ Prepare and present your current and projected climate-related impacts in an organized manner.

Application

☑ Make recommendations on how to limit and prepare for local and global climate change.
☑ Demonstrate an understanding of natural and human factors that influence climate in this region.

UNIT D REVIEW

The following icons indicate the Achievement Chart category addressed by each question.

K/U Knowledge/Understanding T/I Thinking/Investigation C Communication A Application

What Do You Remember?

For each question, select the best answer from the four alternatives.

1. What is the difference between weather and climate? (8.1) K/U
   (a) Weather refers to wind and precipitation, whereas climate refers to the temperature of a region.
   (b) Weather happens on a daily basis, whereas climate is the average of weather over long periods of time.
   (c) Weather changes slowly, whereas climate changes quickly.
   (d) Weather happens only over land, whereas climate happens over both the land and the ocean.
2. Which list includes all the key components of Earth's climate system? (8.4) K/U
   (a) water, land, and living things
   (b) air, water, ice, and land
   (c) air, water, ice, and living things
   (d) air, water, ice, land, and living things
3. The hydrosphere is made up of
   (a) all living things and their habitats.
   (b) all land on Earth's surface.
   (c) all frozen water on Earth.
   (d) all water on Earth. (8.4) K/U
4. Which of the following mechanisms is NOT significant in transferring thermal energy across Earth's surface? (8.8) K/U
   (a) air convection currents
   (b) heat conduction through land masses
   (c) the thermohaline circulation in the ocean
   (d) the Gulf Stream
5. Which gases are the principal contributors to the natural greenhouse effect? (8.6) K/U
   (a) carbon dioxide, methane, water vapour
   (b) carbon dioxide, methane, oxygen
   (c) carbon dioxide, water vapour, oxygen
   (d) argon, carbon dioxide, methane
6. Scientists use proxy records to help them study past climates. Proxy records include
   (a) tree rings.
   (b) atmospheric weather records.
   (c) oceanic temperature measurements.
   (d) satellite observations of polar ice cover. (8.11) K/U
7. Long-term natural changes in Earth's climate over the past 400 000 years have most likely been caused by
   (a) changes in the total amount of ice and water on Earth.
   (b) changes in Earth's orbit and in the angle of its axis.
   (c) a decrease in the number of animals on Earth.
   (d) a decrease in the amount of oxygen in Earth's atmosphere. (8.9) K/U
8. Which of the following is NOT evidence of current climate change? (9.1) K/U
   (a) rising average world temperatures
   (b) rising sea levels
   (c) increasing water pollution
   (d) decreasing Arctic ice cover
9. What is the most likely cause of current climate change? (9.4) K/U
   (a) melting ice at the poles
   (b) changes in solar radiation
   (c) volcanic eruptions
   (d) human emissions of greenhouse gases
10. Which of the following actions is mainly your responsibility and NOT that of your municipality, the province of Ontario, or the government of Canada? (10.5) K/U
    (a) support international treaties to reduce carbon dioxide emissions
    (b) collect methane from landfills for use as energy
    (c) change the power source of electricity-generating stations away from coal
    (d) change transportation habits so that we walk, bicycle, and use public transit more frequently

11. Which of the following statements correctly explains how greenhouse gases affect Earth's temperature? (8.6) K/U
    (a) Greenhouse gases act as a protective layer in Earth's atmosphere by reflecting most of the incoming solar radiation.
    (b) Greenhouse gases keep Earth cool by removing moisture from the atmosphere that would otherwise increase Earth's temperature.
    (c) Greenhouse gases absorb infrared radiation emitted by Earth's surface and emit about half of this radiation back toward Earth's surface.
    (d) Greenhouse gases absorb ultraviolet radiation from the Sun, convert it to infrared radiation, and emit almost all of this radiation back into space.
12. Which of the following is an example of thermal expansion? (9.1, 9.3) K/U
    (a) Water increases in volume as its temperature increases.
    (b) More salt dissolves in sea water as the water travels to the poles.
    (c) Human activity has increased the amount of greenhouse gases in the atmosphere.
    (d) As global temperatures increase, organisms that carry disease may move farther north.
13. Which of the following correctly describes an El Niño event? (8.9) K/U
    (a) Over time, the movement of Earth's landmass produced today's continents from one large supercontinent.
    (b) The vibration of atoms in certain molecules allows these molecules to absorb different types of energy from the Sun.
    (c) The prevailing winds in the Pacific Ocean temporarily switch direction and push warm water east, toward South America.
    (d) Clouds form at low altitudes where they can trap thermal energy near Earth's surface, increasing Earth's temperature.
14. Which statement correctly describes what happens to energy from the Sun once it reaches Earth? (8.3) K/U
    (a) Earth's surface absorbs about half of the total energy and re-radiates it as infrared energy.
    (b) Earth's atmosphere reflects about half of the energy back to space.
    (c) Earth's forests absorb about half of the total energy, converting it into chemical energy.
    (d) Earth's oceans absorb about half of the total energy, causing them to warm up.

Indicate whether each of the statements is TRUE or FALSE. If you think the statement is false, rewrite it to make it true.

15. A location's climate is affected by factors such as distance from the equator, height above sea level, and nearby bodies of water. (8.3, 8.4) K/U
16. Ozone is harmful to life on Earth when it is high up in the stratosphere and helpful when it is at ground level, in the troposphere. (8.4) K/U
17. Earth absorbs much more energy from the Sun than it releases. (8.3) K/U
18. Gases in Earth's atmosphere trap infrared radiation from Earth and thus keep our climate warmer than it would otherwise be. (8.3) K/U
19. Earth's climate changes naturally over very long periods of time. (8.9) K/U
20. Increases in global temperatures are caused by damage to the ozone layer. (8.4) K/U
21. Deforestation is a problem because cutting down forests releases carbon dioxide into the atmosphere and prevents the absorption of carbon dioxide. (9.4) K/U
22. CFCs are greenhouse gases. (9.4) K/U
23. Most greenhouse gases in Canada are produced by household electricity use. (9.5) K/U
24. The leeward side of a mountain range receives less precipitation than the windward side of the mountain range. (8.4) K/U

25. Heating and cooling our homes is the main source of greenhouse gas emissions by individuals. (10.5) K/U

26. Climate change is happening more rapidly near the equator than anywhere else. (10.2) K/U

Copy each of the following statements into your notebook. Fill in the blanks with a word or phrase that correctly completes the sentence.

27. Earth's climate system is made up of the Sun and the __________, __________, lithosphere, and __________. (8.4) K/U

28. Earth absorbs radiation from the Sun in several forms, __________, __________, __________, but re-radiates it back out to space primarily in the form of __________. (8.3) K/U

29. The rise in sea level is probably being caused by two main factors at the present time: __________ and __________. (9.1) K/U

30. The greenhouse gas called __________ is mainly produced by burning fossil fuels. (9.4) K/U

31. The oceans play an important role in balancing Earth's climate because they transfer __________. (8.8) K/U

32. Because of the __________ effect, an ice-covered Earth would reflect more of the Sun's energy than would an Earth with no ice. (8.10) K/U

33. A __________ exists when a cause creates an effect that influences the original cause. (8.10) K/U

34. __________ is the careful management of something you do not own. (10.5) K/U

35. The energy __________ by Earth is equal to the energy radiated by Earth. (8.3) K/U

Write a short answer to each of these questions.

36. Describe the two main ways in which thermal energy is transferred within the climate system. (8.8) K/U

37. Distinguish between the natural greenhouse effect and the anthropogenic greenhouse effect. (8.6, 9.4) K/U

38. Describe the effect of ocean currents on nearby land. (8.8) K/U

39. Describe the main natural and human sources of each of the following greenhouse gases. (8.6, 9.4) K/U
    (a) methane
    (b) nitrous oxide

40. Our climate has changed over the last 50 years. Name two things about the climate in your region that would be different from today if you were growing up in 1960. (10.3) K/U

41. Name two things scientists can learn from examining and testing ice cores. (8.11) K/U

42. Explain how ozone is helpful when it is present in the stratosphere, while it is harmful when it is present in the troposphere. (8.4) K/U

What Do You Understand?

43. The climate impacts our everyday lives. (8.1, 8.2, 10.3) K/U A
    (a) Describe the climate in the region where you live.
    (b) Describe three specific impacts of climate change that you might observe over the next hundred years in the region where you live.

44. Examine Figure 1. (8.6, 9.4) K/U

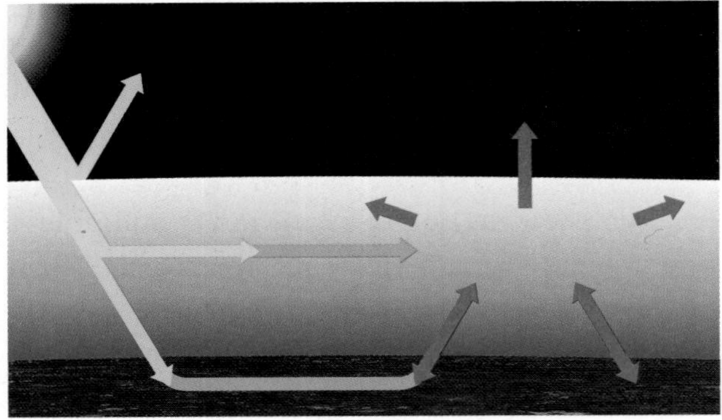

Figure 1

(a) Explain what is happening in Figure 1.
(b) Discuss why the effect shown in the diagram is important for life on Earth.

45. Predict what would happen to temperatures on Earth if Earth absorbed energy from the Sun but did not emit all of it. (8.3) T/I

46. Carbon dioxide is an important natural and anthropogenic greenhouse gas. (8.6, 9.4, 9.5) K/U
    (a) Give one example of a natural source of carbon dioxide.
    (b) Give two examples of human sources of carbon dioxide.
    (c) Give one example of a carbon sink and explain how it works.

47. Look at the two photographs of the same glacier, in the present (Figure 2(a)) and 1939 (Figure 2(b)). (9.1, 10.2) K/U T/I C

Figure 2

    (a) How has the glacier in Figure 2 changed?
    (b) Explain why this change is happening. Include diagrams to illustrate your explanation.
    (c) List the two main human activities that are causing Earth's climate to change.

48. Climate change is occurring as a result of human activities. (9.6, 10.1, 10.2) K/U
    (a) Explain why most scientists think that climate change today is caused by human activities.
    (b) Explain why a few scientists think that most climate change today is not primarily due to human activity.

49. Draw a concept map that shows the methods that scientists use to study past climate. Include the following terms: proxy records, precipitation, temperature, ice, thermometer, written records. (8.11) K/U C

50. Predict how climate change might impact each of the following continents. Include at least two impacts for each. (10.2) T/I
    (a) North America
    (b) Africa
    (c) Australia

51. (a) Describe four negative impacts of climate change that are expected for Ontario.
    (b) Describe two positive impacts of climate change that are expected for Ontario. (10.3) A

52. How does ice influence Earth's climate? (8.10, 10.2) K/U

53. Describe how each of the following natural cycles plays a part in Earth's climate system. (8.4, 8.6, 9.4) K/U C
    (a) the water cycle
    (b) the carbon cycle

54. A friend tells you that we could stop climate change entirely if we immediately reduce greenhouse gas emissions. Use what you have learned in this unit to write a response to this statement. (10.4–10.6) T/I

55. Predict what might happen in each of the following situations. (9.4, 10.2–10.5) T/I
    (a) A non-native plant species migrates into southern Canada.
    (b) More Canadian farmers turn to cattle ranching.
    (c) The Canadian government offers tax incentives for insulation and energy efficiency in new buildings.
    (d) All new buildings were constructed with green roofs.

56. Copy and complete Table 1 in your notebook. (10.2) K/U

Table 1 Potential Costs and Benefits of Higher Temperatures in the Arctic

Potential costs	Potential benefits
1.	1.
2.	2.
3.	3.

Solve a Problem

57. Copy Table 2 into your notebook. List three actions under each heading. (10.4, 10.5) K/U A

Table 2 Acting to Reduce Climate Change

Already being done in Canada	Could be done in Canada	Could be done by you and your family, school, or community

58. Your local government wants to reduce your community's emissions of greenhouse gases. (9.5, 10.4) K/U T/I C
    (a) Identify three major sources of greenhouse gases in your community.
    (b) Come up with a reasonable plan to reduce the amount of greenhouse gases from each source.
    (c) Create a brochure or visual presentation outlining your ideas. Display your presentation at school, at a local community centre, or in a local government building.
59. What colour clothing would you wear to stay cool on a hot, summer day? Use what you know about the albedo effect to explain your answer. (8.10) K/U A
60. Your family needs to buy the items listed below. Explain at least two things your family should consider as they look at each item. (10.5) A
    (a) light bulbs
    (b) a refrigerator
    (c) a car
61. People are cutting down rainforests to clear land for growing crops. (9.4) T/I
    (a) Explain one reason why people would rather grow crops on their land than allow trees to grow on it.
    (b) How do you think you could convince people not to cut down the trees and allow the rainforest to remain?

Create and Evaluate

62. A climate scientist creates a computer model of Earth's climate system. First, she programs the model to show global temperatures over the past 100 years in a world without humans. Next, she programs the model to show global temperatures over the past 100 years in a world with humans. Predict and compare the results from each program. (9.6) T/I A
63. (a) Compare the impacts of climate change that are currently happening (Section 9.1) with the impacts that are expected to happen (Section 10.2).
    (b) Would you agree that the most severe impacts of climate change have already happened? Explain why or why not. (9.1, 10.2) K/U
64. In a school debate, one student argues that Canada's forests are reducing the amount of greenhouse gases Canada emits. The other student argues that Canada's forests are increasing the amount of greenhouse gases emitted by Canada. Discuss how each student could be correct. (9.5) K/U
65. A climate change skeptic argues that scientists are not completely sure how Earth's climate will change in the future. He suggests that scientists should carry out more studies before we start taking action. (10.2) K/U
    (a) Why is it difficult for scientists to accurately predict future climate?
    (b) What is your opinion? Write a short response to this skeptic.

66. In Chapter 8, you learned how the climate system transfers energy around the world. In Chapter 9, you learned that increases in greenhouse gases are causing rising temperatures, in other words, an increase in thermal energy. Use what you know about the way energy is transferred to hypothesize how rising temperatures may affect air and ocean currents. (8.8, 9.4) K/U T/I

67. You want to demonstrate convection currents in water to a class. Describe how you would design this demonstration. (8.8) T/I

68. Contact your local recycling centre to find out which materials can be recycled and which ones cannot. Then, create a poster or other graphic that illustrates what can and cannot be recycled in your area. (10.5) A C

Reflect On Your Learning

69. If you had to describe the most important idea in this unit to a friend, what idea would you choose? Explain why.

70. When you read a text book, you will find some parts more interesting than others.
    (a) What part of this unit was the most interesting to you? Explain why.
    (b) What part of this unit was the least interesting to you? Explain why.
    (c) Find someone in your class who was interested in the section you did not find interesting. Ask your classmate to tell you why it was interesting.

71. In this unit, you learned that greenhouse gases are probably causing rising global temperatures. (9.4)
    (a) What did you know about greenhouse gases before studying this unit?
    (b) How has your understanding of greenhouse gases changed during this unit?

72. How has one of your views on climate change altered after studying this unit? Give a specific example in your explanation.

73. In this unit you learned that individuals can have an impact when they reduce their greenhouse gas emissions. (10.5)
    (a) Describe one way you tried to reduce your greenhouse gas emissions before studying this unit.
    (b) Describe one new way you can reduce your greenhouse gas emissions based on what you learned while studying this unit.

Web Connections 

74. Many people are concerned that food production and transportation may be contributing to climate change. They consider options such as becoming vegetarian, eating only locally grown food, or eating organic. Research the risks and benefits of each of these choices. Consider which might be the best option. Communicate your decision, supported by arguments, in a skit, a blog entry, a documentary video, or some other creative form. (10.4, 10.5) C A

75. Ocean currents around the globe act like a huge conveyor belt, slowly moving thermal energy from the equator to the poles (Figure 3). Some scientists suggest that increases in sea level could disrupt this conveyor belt. Research what impacts can be expected if ocean currents are disrupted and prepare a brief (one-page) summary report. (8.8) T/I C

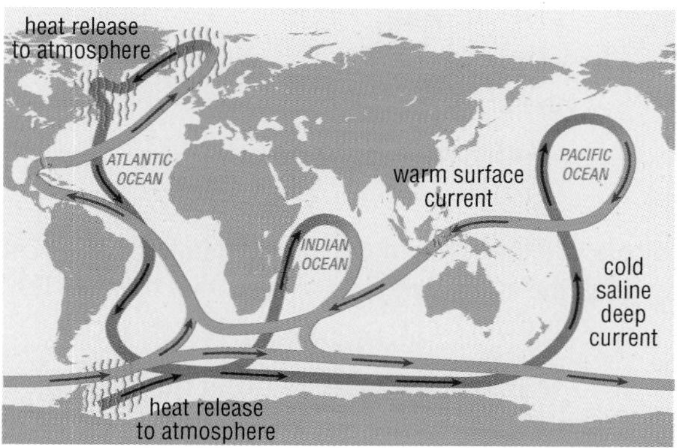

Figure 3

76. Scientists are concerned about changes in climate at the poles. Research climate change at the Arctic and the Antarctic. (10.2) T/I C
    (a) Why is change here so significant?
    (b) Present your findings as a poster or website.

For all Nelson Web Connections,
GO TO NELSON SCIENCE

UNIT D

SELF-QUIZ

The following icons indicate the Achievement Chart category addressed by each question.

K/U Knowledge/Understanding T/I Thinking/Investigation C Communication A Application

For each question, select the best answer from the four alternatives.

1. In which of the following would you expect convection currents to form? (8.8) K/U
   (a) soil
   (b) rock
   (c) air
   (d) ice
2. Which of the following is a factor in causing short-term variations in climate? (8.9) K/U
   (a) continental drift
   (b) volcanic eruptions
   (c) the shape of Earth's orbit
   (d) ozone in the stratosphere
3. Which of the following is an example of a proxy record? (8.11) K/U
   (a) data collected from tree rings
   (b) a recent photo of a glacier in the Arctic
   (c) a graph showing rainfall amounts in 1990
   (d) the weather forecast from yesterday's newspaper
4. Which of the following is a source of methane emissions? (8.6, 9.4) K/U
   (a) nitrogen fertilizers
   (b) aerosol spray cans
   (c) cattle ranching
   (d) photosynthesis in plants

Indicate whether each of the statements is TRUE or FALSE. If you think the statement is false, rewrite it to make it true.

5. Prevailing winds that come from the North Pole will tend to make the regions they pass over receive large amounts of precipitation. (8.8) K/U
6. Melting icebergs will cause a great rise in sea levels. (9.1) K/U
7. The natural greenhouse effect is necessary for life to exist on Earth. (8.6) K/U

Copy each of the following statements into your notebook. Fill in the blanks with a word or phrase that correctly completes the sentence.

8. Grass has a higher albedo than dark-coloured soil. Therefore, grass ________ more sunlight than does dark soil. (8.10) K/U
9. Geothermal energy and hydroelectric power are considered clean energy sources because they do not produce ________. (10.1) K/U

Match each term on the left with the most appropriate description on the right.

10. (a) climate projection
    (b) bioclimate profile
    (c) climate system
    (d) climate

    (i) graphs showing temperature and moisture conditions at a specific location
    (ii) average of the weather in a region over a long period of time
    (iii) global conditions produced by the interactions between Earth's air, land, liquid water, ice, and living things
    (iv) reasonable scientific estimate of a region's future temperature and precipitation conditions
    (8.2, 8.4, 10.1) K/U

11. (a) carbon dioxide
    (b) methane
    (c) water vapour
    (d) nitrous oxide
    (e) ozone

    (i) A gas that forms in the atmosphere as lakes and rivers evaporate.
    (ii) A gas that exists naturally in the stratosphere.
    (iii) A gas released into the atmosphere from certain fertilizers.
    (iv) A gas given off when fossil fuels are burned.
    (v) A gas produced by cattle during their digestive process. (9.4) K/U

Write a short answer to each of these questions.

12. Why would you expect a cloudy, summer night to be warmer than a cloudless, summer night? (8.8) K/U

13. You decide to keep a record of the high and low temperatures in your town for the next two weeks. What kind of research could you do to find how the local temperatures from 50 years ago compare to the data you collect? (8.1) K/U T/I

14. The Sun radiates energy onto all areas of Earth's surface. Explain two reasons why the climate at Earth's poles is colder than the climate at the equator. (8.3, 8.4) K/U

15. Explain how the processes of evaporation and condensation move energy from one place to another. (8.4, 8.8) K/U

16. Predict how Earth's average temperature would change if Earth had no atmosphere. Explain your prediction. (8.4) T/I

17. Describe an example of a positive feedback loop in daily life. (8.10) A

18. Explain how the disappearance of glaciers would affect some people in China. (10.2) T/I

19. Predict at least two ways in which climate change will impact your life. (10.2, 10.3) T/I

20. You are writing an article for your school newspaper about how your community is preparing for the impacts of climate change. To collect information for your article, you have scheduled an interview with a local official. List at least three questions you will ask the official during the interview. C

21. Describe how the shrinking of Arctic ice could affect
    (a) animals such as polar bears and seals that live there
    (b) global climate (10.2) T/I

22. This unit suggested several ways to conserve energy and reduce your greenhouse gas emissions. Write down at least two other items you could add to these suggestions. Explain why these items should be included. (10.5) A

23. Climate affects how we live and what activities we can undertake. (8.1) T/I
    (a) Describe at least two ways in which the climate in your region affects the way you live.
    (b) Choose another region of Canada. How would the climate in that region affect the two things you described in part (a)?

24. Earth's surface is made up of more water than land. Predict what influence it would have on Earth's climate if Earth's surface contained more land than water. (8.4, 8.9) T/I

25. Do you think that Canada should have agreed to join the Kyoto Protocol? Write a short argument explaining your opinion. (10.4) A C

26. This unit introduced many new concepts about climate change. T/I
    (a) Which concept in this unit did you find difficult to understand? Why?
    (b) What further research can you do to help you better understand this concept?

27. In this unit, you learned that ozone can be helpful or harmful to life on Earth, depending on its location in the atmosphere. You also learned that the greenhouse effect is essential to life on Earth, but the anthropogenic greenhouse effect may be harmful. K/U A
    (a) Write a general statement describing the pattern contained in this information.
    (b) Have you ever observed a similar pattern anywhere else in real life?

UNIT E

Light and Geometric Optics

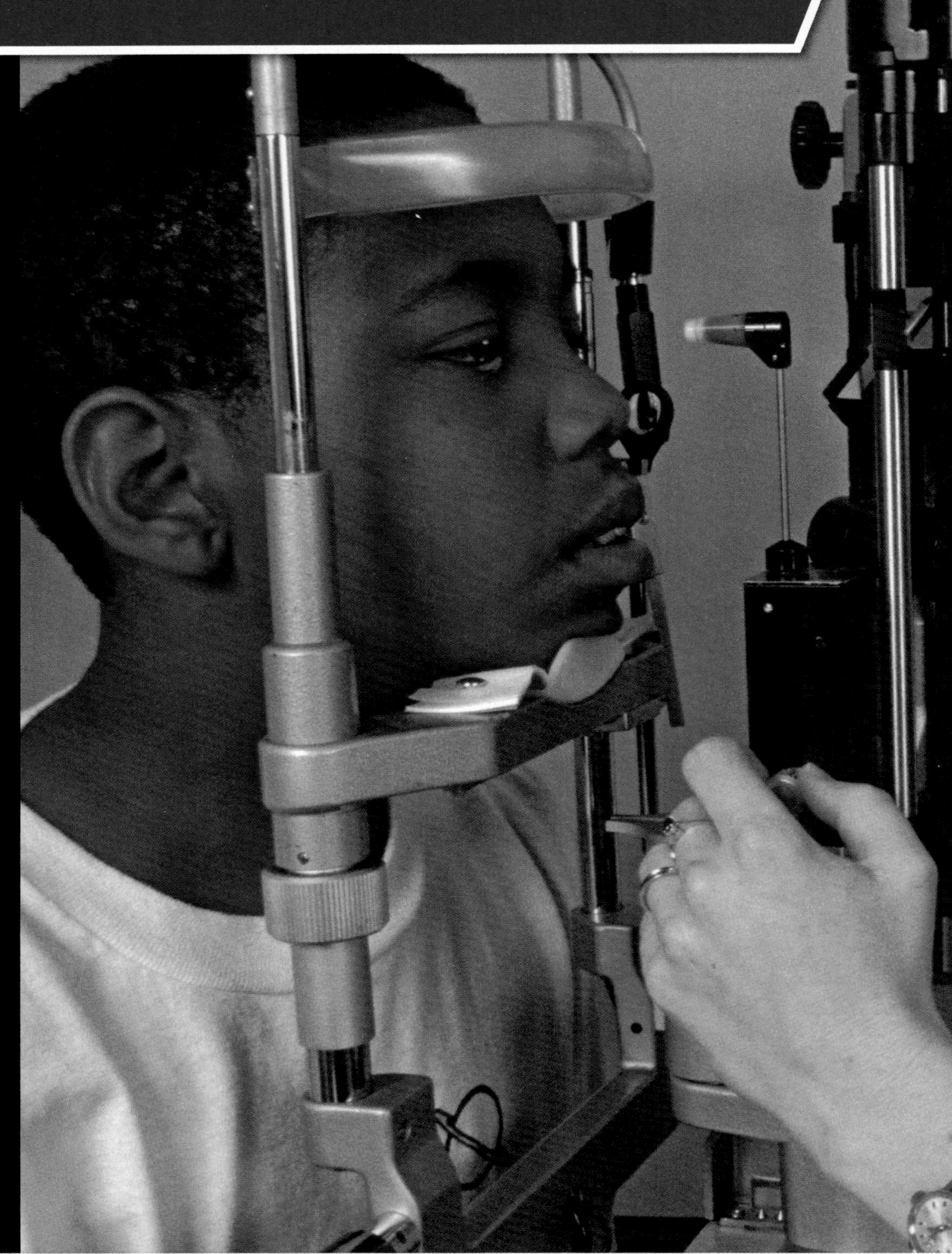

OVERALL Expectations

- evaluate the effectiveness and social benefits of technological devices and procedures that involve light
- investigate the reflection of light in plane and curved mirrors and the refraction of light in converging lenses
- demonstrate an understanding of the characteristics and properties of light related to reflection in mirrors and reflection and refraction in lenses

BIG Ideas

- Light has characteristics and properties that can be manipulated with mirrors and lenses for a range of uses.
- Society has benefitted from the development of a range of optical devices and technologies.

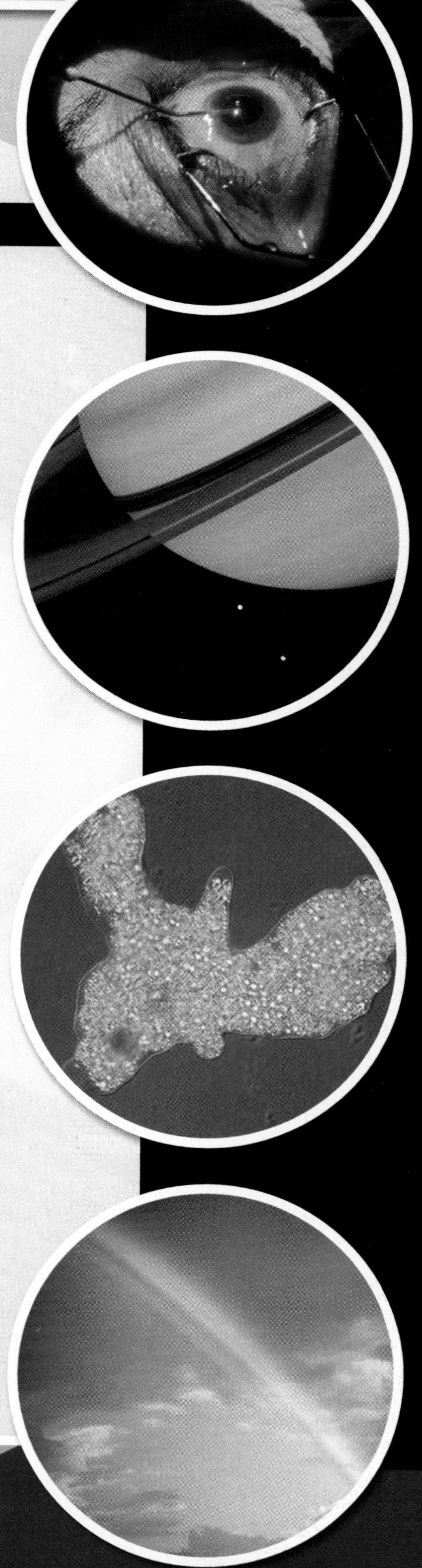

Focus on STSE

WINDOW **ON THE WORLD**

The human eye is the most remarkable of all optical devices. It gathers light and is our main source of information about the world around us.

Vision Problems

Our eyes allow us to see a faint, distant, twinkling star yet also allow us to read a book close up. Sometimes our eyes need help in order to see clearly.

1. How do we help the eye when it has vision problems? A

Unfortunately, current medical technology cannot solve all vision problems. Some people have to learn to cope without using their eyes to see. How do they do this?

2. Write a personal narrative account of how a visually impaired person is able to negotiate the seeing world. What aids, devices, and skills could they use to cope with vision loss? Your narrative should emphasize the social equity and competence of visually impaired persons and how they would like to be treated by others. T/I C A

Optical Devices

The human eye is not the only optical device that is useful to us. A flat mirror lets us see objects behind us, a telescope lets us see distant galaxies, and a microscope allows us to see tiny organisms.

The study of optics allows us to understand how light behaves. This knowledge allows us not only to learn about the universe around us but also to build devices such as cameras that are very useful to society.

3. What are some other optical devices that are very useful to humans? A

UNIT E

LOOKING AHEAD

UNIT E

Light and Geometric Optics

CHAPTER 11

The Production and Reflection of Light

Mirrors are among the most familiar optical devices.

CHAPTER 12

The Refraction of Light

When light travels through different media, besides air, we see some unexpected images.

CHAPTER 13

Lenses and Optical Devices

Lenses can be made of almost any transparent medium, including water.

UNIT TASK Preview

Building an Optical Device

Optical devices such as eyeglasses, microscopes, cameras, or projectors are useful instruments in our society. Optical devices can also be fun. For example, curved amusement park mirrors that seem to distort your body are used for entertainment.

Your task is to design, construct, test, refine, and then evaluate an optical device that satisfies a human need or want. You will use the properties of light presented in this unit. You are to explain how your device works and how it helps or entertains people.

The device should demonstrate that you considered social, health, environmental, and economic factors in designing and constructing it. Try to use environmentally friendly materials that are durable and economically viable. Does the device have any potential negative social or health risks? How will you minimize these?

UNIT TASK Bookmark

The Unit Task is described in detail on page 588. As you work through the unit, look for this bookmark and see how the section relates to the Unit Task.

ASSESSMENT

You will be assessed on how well you

- plan and design your device
- build, test, and improve your prototype
- explain and demonstrate how your device works and why it is useful to society

What Do You Already Know?

PREREQUISITES

Concepts
- Light is a form of energy
- Light travels in straight lines

Skills
- Drawing lines and measuring angles accurately
- Drawing neat, clearly labelled diagrams
- Solving an equation with one unknown
- Using laboratory equipment safely and appropriately
- Communicating scientific ideas appropriately
- Writing lab reports for investigations

1. (a) What is the main source of energy for Earth?
   (b) What evidence supports this? K/U
2. (a) What kind of light is shown in Figure 1?
   (b) What similarities and differences exist between this kind of light and the kind of light in your home? K/U A

Figure 1

3. Hold your hand just above a sheet of paper. T/I
   (a) What happens to the shadow as you move your hand closer to the paper?
   (b) How does the shadow change as you move your hand farther away from the paper? Why does this happen?
4. (a) You are wearing a T-shirt with writing on it. How does the writing appear when you look at the T-shirt in a mirror?
   (b) How is the writing on this ambulance (Figure 2) similar to what you saw looking at your T-shirt in the mirror? K/U

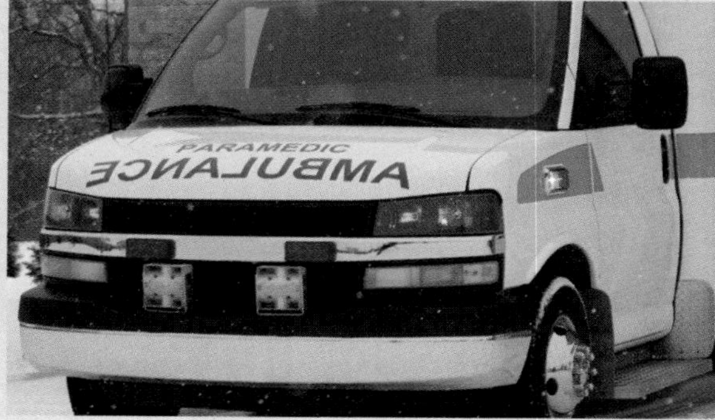
Figure 2

5. Funhouse mirrors are curved mirrors. Describe some of the different shapes your body might have in a funhouse mirror. C
6. (a) What physical shape does a security mirror in a store have (Figure 3)?
   (b) In such a mirror, is your appearance
   - larger or smaller?
   - upside down or right side up?
   - backwards? K/U T/I

Figure 3

7. (a) Describe the appearance of a straw in a glass of water when you look at it from above (Figure 4).
   (b) What do you think causes this unusual appearance? T/I

Figure 4

CHAPTER 11

The Production and Reflection of Light

KEY QUESTION: How do mirrors form images?

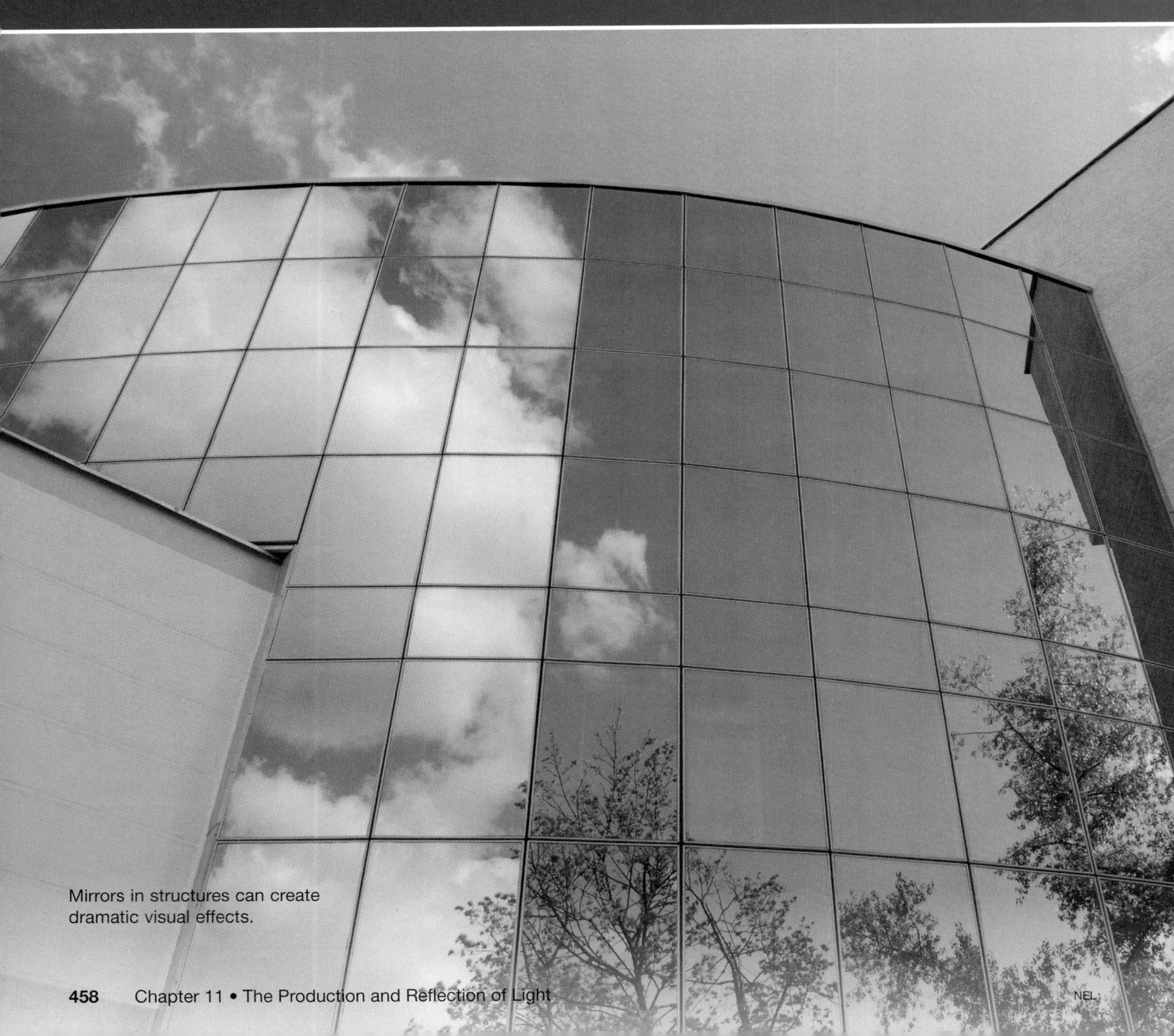

Mirrors in structures can create dramatic visual effects.

UNIT E
Light and Geometric Optics

CHAPTER 11
The Production and Reflection of Light

CHAPTER 12
The Refraction of Light

CHAPTER 13
Lenses and Optical Devices

KEY CONCEPTS

Optical devices benefit our society in many ways.

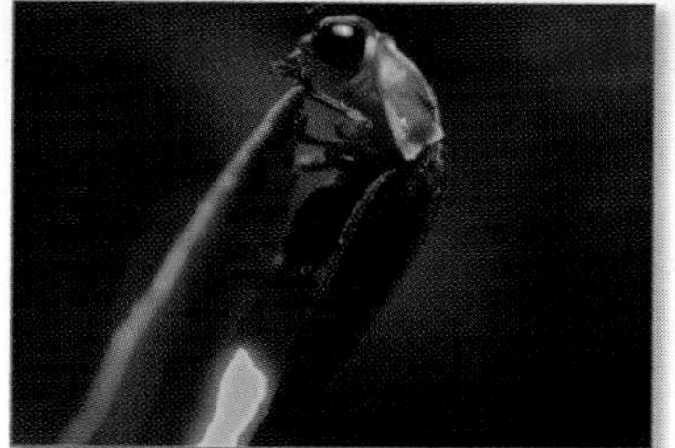

Light is produced by natural and artificial sources.

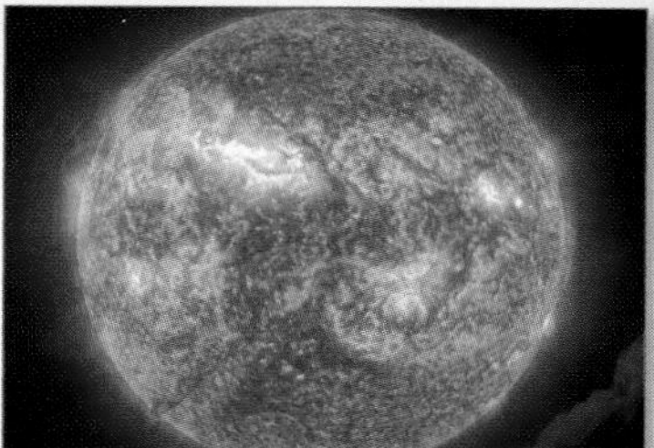

Light is an electromagnetic wave that travels at high speed in a straight line.

When light is reflected off a flat, shiny surface, the image is equal in size to the object and the same distance from the surface.

Images in flat mirrors are located at the point where the backward extensions of reflected rays intersect.

Curved mirrors produce a variety of images.

THE LASER

"It was sweeping round swiftly and steadily, this flaming death, this invisible, inevitable sword of heat. I perceived it coming towards me by the flashing bushes it touched, and was too astounded and stupefied to stir. I heard the crackle of fire in the sand pits and the sudden squeal of a horse that was suddenly stilled. Then it was as if an invisible yet intensely heated finger were drawn through the heather between me and the Martians, and all along a curving line beyond the sand pits the dark ground smoked and crackled."

This passage is from H. G. Wells's *The War of the Worlds*. It describes a "heat-ray" that invading Martians used against helpless humans. Wells's book was published over 100 years ago in 1898. In it, Wells made the first fictional reference to what we today call the laser.

Wells imagined that a laser would be used for destructive purposes. Although this use of a laser does exist today in many Hollywood movies, in reality the laser is a fairly benign invention. After all, the laser is found in CD and DVD players, pointers used in presentations, room-measuring devices used by real estate agents, and scanning devices at the check-out desks of most retail stores. The bright, intense light of a laser beam is used in every field, from manufacturing to entertainment.

Think about applications that you use or experience that make use of a laser. Can you think of other beneficial uses of a laser? Why do you think the image of the laser as a dangerous weapon is persistent even today? Does mass media always portray science and technology in a realistic way?

WHAT DO YOU THINK?

Many of the ideas you will explore in this chapter are ideas that you have already encountered. You may have encountered these ideas in school, at home, or in the world around you. Not all of the following statements are true. Consider each statement and decide whether you agree or disagree with it.

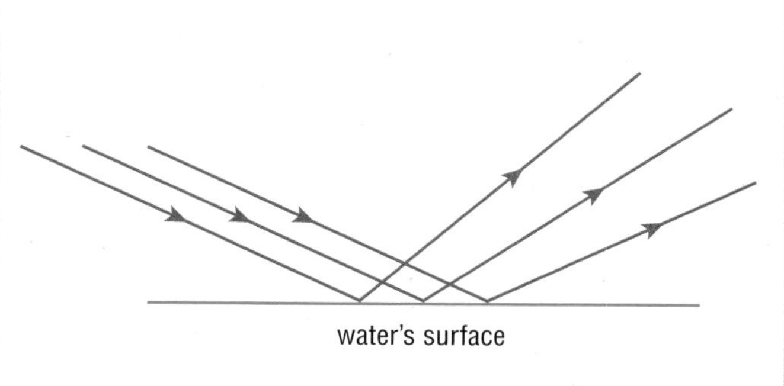

1 This diagram accurately shows light reflecting off the surface of very still water.
Agree/disagree?

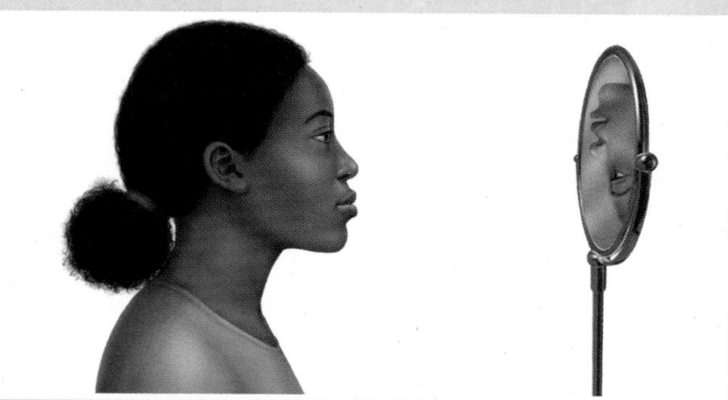

4 This diagram accurately shows how an image appears in a makeup mirror.
Agree/disagree?

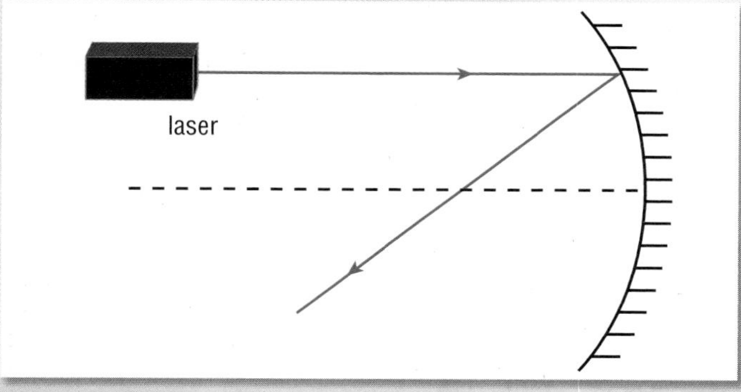

2 This diagram accurately shows a laser beam reflecting off a curved mirror.
Agree/disagree?

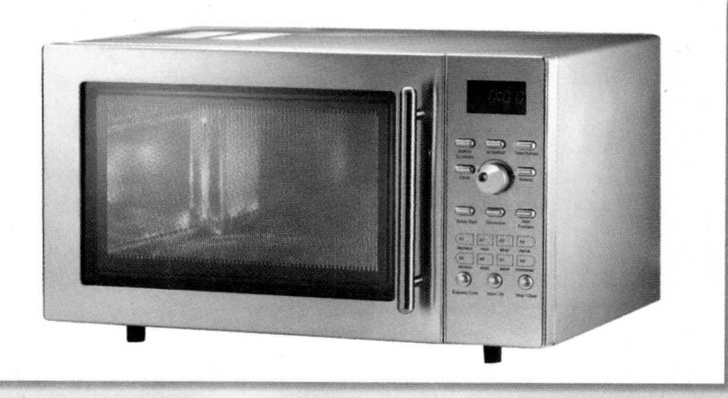

5 Microwaves travel at the speed of light.
Agree/disagree?

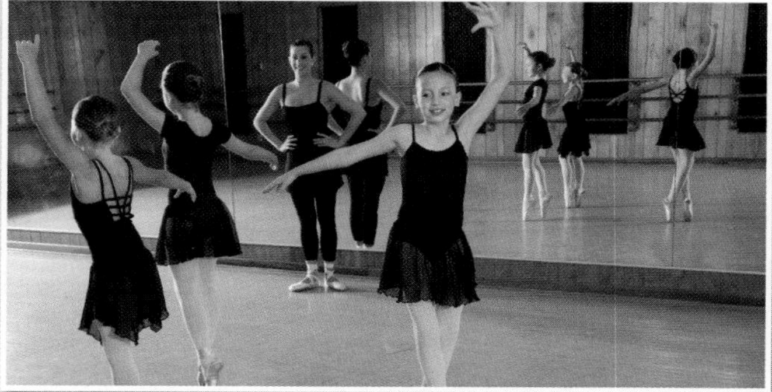

3 A full-length mirror is necessary in order for you to see your whole body in reflection.
Agree/disagree?

6 A luminous object such as a candle radiates light in all directions.
Agree/disagree?

Writing a Persuasive Text

WRITING TIP

As you work through the chapter, look for tips like this. They will help you develop literacy strategies.

One purpose of a persuasive text is to convince an audience to accept an opinion on the basis of powerful reasons and logical thinking. The following is an example of a persuasive text written about lighting sources in Section 11.2. Beside it are the strategies the student used to write the persuasive text effectively.

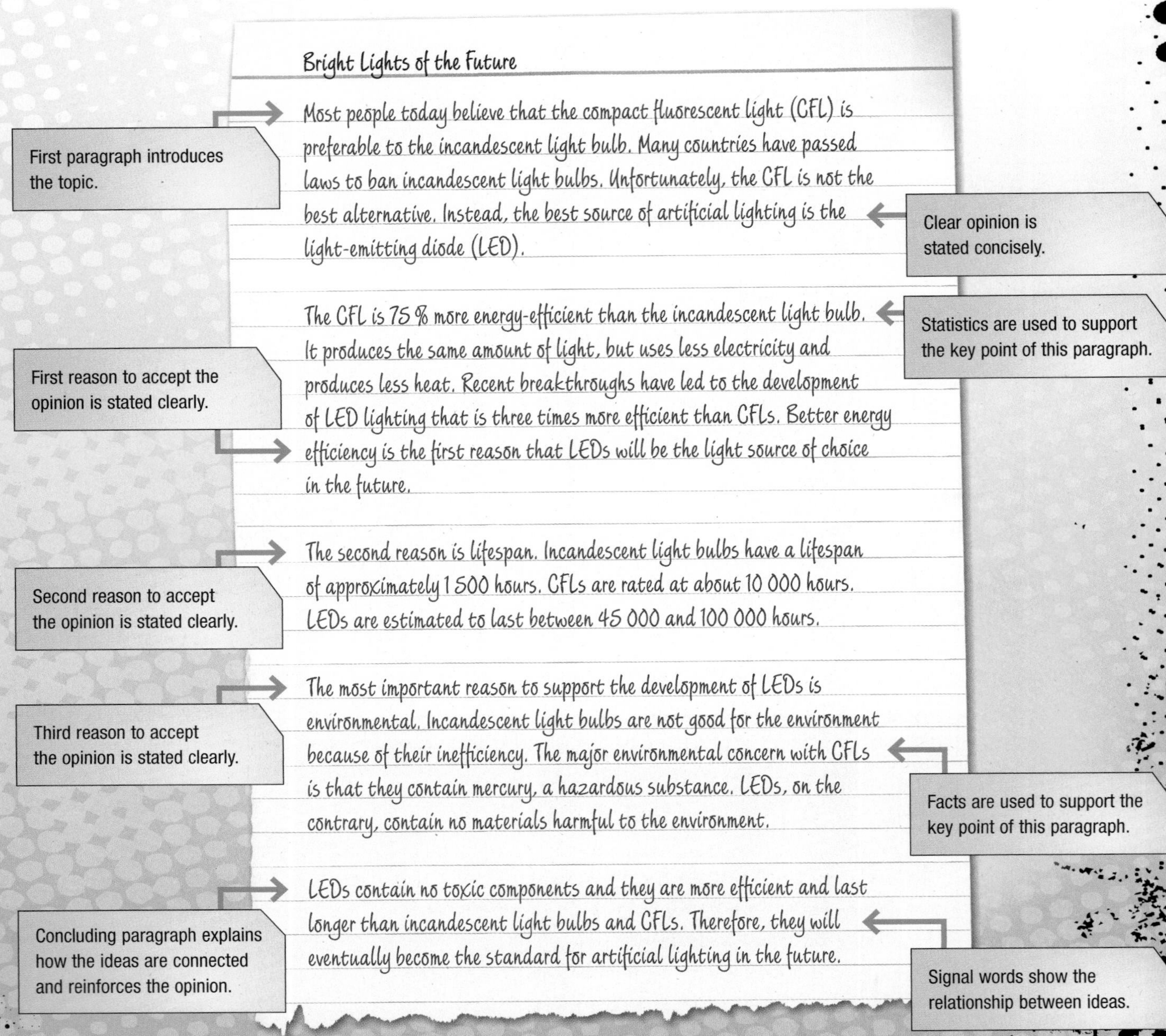

Bright Lights of the Future
First paragraph introduces the topic.
Most people today believe that the compact fluorescent light (CFL) is preferable to the incandescent light bulb. Many countries have passed laws to ban incandescent light bulbs. Unfortunately, the CFL is not the best alternative. Instead, the best source of artificial lighting is the light-emitting diode (LED).
Clear opinion is stated concisely.
The CFL is 75 % more energy-efficient than the incandescent light bulb. It produces the same amount of light, but uses less electricity and produces less heat. Recent breakthroughs have led to the development of LED lighting that is three times more efficient than CFLs. Better energy efficiency is the first reason that LEDs will be the light source of choice in the future.
Statistics are used to support the key point of this paragraph.
First reason to accept the opinion is stated clearly.
The second reason is lifespan. Incandescent light bulbs have a lifespan of approximately 1 500 hours. CFLs are rated at about 10 000 hours. LEDs are estimated to last between 45 000 and 100 000 hours.
Second reason to accept the opinion is stated clearly.
The most important reason to support the development of LEDs is environmental. Incandescent light bulbs are not good for the environment because of their inefficiency. The major environmental concern with CFLs is that they contain mercury, a hazardous substance. LEDs, on the contrary, contain no materials harmful to the environment.
Third reason to accept the opinion is stated clearly.
Facts are used to support the key point of this paragraph.
LEDs contain no toxic components and they are more efficient and last longer than incandescent light bulbs and CFLs. Therefore, they will eventually become the standard for artificial lighting in the future.
Concluding paragraph explains how the ideas are connected and reinforces the opinion.
Signal words show the relationship between ideas.

What Is Light?

What would Earth be like without light from the Sun? Sunlight is the energy that makes life possible on Earth. The Sun is our closest star—about 1.50×10^8 km from Earth. That is nearly 400 times farther than the Moon. The nuclear reactions that occur within the Sun produce tremendous amounts of energy (Figure 1). One form this energy takes is light, which the Sun emits in all directions through the vacuum of outer space.

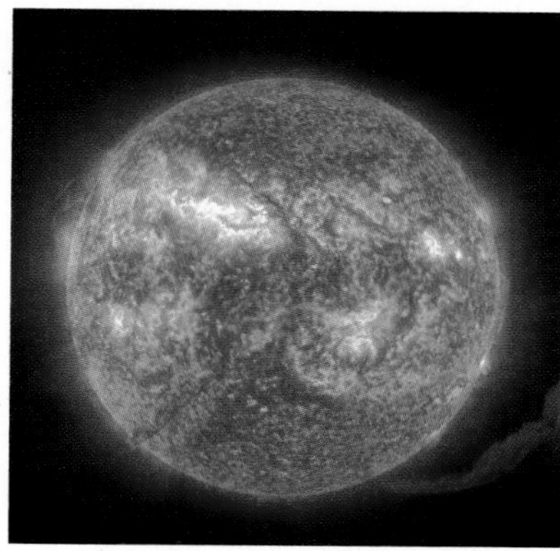

Figure 1 The Sun releases a great deal of energy as a result of its nuclear reactions.

Earth captures only a tiny fraction of the Sun's light. That little fraction, however, provides just enough energy to heat Earth's surface and allow photosynthesis to occur, both in the oceans and on land. From plankton blooms in the ocean to forests on land, plants are the basis of the food chain for almost all organisms on Earth (Figure 2).

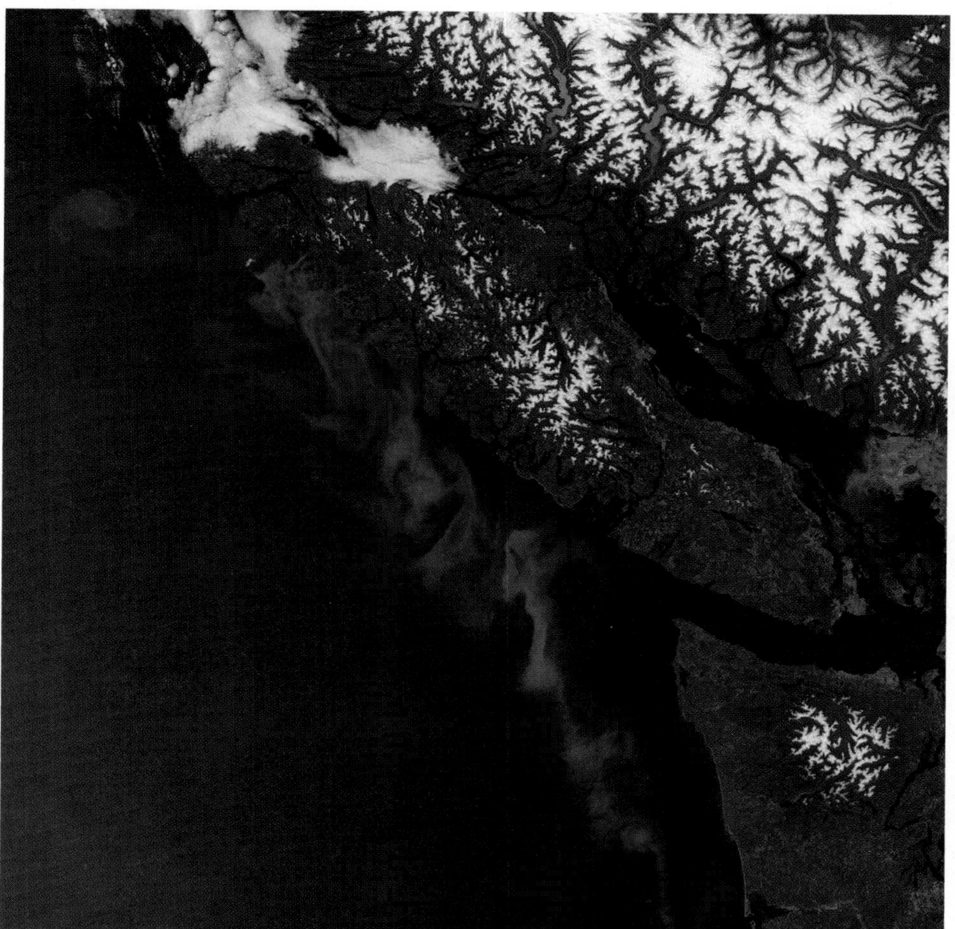

Figure 2 A satellite view of a phytoplankton bloom off the coast of Vancouver Island. Phytoplankton are tiny photosynthesizing organisms that live in the oceans. They produce about half of all the oxygen produced by plants on Earth.

To learn more about the different types of energy produced by the Sun and how the Sun works,

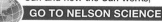

GO TO NELSON SCIENCE

For centuries, scientists have tried to understand the nature of light and its properties. Some of these properties are easily observable. For example, light travels at a very high speed. When you turn on the light switch in a room, the room immediately fills with light. Light travels so fast that something travelling at the speed of light could circle Earth's equator about 7.5 times in just one second.

WRITING TIP

Writing a Persuasive Text

Use the first paragraph to identify the topic and state your main idea/opinion concisely. For example, if you are writing about plants being the basis of the food chain, you could begin: "Light from the Sun enables plants on Earth to grow. Without plants, humans and other animals would starve."

DID YOU KNOW?

The Energy of Light

Scientists call a small packet of light energy a photon. The term "photon" was coined by American chemist Gilbert Lewis in 1926, and was based on the Greek word for "light." Science fiction films have made use of this term (for example, photon torpedoes and photonic cannon).

DID YOU **KNOW?**

The Imaginary Ether

In the 19th century, scientists thought that a luminiferous ("light-carrying") substance, called ether, was necessary for light to travel through outer space. Today, scientists know that light is an electromagnetic wave and does not require a medium to travel through the vacuum of outer space.

Light also travels in straight lines. When you turn on a flashlight in a dark room with dust in the air, you can see a beam of light travelling in a straight line (Figure 3). Sharp shadows around objects such as trees or railings are also evidence of the straight-line nature of light. In fact, that is why you can see Earth's shadow on the Moon during a lunar eclipse. But what exactly is light?

Figure 3 A flashlight shows very clearly that light travels in straight lines.

Light—An Electromagnetic Wave

Recall that heat energy can be transferred by either conduction or convection. Both methods involve particles to transfer heat. That is, conduction and convection require a medium for transmission. A **medium** is any physical substance that acts as a carrier for the transmission of energy. Conduction occurs most often in solids, whereas liquids and gases are often good carriers for convection. Light, however, travels through the vacuum of outer space. This means that light does not require a medium for transmission. Instead, light energy is transferred through **radiation**.

medium any physical substance through which energy can be transferred

radiation a method of energy transfer that does not require a medium; the energy travels at the speed of light

In 1801, English physicist Thomas Young demonstrated that, under certain conditions, light shows wave-like properties. In 1864, James Clerk Maxwell (Figure 4), another English physicist, predicted that electricity and magnetism couple together, in an incredibly fast game of hop-skip-jump, to form a chain travelling through space. He also predicted that the resulting **electromagnetic wave** does not require a medium for transmission and that this wave travels at the speed of light. Unfortunately, Maxwell died at the young age of 48 and did not live to see his prediction confirmed. Proof of the existence of electromagnetic waves came in 1887 when German physicist Heinrich Hertz discovered low-energy electromagnetic waves that we now call radio waves. Additional proof came in 1895 when William Konrad Roentgen, another German scientist, discovered high-energy electromagnetic waves called X-rays.

electromagnetic wave a wave that has both electric and magnetic parts, does not require a medium, and travels at the speed of light

Figure 4 James Clerk Maxwell (1831–1879) predicted the existence of electromagnetic waves.

DID YOU **KNOW?**

X—the Unknown

Roentgen called the rays that he discovered in 1895 "X-rays" because he did not initially know what they were. The "X" stood for unknown. Today, the German word for X-ray is "Roentgen."

Today, scientists have identified many different kinds of electromagnetic waves in addition to radio waves and X-rays. Microwaves, radar, and ultraviolet light are a few examples of other electromagnetic waves. **Visible light** is any electromagnetic wave that the human eye can detect. Scientists classify electromagnetic waves based on the energy of the waves. This classification system is called the **electromagnetic spectrum** (Figure 5).

visible light electromagnetic waves that the human eye can detect

electromagnetic spectrum the classification of electromagnetic waves by energy

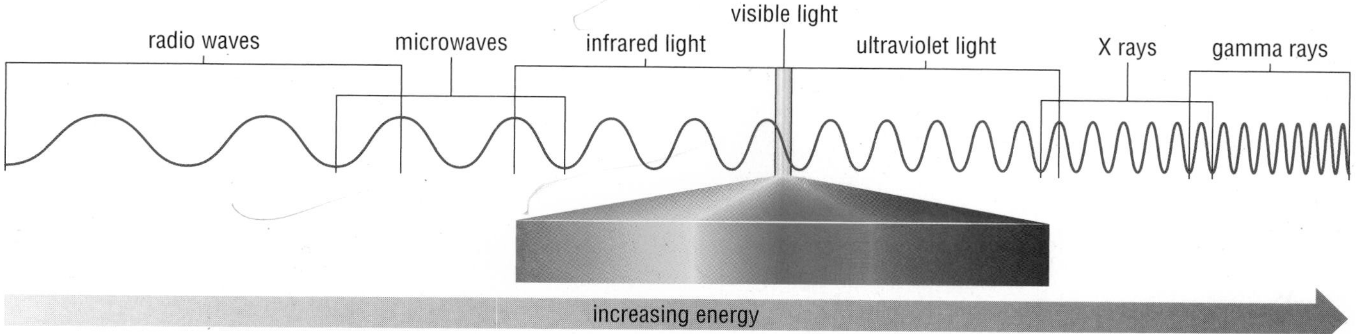

Figure 5 The electromagnetic spectrum. Note the different categories as the energy of the electromagnetic wave increases.

RESEARCH THIS PROTECTING YOURSELF FROM THE SUN

SKILLS: Researching, Analyzing the Issue, Communicating, Evaluating

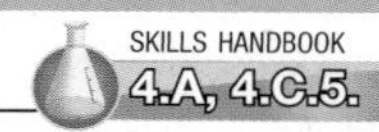

In recent years, newspapers, radio, and TV have been regularly providing updates on how safe it is to go unprotected in the Sun, particularly in the summer. In this activity, you will research why exposure to the Sun and to tanning lamps is now a health concern. You will also investigate the rating scale used to determine safe limits of exposure and how to protect yourself from overexposure (Figure 6).

Figure 6

1. Research why extended sun exposure can be hazardous to your health. Find out some positive benefits of moderate sun exposure.
2. Research the rating scale that has been set up to determine safe exposure limits.
3. Research what you can do to protect yourself from overexposure to the Sun.
4. Research the type of light emitted by tanning lamps. Find out the health concerns associated with exposure to these lamps.

GO TO NELSON SCIENCE

A. Why is some exposure to the Sun necessary? T/I
B. Why can overexposure to the Sun be a health risk? T/I
C. Explain how the UV index has been set up. T/I
D. What is the difference between UVA and UVB? T/I
E. What does SPF mean? T/I
F. How is sunblock different from sunscreen? T/I
G. List the criteria that you would use to choose the best type of sunscreen for yourself. T/I
H. List other things that you can do to protect yourself from overexposure to the Sun. T/I
I. Compare exposure to tanning lamps with Sun exposure. T/I
J. Create an information poster or brochure to educate people about protecting themselves from the Sun. T/I A C

DID YOU KNOW?

Adaptation Affects Detection

A bee's eyes are adapted to detect ultraviolet light when locating flowers. Similarly, the human eye is adapted to detect visible light but not other types of electromagnetic waves. Being unable to see other electromagnetic waves does not mean that they are not there or that you are not affected by them.

WRITING TIP

Writing a Persuasive Text

Use facts, statistics, examples, or reasons to support your main idea. For example, in a persuasive essay about the benefits of electromagnetic waves, you might include supporting details from a variety of sources.

Electromagnetic Waves in Our Society

Electromagnetic waves have many uses. Table 1 shows some of the uses and natural phenomena that involve electromagnetic waves.

Table 1 The Many Uses of Electromagnetic Waves

(increasing energy ↓)

Type of electromagnetic wave	Use/phenomena
radio waves	• AM/FM radio • TV signals • cellphone communication • radar • astronomy (for example, discovery of pulsars)
microwaves	• telecommunications • microwave ovens • astronomy (for example, background radiation associated with the Big Bang)
infrared light	• remote controls (for example, DVD players and game controllers) • lasers • heat detection (for example, leakage from windows, roofs) and remote sensing • keeps food warm (in fast-food restaurants) • astronomy (for example, discovering the chemical composition of celestial bodies) • physical therapy
visible light	• human vision • theatre/concert lighting • rainbows • visible lasers • astronomy (for example, optical telescopes, discovering the chemical composition of celestial bodies)
ultraviolet light	• causes skin to tan and sunburn • increases risk of developing skin cancer • stimulates production of vitamin D • kills bacteria in food and water (sterilization) • "black" lights • ultraviolet lasers • astronomy (for example, discovering the chemical composition of celestial bodies)
X-rays	• medical imaging (for example, of teeth and broken bones) • security equipment (for example, scanning of luggage at airports) • cancer treatment • astronomy (for example, study of binary star systems, black holes, the centres of galaxies)
gamma rays	• cancer treatment • astronomy (for example, study of nuclear processes in the universe) • product of some nuclear decay

The Colours Associated with Visible Light

White visible light is composed of a continuous sequence of colours. This colour sequence is called the **visible spectrum** (Figure 7). Traditionally, seven distinct colours have been identified: red, orange, yellow, green, blue, indigo, and violet.

visible spectrum the continuous sequence of colours that make up white light

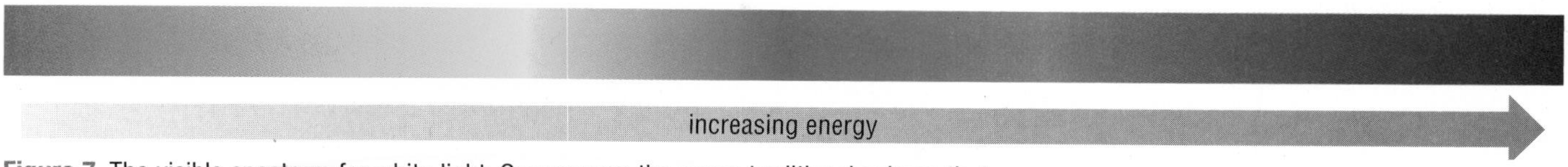

Observe the colours in white light by doing the "Viewing the Visible Spectrum" activity.

increasing energy

Figure 7 The visible spectrum for white light. Can you see the seven traditional colours that Newton identified?

TRY THIS VIEWING THE VISIBLE SPECTRUM

SKILLS MENU: Predicting, Observing, Analyzing, Communicating

Equipment and Materials: a ray box; two triangular prisms; a sheet of white paper

Part A

1. Place one prism on a sheet of paper. Trace its outline.

   When handling the triangular glass prism, take care not to cut your fingers if it is chipped.

2. Shine a single beam of light from the ray box on one side of the prism (Figure 8). Adjust the position of the ray box until a clear spectrum is visible on the other side of the prism. Identify the colours that you can clearly see in the spectrum. Mark the location of each colour on the paper. Note that the spectrum is easier to observe if the room is dark.

When unplugging the ray box, do not pull the electric cord. Pull the plug itself.

Figure 8

A. How many of the seven colours that Newton identified were you able to see? Discuss with your lab partner why all of the seven colours might not be clearly visible. K/U T/I

B. Write down an example in nature where you can see a similar visible spectrum. Share your example with your lab partner. A

Part B

3. Predict whether or not prisms can be used to recombine a spectrum back into a beam of white light. If this is possible, how many prisms do you think would be needed?

4. Test your prediction by using a variety of prisms to first create a spectrum, and then to attempt to reform the beam of white light. Experiment with different numbers and arrangements of prisms. Figure 9 illustrates just one of many possible arrangements.

Figure 9

5. Make careful drawings of any successful prism arrangements.

C. Were your predictions correct? Explain. T/I

D. Make a general statement about the composition of white light. T/I

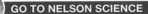

LEARNING TIP

Remembering a Sequence

A mnemonic is a word technique to help remember a more complicated sequence. The fictitious name ROY G. BIV is a mnemonic to remember the order of colours in white visible light: **R**ed, **O**range, **Y**ellow, **G**reen, **B**lue, **I**ndigo, and **V**iolet. Can you think of another way to remember this colour sequence?

To learn more about Newton's life and work,

GO TO NELSON SCIENCE

The previous experiment is similar to one performed by English physicist Isaac Newton (Figure 10) in 1666. As you discovered in the activity, a triangular prism slows down the speed of light. In a vacuum, each colour of the visible spectrum travels at the same high speed: the speed of light. But within the prism, each of these colours travels at a slower speed than the speed of light in a vacuum. Red light, the least energetic, is slowed the least, whereas violet light, the most energetic, is slowed the most. That is why a prism can separate white light into different colours. The behaviour of light entering glass is discussed in more detail in the next chapter.

Figure 10 Isaac Newton (1642–1727) was the first person to separate white light into the visible spectrum.

Newton was the first scientist to state that seven distinct colours were visible in white light. He coined the term "spectrum" to describe them, which means "appearance" in Latin. Some people have difficulty identifying indigo as a separate colour between blue and violet.

DID YOU KNOW?

Invisible Energy

In 1800, German-born astronomer William Herschel placed a thermometer in the path of each colour of the visible spectrum in order to measure the temperature associated with it. He was quite surprised when he placed his thermometer in the dark region beyond the red light and noticed a higher temperature than any temperature in the spectrum. Herschel concluded that there must be an invisible form of energy next to visible red light. He had discovered infrared light. The prefix "infra" means "below."

The Surprising Universe

For centuries, visible light was the only tool that astronomers could use to probe the universe. Today, when astronomers look at an image of a galaxy photographed using X-rays, they see a completely different picture than if they looked at the same galaxy using visible light (Figure 11).

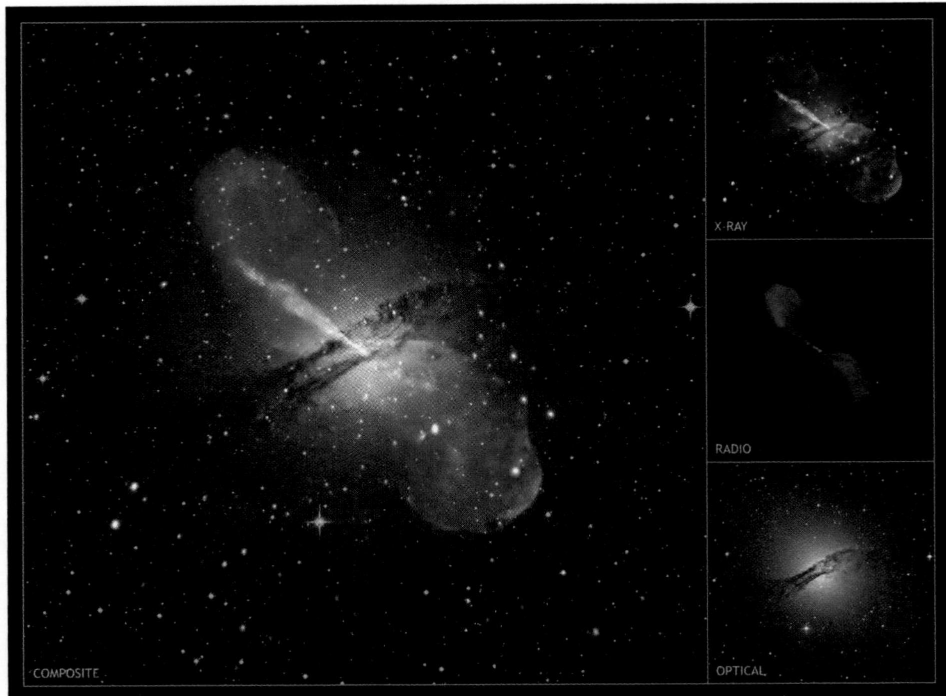

Figure 11 The galaxy Centaurus A appears quite different when examined using X-rays, visible light, or radio waves. A composite photo combining all three electromagnetic rays gives a more complete view of Centaurus A.

Scientists now realize that visible light provides only limited information about the universe. By using other parts of the electromagnetic spectrum to collect and analyze data about stars and galaxies, scientists have discovered that the universe is a far more violent and surprising place than they had imagined.

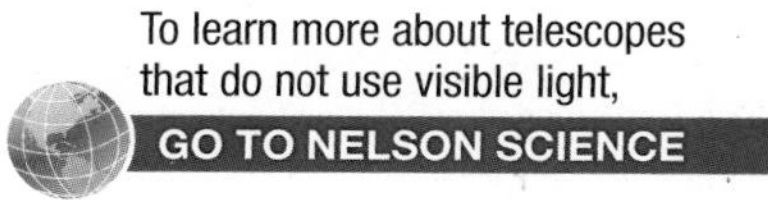
To learn more about telescopes that do not use visible light,
GO TO NELSON SCIENCE

IN SUMMARY

- Electromagnetic waves travel at the speed of light in a vacuum and do not require a medium for transmission.
- Light is an electromagnetic wave.
- The electromagnetic spectrum consists of light listed according to its different energy levels. The order from least energy to most energy is radio waves, microwaves, infrared light, visible light, ultraviolet light, X-rays, and gamma rays.
- White light is composed of a continuous spectrum of colours.

CHECK YOUR LEARNING

1. In terms of heat transfer, how is *radiation* different from *conduction* and *convection*? K/U
2. What two major properties did Maxwell predict that electromagnetic waves would possess? K/U
3. What two discoveries confirmed the existence of electromagnetic waves? K/U
4. Write these electromagnetic waves in order from lowest energy to highest energy: infrared light, X-rays, red light, gamma rays, and microwaves. K/U
5. Sunscreen, if used properly, can protect you from getting a sunburn. From which electromagnetic waves must sunscreen protect the skin? K/U
6. List the seven colours that Newton identified in the visible spectrum of white light. K/U
7. Why is it useful to examine the universe using parts of the electromagnetic spectrum other than visible light? K/U A
8. List some devices that you have used or plan to use today that involve electromagnetic waves. K/U A
9. Match each electromagnetic wave from column A with the term from column B that is most closely related. K/U A

Column A	Column B
(a) X-rays	vitamin D
(b) ultraviolet light	telecommunications
(c) radio waves	cancer treatment
(d) infrared light	radar
(e) microwaves	theatre/concert effects
(f) gamma rays	baggage screening
(g) visible light	DVD player remote control

10. Prepare a mind map listing as many physical properties for light as you can. Base your mind map on Figure 12. T/I C

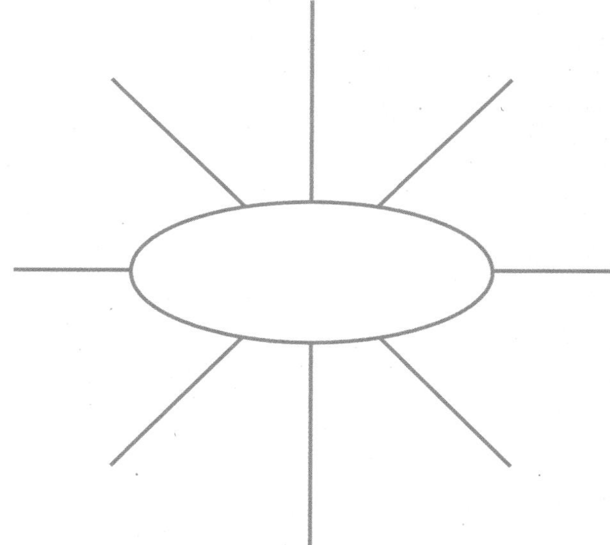

Figure 12

11. Briefly describe how you could demonstrate that white light is composed of many different colours. C

11.2 How Is Light Produced?

Figure 1 The Sun produces its own light, whereas a tree can only reflect light.

luminous produces its own light

non-luminous does not produce its own light

Most people think that light comes only from sources such as the Sun, a light bulb, or a fire. The reality is that light enters your eyes from all objects that you see. You can see a tree in front of you only if light is coming from the tree into your eyes (Figure 1). The difference between light coming from the Sun and light coming from a tree is that the Sun radiates its own light, whereas the tree can only reflect light.

The Sun is **luminous**, which means that it produces its own light. Other examples of luminous sources are a light bulb, a lit match, and a flashlight that is turned on. A tree does not produce its own light, so it is non-luminous. A **non-luminous** source does not produce its own light and can be seen only by using reflected light. Most objects around you are non-luminous: this textbook, a pencil, and a bicycle are just a few. Let's now examine some luminous sources to see exactly how they produce light.

Light from Incandescence

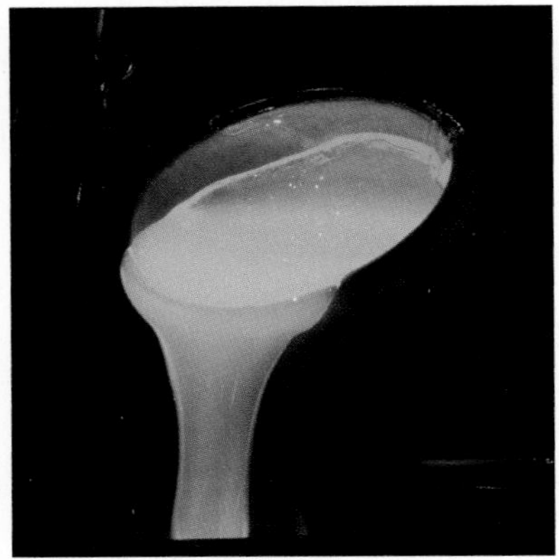
Figure 2 Molten glass glows orange at very high temperatures.

incandescence the production of light as a result of high temperature

When the burner of a stove is set to a high temperature, the filament glows. Any object, as it gets hotter and hotter, will eventually produce light (Figure 2). As the object gets hotter, the colours of light produced change from red, to orange, to yellow, to white, and then to bluish-white. This process of producing light as a result of high temperature is called **incandescence**. Light from a burning candle and the lit sparks flying off a grinder are examples of incandescence.

Incandescence also occurs in an incandescent light bulb (Figure 3). A thin wire filament, usually made of tungsten, glows as electricity passes through it. The filament becomes so hot that it gives off visible light. It also emits infrared light that you feel as heat radiating from the bulb. Depending on the type of bulb, only 5 % to 10 % of the electricity going through the filament is actually converted into visible light. The rest of the energy is converted into infrared light. For this reason, incandescent bulbs are very inefficient light sources.

For an incandescent bulb to work, all the air from the bulb must be removed. It is replaced with non-reactive gases. This way, the tungsten filament cannot combine with oxygen in the air, which would make the filament burst into flame. Even without oxygen present, the filament does eventually disintegrate and break.

American inventor Thomas Edison is credited with producing the first commercially useful incandescent bulb (Figure 4).

Figure 3 The modern incandescent bulb

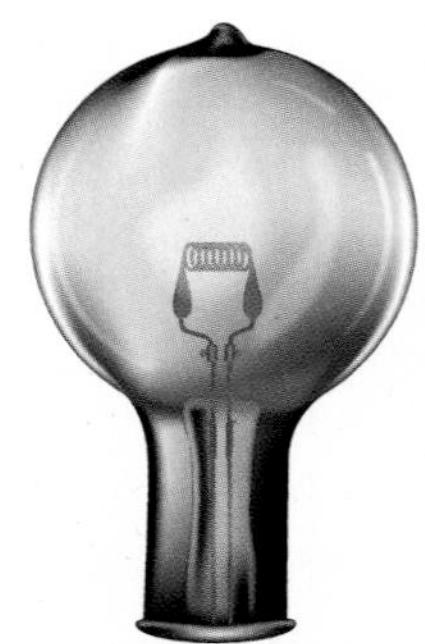
Figure 4 Compare Thomas Edison's patent from 1880 for an incandescent bulb with Figure 3.

Light from Electric Discharge

Every time you see a flash of lightning (Figure 5) or walk past a lit neon sign for a business (Figure 6), you are seeing another form of light production. This light, known as **electric discharge**, is produced by an electric current passing through a gas. The electricity causes the gas to glow. Although the term "neon lighting" is usually used to describe all signs that use this process, many gases other than neon can be used. Neon gas produces the familiar red colour, helium produces a gold-coloured glow, argon a pale violet-blue, and krypton a greyish off-white.

electric discharge the process of producing light by passing an electric current through a gas

Figure 5 Lightning is a dramatic example of an electric discharge through a gas. In this case, the gas is the air in Earth's atmosphere.

Figure 6 Neon gas produces this characteristic red colour as an electric current passes through it.

DID YOU **KNOW?**

Canadian Bright Lights

Many people contributed to the development of the incandescent bulb. Two Torontonians, Henry Woodward and Mathew Evans, filed a Canadian patent in 1874 for an incandescent bulb. Their bulb used a carbon filament and was filled with nitrogen gas. They tried to market their bulb but had no commercial success. Edison recognized the usefulness of their work and purchased the rights to their patent. He incorporated their work into the development of his own incandescent bulb using a carbon filament.

GO TO NELSON SCIENCE

The development of electric discharge in gas tubes comes from the invention in 1855 of a powerful vacuum pump by German physicist Heinrich Geissler. The pump allowed Geissler to evacuate (remove) most of the air from a closed tube. Geissler's colleagues noticed that the remaining air in one of these tubes glowed when an electric current passed through it. Further experiments showed that the colour of the glow depended on which gas was inside the tube. These glowing gas tubes were originally called Geissler tubes (Figure 7). When you look at the different colours of commercial lighting, you are really looking at Geissler tubes.

WRITING **TIP**

Writing Persuasively

Imagine that you are writing an essay promoting the idea that Canadians developed the electric light bulb before Thomas Edison did. Use the body of the essay to state and explain the key points of your argument. For example, state the date of the Canadian patent, reasons for the patent's failure, and so on.

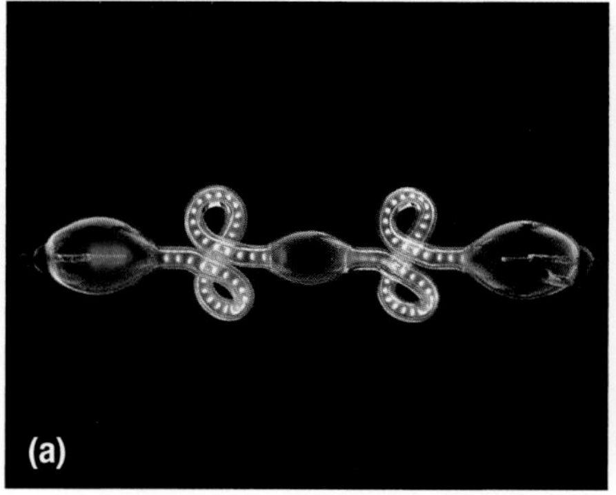

(a)

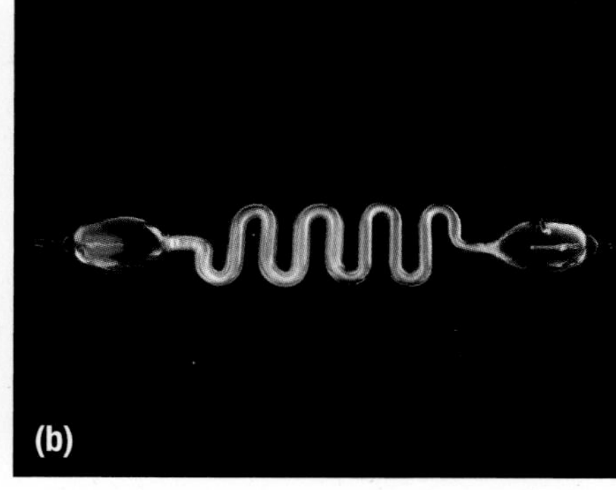

(b)

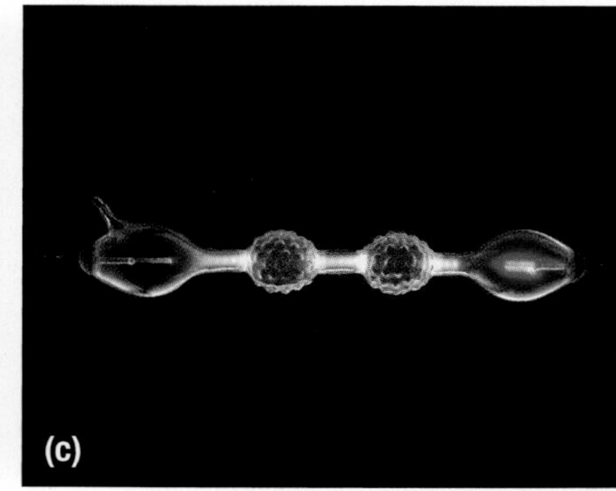

(c)

Figure 7 Because Geissler was also a glass-blower, early Geissler tubes were very elaborate.

DID YOU KNOW?

Watch That Dial!

Watch dials from the early 20th century to shortly after World War II were made luminous through the use of a paint that contained radium. Radium is a radioactive material, and the health effects of radioactivity were not widely understood at that time. As a result, many of the people who worked at those watch factories suffered serious health effects from exposure to radium.

Light from Phosphorescence

People are familiar with objects that glow in the dark, such as the dials on some wristwatches and clocks, and glow-in-the-dark stickers. These glow-in-the-dark materials are coated with phosphors, special materials that give off light through a process called **phosphorescence**. Phosphors absorb light energy, primarily ultraviolet light. These materials keep some of the energy and release visible light of lower energy. However, they do not do so immediately and hold onto this energy for varying periods of time ranging from seconds up to days, depending on the material. Because light is emitted over a period of time, phosphorescent materials are often described as "glow-in-the-dark" (Figure 8).

phosphorescence the process of producing light by the absorption of ultraviolet light resulting in the emission of visible light over an extended period of time

fluorescence the immediate emission of visible light as a result of the absorption of ultraviolet light

Figure 8 Glow-in-the-dark toys are examples of phosphorescence.

Light from Fluorescence

Several laundry detergents claim that they can make clothes brighter. This claim depends on the process of fluorescence. **Fluorescence** occurs when an object absorbs ultraviolet light and immediately releases the energy as visible light.

Detergent manufacturers often add fluorescent dyes to their detergents. Shirts washed in this detergent appear to glow slightly. This process is apparent even in visible light because normal daylight includes a small amount of ultraviolet light. The fluorescent dye on the clothing absorbs UV light and emits visible light. The eye detects both this emitted light and the light normally reflected from the shirt, so the shirt looks brighter. Highlighter pens work on the same principle. The ink in these pens contains a fluorescent dye that causes the ink to glow in the presence of the ultraviolet part of normal daylight.

Fluorescent lights are the most common application of fluorescence. A fluorescent light makes use of both electric discharge and fluorescence. A fluorescent light tube is filled with very low-pressure mercury vapour. The inner surface of the tube is also coated with a fluorescent material. When a fluorescent light is turned on, the electric current causes the mercury atoms to emit ultraviolet light. This ultraviolet light then strikes the fluorescent inner surface of the tube, resulting in the production of visible light (Figure 9).

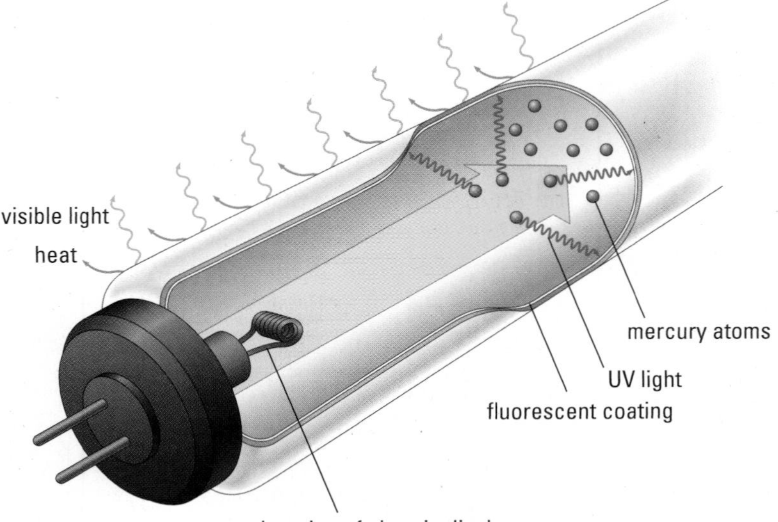

Figure 9 In a fluorescent light, electricity causes the mercury vapour to emit ultraviolet light. This ultraviolet light hits the fluorescent material on the inner surface of the light tube, causing visible light to be emitted.

DID YOU KNOW?

Rock On

The process of fluorescence is named after the mineral fluorite. Fluorite glows when it is illuminated with ultraviolet light. Many other naturally occurring minerals also fluoresce in brilliant colours when exposed to ultraviolet light.

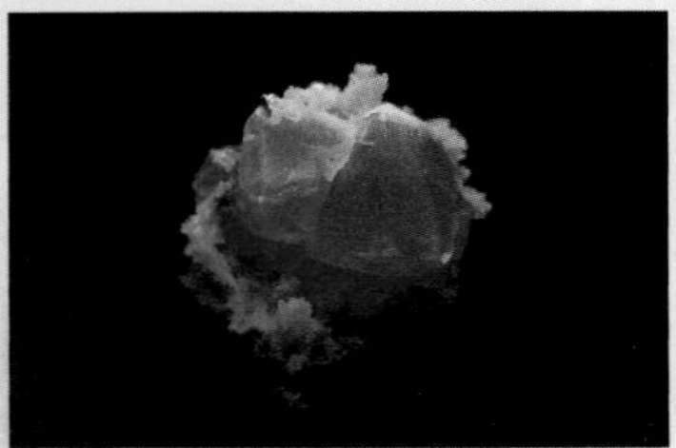

Fluorescent lights are four to five times more energy efficient than incandescent bulbs. A fluorescent light can provide the same light output as an equivalent incandescent bulb but produces much less heat and uses far less electricity. Compact fluorescent lights (CFLs) are recommended for use in homes and businesses because they use less energy to operate (Figure 10).

Widespread use of fluorescent lights could lead to significant energy savings. Although fluorescent lights are more expensive to purchase than incandescent bulbs, they are less expensive to operate and last much longer than incandescent bulbs. There is one slight downside, however, to fluorescent lights. They contain mercury and should *not* be disposed of with regular household waste. Instead, fluorescent lights should be treated like other hazardous household waste, such as paint and batteries. Fluorescent lights need to be taken to appropriate recycling centres, where they will be properly treated.

Figure 10 Compact fluorescent lights (CFLs) can lead to significant energy savings in homes and businesses.

CITIZEN **ACTION**

Thinking for the Future

SKILLS HANDBOOK 4.A.7, 4.C.6.

The Issue

Much of the energy used by our society comes from fossil fuels. These fuels add significant amounts of pollutants and greenhouse gases to Earth's atmosphere. Greenhouses gases are a major factor in global warming.

It is estimated that if each household in Canada replaced just one 60 W incandescent bulb with an equivalent light output 15 W CFL, then there would be an energy saving of $73 million per year. In addition, that one CFL would reduce carbon dioxide emissions by 400 000 t—the equivalent of taking 66 000 cars off the road.

What Can You Do To Help?

Think of how you can personally reduce your contribution to global warming. Examine the lighting in your home, and see where you could use CFLs. Wherever possible, actually change the lights in your home from incandescent to CFL. Estimate how much energy you are saving because of this and the personal effect that you are having on greenhouse gas production. Develop an ad campaign to encourage your community to switch to CFLs. Remember to think globally and act locally.

Light from Chemiluminescence

Have you ever played with a light stick or a glow stick? Have you ever worn a necklace or bracelet at a concert or fair that, when bent and shaken, gave off visible light? If so, you have seen light produced by chemiluminescence. **Chemiluminescence** is the production of light as a direct by-product of a chemical reaction. Almost no heat is produced as a result of this type of reaction. That is why this type of light is often called "cold light".

chemiluminescence the direct production of light as the result of a chemical reaction with little or no heat produced

Light sticks operate by causing two chemicals to mix. The chemicals are originally separate in the light stick. One chemical is in a narrow, small glass vial in the middle of the stick; the second chemical is in the main body of the stick. Bending the light stick in the middle causes the small glass vial to break, allowing the two chemicals to mix in the main body of the stick. The chemical reaction that occurs produces visible light.

Light sticks are inexpensive to manufacture (Figure 11). They are very popular for use in camping, with law enforcement and military personnel, in entertainment venues (concerts, dance halls, amusement parks), and in emergency situations. Light sticks are very durable. Because they have no moving parts and are completely sealed, they are very popular with underwater divers. Light sticks do not require an electric current, so they are useful in hazardous environments where a spark could be quite dangerous. Examine the effect of temperature on a light stick by doing the "Glowing with Light" activity.

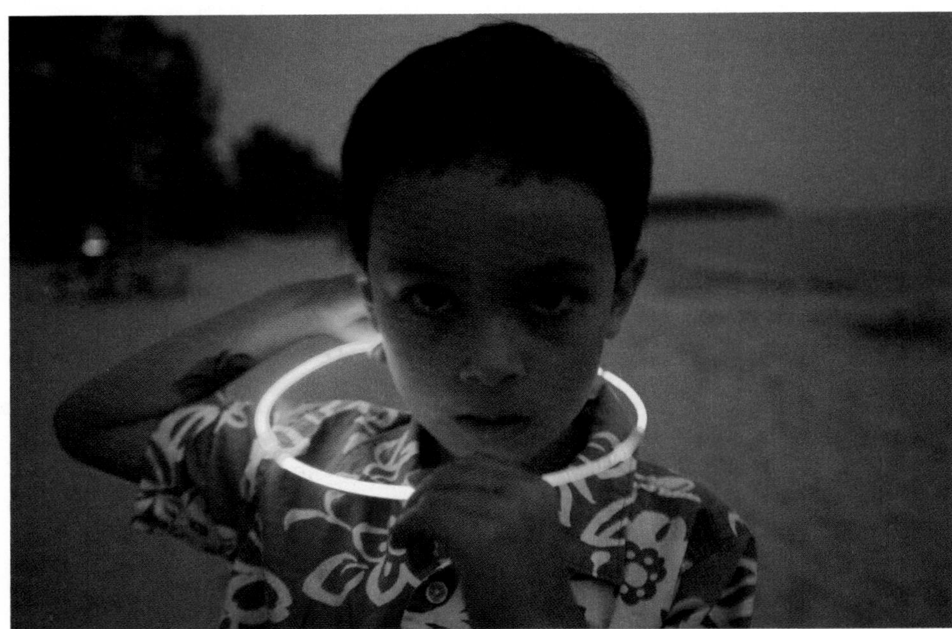

Figure 11 Light sticks, or glow sticks, produce light by chemiluminescence as the result of two chemicals mixing.

TRY THIS GLOWING WITH LIGHT

SKILLS MENU: Predicting, Controlling Variables, Observing

SKILLS HANDBOOK 3.B.

Equipment and Materials: a light stick; freezer; two large beakers or transparent plastic containers; ice cubes; tap water

NOTE: This activity is best done in a darkened room.

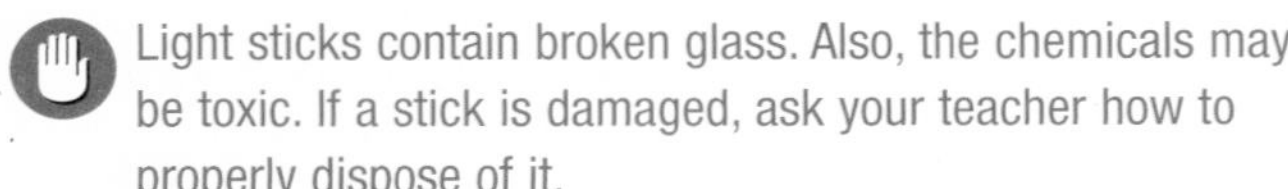

Light sticks contain broken glass. Also, the chemicals may be toxic. If a stick is damaged, ask your teacher how to properly dispose of it.

1. Fill one beaker with warm tap water. Place ice cubes in the other beaker and fill it with cold tap water.
2. Bend the light stick. Observe the amount of light that the light stick emits at room temperature.
3. Predict what will happen when you cool the light stick. Then place the light stick in the ice-water mixture. Observe the amount of light emitted.
4. Predict what will happen if you heat the light stick. Now place the light stick in the beaker with warm water. Again, observe the amount of light emitted.
5. Place the light stick in a freezer for several hours. Predict what will happen. Then take the light stick out of the freezer and observe the amount of light emitted.
6. Let the frozen light stick return to room temperature and make a final observation of the amount of light emitted.

A. Were your predictions correct? T/I

B. What happened to the light output when the light stick was cooled? When it was warmed? T/I

C. Explain the changes in light output. Consult with other classmates. T/I

D. What effect did the freezer have on the light stick? What happened when it returned to room temperature? Explain your observations. T/I

Light from Bioluminescence

When chemiluminescence occurs in living organisms, scientists call it **bioluminescence**. Bioluminescence occurs in a wide variety of organisms, including certain bacteria, fungi, marine invertebrates, fish, and the well-known examples of glow-worms and fireflies (Figure 12). In fact, the glow from a firefly is an excellent example of bioluminescence caused by the chemical reaction of oxygen and luciferin, a substance in the lower abdomen of the insect. The enzyme luciferase is necessary for the chemical reaction to take place. The end result is the production of visible light. Scientists think that living organisms use bioluminescence to protect themselves from predators, to lure prey, or to attract mates.

bioluminescence the production of light in living organisms as the result of a chemical reaction with little or no heat produced

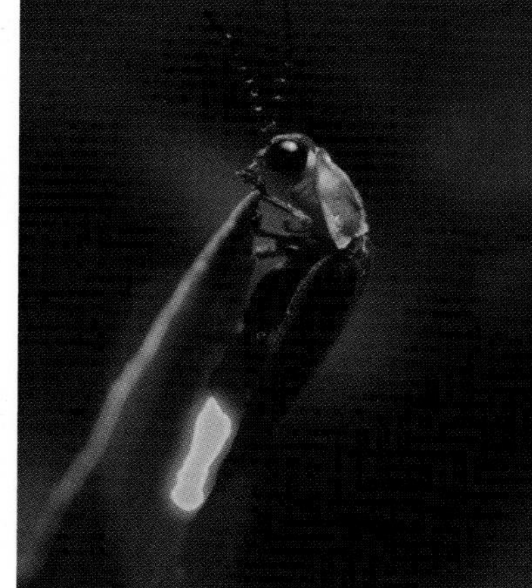

Figure 12 A firefly exhibiting bioluminescence

Light from Triboluminescence

"It is well known that all sugar, whether candied or plain, if it be hard, will sparkle when broken or scraped in the dark." English philosopher Francis Bacon wrote this in *Novum Organum* in 1620, a book in which he argued that science should be based on experimentation. This statement is the first known reference to triboluminescence. **Triboluminescence** is the production of light when certain crystals are scratched, crushed, or rubbed (Figure 13). Unlike other methods of producing light, triboluminescence does not appear to have any practical application at this time. Do the activity called "Eating Candy for the Sake of Science" to observe triboluminescence.

triboluminescence the production of light from friction as a result of scratching, crushing, or rubbing certain crystals

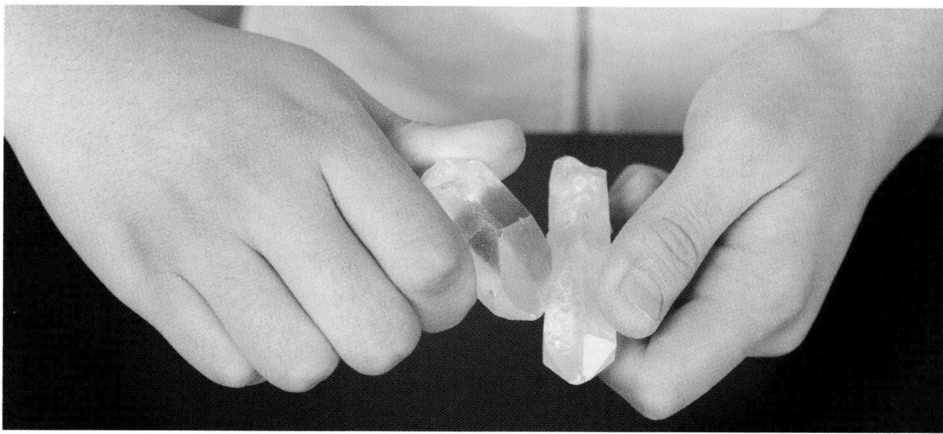

Figure 13 The glow from triboluminescence is visible after rubbing two quartz crystals together.

TRY THIS EATING CANDY FOR THE SAKE OF SCIENCE

SKILLS MENU: Observing

SKILLS HANDBOOK 3.B.6.

Equipment and Materials: two sugar cubes; a wintergreen-flavoured hard candy; a lab partner or a large mirror

1. Go into a completely dark room. Wait at least 5 min until your eyes become adjusted to the dark.
2. Strike the two sugar cubes against one another as if you were striking a match. Observe what happens.
3. Stand in front of a mirror or face your partner. Bite down on a wintergreen candy but keep your mouth open. Observe the inside of your mouth.

A. Describe what you saw when the two sugar cubes were struck. T/I

B. What did you see when biting down on the candy? T/I

light-emitting diode (LED) light produced as a result of an electric current flowing in semiconductors

semiconductor a material that allows an electric current to flow in only one direction

Figure 14 LED Christmas lights use far less electricity than other kinds of lights.

Light from a Light-Emitting Diode (LED)

A **light-emitting diode (LED)** is an electronic device that allows an electric current to flow in only one direction. This is achieved by using special materials called **semiconductors**, such as silicon. Unlike a conductor, which allows current to flow in either direction, semiconductors allow an electric current to flow in only one direction. When an electric current flows in the allowed direction, the LED emits light.

An LED differs in several ways from an incandescent bulb: it does not require a filament; it does not produce much heat as a by-product; and it is more energy efficient. For a long time, the major consumer application of LEDs was as indicator lights in electronic devices (for example, the small red light to show that a radio was on). Recent improvements in LED technology have led to their use in other areas. LEDs are now used in Christmas lights (Figure 14), illuminated signs, and traffic lights. Recent advances in LED manufacturing may reduce the cost of production in the near future. You may well see LEDs being used for street lights and lighting in your home.

IN SUMMARY

- Incandescence is light emitted when a material has been heated.
- Light from an electric discharge is caused by passing an electric current through a gas.
- Phosphorescence and fluorescence are both caused by a material absorbing ultraviolet light. In phosphorescence, visible light is emitted over a period of time, whereas in fluorescence, visible light is emitted immediately.
- Chemiluminescence is light produced from a chemical reaction without an increase in temperature; chemiluminescence in living organisms is called bioluminescence.
- Triboluminescence is light produced from friction with crystals.
- A light-emitting diode (LED) is a special electronic device that produces light when an electric current flows through it.

CHECK YOUR LEARNING

1. In Grade 9 science, you studied the differences among stars, planets, and moons. Which are luminous, and which are non-luminous? Explain why this second group is classified as non-luminous. K/U
2. Why is an incandescent bulb a very inefficient light source? K/U
3. Name the process of producing light by passing an electric current through a gas. K/U
4. What is the main difference between phosphorescence and fluorescence? K/U
5. (a) Do fluorescent brighteners in detergents really make clothes cleaner?
   (b) There is concern that extra additives in detergents can have negative health and environmental impacts. Is it wise to use detergents containing these additives? Explain. K/U T/I A
6. Predict whether or not a fluorescent material would glow if it was illuminated by infrared light. T/I
7. Why is chemiluminescence also called "cold light"? K/U
8. Predict whether or not a light stick would be a good light source in a potentially explosive environment. Explain your prediction. T/I C
9. State several reasons why living organisms might use bioluminescence. K/U C
10. What are two differences between LEDs and incandescent bulbs? K/U C
11. LEDs are considered an even better alternative to CFLs to replace incandescent bulbs. Compare CFLs with LEDs. Are LEDs a better alternative? Be sure to consider environmental, health, and economic factors. Write a brief report to communicate your opinion. K/U T/I A

The Laser—A Special Type of Light

In the previous section, you learned about light from many different sources. Each of those light sources emits electromagnetic radiation of many different energies and in all directions. In this section, you will examine lasers as a light source. You will also find out how laser light is different from other forms of light.

Light from a laser has very special properties. Incandescent bulbs emit electromagnetic waves of many different energy levels. A laser, in contrast, produces electromagnetic waves of exactly the same energy level. This results in visible laser light being of a very pure colour. If you shine red laser light at a triangular prism, the light exiting the triangular prism will still look red. Recall that white light shone at a triangular prism separates into the colours of the visible spectrum.

Laser light is also very intense. This is because the electromagnetic waves travel in exactly the same direction and the waves are exactly in unison (Figure 1). These characteristics are very different from light radiating from an incandescent bulb. That is why laser light is pure in colour, very intense, and concentrated in one narrow beam. For this reason, you should never look directly into a laser beam—it could damage your eyes. These unique properties of lasers make them very useful. A high-energy laser can be used to burn a hole through steel in a manufacturing process. It can also be used by a surveyor to measure distances. Scientists have even used lasers directed at the Moon to measure the Earth–Moon distance 385 000 km with an accuracy of 3 cm.

DID YOU **KNOW?**

Laser Tongue-Twister

"Laser" is an acronym for **L**ight **A**mplification by **S**timulated **E**mission of **R**adiation.

Figure 1 (a) The light bulb emits many different electromagnetic waves.
(b) A laser emits electromagnetic waves that are all exactly the same.

RESEARCH THIS LASERS FOR DIFFERENT USES

SKILLS: Researching, Communicating

SKILLS HANDBOOK 4.A.

There are many different kinds of lasers, and each kind has a specific use (Figure 2). In this activity, you will research some of these lasers and uses.

Figure 2 Lasers are often used in concerts or at dance halls.

1. Research the characteristics and applications of helium-neon lasers.
2. Research at least three practical applications of lasers. Find out what type of laser is used in each case. Identify which type of electromagnetic wave each laser produces.
3. Research the type of laser used to correct vision problems. Identify which type of electromagnetic wave this laser produces.

GO TO NELSON SCIENCE

A. What type of electromagnetic wave is produced by a helium-neon laser? K/U T/I

B. What are three different types of lasers? Where is each of these lasers used? K/U T/I

C. Which laser is used to correct vision problems? What type of electromagnetic wave does this laser produce? T/I

IN SUMMARY

- Laser light consists of electromagnetic waves of exactly the same energy level, travelling in unison in exactly the same direction.
- Laser light is very pure in colour, is intense and concentrated in one narrow beam, and can travel great distances without spreading out.

CHECK YOUR LEARNING

1. List three properties that make laser light different from light emitted by an incandescent bulb or a flashlight. K/U
2. A laser with green light is shone at a triangular prism. What will be the colour of the light leaving the prism? Explain. K/U
3. As you read in this section, scientists have used laser light to accurately measure the distance between Earth and the Moon.
   (a) What properties of laser light allow us to measure such great distances?
   (b) Why could white light produced by a very powerful searchlight not be used for this same measurement? K/U T/I
4. Why should you never look directly into a laser beam? K/U
5. Use a diagram to illustrate how laser light is different from white light. K/U
6. List at least four common applications of lasers that you are familiar with. K/U

The Ray Model of Light

A laser beam clearly shows that light travels in a straight line. This fundamental property of light (Figure 1) can be used to understand how light behaves when it strikes a mirror or a lens.

Figure 1 Evidence that light travels in straight lines is that you never see a flashlight beam go around a corner.

A luminous object such as a candle radiates light in all directions. This is easily seen because a candle will illuminate all objects surrounding it in a complete sphere. You can illustrate this more easily by using the concept of light rays. A **light ray** is a line and arrow representing the direction and straight-line path of light. Because the candle is radiating light in all directions, an infinite number of light rays come from the candle. You need to draw only a few light rays to represent the overall picture (Figure 2).

light ray a line on a diagram representing the direction and path that light is travelling

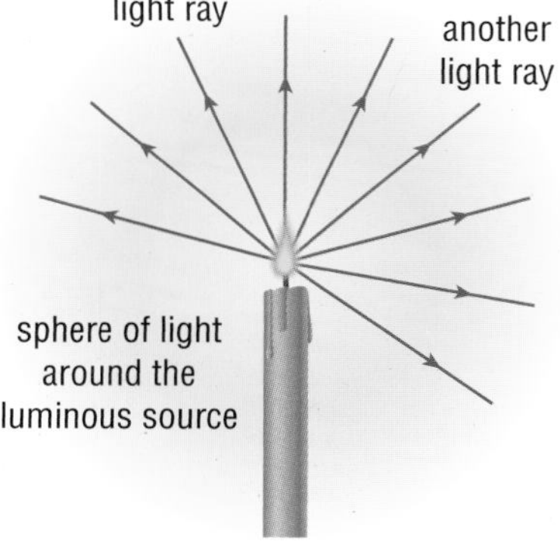

Figure 2 You need to draw only a few light rays to represent the light radiating from a candle.

The use of light rays to determine the path of light when it strikes an object is called **geometric optics**. When light emitted from a source (such as the Sun) strikes an object (such as Earth), the light is called **incident light**. Recall that matter can be classified into three categories, depending on how it behaves when light strikes it. A **transparent** object (such as clear glass) lets light pass through it easily and allows objects behind it to be clearly seen. A **translucent** object (such as frosted glass) allows some light to pass through but does not allow you to clearly see objects behind it. An **opaque** material (such as cardboard) does not allow any light to pass through it. Instead, all incident light is either absorbed or reflected, and it is not possible to see objects behind the opaque material.

geometric optics the use of light rays to determine how light behaves when it strikes objects

incident light light emitted from a source that strikes an object

transparent when a material transmits all or almost all incident light; objects can be clearly seen through the material

translucent when a material transmits some incident light but absorbs or reflects the rest; objects are not clearly seen through the material

opaque when a material does not transmit any incident light; all incident light is either absorbed or reflected; objects behind the material cannot be seen at all

TRY THIS SEEING THE LIGHT

SKILLS MENU: Observing, Evaluating

SKILLS HANDBOOK 3.B.6., 3.B.8.

Equipment and Materials: flashlight; atomizer or similar spray bottle filled with water

1. Place a flashlight on a desk and aim it at the wall. Turn on the flashlight.
2. Turn off the lights in the classroom. Examine the beam on the wall and the air in the path of light from the flashlight to the wall.
3. Place your hand in the path of the beam and observe the beam.
4. Spray water from the spray bottle into the air between the flashlight and the location of the beam on the wall.

A. Could you see the beam of light in the air? T/I

B. What happened to the light beam when you placed your hand in its path? T/I

C. What did you see when you sprayed the water? T/I

D. What is necessary in order for you to see light? T/I

E. Suggest a reason to explain why you could see the light beam only when there was atomized water in the air. T/I

Flat Mirrors

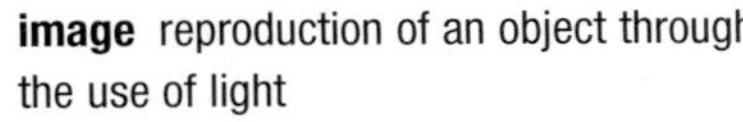

image reproduction of an object through the use of light

mirror any polished surface reflecting an image

reflection the bouncing back of light from a surface

"Mirror, mirror on the wall, who's the fairest one of all?" This famous line from the fairy tale *Snow White* is uttered by the evil queen, who does not want anyone in her land to be prettier than she. The queen is looking at her image in a mirror. An **image** is a reproduction of an original object that is produced through the use of light. A **mirror** is any polished surface that exhibits reflection (Figure 3). **Reflection** is simply the bouncing back of light from any surface. In the previous activity, you noticed that in order to see light, some light rays must reflect off a surface and then enter your eyes. The queen in *Snow White* used the mirror in the way it has been used for thousands of years, as a way to check her physical appearance. This is still the most common use of mirrors today.

Figure 3 On a still day, the water of a lake can be smooth enough to behave like a mirror.

Most mirrors consist of two parts: The front part is a sheet of glass, and the back part is a thin layer of reflective silver or aluminum. This version of the mirror originated in the 12th and 13th centuries. Mirrors existed prior to this time, but they were made of highly polished metals such as bronze (Figure 4), tin, or silver. Mirrors such as those in your home did not become common until the 17th and 18th centuries.

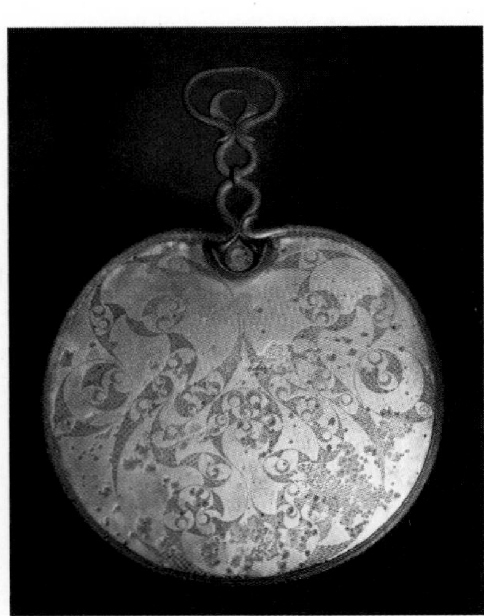

Figure 4 A Celtic bronze mirror that is almost 3000 years old

The reflective part of a mirror is the shiny thin film on the back. The glass protects the thin film and aids in the physical appearance of the mirror. The symbol that is used in physics to represent a mirror refers only to the reflective thin film (Figure 5).

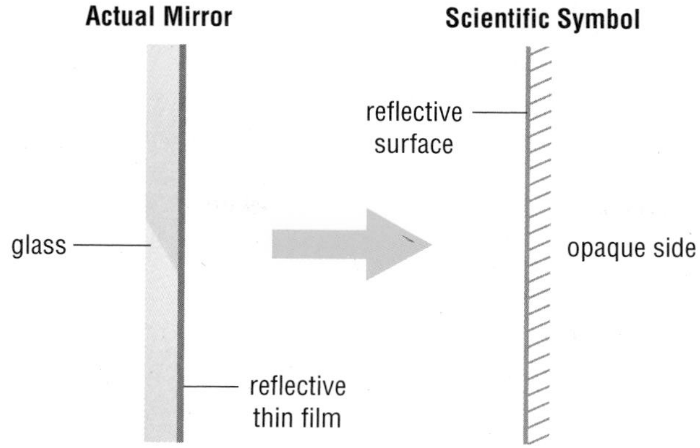

Figure 5 Comparing the side view of an actual mirror with its scientific symbol

The Terminology of Reflection

A **plane** mirror, or a flat mirror, illustrates how predictable the path of light is when it hits the mirror. The original incoming ray is called the **incident ray**. The ray that bounces off the mirror is called the **reflected ray**. The **normal** is the line that is **perpendicular**, or at right angles, to the reflecting mirror surface. The normal is drawn at the point where the incident ray strikes the surface of the mirror. The **angle of incidence** is the angle between the incident ray and the normal. The **angle of reflection** is the angle between the reflected ray and the normal (Figure 6).

You will examine how these terms apply to plane mirrors when you do Activity 11.5: Reflecting Light Off a Plane Mirror.

plane flat

incident ray the incoming ray that strikes a surface

reflected ray the ray that bounces off a reflective surface

normal the perpendicular line to a mirror surface

perpendicular at right angles

angle of incidence the angle between the incident ray and the normal

angle of reflection the angle between the reflected ray and the normal

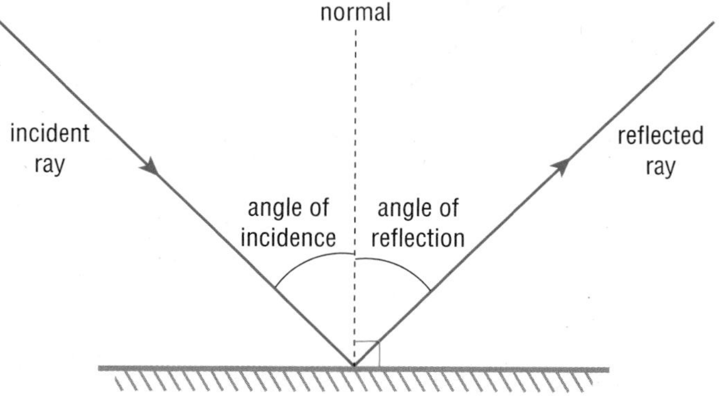

Figure 6 Terminology for reflection off a plane mirror

IN SUMMARY

- Light rays are used to represent the direction and path in which light is travelling.
- Geometric optics uses light rays to determine how light behaves when it strikes objects.

CHECK YOUR LEARNING

1. (a) Name the two parts that make up most mirrors.
   (b) What is the purpose of each part? K/U
2. Clearly explain what is meant by the term *geometric optics*. K/U
3. Classify each of these materials as transparent, translucent, or opaque: a textbook, frosted glass, a single sheet of thin tissue paper, a clean sheet of glass, a rock, clean air, apple juice, sunglasses. K/U
4. What has historically been the main use for plane mirrors? K/U
5. In your own words, clearly distinguish between the terms *normal*, *angle of incidence*, and *angle of reflection*. K/U
6. Brainstorm to create a list of other uses of plane mirrors not mentioned in this section. A

11.5 PERFORM AN ACTIVITY

Reflecting Light Off a Plane Mirror

SKILLS MENU

- Questioning
- Hypothesizing
- Predicting
- Planning
- Controlling Variables
- Performing
- Observing
- Analyzing
- Evaluating
- Communicating

Plane mirrors are all around you. The mirror that you look into when brushing your teeth, the rear-view mirror in a car, and the mirror that the dentist uses to look inside your mouth are all plane mirrors. In this activity, you will examine how light rays behave when they are reflected by a plane mirror.

Purpose

SKILLS HANDBOOK 3.B.

To compare the angle of incidence with the angle of reflection in a plane mirror.

Equipment and Materials

- ray box
- plane mirror and mirror supports
- pencil and ruler
- protractor
- sheet of paper

When unplugging the ray box, do not pull the electric cord. Pull the plug itself.

Procedure

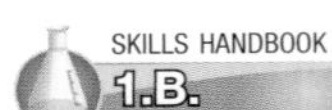

SKILLS HANDBOOK 1.B.

1. Draw a dashed line across the centre of a sheet of paper. Place a mirror on this line. The back of the mirror should be on the line, not the glass part of the mirror. This is because the back or silvered part of the mirror is the reflective part.
2. Place a slit mask on the ray box so that only one ray of light comes out. Aim the incident ray at the mirror.
3. Draw a normal to the mirror (that is, perpendicular to the point where the incident ray strikes the mirror). Label this line "normal" (Figure 1).
4. Place several dots on the page with your pencil showing the path of the incident ray. Then place several dots on the path of the reflected ray.

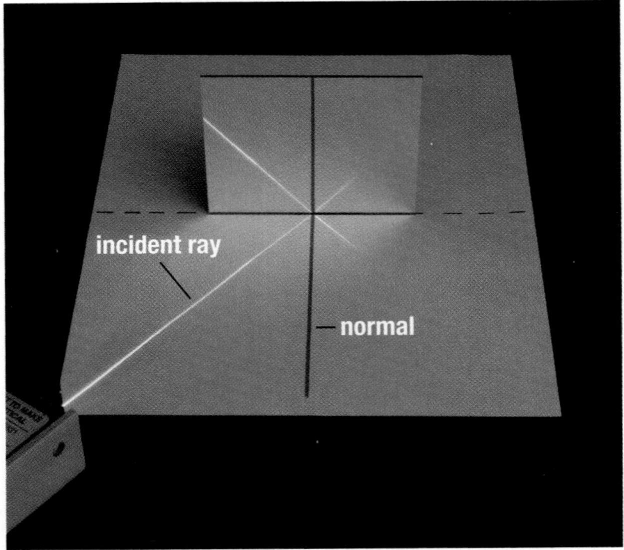

Figure 1

5. Remove the ray box and the mirror. Use your ruler to draw a straight line through the dots until the line hits the top part of the T (where the normal meets the mirror). Label the incident ray as "I1." Repeat this process for the reflected ray. Label the reflected ray as "R1."
6. Use your protractor to measure the angle of incidence and the angle of reflection for trial 1. Remember to measure these angles with respect to the normal. Record your measurements in the "Trial number 1" row in a table similar to Table 1.

Table 1 Observations

Trial number	Angle of incidence	Angle of reflection
1		
2		
3		
4		
5		

7. Repeat steps 1 to 6 three more times, labelling the new incident rays as "I2," "I3," and "I4" and the new reflected rays as "R2," "R3," and "R4." Make sure that the ray box is at a different angle for all of these trials. It must always be pointed at the same intersection of the mirror and the normal. Complete your table for trials 2 to 4.
8. Do a fifth trial, but this time aim the incident ray directly along the normal. Complete your table for trial 5.

Analyze and Evaluate

(a) How did the angle of incidence compare with the angle of reflection? T/I

(b) In trial 5, you aimed the incident ray directly along the normal. Describe the path of the incident and reflected rays for this special case. T/I

(c) Where might errors occur in this activity? T/I

(d) How would these errors affect your conclusion? T/I

Apply and Extend

(e) Billiards is a game that makes use of reflection (Figure 2). How could the results of this activity help you in such a game? T/I A

Figure 2

(f) What other sports or activities make use of the reflection rule that you discovered in this activity? A

11.6 The Laws of Reflection

As you observed in Activity 11.5, when you shine an incident light ray at a plane mirror, the light is reflected off the mirror and forms a reflected ray. Both the incident ray and the reflected ray behave in a predictable way. This predictable behaviour of light leads to the two laws of reflection (Figure 1):

1. The angle of incidence equals the angle of reflection.
2. The incident ray, the reflected ray, and the normal all lie in the same plane.

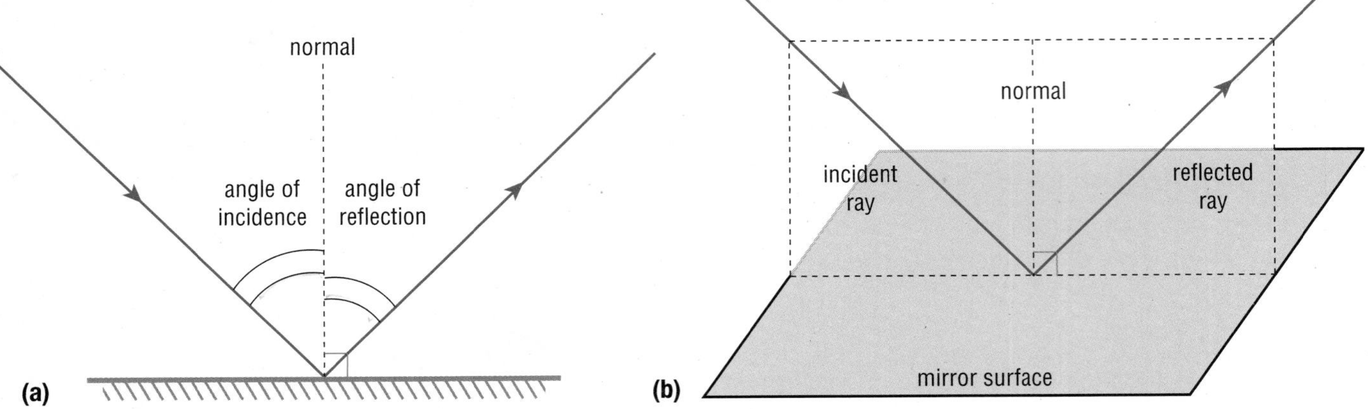

Figure 1 Diagrams illustrating the two laws of reflection

When more than one incident light ray is reflected off a surface, the laws of reflection still hold but the surface affects the way you see the reflected rays. For example, the reflections off a smooth piece of foil are not the same as those off a crumpled piece of foil. Find out by doing the activity "Reflecting Light."

TRY THIS REFLECTING LIGHT

SKILLS MENU: Observing, Evaluating

Equipment and Materials: flashlight (or an unshielded ray box); sheet of aluminum foil (about 30 cm × 30 cm)

1. Place a smooth piece of aluminum foil flat on a table. Turn off the lights in the room and reflect the beam from the flashlight or ray box off the aluminum foil. Aim the light beam so that it is reflected up onto the ceiling. Examine the reflected beam.
2. Now crumple up the aluminum foil. Open up the crumpled piece of foil but do not smooth it out. Again, reflect the light beam off the foil onto the ceiling and examine the reflected beam.

A. Describe the shape of the reflected beam on the ceiling when the aluminum foil was flat and when it was crumpled. T/I

B. Account for the difference in the reflected beam between your two trials. T/I

READING TIP

Using Text Layout

Subheadings are capitalized. This means that there are new concepts or ideas in this section. Write down the main ideas communicated in each subsection.

Reflecting Light Off Surfaces

Assume that a series of parallel incident rays strike a flat reflective surface. The angles of incidence for these rays are all identical. This means that their angles of reflection will also all be identical, and the reflected rays will all be parallel to each other. This is an example of regular or specular reflection.

Specular reflection is the reflection of light off a smooth, shiny surface (Figure 2). Reflection off a plane mirror is specular reflection, as is reflection off the surface of very still water (Figure 3), or a flat piece of aluminum foil.

specular reflection reflection of light off a smooth surface

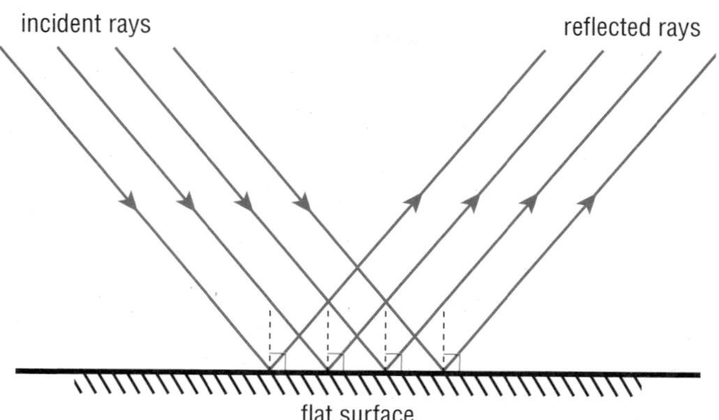

Figure 2 How specular reflection works

Figure 3 A spectacular example of specular reflection. The water acts like a plane mirror.

The mirrored surfaces on a disco mirror ball found in clubs and parties, for example, exhibit specular reflection. The ball consists of a number of plane mirrors on a spherical surface that produce an eye-catching display by reflecting light in all directions (Figure 4).

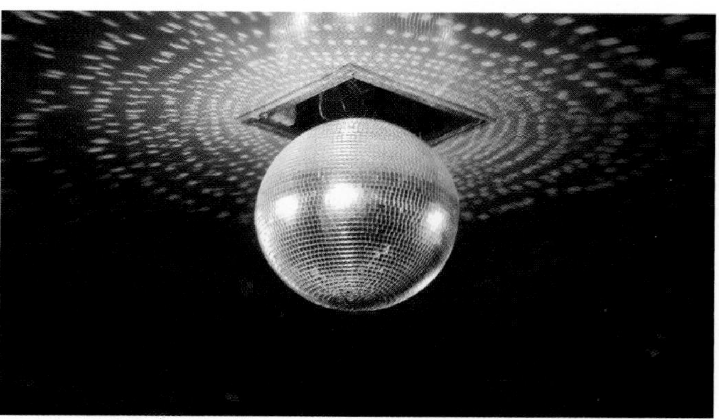

Figure 4 A disco mirror ball is an example of specular reflection from its many surfaces.

What would happen if the parallel incident rays were directed at an irregular surface? Now the incident rays would all have different angles of incidence. This means that their angles of reflection would also be different. The reflected rays would not emerge parallel to each other but would be reflected or scattered, in many different directions. This is called diffuse reflection. **Diffuse reflection** results from the reflection of light off an irregular or dull surface (Figure 5). Examples of diffuse reflection are reflection off a sheet of paper, a water surface with waves (Figure 6), or a crumpled piece of aluminum foil.

diffuse reflection reflection of light off an irregular or dull surface

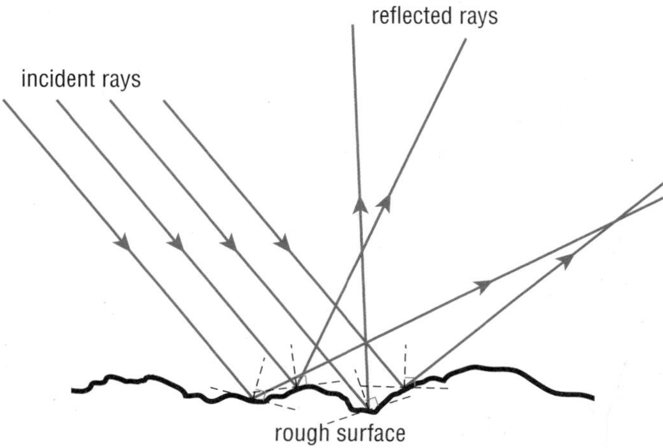

Figure 5 How diffuse reflection works

Figure 6 Diffuse reflection as a result of ripples in the water

TRY THIS RETRO-REFLECTORS

SKILLS MENU: Performing, Observing, Evaluating

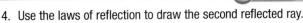

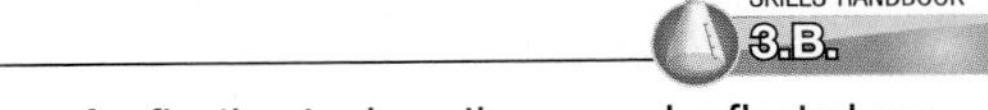

In this activity, you will draw a diagram of a retro-reflector. You will learn more about retro-reflectors in the next chapter.

Equipment and Materials: ruler; protractor; piece of paper; pencil

1. Draw two mirrors meeting at a right angle (forming an L-shape) near the centre of the paper. Pick any orientation that you wish for the mirrors.
2. Use a ruler to draw a ray that strikes one of the mirrors.
3. Use the laws of reflection to determine the angle of incidence. Draw the reflected ray off this mirror. This ray will now act as the incident ray striking the second mirror surface.
4. Use the laws of reflection to draw the second reflected ray.
5. Repeat steps 2 to 4 using a second incident ray that strikes the mirror at a different angle from the first one. Use a different colour for each incident ray.

Optional: Do this experiment with two plane mirrors and a ray box.

A. How did the incident ray compare with the ray that emerged after the two reflections in your first trial? T/I

B. Did this result change for your second trial? T/I

C. Based on your observations, what is the main characteristic of a retro-reflector? T/I

Figure 7 This is how text might appear to a person with dyslexia.

Reflection and Dyslexia

People with dyslexia have difficulty reading print (Figure 7). Many people who are dyslexic complain about the glare off white paper: there is too much reflected light from the paper. The contrast between the white paper and the black text makes reading difficult. For some people, this condition can be helped by the use of coloured filters or glasses that reduce the glare of reflected light from paper. This sometimes makes reading easier.

UNIT TASK Bookmark

You can apply what you learned about optics in this section to the Unit Task described on page 588.

IN SUMMARY

- When light is reflected off a plane mirror, the angle of incidence equals the angle of reflection.
- When a light ray strikes a plane mirror, the incident ray, the reflected ray, and the normal all lie on the same plane.

CHECK YOUR LEARNING

1. Using a diagram, distinguish between the incident ray, the reflected ray, and the normal. K/U C
2. Clearly state the two laws of reflection. K/U
3. (a) What is the difference between specular reflection and diffuse reflection?
   (b) Provide some examples of both specular and diffuse reflection other than those given in this section. K/U A
4. (a) If you were painting the walls in your classroom, would you want the walls to exhibit specular or diffuse reflection? Explain.
   (b) Given your choice, should you use gloss or matte paint? K/U A
5. (a) What would be the angle of reflection for an angle of incidence of 32°?
   (b) What would be the angle of incidence for an angle of reflection of 47°?
   (c) What would the angle of reflection be if the incident ray was 40° from the reflecting mirror surface? K/U
6. Specular reflection and diffuse reflection are concepts that can be applied to every room that you have ever been in. Describe how these concepts can be applied to your kitchen, bathroom, and bedroom. A

TECH CONNECT ✓OSSLT

Cleaning with Light

Imagine that you are a scientist who has been asked to clean a painting or statue that has centuries of dirt and pollution on it. Traditionally, you would do a delicate manual scraping with a scalpel or use messy solvents that are harmful both to you and to the environment. You have to be very slow and careful with the cleaning because you do not want to make a mistake with a priceless work of art!

Today, an alternative is available for restoring artifacts. A modified version of the cosmetic surgery laser is used to heat the dirt on the artifact. The energy in the laser light is absorbed by the dirt, which expands as it is heated. The dirt then falls away from the surface. It sounds almost too good to be true, and the results can be astonishing (Figure 1). Best of all, no delicate, time-consuming scraping with a scalpel is needed, and no toxic or hazardous chemicals are used.

Lasers have many advantages in restoration work. Lasers are accurate, the size and strength of the beam can be easily controlled, they are reliable, there is no physical contact with the surface, and no dangerous chemicals are used. Lasers can be used on artifacts such as statues, pottery, and paintings (Figure 2). Who would have thought that the laser, originally an invention in search of a purpose, would be able to make light of cleaning?

GO TO NELSON SCIENCE

Figure 1 The right half of this statue has been cleaned with a laser. The left half shows what it looked like before cleaning.

Figure 2 (a)–(b) The various stages of restoration of a painting by Italian Renaissance painter Raphael (1483–1520). The painting had shattered into 17 pieces due to a house collapse. (c) It took 10 years and the use of a UV laser to restore the painting to its original state.

11.7 Images in Plane Mirrors

READING TIP

Using What You Know
If a question is posed in a paragraph, stop reading immediately after the question. Continue reading only after you tried to answer the question.

You are standing in front of a mirror in the bathroom while brushing your teeth. You notice that the writing on your T-shirt seems backwards. You have always wondered why the writing on the hood of a police car or ambulance appears backwards. What exactly is going on here? Let's begin by doing the activity "Writing Reflectively." Can you write a message that can be read in a mirror? Here is your chance to try!

TRY THIS WRITING REFLECTIVELY

SKILLS MENU: Questioning, Performing, Observing, Evaluating, Communicating

SKILLS HANDBOOK
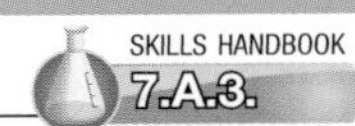

Equipment and Materials: plane mirror; mirror supports (optional); a sheet of paper; pencil

1. Place the mirror on the upper half of the sheet of paper. You can use mirror supports or your hand to hold the mirror.
2. While looking into the mirror, carefully print your name on the piece of paper so that it appears correctly in the mirror, not on the page. You may have to practise a few times until you are able to do this successfully. Also, try writing with the hand that you do not usually use.
3. Once you are comfortable with writing using a mirror, use the mirror to carefully print a short message.
4. Exchange your message with a partner. Try to decipher the message that you have received from your partner. Check the accuracy of your translation of your partner's message by using a mirror.

A. Describe the appearance of your written name on the paper compared with its appearance in the mirror. T/I
B. Was writing while looking into a mirror difficult? Why? Which letters did you find more difficult to do? T/I
C. Did you find it easier writing with one hand than with the other? If so, suggest a reason why. T/I
D. From this activity, what general conclusion can you make about how an object and its image in a plane mirror are related? T/I
E. Leonardo da Vinci was a left-handed Italian artist and scientist born in the 15th century. He used mirror writing in his notebooks when he was writing about his inventions and other ideas (Figures 1 and 2). Why do you think he did this? T/I

GO TO NELSON SCIENCE

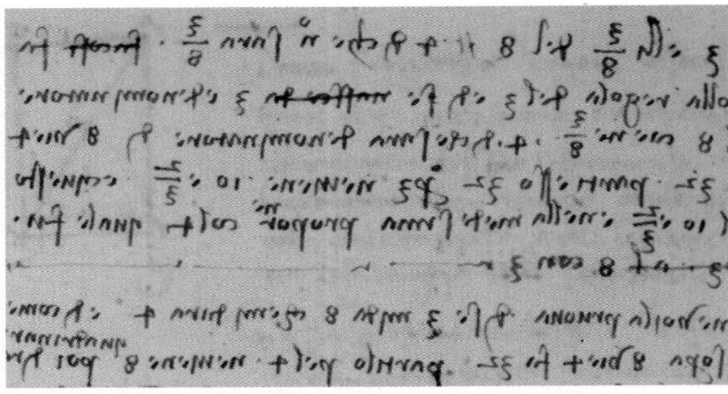

Figure 1 A section from one of Leonardo da Vinci's notebooks showing his backwards writing

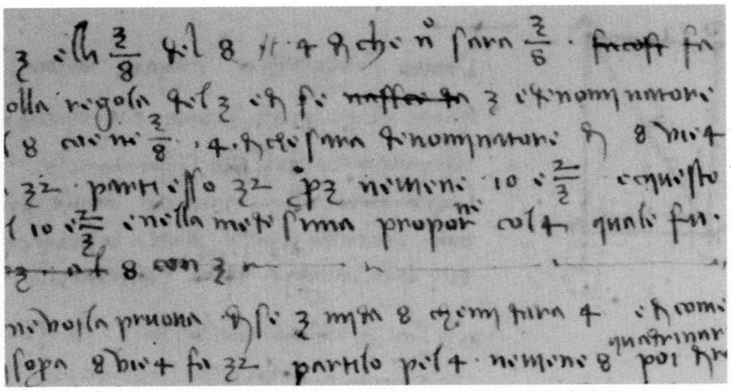

Figure 2 The same section from Figure 1 now reflected in a mirror. Note that the numbers are now readable.

Using Light Rays to Locate an Image

Light rays and the laws of reflection help determine how and where an image is formed in a plane mirror. A light source radiates millions of light rays in all directions, but you are only concerned with the rays that actually strike the mirror and are reflected into your eyes. These rays are reflected off the mirror, with the angle of incidence being equal to the angle of reflection.

To learn more about producing multiple images of an object in plane mirrors, try the activity below.

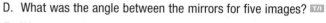

TRY THIS PRODUCING IMAGES, AND MORE IMAGES, AND MORE IMAGES ...

SKILLS MENU: Predicting, Observing, Analyzing

SKILLS HANDBOOK 3.B.

Equipment and Materials: two plane mirrors; two mirror supports; ruler; protractor; a die; paper; pencil

1. Place the two mirrors at right angles to each other at the top of the sheet of paper. Place the die directly in front of the right angle formed by the mirrors (Figure 3). Record how many images you see in the mirrors.

Figure 3

2. Gently move one of the mirrors, changing the angle between the two mirrors, until you see four complete images. Draw lines on the paper at the base of the two mirrors. Measure and record the angle between them.
3. Now gently move one of the mirrors until you see five images. Again, draw lines on the paper at the base of the two mirrors. Measure and record the angle between them.
4. Based on your previous results, predict what angle between the mirrors would produce six images, then seven, eight, nine, and so on.
5. Continue moving the mirrors, counting the total number of images, and measuring the angle between the mirrors as long as you are able to.

A. How many images were visible when the mirrors were at right angles to each other? T/I
B. Use your knowledge of light rays to explain why this number of images was formed. K/U T/I
C. What was the angle between the mirrors for four images? T/I
D. What was the angle between the mirrors for five images? T/I
E. Were your angle predictions correct for six, seven, eight, and nine images? If not, explain why. T/I
F. What was the total number of images that you were able to count? Why were you not able to exceed this value? T/I
G. A hall of mirrors in an amusement park seems to produce an infinite number of images when you look into it. This effect is also commonly seen in elevators that have two plane mirrors on opposite walls (Figure 4). T/I C A
    (a) Suggest a reason why elevator designers use this effect.
    (b) On a piece of paper, draw two plane mirrors that are parallel to each other. Add light rays to show how this set-up can produce multiple images.

Figure 4 Multiple images produced by parallel plane mirrors

DID YOU **KNOW?**

Virtual Reality

The word "virtual" in virtual reality also refers to something that does not really exist, or something that is imaginary.

virtual image an image formed by light coming from an apparent light source; light is not arriving at or coming from the actual image location

From everyday experience, you know that light travels in a straight line. This belief is so strong that when your eyes detect reflected light from a plane mirror, your brain projects these light rays backwards in a straight line. This results in your brain thinking that there is a light source *behind* the mirror and that this is where the light rays originate (Figure 5). It is this apparent light source behind the mirror that results in you seeing an image behind the mirror. There is, of course, no real light source behind the mirror because the mirror is opaque. This kind of image is called a virtual image. A **virtual image** is an image in which light does not actually arrive at or come from the image location. The light only *appears* to come from the image (Figure 6). Your eyes detect the light rays, but your brain determines where the image is located. You will learn more about the eye in Chapter 13.

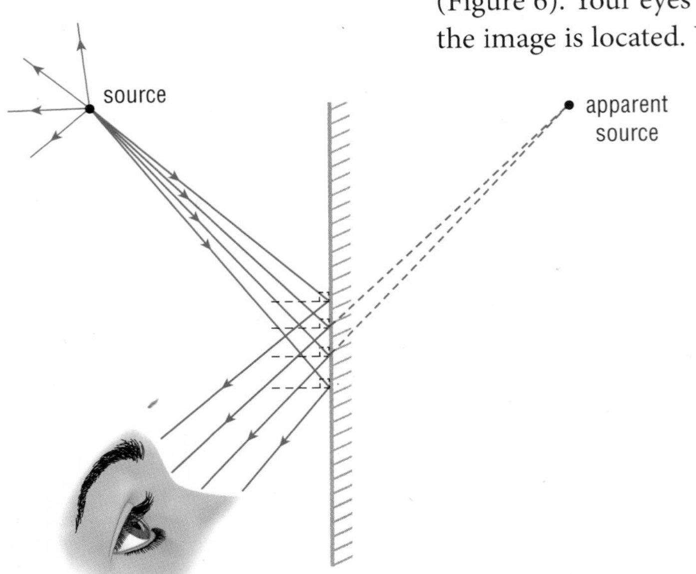

Figure 5 Light rays and the laws of reflection can be used to explain how the eye forms an image of the light source behind the opaque mirror. Note that only light rays that are reflected off the mirror and into your eyes contribute to the location of the apparent source.

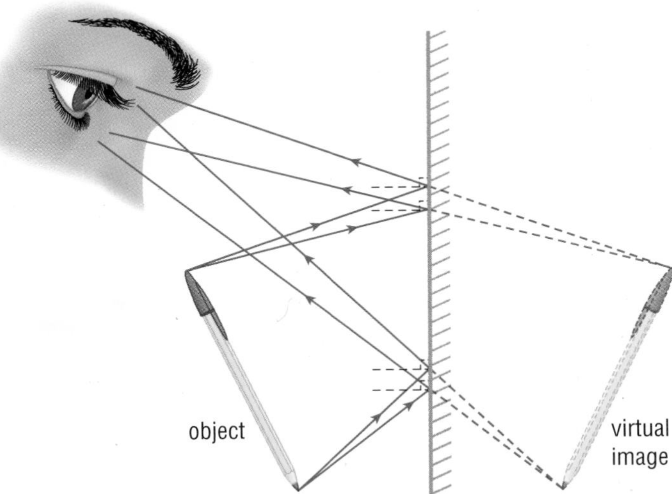

Figure 6 Note that light rays behind the mirror are drawn as dashed lines. This indicates that these rays do not really exist. Your brain projects these rays behind the mirror and forms a virtual image behind the mirror.

DID YOU **KNOW?**

Playing With Light

A kaleidoscope is a toy that consists of three mirrors that reflect multiple images of an object. The mirrors are mounted in a tube at 60° to one another, with the viewer looking in one end and light coming in from the other end through a screen. Many pleasing hexagonal images are produced as the tube is rotated. The kaleidoscope was invented in England in 1816 by David Brewster.

Using Equal Perpendicular Lines to Locate an Image

You can use light rays and the laws of reflection to show how a plane mirror produces a virtual image and where that is located. The use of light rays also demonstrates another interesting property that allows you to locate the image more directly. Consider again how light rays and the laws of reflection can be used to show how the human eye forms an apparent light source behind an opaque mirror (Figure 5). This time, however, draw a line between the original object and the location of the image. This is the object–image line. There are two interesting observations to make about an object and its image in a plane mirror:

1. The distance from the object to the mirror is exactly the same as the distance from the image to the mirror. In other words, the image appears to be located the same distance behind the mirror as the object is in front of the mirror.
2. The object–image line is perpendicular to the mirror surface.

Both observations can be stated like this: A plane mirror divides the object–image line in half and is perpendicular to that line (Figure 7). The use of equal perpendicular lines allows you to easily locate the image of an object. All you need to do is pick several points on the object and then use the object–image lines and the mirror to locate the image. When you have enough points, you can draw the virtual image (Figure 8). This method does not require you to draw light rays or to measure any angles of incidence and reflection.

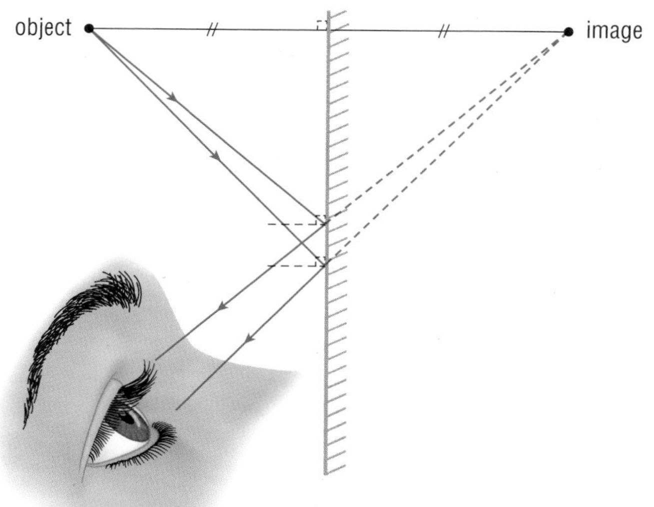

Figure 7 A plane mirror divides the object-image line in half and is perpendicular to that line. This method can be used to locate the image of an object without using light rays.

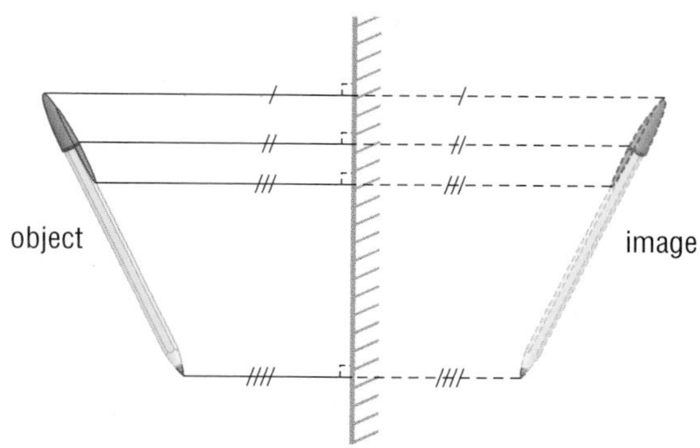

Figure 8 By choosing enough points on an object and drawing a series of object–image lines and lines of equal length that are perpendicular to the mirror, you can accurately locate the virtual image of the object. Note that the image is drawn with dashed lines to indicate that it is a virtual image.

Looking into a Plane Mirror

If you look at a printed word, such as the word SCIENCE!, then turn your head to view the same word in a plane mirror, the letters appear to be flipped from left to right (Figure (9). The image in the mirror appears to be backwards compared to how we view the object directly. This is why the word on the front of an ambulance is written backwards: so that it can be read when seen in a rear-view mirror (Figure 10).

Figure 9 Imagine a word written on thin paper and held in front of a plane mirror. Now imagine how the *object* would appear if you were standing between the object and the mirror.

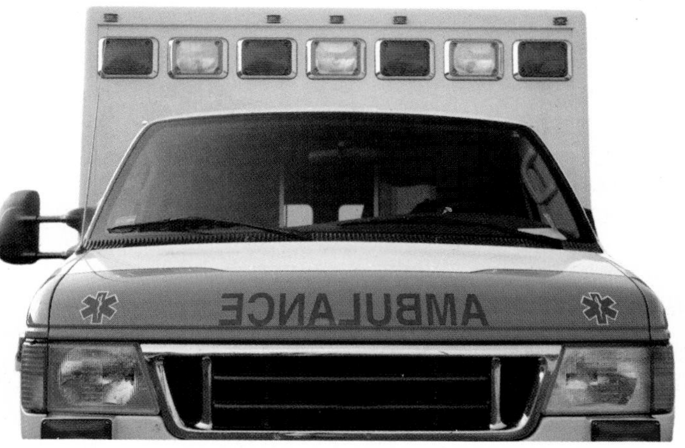

Figure 10 The writing on the hood of an ambulance is reversed.

Saving Money With Mirrors

The use of large plane mirrors in interior designs and movie productions saves money. The mirrors are placed appropriately so that a corridor seems twice as long, when, in reality, it is just a reflection of a shorter corridor in front. Science fiction movies often use this technique when filming images of futuristic machinery stretching off into the distance.

To learn more about the work of interior designers and how they use mirrors,

GO TO NELSON SCIENCE

The Acronym SALT

When you describe the properties of an image, you need to examine four characteristics:

1. size of image (compared to the object: same size, smaller, or larger)
2. attitude of image (which way the image is oriented compared to the object: upright or inverted)
3. location of image
4. type of image (real or virtual) A real image is an image formed when light is actually arriving at the image location. You will learn more about real images in Section 11.9.

Use the acronym **SALT**, for **S**ize, **A**ttitude, **L**ocation, and **T**ype, to remember these four image characteristics (Figure 11).

An image in a plane mirror is always the same size as the object (**S**ize), upright (**A**ttitude), behind the mirror (**L**ocation: the same distance behind as the object is in front), and virtual (**T**ype).

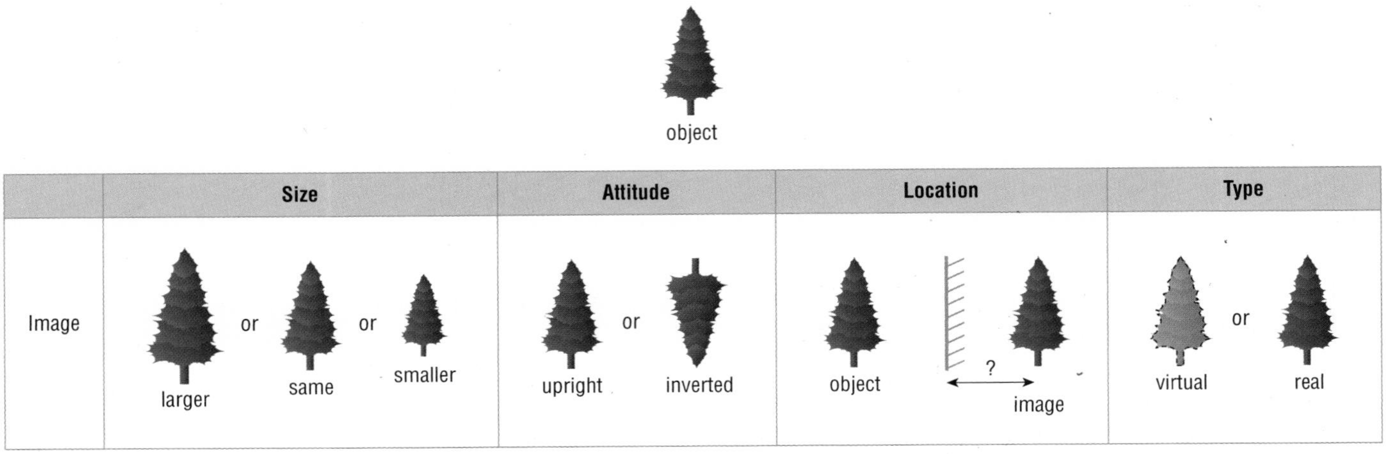

	Size	Attitude	Location	Type
Image	larger or same or smaller	upright or inverted	object ? image	virtual or real

Figure 11

UNIT TASK Bookmark

Think about how you can use ray diagrams and the laws of reflection as you work on the Unit Task described on page 588.

IN SUMMARY

- When reflected light off a plane mirror enters your eyes, your brain projects these rays backwards to form an apparent light source located behind the mirror.
- A virtual image is formed by the apparent light source because no light rays are actually arriving at or coming from the image location.
- A plane mirror divides the object–image line in half and is perpendicular to that line.
- The acronym **SALT** (for **S**ize, **A**ttitude, **L**ocation, and **T**ype) can be used to remember the four image characteristics.
- An image in a plane mirror is always the same size as the object, upright, behind the mirror, and virtual.

CHECK YOUR LEARNING

1. In your own words, describe what is meant by the term "virtual image." K/U
2. You stand 1.8 m in front of a plane mirror as you are brushing your teeth. Use SALT to describe the characteristics of the image. K/U
3. You are wearing a T-shirt that has the word "OPTICS" on it. You stand in front of a plane mirror. Write in your notebook how this word appears to you as you look in the mirror. K/U
4. Copy Figure 12 into your notebook. Use a ruler and a protractor to draw normals and reflected rays for the two incident rays. Then project these reflected rays to locate the apparent source behind the mirror. (Refer to Figure 5 in this section for help with this.) Verify your answer using an object–image line and lines of equal length that are perpendicular to the mirror. T/I C

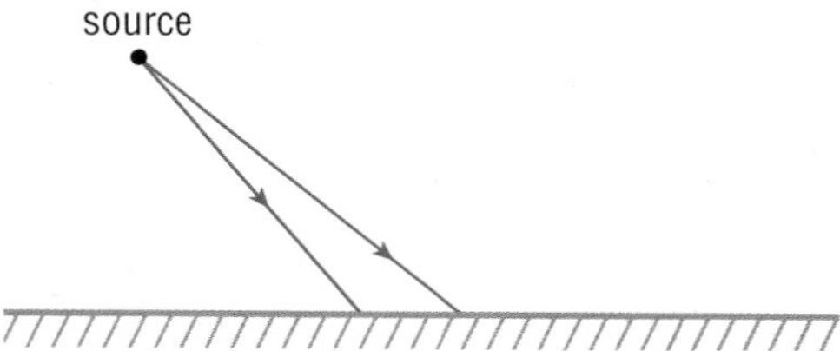

Figure 12

5. Copy all three parts of Figure 13 into your notebook, leaving plenty of space around each part. Draw object–image lines and lines of equal length that are perpendicular to the mirror to determine the image of each object. Use SALT to describe the characteristics of each image. T/I C

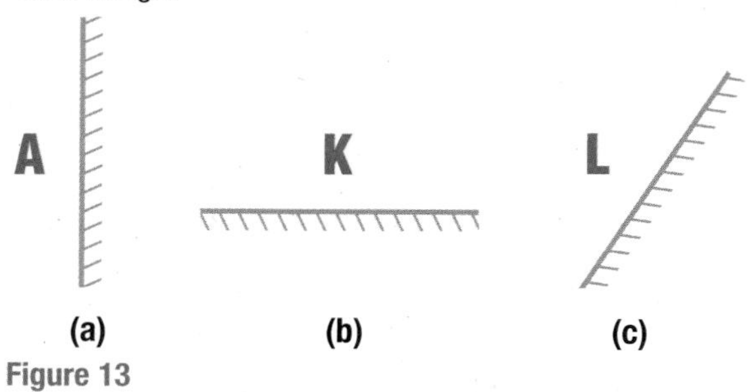

Figure 13

6. (a) What does the acronym SALT stand for?
   (b) In your own words, write a brief explanation of each of these four terms. K/U
7. Emergency vehicles often have words painted backwards on their hoods. Why do you think this is so? Write a brief explanation, including examples of how this is used in your community. K/U A
8. Explain how the backwards writing in a mirror in the activity "Writing Reflectively" demonstrates the properties of an image in a plane mirror. K/U A
9. Your parents have bought a new mirror for your bedroom. At first, you are dismayed because the mirror is only half your height and you do not think that you will be able to see an image of your entire body. You immediately notice, however, that the mirror does allow you to see your entire body. Copy Figure 14 into your notebook and use light rays to show that you really can see your feet in the mirror the way it is set up. T/I C

Figure 14

10. Brainstorm to create a list of effects interior designers might create using mirrors. Explain each effect using what you have learned about light and reflection. K/U A
11. A periscope is a device that is used to see around corners, over a wall, or above water. Simple periscopes contain two plane mirrors.
    (a) Predict how these mirrors are arranged.
    (b) Draw a diagram to illustrate how such a periscope would work. T/I C
12. Were you surprised to learn that your brain can be "fooled" into thinking that an apparent source (virtual image) can be located behind an opaque plane mirror? Discuss this with a partner. C

Locating Images in a Plane Mirror

You use plane mirrors every day. In this activity, you will examine the image in a plane mirror using the four characteristics: size, attitude, location, and type.

SKILLS MENU

- Questioning
- Hypothesizing
- Predicting
- Planning
- Controlling Variables
- Performing
- Observing
- Analyzing
- Evaluating
- Communicating

Testable Question

What are the characteristics of the image in a plane mirror?

Hypothesis/Prediction

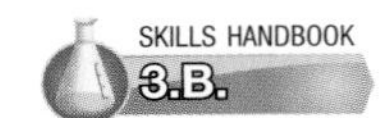

Predict the size, attitude, location, and type of image that will be produced. Give reasons for your hypothesis.

Experimental Design

You will use a ray box to test your predictions regarding the characteristics of an image in a plane mirror.

Equipment and Materials

- ray box
- plane mirror and mirror supports
- pencil and ruler
- sheet of paper

When unplugging the ray box, do not pull the electric cord. Pull the plug itself.

Procedure

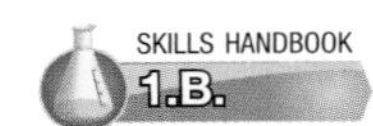

1. Draw a horizontal line across the centre of the sheet of paper.
2. Mount a plane mirror on this line. Remember to place the back of the mirror on the line.
3. Draw an arrow in front of the mirror that is at an angle to the mirror. Place a small circle over the arrow head and label it O1. Place a small circle over the tail of the arrow and label it O2 (Figure 1).

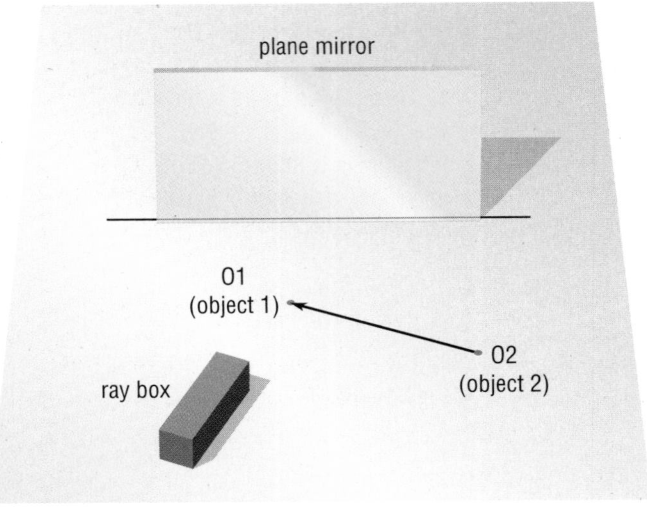

Figure 1

4. Use the ray box to send a ray through O1 that is reflected off the mirror. Use your pencil to mark at least three points on the incident ray and at least three points on the reflected ray. Use your ruler to connect the points and draw both the incident and reflected rays on the paper.
5. Send a second ray through the arrow head, O1, but in a different direction. Repeat the procedure in step 4 to trace the incident and reflected rays on the paper.
6. Send two different light rays through the arrow tail, O2, and trace the incident and reflected rays.
7. Remove the mirror and, using dashed lines, extend the two *reflected rays* from O1 back behind the mirror to where they intersect. Label the point where they intersect behind the mirror as I1.
8. Repeat step 7 for the two *reflected rays* from O2 to locate I2.
9. Use the locations of I1 and I2 to sketch the image behind the mirror using a dashed line (----).

Table 1 Observations

		Arrow head (01)	Arrow tail (02)
length	object–mirror		
	image–mirror		
angle	object–mirror		
	image–mirror		

10. Draw a line from O1 to I1. Measure the length of this line from the object to the mirror and from the image to the mirror. Measure the angle that these lines make with the mirror. Record your results in a table in your notebook similar to Table 1.
11. Repeat step 10 for the line from O2 to I2. Record your results in your table.

Analyze and Evaluate

(a) How did the size (length) of the image compare to the original object? Measure the object and the image with a ruler. T/I

(b) Describe the attitude of the image compared to the original object. C

(c) How does the object–mirror distance compare with the image–mirror distance in a plane mirror? T/I

(d) What angle is formed between the mirror and the object–image line? T/I

(e) What type of image is formed by a plane mirror? How do you know? T/I

(f) Answer the Testable Question at the beginning of this investigation. K/U

(g) Was your prediction correct? Why or why not? T/I

(h) Was your equipment adequate to allow you to answer the Testable Question? Explain. T/I

Apply and Extend

(i) What is a simple way to locate the image in a plane mirror without using light rays? T/I

(j) Use this method to locate the images of the objects in Figure 2 by tracing them into your notebook. C

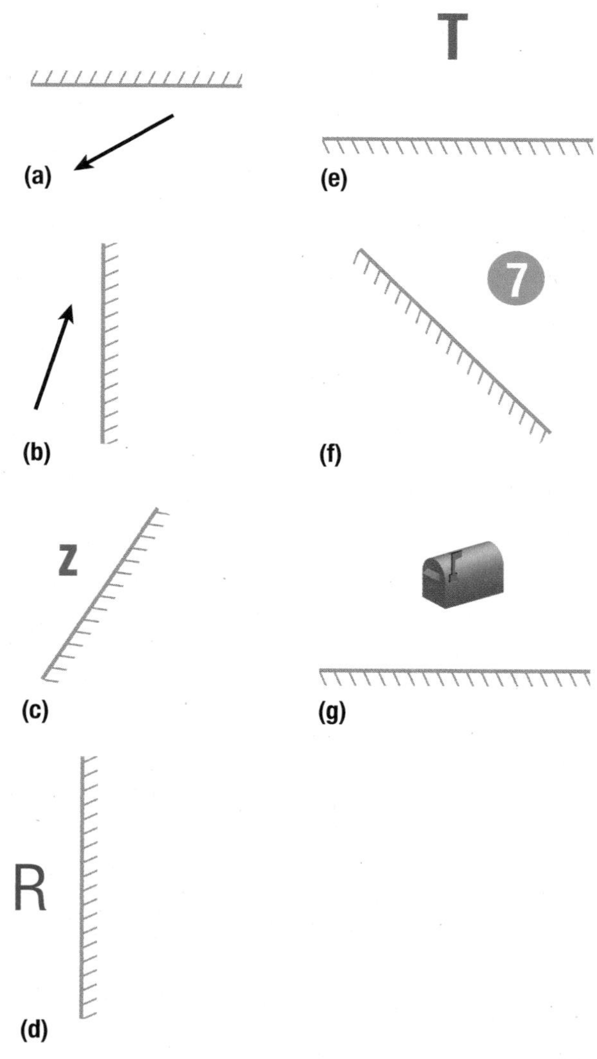

Figure 2

11.9

Images in Curved Mirrors

concave (converging) mirror a mirror shaped like part of the surface of a sphere in which the inner surface is reflective

convex (diverging) mirror a mirror shaped like part of the surface of a sphere in which the outer surface is reflective

Every time you use a flashlight or a makeup mirror, or look into a security mirror at a store, you are using a curved mirror. Curved mirrors are created when you make part of the surface of a sphere reflective. If the reflection is from the inner surface of the sphere, the mirror is **concave**. The centre of a concave mirror bulges away from you. In a **convex mirror**, the reflection is from the outer surface of the sphere. The centre of a convex mirror bulges toward you (Figure 1).

The Terminology of Concave Mirrors

centre of curvature the centre of the sphere whose surface has been used to make the mirror

principal axis the line through the centre of curvature to the midpoint of the mirror

vertex the point where the principal axis meets the mirror

Concave and convex mirrors are described using similar terms. The **centre of curvature** of a mirror is the centre of the sphere, part of whose surface forms the curved mirror. It is labelled as *C*. The **principal axis** of the mirror is the line going through the centre of curvature and the centre of the mirror. Figure 2 shows the side view of a concave mirror. Notice that, in two dimensions, the mirror forms part of a circle. Because the principal axis goes through the centre of the circle, this axis is a radius of the circle. This means that the principal axis intersects the mirror at 90° and is normal to the surface. The **vertex** is the point where the principal axis intersects the mirror. It is labelled as *V*.

READING TIP

Relate It to Yourself

Relate what you have just read about curved mirrors to your own personal life. Think of examples of concave and convex mirrors that you either personally use or are familiar with.

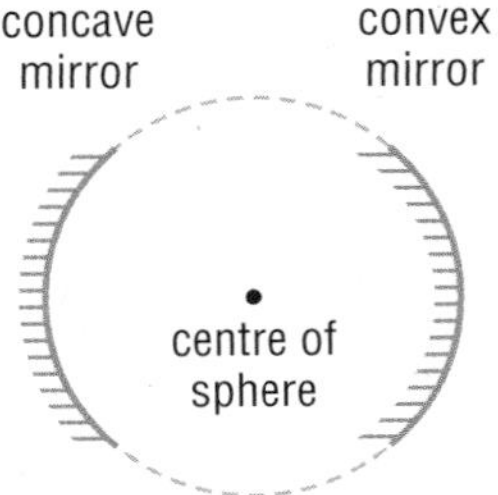

Figure 1 If you make part of the inner surface of a sphere reflective, you get a concave mirror. If you make part of the outer surface reflective, you get a convex mirror.

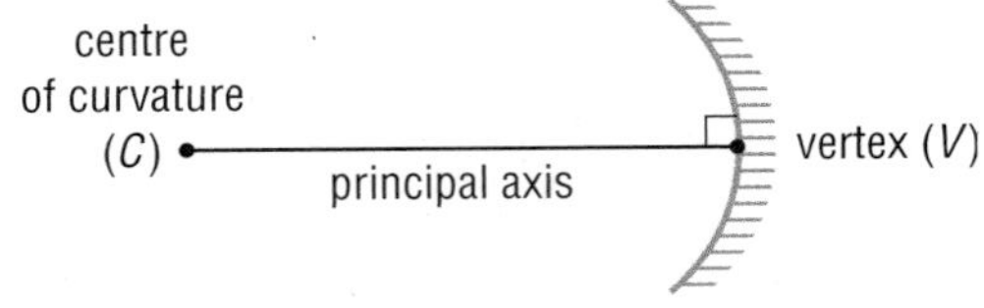

Figure 2 The side view of a concave mirror

converge to meet at a common point

focus the point at which light rays parallel to the principal axis converge when they are reflected off a concave mirror

Any light rays that are parallel to the principal axis will be reflected off the mirror through a single point. This point, where parallel light rays come together, or **converge**, is called the **focus**. It is labelled as *F* (Figure 3). Because a concave mirror focuses parallel rays at *F*, this type of mirror is also called a converging mirror (Figure 4).

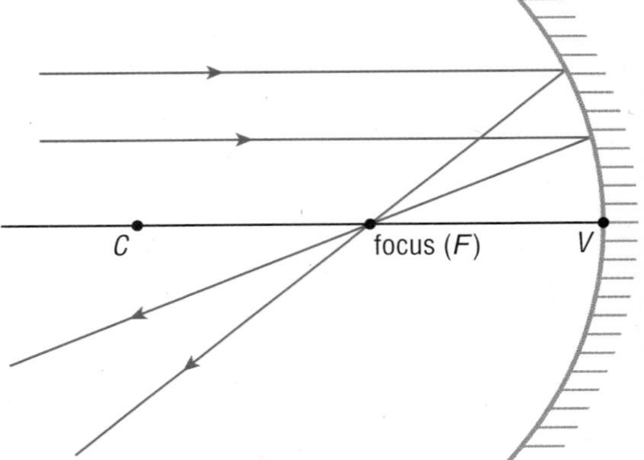

Figure 3 The focus is the point where all incident rays that are parallel to the principal axis converge when they are reflected off the mirror surface.

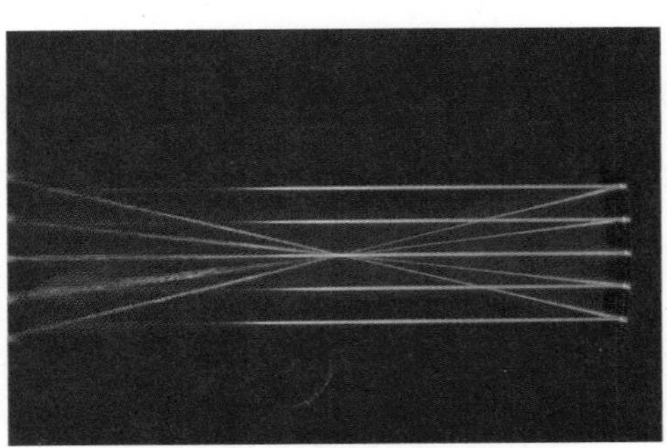

Figure 4 A concave mirror showing the convergence of parallel light rays

How to Locate the Image in a Converging (Concave) Mirror

To determine the image of an object in front of a concave mirror, you need to draw at least two incident rays from the top of the object. These rays will be reflected off the mirror and may or may not cross to form an image. Figure 5 shows several rules that you can use to draw the incident and reflected rays.

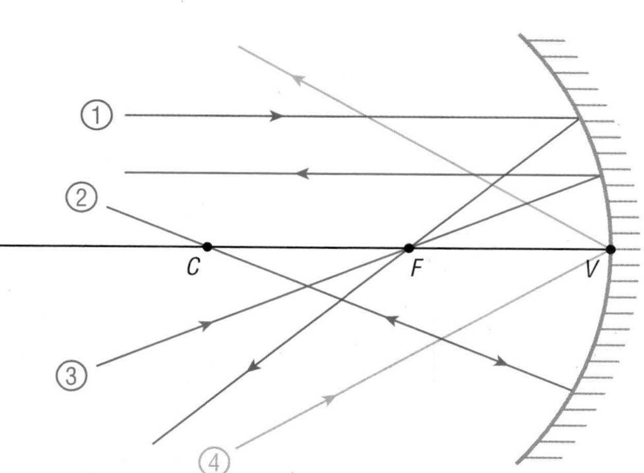

Figure 5 Imaging rules for a concave mirror

1. A light ray parallel to the principal axis is reflected through the focus. This is how the focus is defined.
2. A light ray through the centre of curvature is reflected back onto itself. This rule makes sense because any line through the centre of curvature is a radius of the circle formed by the mirror. A radius is always at 90° to the mirror. A ray along the normal has an angle of incidence of 0°. This means that the angle of reflection is also 0°. The reflected ray will return back on the same path.
3. A ray through *F* will reflect parallel to the principal axis. This rule uses the fact that the angle of incidence is always equal to the angle of reflection. Even if you switch the incident and reflected rays, the light will still follow the same path; only the direction will change. This principle is called the reversibility of light.
4. A ray aimed at the vertex will follow the law of reflection. Because the principal axis is perpendicular to the surface of the mirror, the angle of incidence can be easily measured.

You can use the imaging rules for a concave mirror to find the characteristics of images at a variety of object locations (Figure 6).

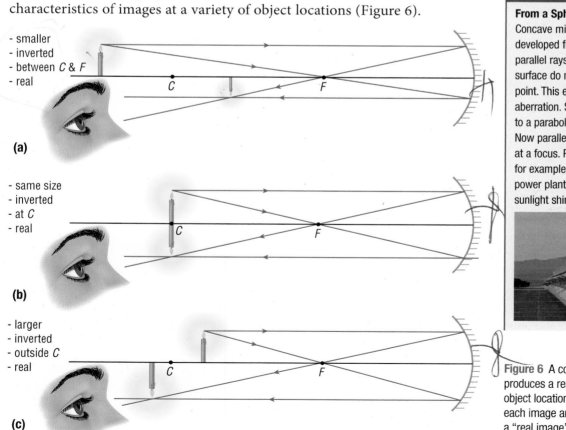

- smaller
- inverted
- between *C* & *F*
- real

C F

(a)

- same size
- inverted
- at *C*
- real

C F

(b)

- larger
- inverted
- outside *C*
- real

C F

(c)

Figure 6 A converging (concave) mirror produces a real image at these three object locations. The characteristics of each image are shown. (An explanation of a "real image" is given on the next page.)

DID YOU **KNOW?**

From a Sphere to a Parabola

Concave mirrors were originally developed from a spherical surface. But parallel rays reflecting off a spherical surface do not all go through a single point. This effect is called spherical aberration. Slightly modifying the shape to a parabolic one corrects this problem. Now parallel rays do, in fact, converge at a focus. Parabolic solar collectors, for example, are used in many solar power plants to increase the intensity of sunlight shining at the focus.

READING TIP

Take Time with Diagrams
Examine diagrams carefully. Take time to understand each part of a diagram. When you have a proper understanding of diagrams, it will help you understand concepts and applications.

real image an image that can be seen on a screen as a result of light rays actually arriving at the image location

Images in a Converging Mirror

If you place a luminous source at a distance greater than *C*, you can locate an image of this source by moving a paper screen back and forth in front of the mirror. The image is smaller, inverted, and somewhere between *C* and *F*. In this case, light is actually arriving at the image location. This type of image is called a **real image**. Any image that can be formed on a screen is a real image because light rays are actually arriving at the image location.

When an object is beyond *C*, at *C*, or between *C* and *F*, the reflected rays actually meet in front of the mirror, forming an inverted, real image each time.

Many devices make use of the properties of concave mirrors. A car headlight, a flashlight, and a searchlight all use concave mirrors. In a searchlight, the light source (the filament) is at the focus, and the reflected rays form a parallel beam (Figure 7). Car headlights and most flashlights are designed in a similar way, except that the filament is slightly inside the focus so that the reflected rays spread slightly apart and illuminate a greater area.

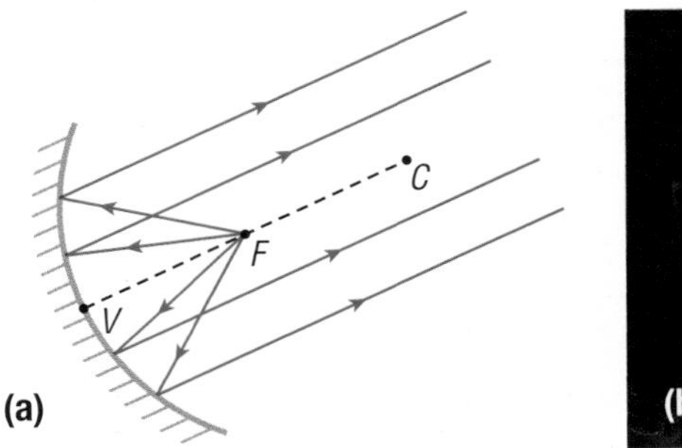

Figure 7 A searchlight radiates light rays from the focus (*F*) so that the reflected rays are parallel.

The reverse of the searchlight application occurs when a concave mirror is used in a reflecting telescope. In this situation, parallel light rays are focused to a sharp image after reflecting off the concave mirror in the telescope. A parabolic solar cooker operates in a similar way. Parallel rays from the Sun converge at the focus where a pot is located. The energy absorbed at the focus in a solar cooker can heat water to boiling. Radio telescopes and satellite dishes are devices that also cause parallel rays to converge. The difference here is that electromagnetic waves other than visible light are used (Figure 8). These other kinds of electromagnetic waves act just like parallel light rays; they converge at the focus of a concave mirror.

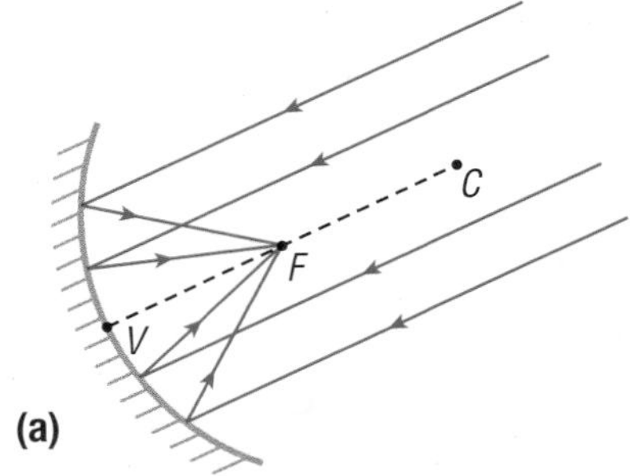

Figure 8 A TV satellite dish receives parallel rays and reflects them to the focus (*F*) where a detector is located.

DID YOU KNOW?

Costly Error
The Hubble space telescope is an important and expensive instrument used for astronomy. This \$2.5 billion telescope was launched in 1990. There had been a small error in the grinding of Hubble's concave mirror. This 0.0023 mm error (about $\frac{1}{50}$th the thickness of a human hair) resulted in Hubble producing fuzzy images (left). The error was corrected in a shuttle servicing mission in 1993, and astronomers around the world breathed a huge sigh of relief. Since that time, Hubble has produced stunning, razor-sharp images of distant celestial objects (right).

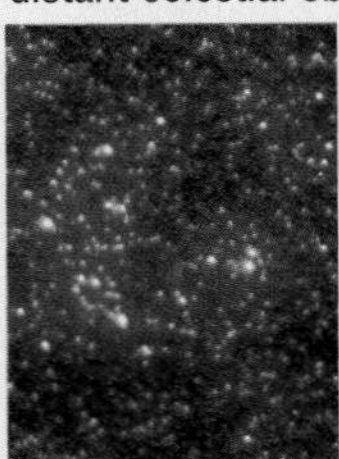

To learn more about how the mirror of the Hubble space telescope was repaired,

GO TO NELSON SCIENCE

No real image is produced when an object is located at *F* in front of a concave mirror. The reflected rays are parallel and do not intersect to form an image (Figure 9). If you extend the rays behind the mirror using dashed lines, you cannot even see a virtual image.

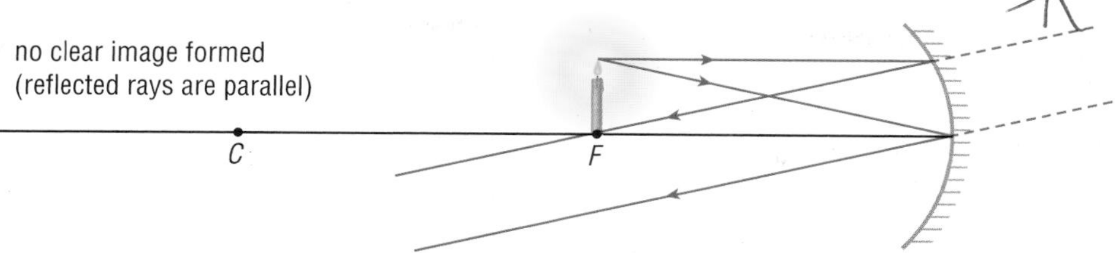

Figure 9 No image is formed when the object is at *F* because the reflected rays are parallel.

WRITING TIP

Using Signal Words

When writing, use signal words such as "moreover" and "also" to show similarities between ideas. Use "however" and "although" to indicate differences between ideas. Use "for instance" to signal examples, and "therefore" to indicate a conclusion.

No real image is produced when an object is between *F* and the concave mirror. The reflected rays spread apart, or **diverge** (Figure 10). The human brain, however, extrapolates the diverging rays backwards to where they appear to originate, which in this case is behind the mirror. This results in a virtual image because the concave mirror is opaque. These extrapolated rays cannot actually come from behind the mirror.

diverge to spread apart

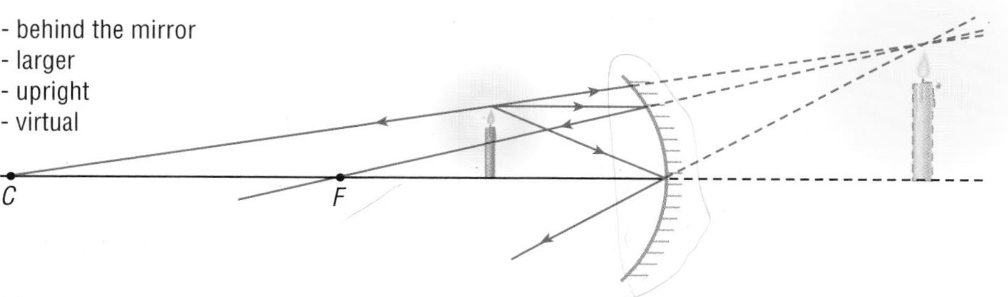

Figure 10 A virtual image behind a concave mirror is formed when an object is between *F* and the mirror.

A virtual image behind a concave mirror is always larger and upright. A shaving mirror and a makeup mirror are two common examples that make use of this property (Figure 11).

Table 1 summarizes the image characteristics in a converging (concave) mirror.

Figure 11 The large virtual image behind the concave mirror allows makeup to be put on more easily.

Table 1 The Imaging Properties of a Converging Mirror

OBJECT	IMAGE			
Location	**Size**	**Attitude**	**Location**	**Type**
beyond *C*	smaller	inverted	between *C* and *F*	real
at *C*	same size	inverted	at *C*	real
between *C* and *F*	larger	inverted	beyond *C*	real
at *F*	no clear image			
inside *F*	larger	upright	behind mirror	virtual

How to Locate the Image in a Diverging (Convex) Mirror

The parts of a convex mirror and the imaging rules for a convex mirror are similar to those for a concave mirror. The difference is that *F* (now called a virtual focus) and *C* are *behind* the mirror and light rays seem to come from an apparent light source behind the mirror (Figure 12).

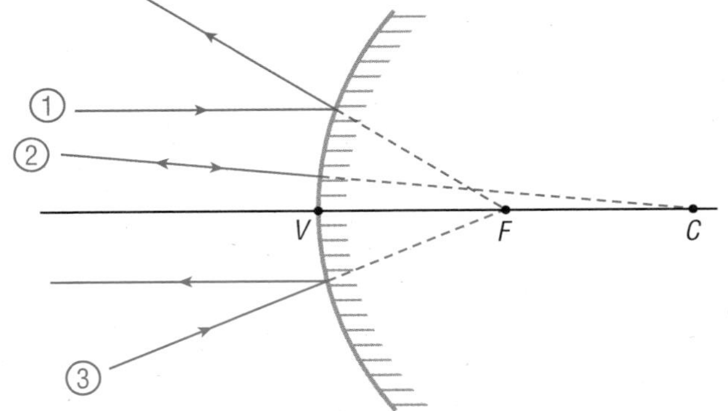

① A ray parallel to the principal axis is reflected as if it had come through the focus (*F*).

② A ray aimed at the centre of curvature (*C*) is reflected back upon itself.

③ A ray aimed at the focus (*F*) is reflected parallel to the principal axis.

Figure 12 Imaging rules for a convex mirror

Images in a Diverging Mirror

The rays reflected off a convex mirror always diverge. For this reason, a convex mirror is also called a diverging mirror. Reflected rays from an object never cross in front of the mirror to form a real image. The human brain, again, extrapolates these rays behind the mirror to where they appear to converge. This results in the diverging (convex) mirror producing a smaller, upright, virtual image. This property makes convex mirrors very useful as security mirrors in stores (Figure 13). Convex mirrors show a wide range of view with their smaller virtual image. Convex mirrors are also used as the side-view mirror on vehicles. "Objects in mirror are closer than they appear" is frequently printed at the bottom of side-view mirrors to remind viewers that they are seeing a smaller image.

For computer simulations involving curved mirrors,

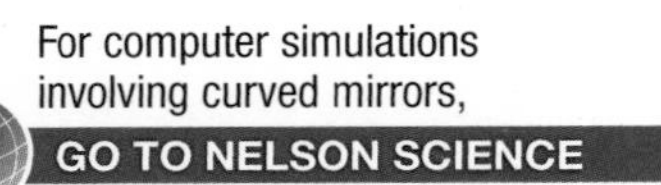

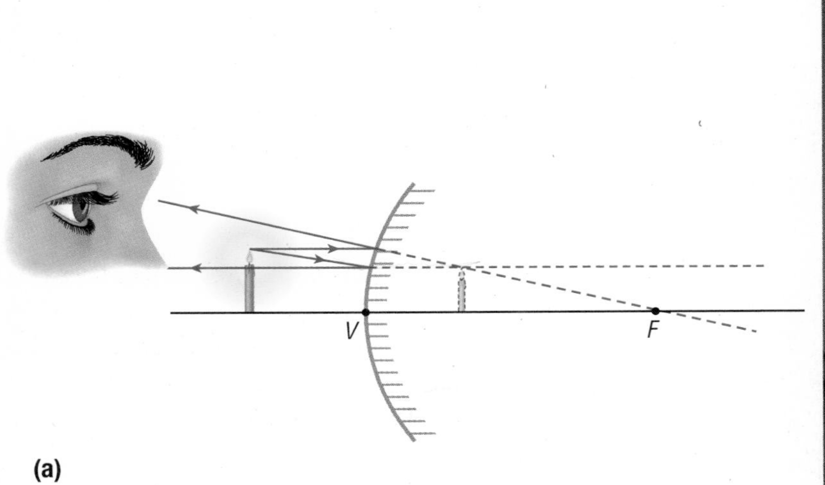

Figure 13 (a) A convex mirror always produces a smaller virtual image. (b) A security mirror in a store illustrates this property.

UNIT TASK Bookmark

You can apply what you learned about optics in this section to the Unit Task described on page 588.

WRITING TIP

Writing a Persuasive Text
Conclude by connecting the main idea and key points. For example, for a persuasive text on images in a converging mirror, you might conclude by saying that a converging mirror produces an inverted, real image if the object is beyond *F*.

IN SUMMARY

- A converging (concave) mirror has its focus on the same side as the object; a diverging (convex) mirror has its focus *behind* the mirror.
- A light ray that is parallel to the principal axis of a curved mirror is reflected through the focus (*F*); if the mirror is diverging (convex), parallel rays are reflected away from the virtual focus, which is behind the mirror.
- At least two incident rays are drawn to determine whether or not an image is formed and, if so, its characteristics. These rays usually originate from the top of the object.
- A converging (concave) mirror produces an inverted, real image if the object is beyond *F*; if the object is at *F*, no image is formed; and if the object is between *F* and the mirror, a larger, upright, virtual image is formed.
- A diverging (convex) mirror always produces a smaller, upright, virtual image.

CHECK YOUR LEARNING

1. List examples of how concave or convex mirrors might be used at your school. A
2. Describe the difference between a real image and a virtual image. K/U
3. Use a diagram to show how to locate the focus in a concave mirror. K/U C
4. In your own words, state the imaging rules for concave mirrors. K/U
5. You are looking at your image in a makeup or shaving mirror. Where is your head located with respect to the focus (*F*)? T/I
6. Why will a diverging (convex) mirror never produce a real image? Include a diagram in your explanation. K/U C
7. Examine the image formed by the mirror in Figure 14.
   (a) What kind of mirror is this?
   (b) Where is this image located?
   (c) What type of image is it? K/U

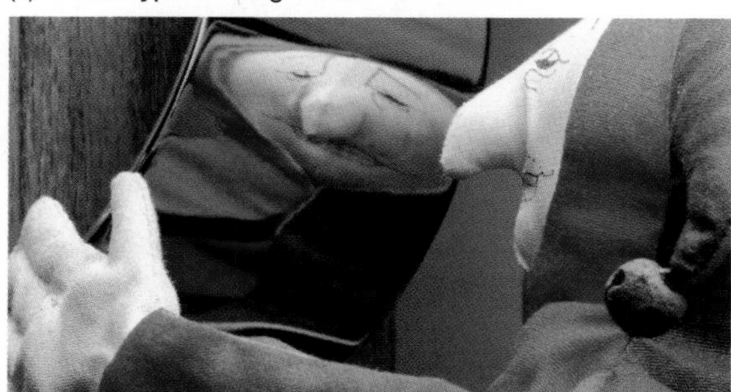

Figure 14

8. Copy Figure 15 into your notebook. Locate the image for each object and state its characteristics. T/I C

(a)

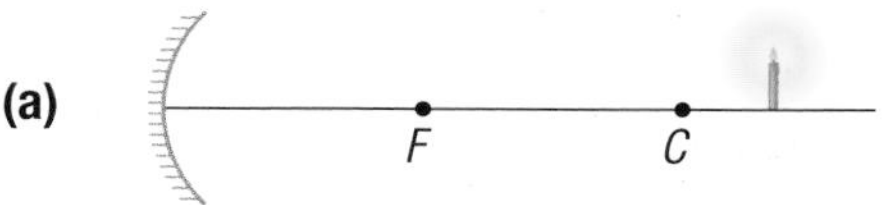

(b)

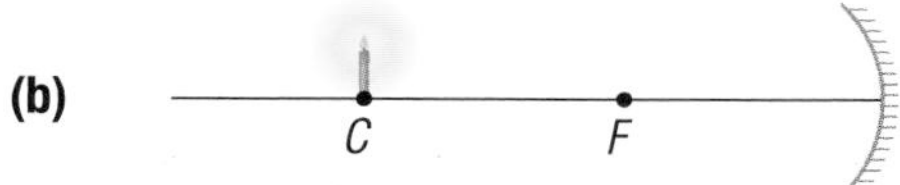

(c)

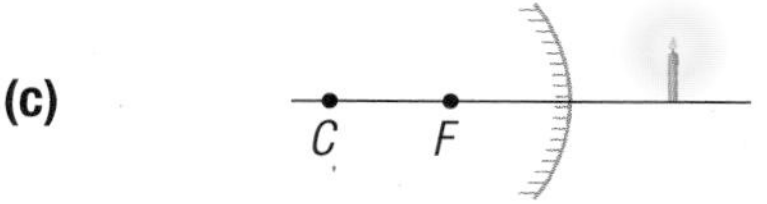

Figure 15

9. What is the relationship between the type and the attitude of an image? K/U
10. (a) Why are convex mirrors placed on sharp turns in parking garages?
    (b) State other uses for convex mirrors. A

11.10 CONDUCT AN INVESTIGATION

Locating Images in Curved Mirrors

Curved mirrors are all around you. These mirrors are found in makeup mirrors, in flashlights, on cars, and as security mirrors in many stores. In this investigation, you will examine the imaging properties of curved mirrors.

SKILLS MENU

- Questioning
- Hypothesizing
- Predicting
- Planning
- Controlling Variables
- Performing
- Observing
- Analyzing
- Evaluating
- Communicating

Testable Question

What are the characteristics (size, attitude, location, and type) of the images in a converging (concave) mirror and in a diverging (convex) mirror?

Hypothesis/Prediction

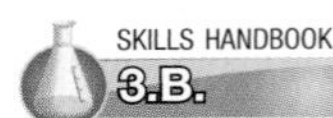

Use your understanding of the properties of light to write a hypothesis predicting the characteristics of images in converging and diverging mirrors.

Experimental Design

You will use a candle to test your predictions regarding the characteristics of images in converging and diverging mirrors.

Equipment and Materials

- converging mirror
- diverging mirror
- mirror support
- metre stick with two supports
- candle with holder
- paper screen
- chalk that can be easily erased

Procedure

Part A: Locating *F* and *C*

1. Place the two stick supports under the ends of the metre stick.
2. Place the converging mirror in the mirror support and place this assembly near the end of the metre stick.
3. Find the focus, *F*, of the mirror. To do this, select a relatively distant object (such as the open slats of window blinds) that is visible when the lights have been turned off. Make sure that you are as far away from this distant object as possible. Aim the metre stick–mirror assembly at the distant object, which should be reflected in the mirror. Move the sheet of paper back and forth in front of the mirror until you see an image of the distant object on the paper. Adjust the paper to get the image as sharp as possible. (Do not completely cover the mirror with the paper.) The distance from the paper to the mirror gives the position of the focus. Mark *F* on the ruler with the chalk.
4. The centre of curvature (*C*) is located at a distance that is twice the distance from the mirror to *F*. For example, if *F* is 30 cm, *C* would be 60 cm from the mirror. Mark *C* on the ruler with the chalk.

Part B: Locating Images in the Mirrors

5. Prepare a table similar to Table 1 in which to record your observations.

Table 1 Observations of Candle Images

Object location	Size of image	Attitude of image	Location of image	Type of image
beyond *C*				
at *C*				
between *C* and *F*				
at *F*				
inside *F*				

6. Place a lit candle beyond *C*, as shown in Figure 1. Move the paper screen back and forth to locate a focused real image. Record the characteristics of each image (size, attitude, location, and type) in your table. Use *C* and *F* as reference points when describing the image location.

When using a candle, tie back long hair and loose clothing. Place a piece of paper under the candle to catch any falling wax. Be careful when moving the candle—the wax is hot.

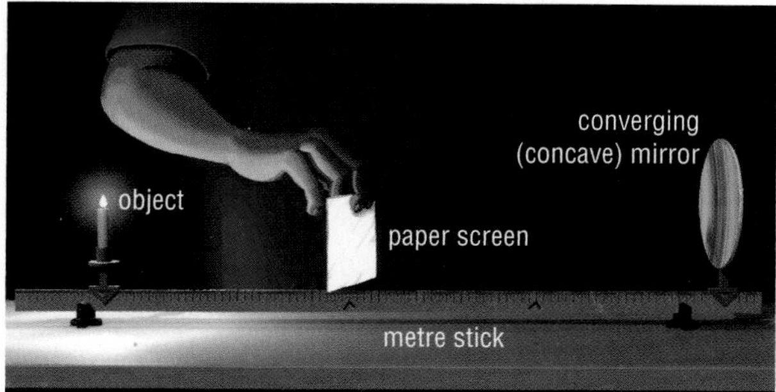

Figure 1

7. Next, place a lit candle at *C*. Again, move the screen to find the image and record its characteristics.
8. Repeat step 7 with the candle at three other positions:
   - between *C* and *F*
   - at *F*
   - between *F* and the mirror

   Note that you may need assistance from your teacher for the last two object locations: at *F* and between *F* and the mirror.
9. Being careful of the candle flame, replace the converging mirror with a diverging one. Attempt to find an image on the paper screen. Record your observations.
10. Now, look *into* the convex mirror and locate the image of the candle. Observe and record the image characteristics.
11. Move the candle back and forth. Observe and record any changes in the image.

Analyze and Evaluate

(a) What type of image(s) does a converging mirror produce? T/I

(b) Where must an object be located for a converging mirror to produce a real image? T/I

(c) What happens to the size of the real image as the object is slowly moved from its original position beyond *C* toward *F*? T/I

(d) What is the only location where a converging mirror will not produce an image? T/I

(e) Where must an object be located for a converging mirror to produce a virtual image? T/I

(f) When you used the diverging mirror, why did you not have to follow a procedure similar to that used for the converging mirror? T/I

(g) Answer the Testable Question at the beginning of this investigation. K/U

(h) Was your prediction correct? Why or why not? T/I

Apply and Extend

(i) What kind of mirror is used as a security mirror in almost all stores? What image characteristics make this mirror so useful? A

(j) Flashlights and car headlights all have converging mirrors. Why would it not be a good idea to use a diverging mirror instead? A

(k) Early settlers often placed a diverging reflective surface behind oil lamps that were mounted on walls. What was the purpose of this mirror? Why was a converging mirror not used instead? A

(l) A converging mirror has a focus 12 cm from the mirror. Predict the characteristics of the image produced if a candle is placed: T/I C

- 30 cm from the mirror
- 18 cm from the mirror
- 9 cm from the mirror

Check your predictions by using ray diagrams.

CHAPTER 11 LOOKING BACK

KEY CONCEPTS SUMMARY

Optical devices benefit our society in many ways.

- Plane mirrors are widely used for personal hygiene and grooming. (11.4)
- Converging mirrors are used in car headlights, searchlights, reflecting telescopes, solar cookers, and shaving and makeup mirrors. (11.9)
- Some examples of diverging mirrors are side-view mirrors on cars and security mirrors in stores. (11.9)

Light is produced by natural and artificial sources.

- Incandescence is light emitted when a material has been heated. (11.2)
- Light from an electric discharge is caused by passing an electric current through a gas. (11.2)
- Phosphorescence and fluorescence are both caused by a material absorbing ultraviolet light. (11.2)
- Chemiluminescence and bioluminescence are the production of light from a chemical reaction without an increase in temperature. (11.2)
- Triboluminescence is light produced from friction with crystals. (11.2)

Light is an electromagnetic wave that travels at high speed in a straight line.

- Electromagnetic waves travel at the speed of light in a vacuum and do not require a medium for transmission. (11.1)
- The electromagnetic spectrum consists of light with a variety of energies: radio waves (least energy), microwaves, infrared light, visible light, ultraviolet light, X-rays, and gamma rays (most energy). (11.1)
- White light is composed of a continuous spectrum of colours. (11.1)

When light is reflected off a flat, shiny surface, the image is equal in size to the object and the same distance from the surface.

- Light rays are used to represent the direction in which light is travelling. (11.4)
- When light is reflected off a plane mirror, the angle of incidence equals the angle of reflection. (11.6)
- When a light ray strikes a plane mirror, the incident ray, the reflected ray, and the normal all lie in the same plane. (11.6)

Images in flat mirrors are located at the point where the backward extensions of reflected rays intersect.

- A plane mirror divides the object–image line in half and is perpendicular to that line. (11.7)
- The four image characteristics are **S**ize, **A**ttitude, **L**ocation, and **T**ype (**SALT**). (11.7)
- An image in a plane mirror is always the same size as the object, upright, behind the mirror, and virtual. (11.7)

Curved mirrors produce a variety of images.

- At least two incident rays from an object are needed to determine whether or not an image is formed and, if so, its characteristics. (11.7, 11.9)
- A converging (concave) mirror produces an inverted, real image if the object is anywhere beyond *F*; if the object is at *F*, no image is formed; and if the object is between *F* and the mirror, a larger, upright, virtual image is formed. (11.9)
- A diverging (convex) mirror always produces a smaller, upright, virtual image. (11.9)

WHAT DO YOU THINK NOW?

You thought about the following statements at the beginning of the chapter. You may have encountered these ideas in school, at home, or in the world around you. Consider them again and decide whether you agree or disagree with each one.

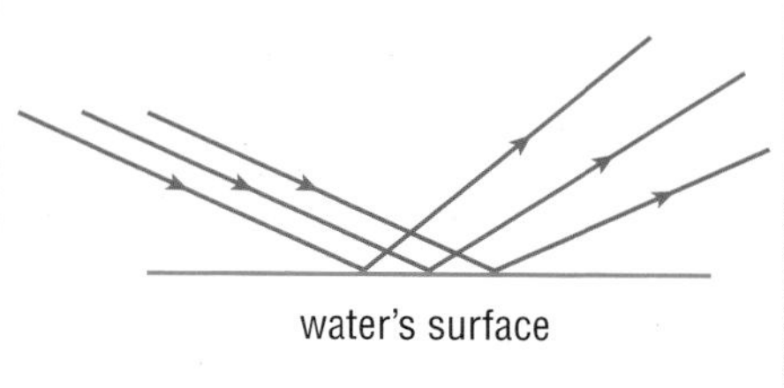

1 This diagram accurately shows light reflecting off the surface of very still water.
Agree/disagree?

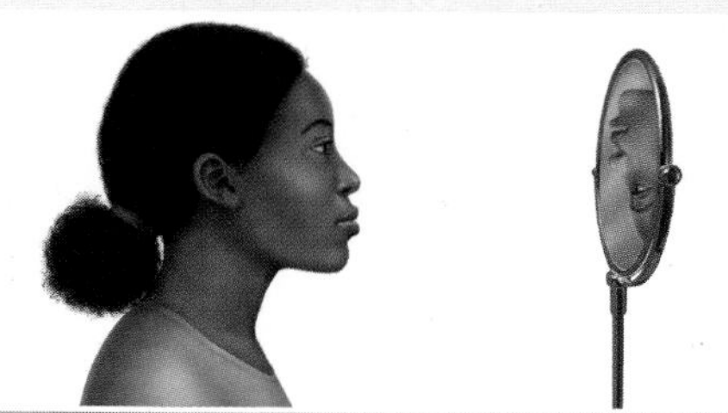

4 This diagram accurately shows how an image appears in a makeup mirror.
Agree/disagree?

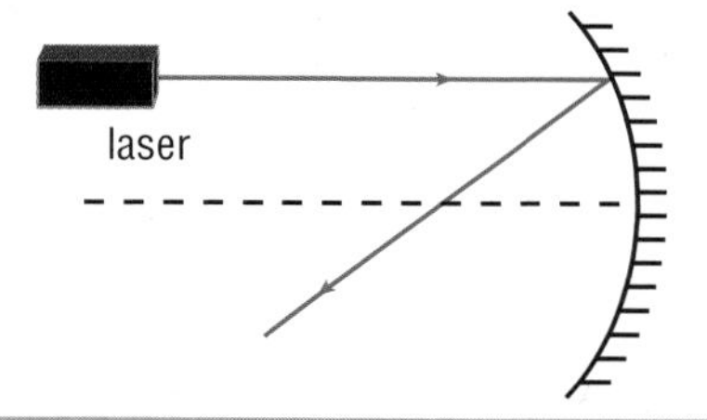

2 This diagram accurately shows a laser beam reflecting off a curved mirror.
Agree/disagree?

5 Microwaves travel at the speed of light.
Agree/disagree?

3 A full-length mirror is necessary in order for you to see your whole body in reflection.
Agree/disagree?

6 A luminous object such as a candle radiates light in all directions.
Agree/disagree?

How have your answers changed?
What new understanding do you have?

Vocabulary

electromagnetic wave (p. 464)
visible light (p. 465)
electromagnetic spectrum (p. 465)
visible spectrum (p. 467)
luminous (p. 470)
non-luminous (p. 470)
light ray (p. 479)
incident light (p.479)
transparent (p. 479)
translucent (p. 479)
opaque (p. 479)
image (p. 480)
mirror (p. 480)
reflection (p. 480)
plane (p. 481)
incident ray (p. 481)
reflected ray (p. 481)
normal (p. 481)
perpendicular (p. 481)
angle of incidence (p. 481)
angle of reflection (p. 481)
virtual image (p. 490)
concave (converging) mirror (p. 496)
convex (diverging) mirror (p. 496)
centre of curvature (p. 496)
principal axis (p. 496)
vertex (p. 496)
converge (p. 496)
focus (p. 496)
real image (p. 498)
diverge (p. 499)

BIG Ideas

- ✓ Light has characteristics and properties that can be manipulated with mirrors and lenses for a range of uses.
- ✓ Society has benefitted from the development of a range of optical devices and technologies.

CHAPTER 11

REVIEW

The following icons indicate the Achievement Chart category addressed by each question.

K/U Knowledge/Understanding T/I Thinking/Investigation
C Communication A Application

What Do You Remember?

1. In your notebook, match each item in column A with the most appropriate phrase from column B. (11.1, 11.2, 11.4, 11.9) K/U

Column A	Column B
luminous	visible electromagnetic waves
transparent	90° to a surface
white light	produces its own light
concave mirror	from an apparent light source
real image	diverging mirror
normal	transmits all incident light
virtual image	converging mirror
convex mirror	seen on a screen

2. What two properties do all electromagnetic waves have in common? (11.1) K/U

3. Describe the characteristics of an image in a plane mirror. (11.7) K/U

4. In your own words, describe the two laws of reflection. (11.6) K/U

5. Briefly describe how light is produced by
   (a) phosphorescence
   (b) electric discharge
   (c) triboluminescence (11.2) K/U

6. List these electromagnetic waves in order from lowest energy to highest energy: green light, microwaves, X-rays, ultraviolet light, infrared light, red light, and radio waves. (11.1) K/U

What Do You Understand?

7. How will the word **PHYSICS** appear when viewed in a plane mirror?(11.7) K/U

8. Why is the light ray concept a useful model to use when determining the behaviour of light? (11.4) T/I

9. Why will a convex mirror never form a real image? (11.9) K/U

10. Where must an object be located in order for a concave mirror to form
    (a) a real image?
    (b) a virtual image? (11.9) K/U

11. Classify each object as exhibiting specular or diffuse reflection. Justify each of your answers. (11.6) K/U
    (a) dry asphalt
    (b) a car windshield producing glare in your eyes
    (c) a sweater
    (d) high-gloss paint

12. Copy and complete Table 1 for each initial condition. (11.6) K/U T/I

Table 1 Ray Angles

Initial condition	Angle of incidence	Angle of reflection
angle between the reflected ray and the normal is 47°		
angle between the incident ray and the normal is 52°		
angle between the incident ray and a plane mirror is 14°		
the incident ray comes in along the normal		

13. Copy Figure 1 into your notebook. Draw light rays that follow the imaging rules for curved mirrors to determine the image characteristics for each object (11.9). T/I C

(a)
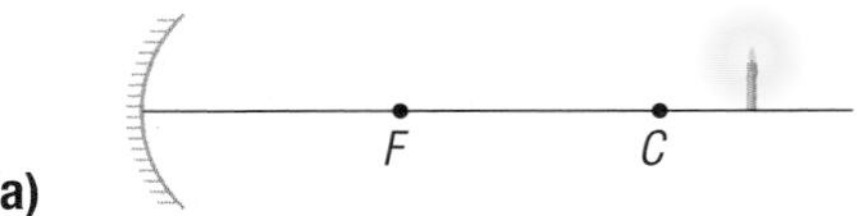

(b)
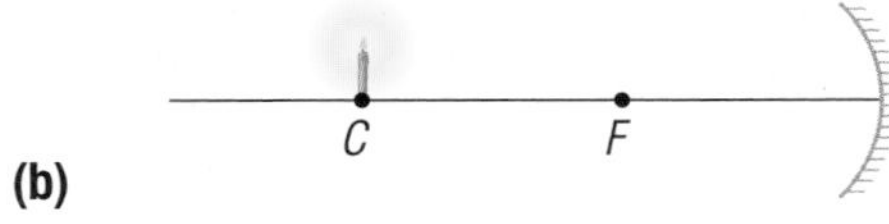

(c)
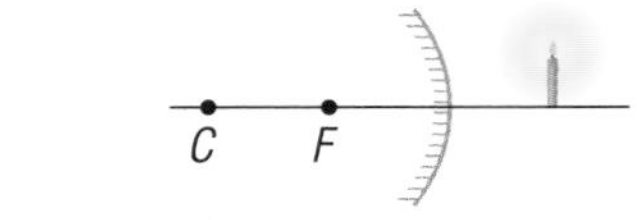

Figure 1

Solve a Problem

14. Your younger brother does not think that light is a form of energy. Write a short script to show how you would convince him otherwise. (11.1) T/I C

15. You walk into a darkened room and turn on a flashlight. You see an image of the flashlight reflecting off a plane mirror in front of you. The image is 8.4 m away. How far away is the plane mirror? Explain. (11.7) K/U C

16. Copy Figure 2 into your notebook. Use light rays to determine which of the object(s) would be visible by looking into the mirror from the eye location. (11.7) T/I C

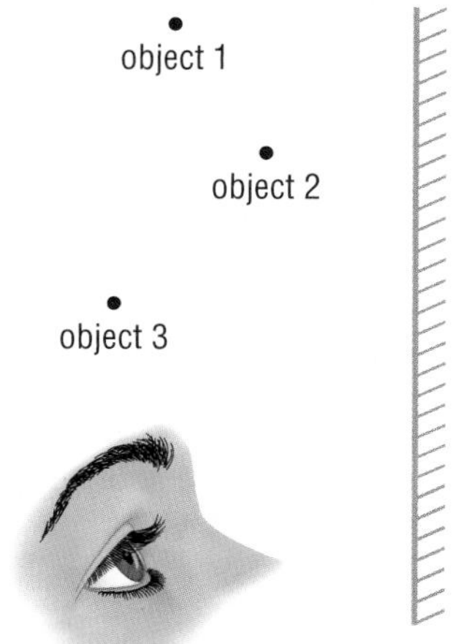

Figure 2

17. A photograph is covered with non-glare glass. The surface of this glass is rougher than regular glass. How would this feature contribute to it being non-glare glass? (11.6) A

18. A concave mirror has a focus of 75 cm. An object is placed 60 cm in front of the mirror. Describe the image characteristics of this object. (11.9) T/I C

19. Copy Figure 3 into your notebook. Use light rays to locate the focus (F) and the centre of curvature (C) of this mirror. (11.9) T/I C

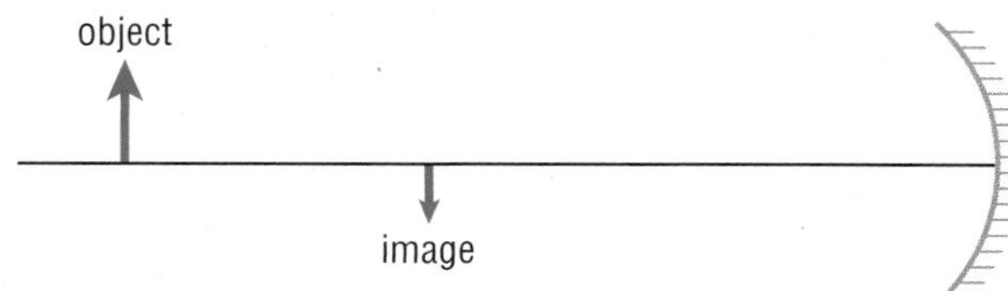

Figure 3

Create and Evaluate

20. Compare and contrast incandescent lighting and fluorescent lighting in these categories: (11.2) K/U A
    (a) method of producing light
    (b) efficiency in producing light
    (c) initial cost
    (d) long-term cost
    (e) environmental consequences

Reflect on Your Learning

21. In this chapter, you learned that images in a mirror are either behind the mirror or visible in front on a screen.
    (a) Were you surprised to learn this? Explain.
    (b) Many people think that the image in a plane mirror is on the mirror surface. How would you explain to them that it really is not there but *behind* the mirror?

Web Connections 

22. Research why the incandescent bulb was so popular for so long. Express your opinion in two paragraphs. (11.2) T/I A C

23. Figure 4 shows the world's largest blast furnace powered by the Sun. It is located in Odeillo, France. (11.9) K/U T/I A

Figure 4

    (a) What kind of mirror is built into the wall of the building?
    (b) Where do you think the focus (that is, the site of the blast furnace) of this large mirror is located?
    (c) What are some advantages and disadvantages of using light as the energy source for this blast furnace?

To do an online self-quiz and for all Nelson Web Connections,
GO TO NELSON SCIENCE

CHAPTER 11

SELF-QUIZ

The following icons indicate the Achievement Chart category addressed by each question.

K/U Knowledge/Understanding T/I Thinking/Investigation
C Communication A Application

For each question, select the best answer from the four alternatives.

1. By which of these methods is light energy transmitted? (11.1) K/U
   (a) radiation
   (b) inversion
   (c) conduction
   (d) emission

2. Which of these objects is considered luminous? (11.2) K/U
   (a) a tree
   (b) a mirror
   (c) a window
   (d) a lit match

3. Figure 1 shows a light bulb positioned in front of a concave mirror. Which statement correctly describes the properties of the resulting image? (11.9) K/U

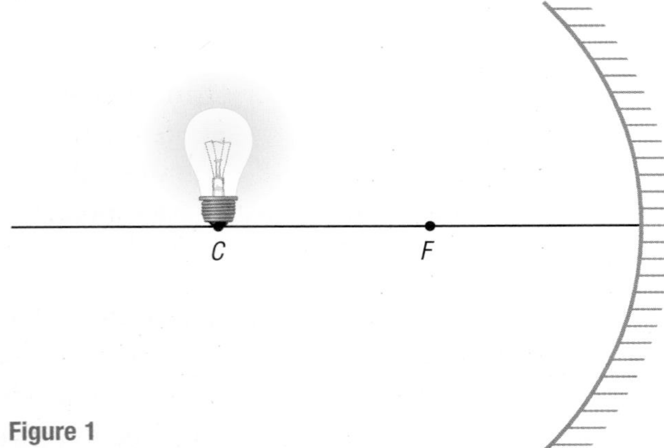

Figure 1

   (a) The image will be an inverted, virtual image that is smaller than the light bulb.
   (b) The image will be an inverted, real image that is the same size as the light bulb.
   (c) The image will be an upright, virtual image that is larger than the light bulb.
   (d) The image will be an upright, real image that is the same size as the light bulb.

4. Which of these correctly shows how the word **LIGHT** would look when viewed in a plane mirror? (11.7) K/U
   (a) THƎI⅃
   (b) LIGHT
   (c) ⅃IƎHT
   (d) THGIL

Indicate whether each statement is TRUE or FALSE. If you think the statement is false, rewrite it to make it true.

5. In order to see an object, light must come from that object into your eyes. (11.2) K/U

6. The four characteristics of an image in a mirror are size, attitude, luminosity, and type. (11.7) K/U

Copy each of the following statements into your notebook. Fill in the blanks with a word or phrase that correctly completes the sentence.

7. A ______________ mirror is a curved mirror shaped like part of the outer surface of a sphere in which the outer surface is reflective. (11.9) K/U

8. The image of an object you see in a plane mirror is called a ______________ image. (11.7) K/U

Match each term on the left with the appropriate example on the right.

9. (a) incandescence — (i) lightning
   (b) chemiluminescence — (ii) fireflies
   (c) bioluminescence — (iii) traditional light bulbs
   (d) fluorescence — (iv) glow sticks
   (e) electric discharge — (v) energy-efficient light bulbs

   (11.2) K/U

Write a short answer to each of these questions.

10. Describe two properties of light. (11.1) K/U

11. You have learned that light rays travel in a straight line. Does this also mean that a light ray cannot change direction? Why or why not? (11.4) K/U

12. Give at least two examples of convex mirrors other than those mentioned in the text. (11.9) K/U A

13. What property of radio waves allows you to send and receive calls on a cell phone? Explain your answer. (11.1) T/I A

14. Name three types of electromagnetic waves and their uses that you might encounter in your daily life. (11.1) A

15. Your family plans to buy new Christmas lights this year. What two reasons would you give to convince them to buy LED lights instead of incandescent lights? (11.2) C

16. Laser light has special properties. Use what you know about the properties of lasers to answer these questions.
    (a) Why does visible light from a laser have a very pure colour?
    (b) Why can some forms of laser light be used to burn holes through steel? (11.3) K/U

17. (a) Give an example of a situation in which it would be better to use a translucent material instead of a transparent one. Explain your answer.
    (b) Give an example of a situation in which it would be better to use a transparent material instead of a translucent one. Explain your answer. (11.4) T/I A

18. Consider the uses and properties of mirrors. How could a mirror be useful to people in each situation?
    (a) a hiker in a wilderness area
    (b) a dentist examining teeth
    (c) a security guard in a store (11.9) T/I

19. Your friend shines a flashlight onto the open palm of your hand. Would you expect the light reflected from your hand to be specular reflection or diffuse reflection? Explain. (11.6) T/I

20. (a) What do you think the term "mirror image" means?
    (b) Give an example of a mirror image. (11.7) T/I

21. In this chapter, you learned the mnemonic ROY G. BIV to help you remember the order of the colours in the visible spectrum. Devise a mnemonic to help you remember the seven types of electromagnetic waves in the electromagnetic spectrum in order, from those with the lowest energy to those with the highest energy. (11.1) C

22. Do you think it is false advertising when a company claims that a laundry detergent cleans clothes better because the clothes become brighter when washed? Justify your answer. (11.2) T/I

23. Your friend explains why light reflected off a bumpy surface produces a fuzzy image. He says that a light ray striking a rough surface produces a reflected ray with an angle of reflection that is not equal to the angle of incidence. What do you think of your friend's explanation? Explain your answer. (11.6) T/I

24. Some restaurants cover an entire wall with a large, plane mirror. What effect does this have on the appearance of the restaurant? Why does this happen? (11.7) K/U A

CHAPTER

12 The Refraction of Light

KEY QUESTION: How does light behave as it travels from one medium into another?

Light travelling through different media can create unusual images.

UNIT E

Light and Geometric Optics

CHAPTER 11

The Production and Reflection of Light

CHAPTER 12

The Refraction of Light

CHAPTER 13

Lenses and Optical Devices

KEY CONCEPTS

Light changes direction predictably as it travels through different transparent media.

Light bends toward the normal when it slows down in a medium with a higher index of refraction.

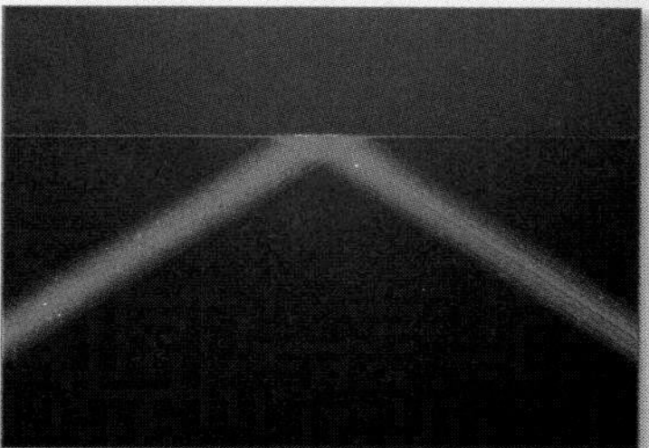

Total internal reflection may occur when an incident ray is aimed at a medium with a lower index of refraction.

Many optical devices make use of the refraction and reflection of light.

The refraction and reflection of light can be used to explain natural phenomena.

Understanding the behaviour of light is key to many careers.

ENGAGE IN SCIENCE

FROM EARTH TO THE MOON AND BACK!

The following conversation was recorded on July 21, 1969.

Armstrong: The top of that next little ridge there. Wouldn't that be a pretty good place?

Aldrin: All right. Should I put the LR-cubed right about here?

Armstrong: All right.

Aldrin: I'm going to have to get on the other side of this rock here.

Armstrong: I would go right around that crater to the left there. Isn't that a level spot there?

Aldrin: I think this right here is just as level.

There are two remarkable things about this conversation. The first is that this conversation took place on the surface of the Moon in a location called the Sea of Tranquility. It took place between two NASA astronauts from the Apollo 11 mission: Neil Armstrong, who was the first human on the Moon, and Edwin "Buzz" Aldrin, who was the second human on the Moon. The second remarkable thing is that they were talking about the placement of a special device called the "LR-cubed."

The "LR-cubed" stands for Laser Ranging Retro-Reflector. This amazing device was part of an experiment that is still working today, decades after its original installation. What exactly is the LR^3 (LR-cubed), and how does it work? In a small group, discuss what you think this device might be. You will learn the answers in this chapter.

WHAT DO YOU THINK?

Many of the ideas you will explore in this chapter are ideas that you have already encountered. You may have encountered these ideas in school, at home, or in the world around you. Not all of the following statements are true. Consider each statement and decide whether you agree or disagree with it.

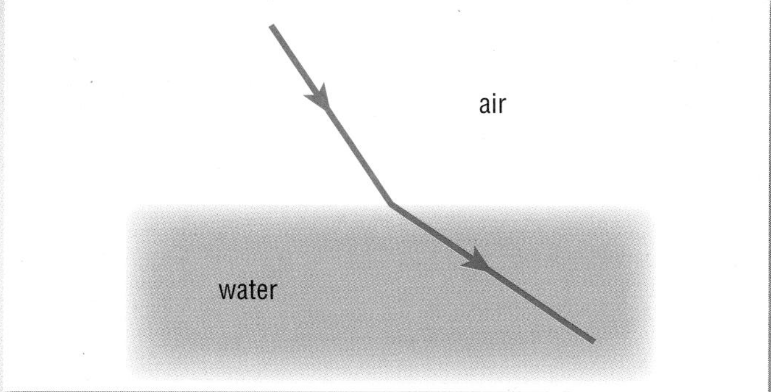

1 This diagram accurately shows light passing from air into water.
Agree/disagree?

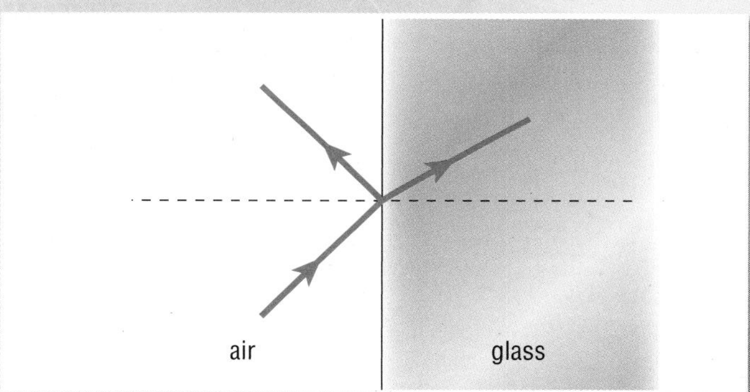

2 Light can be both reflected and transmitted when it goes from air into glass.
Agree/disagree?

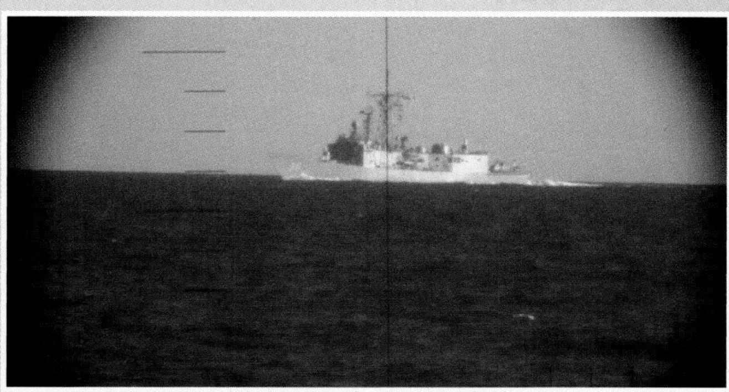

3 A periscope is an optical device that allows you to see around corners.
Agree/disagree?

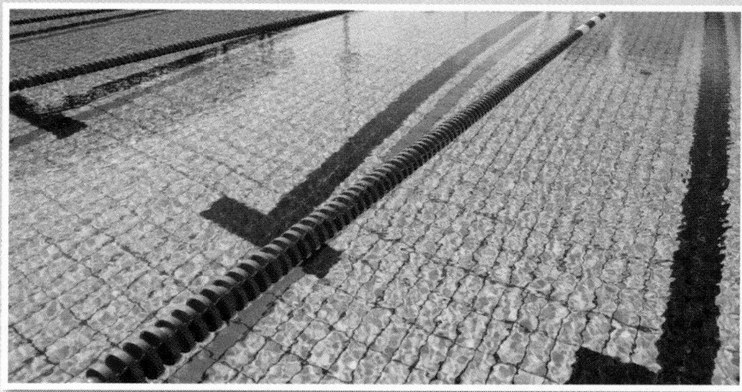

4 A swimming pool is always deeper than it appears.
Agree/disagree?

5 In order to see a rainbow, the Sun must be shining in one part of the sky while rain or clouds are in the opposite part of the sky.
Agree/disagree?

6 When you see a mirage, you are really looking at an image of the sky.
Agree/disagree?

Evaluating

When you evaluate a text, you apply your prior knowledge and analysis of the content to make judgments about its ideas, information, and reliability. Support your judgments with convincing reasons and specific details or quotations from the text. Ask yourself the following questions when evaluating texts:

- Does the main idea seem reasonable? Can I accept it?
- Is the information accurate and credible?
- Do I agree with the author's point of view on the subject?
- Can I find clues about the text's biases and the values it supports?

READING TIP

As you work through the chapter, look for tips like this. They will help you develop literacy strategies.

Figure 1 These copper rings are "invisible" to microwaves.

Hiding in Plain Sight—The Invisibility Cloak

Surprisingly, scientists are paying serious attention to the idea of some kind of invisibility device. They recently demonstrated a device that could bend microwaves around copper rings, in effect, "cloaking" the copper rings. This means that if you subjected the device to microwaves, you could not detect the rings (Figure 1). This result astounded physicists around the world. Light is also an electromagnetic wave, just like microwaves. Could this concept be applied to light?

Evaluating *in Action*

Evaluating is a high-level thinking process in which you question the information in a text so as not to take it for granted. Here is how one student used the strategies to evaluate the selection about the invisibility cloak.

Questions I Asked of the Text	Clues in the Text	My Final Opinion/Judgment
Is this idea reasonable?	The technology is being developed and tested.	In the future, an invisibility device may exist.
Is the information credible?	It sounds credible but no sources are given.	I would have to read about this from different sources before I believed it.
Do I agree with the author's point of view?	The author appears to support this device.	The information is logically presented. I agree with the author but need more information to make a final decision.
Is the text biased?	The text says nothing negative.	I worry that this technology will be used by the military. I want to know more about possible applications.

What Is Refraction?

12.1

Light travels in straight lines through air. But what happens when it travels from one material into another? You likely have noticed the strange phenomenon of a spoon or stir stick, when placed in a glass of water, looking somewhat disconnected at the surface of the water. The spoon is not really disconnected. It is made of a solid material. So what is happening? Why does an object appear disconnected, or broken, at the water's surface? Find out by doing the "Exploring With Light" activity.

TRY THIS EXPLORING WITH LIGHT

SKILLS: Observing, Analyzing

SKILLS HANDBOOK 3.B.

Equipment and Materials: beaker or other transparent container; stir stick; coin; water

1. Place a coin in the middle of a beaker. Fill the beaker with water.
2. Look at the coin from the edge of the beaker. Be sure that you are looking above the beaker through the surface of the water as in Figure 1.
3. While looking above the beaker, aim a stir stick just inside the outer edge of the coin so that it looks as though you are going to just touch the coin. Place the stir stick in the water and attempt to touch the coin.
4. Now look at the beaker from the side and notice the position of the stir stick and the position of the coin.

A. Did you touch the outer edge of the coin when you placed the stir stick in the water? T/I

B. Describe where the stir stick actually went. T/I

C. Why do you think the stir stick missed the coin even though you had aimed the stick directly at the coin while looking above the beaker? Write a brief explanation. T/I

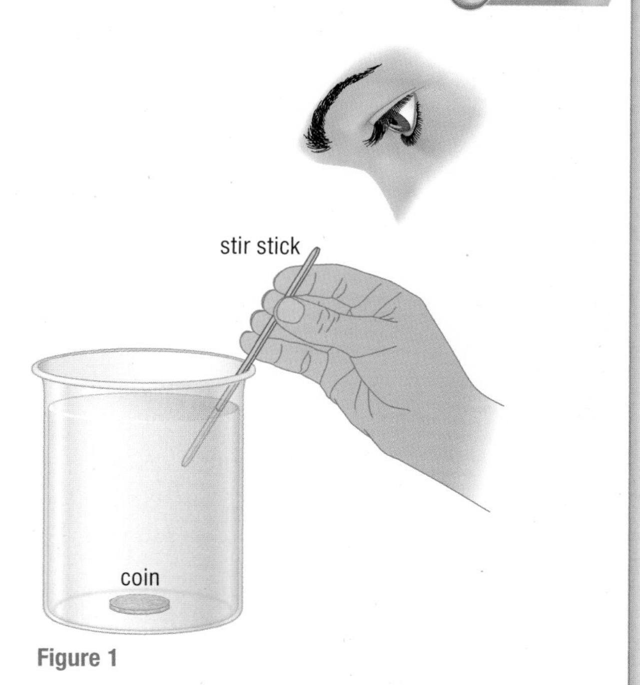

Figure 1

Bending Light

From the previous activity, you noticed that, even though you had aimed the stir stick at the coin, you missed touching the coin. The light coming from the coin to your eyes became bent. The light from the coin went through the water and then through the air into your eyes. Along this path, the light changed direction as it went from water into air. This bending of light when it travels from one material (medium) into another is called **refraction**. Refraction causes interesting effects whenever light travels from one medium to another.

refraction the bending or change in direction of light when it travels from one medium into another

LEARNING TIP

Using an Analogy

An analogy allows you to make a comparison by using as your example something that is simpler and easier to understand. Although a wagon moving from pavement onto sand is not a beam of light, this scenario allows you to understand the principle of refraction using everyday objects.

What Causes Refraction?

The refraction of light becomes obvious when you shine a powerful beam of light into water at an angle (Figure 2). Why does the direction change? A useful analogy is to think of a wagon travelling at an angle from pavement onto sand (Figure 3). When the right front wheel hits the sand, it slows down. The left front wheel, however, does not slow down because it is still on the pavement. This results in the wagon pivoting about the slower right front wheel. So the direction of the wagon changes as it moves from pavement onto sand. You might notice a similar effect when a car travels at an angle from pavement onto mud or snow.

Figure 2 The refraction of light going from air into water

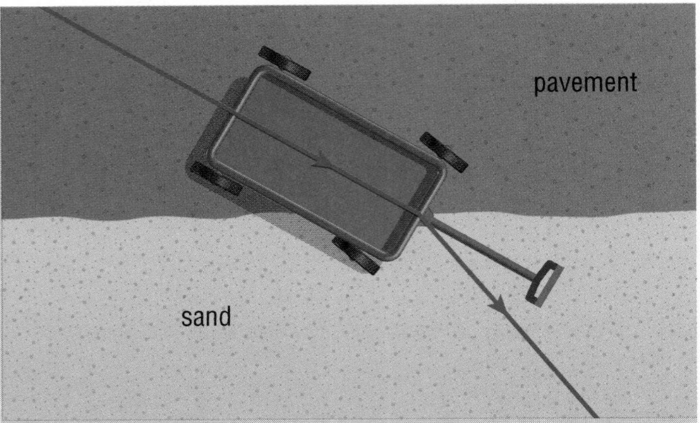

Figure 3 A wagon changes direction when travelling at an angle from pavement onto sand because one front wheel slows down while the other wheels continue moving at a higher speed.

DID YOU KNOW?

Measuring the Speed of Light

The first calculation of the speed of light that was close to today's accepted value of 3.00×10^8 m/s was done by Dutch mathematician Christian Huygens. He based his calculations on astronomical observations of Io, one of Jupiter's moons, collected by Danish astronomer Olaus Roemer in 1676.

GO TO NELSON SCIENCE

How Fast Is the Speed of Light?

The analogy of refraction using the wagon involves one of the front wheels slowing down as the wagon travels from one surface onto another. Is this the case with light? That is, is the speed of light in water different from the speed of light in air? Measurements of the speed of light clearly indicate that this is indeed the case. Light travels at a speed of 3.00×10^8 m/s in a vacuum, at a speed of 2.26×10^8 m/s in water, and at a speed of 1.76×10^8 m/s in acrylic. (Note that the speed of light in air is slightly less than the speed of light in a vacuum. This difference, however, is so small that it is not significant. Therefore, we use the same value for the speed of light in a vacuum and the speed of light in air.)

The Rules for Refraction

angle of refraction the angle between the refracted ray and the normal

Because the speed of light changes depending on the medium through which it is travelling, it is possible to make two statements about refraction in general and about the **angle of refraction** in particular. The angle of refraction is the angle between the refracted light ray and the normal.

1. The incident ray, the refracted ray, and the normal all lie in the same plane. The incident ray and the refracted ray are on opposite sides of the line that separates the two media.

2. Light bends *toward* the normal when the speed of light in the second medium is less than the speed of light in the first medium (Figure 4). Light bends *away* from the normal when the speed of light in the second medium is greater. (This second statement can also be predicted from the principle of the reversibility of light; light still follows the same path even if you switch its original direction. Recall that this principle also applies to the reflection of light.)

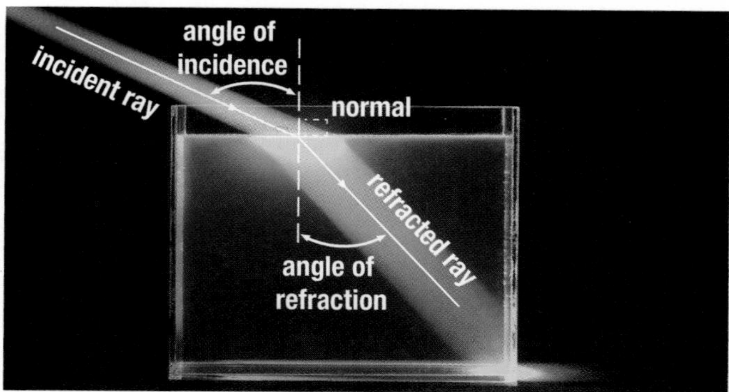

Figure 4 Light bends toward the normal when its speed decreases in a material.

The Bent Spoon

Using the concept of refraction, it is possible to explain why a spoon in a glass of water appears bent (Figure 5). Light coming from the part of the spoon below the water's surface must travel through the water into air. The speed of light increases as light goes from water into air so, if it hits the water–air boundary at an angle, light will bend away from the normal. The human brain perceives light to travel in a straight line, so it will project these light rays backwards to a virtual light source behind the real spoon (Figure 6). This is similar to how the brain projects light rays to form a virtual image in a plane mirror.

READING TIP

Evaluating

To determine whether a main idea seems reasonable, think about examples that you know. For instance, you may have seen a fallen tree branch appear to be bent in a river. By applying the explanation of refraction to a similar example from your own experience, you might determine if the explanation is reasonable.

Figure 5 The familiar "bending" of a spoon in a glass of water

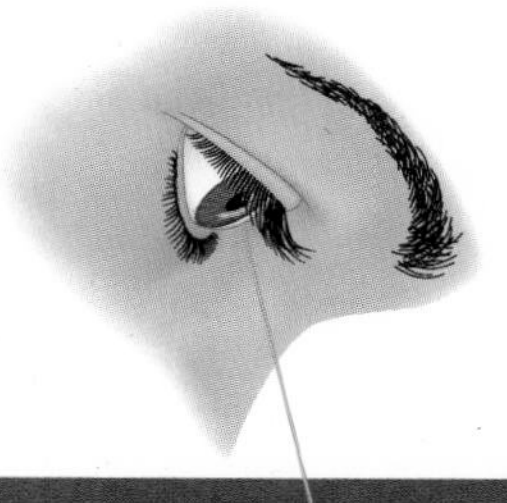

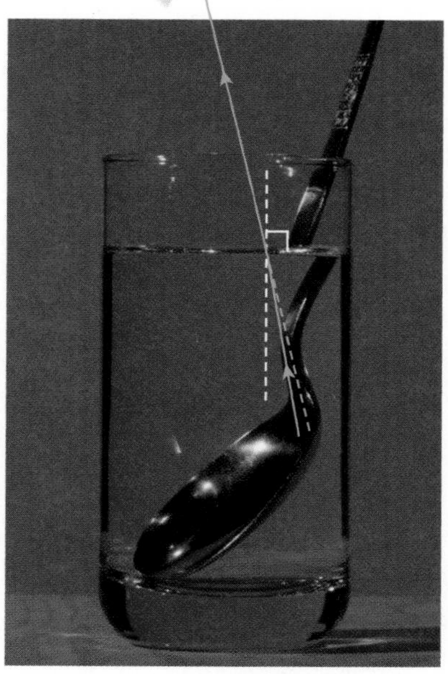

Figure 6 The brain thinks that the spoon is behind where it really is because light is refracted away from the normal when travelling from water into air.

Partial Reflection and Refraction

Light has some very interesting and unique properties that are not observed with most other matter in nature. Do the "Examining Light in a Window" activity on the next page to observe one of these properties.

TRY THIS EXAMINING LIGHT IN A WINDOW

SKILLS: Questioning, Observing

SKILLS HANDBOOK 3.B.

1. Stand directly in front of a clean window and look through it. This experiment works better if it is dimmer outside or the window is in shadow. Look at a tree, a person, a building, or some other object outside.
2. Continue to stand directly in front of the window, but this time focus on the glass directly in front of you. You should be able to see a faint reflection of yourself in the window. (Try turning the light off or on in the room if you have trouble seeing your reflection.)

A. During late dawn or early dusk, someone standing outside the window would be able to see you clearly through the window. What does this tell you about how light has travelled from you toward the outside observer? T/I

B. You are also able to see your reflection in the glass. What does this tell you about the light that travels from you toward the glass surface directly in front of you? T/I

C. Light coming from you appears to have two different behaviours based on questions A and B. Do you find this to be surprising? Explain.

D. A thought experiment is one that you do in your head; it is not one that you actually do. Let us do a brief thought experiment. Imagine that you were to throw a tennis ball at a window. Predict the two possible outcomes when the ball hits the glass. Is it possible that these two outcomes could both occur at the *same time*? T/I

E. Unlike the ball, light travelled through the window (to the outside observer) and also bounced back (allowing you to see your reflection) at the same time. This property does not appear to happen with most other things (such as balls). Does this imply that light has at least one property that might make it special or unique? Explain. T/I

Refraction is often accompanied by reflection. Some of the light that strikes water is reflected off the water, but a great deal of light is also refracted as it enters the water and illuminates the water below the surface (Figure 7). A transparent window exhibits the same property that light can be both reflected and refracted at the same time. This is called partial reflection and refraction. This effect is enhanced if glass has a special film coating behind it that allows some of the incident light to be refracted but that also reflects much of the incident light. This results in a mirrored surface that you can see through, but others cannot. This type of surface is called a silvered two-way mirror, and it is exactly how mirrored sunglasses are made (Figure 8).

To see some interesting simulations involving refraction,

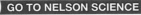

Figure 7 A beam of light is both reflected and refracted when it strikes water. The tree is visible as a result of reflection, whereas the fish is visible due to refraction.

Figure 8 Mirrored sunglasses allow you to see out but do not allow others to see in.

Two-way mirrors are used in the windows of many buildings (Figure 9). In the summer time, these windows reflect some of the incident sunlight, which reduces air-conditioning costs. The reflection of clouds and the blue sky off these windows also makes them very visually appealing.

Figure 9 Many modern buildings use windows that illustrate the partial transmission and reflection of light.

UNIT TASK Bookmark

You can apply what you learned about refraction at the boundary of two media as you think about the Unit Task described on page 588.

IN SUMMARY

- Refraction is the bending or change in direction of light when it travels from one medium into another.
- The speed of light depends on the medium that it is passing through.
- Light bends toward the normal when it slows down in a medium and away from the normal when it speeds up in a medium.
- Light can undergo partial reflection and refraction at the same time at a surface.

CHECK YOUR LEARNING

1. Clearly explain what is meant by the term "refraction." K/U
2. (a) Explain why refraction takes place.
   (b) What conditions must be present for refraction to take place?
   (c) From your answers to (a) and (b), make a prediction about the speed of light in water as compared to the speed of light in air. K/U
3. Figure 10 represents a beam of light going from one medium into another. T/I

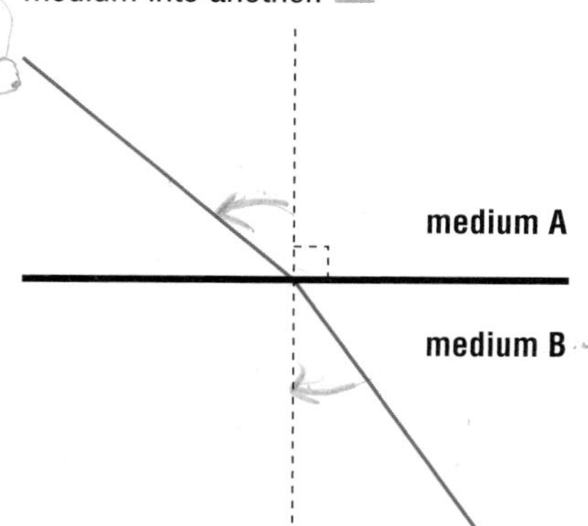

Figure 10

   (a) One medium is air, in which light has a speed of 3.00×10^8 m/s. The other medium is ice, in which light has a speed of 2.29×10^8 m/s. Use this information to identify medium A and medium B. Explain.
   (b) Do you know in which direction the light beam is travelling? Does it matter? Explain.
4. Which way will light bend if it is travelling
   (a) faster in a medium?
   (b) slower in a medium? K/U
5. What property of light is illustrated in Figure 11? K/U

Figure 11

6. Give some examples of where you have seen two-way mirrors used. A
7. In your own words, explain some practical applications of partial reflection and refraction. A C

The Path of Light—From Air into Acrylic

SKILLS MENU

- Questioning
- Hypothesizing
- Predicting
- Planning
- Controlling Variables
- Performing
- Observing
- Analyzing
- Evaluating
- Communicating

In Chapter 11, you learned that light travels in straight lines. Mirrors and their applications depend on this property. But what happens when light travels from one medium into another? In this activity, you will explore the path of light as it travels through two transparent media.

Purpose

To explore the path of light as it travels from one transparent medium into another.

Equipment and Materials

- ray box with single slit
- semicircular acrylic block
- polar graph paper (or ruler and protractor)

Procedure

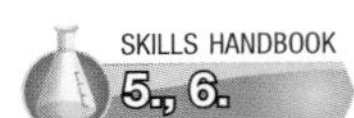

SKILLS HANDBOOK 5., 6.

1. Place the acrylic block at the centre of the polar graph paper with its flat edge along the horizontal centre line. Centre the acrylic block on the paper. The 0–180° line on the polar graph paper now acts as the normal and passes through the centre of the acrylic surface. You will be projecting light rays at the exact centre of the polar graph paper (the origin) (Figure 1).

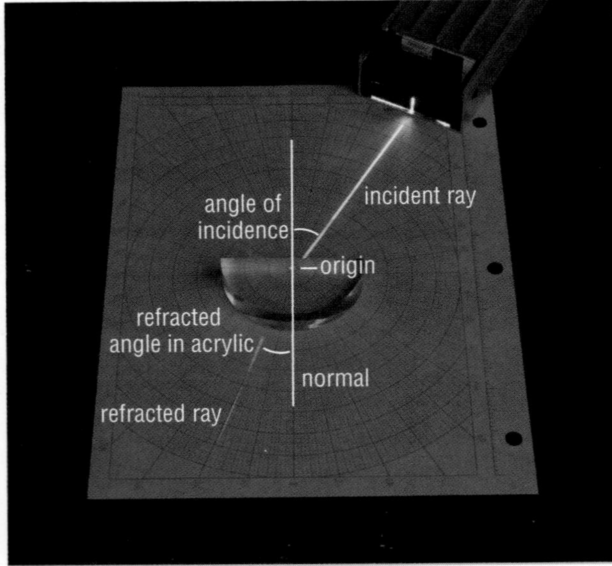

Figure 1

2. Project a single ray of light at the centre of the flat edge of the block at an angle of incidence of 0° (that is, you will be projecting along the normal). Measure the corresponding angle in the acrylic block. (Note that if polar graph paper is not available, you will have to mark the path of the light rays with a pencil and measure the appropriate angles later with a protractor.)
3. Repeat step 2, but for angles of incidence in air of 10°, 20°, 30°, 40°, 50°, and 60°. Always measure the angle of the refracted ray in the acrylic block with respect to the normal. Record your results in a table similar to Table 1 at the top of the next page. (Note that at some point you should also notice a reflected ray going back into the air from the surface of the acrylic. You can actually see the reflected ray in Figure 1. This ray follows the laws of reflection that you learned about in Chapter 11. Please ignore this reflected ray for now because in this experiment, we are concentrating on the refracted ray in the acrylic block. We will return to this reflected ray later in the chapter.)

When unplugging the ray box, do not pull the electric cord. Pull the plug itself.

Table 1 Refraction of Light as it Passes from Air into Acrylic

Angle of incidence in air ($\angle i$)	Angle of refracted ray in acrylic block ($\angle R$)	$\frac{\angle i}{\angle R}$	sin $\angle i$	sin $\angle R$	$\frac{\sin \angle i}{\sin \angle R}$
0°					
10°					
20°					
30°					
40°					
50°					
60°					

Analyze and Reflect

(a) What was the angle of the refracted ray in acrylic for the initial angle in air of 0°? Does this answer make sense to you? Explain. T/I

(b) How did the value of the angle in air (angle of incidence) compare with the angle in acrylic for the remaining measurements? T/I

(c) Which way did the refracted ray in the acrylic bend when compared with the normal? T/I

(d) Why do think you used a semi-circular medium in this experiment? T/I

Apply and Extend

(e) What trend do you notice in the angle of the refracted ray in acrylic as the angle of incidence in air steadily increases? T/I

(f) With the exception of the first measurement, what do you notice about the ratio $\frac{\angle i}{\angle R}$? T/I

(g) Again, with the exception of the first measurement, what do you notice about the ratio $\frac{\sin \angle i}{\sin \angle R}$? T/I

(h) Why were you not able to calculate $\frac{\angle i}{\angle R}$ or $\frac{\sin \angle i}{\sin \angle R}$ for an angle of incidence of 0°? T/I

(i) Which ratio, $\frac{\angle i}{\angle R}$ or $\frac{\sin \angle i}{\sin \angle R}$, is nearly constant for light travelling from air into acrylic? T/I

The Refraction of Light through Different Media

SKILLS MENU
- Questioning
- Hypothesizing
- Predicting
- Planning
- Controlling Variables
- Performing
- Observing
- Analyzing
- Evaluating
- Communicating

In Activity 12.2, you examined the refraction of light when it travelled from air into a semicircular acrylic block. You also learned that light bends toward the normal when it slows down in a medium. What happens when light goes through different media? What can you learn by examining and comparing refraction in different media?

Purpose

To explore how the angle of refraction changes in different media.

Equipment and Materials

- semicircular plastic dish
- polar graph paper (or ruler and protractor)
- ray box with single slit
- water
- vegetable oil
- glycerol (optional)
- dish detergent (to clean vegetable oil residue on plastic dish)

Procedure

1. Place the semicircular dish at the centre of the polar graph paper with its flat edge along the horizontal centre line. Centre the semicircular dish on the paper. The 0–180° line on the polar graph paper now acts as the normal and passes through the centre of the semicircular dish. You will be projecting light rays at the exact centre of the polar graph paper (the origin). This experimental setup is identical to that in Activity 12.2 (See Figure 1 on page 520).
2. Fill the dish with water.
3. Project a single ray of light at the centre of the semicircular dish at an angle of incidence of 0° (that is, you will be projecting along the normal). Measure the corresponding angle that light follows in the water in the semicircular dish. (If polar graph paper is not available, you will have to mark the path of the light rays with a pencil and measure the appropriate angles later with a protractor.)
4. Repeat step 3, but for angles of incidence in air of 10°, 20°, 30°, 40°, 50°, and 60°. Always measure the angle of the refracted ray in the semicircular dish with respect to the normal. Record your results in a table similar to Table 1. Accuracy is extremely important in this experiment.

When unplugging the ray box, do not pull the electric cord. Pull the plug itself.

5. Repeat steps 1 to 4, but this time put vegetable oil in the semicircular dish. Enter your data in the last row of your table. (Make sure that you use the dish detergent to carefully clean the vegetable oil from the semicircular dish after completing step 5.)

Table 1 Refraction of Light in Different Media

Angle of incidence in air	0°	10°	20°	30°	40°	50°	60°
Angle of refraction in water							
Angle of refraction in vegetable oil							

Analyze and Reflect

(a) Did the refracted ray bend toward the normal
- in the water?
- in the vegetable oil? T/I

(b) With the exception of an angle of incidence of 0°, compare the angles of refraction for water and for vegetable oil for the same angles of incidence. T/I

(c) What do your results tell you about the speed of light in water or in vegetable oil compared with the speed of light in air? T/I

(d) In which medium, water or vegetable oil, did the speed of light slow down more? T/I

Apply and Extend

(e) As in Activity 12.2, calculate $\frac{\sin \angle i}{\sin \angle R}$ using your data for vegetable oil. T/I

(f) With the exception of $\angle i = 0°$, how does your value of $\frac{\sin \angle i}{\sin \angle R}$ for vegetable oil in compare with the value for acrylic in Activity 12.2? T/I

(g) Could the method used in this activity be used to distinguish between two liquids that appear to be identical? Explain. T/I

(h) Check your answer from part (g) by repeating steps 1 or 4 of the Procedure using glycerol. First check the MSDS for glycerol and follow any necessary safety precautions. Record and analyze your data as before. T/I

12.4 The Index of Refraction

French physicist Jean Foucault made the first measurement of the speed of light in a medium (other than a vacuum or air), in 1862. He measured the speed of light in water to be 2.25×10^8 m/s. Since then, the speed of light has been measured for a variety of media. The speed of light is different for each medium, but is always less than the speed of light in a vacuum. The change in the speed of light at the boundary of a substance causes refraction. The speed of light in a medium is a distinctive optical property of that medium.

To learn more about Foucault's experiments to measure the speed of light in various media,

GO TO NELSON SCIENCE

The **index of refraction** for a medium is defined as the ratio of the speed of light in a vacuum to the speed of light in that medium. Mathematically, the index of refraction is written as

index of refraction the ratio of the speed of light in a vacuum to the speed of light in a medium, $n = \frac{c}{v}$; this value is also equal to the ratio of the sine of the angle of incidence in a vacuum to the sine of the refracted ray in a medium, $n = \frac{\sin \angle i}{\sin \angle R}$

$$n = \frac{c}{v}$$

where n is the index of refraction, c is the speed of light in a vacuum, and v is the speed of light in a given medium.

Because c and v are both speeds and can be expressed in the same units of m/s, the units in the equation divide out, so n has no units. This means that the index of refraction (n) is a dimensionless quantity.

The same index of refraction values can be calculated using the sines of angles. Imagine an incident ray passing from a vacuum into a transparent medium. The index of refraction for the medium can also be written as

$$n = \frac{\sin \angle i}{\sin \angle R}$$

where $\angle i$ is the angle of incidence and $\angle R$ is the angle of refraction.

Table 1 shows the indices of refraction for various media. Compare your values for n for acrylic and vegetable oil from Activities 12.2 and 12.3 with the values in Table 1.

Table 1 The Index of Refraction for Various Media

Medium	Index of refraction (n)
air/vacuum	1.00
ice	1.31
pure water	1.33
ethyl alcohol	1.36
quartz	1.46
vegetable oil	1.47
olive oil	1.48
acrylic	1.49
glass	1.52
zircon	1.92
diamond	2.42

SAMPLE PROBLEM 1 Calculating the Index of Refraction

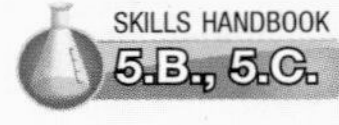

The speed of light in sodium chloride (salt) is 1.96×10^8 m/s. Calculate the index of refraction for sodium chloride (Figure 1).

Given: $c = 3.00 \times 10^8$ m/s

$v_{\text{sodium chloride}} = 1.96 \times 10^8$ m/s

Required: $n = ?$

Analysis and Solution: $n = \frac{c}{v}$

$$= \frac{3.00 \times 10^8 \text{ m/s}}{1.96 \times 10^8 \text{ m/s}}$$

$$\doteq 1.53$$

Statement: The index of refraction for sodium chloride (salt) is about 1.53.

Figure 1

You have just seen how the formula $n = \frac{c}{v}$ can be used to determine the index of refraction, n, for a medium. The same formula can also be used to calculate the speed at which light travels in that medium.

SAMPLE PROBLEM 2 Calculating the Speed of Light

Calculate the speed of light in olive oil.

Given: $c = 3.00 \times 10^8$ m/s

$n_{\text{olive oil}} = 1.48$ (from Table 1)

Required: $v = ?$

Analysis and Solution: $n = \frac{c}{v}$

$$v \times n = c$$

$$v = \frac{c}{n}$$

$$= \frac{3.00 \times 10^8 \text{ m/s}}{1.48}$$

$$\doteq 2.03 \times 10^8 \text{ m/s}$$

Statement: The speed of light in olive oil is about 2.03×10^8 m/s.

LEARNING TIP

Word Origins

The symbol "c" for the speed of light may seem strange because there is no letter "c" in the word light. The "c" comes from the Latin word "celeritas," meaning "velocity."

DID YOU KNOW?

The Real Value

The real value for the index of refraction for air is 1.000 293. None of the calculations in this text go to this degree of precision, so, for this course, you can treat the indices of refraction for air and vacuum the same.

IN SUMMARY

- The index of refraction for a medium is defined as the ratio of the speed of light in a vacuum to the speed of light in that medium; it is a dimensionless quantity.
- Mathematically, the index of refraction is defined as $n = \frac{c}{v}$ or $n = \frac{\sin \angle i}{\sin \angle R}$.

CHECK YOUR LEARNING

1. (a) What is meant by the term "index of refraction"?
   (b) Why is it a dimensionless quantity? K/U
2. The speed of light in vinegar is 2.30×10^8 m/s. Determine the index of refraction for vinegar. T/I
3. The speed of light in sapphire is 1.69×10^8 m/s. What is the index of refraction for sapphire (Figure 2)? T/I

Figure 2

4. Use Table 1 to calculate the speeds of light in
   (a) quartz
   (b) diamond T/I
5. An 80 % sugar solution has an index of refraction of 1.49. Calculate the speed of light in this solution. T/I
6. The index of refraction for acetone is 1.36. What is the speed of light in acetone? T/I
7. The speed of light in an unknown substance is 2.20×10^8 m/s.
   (a) Calculate the index of refraction for this substance.
   (b) Use Table 1 to determine a possible identity of the unknown substance. T/I
8. Suppose you calculated the speed of light in an unknown substance to be 4.00×10^8 m/s. How could you tell if you made an error in your calculations? K/U
9. A light ray travelling from diamond into air has an angle of refraction of 56°. A piece of glass is then placed right next to the diamond.
   (a) How will the angle of refraction change?
   (b) Justify your answer by considering any changes to the speed of light. Include a ray diagram. T/I C
10. Why is the index of refraction a unique property of a medium? K/U

12.5

Total Internal Reflection

When light travels from one medium into another, some of the light is reflected and some is refracted. As you know, light slows down when it travels from air into acrylic or water. This results in the light bending toward the normal.

Light, however, bends away from the normal when it speeds up at the boundary of two media. (An example of this is when light travels from acrylic into air.) In this situation, the angle of refraction is always larger than the angle of incidence (Figure 1). In fact, the angle of refraction continues to increase as the angle of incidence increases. Eventually, the angle of refraction will become 90°. The angle of incidence at this point is called the **critical angle**. The critical angle is the angle of incidence that produces a refracted angle of 90°.

critical angle the angle of incidence that results in an angle of refraction of 90°

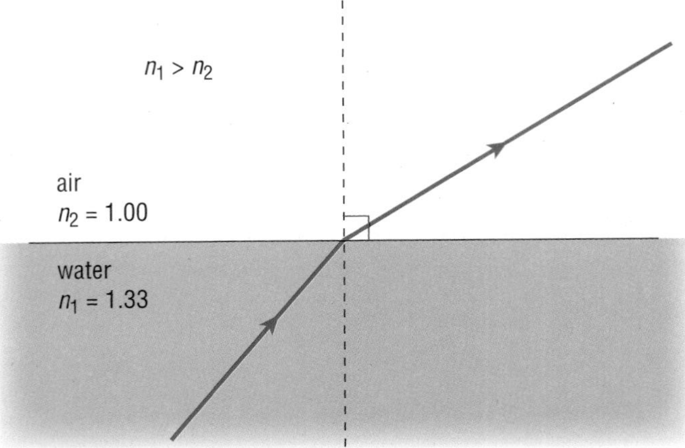

Figure 1 Medium 1 (water) has an index of refraction that is greater than that of medium 2 (air). So an incident ray in water speeds up as it goes into air.

If you increase the angle of incidence past the critical angle, the refracted ray will no longer exit the medium. Instead, it will reflect back into the medium. In other words, the refracted ray disappears; only a reflected ray is visible. This phenomenon is called **total internal reflection** (Figure 2).

total internal reflection the situation when the angle of incidence is greater than the critical angle

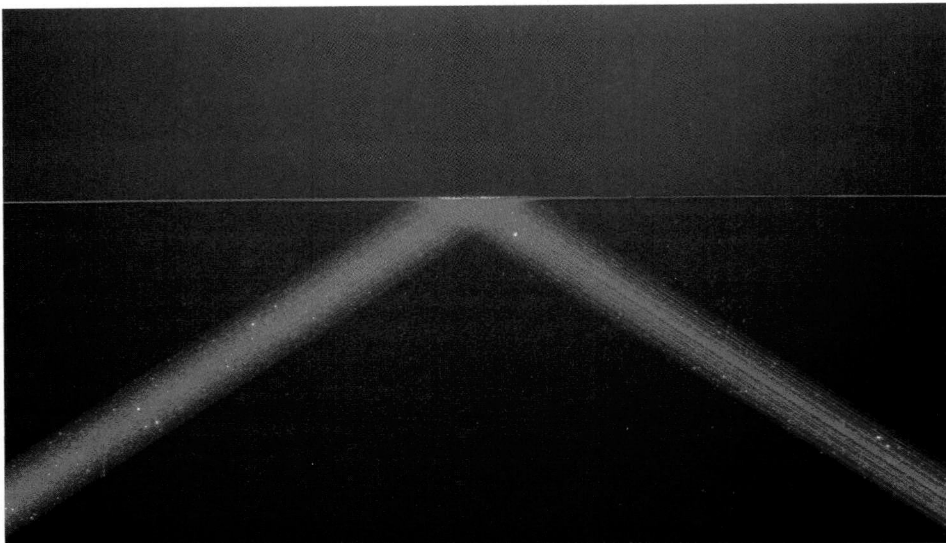

Figure 2 Total internal reflection of laser light in water

LEARNING TIP

Understanding Diagrams

It is often easier to understand a concept from a diagram than from words. Total internal reflection is a perfect example of this. After looking at Figure 2, draw your own diagrams to illustrate total internal reflection. Show your diagrams to a classmate and explain them.

Total internal reflection occurs when these two conditions are met:

1. Light is travelling more slowly in the first medium than in the second.
2. The angle of incidence is large enough that no refraction occurs in the second medium. Instead, the ray is reflected back into the first medium (Figure 3).

To see some interesting simulations for total internal reflection,

GO TO NELSON SCIENCE

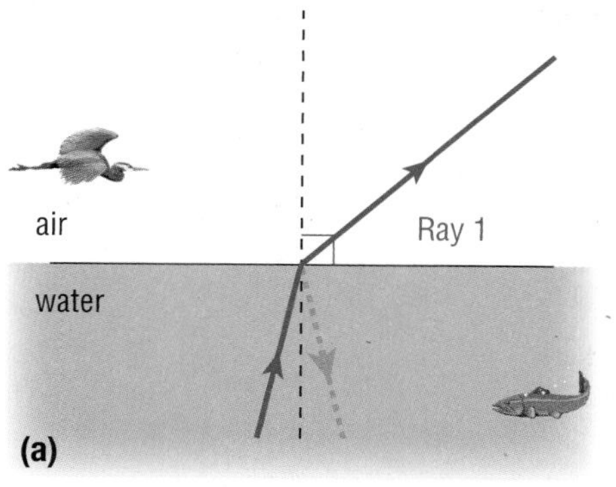

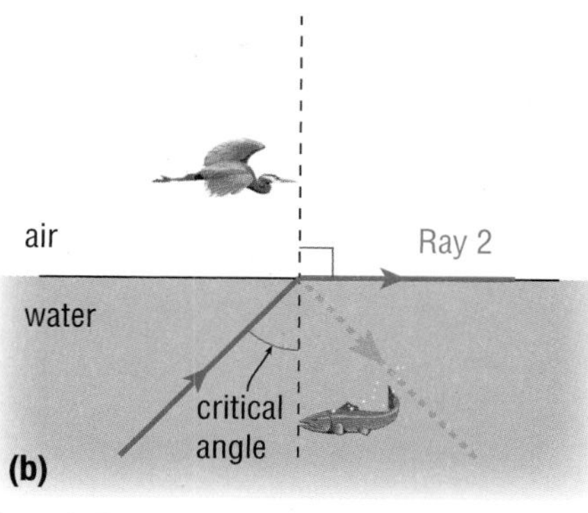

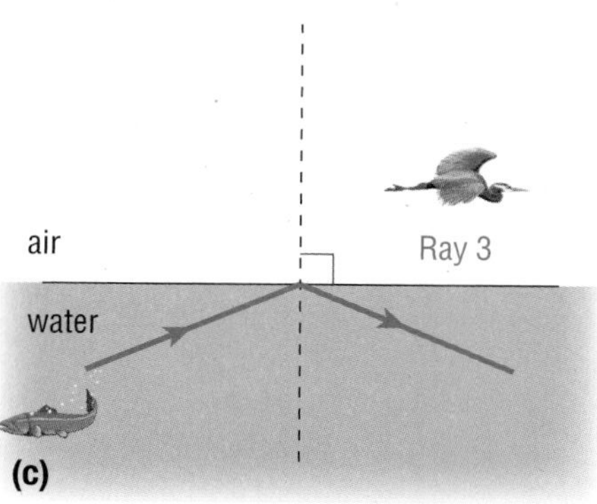

Figure 3 Ray 1 is refracted as it passes from water into air. Ray 2 has an angle of refraction of 90°; the angle of incidence is equal to the critical angle. Ray 3 is reflected internally back into water and does not go into air. Note that in (a) and (b), some light is reflected internally but not as strongly as in (c).

Water has a critical angle of 48.8°. This means that an angle of incidence greater than 48.8° would result in total internal reflection in the water. The critical angle is a physical property of a medium. (Recall that the index of refraction is another physical property.)

Diamonds Are Forever

One of the features that make diamonds so attractive in jewellery is the fact that they sparkle. This "sparkling" is due to the cut of the diamond faces, which, combined with the high index of refraction for diamond ($n = 2.42$), results in the total internal reflection of light. The high refractive index means that diamonds have a very small critical angle: 24.4°. So a great deal of incident light undergoes total internal reflection inside the diamond. A light ray can bounce around several times inside the diamond before eventually exiting through a top face of the gemstone (Figure 4(a)). This causes the "sparkling" effect that makes diamonds so appealing (Figure 4(b)).

To learn more about how jewellers and gem cutters use optics in their work,

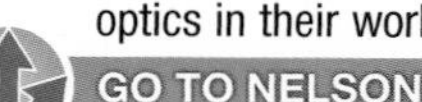

GO TO NELSON SCIENCE

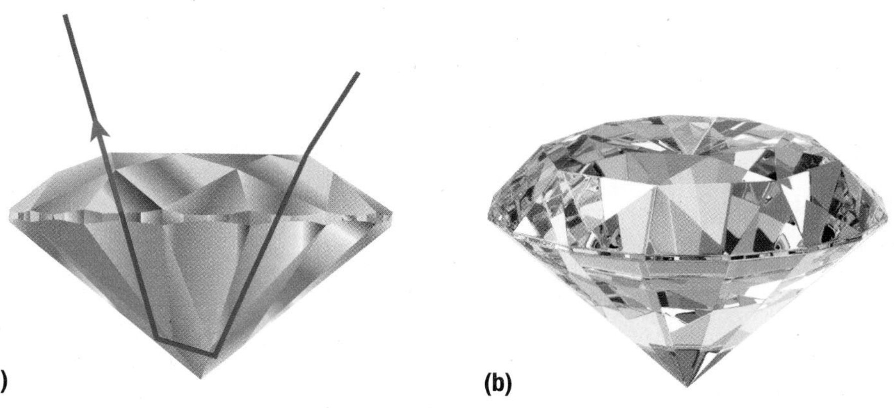

Figure 4 (a) Many light rays undergo two total internal reflections inside a diamond. (b) This is what makes a diamond "sparkle."

READING TIP

Evaluating

Examine illustrations and captions carefully to determine how they increase your understanding of a text. Make connections to what you already know about the topic. For example, you might have seen pictures of diamonds that appear to support the explanation for why diamonds sparkle.

Fibre Optics

Fibre optics is a technology that uses light to transmit information along a glass cable. The light must not escape as it travels along the cable. To achieve this, the cable must have a small critical angle so that light entering it will have an angle of incidence greater than the critical angle. Substances that have a small critical angle include high-purity glass and special types of plastics, such as Lucite (Figure 5).

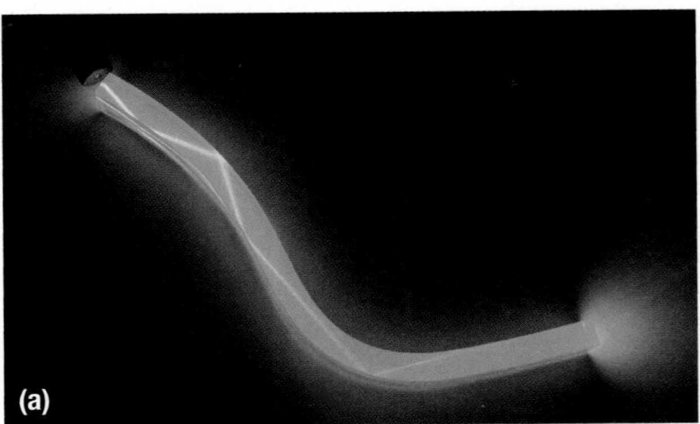

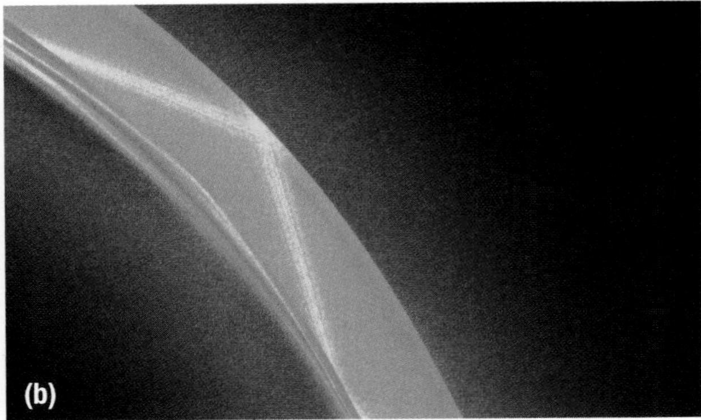

Figure 5 (a) A laser beam undergoes total internal reflection in a Lucite rod. (b) In close-up, you can see the point at which total internal reflection occurs.

READING TIP

Evaluating

Check a text for clues of bias. Are advantages and disadvantages described? If only advantages are given, can you think of any disadvantages? For example, in a text about the applications of fibre optics, does the author treat the topic in a balanced way? Are both pros and cons described? If not, does the text show a bias toward one side of the issue?

To learn more about careers in fibre optics,

GO TO NELSON SCIENCE

Fibre optics is used extensively in the communications industry for phones, computers, and TVs. Fibre optics also plays a major role in the movie industry. Science fiction films sometimes make use of fibre-optic cables to represent small windows in "giant" spaceships (Figure 6). The automotive industry uses optical fibres to transmit light to the instrument panel in cars. As you learned in Chapter 2, medical professionals use fibre-optic technologies to see into parts of the human body that would otherwise be inaccessible. An endoscope is a fibre-optic device that allows doctors to check the health of various internal organs (Figure 7). The endoscope consists of two separate fibre-optic bundles. One bundle shines light into the body. The second bundle carries the reflected light back to the instrument. A colonoscopy, for example, is now a very common procedure among males over 50 years of age. In this procedure, doctors use an endoscope to check for growths that could develop into colon cancer.

Figure 6 Light travels along optical fibres and then emerges at the ends.

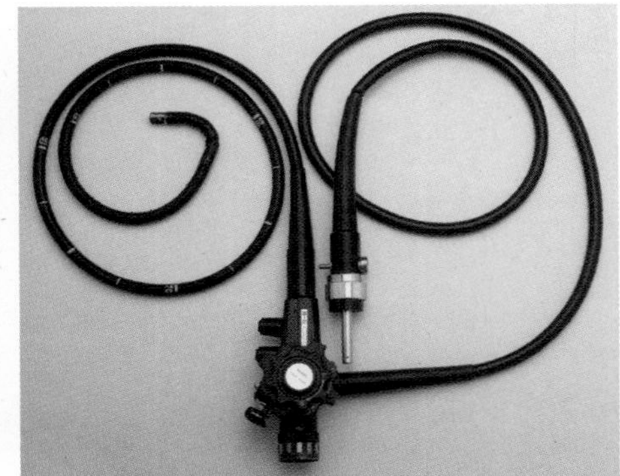

Figure 7 An endoscope is an important fibre-optic device used for medical diagnoses.

The Triangular Prism

A triangular prism also exhibits total internal reflection. The critical angle for glass is about 41.1°. If a prism is oriented in such a way that the angle of incidence is greater than 41.1°, total internal reflection will result. Prisms are much more useful to reflect light than mirrors because a prism reflects almost 100 % of the light internally. Mirrors reflect most incident light but lose a little through absorption. Also, the silvered surface of mirrors deteriorates over time. For these reasons, most optical devices, such as cameras and binoculars, use prisms instead of mirrors. The emergent ray can be either 90° or 180° relative to the incident ray, depending on the placement of the prism (Figure 8).

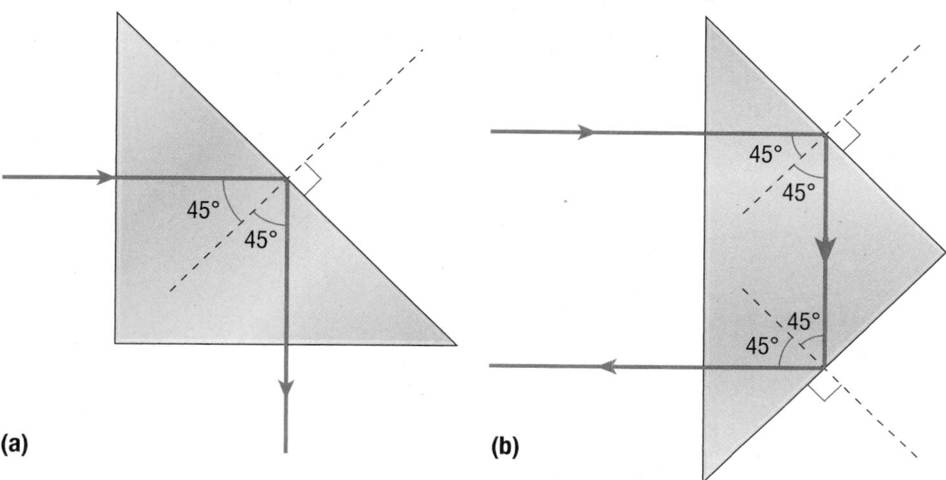

Figure 8 By changing the orientation of the prism, you can change the direction of the emergent ray by either 90° or 180°. In (a) the light ray goes through just one reflection. In (b) it goes through two reflections.

In Chapter 11, you learned that plane mirrors could be used to make a simple periscope. A more complex periscope uses triangular prisms to change the direction of light by 90° (Figure 9). Each triangular face has angles of 45°, 45°, and 90°. A pair of binoculars uses two such prisms to change the direction of light by 180° (Figure 10).

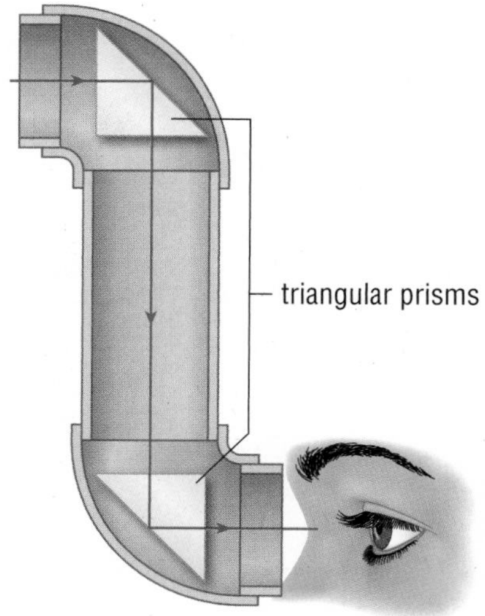

Figure 9 A periscope uses triangular prisms to change the direction of light by 90° twice.

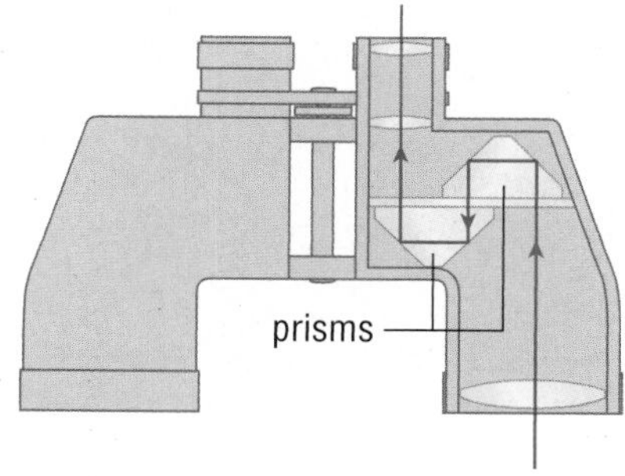

Figure 10 Binoculars use two triangular prisms to change the path of light.

Retro-reflectors and Prisms

retro-reflector an optical device in which the emergent ray is parallel to the incident ray

A **retro-reflector** is an optical device that returns any incident light back in exactly the same direction from which it came. The prism orientation in Figure 8(b) on the previous page is an example of a retro-reflector because the emergent ray is parallel to the incident ray as a result of two total internal reflections.

If you cut off the corner of a glass cube, you would produce a corner cube retro-reflector. This type of retro-reflector has three perpendicular faces (like the corner of a room) (Figure 11). It will reflect an incident ray coming from any direction back along its original direction.

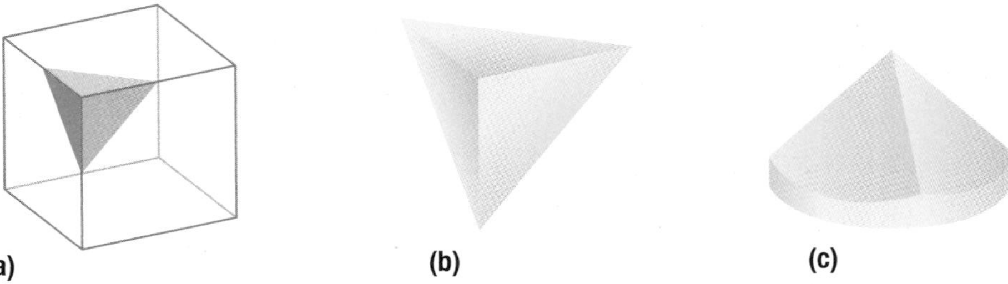

Figure 11 (a) and (b) A corner cube retro-reflector is created by cutting off the corner of a cube. (c) The resulting corner cube retro-reflector after three corners have been ground down.

The Laser Ranging Retro-Reflector (LR^3 or "LR-cubed") left on the Moon by the Apollo 11 astronauts is an example of this type of retro-reflector. The LR^3 is an array of 100 corner cube retro-reflectors set up in a 10 × 10 grid mounted on a square aluminum panel that is 46 cm long, about the size of a large pizza box. These 100 corner-cubed prisms on the Moon are made of quartz. With this device, scientists on Earth have been able to shine a very powerful laser beam at the Moon and bounce it off the LR^3. This enabled scientists to determine the Earth–Moon distance with an accuracy of 3 cm. The LR^3 is still working, and recent laser measurements have refined the Earth–Moon distance even further, down to within a few millimetres.

You may not have heard of retro-reflectors before, but you have almost certainly seen them. They are built into bike reflectors and the reflective strips on clothing and helmets. Road signs also contain tiny retro-reflectors in the paint so that you can see the signs at night (Figure 12).

To learn more about the LR^3 on the Moon and retro-reflectors in general,

GO TO NELSON SCIENCE

Figure 12 Retro-reflectors on road signs help you see the signs at night.

UNIT TASK Bookmark

You can apply what you learned about total internal reflection as you plan what optical device you will construct for the Unit Task described on page 588.

IN SUMMARY

- The critical angle is the angle of incidence for which the angle of refraction is 90°. This occurs only when light passes from one medium into another with a lower index of refraction.
- Total internal reflection occurs if the angle of incidence is greater that the critical angle.
- Optical devices such as periscopes, binoculars, and fibre-optic cables make use of total internal reflection.
- A triangular prism, depending on its orientation, can change the direction of light by 90° (one total internal reflection) or 180° (two total internal reflections).

CHECK YOUR LEARNING

1. What two conditions must be satisfied in order for total internal reflection to occur? K/U
2. Why does total internal reflection occur only when light travels more slowly through the first medium than in the second and not the other way around? Include a ray diagram with your answer. T/I
3. The critical angle for sapphire is 34.4°. For each angle of incidence, determine if it would result in total internal reflection in a sapphire. K/U

   (a) 23.7° (b) 34.7° (c) 53.4° (d) 31.5°
4. What is the advantage of using triangular prisms over plane mirrors in optical devices requiring the reflection of light? K/U A
5. Will you get more total internal reflection with a medium that has a small critical angle or with one that has a large critical angle? Explain. K/U
6. Brainstorm to create a list of suggestions for using retro-reflectors to improve road safety on a winding, dark, country road. A
7. Briefly describe three applications that make use of the total internal reflection of light. A
8. Figure 13 shows light travelling through two different media. In which diagrams would total internal reflection be possible if the angle of incidence were increased? T/I
9. Look again at Figure 13. For each diagram, in which medium would total internal reflection occur? Explain your answers. T/I

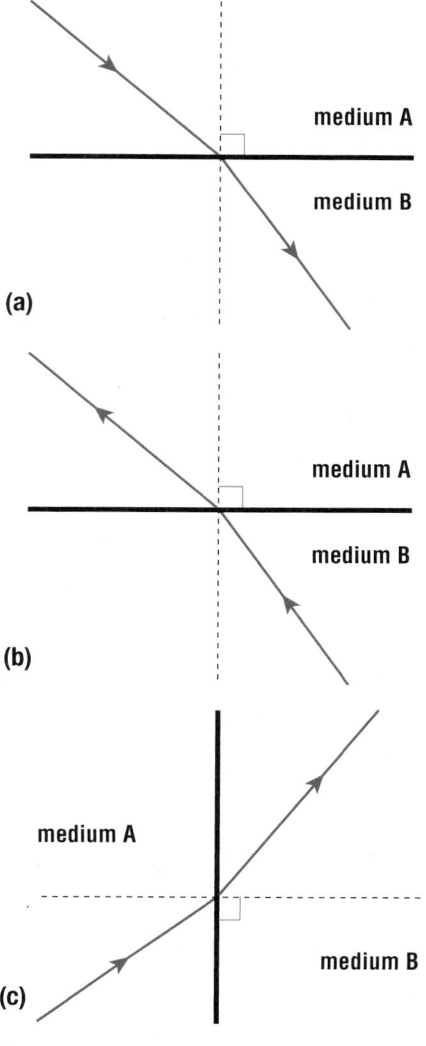

Figure 13

Measuring the Critical Angle for Various Media

Light bends away from the normal when it travels more slowly in the first medium than in the second. Total internal reflection in the first medium will occur when light has an angle of incidence greater than the critical angle for that medium.

SKILLS MENU

- Questioning
- Hypothesizing
- Predicting
- Planning
- Controlling Variables
- Performing
- Observing
- Analyzing
- Evaluating
- Communicating

Purpose

To explore the critical angle for various media.

Equipment and Materials

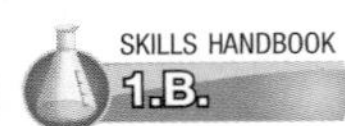

- ray box with single slit
- semicircular acrylic block
- semicircular plastic dish
- water
- glycerol
- vegetable oil
- polar graph paper (or ruler and protractor)
- dish detergent (to clean vegetable oil residue on plastic dish)

Procedure

1. Place the acrylic block at the centre of the polar graph paper. The flat edge of the block should lie along the horizontal centre line of the polar graph paper. Centre the glass block carefully on the paper. The 0–180° line on the polar graph paper now acts as the normal. This line passes through the centre of the acrylic surface. You will be projecting light rays at the exact centre of the polar graph paper (the origin). The setup is similar to that for previous investigations in this chapter with one exception: *You will now be aiming light rays at the curved part of the block.* You will aim the ray box so that the incident ray goes through the centre of the block (that is, the origin on the polar graph paper). Position the ray box so that you produce a refracted angle in air that is larger than the angle of incidence in the acrylic block (Figure 1).

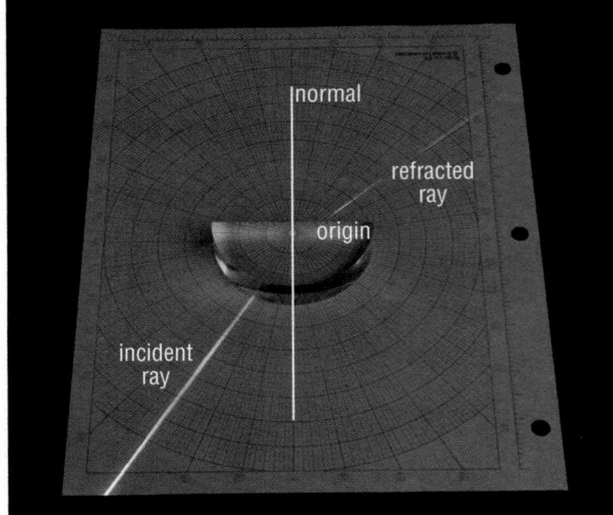

Figure 1

2. Slowly move the ray box to increase the angle of incidence in the acrylic block. Eventually, the refracted ray leaving the block will disappear. Measure the angle of incidence at this point. Enter your data in the first row of a table similar to Table 1 shown below. (If you are not using polar graph paper, mark the path of the light rays with a pencil. Measure the angles later with a protractor.)

Table 1 Critical Angles for Different Media

Medium	Critical angle
acrylic	
water	
glycerol	
vegetable oil	

3. Repeat steps 1 and 2 using the semicircular plastic dish. Use water first, then glycerol, and finally vegetable oil. (Make sure that you use the dish detergent to clean the glycerol and the vegetable oil carefully from the semicircular dish after each use.)

When unplugging the ray box, do not pull the electric cord. Pull the plug itself.

Analyze and Reflect

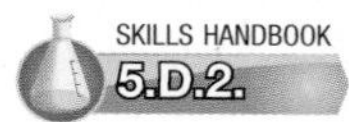

(a) The accepted values for the critical angles for acrylic, water, and vegetable oil are 42.2°, 48.8°, and 42.9° respectively. Compare your measured values with the accepted values. T/I

(b) What sources of error could account for any differences between your measured values for the critical angles and the accepted values? T/I

(c) Did partial reflection and refraction still occur when light was travelling more slowly in the first medium than in the second? Explain. T/I

Apply and Extend

(d) The index of refraction for acrylic is 1.49. For water, it is 1.33. For vegetable oil, it is 1.47. Given these values and your measured critical angles, what trend do you think exists between the index of refraction and the critical angle? T/I

(e) Research to find out the critical angle and index of refraction for glycerol. How close were your measured values to the accepted values? T/I

GO TO NELSON SCIENCE

SCIENCE WORKS

Hiding in Plain Sight—The Invisibility Cloak

"... if its refractive index could be made the same as that of air; for then there would be no refraction or reflection as the light passed from glass to air." With this statement, the central character in H. G. Wells's novel *The Invisible Man* attempts to explain how he could become invisible. Many works of entertainment have used invisibility as a plot device, from Wells's 1897 novel to Harry Potter's invisibility cloak. But is it just fiction, or is there something more to the idea?

Surprisingly, scientists are paying serious attention to the idea of some kind of invisibility device. They recently demonstrated a device that could bend microwaves around copper rings, in effect, "cloaking" the copper rings. This means that if you subjected the device to microwaves, you could not detect the rings (Figure 1). This result astounded physicists around the world. Light is also an electromagnetic wave, just like microwaves. Could this concept be applied to light?

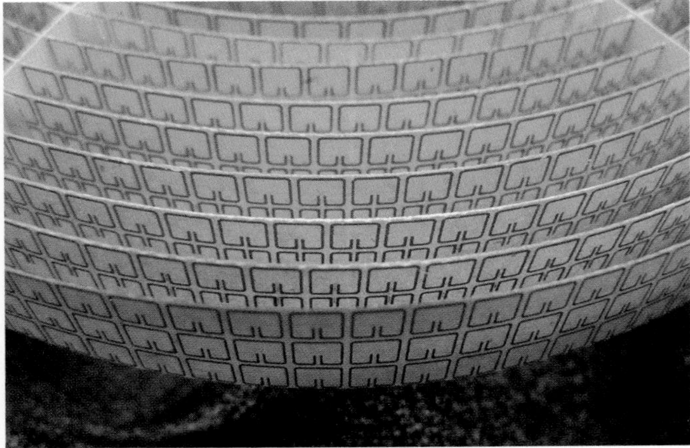

Figure 1 These copper rings are "invisible" to microwaves. Not quite Harry Potter's cloak yet.

Scientists are now working with special materials called metamaterials in an attempt to mimic the microwave experiment using light. Metamaterials have extraordinary properties that are not found in natural materials; they respond to electromagnetic waves in entirely new ways. These new materials have a negative index of refraction. A negative index of refraction means that the material still refracts light, but the refracted ray is on the *same side* of the normal in the second material (Figure 2). (In all naturally occurring substances, light is refracted on the opposite side of the normal.) This unusual property allows scientists to bend light in unexpected ways. Metamaterials are extremely difficult to construct, and adapting them to bend light around a substance still involves many practical difficulties.

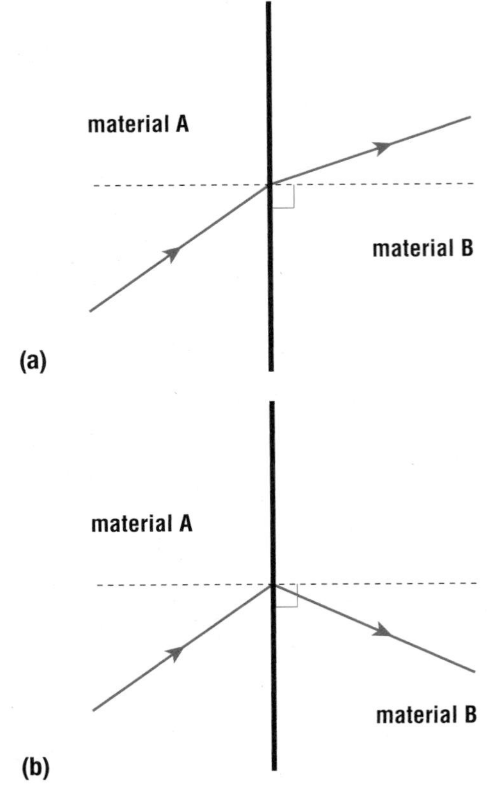

Figure 2 (a) In regular refraction, the light ray is on the opposite side of the normal as compared to the incident ray. (b) Negative refraction occurs when the refracted light ray is on the *same side* of the normal as the incident ray.

The physics of cloaking, however, is valid. It has already been successfully demonstrated with microwaves. True invisibility may well turn out to be not just fiction (Figure 3).

Figure 3 What an invisibility cloak might look like

Phenomena Related to Refraction

Nature has many interesting phenomena involving light. Geometric optics is a useful tool that we can use to explain many of these phenomena.

Apparent Depth

A pencil partly under water looks bent when viewed from above (Figure 1). We can explain this using the concept of refraction and the knowledge that our brains perceive light rays to always travel in a straight line. Light from the submerged pencil tip reaches your eyes. Your brain then projects the rays backwards in a straight line to create a virtual image in the water. This virtual image is higher than the actual pencil tip, resulting in the pencil appearing to be bent. The pencil tip appears to be at a shallower depth than it really is. The distance from the surface of the water to where the object appears to be (the virtual image) is called the **apparent depth** (Figure 2). A paddle in the water also appears to be closer to the surface for the same reason.

Figure 1 A "bent" pencil in water

apparent depth the depth that an object appears to be at due to the refraction of light in a transparent medium

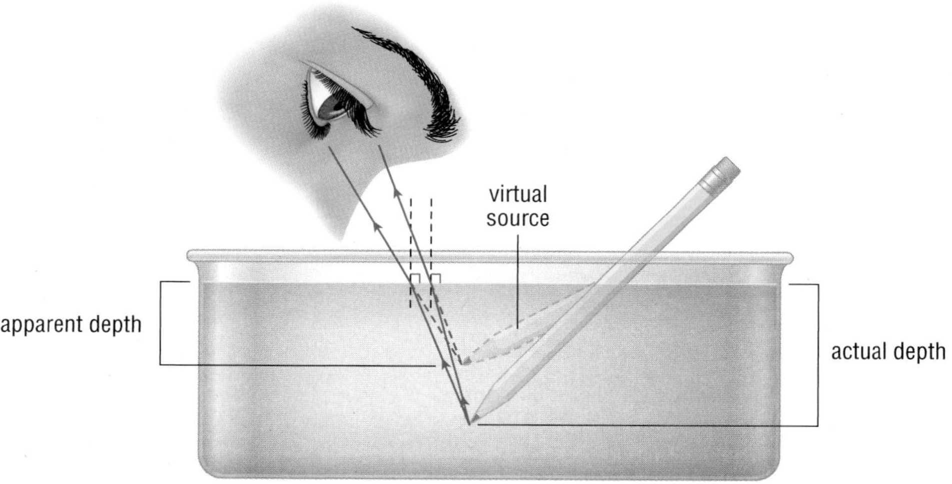

Figure 2 Refraction causes the pencil to appear closer to the surface than it actually is.

Objects under water always appear to be nearer to the surface than they actually are. Apparent depth is an optical illusion. This is what makes fish in water appear to be closer to the surface than they actually are (Figure 3).

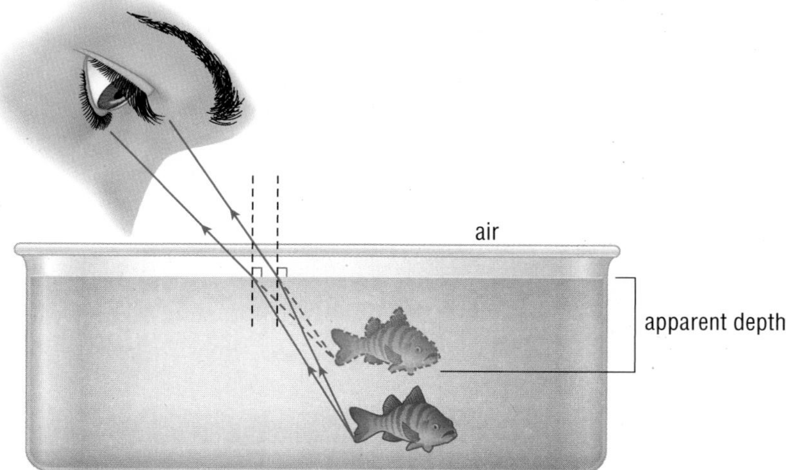

Figure 3 The illusion of apparent depth

For the same reason, the legs of someone standing in water appear to be shorter. Figure 4 is an illustration of refraction and apparent depth.

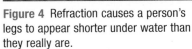

Figure 4 Refraction causes a person's legs to appear shorter under water than they really are.

The "Flattened" Sun

Sunsets offer a unique opportunity to see an unusual image due to refraction. People notice that when the Sun is near the horizon during sunset, it appears to be flattened. The Sun, of course, is not really flattened (Figure 5). When the Sun is close to the horizon, light from the bottom of the Sun is refracted more than light from the top of the Sun. Part of the reason is that air is more dense near Earth's surface than higher up in the atmosphere. So the increased density of air closer to Earth results in greater bending of the Sun's rays. In addition, the light rays from the bottom of the Sun have a greater angle of incidence than the light rays from the top of the Sun. This results in the Sun having a flattened appearance rather than its familiar round shape.

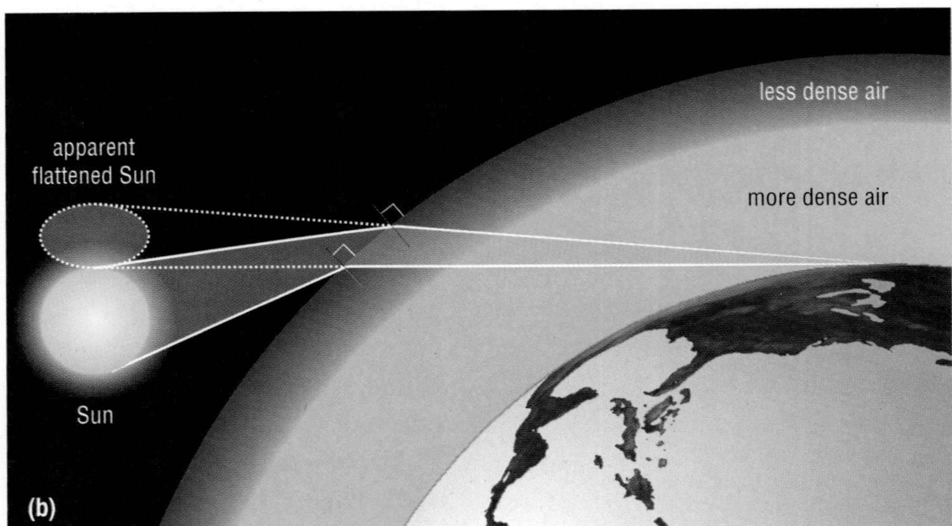

Figure 5 The flattening of the Sun is the result of refraction in Earth's atmosphere.

Water on Pavement—The Mirage

Many people have noticed what appears to be a pool of water in front of them as they drive along a highway (Figure 6(a)). This pool of water seems to be only a short distance away, yet the car never seems to reach it. The pool appears to be constantly moving away.

The pool of water on a highway is a **mirage**. A mirage can appear when light is travelling from cool air into warmer air. The index of refraction for air decreases as the air gets warmer. This results in light bending farther away from the normal as the air temperature continues to increase. Eventually, total internal reflection occurs in the lowest (hottest) air layer (Figure 6(b)). The light ray now travels up from the hottest layer to the cooler layer above and is gradually refracted toward the normal as the air temperature decreases. This light ray eventually enters your eyes. A motorist who sees this curved light forms a virtual image on the highway.

mirage a virtual image that forms as a result of refraction and total internal reflection in Earth's atmosphere

In reality, the pool of water is a virtual image of the sky on the highway. Because the human brain perceives light to travel in a straight line, the motorist projects the image of the sky onto the highway.

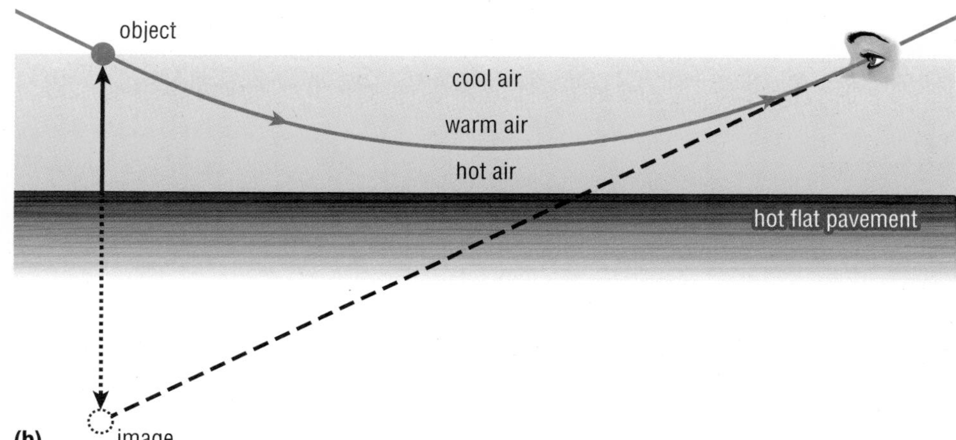

Figure 6 There appears to be a pool of water on the highway. This illusion is caused by the refraction and reflection of light as it goes through air of different temperatures.

Shimmering

You may have noticed that when the Moon is out at night above a lake, you can see a shimmering image of the Moon on the water's surface (Figure 7). As with a mirage, shimmering is caused by light being refracted as it passes through air of different temperatures.

At night, the air just above a lake is much warmer than air farther away from the water's surface. Moonlight passes through layers of air that have different temperatures. In the coldest air layer, light travels more slowly so a light ray going through this layer bends toward the normal. As the light ray continues travelling downward toward the warmest air layer (just above the lake), its speed increases, so the light ray bends farther and farther away from the normal. Eventually, total internal reflection occurs in the lowest warm air layer. This results in multiple virtual images of the Moon on the water's surface.

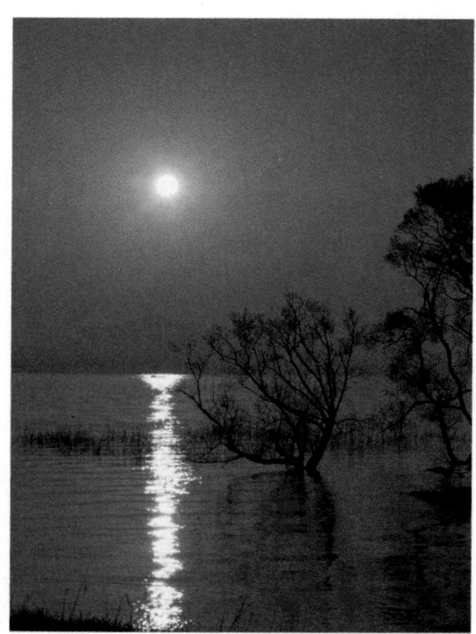

Figure 7 Shimmering on a lake is caused by light travelling at slightly different speeds through air layers of different temperatures.

The Rainbow

dispersion the separation of white light into its constituent colours

In the previous chapter, you learned how Isaac Newton used a triangular prism to separate white light into a continuous sequence of colours. The separation of white light into its spectrum is called **dispersion**. Dispersion occurs because each colour of visible light travels at a slightly different speed when it goes through the glass prism. Violet light slows down more than red light when it enters the prism. That is why you see violet light being refracted more than any other colour (that is, bending more toward the normal than any other colour). Red light is refracted the least (Figure 8).

READING TIP

Challenging Beliefs

The explanation of a natural phenomenon or an optical illusion may challenge a belief that you hold about something you have experienced. For example, the explanation of a rainbow or a mirage may convince you that you cannot always believe what you see, or think you see.

Figure 8 The colours in visible light travel at different speeds through a triangular glass prism.

The rainbow is an optical phenomenon that is produced by water droplets in Earth's atmosphere (Figure 9). The first step in the process involves refraction as light enters the raindrop (going from air into water), resulting in dispersion. The second step is partial internal reflection when this light hits the back of the raindrop. The third step is refraction as the light now exits the raindrop (going from water into air). This is the light that your eyes see, which you perceive as a rainbow. Your brain projects these light rays backwards and forms a virtual image of the spectrum: a rainbow (Figure 10). You can only see a rainbow when the Sun is behind you.

DID YOU KNOW?

Over the Rainbow

The geometry of a rainbow is very precise. The angle between the rainbow and the viewer is always the same: about 42°. All rainbows would form complete circles if the ground were not in the way.

Figure 9 As long as the Sun remains behind you, a rainbow moves as you move.

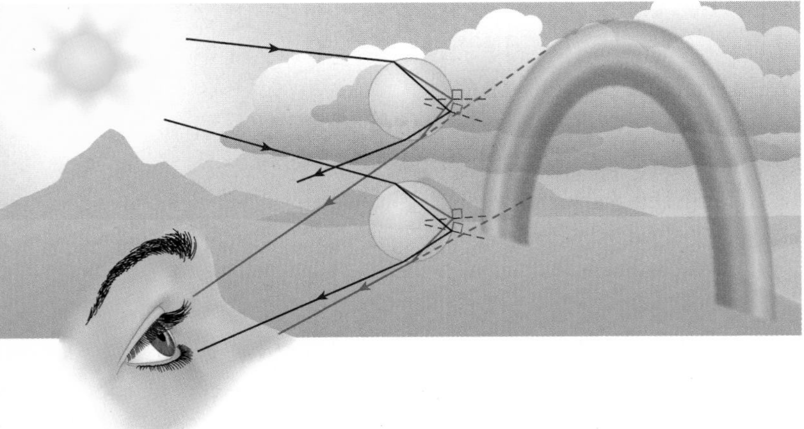

Figure 10 The rainbow is caused by a combination of dispersion and partial internal reflection in water droplets in the atmosphere. Millions of raindrops are necessary to produce a rainbow.

RESEARCH THIS OTHER ATMOSPHERIC OPTICAL PHENOMENA

SKILLS: Researching, Communicating

SKILLS HANDBOOK 4.A., 4.B.

Light produces far more atmospheric phenomena than the few mentioned in this section (Figure 11). In this activity you will research one of these phenomena.

Figure 11 The belt of Venus (the dark blue band above the horizon) is another natural optical phenomena.

Complete *one* of these research questions.

1. Research how a sun dog or a moon dog is produced.
2. Research how an icebow is formed.
3. Research how a "green flash" is produced.
4. Research another interesting atmospheric light phenomenon that has not been mentioned in this section.

Answer the question that matches the phenomenon you chose.

A. What conditions are necessary to produce a sun dog or a moon dog? T/I

B. What is the primary difference between a rainbow and an icebow? T/I

C. Explain how a "green flash" is produced in the atmosphere. Why is the "green flash" so difficult to see? T/I

D. Briefly explain how the other light phenomenon you researched is produced in the atmosphere. T/I

IN SUMMARY

- Objects in water appear to be at a shallower depth (an apparent depth) than they really are as a result of the refraction of light.
- The Sun appears flattened near the horizon because light from the bottom of the Sun is refracted more through Earth's atmosphere than light from the top of the Sun.
- Shimmering is caused by light travelling at slightly different speeds through air layers of different temperatures.
- A mirage is the result of refraction and total internal reflection in layers of air of different temperatures.
- A rainbow is caused by refraction of sunlight and partial internal reflection in water droplets in Earth's atmosphere.

CHECK YOUR LEARNING

1. (a) Explain what is meant by the term "apparent depth."
   (b) What causes this phenomenon? K/U
2. You want to scoop a fish out of water. Where should you aim relative to the fish image in order to really capture it? Explain. K/U
3. In the explanation of a mirage, the text mentioned that the index of refraction for air decreases as the air gets warmer. What does this tell you about the speed of light in cold air compared with the speed of light in warm air? Explain. K/U
4. What are you really looking at when you see a pool of water on a highway that you know is dry? Explain. K/U
5. In dispersion, violet light is refracted more than red light. What does this tell you about how the index of refraction for violet light compares with that for red light? Explain. K/U
6. What three changes in direction does a light ray undergo when it interacts with a raindrop to form a rainbow? K/U
7. Would a rainbow still form if the speed of light did not change for different colours in a raindrop? Explain. K/U

CHAPTER 12 LOOKING BACK

KEY CONCEPTS SUMMARY

Light changes direction predictably as it travels through different transparent media.

- The speed of light depends on the medium that it is passing through. (12.1)
- Light bends toward the normal when it slows down in a medium and away from the normal when it speeds up in a medium. (12.1)
- Light can undergo partial reflection and refraction at the same time at a surface. (12.1)

Light bends toward the normal when it slows down in a medium with a higher index of refraction.

- The index of refraction for a medium is defined as the ratio of the speed of light in a vacuum to the speed of light in that medium; it is a dimensionless quantity. (12.4)
- Mathematically, the index of refraction is defined as $n = \frac{c}{v}$ or $n = \frac{\sin \angle i}{\sin \angle R}$. (12.4)

Total internal reflection may occur when an incident ray is aimed at a medium with a lower index of refraction.

- The critical angle is the angle of incidence for which the angle of refraction is 90°; this occurs only when light passes from one medium into another with a lower index of refraction. (12.5)
- Total internal reflection occurs if the angle of incidence is greater that the critical angle. (12.5)

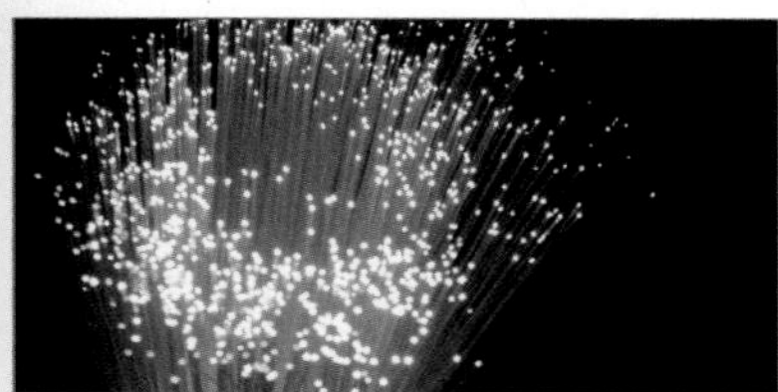

Many optical devices make use of the refraction and reflection of light.

- Optical devices such as periscopes, binoculars, retro-reflectors, and fibre-optic cables make use of total internal reflection. (12.5)
- A triangular prism, depending on its orientation, can change the direction of light by 90° (one total internal reflection) or 180° (two total internal reflections). (12.5)

The refraction and reflection of light can be used to explain natural phenomena.

- Objects in water appear to be at a shallower depth (an apparent depth) than they really are as a result of the refraction of light. (12.7)
- Shimmering and mirages are the result of refraction and total internal reflection in layers of air of different temperatures. (12.7)
- A rainbow is caused by refraction and partial internal reflection of sunlight in water droplets in Earth's atmosphere. (12.7)

Understanding the behaviour of light is key to many careers.

- Jewellers apply the total internal reflection of light in precious and semiprecious stones to create very appealing designs. (12.5)
- Manufacturers of glass prisms apply the refraction and total internal reflection of light to manufacture prisms that can accurately change the direction of light by 90° or 180°. (12.5)

WHAT DO YOU THINK NOW?

You thought about the following statements at the beginning of the chapter. You may have encountered these ideas in school, at home, or in the world around you. Consider them again and decide whether you agree or disagree with each one.

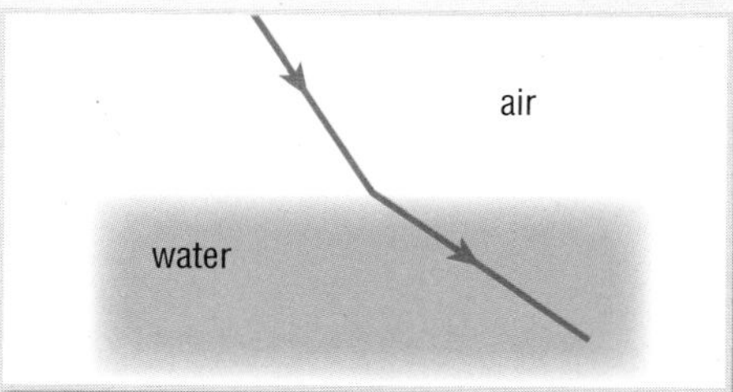

1 This diagram accurately shows light passing from air into water.
Agree/disagree?

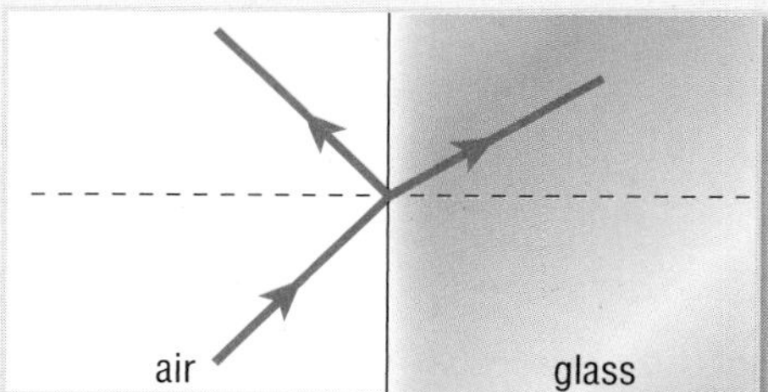

2 Light can be both reflected and transmitted when it travels from air into glass.
Agree/disagree?

3 A periscope is an optical device that allows you to see around corners.
Agree/disagree?

4 A swimming pool is always deeper than it appears.
Agree/disagree?

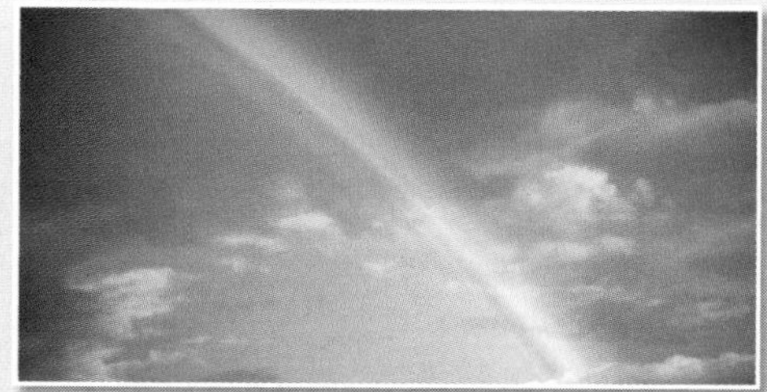

5 In order to see a rainbow, the Sun must be shining in one part of the sky while rain or clouds are in the opposite part of the sky.
Agree/disagree?

6 When you see a mirage, you are really looking at an image of the sky.
Agree/disagree?

How have your answers changed?
What new understanding do you have?

Vocabulary

refraction (p. 515)
angle of refraction (p. 516)
index of refraction (p. 524)
critical angle (p. 526)
total internal reflection (p. 526)
retro-reflector (p. 530)
apparent depth (p. 535)
mirage (p. 537)
dispersion (p. 538)

BIG Ideas

- ✓ **Light has characteristics and properties that can be manipulated with mirrors and lenses for a range of uses.**
- ✓ **Society has benefitted from the development of a range of optical devices and technologies.**

CHAPTER 12 REVIEW

The following icons indicate the Achievement Chart category addressed by each question.

K/U Knowledge/Understanding T/I Thinking/Investigation
C Communication A Application

What Do You Remember?

1. What is the procedure for measuring the angle of refraction? Include a diagram in your answer. (12.1) K/U C
2. Light bends toward the normal when going from material A into material B. Compare the speed of light in the two materials. (12.1) K/U
3. Light slows down as it travels into a second medium. Which way is the light bending, if at all? (12.1) K/U
4. Four materials have indices of refraction of 1.72, 1.00, 2.30, and 1.50. Which material will refract light the most? Why? (12.4) K/U
5. What is the angle of refraction at the critical angle? (12.5) K/U
6. A certain substance has a critical angle of 24.5°. What minimum angle of incidence, to the nearest tenth of a degree, will result in total internal reflection? (12.5) K/U
7. When you see an object under water, how does the apparent depth of the object compare with the actual position of the object? (12.7) K/U
8. In your own words, describe how a rainbow forms. (12.7) K/U
9. Figure 1 shows a light ray striking a window. (12.1, 12.5) K/U T/I C
   (a) Copy the diagram into your notebook. Draw the path of the light ray through the window.
   (b) Explain why the light ray behaves as it does in the glass.
   (c) Will the angle of refraction ever reach 90°? Explain.

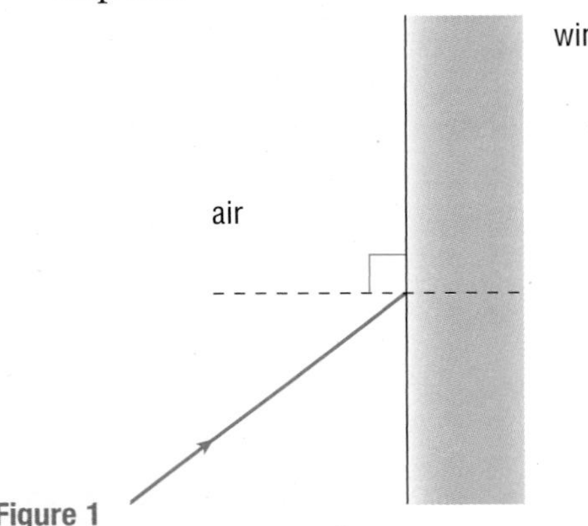

Figure 1

What Do You Understand?

10. Figure 2 represents a beam of light travelling in two different media. (12.4) K/U
    (a) Which medium has the higher index of refraction?
    (b) In which medium will light travel slower?
    (c) Why did you not need to know in which direction the light was travelling to answer the first two questions?

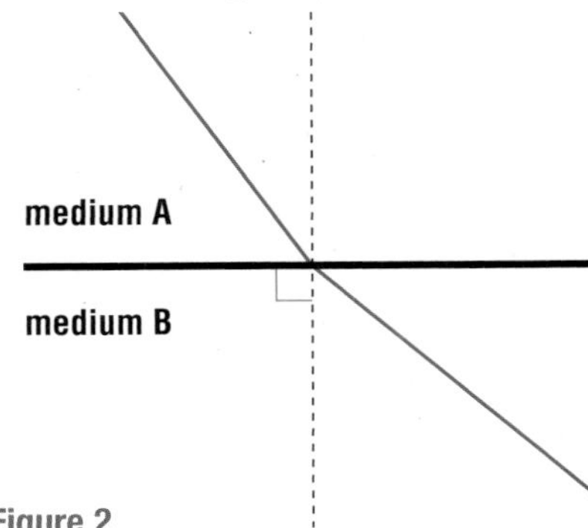

Figure 2

11. A beam of light is travelling from air into glass. Would refraction still occur if light travelled at the same speed in air *and* in glass? Explain. (12.1) K/U
12. Periscopes could be made with mirrors. They are, however, usually made with right-angled prisms. (12.5) K/U A
    (a) How can prisms behave like mirrors?
    (b) Why are prisms usually used instead of mirrors?
13. What type of image is a rainbow? Explain. (12.7) K/U
14. A shimmering pool of water on the road is often visible just after hot asphalt has been laid. With the aid of a diagram, explain why this is so. (12.7) A C

Solve a Problem

15. The speed of light in carbon disulfide is 1.84×10^8 m/s. Calculate the index of refraction for carbon disulfide. (12.4) T/I
16. The speed of light in arsenic trisulfide glass is 1.47×10^8 m/s. What is the index of refraction for this medium? (12.4) T/I

17. What is the index of refraction for fluorite if light travels at a speed of 2.10×10^8 m/s in fluorite? (12.4) T/I
18. The index of refraction for vegetable oil is 1.47. Determine the speed of light in vegetable oil. (12.4) T/I
19. Heavy flint glass has an index of refraction of 1.65. What is the speed of light in this medium? (12.4) T/I
20. Zircon has an index of refraction of 1.92. Determine the speed of light in zircon. (12.4) T/I

Create and Evaluate

21. The direction of light in a fibre-optic cable is constantly changing throughout the entire length of the cable. (12.5) T/I A
    (a) How can this be when you know that light travels in straight lines?
    (b) Research the advantages of fibre-optic cable over copper cable.
22. Create a t-chart similar to Figure 3. In your chart, compare a mirage with a rainbow. (12.7) K/U

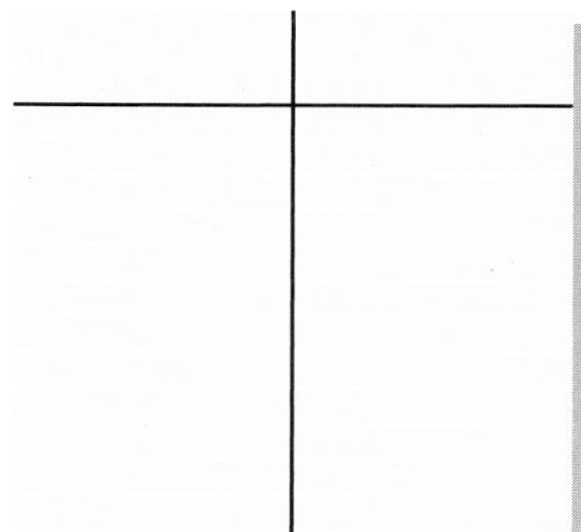

Figure 3

23. You are having a disagreement with your brother at a swimming pool. A puck has been thrown in the water and is resting at the bottom of the deep end. Your brother wants to dive directly at the puck in order to retrieve it. You tell him that the puck is not really where it appears to be. He wonders how this can possibly be. Defend your statement with an explanation of why the puck is not where it appears to be. Your explanation should state where your brother should dive. Present your explanation orally, in written form, or as a diagram.(12.7) A C
24. In this chapter, you learned about invisibility cloaks.
    (a) Form an opinion on whether you think science and technology always produce beneficial products.
    (b) What are some potential applications or uses of an invisibility cloak? Evaluate each use in terms of its benefit to society.
    (c) Do you think the creation of an invisibility cloak is a good idea? Explain.(12.6) A

Reflect on Your Learning

25. A knowledge of science can lead to a greater understanding of nature.
    (a) How did your knowledge of refraction give you greater insight into the production of certain atmospheric phenomena?
    (b) Did this extra knowledge enhance or change the way you look at fairly common events such as shimmering or the formation of mirages or rainbows? Explain.

Web Connections 

26. Research the historical methods used to measure the speed of light by Foucault and others. Present your research orally or in written form. T/I C
27. It is often possible to see two rainbows formed at the same time. Sometimes it is even possible to see three rainbows on top of one another. Research how secondary and tertiary (the third bow) rainbows are formed. Prepare an illustrated presentation of your findings. T/I C
28. The rear-view mirror inside a vehicle can be used for both daytime and nighttime driving. Research how this mirror works so that shifting its position for night driving still allows images to be formed, but with a great reduction in glare. Write a short article, with diagrams, for the motoring section of your local paper. T/I C A

To do an online self-quiz and for all Nelson Web Connections, GO TO NELSON SCIENCE

SELF-QUIZ

The following icons indicate the Achievement Chart category addressed by each question.

K/U Knowledge/Understanding T/I Thinking/Investigation
C Communication A Application

For each question, select the best answer from the four alternatives.

1. As light passes from air into water,
   (a) the light bends toward the normal.
   (b) the light bends away from the normal.
   (c) the light continues on a straight path and does not bend.
   (d) the light is completely reflected off the surface of the water. (12.1) K/U

2. The index of refraction for glass is 1.52. What is the speed of light in glass? (12.4) K/U
   (a) 1.52×10^8 m/s
   (b) 1.97×10^8 m/s
   (c) 3.00×10^8 m/s
   (d) 6.57×10^8 m/s

3. You see a submerged seashell that you want to bring home for your collection. Which statement correctly describes where you should place your net in order to pick up the shell? (12.7) K/U
   (a) Place the net directly above the image of the shell in the water.
   (b) Place the net directly behind the image of the shell in the water.
   (c) Place the net directly in front of the image of the shell in the water.
   (d) Place the net directly below the image of the shell in the water.

4. Which condition must be true for total internal reflection to occur? (12.5) K/U
   (a) The light ray must bend toward the normal.
   (b) The angle of reflection must equal the angle of incidence.
   (c) The angle of incidence must be greater than the critical angle.
   (d) Light must be travelling faster in the first medium than in the second.

Indicate whether each statement is TRUE or FALSE. If you think the statement is false, rewrite it to make it true.

5. Light travels faster in cool air than in warm air. (12.7) K/U

6. The legs of someone standing in water appear to be longer than they actually are. (12.7) K/U

7. When light refracts, it changes direction as it travels from one medium into another. (12.1) K/U

Copy each of the following statements into your notebook. Fill in the blanks with a word or phrase that correctly completes the sentence.

8. If the __________ of a light ray passing through water is greater than the critical angle for water, the resulting light ray will reflect back into the water. (12.5) K/U

9. Triangular prisms are more useful in optical devices than mirrors because these prisms __________ almost 100 % of the light internally. (12.5) K/U

10. The index of refraction for a given medium is a __________ quantity because it has no units. (12.4) K/U

Match each term on the left with the most appropriate description on the right.

11. (a) index of refraction
    (b) mirage
    (c) dispersion
    (d) total internal reflection
    (e) retro-reflector

    (i) the result of white light passing through a prism
    (ii) the process that makes a diamond sparkle
    (iii) the ratio of the speed of light in a vacuum to its speed in a medium
    (iv) a device used on road signs to make them visible at night
    (v) an image that appears to be a puddle of water on the road at a distance

    (12.4, 12.5, 12.7) K/U

Write a short answer to each of these questions.

12. Give an example in which light undergoes partial reflection and partial refraction at the same time. (12.1) K/U

13. You have learned that light travels in straight lines. You have also learned that light bends. Explain why these two statements do not contradict each other. (12.1) K/U T/I

14. Look at the indices of refraction in Table 1.

Table 1 Indices of Refraction for Three Media

Medium	Index of Refraction
A	2.30
B	1.76
C	1.98

Through which medium, listed in Table 1, does light pass
(a) fastest?
(b) slowest? (12.4) K/U

15. The speed of light in sulfuric acid at room temperature is 2.11×10^8 m/s. What is the index of refraction for sulfuric acid? Show your calculations. (12.4) K/U

16. (a) How does the technology of fibre optics make use of light?
(b) Why is it important that a material used to produce fibre optic cables have a very small critical angle? (12.5) K/U

17. Suggest at least two uses for retro-reflectors in your home. (12.5) A

18. Consider how your eyes and brain interpret information to allow you to see the image of an object in a plane mirror. Compare this process to the way your eyes and brain interpret information to allow you to see an image that is underwater. (12.7, 11.7) T/I

19. (a) White light slows down when it passes through a prism. Use what you know about dispersion, refraction, and the colours of the visible spectrum to explain why the colours that make up white light appear in a specific order as they exit the prism.
(b) Would red laser light be dispersed if it passed through a prism? Why or why not? (12.7, 11.3) T/I

20. Would you be able to see a rainbow in the sky during a rainstorm if there were clouds between you and the Sun? Why or why not? (12.7) T/I

21. (a) Name three devices that make use of total internal reflection.
(b) Draw a diagram that illustrates the phenomenon of total internal reflection. Include arrows in your diagram that indicate the direction in which the light is travelling. Also show the critical angle with a dashed line. (12.5) K/U C

22. Write a definition of "apparent depth" in your own words. (12.7) K/U C

23. Your friend claims that mirages do not really exist, and that if you try to take a picture of one, the mirage will not show up in the photo. Do you think your friend is correct? Explain your answer. (12.7) T/I

24. A company that installs a special coating on car windows has hired you to market this product. This special coating allows the driver to see out, but limits what others see when they look into the car. Design an ad for this product showing how the coating works. (12.6) C A

CHAPTER 13

Lenses and Optical Devices

KEY QUESTION: How do lenses produce images, and how do lenses benefit humans?

Water droplets can act as natural lenses.

UNIT E

Light and Geometric Optics

CHAPTER 11
The Production and Reflection of Light

CHAPTER 12
The Refraction of Light

CHAPTER 13
Lenses and Optical Devices

KEY CONCEPTS

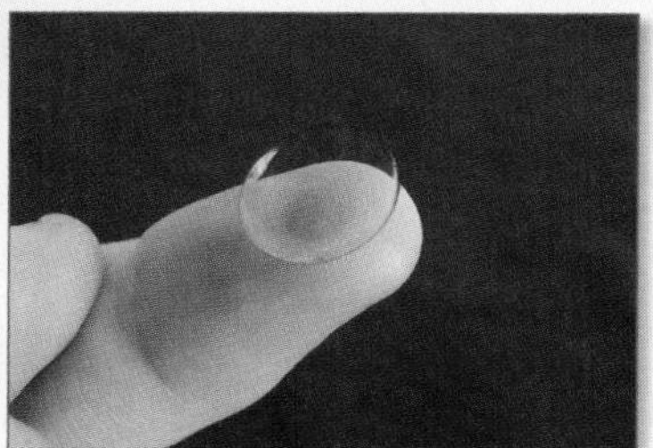

A lens is a transparent object used to change the path of light.

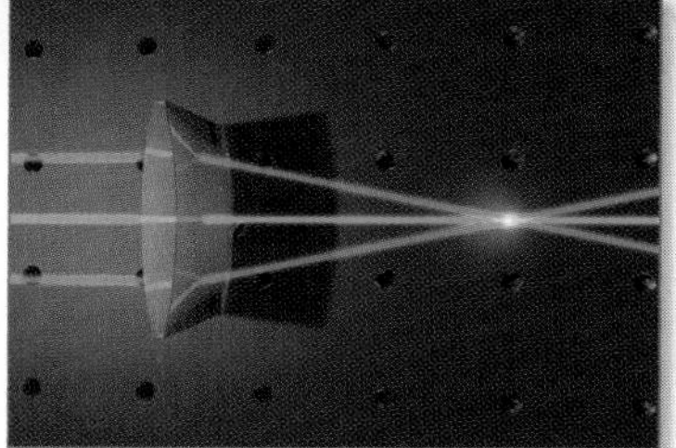

Parallel light rays are refracted through a focus when they pass through a converging lens.

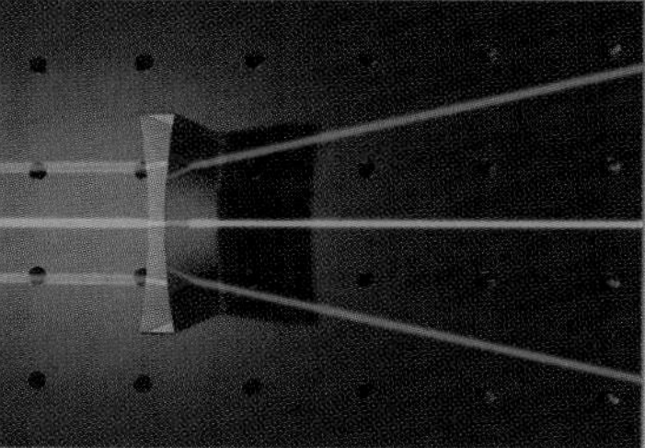

Geometric optics can be used to determine the path of light rays through lenses.

Both ray diagrams and algebraic equations can be used to determine the characteristics of an image in a lens.

Lenses have many technological uses that benefit humans.

The eye can be treated as a lens, and vision problems can be corrected with other lenses.

CHECKING THE FACTS in Fiction

Lord of the Flies is a well-known novel by William Golding. The story describes the actions of a group of British schoolboys who have been stranded on an island. A key part of the story concerns a boy named Piggy. Piggy is near-sighted, and wears glasses to correct his vision. He uses the lenses of his glasses to start a fire. The schoolboys hope that the fire will be noticed and that they will be rescued.

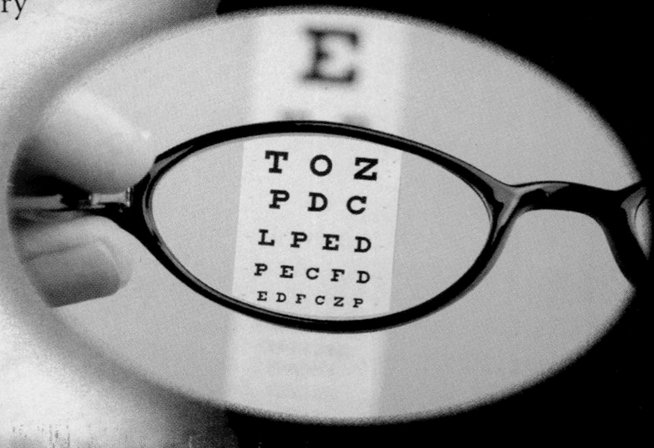

Piggy's use of his glasses to concentrate light is an application of lenses. But is Piggy's use of his glasses scientifically correct? Would he really be able to start a fire with glasses designed to correct near-sightedness?

WHAT DO YOU THINK?

Many of the ideas you will explore in this chapter are ideas that you have already encountered. You may have encountered these ideas in school, at home, or in the world around you. Not all of the following statements are true. Consider each statement and decide whether you agree or disagree with it.

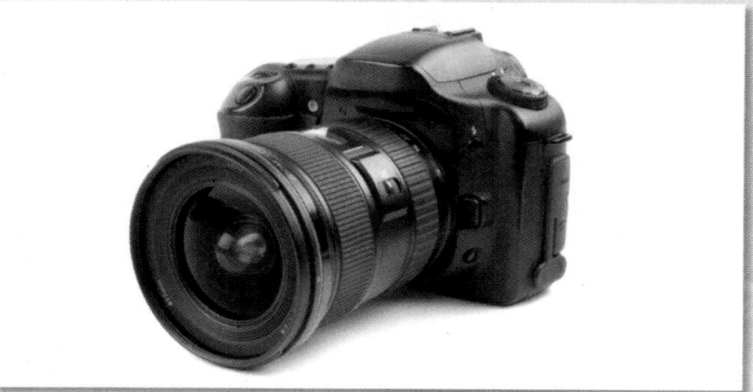

1 A camera lens produces images that are upside down and flipped horizontally.
Agree/disagree?

2 If you are near-sighted, then thick lenses will help you see better.
Agree/disagree?

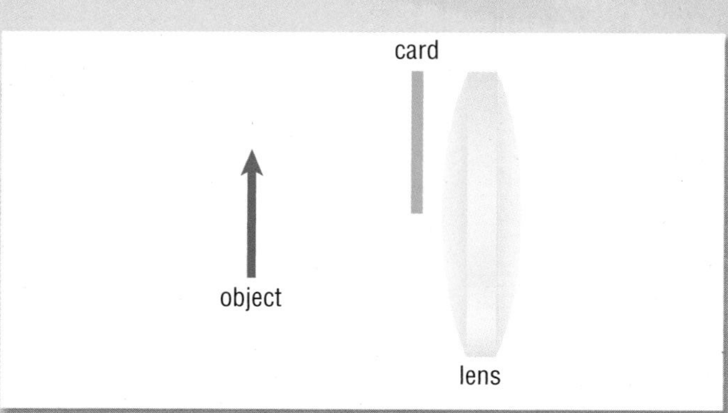

3 If you cover up half of a lens, then the image that you will see through the lens will be only half of the object.
Agree/disagree?

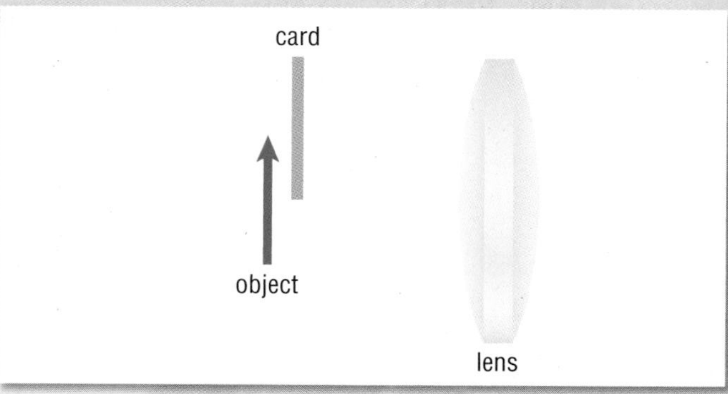

4 If you cover up half of an object, then the image that you will see through a lens will be only half of the object.
Agree/disagree?

5 A magnifying glass is very similar to a microscope in how it forms an image.
Agree/disagree?

6 Many people require glasses for reading as they age because the ability of the eye to focus decreases with age.
Agree/disagree?

FOCUS ON
WRITING

Writing a Critical Analysis

When you write a critical analysis, you examine an issue in depth by raising questions about the accuracy and consistency of the information and the reliability of the sources, and then make a judgment or recommend a course of action. Use the strategies listed next to the text to improve your critical writing.

WRITING TIP

As you work through the chapter, look for tips like this. They will help you develop literacy strategies.

Is Laser Eye Surgery Appropriate For My Friend?

My friend is constantly mentioning how the look of her glasses bothers her and that they are inconvenient to wear when she plays sports.

First paragraph establishes the context of your critical analysis.

With laser eye surgery, my friend would no longer have to wear glasses. She could stop worrying about how she looks wearing glasses and would be able to play sports more comfortably.

Second paragraph narrows the topic to a specific detail.

During my research, I discovered several different companies that claim histories of successful laser eye surgeries. The companies advertising "great deals" caused me to worry that they are having trouble finding clients.

Conduct research to collect a range of viewpoints on the issue.

Ask questions about important information that is omitted or is inconsistent.

I also learned from my research that being able to see upon waking, not having to wear glasses or contact lenses, and improved personal safety are all benefits of the surgery. These benefits all contributed to a majority of people saying their quality of life had improved because of the surgery.

Loss of night vision seems to be a concern with my parents' friends. One neighbour says she has trouble seeing at night since her surgery. This concern may no longer be valid because she had the surgery 10 years ago when it first started to become popular. Technology has come a long way since then.

Determine your position on the issue and whether it is consistent with other people's perspectives.

After researching the issue, I have come to a decision that if my friend finds a reputable company to perform the surgery, laser eye surgery is appropriate for her.

Make a judgment or propose possible strategies, alternatives, or solutions.

Lenses and the Formation of Images

13.1

We see the world through lenses. This is certainly true for those of us who wear glasses or contact lenses. Even if you do not require any vision aids, you still see the world through the lenses in your eyes. In this chapter, you will explore how lenses form images and how lenses are used in our society.

Basic Lens Shapes

Lenses consist of two basic shapes. The first is a **converging lens**, so named because parallel light rays converge through a single point after refraction through the lens (Figure 1). The converging lens is thickest in the middle and thinnest at the edge.

converging lens a lens that is thickest in the middle and that causes incident parallel light rays to converge through a single point after refraction

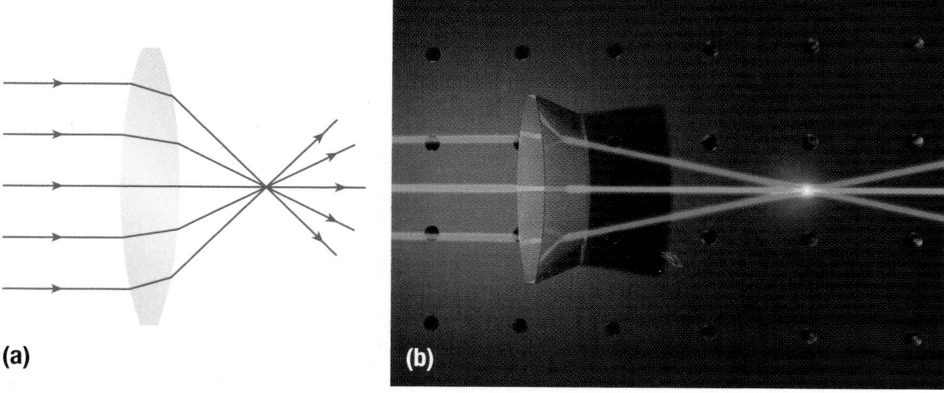

Figure 1 A converging lens brings refracted rays together through a single point.

The second kind of lens is a **diverging lens**. In a diverging lens, parallel light rays diverge after refraction from the lens (Figure 2). A diverging lens is thinnest in the middle and thickest at the edge.

diverging lens a lens that is thinnest in the middle and that causes incident parallel light rays to spread apart after refraction

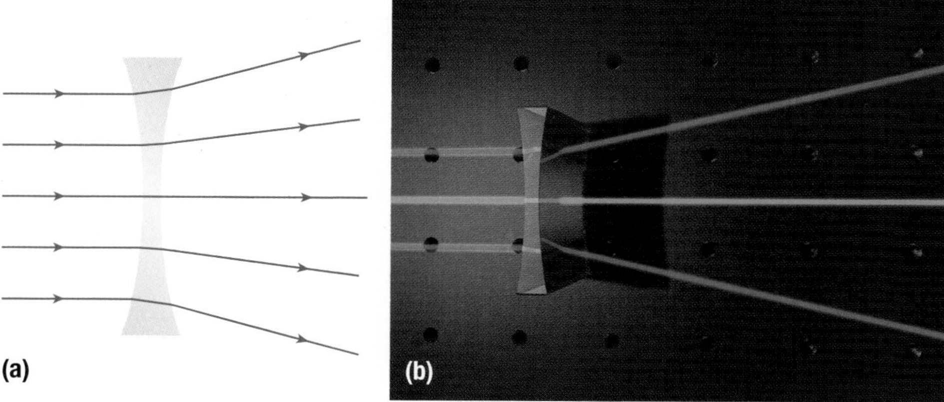

Figure 2 Light rays spread apart after refraction in a diverging lens.

Simplifying the Path of Light Rays Through a Lens

In a lens, light is refracted at the first air to glass surface. Light then travels through the glass of the lens and is refracted again at the glass to air surface on the other side. This means that there are always two refractions in a lens. We are, however, concerned only with the direction of the incident ray entering the lens and the ray leaving the lens. Ray diagrams can be greatly simplified by drawing a dashed vertical line through the centre of the lens and showing refraction occurring at this line. The central line is a reference point and shows light being refracted only once (Figure 3). It is used as a shortcut for both converging and diverging lenses.

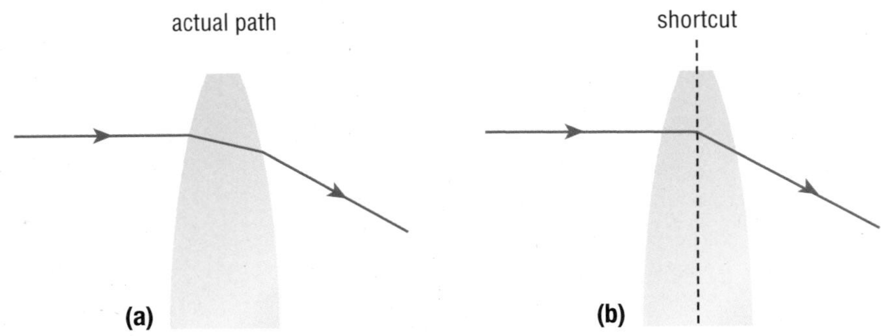

Figure 3 By drawing one refracted ray at the central dashed line of a lens, you can greatly simplify ray diagrams.

The Terminology of Converging Lenses

optical centre point at the exact centre of the lens

principal focus the point on the principal axis of a lens where light rays parallel to the principal axis converge after refraction

The centre of the lens is called the **optical centre** (O). The line through the optical centre that is perpendicular to the central dashed line of the lens is the principal axis (similar in purpose to the principal axis in curved mirrors). Light rays parallel to the principal axis converge through a single point on the principal axis called the **principal focus** (F). Light can strike the lens from either side, and both sides of the lens can focus parallel light rays. To tell them apart, the focus that is on the same side of the lens relative to the incident rays is usually labelled as the secondary principal focus (F') (Figure 4).

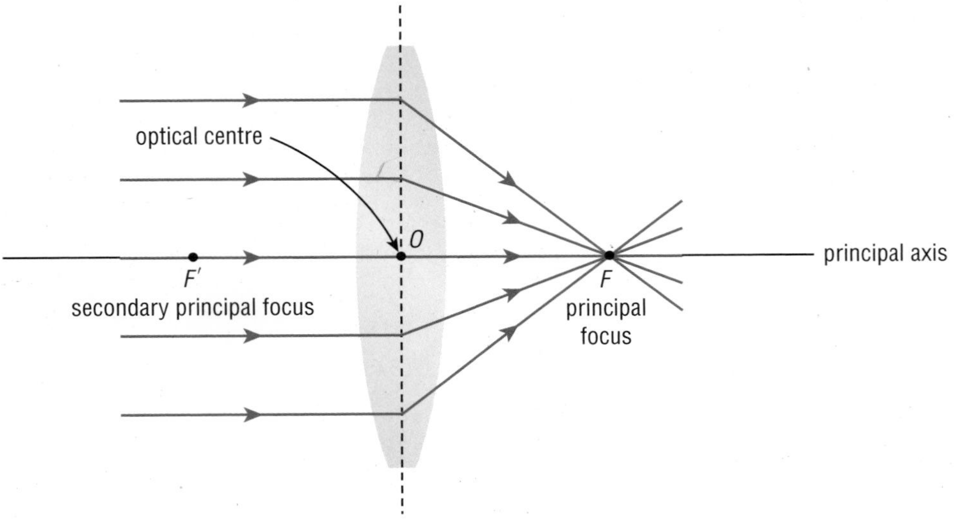

Figure 4 Terminology for a converging lens

The Terminology of Diverging Lenses

Light rays parallel to the principal axis of a diverging lens do not converge. Instead, the refracted rays spread apart. If you project these diverging rays backwards, it looks as if they come from a virtual focus. This point is now the principal focus (F). The secondary principal focus (F') is now on the other side of the lens, where the rays actually diverge (Figure 5). Note that F and F' are equally far apart from the optical centre of both a converging lens and a diverging lens.

You will examine how these terms relate to images in these two kinds of lenses when you do Activity 13.2.

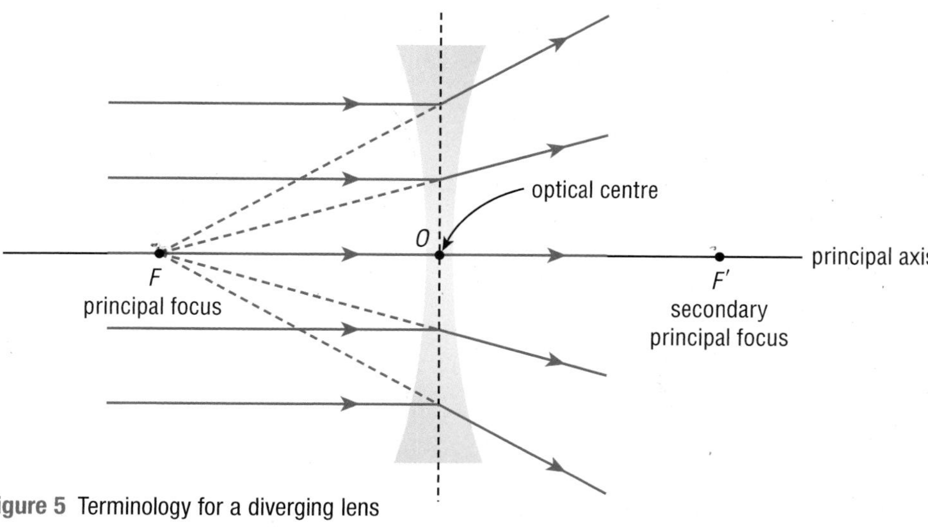

Figure 5 Terminology for a diverging lens

IN SUMMARY

- A converging lens brings parallel light rays together through a focus after refraction.
- A diverging lens spreads parallel light rays apart after refraction so that it looks as if they have come from a virtual focus.
- The principal focus of a *converging lens* is on the *opposite* side of the lens as the incident rays.
- The principal focus of a *diverging lens* is on the *same* side of the lens as the incident rays.

CHECK YOUR LEARNING

1. Why is a knowledge of lenses important even if you do not require glasses? A
2. What is the difference between a converging lens and a diverging lens? Mention the paths of light rays in your explanation. K/U
3. (a) How many refractions actually occur as a light ray travels through a lens? Identify the locations of these refractions on a diagram.
   (b) Why is it possible to simplify the number of actual refractions in a lens down to one refraction at a central line through the optical centre? K/U
4. Can a converging lens have more than one focus? Explain. K/U
5. You are given two lenses, a converging lens and a diverging lens. Can you tell them apart just by feeling their shape? Explain. K/U
6. (a) On what side of a converging lens is the principal focus located? Explain.
   (b) Where is the principal focus of a diverging lens located?
   (c) Why is a diverging lens different from a converging lens? K/U

13.2 PERFORM AN ACTIVITY

Locating Images in Lenses

Lenses are used in many optical devices such as cameras and eyeglasses. In this activity, you will examine the images produced in converging and diverging lenses. Remember to pay particular attention to the four characteristics of images: size, attitude, location, and type (SALT).

SKILLS MENU

- Questioning
- Hypothesizing
- Predicting
- Planning
- Controlling Variables
- Performing
- Observing
- Analyzing
- Evaluating
- Communicating

Purpose

To explore the characteristics of images produced by converging and diverging lenses.

Equipment and Materials

- converging lens with support
- diverging lens
- metre stick with two supports
- candle with holder
- paper screen and holder
- second sheet of paper or a small piece of cardboard
- chalk that can be easily erased

Procedure

Part A: Locating Reference Positions for a Converging Lens

1. Place the two metre stick supports under the ends of the ruler.
2. Place the converging lens in the lens support and place the lens and lens support in the middle of the ruler (at the 50 cm mark).
3. Aim the metre stick–lens assembly at a relatively distant object that is transmitting external light in the classroom when the lights have been turned off. Suitable objects are the slats of an open window blind, a window frame, or a door frame in a room with a window. Make sure that you are as far away as possible from this distant object. Move the sheet of paper back and forth behind the lens until you see as sharp an image of the distant object as possible. Mark this location on the ruler as *F* (principal focus). Also, mark in twice this distance (2*F*) from the lens.
4. Mark these same positions on the opposite side of the lens, but mark them as *F′* (secondary principal focus) and 2*F′* respectively.

Part B: Locating Images in a Converging Lens

5. Place a lit candle at these five positions: beyond 2*F′*, at 2*F′*, between 2*F′* and *F′*, at *F′*, and between *F′* and the lens. Move the paper screen back and forth until you locate an image. Figure 1 on the next page shows the setup for this procedure. Describe the characteristics of each image (size, attitude, location, and type) that you were able to locate. Use 2*F′* and *F′* as reference points when describing the image location. Record your observations in a table similar to Table 1.

   Note that you may need assistance from your teacher for the last two object locations: at *F′* and inside *F′*.

When using a candle, tie back long hair and loose clothing.

Place a piece of paper under the candle to catch any falling wax.

Be careful when moving the candle—the wax is hot.

Table 1 Image Characteristics in Lenses

Object location	Size of image	Attitude of image	Location of image	Type of image
beyond 2*F′*				
at 2*F′*				
between 2*F′* and *F′*				
at *F′*				
inside *F′*				

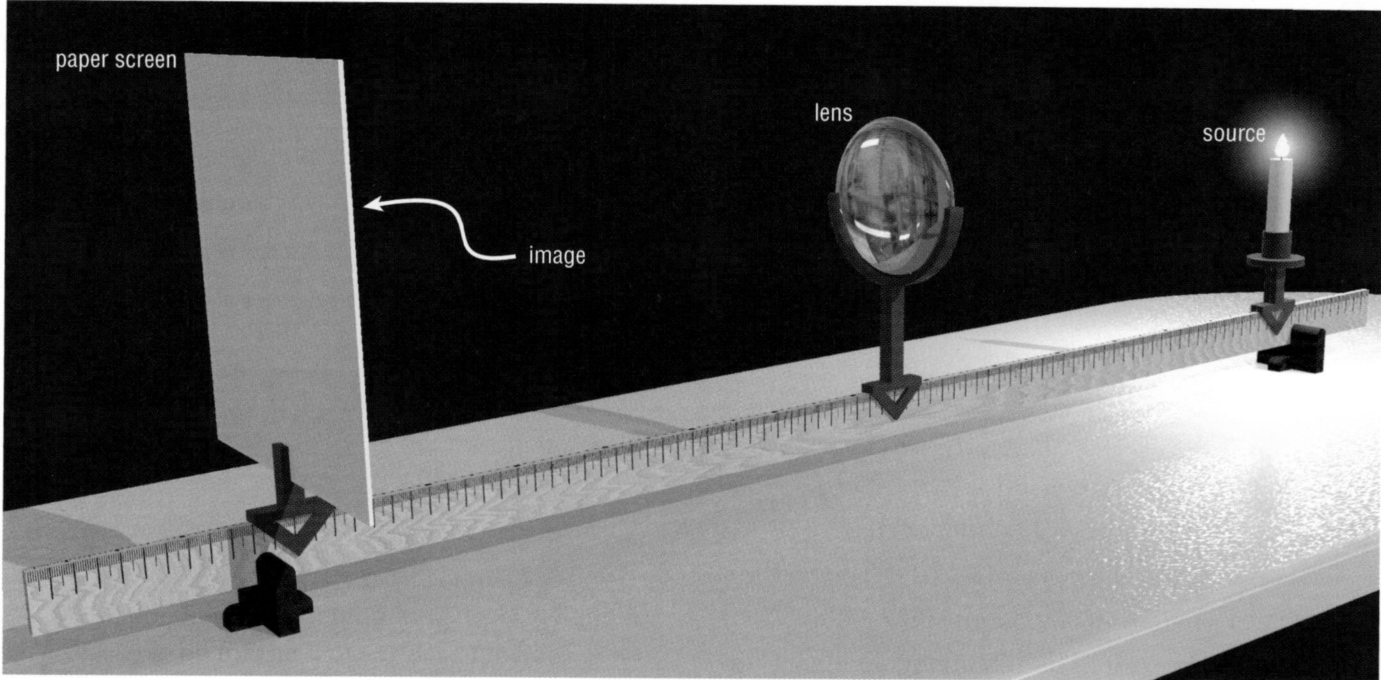

Figure 1

6. Move the candle back to its original position beyond $2F'$. Now cover half of the lens with the second piece of paper or cardboard. Locate and describe the image.
7. Move the second piece of paper or cardboard to cover half of the flame. Locate and describe the image.

Part C: Locating Images in a Diverging Lens

8. Replace the converging lens with a diverging lens. Attempt to find an image on the screen. Now look *into* the diverging lens, locate the image of the candle, and describe its characteristics. Move the lens back and forth to see if there is any change in image characteristics. Record your observations.

Analyze and Evaluate

(a) Where must an object be located for a converging lens to produce a real image? T/I

(b) What happened to the size of the real image as the object was slowly moved toward the lens from its original position beyond $2F'$? T/I

(c) What was the only location where the converging lens did not produce an image? T/I

(d) Where must an object be located for a converging lens to produce a virtual image? T/I

(e) What were the characteristics of the image in the diverging lens for all object locations? T/I

(f) Why did you not have to follow the same procedure for the diverging lens as you did for the converging lens? T/I

(g) Why were you still able to see the object when half of the lens was covered? Why was the brightness of the image reduced? T/I

(h) Why did you lose half of the image when you covered half of the object? T/I

Apply and Extend

(i) List some optical devices that use a lens to produce a real image. A

(j) Name an optical device that uses a lens to produce a larger, virtual image. A

(k) Suppose F for a converging lens is 23 cm, and a luminous source is placed at different positions in front of the lens. Predict the image characteristics for each position. T/I

- 64 cm from the lens
- 40 cm from the lens
- 10 cm from the lens

13.3 Images in Lenses

In Activity 13.2, you noticed that two things affect the characteristics of the image formed: the kind of lens (converging or diverging) and the location of the object. You can determine the image characteristics by drawing ray diagrams, just as you did with converging and diverging mirrors. As before, you need to draw only two light rays to locate the image. The only difference is that with mirrors you considered reflected rays, but with lenses you consider *refracted* rays.

In order to understand how to draw ray diagrams for lenses, it is important to understand how the incident and emergent rays are related to each other. The **emergent ray** is the ray that leaves the lens, being refracted as it goes from the lens back into air. Find out about their relationship by doing the activity "Exploring the Rectangular Prism."

emergent ray the light ray that leaves a lens after refraction

TRY THIS EXPLORING THE RECTANGULAR PRISM

SKILLS: Predicting, Performing, Observing, Analyzing

Equipment and Materials: ray box; single-slit mask; rectangular prism; blank sheet of paper

1. Lay the rectangular prism on its large flat face in the middle of the sheet of paper.

   When handling the glass prism, take care not to cut your fingers if it is chipped.

2. Use the ray box and the single-slit mask to produce a light ray.

   When unplugging the ray box, do not pull the electric cord. Pull the plug itself.

3. Aim the light ray at the prism so that you can see an emergent ray on the other side of the prism. Carefully examine how the incident ray and the refracted ray are positioned relative to each other. (If you notice another ray going between the prism and the paper, press down on the prism slightly.)
4. Now place the prism on its thin side and repeat step 3.

A. How did the rectangular prism on its large flat surface affect the emergent ray? T/I

B. What changed when you laid the prism on its thin side? T/I

C. How would the incident ray and the emergent ray compare if you had used an extremely thin rectangular prism? T/I

D. If a very thin rectangular prism is available, test your prediction by repeating this experiment. T/I

You can use a rectangular prism to understand how a lens works. An incident ray directed at a rectangular glass prism undergoes two refractions. The first is at the air–glass boundary as the ray enters the prism. The second is at the glass–air boundary when the ray emerges from the prism. In a rectangular prism, these two boundaries are parallel because the two surfaces of the prism are parallel to each other. So the emergent ray is parallel to the incident ray but displaced sideways. The amount of sideways displacement depends on the thickness of the prism (Figure 1, next page).

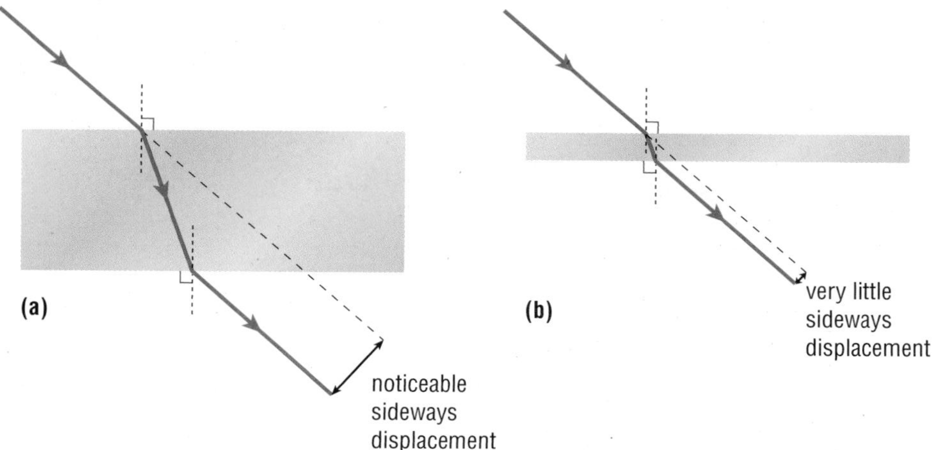

Figure 1 The emergent ray through a rectangular prism is parallel to the incident ray but displaced sideways. Reducing the width of the prism greatly reduces the amount of sideways displacement.

A very thin rectangular prism results in very little displacement of the emergent ray (Figure 1(b)). If the prism is thin enough, the emergent ray appears to be almost unaffected by the presence of the prism. This fact is important when considering images formed by converging lenses.

How to Locate the Image in a Converging Lens

The three imaging rules for converging lenses are shown in Figure 2.

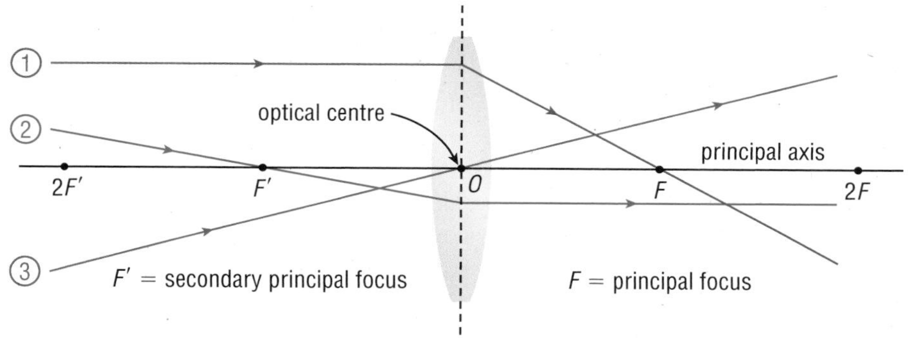

Figure 2 Imaging rules for a converging lens

1. A ray parallel to the principal axis is refracted through the principal focus (*F*).
2. A ray through the secondary principal focus (*F*′) is refracted parallel to the principal axis. This rule comes from the reversibility of light.
3. A ray through the optical centre (*O*) continues straight through without being refracted. This is true because the middle part of the lens acts like a very thin rectangular prism with no noticeable sideways displacement.

Note that these rules are true only for thin lenses. We will only discuss thin lenses in this chapter.

Images in a Converging Lens

You can investigate the images formed by a converging lens. If you place a luminous source at a distance greater than $2F'$, you can locate an image of this source by moving a paper screen back and forth on the other side of the lens. The image is smaller, inverted, and located somewhere between F and $2F$. This image is real. Light is actually arriving at the image location, and you can see the image on a paper screen.

WRITING TIP

Writing a Critical Analysis

Imagine that you are writing a critical analysis of a camera that has a lens with a fixed focal length. You might refer to the imaging rules for converging lenses in the first paragraph to set the context for your analysis.

Using the imaging rules for a converging lens, you can show how the lens produced this image. You can also predict how the lens produces images for other object locations (Figure 3).

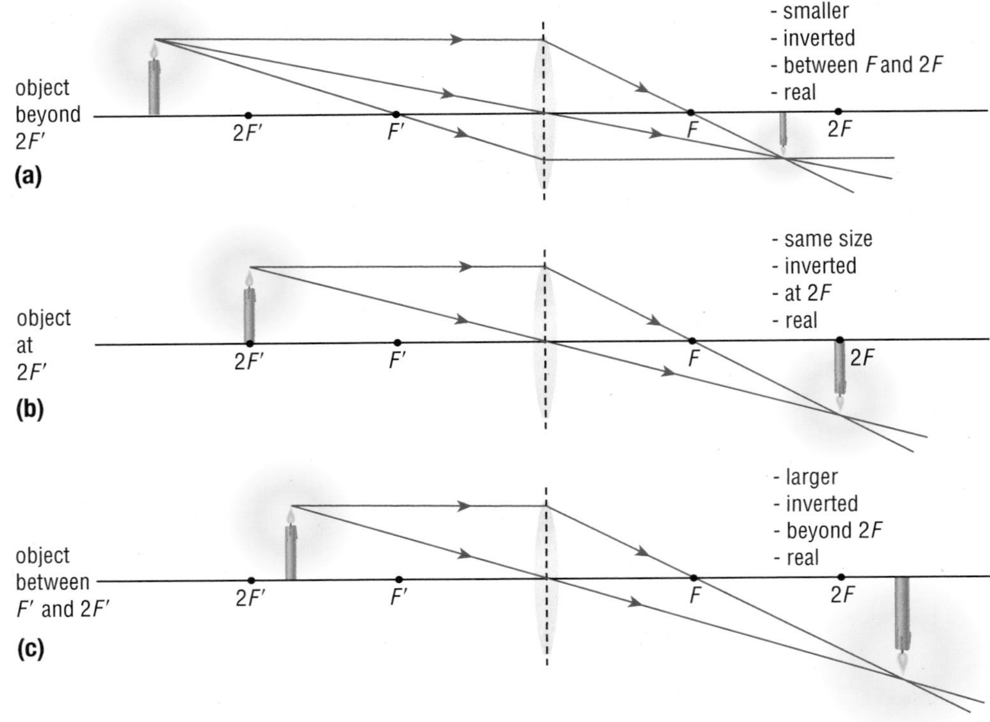

Figure 3 A converging lens produces a real image for these three object locations.

READING TIP

Making Connections

Compare Figure 4 on this page with the three diagrams for the concave mirror in Figure 6 in Section 11.9 on page 497. Examine how the image characteristics are related.

When an object is located beyond $2F'$, the image is smaller than the object and is between $2F$ and F. As you slowly move the object toward the lens, the image gets larger and larger. Eventually, the image and the object are the same size when the object is located at $2F'$; the image is now at $2F$. If you continue moving the object between $2F'$ and F', you get a larger image than the object; the image is now outside $2F$. Note that for all these image positions, the image is always inverted and real.

When you move the object to the secondary principal focus (F'), no image is produced. The refracted rays are parallel and do not cross to form an image (Figure 4). Even if you extend the rays backwards, there is no virtual image. The reason is that the rays are parallel and do not form a virtual source.

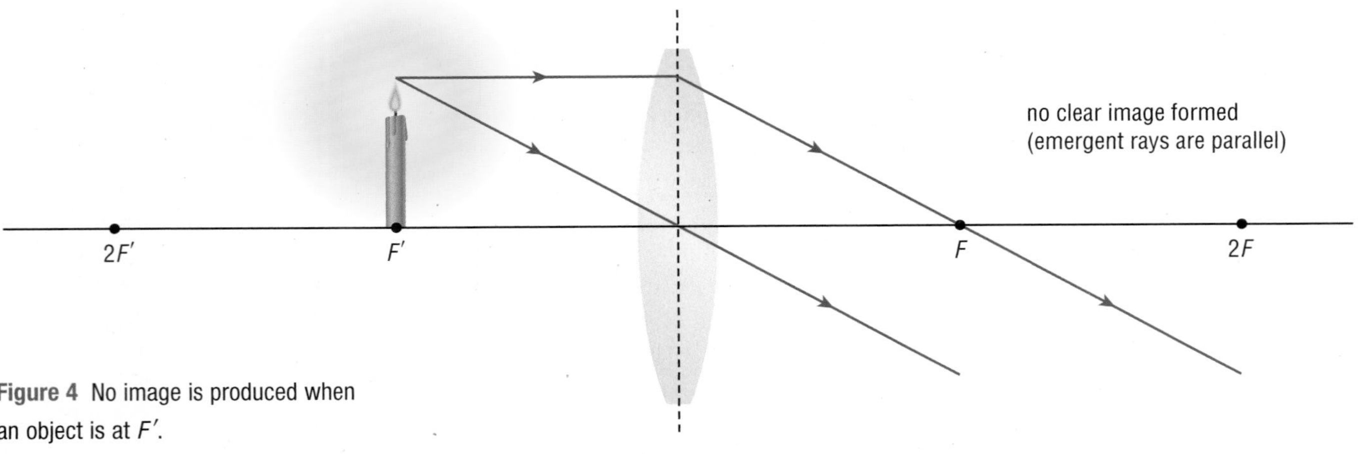

Figure 4 No image is produced when an object is at F'.

No real image is produced when the object is between F' and the lens. The refracted rays spread apart or diverge. However, the human brain projects these rays backwards and produces a virtual image behind the object (Figure 5). (Note that virtual images are often described as being behind the lens because light rays do not actually arrive at the image location; they only appear to.)

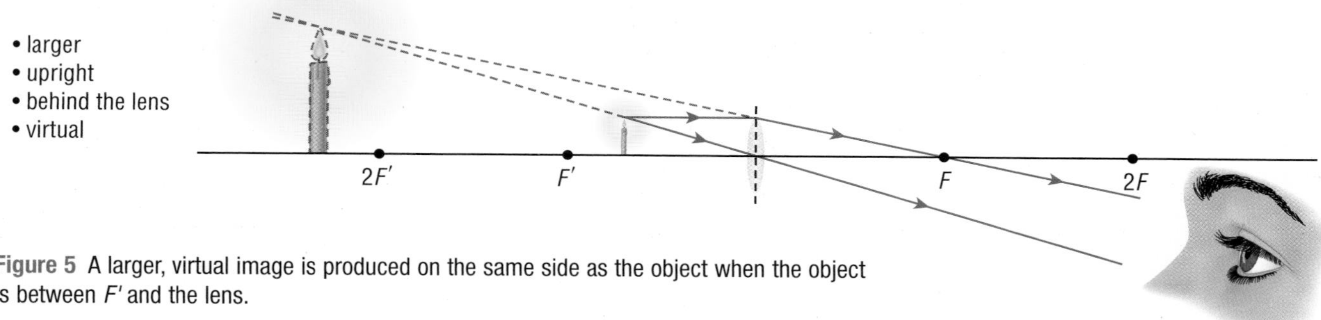

Figure 5 A larger, virtual image is produced on the same side as the object when the object is between F' and the lens.

Table 1 summarizes the image characteristics in a converging lens.

Table 1 The Imaging Properties of a Converging Lens

OBJECT	IMAGE			
Location	**Size**	**Attitude**	**Location**	**Type**
beyond $2F'$	smaller	inverted	between $2F$ and F	real
at $2F'$	same size	inverted	at $2F$	real
between $2F'$ and F'	larger	inverted	beyond $2F$	real
at F'	no clear image			
inside F'	larger	upright	same side as object (behind lens)	virtual

How to Locate the Image in a Diverging Lens

The imaging rules for a diverging lens are similar to those for a converging lens. The only difference is that light rays do not actually come from the principal focus (F); they only appear to (Figure 6).

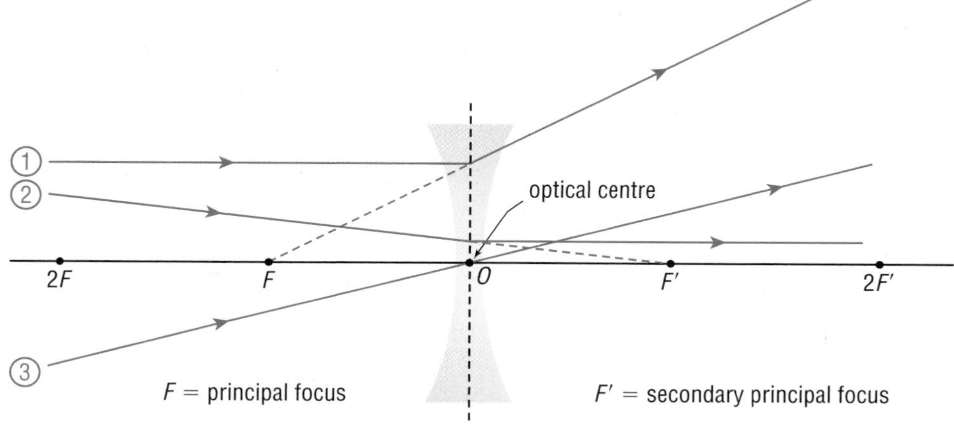

① A ray parallel to the principal axis is refracted as if it had come through the principal focus (F).

② A ray that appears to pass through the secondary principal focus (F') is refracted parallel to the principal axis.

③ A ray through the optical centre (O) continues straight through on its path.

Figure 6 Imaging rules for a diverging lens

Images in a Diverging Lens

A diverging lens always produces the same image characteristics no matter where the object is. The image is always smaller, upright, virtual and on the same side of the lens as the object (Figure 7). The human brain perceives this virtual image by extending the diverging rays backwards to a virtual source.

To investigate lenses by using computer simulations,

GO TO NELSON SCIENCE

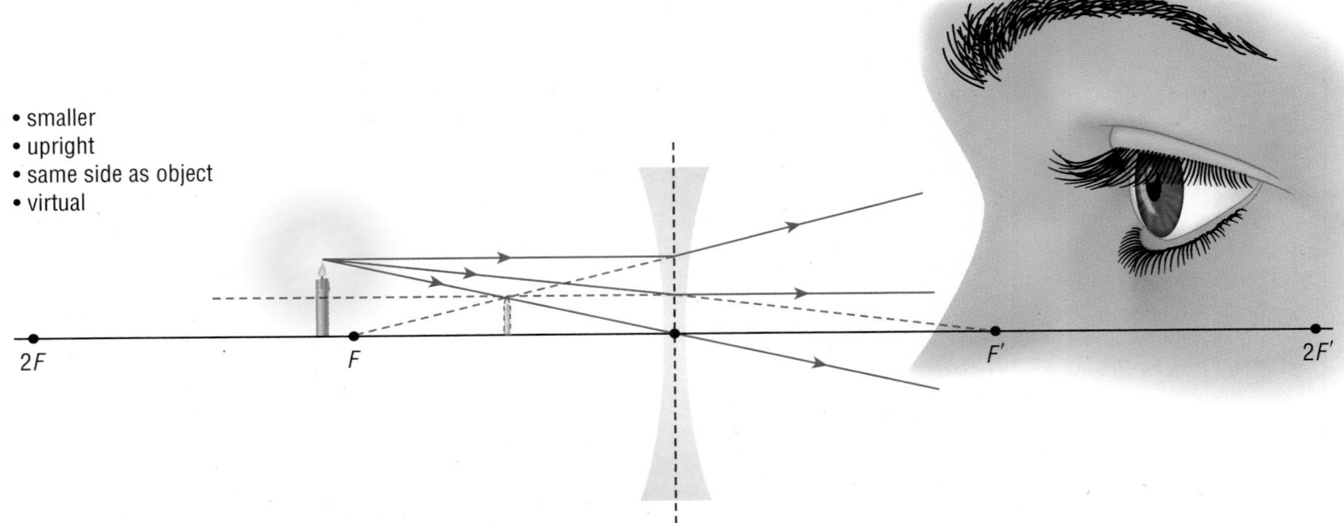

Figure 7 A diverging lens always forms a smaller, upright, virtual image that is on the same side of the lens as the object.

UNIT TASK Bookmark

How can you use the image characteristics of lenses as you plan your optical device for the Unit Task described on page 588?

IN SUMMARY

- A converging lens produces both real and virtual images. The image size and attitude will vary depending on the location of the object.
- A diverging lens always produces a smaller, upright, virtual image.

CHECK YOUR LEARNING

1. (a) In your own words, state the imaging rules for converging lenses.
   (b) How are these rules slightly different for diverging lenses? K/U
2. Copy Figure 8 into your notebook. T/I C
   (a) Add light rays to the diagrams to locate the image for each object.
   (b) Describe the image characteristics for each object.

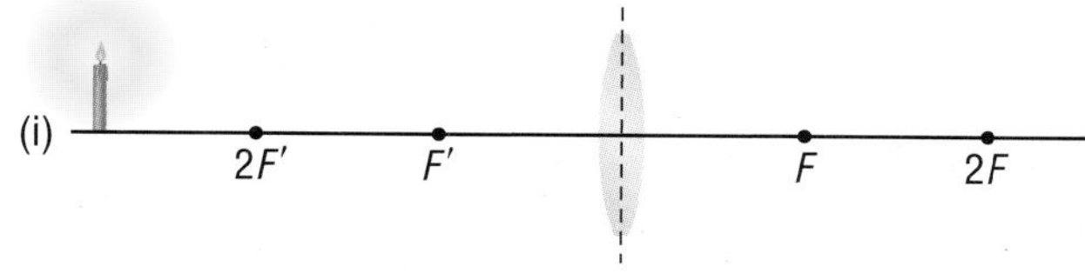

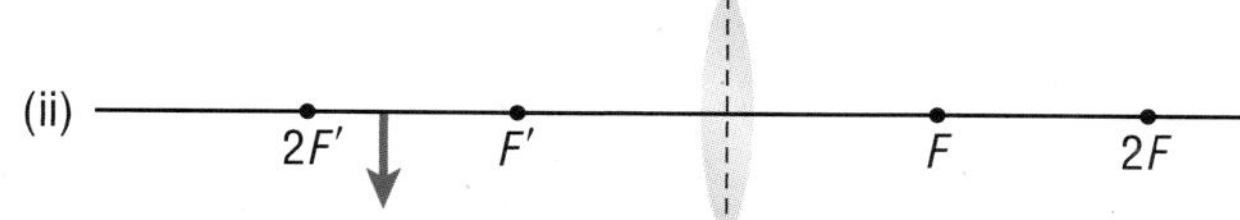

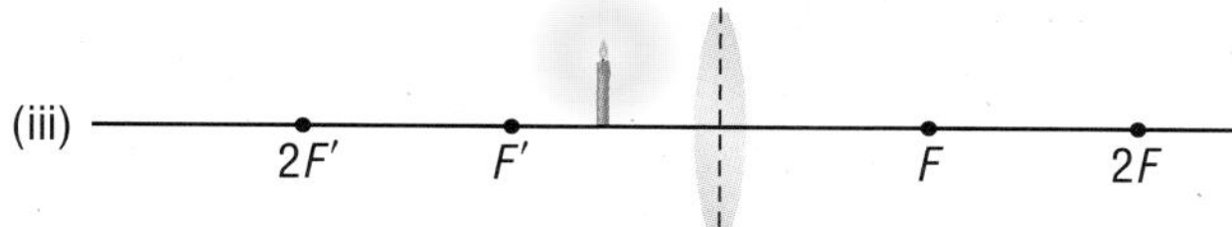

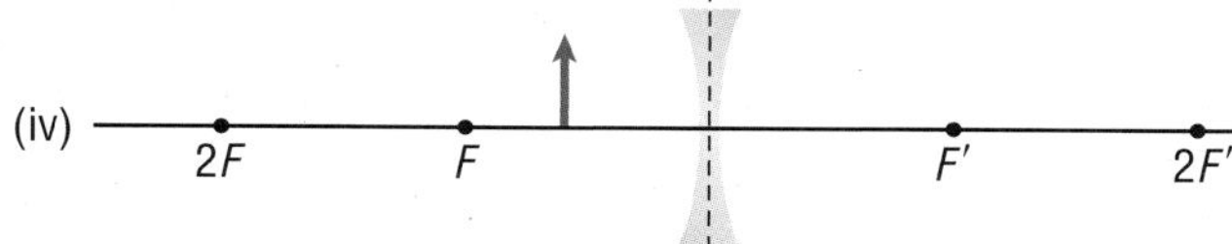

Figure 8

3. Copy Figure 9 into your notebook. Use light rays to locate F. T/I C

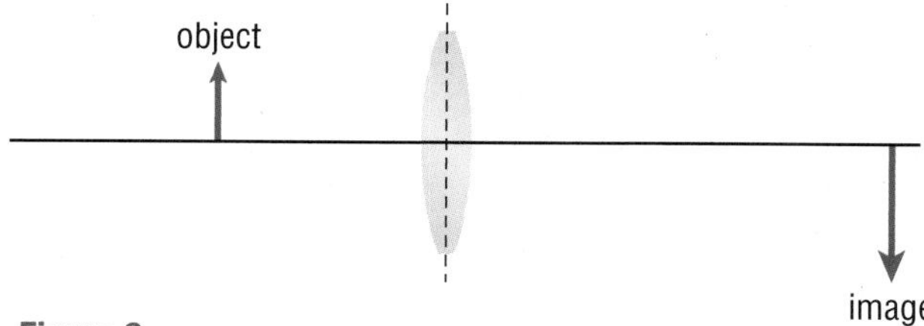

Figure 9

4. Copy Figure 10 into your notebook. T/I C
   (a) A screen is used to cover half of the lens (Figure 10(i)). Use light rays to locate the image on the diagram.
   (b) A screen is used to cover half of the object (Figure 10(ii)). Use light rays to locate the image on the diagram.

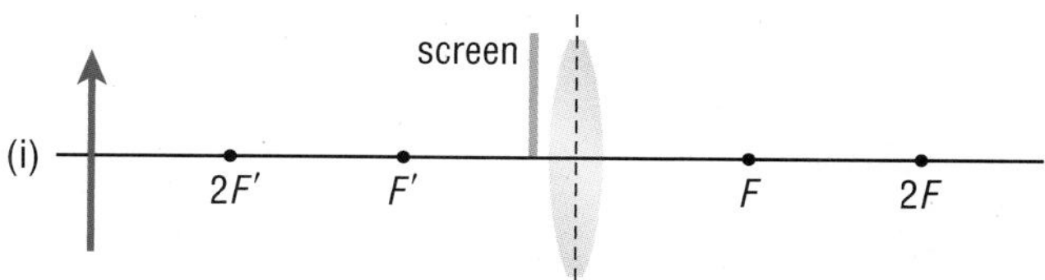

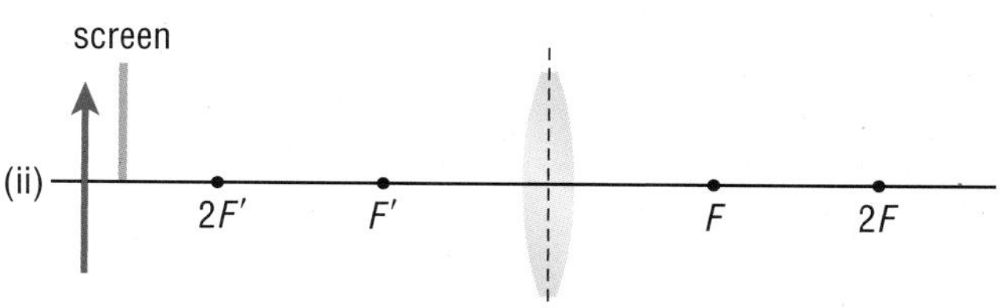

Figure 10

5. Why does a diverging lens never produce a real image? K/U
6. How is the virtual image produced by a converging lens different from the virtual image produced by a diverging lens? K/U
7. Write a general statement that is valid for both kinds of lenses that summarizes the relationship between the type and the attitude of the image. K/U T/I
8. When you watch a movie projected onto a screen, you are seeing an image. Traditional-style movie projectors include a light and a lens to project the picture onto the screen. T/I C
   (a) What type of lens is used in the projector? Explain.
   (b) Draw a ray diagram that includes the film (the object), the lens, and the image on the screen.
   (c) Describe the characteristics of this image.

13.4 The Lens Equations

There are two ways to determine the characteristics of images formed by lenses. You can use either ray diagrams or algebra. You used ray diagrams in the previous section. In this section, you will use algebra to determine the image characteristics.

Lens Terminology

Figure 1 illustrates some variables that we must first define:

d_o = distance from the object to the optical centre
d_i = distance from the image to the optical centre
h_o = height of the object
h_i = height of the image
f = focal length of the lens; distance from the optical centre to the principal focus (F)

Note that the focal length (f) is the same distance whether it goes to F or F'.

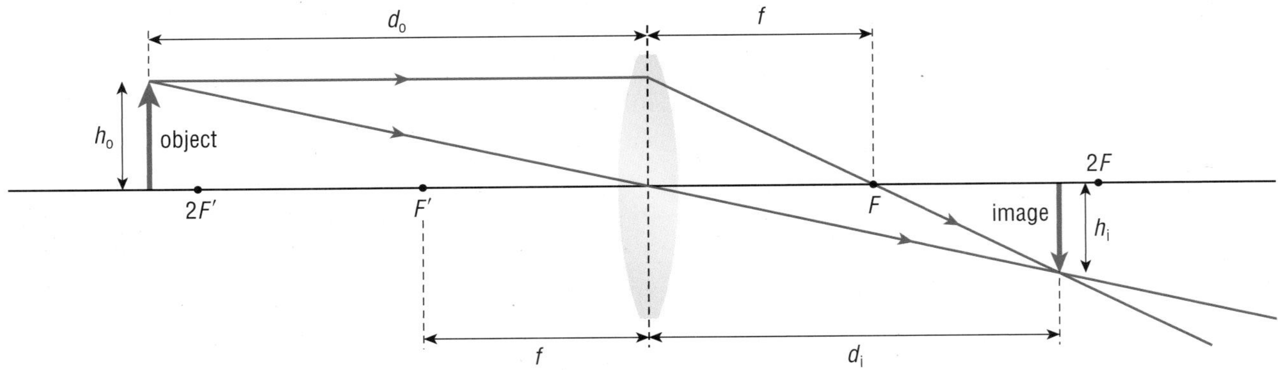

Figure 1 An illustration of the variables d_o, d_i, h_o, h_i, and f

The Thin Lens Equation

thin lens equation the mathematical relationship among d_o, d_i, and f; $\frac{1}{d_o} + \frac{1}{d_i} = \frac{1}{f}$

There is a very useful equation that relates the focal length (f), the object distance (d_o), and the image distance (d_i).

The equation $\frac{1}{d_o} + \frac{1}{d_i} = \frac{1}{f}$ is called the **thin lens equation.**

To use the thin lens equation, you need to follow this sign convention:

- Object distances (d_o) are always positive.
- Image distances (d_i) are positive for real images (when the image is on the opposite side of the lens as the object) and negative for virtual images (when the image is on the same side of the lens as the object).
- The focal length (f) is positive for converging lenses and negative for diverging lenses.

The thin lens equation was originally derived with the help of a diagram similar to Figure 2 on the next page. You do not need to be able to derive the equation yourself, but you might be interested in how it was done.

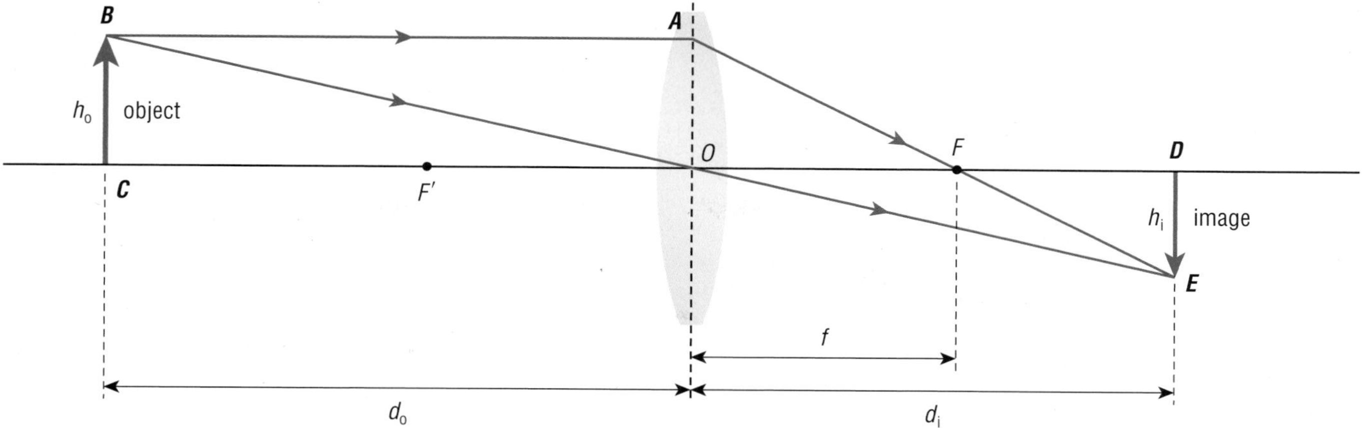

Figure 2 Diagram to derive the thin lens equation

$$\frac{ED}{DF} = \frac{AO}{OF}$$ ΔEDF and ΔAOF are similar (angle–angle similarity).

$$\frac{h_i}{d_i - f} = \frac{h_o}{f}$$

$$h_i = \frac{h_o}{f}(d_i - f)$$ Rearrange the equation to get h_i on the left side.

$$\frac{h_i}{h_o} = \frac{d_i - f}{f}$$

Since $\frac{h_i}{h_o} = \frac{d_i}{d_o}$ ΔEDO and ΔBCO are similar (angle–angle similarity).

Therefore, $\frac{d_i}{d_o} = \frac{d_i - f}{f}$ Substitute $\frac{d_i}{d_o}$ for $\frac{h_i}{h_o}$.

$$\frac{d_i}{d_o} = \frac{d_i}{f} - \frac{f}{f}$$

$$\frac{1}{d_o} = \frac{1}{f} - \frac{1}{d_i}$$ Now divide both sides by d_i and rearrange the equation.

$$\frac{1}{d_o} + \frac{1}{d_i} = \frac{1}{f}$$

Let's look at some problems that can be solved by using the thin lens equation.

SAMPLE PROBLEM 1 Using the Thin Lens Equation for a Converging Lens

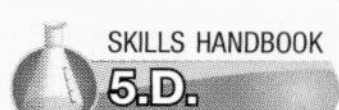

A converging lens has a focal length of 17 cm. A candle is located 48 cm from the lens (Figure 3). What type of image will be formed, and where will it be located?

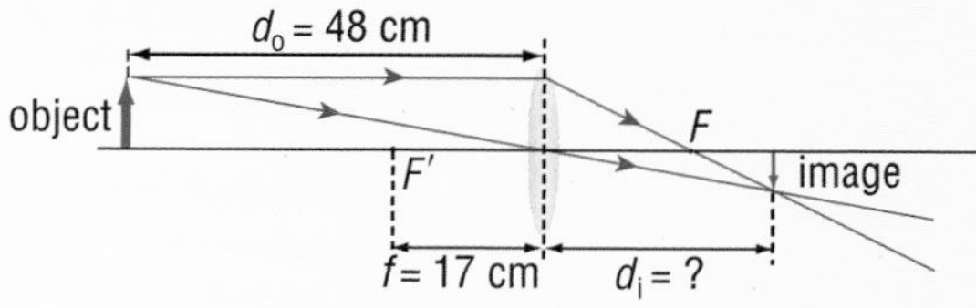

Figure 3

Given: $f = 17$ cm

$d_o = 48$ cm

Required: $d_i = ?$

Analysis and Solution: $\frac{1}{d_o} + \frac{1}{d_i} = \frac{1}{f}$

$$\frac{1}{d_i} = \frac{1}{f} - \frac{1}{d_o}$$

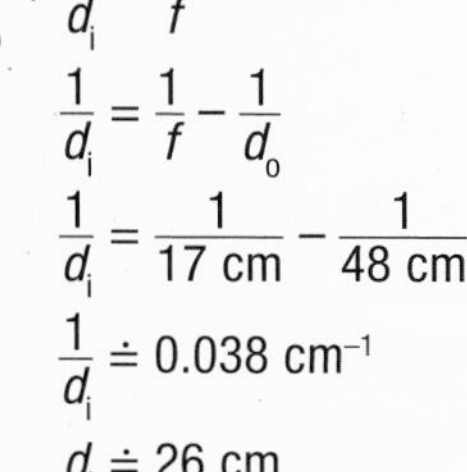

$$\frac{1}{d_i} = \frac{1}{17\text{ cm}} - \frac{1}{48\text{ cm}}$$

$$\frac{1}{d_i} \doteq 0.038\text{ cm}^{-1}$$

$$d_i \doteq 26\text{ cm}$$

Statement: The image of the candle is real and will be about 26 cm from the lens, opposite the object.

The thin lens equation applies equally well to a diverging lens (Figure 4).

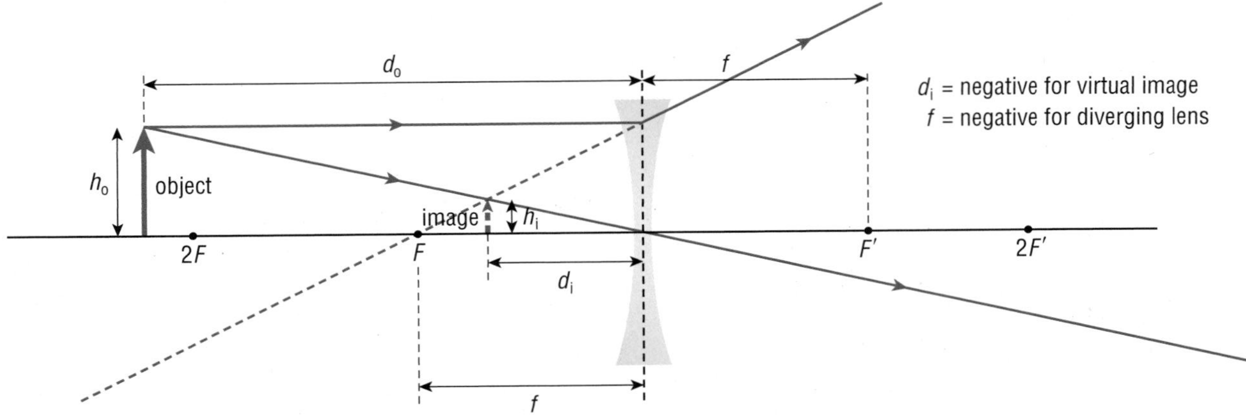

Figure 4 Lens equation variables for a diverging lens

SAMPLE PROBLEM 2 Using the Thin Lens Equation for a Diverging Lens

A diverging lens has a focal length of 29 cm. A virtual image of a marble is located 13 cm in front of the lens (Figure 5). Where is the marble located?

Given: $f = -29\text{ cm}$

$d_i = -13\text{ cm}$

Required: $d_o = ?$

Analysis and Solution:

$$\frac{1}{d_o} + \frac{1}{d_i} = \frac{1}{f}$$

$$\frac{1}{d_o} = \frac{1}{f} - \frac{1}{d_i}$$

$$\frac{1}{d_o} = \frac{1}{-29\text{ cm}} - \frac{1}{-13\text{ cm}}$$

$$\frac{1}{d_o} = 0.043\text{ cm}^{-1}$$

$$d_o \doteq 23\text{ cm}$$

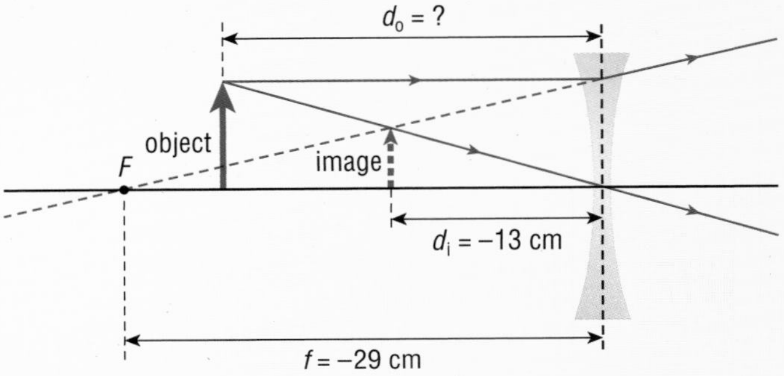

d_o = ?
F
object
image
d_i = −13 cm
f = −29 cm

Figure 5

Statement: The marble is located 23 cm from the lens, on the same side as the image.

LEARNING TIP

Does It Make Sense?

A problem is not completed once you have an answer. Stop for a moment and reflect on your answer. Does it make sense given the situation in the problem? Is the numerical value appropriate? Do you have the correct units?

The Magnification Equation

When you compare the size of the image with the size of the object, you are determining the magnification of the lens. To derive the thin lens equation, it is important to use the relationship $\frac{h_i}{h_o} = \frac{d_i}{d_o}$. This relationship is used to obtain the magnification equation.

> The magnification equation can be stated as $M = \frac{h_i}{h_o} = -\frac{d_i}{d_o}$.

The sign convention is the same as before, with two additions:

- Object (h_o) and image (h_i) heights are positive when measured upward from the principal axis and negative when measured downward.
- Magnification (M) is positive for an upright image and negative for an inverted image.

The magnification (M) is a dimensionless quantity because the units divide out.

Let's see how the magnification equation can be used to solve problems.

SAMPLE PROBLEM 3 Finding the Magnification of a Converging Lens

A toy of height 8.4 cm is balanced in front of a converging lens. An inverted, real image of height 23 cm is noticed on the other side of the lens (Figure 6). What is the magnification of the lens?

Given: $h_o = 8.4$ cm

$h_i = -23$ cm

Required: $M = ?$

Analysis and Solution: $M = \dfrac{h_i}{h_o}$

$M = \dfrac{-23 \text{ cm}}{8.4 \text{ cm}}$

$M \doteq -2.7$

Statement: The lens has a magnification of –2.7.

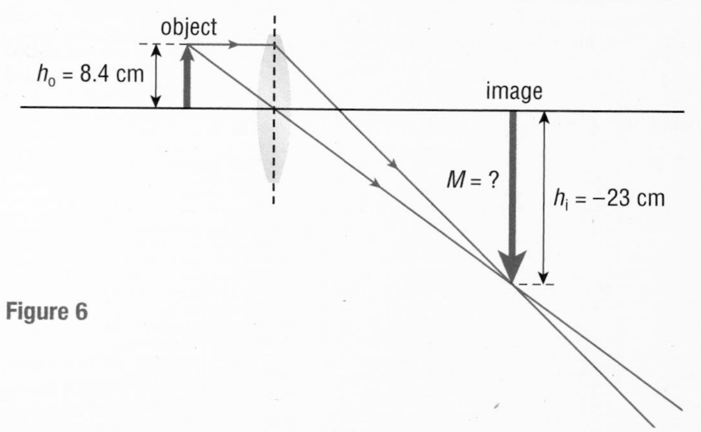

Figure 6

SAMPLE PROBLEM 4 Locating the Image

A small toy building block is placed 7.2 cm in front of a lens. An upright, virtual image of magnification 3.2 is noticed (Figure 7). Where is the image located?

Given: $d_o = 7.2$ cm

$M = 3.2$

Required: $d_i = ?$

Analysis and Solution: $M = -\dfrac{d_i}{d_o}$

$-Md_o = d_i$

$d_i = -Md_o$

$d_i = -(3.2)(7.2 \text{ cm})$

$d_i \doteq -23$ cm

Statement: The image of the toy block is located 23 cm from the lens, on the same side as the object.

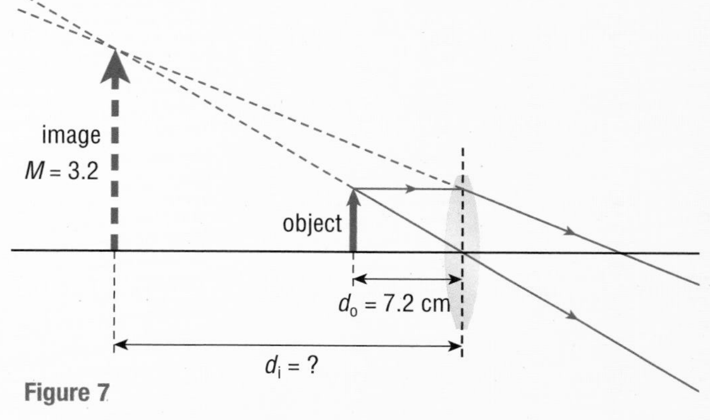

Figure 7

The magnification equation also applies to diverging lenses.

SAMPLE PROBLEM 5 Finding the Magnification of a Diverging Lens

A coin of height 2.4 cm is placed in front of a diverging lens. An upright, virtual image of height 1.7 cm is noticed on the same side of the lens as the coin (Figure 8). What is the magnification of the lens?

Given: $h_o = 2.4$ cm

$h_i = 1.7$ cm

Required: $M = ?$

Analysis and Solution: $M = \dfrac{h_i}{h_o}$

$M = \dfrac{1.7 \text{ cm}}{2.4 \text{ cm}}$

$M \doteq 0.71$

Statement: The lens has a magnification of 0.71.

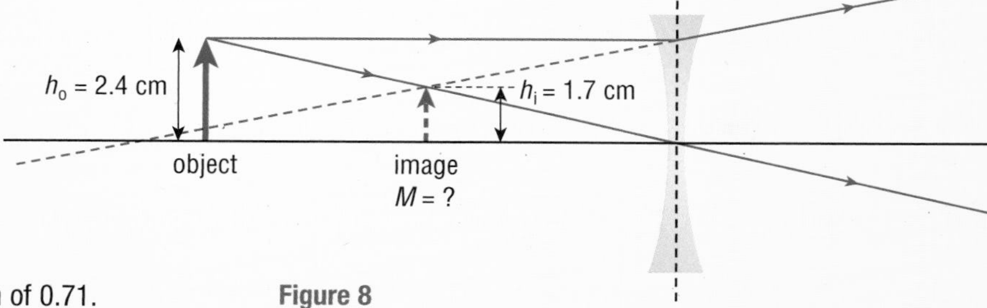

Figure 8

Table 1 summarizes the sign conventions for lenses.

Table 1 Sign Conventions for Lenses

Variable	Positive	Negative
(object distance) d_o	always	never
(image distance) d_i	real image (image is on opposite side of lens as object)	virtual image (image is on same side of lens as object)
(height of object) h_o	when measured upward	when measured downward
(height of image) h_i	when measured upward	when measured downward
(focal length) f	converging lens	diverging lens
(magnification) M	upright image	inverted image

IN SUMMARY

- Thin lens equation: $\frac{1}{d_o} + \frac{1}{d_i} = \frac{1}{f}$
- Magnification equation: $M = \frac{h_i}{h_o} = -\frac{d_i}{d_o}$

CHECK YOUR LEARNING

For each of these questions, draw a ray diagram to check your answer.

1. A converging lens has a focal length of 23 cm. A frog is 32 cm from the lens. Use the thin lens equation to calculate where the image of the frog will be located. T/I C
2. A pencil is located 53 cm from a diverging lens. An upright, virtual image of the pencil is observed 18 cm from the lens. Use the thin lens equation to calculate the focal length of this lens. T/I C
3. A diverging lens has a focal length of 34 cm. An upright, virtual image of a small booklet is located 13 cm behind the lens. Where is the booklet located? T/I C
4. A converging lens has a focal length of 16 cm. An insect is located 11 cm from the lens. Where will the image of the insect be located? T/I C
5. A vase of height 12 cm is placed in front of a converging lens. An inverted image of height 35 cm is noticed on the other side of the lens. T/I C
   (a) Use the magnification equation to calculate the magnification of the lens.
   (b) What type of image is it?
6. A playing card of height 14 cm is placed in front of a converging lens. An inverted, real image of height 7.9 cm is noticed on the other side of the lens. What is the magnification of the lens? T/I C
7. A postage stamp of height 2.8 cm is placed in front of a diverging lens. A virtual image of height 1.3 cm is noticed on the same side of the lens as the stamp. T/I C
   (a) What is the magnification of the lens?
   (b) What is the attitude of the image?
8. A small fork is placed 9.4 cm in front of a lens. An upright, virtual image of the fork with a magnification of 5.6 times is observed. K/U T/I C
   (a) Where is the image located?
   (b) What is the focal length for this lens?
   (c) What kind of lens is this? Explain.

Lens Applications

Lenses make use of the phenomenon of refraction. Optical devices that involve lenses have been used for centuries. These devices have benefitted humans greatly and have often led to great advances in knowledge. In this section, you will examine some of these optical devices.

The Camera

A converging lens produces an inverted, real image as long as the object is at a distance greater than F' (the secondary principal focus). The camera is a good example of a device that makes use of this fact.

A camera takes light from large, distant objects and forms smaller, real images on either film in a traditional camera or the sensor in a digital camera (Figure 1). This means that the object must be located at more than twice the focal length of the lens (that is, beyond $2F'$). As the object changes position, its image will change location. The location of the real image, however, will be somewhere between F and $2F$. You cannot move the film in a camera back and forth to create a sharp image. So to compensate for the fixed position of the film, you move the lens in and out. That way, a sharp image always falls on the film. This is what is called "focusing."

WRITING TIP

Writing a Critical Analysis

Imagine that you are writing a critical analysis of a camera that has a lens with a fixed focal length. You can conduct research on the advantages and disadvantages of various types of lenses: fixed focal length, zoom, interchangeable, and converter. Check for information that is missing or inconsistent. Ask questions such as, "What is the key difference between digital and optical zoom lenses?"

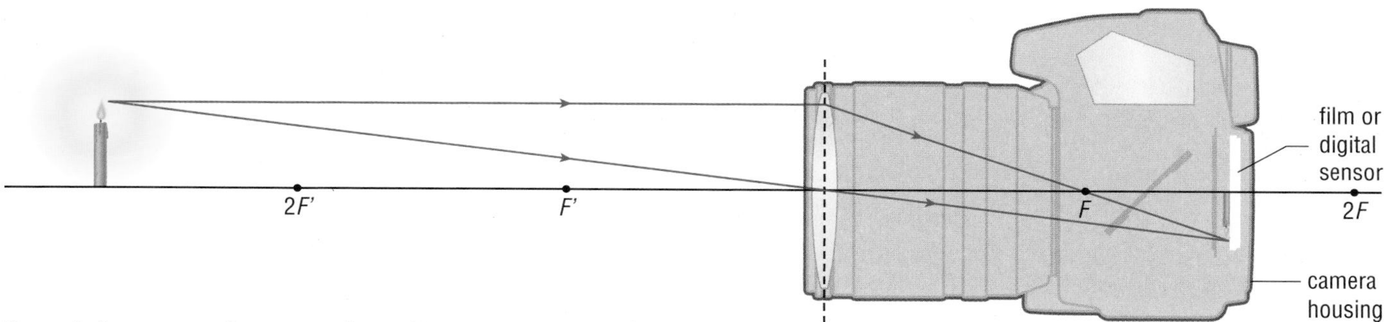

Figure 1 A camera produces a smaller, real image.

A traditional film camera uses flexible roll film to capture images. Flexible roll film was invented by George Eastman in 1884. Until that time, photographers had used glass plates coated with light-sensitive chemicals. George Eastman may not be a familiar name today, but the company that he founded to sell his products, Kodak, certainly is.

Today, digital cameras are far more popular than film cameras. Instead of film, a digital camera uses a light-sensitive device made of silicon called a charge-coupled device (CCD) (Figure 2). CCDs are at the heart of many other modern optical devices such as TV cameras, medical imaging systems, telescopes, and certain kinds of microscopes.

Figure 2 In a digital camera, a charge-coupled device (CCD) replaces the film found in a traditional camera.

The Movie Projector

A movie projector is, in a sense, the opposite of a camera. A projector takes a small object (the film) and projects a large, inverted, real image on a screen (Figure 3). Because the image is larger than the object, the film must be located between F' and $2F'$. Also, because the image is inverted, the film must be loaded into the projector upside down so that what you see on the movie screen is upright. An overhead projector works in a similar way.

DID YOU KNOW?

Size Matters

The IMAX film projection system was developed in Canada. IMAX film is 10 times larger than regular 35 mm film and feeds through the projector at three times the speed of other systems. The film canisters are so large and heavy that the film runs through the projector horizontally rather than vertically.

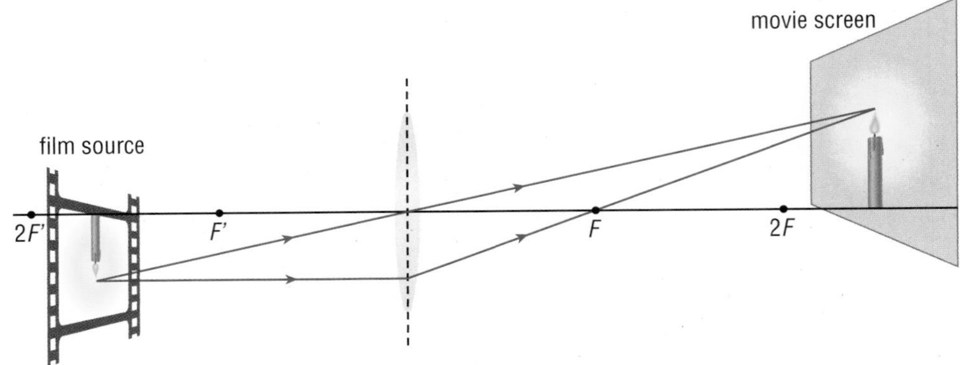

Figure 3 A projector produces a larger, inverted, real image.

Many theatres have upgraded to digital projection. The movies that these digital cinemas show are no longer stored on film; they are instead stored on DVDs or hard drives or are distributed to the theatre by direct satellite transmission.

The Magnifying Glass

Probably the simplest optical device is the magnifying glass. It is a simple converging lens in which the object is located between F' and the lens. No real image is produced at this object location. The refracted rays spread apart or diverge. However, the human brain extends these rays backwards and produces an enlarged, virtual image located on the same side of the lens as the object (Figure 4).

DID YOU KNOW?

The Oldest Magnifying Glass

The earliest recorded example of using a magnifying glass to produce a magnified image comes from the *Book of Optics* published by Ibn al-Haytham in 1021.

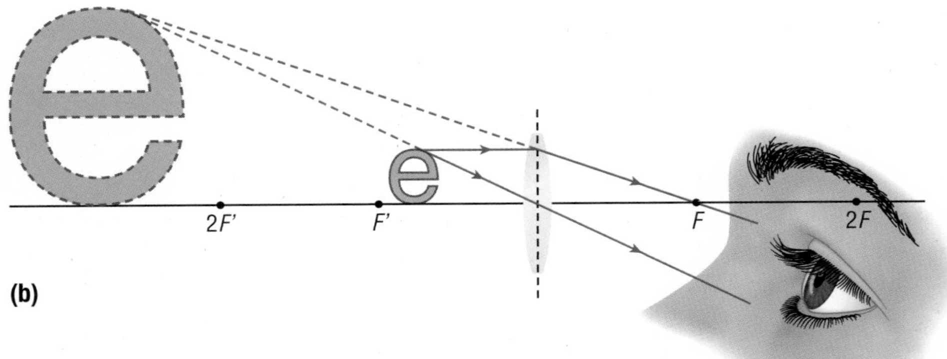

Figure 4 A magnifying glass is a converging lens that produces a larger, upright, virtual image on the same side of the lens as the object.

The magnifying glass is also called a simple microscope. Dutch scientist Antonie van Leeuwenhoek became famous for his observations using simple microscopes (Figure 5). Leeuwenhoek's simple microscopes, however, had one serious shortcoming. The converging lenses had very short focal lengths and so had to be held very close to the eye, which led to a lot of eye strain. The design of this simple microscope was improved and eventually led to the compound microscope.

Figure 5 Using a microscope like this, van Leeuwenhoek was the first person to observe bacteria and blood cells. The specimen was mounted on the tip of the specimen holder in front of the lens.

The Compound Microscope

The compound microscope is an arrangement of two converging lenses. It produces two enlarged, inverted images: one real and one virtual. The real image is formed by the objective lens. You do not see this image because it is in the body tube of the microscope, between the objective and the eyepiece lenses. The virtual image is formed by the eyepiece lens. This larger, virtual image is the image that you actually see (Figure 6).

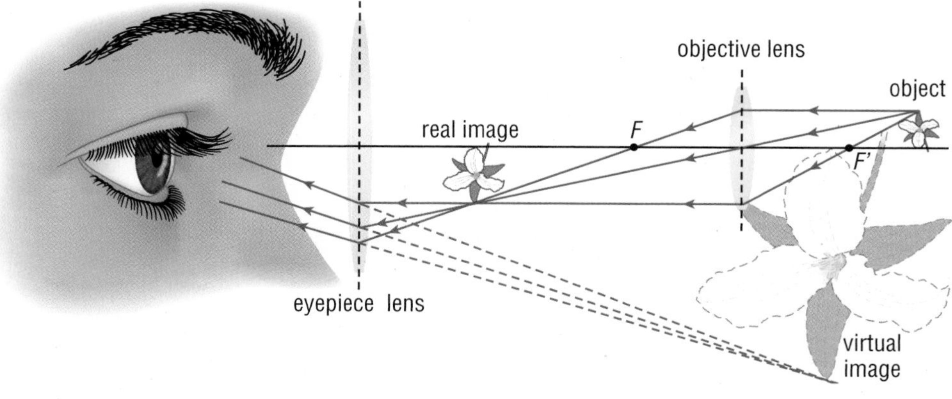

Figure 6 A compound microscope produces two images. You see only the larger, virtual image.

The Refracting Telescope

A refracting telescope operates on the same principle as a compound microscope. The difference is that the object is much farther away. In fact, the object in a refracting telescope is so far beyond $2F'$ that incident rays passing through the objective lens are considered to be parallel. A refracting telescope, like a compound microscope, produces two enlarged, inverted images: one real image that you do not see (inside the tube of the telescope) and one larger, virtual image that you do see (Figure 7).

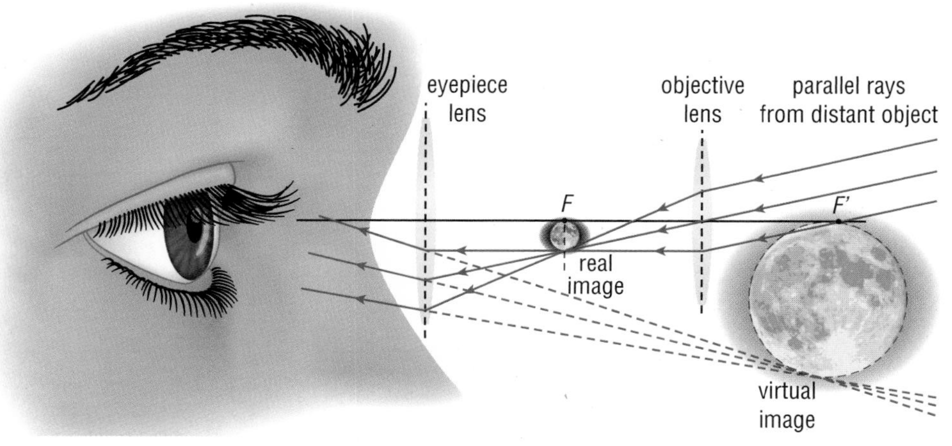

Figure 7 A refracting telescope also produces two images. You see only the larger, virtual image.

Figure 8 This 1.02 m refracting telescope at the Yerkes Observatory is the largest refracting telescope in the world.

Because of gravity, there is a size limit on refracting telescopes. If the objective lens is too large, it will start to sag slightly under its own weight, resulting in distorted images. The largest refractor in the world is the Yerkes telescope (Figure 8). Its objective lens has a diameter of 1.02 m.

The image in a refracting telescope is inverted, which is not a problem when viewing distant stellar objects. This can, however, be a problem when you view objects on Earth. For this application, you need to use a terrestrial telescope that has a third converging lens placed between the objective lens and the eyepiece. This extra lens corrects the inverted image so that you always see an upright image in the eyepiece.

UNIT TASK Bookmark

You can apply what you learned about applications of lenses in this section to the Unit Task described on page 588.

IN SUMMARY

- A camera uses a converging lens to produce a smaller, inverted, real image of a large object; the object is beyond $2F'$, and the real image is located between F and $2F$ in the camera body.
- A movie projector uses a converging lens to produce a larger, inverted, real image of a small object; the object (the film strip) lies between F' and $2F'$, and the image is located beyond $2F$.
- A magnifying glass, or simple microscope, is a converging lens in which the object is located between the lens and F'. A larger, upright, virtual image is formed on the same side of the lens as the object.
- A compound microscope consists of two converging lenses and produces a larger, inverted, virtual image. The object is located close to the objective lens.
- A refracting telescope consists of two converging lenses and produces a larger, inverted, virtual image. The object is so far away from the objective lens that incident rays that pass through the lens are essentially parallel.

CHECK YOUR LEARNING

1. What is the purpose of focusing in a camera? K/U
2. Briefly describe three technologies that have been used to capture images in cameras. K/U
3. (a) Why would the film strip in a movie projector never be located beyond $2F'$?
   (b) Where is the film strip actually located? Why? T/I
4. Why is the image that you see on the screen in a movie theatre not inverted? K/U
5. Why must the object in a magnifying glass be located between F' and the lens? K/U
6. Why can you not see a real image in either a compound microscope or a refracting telescope? K/U
7. Use a t-chart to summarize the similarities and differences between a compound microscope and a refracting telescope. K/U C
8. (a) Why is the refracting telescope not suitable for viewing objects on Earth?
   (b) How has this problem been overcome in the design of the terrestrial telescope? K/U

The Einstein Ring

One of Albert Einstein's predictions was that mass bends space. A large mass, such as a galaxy, should distort the space around it so much that it changes the path of light from galaxies behind it. The large mass of the galaxy would, in effect, act like a large converging lens. The bending of light by a large mass is called gravitational lensing (Figure 1).

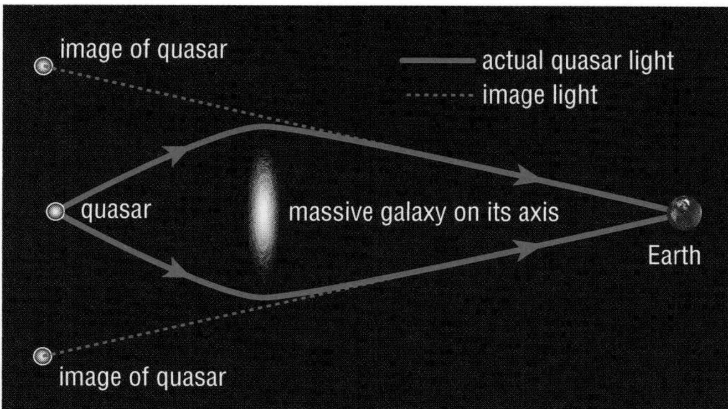

Figure 1 The massive galaxy in front acts like a gravitational lens and bends light from the distant galaxy (quasar) behind it.

For a long time, scientists did not know whether gravitational lenses actually existed. Recent advances in imaging technology have now provided conclusive evidence that gravitational lenses do exist. The large smears and arcs of light in Figure 2 are examples of gravitational lensing. The dense cluster of galaxies in the foreground causes the smearing of light. The effect is identical to the smearing of a light source caused by an off-centre, irregularly shaped converging lens.

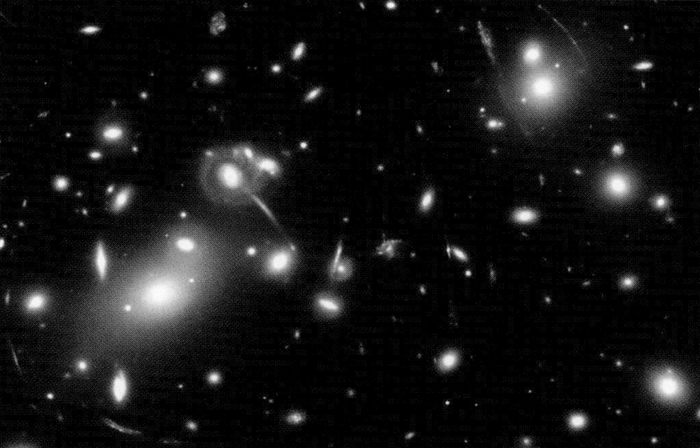

Figure 2 The massive galaxy cluster in the foreground acts like a converging lens. The result is that light from galaxies behind it is bent into arcs.

If the central gravitational lens is symmetrical, and if it is directly in front of another galaxy, then it acts like a perfect converging lens. A complete circle of light from the distant galaxy is visible instead of a smeared arc. This is called an Einstein ring (Figure 3). Several have now been observed.

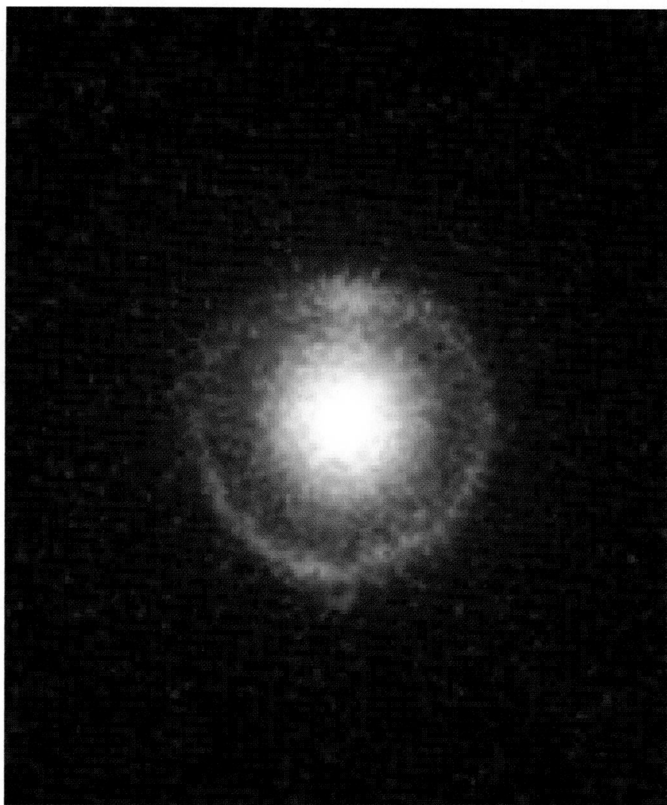

Figure 3 An Einstein ring photographed by the Hubble space telescope

The universe is an astonishing and amazing place. It is just as astonishing that concepts from optics can lead to an understanding of some of these unusual physical phenomena!

GO TO NELSON SCIENCE

13.6

The Human Eye

The human eye is the optical instrument that helps most of us learn about the external world; it is what you are using to read this sentence. It is a remarkable apparatus that acts as our window on the universe.

Parts of the Human Eye

"I am a camera with its shutter open, quite passive, recording, not thinking." This is the opening line from Christopher Isherwood's 1939 novel *Goodbye to Berlin*. Isherwood's opening line compares the eye to a camera as a recorder of events. Isherwood used the eye as an analogy, and the example of the eye as a camera is a good comparison.

The human eye is an amazing optical device that in many ways is very similar to a camera. A camera has a diaphragm that controls the amount of light entering it. The diaphragm on a compound microscope has exactly the same function. The iris in the eye has this function. The iris is the coloured part of the eye, and it opens and closes around a central hole to let in more or less light. The hole in the iris is called the pupil, comparable to the aperture in a camera, and is where light enters the eye. A camera has a converging lens to refract light to form a sharp image. The eye also has structures (the lens and cornea) that cause light to converge. The cornea is the transparent bulge on top of the pupil that focuses light. Light is refracted more through the cornea than through the lens.

In a camera, the image is focused on film or on a digital sensor. Light-sensitive cells in the retina at the back of each eye cavity accomplish the same task. The retina converts the light signal into an electrical signal that is transmitted to the brain through the optic nerve (Figure 1). The optic nerve creates a blind spot at the back of each eye because there are no light-sensitive cells in this small area. You do not notice the blind spot in normal vision because each eye compensates for the blind spot of the other eye. That is, your left eye can see what is in the blind spot of your right eye and vice versa.

DID YOU KNOW?

Retina Cells

The retina contains basically two kinds of light-receptive cells: rods and cones. These cells are identified by their basic shape. The cone cells allow you to see colour. The rod cells are much more sensitive to light and are predominantly for black and white vision.

WRITING TIP

Writing a Critical Analysis

In a critical analysis of the human eye, you might comment on similarities between its parts and those of a camera: pupil/aperture, iris/diaphragm, retina/digital sensor. You might form an opinion that the development of the camera as an optical instrument was based on the parts of the human eye and how they function. You can do further research to find out whether anyone else agrees with your opinion.

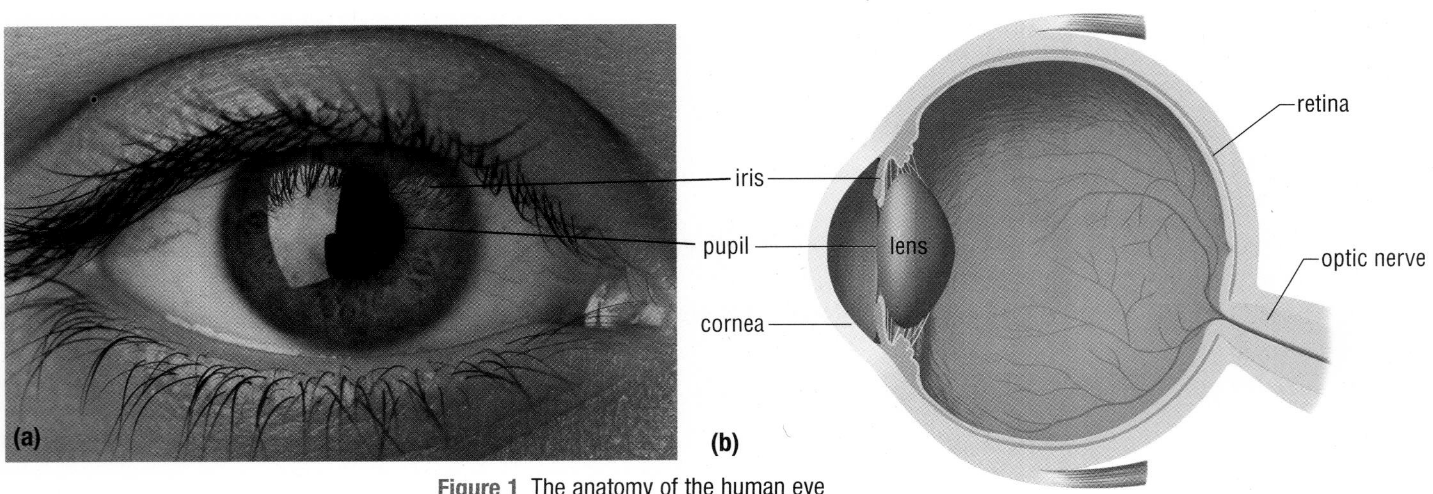

Figure 1 The anatomy of the human eye

TRY THIS DISCOVERING YOUR BLIND SPOT

SKILLS: Performing, Observing, Communicating

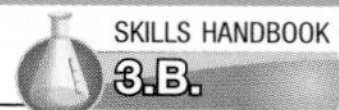

Equipment and Materials: pencil; blank sheet of paper

1. Place a dot with a diameter of 2–3 cm on the sheet of paper. Place an "X" of similar size so that it is 6 cm to the right of the dot.
2. Hold the paper at arm's length in your right hand. Close your right eye and look at the "X" with your left eye. You should also be able to see the small dot out of the corner of your eye. This is called "peripheral vision."
3. Keep your eye focused on the "X" and slowly bring the paper straight toward you. At a certain point, the dot should disappear.
4. Keep moving the paper until the dot reappears.

A. Why did the dot disappear? T/I

B. Why did the dot reappear when you continued moving the paper? T/I

C. Why do you not normally notice this "hole" in your vision? T/I

Most people think that they see with their eyes. In reality, the eye acts as a light gathering instrument. We actually "see" with our brain. The cornea–lens combination of the eye acts like a converging lens and produces a smaller, real, inverted image on the retina (Figure 2). Electrical impulses from the retina travel through the optic nerve to the brain where we "see" the image. The brain takes the inverted image from the retina and flips it so that the image we "see" appears upright.

DID YOU KNOW?

Eye See
The average person blinks about 15 000 times a day.

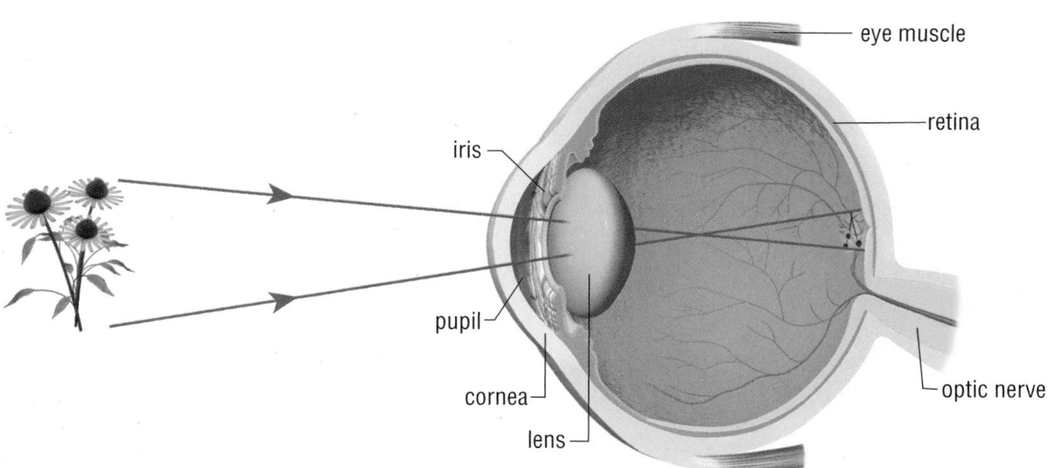

Figure 2 The eye acts like a converging lens and produces a smaller, inverted, real image on the retina.

Eye Accommodation

A camera focuses by moving the lens in and out because the plane of the film or digital sensor is fixed. The human eye, however, cannot move the lens in and out like a camera. Eyes have evolved a different way of producing a clear image.

In humans, eye muscles called ciliary muscles help the eye focus on distant and nearby objects by slightly changing the shape of the eye lens. This change in shape of the eye lens changes the focal length of the lens to allow focusing of the image on the retina. This process is called **accommodation**. A healthy eye can accommodate itself to view distant and nearby objects (Figure 3).

accommodation the changing of shape of the eye lens by eye muscles to allow a sharply focused image to form on the retina

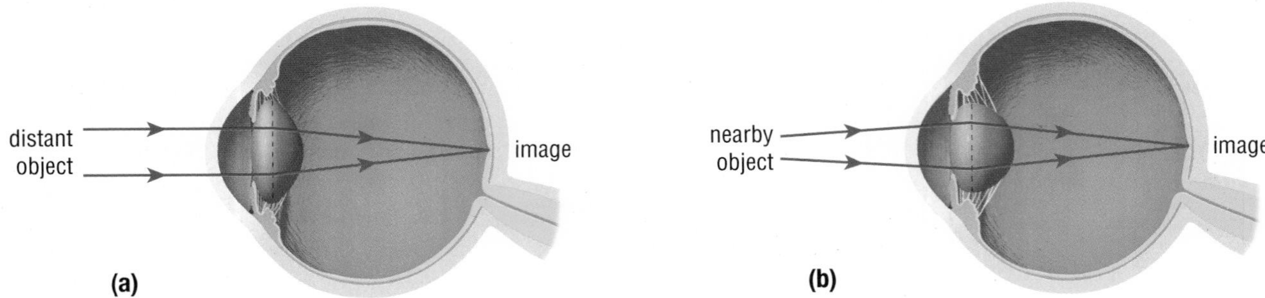

Figure 3 A healthy eye can focus light from both distant objects (a) and nearby objects (b) on the retina. Notice that the lens is slightly fatter when focused on nearby objects.

Focusing Problems

For some people, the process of accommodation does not work as well as it should. These people's eyes cannot focus on objects at every distance. This can result in blurred vision. The difficulty might be with focusing on nearby objects or on distant objects.

Hyperopia (Far-sightedness)

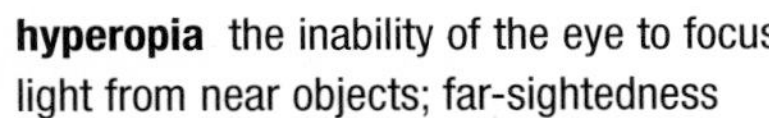

hyperopia the inability of the eye to focus light from near objects; far-sightedness

A person who has **hyperopia** is far-sighted. This means that the person has no difficulty seeing distant objects. Seeing nearby objects is a problem, however, because the eye cannot refract light well enough to form an image on the retina. Far-sightedness usually occurs because the distance between the lens and the retina is too small or because the cornea–lens combination is too weak. Instead, light from all nearby objects focuses *behind* the retina (Figure 4).

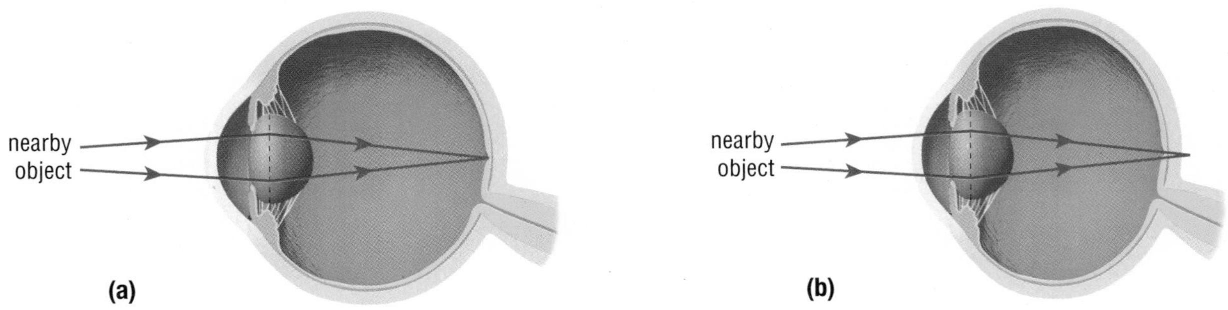

Figure 4 (a) A normal, healthy eye focuses light from a nearby object onto the retina. (b) A far-sighted eye focuses light from a nearby object behind the retina.

The far-sighted eye needs help in refracting light. A converging lens will do the trick. The actual lens shape is modified from the basic converging lens shape and is called a **positive meniscus**. A positive meniscus lens is a converging lens because the middle part of the lens is still thicker than the edge (Figure 5). A positive meniscus is much more cosmetically appealing than the thick shape of a basic converging lens.

positive meniscus a modified form of the converging lens shape

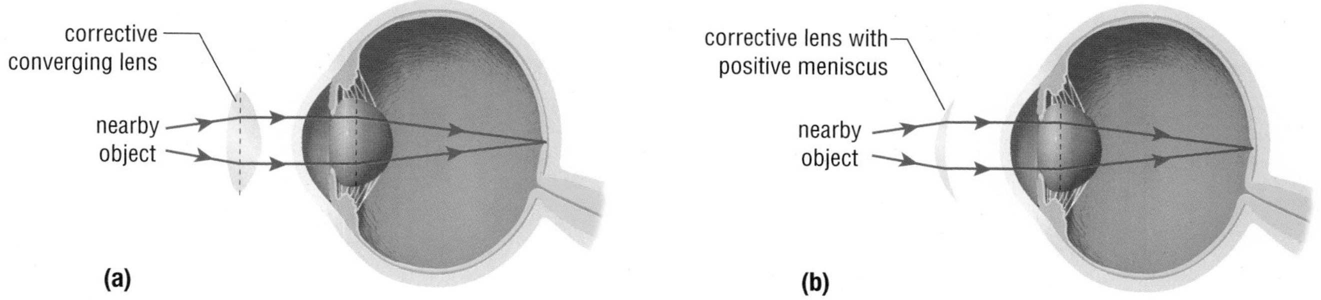

Figure 5 (a) A corrective converging lens will correct far-sightedness. (b) A corrective lens with a positive meniscus has the same effect because it, too, is thickest in the middle.

Presbyopia

Many people find it harder to read small print as they get older. The reason is that the eye lens loses its elasticity. This loss of accommodation results in a form of far-sightedness called **presbyopia**. Presbyopia is an age-related vision condition and, unlike hyperopia, is not a result of the eyeball being too short for focusing. Presbyopia can also be corrected by glasses with converging lenses.

presbyopia a form of far-sightedness caused by a loss of accommodation as a person ages

Myopia (Near-sightedness)

A person who has **myopia** is near-sighted. This means that the eye can focus light rays from nearby objects on the retina; this person can see close up quite clearly. Distant objects, however, are a problem. Myopia usually occurs because the distance between the lens and the retina is too large or because the cornea–lens combination converges light too strongly. In the near-sighted eye, light from distant objects is brought to a focus *in front* of the retina (Figure 6).

myopia the inability of the eye to focus light from distant objects; near-sightedness

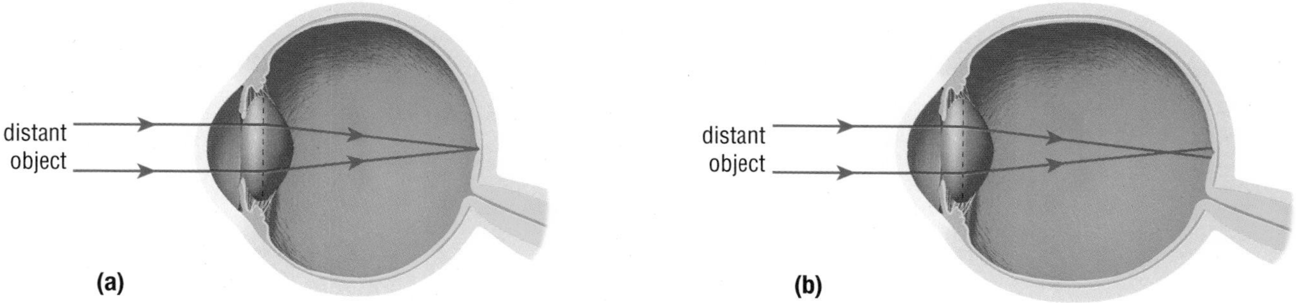

Figure 6 (a) A normal, healthy eye focuses light from a distant object onto the retina. (b) A near-sighted eye focuses light from a distant object in front of the retina.

DID YOU **KNOW?**

A Common Sight

Myopia is a common vision condition that affects nearly 30 % of Canadians.

negative meniscus a modified form of the diverging lens shape

The near-sighted eye can focus the image if incoming light rays diverge a little. A diverging lens achieves this. The actual lens shape is modified from the basic diverging lens shape and is called a **negative meniscus**. A negative meniscus lens is a diverging lens because the edge of the lens is still thicker than the middle part (Figure 7). A negative meniscus is, again, much more cosmetically appealing than the thick edge of a basic diverging lens.

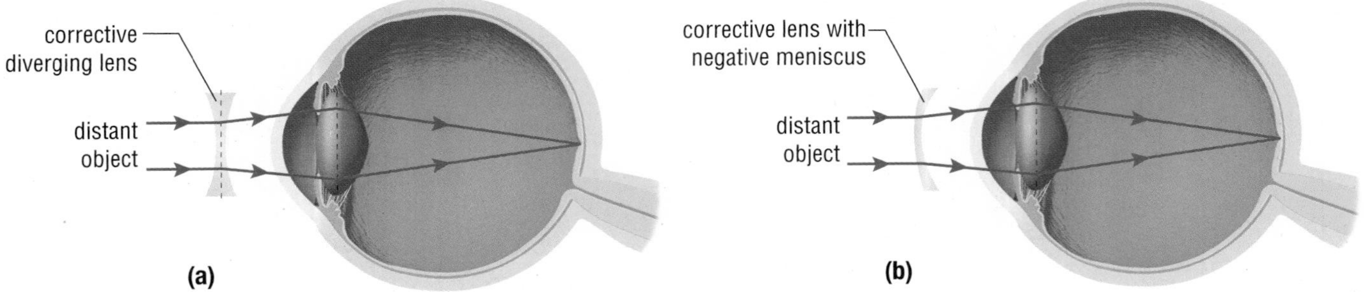

Figure 7 (a) A corrective diverging lens will correct near-sightedness. (b) A corrective lens with a negative meniscus has the same effect because it, too, is thinnest in the middle.

Contact Lenses

contact lens a lens that is placed directly on the cornea of the eye

A **contact lens** is a lens that is placed directly on the cornea of the eye. Contact lenses serve the same purpose as glasses. A contact lens can be shaped so that it can be used for correcting far-sightedness or near-sightedness (Figure 8). A contact lens is usually invisible when placed on the cornea.

Contact lenses can also be used for strictly cosmetic purposes when they are used to change the colour of the eye. Movie makeup artists make use of contact lenses in this way to transform actors into zombies or demons (Figure 9).

WRITING **TIP**

Writing a Critical Analysis

In a critical analysis of contact lenses, you might focus on the problems that some people experience. These problems could include dry eyes, irritation, and blurred vision. You might conclude by suggesting possible solutions to these problems such as better contact lens care, nocturnal contact lenses, permanent implantable contact lenses, or laser eye surgery.

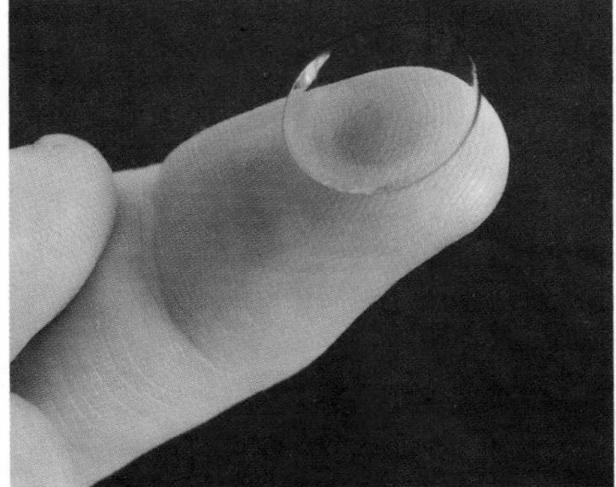

Figure 8 A contact lens can be used to improve vision.

Figure 9 A contact lens creates this "cat's eye" effect.

RESEARCH THIS OTHER VISION PROBLEMS

SKILLS: Researching, Communicating

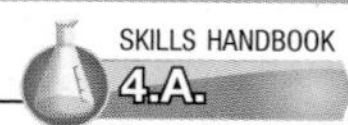

The human eye can suffer from vision problems for many other reasons than the ones mentioned here. Just as you have regular health check-ups with your family doctor, it is also important to take care of the health of your eyes. Regular visits to an optometrist will alert you to eye problems.

1. Research the cause, development, and treatment of each of these common eye problems.
   - astigmatism
   - glaucoma
   - cataracts

GO TO NELSON SCIENCE

A. What causes astigmatism? How is it treated? T/I

B. How does an optometrist check if someone has glaucoma? T/I

C. What factors contribute to the formation of cataracts? T/I

D. Prepare a visual presentation that summarizes these vision problems, their causes, and how they are treated. T/I C

UNIT TASK Bookmark

In this section, you learned about vision and vision problems. How can you apply your knowledge to the Unit Task described on page 588?

IN SUMMARY

- The cornea–lens combination in the eye acts like a converging lens; the brain flips the inverted image that it receives from the eye so that what you see is upright.
- The eye focuses through accommodation; the shape of the eye lens is changed slightly by eye muscles.
- Hyperopia means that a person is far-sighted; near vision is corrected with a converging lens.
- Presbyopia is an age-related condition of far-sightedness that is caused by a loss of accommodation.
- Myopia refers to a person who is near-sighted; distant vision is corrected with a diverging lens.

CHECK YOUR LEARNING

1. Describe at least three similarities between a camera and the human eye. A
2. The text states that we actually "see with our brain." What is meant by this? K/U
3. (a) What is the difference between far-sightedness and near-sightedness?
   (b) What simple lens shape would correct each of these problems? K/U
4. The actual lens shapes to correct the two vision problems of far-sightedness and near-sightedness have been modified. K/U A
   (a) What are these new lens shapes called? Draw an example of each new shape.
   (b) Why have they been changed from the basic lens shapes?
5. (a) People often require reading glasses as they get older. What vision problem do these people usually have, and what causes it?
   (b) Which corrective lens shape corrects this problem, a positive meniscus or a negative meniscus? Explain. K/U A
6. The introduction to this chapter described how glasses were used to start a fire in *Lord of the Flies.* Would eyeglasses to correct near-sightedness be able to do this? Explain, with the aid of a diagram. K/U C A

Laser Eye Surgery

SKILLS MENU
- Defining the Issue
- Researching
- Identifying Alternatives
- Analyzing the Issue
- Defending a Decision
- Communicating
- Evaluating

Laser eye surgery is prominently advertised as providing an alternative to glasses and contact lenses. Many laser surgery clinics have opened throughout the country (Figure 1). Laser eye surgery involves using a laser to reshape the cornea of the eye in order to improve vision. It is a technique that is becoming more widely used.

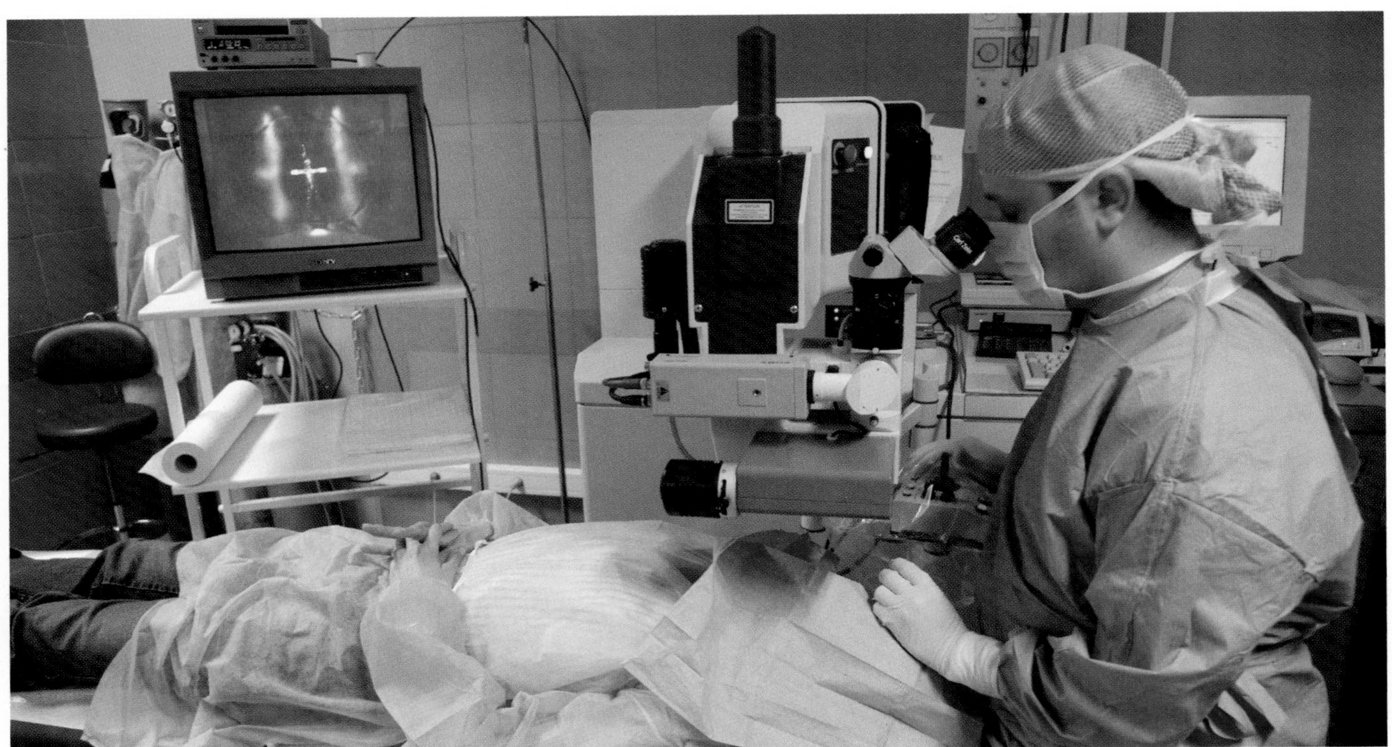

Figure 1 Laser eye surgery is now so routine that it is performed at clinics in shopping malls.

The Issue

Your best friend suffers from myopia (near-sightedness). Your friend presently wears glasses to correct this vision problem. Every day on the bus to school, you and your friend see ads from laser eye surgery clinics. They promise to provide you "Freedom from glasses and contacts. Become the person you always wanted to be!" Your friend eventually confides in you. She is wondering if laser surgery is right for her.

Goal

Your friend has asked you for advice about laser eye surgery. You need to become informed about the issues relating to laser surgery. You will need to come to a decision about whether or not you think laser surgery is appropriate for your friend.

Gather Information

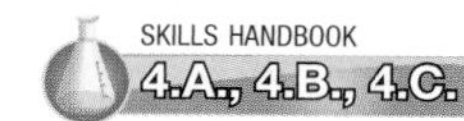

Work in pairs or in small groups to learn about the following issues associated with laser eye surgery.

- What main procedures are used?
- How long does it take, and how expensive is it?
- How long does recovery take?
- What is the success rate?
- How safe is it?
- What are some possible drawbacks to laser surgery?
- How big a factor is age and occupation in your decision?

Consider how to find reliable, unbiased information. Think about the purpose of commercial websites, advertisements, government publications, and medical reports. Which sources are likely to be most reliable?

Identify Solutions

Think about these questions as you collect and analyze information. T/I A

- What are the potential positive benefits of laser eye surgery?
- Does your friend meet the criteria for a candidate for laser eye surgery?
- What negative concerns does your friend need to be aware of?
- How big a factor should the cosmetic/appearance issue be?
- What are the alternatives to laser surgery for your friend?
- Should your friend wait several more years before making a decision?

Make a Decision

Make a clear recommendation to your friend about whether she should pursue laser surgery. Clearly state the criteria you used in arriving at your decision. T/I

Communicate

Prepare a consumer news report that clearly identifies all the concerns, issues, and possible solutions related to laser eye surgery. The report could be either an article that would appear in a popular science magazine or a Health Unit pamphlet. Your report should indicate the factors you used to analyze the issue and should make use of graphic organizers. Your report should end with a clear recommendation and how you arrived at this recommendation. T/I C A

GO TO NELSON SCIENCE

CHAPTER 13 LOOKING BACK

KEY CONCEPTS SUMMARY

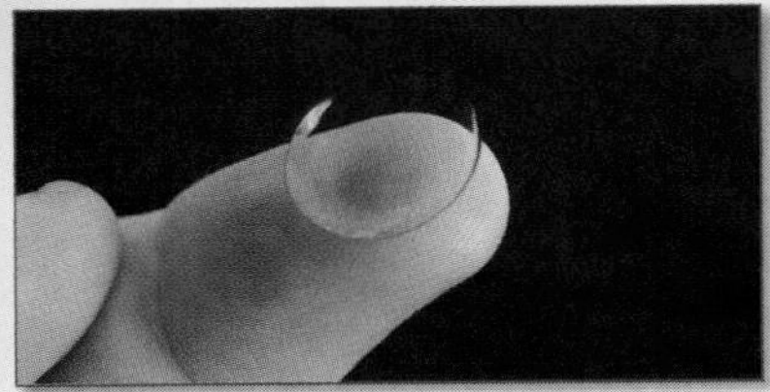

A lens is a transparent object used to change the path of light.

- Light is able to pass through a lens but is refracted (bent) by the lens. (13.1)
- The shape and the material of the lens affect how light is refracted. (13.1)
- Lenses with specific characteristics are created for use in optical devices. (13.5, 13.6)

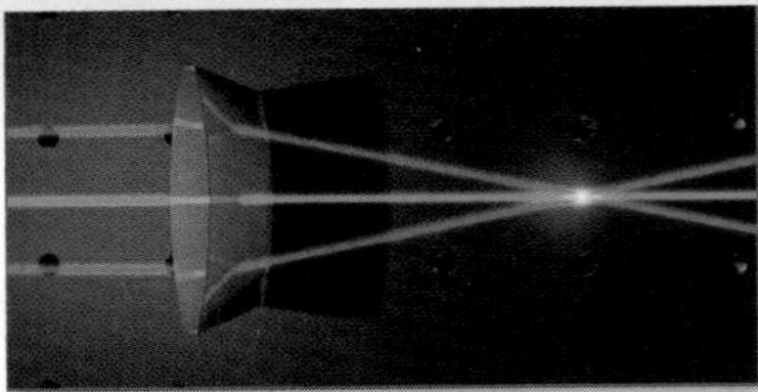

Parallel light rays are refracted through a focus when they pass through a converging lens.

- A converging lens brings parallel light rays together through a focus after refraction. (13.1)
- A diverging lens spreads light rays apart after refraction so that it looks as if they have come from a virtual focus. (13.1)
- The principal focus of a *converging lens* is on the *opposite* side of the lens as the incoming light rays. (13.1)
- The principal focus of a *diverging lens* is on the *same* side of the lens as the incoming light rays. (13.1)

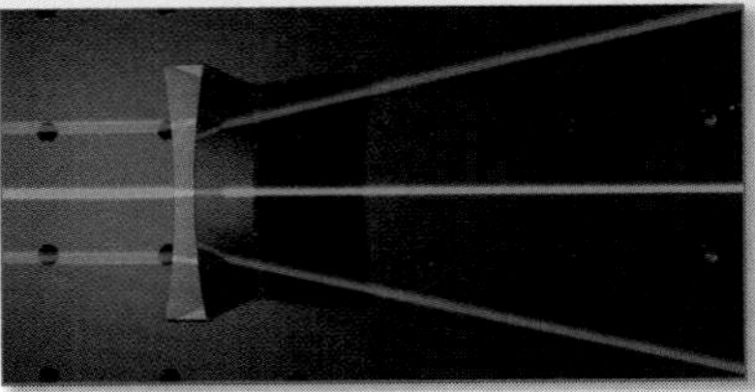

Geometric optics can be used to determine the path of light rays through lenses.

- A converging lens produces an inverted, real image when the object is beyond F'. The image is smaller if the object is beyond $2F'$, the same size at $2F'$, and larger if the object is between $2F'$ and F'. (13.3)
- A converging lens produces a virtual image when the object is between F' and the lens. The image is always larger, upright, and on the same side of the lens as the object. (13.3)
- A diverging lens always produces a smaller, upright, virtual image that is on the same side of the lens as the object. (13.3)

Both ray diagrams and algebraic equations can be used to determine the characteristics of an image in a lens.

- Thin lens equation:

$$\frac{1}{d_o} + \frac{1}{d_i} = \frac{1}{f} \quad (13.4)$$

- Magnification equation:

$$M = \frac{h_i}{h_o} = -\frac{d_i}{d_o} \quad (13.4)$$

Lenses have many technological uses that benefit humans.

- A camera produces a smaller, inverted, real image of a large, distant object. A film projector produces a larger, inverted, real image of a small, nearby object. (13.5)
- A compound microscope and a refracting telescope both contain two converging lenses and both produce a larger, inverted, virtual image. (13.5)

The eye can be treated as a lens, and vision problems can be corrected with other lenses.

- The cornea–lens combination in the eye acts like a converging lens. Eye muscles change the shape of the eye lens to allow you to focus on objects that are distant and nearby. (13.6)
- Converging lenses help a far-sighted person see nearby objects. (13.6)
- Diverging lenses help a near-sighted person to see distant objects. (13.6)

WHAT DO YOU THINK NOW?

You thought about the following statements at the beginning of the chapter. You may have encountered these ideas in school, at home, or in the world around you. Consider them again and decide whether you agree or disagree with each one.

1 A camera lens produces images that are upside down and flipped horizontally.
Agree/disagree?

2 If you are near-sighted, then thick lenses will help you see better.
Agree/disagree?

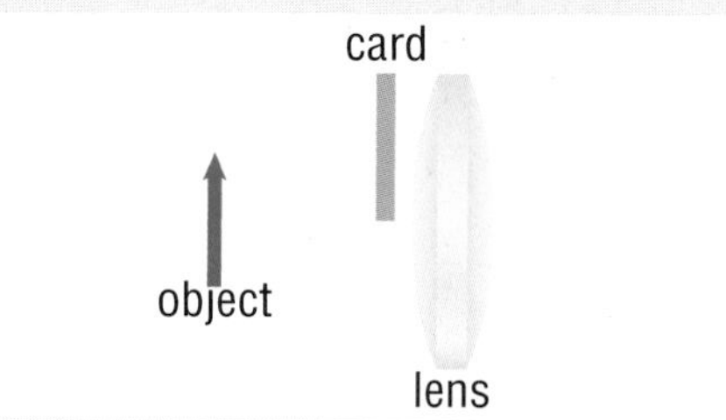

3 If you cover up half of a lens, then the image that you will see through the lens will be only half of the object.
Agree/disagree?

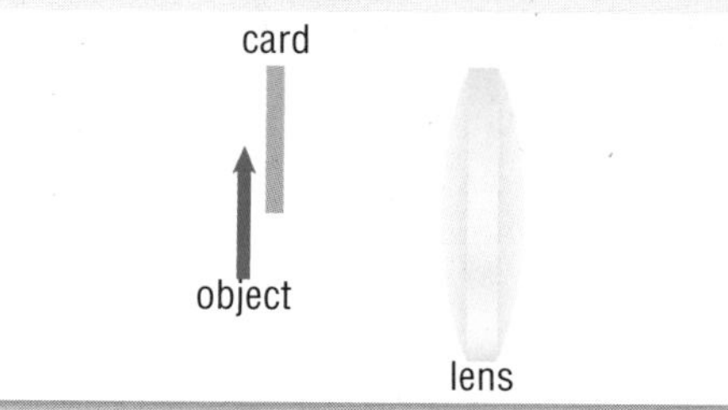

4 If you cover up half of an object, then the image that you will see through a lens will be only half of the object.
Agree/disagree?

5 A magnifying glass is very similar to a microscope in how it forms an image.
Agree/disagree?

6 Many people require glasses for reading as they age because the ability of the eye to focus decreases with age.
Agree/disagree?

How have your answers changed?
What new understanding do you have?

Vocabulary

converging lens (p. 551)
diverging lens (p. 551)
optical centre (p. 552)
principal focus (p. 552)
emergent ray (p. 556)
thin lens equation (p. 562)
accommodation (p. 574)
hyperopia (p. 574)
positive meniscus (p. 575)
presbyopia (p. 575)
myopia (p. 575)
negative meniscus (p. 576)
contact lens (p. 576)

BIG Ideas

- ✓ **Light has characteristics and properties that can be manipulated with mirrors and lenses for a range of uses.**
- ✓ **Society has benefitted from the development of a range of optical devices and technologies.**

The following icons indicate the Achievement Chart category addressed by each question.

K/U Knowledge/Understanding T/I Thinking/Investigation C Communication A Application

What Do You Remember?

1. What type(s) of image does a converging lens produce? (13.3) K/U
2. What type(s) of image does a diverging lens produce? (13.3) K/U
3. Name the centre of a lens. (13.1) K/U
4. List at least three lens applications that produce a virtual image. (13.5) K/U
5. What type of image does a compound microscope produce? If more than one image is produced, state the type of each. (13.5) K/U
6. State the location of an object when viewed through a refracting telescope. (13.5) K/U
7. Which eye condition is treated with a negative meniscus lens? (13.6) K/U
8. What do your eye muscles adjust for when they cause the lens to change shape slightly? (13.6) K/U

What Do You Understand?

9. A converging lens has a focal length of 17 cm. Describe the characteristics of the image of a candle that is placed at these distances from the lens: (13.3, 13.4) T/I
   (a) 34 cm
   (b) 52 cm
   (c) 17 cm
   (d) 25 cm
   (e) 12 cm
10. Why will a diverging lens never produce a real image? (13.3) K/U
11. What is the relationship between the type (virtual or real) and attitude (inverted or upright) of the image for:
    (a) a converging lens?
    (b) a diverging lens? (13.3) K/U
12. What are the characteristics of the image formed on the retina of the eye? Explain. (13.6) K/U
13. (a) Copy Figure 1 into your notebook. Add light rays to locate the image of each object.
    (b) Describe the characteristics of each image. (13.3) T/I C

(i)

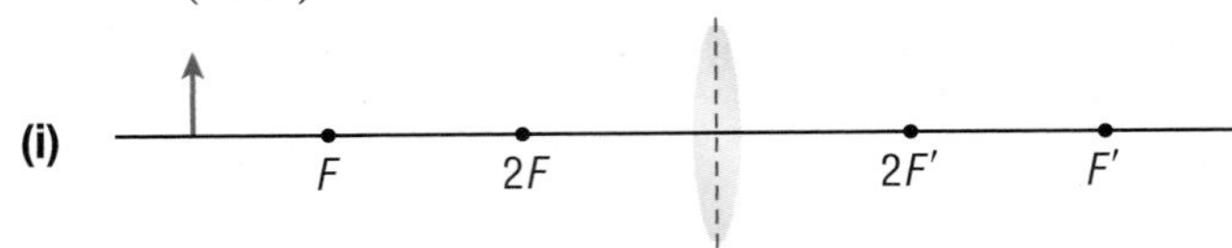

(ii)

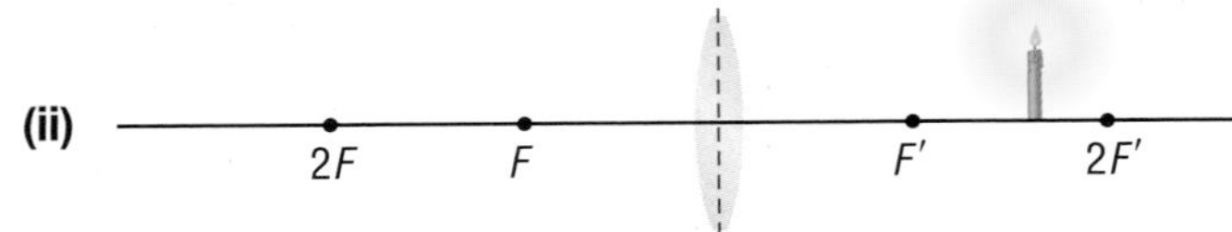

(iii)

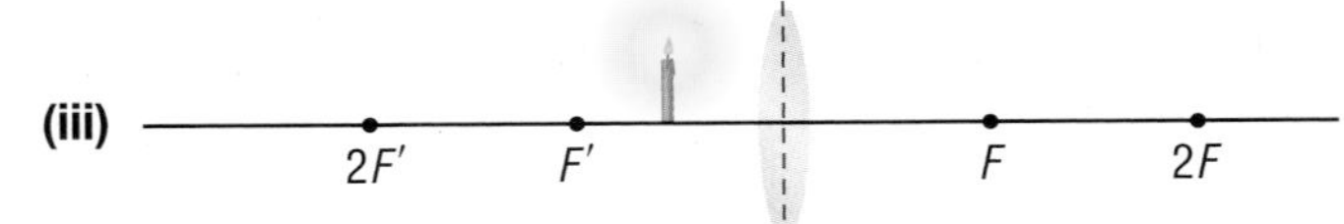

(iv)

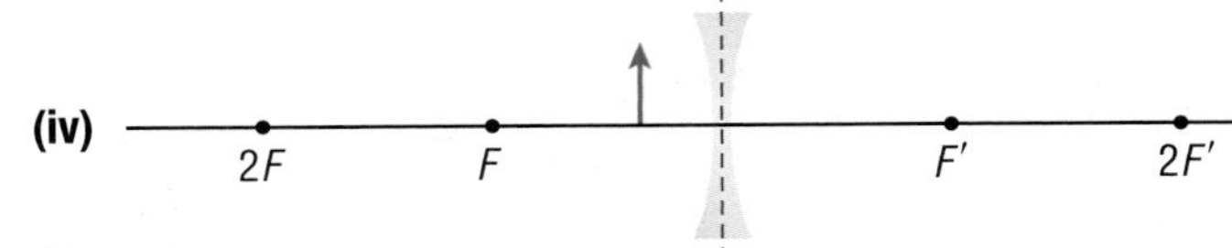

Figure 1

14. Distinguish between hyperopia and presbyopia. (13.6) K/U
15. Using diagrams, explain how
    (a) a diverging lens corrects near-sightedness.
    (b) a converging lens corrects far-sightedness. (13.6) K/U C
16. Distinguish between astigmatism, glaucoma, and cataracts. (13.6) K/U

Solve A Problem

17. A converging lens has a focal length of 21 cm. A candle is located 57 cm from the lens. (13.3, 13.4) T/I C
    (a) What type of image will be formed?
    (b) Where will the image be located?
18. A magnifying glass has a focal length of 18 cm. A leaf is located 13 cm from the lens. Where will the image of the leaf be located? (13.3, 13.4) T/I C

19. A rose of height 14 cm is placed in front of a converging lens. An inverted, real image of height 43 cm is noticed on the other side of the lens. What is the magnification of the lens? (13.3, 13.4) T/I C

20. A model ship of height 18 cm is placed in front of a diverging lens. A smaller, virtual image of height 12 cm is noticed. (13.3, 13.4) K/U T/I C
    (a) What is the magnification of the lens?
    (b) On what side of the lens would this virtual image be located?
    (c) Would it ever be possible for this lens to produce an image that is larger than the object? Explain.

21. A camera is focused to take a photo of a bee at a flower. The film in the camera is 11 cm from the centre of a lens. The bee is 53 cm from the camera. What is the focal length of the camera lens? (13.3, 13.4) T/I C

22. An upright object is placed 17 cm in front of a lens. This is less than the focal length of the lens. An upright, virtual image is observed that has a magnification of 2.8 times the object. (13.3, 13.4) T/I C
    (a) What kind of lens must this be? Explain.
    (b) Where is the image located?

23. Figure 2 shows the location of the *image*. Copy the diagram into your notebook and add light rays to locate the object. (13.3) T/I C

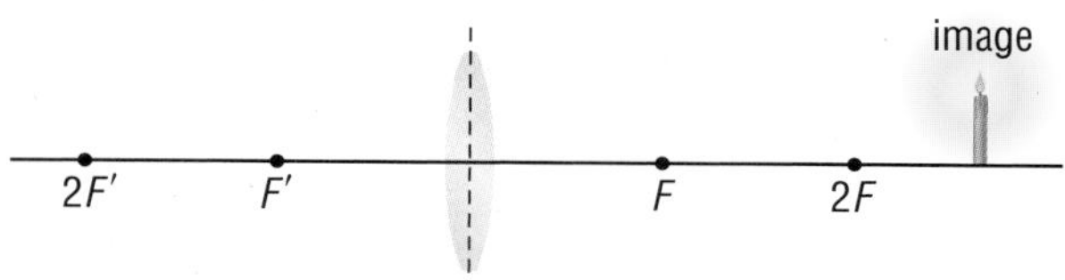

Figure 2

24. A student previously had to sit at the front of the class to clearly see writing on the board. The student now wears glasses and can read the board quite comfortably from the back of the classroom. (13.6) K/U C A
    (a) What kind of vision problem does this student have?
    (b) Do the student's eyeglasses have lenses that have a positive meniscus or a negative meniscus? Explain.

Create and Evaluate

25. Compare the similarities between a camera and the human eye. Contrast the differences between them. (13.5, 13.6) A

26. Assume for the moment that an invisibility cloak is actually possible. What kind of vision would somebody under this cloak experience? Explain by making reference to how the eye forms images. (13.6, 12.6) A

Reflect On Your Learning

27. In this chapter, you learned that lenses have many practical applications in our society. Now imagine that all devices using lenses just disappeared. How would this affect your life?

Web Connections 

28. Research the major developments in the invention of eyeglasses. Present your findings as a timeline. (13.6) T/I C

29. Find out how night vision goggles work. Draw a diagram to illustrate your discovery. (13.6) T/I C

30. How do 3-D glasses allow the eye to perceive images in 3-D? Create a pamphlet that advertises this technology to movie theatres. (13.6) T/I C A

31. Research these lens imperfections and how they can be reduced. Write a brief report on your findings. (13.5) T/I C A
    (a) chromatic aberration
    (b) spherical aberration

32. Research these three occupations and how they are different from one another. Present your findings in a table. (13.6) T/I C
    (a) optician
    (b) optometrist
    (c) ophthalmologist

To do an online self-quiz or for all Nelson Web Connections,
GO TO NELSON SCIENCE

SELF-QUIZ

The following icons indicate the Achievement Chart category addressed by each question.

K/U Knowledge/Understanding T/I Thinking/Investigation C Communication A Application

For each question, select the best answer from the four alternatives.

1. Which statement correctly describes the path of a light ray when it reaches the optical centre of a converging lens? (13.1) K/U
   (a) The ray is refracted parallel to the principal axis.
   (b) The ray is refracted through the principal focus.
   (c) The ray is reflected back in the direction from which it came.
   (d) The ray continues on its original path without changing direction.
2. A compound microscope produces
   (a) a small, inverted image and a larger, upright, real image.
   (b) an inverted, real image and a larger, inverted, virtual image.
   (c) an upright, virtual image and a larger, upright, virtual image.
   (d) a small, inverted image and a larger, upright, virtual image. (13.5) K/U
3. Where should you place an object in front of a converging lens to produce an inverted, real image that is the same size as the object? (13.3) K/U
   (a) at F'
   (b) at $2F'$
   (c) beyond $2F'$
   (d) between $2F'$ and F'
4. A lens that has a magnification of 1.0 will
   (a) not produce an image.
   (b) produce an image that is larger than the original object.
   (c) produce an image that is smaller than the original object.
   (d) produce an image that is the same size as the original object. (13.4) K/U

Indicate whether each statement is TRUE or FALSE. If you think the statement is false, rewrite it to make it true.

5. A converging lens produces an image by reflecting light rays that strike it. (13.1) K/U
6. When using the thin lens equation, the distance from the object to the optical centre of the lens must be expressed as a positive value. (13.4) K/U

Copy each of the following statements into your notebook. Fill in the blanks with a word or phrase that correctly completes the sentence.

7. A ___________ lens always forms an image on the same side of the lens as the object. (13.1) K/U
8. A magnifying glass produces a larger, ___________ image of an object. (13.5) K/U
9. ___________ occurs when the eye lens changes shape to allow an image to focus on the retina. (13.6) K/U

Figure 1 shows a convex lens. Match the term on the left with the appropriate location on the right.

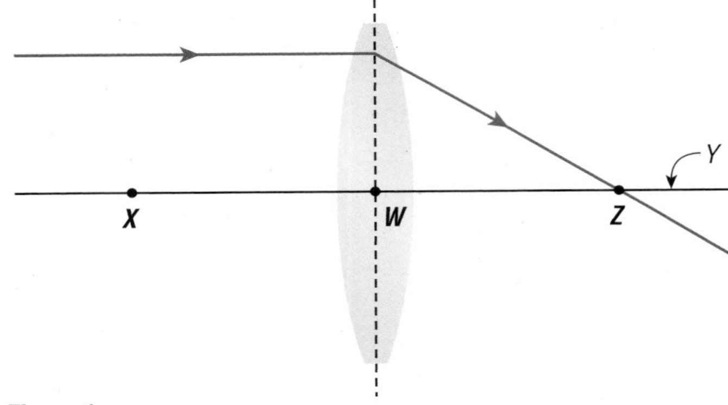

Figure 1

10. (a) principal focus (i) W
    (b) optical centre (ii) X
    (c) principal axis (iii) Y
    (d) secondary principal focus (iv) Z (13.1) K/U

Write a short answer to each of these questions.

11. Is it possible to position an object in front of a converging lens so that no real image is produced by the lens? Explain why or why not. (13.3) K/U

12. A converging lens has a focal length of 15 cm. A ring is placed 32 cm from the lens. How far from the lens is the image of the ring? Show your work. (13.3, 13.4) T/I

13. A small figurine is placed 6.5 cm in front of a lens. The lens produces a virtual image with a magnification of 2.9. (13.3, 13.4) T/I
    (a) Is the image located on the same side of the lens as the figurine?
    (b) How far from the lens is the image located? Show your work.

14. A flower 10.4 cm high is placed in front of a diverging lens. An upright, virtual image that is 4.1 cm high is formed on the same side of the lens as the flower. What is the magnification of the lens? Show your work. (13.3, 13.4) T/I

15. A converging lens has a focal length of 22 cm. An object placed 63 cm from the lens has an image that is 34 cm from the lens. If the distance between the object and the lens is doubled, will the distance between the image and the lens also double? Show your work. (13.3, 13.4) T/I

16. (a) Is the film in a movie projector loaded into the projector right side up or upside down?
    (b) Explain why the film is loaded the way you stated in part (a). (13.5) K/U

17. (a) Some people with presbyopia hold a book farther than a normal distance from their eyes so that they can read the print. How does this help their eyes focus?
    (b) What type of glasses would help a person with presbyopia? (13.6) K/U A

18. Your friend does not understand how a distance used in the thin lens equation can be negative. Explain why there is a negative sign in front of a distance representing a virtual image. (13.4) T/I C

19. You want to take a photograph of a large group of people. At first, everyone does not fit into the picture. When you move farther away from the group, everyone fits into the picture, but the people appear smaller than before. Use what you know about lenses to explain why this happens. (13.3, 13.5) A

20. Imagine you worked for a newspaper at the time when Antonie van Leeuwenhoek invented his microscope. You have just attended a demonstration of the device. Write a paragraph for your newspaper describing the event and what you saw. (13.5) C

21. The magnification equation is expressed mathematically as $M = \frac{h_i}{h_o} = -\frac{d_i}{d_o}$. Express the magnification equation in words. (13.4) C

22. Explain which type of telescope would be most appropriate in each situation. (13.5) K/U T/I
    (a) A sailor on open water needs to see a distant landmass.
    (b) An astronomer wants to observe a star in another galaxy.

23. In this chapter, you learned that the human eye is similar to a camera in many ways. What part of a camera has a function similar to the function of the human retina when you see an object? Explain your answer. (13.5, 13.6) T/I

24. For a converging lens, why is the distance from the optical centre to the principal focus the same as the distance from the optical centre to the secondary principal focus? (13.1) T/I

UNIT E

Light and Geometric Optics

CHAPTER 11
The Production and Reflection of Light

KEY CONCEPTS

Optical devices benefit our society in many ways.

Light is produced by natural and artificial sources.

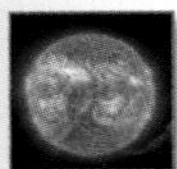
Light is an electromagnetic wave that travels at high speed in a straight line.

When light is reflected off a flat, shiny surface, the image is equal in size to the object and the same distance from the surface.

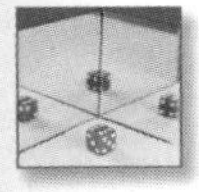
Images in flat mirrors are located at the point where the backward extensions of reflected rays intersect.

Curved mirrors produce a variety of images.

CHAPTER 12
The Refraction of Light

KEY CONCEPTS

Light changes direction predictably as it travels through different transparent media.

Light bends toward the normal when it slows down in a medium with a higher index of refraction.

Total internal reflection may occur when an incident ray is aimed at a medium with a lower index of refraction.

Many optical devices make use of the refraction and reflection of light.

The refraction and reflection of light can be used to explain natural phenomena.

Understanding the behaviour of light is key to many careers.

CHAPTER 13
Lenses and Optical Devices

KEY CONCEPTS

A lens is a transparent object used to change the path of light.

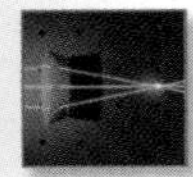
Parallel light rays are refracted through a focus when they pass through a converging lens.

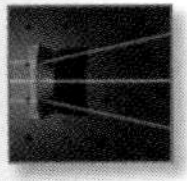
Geometric optics can be used to determine the path of light rays through lenses.

Both ray diagrams and algebraic equations can be used to determine the characteristics of an image in a lens.

Lenses have many technological uses that benefit humans.

The eye can be treated as a lens, and vision problems can be corrected with other lenses.

MAKE A SUMMARY

This summary activity will help you consolidate your understanding of important vocabulary in this unit.

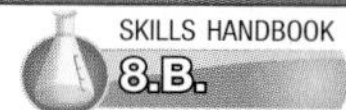

Part A: Word Mixer Activity

1. Your class is divided evenly in two groups. Each person in Group 1 receives a card on which a word or concept from the unit is printed (word wall cards).
2. Each person in Group 2 receives a card with a definition on it.
3. Match the words or concepts with the appropriate definitions by finding the person in the other group with a match for your card. Once you find a match, you and your partner can continue with Part B. K/U

Part B: Word Development Guide

4. You and your partner each fill out a Word Development Guide based on Figure 1. K/U C
5. When you have finished, compare your Word Development Guide with the one completed by your partner. Note how your entries in the boxes were similar or different. On your original sheet, add any good suggestions or entries that your partner has written. Keep this guide for future reference.
6. Repeat Parts A and B with other words or concepts, according to your teacher's directions.

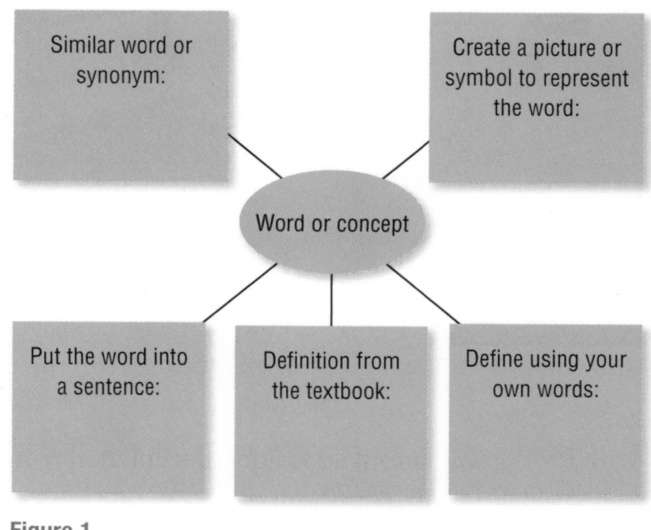

Figure 1

CAREER LINKS

List the careers mentioned in this unit. Choose two of the careers that interest you or choose two other careers that relate to light and geometric optics. For each of these careers, research the following information:

- educational requirements (secondary and post-secondary)
- skill/personality/aptitude requirements
- potential salary
- duties/responsibilities

Assemble the information you have discovered into a chart. Your chart should compare your two chosen careers, and explain how they use light and geometric optics.

GO TO NELSON SCIENCE

Building an Optical Device

Optical devices have had a great impact on our society and on our knowledge and understanding of the physical world. The invention of the compound microscope, for example, opened up the entire field of microbiology (Figure 1). The study of diseases, their causes, and how to prevent them was made possible through the use of the microscope. The telescope in the last two decades has totally revolutionized our understanding of astronomy. Modern telescopes provide incredibly detailed pictures of the planets and moons in our solar system, comets, exploding stars, Einstein rings, and other phenomena.

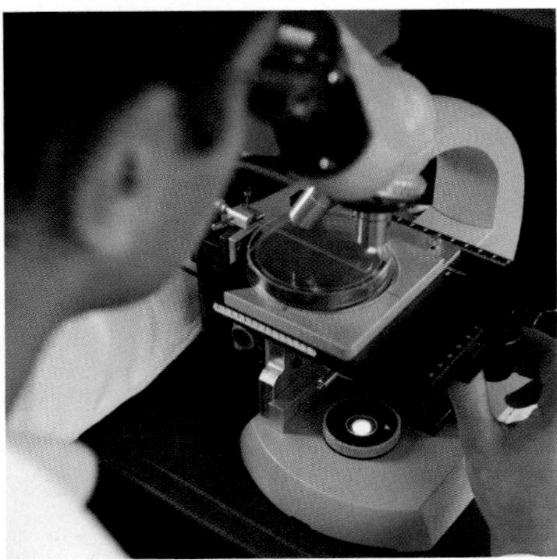

Figure 1 Microscopes are important tools for diagnosing diseases, especially those caused by micro-organisms.

In addition to their use as tools for scientific exploration, optical devices are often used for entertainment. Movie projectors are used to play the latest action movie or breathtaking documentary. Many people enjoy photography as a leisure activity and use cameras to record memories of trips (Figure 2).

Figure 2 We use cameras so often that we barely think about the technology involved.

SKILLS MENU

Questioning	Performing
Hypothesizing	Observing
Predicting	Analyzing
Planning	Evaluating
Controlling Variables	Communicating

For this Unit Task you will design and build a prototype of an optical device.

Purpose

To build one of these optical devices:

- a reflecting telescope
- a microscope
- a projector
- a device that enables people to see spaces normally hidden from view (for example, the back of one's head, hidden corners of a store, under or around furniture)
- a device of your own design (either for a serious purpose or for entertainment)

Equipment and Materials

Your optical device must use at least two components from this list:

- plane mirror
- converging mirror
- diverging mirror
- prism
- converging lens
- diverging lens
- ray box

These components will be available from the teacher. You may use more than two if your school's supply of equipment allows this.

Your device will also require some materials to hold your optical components together. The school may provide some of these materials (for example, glue, cardboard, tape).

Procedure

Your school may require that school materials stay on school property. If this is the case, your teacher will allow you time to construct, test, and modify your device in the Science classroom or lab.

1. Select a device that you will construct.
2. Design your device. Your design should include a neat, scale drawing.
3. Prepare a list of materials that you will need. Have your teacher review this list.
4. Construct a prototype of your device.
5. Test your prototype to see how well it works.
6. Modify your prototype until it produces a result that you are satisfied with. This final version of the prototype will be your finished device.

Analyze and Evaluate

(a) How well does your device do what you want it to do?

(b) What problems did you run into in construction?

(c) How did you solve these construction problems?

(d) Describe the image that your device produces under these headings:

- magnification
- quality of image
- image characteristics

(e) How easy is your device to use?

(f) How would you do this activity differently if you were to do it all over again?

Apply and Extend

(g) Draw a neat ray diagram to illustrate how your device produces an image.

(h) Demonstrate your device to your class.

(i) Research real-life examples of the device that you designed.

ASSESSMENT CHECKLIST

Your completed Performance Task will be assessed according to these criteria:

Knowledge/Understanding

- ☑ Demonstrate a knowledge of geometric optics.
- ☑ Demonstrate a knowledge of the imaging properties of optical components.

Thinking/Investigation

- ☑ Develop a plan for constructing the optical device.
- ☑ Incorporate safety principles.
- ☑ Incorporate environmental considerations.
- ☑ Perform the task in an organized manner.
- ☑ Analyze the results.
- ☑ Evaluate the design and modify it as necessary.

Communication

- ☑ Prepare an appropriate lab report addressing the finalized design, the materials used, a scale diagram, and an evaluation of the device's effectiveness.
- ☑ Communicate clearly during your design demonstration.
- ☑ Demonstrate an understanding of ray diagrams and the proper symbols of optical components.

Application

- ☑ Use two or more optical components together.
- ☑ Achieve the stated goal with the optical device.
- ☑ Build a mechanically sound, environmentally friendly, and safe device.
- ☑ Pay attention to the overall appearance and aesthetics of the device.

The following icons indicate the Achievement Chart category addressed by each question.

K/U Knowledge/Understanding T/I Thinking/Investigation C Communication A Application

What Do You Remember?

For each question, select the best answer from the four alternatives.

1. The angle of incidence is the angle between the
   (a) incident ray and the surface of a material
   (b) refracted ray and the normal
   (c) incident ray and the normal
   (d) refracted ray and the surface of a material (11.4) K/U

2. Light reflecting off a plane mirror is an example of
   (a) specular reflection
   (b) diffuse reflection
   (c) total internal reflection
   (d) refraction (11.6) K/U

3. Which of these best describes the index of refraction for a medium? (12.4) K/U
   (a) the speed of light in a vacuum times the speed of light in the medium
   (b) the speed of light in a vacuum minus the speed of light in the medium
   (c) the speed of light in a vacuum divided by the speed of light in the medium
   (d) the speed of light in a vacuum plus the speed of light in the medium

4. Which of these best describes the angle of refraction? (12.1) K/U
   (a) the angle between the refracted ray and the normal
   (b) the angle between the refracted ray and a reflective surface
   (c) the angle between the refracted ray and the reflected ray
   (d) the angle between the refracted ray and the incident ray

5. Which medium has an index of refraction of exactly 1.0? (12.4) K/U
   (a) vegetable oil
   (b) glass
   (c) vacuum
   (d) water

6. Which material will refract light the least?
   (a) material A with $n = 1.72$
   (b) material B with $n = 2.34$
   (c) material C with $n = 1.58$
   (d) material D with $n = 1.92$ (12.4) K/U

7. A larger, upright, virtual image is noticed in a converging lens. The object must be located
   (a) beyond $2F'$
   (b) between $2F'$ and F'
   (c) at F'
   (d) between F' and the lens (13.3) K/U

8. A person who is far-sighted can clearly see
   (a) distant objects but not nearby objects
   (b) both distant and nearby objects
   (c) nearby objects but not distant objects
   (d) none of the above (13.6) K/U

Indicate whether each statement is TRUE or FALSE. If you think the statement is false, rewrite it to make it true.

9. Light is a form of energy. (11.1) K/U
10. Light is visible electromagnetic waves. (11.1) K/U
11. Light can only travel through a medium. (11.1) K/U
12. A light ray is either reflected or transmitted when it strikes a surface such as a piece of glass. (12.1) K/U
13. The centre of a lens is called the centre of curvature. (13.1) K/U
14. A converging lens spreads parallel light rays apart after refraction. (13.1) K/U
15. A positive meniscus lens is a converging lens. (13.6) K/U
16. A light ray that is parallel to the principal axis of a lens is not refracted. (13.3) K/U
17. Einstein predicted that gravitational forces can change the path of light. (13.5) K/U
18. The function of film in a camera is most like the function of the iris of a human eye. (13.5) K/U

19. An object located at the secondary principal focus, F', of a converging lens has an image that is exactly the same size as the object. (13.3) K/U

Copy each of the following statements into your notebook. Fill in the blanks with a word or phrase that correctly completes the sentence.

20. A light source that produces its own light is called _________. (11.2) K/U

21. All electromagnetic waves travel at the speed of _________. (11.1) K/U

22. _________ rays are the type of electromagnetic radiation with the greatest energy. (11.1) K/U

23. A diverging (convex) mirror can only produce images of a _________ size and a(n) _________ attitude. (11.9) K/U

24. A magnifying glass is an example of a _________ lens. (13.5) K/U

25. At the critical angle, the angle of refraction is _________. (12.5) K/U

26. A lens that is thinnest in the middle and thickest at the edge is called a _________ lens. (13.1) K/U

27. Light rays that are parallel to the principal axis of a converging lens meet at the principal _________. (13.1) K/U

28. Total internal reflection will occur when the angle of incidence is greater than the _________ angle. (12.5) K/U

29. The term for near-sightedness is _________. (13.6) K/U

Match each type of electromagnetic radiation on the left with its application on the right.

30. (a) ultraviolet light (b) infrared light (c) visible light (d) radio waves (e) X rays

(i) human vision (ii) "black" lights (iii) DVD player remote controls (iv) dental imaging (v) radar (11.1) K/U

Write a short answer to each question.

31. Classify each of these materials as transparent, translucent, or opaque: (11.4) K/U
    (a) a clean sheet of glass
    (b) a clear, cold winter sky
    (c) frosted glass
    (d) a brick
    (e) a tree
    (f) hazy, polluted air
    (g) a glass of water

32. Describe the shape of a converging lens as compared to the shape of a diverging lens. (13.1) K/U

33. Define each of these terms: (11.4, 11.7, 11.9, 12.1, 12.7, 13.4) K/U
    (a) angle of incidence
    (b) angle of refraction
    (c) focus
    (d) magnification
    (e) mirage
    (f) virtual image

What Do You Understand?

34. Give an example of each of these objects. (11.4) K/U A
    (a) an opaque object
    (b) a transparent object
    (c) a translucent object

35. (a) Use capital letters to write a three-letter word that appears exactly the same when viewed in a plane mirror.
    (b) Use capital letters to write a three-letter word that appears as a different word when viewed in a plane mirror. (11.7) A

36. Describe how each of these processes produces light. (11.2) K/U
    (a) electric discharge
    (b) bioluminescence
    (c) chemiluminescence

37. Sketch how the word **OPTICS** would appear when viewed in a plane mirror. (11.7) K/U C

38. What two properties do all electromagnetic waves possess? (11.1) K/U

39. Compare the process of light emission in
   (a) incandescence and fluorescence
   (b) triboluminescence and phosphorescence (11.2) K/U

40. Classify each material as exhibiting specular or diffuse reflection. Justify your answer in each case. (11.6) K/U C
   (a) a highly waxed floor
   (b) a crumpled sheet of aluminum foil
   (c) a dense carpet
   (d) a flat sheet of aluminum foil, shiny side up

41. Explain why a laser would not be a useful tool for illuminating a dark room. (11.3) T/I

42. Clearly explain why diverging mirrors and diverging lenses can never form a real image. Illustrate your answer with diagrams. (11.9, (13.3) K/U C

43. Copy and complete Table 1 in your notebook. (11.6) K/U T/I

Table 1 Angles of Incidence and Reflection

Description	Angle of incidence	Angle of reflection
angle between the incident ray and the normal is 38°		
	12°	
angle between the reflected ray and the flat mirror surface is 43°		
angle between the reflected ray and the normal is 23°		
	0°	0°

44. Copy Figure 1 into your notebook.
   (a) Use at least three object–image lines and lines of equal length that are perpendicular to the mirror to determine the image for each object.
   (b) Describe the four characteristics of each image. (11.7) T/I C

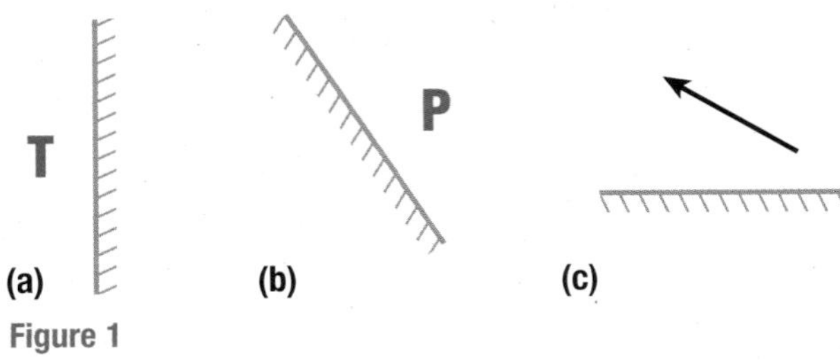

Figure 1

45. Copy Figure 2 into your notebook.
   (a) Locate the image of each object
   (b) State the four characteristics for each image. (11.9) T/I C

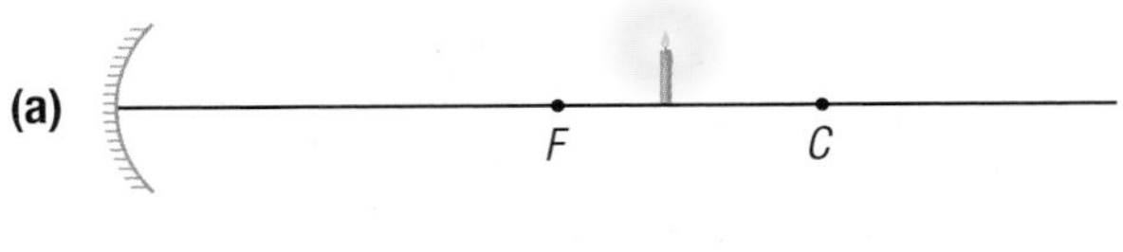

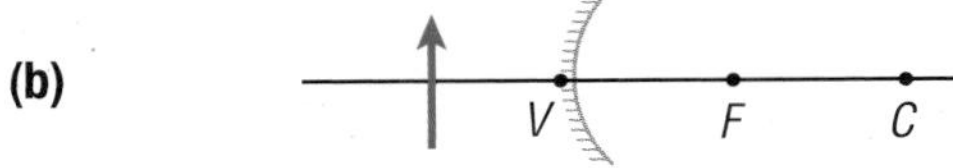

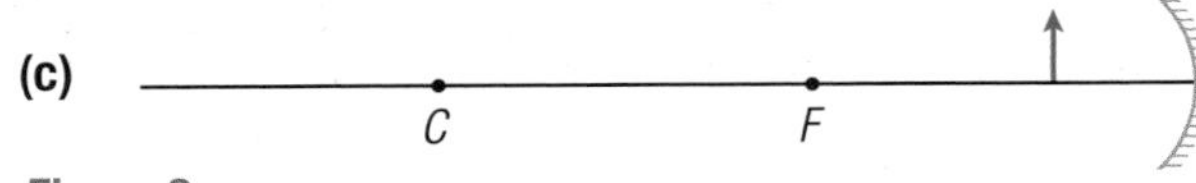

Figure 2

46. Figure 3 represents a beam of light travelling through two different media.
   (a) Which medium has the greater index of refraction?
   (b) In which medium will light travel slower? (12.4) K/U

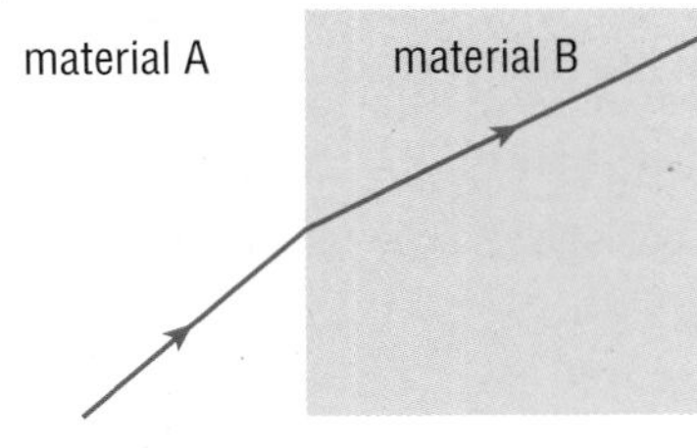

Figure 3

47. Copy Figure 4 into your notebook. (13.3) T/I C
(a) Locate the image for each object.
(b) State the four characteristics for each image.

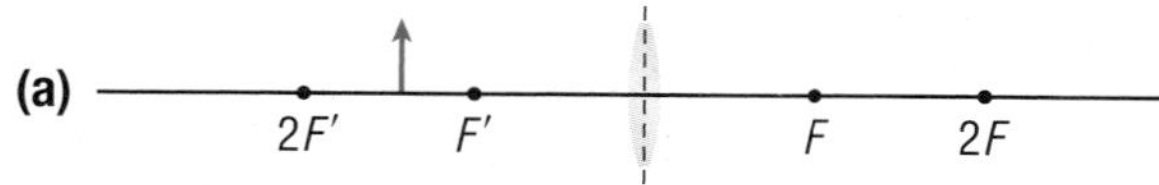

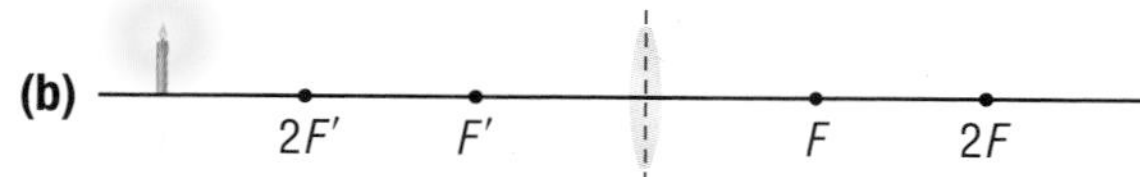

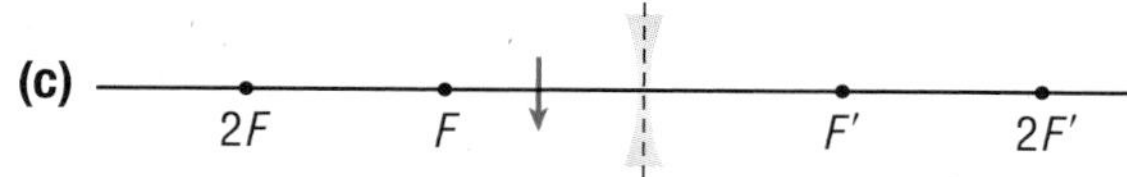

Figure 4

Solve a Problem

48. You like wearing white shirts. Every time you wash your shirt, however, you notice that it never seems to come out as white as it was when you first bought it. The next time you are in the supermarket you notice a new detergent that claims it can make your shirts "whiter than white." This would seem to solve your problem, if the claim were true. Analyze and assess the merits of this claim. (11.2) T/I A

49. Australia has one of the highest rates of skin cancer in the world. (11.2) K/U A
(a) Which part of the electromagnetic spectrum is damaging to the skin?
(b) What could you do to prevent some of this damage, and why would these techniques be useful?

50. Your cat likes playing in front of a large mirror on the wall. She is constantly trying to reach her image in the mirror, with little success. She moves back and forth in front of the mirror, but nothing seems to help. Can you account for her frustration? Use a diagram to demonstrate why she will never be able to touch her image. (11.7) K/U C

51. A converging (concave) mirror has a focus (F) at 27 cm.
(a) An object is placed 41 cm away from the mirror.
- Where will the image of this object be located?
- Is the image actually there even if you cannot see it?
- What could you do to see the image?

(b) The object is now placed 20 cm away from the mirror. Where is the image located, and how can you see it? (11.9) T/I C

52. Determine the speed of light in fused quartz if fused quartz has an index of refraction of 1.46. (12.4) T/I C

53. The speed of light in turpentine is 2.04×10^8 m/s. What is the index of refraction for turpentine? (12.4) T/I C

54. As a beam of light passes from air into a transparent medium, the angle between the light ray and the normal changes from 45° to 30°. What is the speed of light in the medium? Show your work. (12.4) T/I

55. A diverging lens has a focal length of 30 cm. A golf ball is located 23 cm in front of the lens. Where is the image located? Show your work. (13.3, 13.4) T/I

56. A converging lens has a focal length of 34 cm. A tree is located 45 cm from the lens. Calculate the location of the image and state its characteristics. (13.3, 13.4) T/I C

57. An apple is located 34 cm from a converging lens. A real image of the apple is observed 21 cm from the lens. Calculate the distance from the lens to F. (13.3, 13.4) T/I C

58. For a magnifying glass, F is 24 cm from the lens. The magnifying glass is held 17 cm in front of a sea shell.
(a) Calculate the location of the image and state its characteristics.
(b) Determine the magnification of the magnifying glass. (13.3, 13.4) T/I C

59. A diverging lens is placed 13 cm above a sculpture. The image is located 5.0 cm from the lens, on the same side of the lens as the sculpture. (13.3, 13.4) K/U T/I C
   (a) What kind of image is produced by the lens?
   (b) Determine the focal length of this lens.

60. *F* is 27 cm from a diverging lens. The lens is placed in front of a fruit bowl. A virtual, upright image is located 12 cm on the same side of the lens as the fruit bowl. What is the distance between the lens and the fruit bowl? (13.3, 13.4) T/I C

61. A cat of height 19 cm is placed in front of a converging lens. An inverted, real image of height 58 cm is located on the other side of the lens. Calculate the magnification of the lens. (13.3, 13.4) T/I C

Create and Evaluate

62. Assess which form of lighting would be most useful in saving energy in your home. Summarize your findings in a brief report, including your reasoning. (11.2) C A

63. You are driving with your mother on a winding road. You notice, on a particularly sharp turn, that convex mirrors are placed on poles at the side of the road. What is the purpose of these mirrors? (11.9) A

64. The Sun's energy can be both harmful and beneficial for humans.
   (a) Explain why the Sun is considered to be the original source of almost all energy used by organisms on Earth.
   (b) Create a public notice for a local pool explaining the risks of too much Sun exposure. Include a list of ways that people can protect themselves from the Sun's harmful rays. (11.1) C A

65. As a light ray passes from one transparent medium into another, it changes both direction and speed. Draw a diagram showing how the change in direction is related to the change in speed. (12.1) C

66. Compare using a t-chart, the functioning of the human eye with that of a camera. (13.5, 13.6) C A

67. The word "radiation" has a negative meaning for many people. Devise an argument to illustrate the positive aspects of radiation. (11.1) C A

68. Your aunt's vision has become increasingly poor. She is now having trouble reading the newspaper. (13.6) T/I C
   (a) What is the name of her vision problem?
   (b) Draw a diagram with light rays to illustrate her vision problem.
   (c) What kind of lens would an optometrist likely recommend for her? Draw a diagram to illustrate how this lens would correct her vision problem.

69. Compare the image characteristics of these pairs of optical instruments: (11.9, 13.3) K/U
   (a) a converging lens and a converging mirror
   (b) a diverging lens and a diverging mirror

Reflect on Your Learning

70. Assume that all forms of electromagnetic waves, with the exception of visible light, disappeared. Would this make an impact on your life? Explain.

71. Other than the fact that your eyes can see visible light, there is nothing particularly unique about this narrow band in the electromagnetic spectrum. Some animal species can detect light outside the visible range. Reflect on how your life would be different if you could see *all* types of electromagnetic radiation.

72. Several unusual properties and applications of light were described in this unit. Select two properties or applications that you found to be most surprising and explain your reasoning.

73. (a) Prepare a concept map (based on Figure 5) with you at the centre showing all the ways that mirrors affect your life.
(b) Prepare a second concept map showing all the ways that lenses affect your life.

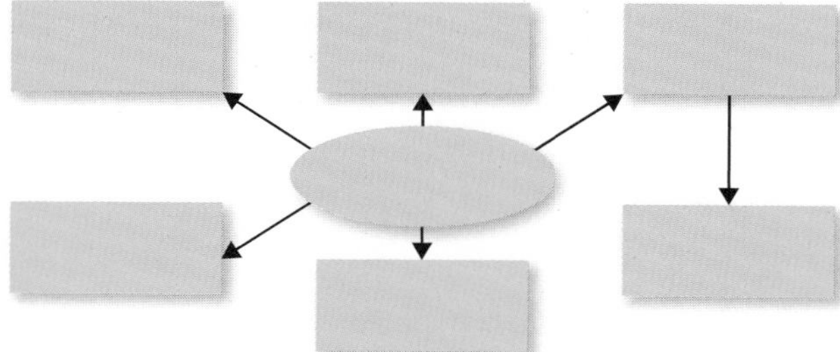

Figure 5

Web Connections 

74. Reflecting telescopes are a major tool for astronomical research involving visible light. Research the different kinds of reflecting telescopes. Present your findings as a timeline of their historical development. (11.9) T/I

75. The laser has many applications in medicine. (11.3) T/I A C
(a) One application is laser eye surgery, which you examined in Activity 13.7. Research other medical uses of the laser. Present your findings in a table.
(b) The laser is also often used for cosmetic purposes. Research some of these applications of the laser. Write a paragraph to summarize your findings.

76. Research the kind of telescope that Galileo built, and the major discoveries that he made with it. Write two paragraphs to summarize your findings. (11.9, 13.5) T/I C

77. Canada's first space telescope is known as MOST. Research MOST. How is it constructed, and what are its research goals? (11.9, 13.5) T/I A

78. (a) Research the history and the development of the compound microscope. Present your findings as a timeline.
(b) How is an electron microscope different from a compound microscope? (13.5) T/I A

79. (a) Optics has made many positive contributions to society. Summarize the applications that have been discussed in this unit. Part of your summary should include why this application has such a beneficial effect on society.
(b) Use the Internet and other information resources to research interesting applications of optics. Focus on applications that are either not discussed in great detail or not at all in this unit. This second summary should include a discussion of why these applications are beneficial to society. (11.9, 12.5, 13.5) T/I A C

80. Research the development of movie cameras from 1900 to the present. (13.5) T/I A C
(a) Explain how improvements in technology have enabled filmmakers to film an increasingly wide range of subjects.
(b) Choose two or three specialty filming technologies (e.g., underwater, microscopic, stop-motion) to illustrate how advances in movie making have advanced our understanding of the natural world.

For all Nelson Web Connections,
GO TO NELSON SCIENCE

SELF-QUIZ

The following icons indicate the Achievement Chart category addressed by each question.

K/U Knowledge/Understanding | C Communication | T/I Thinking/Investigation | A Application

For each question, select the best answer from the four alternatives.

1. Specular reflection refers to reflection from
   (a) a flat mirror.
   (b) a curved mirror.
   (c) an uneven surface.
   (d) a transparent surface. (11.6) K/U
2. A person who can read fine print without glasses but who must wear glasses to drive most likely has
   (a) astigmatism.
   (b) hyperopia.
   (c) myopia.
   (d) presbyopia. (13.6) K/U
3. After striking a converging lens, light rays that are parallel to the principal axis pass through
   (a) the optical center.
   (b) the principal axis.
   (c) the principal focus.
   (d) the secondary principal focus. (13.1) K/U
4. What is the process by which materials that absorb light energy release light at a later time? (11.2) K/U
   (a) chemiluminescence
   (b) fluorescence
   (c) incandescence
   (d) phosphorescence
5. In which case does light bend toward the normal? (12.1) K/U
   (a) light passing through glass
   (b) light passing from glass into air
   (c) light passing from air into water
   (d) light passing from cold air into warm air

Indicate whether each statement is TRUE or FALSE. If you think the statement is false, rewrite it to make it true.

6. The higher the index of refraction for a medium, the faster light will travel through that medium. (12.4) K/U
7. Visible light is a very narrow band of energy in the electromagnetic spectrum. (11.1) K/U
8. To see the mirror image of an object printed on a page, turn the page upside-down. (11.6) K/U
9. Loss of accommodation can lead to presbyopia. (13.6) K/U

Copy each of the following statements into your notebook. Fill in the blanks with a word or phrase that correctly completes the sentence.

10. The angle between a light ray striking a surface and the normal is called the angle of __________. (11.4) K/U
11. The heating wires in a toaster glow red when electricity passes through them. This process of producing light using an object at high temperature is called __________. (11.2) K/U

Match each term on the left with the most appropriate definition on the right.

12. (a) transparent — (i) produces its own light
    (b) translucent — (ii) allows light to pass through easily
    (c) luminous — (iii) allows some light to pass through
    (d) opaque — (iv) does not allow any light to pass through
    (11.2, 11.4) K/U

Write a short answer to each of these questions.

13. An object is located between F' and $2F'$ of a converging lens. Describe the size and attitude of the resulting image. (13.3) K/U
14. Human eyes can change focal length in order to focus on objects over a range of distances. Briefly describe the mechanism by which eyes change their focal length. (13.6) K/U

15. Figure 1 shows someone looking through a periscope. Copy the diagram into your notebook. Add mirrors to complete the diagram. Show the path of light through the periscope with arrows. (11.7) C

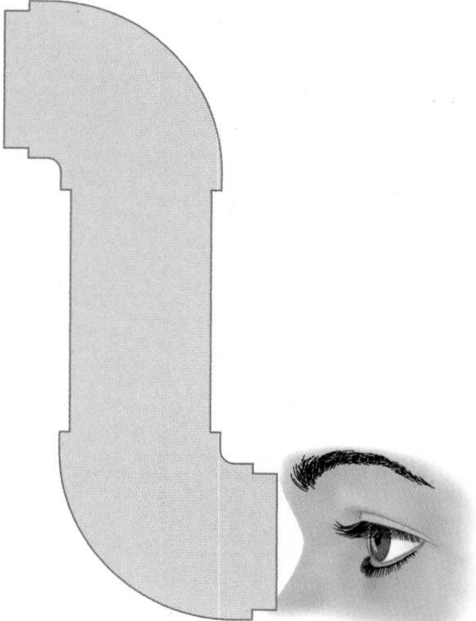

Figure 1

16. You want to position a light so that it shines through the water of an aquarium at an angle of 45°. The index of refraction for air is 1.00, and for water it is 1.33. To correctly position the light, what would the angle of incidence have to be? (12.4) T/I

17. A scientist needs to determine the speed of light in hexane, a clear liquid hydrocarbon. (13.3) C
    (a) Which two pieces of data are needed in order to calculate the speed?
    (b) Which equation should the scientist use? (12.4) T/I

18. (a) Copy Figure 2 into your notebook. Use light rays to locate the image.
    (b) Describe the image characteristics.

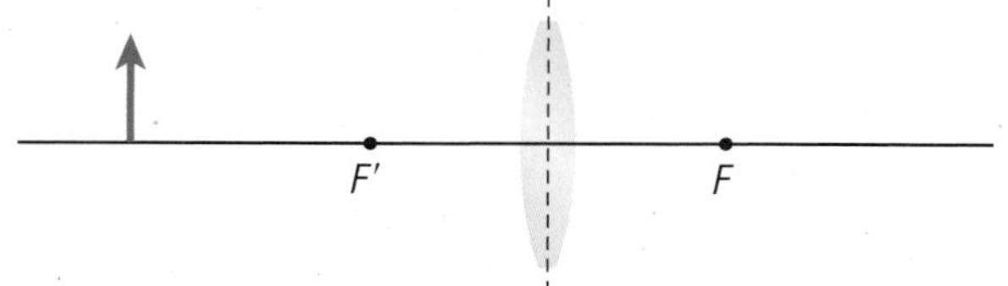

Figure 2

19. (a) Compare incandescent light bulbs with fluorescent light bulbs in terms of light production.
    (b) Your parents are considering using fluorescent bulbs instead of incandescent bulbs in your house. Describe two advantages of replacing incandescent bulbs with compact fluorescent lights. (11.2) K/U C

20. Two marine biologists are discussing the best way to shoot tranquilizer darts into sharks so that they can attach tracking devices onto them. One biologist wants to dive underwater to shoot the sharks, whereas the other wants to shoot them from a boat. Although shooting from the boat is safer, it presents another problem. (12.7) T/I
    (a) Describe the problem.
    (b) Explain how to solve the problem in part (a).

21. A student bought a magnifying glass that was advertised as having a magnification of 4. (13.4) T/I A
    (a) How can the student check the magnification of the magnifying glass?
    (b) Which equation should the student use?

22. A converging lens has a focal length of 20 cm. A cup is located 50 cm from the lens. (13.3, 13.4) A
    (a) What type of image will be formed?
    (b) Where will the image be located? Show your calculations.

23. (a) Identify three devices you have used that make use of electromagnetic radiation. Choose devices that use radiation from three different parts of the electromagnetic spectrum.
    (b) List the devices in order of increasing energy. (11.1) A

24. (a) Explain how a wax coating changes a surface in terms of specular and diffuse reflection.
    (b) Describe a process that would have the opposite effect on a surface. (11.6) A

APPENDIX A Skills Handbook

CONTENTS

1.A. Having a Safe Attitude

Science investigations can be a lot of fun, but certain safety hazards exist in any laboratory. You should know about them and about the precautions you must take to reduce the risk of an accident.

Why is safety so important? Think about the safety measures you already take in your daily life. Your school laboratory, like your kitchen, need not be a dangerous place. In any situation, you can avoid accidents when you understand how to use equipment and materials and follow proper procedures. For example, you can take a hot pizza out of the oven safely if you take a few common-sense safety precautions. Similarly, corrosive acids can be used safely if appropriate safety precautions are taken.

Safety in the laboratory combines common sense with the foresight to consider the worst-case scenario. The activities and investigations in this textbook are safe, as long as you follow proper safety precautions. General laboratory safety rules are outlined below (Table 1). Your teacher may provide you with additional safety rules for specific tasks.

Table 1 Practise Safe Science in the Classroom

Be science ready	Follow instructions	Act responsibly
• Come prepared with your textbook, notebook, pencil, and anything else you need. • Tell your teacher about any allergies or medical problems. • Keep yourself and your work area tidy and clean. Keep aisles clear. • Keep your clothing and hair out of the way. Roll up your sleeves, tuck in loose clothing, and tie back loose hair. Remove any loose jewellery. • Wear closed shoes (not sandals). • Do not wear contact lenses while doing investigations. • Read all written instructions carefully before you start an activity or investigation.	• Do not enter a laboratory unless a teacher is present or you have permission to do so. • Listen to your teacher's directions. Read written instructions. Follow them carefully. • Ask your teacher for directions if you are not sure what to do. • Wear eye protection or other safety equipment when instructed by your teacher. • Never change anything, or start an activity or investigation on your own, without your teacher's approval. • Get your teacher's approval before you start an investigation that you have designed yourself.	• Pay attention to your own safety and the safety of others. • Know the location of MSDS (Material Safety Data Sheet) information, exits, and all safety equipment, such as the first-aid kit, fire blanket, fire extinguisher, and eyewash station. • Alert your teacher immediately if you see a safety hazard, such as broken glass, a spill, or unsafe behaviour. • Stand while handling equipment and materials. • Avoid sudden or rapid motion in the laboratory, especially near chemicals or sharp instruments. • Never eat, drink, or chew gum in the laboratory. • Do not taste, touch, or smell any substance in the laboratory unless your teacher asks you to do so. • Clean up and put away any equipment after you are finished. • Wash your hands with soap and water at the end of each activity or investigation.

1.B. Specific Safety Hazards

Follow these instructions to use materials and equipment safely in the science classroom.

1.B.1. Chemicals

Some of the chemicals recommended for use in this course are dangerous if used incorrectly. Be sure to follow these rules to avoid accidents.

- Assume that any unknown chemicals are hazardous.
- Reduce exposure to chemicals to the absolute minimum. Avoid direct skin contact, if possible.
- When taking a chemical from a container, first check the label to be sure you are taking the correct substance. Replace the lid securely when you have taken what you need.
- Never use the contents of a container that has no label or has an illegible label. Give any such containers to your teacher.
- Place test tubes in a rack before pouring liquids into them. If you must hold a test tube, tilt it away from you, and others, before pouring in a liquid.
- Pour liquid chemicals carefully (down the side of the receiving container or down a stirring rod) to avoid splashing. Always pour from the side opposite the label so that drips will occur only on the side away from your hand.
- When you are instructed to smell a chemical by your teacher, first take a deep breath to fill your lungs with just air, then waft or fan the vapours toward your nose.
- Do not return surplus chemicals to stock bottles and do not pour them down the drain. Dispose of excess chemicals as directed by your teacher.
- If any part of your body comes in contact with a chemical, wash the area immediately and thoroughly with cool water. Rinse affected eyes for at least 15 minutes. Alert your teacher.

1.B.2. Heat Sources

Heat sources, such as hot plates, light bulbs, and Bunsen burners, can cause painful burns. Use caution when there are hot objects around.

- Secure your Bunsen burner to a retort stand with a clamp before lighting it.
- Your teacher will show you the proper method of lighting and adjusting the Bunsen burner. Always follow this method.
- Never leave a lighted burner unattended as a clean blue flame is almost invisible.
- Never heat a flammable material over a Bunsen burner. Make sure there are no flammable materials nearby.
- Never look down the barrel of a burner.
- When heating liquids in glass containers, use only heat-resistant glassware. If a liquid is to be heated to boiling, use boiling chips to prevent "bumping." Keep the open end of the container away from yourself and others. Never allow a container to boil dry.
- When heating a test tube over a Bunsen burner, use a test-tube holder and a spurt cap. Hold the test tube at an angle, with the opening facing away from yourself and others. Heat the upper half of the liquid first and then move it gently into the flame to distribute the heat evenly.
- Always turn off the gas at the valve, not using the gas-adjustment screw of the Bunsen burner.
- If you burn yourself, immediately apply cool water to the affected area and inform your teacher.

1.B.3. Glass and Sharp Objects

Handle glass carefully: it can break, leaving sharp edges and splinters.

- Never use glassware that is broken, cracked, or chipped.
- Never pick up broken glass with your fingers. Use gloves as well as a broom and a dustpan to remove glass from the area.
- Dispose of glass fragments in special containers marked "Broken Glass."
- If you cut yourself, inform your teacher immediately. Embedded glass or continued bleeding requires medical attention.
- Select the appropriate instrument for the task. Never use a knife when scissors would work better.
- Never carry a scalpel in the laboratory with an exposed blade; transport it in a dissection case or box or on a dissection tray.

- Make sure your cutting instruments are sharp. Dull cutting instruments require more pressure than sharp instruments and are, therefore, much more likely to slip. If your scalpel blade needs to be changed, ask your teacher to change it for you.
- Always cut away from yourself and others. Cut downward, on a tray, cutting board, or paper towel.

1.B.4. Electricity and Light

- Never touch an electrical device, electrical cord, or outlet with wet hands.
- Keep water away from electrical equipment.
- Do not use the equipment if wires or plugs are damaged or if the ground pin has been removed.
- If using a light source, check that the wires of the light fixture are not frayed and that the bulb socket is in good shape and is well secured to a stand.
- Make sure that electrical cords are not a tripping hazard.
- When unplugging equipment, hold the plug to remove it from the socket. Do not pull on the cord.
- When using variable power supplies, start at low voltage and increase slowly.
- Do not look directly at any bright source of light. You cannot rely on the sensation of pain to protect your eyes.
- Never point a laser beam (directly or after reflection) at anybody's eyes.

1.B.5. Living Things

- Treat all living things with care and respect. Never treat an animal in a way that would cause it injury or harm.
- Animals that live in the classroom should be kept in a clean, healthy environment.
- Wear gloves and wash your hands before and after feeding or handling an animal, touching materials from the animal's cage or tank, or handling bacterial cultures.
- Human blood, urine, or saliva samples should not be tested or used in investigations due to the risk of contracting a disease from these fluids.

1.C. Accidents Can Happen

The following guidelines apply if an injury occurs, such as a burn, a cut, a chemical spill, ingestion, inhalation, an electrical accident, or a splash in the eyes.

- If an injury occurs, inform your teacher immediately.
- If the injury is from contact with a chemical, wash the affected area with a continuous flow of cool water for at least 15 minutes. Remove contaminated clothing. Consult the Material Safety Data Sheet (MSDS) for the chemical; this sheet provides information about the first-aid requirements. If the chemical is splashed into your eyes, have another student assist you in getting to the eyewash station immediately. Rinse with your eyes open for at least 15 minutes.
- If you have ingested or inhaled a hazardous substance, inform your teacher immediately. Consult the MSDS for the first-aid requirements for the substance.
- If the injury is from a burn, immediately immerse the affected area in cold water. This will reduce the temperature and prevent further tissue damage.
- In the event of electrical shock, do not touch the affected person or the equipment the person was using. Break contact by switching off the source of electricity or by removing the plug.

1.D. Safety Conventions and Symbols

The activities and investigations in *Nelson Science Perspectives 10* are safe to perform provided precautions are taken. This is why general safety hazards are identified with caution symbols (Figure 1).

Figure 1 Potential safety hazards are identified with caution symbols.

More specific hazards, related to dangerous chemicals, are indicated with the appropriate WHMIS symbol. Make sure that you read the information and associated instructions (in red type) carefully. You must understand and follow these instructions in order to perform the activity or investigation safely. Check with your teacher if you are unsure.

1.D.1. Workplace Hazardous Materials Information System (WHMIS) Symbols

The Workplace Hazardous Materials Information System (WHMIS) provides workers and students with complete and accurate information about hazardous products. Clear and standardized labels must be present on the product's original container or must be added to other containers if the product is transferred. If the material is hazardous, the label will include one or more of the WHMIS symbols (Figure 2).

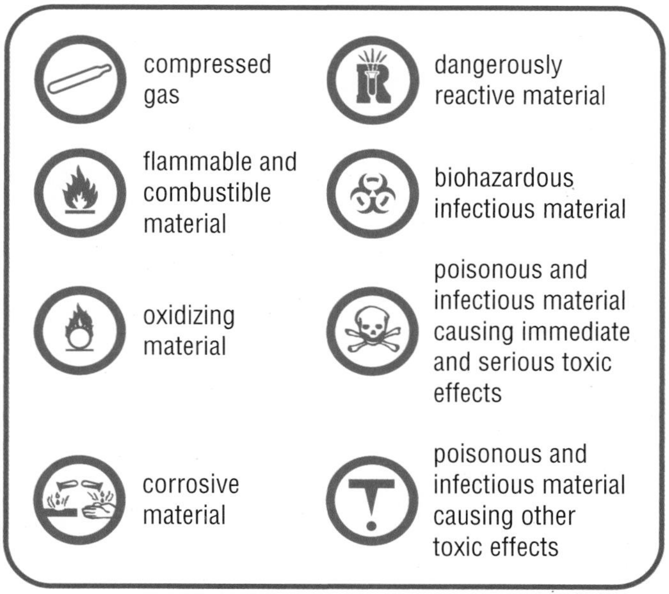

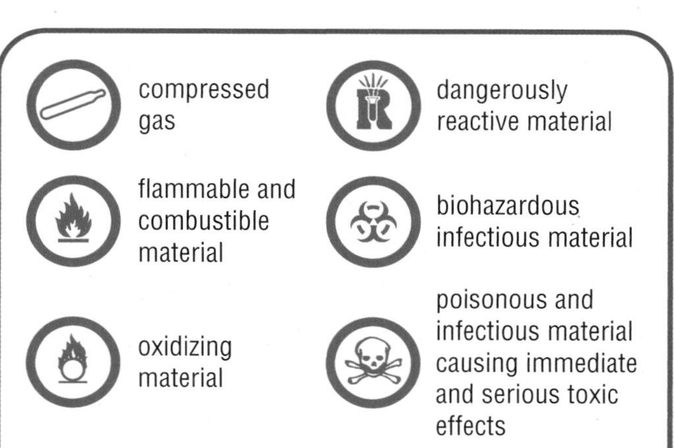

Figure 2 WHMIS symbols identify dangerous materials that are used in all workplaces, including schools.

1.D.2. Hazardous Household Product Symbols (HHPS)

The *Canadian Hazardous Products Act* requires manufacturers of consumer products to include a symbol that specifies both the nature and the degree of any hazard. The Hazardous Household Products Symbols (HHPS) were designed to do this. The symbol is made up of a picture and a frame. The picture tells you the type of danger. The frame tells you whether it is the contents or the container that poses the hazard (Figure 3).

Symbol	Danger
	**Explosive** This container can explode if it is heated or punctured.
	**Corrosive** This product will burn skin or eyes on contact, or throat and stomach if swallowed.
	**Flammable** This product, or its fumes, will catch fire easily if exposed to heat, flames, or sparks.
	**Poisonous** Licking, eating, drinking, or sometimes smelling, this product is likely to cause illness or death.

Figure 3 Household Hazardous Products Symbols (HHPS) appear on many products that are used in the home. A triangular frame indicates that the container is potentially dangerous. An octagonal frame indicates that the product inside the container poses a hazard.

Selecting the correct tools and equipment and using them properly are essential for your safety and the safety of your classmates.

2.A. Working with Dissecting Equipment

Dissection tools are very sharp and must be handled carefully. Read through the following safety tips before you begin any dissections (Figure 1).

- Always hold the dissection tool by the handle.
- To move dissection equipment to and from your workstation, place all tools in the dissection tray and carry the tray with both hands. Do not carry tools by hand. If you forgot a piece of equipment, carry the tray back to get what you need.
- At your workstation, imagine your dissection tray sitting in the middle of a 30 cm circle (about the size of a large dinner plate). The tools never leave this circle. If you leave your workstation or turn to answer a question, place the tools in the dissection tray. Tools must always lie flat on the inside or outside of the dissecting tray, not propped up on the side of the tray.
- Always cut away from yourself and others. Cut downward onto a tray.
- If you are asked to rinse a tool after you have used it, remember that the tool is sharp. Hold it by the handle and rinse it under running water; do not wipe sharp edges or points.

Figure 1 Use safety equipment and always work within a 30 cm circle at your workstation when dissecting.

2.B. Testing for Electrical Conductivity

Before you start any conductivity testing, ask your teacher for specific operating instructions on your school's equipment (Figure 2). Devices used to test electrical conductivity vary considerably. Two metal electrodes are inserted into the sample to be tested.

Figure 2 You should use only low-voltage (battery-powered) conductivity testers.

In many cases, if the sample conducts electricity, the conductivity tester gives a positive result (e.g., a light bulb turns on). Some conductivity testers give variable results (Table 1).

Table 1 Results of Conductivity Testing

Observation	Sample
bright glow	good conductor
faint glow	poor conductor
no glow	nonconductor (insulator)

2.C. Using the pH Meter

pH meters are used to measure acidity or alkalinity (Figure 3). Your teacher will provide specific operating instructions for your school's pH meters. Most pH meters need to be calibrated by placing the probe into a solution of a specific pH and then adjusting the meter so that it displays this pH value.

Figure 3 A pH meter

2.D. Using the Microscope

To view cells and other small objects closely, you will use a microscope (Figure 4). You will most likely use a compound light microscope that uses more than one lens and a light source to make an object appear larger. The object is magnified first by the lens just above the object: the objective lens. Then that image is magnified by the eyepiece: the ocular lens. The comparison of the actual size of the object with the size of its image is called the magnification. The parts of a compound light microscope and their functions are listed in Table 2.

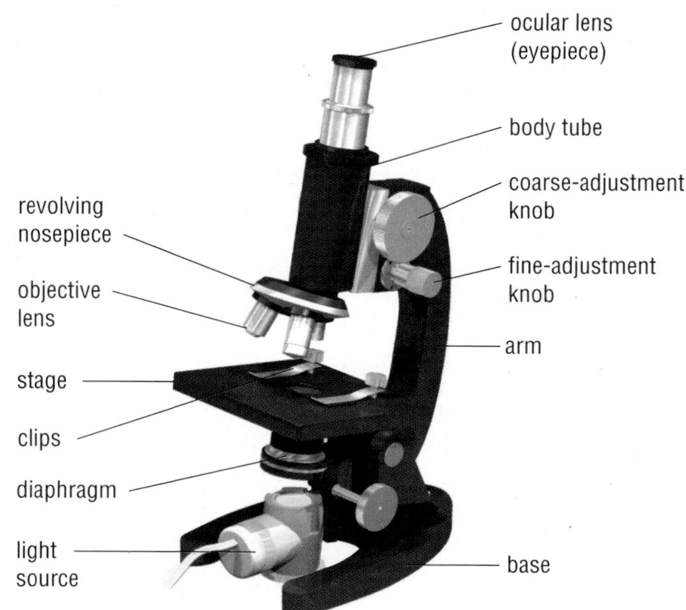

Figure 4 A compound light microscope

Table 2 Parts of a Light Microscope

Part	Function
stage	• supports the microscope slide • has a central opening that allows light to pass through the slide
clips	• holds the slide in position on the stage
diaphragm	• controls the amount of light that reaches the object being viewed
objective lenses	• magnify the object • have three possible magnifications: low power (4×), medium power (10×), and high power (40×)
revolving nosepiece	• holds the objective lenses • rotates, allowing the objective lenses to be changed
body tube	• contains the eyepiece (ocular lens) • supports the objective lenses
eyepiece (ocular lens)	• is the part you look through to view the object • magnifies the image of the object, usually by 10×
coarse-adjustment knob	• moves the body tube up or down to get the object into focus • is used with the low-power objective lens only
fine-adjustment knob	• moves the tube to get the object into sharp focus • is used with medium-power and high-power magnification • is used only after the object has been located, centred, and focused under lower power magnification using the coarse-adjustment knob
light source	• may be an electric light bulb or a mirror that can be angled to direct light through the object being viewed

2.D.1. Microscope Skills

The basic microscope skills are presented as instructions. This allows you to practise these skills before you need to use them in the activities in *Nelson Science Perspectives 10*.

MATERIALS

- scissors
- microscope slide
- cover slip
- medicine dropper
- compound microscope
- compass or Petri dish
- pencil
- transparent ruler
- small piece of newspaper with writing
- small piece of onion membrane
- water
- two pieces of thread, different colours

PREPARING A DRY MOUNT

A dry mount is a way to prepare a microscope slide that does not use water.

1. Find a small, flat object, such as a single letter cut from a newspaper article. A letter "e" works well.
2. Place the object in the centre of a microscope slide.
3. Hold a cover slip between your thumb and forefinger. Place the edge of the cover slip to one side of the object (Figure 5). Gently lower the cover slip onto the slide so that it covers the object.

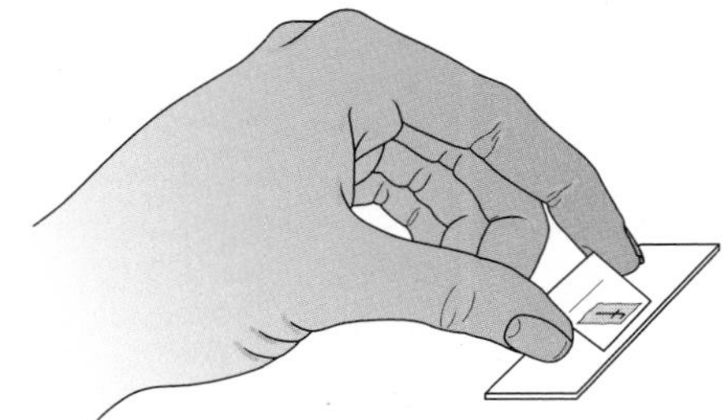

Figure 5

PREPARING A WET MOUNT

A wet mount is a method of preparing a microscope slide that uses a drop of water.

1. Find a small, flat object such as a piece of onion membrane.
2. Place the object in the centre of a microscope slide.
3. Place one or two drops of water on the object (Figure 6).

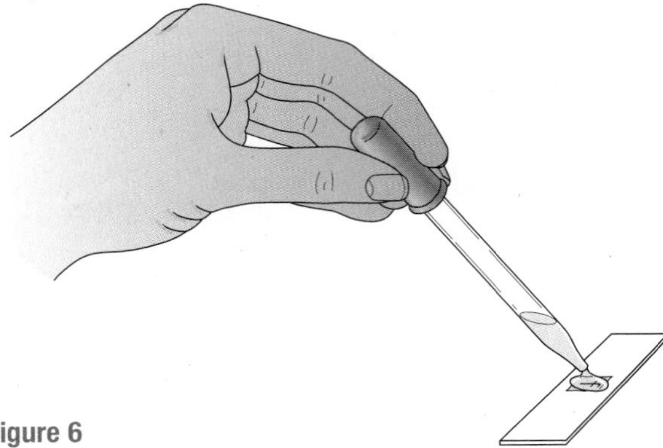

Figure 6

4. Holding the cover slip with your thumb and forefinger at a 45° angle, place one edge of the cover slip on the slide (Figure 7). Gently lower the cover slip, allowing the air to escape.

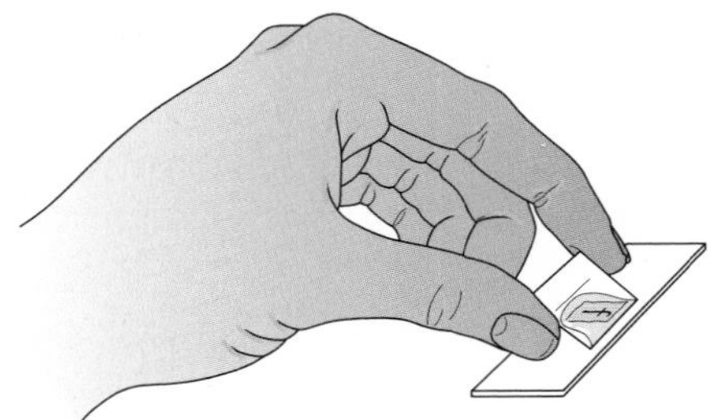

Figure 7

VIEWING OBJECTS UNDER THE MICROSCOPE

1. Make sure that the low-power objective lens is positioned over the diaphragm. Either raise the objective lens or lower the microscope stage as far as possible. Place your dry-mount prepared slide in the centre of the stage. Use the stage clips to hold the slide in position. Turn on the light source (Figure 8).

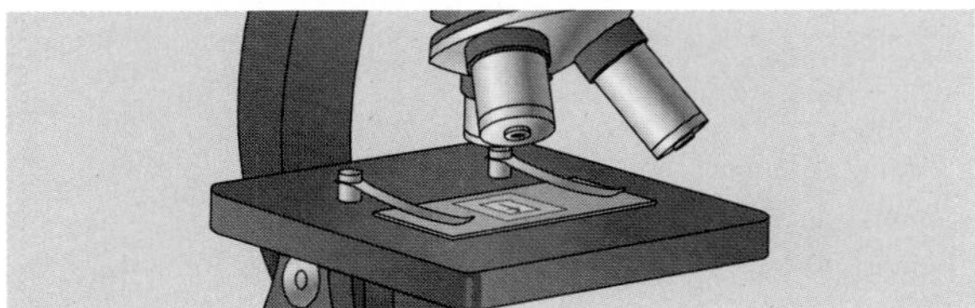

Figure 8

2. View the microscope stage from the side. Then, using the coarse-adjustment knob, lower the low-power objective lens until it is close to the object. (Some microscopes have moveable stages, rather than moveable lenses.) Do not allow the lens to touch the cover slip (Figure 9). Make sure that you know which way to turn the knob to raise the objective lens.

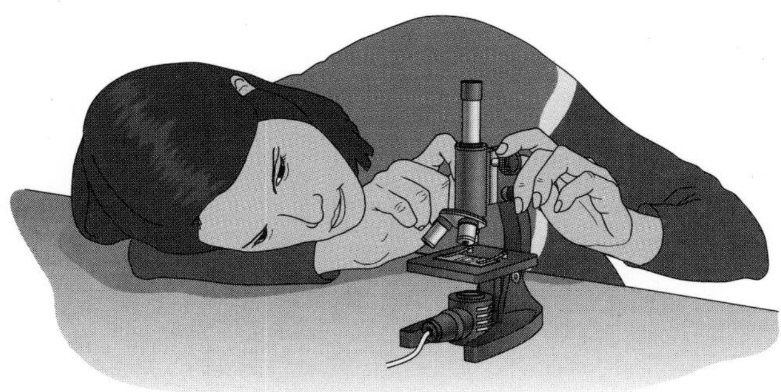

Figure 9

3. Look through the eyepiece. Slowly raise the objective lens using the coarse-adjustment knob until the image is in focus. Note that the object is facing the "wrong" way and is upside down. The area you can see is called the *field of view*.
4. Draw a circle in your notebook to represent the field of view. Look through the microscope and draw what you see. Make the object fill the same proportion of area in your drawing as it does in the microscope.
5. While you look through the microscope, slowly move the slide horizontally away from you. Note that the object appears to move toward you. Now move the slide to the left. Notice that the object appears to move to the right.
6. Rotate the nosepiece to the medium-power objective lens. Use the fine-adjustment knob to bring the object into focus. Notice that the object appears larger. Always use the fine adjustment when the medium- or high-power objectives are in place; the coarse adjustment may damage the slide or lenses.
7. Adjust the position of the object so that it is directly in the centre of the field of view. Rotate the high-power objective lens into place. Again, use the fine-adjustment knob to focus the image. Notice that you see less of the object than you did under medium-power magnification. Also notice that the object appears larger.

DETERMINING THE FIELD OF VIEW

The field of view is the area that you observe when you look through the microscope.

1. Put the low-power objective lens in place. Place a transparent ruler on the stage. Position the millimetre marks of the ruler immediately below the objective lens.
2. Use the coarse-adjustment knob to focus on the marks of the ruler.
3. Move the ruler so that one of the millimetre markings is just at the edge of the field of view. Note the diameter of the field of view, in millimetres, under the low-power objective lens (Figure 10).

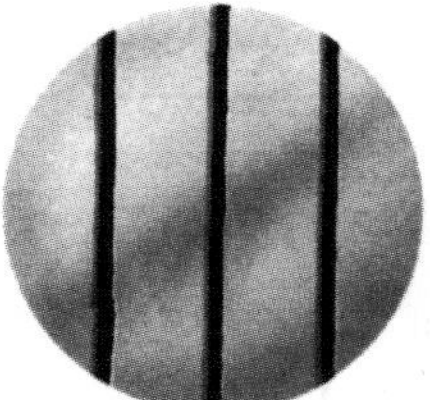

Figure 10

4. Rotate to the medium-power objective lens. Repeat steps 2 and 3 to measure the field of view for this lens.
5. Most high-power objective lenses provide a field of view that is less than 1 mm in diameter, so it cannot be measured with a ruler.

The following steps can be used to calculate the field of view of a high-power lens.

- Calculate the ratio of the magnification of the high-power objective lens to the magnification of the low-power objective lens.

$$\text{ratio} = \frac{\text{magnification of high-power lens}}{\text{magnification of low-power lens}}$$

For example, if the low-power lens is 4× magnification and the high-power lens is 40× magnification, then

$$\text{ratio} = \frac{40\times}{4\times} = 10$$

- Use the ratio to determine the field of diameter (diameter of the field of view) under high-power magnification.

$$\text{field of view diameter (high power)} = \frac{\text{field of view diameter (low power)}}{\text{ratio}}$$

For example, if the diameter of the low-power field of view is 2.5 mm, then

$$\text{field of view diameter (high power)} = \frac{2.5\text{ mm}}{10} = 0.25\text{ mm}$$

ESTIMATING SIZE

1. Measure the field of view, in millimetres, as shown above.
2. Remove the ruler and replace it with the object under investigation.
3. Estimate the number of times the object could fit across the field of view.
4. Calculate the width of the object:

$$\text{width of object} = \frac{\text{width of field of view}}{\text{number of objects across field}}$$

Remember to include units.

PUTTING AWAY THE MICROSCOPE

After you have completed an activity using a microscope, follow these steps:

1. Rotate the nosepiece to the low-power objective lens.
2. Raise the lenses (or lower the stage) as far as possible.
3. Remove the slide and the cover slip (if applicable).
4. Clean the slide and the cover slip and return them to their appropriate location.
5. Return the microscope, using two hands, to the storage area.

2.E. Using Other Scientific Equipment

For some labs in this textbook, a specific list of required materials and equipment is provided. In others, you have to decide what equipment and materials are necessary. Always keep safety in mind as you make your selections. Be sure to include appropriate safety equipment, such as eye protection, gloves, or lab aprons, in your planning. For more information on safety, refer to the Safe Science section of the Skills Handbook (page 600).

Figure 11 shows some of the equipment required for the labs in this text.

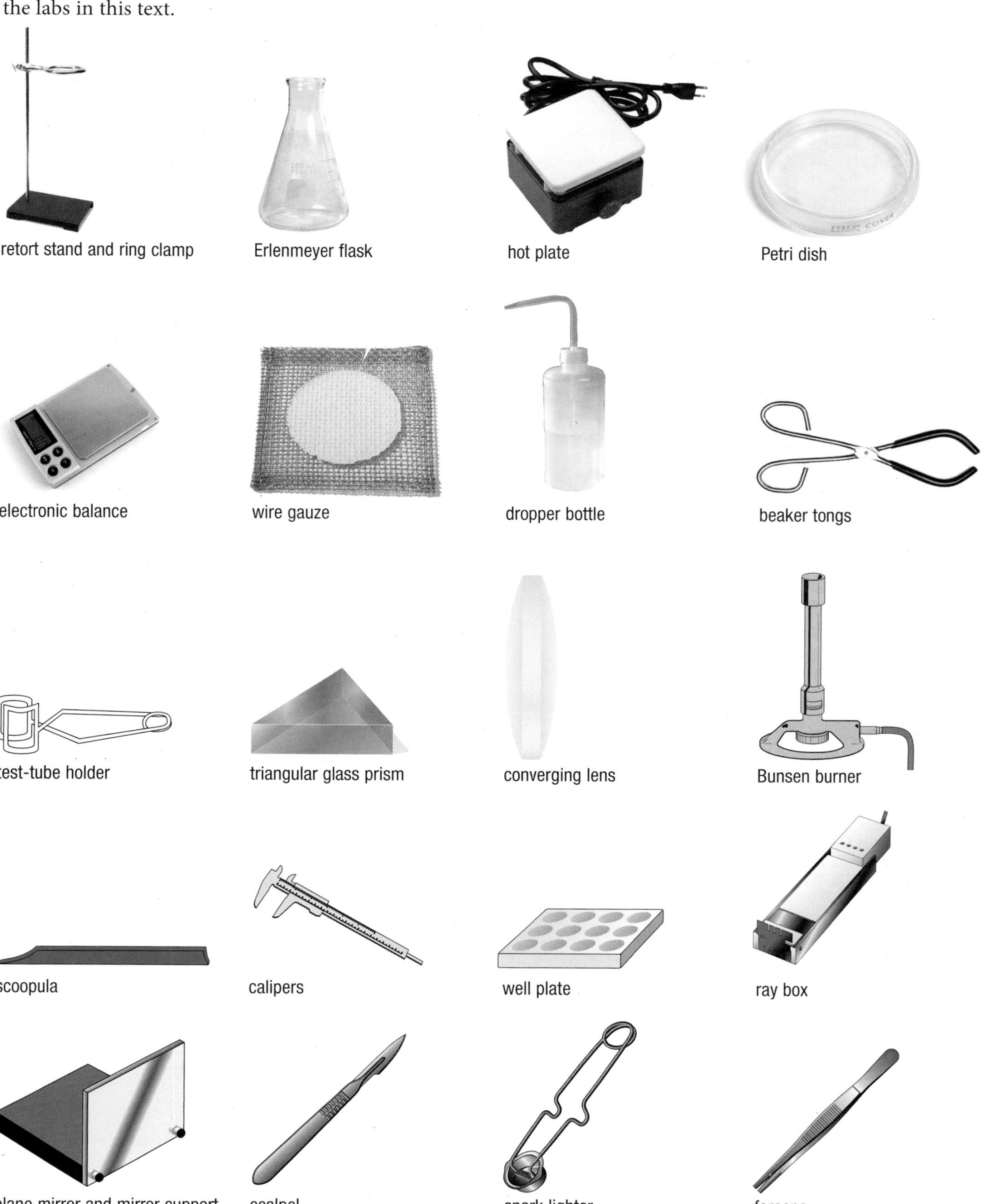

Figure 11 Some typical laboratory equipment

3 SCIENTIFIC INQUIRY SKILLS

3.A. Thinking as a Scientist

Imagine that you are planning to buy a new personal electronic device. First, you write a list of questions. Then you check print and Internet sources, visit stores, and talk to your friends to find out which is the best purchase. When you try to solve problems in this way, you are conducting an investigation and thinking like a scientist.

- Scientists investigate the natural world to describe it. For example, climatologists study the growth rings of trees to make inferences about what the climate was like in the past (Figure 1).

Figure 1 Looking at and measuring tree rings tell scientists how climate affects the growth of trees.

- Scientists investigate objects to classify them. For example, chemists classify substances according to their properties (Figure 2).

Figure 2 These compounds are similar because they are insoluble in water.

- Scientists investigate the natural world to test their ideas about it. For example, biologists ask cause-and-effect questions about the impact of climate change on water in the Arctic (Figure 3). They also propose hypotheses to answer their questions. Then they design investigations to test their hypotheses. This process leads scientists to come up with new ideas to be tested and more questions that need answers.

Figure 3 Water samples can be tested for pH, dissolved substances, and other impurities.

3.B. Scientific Inquiry

You need to use a variety of skills when you do scientific inquiry and design or carry out an experiment. Refer to this section when you have questions about any of the following skills and processes:

- questioning
- hypothesizing
- predicting
- planning
- controlling variables
- performing
- observing (Figure 4, next page)
- analyzing
- evaluating
- communicating

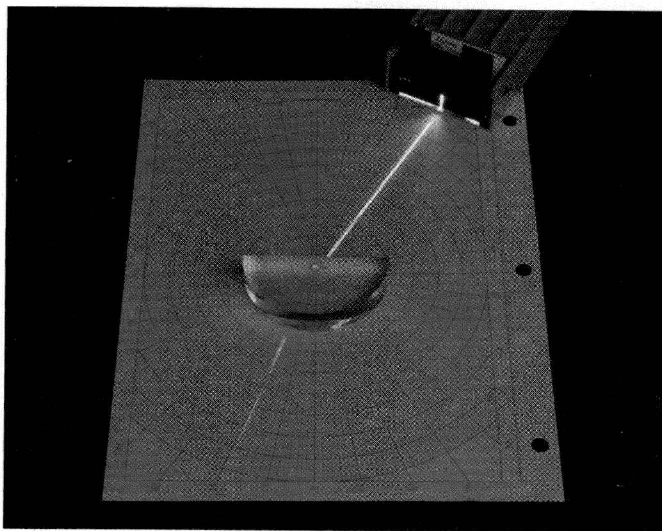

Figure 4 In each unit you will have an opportunity to develop new skills of scientific inquiry.

3.B.1. Questioning

Scientific investigations result from our curiosity about the natural world. Our observations lead us to wonder "Why does that happen?" or "What would happen if I …?" Before you can conduct an investigation, you must have a clear idea of what you want to know. This will help you develop a question that will lead you to the information you want. Try to avoid questions that lead simply to "yes" and "no" answers. Instead, develop questions that are testable and lead to an investigation.

Sometimes, an investigation starts with a special type of question called a "cause-and-effect" question. A cause-and-effect question asks whether something is causing something else to happen. It might start in one of the following ways: What causes …? How does … affect …?

3.B.2. Controlling Variables

Considering the variables involved is an important step in designing an effective investigation. Variables are any factors that could affect the outcome of an investigation.

There are three kinds of variables in an investigation:

1. The variable that is changed is called the independent variable, or cause variable.
2. The variable that is affected by a change is called the dependent variable, or effect variable. This is the variable you measure to see how it was affected by the independent variable.
3. All the other conditions that remain unchanged and did not affect the outcome of the investigation are called the controlled variables.

For example, consider the variables involved in an investigation to determine the effects of the mass of dissolved salt on the boiling point of water (Figure 5). In this investigation,

- the mass of dissolved salt is the independent variable
- the boiling point is the dependent variable
- the amount of water in the beaker and the type of salt used are two controlled variables

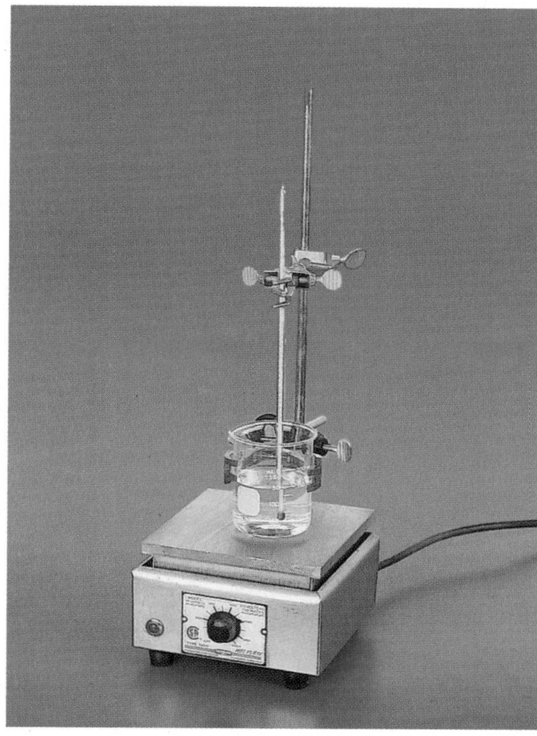

Figure 5 Measuring the temperature at which water, containing a known mass of salt, boils.

When scientists conduct controlled experiments (described in Section 1.1), they make sure that they change only one independent variable at a time. This way, they may assume that their results are caused by the variable they changed and not by any of the other variables they identified.

3.B.3. Predicting and Hypothesizing

A prediction states what is likely to happen as the result of a controlled experiment. Scientists base predictions on their observations and knowledge. They look for patterns in the data they gather to help them understand what might happen next or in a similar situation (Figure 6). A prediction may be written as an "if ... then ..." statement. For this investigation, your prediction might be, "If the amount of dissolved salt is increased, then the boiling point will also increase."

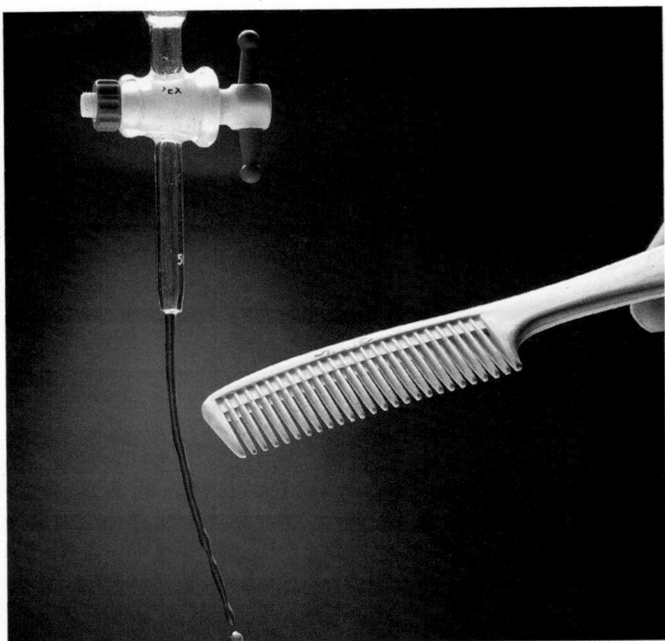

Figure 6 A positively charged piece of plastic causes a stream of tap water to bend. What would you predict might happen if a negatively charged object is brought near the stream of water?

In summary, a prediction states what you think will happen. Remember, however, that predictions are not guesses. They are suggestions based on prior knowledge and logical reasoning.

A prediction can be used to generate a hypothesis. A hypothesis is a tentative answer about the outcome of a controlled experiment along with an explanation for the outcome. A hypothesis may be written in the form of an "if ... then ... because ..." statement. *If* the cause variable is changed in a particular way, *then* the effect variable will change in a particular way, and this change occurs *because* of certain reasons. For example, "If the amount of salt is increased, then the boiling point will also increase because salt forms attractions with water molecules and prevents them from changing into gas." If your observations confirm your prediction, then they support your hypothesis. You can create more than one hypothesis from the same question or prediction. Another student might test the hypothesis, "If the amount of salt is increased, then the boiling point will not change because the attractions that salt forms with water are very weak." Of course, both of you cannot be correct. When you conduct an investigation, your observations do not always confirm your prediction. Sometimes, they show that your hypothesis is incorrect. An investigation that does not support your hypothesis is not a bad investigation or a waste of time. It has contributed to your scientific knowledge. You can re-evaluate your hypothesis and design a new investigation.

3.B.4. Planning

You have been asked to design and carry out the boiling point investigation described in 3.B.3. First, you create an experimental design. To conduct a controlled experiment that tests your hypothesis, you decide to change only one variable—the amount of dissolved salt. This is your independent variable. You dissolve 5.0 g, 10.0 g, 15.0 g, and 20.0 g samples of salt in identical beakers containing exactly 100 mL of water. You also prepare a beaker containing water with no salt. Each solution is then heated using the same hot plate until it boils. A thermometer that is suspended in each solution measures the temperature at which boiling occurs—the dependent variable.

Now consider the equipment and materials you need. Be sure to include any safety equipment, such as an apron or eye protection, in your equipment and materials list. How will you secure the beaker so that it doesn't accidentally fall off the hot plate as it boils? Will you use tap water or distilled water?

You need to write a procedure—a step-by-step description of how you will perform your investigation. A procedure should be written as a series of numbered steps, with only one instruction per step. For example:

1. Wear safety goggles, a protective apron, and heat-resistant gloves.
2. Set up the equipment as shown in the diagram. (Include a clearly labelled diagram.)
3. Add 100 mL of distilled water to the beaker.
4. Add one boiling chip to the beaker to ensure that the water boils gently and does not bump.
5. Dissolve 5.0 g of sodium chloride in the water.
6. Set the heating control on the hot plate to 50 %.
7. Wait for the mixture to boil.
8. Turn off the heat and allow the beaker to cool.
9. Repeat steps 3–8 using 10.0 g, 15.0 g, and 20.0 g of sodium chloride.
10. Allow the apparatus to cool before dismantling it.
11. Once cool, return all equipment to the storage cabinet.

Your procedure must be clear enough for someone else to follow, and it must explain how you will deal with each of the variables in your investigation. The first step in a procedure usually refers to any safety precautions that need to be addressed, and the last step relates to any cleanup that needs to be done. Your teacher must approve your procedure and list of equipment and materials before you begin.

3.B.5. Performing

As you carry out an investigation, be sure to follow the steps in the procedure carefully and thoroughly. Check with your teacher if you find that significant alterations to your procedure are required. Use all equipment and materials safely, accurately, and with precision. Be sure to take detailed, careful notes and to record all of your observations. Record numerical data in a table.

3.B.6. Observing

When you observe something, you use your senses to learn about it. You can also use tools, such as a balance or a microscope. Observations of measured quantities such as temperature, volume, and mass are called quantitative observations. Numerical data from quantitative observations are usually recorded in data tables or graphs.

Other observations describe characteristics that cannot be expressed in numbers. These are called qualitative observations (Figure 7). Colour, smell, clarity, and state of matter are common examples of qualitative observations. Qualitative observations can be recorded using words, pictures, or labelled diagrams.

Figure 7 Qualitative observations such as a colour change, bubbles, an irritating odour, or a sizzling sound suggest that a chemical change is occurring when certain chemicals are mixed.

As you work on an investigation, be sure to record all of your observations, both qualitative and quantitative, clearly and carefully. If a data table is appropriate for your investigation, use it to organize your observations and measurements. (See Data Tables and Graphs, page 634.) Include all observations and measurements in your final lab report or presentation. It is important to remain impartial when recording observations. Record exactly what you observe. Observations from an experiment may not always be what you expect them to be.

SCIENTIFIC DRAWINGS

Scientific drawings are done to record observations as accurately as possible. They are also used to communicate, which means that they must be clear, well labelled, and easy to understand. Below are some tips that will help you produce useful scientific drawings.

Getting Started

The following materials and ideas will help you get started:

- Use plain, blank paper. Lines might obscure your drawing or make your labels confusing.
- Use a sharp pencil rather than a pen or marker as you will probably need to erase parts of your drawing and do them over again (Figure 8).
- Observe and study your specimen or equipment carefully, noting details and proportions, before you begin the drawing.

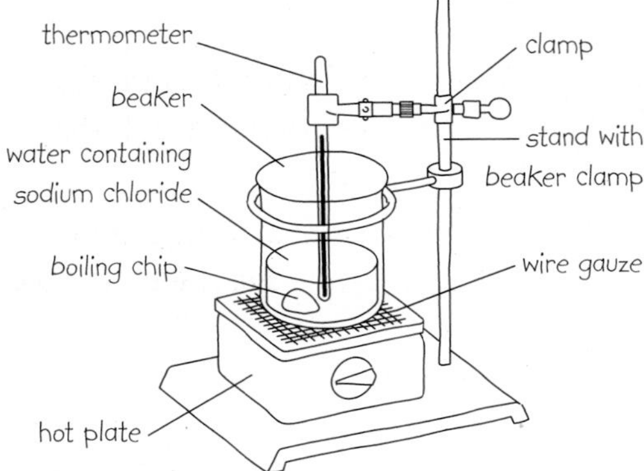

Figure 8 Your drawing of an experimental setup should show how the equipment was assembled.

- Create drawings large enough to show details. For example, a third of a page might be appropriate for a diagram of a single cell or a unicellular organism. When drawing lab equipment, do not include unnecessary details.
- Label your diagram clearly. Use a ruler to draw the label lines.

Scale Ratio

You may want to indicate the actual size of your object on your drawing. To do this, use a ratio called the scale ratio.

- If your diagram is 10 times larger than the real object (e.g., a tiny organism), your scale ratio is 10×.
- In general, scale ratio = $\frac{\text{size of drawing}}{\text{actual size of object}}$

You can also show the actual size of the object on your diagram (Figure 9).

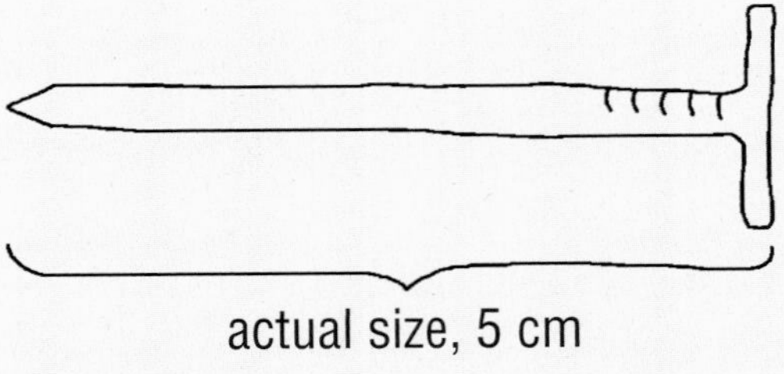

Figure 9 This is an example of a scientific drawing that indicates the actual size of the object.

Checklist for Scientific Drawing

- ✔ Use plain, blank (unlined) paper and a sharp, hard pencil.
- ✔ Draw as large as necessary to show details clearly.
- ✔ Do not shade or colour.
- ✔ Draw label lines that are straight and run outside your drawing. Use a ruler.
- ✔ Include labels, a title or caption, and, if appropriate, the magnification of the microscope you are using.

3.B.7. Analyzing

You analyze data from an investigation to make sense of it. You examine and compare the measurements you have made. You look for patterns and relationships that will help you explain your results and give you new information about the question you are investigating.

Once you have analyzed your data, you can tell whether your prediction or hypothesis is correct. You can also write a conclusion that indicates whether or not the data supports your hypothesis (Figure 10). You may even come up with a new hypothesis that can be tested in a new investigation.

Figure 10 Students analyze data and make notes about their observations in order to find any patterns.

3.B.8. Evaluating

How useful is the evidence from an investigation? You need quality evidence before you can evaluate your prediction or hypothesis. If the evidence is poor or unreliable, you can identify areas of improvement for when the investigation is repeated.

Below are some things to consider when evaluating the results of an investigation:

- *Plan:* Were there any problems with the way you planned your experiment or your procedure? Did you control for all the variables, except the independent variable?
- *Equipment and Materials:* Could better or more accurate equipment have been used? Was something used incorrectly? Did you have difficulty with a piece of equipment?
- *Observations:* Did you record all the observations that you could have? Or did you ignore or overlook some observations that might have been important?
- *Skills:* Did you have the appropriate skills for the investigation? Did you have to use a skill that you were just beginning to learn?

Once you have identified areas in which errors could have been made, you can judge the quality of your evidence.

3.B.9. Communicating

When you plan and carry out your own investigation, it is important to share both your process and your findings. Other people may want to repeat your investigation, or they may want to use or apply your findings in another situation. Your write-up or report should reflect the process of scientific inquiry that you used in your investigation.

SAMPLE LAB REPORT

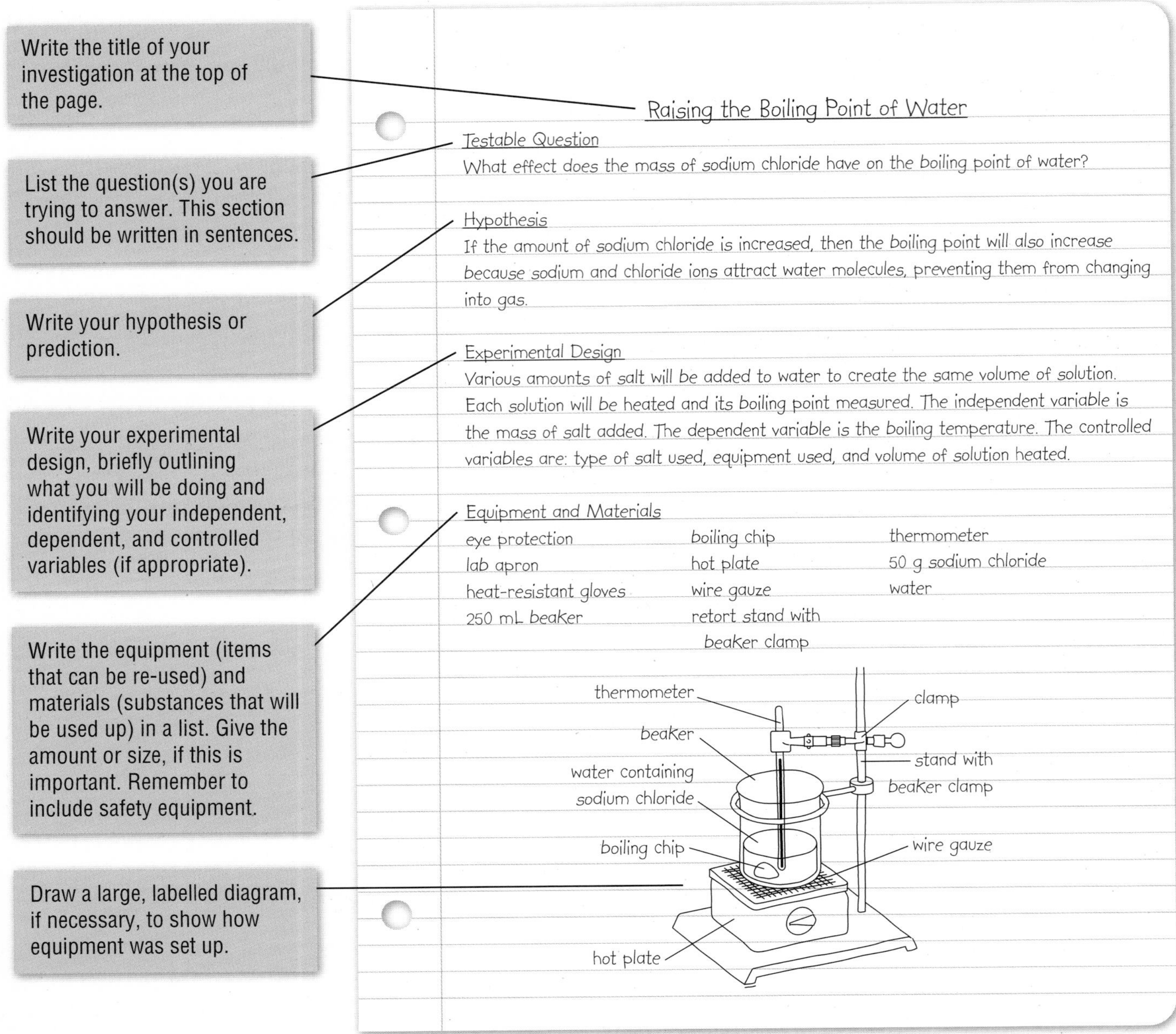

Write the title of your investigation at the top of the page.
List the question(s) you are trying to answer. This section should be written in sentences.
Write your hypothesis or prediction.
Write your experimental design, briefly outlining what you will be doing and identifying your independent, dependent, and controlled variables (if appropriate).
Write the equipment (items that can be re-used) and materials (substances that will be used up) in a list. Give the amount or size, if this is important. Remember to include safety equipment.
Draw a large, labelled diagram, if necessary, to show how equipment was set up.
Raising the Boiling Point of Water
Testable Question
What effect does the mass of sodium chloride have on the boiling point of water?
Hypothesis
If the amount of sodium chloride is increased, then the boiling point will also increase because sodium and chloride ions attract water molecules, preventing them from changing into gas.
Experimental Design
Various amounts of salt will be added to water to create the same volume of solution. Each solution will be heated and its boiling point measured. The independent variable is the mass of salt added. The dependent variable is the boiling temperature. The controlled variables are: type of salt used, equipment used, and volume of solution heated.
Equipment and Materials
eye protection
lab apron
heat-resistant gloves
250 mL beaker
boiling chip
hot plate
wire gauze
retort stand with beaker clamp
thermometer
50 g sodium chloride
water
thermometer
clamp
beaker
water containing sodium chloride
stand with beaker clamp
boiling chip
wire gauze
hot plate

Procedure

1. Safety goggles, a protective apron, and heat resistant gloves were obtained.
2. The equipment was set up as shown in the diagram.
3. 100 mL distilled water was added to the beaker.
4. One boiling chip was added to the beaker to prevent the solution from bumping.
5. 5.0 g sodium chloride was dissolved in the water.
6. The hot plate heater control was set to 50 %.
7. The temperature at which the water boiled was recorded.
8. The heat was turned off and the beaker was allowed to cool.
9. Steps 3–8 were repeated using 10.0 g, 15.0 g, and 20.0 g of sodium chloride.
10. The apparatus was allowed to cool before it was returned.

Describe the procedure using numbered steps. Each step should start on a new line. Write the steps as they occurred, using past tense and passive voice. Make sure that your steps are clear so that someone else could repeat your investigation. Include any safety precautions.

Observations

Table of Observations

Mass of sodium chloride (g)	Boiling point (°C)
5	100.4
10	100.8
15	101.4
20	101.8

In each case, the water boiled gently. Small bubbles formed throughout the solution as it boiled.

Present your observations in a form that is easily understood. Quantitative observations should be recorded in one or more tables, with units included. Qualitative observations can be recorded in words or drawings.

Analysis and Evaluation

The boiling point of the solution increased as more salt was dissolved in it. Water condensing on the thermometer made some of the temperatures difficult to read. This problem could be overcome by using a temperature probe instead of a thermometer. The data clearly support the hypothesis that the boiling point of water increases as the mass of dissolved sodium chloride increases.

Analyze your results and evaluate your procedure. If you have created graphs, refer to them here. If necessary, include them on a separate piece of graph paper. Write a conclusion indicating whether or not your results support your hypothesis or prediction. Answer any Analyze and Evaluate questions here.

Applications and Extensions

Adding table salt (sodium chloride) to boiling water does not increase the boiling point of the water a great deal. Therefore, adding a little salt to boiling water when cooking will not speed up the cooking process.

Describe how the knowledge you gained from your investigation relates to real-life situations. How can this knowledge be used? Answer any Apply and Extend questions here.

In modern society, you are constantly bombarded with information. Some of this information is reliable, and some is not. Trying to find the "right" information to conduct scientific research may seem overwhelming. However, the task is less daunting when you learn how to search efficiently for the information you need. Then you must know how to assess its credibility. Here are some tips that will help you in your research.

4.A. General Research Skills

4.A.1. Identify the Information You Need

- Identify your research topic.
- Identify the purpose of your research.
- Identify what you already know about the topic.
- Identify what you do not yet know.
- Develop a list of key questions that you need to answer.
- Identify categories based on your key questions.
- Use these categories to identify key search words.

4.A.2. Identify Sources of Information

Identify places where you could look for information about your topic. These places might include programs on television, people in your community, print sources (Figure 1), and electronic sources (such as CD-ROMs and Internet sites).

Figure 1 Your school library and local public library are both excellent sources of information.

Refer to "Using the Internet" on page 619. Remember, gathering information from a variety of sources will improve the quality of your research.

4.A.3. Evaluate the Sources of Information

Read through your sources of information and decide whether they are useful and reliable. Here are five things to consider:

- *Authority:* Who wrote or developed the information or who sponsors the website? What are the qualifications of this person or group?
- *Accuracy:* Are there any obvious errors or inconsistencies in the information? Does the information agree with that of other reliable sources?
- *Currency:* Is the information up to date? Has recent scientific information been included?
- *Suitability:* Does the information make sense to someone with your experience or of your age? Do you understand it? Is the information well organized?
- *Bias:* Are facts reported fairly? Are there reasons why your sources might express some bias? Are facts deliberately left out?

4.A.4. Record and Organize the Information

After you have gathered and evaluated your sources of information, you can start organizing your research. Identify categories or headings for note taking. In your notebook, use point-form notes to record information in your own words under each heading. You must be careful not to copy information directly from your sources. If you quote a source, use quotation marks. Record the title, author, publisher, page number, and date for each of your sources. For websites, record the URL (website address). All of these details are necessary to help you keep track of your sources of information so that you can go back to them to clarify any points in the future. You will also need this information to create a bibliography. If necessary, add to your list of questions as you find new information.

To help you organize the information further, you may want to use pictures, graphic organizers, and diagrams. (See the Literacy section on page 641.)

4.A.5. Make a Conclusion

Look at your original research question. What did you learn from the information you gathered? Can you state and explain a conclusion based on that information? Do you need further information? If so, where would you look for that information? Do you have an informed opinion on the research topic that you did not have before you started? If not, what additional information do you need to reach such an opinion?

4.A.6. Evaluate Your Research

Now that your research is complete, reflect on how you gathered and organized the information (Figure 2). Can you think of ways to improve the research process for next time? How valuable were the sources of information you selected?

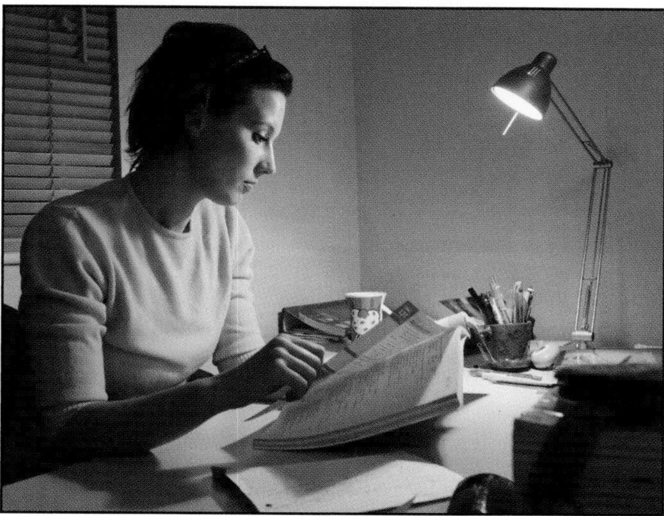

Figure 2 Keep track of your sources so you do not forget where you found your information.

4.A.7. Communicate Your Conclusions

Choose a format for communication that suits your audience, your purpose, and the information you gathered. Are labelled diagrams, graphs, or charts appropriate?

4.B. Using the Internet

The Internet is a vast and constantly growing network of information. You can use search engines to help you find what you need, but keep in mind that not everything you find will be useful, reliable, or true.

4.B.1. Search Results

Once you have entered your search word or phrase, a list of web page "matches" will appear. If your keywords are general, you are likely to get a high number of matches. Therefore, you need to refine your search. Most search engines provide online help and search tips. Look at these to find ideas for better searching.

Every web page has a URL (universal resource locator). The URL may tell you the name of the organization hosting the web page, or it may indicate that you are looking at a personal page (often indicated by the ~ character in the URL). The URL also includes a domain name, which provides clues about the organization hosting the web page (Table 1). For example, a URL that includes "ec.gc.ca" indicates that the content is hosted by Environment Canada—a reliable source.

Table 1 Common URL Codes and Organizations

Code	Organization
ca	Canada
com or co	commercial
edu or ac	educational
org	nonprofit
net	networking provider
mil	military
gov	government
int	international organization

4.B.2. Evaluating Internet Resources

Anyone can post information on the Internet without verifying its accuracy. Therefore, you must learn to evaluate the information that you find on the Internet as coming from dependable and legitimate sources.

Use the following questions to help determine the quality of an Internet source. The greater the number of questions answered "yes," the more likely it is that the source is of high quality.

- Is it clear who is sponsoring the page? Does the site seem to be permanent or sponsored by a reputable organization?
- Is there information about the sponsoring organization? For example, is a telephone number or address given to contact for more information?
- Is it clear who developed and wrote the information? Are the author's qualifications provided?
- Are the sources for factual information given so that they can be checked?
- Are there dates to indicate when the page was written, placed online, or last revised?
- Is the page presented as a public service? Does it present balanced points of view?

4.B.3. Using School Library Resources

Many schools and school boards have access to online encyclopedias with science sections in them. Find out if your school or board has a website where you can access these resources. You may need a password.

4.B.4. Using the Nelson Website

When you see the Nelson Science icon in your textbook, you can go to the Nelson website and find links to useful sources of information.

4.C. Exploring an Issue Critically

An issue is a situation in which several points of view need to be considered in order to make a decision. It is often difficult to come to a decision that everyone agrees with. When a decision affects many people or the environment, it is important to explore the issue critically. Think about all the possible solutions and try to understand all the different perspectives—not just your own point of view. Consider the risks and benefits of each possible solution. Put yourself in the place of several of the stakeholders, to try to understand their positions.

Exploring an issue critically also means researching and investigating your ideas and communicating with others. Figure 3 shows all the steps in the process.

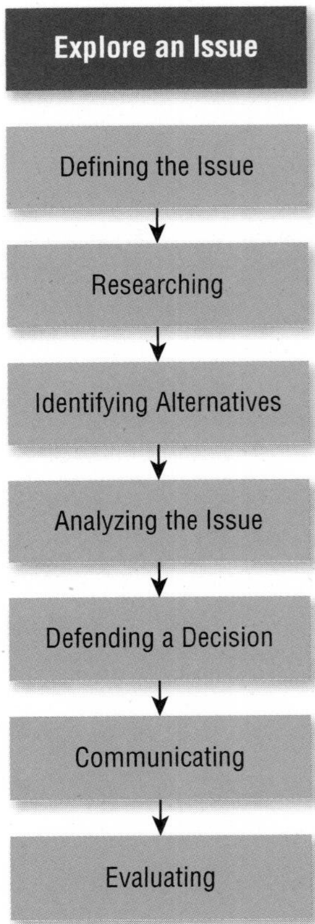

Figure 3 You may perform some or all of these steps as you explore an issue critically.

4.C.1. Defining the Issue

To explore an issue, first identify what the issue is. An issue has more than one solution, and there are different points of view about which solution is the best. Rephrase the issue as a question: "What should …?" The issue can also include information about the *role* a person takes when thinking about an issue. For instance, you may think about the issue from someone else's point of view—you may take the role of a landowner, a government worker, or a tour guide. The issue can also include a description of who your audience will be—will it be other students, a meeting of government officials, or your parents? Be sure to take into account your role and audience when defining your issue.

4.C.2. Researching

Ensure that the decision you reach is based on a good understanding of the issue. You must be in a position to choose the most appropriate solution. To do this, you need to gather factual information that represents all the different points of view. Develop good questions and a plan for your research. Your research may include talking to people, reading about the topic in newspapers and magazines, and doing Internet research.

As you collect information, make sure that it is reliable, accurate, and current. Avoid biased information that favours only one side of the issue. It is important to ensure that the information you have gathered represents all aspects of an issue. Are the sources valuable? Could you find better information elsewhere?

4.C.3. Identifying Alternatives

Consider possible solutions to the issue. Different stakeholders may have different ideas on this. Consider all reasonable options. Be creative about combining the suggestions. For example, suppose that your municipal council is trying to decide how to use some vacant land next to your school. You and other students have asked the council to use the land as a nature park. Another group is proposing that the land be used to build a seniors' home because there is a shortage of this kind of housing. The school board would like to use the land to build a track for sporting events.

After defining the issue and researching, you can now generate a list of possible solutions. You might, for example, come up with the following choices for the land-use issue:

- Turn the land into a nature park for the community and the school.
- Use the land as a playing field and track for the community and the school.
- Create a combination of a nature park and a playing field.
- Use the land to build a seniors' home with a nature park.

4.C.4. Analyzing the Issue

Develop criteria to evaluate each possibility. For example, should the solution be the one that has the most community support, or should it be the one that best protects the environment? Should it be the least costly, financially, or the one that creates the most jobs? You need to decide which criteria you will use to evaluate the alternatives so that you can decide which solution is the best.

4.C.5. Defending a Decision

This is the stage where everyone gets a chance to share ideas and information gathered about the issue. Then the group needs to evaluate all the possible alternatives and decide on one solution based on the criteria.

COST–BENEFIT ANALYSIS

A cost–benefit analysis can help you determine the best solution to a complex problem when a number of solutions are possible. First, research possible costs and benefits associated with a proposed solution. Costs are not always financial. You may be comparing advantages and disadvantages. Then, based on your research, try to decide the level of importance of each cost and benefit. This is often a matter of opinion. However, your opinions should be informed by researched facts.

Once you have completed your research and identified costs and benefits, you may conduct the cost–benefit analysis as follows:

1. Create a table similar to Table 2.
2. List costs and benefits.
3. Rate each cost and benefit on a scale from 1 to 5, where 1 represents the least important cost or benefit and 5 represents the most important cost or benefit.
4. After rating each cost and benefit, add up the results to obtain totals. If the total benefits outweigh the total costs, you may decide to go ahead with the proposed solution.

Table 2 Cost–Benefit Analysis of Using Land to Build a Seniors' Home with a Nature Park

Costs		Benefits	
Possible result	**Cost of result (rate 1 to 5)**	**Possible result**	**Benefit of result (rate 1 to 5)**
land cannot be used for sports	2	seniors' home provides necessary housing	5
expensive to maintain	4	park preserves some habitat for plants and animals	4
nature park will be very small	3	park increases value of seniors home	3
Total cost value	**9**	**Total benefit value**	**12**

4.C.6. Communicating

You might be told how you will communicate your decision. For example, your class might hold a formal debate. Alternatively, you might be free to choose your own method of communication.

You could choose one of the following methods of communicating your decision:

- Write a report.
- Give an oral presentation.
- Make a poster.
- Prepare a slide show.
- Create a video (Figure 4).
- Organize a town hall or panel discussion.
- Create a blog or webcast.
- Write a newspaper article.

Figure 4 Creating a video is an effective way to communicate information about an issue.

Choose a type of presentation that will share your decision or recommendation in a way that is suitable for your audience. For example, if your audience is small, it might be easiest to present your decision in person. An oral presentation is a good way to present your decision to many people at one time. If your presentation includes visuals, ensure that they are large and clear enough for your audience to see. Creating a poster or a blog allows people to read your recommendation on their own, but you must find a way to let others know where to find this information.

Whatever means you use, however, you should

- state your position clearly, considering your audience;
- support it with objective data if possible, and with a persuasive argument; and
- be prepared to defend your position against opposition (Figure 5).

Figure 5 Share your information in a way that makes it easy for your audience to understand.

4.C.7. Evaluating

The final step of the decision-making process includes evaluating the decision itself and the process used to reach the decision. After you have made a decision, carefully examine the thinking that led to this decision. Some questions to guide your evaluation include the following:

- What was my initial perspective on the issue? How has my perspective changed since I first began to explore the issue?
- How did I gather information about the issue? What criteria did I use to evaluate the information? How satisfied am I with the quality of my information?
- What information did I consider to be the most important when making my decision?
- How did I make my decision? What process did I use? What steps did I follow?
- To what extent were my arguments factually accurate and persuasively made? (Figure 6)
- In what ways does my decision resolve the issue?
- What are the likely short-term and long-term effects of my decision?
- How might my decision affect the various stakeholders?
- To what extent am I satisfied with my decision?
- If I had to make this decision again, what would I do differently?

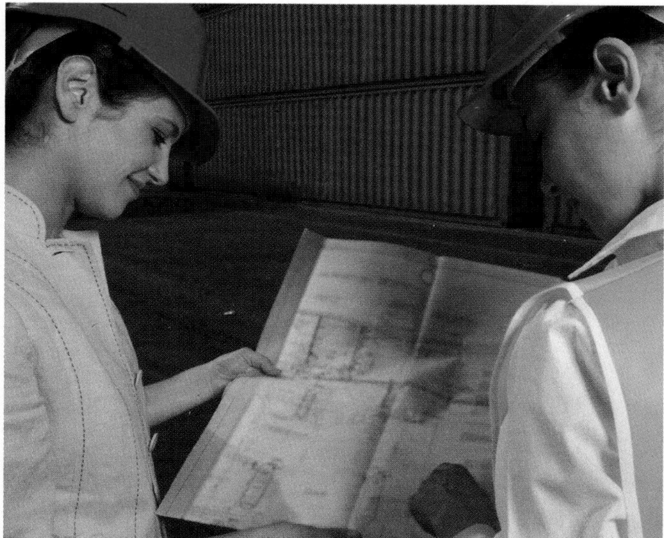

Figure 6 Were the arguments clearly presented and backed up by evidence?

Effective communication of experimental data is an important part of science. To avoid confusion when reporting measurements or using measurements in calculations, there are a few accepted conventions and practices that should be followed.

5.A. SI Units

The scientific communities of many countries, including Canada, have agreed on a system of measurement called SI (Système international d'unités). This system consists of the seven fundamental SI units, called base units, shown in Table 1.

Table 1 The Seven SI Base Units

Quantity name	Unit name	Unit symbol
length	metre	m
mass	kilogram	kg*
time	second	s
electric current	ampere	A
temperature	kelvin	K**
amount of substance	mole	mol
light intensity	candela	cd

*The kilogram is the only base unit that contains a prefix.
**Although the base unit for temperature is a kelvin (K), the common unit for temperature is a degree Celsius (°C).

All other physical quantities can be expressed as a combination of these seven SI base units. For example, the speed of an object is determined by the distance it travels in a specified time period. Therefore, the unit for speed is metres (distance) per second (time) or m/s. Units that are formed using two or more base units are called derived units. Some derived units have special names and symbols. For example, the unit of force that causes a mass of 1 kg to accelerate at a rate of 1 m/s^2 (metre per second per second) is known as a newton (N). In base units, the newton is $m{\cdot}kg/s^2$. The dot between m and kg means "multiplied by," but m·kg is simply read as "metre kilogram." The slash means "divided by" and is read "per." The whole unit is read "metre kilogram per second squared." You can see why a special name and symbol are given to some derived units.

Some common quantities and their units are listed in Table 2. Note that the symbols representing the quantities are italicized, whereas the unit symbols are not.

Table 2 Common Quantities and Units

Quantity name	Quantity symbol	Unit name	Unit symbol
distance	*d*	metre	m
area	*A*	square metre	m^2
volume	*V*	cubic metre	m^3
		litre	L
speed	*v*	metre per second	m/s
acceleration	*a*	metre per second per second	m/s^2
concentration	*c*	gram per litre	g/L
temperature	*t*	degree Celsius	°C
pressure	*p*	pascal	Pa
energy	*E*	joule	J
work	*W*	joule	J
power	*P*	watt	W
electric potential	*V*	volt	V
electrical resistance	*R*	ohm	Ω
current	*I*	ampere	A

5.A.1 Converting Units

An important feature of SI is the use of prefixes to express small or large sizes of any quantity conveniently. SI prefixes act as multipliers to increase or decrease the value of a number in multiples of 10 (Table 3, next page). The most common prefixes change the size in multiples of 1000 (10^3 or 10^{-3}), except for *centi* (10^{-2}), as in centimetre.

SI prefixes are also used to create conversion factors (ratios) to convert between larger or smaller values of a unit. For example,

1 km = 1000 m

Therefore, $\frac{1\text{kg}}{1\ 000\text{ g}} = \frac{1\ 000\text{ g}}{1\text{ kg}} = 1$

Multiplying by a conversion factor is like multiplying by 1: it does not change the quantity, only the unit in which it is expressed. Let's see how to convert from one unit to another.

Table 3 Common SI Prefixes

Prefix	Symbol	Factor by which unit is multiplied	Example
giga	G	1 000 000 000	1 000 000 000 m = 1 Gm
mega	M	1 000 000	1 000 000 m = 1 Mm
kilo	k	1 000	1 000 m = 1 km
hecto	h	100	100 m = 1 hm
deca	da	10	10 m = 1 dam
		1	
deci	d	0.1	0.1 m = 1 dm
centi	c	0.01	0.01 m = 1 cm
milli	m	0.001	0.001 m = 1 mm
micro	μ	0.000 001	0.000 001 m = 1 μm
nano	n	0.000 000 001	0.000 000 001 m = 1 nm

SAMPLE PROBLEM 1 Using Conversion Factors

A block of cheese at a grocery store has a mass of 1 256 g. Its price is $15.00/kg. What is the price of the block of cheese?

First, convert the mass from grams to kilograms.

There are two possible conversion factors between g and kg, as shown above.

Always choose the conversion factor that cancels the original unit. In this case, the original unit is g, so the correct conversion factor is $\frac{1\text{ kg}}{1\ 000\text{ g}}$

$$1\ 256\ \cancel{\text{g}} = \frac{1\text{ kg}}{1\ 000\ \cancel{\text{g}}} = 1.256\text{ kg}$$

The original units, g, cancel (divide to give 1), leaving kg as the new unit.

Now that you have determined the mass in kg, multiply the price per kg by the mass in kg.

$$\$15.00/\text{kg} \times 1.256\text{ kg} = \$18.84$$

The price of the block of cheese is $18.84.

PRACTICE

Conversions

Make the following conversions. Refer to Table 3 if necessary.

(a) Write 3.5 s in ms.

(b) Change 5.2 A to mA.

(c) Convert 7.5 μg to ng.

Convenient conversion factors to convert between millimetres and metres are

$$\frac{1\ 000\text{ mm}}{1\text{ m}} \text{ and } \frac{1\text{ m}}{1\ 000\text{ mm}}$$

Conversion factors can be used for any unit equality, such as 1 h = 60 min and 1 min = 60 s.

5.B. Solving Numerical Problems Using the GRASS Method

In science and technology, you sometimes have problems that involve quantities (numbers), units, and mathematical equations. An effective method for solving these problems is the GRASS method. This always involves five steps: Given, Required, Analysis, Solution, and Statement.

Given: Read the problem carefully and list all of the values that are given. Remember to include units.

Required: Read the problem again and identify the value that the question is asking you to find.

Analysis: Read the problem again and think about the relationship between the given values and the required value. There may be a mathematical equation you could use to calculate the required value using the given values. If so, write the equation down in this step. Sometimes it helps to sketch a diagram of the problem.

Solution: Use the equation you identified in the "Analysis" step to solve the problem. Usually, you substitute the given values into the equation and calculate the required value. Do not forget to include units and to round off your answer to an appropriate number of digits. (See Sections 5.C. and 5.D. on significant digits and scientific notation for help.)

Statement: Write a sentence that describes your answer to the question you identified in the "Required" step.

Sometimes two of these steps can be addressed together. (See Sample Problem 2 below).

5.C. Scientific Notation

Scientists often work with very large or very small numbers. Such numbers are difficult to work with when they are written in common decimal notation. For example, the speed of light is about 300 000 000 m/s. There are many zeros to keep track of if you have to multiply or divide this number by another number.

Sometimes it is possible to change a very large or very small number, so that the number falls between 0.1 and 1000, by changing the SI prefix. For example, 237 000 000 mm can be converted to 237 km, and 0.000 895 kg can be expressed as 895 mg.

Alternatively, very large or very small numbers can be written using scientific notation. Scientific notation expresses a number by writing it in the form $a \times 10n$, where the letter a, referred to as the coefficient, is a value that is at least 1 and less than 10. The number 10 is the base, and n represents the exponent. The base and the exponent are read as "10 to the power of n." Powers of 10 and their decimal equivalents are shown in Table 4 on page 627.

SAMPLE PROBLEM 2 Locating the Image

A small toy building block is placed 7.2 cm in front of a lens. An upright, virtual image of magnification 3.2 is noticed. Where is the image located?

Given: $d_0 = 7.2$ cm

$M = 3.2$

Required: $d_1 = ?$

Analysis and Solution:

$$M = \frac{-d_1}{d_0}$$

$$-Md_0 = d_1$$

$$d_1 = -Md_1$$

$$= -(3.2)(7.2\text{ cm})$$

$$d_1 = -23\text{ cm}$$

Statement: The image of the toy block is located 23 cm from the lens, on the same side as the object.

Table 4 Powers of 10 and Decimal Equivalents

Power of 10	Decimal equivalent
10^9	1 000 000 000
10^8	100 000 000
10^7	10 000 000
10^6	1 000 000
10^5	100 000
10^4	10 000
10^3	1 000
10^2	100
10^1	10
10^0	1
10^{-1}	0.1
10^{-2}	0.01
10^{-3}	0.001
10^{-4}	0.000 1
10^{-5}	0.000 01
10^{-6}	0.000 001
10^{-7}	0.000 000 1
10^{-8}	0.000 000 01
10^{-9}	0.000 000 001

To write a large number in scientific notation, follow these steps:

1. To determine the exponent, count the number of places you have to move the decimal point to the left, to give a number between 1 and 10. For example, when writing the speed of light (300 000 000 m/s) in scientific notation, you have to move the decimal point eight places to the left. The exponent is therefore 8.
2. To form the coefficient, place the decimal point after the first digit. Now drop all the trailing zeros *unless* all the numbers after the decimal are zeros, in which case, keep one zero. In our example, the coefficient is 3.0.
3. Combine the coefficient with the base, 10, and the exponent. For example, the speed of light is 3.0×10^8 m/s.

Very small numbers (less than 1) can also be expressed in scientific notation. To find the exponent, count the number of places you move the decimal point to the right, to give a coefficient between 1 and 10. For very small numbers, the base (10) must be given a negative exponent.

For example, a millionth of a second, 0.000 001 s, can be written in scientific notation as 1×10^{-6} s. Table 5 shows several examples of large and small numbers expressed in scientific notation.

To multiply numbers in scientific notation, multiply the coefficients and add the exponents. For example,

$$\begin{aligned}(3 \times 10^3)(5 \times 10^4) &= 15 \times 10^{3+4} \\ &= 15 \times 10^7 \\ &= 1.5 \times 10^8\end{aligned}$$

Note that when writing a number in scientific notation, the coefficient should be at least 1 and less than 10.

When dividing numbers in scientific notation, divide the coefficients and subtract the exponents. For example,

$$\begin{aligned}\frac{8 \times 10^6}{2 \times 10^4} &= 4 \times 10^{6-4} \\ &= 4 \times 10^2\end{aligned}$$

Table 5 Numbers Expressed in Scientific Notation

Large or small number	Common decimal notation	Scientific notation
124.5 million km	124 500 000 km	1.245×10^8 km
154 thousand nm	154 000 nm	1.54×10^5 nm
753 trillionths of a kg	0.000 000 000 753 kg	7.53×10^{-10} kg
315 billionths of a m	0.000 000 315 m	3.15×10^{-7} m

5.D. Uncertainty in Measurement

There are two types of quantities used in science: exact values and measurements. Exact values include defined quantities: those obtained from SI prefix definitions (such as 1 km = 1000 m) and those obtained from other definitions (such as 1 h = 60 min).

Exact values also include counted values, such as 5 beakers or 10 cells. All exact values are considered completely certain. In other words, 1 km is exactly 1000 m, not 999.9 m or 1000.2 m. Similarly, 5 beakers could not be 4.9 or 5.1 beakers; 5 beakers are exactly 5 beakers.

Measurements, however, have some uncertainty. The uncertainty depends on the limitations of the particular measuring instrument used and the skill of the measurer.

5.D.1. Significant Digits

The certainty of any measurement is communicated by the number of significant digits in the measurement. In a measured or calculated value, significant digits are the digits that are certain, plus one estimated (uncertain) digit. Significant digits include all the digits that are correctly reported from a measurement.

For example, 10 different people independently reading the water volume in the graduated cylinder in Figure 1 would all agree that the volume is at least 50 mL. In other words, the "5" is certain. However, there is some uncertainty in the next digit. The observers might report the volume as being 56 mL, 57 mL, or 58 mL. The only way to know for sure is to measure with a more precise measuring device. Therefore, we say that a measurement such as 57 mL has two significant digits: one that is certain (5) and the other that is uncertain (7). The last digit in a measurement is always the uncertain digit. For example, 115.6 g contains three certain digits (115) and one uncertain digit (6).

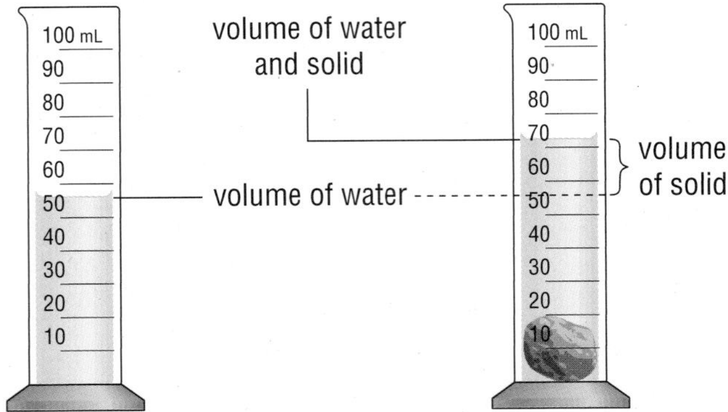

Figure 1 This figure shows the difficulty of making observations using an imprecise measuring device.

Table 6 provides the guidelines for determining the number of significant digits, along with examples to illustrate each guideline.

Table 6 Guidelines for Determining Significant Digits

Guideline	Example	
	Number	**Number of significant digits**
Count from left to right, beginning with the first non-zero digit.	345	3
	457.35	5
Zeros at the beginning of a number are never significant.	0.235	3
	0.003	1
All non-zero digits in a number are significant.	1.1223	4
	76.2	2
Zeros between digits are significant.	107.05	5
	0.02094	4
Zeros at the end of a number with a decimal point are significant.	10.0	3
	303.0	4
Zeros at the end of a number without a decimal point are ambiguous.	5400	at least 2
	200 000	at least 1
All digits in the coefficient of a number written in scientific notation are significant.	5.4×10^3	2
	5.40×10^3	3
	5.400×10^3	4

ROUNDING

Use these rules when rounding answers to the correct number of significant digits:

1. When the first digit discarded is less than 5, the last digit kept (i.e., the one before the discarded digit) should not be changed.

   Example:

   3.141 326 rounded to four digits is 3.141.

2. When the first digit discarded is greater than 5, or when it is 5 followed by at least one digit other than zero, the last digit kept is increased by one unit.

   Examples:

   2.221 372 rounded to five digits is 2.2214.
   4.168 501 rounded to four digits is 4.169.

3. When the first digit discarded is 5 followed by only zeros, the last digit kept is increased by 1 if it is odd but not changed if it is even. Note that when this rule is followed, the last digit in the final number is always even.

   Examples:

   2.35 rounded to two digits is 2.4
   2.45 rounded to two digits is 2.4
   6.75 rounded to two digits is 6.8

4. When adding or subtracting measured quantities, look for the quantity with the fewest number of digits to the right of the decimal point. The answer can have no more digits to the right of the decimal point than this quantity has. In other words, the answer cannot be more precise than the least precise value.

   Example:

   $$\begin{array}{r} 12.52\ \text{g} \\ +\ 349.0\ \text{g} \\ +\ 8.24\ \text{g} \\ \hline 369.76\ \text{g} \end{array}$$

   Because 349.0 g is the quantity with the fewest digits to the right of the decimal point, the answer must be rounded to 369.8 g.

   Example:

   $$\begin{array}{r} 157.85\ \text{mL} \\ -\ 32.4\ \text{mL} \\ \hline 125.45\ \text{mL} \end{array}$$

   Because 32.4 mL has the fewest decimal places, the answer must be rounded to 125.4 mL. Note that Rule 3 applies.

5. When multiplying or dividing, the answer must contain no more significant digits than the quantity with the fewest number of significant digits.

   Examples:

   $$m = \frac{1.15\ \text{g}}{\cancel{\text{cm}^3}} \times 16\ \cancel{\text{cm}^3} = 18\ \text{g}$$

   $$\Delta t = 1.25\ \cancel{\text{h}} \times \frac{60\ \text{min}}{1\ \cancel{\text{h}}} = 75.0\ \text{min}$$

In other words, the answer cannot be more certain than the least certain value.

Note, in the second example, 1.25 h is a measurement and, as a result, contains uncertainty. However, 60 min/h is an exact quantity. Since it has no uncertainty, the number of significant digits in the final answer is based only on the measured value of 1.25 h.

Rule 5 also applies when multiplying or dividing measurements expressed in scientific notation. For example,

$$\begin{aligned}(3.5 \times 10^3\ \text{km})(7.4 \times 10^2\ \text{km}) &= 25.9 \times 10^5\ \text{km}^2 \\ &= 2.59 \times 10^6\ \text{km}^2 \\ &= 2.6 \times 10^6\ \text{km}^2\end{aligned}$$

The coefficient should be rounded to the same number of significant digits as the measurement with the least number of significant digits (the least certain value). In this example, both measurements have only two significant digits, so the coefficient 2.59 should be rounded to 2.6 to give a final answer of $2.6 \times 10^6\ \text{km}^2$.

Similarly,

$$\begin{aligned}\frac{3.9 \times 10^6\ \text{m}}{5.3 \times 10^3\ \text{s}} &= 0.7377 \times 10^3\ \text{m/s} \\ &= 7.377 \times 10^2\ \text{m/s} \\ &= 7.4 \times 10^2\ \text{m/s}\end{aligned}$$

5.D.2. Measurement Errors

There are two types of error that can occur when measurements are taken: random and systematic. Random error results when an estimate is made to obtain the last significant digit for a measurement. The size of the random error is determined by the precision of the measuring instrument. For example, when measuring length, it is necessary to estimate between the marks on the measuring tape. If these marks are 1 cm apart, the random error is greater and the precision is less than if the marks were 1 mm apart. Systematic error is caused by a problem with the measuring system itself, such as equipment not set up correctly. For example, if a balance is not tared (re-set to zero) at the beginning, all the measurements taken with the balance will have a systematic error.

The precision of measurements depends on the markings (gradations) of the measuring device. Precision is the place value of the last measurable digit. For example, a measurement of 12.74 cm is more precise than a measurement of 127.4 cm because 12.74 was measured to hundredths of a centimetre, whereas 127.4 was measured to tenths of a centimetre.

When adding or subtracting measurements with different precisions, round the answer to the same precision as the least precise measurement. Consider the following:

$$\begin{array}{r} 11.7\text{ cm} \\ 3.29\text{ cm} \\ +\ 0.542\text{ cm} \\ \hline 15.532\text{ cm} \end{array}$$

The first measurement, 11.7 cm, is measured to one decimal place and is the least precise. The answer must be rounded to one decimal place, or 15.5 cm.

No matter how precise a measurement is, it still may not be accurate. Accuracy refers to how close a value is to its accepted value. Figure 2 uses the results of a horseshoe game to explain precision and accuracy.

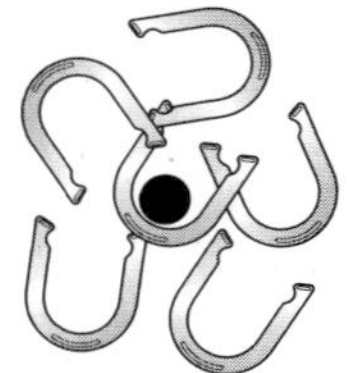

(a) precise and accurate

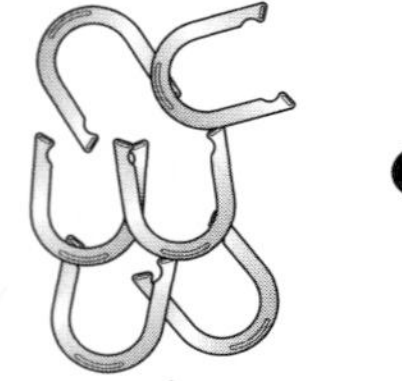

(b) precise but not accurate

(c) accurate but not precise

(d) neither accurate nor precise

Figure 2 The patterns of the horseshoes illustrate the comparison between accuracy and precision.

How certain you are about a measurement depends on two factors: the precision of the instrument and the size of the measured quantity. Instruments that are more precise give more certain values. For example, a measurement of 13 g is less precise than one of 12.76 g because the second measurement has more decimal places than the first. Certainty also depends on the size of the measurement. For example, consider the measurements 0.4 cm and 15.9 cm. Both have the same precision (number of decimal places): both are measured to the nearest tenth of a centimetre. Imagine that the measuring instrument is precise to ± 0.1 cm, however. An error of 0.1 cm is much more significant for the 0.4 cm measurement than it is for the 15.9 cm measurement because the second measurement is much larger than the first. For both factors—the precision of the instrument used and the value of the measured quantity—the more digits there are in a measurement, the more certain you are about the measurement.

5.E. Using the Calculator

A calculator is a very useful device that makes calculations easier, faster, and probably more accurate. However, like any other electronic instrument, you need to learn how to use it. These guidelines apply to a basic scientific calculator. If your calculator is different, such as a graphing calculator, some of the instructions and operations may use different keys or sequences, so always check the manual.

GENERAL POINTS

- Most calculators follow the usual mathematical rules for order of operations—multiplication/division before addition/subtraction. For example, if you are calculating y using $y = mx + b$, you can enter the values of m times x plus b in one sequence. The calculator will "know" that m and x must be multiplied first before b is added.
- Calculators do not keep track of significant digits. For example, 12.0 is the same as 12 for a calculator.
- Some calculator keys such as [$\frac{1}{x}$] [+/−], [x^2], (and its second function, [$\sqrt{x}$]) apply the operation only to the value in the display regardless of other operations in progress. This means you can quickly change the sign of the number, convert to the reciprocal, square the number, or determine the square root while inputting a sequence of calculations.
- Do not clear all numbers from the calculator until you are completely finished with a question: the result of one calculation can be reused to start the next one.
- All scientific calculators have at least one memory location where you can store a number (M+ and STO are common keys) and recall it later (usually with MR or RCL). Use it to avoid re-entering many digits.

MULTIPLICATION AND DIVISION

- Division is the inverse of multiplication. This means that dividing by a number is the equivalent to multiplying by the inverse of the same number. For example,

  $\frac{12\text{ km}}{0.75\text{ h}}$ is the same as $12\text{ km} \times \frac{1}{0.75\text{ h}}$

  and equals $16\,\frac{\text{km}}{\text{h}}$.

This is particularly useful when you want to divide by a number that is currently in the display of your calculator. For example, you have just finished converting 45 min to 0.75 h in your calculator and now you want to calculate the speed.

Display: 0.75

Press:

New display: 16

- Brackets, ( ), are useful to force the calculator to perform the operation(s) inside the brackets first, before continuing with the calculation. The calculation of the slope of a line is a good example.

$$\begin{aligned} \text{slope} &= \frac{\Delta d}{\Delta t} \\ &= \frac{(15.2 - 4.1)\text{ m}}{(6.5 - 3.6)\text{ s}} \\ &= 3.8\,\frac{\text{m}}{\text{s}} \end{aligned}$$

  If you do not use the brackets on your calculator, you will have to calculate the numerator and denominator separately and then divide.

SCIENTIFIC NOTATION

On many calculators, scientific notation is entered using a special key, labelled EXP or EE. This key includes "×10" from the scientific notation, and you need to enter only the exponent. For example, to enter

7.5×10^4 press 7 . 5 EXP 4

3.6×10^{-3} press 3 . 6 EXP +/− 3

CALCULATING A MEAN

There are many statistical methods of analyzing experimental evidence. One of the most common and important is calculating an arithmetic mean, or simply a mean. The mean of a set of values is the sum of all reasonable values divided by the total number of values. (This is also commonly known as the average of a set of values, but this term is not recommended because it is too vague and open to different interpretations.)

Suppose you measure the root growth of five seedlings. Your root measurements after three days are 1.7 mm, 1.6 mm, 1.8 mm, 0.4 mm, and 1.6 mm. What is the mean root growth? Inspection of the measurements shows that the 0.4 mm measurement clearly does not fit with the rest. Perhaps this seedling was infected with a fungus, or some other problem occurred. You should not include this result in your mean but leave it in your evidence table so everyone can see the decision that you made. Using only reasonable values,

$$\text{mean root growth} = \frac{1.7\text{ mm} + 1.6\text{ mm} + 1.8\text{ mm} + 1.6\text{ mm}}{4}$$
$$= 1.7\text{ mm}$$

Means are important in all areas of science because multiple measurements or trials are widely used to increase the reliability of the results.

EQUATIONS

Algebra is a set of rules and procedures for working with mathematical equations. In general, your equations will contain one unknown quantity. Whatever mathematical operation is performed to one side of an equation must be performed to the other side. To solve for an unknown value, you need to isolate it on one side of the equal sign. To accomplish this, you should follow three rules:

1. The same quantity can be added or subtracted from both sides of the equation without changing the equality.

The following examples illustrate this rule:

$100\text{ m} = 100\text{ m}$	$x + b = y$
$100\text{ m} - 5\text{ m} = 100\text{ m} - 5\text{ m}$	$x + b - b = y - b$
$95\text{ m} = 95\text{ m}$	$x = y - b$

The example on the left shows the application of this rule using quantities with numbers and units. The rule works equally well with quantity symbols. The example on the right shows how to isolate x.

2. The same quantity can be multiplied or divided on both sides of the equation without changing the equality.

The following examples illustrate this rule. The example on the left shows the use of this rule with known quantities. Use the same rule with an equation containing quantity symbols to isolate one of the quantities. To solve for d in the example on the right, multiply both sides by t. Notice that t divided by t equals 1. Multiplying or dividing any quantity by 1 does not change the quantity; therefore, $d \times 1 = d$.

$120\text{ m} = 120\text{ m}$	$v = \frac{d}{t}$
$\frac{120\text{ m}}{8.0\text{ s}} = \frac{120\text{ m}}{8.0\text{ s}}$	$v \times t = \frac{d}{\cancel{t}} \times \cancel{t}$
$15\text{ m/s} = 15\text{ m/s}$	$vt = d$ or $d = vt$

3. The same power (such as square or square root) can be applied to both sides of the equation without changing the equality.

The following examples illustrate this rule:

$$25\ \text{s}^2 = 25\ \text{s}^2 \qquad b^2 = A$$

$$\sqrt{25\ \text{s}^2} = \sqrt{25\ \text{s}^2} \qquad \sqrt{b^2} = \sqrt{A}$$

$$5.0\ \text{s} = 5.0\ \text{s} \qquad b = \sqrt{A}$$

If several of the rules listed above are required to isolate an unknown quantity, then you should apply Rule 1 first, whenever possible, and then apply Rule 2. In general, Rule 3 should be used last.

5.F. Working with Angles

Measuring angles is an important skill in the study of optics. The angle at which a light ray strikes a reflecting surface is usually measured from a perpendicular reference line called a normal. A protractor is used to measure the angle that an incoming (or incident) light ray makes relative to the normal (Figure 3).

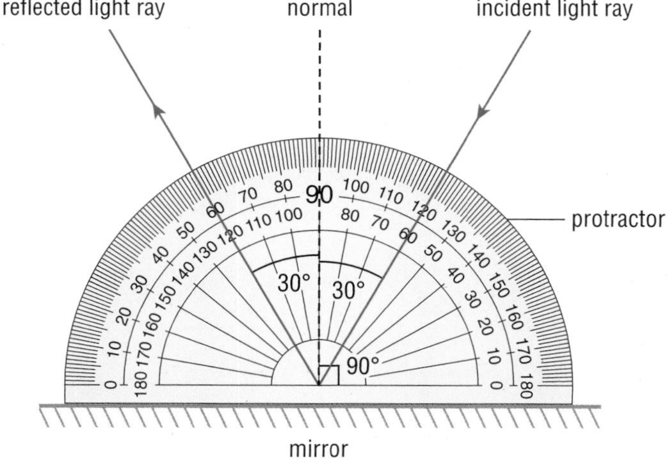

Figure 3 Angles are measured from the normal using a protractor.

SINE OF AN ANGLE

You may be required to determine the sine of an angle. Sine is a specific mathematical function that you will learn more about in future science and math courses. For now, all you need to know is how to calculate the sine of an angle using your calculator. The sequence of buttons to press to determine the sine of an angle varies with the model of the calculator. However, these steps work for most common calculators:

- Make sure that the calculator is in "degree" mode. (Check the manual if you are unsure.)
- Press the "sin" button.
- Enter the angle (e.g., 30).
- Press the equal button.

The sine of 30° is 0.5.

PRACTICE

Sines of Angles

Determine the sine of the following angles:

(a) 45° (b) 60° (c) 64° (d) 90°

INVERSE SINE

Sometimes you will be given the sine of an angle and be asked to determine the angle.

Given the sine of an angle, you can also solve for the angle. Follow these steps:

- Make sure that the calculator is in "degree" mode.
- Press the shift "$\sin^{-1}$" buttons.
- Enter the value (e.g., 0.5).
- Press the equal button.

The angle whose sine equals 0.5 is 30°.

PRACTICE

Angles of Sines

Determine the angles that have the following sines:

(e) 0.7071 (f) 0.8660 (g) 0.8988 (h) 1

Data tables are an effective means of recording both qualitative and quantitative observations. Making a data table should be one of your first steps as you prepare to conduct an investigation. It is particularly useful for recording the values of the independent variable (the cause) and the dependent variables (the effects), as shown in Table 1.

Table 1 A Running White-Tailed Deer

Time (s)	Distance (m)
0	0
1.0	13
2.0	25
3.0	40
4.0	51
5.0	66
6.0	78

Follow these guidelines to make a data table:

- Use a ruler to make your table.
- Write a title that precisely describes your data.
- Include the units of measurement for each variable, when appropriate.
- List the values of the independent variable in the left-hand column.
- List the values of the dependent variable(s) in the column(s) to the right.

6.A. Graphing Data

A graph is a visual representation of quantitative data. Graphing data often makes it easier to identify a trend or pattern in the data, which indicates a relationship between the variables. There are many types of graphs that can be used to organize data. Three of the most useful kinds of graphs are bar graphs, circle graphs, and point-and-line graphs. Each kind of graph has its own special uses. You need to identify which type of graph is most appropriate for the data you have collected. Then you can construct the graph.

BAR GRAPHS

A bar graph helps you make comparisons when one variable is in numbers (e.g., rainfall) and the other variable is not (e.g., month of the year). Figure 1 shows a bar graph of the distribution of rainfall over a 12-month period in Ottawa. Each bar stands for a different month. This graph clearly shows that rainfall is higher in the summer months and is lower in the winter months.

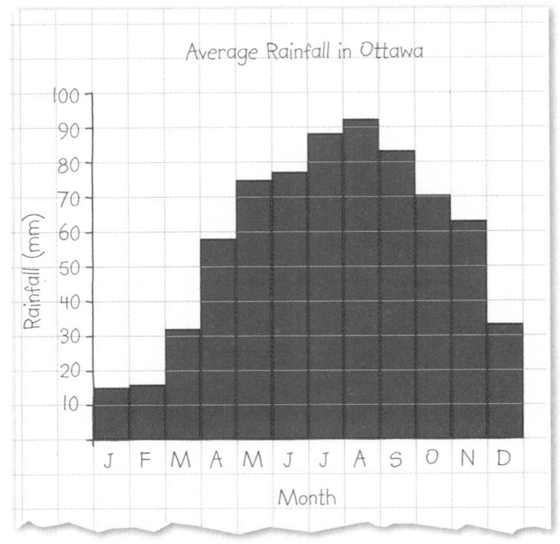

Figure 1 A bar graph

CIRCLE GRAPHS

Circle graphs (pie charts) are useful to show how the whole of something is divided into many parts. For example, the circle graph in Figure 2 shows that transportation is the largest source of greenhouse gas emissions by individuals.

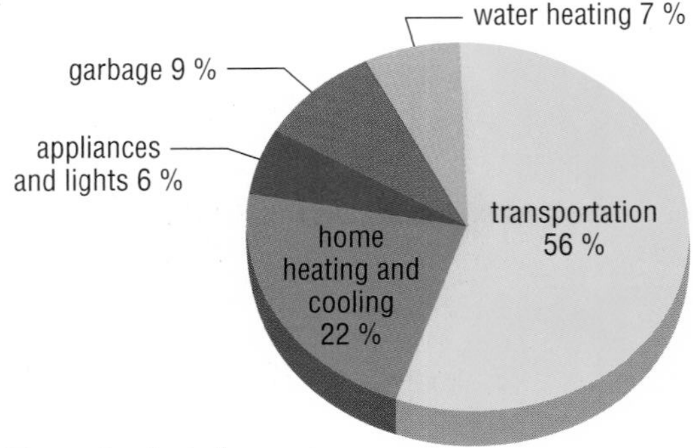

Figure 2 A circle graph

POINT-AND-LINE GRAPHS

When both variables are quantitative, use a point-and-line graph. This format shows all the data points and a line of best fit indicating any relationship between the variables. For example, we can use the following guidelines and the data in Table 1 to construct the line graph in Figure 3.

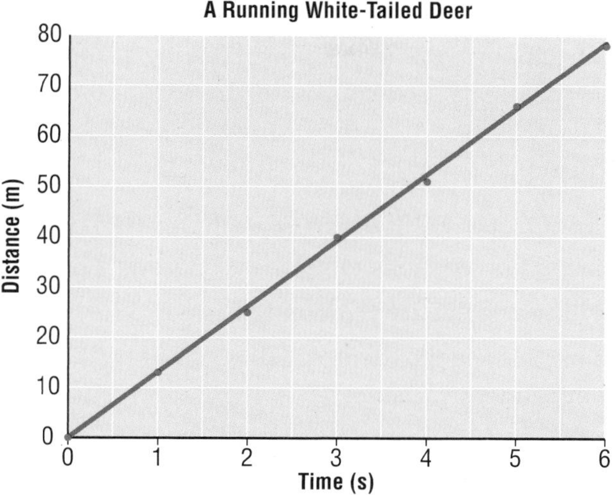

Figure 3 A point-and-line graph of the data from Table 1

Making Point-and-Line Graphs

1. Use graph paper. Draw a horizontal line close to the bottom of the paper as the x-axis and a vertical line close to the left edge as the y-axis.
2. You will generally plot the independent variable along the x-axis and the dependent variable along the y-axis. The exception is when one variable is time: always plot time on the x-axis. The slope of the graph then always represents a rate. Label each axis, including the units.
3. Give your graph a title: a short, accurate description of the data represented by the graph.
4. Determine the range of values for each variable. The range is the difference between the greatest and least values. Graphs often include a little extra length on each axis.
5. Choose a scale for each axis. The scale will depend on how much space you have and the range of values for each axis. Spaces on the grid usually represent equal increments, such as 1, 2, 5, 10, or 100.
6. Plot the points. In Figure 3, the first set of points is 0 on the x-axis and 0 on the y-axis: the origin.
7. After all the points are plotted, try to visualize a line through the points to show the relationship between the variables. Not all points may lie exactly on a line. Draw the line of best fit—a straight or smoothly curving line through the points so that there is approximately the same number of points on each side of the line. The line's purpose is to show the overall pattern of the data.
8. If you are plotting more than one set of data on the same graph, use different colours or symbols for each. Provide a legend.

CALCULATING SLOPES

If the line of best fit on a graph is a straight line, there is a simple relationship between the two variables. You can represent this linear (straight-line) relationship with this mathematical equation:

$$y = mx + b$$

where y is the dependent variable (on the y-axis), x is the independent variable (on the x-axis), m is the slope of the line, and b is the y-intercept (the point where the line touches the y-axis). To determine the slope, choose two points on the line, (x_1, y_1) and (x_2, y_2). (Note that these are not necessarily data points: just two points on the line.) The slope is equal to

$$m = \frac{\text{rise}}{\text{run}} = \frac{y_2 - y_1}{x_2 - x_1}$$

For the graph in Figure 4, suppose you choose the two points (1.5, 20) and (5.5, 72). You can use them to calculate the slope.

$$\begin{aligned} m &= \frac{y_2 - y_1}{x_2 - x_1} \\ &= \frac{72\text{ m} - 20\text{ m}}{5.5\text{ s} - 1.5\text{ s}} \\ &= \frac{52\text{ m}}{4.0\text{ s}} \\ &= 13\text{ m/s} \end{aligned}$$

The slope of the line is 13 m/s. The positive number indicates a direct relationship.

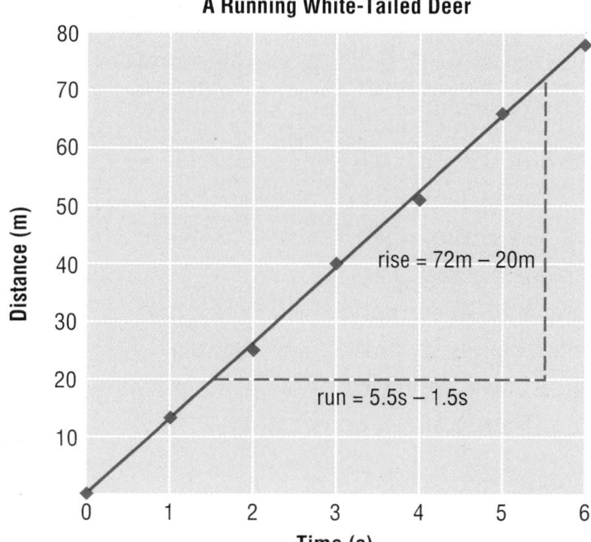

Figure 4 Calculating the slope of a line of best fit

6.B. Using Computers for Graphing

You can use spreadsheet or graphing programs on your computer to construct bar, circle, and point-and-line graphs. In addition, such programs can use statistical analysis to compute the line of best fit. The following instructions guide you to produce best-fit line graph values for two variables on a spreadsheet.

STEPS FOR CREATING A GRAPH IN A SPREADSHEET PROGRAM

1. Start the spreadsheet program. Enter the data with the values for x (the first column in your data table) that start in cell A2. Now enter the values for y from the other column that start in B2.
2. Highlight the two columns containing your data (A2 … B6). From the toolbar, select the button for creating graphs. The cursor may change to a symbol, which allows you to "click and drag" to choose the size of your graph.
3. A series of choices will now be presented to you, so you can specify what kind of graph you want to create. A scatter graph is most appropriate for our data.
   - Click in the "Title" box and type the title of the graph.
   - Click in "Value (X)" box and type the label and units for the x-axis.
   - Click the "Finish" button.
4. To add a line of best fit, point your cursor to one of the highlighted data points and right click once. Select the "Trendline" option. Make sure that the "Linear" box is highlighted under Type.
5. To find the slope or y-intercept, click on the "Options" tab and select the box beside "Display Equation on Chart." This will put the values for $y = mx + b$ on the graph, giving you the slope and y-intercept value.
6. Save your spreadsheet before you close the program.

6.C. Interpreting Graphs

When data from an investigation are plotted on an appropriate graph, patterns and relationships become easier to see and interpret. You can more easily tell if the data support your hypothesis. Looking at the data in the graph may also lead you to a new hypothesis. Or you can extract meaning from other people's investigations.

WHAT TO LOOK FOR WHEN READING GRAPHS

Here are some questions to help you interpret a graph:

- What variables are represented?
- What is the dependent variable? What is the independent variable?
- Are the variables quantitative or qualitative?
- If the data are quantitative, what are the units of measurement?
- What do the highest and lowest values represent on the graph?
- What is the range between the highest and the lowest values on each axis?
- Are the axes continuous? Do they start at zero?
- What patterns or trends exist between the variables?
- If there is a linear relationship, what might the slope of the line tell you?

USING GRAPHS FOR PREDICTING

If a graph shows a regular pattern, you can use it to make predictions. For example, you could use the graph in Figure 3 on page 635 to predict the distance travelled by the deer in 8.0 s. To do this, you extrapolate the graph (extended beyond the measured points), assuming that the observed trend would continue. You should be careful when predicting values outside of your measured range. The farther you are from the known values, the less reliable will be your prediction.

7.A. Working Together

Teamwork is just as important in science as it is on the playing field or in the gym. Scientific investigations are usually carried out by teams of people working together. Ideas are shared, experiments are designed, data are analyzed, and results are evaluated and shared with other investigators. Group work is necessary and is usually more productive than working alone.

Several times throughout the year, you may be asked to work with one or more of your classmates. Whatever the task that your group is assigned, you need to follow a few guidelines to ensure a productive and successful experience.

7.A.1. General Guidelines for Effective Teamwork

- Keep an open mind. Everyone's ideas deserve consideration.
- Divide the task among all the group members. Choose a role best suited to your particular strengths.
- Work together and take turns. Encourage, listen, clarify, help, and trust one another.
- Remember that the success of the team is everyone's responsibility. Every member needs to be able to demonstrate what the team has learned and support the team's final decision.

7.A.2. Exchanging Information Orally

You will be involved in a number of different activities during your science course. These activities help you express your ideas and learn about new ones. Here are three of the more successful discussion formats that your group may use to share ideas:

- In a **think-pair-share activity**, you and your partner are given a problem. Each of you develops a response (usually within a time limit). Then share your ideas with each other to resolve the problem. You may also be asked to share your results with a larger group or the class.
- In a **jigsaw activity**, you are an active member in two teams: your home team and your expert team. Each member of the home team chooses or is assigned to a particular area of research. Each home team member then meets with an expert team in which everyone is working on the same area of research. In your expert team, you may work together to come up with answers to questions in your area of research. Once you have accomplished your task as a team, you return to your home team, and it is your responsibility to teach what you have learned to the members of your home team. Each person on the home team will do the same thing.

  The guidelines of teamwork are important with the jigsaw, so try to keep them in mind as you work with your expert and home teams.
- A **round table activity** can be used to give your group an opportunity to review what they know. Your group is given a pen, paper, and a question or questions. Pass the pen and paper around and take turns writing one line of the solution. You can pass on your turn if you wish. Keep working until the solution is complete. Check to ensure that everyone understands the solution. Finally, working as a team, review the steps to the solution.

7.A.3. Investigations and Activities

This kind of work is most effective when completed by small groups. Here are some suggestions for effective group performance during investigations and activities:

- Make sure that each group member understands and agrees to the role assigned to him or her.
- Take turns doing the work during similar and repeated activities.
- Safety must come first. Be aware of where other group members are and what they are doing. Ask yourself, what potential hazards are involved in this activity? What are the safety procedures?

- Take responsibility for your own learning by making notes and contributing to any discussions about the activities. Make your own observations and compare them with the observations of other group members (Figure 1).

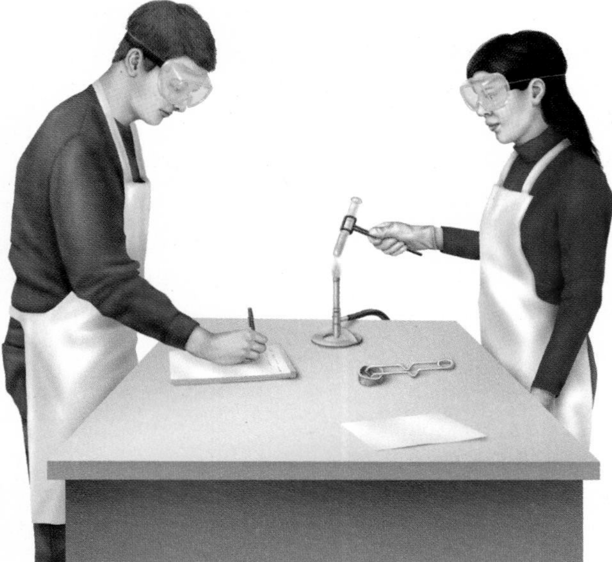

Figure 1 When you are working with one student or several, you can share and compare your results.

7.A.4. Explore an Issue

Follow these guidelines when you are conducting research with a group:

- Divide the topic into several topics and assign one topic to each group member.
- Keep records of the sources used by each group member.
- Decide on a format for exchanging information (such as photocopies of notes, oral discussion, electronically).
- When the time comes to make a decision and take a position on an issue, allow for the contributions of each group member. Make decisions by compromise and consensus.
- Communicating your position should also be a group effort.

7.A.5. Evaluating Teamwork

After you have completed a task with your group, evaluate your team's effectiveness using these criteria: strengths and weaknesses, opportunities, and challenges.

Reflect on your experience by asking yourself the following questions:

- What were the strengths of your teamwork?
- What were the weaknesses of your teamwork?
- What opportunities were provided by working with your group?
- What challenges did you encounter as a member of a group?

7.B. Setting Goals and Monitoring Progress

Think back to your last school year. What classes did you do well in? Why do you think you were successful? In which classes did you have difficulty? Why do you think you had difficulty? What could you do differently this year to improve your performance? Use your answers to these questions to reflect on your past experiences to make new and positive changes. Things that you want to accomplish today, this week, and this year are all called goals. Learning to set goals and to make a plan to achieve them takes skill, patience, and practice.

7.B.1. Setting Goals

ASSESS YOUR STRENGTHS AND WEAKNESSES

The process of setting goals starts with honest reflection. Maybe you have noticed that you do better on projects than you do on tests and exams. You may perform better when you are not pressured by time. Inattention in class and poor study habits may be weaknesses that result in poor performance.

REALISTIC GOALS THAT YOU CAN MEASURE

Do not set yourself up to fail by setting goals that you cannot possibly achieve. Saying, "I will have the best mark in the class at the end of the semester" may not be realistic. Setting a goal to increase your test marks by 10 % this semester may be achievable, however. You will find it easier to reach your goals if you can tell whether you are getting closer to them. A goal to increase your test marks by 10 % this semester is easy to measure. When you are thinking of setting goals, remember the acronym SMART: Specific, Measurable, Attainable, Realistic, and Time-limited.

SHARE YOUR GOALS

People whom you respect can often help you set and clarify goals. Someone who knows your strengths and weaknesses may be able to think of possibilities you may not have considered. Sharing your goals with a trusted friend or adult will often provide needed support to help you reach your goals.

PLANNING TO MEET YOUR GOALS

Once you have made a list of realistic goals, create a plan to achieve them. A successful plan usually consists of two parts: an action plan and target dates.

THE ACTION PLAN

First, make a list of the actions or behaviours that might help you reach your goals (Figure 2).

Goal: To increase my grades by 10 % by the end of the term

Possible actions:

- Arrange to work with a partner, or other students in the class, to list things we must know for tests or assignments.
- Ask the teacher to explain anything on the list that I cannot explain myself or do not understand.
- Choose a study area at home or in a public place. Use it.
- Use an organizer to make a weekly schedule and to keep a record of all evaluations.

Figure 2 One student's goal and action plan to improve test results

If you have made an honest assessment of your strengths and weaknesses, then you know what you have to do to improve. If you want to improve your test marks, you could try to work with others to prepare for tests. You could also use a weekly planner or reorganize your study area at home.

Identify what is preventing you from achieving your goal. Think of ways to overcome these obstacles. Ask friends for tips on the different ways that they maintain good study habits and improve test results.

SETTING TARGET DATES

Suppose you want to improve test results by 10 % by the end of the semester. How much time do you have? How many tests are scheduled between now and then? Work back from your target date at the end of the semester. Determine the dates of all the tests between now and the end of the semester. These dates will give you short-term targets that, if you hit them, will make it easier for you to reach your overall target. Figure 3 shows an example of a working schedule.

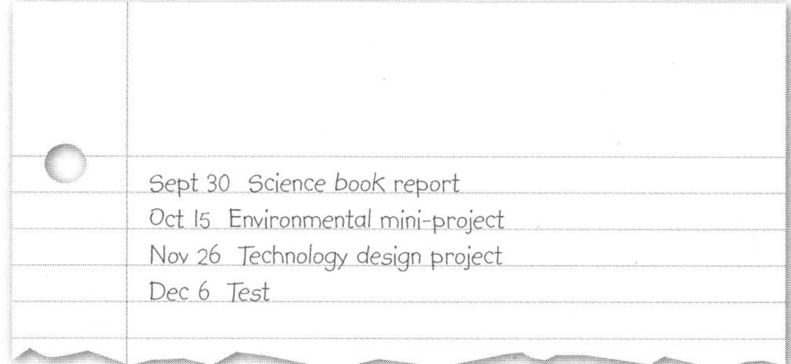

Sept 30 Science book report
Oct 15 Environmental mini-project
Nov 26 Technology design project
Dec 6 Test

Figure 3 Test dates

Once you complete your schedule in your planner, transfer it to a calendar in your study area. Refer to either your planner or your calendar every day.

7.B.2. Monitoring Progress

Remember to measure your progress along the way. It is always important to look at and monitor the results of your tests and activities during the school year rather than just at the end of it. You might decide, for example, to check your progress after the first test. Did these results meet your short-term target to improve test results by 10 % by the end of the semester? If you do not seem to be on track to meet your goal, then you may need to change your plan. For example, perhaps you need to study by yourself or with one friend rather than with a group of friends.

It is always possible to change your plan or even adjust your goal. The most important thing is to keep moving forward and to remain committed to improvement.

7.C. Good Study Habits

Studying takes many forms. Developing good study skills can help you study and learn more successfully. Below are some tips to help you achieve better study habits.

7.C.1. Your Study Space

- *Organize your work area.* The place where you study should be tidy and organized. Place all papers, books, magazines, and pictures in appropriate areas of your study area (e.g., keep books in a bookcase or crate and magazines in a stack on the floor). This will make it easier for you to focus on your school work.
- *Maintain a quiet work area.* Where possible, make sure that your work area is free from distractions—telephone, music, television, and other family members. If there are too many distractions at home, you can usually find an appropriate space at the school or public library. Any quiet space, free from interruptions, can be a productive work area.
- *Make sure that you are comfortable in your work area.* If possible, personalize your work area. For example, make sure that the light is right for your needs. Decide what works best for you and create a study area that provides a productive and positive environment in which you feel comfortable.
- *Be prepared—bring everything you need.* It is important to have all the necessary materials and equipment that you will need when you begin to study. You can easily increase your productivity by gathering materials such as pens, pencils, notebooks, or textbooks in one place near your computer or on your desk. Continually getting up to find something you need decreases your ability to stay on task and be productive.

7.C.2. Study Habits

- *Take notes.* Take notes during class. Outside class, review the appropriate section of the textbook. Read or view additional material on the topic from other sources, such as newspapers, magazines, the Internet, and television. Ask a friend to share notes with you.
- *Use graphic organizers.* You can use a variety of graphic organizers to help you summarize a concept or unit (page 642). They also help you more easily connect different concepts.
- *Schedule your study time.* Use a daily planner and take it with you to class. Write any homework assignments, tests, or projects in it. Use it to create a daily "to do" list. This will help you complete work to hand in and avoid last-minute panic. Also, jot down in your planner when, where, and with whom you plan to study for certain topics.
- *Take study breaks.* It is important to schedule breaks into your study time. For example, you could decide to take a study break after completing one or two items on your "to do" list. Taking breaks allows you to relax and to recharge your brain so that you can keep focused when you return to your assignments.

8.A. Reading Strategies

The skills and strategies that you use to help you read depend on the type of material you are reading. Reading a science book is different from reading a novel. When you are reading a science book, you are reading for information. Here are some strategies to help you read for information.

8.A.1. Before Reading

Skim the section you are going to read. Look at the illustrations, headings, and subheadings.

- *Preview.* What is this section about? How is it organized?
- *Make connections.* What do I already know about the topic? How is it connected to other topics I already know about?
- *Predict.* What information will I find in this section? Which parts provide the most information?
- *Set a purpose.* What questions do I have about the topic?

8.A.2. During Reading

Pause and think as you read. Spend time on the photographs, illustrations, tables, and graphs, as well as on the words.

- *Check your understanding.* What are the main ideas in this section? How would I state them in my own words? What questions do I still have? Should I reread? Do I need to read more slowly, or can I read more quickly?
- *Determine the meanings of key science terms.* Can I figure out the meanings of terms from context clues in the words or illustrations? Do I understand the definitions of terms in bold type? Is there something about the structure of a new term that will help me remember its meaning? Which terms should I look up in the glossary?
- *Make inferences.* What conclusions can I make from what I am reading? Can I make any conclusions by "reading between the lines"?
- *Visualize.* What mental pictures can I make to help me understand and remember what I am reading? Should I make a sketch?
- *Make connections.* How is the information in this section like information I already know?
- *Interpret visuals and graphics.* What additional information can I get from the photographs, illustrations, tables, or graphs?

8.A.3. After Reading

Many of the strategies you use during reading can also be used after reading. For example, this textbook provides summaries and questions at the ends of sections. These questions will help you check your understanding and make connections to information that you have just read or to other parts in the textbook.

At the end of each chapter are a Key Concepts Summary and a Vocabulary list, followed by a Chapter Review and Chapter Self-Quiz.

- *Locate needed information.* Where can I find the information I need to answer the questions? Under what heading might I find the information? What terms in bold type should I look for? What details do I need to include in my answers?
- *Synthesize.* How can I organize the information? What graphic organizer could I use? What headings or categories could I use?
- *React.* What are my opinions about this information? How does it, or might it, affect my life or my community? Do other students agree with my reactions? Why or why not?
- *Evaluate information.* What do I know now that I did not know before? Have any of my ideas changed because of what I have read? What questions do I still have?

8.B. Graphic Organizers

Diagrams that are used to organize and display ideas visually are called graphic organizers. Graphic organizers are especially useful in science and technology studies when you are trying to connect together different concepts, ideas, and data. Different graphic organizers have different purposes. They can be used to

- show processes
- organize ideas and thinking
- compare and contrast
- show properties or characteristics
- review words and terms
- collaborate and share ideas

TO SHOW PROCESSES

Graphic organizers can show the stages in a process (Figure 1).

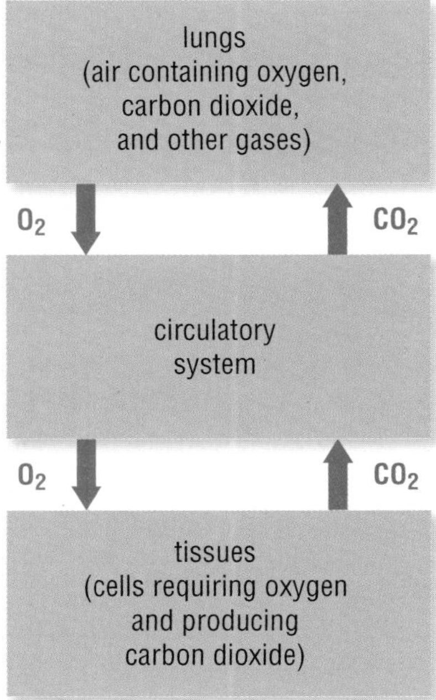

Figure 1 This organizer shows that oxygen and carbon dioxide are transported around the body.

TO ORGANIZE IDEAS AND THINKING

A **concept map** is a diagram showing the relationships between ideas (Figure 2). Words or pictures representing the ideas are connected by arrows and words or expressions that explain the connections. You can use a concept map to brainstorm what you already know, to map your thinking, or to summarize what you have learned.

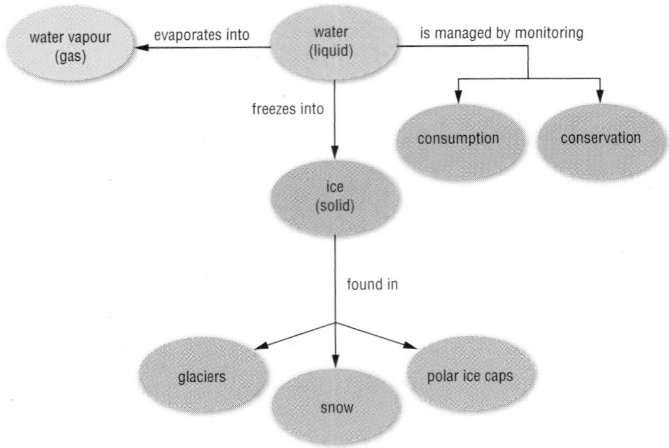

Figure 2 Concept maps help show the relationships among ideas.

Mind maps are similar to concept maps, but they do not have explanations for the connections between ideas.

You can use a **tree diagram** to show concepts that can be broken down into smaller categories (Figure 3).

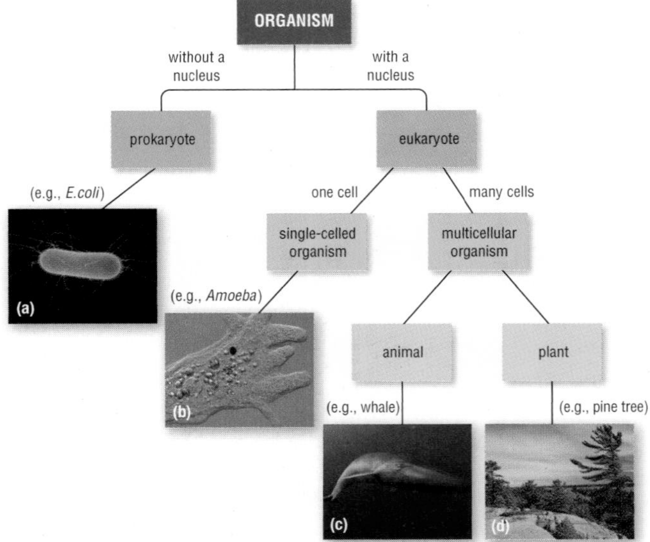

Figure 3 Tree diagrams are very useful for classification.

You can use a **fishbone diagram** to organize the important ideas under the major concepts of a topic you are studying (Figure 4).

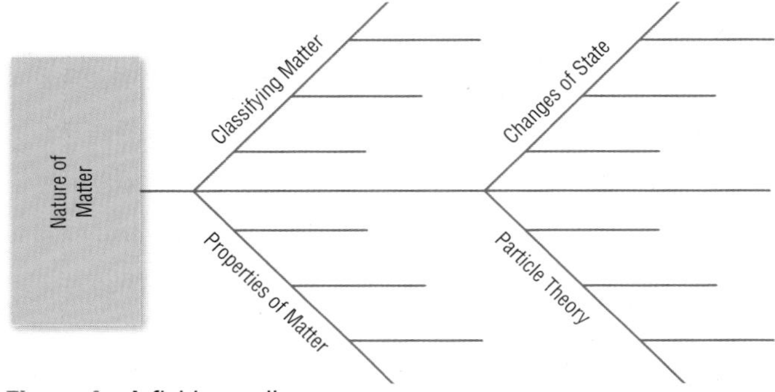

Figure 4 A fishbone diagram

What do we **Know**?	What do we **Want** to find out?	What did we **Learn**?
Carbon dioxide is a greenhouse gas.	Are there any other important greenhouse gases? If so, where do they come from?	Methane, water, and nitrous oxide are other common greenhouse gases. Methane is released from decaying organic matter and from the digestive tracts of grazing animals. Nitrous oxide is released in automobile emissions.
The greenhouse effect traps solar energy in the atmosphere.	If the energy of the Sun can get through the atmosphere and warm Earth's surface, how is the energy trapped by greenhouse gases?	The atmosphere is transparent to light, allowing light rays from the Sun to strike Earth's surface. This energy is absorbed and then released as infrared waves. Because the atmosphere is not transparent to infrared waves, energy is trapped in the atmosphere and warms Earth.

Figure 5 A K-W-L chart

You can use a **K-W-L** chart to write down what you know (K), what you want (W) to find out, and, afterwards, what you have learned (L) (Figure 5).

TO COMPARE AND CONTRAST

You can use a **comparison matrix** (a type of table) to compare related concepts (Table 1).

Table 1 Subatomic Particles

	Proton	Neutron	Electron
electrical charge	positive	neutral	negative
symbol	p^+	n^0	e^-
location	nucleus	nucleus	orbit around the nucleus

You can use a **Venn diagram** to show similarities and differences (Figure 6).

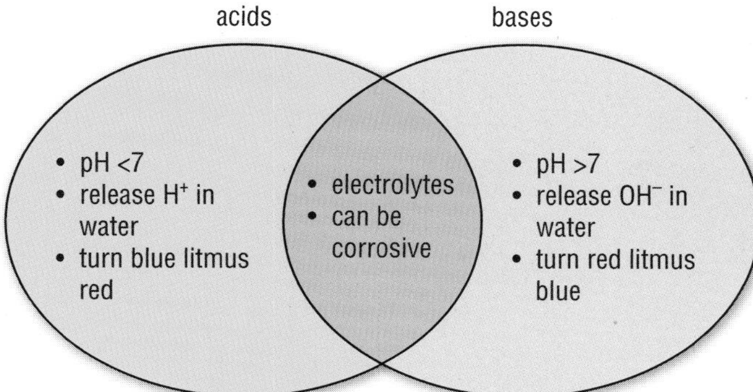

Figure 6 A Venn diagram

You can use a **compare-and-contrast chart** to show similarities and differences between two substances, actions, ideas, and so on (Figure 7).

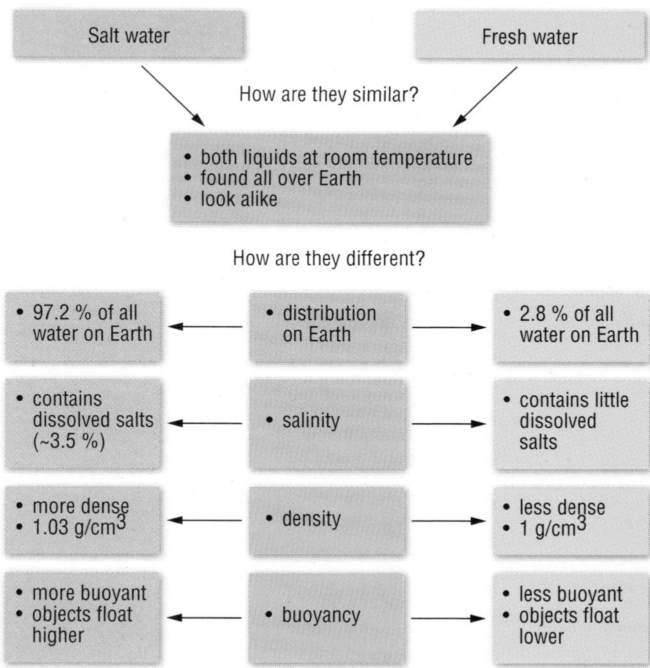

Figure 7 A compare-and-contrast chart

TO SHOW PROPERTIES OR CHARACTERISTICS

You can use a **bubble map** to show properties or characteristics (Figure 8).

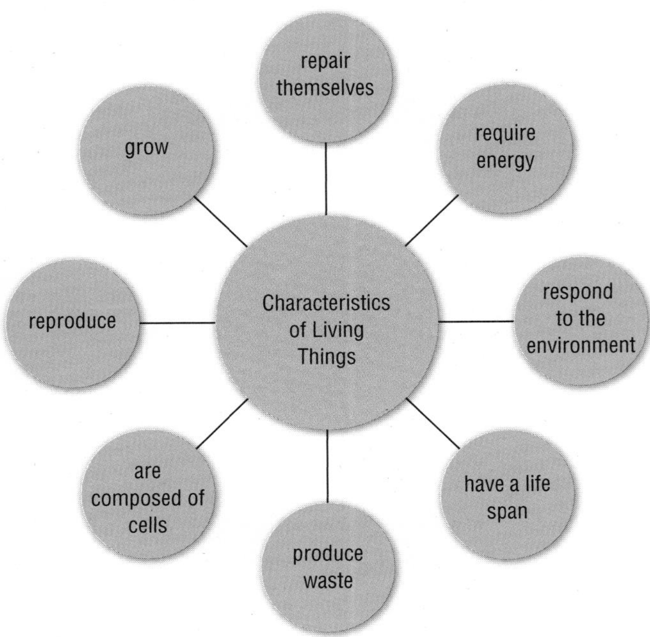

Figure 8 A bubble map

TO REVIEW WORDS AND TERMS

You can use a **word wall** to list, in no particular order, the key words and concepts for a topic (Figure 9).

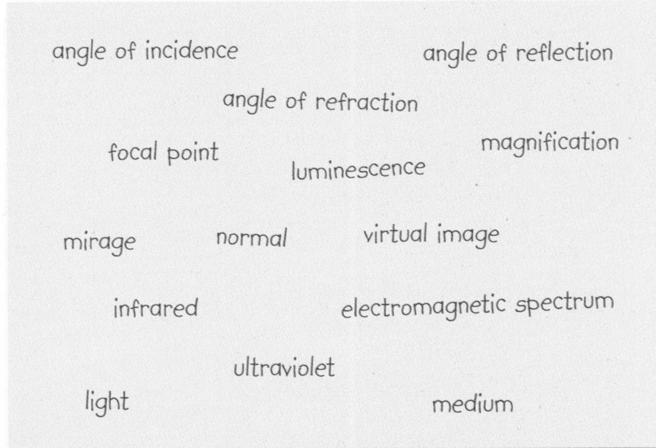

Figure 9 A word wall

TO COLLABORATE AND SHARE IDEAS

A **placemat organizer** gives students in a small group a space to write down what they know about a certain topic. Then group members discuss their answers and write in the middle section what they have in common (Figure 10).

Before:

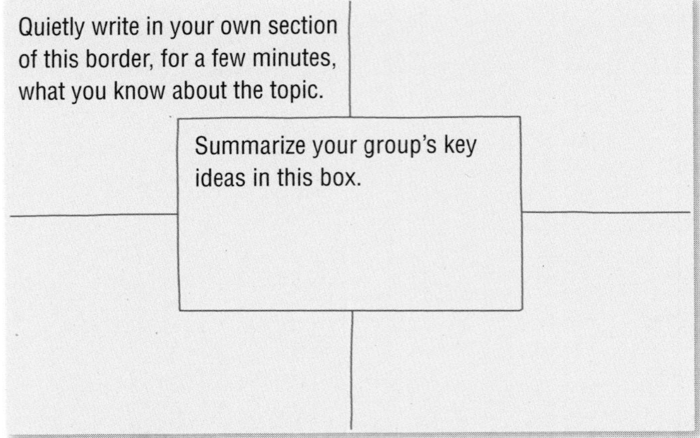

After:

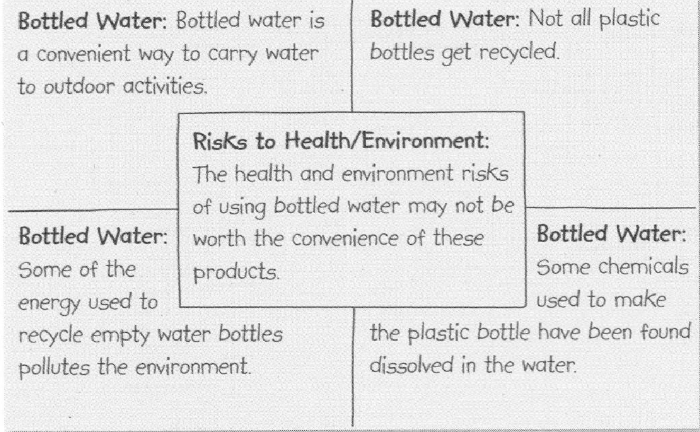

Figure 10 A placemat organizer

Prefix	Latin or Greek	Meaning	Example
ant-	Greek	opposing	antacid; Antarctic
anthrop-	Greek	related to people	anthropogenic
aqu-	Latin	water	aqueous solution; aquatic ecosystem
bio-	Greek	life	biology; biosphere
cardio-	Greek	related to the heart	cardiology
co-	Latin	with or together	covalent bond; converging lens
di-	Greek	two	diatomic molecule
epi-	Greek	upon or over	epidermis
hal-	Greek	sea salt	halogen; halophile
hemo- (haemo-)	Greek	related to blood	hemoglobin
hydro-	Greek	related to water	hydrosphere; hydroelectric
hyper-	Greek	more or excessively	hyperopia; hyperactive
infra-	Latin	below or lower	infrared radiation
inter-	Latin	between	interphase; interglacial period
litho-	Greek	rock	lithosphere
lum-	Latin	related to light	luminous
meta-	Greek	beyond	metamorphosis
micro-	Greek	small	micro-organism
mono-	Greek	one	carbon monoxide
peri-	Greek	around	peripheral nervous system; periderm
poly-	Greek	many	polyatomic ion
pro-	Greek and Latin	before	prophase; propagate
pseudo-	Greek	false	pseudoscience
retro-	Latin	turned back	retro-reflector; retrograde
sal-	Latin	related to salt	saline solution
therm-	Greek	heat	thermometer; thermocline
trans-	Latin	through or across	transparent; transistor
ultra-	Latin	above or beyond	ultraviolet radiation
xeno-	Greek	foreign or other	xenotransplantation
Suffix	**Latin or Greek**	**Meaning**	**Example**
-gen	Greek	to make or generate	carcinogen; halogen
-meter	Greek	measure	conductivity meter
-ology	Greek	study of	geology
-stasis	Greek	location	metastasis

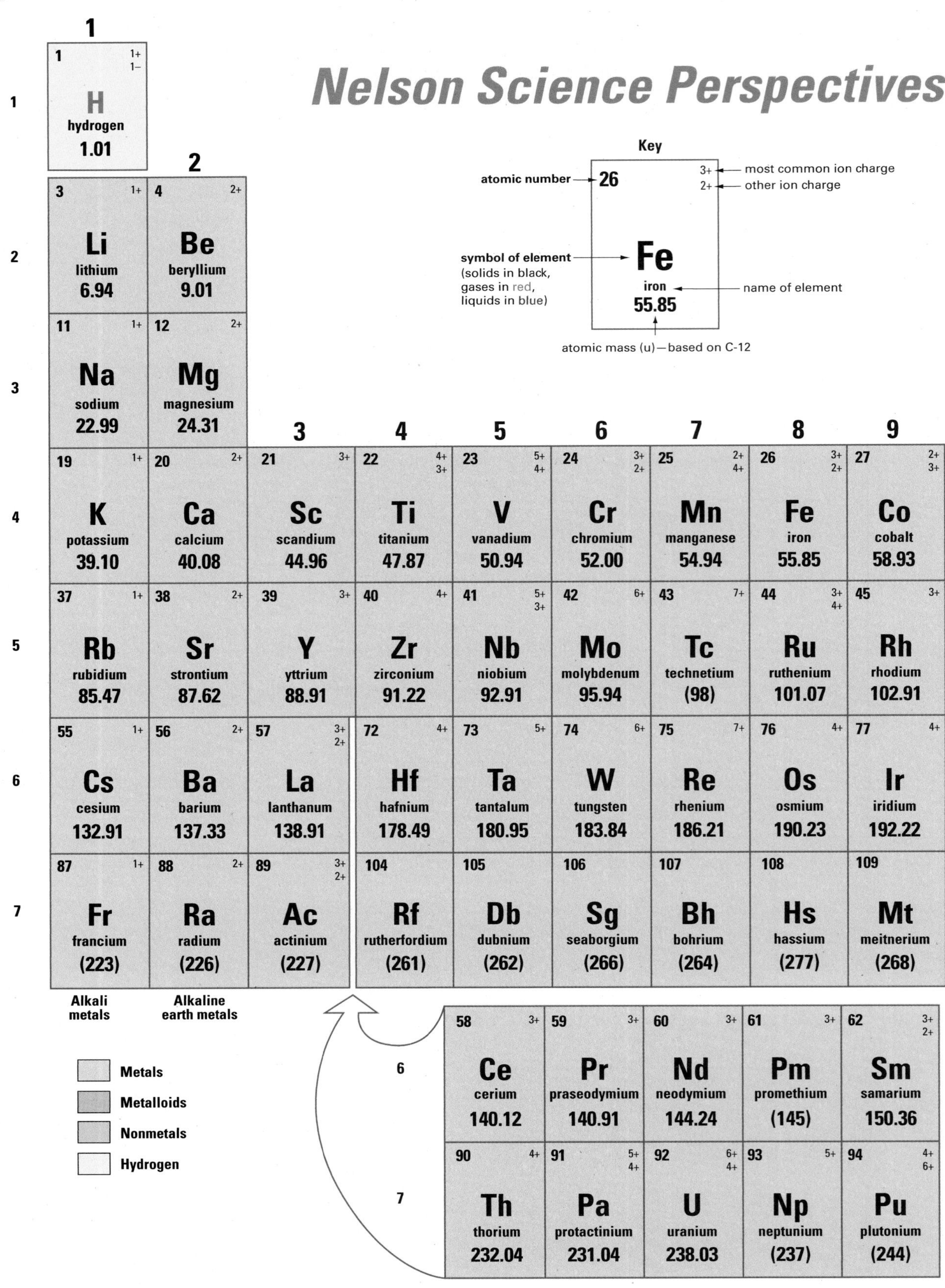
Nelson Science Perspectives
Key
atomic number 26
3+ most common ion charge
2+ other ion charge
symbol of element Fe
(solids in black, gases in red, liquids in blue)
iron name of element
55.85
atomic mass (u)—based on C-12
1
2
3
4
5
6
7
8
9
1
2
3
4
5
6
7
1 1+ 1– H hydrogen 1.01
3 1+ Li lithium 6.94
4 2+ Be beryllium 9.01
11 1+ Na sodium 22.99
12 2+ Mg magnesium 24.31
19 1+ K potassium 39.10
20 2+ Ca calcium 40.08
21 3+ Sc scandium 44.96
22 4+ 3+ Ti titanium 47.87
23 5+ 4+ V vanadium 50.94
24 3+ 2+ Cr chromium 52.00
25 2+ 4+ Mn manganese 54.94
26 3+ 2+ Fe iron 55.85
27 2+ 3+ Co cobalt 58.93
37 1+ Rb rubidium 85.47
38 2+ Sr strontium 87.62
39 3+ Y yttrium 88.91
40 4+ Zr zirconium 91.22
41 5+ 3+ Nb niobium 92.91
42 6+ Mo molybdenum 95.94
43 7+ Tc technetium (98)
44 3+ 4+ Ru ruthenium 101.07
45 3+ Rh rhodium 102.91
55 1+ Cs cesium 132.91
56 2+ Ba barium 137.33
57 3+ 2+ La lanthanum 138.91
72 4+ Hf hafnium 178.49
73 5+ Ta tantalum 180.95
74 6+ W tungsten 183.84
75 7+ Re rhenium 186.21
76 4+ Os osmium 190.23
77 4+ Ir iridium 192.22
87 1+ Fr francium (223)
88 2+ Ra radium (226)
89 3+ 2+ Ac actinium (227)
104 Rf rutherfordium (261)
105 Db dubnium (262)
106 Sg seaborgium (266)
107 Bh bohrium (264)
108 Hs hassium (277)
109 Mt meitnerium (268)
Alkali metals
Alkaline earth metals
6
58 3+ Ce cerium 140.12
59 3+ Pr praseodymium 140.91
60 3+ Nd neodymium 144.24
61 3+ Pm promethium (145)
62 3+ 2+ Sm samarium 150.36
7
90 4+ Th thorium 232.04
91 5+ 4+ Pa protactinium 231.04
92 6+ 4+ U uranium 238.03
93 5+ Np neptunium (237)
94 4+ 6+ Pu plutonium (244)
Metals
Metalloids
Nonmetals
Hydrogen

Periodic Table of the Elements

Measured values are subject to change as experimental techniques improve. Atomic molar mass values in this table are based on IUPAC Web site values (2005).

10	11	12	13	14	15	16	17	18	
								2 — **He** helium 4.00	1
			5 — **B** boron 10.81	6 — **C** carbon 12.01	7 3– **N** nitrogen 14.01	8 2– **O** oxygen 16.00	9 1– **F** fluorine 19.00	10 — **Ne** neon 20.18	2
			13 3+ **Al** aluminium 26.98	14 — **Si** silicon 28.09	15 3– **P** phosphorus 30.97	16 2– **S** sulfur 32.07	17 1– **Cl** chlorine 35.45	18 — **Ar** argon 39.95	3
28 2+ 3+ **Ni** nickel 58.69	29 2+ 1+ **Cu** copper 63.55	30 2+ **Zn** zinc 65.41	31 3+ **Ga** gallium 69.72	32 4+ **Ge** germanium 72.64	33 3– **As** arsenic 74.92	34 2– **Se** selenium 78.96	35 1– **Br** bromine 79.90	36 — **Kr** krypton 83.80	4
46 2+ 3+ **Pd** palladium 106.42	47 1+ **Ag** silver 107.87	48 2+ **Cd** cadmium 112.41	49 3+ **In** indium 114.82	50 4+ 2+ **Sn** tin 118.71	51 3+ 5+ **Sb** antimony 121.76	52 2– **Te** tellurium 127.60	53 1– **I** iodine 126.90	54 — **Xe** xenon 131.29	5
78 4+ 2+ **Pt** platinum 195.08	79 3+ 1+ **Au** gold 196.97	80 2+ 1+ **Hg** mercury 200.59	81 1+ 3+ **Tl** thallium 204.38	82 2+ 4+ **Pb** lead 207.2	83 3+ 5+ **Bi** bismuth 208.98	84 2+ 4+ **Po** polonium (209)	85 1– **At** astatine (210)	86 — **Rn** radon (222)	6
110 **Ds** darmstadtium (281)	111 **Rg** roentgenium (272)	112 **Uub** ununbium (285)	113 **Uut** ununtrium (284)	114 **Uuq** ununquadium (289)	115 **Uup** ununpentium (288)	116 **Uuh** ununhexium (291)	117 **Uus** ununseptium	118 **Uuo** ununoctium (294)	7
							Halogens	Noble gases	

63 3+ 2+ **Eu** europium 151.96	64 3+ **Gd** gadolinium 157.25	65 3+ **Tb** terbium 158.93	66 3+ **Dy** dysprosium 162.50	67 3+ **Ho** holmium 164.93	68 3+ **Er** erbium 167.26	69 3+ **Tm** thulium 168.93	70 3+ 2+ **Yb** ytterbium 173.04	71 2+ **Lu** lutetium 174.97	6
95 3+ 4+ **Am** americium (243)	96 3+ **Cm** curium (247)	97 3+ 4+ **Bk** berkelium (247)	98 3+ **Cf** californium (251)	99 3+ **Es** einsteinium (252)	100 3+ **Fm** fermium (257)	101 2+ 3+ **Md** mendelevium (258)	102 2+ 3+ **No** nobelium (259)	103 3+ **Lr** lawrencium (262)	7

APPENDIX B What Is Science?

Characteristics of Science

Science is both a collection of what we know and how we know about the world around us—both a body of knowledge and a process for acquiring knowledge.

- *Science starts from observations that lead to questions.* The natural world is full of fascinating phenomena that raise questions in the curious mind. These questions form the basis of scientific investigation.
- *Scientific knowledge can come from observation.* Observations that are made and recorded in a structured, organized fashion can provide evidence that leads to knowledge and understanding. Empirical knowledge includes knowledge gained in the process of scientific inquiry, as well as observations and knowledge gained from Aboriginal peoples, as part of their traditional ecological knowledge and wisdom (TEKW).
- *Science is done differently in different cultures.* The methods used to conduct scientific investigations may vary depending on social and cultural traditions and practices. However, the purpose of scientific inquiry remains the same: to understand and explain the natural world.
- *Scientific knowledge is tentative but reliable.* Scientific hypotheses are ideas to be tested. Repeated testing that produces consistent results can lead to laws and theories that help to describe and explain the natural world. Current laws and theories are reliable because of extensive testing, but they can change if new evidence suggests that a change is necessary. In science, a conclusion is never regarded as final.
- *Science is progressive.* Scientific knowledge builds on existing knowledge. Our understanding of the natural world grows as new knowledge is acquired and as new technologies enable further scientific investigation.
- *Science is repeatable, self-correcting, and not based on authority.* A hypothesis should be tested in such a way that the same test can be repeated by different people. This way, inconsistencies can be identified and theories can be updated to explain new observations. Modern science does not accept the proclamations of famous individuals or people in positions of political or social authority unless they can provide credible evidence to support their claims.

Misconceptions about Science

Many people misunderstand what science is and what it can do. Some of the most common misconceptions are briefly described here.

MISCONCEPTION #1: ALL SCIENTISTS FOLLOW A SINGLE SCIENTIFIC METHOD.

Many people think that there is one "scientific method" that all scientists follow to conduct research. This method is usually described as a series of steps:

1. A scientist asks a testable question and develops a hypothesis.
2. The scientist designs and carries out an experiment, makes observations, and analyzes them.
3. The scientist draws a conclusion based on the evidence and compares the conclusion with the hypothesis to determine whether the evidence supports the hypothesis.

This is a valid research method, but different scientists use different skills, technologies, and methods. There are similarities, however. Science follows procedures that generally lead to a logical conclusion, which addresses the goal of the investigation. However, there is no single scientific method that all scientists follow, step by step. The term "scientific method" refers to the general types of mental and physical activities that scientists use to create, refine, extend, and apply knowledge.

MISCONCEPTION #2: SCIENCE ALWAYS INVOLVES EXPERIMENTATION.

Experimentation is not the only approach to conducting scientific investigations, nor is it the only way of building scientific knowledge. Many scientific investigations are not experiments. Some sciences, such as astronomy and environmental science, do not lend themselves to experimentation because scientists cannot control the conditions in which a phenomenon occurs. Other types of scientific investigations are equally valid for producing valuable scientific knowledge. For example, most of our understanding about climate change is based on extensive observations and analysis of naturally occurring phenomena.

MISCONCEPTION #3: SCIENCE INVESTIGATIONS PROVIDE PROOF.

Although scientific investigations can result in scientific knowledge, they cannot provide proof. Empirical evidence can support or validate a law or theory but can never prove a law or theory to be true. Science can only show that an idea is false or disproven. Consider the law of gravity. The evidence collected worldwide leads scientists to conclude that objects denser than air always fall downward, or toward Earth's centre. An observation in which an object denser than air falls upward (away from Earth's centre) would seriously challenge our current understanding of the law of gravity.

Scientific laws are reliable. Because they are based on such vast numbers of observations, it is very unlikely that evidence will be found to prove them false. In fact, scientific laws are seldom proven false. The law of gravity and other scientific laws are probably as close to scientific "truth" as we may ever come.

MISCONCEPTION #4: SCIENCE IS NOT VERY SUCCESSFUL.

Science is often criticized because of what it has not done—for example, it has not found a cure for cancer or the common cold. However, when we consider the outstanding achievements of science, we can see that it has been very successful in helping us understand the structure of the natural world and how it functions. For example, we are confident in our knowledge that matter is made of non-visible atoms, that living things are made of cells that pass on information in the genetic material DNA, and that the continents are slowly moving across the surface of Earth.

Scientific and technological knowledge has allowed us to land on the Moon, communicate at the speed of light, and perform open-heart surgery. Recent advances in the diagnosis and treatment of cancer are possible because scientists are learning more and more about what causes cancer and how different types of cancers behave. A cure for all cancers has not yet been found, but the knowledge that is being gathered may one day provide this cure. Science is not perfect, but it is the most reliable way we have to explore and make sense of the natural world.

MISCONCEPTION #5: SCIENCE CAN PROVIDE THE ANSWERS TO ALL QUESTIONS.

Although scientific inquiry is a highly efficient way to learn about the structures and functions of the natural world, it cannot answer moral, ethical, and social questions. Should we allow mining in environmentally sensitive areas? Which of 10 patients should receive a donor kidney? Questions such as these cannot be answered by science. Scientific knowledge can, however, provide information to help individuals and groups make important decisions on such issues.

What Is Not science?

It is important to understand what science is about. It is also important to recognize what is not science. Information can be presented, intentionally or unintentionally, as science even though it is not really science. Examples abound on television, in magazines, and on the Internet.

Pseudoscience

Pseudoscience (*pseudo* means "false") is the practice of presenting claims so that they appear scientific, even though they have not been scientifically tested and are not supported by sound scientific evidence. For example, the alternative medical practice of magnetic healing (Figure 1) relies on the scientific concept of magnetic fields to give it credibility. Supporters of magnetic healing claim that magnetic fields promote the healing of bones and the improvement of blood circulation. These supporters, including some apparently credible scientists, provide testimonials about the healing effects of magnetic therapy. However, the little scientific testing done on magnetic healing has found no evidence of healing effects.

Figure 1 The magnetic field in most magnetic therapy devices is not strong enough to penetrate the skin.

Faulty Science

Faulty science results when scientists do not follow the established standards and practices that science requires. Faulty science often results when scientific methods are affected by bias. Bias occurs when a scientist allows emotions or personal values to affect the analysis of data. In general, scientists strive to be objective and unbiased when conducting scientific investigations. However, because they are human, the investigators may believe something to be true before they have evidence to support it. Because of their beliefs, and possibly a desire to be accepted by other scientists, investigators may design and conduct experiments that are guaranteed to obtain results that support their position. Valid evidence may also be incorrectly analyzed or interpreted.

Hoaxes and Frauds

Hoaxes and frauds are intentional attempts to mislead people with false claims or with information that is misrepresented. Scientific hoaxes and frauds prey upon people's lack of scientific knowledge.

Hoaxes may be carried out as a joke, but they may also be motivated by greed, desire for fame, or pressure to announce a significant scientific discovery. Most hoaxes are pranks, and the only harm done is the embarrassment of falling for the trick.

Would you fall for a hoax? How can you guard against being deceived? One way is to look closely at the claim and compare it with what you already know. Figure 2 is a touched-up photo of a tsunami about to crash over a city. The wave appears to be higher than the 20-storey building. An average storey is 3 m, so this wave would be higher than 60 m. This is unlikely since tsunami waves generally range in height from 1 m to 20 m.

Figure 2 Is this tsunami real?

Myths and Urban Legends

There are many interesting stories or claims that have circulated so widely, and for so long, that they are generally accepted as true. Such stories, referred to as urban legends, are often presented as being scientific, and many people never question their validity. The problem with urban legends is that they are usually far-fetched but somewhat believable at the same time. Although most of these stories are false, some are actually true. For others, it might be impossible to determine whether they are true or false. Can you identify those that are true?

- A penny placed on train tracks will derail a train.
- Sharks do not get cancer, so eating shark cartilage can prevent cancer in humans.
- The average person accidentally swallows eight spiders per year.
- Water that is boiled in a microwave can explode.
- Lemmings jump off cliffs to commit suicide.
- Hair grows back thicker and longer after it is shaved.
- Swimming after eating may result in a cramp that causes a person to drown.

Science in Marketing

How many advertisements do you see in an average day? Ten? One hundred? Have you ever thought about why companies advertise and whether advertising works? What strategies do advertisers use to catch our attention and to convince us to buy their products?

A common advertising strategy is to claim that a product was scientifically developed. Examples include "scientifically formulated" facial cleansers or "clinically proven" weight loss supplements. Advertisers believe that the positive public image of science will convince consumers that a product is effective, of good quality, or safe. Advertisements may feature images of researchers in white lab coats in a laboratory, or wording that conjures up this image in people's minds. The problem with this type of claim is that consumers are expected to believe that credible scientific investigations were used to develop and test the product.

Advertisers sometimes use scientific language to make the product seem authentic. For example, a common advertising phrase in today's environmentally sensitive world is "chemical free," as in "chemical-free cleaning products." Although this may sound convincing at first, after a little thought it is obvious that no product can be chemical free. All products are made of matter, and all matter is made of chemicals. So, the "chemical free" claim is meaningless.

TRY THIS SCIENTIFIC OR NON-SCIENTIFIC?

SKILLS: Analyzing, Evaluating

To be considered scientific, evidence or information must be both falsifiable and reproducible. In other words, we must be able to conduct a fair test to determine whether the information is false, and we must be able to reproduce the research to produce the same information. Using these two criteria and the information you acquired about the nature of science and about what is *not* science, analyze each of the following statements and decide whether it is scientific or non-scientific. Explain your answer. (Remember, the aim is not to decide whether these statements are true or false but rather whether they are scientific or non-scientific.)

(a) The pyramids were built by extraterrestrial beings.

(b) Wearing a magnetic bracelet will improve your golf game.

(c) Ancient Egyptian kings and queens were mummified so that their soul could go to the afterlife.

(d) The speed of an object falling due to gravity will increase by 9.8 m/s each second.

(e) Adding fertilizer to a lawn will make the grass greener.

(f) There is a creator who created and controls the universe.

(g) The more you study, the lower your test scores will be.

(h) There is life elsewhere in the universe.

(i) If you turn a potted plant upside down, the plant stem will continue to grow toward the ground.

(j) Plants are aware of their surroundings.

(k) Your horoscope can be used to predict your future.

Although many products and foods are indeed scientifically formulated and clinically proven, it is impossible to judge from the advertisement whether the quality of the scientific investigation met acceptable standards. In many cases, a thorough investigation of the product reveals that there is no accepted scientific basis for the claims that are made.

Being Skeptical

Skepticism is an important part of being scientifically literate. Being skeptical does not mean that you dismiss all claims that you encounter; it means that you must question all claims, including your own beliefs and conclusions, until sufficient evidence is available for you to accept or reject the claims. A skeptical person must be open-minded and be willing to change his or her understanding about nature when faced with sound scientific evidence and logical arguments.

How can you tell what is scientific?

You encounter bizarre and outrageous claims daily about science-related matters. You need a way to decide what is true and what is pseudoscientific, faulty science, or an urban legend. The methods of science combined with ordinary critical thinking will help you make such decisions. You should ask yourself a few questions when faced with a claim:

- Does the information source (person and/or organization) making the claim have appropriate credentials or qualifications?
- Is there evidence of bias or a conflict of interest?
- Were valid and appropriate scientific methods used to produce the information?
- Were the results reviewed by other scientists? Were the other scientists able to replicate the research and get the same results?
- Do the conclusions make sense? Are the conclusions supported by evidence?
- Does the phenomenon or product work as claimed?

As you encounter the topics and issues in this textbook, and claims to scientific knowledge in your everyday life, analyze the available information to determine what is and what is not scientific. Knowing that people will provide arguments to support their opinion or position, think critically about the claims of those representing different perspectives. Above all, use empirical evidence, ordinary common sense, and your personal values to help you accept or reject an argument and to make and defend your own position on an issue.

Ethical Behaviour in Science

Ethics is the study of what is right and wrong and how this affects the way we behave as individuals or as a group. Ethics provides us with a guide to which behaviours are acceptable and which are unacceptable.

There are acceptable and unacceptable behaviours in science. A scientist should act in an unbiased manner. We know, however, that scientists are sometimes affected by their emotions and their values. There is a set of general rules, or guiding principles, that all scientists should follow when conducting scientific research.

- *The principle of scientific honesty:* Do not commit scientific fraud. In other words, do not fabricate, trim, destroy, or misrepresent valid data. Report all results accurately.
- *The principle of carefulness:* Make every effort to avoid careless errors or sloppiness in all aspects of scientific work. Think about the potential consequences of careless errors.
- *The principle of intellectual freedom:* Do not be afraid to pursue new ideas and criticize all ideas, old and new. Feel free to conduct research in an area that interests you.
- *The principle of openness:* Share data, results, methods, theories, equipment, and so on. Allow others to see your work and be open to criticism and suggestions.
- *The principle of credit:* Do not claim other people's work as your own. Give credit where it is due.
- *The principle of public responsibility:* Report the results of scientific research to the public if it has important implications for society, but only after the results has been validated and verified by other scientists.

Science and technology are very different activities, but we often hear about them together. This is because they are closely related and often go hand in hand. Scientists rely on technologies to further their research, and technologists rely on scientists to understand the scientific basis of technological developments.

A Partnership: Science and Technology

Scientific research produces knowledge, or understanding of natural phenomena. Technologists and engineers look for ways to apply this knowledge in the development of practical products and processes. For example, scientists want to know how different materials affect the transmission of X-rays (Figure 3). Technologists and engineers focus on using that knowledge to explore inside solid objects, including the human body. Scientists use technologies in their research. Engineers and technologists may use scientific knowledge and principles when designing and developing new technologies.

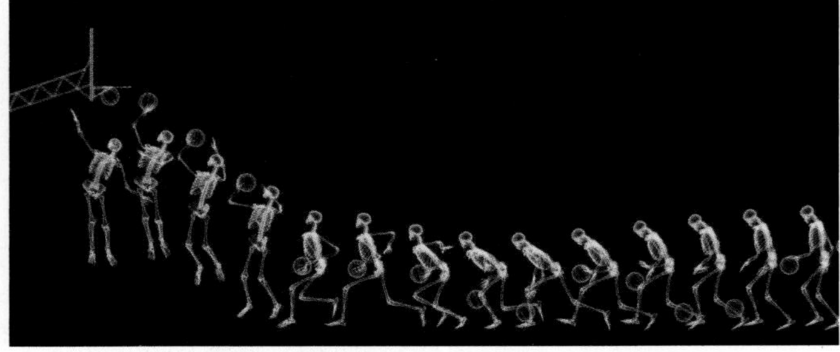

Figure 3 Knowing that X-rays do not pass through dense material such as bones, technologists designed machines that use X-rays and film or a computer to provide an internal view of the human body.

In some cases, technological inventions and innovations occur before the scientific principles are known. For example, practitioners of early traditional medicine used specific plants to treat specific conditions long before modern medical professionals understood how the chemicals in these plants affected human body functions. In other cases, scientific discoveries are made because of technological inventions. For example, the invention of glass lenses led to the development of early microscopes (Figure 4), which allowed scientists to see micro-organisms that cause diseases.

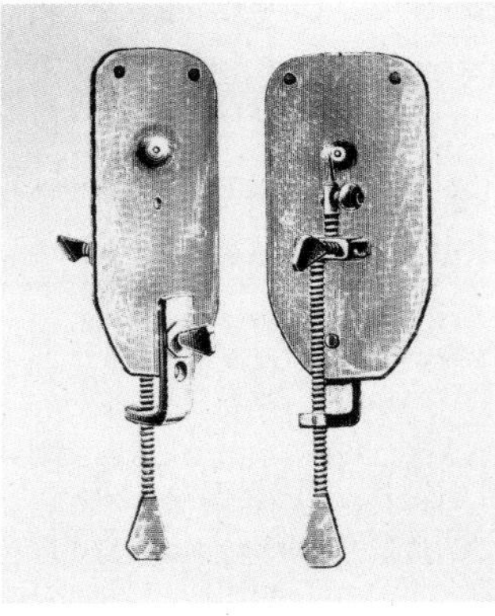

Figure 4 The development of the light microscope followed the understanding of the structure and behaviour of glass lenses.

DID YOU KNOW?

Discovery of Bacteria

Antonie van Leeuwenhoek (1632–1723), a Dutch scientist with no formal education, designed many simple microscopes, which enabled him to discover bacteria. He described them as "incredibly small, nay so small, in my sight, that I judged that even if 100 of these very wee animals lay stretched out one against another, they could not reach to the length of a grain of coarse sand."

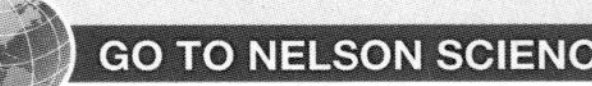

GO TO NELSON SCIENCE

The invention of glass lenses also led to the development of telescopes, which allowed astronomers to observe our solar system and the universe. The telescope, in turn, led to more accurate astronomical observations and measurements, which contributed to the change from an Earth-centred scientific model of the universe to a Sun-centred model. Many other technologies, including the thermometer and the computer, have greatly helped the advancement of science. You can see how science and technology often support each other.

Sometimes, technological inventions follow scientific discoveries (Table 1 on page 655). For example, the television was invented after scientists had created theories to explain the structure of the atom and understood electrons, current electricity, and electromagnetism. The relationship between science and technology is mutually beneficial; scientific discoveries lead to technological advances, which lead to further scientific discoveries, and so on.

You may have heard in the media that some disease-causing bacteria are becoming resistant to antibiotics. Viruses are another form of micro-organism (Figure 5). Scientists recently discovered more about how viruses make, package, and insert genetic information into living cells. With this knowledge, technologists can design a virus that will seek out and kill disease-causing bacteria.

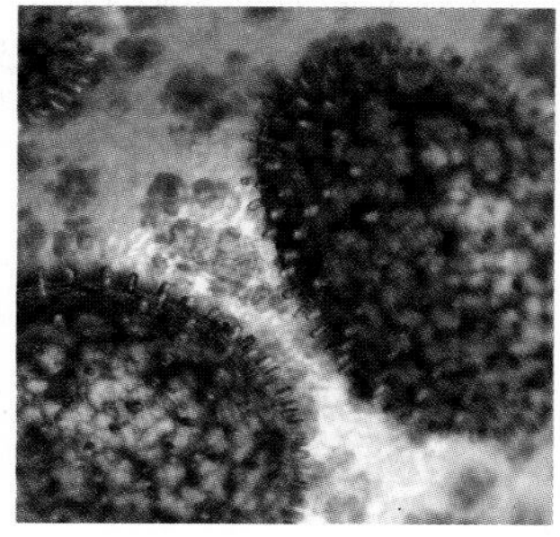

Figure 5 The Influenza-A virus

Table 1 Examples of the Science–Technology Relationship

Science	Technology	Example
Biologists learn how the heart functions.	Engineers and technologists design replacement valves and artificial hearts.	Replacement heart valve
Chemists learn about the structure of materials.	Engineers and technologists design useful products, using materials with suitable properties.	New materials such as Kevlar or Zylon, used in protective gear
Earth scientists observe how radio waves are reflected off different surfaces, such as snow and rock.	Technologists use satellite radar imaging to view Earth from space.	Satellite image of a coastline
Physicists discover how light behaves when it passes through or reflects off different materials.	Technologists design lenses and mirrors for telescopes, microscopes, and other optical devices.	Optical telescope

Science, Technology, and Society

You do not have to look far to find evidence of the impact of science and technology on society. Many important discoveries and inventions have occurred within the last century—vaccines, antibiotics, organ transplants, reproductive technologies, genetic engineering, pesticides, atomic weapons, computers, lasers, plastics, televisions, communication satellites, and the Internet. The homes we live in, the food we eat, the vehicles we drive, and the electronic gadgets we use are all products of scientific and technological achievement (Figure 6).

Figure 6 Examples of science and technology in our daily lives: (a) gaming console (b) hybrid car engine.

There are obvious influences of science and technology on society. But science and technology are influenced by society as well. The values and priorities of society at a particular time can influence the direction and progress of developments in both science and technology. For example, our desire to reduce greenhouse gas emissions is encouraging the rapid development of vehicles and machinery that use alternative fuels.

Scientific and technological research is very expensive. Research facilities employ highly paid and highly skilled professionals. The facilities consume large amounts of energy and require expensive and sophisticated tools and equipment. Funding for research comes from both private and public sources. It may be a long time between the beginning of research and development and the release of a new product or process.

Basic Research

Basic research helps people learn more about how the natural world works and is essentially the same as scientific investigation. Basic research—in areas such as biochemistry, particle physics, astronomy, and geology—usually receives funding from government grant agencies.

This type of research often produces knowledge that engineers and technologists can use to develop practical solutions to everyday problems. The priorities of government, representing the priorities of the public, determine which areas of research are funded.

Applied Research

Applied research is primarily focused on developing new and better solutions to practical problems. Applied research can be equated with the development of technology. Research into the development of technology—such as new cosmetics, sports equipment, telephones, automobiles, computer software and hardware, medical equipment, and pharmaceuticals—is usually carried out by privately owned companies. The marketplace, or the public demand for new products, will obviously influence which areas of research private companies fund. If market analysis shows that there is a demand for better cellphones or more fuel-efficient cars, research and development in these areas will be supported.

There are risks and benefits associated with many scientific and technological developments. Although most technologies are developed with the intention of solving problems, there are often unintended consequences associated with their use. Social networking sites, and other Internet technologies, can greatly facilitate communication; however, the same technology is also used to commit crimes.

It is difficult, if not impossible, to foresee all of the consequences of technologies. Often, new applications of scientific and technological knowledge are found long after the knowledge was first acquired. An important question remains up for debate. Who is responsible for the negative impacts of science and technology on society: scientists, science and technology, or human use of science and technology?

Science, Technology, and the Environment

Since the beginning of the Industrial Revolution in the late 1700s, the industrialized world has used natural resources at an increasing rate. With the use of resources, the development of technology, and the human population all increasing, we are producing waste faster than ever.

TRY THIS DISTINGUISHING BETWEEN BASIC AND APPLIED RESEARCH

SKILLS: Analyzing, Evaluating, Communicating

Science and technology are so intertwined and dependent on each other that it is sometimes difficult to distinguish between the two or to determine where one ends and the other begins.

1. In a small group, consider each of the following descriptions. Discuss each description and decide whether it refers to basic research, applied research, or both. Explain your choice for each statement.
   - Researchers have designed a virus that will carry genetic information to a specific type of bacterial cell, causing the bacterium to self-destruct.
   - Eliminating the oxygen from a package containing fresh produce extends its refrigerator life.
   - NASA astronauts observed that crystals grown in the International Space Station are perfectly formed.
   - A group of researchers has discovered how a virus delivers DNA to a living cell.
   - Certain materials, known as superconductors, can conduct electrical energy very efficiently, with practically no loss of energy as heat.
   - Government researchers are trying to determine what causes some maple trees to lose all of their leaves in midsummer.
   - Researchers have designed a flexible lens that changes shape in response to electrical stimuli.
   - Superconductor materials are used in new computers.
   - A car manufacturer's research lab has built a battery for its electric car that can store 50 % more energy than other batteries.
   - Chemical researchers have found a new chemical process that absorbs carbon dioxide.

Many of the by-products of human industry and technology become pollutants. For example, there are hundreds of millions of discarded cellphones in North America, most of which end up in landfills (Figure 7).

Figure 7 Cellphones contain hazardous substances such as lead, mercury, and arsenic. These metals can leach into groundwater if they are dumped into landfills.

The impact of science and technology on the environment is not always negative. There are many positive effects of science and technology. New knowledge improves our understanding of the natural world, and new technologies allow us to live with the environment in a more sustainable way.

Figure 8 illustrates how technology can transform the energy of the Sun into electrical energy. This technology is generally more environmentally friendly than the use of fossil fuels. Researchers in the auto industry are busy developing vehicles that use alternative sources of energy and are more fuel efficient. The intention is that these vehicles will have less of an impact on the environment.

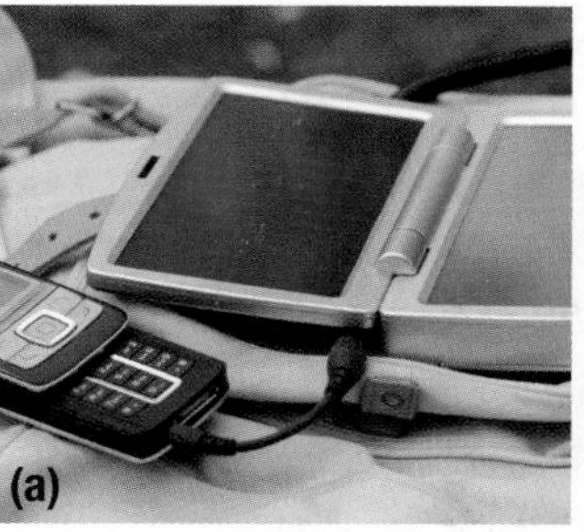

Figure 8 (a) Small solar panels can charge your cellphone. (b) Large panels can heat water or power your home.

The relationships among science, technology, society, and the environment are complex. To make wise personal decisions and to act as responsible citizens we must think carefully about the influences of each of these elements on the other. We each have a responsibility to ourselves, to society, and to future generations to become scientifically and technologically literate.

Numerical and Short Answers

This section includes numerical and short answers to questions in Check Your Learning, Chapter Review, Chapter Self-Quiz, Unit Review, and Unit Self-Quiz.

Unit B

Section 2.1, p. 32

3. size
7. structure and protection

Section 2.3, p. 37

1. growth, reproduction, repair
4. osmosis and diffusion
6. mitosis

Section 2.5, p. 44

1. interphase
5. (a) cytokinesis
   (b) metaphase
   (c) anaphase
   (d) cytokinesis
   (e) telophase
   (f) interphase

Section 2.7, p. 55

1. cancer cells do not stop dividing
2. (b) no

Section 2.9, p. 60

2. e.g., breathe, eat, eliminate waste, keep constant body temperature

Chapter 2 Review, pp. 64–65

1. interphase, mitosis, cytokinesis
2. interphase
4. (a) (i) plant; (ii) plant; (iii) animal; (iv) animal
   (b) (i) metaphase; (ii) end of interphase; (iii) metaphase; (iv) telophase
6. diffusion
10. spindle fibres
14. root tip
16. (a) A: interphase: 85%; prophase: 3%; metaphase: 1%; anaphase: 0.6%; telophase: 0.3%, cytokinesis: 0.3%.
    B: interphase: 77%; prophase: 14%; metaphase: 3%; anaphase: 2%; telophase: 1%; cytokinesis: 1%.

Chapter 2 Self-Quiz, pp. 66–67

1. (c)
2. (a)
3. (a)
4. (d)
5. F
6. F
7. T
8. asexual
9. carcinogens
10. (a) (iii)
    (b) (iv)
    (c) (i)
    (d) (ii)
    (e) (v)
11. 60 chromosomes

Section 3.1, p. 76

2. (a) heart
   (b) lungs

Section 3.3, p. 82

3. e.g., saliva, stomach acid, enzymes, bile, mucous
4. muscle

Section 3.4, p. 87

2. e.g., blood, oxygen, food, carbon dioxide
5. (a) artery
   (b) capillary
   (c) vein

Section 3.6, p. 95

1. nasal cavity, mouth, trachea, bronchi, lungs, bronchioles, alveoli
5. (b) stomach and lunch secretion

Section 3.7, p. 98

4. rejection, risk of infection

Section 3.8, p. 101

1. movement, support, protection

Section 3.10, p. 107

4. e.g., breathing during exercise, digestion, appetite
6. e.g., CT scan

Section 3.11, p. 111

2. (a) circulatory system
3. skin, respiratory system

Chapter 3 Review, pp. 116–117

1. organism
5. (a) smooth
6. (a) musculoskeletal
   (b) movement
   (c) connective, muscle
7. (a) circulatory system
   (b) heart
10. (a) move, grasp, throw, catch
    (b) flap, fly

Chapter 3 Self-Quiz, pp. 118–119

1. (d)
2. (a)
3. (b)
4. (c)
5. F
6. F
7. central; peripheral
8. alveoli
9. (a) (iv)
   (b) (v)
   (c) (i)
   (d) (ii)
   (e) (iii)
10. e.g., respiratory and urinary systems
12. lungs

Section 4.1, p. 128

1. root and shoot

Section 4.2, p. 133

1. growth and differentiation

Section 4.6, p. 147

3. apical, lateral
5. apical

Chapter 4 Review, pp. 150–151

1. (a) root, shoot
   (b) dermal, vascular, ground
   (c) roots, stem, leaves, flower
   (d) xylem, phloem
6. e.g., respiration, reproduction, growth
10. grana
19. vascular tissue
21. oxygen, sugar
23. (a) grana
    (b) stem
    (c) leaf
    (d) apical meristem of shoot

Chapter 4 Self-Quiz, pp. 152–153

1. (a)
2. (b)
3. (c)
4. (d)
5. (d)
6. (a)
7. F
8. T
9. T
10. chloroplast
11. grana
12. pollen
13. (a) (iii)
    (b) (i)
    (c) (iv)
    (d) (ii)
16. (a) sugar and oxygen
18. tree with diameter of 3 m

Unit B Review, pp. 158–163

1. (a)
2. (b)
3. (b)
4. (d)
5. (b)
6. (c)
7. (a)
8. (c)
9. (a)
10. (b)
11. (d)
12. T
13. F
14. F
15. F
16. F
17. F
18. F
19. F
20. F
21. bone marrow, umbilical cord blood
22. nerve, muscle
23. urinary, digestive, circulatory, respiratory
24. nervous
25. water
26. pairs, contract, relax
27. higher, lower
28. cellular differentiation
29. (a) (iii)
    (b) (v)
    (c) (iv)
    (d) (ii)
    (e) (i)
50. desert

Unit B Self-Quiz, pp. 164–165

1. (b)
2. (d)
3. (d)
4. (a)
5. T
6. F
7. T
8. lateral
9. circulatory
10. (a) (iii)
    (b) (iv)
    (c) (i)
    (d) (v)
    (e) (ii)
14. no

Unit C

Section 5.1, p. 178

2. (a) physical
   (b) chemical
   (c) chemical
   (d) physical
   (e) chemical
   (f) chemical
3. (a) physical
   (b) physical
   (c) chemical
   (d) chemical
   (e) physical
   (f) chemical
5. chemical
8. chemical
9. e.g., chemical

Section 5.3, p. 183

6. workplace label
7. e.g., hydrochloric acid
8. e.g., chlorine bleach

Section 5.4, p. 187

3. (a) fluorine
   (b) strontium
   (c) helium
   (d) iodine
   (e) potassium
   (f) aluminum
   (g) neon
5. (a) alkali metal
   (b) 1
6. (a) (i) non-metal; (ii) non-metal; (iii) metal; (iv) metal

Section 5.5, p. 191

4. (a) magnesium plus 2
   (b) sulfide minus 2
   (c) iron plus 3
   (d) bromide minus 1
   (e) nitride minus 3
5. (a) P^{3-}, Cl^{-}, Ar
   (b) Na^{+}, F^{-}, Ne
   (c) S^{2-}, Cl^{-}, Ar
   (d) Se^{2-}, Br^{-}, As^{3-}
   (e) Ba^{2+}, I^{-}, Xe
6. (a) 2
   (b) 2^{+}

Section 5.6, p. 195

1. metals and non-metals
6. (a) Na^{+}, F^{-}; 1:1
   (b) Li^{+}, N^{3-}; 3:1
   (c) F^{3+}, Cl^{-}; 1:3
   (d) K^{+}, O^{2-}; 2:1
7. (a) X: metal; Y: non-metal
   (b) XY_3
9. silver sulphide

Section 5.7, p. 200

2. (a) calcium fluoride
   (b) potassium sulfide
   (c) aluminum oxide
   (d) lithium bromide
   (e) calcium phosphide
3. (a) KBr
   (b) CaO
   (c) Na_2S
4. SnO_2
5. copper(I) bromide: CuBr; copper(II) bromide: $CuBr_2$
7. (a) $CaCl_2$
   (b) $AlBr_3$
   (c) MgS
   (d) Li_3N
   (e) Ca_3N_2
9. (a) $FeBr_2$
   (b) MnO_2
   (c) $SnCl_4$
   (d) Cu_2S
   (e) FeN
   (f) CuO
   (g) lead(II) chloride
   (h) iron(III) oxide
   (i) tin(II) sulfide
   (j) copper(II) phosphide
   (k) calcium bromide
   (l) copper(II) fluoride
   (m) potassium phosphide
   (n) copper(I) phosphide

Section 5.9, p. 205

1. (a) nitrate ion, potassium nitrate
   (b) hydroxide ion, calcium hydroxide
   (c) carbonate ion, calcium carbonate
   (d) sulfate ion, copper(II) sulfate
   (e) hydroxide ion, potassium hydroxide
   (f) nitrate ion, iron(III) nitrate
   (g) chlorate ion, copper(II) chlorate
   (h) phosphate ion, ammonium phosphate
2. (a) KNO_3
   (b) $BaSO_4$
   (c) NH_4NO_3
   (d) $(NH_4)_2SO_4$
   (e) $KClO_3$
   (f) $Cu(NO_3)_2$
   (g) $PbSO_4$
   (h) $Sn_3(PO_4)_2$
3. (a) -ate
   (b) -ide
4. fertilizer
5. (a) tin(II) carbonate
   (b) calcium chloride
   (c) iron(III) hydroxide
   (d) manganese(IV) oxide
   (e) potassium sulfide
   (f) ammonium sulfate
   (g) manganese(II) chlorate
   (h) lead(II) iodide
6. (a) $CaSO_4$
   (b) NH_4Cl
   (c) Cu_2CO_3
   (d) BaS
   (e) $Ca(ClO_3)_2$
   (f) $Sn(OH)_2$
   (g) $Fe_3(PO_4)_4$
   (h) AlN
8. when the cation is ammonium
9. cation
11. (a) NaCl, $NaClO_3$
    (b) Cl^-, ClO_3^-
    (c) $CaCl_2$, $Ca(ClO_3)_2$

Section 5.10, p. 212

2. (a) CO
   (b) SF_4
   (c) N_2O_4
   (d) NBr_3
   (e) CS_2
4. (a) 1 H, 6 O
   (b) 1 H, 2 O

Chapter 5 Review, pp. 216–217

5. (a) CO_2: 1 C, 2 O
   (b) N_2: 2 N
   (c) CCl_4: 1 C, 4 Cl
   (d) HBr: 1 H, 1 Br
6. (a) iron(III) chloride, ionic
   (b) copper(II) sulfate, ionic
   (c) nitrogen tri-iodide, molecular
   (d) lead(IV) oxide, ionic
   (e) diphosphorous trioxide, molecular
   (f) tin(II) nitrate, ionic
   (g) CBr_4, molecular
   (h) $CaCO_3$, ionic
   (i) NO, molecular
   (j) H_2S, molecular
7. (a) 10
   (b) not the same
   (c) 18
   (d) not the same
8. (a) KCl, ionic
   (b) CO, molecular
   (c) CCl_4, molecular
   (d) CaI_2, ionic
   (e) SO_2, molecular
   (f) Li_2O, ionic
10. (a) CaS, calcium sulfide
    (b) $AlCl_3$, aluminum chloride
    (c) Na_3P, sodium phosphide
    (d) Al_2S_3, aluminum sulfide
11. (a) $Ca(NO_3)_2$
    (b) Ag_2CO_3
    (c) iron(III) hydroxide
    (d) copper(II) chlorate
    (e) $Pb_3(PO_4)_2$
12. (a) 1, 1
    (b) 1, 1
    (c) 2, 1

Chapter 5 Self-Quiz, pp. 218–219

1. (a)
2. (c)
3. (d)
4. (b)
5. T
6. F
7. period
8. nucleus
9. (a) (iv)
   (b) (i)
   (c) (iii)
   (d) (ii)
10. (a) (ii)
    (b) (iii)
    (c) (i)
    (d) (iv)

Section 6.1, p. 227

4. (a) reactants: $AgNO_3$, NaCl; products: AgCl, $NaNO_3$
   (b) $AgNO_3$, NaCl, $NaNO_3$
   (c) AgCl
5. (a) H_2, $ZnSO_4$, energy
   (b) water
7. (a) sugar + energy $\rightarrow$ carbon + water
8. (a) glucose $\rightarrow$ ethanol + carbon dioxide
   (b) chemical
9. (a) $CO_2 + H_2O \rightarrow H_2CO_3$

Section 6.3, p. 232

6. (a) 10 g

Section 6.4, p. 236

5. (a) $2KI \rightarrow 2K + I_2$
   (b) $Mg + 2AgNO_3 \rightarrow 2Ag + Mg(NO_3)_2$
   (c) $Na + 2H_2O \rightarrow H_2 + 2\ NaOH$
   (d) $Pb(NO_3)_2 + 2NaCl \rightarrow PbCl_2 + 2NaNO_3$
6. (a) $2C_8H_{18} + 25\ O_2 \rightarrow 16CO_2 + 18H_2O$
   (b) 8 molecules of CO_2
7. (a) balanced as written
   (b) $2K + Br_2 \rightarrow 2KBr$
   (c) $2H_2O_2 \rightarrow 2H_2O + O_2$
   (d) $4Na + O_2 \rightarrow 2Na_2O$
   (e) $N_2 + 3H_2 \rightarrow 2NH_3$
   (f) balanced as written
   (g) $CaSO_4 + 2KOH \rightarrow Ca(OH)_2 + K_2SO_4$
   (h) $Ba + 2HNO_3 \rightarrow H_2 + Ba(NO_3)_2$
   (i) $H_3PO_4 + 3NaOH \rightarrow 3H_2O + Na_3PO_4$
   (j) $C_3H_8 + 5O_2 \rightarrow 3CO_2 + 4H_2O$
   (k) $Al_4C_3 + 12H_2O \rightarrow 3CH_4 + 4Al(OH)_3$
   (l) $FeBr_3 + 3Na \rightarrow Fe + 3NaBr$
   (m) $2Fe + 3H_2SO_4 \rightarrow 3H_2 + Fe_2(SO_4)_3$
   (n) $2C_2H_6 + 7O_2 \rightarrow 4CO_2 + 6H_2O$
8. (a) ammonium dichromate → nitrogen gas + water + chromium oxide
   (b) 1.5 g

Section 6.5, p. 239

1. (a) decomposition
   (b) synthesis
   (c) synthesis
   (d) decomposition
2. (a) $ZnCl_2 \rightarrow Zn + Cl_2$
   (b) $2K + I_2 \rightarrow 2KI$
   (c) $K_2O + H_2O \rightarrow 2KOH$
   (d) $CaCO_3 \rightarrow CaO + CO_2$
3. (a) copper (II) oxide → copper + oxygen
   (b) decomposition
   (c) $2CuO \rightarrow 2Cu + O_2$
4. (a) $H_2\ (g) + Cl_2\ (g) \rightarrow 2HCl\ (g)$; synthesis
   (b) $2H_2O_2\ (l) \rightarrow 2H_2O\ (l) + O_2(g)$; decomposition
   (c) $2\ KClO_3\ (s) \rightarrow 2KCl\ (s) + 3O_2\ (g)$; decomposition
   (d) $3H_2 + N_2 \rightarrow 2\ NH_3$; synthesis
   (e) $4Al + 3O_2 \rightarrow Al_2O_3$; decomposition

Section 6.6, p. 243

2. (a) an element and a compound
   (b) two compounds
3. (a) single
   (b) double
   (c) single
   (d) double
   (e) single
4. (a) $2Al + Fe_2O_3 \rightarrow Al_2O_3 + 2Fe$
   (b) $BaCl_2 + Na_2SO_4 \rightarrow BaSO_4 + 2NaCl$
   (c) $Zn + CuSO_4 \rightarrow ZnSO_4 + Cu$
   (d) $AgNO_3 + Na_3PO_4 \rightarrow Ag_3PO_4 + NaNO_3$
   (e) $2Ca + 2H_2O \rightarrow H_2 + 2Ca(OH)_2$
5. (a) single displacement
7. (a) single displacement
8. (a) $Ag + 2HNO_3 \rightarrow AgNO_3 + NO_2 + H_2O$

Section 6.9, p. 251

2. (a) $S(s) + O_2(g) \rightarrow SO_2(g)$ + energy
   (b) $2Ca(s) + O_2(g) \rightarrow 2CaO(s)$ + energy
   (c) $C_3H_8(g) + 5O_2 \rightarrow 3O_2 + 4H_2O$ + energy
   (d) $C_2H_4(g) + 3O_2 \rightarrow 2CO_2 + 2H_2O$ + energy
3. (a) hydrocarbon + oxygen → carbon dioxide + water + energy
   (b) $C_3H_8 + 5O_2 \rightarrow 3CO_2 + 4H_2O$ + energy
6. synthesis reactions; decomposition reactions; single displacement; double displacement; combustion

Section 6.10, p. 254

3. (a) water, oxygen
   (b) electrolytes

Chapter 6 Review, pp. 258–259

1. (a) 2
   (b) 1
   (c) 3
   (d) 2
   (e) 5
   (f) ZnS: solid; oxygen: gas; ZnO: solid; SO_2: gas
3. (a) synthesis reaction
   (b) double-replacement reaction
   (c) combustion reaction
   (d) decomposition reaction
   (e) single-displacement reaction
11. (a) 0.3 g

Chapter 6 Self-Quiz, pp. 260–261

1. (b)
2. (d)
3. (b)
4. (b)
5. T
6. F
7. coefficient
8. product
9. (a) (iv)
   (b) (iii)
   (c) (ii)
   (d) (i)
   (e) (v)
13. products and reactants, ratios, states

Section 7.2, p. 271

2. (a) basic
   (b) acidic
   (c) basic
   (d) basic
   (e) basic
3. (a) potassium hydroxide
   (b) nitric acid
   (c) $Ba(OH)_2(aq)$
   (d) potassium hydrogen carbonate
   (e) $NaHCO_3(aq)$
4. H
5. hydroxide ions

Section 7.3, p. 275

2. (a) 2
   (b) 10
   (c) 9

3. (a) highly basic
   (b) slightly acidic
   (c) highly acidic
   (d) highly acidic
   (e) slightly basic

Section 7.5, p. 281

2. (a) $HCl(aq) + KOH(aq) \rightarrow KCl(aq) + H_2O(l)$
   (b) $H_2SO_4(aq) + 2KOH(aq) \rightarrow K_2SO_4(aq) + 2H_2O(l)$
3. (a) $H_2CO_3(aq) + 2KOH(aq) \rightarrow K_2CO_3(aq) + 2H_2O(l)$
7. (a) CO_2, H_2O, and a Ca salt form
   (b) $H_2SO_4(aq) + CaCO_3(s) \rightarrow CaSO_4(aq) + H_2O(l) + CO_2(g)$

Section 7.8, p. 290

4. (a) sulfur dioxide, nitrogen oxides
   (b) e.g., $3NO_2(g) + H_2O(l) \rightarrow 2HNO_3(aq) + NO(g)$
5. (a) sulfur dioxide: <3 %; nitrogen oxide: 51 %
6. (a) water birds, other large organisms
   (b) fish, other small organisms
11. (a) $CaO(s) + H_2O(l) \rightarrow Ca(OH)_2(aq)$

Chapter 7 Review, pp. 294–295

1. (a) phosphoric acid
   (b) hydrobromic acid
   (c) iron(III) hydroxide
   (d) sulfuric acid
   (e) calcium hydrogen carbonate
   (f) potassium nitrate
3. (a) hydrogen
   (b) hydrogen
   (c) acid: yellow; base: blue
   (d) water and a salt
4. (a) sulfur dioxide, nitrogen oxides
   (b) burning fossil fuels, mining and refining metals
   (c) e.g., scrubbers in smokestacks, catalytic converters in vehicles
5. (a) $H_3PO_4(aq)$, $HBr(aq)$, $H_2SO_4(aq)$
   (b) $Fe(OH)_3$, $Ca(HCO_3)_2$
   (c) $KNO_3(aq)$
6. (a) e.g., vinegar
   (b) e.g., drain cleaner
8. (a) potassium hydroxide
   (b) water, calcium sulfate
   (c) sodium phosphate
9. (a) $HCl(aq) + KOH(aq) \rightarrow H_2O(l) + KCl(aq)$
   (b) $H_2SO_4(aq) + Ca(OH)_2(aq) \rightarrow 2H_2O(l) + CaSO_4(s)$
   (c) $H_3PO_4(aq) + 3NaOH(aq) \rightarrow 3H_2O(l) + Na_3PO_4(aq)$
15. (a) 7.6–7.8
16. A: hydrochloric acid; B: potassium hydroxide; C: sugar

Chapter 7 Self-Quiz, pp. 296–297

1. (c)
2. (a)
3. (b)
4. (c)
5. F
6. F
7. F
8. water
9. pH
10. (a) (iii)
    (b) (i)
    (c) (iv)
    (d) (ii)
13. (a) $HNO_3 + KOH \rightarrow KNO_3 + H_2O$
    (b) $Ca(OH)_2 + H_2SO_4 \rightarrow CaCO_3 + 2H_2O$
    (c) $H_2SO_4 + NaOH \rightarrow Na_2SO_4 + 2H_2O$
15. (a) hydrogen ions, hydroxide ions
    (b) water
    (c) neutralization reaction
16. (a) basic
18. (a) either end of the pH scale

Unit C Review, pp. 302–307

1. (d)
2. (a)
3. (b)
4. (c)
5. (c)
6. (d)
7. (a)
8. (b)
9. (c)
10. F
11. T
12. F
13. F
14. F
15. T
16. F
17. F
18. F
19. F
20. F
21. T
22. F
23. T
24. chemical
25. products
26. element
27. noble gas
28. sulfur dioxide
29. rust
30. indicator
31. metals
32. physical
33. compound
34. 18
35. hydrogen
36. $K_2SO_4(aq) + BaCl_2(aq) \rightarrow BaSO_4(s) + 2KCl(aq)$
37. synthesis
40. synthesis
41. $2C_2H_2 + 5O_2 \rightarrow 4CO_2 + 2H_2O$
43. e.g., Ar, Cl^-, P^{3-}
44. (a) $MgCl_2$
    (b) Al_2S_3
    (c) $SnSO_4$
    (d) Fe_2O_3
    (e) $Pb(NO_3)_2$
    (f) Ag_3PO_4
    (g) H_2SO_4
    (h) HCl
    (i) ClO_2
    (j) N_2O
45. (a) potassium oxide
    (b) copper(II) sulfide
    (c) sodium phosphate
    (d) lead(II) hydroxide
    (e) nitric acid
    (f) carbon monoxide
    (g) nitrogen monoxide

47. (a) synthesis
(b) single displacement
(c) double displacement
(d) decomposition
(e) combustion

48. (a) $2NH_3(g) + H_2SO_4(aq) \rightarrow (NH_4)_2SO_4(aq)$
(b) $2Al(s) + 3CuCl_2(aq) \rightarrow 2AlCl_3(aq) + 3Cu(s)$
(c) $H_3PO_4(aq) + 3NaOH(aq) \rightarrow 3H_2O(l) + Na_3PO_4(aq)$
(d) $Al_2(SO_4)_3(s) \rightarrow Al_2O_3(s) + 3SO_3(g)$
(e) $2C_2H_6(g) + 7O_2(g) \rightarrow 4CO_2(g) + 6H_2O(l)$

49. (a) carbon + iron(III) oxide $\rightarrow$ iron + carbon dioxide
(b) $3C(s) + 2Fe_2O_3(s) \rightarrow 4Fe(l) + 3CO_2(g)$
(c) single displacement

53. (a) potassium sulfide
(b) carbon tetrabromide
(c) propane
(d) copper sulfate
(e) silver nitrate
(f) lead dioxide
(g) nitrogen(I) oxide

54. $2FeCl_3 + SnCl_2 \rightarrow 2FeCl_2 + SnCl_4$

56. 43 g

59. 22 g

60. XH_3

Unit C Self-Quiz, pp. 308–309

1. (b)
2. (a)
3. (b)
4. (c)
5. (a)
6. (c)
7. T
8. F
9. T
10. double displacement
11. acid
12. potassium nitrate
13. (a) (iii)
(b) (ii)
(c) (v)
(d) (i)
(e) (iv)

14. $HCl + KOH \rightarrow KCl + H_2O$

16. (a) Al_2O_3
(b) $FeCl_2$
(c) $(NH_4)_2SO_4$

18. CaX

20. (a) hydrogen
(b) $Zn + H_2SO_4 \rightarrow ZnSO_4 + H_2$

21. (b) $2H_2O_2 \rightarrow 2H_2O + O_2$

Unit D

Section 8.1, p. 321

1. (a) weather
(b) weather
(c) climate
(d) weather
(e) weather

Section 8.2, p. 324

1. (a) temperature, precipitation, plant communities
(b) landform, soil, animals, human factors

2. (a) dry arid
(b) yes

4. e.g., 11

Section 8.4, p. 335

1. (a) atmosphere, litmosphere, hydrosphere, living things

2. troposphere, stratosphere, mesosphere, thermosphere, exosphere

Section 8.6, p. 342

3. (b) water vapour, carbon dioxide

6. (a) e.g., respiration
(b) e.g., digestive systems of some animals
(c) e.g., volcanic eruptions
(d) e.g., ice sheets that melt

Chapter 8 Review, pp. 364–365

3. atmosphere, hydrosphere, lithosphere, living things

4. (b) e.g., carbon dioxide, methane, nitrous oxide

7. high cloud, low cloud, albedo

Chapter 8 Self-Quiz, pp. 366–367

1. (b)
2. (c)
3. (c)
4. (c)
5. (d)
6. (b)
7. T
8. F
9. F
10. T
11. climate
12. weather
13. proxy records
14. greenhouse gases; thermal energy
15. ocean currents; temperature; salinity
16. (a) (iv)
(b) (iii)
(c) (ii)
(d) (i)
(e) (v)

18. (b) photosynthesis, respiration

Chapter 9 Self-Quiz, pp. 400–401

1. (c)
2. (c)
3. (a)
4. (d)
5. F
6. T
7. T
8. T
9. sinks
10. fossil fuels
11. 400 000
12. (a) (ii)
(b) (i)
(c) (iv)
(d) (iii)

Chapter 10 Self-Quiz, pp. 440–441

1. (c)
2. (b)
3. (a)
4. (d)
5. (b)
6. (c)
7. F
8. T
9. Intergovernmental Panel on Climate Change (IPCC)

10. permafrost
11. predictions; models
12. (a) (i)
    (b) (iv)
    (c) (ii)
    (d) (iii)
    (e) (v)

Unit D Review, pp. 446–451

1. (b)
2. (d)
3. (d)
4. (b)
5. (a)
6. (a)
7. (b)
8. (c)
9. (d)
10. (d)
11. (c)
12. (a)
13. (c)
14. (d)
15. T
16. F
17. F
18. T
19. T
20. T
21. T
22. F
23. F
24. T
25. T
26. F
27. atmosphere; living things; hydrosphere
28. ultraviolet radiation; infrared radiation; visible light; lower level infrared radiation
29. melting ice; thermal expansion of water
30. carbon dioxide
31. thermal energy
32. albedo
33. feedback loop
34. stewardship
35. absorbed

Unit D Self-Quiz, pp. 452–453

1. (c)
2. (b)
3. (a)
4. (c)
5. T
6. F
7. T
8. reflects
9. greenhouse gases
10. (a) (iv)
    (b) (i)
    (c) (iii)
    (d) (ii)
11. (a) (iv)
    (b) (v)
    (c) (i)
    (d) (iii)
    (e) (ii)

Unit E

Section 11.1, p. 469

9. (a) baggage screening
   (b) vitamin D
   (c) radar
   (d) DVD player remote control
   (e) telecommunications
   (f) cancer treatment
   (g) theatre/concert lights

Section 11.2, p. 476

3. electric discharge
5. (a) no

Section 11.3, p. 478

2. green

Section 11.4, p. 481

1. (a) glass; shiny film on back surface of glass

Section 11.6, p. 486

4. (a) diffuse reflection
5. (a) 32°
   (b) 47°
   (c) 50°

Section 11.7, p. 493

6. (a) S = size; A = attitude; L = location; T = type

Section 11.9, p. 501

7. (a) convex
   (b) behind the mirror
   (c) virtual
8. (a) smaller, inverted, between *C* and *F*, real
   (b) same size, inverted, at *C*, real
   (c) smaller, upright, on the other side of the mirror between the mirror and *F*, virtual

Chapter 11 Review, pp. 506–507

7. laterally inverted
10. (a) farther away from the mirror than its focus
    (b) between the mirror's surface and its focus
11. (a) diffuse
    (b) specular
    (c) diffuse
    (d) specular
15. 4.2 m
16. All objects are visible.
18. larger, upright, other side of the mirror from the object, virtual

Chapter 11 Self-Quiz, pp. 508–509

1. (a)
2. (d)
3. (b)
4. (a)
5. T
6. F
7. convex
8. virtual
9. (a) (iii)
   (b) (iv)
   (c) (ii)
   (d) (v)
   (e) (i)
19. diffuse

Section 12.1, p. 519

3. (a) medium A: air; medium B: ice
   (b) no; no
4. (a) away from the normal
   (b) toward the normal
5. partial reflection and refraction

Section 12.4, p. 525

2. $n_{vinegar} = 1.30$
3. $n_{sapphire} = 1.78$
4. (a) $v_{quartz} = 2.05 \times 10^8$ m/s
   (b) $v_{diamond} = 1.24 \times 10^8$ m/s
5. $v_{solution} = 2.01 \times 10^8$ m/s
6. $v_{acetone} = 2.21 \times 10^8$ m/s
7. (a) $n = 1.36$
   (b) ethyl alcohol
9. (a) decrease to 33°

Section 12.5, p. 531

3. (a) no
   (b) yes
   (c) yes
   (d) no
8. (b) and (c)
9. (a) medium B
   (b) medium B
   (c) medium A

Section 12.7, p. 539

2. below the apparent position
7. no

Chapter 12 Review, pp. 542–543

2. Light travels slower in material B.
4. 2.30
5. 90°
6. 24.6°
9. (c) no
10. (a) material A
    (b) material A
11. no
13. virtual
15. $n_{carbon\ disulfide} = 1.63$
16. $n_{atg} = 2.04$
17. $n_{fluorite} = 1.43$
18. $v_{veg.\ oil} = 2.04 \times 10^8$ m/s
19. $v_{flint\ glass} = 1.82 \times 10^8$ m/s
20. $v_{zircon} = 1.56 \times 10^8$ m/s

Chapter 12 Self-Quiz, pp. 544–545

1. (a)
2. (b)
3. (d)
4. (c)
5. F
6. F
7. T
8. angle of incidence
9. reflect
10. dimensionless
11. (a) (iii)
    (b) (v)
    (c) (i)
    (d) (ii)
    (e) (iv)
14. (a) medium B
    (b) medium A
15. $n = 1.42$
19. (b) no
21. (a) e.g., periscopes, binoculars, retro-reflectors

Section 13.1, p. 553

3. (a) two
4. yes
5. yes
6. (a) opposite the incident light
   (b) same side as the incident light
   (c) one curves in, one curves out

Section 13.3, p. 561

8. (a) converging lens

Section 13.4, p. 566

1. $d_i = 82$ cm
2. $f = -27$ cm
3. $d_o = 21$ cm
4. $d_i = -35$ cm
5. (a) $M = -2.9$
   (b) real
6. $M = -0.56$
7. (a) $M = 0.46$
   (b) upright
8. (a) $d_i = -53$ cm
   (b) $f = 11$ cm
   (c) converging lens

Section 13.6, p. 577

3. (b) far-sightedness: converging lenses; near-sightedness: diverging lenses
4. (a) positive meniscus; negative meniscus
5. (a) presbyopia
   (b) positive
6. no

Chapter 13 Review, pp. 582–583

1. real and virtual
2. virtual
3. optical centre
4. e.g., magnifying glass, refracting telescope, compound microscope
5. objective lens: enlarged, inverted, real; eyepiece: enlarged, virtual
6. beyond $2F'$ for objective lens
7. near-sightedness
8. varying distances from objects
17. (a) real
    (b) 33 cm from centre of lens, opposite side from candle
18. $d_i = -47$ cm
19. $M = -3.1$
20. (a) $M = 0.67$
    (b) same side as the ship
    (c) no
21. $f = 9.1$ cm
22. (a) converging
    (b) 48 cm from centre of lens; same side as object
24. (a) near-sightedness
    (b) negative meniscus

Chapter 13 Self-Quiz, pp. 584–585

1. (d)
2. (b)
3. (b)
4. (d)
5. F
6. T
7. diverging
8. virtual
9. accommodation
10. (a) (iv)
    (b) (i)
    (c) (iii)
    (d) (ii)
12. $d_i = 28$ cm
13. (a) yes
    (b) 18.9 cm
14. $M = 0.39$
16. (a) upside down
17. (b) glasses with converging lenses
23. the film

Unit E Review, pp. 590–595

1. (c)
2. (a)
3. (c)
4. (a)
5. (c)
6. (c)
7. (b)
8. (a)
9. T
10. T
11. F
12. F
13. F
14. F
15. T
16. F
17. T
18. F
19. F
20. luminous
21. light
22. X
23. smaller, upright
24. converging
25. 90°
26. diverging
27. focus
28. critical
29. myopia
30. (a) (ii)
    (b) (iii)
    (c) (i)
    (d) (v)
    (e) (iv)
31. (a) transparent
    (b) transparent
    (c) translucent
    (d) opaque
    (e) opaque
    (f) translucent
    (g) transparent
40. (a) specular
    (b) diffuse
    (c) diffuse
    (d) specular
46. (a) material B
    (b) material B
49. (a) ultraviolet
52. $c_{quartz} = 2.05 \times 10^8$ m/s
53. $n_{turpentine} = 1.47$
54. $c = 2.12 \times 10^8$ m/s
55. $d_i = -13$ cm
56. $d_i = 139$ cm, larger, inverted, beyond 2F, real
57. 13 cm
58. (a) $d_i = -58$ cm; larger, upright, beyond 2F', virtual
    (b) $M = 3.4$
59. (a) virtual
    (b) $f = -8.1$ cm
60. $d_o = 22$ cm
61. $M = -3.1$
68. (a) presbyopia
    (c) converging

Unit E Self-Quiz, pp. 596–597

1. (a)
2. (c)
3. (c)
4. (d)
5. (c)
6. F
7. T
8. F
9. T
10. incidence
11. incandescence
12. (a) (ii)
    (b) (iii)
    (c) (i)
    (d) (iv)
13. larger, inverted
16. 70°
17. (a) c and v_{hexane}
    (b) $n = \frac{c}{v_{hexane}}$
21. (b) $M = \frac{h_i}{h_o}$
22. (a) real
    (b) 33 cm from lens, opposite the object

Appendix A

Section 5.A, p. 625

(a) 3500 ms
(b) 5200 mA
(c) 7500 ng

Section 5.F., p. 633

(a) 0.7071
(b) 0.8660
(c) 0.8988
(d) 1
(e) 45°
(f) 60°
(g) 64°
(h) 90°

Glossary

A

accommodation the changing of shape of the eye lens by eye muscles to allow a sharply focused image to form on the retina (p. 574)

acid an aqueous solution that conducts electricity, tastes sour, turns blue litmus red, and neutralizes bases (p. 268)

acid–base indicator a substance that changes colour depending on whether it is in an acid or a base (p. 270)

acid leaching the process of removing heavy metals from contaminated soils by adding an acid solution to the soil and catching the solution that drains through (p. 274)

acid precipitation any precipitation (e.g., rain, dew, hail) with a pH less than the normal pH of rain, which is approximately 5.6 (p. 285)

albedo [al-BEE-do] a measure of how much of the Sun's radiation is reflected by a surface (p. 355)

albedo effect [al-BEE-do uh-fekt] the positive feedback loop in which an increase in Earth's temperature causes ice to melt, so more radiation is absorbed by Earth's surface, leading to further increases in temperature (p. 356)

alkali metals the elements (except hydrogen) in the first column of the periodic table (Group 1) (p. 184)

alkaline earth metals the elements in the second column of the periodic table (Group 2) (p. 184)

alveolus [al-vee-O-luhs] (plural: alveoli [al-vee-O-lye]) tiny sac of air in the lungs that is surrounded by a network of capillaries; where gas exchange takes place between air and blood (p. 92)

anaphase [AN-uh-fayz] the third phase of mitosis, in which the centromere splits and sister chromatids separate into daughter chromosomes, and each moves toward opposite ends of the cell (p. 42)

angle of incidence the angle between the incident ray and the normal (p. 481)

angle of reflection the angle between the reflected ray and the normal (p. 481)

angle of refraction the angle between the refracted ray and the normal (p. 516)

anion [AN-eye-awn] a negatively charged ion (p. 190)

anthropogenic [AN-thruh-puh-JEN-ik] resulting from a human influence (p. 384)

anthropogenic greenhouse effect the increase in the amount of lower-energy infrared radiation trapped by the atmosphere as a result of higher levels of greenhouse gases in the atmosphere due to human activities, which is leading to an increase in Earth's average global temperature (p. 387)

apical meristem [AY-puh-kuhl MEH-ruh-stem] undifferentiated cells at the tips of plant roots and shoots; cells that divide, enabling the plant to grow longer and develop specialized tissues (p. 143)

apparent depth the depth that an object appears to be at, due to the refraction of light in a transparent medium (p. 535)

artery a thick-walled blood vessel that carries blood away from the heart (p. 84)

asexual reproduction the process of producing offspring from only one parent; the production of offspring that are genetically identical to the parent (p. 36)

atmosphere the layers of gases surrounding Earth (p. 330)

B

base an aqueous solution that conducts electricity and turns red litmus blue (p. 270)

benign tumour a tumour that does not affect surrounding tissues other than by physically crowding them (p. 48)

bioclimate profile a graphical representation of current and future climate data from a specific location (p. 323)

bioluminescence [BYE-o-loo-muh-NES-ens] the production of light in living organisms as the result of a chemical reaction with little or no heat produced (p. 475)

biophotonics [BYE-o-fo-TAW-niks] the technology of using light energy to diagnose, monitor, and treat living cells and organisms (p. 55)

Bohr–Rutherford diagram a model representing the arrangement of electrons in orbits around the nucleus of an atom (p. 185)

buffering capacity the ability of a substance to resist changes in pH (p. 288)

C

cancer a broad group of diseases that result in uncontrolled cell division (p. 48)

capillary a tiny, thin-walled blood vessel that enables the exchange of gases, nutrients, and wastes between the blood and the body tissues (p. 84)

carbon sink a reservoir, such as an ocean or a forest, that absorbs carbon dioxide from the atmosphere and stores the carbon in another form (p. 339)

carcinogen [kahr-SIN-uh-juhn] any environmental factor that causes cancer (p. 49)

cation [KAT-eye-awn] a positively charged ion (p. 190)

cell cycle the three stages (interphase, mitosis, and cytokinesis) through which a cell passes as it grows and divides (p. 40)

cell theory a theory that all living things are made up of one or more cells, that cells are the basic unit of life, and that all cells come from pre-existing cells (p. 29)

cellular differentiation the process by which a cell becomes specialized to perform a specific function (p. 77)

central nervous system the part of the nervous system consisting of the brain and the spinal cord (p. 104)

centre of curvature the centre of the sphere whose surface has been used to make the mirror (p. 496)

centromere [SEN-truh-meer] the structure that holds chromatids together as chromosomes (p. 41)

chemical equation [KEM-uh-kuhl ee-KWAY-zhuhn] a way of describing a chemical reaction using the chemical formulas of the reactants and products (p. 225)

chemical property a description of what a substance does as it changes into one or more new substances (p. 175)

chemical reaction a process in which substances interact, causing the formation of new substances with new properties (p. 225)

chemiluminescence [KEM-uh-loo-muh-NES-ens] the direct production of light as the result of a chemical reaction with little or no heat produced (p. 473)

chromatid [KRO-muh-tid] one of two identical strands of DNA that make up a chromosome (p. 41)

chromosome [KRO-muh-zom] a structure in the cell nucleus made up of a portion of the cell's DNA, condensed into a structure that is visible under a light microscope (p. 41)

circulatory system the organ system that is made up of the heart, the blood, and the blood vessels; the system that transports oxygen and nutrients throughout the body and carries away wastes (p. 83)

clean energy source a source of energy that produces no significant greenhouse gases (p. 407)

climate the average of the weather in a region over a long period of time (p. 320)

climate projection a scientific forecast of future climate based on observations and computer models (p. 408)

climate system the complex set of components that interact with each other to produce Earth's climate (p. 325)

combustion the rapid reaction of a substance with oxygen to produce oxides and energy; burning (p. 248)

complete combustion a combustion reaction of hydrocarbons that uses all the available fuel and produces only carbon dioxide, water, and energy; occurs when the supply of oxygen is plentiful (p. 248)

compound a pure substance composed of two or more elements in a fixed ratio (p. 186)

concave (converging) mirror a mirror shaped like part of the surface of a sphere in which the inner surface is reflective (p. 496)

concentration the amount of a substance (solute) present in a given volume of solution (p. 37)

connective tissue a specialized tissue that provides support and protection for various parts of the body (p. 75)

contact lens a lens that is placed directly on the cornea of the eye (p. 576)

continental drift the theory that Earth's continents used to be one supercontinent named Pangaea (p. 348)

controlled experiment an experiment in which the independent variable is purposely changed to find out what change, if any, occurs in the dependent variable (p. 8)

convection current a circular current in air and other fluids caused by the rising of warm fluid as cold fluid sinks (p. 345)

converge to meet at a common point (p. 496)

converging lens a lens that is thickest in the middle and that causes incident parallel light rays to converge through a single point after refraction (p. 551)

convex (diverging) mirror a mirror shaped like part of the surface of a sphere in which the outer surface is reflective (p. 496)

correlational study a study in which an investigator looks at the relationship between two variables (p. 8)

corrosion the breakdown of a metal resulting from reactions with chemicals in its environment (p. 252)

covalent bond a bond that results from the sharing of outer electrons between non-metal atoms (p. 207)

critical angle the angle of incidence that results in an angle of refraction of 90° (p. 526)

cuticle a layer of wax on the upper and lower surfaces of a leaf that blocks the diffusion of water and gases (p. 137)

cytokinesis [SYE-to-kuhn-EE-suhs] the stage in the cell cycle when the cytoplasm divides to form two identical cells; the final part of cell division (p. 40)

D

daughter cell one of two genetically identical, new cells that result from the division of one parent cell (p. 40)

decomposition reaction a reaction in which a large or more complex molecule breaks down to form two (or more) simpler products; general pattern: AB → A + B (p. 238)

dependent variable a variable that changes in response to the change in the independent variable (p. 8)

dermal tissue system the tissues covering the outer surface of the plant (p. 126)

diatomic molecule a molecule consisting of only two atoms of either the same or different elements (p. 207)

diffuse reflection reflection of light off an irregular or dull surface (p. 485)

diffusion a transport mechanism for moving chemicals into and out of the cell, from an area of higher concentration to an area of lower concentration (p.37)

digestive system the organ system that is made up of the mouth, esophagus, stomach, intestines, liver, pancreas, and gall bladder; the system that takes in, breaks up, and digests food and then excretes the waste (p. 80)

dispersion [dis-PUR-shuhn] the separation of white light into its constituent colours (p. 538)

diverge to spread apart (p. 499)

diverging lens a lens that is thinnest in the middle and that causes incident parallel light rays to spread apart after refraction (p. 551)

DNA (deoxyribonucleic acid) the material in the nucleus of a cell that contains all of the cell's genetic information (p. 30)

double displacement reaction a reaction that occurs when elements in different compounds displace each other or exchange places, producing two new compounds (p. 242)

dry deposition the process in which acid-forming pollutants fall directly to Earth in the dry state (p. 286)

E

El Niño [el NEEN-yo] a recurring change in the Pacific winds and ocean currents that brings warm, moist air to the west coast of South America (p. 352)

electric discharge the process of producing light by passing an electric current through a gas (p. 471)

electrocardiogram (ECG) a diagnostic test that measures the electrical activity pattern of the heart through its beat cycle (p. 86)

electrolyte [e-LEK-tro-lyet] a compound that separates into ions when it dissolves in water, producing a solution that conducts electricity (p. 194)

electromagnetic spectrum [e-LEK-tro-mag-NET-ik SPEK-truhm] the classification of electromagnetic waves by energy (p. 465)

electromagnetic wave a wave that has both electric and magnetic parts, does not require a medium, and travels at the speed of light (p. 464)

element a pure substance that cannot be broken down into simpler substances (p. 184)

emergent ray the light ray that leaves a lens after refraction (p. 556)

epidermal tissue [ep-uh-DUHR-muhl TISH-yoo] (epidermis [ep-uh-DUHR-mis]) a thin layer of cells covering all non-woody surfaces of the plant (p. 131)

epithelial tissue [e-puh-THEE-lee-uhl TISH-yoo] (or epithelium) a thin sheet of tightly packed cells that covers body surfaces and lines internal organs and body cavities (p. 75)

eukaryote [yoo-KEHR-ee-uht] a cell that contains a nucleus and other organelles, each surrounded by a thin membrane (p. 29)

experimental design a brief description of the procedure in which the hypothesis is tested (p. 11)

F

feedback loop a process in which the result acts to influence the original process (p. 340)

fluorescence [flo-RES-ens] the immediate emission of visible light as a result of the absorption of ultraviolet light (p. 472)

focus the point at which light rays parallel to the principal axis converge when they are reflected off a concave mirror (p. 496)

G

galvanized steel steel that has been coated with a protective layer of zinc, which forms a hard, insoluble oxide (p. 254)

geometric optics the use of light rays to determine how light behaves when it strikes objects (p. 479)

greenhouse effect a natural process whereby gases and clouds absorb infrared radiation emitted from Earth's surface and radiate it, heating the atmosphere and Earth's surface (p. 338)

greenhouse gas any gas in the atmosphere (such as water vapour, carbon dioxide, and methane) that absorbs lower-energy infrared radiation (p. 339)

ground tissue system all plant tissues other than those that make up the dermal and vascular tissue systems (p. 126)

group a column of elements in the periodic table with similar properties (p. 184)

guard cell one of a pair of special cells in the epidermis that surround and control the opening and closing of each stomate (p. 137)

H

halogens [HA-luh-jenz] the elements in the seventeenth column of the periodic table (Group 17) (p. 184)

heat sink a reservoir, such as the ocean, that absorbs and stores thermal energy (p. 344)

hierarchy [HYE-ur-ahr-kee] an organizational structure, with more complex or important things at the top and simpler or less important things below it (p. 73)

hydrosphere [HYE-druhs-feer] the part of the climate system that includes all water on and around Earth (p. 333)

hyperopia [hye-per-O-pee-uh] the inability of the eye to focus light from near objects; far-sightedness (p. 574)

hypothesis a possible answer or untested explanation that relates to the initial question in an experiment (p. 11)

I

ice age a time in Earth's history when Earth is colder and much of the planet is covered in ice (p. 348)

image reproduction of an object through the use of light (p. 480)

impacts of climate change effects on human society and our natural environment that are caused by changes in climate, such as rises in Earth's global temperature (p. 412)

incandescence [in-kan-DES-ens] the production of light as a result of high temperature (p. 470)

incident light light emitted from a source that strikes an object (p. 479)

incident ray the incoming ray that strikes a surface (p. 481)

incomplete combustion a combustion reaction of hydrocarbons that may produce carbon monoxide, carbon, carbon dioxide, soot, water, and energy; occurs when the oxygen supply is limited (p. 249)

independent variable a variable that is changed by the investigator (p. 8)

index of refraction the ratio of the speed of light in a vacuum to the speed of light in a medium, $n = \frac{c}{v}$; this value is also equal to the ratio of the sine of the angle of incidence in a vacuum to the sine of the refracted ray in a medium, $n = \frac{\sin \angle i}{\sin \angle R}$ (p. 524)

infrared radiation a form of invisible lower-energy radiation (p. 325)

interglacial period a time between ice ages when Earth warms up (p. 349)

Intergovernmental Panel on Climate Change (IPCC) a group of several thousand climate scientists who have summarized the latest scientific research on climate change (p. 412)

interphase the phase of the cell cycle during which the cell performs its normal functions and its genetic material is copied in preparation for cell division (p. 40)

ion [EYE-awn] a charged particle that results when an atom gains or loses one or more electrons (p. 188)

ionic bond the simultaneous strong attraction of positive and negative ions in an ionic compound (p. 192)

ionic compound a compound made up of one or more positive metal ions (cations) and one or more negative non-metal ions (anions) (p. 192)

K

Kyoto Protocol a plan within the United Nations for controlling greenhouse gas emissions (p. 423)

L

lateral meristem undifferentiated cells under the bark in the stems and roots of woody plants; cells that divide, enabling the plant to grow wider and develop specialized tissues in the stem (p. 143)

law of conservation of mass the statement that, in any given chemical reaction, the total mass of the reactants equals the total mass of the products (p. 230)

light-emitting diode (LED) light produced as a result of an electric current flowing in semiconductors (p. 476)

light ray a line on a diagram representing the direction and path that light is travelling (p. 479)

lithosphere [LI-thuhs-feer] the part of the climate system made up of the solid rock, soil, and minerals of Earth's crust (p. 334)

luminous produces its own light (p. 470)

M

malignant tumour [muh-LIG-nuhnt TOO-mur] a tumour that interferes with the functioning of surrounding cells; a cancerous tumour (p. 48)

medium any physical substance through which energy can be transferred (p. 464)

meristematic cell an undifferentiated plant cell that can divide and differentiate to form specialized cells (p. 129)

metaphase [MET-uh-fayz] the second stage of mitosis, in which the chromosomes line up in the middle of the cell (p. 41)

metastasis [muh-TAS-tuh-sis] the process of cancer cells breaking away from the original (primary) tumour and establishing another (secondary) tumour elsewhere in the body (p. 48)

mirage [muh-RAWJ] a virtual image that forms as a result of refraction and total internal reflection in Earth's atmosphere (p. 537)

mirror any polished surface reflecting an image (p. 480)

mitigation [mi-ti-GAY-shuhn] reducing an unwanted change by deliberate decisions and actions (p. 423)

mitosis [mye-TO-suhs] the stage of the cell cycle in which the DNA in the nucleus is divided; the first part of cell division (p. 40)

molecular compound a pure substance formed from two or more non-metals (p. 206)

molecule a particle in which atoms are joined by covalent bonds (p. 207)

muscle tissue a group of specialized tissues containing proteins that can contract and enable the body to move (p. 75)

musculoskeletal system the organ system that is made up of bones and skeletal muscle; the system that supports the body, protects delicate organs, and makes movement possible (p. 99)

mutation a random change in the DNA (p. 49)

myopia [mye-O-pee-uh] the inability of the eye to focus light from distant objects; near-sightedness (p. 575)

N

negative meniscus a modified form of the diverging lens shape (p. 576)

nerve tissue specialized tissue that conducts electrical signals from one part of the body to another (p. 75)

nervous system the organ system that is made up of the brain, the spinal cord, and the peripheral nerves; the system that senses the environment and coordinates appropriate responses (p. 104)

neuron [NOO-rawn] a nerve cell (p. 104)

neutral [NOO-truhl] neither acidic nor basic; with a pH of 7 (p. 272)

neutralization reaction a chemical reaction in which an acid and a base react to form an ionic compound (a salt) and water; the resulting pH is closer to 7 (p. 278)

noble gases the elements in the eighteenth column of the periodic table (Group 18) (p. 184)

non-luminous does not produce its own light (p. 470)

normal the perpendicular line to a mirror surface (p. 481)

O

observational study the careful watching and recording of a subject or phenomenon to gather scientific information to answer a question (p. 8)

opaque [o-PAYK] when a material does not transmit any incident light; all incident light is either absorbed or reflected; objects behind the material cannot be seen at all (p. 479)

optical centre point at the exact centre of the lens (p. 552)

organ a structure composed of different tissues working together to perform a complex body function (p. 74)

organ system a system of one or more organs and structures that work together to perform a major vital body function such as digestion or reproduction (p. 74)

organelle [OR-guh-nel] a cell structure that performs a specific function for the cell (p. 29)

osmosis [awz-MO-suhs] the movement of a fluid, usually water, across a membrane toward an area of high solute concentration (p. 37)

P

palisade layer a layer of tall, closely packed cells containing chloroplasts, just below the upper surface of a leaf; a type of ground tissue (p. 137)

Pap test a test that involves taking a sample of cervical cells to determine if they are growing abnormally (p. 50)

periderm tissue tissue on the surface of a plant that produces bark on stems and roots (p. 131)

period a row of elements in the periodic table (p. 184)

peripheral nervous system [puh-RIF-ruhl NUR-vuhs SIS-tuhm] the part of the nervous system consisting of the nerves that connect the body to the central nervous system (p. 104)

perpendicular at right angles (p. 481)

pH a measure of how acidic or basic a solution is (p. 272)

pH scale a numerical scale ranging from 0 to 14 that is used to compare the acidity of solutions (p. 272)

phloem [FLO-uhm] vascular tissue in plants that transports dissolved food materials and hormones throughout the plant (p. 132)

phosphorescence [FAWS-fuh-RES-ens] the process of producing light by the absorption of ultraviolet light resulting in the emission of visible light over an extended period of time (p. 472)

physical property a description of a substance that does not involve forming a new substance; for example, colour, texture, density, smell, solubility, taste, melting point, and physical state (p. 175)

plane flat (p. 481)

plate tectonics the theory explaining the slow movement of the large plates of Earth's crust (p. 348)

polyatomic ion [paw-lee-uh-TAW-mik EYE-awn] an ion made up of more than one atom that acts as a single particle (p. 202)

positive meniscus a modified form of the converging lens shape (p. 575)

precipitate [pruh-SI-puh-tayt] a solid formed from the reaction of two solutions (p. 242)

prediction a statement that predicts the outcome of a controlled experiment (p. 11)

presbyopia [PREZ-bye-O-pee-uh] a form of far-sightedness caused by a loss of accommodation as a person ages (p. 575)

principal axis the line through the centre of curvature to the midpoint of the mirror (p. 496)

principal focus the point on the principal axis of a lens where light rays parallel to the principal axis converge after refraction (p. 552)

product a chemical that is produced during a chemical reaction (p. 225)

prokaryote [pro-KEHR-ee-uht] a cell that does not contain a nucleus or other membrane-bound organelles (p. 29)

prophase [PRO-fayz] the first stage of mitosis, in which the chromosomes become visible and the nuclear membrane dissolves (p. 41)

proxy record stores of information in tree rings, ice cores, and fossils that can be measured to give clues to what the climate was like in the past (p. 358)

Q

qualitative observation a non-numerical observation that describes the qualities of objects or events (p. 12)

quantitative observation a numerical observation based on measurements or counting (p. 12)

R

radiation a method of energy transfer that does not require a medium; the energy travels at the speed of light (p. 464)

reactant a chemical, present at the start of a chemical reaction, that is used up during the reaction (p. 225)

real image an image that can be seen on a screen as a result of light rays actually arriving at the image location (p. 498)

reflected ray the ray that bounces off a reflective surface (p. 481)

reflection the bouncing back of light from a surface (p. 480)

refraction the bending or change in direction of light when it travels from one medium into another (p. 515)

respiratory system the organ system that is made up of the nose, mouth, trachea, bronchi, and lungs; the system that provides oxygen for the body and allows carbon dioxide to leave the body (p. 91)

retro-reflector an optical device in which the emergent ray is parallel to the incident ray (p. 530)

root system the system in a flowering plant, fern, or conifer that anchors the plant, absorbs water and minerals, and stores food (p. 126)

S

semiconductor a material that allows an electric current to flow in only one direction (p. 476)

sexual reproduction the process of producing offspring by the fusion of two gametes; the production of offspring that have genetic information from each parent (p. 36)

shoot system the system in a flowering plant that is specialized to conduct photosynthesis and reproduce sexually; it consists of the leaf, the flower, and the stem (p. 127)

single displacement reaction a reaction in which an element displaces another element in a compound, producing a new compound and a new element (p. 240)

specialized cell a cell that can perform a specific function (p. 58)

specular reflection reflection of light off a smooth surface (p. 485)

spongy mesophyll [spun-jee MES-uh-fil] a region of loosely packed cells containing chloroplasts, in the middle of a leaf; a type of ground tissue (p. 137)

state symbol a symbol indicating the physical state of the chemical at room temperature (i.e., solid (s), liquid (l), gas (g), or aqueous (aq)) (p. 226)

stem cell an undifferentiated cell that can divide to form specialized cells (p. 77)

stomate [STO-mayt] (plural: stomata [sto-MAH-tah]) an opening in the surface of a leaf that allows the exchange of gases (p. 137)

synthesis reaction a reaction in which two reactants combine to make a larger or more complex product; general pattern: $A + B \rightarrow AB$ (p. 237)

T

telophase [TEL-uh-fayz] the final phase of mitosis, in which the chromatids unwind and a nuclear membrane reforms around the chromosomes at each end of the cell (p. 42)

thermal energy the energy present in the motion of particles at a particular temperature (p. 327)

thermal expansion the increase in the volume of matter as its temperature increases (p. 376)

thermohaline circulation the continuous flow of water around the world's oceans driven by differences in water temperatures and salinity (p. 346)

thin lens equation the mathematical relationship among d_o, d_i, and f; $\frac{1}{d_o} + \frac{1}{d_i} = \frac{1}{f}$ (p. 562)

tissue a collection of similar cells that perform a particular, but limited, function (p. 74)

tissue culture propagation a method of growing many identical offspring by obtaining individual plant cells from one parent plant, growing these cells into calluses, and then into whole plants (p. 146)

total internal reflection the situation when the angle of incidence is greater than the critical angle (p. 526)

translucent when a material transmits some incident light but absorbs or reflects the rest; objects are not clearly seen through the material (p. 479)

transparent when a material transmits all or almost all incident light; objects can be clearly seen through the material (p. 479)

triboluminescence [TRYE-bo-loo-muh-NES-ens] the production of light from friction as a result of scratching, crushing, or rubbing certain crystals (p. 475)

tumour a mass of cells that continue to grow and divide without any obvious function in the body (p. 48)

ultraviolet radiation a form of invisible higher-energy radiation (p. 325)

variable any condition that changes or varies the outcome of a scientific inquiry (p. 8)

vascular tissue system the tissues responsible for conducting materials within a plant (p. 126)

vegetative reproduction the process in which a plant produces genetically identical offspring from its roots or shoots (p. 145)

vein a blood vessel that returns blood to the heart (p. 84)

vertex [VUR-tex] the point where the principal axis meets the mirror (p. 496)

virtual image an image formed by light coming from an apparent light source; light is not arriving at or coming from the actual image location (p. 490)

visible light electromagnetic waves that the human eye can detect (p. 465)

visible spectrum the continuous sequence of colours that make up white light (p. 467)

weather atmospheric conditions, including temperature, precipitation, wind, and humidity, in a particular location over a short period of time, such as a day or a week (p. 319)

word equation a way of describing a chemical reaction using the names of the reactants and products (p. 225)

xenotransplantation [ZEE-no-tranz-plan-TAY-shuhn] the process of transplanting an organ or tissue from one species to another (p. 97)

xylem [ZYE-luhm] vascular tissue in plants that transports water and dissolved minerals from the roots to the leaves and stems of the plant (p. 132)

Index

D

E

M

N

O

S

T

U

V

W

X

Y

Z

Credits

This page constitutes an extension of the copyright page. We have made every effort to trace the ownership of all copyrighted material and to secure permission from copyright holders. In the event of any question arising as to the use of any material, we will be pleased to make the necessary corrections in future printings. Thanks are due to the following authors, publishers, and agents for permission to use the material indicated. **Legend: b = bottom, c = centre, l = left, r = right, t = top, [CAPITAL] initials correspond to figure numbers.**

Unit A

Chapter 1. 2: NASA **3:** [b] © Park Street/PhotoEdit; [fire] The Toronto Star/ The Canadian Press (Rene Johnston); [polar bear] NORBERT ROSING/National Geographic Image Collection; [t] Don Farrall/Digital Vision/Getty **4:** © Jenny E. Ross/Corbis **5:** [bc] Michael Blann/Lifesize/Getty; [bl] © ARCTIC IMAGES/Alamy; [br] © Jim West/Alamy; [tc] BRITISH ANTARCTIC SURVEY/SCIENCE PHOTO LIBRARY; [tl] Monkey Business Images/Shutterstock; [tr] © 2009 Jupiterimages **6:** Lucas Oleniuk/The Toronto Star **7:** [bkgd] rahulred/Shutterstock **8:** [tl] Monkey Business Images/Shutterstock **10:** [b] douglas knight/Shutterstock **11:** © Ian Shaw/ Alamy **12:** [bl] Photo by Andy Bourne; [br] © 2009 Jupiterimages; [tl] BRITISH ANTARCTIC SURVEY/SCIENCE PHOTO LIBRARY **13:** Graph adapted from: Scenarios for GHG emissions from 2000 to 2100 (in the absence of additional climate policies) and projections of surface temperatures (Fig. SPM.5, right panel, page 7). IPCC, 2007: Summary for Policymakers. In: *Climate Change 2007: The Physical Science Basis.* [Solomon, S., D. Qin et. al. (eds.)]. Cambridge University Press, Cambridge, UK and New York, NY, USA. p. 7 **14:** Michael Blann/Lifesize/Getty **16:** Excerpt from: Sagan, Carl and Ann Druyan. *A Demon-Haunted World: Science as a Candle.* New York: The Random House Publishing Group, A Ballantine Book, 1996; [bc] © PHOTOTAKE/Alamy; [br] © Daniel J. Cox/Corbis; [l] ©iStockphoto/Andreas Reh **17:** Brand X Pictures/Photolibrary **19:** [bc] Michael Blann/Lifesize/Getty; [bl] © ARCTIC IMAGES/Alamy; [br] © Jim West/Alamy; [tc] BRITISH ANTARCTIC SURVEY/SCIENCE PHOTO, LIBRARY; [tl] Monkey Business Images/Shutterstock; [tr] © 2009 Jupiterimages

Unit B

Chapter 2. 20: Don Farrall/Digital Vision/Getty **21:** [b] ©iStockphoto/Kiyoshi Takahase Segundo; [dialysis] © Helene Rogers/Alamy; [surgery] © 2009 Jupiterimages; [X-ray] ©iStockphoto/ksass **22:** [c] © MedicalRF/Corbis; [l] STEVE GSCHMEISSNER/SCIENCE PHOTO LIBRARY; [r] BJANKA KADIC/SCIENCE PHOTO LIBRARY **24:** STEVE GSCHMEISSNER/SCIENCE PHOTO LIBRARY **25:** [tc] Biophoto Associates/Photo Researchers; [br] SMC Images/Photodisc/Getty; [bl] P&R Fotos/age fotostock/Photolibrary; [tl] Wim van Egmond/Visuals Unlimited/ Getty **26:** [b, t] Peterborough Examiner/The Canadian Press (Clifford Skarstedt) **27:** [bl] Dorling Kindersley/Getty; [br] © 2009 Creatas Images/Jupiterimages; [cl] Ed Reschke/Peter Arnold; [cr] ANDREW LAMBERT PHOTOGRAPHY/SCIENCE PHOTO LIBRARY; [tl] STEVE GSCHMEISSNER/SCIENCE PHOTO LIBRARY; [tr] © Visuals Unlimited/Corbis **28:** [paper] Robyn Mackenzie/Shutterstock; [bkgd] rahulred/Shutterstock **29:** [A] © Eraxion/Dreamstime; [B] Wim van Egmond/Visuals Unlimited/Getty; [C] © 2009 David B Fleetham/Oxford Scientific/Jupiterimages; [D] © All Canada Photos/Alamy **30:** [bl] Ed Reschke/Peter Arnold; [cl] Scimat/ Photo Researchers **31:** [Fig. 5] Keith R. Porter/Photo Researchers; [Fig. 6] Steve Gschmeissner/Photo Researchers; [Fig. 7] SCIENCE PHOTO LIBRARY; [Fig. 8A] Biophoto Associates/Photo Researchers; [Fig. 8B] Steve Gschmeissner/Photo Researchers **32:** [Fig. 9] Ed Reschke/Peter Arnold **33:** [l] Joseph T. Collins/ Photo Researchers, Inc., [r] Steve Gschmeissner/Photo Researchers, Inc. **36:** [bl] Suzie Gibbons/Garden Picture Library/Photolibrary; [r] Radius Images/ Photolibrary; [t] Kwangshin Kim/Photo Researchers **41:** [A, B] P&R Fotos/age fotostock/Photolibrary **42:** [bl] Dr. Robert Calentine/Visuals Unlimited/Getty; [br] Ed Reschke/Peter Arnold; [cl, cr] P&R Fotos/age fotostock/Photolibrary **44:** Michael Abbey/Photo Researchers **45:** [bkgd] Svetlana Privezentseva/Shutterstock; [l] © 2009 Jupiterimages; [r] © 2009 Creative Concept/Jupiterimages **46:** [l] Biology Media/ Photo Researchers; [r] Ed Reschke/Peter Arnold **47:** [A, B, C, D, E] Ed Reschke/Peter Arnold **50:** [Benign Asymmetry] © Don Garbera/Phototake—All rights reserved.; [Benign Border, Benign Colour, Benign Diameter] © ISM/Phototake—All rights reserved.; [Malignant Asymmetry] © Pulse Picture Library/CMP Images/Phototake; [Malignant Border, Malignant Diameter, Malignant Colour] NMSB/CMSP **52:** [bl] SMC Images/Photodisc/Getty; [t] © Bettmann/CORBIS; [tl] Dave King © Dorling Kindersley, Courtesy of The Science Museum, London **53:** [Fig. 10] Simon Fraser/ Photo Researchers; [Fig. 8] SIMON FRASER/SCIENCE PHOTO LIBRARY; [Fig. 9] SIMON FRASER/FREEMAN HOSPITAL, NEWCASTLE UPON TYNE/SCIENCE PHOTO LIBRARY **54:** © Luca Medical/Alamy **56:** [l] © 2009 Image Source/ Jupiterimages; [r] Biodisc/Visuals Unlimited/Getty **58:** PROF. P. MOTTA/DEPT. OF ANATOMY/UNIVERSITY "LA SAPIENZA," ROME/SCIENCE PHOTO LIBRARY **59:** [A] PROFESSORS P.M. MOTTA & S. CORRER/SCIENCE PHOTO LIBRARY; [B] EYE OF SCIENCE/SCIENCE PHOTO LIBRARY; [C] STEVE GSCHMEISSNER/ SCIENCE PHOTO LIBRARY; [D] ANDREW SYRED/SCIENCE PHOTO LIBRARY; [E] Dr. David M. Phillips/Visuals Unlimited/Getty; [F] EYE OF SCIENCE/SCIENCE PHOTO LIBRARY; [G] STEVE GSCHMEISSNER/SCIENCE PHOTO LIBRARY; [H] SUSUMU NISHINAGA/SCIENCE PHOTO LIBRARY; [I] PAUL ZAHL/National Geographic Stock **60:** [A] Educational Images Ltd./CMSP; [B] Dr. Richard Kessel & Dr. Gene Shih/Visuals Unlimited/Getty; [C] STEVE GSCHMEISSNER/SCIENCE PHOTO LIBRARY; [D] Dr. Richard Kessel & Dr. Gene Shih/Visuals Unlimited/ Getty; [E] © Visuals Unlimited/Corbis; [F] Dr. Dennis Kunkel/Visuals Unlimited/ Getty **61:** Dr. Gopal Murti/Visuals Unlimited/Getty **62, 154:** [bl] P&R Fotos/age fotostock/Photolibrary; [br] SMC Images/Photodisc/Getty; [tc] Biophoto Associates/ Photo Researchers; [tl] Wim van Egmond/Visuals Unlimited/Getty **63:** [bl] Dorling Kindersley/Getty; [br] © 2009 Creatas Images/Jupiterimages; [cl] Ed Reschke/Peter Arnold; [cr] ANDREW LAMBERT PHOTOGRAPHY/SCIENCE PHOTO LIBRARY; [tl] STEVE GSCHMEISSNER/SCIENCE PHOTO LIBRARY; [tr] © Visuals Unlimited/Corbis

Chapter 3. 68: © MedicalRF/Corbis **69:** [br] PHANIE/Photo Researchers; [tl] ©iStockphoto/Kevin Snair **70:** [t] AP Photo/Hospital Clinic of Barcelona, HO **71:** [bl] NANCY KEDERSHA/SCIENCE PHOTO LIBRARY; [br] R. Andrew Odum/ Peter Arnold;: [cl] Michael Abbey/Photo Researchers; [cr] ©iStockphoto/Kevin Snair; [tl] urbanraven/BigStockPhoto; [tr] © Photodisc/Alamy **73:** [c] John A. Anderson/ Shutterstock; [l] Olga van de Veer/BigStockPhoto; [r] Gerald A. DeBoer/Shutterstock **77:** [b] © Collection CNRI/Phototake—All rights reserved. **78:** PHANIE/Photo Researchers **79:** Reinhard Dirscherl/Visuals Unlimited/Getty **83:** [b] NATIONAL CANCER INSTITUTE/SCIENCE PHOTO LIBRARY **84:** [bl] Ed Reschke/Peter Arnold; [br] Ed Reschke/Peter Arnold; [t] Photo by Colin Rowe **85:** © BSIP Cavallini James **86:** [c] Medicimage/Photolibrary **86:** [electrocardiogram] Oscar Ruben Calero de Diago/BigStockPhoto; [t] © Dr. David M. Phillips/Visuals Unlimited/Alamy **87:** [b] Dr. Gladden Willis/Visuals Unlimited/Getty; [c] STEVE GSCHMEISSNER/ SCIENCE PHOTO LIBRARY; [t] © Visuals Unlimited/Corbis **88:** [l] ©iStockphoto/ Kevin Snair; [r] CMSP **89:** [l] Educational Images Ltd./CMSP **89:** [r] Educational Images Ltd./CMSP **90:** [bkgd] Terrance Emerson/Shutterstock; [br] Lydia Bilby-Sparling/Shutterstock; [l] ©iStockphoto/Douglas Allen; [tr] Bernard Weil/GetStock **91:** [c] SUSUMU NISHINAGA/SCIENCE PHOTO LIBRARY **94:** GUSTOIMAGES/ SCIENCE PHOTO LIBRARY **95:** right, Toronto Globe and Mail/The Canadian Press (Tibor Kolley) **97:** [card] Courtesy Lynn McLeod **99:** [tl] ANDREW SYRED/ SCIENCE PHOTO LIBRARY **100:** [tr] © Visuals Unlimited/Corbis **101:** [b] Igor Gorelchenkov/Shutterstock **106:** [bl] AP Photo/Mark Gilliland **107:** DU CANE MEDICAL IMAGING LTD/SCIENCE PHOTO LIBRARY **109:** [b] MedicalRF/ Photo Researchers; [t] Susumu Nishinaga/Photo Researchers **110:** [bc] © 2009 Jupiterimages; [bl] Ffion/BigStockPhoto; [br] ©iStockphoto/Les McGlasson; [cl] James H. Robinson/Photo Researchers; [t] Cathy Keifer/BigStockPhoto **111:** [lynx rabbit] © 2009 Creatas Images/Jupiterimages **112:** [bkgd] dwphotos/Shutterstock; [l] © Nic Cleave Photography/Alamy **113:** AP Photo/John Amis **114, 154:** [br] PHANIE/Photo Researchers; [tl] ©iStockphoto/Kevin Snair **115:** [bl] NANCY KEDERSHA/SCIENCE PHOTO LIBRARY; [br] R. Andrew Odum/Peter Arnold; [cl] Michael Abbey/Photo Researchers; [cr] ©iStockphoto/Kevin Snair; [tl] urbanraven/ BigStockPhoto; [tr] © Photodisc/Alamy

Chapter 4. 120: BJANKA KADIC/SCIENCE PHOTO LIBRARY **121:** [bc] DR. KEITH WHEELER/SCIENCE PHOTO LIBRARY; [bl] From Ow, David W. et al. "Transient and Stable Expression of the Firefly Luciferase Gene in Plant Cells and Transgenic Plants." *Science*, Nov 1, 1986, published by The American Association for the Advancement of Science. Reprinted with permission from AAAS.; [br] © Biodisc/ Visuals Unlimited/Alamy; [tc] © Graham Oliver/Alamy; [tl] prism68/Shutterstock; [tr] Ed Reschke/Peter Arnold **122:** [b] Mike Flippo/Shutterstock; [bkgd] Christophe Testi/Shutterstock; [t] © Macduff Everton/CORBIS **123:** [bl] Filipe B. Varela/ Shutterstock; [br] Fredrik Ehrenstrom/Oxford Scientific/Photolibrary; [cl] © Alaska Stock LLC/Alamy; [cr] Jasmina007/Shutterstock; [tl] Maksym Gorpenyuk/ Shutterstock; [tr] David M Dennis/Oxford Scientific/Photolibrary **124:** [paper] Robyn Mackenzie/Shutterstock; [bkgd] rahulred/Shutterstock; Pippin Lee/fotoboof **126:** [bc] GEOFF TOMPKINSON/SCIENCE PHOTO LIBRARY; [bl] Japan Travel Bureau/Photolibrary; [br] Yuji Sakai/Digital Vision/Getty; [t] Angelo Cavalli/ Photodisc/Getty **127:** [bc] Walid Nohra/Shutterstock; [br] © Fackler Poinsettias/ Alamy; [l] © Custom Life Science Images/Alamy; [t] DR. KARI LOUNATMAA/ SCIENCE PHOTO LIBRARY **128:** [b] Ed Reschke/Peter Arnold; [c] Christian Musat/Shutterstock; [t] © blickwinkel/Alamy **129:** [l] prism68/Shutterstock; [r] © Biodisc/Visuals Unlimited/Alamy **131:** [c] © Nigel Cattlin/Alamy; [l] © BrazilPhotos/Alamy; [r] Michael Coyne/Lonely Planet **132:** [b] Ed Reschke/ Peter Arnold **134:** [br] From Ow, David W. et al. "Transient and Stable Expression of the Firefly Luciferase Gene in Plant Cells and Transgenic Plants." *Science*, Nov 1, 1986, published by The American Association for the Advancement of Science. Reprinted with permission from AAAS.; [l] Science VU/Visuals Unlimited **135:** Anna Jurkovska/Shutterstock **136:** [l] © Graham Oliver/Alamy; [r] © Matthew Mawson/ Alamy **137:** [b] Ed Reschke/Peter Arnold **139:** [t] DR. KEITH WHEELER/SCIENCE PHOTO LIBRARY **140:** Eye of Science/Photo Researchers **142:** [bkgd] Terrance

Emerson/Shutterstock; [bl] © inga spence/Alamy; [r] Pippin Lee/fotoboof; [tl] © Wayne Higgins/Alamy **144:** [bl] Doug Fraser; [t] © PHOTOTAKE/Alamy **145:** [l] Mel McIntyre/Shutterstock; [r] James Randklev/Photographer's Choice/Photolibrary **146:** [bl] © P. Dumas/Phototake—All rights reserved.; [br] SINCLAIR STAMMERS/ SCIENCE PHOTO LIBRARY.; [tl] Wally Eberhart/Visuals Unlimited/Getty; [tr] Michael Howes/Garden Picture Library/Photolibrary **147:** Heinrich van den Berg/ Gallo Images/Getty **148, 154:** [bc] DR. KEITH WHEELER/SCIENCE PHOTO LIBRARY; [bl] From Ow, David W. et al. "Transient and Stable Expression of the Firefly Luciferase Gene in Plant Cells and Transgenic Plants." *Science*, Nov 1, 1986, published by The American Association for the Advancement of Science. Reprinted with permission from AAAS.; [br] © Biodisc/Visuals Unlimited/Alamy; [tc] © Graham Oliver/Alamy; [tl] prism68/Shutterstock; [tr] Ed Reschke/Peter Arnold **149:** [bl] Filipe B. Varela/Shutterstock; [br] Fredrik Ehrenstrom/Oxford Scientific/ Photolibrary; [cl] © Alaska Stock LLC/Alamy; [cr] Jasmina007/Shutterstock; [tl] Maksym Gorpenyuk/Shutterstock; [tr] David M. Dennis/Oxford Scientific/ Photolibrary **150:** [bl] STEVE GSCHMEISSNER/SCIENCE PHOTO LIBRARY; [br] Ed Reschke/Peter Arnold; [tl] DR. DAVID FURNESS, KEELE UNIVERSITY/ SCIENCE PHOTO LIBRARY; [tr] EYE OF SCIENCE/SCIENCE PHOTO LIBRARY **151:** Jerry Horbert/Shutterstock **153:** [c] DR. KEITH WHEELER/SCIENCE PHOTO LIBRARY

Unit C

Chapter 5. 166: The Toronto Star/The Canadian Press (Rene Johnston) **167:** [b] Steve Winter/National Geographic/Getty; [rust] ©iStockphoto/Ewald Froech; [t] Courtesy of David Richardson; [truck] © 2009 Jupiterimages **168:** [c] SVEN DILLEN/AFP/ Getty; [l] NASA Marshall Space Flight; [r] T. Kitchin & V. Hurst/All Canada Photos/ Photolibrary **169:** [billiard] ©iStockphoto/chris scredon; [bl] © Jewellery specialist/ Alamy; [girl] WILLIAM WEST/AFP/Getty; [kettle] © Darren Matthews/Alamy **170:** NASA Marshall Space Flight Center **171:** [bc] Creatas/Photolibrary; [br] Feng Yu/Shutterstock; [tc] Charles D. Winters/Photo Researchers/Photolibrary; [tl] © Michael Newman/PhotoEdit; [tr] © Leslie Garland Picture Library/Alamy **172:** Christian Darkin/Photo Researchers **173:** [br] © 2009 Stock Connection/ Jupiterimages; [cl] ©iStockphoto/Skip O'Donnell; [cr] © David Young-Wolff/ PhotoEdit; [tr] Charles Winters/Photo Researchers **174:** [bkgd] rahulred/Shutterstock **175:** Phanie/First Light **176:** JoLin/Shutterstock **177:** [b] © 2009 BananaStock/ Jupiterimages; [candle] South12th Photography/Shutterstock; [firefly] Photo Researchers/First Light; [lb] Lambert Photography/Photolibrary; [tl] Martyn F. Chillmaid/SPL/photolibrary; [tr] © sciencephotos/Alamy **179:** [b] © 2009 Jupiterimages; [bkgd] Terrance Emerson/Shutterstock; [l] Visions LLC/Photolibrary **184:** [bl] Klaus Guldbrandsen/Photo Researchers; [br] © Leslie Garland Picture Library/Alamy; [tl] Charles D. Winters/Photo Researchers; [tr] Charles D. Winters/ Photo Researchers **185:** [tc] Charles D. Winters/Photo Researchers/Photolibrary; [tl] Charles D. Winters/Photo Researchers/Photolibrary; [tr] Charles D. Winters/ Photo Researchers/Photolibrary **187:** [bl] Don Farrall/Photodisc/Getty; [br] Photo Researchers/Photolibrary; [tl] © Leslie Garland Picture Library/Alamy; [tr] © 1989 Richard Megna/Fundamental Photographs **188:** Novastock/Photolibrary **192:** [B] Charles D. Winters/Photo Researchers; [C] Charles D. Winters/Photo Researchers; [D] Yoav Levy/Photolibrary; [E] Spencer Jones/Picturearts/Photolibrary; [A] Charles D. Winters/Photo Researchers **193:** [tl] Dr. Jeremy Burgess/Photo Researchers **196:** [b] D.Harms/WILDLIFE/Peter Arnold; [cl] Charles D. Winters/Photo Researchers; [t] Dusan Po/Shutterstock **197:** [b] © 2009 Jupiterimages **198:** [l] © 1994 Richard Megna/Fundamental Photographs **200:** Visuals Unlimited/Getty **202:** [t] Getty **206:** [bc] cappi thompson/Shutterstock; [bl] Radu Razvan/ Shutterstock; [br] Creatas/Photolibrary; [tl] © Michael Newman/PhotoEdit; [tr] © Didier Dorval/Masterfile **208:** [A] © Carolinasm/Dreamstime; [B] Dr. Tim Evans/ Photo Researchers; [C] © Scott Camazine/Phototake—All rights reserved. **210:** [tl] Feng Yu/Shutterstock **211:** The Canadian Press (Sam Leung) **214, 296:** [bc] Creatas/ Photolibrary; [br] Feng Yu/Shutterstock; [tc] Charles D. Winters/Photo Researchers/ Photolibrary; [tl] © Michael Newman/PhotoEdit; [tr] © Leslie Garland Picture Library/Alamy **215:** [br] © 2009 Stock Connection/Jupiterimages; [cl] ©iStockphoto/ Skip O'Donnell; [cr] © David Young-Wolff/PhotoEdit; [tr] Charles Winters/Photo Researchers **217:** [bl] © Mallory Morrison Lann/Alamy

Chapter 6. 220: SVEN DILLEN/AFP/Getty **221:** [bl] Andrew Lambert Photography/ Photo Researchers; [bc] The Canadian Press (Andrew Vaughan); [br] Boris Spremo/ Getstock; [tl] University of Puget Sound, Tacoma, WA, U.S.A.; [tr] © Rob Walls/ Alamy **222:** [r] Edward Kinsman/Photo Researchers; [torn paper bkgd] Robyn Mackenzie/Shutterstock **223:** [bl] Yuri Arcurs/Shutterstock; [br] © Dino Fracchia/ Alamy; [cl] © 2009 Design Pics/Jupiterimages; [cr] ©iStockphoto/Eileen Hart; [tl] Michael Stokes/Shutterstock; [tr] Simone van den Berg/Shutterstock **224:** [bkgd] rahulred/Shutterstock; [paper] Robyn Mackenzie/Shutterstock; Rene Johnston/ The Toronto Star **225:** [bc] Andrew Lambert Photography/Photo Researchers; [bl] Martyn F. Chillmaid/Photo Researchers; [br] Martyn F. Chillmaid/Photo Researchers; [t] ©iStockphoto/Johann Helgason **226:** [bl] Andrew Lambert Photography/Photo Researchers; [br] Charles D. Winters/Photo Researchers **227:** [b] Charles D. Winters/ Photo Researchers **230:** [t] Shironina Lidiya Alexandrovna/Shutterstock **234:** Creatas/ Photolibrary **237:** [c] Andrew Lambert Photography/Photo Researchers; [l] Charles D. Winters/Photo Researchers; [r] Charles D. Winters/Photo Researchers **238:** [b] © Rick Fischer/Masterfile **239:** [t] Toronto Star Syndicate/The Canadian Press **240:** [bl] Charles D. Winters/Photo Researchers; [br] Charles D. Winters/Photo Researchers; [t] University of Puget Sound, Tacoma, WA, U.S.A. **241:** [b] © Paul A. Souders/CORBIS; [t] Charles D. Winters/Photo Researchers **242:** [bl] Charles D. Winters/Photo Researchers; [r] Charles D. Winters/Photo Researchers **243:** [b] © Tony Freeman/PhotoEdit; [t] David Taylor/Photo Researchers **244:** Charles D. Winters/Photo Researchers **248:** Rene Johnston/The Toronto Star **249:** [b] Corbis/ Photolibrary; [t] Boris Spremo/Getstock **250:** [l] © Charles D. Winters/Photo Researchers; [r] © Rob Walls/Alamy **251:** Courtesy Coleman **252:** © Howard Sandler/iStockphoto **253:** [br] 3D4Medical/Getty; [l] The Canadian Press (Andrew Vaughan); [tr] DCD/Shutterstock **254:** Chad McDermott/Shutterstock **255:** [bkgd] Svetlana Privezentseva/Shutterstock; [l] © Petra Johansson/Alamy; [r] Duard van der Westhuizen/Shutterstock **256, 298:** [bc] The Canadian Press (Andrew Vaughan); [bl] Andrew Lambert Photography/Photo Researchers; [br] Boris Spremo/Getstock; [tl] University of Puget Sound, Tacoma, WA, U.S.A.; [tr] © Rob Walls/Alamy **257:** [bl] Yuri Arcurs/Shutterstock; [br] © Dino Fracchia/Alamy; [cl] © 2009 Design Pics/Jupiterimages; [cr] ©iStockphoto/Eileen Hart; [tl] Michael Stokes/Shutterstock; [tr] Simone van den Berg/Shutterstock

Chapter 7. 262: T. Kitchin & V. Hurst/All Canada Photos/Photolibrary **263:** [draino] Pippin Lee/fotoboof; [bl] © Leslie Garland Picture Library/Alamy; [br] Mike Grandmaison; [tl] ©iStockphoto/Jakub Semeniuk; [tr] North Bay Nugget/The Canadian Press (Rick Owen/Osprey Media) **264:** [bkgd] Iakov Kalinin/Shutterstock; [l] © Richard Hamilton Smith/CORBIS; [r] Mike Grandmaison **265:** [bl] Martyn Chillmaid/Oxford Scientific/Photolibrary; [cl] ©Stefan Sollfors/Alamy; [cr] Peter Doomen/Shutterstock; [tl] Biophoto Associates/Photo Researchers; [tr] © 2008 Jupiterimages **266:** [bkgd] rahulred/Shutterstock **268:** [tl] Biophoto Associates/Photo Researchers **269:** [cola] Pippin Lee/fotoboof **270:** © Leslie Garland Picture Library/ Alamy **271:** Edward H. Gill/CMSP **272:** [t] Dr. Kari Lounatmaa/Photo Researchers **273:** © Mark Spowart/Alamy **274:** [draino] Pippin Lee/fotoboof; [r] © Rachel Weill/ Botanica/Jupiterimages **275:** Science Faction **279:** North Bay Nugget/The Canadian Press (Rick Owen/Osprey Media) **280:** [b] David R. Frazier/Photo Researchers; [t] © Ted Spiegel/CORBIS **281:** [l] © Westend61/Alamy; [r] © Stockphoto/Jakub Semeniuk **282:** [b] Charles D. Winters/Photo Researchers; [t] © PHOTOTAKE/ Alamy **283:** Simon Fraser/Photo Researchers **284:** [b] AP Photo/The Canadian Press (Marianna Bertagnolli); [bkgd] Svetlana Privezentseva/Shutterstock **286:** [pie chart] Source: US and Canadian National Emissions by Sector for Selected Pollutants. Fig. 26. © Her Majesty The Queen in Right of Canada, Environment Canada, 2006. Reproduced with the permission of the Minister of Public Works and Government Services Canada. **287:** Graph adapted from: Environmental Protection Agency, Washington, DC; [bass] ©iStockphoto/George Peters; [clam] ©iStockphoto/ ROBERTO ADRIAN; [crayfish] Andrei Nekrassov/Shutterstock; [frog] ©iStockphoto/Kevin Snair; [loon] Pavel Cheiko/Shutterstock; [mayfly] David Dohnal/Shutterstock; [perch] ©iStockphoto/David Dohnal; [salamander] © Ingram Publishing (Superstock Limited)/Alamy; [snail] Sasha Radosavljevich/ Shutterstock; [t] Source: US and Canadian National Emissions by Sector for Selected Pollutants. Fig. 26. Environment Canada, 2004. Reproduced with the permission of the Minister of Public Works and Government Services Canada, 2009.; [trout] ©iStockphoto/arne thaysen **288:** Mike Grandmaison **289:** [after] © 1994 Kristen Brochmann/Fundamental Photographs; [before] © 1994 NYC Parks Photo Archive/ Fundamental Photographs **292, 298:** [bl] © Leslie Garland Picture Library/ Alamy; [draino] Pippin Lee/fotoboof; [br] Mike Grandmaison; [tl] ©iStockphoto/ Jakub Semeniuk; [tr] North Bay Nugget/The Canadian Press (Rick Owen/Osprey Media) **293:** [bl] Martyn Chillmaid/Oxford Scientific/Photolibrary; [cl] ©Stefan Sollfors/Alamy; [cr] Peter Doomen/Shutterstock; [tl] Biophoto Associates/Photo Researchers; [tr] © 2008 Jupiterimages **295:** © Swerve/Alamy **299:** [antacid] Linda Muir/Shutterstock; [draino] Pippin Lee/fotoboof; [baking soda] © mediablitzimages (uk) Limited/Alamy; [battery] Egidijus Skiparis/Shutterstock; [cola] ©iStockphoto/ malcolm romain; [hydrogen peroxide] © BroadSpektrum/Alamy; [jewellery] © Matt Brasier/Masterfile; [salt] Spencer Jones/Picturearts/Photolibrary; ©iStockphoto/Steve Gray **300:** [l] © Patrick Robert/Corbis; [r] Harry Rogers/ Photo Researchers **301:** © Tony Freeman/PhotoEdit **304:** Soleil Noir/Photonons/ Photolibrary **306:** Stephen Finn/Shutterstock

Unit D

Chapter 8. 310: NORBERT ROSING/National Geographic Image Collection **311:** [b] Mirko Iannace/age fotostock/Photolibrary; [refinery] Grant Faint/ Photographer's Choice/Photolibrary; [t] © Bob Daemmrich/PhotoEdit; [tree] ©iStockphoto/Robert Koopmans **312:** [l] Bildagentur RM/Tips Italia/Photolibrary; [c] Jasper Yellowhead Museum and Archives, 89.36.263, Leonard Jeck fonds; [r] The Canadian Press (Dave Chidley) **313:** [b] Radius Images/Photolibrary; [t] © Michael P. Gadomski/SuperStock **314:** Bildagentur RM/Tips Italia/Photolibrary **315:** [bc] InterNetwork Media/Photodisc/Getty; [bl] Linux Patrol/Shutterstock; [br] © Dietrich Rose/zefa/Corbis; [tc] Thomas Kitchin & Victoria Hurst/First Light/Getty; [tl] © Dale Wilson/Masterfile; [tr] © 2009 Doug Allan/Jupiterimages **316:** [grass] digitalife/Shutterstock; [mammoth] © Mary Evans Picture Library/ Alamy; [snow] nfaustino/Shutterstock **317:** [br] Ethan Meleg/All Canada Photos/ Getty; [cl] Michal Szczepaniak/Shutterstock; [cr] Photodisc/Photolibrary; [tl]

Courtesy of SOHO/EIT consortium. SOHO is a project of international cooperation between ESA and NASA.; [tr] © Richard Broadwell/Alamy **318:** [paper] Robyn Mackenzie/Shutterstock; [bkgd] rahulred/Shutterstock **319:** [l] Graca Victoria/Shutterstock; [r] Toronto Star/The Canadian Press (Rick Madonik) **320:** British Antarctic Survey/Photo Researchers **321:** [l] © John Foster/Masterfile; [r] Mike Grandmaison/First Light **322:** [l] © Masterfile; [r] Sergey I/Shutterstock **325:** © Dale Wilson/Masterfile **328:** [b] © Gary Cook/Alamy **329:** [r] © 2009 Doug Allan/Jupiterimages **330:** [atmosphere] Michal Szczepaniak/Shutterstock; [hydrosphere] Linux Patrol/Shutterstock; [lithosphere] Eryk Jaegermann/Index Stock Imagery/Photolibrary; [living things] ©iStockphoto/Neta Degany **331:** [bl] ©NASA/Photo Researchers **332:** The Canadian Press (Nathan Denette) **334:** [t] Jerry Kobalenko/Photographer's Choice/Getty **335:** Thomas Kitchin & Victoria Hurst/First Light/Getty **336:** © Ray A. Akey/Alamy **339:** [t] Science Source/USGS/Photo Researchers, Inc **340:** Wave RF/Photolibrary **341:** © Amazon-Images/Alamy **344:** [t] UncleGenePhoto/Shutterstock **345:** Adapted from: Federation of American Scientists and Drake, Dr. John and Dr. Philip Jones, "Developing Models for Predictive Climate Science," Scientific Discovery through Advanced Computing: IOP Publishing. **346:** Adapted from: NASA Earth Observatory, Explaining Rapid Climate Change: Tales from the Ice and Union of Concerned Scientists: Abrupt Climate Change. **347:** NASA Goddard Space Flight Center (NASA-GSFC) **348:** [t] Henry, P./Peter Arnold **349, 357:** [graph] source: "Temperature and CO_2 Concentration in the Atmosphere Over the Past 400 000 Years (From the Vostok Ice Core) - UNEP/GRID-Arendal." A Climate Change Primer. **350:** [b] Scuddy/Shutterstock **351:** [l] Photodisc/Getty; [r] NASA/SCIENCE PHOTO LIBRARY **354:** [b] Courtesy of James Teller; [bkgd] Terrance Emerson/Shutterstock; [tr] Jon Nelson **355:** [l] © 2009 Roy Hsu/Workbook Stock/Jupiterimages **358:** [b] BRITISH ANTARCTIC SURVEY/SCIENCE PHOTO LIBRARY **358:** [t] ©Ann Ronan Picture Library/Heritage-Images/The Image Works **359:** [b] Stockbyte/Photolibrary; [tl] © Dietrich Rose/zefa/Corbis; [tr] Peter Kelly **360:** [bl] © Lowell Georgia/Corbis; [br] © Garry Black/Masterfile; [t] © Dennis Kunkel/Phototake **362:** [bc] InterNetwork Media/Photodisc/Getty; [bl] Linux Patrol/Shutterstock; [br] © Dietrich Rose/zefa/Corbis; [tc] Thomas Kitchin & Victoria Hurst/First Light/Getty; [tl] © Dale Wilson/Masterfile; [tr] © 2009 Doug Allan/Jupiterimages **363:** [br] Ethan Meleg/All Canada Photos/Getty; [cl] Michal Szczepaniak/Shutterstock; [cr] Photodisc/Photolibrary; [tl] Courtesy of SOHO/EIT consortium. SOHO is a project of international cooperation between ESA and NASA.; [tr] © Richard Broadwell/Alamy **364:** Walter Bibikow/age fotostock/Photolibrary

Chapter 9. 368: [inset] Jasper Yellowhead Museum and Archives, 89.36.263, Leonard Jeck fonds; © Chunli Li/Dreamstime **369:** [bc] The Canadian Press (Larry MacDougal); [bl] NASA/Goddard Institute for Space Studies; [br] Adapted from: Graph, "Global and Continental Temperature Change, Global." IPCC, 2007: *Summary for Policymakers.* In: *Climate Change 2007: The Physical Science Basis.* [Solomon, S., D. Qin et. al. (eds.)]. Cambridge University Press, Cambridge, UK and New York, NY, USA. p. 6; [tc] © Dennis MacDonald/Alamy; [tl] NASA/Goddard Space Flight Center Scientific Visualization Studio. Thanks to Rob Gerston (GSFC) for providing the data.; [tr] British Antarctic Survey/Photo Researchers **370:** [l] Graham Ashford, International Institute for Sustainable Development; [r] Graham Ashford, International Institute for Sustainable Development **371:** [bl] © Dwayne Newton/PhotoEdit; [br] Radius Images/Photolibrary; [cl] ©iStockphoto/Jason Lugo; [cr] © Mike Booth/Alamy; [tl] Hulton Archive/American Stock/Getty; [tr] REUTERS/Mathieu Belanger/Landov **372:** [bkgd] rahulred/Shutterstock; [paper] Robyn Mackenzie/Shutterstock **372, 374:** [map] "Increases in annual temperatures for a recent five-year period, relative to 1951–1980." Hugo Ahlenius, UNEP/GRID-Arendal Maps and Graphics Library. June 2007. Hansen, J., Sato, M., Ruedy, R., Lo, K., Lea, D.W. and Medina-Elizade, M. (2006). Global temperature change. Proc. Natl. Acad. Sci., 103, 14288-14293. http://www.unep.org/geo/ice_snow **373:** [bl] NASA/Goddard Institute for Space Studies; [t] Maxppp/Landov **375:** Graphs adapted from: Graph, Changes in Temperature, Sea Level and Northern Hemisphere Snow Cover (parts a and b) (Fig. SPM.1). IPCC, 2007: *Summary for Policymakers.* In: *Climate Change 2007: The Physical Science Basis.* [Solomon, S., D. Qin et. al. (eds.)]. Cambridge University Press, Cambridge, UK and New York, NY, USA. p. 3; [tl] NASA/Photo Researchers.; [tr] NASA/Goddard Space Flight Center Scientific Visualization Studio. Thanks to Rob Gerston (GSFC) for providing the data. **376:** [b] Courtesy of Richard Peltier; [t] Provided by the SeaWiFS Project, NASA/Goddard Space Flight Center, and ORBIMAGE **377:** [l] Xinhua/Landov; [r] © Newspix/Calum Robertson **379:** [bkgd] dwphotos/Shutterstock; [r] RADARSAT-2 Data and Products © MacDonald, Dettwiler and Associates Ltd. (2008)—All Rights Reserved. RADARSAT is an official mark of the Canadian Space Agency.; [tl] © Canadian Space Agency **380:** [maps] National Snow and Ice Data Center; [tr] Manfred Thonig/Picture Press/Photolibrary **381:** Doug Fraser **384:** Graph adapted from: Concentrations of Greenhouse Gases, from 0 to 2005 (FAQ 2.1, Fig. 1). *Intergovernmental Panel on Climate Change, 2007: Working Group I Report "The Physical Science Basis."* [Solomon, S., D. Qin et. al.] Cambridge University Press, Cambridge, UK and New York, NY, USA. p. 135. **385:** [bl] Radius Images/Photolibrary; [br] TED MEAD/Photolibrary; [tl] ©iStockphoto/Imre Cikajlo; [tr] © Dennis MacDonald/Alamy **386:** Table, sources: "Global and Continental Temperature Change, Global." IPCC, 2007: Summary for Policymakers. In: *Climate Change 2007: The Physical Science Basis.* [Solomon, S., D. Qin et. al. (eds.)]. Cambridge University Press, Cambridge, UK and New York, NY, USA. Also, The NOAA Annual Greenhouse Gas Index. National Oceanic and Atmospheric Administration, Earth Systems Research Laboratory, Global Monitoring Division.; [b] British Antarctic Survey/Photo Researchers **388:** © Steve Bloom Images/Alamy **390:** [b] The Canadian Press (Larry MacDougal); [r] © Gloria H. Chomica/Masterfile; [t] © 2009 Jupiterimages **391:** [A] © Peter Christopher/Masterfile; [B] Larry Lee/Photolibrary; [C] Noel Hendrickson/Photographer's Choice/Getty; [D] ©iStockphoto/Amy Walters; [b] Yvan/Shutterstock **392:** [graph] Is Canada's forest a carbon sink or source? October 2007. Natural Resources Canada, Canadian Forest Service, Ottawa. Canadian Forest Service Science-Policy Notes. 2 p. Reproduced with the permission of Natural Resources Canada, Canadian Forest Service, 2009. **393:** [l] Photodisc/Photolibrary; [r] SCIENCE PHOTO LIBRARY **394:** Graph adapted from: "Global and Continental Temperature Change, Global." IPCC, 2007: Summary for Policymakers. In: *Climate Change 2007: The Physical Science Basis.* [Solomon, S., D. Qin et. al. (eds.)]. Cambridge University Press, Cambridge, UK and New York, NY, USA.; [b] INSADCO Photography/Doc-Stock/Photolibrary; [t] DR. KEITH WHEELER/SCIENCE PHOTO LIBRARY **396:** [bc] The Canadian Press (Larry MacDougal); [bl] NASA/Goddard Institute for Space Studies; [br] Graph adapted from: "Global and Continental Temperature Change, Global." IPCC, 2007: Summary for Policymakers. In: *Climate Change 2007: The Physical Science Basis.* [Solomon, S., D. Qin et. al. (eds.)]. Cambridge University Press, Cambridge, UK and New York, NY, USA.; [tc] © Dennis MacDonald/Alamy; [tl] NASA/Goddard Space Flight Center Scientific Visualization Studio. Thanks to Rob Gerston (GSFC) for providing the data.; [tr] British Antarctic Survey/Photo Researchers **397:** [bl] © Dwayne Newton/PhotoEdit; [br] Radius Images/Photolibrary; [cl] ©iStockphoto/Jason Lugo; [cr] © Mike Booth/Alamy; [tl] Hulton Archive/American Stock/Getty; [tr] REUTERS/Mathieu Belanger/Landov **399:** [t] NASA/Goddard Institute for Space Studies

Chapter 10. 402: The Canadian Press (Dave Chidley) **403:** [bc] Gary Strand, National Center for Atmospheric Research; [bl] ©iStockphoto/YinYang; [br] Manfred Steinbach/Shutterstock; [tc] © infocusphotos/Alamy; [tl] REUTERS/Rafiqur Rahman/Landov; [tr] Simon Hayter/GetStock **404:** [all r] Courtesy of the Vancouver Convention Centre; [grass] digitalife/Shutterstock **405:** [bl] Eky Chan/Shutterstock; [br] Albert H. Teich/Shutterstock; [cl] REUTERS/Heino Kalis/Landov; [cr] ©iStockphoto/Don Wilkie; [tl] ©iStockphoto/Andrew Penner; [tr] Radius Images/Photolibrary **406:** [paper] Robyn Mackenzie/Shutterstock; [bkgd] rahulred/Shutterstock; © Rick Friedman/Corbis **407:** [r] Lothar Schulz/fStop/Photolibrary **408:** Gary Strand, National Center for Atmospheric Research **409:** Adapted from: Scenarios for GHG emissions from 2000 to 2100 (in the absence of additional climate policies) and projections of surface temperatures (Fig. SPM.5, right panel, page 7). IPCC, 2007: Summary for Policymakers. In: *Climate Change 2007: The Physical Science Basis.* [Solomon, S., D. Qin et. al. (eds.)]. Cambridge University Press, Cambridge, UK and New York, NY, USA. p 7 **410:** [biofuels] Fesus Robert/Shutterstock; [geothermal] © Mark Boulton/Alamy; [hydro] Omni Photo Communications/Index Stock Imagery/photolibrary; [nuclear] John Edwards/Stone/Getty; [solar] Manfred Steinbach/Shutterstock; [wind] © Lloyd Sutton/Masterfile **411:** Javier Larrea/age fotostock/Photolibrary **412:** [bl] AP Photo/The Canadian Press (John McConnico); [r] Courtesy Ryan Danby; [tl] TORSTEN BLACKWOOD/AFP/Getty **413:** [bl] © Gideon Mendel/ActionAid/Corbis; [br] Gerard Soury/Oxford Scientific (OSF)/Photolibrary; [cr] ALEXANDER JOE/AFP/Getty; [tr] REUTERS/Rafiqur Rahman/Landov **415:** [b] The Canadian Press (Sam Soja); [t] © infocusphotos/Alamy **416:** [b] Richard Olsenius/National Geographic/Getty **417:** Toronto Star/The Canadian Press (Tony Bock) **418:** [b] © Hasse Schroder/Jupiterimages; [bkgd] Svetlana Privezentseva/Shutterstock; [tr] © Ashley Cooper/Alamy **419:** [tl] © Garry Black/Masterfile; [br] Simon Hayter/GetStock **420:** [b] Melissa Farlow/National Geographic/Getty **421:** [l] © Cobretti/Dreamstime; [r] John Czenke/Shutterstock **422:** © Rick Friedman/Corbis **423:** Graph source: "Canada's GHG Emissions 1990–2007" from *Information on Greenhouse Gas Sources and Sinks: Canada's 2007 Greenhouse Gas Inventory—A Summary of Trends*, p. 1 © Her Majesty The Queen in Right of Canada, Environment Canada, 2009. Reproduced with the permission of the Minister of Public Works and Government Services Canada. **424:** [l] © Dan Lamont/Corbis; [r] The Canadian Press (Tom Hanson) **425:** Graph source: "Table A11-12: 1990–2006 GHG Emission Summary for Ontario" from National Inventory Report, 1990–2006. p 578. © Her Majesty The Queen in Right of Canada, Environment Canada, 2008. Reproduced with the permission of the Minister of Public Works and Government Services Canada.; [b] Courtesy Hatch **426:** [b] © Ilene MacDonald/Alamy; [c] ©iStockphoto/archives; [t] ©iStockphoto/YinYang **427:** [Fig. 10] Photo © Toronto and Region Conservation. All Rights Reserved.; [Fig. 11] Wally Stemberger/Shutterstock; [Fig. 12] Alt-6/First Light; [Fig. 9] © john t. fowler/Alamy **429:** Source: "Canada's GHG Emissions by Sector, End-Use and Sub-Sector—Including Electricity-Related Emissions." *Natural Resources Canada Energy Use Data Handbook 1990 and 1998 to 2004.* Natural Resources Canada, 2006, pp 8–10. Reproduced with the permission of the Minister of Public Works and Government Services Canada, courtesy of Natural Resources Canada, 2009. **430:** [l] © 2009 Jupiterimages; [r] © Imageplus/Corbis **431:** Courtesy of the Ministry of Natural Resources **432:** [Fig. 4] © 2009 Jupiterimages; [Fig. 5] Joel Blit/Shutterstock; [Fig. 6] © Michael Mahovlich/Masterfile; [Fig. 7] © Bryan & Cherry Alexander Photography/Alamy; [Fig. 8] © J. David Andrews/Masterfile **434:** Paul Souders/The Image Bank/Getty **435:** © Simon Jarratt/Corbis/Jupiterimages

436, 442: [bc] Gary Strand, National Center for Atmospheric Research; [bl] ©iStockphoto/YinYang; [br] Manfred Steinbach/Shutterstock; [tc] © infocusphotos/ Alamy; [tl] REUTERS/Rafiqur Rahman/Landov; [tr] Simon Hayter/GetStock **437:** [bl] Eky Chan/Shutterstock; [br] Albert H. Teich/Shutterstock; [cl] REUTERS/ Heino Kalis/Landov; [cr] ©iStockphoto/Don Wilkie; [tl] ©iStockphoto/Andrew Penner; [tr] Radius Images/Photolibrary **439:** Panoramic Images/Getty **441:** ©iStockphoto/Ashok Rodrigues **443:** © 2007. Dan Piraro. King Features Syndicate **444:** [l] © Brian Sytnyk/Masterfile; [r] ©iStockphoto/Klaas Lingbeek van Kranen **449:** [b] Jasper Yellowhead Museum and Archives, 89.36.263, Leonard Jeck fonds; [t] © Chunli Li/Dreamstime **451:** Adapted from: NASA Earth Observatory, Explaining Rapid Climate Change: Tales from the Ice and Union of Concerned Scientists: Abrupt Climate Change.

Unit E

Chapter 11. 454: © Park Street/PhotoEdit **455:** [Saturn] NASA Jet Propulsion Laboratory; [amoeba] Michael Abbey/Photo Researchers; [b] 2happy/Shutterstock; [t] Medicimage/Photolibrary **456:** [c] Photo by George Silk/Time Life Pictures/Getty; [l] Vitaliy Minsk/Shutterstock; [r] Novastock/Photolibrary **457:** [bl] ©iStockphoto/ David Wilson; [br] Jozsef Szasz-Fabian/Shutterstock; [tl] Martin Heitner/ Photolibrary; [tr] © Helen King/Corbis **458:** Vitaliy Minsk/Shutterstock **459:** [bl] Pavel Cheiko/Shutterstock; [br] © Michael Newman/PhotoEdit; [tc] Kim Steele/Photonica/Getty; [tl] Chris Hill/Shutterstock; [tr] Courtesy of SOHO (ESA & NASA) **460:** THE KOBAL COLLECTION/DREAMWORKS/PARAMOUNT **461:** [bl] Kablonk! Kablonk!/Photolibrary; [br] Gary Paul Lewis/Shutterstock; [cr] trailexplorers/Shutterstock **462:** [bkgd] rahulred/Shutterstock **463:** [l] Jeff Schmaltz, MODIS Land Rapid Response Team at NASA GSFC; [tr] Courtesy of SOHO (ESA & NASA) **464:** [b] © Baldwin H. Ward & Kathryn C. Ward/Corbis; [t] Terry Underwood Evans/Shutterstock **465:** [b] Yuri Arcurs/Shutterstock **466:** [X-ray] Scott Camazine/Photo Researchers; [gamma] NASA/SCIENCE PHOTO LIBRARY; [infrared] © 2009 Daisy Rae/FoodPix/Jupiterimages; [microwave] trailexplorers/ Shutterstock; [radio] Julián Rovagnati/Shutterstock; [sunburn] Ronald Sumners/ Shutterstock; [visible] © 2009 Jupiterimages **468:** [b] X-ray: NASA/CXC/CfA/R. Kraft et al; Radio: Courtesy of Dr. Martin Hardcastle/U.S. National Radio Astronomy Observatory; Optical: European Organization for Astronomical Research in the Southern Hemisphere/WFI/M.Rejkuba et. al.; [t] Science Source/Photo Researchers **470:** [cl] © imagebroker/Alamy; [modern bulb] graphyx/Shutterstock; [tl] © 2009 Jupiterimages **471:** [all b] ©SSPL/The Image Works; [tl] © Craig Aurness/Corbis; [tr] michael ledray/Shutterstock **472:** [bl] Mark A. Schneider/Photo Researchers; [tl] Imagestate RM/Pictor/Photolibrary **473:** Chris Hill/Shutterstock **474:** © Pierre Arsenault/Alamy **475:** [tr] Kim Steele/Photonica/Getty **476:** Adam Filipowicz/ Shutterstock **478:** © Julian Smith/Corbis **480:** [l] © Werner Forman/Corbis; [r] ©iStockphoto/Gord Horne **483:** [r] Bridget McPherson/Shutterstock **485:** [b] Pavel Cheiko/Shutterstock; [c] semenovp/Shutterstock; [t] Pavel Cheiko/Shutterstock **486:** [dyslexia] Joanne Meredith, vrse design inc. **487:** [b] Reuters/Landov; [bkgd] dwphotos/Shutterstock; [t] © ANNEBICQUE BERNARD/Corbis SYGMA **488:** [b] The British Library/Imagestate RM/Photolibrary **489:** [br] © FABRIZIO BENSCH/Reuters/Corbis **490:** [b] ©iStockphoto/Loretta Hostettler **491:** [bl] Doug Fraser; [br] © Zimmytws/Dreamstime **496:** [br] © Visuals Unlimited/Corbis **497:** [br] © 2009 Brakefield Photo/Jupiterimages **498:** [bl] Dr. R. Jedrzejewski (STScI) NASA, ESA; [br] Alexey Gostev/Shutterstock; [tr] Oleg Kozlov, Sophy Kozlova/ Shutterstock **499:** [br] © Michael Newman/PhotoEdit **500:** [br] © Helen King/Corbis **501:** [bl] © Richard Megna/Fundamental Photographs **504, 586:** [bl] Pavel Cheiko/ Shutterstock; [br] © Michael Newman/PhotoEdit; [tc] Kim Steele/Photonica/ Getty; [tl] Chris Hill/Shutterstock; [tr] Courtesy of SOHO (ESA & NASA) **505:** [bl] Kablonk! Kablonk!/Photolibrary; [br] Gary Paul Lewis/Shutterstock; [cr] trailexplorers/Shutterstock **507:** [br] Factoria singular fotografia/Shutterstock

Chapter 12. 510: Photo by George Silk/Time Life Pictures/Getty **511:** [bc] ©iStockphoto/Greg Nicholas; [bl] Macs Peter/Shutterstock; [br] ©iStockphoto/ Evgeny Terentev; [tc] © Richard Megna/Fundamental Photographs; [tl] © Clayton J. Price/Corbis; [tr] GIPhotoStock/Photo Researchers **512:** Source of astronaut conversation: from the transcription of the Technical Air-to-Ground Voice Transmission (GOSS NET 1) from the Apollo 11 mission. Courtesy NASA.; [bkgd] beaucroft/Shutterstock; [inset] NASA **513:** [bl] Chris Anderson/Aurora/Getty; [br] Stephen St. John/National Geographic/Getty; [cr] 2happy/Shutterstock; [tr] © 2009 Jupiterimages **514:** [bkgd] rahulred/Shutterstock; [copper] David Schurig; [paper] Robyn Mackenzie/Shutterstock **516:** [l] © Richard Megna/Fundamental Photographs **517:** [b] Jerome Wexler/Photo Researchers; [t] © Richard Megna/ Fundamental Photographs **518:** [l] Albert Cheng/Shutterstock; [r] ©iStockphoto/ Adam Neiland **519:** [b] Joe Henderson/Visuals Unlimited; [t] Tyler Fox/Shutterstock **524:** © Visuals Unlimited/Corbis **525:** Wayne Scherr/Photo Researchers **526:** [b] GIPhotoStock/Photo Researchers **527:** [b] ©iStockphoto/Evgeny Terentev **528:** [bl] Macs Peter/Shutterstock; [br] Dave King © Dorling Kindersley, Courtesy of The Science Museum, London; [t] Edward Kinsman/Photo Researchers **530:** [b] © 2009 Plainpicture/Jupiterimages **534:** [bkgd] Terrance Emerson/ Shutterstock; [l] David Schurig; [r] AP Photo/Shizuo Kambayashi **535:** [t] Southern Illinois University/Photo Researchers **536:** [bl] © Mu Xiang Bin/Redlink/Corbis; [tl] © Lawcain/Dreamstime **537:** [br] © YOSHITSUGU NISHIGAKI/amanaimages/ Corbis; [tl] Stephen St. John/National Geographic/Getty **538:** [bl] ©iStockphoto/ Greg Nicholas; [t] © Clayton J. Price/Corbis **539:** Bill Hatcher/National Geographic/ Getty **540, 586:** [bc] ©iStockphoto/Greg Nicholas; [bl] Macs Peter/Shutterstock; [br] ©iStockphoto/Evgeny Terentev; [tc] © Richard Megna/Fundamental Photographs; [tl] © Clayton J. Price/Corbis; [tr] GIPhotoStock/Photo Researchers **541:** [bl] Chris Anderson/Aurora/Getty; [br] Stephen St. John/National Geographic/Getty; [cr] 2happy/Shutterstock; [tr] © 2009 Jupiterimages

Chapter 13. 546: Novastock/Photolibrary **547:** [bc] Romanchuck Dimitry/ Shutterstock; [bl] Lein de León Yong/Shutterstock; [br] © 2009 Jupiterimages; [tc] David Parker/Photo Researchers; [tl] Robert St-Coeur/Shutterstock; [tr] David Parker/Photo Researchers **548:** [inset] ©iStockphoto/Gene Chutka; [paper] javarman/ Shutterstock; ©Columbia Pictures/Courtesy Everett Collection **549:** [br] © Iofoto/ Dreamstime; [cl] © 2009 Jupiterimages; [cr] Lein de León Yong/Shutterstock; [tl] Romanchuck Dimitry/Shutterstock **550:** [bkgd] rahulred/Shutterstock **551:** [b] David Parker/Photo Researchers; [t] David Parker/Photo Researchers **567:** [b] ©iStockphoto/Sergii Shcherbakov **568:** [bl] Lein de León Yong/Shutterstock **569:** [tr] Tetra Images/Photolibrary **570:** [t] © Roger Ressmeyer/Corbis **571:** top right, [Fig. 1] Source: NASA; [bkgd] Svetlana Privezentseva/Shutterstock; [bl] NASA, ESA, Richard Ellis (Caltech) and Jean-Paul Kneib (Observatoire Midi-Pyrenees, France); [r] NASA, ESA, A. Bolton (Harvard-Smithsonian CfA) and the SLACS Team, STScI **572:** [bl] Chepe Nicoli/Shutterstock **576:** [l] Robert St-Coeur/Shutterstock; [r] Ciaran Griffin/Stockbyte/Photolibrary **578:** Olivier Voisin/ Photo Researchers **580, 586:** [bc] Romanchuck Dimitry/Shutterstock; [bl] Lein de León Yong/Shutterstock; [br] © 2009 Jupiterimages; [tc] David Parker/Photo Researchers; [tl] Robert St-Coeur/Shutterstock; [tr] David Parker/Photo Researchers **581:** [br] © Iofoto/Dreamstime; [cl] © 2009 Jupiterimages; [cr] Lein de León Yong/ Shutterstock; [tl] Romanchuck Dimitry/Shutterstock **588:** [b] ©iStockphoto/Ulina Tauer; [t] © Fancy/Veer/Corbis

Appendix A

598: © Monkeybusinessimages/Dreamstime **599:** [bulb] © Corbis Premium RF/ Alamy; [glassware] © Steve Allen/Brand X/Corbis; [microscope] © sciencephotos/ Alamy; [purple slide] © Biodisc/Visuals Unlimited/Alamy **604:** [conductivity tester] tomek_/Shutterstock; [dissection] corbis/First Light; [pH meter] Charles D. Winters/ Photo Researchers **609:** [Erlenmeyer] From GIUSEPPI/HAMMILL. *Nelson Science & Tech Perspectives 8*. © 2009 Nelson Education Ltd. p.386. Photo by Dave Starrett; [Petri dish] From GIUSEPPI/HAMMILL. *Nelson Science & Tech Perspectives 8*. © 2009 Nelson Education Ltd. p.386. Photo by Dave Starrett; [balance] Coprid/ Shutterstock; [dropper bottle] From GIUSEPPI/HAMMILL. *Nelson Science & Tech Perspectives 8*. © 2009 Nelson Education Ltd. p.130. Photo by Dave Starrett; [hot plate] From GIUSEPPI/HAMMILL. *Nelson Science & Tech Perspectives 8*. © 2009 Nelson Education Ltd. p.386. Photo by Dave Starrett; [retort stand and ring clamp] From GIUSEPPI/HAMMILL. *Nelson Science & Tech Perspectives 8*. © 2009 Nelson Education Ltd. p.386. Photo by Dave Starrett **610:** [scientist] © Patrick Robert/ Corbis; [tree] S.J. Krasemann/Peter Arnold **611:** [boiling water] From GIUSEPPI/ HAMMILL. *Nelson Science & Tech Perspectives 8*. © 2009 Nelson Education Ltd. p. 393. Photo by Dave Starrett **612:** [comb] Charles D. Winters/Photo Researchers **613:** [reaction] Martyn Chillmaid/Oxford Scientific/Photolibrary **615:** [students] Laurence Gough/Shutterstock **619:** [homework] Sofos Design/Shutterstock **622:** [reporter] © Zeffss/Dreamstime **623:** [workers] Denkou Images/Alamy; [student presentation] ©iStockphoto/Chris Schmidt

Appendix B

648: Ulrich Mueller/Shutterstock **651:** [bracelet] © Canadafirst/Dreamstime; [tsunami] © Christophe Fouquin/Fotolia **654:** [microscope] © Bettmann/Corbis; [X-ray] riccardocova/BigStockPhoto **655:** [satellite] Corbis/Photolibrary; [telescope] © Roger Ressmeyer/Corbis; [valve] © 2009 Creatas Images/Jupiterimages; [vest] © 2009 Hemera Technologies/PhotoObjects/Jupiterimages; [virus] MedicalRF/ The Medical File/Peter Arnold **656:** [WiiFit] The Canadian Press (Nathan Denette); [engine] Mario Beauregard/age fotostock/Photolibrary **657:** [house] Tom Uhlenberg/ Shutterstock; [phone] © David Burton/Beateworks/Corbis; [waste] Peter Grosch/ Shutterstock

Studio Photographer: Dave Starrett

Nelson Science Perspectives

Key

atomic number	26
most common ion charge	3+
other ion charge	2+
symbol of element (solids in black, gases in red, liquids in blue)	Fe
name of element	iron
atomic mass (u)—based on C-12	55.85

Period	1	2	3	4	5	6	7	8	9
1	1 (1+, 1−) **H** hydrogen 1.01								
2	3 (1+) **Li** lithium 6.94	4 (2+) **Be** beryllium 9.01							
3	11 (1+) **Na** sodium 22.99	12 (2+) **Mg** magnesium 24.31							
4	19 (1+) **K** potassium 39.10	20 (2+) **Ca** calcium 40.08	21 (3+) **Sc** scandium 44.96	22 (4+, 3+) **Ti** titanium 47.87	23 (5+, 4+) **V** vanadium 50.94	24 (3+, 2+) **Cr** chromium 52.00	25 (2+, 4+) **Mn** manganese 54.94	26 (3+, 2+) **Fe** iron 55.85	27 (2+, 3+) **Co** cobalt 58.93
5	37 (1+) **Rb** rubidium 85.47	38 (2+) **Sr** strontium 87.62	39 (3+) **Y** yttrium 88.91	40 (4+) **Zr** zirconium 91.22	41 (5+, 3+) **Nb** niobium 92.91	42 (6+) **Mo** molybdenum 95.94	43 (7+) **Tc** technetium (98)	44 (3+, 4+) **Ru** ruthenium 101.07	45 (3+) **Rh** rhodium 102.91
6	55 (1+) **Cs** cesium 132.91	56 (2+) **Ba** barium 137.33	57 (3+, 2+) **La** lanthanum 138.91	72 (4+) **Hf** hafnium 178.49	73 (5+) **Ta** tantalum 180.95	74 (6+) **W** tungsten 183.84	75 (7+) **Re** rhenium 186.21	76 (4+) **Os** osmium 190.23	77 (4+) **Ir** iridium 192.22
7	87 (1+) **Fr** francium (223)	88 (2+) **Ra** radium (226)	89 (3+, 2+) **Ac** actinium (227)	104 **Rf** rutherfordium (261)	105 **Db** dubnium (262)	106 **Sg** seaborgium (266)	107 **Bh** bohrium (264)	108 **Hs** hassium (277)	109 **Mt** meitnerium (268)
	Alkali metals	Alkaline earth metals							

Period					
6	58 (3+) **Ce** cerium 140.12	59 (3+) **Pr** praseodymium 140.91	60 (3+) **Nd** neodymium 144.24	61 (3+) **Pm** promethium (145)	62 (3+, 2+) **Sm** samarium 150.36
7	90 (4+) **Th** thorium 232.04	91 (5+, 4+) **Pa** protactinium 231.04	92 (6+, 4+) **U** uranium 238.03	93 (5+) **Np** neptunium (237)	94 (4+, 6+) **Pu** plutonium (244)

- Metals
- Metalloids
- Nonmetals
- Hydrogen